BIODIVERSITY OF FUNGI

Inventory and Monitoring Methods

BIODIVERSITY OF FUNGI

Inventory and Monitoring Methods

GREGORY M. MUELLER

Field Museum of Natural History

GERALD F. BILLS

Merck Sharp & Dohme de España

MERCEDES S. FOSTER

USGS Patuxent Wildlife Research Center

ELSEVIER
ACADEMIC
PRESS

Amsterdam • Boston • Heidelberg • London
New York • Oxford • Paris • San Diego
San Francisco • Singapore • Sydney • Tokyo

Acquisitions Editor: David Cella
Project Manager: Brandy Palacios
Associate Editor: Kelly Sonnack
Marketing Manager: Linda Beattie
Cover Design: Eric Decicco
Full Service Provider: Graphic World, Inc.
Composition: SNP Best-set Typesetter Ltd., Hong Kong
Printer: CTPS

Elsevier Academic Press
200 Wheeler Road, Burlington, MA 01803, USA
525 B Street, Suite 1900, San Diego, California 92101-4495, USA
84 Theobald's Road, London WC1X 8RR, UK

This book is printed on acid-free paper. ∞

Library of Congress Cataloging-in-Publication Data
Application submitted

British Library Cataloguing in Publication Data
A catalogue record for this book is available from the British Library

ISBN: 0-12-509551-1

For all information on all Elsevier Academic Press publications
visit our Web site at www.books.elsevier.com

Front cover: The slime mold *Physarum roseum* on a decaying leaf. Slime molds are fungal-like organisms traditionally studied by mycologists. Photo by Ray Simons.

Back cover: A species of *Mycena* found in Yunnan, China. Species of the mushroom genus *Mycena* are commonly encountered throughout the world. Photo by Gregory M. Mueller.

CONTENTS

FOREWORD

DAVID L. HAWKSWORTH, CBE

Past-President, International Union of Biological Sciences; Honorary President, International Mycological Association

The ALL SPECIES Summit, convened by Edward O. Wilson at Harvard University from October 13–15, 2001, viewed the discovery and description of the unknown species on Earth as the major challenge facing scientists today. Such an effort was deemed comparable to the Human Genome Project in the scale of resources and level of international cooperation needed to accomplish the task, or at least make major inroads into it, within 25 years. The extent of the undiscovered organismal diversity with which we share the planet is far from clear, but it is now generally accepted that only 5–10% of the extant species have been discovered and named. Fungi constitute a significant proportion of the yet-undiscovered biota that is crucial to the maintenance of ecological processes and human well-being. Without fungi, major problems in nutrient cycling would occur, plants would suffer without the nutrients that fungi secure for them from the soil, many animals would be without food, woody materials would not be broken down, some insects and other animals would not be able to digest plant materials, and even soil structures would differ. In addition, many fungi are of direct benefit to humankind—for instance, as sources of antibiotics and other pharmaceutical products, food and drink, industrial chemicals and enzymes, and biological agents that control weeds and pests. Fungi have particular promise as a component of sustainable development, and in some developing countries, a Non-Green Revolution is already a topic of conversation. However, other fungi are of concern because of their adverse effects—for example, as plant pathogens, toxin producers, causes of skin and other diseases in humans, food-spoilage agents, and biodeteriogens of manufactured materials and even buildings.

The studies by mycologists embrace filamentous fungi, lichen fungi, molds, mushrooms, slime molds, some chromists (oomycetes), and yeasts, although some now are classified in different kingdoms. At least 74,000, but possibly as many as 120,000, fungal species so far have been named. Back in 1990, the number of fungi actually present on Earth was estimated conservatively at 1,500,000 species. That figure has been widely cited and generally accepted as a working hypothesis. Evidence accumulated since that time, especially from studies on plants in the tropics but also from critical molecular investigations, suggests that, if anything, the figure is too low. That suggestion implies that we know at most only about 5% of the fungal species on Earth.

How are the undiscovered fungi to be found, and how are those present in a site to be inventoried? No single site in the world has been comprehensively surveyed for fungi. We know that even in temperate areas we can expect about six times as many fungi as native plants to occur in a site and that 2,500 to 3,000 species can be found in about 200 hectares if a variety of specialists study the site for more than 25 years. Fungi cannot be trapped or fogged and collected *en masse* in the same way as insects. They may produce fruit bodies only at intervals of many years, and those structures then may persist for only a few hours. In contrast, other fungi are perennial and can be found at any time of the year, as in the case of lichen-forming fungi and many polypores. Some fungi live inside plants or arthropods, and the same plant may support different fungi during different stages of its life cycle from flowering to decay. Many fungi are found only by examining leaves and woody surfaces microscopically, or they can be detected only by isolating and culturing the fungi until they sporulate. Furthermore, some fungi have more than one single sporing stage in their life cycles, and in those so-called pleomorphic fungi, only one stage may be found. In some instances, the different stages are given separate scientific names, which can be confusing.

Providing quantitative data poses a separate raft of problems. A single underground mycelium—that is, one individual—may produce numerous above-ground fruit bodies over a wide area. Similarly, a twig may have numerous fruit bodies of some ascomycete that we now know may arise from discrete but adjacent individuals in some cases and from a single wide-spreading mycelium in others. Isolation into culture is not the answer because the species that are isolated can depend on the methods used. Furthermore, many fungi do not grow, or do not grow readily, in culture—and when they do, they often remain sterile, defying identification by all but molecular methods. In addition, many fungi isolated from soil may not have been actively growing in the soil but present only as resting spores awaiting suitable substrata on which to grow. Major discrepancies thus can occur between fungi isolated and those known to be present either by molecular methods or because their fruit bodies were present.

Because of these multifarious complications, the development of standardized approaches to fungal inventorying and the selection of groups that might be bioindicators (surrogates) for fungal biodiversity are critical to facilitate comparisons between different sites. The desirability of having a suite of standard methods and bioindicators for sampling diverse habitats often has been discussed, but no broad consensus has been reached. In the future, new methods for describing and communicating information on fungi that are discovered will have to be developed; the numbers of new species to be discovered and described are such that traditional publication methods will be unable to cope—Internet publishing linked to an internationally recognized registration and archival service may be the way ahead.

Fungi are so diverse that various kinds of fungi demand different procedures for their detection and identification. The techniques that mycologists specializing in different kinds of fungi or habitats use can often appear mysterious to other biologists, and they rarely are brought together in one place. The present work is the most comprehensive attempt to date to take the mystique out of surveying fungi. The major compilations edited by C. Booth (1971, *Methods in Microbiology*, vol. 4, Academic Press, London) and R. B. Stevens (1981, *Mycology Guidebook*, University of Washington Press, Seattle) are less comprehensive and now inevitably somewhat dated. Methods hardly are addressed in the multivolume *The Mycota* (1994 on, Springer-Verlag, Berlin). Protocols for many groups, however, were developed carefully by A. Y. Rossman and colleagues (1998, *Protocols for an All Taxa Biodiversity Inventory of Fungi in a Costa Rican Conservation Area*, Parkway Publishers, Boone, North Carolina); that work is a useful adjunct to this volume in exploring how a comprehensive fungal inventory might be attempted.

Biodiversity of Fungi is particularly timely as the concept of a global project to describe all Life on Earth starts to catch hold and roll. Here are authoritative accounts of how to proceed in, and the difficulties associated with, inventorying fungi of diverse groups and habitats. Leading specialists have distilled decades of hands-on experience into their chapters, imparting knowledge that will be of benefit to all future generations of mycologists. Every mycologist is almost sure to learn something here, even in his or her own specialty, and no mycology laboratory should be without a copy. It is of great credit to the foresight, determination, and skill of the editors that they brought such a major work of outstanding academic quality to fruition—and at a time when it promises to be needed more than ever before.

PREFACE

This volume is particularly important because, with the exception of clinical diagnoses for human pathogens and quality-control procedures for detection of contamination in foods, cosmetics, and drugs, standardized protocols for the qualitative or quantitative sampling of fungi do not exist. Stevens (1974) compiled a variety of information on collecting, isolating, and preserving many groups of fungi in his *Mycological Guidebook*. Although this now out-of-print book is full of useful information, it was developed as a teaching tool to complement courses in mycology, not as a guide to developing methods for inventory and monitoring. Rossman and colleagues (1998) provided, in outline form, a series of recommendations for sampling fungi as part of the planned All Taxa Biodiversity Inventory of the Guanacaste Conservation Area in Costa Rica (see Allen 2001 and Appendix III). Their treatments, by design, are site-specific and lack the detail necessary to facilitate modification of the protocols for use in other studies or at other sites. Thus, generalized protocols for sampling fungal diversity or for documenting changes in fungal diversity and distributions over time do not exist.

Amy Rossman approached Mercedes Foster about the possibility of publishing a much-needed protocols book on fungi, an idea that Foster enthusiastically endorsed. Planning for the book began early in 1994 when Rossman recruited Hal Burdsall, Gerald Bills, and Gregory Mueller to help carry out the ambitious project. The team discussed at length how best to present the information that investigators would need to inventory and monitor the taxonomically and ecologically diverse group of organisms treated as fungi. Ultimately, the group decided that a book organized according to sampling methodology rather than along taxonomic lines would be the most useful and least repetitive. They then identified primary authors for each chapter.

A workshop was held from October 15–19, 1995, at the Systematic Mycology Laboratory of the U.S. Department of Agriculture in Beltsville, Maryland, to discuss the contents of the chapters. The lead author for each chapter provided a draft overview of his or her chapter to the more than 65 workshop participants. The discussions following those oral presentations served as preliminary reviews of the chapters' contents. They also provided an opportunity for authors to reach consensus on the main points to be included in each chapter, thereby providing a degree of uniformity to the volume and ensuring that each chapter would focus on the primary goals of the endeavor—to provide standardized methods for quantitative inventory and monitoring of fungi and related groups.

Manuscripts were completed and submitted to Mueller. Each manuscript was reviewed by one or two specialists and edited by Mueller or Bills before being sent to Foster for additional editing. Foster returned the manuscripts to Mueller and Bills, who reviewed and incorporated her suggestions. Each fully edited manuscript then was returned to the appropriate authors for their approval and any necessary updating. The final version comprised approximately 2,150 pages; Foster reviewed the entire work for overall content, format, and usability.

Throughout this endeavor, we worked to involve as many mycologists as possible who actively study fungal biodiversity, regardless of country or institution. Ultimately, more than 88 mycologists and colleagues, with an overwhelming aggregate of knowledge and experience, contributed to the volume by authoring or coauthoring chapters. Consequently, at this writing, the volume represents state-of-the-art knowledge about sampling methodologies and data analyses for fungi. As such, we believe that it will make a major contribution to the study of fungi and biodiversity. It will become apparent to the user, however, that the state of knowledge regarding fungal sampling is uneven from habitat to habitat and taxon to taxon and that few quantitative procedures have been developed. Nevertheless, this volume represents important early, fundamental steps in the study of fungal biodiversity. We hope that it will stimulate those investigators interested in fungi, and in biodiversity in general, to develop new techniques and to refine and improve those presented here.

ACKNOWLEDGMENTS. Many people worked very hard to bring this ambitious project to completion. First we thank Amy Rossman for developing the idea and then "selling" it to Mercedes Foster (who responded enthusiastically from the beginning). Without Amy's vision this project would not have become a reality. Hal Burdsall, David Malloch, and Meredith Blackwell were instrumental in obtaining the initial manuscript submissions. Ken Kuehl (Homewood, Illinois) edited many of the chapters. Susan Hamnik and Betty Strack (both from The Field Museum) provided invaluable editorial and organizational assistance. Fiona Wilkinson (U.S. Geological Survey, Patuxent Wildlife Research Center) undertook the monumental task of checking each literature reference for accuracy and consistency. She also assisted with the preparation of the figures and helped Foster with a myriad of miscellaneous tasks related to the preparation of the manuscript. Martha A. Rosen, David T. Steere Jr., Wanda L. West, Clare Catron, and Ruth F. Schaller of the Smithsonian Institution, National Museum of Natural History Branch Library, worked endlessly to obtain copies of publications on interlibrary loan. Finally, the editors wish to acknowledge the authors and contributors for their outstanding contributions and patience during this project. Financial support, for which we are most grateful, was supplied to the project by the United States Department of Agriculture (USDA/Forest Service, Forest Products Laboratory, Madison, Wisconsin [Research Joint Venture Agreement 95-RJVA-2610]; USDA/Forest Service, Pacific Northwest Research Station, Portland, Oregon [Grant #PNW 95-0786]; and the U.S. Department of the Interior (The National Biological Service, Biological Survey Project, Washington, DC; USGS Patuxent Wildlife Research Center, Laurel, Maryland; and the USGS Biological Resources Division Office of Information and Informatics, Reston, Virginia). Mueller acknowledges the support of NSF (DEB 9972027).

Greg Mueller, Gerald Bills, and *Mercedes Foster*

ADDRESSES OF AUTHORS AND CONTRIBUTORS

Joseph F. Ammirati, Jr.
Department of Biology
University of Washington
Box 351800
Seattle, Washington 98195-1800
USA

George L. Barron
Department of Environmental Biology
Ontario Agriculture College
University of Guelph
Guelph, Ontario N1G 2W1
Canada

Richard K. Benjamin
Rancho Santa Ana Botanic Garden
1500 North College Avenue
Claremont, California 91711-3157
USA

Gerald L. Benny
Department of Plant Pathology
University of Florida
1453 Fifield Hall
Gainesville, Florida 32611-0680
USA

Stephen P. Bentivenga
Department of Biology and Microbiology
University of Wisconsin at Oshkosh
800 Algoma Boulevard
Oshkosh, Wisconsin 54901-8640
USA

Gerald F. Bills
Centro de Investigación Básica
Merck Sharp & Dohme de España, S. A.
Josefa Valcarcel 38
E-28027 Madrid
Spain

Meredith Blackwell
Department of Biological Sciences
Louisiana State University
380 Life Sciences Building
Baton Rouge, Louisiana 70803-1715
USA

Brenda E. Callan
Pacific Forestry Centre
Canadian Forest Service, Natural Resources Canada
506 West Burnside Road
Victoria, British Columbia V8Z 1M5
Canada

Paul F. Cannon
CABI Bioscience
Bakeham Lane
Egham, Surrey TW20 9TY
United Kingdom

Lori M. Carris
Department of Plant Pathology
Washington State University
P.O. Box 646430
329 Johnson Hall
Pullman, Washington 99163-6430
USA

Michael A. Castellano
USDA Forest Service
Forestry Sciences Laboratory
3200 Jefferson Way
Corvallis, Oregon 97331
USA

James C. Cavender
Department of Environmental and Plant Biology
Ohio University
Porter Hall, Room 309A
Athens, Ohio 45701
USA

Ignacio H. Chapela
Department of Environmental Science, Policy, and Management
University of California at Berkeley
1385 Francisco Street
Berkeley, California 94720-3110
USA

Martha Christensen
Botany Department
University of Wyoming
P.O. Box 3165
Laramie, Wyoming 82071-3165
USA

Daniel Cullen
USDA Forest Service
Forest Products Laboratory
Institute for Microbial and Biochemical Technology
One Gifford Pinchot Drive
Madison, Wisconsin 53705-2898
USA

Daniel L. Czederpiltz
USDA Forest Service
Forest Products Laboratory
Center for Forest Mycology Research
One Gifford Pinchot Drive
Madison, Wisconsin 53705-2398
USA

Paul Diederich
Musée National d'Histoire Naturelle
Marché-aux-Poissons L-2345
Luxembourg

Frank M. Dugan
USDA Agricultural Research Service
Western Regional Plant Introduction Station
Washington State University

59 Johnson Hall
Pullman, Washington 99164-6402
USA

David F. Farr
USDA Agricultural Research Service
Systematic Botany and Mycology Laboratory
Room 304, Building 011A
Beltsville, Maryland 20705
USA

Ellen R. Farr
Department of Botany, MRC-166
Smithsonian Institution
Washington, DC 20560-0166
USA

Jack W. Fell
Rosenstiel School of Marine and Atmospheric Science
University of Miami
4600 Rickenbacker Causeway
Miami, Florida 33149
USA

Mercedes S. Foster
USGS Patuxent Wildlife Research Center
National Museum of Natural History
10th and Constitution, N.W.
Washington, D.C. 20560-0111
USA

Walter Gams
Centraalbureau voor Schimmelcultures
P.O. Box 273
3740 AG Baarn
The Netherlands

David L. Hawksworth
Departamento de Biología Vegetal
II, Facultad de Farmacia, Universidad Complutense
Plaza de Ramon y Cajal, Ciudad Universitaria
E-28040 Madrid
Spain

Dwayne D. Hegedus
Molecular Genetics Section
Agriculture and Agri-Food Canada
Saskatoon, Saskatchewan 57N OX2
Canada

Sabine M. Huhndorf
Department of Botany
Field Museum of Natural History
1400 South Lake Shore Drive

Chicago, Illinois 60605-2496
USA

Richard A. Humber
USDA Agricultural Research Service
Plant Protection Research Unit
Plant, Soil and Nutrition Laboratory
Tower Road
Ithaca, New York 14853
USA

Kevin G. Jones
Department of Biological Sciences
Louisiana State University
380 Life Sciences Building
Baton Rouge, Louisiana 70803-1715
USA

Shung-Chang Jong
American Type Culture Collection
10801 University Boulevard
Manassas, Virginia 20110-2209
USA

Harold W. Keller
Department of Biology
Central Missouri State University
Warrensburg, Missouri 64093
USA

George G. Khachatourians
Department of Applied Microbiology
College of Agriculture
University of Saskatchewan
Saskatoon, Saskatchewan S7N 5A8
Canada

Kier D. Klepzig
USDA Forest Service
Forest Insect Research Project
Southern Research Station
2500 Shreveport Highway
Pineville, Louisiana 71360
USA

Jan Kohlmeyer
Institute of Marine Sciences
University of North Carolina at Chapel Hill
Morehead City, North Carolina 28557
USA

Richard E. Koske
Department of Biological Sciences
University of Rhode Island

Kingston, Rhode Island 02881-0812
USA

John C. Krug
Department of Botany
University of Toronto
25 Willcocks Street
Toronto, Ontario M5S 3B2
Canada

Cletus P. Kurtzman
USDA Agricultural Research Service
National Center for Agricultural Utilization Research
1815 N. University Street
Peoria, Illinois 61604
USA

Deborah M. Langsam
Biology Department, McEniry Building, Room 306
University of North Carolina at Charlotte
9201 University City Boulevard
Charlotte, North Carolina 28223-0001
USA

Patrick R. Leacock
Department of Botany
Field Museum of Natural History
1400 South Lake Shore Drive
Chicago, Illinois 60605-2496
USA

Robert W. Lichtwardt
Department of Botany and Evolutionary Biology
University of Kansas
Lawrence, Kansas 66045-2106
USA

D. Jean Lodge
USDA Forest Service
Forest Products Laboratory
Center for Forest Mycology Research
P.O. Box 1377
Luquillo, Puerto Rico 00773-1377
USA

Joyce E. Longcore
Department of Biological Sciences
University of Maine
5722 Deering Hall
Orono, Maine 04469-5722
USA

Daniel L. Luoma
Department of Forest Science

321 Richardson Hall
Oregon State University
Corvallis, Oregon 97331-5752
USA

David Malloch
Department of Botany
University of Toronto
25 Willcocks Street
Toronto, Ontario M5S 3B2
Canada

Milagro Mata
Costa Rica National Institute of Biodiversity (INBio)
Santo Domingo, Heredia
Costa Rica

Bruce McCune
Department of Botany and Plant Pathology
Cordley Hall 2082
Oregon State University
Corvallis, Oregon 97331-2902
USA

Donna L. Moore
Division of Biology and Chemistry
Corning Community College
Corning, New York 14830
USA

Joseph B. Morton
Division of Plant and Soil Sciences
West Virginia University
401 Brooks Hall
P.O. Box 6057
Morgantown, West Virginia 26506-6057
USA

Gregory M. Mueller
Department of Botany
Field Museum of Natural History
1400 South Lake Shore Drive
Chicago, Illinois 60605-2496
USA

Karen K. Nakasone
USDA Forest Service
Forest Products Laboratory
Center for Forest Mycology Research
One Gifford Pinchot Drive
Madison, Wisconsin 53705-2398
USA

Steven Y. Newell
Marine Institute

University of Georgia
Sapelo Island, Georgia 31327
USA

Hiroaki Noda
National Institute of Sericultural and Entomological
Science
Owashi, Tsukuba 305
Japan

Thomas E. O'Dell
National Park Service
Grand Staircase—Escalante National Monument
190 E. Center Street
Kanab, UT 84741
USA

Stephen W. Peterson
USDA Agricultural Research Service
National Center for Agricultural Utilization Research
1815 N. University Street
Peoria, Illinois 61604
USA

Donald H. Pfister
Farlow Herbarium
Harvard University
Cambridge, Massachusetts 02138
USA

Kadri Põldmaa
Institute of Zoology and Botany
Riia 181
EE-2400 Tartu
Estonia

Jon D. Polishook
Merck Research Laboratories
P.O. Box 2000
RY80Y-120
Rahway, New Jersey 07065-0900
USA

Martha J. Powell
Department of Biological Sciences
Box 870344
University of Alabama
Tuscaloosa, Alabama 35487
USA

Thomas G. Rand
Department of Biology
St. Mary's University
Halifax, Nova Scotia B3H 3C3
Canada

Russell J. Rodriguez
USGS Western Fisheries Research Center
6506 N.E. 65th Street
Seattle, Washington 98115
USA

Richard A. Roeper
Department of Biology
Alma College
Alma, Michigan 48801
USA

Roger Rosentreter
Bureau of Land Management
1387 Vinnel Way
Boise, Idaho 83709
USA

Leif Ryvarden
Botany Department, Biological Institute
University of Oslo
Box 1045, Blindern N-0316
Norway

John Paul Schmit
Department of Plant Biology
University of Illinois at Champaign-Urbana
265 Morrill Hall
505 S Goodwin Avenue
Urbana, Illinois 61801
USA

Carol A. Shearer
Department of Plant Biology
University of Illinois at Champaign-Urbana
265 Morrill Hall
505 S Goodwin Avenue
Urbana, Illinois 61801
USA

Harrie J. M. Sipman
Freie Universität Berlin
ZE Botanischer Garten & Botanisches Museum
Königin-Luise-Strasse 6-8
D-14191 Berlin
Federal Republic of Germany

Joseph W. Spatafora
Department of Botany and Plant Pathology
Cordley Hall 2082
Oregon State University
Corvallis, Oregon 97331-2902
USA

Frederick W. Spiegel
Department of Biological Sciences
Science Center 601
University of Arkansas
Fayetteville, Arkansas 72701
USA

Steven L. Stephenson
Department of Biological Sciences
Science Center 632
University of Arkansas
Fayetteville, Arkansas 72701
USA

Jogeir N. Stokland
Norwegian Institute of Land Inventory
Raveien 9, N-1430 Ås
Norway

Jeffrey K. Stone
Department of Botany and Plant Pathology
Cordley Hall 2082
Oregon State University
Corvallis, Oregon 97331-2902
USA

Sidney L. Stürmer
Dept. de Ciências Naturais
Universidade Regional de Blumenau—FURB
Cx.P. 1507 Blumenau
89010-971 Santa Catarina
Brazil

Richard C. Summerbell
Centraalbureau voor Schimmelcultures
P.O. Box 273
3740 AG Baarn
The Netherlands

Brian C. Sutton
Apple Tree Cottage, Blackheath
Wenhaston
Suffolk IP 19 9HD
United Kingdom

Barbara M. Thiers
The New York Botanical Garden
200th Street at Southern Boulevard
Bronx, New York 10458-5126
USA

Greg Thorn
Department of Plant Sciences
University of Western Ontario
P.O. Box 3165, University Station

London, Ontario N6A 5B7
Canada

James M. Trappe
Department of Forest Science
Oregon State University
Corvallis, Oregon 97331-7501
USA

Loengrin Umaña
National Institute of Biodiversity (INBio)
Santo Domingo, Heredia
Costa Rica

Brigitte Volkmann-Kohlmeyer
Institute of Marine Sciences
University of North Carolina at Chapel Hill
Morehead City, North Carolina 28577
USA

Chun-Juan K. Wang
College of Environmental Sciences and Forestry
1 Forestry Drive
State University of New York at Syracuse
Syracuse, New York 13210-2788
USA

Alexander Weir
Faculty of Environment and Forest Biology
350 Illick Hall, 1 Forestry Drive
State University of New York at Syracuse
Syracuse, New York 13210-2788
USA

James F. White, Jr.
Department of Plant Pathology
Foran Hall, Cook College
Rutgers University
New Brunswick, New Jersey 08903
USA

Howard G. Wildman
Cerylid Biosciences
576 Swan Street

Richmond, Victoria 3121
Australia

Fiona A. Wilkinson
USGS Patuxent Wildlife Research Center
National Museum of Natural History
10th and Constitution, N.W.
Washington, D.C. 20560-0111
USA

Michael R. Willig
Department of Biological Sciences
Texas Technical University
P.O. Box 43131
Lubbock, Texas 79409-3131
USA

Susan Will-Wolf
Department of Botany, 317 Birge Hall
University of Wisconsin at Madison
430 Lincoln Drive
Madison, Wisconsin 53706-1381
USA

Qiuxin (Florence) Wu
AEMTEK, Inc.
46309 Warm Springs Boulevard
Suite A
Fremont, California 94539
USA

Daniel A. Wubah
College of Science and Mathematics
MSC 7502/112 Sheldon Hall
James Madison University
Harrisonburg, Virginia 22807
USA

John C. Zak
Ecology Program
Department of Biological Sciences
Texas Technical University
Lubbock, Texas 79409
USA

INTRODUCTION

GREGORY M. MUELLER AND GERALD F. BILLS

Fungi are among the most important organisms in the world, not only because of their vital roles in ecosystem functions but also because of their influence on humans and human-related activities. Fungi are essential to such crucial activities as decomposition, nutrient cycling, and nutrient transport and are indispensable for achieving sustainable development (Palm and Chapela 1998). Some species are important plant and animal pathogens; others form obligate mutualistic symbioses with sundry species of plants, algae, cyanobacteria, and animals. Fungi are also of great economic importance, having both positive and negative effects on human activities. They have been domesticated for use in the brewing, baking, industrial fermentation, pharmaceutical, and biotechnical industries, and other species are cultivated or collected for use as food. At the same time, fungi cause many millions of dollars in damage each year through food spoilage, destruction or degradation of materials used by humans, and diseases of plants and animals (including humans). Because of the diversity and abundance of fungi, and the vital roles they play in both natural and altered ecosystems as well as human-related activities, we believe that fungi must be included in considerations of biodiversity conservation, land-use planning and management, and related subjects. They frequently have been ignored, however, because of their perceived intractableness.

Fungi (Kingdom Fungi) and fungus-like groups (e.g., water molds, Kingdom Straminipila; slime molds and relatives, Kingdom Protista) encompass an astounding breadth of taxa, morphologies (ranging from amoebalike protists and single-celled aquatic chytridiomycetes to large basidiomycete mushrooms), ecologies, and life history strategies, yet only limited and incomplete information is available for most species. Current estimates of species numbers for fungi differ significantly (e.g., Hawksworth 1991; May 1991), but the 1.5 million species hypothesized by Hawksworth is a commonly used figure. If his estimate is correct, then less than 5%

of the fungi have been described. To further complicate matters, fungi tend to form symbiotic relationships with plants and animals, necessitating consideration of the plant or animal host when treating obligate plant mutualists, such as mycorrhizal fungi and endophytes, plant pathogens, and animal-associated fungi.

Although the diversity of fungi and fungus-like organisms is daunting, we believe that their variety, numbers, and importance mandate their inclusion in conservation/development dialogues and biodiversity projects. We also believe that they are tractable in a meaningful way. Participants at a workshop held in 1995 to explore the feasibility of preparing this book estimated that as many as 80% of the species in some species-rich groups of fungi (e.g., endophytes, soil fungi) are widely distributed and commonly encountered, leaving only 20% with more limited distributions. If that estimate is correct, then sites share enough species to make broad-scale inventory work possible yet harbor sufficient numbers of unique species to make valuable contributions to our understanding of fungal biodiversity and the ecological (e.g., competition and beneficial collaborations), evolutionary (e.g., speciation), and genetic processes of these fungi and their associated organisms.

The contributors to this volume have provided protocols for assessing the diversity of all groups of true fungi. They also have provided protocols for the phylogenetically unrelated but fungus-like organisms that traditionally have been studied by mycologists, to ensure their coverage in biodiversity surveys. Overall, taxa in three kingdoms comprising 11 phyla are treated (Fungi: Chytridiomycota, Zygomycota, Ascomycota, Basidiomycota; Straminipila: Oomycota, Hyphochytriomycota, Labyrinthulomycota; and Protista: Plasmodiophoromycota, Dictyosteliomycota, Acrasiomycota, Myxomycota). The incredible diversity of forms and habits represented among those taxa presented major challenges to the 88 contributing authors. Because of the skill and ingenuity with which they met those challenges, however, it is now possible to inventory fungal biodiversity and to include them in monitoring programs.

Although the contributors to this volume are the leading experts in their areas, the survey methods they propose here have not necessarily been tested. Techniques presented for some groups, such as the obvious and well-studied terrestrial macrofungi and lichens, have been vetted thoroughly, although there is always room for further refinement. For most groups, however, the authors have developed protocols based on their understandings of and experiences with the groups. Those protocols require additional testing and likely some modification. It is essential that workers fully document and justify the techniques that they use and, when possible, discuss their techniques in relation to the ones presented in this book. That will allow for evaluation of methods so that mycologists can designate standard protocols and embrace them. Standardized methods for each operational group of fungi are necessary to facilitate and promote quantitative sampling and comparisons of survey results across sites, times, and studies. Our hope is that the next edition of this book will demonstrate the achievement of that goal.

SCIENTIFIC CONTRIBUTIONS OF FUNGAL SURVEYS

The most basic data obtained from fungal surveys contribute to our knowledge of the numbers of species of fungi and fungus-like organisms that exist and their distributions and abundances across habitats and landscapes and around the globe. Such information constitutes the baseline against which to measure changes in the presence and abundances of species at particular sites in response to natural or, especially, human-induced (e.g., global warming; air, water, and soil pollution; forest fragmentation) environmental perturbations. During the past 20–30 years, for example, both species compositions and abundances of macrofungi at sites in Europe have changed markedly. Several previously common species are no longer encountered, and a number of other species have been designated as rare or endangered (Arnolds 1988, 1991; Pegler et al. 1993). Without baseline data, those changes could not have been detected.

Because of the pivotal roles played by fungi in all ecosystems, species composition and/or abundance often are used as indicators of ecosystem health and as indicators of the effects of alternative management and use schemes, pollution, and natural environmental perturbations on those ecosystems. Monitoring efforts may focus on entire fungal communities or on functionally or taxonomically related fungal groups. Fungi are a large and important component of the program monitoring the health, management, and use of old growth forests in the Pacific Northwest region of the United States (Forest Ecosystem Management Assessment Team [FEMAT] 1993). A resurgence of interest in determining the diversity of Chytridiomycota and the distributions of those species has followed the discovery that chytrids may be an important factor in the global decline of amphibian populations (Daszak et al. 1999, 2000; Longcore et al. 1999).

Rigorous fungal surveys also produce data that help to answer fundamental scientific questions being posed by the research community. Those questions include the following: What do fungal distribution data tell us about the effects that historical events such as glaciations, orogenies, and plate tectonics may have had on the distributions of plants, animals, and fungi? What levels of host, substratum, habitat, and other environmental specificities do fungi exhibit, and how and why? How do fungi influence the success or failure of land management practices?

Finally, inventories result in the accumulation of herbarium specimens and live fungal cultures as well as the discovery of many new species. Such materials serve as the foundation for many different studies by researchers in the systematics, ecological, physiological, genetic, and conservation communities.

DISSEMINATION OF RESULTS

Inventory results generally are reported in the scientific literature or in reports pertaining to particular development, management, or conservation projects. Such information also can be used in education programs appropriate for a diverse array of audiences. Information on the ecology and life histories of fungi as well as their roles in ecosystems' functions, for example, is valuable for traditional classroom teaching and can be incorporated into curricula at levels from primary school through university and into graduate training. Survey data also form a basis for the production of field guides for anyone with an interest in fungi (e.g., naturalists, ecotourists, citizen scientists, mushroom hunters looking for edibles) and can be used for presentations and programs at parks, recreation centers, and summer camps for lay audiences. It is also vital that data derived from surveys and monitoring programs reach citizens, land managers, and politicians in a format that will help them to make informed decisions about zoning, designation of parklands, farming and logging practices, and other activities under their jurisdictions.

Rossman and colleagues (1998) provided lists of potential economic, educational, and scientific products that they thought might be developed based on the results of a fungal survey of the Guanacaste Conservation Area in Costa Rica. Similar lists could be drawn up for most inventory projects. Any resulting products would increase the value of such studies.

Because fungal survey data are used in multiple ways by diverse audiences, information resulting from those studies must be provided not only in traditional scientific publications and reports but also in formats accessible to broad audiences. In addition to fully illustrated and colorful natural history articles and books and field guides, the use of electronic venues, such as CD-ROMs and websites, to disseminate data is imperative. Scientists also should be encouraged to interact with lay audiences through presentations and workshops. Finally, incorporating volunteers into field and laboratory projects is a good way to educate and excite citizens about fungi and to obtain excellent extra help for a project.

RESOURCES

The technical literature for identifying fungi is widely scattered and voluminous. An additional complication is the limited availability of many important works that are no longer in print. Collaboration with long-established libraries in North America and Europe can facilitate access to most of the out-of-print publications.

In addition to traditional printed sources of information, numerous resources are available over the Internet. Fungus-related websites are increasing rapidly in number and content, and new CD-ROMs, interactive identification keys, and simple field guides are being developed constantly. The Mycological Resources website (Appendix III) is particularly comprehensive. Easily accessible information banks that can be produced rapidly using new technologies are vital resources for the initiation and expansion of fungal inventory and monitoring projects. Such repositories should serve as models for management and dissemination of data from inventories of other taxa.

Material housed in herbaria are a critical source of information for systematists and also are an invaluable resource for all individuals and organizations interested in biodiversity conservation, land-use planning and management, and related topics. Collections maintained in herbaria and their associated data (some of them having originated more than 100 years ago) document species distribution patterns, fruiting phenologies, host associations, and species composition data for specific sites.

They also provide the critical comparative material essential for identification of newly collected specimens, for defining and understanding morphological variation within particular species, and for arriving at taxonomic decisions. Herbarium collections also provide materials for chemical and molecular analyses (e.g., Arugete et al. 1998; Mueller 1999). Herbaria contain the historical information necessary to examine changes in species compositions and abundances, fruiting patterns, host associations, and so forth caused by natural events and human activities.

ORGANIZATION OF THE VOLUME

This book is organized by technique and by functional ecological group rather than by taxonomic unit; no one section, for example, treats all of the ascomycetes. Rather, sections focus on fungi in freshwater habitats, fungi associated with insects, terrestrial macrofungi, and so forth; each section deals with a subset of ascomycetes, basidiomycetes, and other fungi and fungus-like organisms. We adopted this approach because the organisms covered in this volume fall into ecological groups whose species have very different life histories and require very different sampling techniques, independent of their taxonomy. For example, distinct protocols are necessary for sampling (collecting and identifying) each of the following groups: single-celled aquatic species, other aquatic fungi, soil-associated species, obligate root mutualists, endophytes, plant pathogens, lichens, wood-inhabiting macrofungi and microfungi, terrestrial macrofungi, and others. For that reason, fungal inventories typically are designed around operational groups based on habitat or substratum use, ecological habits, or whatever, rather than on strict taxonomic criteria. In addition, treatment by taxonomic group would lead to significant duplication of information. Protocols for sampling aquatic habitats, endophytes, leaf parasites, animal associates, wood inhabitants, and terrestrial forms, for example, would have to be presented in chapters on both ascomycetes and basidiomycetes as well as in chapters dealing with some other fungal groups.

The volume begins with an overview of fungal taxa, a consideration of the analysis and treatment of fungal biodiversity data, the use of molecular methods in inventory and monitoring, and information on preparation, storage, and use of herbarium specimens and live cultures. Those chapters should be read carefully by every investigator, regardless of the fungal group with which she or he works. The volume ends with appendices that provide sources of supplies, information on culture media, and so forth; a glossary relevant to all groups; and an extensive bibliography of publications related to fungal biodiversity.

The bulk of the book is devoted to sampling protocols (*sensu lato*). We envision that these chapters will be used differently in that different projects, or at least discrete teams within a project, will focus on fungi in only one, or a few, ecological groups. Consequently, different teams of specialists and parataxonomists participating in ongoing fungal inventories such as those being carried out as part of the Costa Rican National Biodiversity Inventory (Appendix III) and as part of the Great Smoky Mountains National Park All Taxa Biodiversity Inventory (Appendix III) mostly will use different chapters. To facilitate such usage, each chapter is designed to be somewhat autonomous. However, we have tried to present the protocols in a way that will make them accessible to anyone with university-level training so that they can be used by individuals conducting environmental impact studies or preliminary surveys of unknown areas, individuals preparing checklists for parks and nature areas, students, and other nonprofessional mycologists.

In addition to sampling protocols, chapters on fungal groups contain considerable information on their biology (e.g., life history, ecology, morphology). Such information is provided to assist investigators designing inventory programs to make informed decisions about temporal and spatial scales appropriate for sampling the particular target group. Scale is an important, primary consideration when developing sampling protocols (e.g., Colwell and Coddington 1994; see also "Spatial Scale of Biodiversity," in Chapter 5, this volume).

Fungi and their allies comprise one of the largest and least-known groups of organisms despite their ubiquity and importance to nature and to humans. As habitats are degraded and destroyed each year, hundreds or perhaps thousands of species are being lost even before they are known. We hope that this volume will stimulate and facilitate large-scale inventory and monitoring programs to detect, identify, and protect the biodiversity of this extremely heterogeneous and fascinating group of organisms.

I

GENERAL ISSUES

FUNGI AND THEIR ALLIES

MEREDITH BLACKWELL AND JOSEPH W. SPATAFORA

Fungi are heterotrophic organisms that permeate our environment. With few exceptions fungi have filamentous bodies enclosed by cell walls, are nonmotile, and reproduce both sexually and asexually by spores. During the last decade, mycologists have made unprecedented progress toward producing a phylogenetic classification of fungi; a skeleton phylogeny based on analyses of DNA characters was developed relatively early on (Bruns et al. 1991; Alexopoulos et al. 1996; McLaughlin et al. 2001). Such a phylogeny serves as a basis on which hypotheses of fungal evolution can be developed. One important finding has been that "fungi" are polyphyletic (Table 1.1); their morphologies are convergent, having been derived independently from among several independent eukaryotic lineages (Fig. 1.1). A monophyletic group, exclusive of slime molds and oomycetes, is well defined and supported as "true fungi," a kingdom-level taxon (Barr 1992; Bruns et al. 1993; Baldauf et al. 2000; Keeling et al. 2000).

Historically fungi have been compared with plants and included in the study of botany. Contemporary studies, however, indicate that members of Kingdom Fungi are most closely related to animals, not plants, possibly through a choanoflagellate-like ancestor (Barr 1992; Baldauf and Palmer 1993; Wainright et al. 1993; Ragan et al. 1996; Baldauf et al. 2000; Cavalier-Smith 2001). The phylogenetic positions of other groups once considered as fungi are not always well defined among eukaryotes (see "Kingdom Straminipila" and "Slime Molds," later in this chapter).

Thus far, fewer than 800 fungi or fungus-like taxa have been included together in any single phylogenetic reconstruction (Tehler et al. 2000), a number representing less than 1% of the 80,000 currently listed species (see Hawksworth et al. 1995). Estimates of up to 1.5 million species of fungi (six times more than the number of recognized species of land plants) are daunting; the missing fungi must be discovered, identified, and incorporated into the taxonomic framework. In addition to the relatively low numbers of taxa included in the trees, taxonomic coverage across higher taxa is limited. Until now only molecular characters from ribosomal RNA genes (rDNA) have been used widely in phylogenetic studies of fungi. The recent increase in use of other DNA regions and incorporation of phenotypic characters to resolve conflicting trees based on single genes and analytical approaches is encouraging.

Fungal classification, as opposed to phylogeny, is in a state of flux. Ideally, classifications should reflect phylogenetic relationships, but the relationships of all the groups covered in this volume have not been resolved nor have all of the groups been represented in analyses. Thus, the formal and complex classification systems that have been devised (e.g., Cavalier-Smith 2001) may vary in their inclusiveness. Systematists who study many types of organisms face this situation. Nevertheless, one reactionary suggestion called for a moratorium on publication of higher-level taxonomic systems until 2050 (Karatygin 1999). Hibbett and Donoghue (1998) suggested a more practical solution—the use of clade names rather than formal taxon designations. We use this approach, especially for higher taxa, in the following discussion.

Nowhere are examples of convergent evolution more evident than among organisms occupying the same habitat, which are grouped together for treatment in this volume. In this chapter we review the current state of knowledge of the phylogeny of true fungi (Kingdom Fungi) and fungus-like organisms placed in several additional distinct clades (Table 1.1). The two-volume work edited by McLaughlin et al. (2001) provides detailed information on fungi and fungus-like organisms; only a few of the chapters are cited here.

A BROAD VIEW OF EUKARYOTES

Previous studies based on small subunit rRNA genes (SSU rDNA) suggest that it may not be possible to reconstruct patterns of an ancient rapid radiation of the so-called "crown eukaryotes" (see "Kingdom Fungi,"

TABLE 1.1

Fungi and Fungus-like Organisms Previously Classified as Fungi*

Fungi
Chytridiomycota
Zygomycota
Ascomycota
Basidiomycota
Straminipila
Oomycota
Hyphochytriomycota
Labyrinthulales
Thraustochytriales
Slime molds
Plasmodiophorales
Myxomycota
Dictyosteliomycota
Acrasiomycota

* Taxonomic ranks of the clades are not indicated (see text).

later in this chapter), especially when based solely on characters from rDNA. Baldauf and colleagues (Baldauf et al. 2000), in contrast, used multiple genes and partitioning of data to develop a hypothesis of the relationships of the large diverse group of eukaryotes, which includes fungi and fungus-like organisms (Fig. 1.1). Their study differed from others in its approach because Baldauf and colleagues (2000) analyzed deduced amino acid sequences from four protein-coding regions (α-tubulin, β-tubulin, actin, and elongation factor 1-alpha [EF-1α]) in addition to rDNA. Consequently, they obtained better resolution of the deep branches in the tree (that is, those that represent the earliest divergences over a great period).

Many of the relationships proposed in Baldauf and colleagues' (2000) study were not entirely new; however, the large data set they used supported several controversial or conflicting tree branches, depending on their use of either rDNA or protein-coding genes. Relationships identified or strengthened in their phylogenetic analyses included the Fungi-Microsporidia link, the grouping of dictyostelid and plasmodial slime molds, and the remote basal position of the acrasid slime molds. In some instances, correspondence of non-molecular traits (e.g., flagellum type) and arrangement of mitochondrial cristae provided additional support. Taking a broad view of fungi and fungus-like organisms and using a wide variety of characters, Baldauf and colleagues (2000) derived the following relationships (Fig. 1.1):

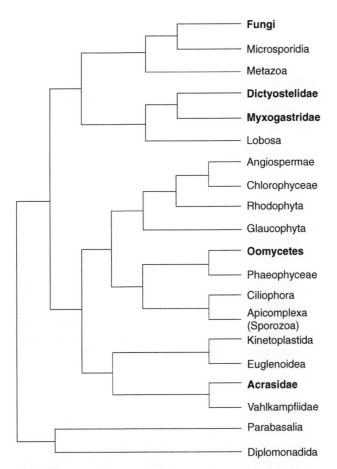

FIGURE 1.1 Phylogenetic positions of organisms studied by mycologists (*bold text*). The tree is unrooted, and all organisms are eukaryotic (i.e., they possess nuclei and other membrane-bounded organelles). The tree is based on phylogenetic analyses of sequences of up to four protein-coding (α-tubulin, β-tubulin, actin, and elongation factor 1-alpha [EF-1α]) regions in addition to rDNA. (Based on Baldauf et al. 2000.)

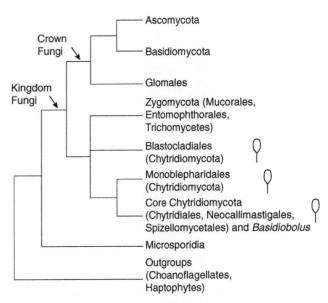

FIGURE 1.2 Members of the Kingdom Fungi showing Microsporidia as a basal lineage. Note the monophyletic grouping of the crown, or terrestrial, fungi (Glomales, Ascomycota, Basidiomycota). Resolution of the traditional Chytridiomycota (flagellated-cell symbol) is incomplete; some trees place *Basidiobolus* (Zygomycota) among the core chytrids, perhaps an artifact of trees based on rDNA analyses. The Blastocladiales may represent a lineage of flagellated fungi separate from the main body of chytrids. (Based on James et al. 2000; Keeling et al. 2000.)

- Kingdom Fungi, (traditionally the phyla Ascomycota, Basidiomycota, Zygomycota, and Chytridiomycota; see "Kingdom Fungi," later in this chapter) comprises a monophyletic group of "true" Fungi.
- Microsporidia are a sister group to Fungi, and these two taxa form a monophyletic grouping with the Metazoa (animals) linked by a choanoflagellate-like common ancestor.
- Although still scantily sampled, dictyostelid and myxomycete (myxogastrid) slime molds form a monophyletic group that, together with certain amoebae, is the sister group to Fungi, Microsporidia, and Metazoa.
- Oomycetes and other groups with a heterokont or derivative flagellar condition (e.g., hyphochytrids, brown algae, diatoms, chrysophytes) form a sister

clade to the clade made up of Ciliophora (ciliates) and Apicomplexa (sporozoa).
- Myxomycete and dictyostelid slime molds may comprise a monophyletic group; however, broader taxon sampling is needed.
- Acrasid slime molds occur in a clade with predaceous vahlkampfiid amoebae, remote from other slime-mold groups.
- One other group of fungus-like organisms, Plasmodiophorales, has been placed in a clade of protozoans near alveolates (Castlebury and Domier 1998).

KINGDOM FUNGI

Historically the monophyletic Fungi have been classified in four phyla: Chytridiomycota, Zygomycota, Basidiomycota, and Ascomycota (Barr 1992; Hawksworth et al. 1995; Alexopoulos et al. 1996); however, current studies have shown that such a simple classification does not represent the phylogeny of those organisms accurately (Fig. 1.2). The phyla Chytridiomycota and Zygomycota do not emerge as monophyletic groups in modern analyses, and the two groups intergrade at several points based on analyses of SSU rDNA (Nagahama et al. 1995; James et al. 2000). The traditional chytrids bear flagellated cells (zoospores or gametes) at some stage in their life cycles, which is a plesiomorphic or ancestral trait. Phylogenetic analyses indicate that some chytrid lineage occupies the most basal branch of Kingdom Fungi—a finding consistent with a choanoflagellate-like ancestor. Current information indicates that the grouping of 800 species of Chytridiomycota, largely defined by the ancestral character state of a smooth posterior flagellum, is not monophyletic.

Zygomycota, a group of more than 1000 species, is defined by the presence of a meiospore, known as the zygospore, and the absence of a flagellum. As defined, the zygomycetes are polyphyletic. The genus *Basidiobolus* is considered to have been derived from within the core chytrids in trees based on rDNA (Nagahama et al. 1995; James et al. 2000; O'Donnell et al. 2001). In trees based on the analysis of protein coding genes, however, *Basidiobolus* is retained in the zygomycetes (see "Zygomycota," later in this chapter; Keeling et al. 2000). A second traditionally Zygomycete group, the Glomales, is a sister clade to the Ascomycota/Basidiomycota clade (Simon et al. 1993; Nagahama et al. 1995; Gehrig et al. 1996; Redecker et al. 2000b; Platt and Spatafora 2000; Gernandt et al. 2001), and together the three groups constitute the monophyletic "crown" fungi. The crown-fungus clade represents the most derived, higher monophyletic grouping within the Kingdom Fungi. The timing of the appearance of this clade on Earth probably coincided with the origin and diversification of land plants (Simon et al. 1993; Blackwell 2000; Redecker et al. 2000a; Berbee and Taylor 2001). An estimated 80% of all plant species are associated with Glomales, indicating the importance of these arbuscular mycorrhiza-forming fungi for life on land. The Ascomycota and Basidiomycota are the two largest phyla of the Kingdom Fungi and together comprise more than 95% of all known fungal taxa (Hawksworth et al. 1995). Today the crown groups predominate in terrestrial environments and make life on Earth possible.

Recent studies of the true Fungi indicate the following:

- The Chytridiomycota is not monophyletic, although some clade of the group is the basal member of the Kingdom Fungi.
- A monophyletic clade of core zygomycetes does not include Glomales or, perhaps, *Basidiobolus* species.
- Ascomycota and Basidiomycota are monophyletic sister taxa.
- Glomales, Ascomycota, and Basidiomycota comprise a large monophyletic clade of crown fungi.

PHYLUM CHYTRIDIOMYCOTA (ZOOSPORIC FUNGI)

As mentioned earlier, the Chytridiomycota (Fig. 1.2) has been defined traditionally on the basis of the presence of a single posteriorly inserted smooth flagellum (Barr 2001). At the time molecular methods came into use, some mycologists doubted that the Chytridiomycota were true fungi, but their inclusion in that group was confirmed based on rDNA evidence (Förster et al. 1990; Bowman et al. 1992; Bruns et al. 1993). While subsequent studies, including those with increased taxon sampling, have shown that the Chytridiomycota are not monophyletic, chytrid groups are basal in phylogenetic trees (James et al. 2000; Tanabe et al. 2000).

Barr and Desaulniers (1988) established five orders of chytrids (Blastocladiales, Monoblepharidales, Neocallimastigales, Spizellomycetales, and Chytridiales) based on the ultrastructure of zoospores. Their implicit evolutionary hypothesis is congruent with phylogenies based on mitochondrial DNA (Paquin et al. 1997). The relationships among clades were tested in 2000 using SSU rDNA characters (James et al. 2000). That phylogenetic analysis distinguished four monophyletic clades, consisting of the orders Blastocladiales, Monoblepharidales, Neocallimastigales, and Spizellomycetales, and four small independent groups of chytrids, results consistent with the groupings based on zoospore ultrastructure. The relationships among the orders and chytrid clades, however, were not resolved.

Some of the chytrids with unique zoospore morphologies have not been classified in previously existing orders; several of those, in fact, including *Batrachochytrium dendrobatidis*, the chytrid associated with amphibian decline, do not fall within any of the groups defined on the basis of DNA characters. James and his colleagues (2000) predicted that additional chytrid diversity will be uncovered when more chytrids with unique zoospore types are included in phylogenetic analyses. Based on data presently available, we can conclude the following:

- The orders of Chytridiomycota recognized on the basis of zoospore morphology generally are well supported by molecular characters, although the relationships among the individual groups have not been resolved.
- A core group of Chytridiomycota forms a clade.
- The best sampled members of the phylum, the Chytridiales, can be grouped with rDNA; the clades are consistent with zoospore ultrastructure.
- The Blastocladiales may represent a lineage independent of the other orders of Chytridiomycota.

PHYLUM ZYGOMYCOTA

Zygomycetes comprise a monophyletic group if the Glomales, and perhaps *Basidiobolus ranarum* (Entomophthorales), are excluded. *Basidiobolus ranarum* commonly is found in the same terrestrial habitats with insects and amphibians. Although it is not flagellated, this species has been placed within the core chytrid group in analyses based on rDNA evidence (Nagahama et al. 1995; Jensen et al. 1998). In trees based on analyses of α- and β-tubulin genes, however, *B. ranarum* falls within the zygomycetes (Fig. 1.2; Keeling et al. 2000). The discrepancy may result from an accelerated rate of sequence evolution in zygomycete SSU rDNA compared to other fungi and the resulting long-branch attraction in phylogenetic analyses using that gene (James et al. 2000). Unlike members of the Entomophthorales, *Basidiobolus* has a distinctive spindle-pole body (nucleus-associated organelle) with a microtubular structure reminiscent of centrioles, which are known only among chytrids in the fungi (McKerracher and Heath 1985; Alexopoulos et al. 1996; Benny et al. 2001). In addition, species of *Basidiobolus* have a rocket-like forcible spore-release mechanism that differs from the cannon-like mechanisms of the remaining species of Entomophthorales. However, most species of *Basidiobolus* share capilliconidia (distinctive animal-dispersed diaspores) with certain species in the Entomophthorales. The *Basidiobolus* example highlights problems systematists encounter when trees based on different characters, in this case genetic and morphological ones, conflict.

Several traditional zygomycete lineages are considered to be monophyletic. Those include Entomophthorales, Trichomycetes (Harpellales, excluding Amoebidiales), Kickxellales (excluding *Spiromyces*), Mucorales, Mortierellales, Dimargaritales, and Zoopagales (James et al. 2000; Tanabe et al. 2000; Benny et al. 2001; O'Donnell et al. 2001). The relationships among the monophyletic lineages, however, have not been well supported in most cases (Fig. 1.3).

One of the few relationships that is well supported, based on analysis of SSU rDNA, is that between the Harpellales and Kickxellales. Those groups also share the ultrastructural feature of plugged, flared septal pores. A suggestion that the Harpellales were derived from the Kickxellales was rejected by O'Donnell and colleagues (1998). The enlarged Harpellales-Kickxellales clade also contains Zoopagales and *Spiromyces*; most species of both of these groups are parasites of fungi or invertebrate animals (Tanabe et al. 2000).

O'Donnell and colleagues (2001) included a greater number of species and greater taxon diversity (54 genera) in their phylogenetic analyses of Mucorales. They also analyzed a greatly expanded molecular data set of SSU

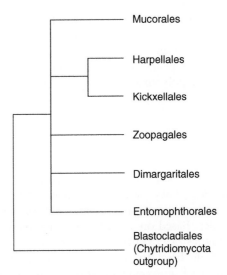

FIGURE 1.3 Major clades of the phylum Zygomycota, excluding the Glomales, based on analyses of nuclear rDNA and elongation factor 1-alpha (EF 1-α) data. The terminal clades (orders) are generally well supported, but their interrelationships are poorly resolved. As a group, they represent the loss of the flagellated stage, possibly from a blastocladialean ancestor. (Based on O'Donnell et al. 1998, 2001; Benny et al. 2001.)

rDNA, large subunit rDNA (LSU rDNA), and EF-1α gene exons and 11 morphological traits. Their new tree for the Mucorales is at odds with the phylogenetic hypotheses based on morphology. It supports a monophyletic grouping of Mucorales and identifies several members of the Mortierellales as a basal outgroup. The striking result of the study, however, is the lack of support for the traditionally recognized families of the order. In particular the larger families Mucoraceae, Thamnidiaceae, and Pilobolaceae were deemed polyphyletic.

Only the Harpellales and the Amoebidiales, the two groups of Trichomycetes that have been cultured, have been included in phylogenetic studies. The Amoebidiales has been determined to represent a protozoan lineage and thus has been excluded from the Trichomycetes (Benny and O'Donnell 2000; Benny 2001). Our present understanding of the phylum Zygomycota suggests the following:

- It includes several monophyletic groups, including Entomophthorales, Harpellales, Kickxellales (excluding *Spiromyces*), Mucorales, Mortierellales, Dimargaritales, and Zoopagales.
- Amoebidiales is not related to harpellalean trichomycetes; it groups with protozoans.
- Harpellales and Kickxellales are sister taxa.
- Members of the Mortierellaceae are a basal sister group of the Mucorales.

- Some of the established family-level taxa in Mucorales are artificial, including Mucoraceae, Thamnidiaceae, and Pilobolaceae.
- Several large genera of the Mucorales (e.g., *Mucor*, *Absida*) are polyphyletic.

CLADE GLOMALES

The Glomales comprises fungi with an arbuscular mycorrhizal (AM) habit, and as mentioned previously, the group is associated with an estimated 80% of the world's plant species. Based on an early molecular study, Simon and colleagues (1993) placed the Glomales in a monophyletic clade basal to the Ascomycota-Basidiomycota clade (Fig. 1.4), a placement that has been supported by subsequent studies (Nagahama et al. 1995; Gehrig et al. 1996; Redecker et al. 2000b). Glomales includes two major lineages, Glomaceae and its sister group Gigasporaceae (Redecker et al. 2000b). It is surprising that two previously unrecognized families, Archaeosporaceae and Paraglomaceae (Redecker et al. 2000b; Morton and Redecker 2001), despite sharing many superficial morphological similarities with Glomales, have evolved from outside of the Glomalean lineages. Differences in mycorrhizal morphology, fatty acid profiles, and immunological reactions against monoclonal antibodies all agree with the molecular evidence. Several investigators (Redecker et al. 2000b; Morton and Redecker 2001) have suggested that *Geosiphon*, a symbiont of cyanobacteria, may be derived from within, rather than being

ancestral to, the AM fungi as Gehrig and colleagues (1996) suggested.

- The Glomales are best understood as a group of terrestrial crown fungi.
- Earlier studies of morphological characters did not provide an accurate view of the genetic diversity of the group.
- The ancestor of the Glomales and the other crown fungi among the more basal lineages of Zygomycota and Chytridiomycota is yet to be identified.

PHYLUM ASCOMYCOTA

The Ascomycota is the largest phylum of the Kingdom Fungi, with approximately 32,000 species (Hawksworth et al. 1995). It is characterized by the production of meiospores (ascospores) within sac-shaped cells (asci) and includes many commonly encountered fungi. Some of those species have had major impacts on human civilization as food (e.g., *Saccharomyces cerevisiae*, a yeast), medicine (e.g., *Penicillium chrysogenum*, the source of penicillin), and disease-causing microbes (e.g., *Pneumocystis jiroveci*, an agent of pneumonia).

Three major groups or classes of Ascomycota, including Euascomycetes (mostly filamentous, sporocarp-producing as well as mitosporic or conidial forms), Saccharomycetes (the true yeasts), and Archiascomycetes (a paraphyletic assemblage of basal taxa) generally are recognized (Nishida and Sugiyama 1994; Taylor et al. 1994, but see Eriksson and colleagues 2003 for an alternative classification). Relationships of the major groups within each of those classes can be characterized best as tenuous because most are represented by well-supported terminal clades (Fig. 1.5) that are linked by poorly supported basal nodes (Berbee and Taylor 1993b; Gargas et al. 1995; Spatafora 1995; Tehler et al. 2000).

CLASS ARCHIASCOMYCETES

The Archiascomycetes is recognized based on phylogenetic analyses of rDNA sequence data (Nishida and Sugiyama 1993, 1994; Taylor et al. 1994) and includes yeastlike, filamentous, and possibly sporocarp-producing species. The largest order of the Archiascomycetes is the Taphrinales, which includes approximately 100 species of plant-pathogenic fungi (Kramer 1987). Taphrinales lack both sporocarps and ascogenous hyphae. Species in the order are dimorphic, having a saprobic yeast phase and a filamentous pathogenic phase; that life-history trait is shared with certain basidiomycetes (e.g., rusts), which has led some mycologists to hypothesize that the order

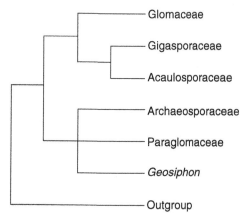

FIGURE 1.4 Major clades of the arbuscular-mycorrhizae–forming Glomales, including the traditional families Glomaceae, Gigasporaceae, and Acaulosporaceae, as well as the newly discovered lineages of Archaeosporaceae and Paraglomaceae and the *Nostoc*-symbiont *Geosiphon*. The ancestor of the Glomales and the other crown fungi (see Fig. 1.2) is yet to be identified among the Zygomycota and Chytridiomycota. (Based on Redecker et al. 2000b.)

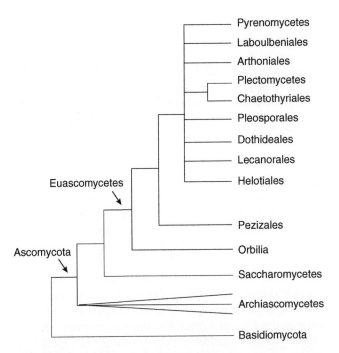

FIGURE 1.5 Major clades of the phylum Ascomycota, including the three classes Archiascomycetes, Saccharomycetes, and Euascomycetes, and the major lineages of the Euascomycetes polytomy, from analyses of nuclear rDNA and RNA polymerase II. (Based on Berbee and Taylor 1993b; Nishida and Sugiyama 1994; Gargas et al. 1995; Spatafora 1995; Berbee 1996; Landvik 1996; Liu et al. 1999; Berbee et al. 2000; Platt 2000; Tehler et al. 2000; Gernandt et al. 2001.)

was the primitive lineage of the Ascomycota (Savile 1968). The Archiascomycetes includes only a few other genera, which are exemplified by *Pneumocystis jiroveci*, the causal agent of pneumocystis pneumonia; species of *Schizosaccharomyces*, the fission yeasts; and *Saitoella complicata*, an asexual soil-dwelling yeast. Additional evidence suggests that the ascoma-producing genus *Neolecta* is also a member of the Archiascomycetes (Landvik 1996). *Neolecta* produces stipitate, tonguelike sporocarps with asci arranged in a hymenial layer that lacks sterile cells (paraphyses). As previously mentioned, the Archiascomycetes is probably not monophyletic but likely encompasses a series of basal lineages in phylogenetic trees based on nuclear SSU rDNA. Studies involving multiple independent loci are needed to test the validity of the grouping.

CLASS SACCHAROMYCETES

Most fungi that biologists consider to be "true yeasts" are members of the Saccharomycetes. Species of the class can be found in virtually all environments and habitats. They occupy a paramount position in food production for humans (e.g., *Saccharomyces cerevisiae*, bakers' and

brewers' yeast) and associate with mammals as pathogens (e.g., *Candida albicans*, a facultative human pathogen) and with various animal groups as mutualists (e.g., numerous species of endosymbionts of arthropods; Kurtzman and Fell 1998). The Saccharomycetes is a monophyletic clade (Fig. 1.5) that probably shares a most recent common ancestor with the Euascomyetes (Berbee and Taylor 1993b; Kurtzman and Robnett 1995). No true yeast produces a sporocarp, and in general the class is comprised of fungi that lack ascogenous hyphae but whose life histories include a budding growth phase. Those generalizations do not apply to numerous species that display filamentous growth (e.g., *Ascoidea* species; Batra 1987) and species of *Cephaloascus* that produce ascophores, which have been interpreted as ascogenous hyphae (von Arx and van der Walt 1987). Members of the group also share a number of ultrastructural features, including details of nuclear division and ascospore delimitation.

CLASS EUASCOMYCETES

The Euascomycetes is the largest class of Ascomycota and includes the major lineages of filamentous, sporocarp-producing taxa and their equally diverse asexual relatives (anamorphs). It is arguably the most successful group of fungi, including parasites, pathogens, and mutualists of plants, algae, and animals, as well as saprobes able to decompose virtually all known organic substrata. Mycologists have relied heavily on the morphologies of sporocarps (ascomata) and asci as bases for traditional classifications (Nannfeldt 1932; Seaver 1942, 1951; Luttrell 1951, 1955; Ainsworth et al. 1973b). More recent molecular studies have shown that such groups are not natural. Rather, morphological traits of the ascomata repeatedly have been lost and gained during the evolutionary history of the group. Convergent evolution in ascus morphology is also apparently common, especially among modes of dehiscence (Berbee and Taylor 1992a; Spatafora and Blackwell 1994a, 1994b; Blackwell 1994). As with many higher groups of fungi, the Euascomycetes is characterized by numerous well-supported terminal clades (Fig. 1.5), although the relationships among those clades cannot be resolved confidently with current data and analyses. Several authors have postulated that the poorly resolved base of the euascomycete clade may represent a radiation event in which many of the major clades of the class originated over a relatively short period (Berbee and Taylor 1993a; Spatafora 1995; Berbee et al. 2000). However, their hypotheses are based on analyses of the single nuclear SSU rDNA gene, which led to a similar hypothesis for the major groups of eukaryotes (Patterson and Sogin 1992). That latter hypothesis proved erroneous or at least resolvable by analyses involving numerous independent loci (Baldauf et al. 2000).

The nomenclature for the major groups within the Euascomycetes is in a state of flux. Various classifications have been proposed (Eriksson et al. 2003) with the ultimate goal of accurately reflecting monophyletic clades within that group. The names used here represent a composite of formal and informal designators that refer to major groups (Alexopoulos et al. 1996). The primitive sporocarp morphology of the Euascomycetes appears to be that of the apothecium (Berbee and Taylor 1993b; Gargas et al. 1995; Tehler et al. 2000; Platt and Spatafora 2000; Gernandt et al. 2001), which is characterized by an exposed fertile layer of asci (hymenium). The most basal lineage of the apothecial fungi (Fig. 1.5), and of the Euascomycetes, includes the genus *Orbilia*, a poorly known clade of nematophagous ascomycetes (Pfister 1997; Platt 2000). The Pezizales is the next basal lineage of the class (Berbee and Taylor 1993b; Gargas et al. 1995; Platt 2000). It also may be the best-known and largest group of apothecial fungi and includes numerous macroscopic forest species (e.g., *Morchella* species, *Gyromitra* species, *Helvella* species). The Pezizales is characterized by operculate asci. The monophyly of the order is not strongly supported, however, and numerous familial revisions have been proposed (O'Donnell et al. 1996). Moreover, the group includes numerous, independent lineages of truffles (O'Donnell et al. 1996) and, as an order, probably contains the majority of ectomycorrhizal species of ascomycetes. In most phylogenetic analyses of the SSU rDNA, the remainder of the euascomycetes form a large polytomy (Berbee et al. 2000). Analyses of nucleotide and amino acid data from RNA polymerase II along with other nucleotide sequence data (e.g., β-tubulin, EF-1α) hold great promise for resolving the components of that polytomy (Liu et al. 1999; Baldauf et al. 2000; O'Donnell et al. 2001).

The Helotiales is another major group of apothecial fungi. The group includes endophytes, mycorrhizal fungi, plant pathogens, and saprobes of litter and woody debris. Those fungi often are referred to as the inoperculate discomycetes because their asci lack the operculum of the Pezizales. The general morphology of the apothecium varies from cup-shaped to spathulate to hysteriform. The order is grossly polyphyletic with representatives found scattered throughout the basal Euascomycetes (Holst-Jensen et al. 1997; Platt 2000; Gernandt et al. 2001). The name Leotiales also has been used as the valid name for the order, but it has been replaced by Helotiales (Korf and Lizon 2000). Regardless of its nomenclature, the order is not monophyletic and will require significant phylogenetic and monographic research before an accurate classification can be devised.

The other major group of apothecial ascomycetes includes the lichenized species of the Lecanorales *sensu lato* (Lecanorales, Peltigerales, and related taxa). This is a large group of fungi, including more than 7000 species, and in many ways it epitomizes the modern view of symbiosis. The phylogeny of lichenized ascomycetes has been studied less than that of nonlichenized species, and the integration of the two groups in phylogenetics and classifications is relatively recent. In analyses of nuclear rDNA, two main groups of lichenized ascomycetes are resolved (Gargas et al. 1995; Lutzoni et al. 2002): One contains the apothecial forms of Lecanorales *s.l.*, and the second clade comprises the Arthoniales, which is possibly more closely related to the Pleosporales and pyrenomycetes than it is to the Lecanorales. The work of Lutzoni and colleagues (2002) supports the hypotheses that the gain of lichenization occurred early and infrequently during the evolution of the Euascomycetes and that some major lineages of nonlichenized ascomycetes, such as the Plectomycetes and the Chaetothyriales, may be derived from the loss of lichenization. The loss of lichenization was not predicted from previous phylogenetic studies and represents a novel hypothesis for the evolution of the Ascomycota.

Loculoascomycetes is a loose term that refers to all taxa that produce ascogenous hyphae in preformed locules within a stroma, a process known as ascostromatic development (Luttrell 1955). Many of those species possess bitunicate asci that display a "jack-in-the-box" mode of dehiscence (Reynolds 1989). Different classifications have led to the recognition of more than 10 orders in this group (M. E. Barr 1979, 1987; Alexopoulos et al. 1996). Here we focus on three orders, Pleosporales, Dothideales, and Chaetothyriales, which include the majority of the ascostromatic species. Most molecular phylogenetic analyses have been performed with these taxa. The Pleosporales is supported as a monophyletic clade (Berbee 1996), consistent with the synapomorphy of sterile cells (pseudoparaphyses) interspersed among the asci in members of the group (Luttrell 1955, 1965).

The Dothideales is defined largely by a lack of paraphysoids (Luttrell 1955, 1965). Its position as a monophyletic group is weakly supported and is sensitive to taxon sampling (Berbee 1996). The Chaetothyriales is characterized by evanescent apical pseudoparaphyses (M. E. Barr 1979, 1987) and represents a monophyletic group. It is not supported as a member of the Dothideales (Winka et al. 1998); rather, it is related more closely to the plectomycetes based on analyses of both SSU rDNA (Spatafora 1995) and chitin synthetases (Bowen et al. 1992). Inclusion of all three orders in a monophyletic Loculoascomycetes is not supported by the data, which suggest at least two independent origins of ascostromatic development (Spatafora 1995; Berbee 1996; Winka et al. 1998).

Plectomycetes (Eurotiomycetes of Eriksson and Winka 1998) is a term with a long and confused nomenclatural history (Luttrell 1951; Fennell 1973; Berbee and Taylor 1992b; Geiser and LoBuglio 2001). Here we use it to denote the clade of Eurotiales and Onygenales, which contain the majority of taxa that have been considered as plectomycetes. The Eurotiales and Onygenales include numerous species that produce closed ascomata (cleistothecia) and relatively simple, evanescent asci (protunicate asci). The clade also encompasses numerous medically important anamorphic fungi that are both the source of life-saving compounds (e.g., penicillin isolated from *Penicillium chrysogenum*) and life-threatening diseases (e.g., valley fever caused by *Coccidioides immitis*). Phylogenetic analyses of nuclear SSU rDNA (Berbee and Taylor 1992b), chitin synthetases (Bowen et al. 1992), and RNA polymerase II (Liu et al. 1999) provide strong support for a Eurotiales/Onygenales relationship. Data also support inclusion of *Elaphomyces*, a genus of ectomycorrhizal truffles, as a member of the plectomycete clade, which would make it the only ascomycete truffle lineage known outside of the Pezizales (Geiser and LoBuglio 2001).

Pyrenomycetes (Sordariomycetes of Eriksson and Winka 1998) is a class name formerly used to refer to a group of species with a particular shared morphology. It now is used to designate a clade that includes the orders Diaporthales, Halosphaeriales, Hypocreales, Lulworthiales, Microascales, Ophiostomatales, Phyllachorales, Sordariales, and Xylariales (Berbee and Taylor 1992a, 1992b; Hausner et al. 1992; Spatafora and Blackwell 1993, 1994a, 1994b; Spatafora et al. 1998; Kohlmeyer et al. 2000). Although the taxa in the clade encompass a wide range of macromorphologies and micromorphologies, the vast majority have a flask-shaped ascomata or perithecia. The Pyrenomycetes clade also includes numerous lineages of cleistothecial fungi, which represent multiple and repeated losses of the perithecial neck and ostiole (Berbee and Taylor 1992a; Spatafora and Blackwell 1994a; Suh and Blackwell 1999), and yeastlike endosymbionts (Suh et al. 2001). Although almost all perithecial ascomycetes are included in the pyrenomycetes clade, there are a few exceptions. *Pyxidiophora* species and the Laboulbeniales are arthropod-associated ascomycetes comprising a separate clade based on SSU rDNA analyses (Blackwell 1994). The Laboulbeniales are a derived group of fungi that produce unusual reproductive structures (Thaxter 1924, 1926, 1932; Tavares 1985, Blackwell 1994). *Pyxidiophora* species produce more standard perithecia that are morphologically similar to those of the Ophiostomatales (Blackwell 1994). These results are consistent with the hypotheses that long-necked perithecia have evolved repeatedly in the ascomycetes in conjunction with arthropod dispersal of ascospores (Cain 1972; Malloch and Blackwell 1993b; Blackwell 1994).

In summary, our present knowledge of the Ascomycota supports the following conclusions:

- There are three major groups of ascomycetes: the paraphyletic grouping of yeasts and yeastlike fungi based largely on characteristics of rDNA (Archiascomycetes), the true yeasts (Saccharomycetes), and the filamentous or ascohymenial ascomycetes (Euascomycetes).
- Many monophyletic lineages at about the level of order have been discerned among the Euascomycetes, but they do not correspond with the traditional groupings of plectomycetes, pyrenomycetes, discomycetes, and loculoascomycetes based on sporocarp morphology.
- Relationships among the monophyletic groups of Euascomycetes have not been resolved in most cases.
- Convergent morphological features are common throughout the Euascomycetes, which has led to the formal description of a number of polyphyletic taxa.
- Lichenization occurred early during the evolution of the Euascomycetes, and some lineages of nonlichenized ascomycetes may be derived via the loss of lichenization.

PHYLUM BASIDIOMYCOTA

The Basidiomycota is the second largest phylum of Kingdom Fungi, with approximately 23,000 species (Hawksworth et al. 1995), including many of the common macroscopic forest fungi (e.g., mushrooms, shelf fungi). The group is characterized by having meiospores (basidiospores) that are produced on club-shaped cells or basidia. The morphology of the basidium has played a central role in past classifications of the Basidiomycota, with fungi possessing septate basidia assigned to the Phragmobasidiomycetes (Heterobasidiomycetes) and fungi with nonseptate basidia classified in the Homobasidiomycetes (Holobasidiomycetes) (Patouillard 1900; Donk 1972, reviewed in Swann and Taylor 1993). Recent phylogenetic studies have shown that the Phragmobasidiomycetes is not a monophyletic taxon and that the septate basidium is probably an ancestral character state for the Basidiomycota (Swann and Taylor 1993, 1995b; Wells 1994; Hibbett and Thorn 2001). Today the Basidiomycota generally is considered to include the classes Urediniomycetes (rusts and relatives), Ustilagniomycetes (smuts), and Hymenomycetes (mushrooms and relatives) (Fig. 1.6), with molecular studies indicating a greater degree of convergent evolution among macromorphological and micromorphological traits than previously appreciated (Wells 1994; Swann and Taylor 1995a, 1995b). The relationships of the three classes within the Basidiomycota are controversial, as are

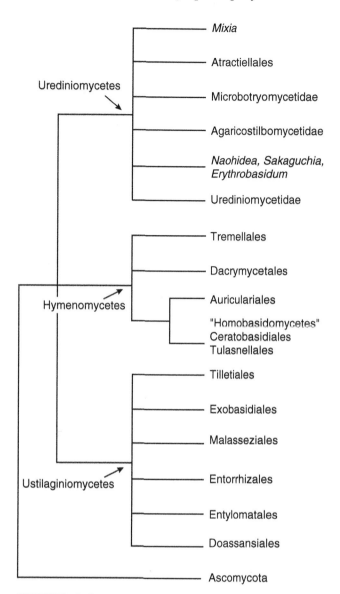

FIGURE 1.6 Major clades of the phylum Basidiomycota, including the classes Ustilaginiomycetes, Urediniomycetes, and Hymenomycetes and their respective major subclass clades based on analyses of small subunit (SSU) and large subunit (LSU) rDNA sequences. (Based on Labyrinthulales and Thraustochytriales, Hibbett and Thorn 2001; Swann et al. 2001; D. S. Hibbett, unpublished data.)

many of the ordinal relationships within each class (McLaughlin et al. 1995; Swann and Taylor 1995b; Swann et al. 1999; Moncalvo et al. 2000; Hibbett and Thorn 2001).

CLASS USTILAGINIOMYCETES

Fungi of the Ustilaginiomycetes are characterized by the production of teliospores (diploid overwintering spores)

and dimorphic life histories that include a saprobic yeast state and a pathogenic filamentous state. Because of certain morphological similarities, members of this group often have been confused with some members of the Urediniomycetes (see "Urediniomycetes," later in this chapter), which has led to controversial and conflicting classifications (reviewed in Swann and Taylor 1993). Traditional classifications placed the two groups in the Teliomycetes, based on possession of the teliospore (Talbot 1968), or in the Phragmobasidiomycetes, based largely on the presence of septate basidia (Lowy 1968; Donk 1972). Phylogenetic analyses of molecular characters do not support monophyly of either group (Swann and Taylor 1993; Swann et al. 1999, 2001). Furthermore, the teliospore and the overall "smut" morphology (i.e., basidiospores produced in a darkly pigmented sooty mass) are other examples of convergent evolution in morphology among fungi (Swann et al. 1999). Some classifications have considered the Ustilaginiomycetes to represent a distinct phylum of fungi, the Ustomycota (Moore 1996), whereas others have included a broad grouping of smutlike taxa in the Ustomycetes of the Basidiomycota (Hawksworth et al. 1995). Current studies support designation of the Ustilaginiomycetes as a member of the Basidiomycota, rather than as a separate phylum (Swann and Taylor 1993, 1995a). Ultrastructural studies and molecular phylogenetic analyses support the hypothesis that Ustilagniomycetes includes three major clades, or subclasses, each containing one to several orders: Entorrhizomycetidae, including Entorrhizales; Ustilaginomycetidae, including Ustilaginales and Urocystales; and Exobasidiomycetidae, including Doassansiales, Entylomatales, Exobasidiales, Georgefisherales, Graphiolales, Malasseziales, Microstromatales, and Tilletiales (Bauer et al. 1997, 2001). Numerous taxa that once were assigned to the Ustilaginales clade (e.g., *Microbotryum* species) now have been demonstrated convincingly to be derived members of the Urediniomycetes (Swann et al. 1999, 2001).

CLASS UREDINIOMYCETES

The Urediniomycetes is a large group of dimorphic, yeastlike fungi that includes the subclasses Urediniomycetidae (encompassing the orders Uredinales and Septobasidiales), Microbotryomycetidae, and Agaricostilbomycetidae, as well as the Atractiellales, *Erythrobasidium*-clade, and the genus *Mixia* (Swann et al. 1999, 2001; Fell et al. 2000). Filamentous Urediniomycetes all have septate mycelia with simple pores that are neither surrounded by flared cell walls (dolipore septum) nor associated with membranes (parenthesomes) (Swann and Taylor 1993).

The Uredinales contains plant pathogens that display some of the more complex life cycles among fungi. In the most extreme examples the fungi are heteroecious, requiring two distantly related hosts to complete their life cycles, and macrocyclic, possessing up to five distinct spore-producing states. More derived autoecious species with a reduced number of distinct stages also exist. The Septobasidiales are of interest because of the extreme specialization of the group on scale insects. The Microbotryomycetidae is a heterogenous group of species that is recognized mostly by molecular characters and includes smutlike species of *Microbotryum* and yeastlike species of *Rhodotorula*, *Rhodosporidium*, and *Sporidiobolus* (Swann et al. 1999; Fell et al. 2000). Basidiomycetous yeasts, however, are a polyphyletic group; the order Sporidiales (Moore 1980), for example, includes members of no fewer than three clades of the Urediniomycetes, including the *Erythrobasidium* and *Agaricostilbum* clades (Fell et al. 2000). The genus *Mixia* is a fern parasite and is probably the most enigmatic member of the Urediniomycetes; it previously was classified in the Taphrinales of the Ascomycota (Nishida et al. 1995).

CLASS HYMENOMYCETES

The Hymenomycetes consists of the fleshy forest fungi (e.g., mushrooms, jelly fungi, shelf fungi) with which biologists, naturalists, and nonmycologists are most familiar. A considerable amount of sequence data—again rDNA—has been collected on numerous groups of Hymenomycetes, and robust and testable phylogenetic hypotheses are beginning to emerge (reviewed in Hibbett and Thorn 2001). The clade is united by a unique mycelial structure, the dolipore septum, in which the cell walls near the pore of the septum flare, and a membrane structure, the parenthesome, occurs on either side of the pore (Moore 1985). The parenthesome may be perforated or not, depending on the clade, with the imperforated form being ancestral for the class. Reconstruction of the character states is not, however, without homoplasy (Hibbett and Thorn 2001). The orders Ceratobasidiales and Tulasnellales, which include plant pathogenic (e.g., *Rhizoctonia*) and saprobic species, are also members of the Hymenomycetes and possibly closely related to the Auriculariales (Swann and Taylor 1995a; Swann et al. 2001). Many of the taxa in these orders possess septate or deeply divided basidia and exhibit a dimorphic life cycle that includes yeast phases, a finding consistent with those traits being ancestral for the Basidiomycota.

The most derived clade of the Hymenomycetes includes homobasidiomycetous taxa with nonseptate basidia that lack a yeast phase in their life cycles. This clade includes the mushrooms, shelf fungi, and stinkhorns, all of which may produce mycorrhizae, decay litter and wood, or act as plant pathogens and insect symbionts. Traditional classifications of the homobasidiomycetous fungi were based largely on basidiocarp morphology with particular emphasis on the spore-producing region or hymenophore. For example, all of the mushrooms and their gilled relatives were grouped in the Agaricales, and all of the poroid forms and their relatives were grouped in the Aphyllophorales (for more complete reviews see Ainsworth et al. 1973b; McLaughlin et al. 1995; Alexopoulos et al. 1996). Phylogenetic analyses of molecular data from both nuclear and mitochondrial genomes do not support those classifications (Hibbett and Donoghue 1995; Hibbett et al. 1997; Moncalvo et al. 2000, 2002). The consensus arising from those studies is that overall, basidiocarp morphology it is not a phylogenetically informative trait at higher taxonomic levels because of repeated episodes of convergent and divergent evolution. The Hymenomycetes includes an estimated eight major clades (polyporoid, euagaric, bolete, thelephoroid, russuloid, hymenochaetoid, cantharelloid, and gomphoid-phalloid), each of which encompasses multiple basidiocarp and hymenophore morphologies. Cantharelloids, for example, include species with any one of four basidiocarp types, the fewest in any of the clades; all eight clades, however, include species with a crust or corticioid morphology (Hibbett and Thorn 2001).

The current resolution of the basal nodes, or backbone, of the Hymenomycetes is not strongly supported by current data and in many ways mirrors our poor understanding of the Euascomycetes. The imperforated parenthesome appears to be an ancestral character for the class, but again the character is homoplasious (Hibbett and Thorn 2001). The lack of support for basal nodes and the presence of the numerous basidiocarp morphologies in each major clade make identification of the ancestral basidiocarp morphology difficult. The most basal lineage of the homobasidiomycetous Hymenomycetes may be the gomphoid-phalloid clade, a major clade of the class that was not anticipated based on gross morphology but is strongly supported by molecular data (Colgan et al. 1997; Hibbett et al. 1997; Humpert et al. 2001). That clade contains six of the seven major basidiocarp types, suggesting that most, if not all, major basidiocarp morphologies have been present since early in the evolutionary history of the homobasidiomycetes.

The Agaricales (sensu Moncalvo et al. 2000, 2002; Euagarics clade sensu Hibbett et al. 1997) is currently the best-studied group of basidiomycetes and the best-known group phylogenetically (see Singer 1986b,

Moncalvo et al. 2000, and Hibbett and Thorn 2001 for a complete list of citations). Traditionally the order included all mushroom-forming species with a gilled or lamellate hymenophore plus the fleshy poroid mushrooms of the Boletaceae (Singer 1986b). Numerous more recent phylogenetic investigations do not support such a classification, suggesting the existence of five to six separate clades of gilled mushrooms among the homobasidiomycetous Hymenomycetes alone (Hibbett et al. 1997). For example, *Russula* and *Lactarius*, two common mycorrhizal genera of the Russulales, are not members of the Agaricales but rather are part of a large and morphologically heterogeneous clade that includes poroid, toothed, corticioid, and gasteroid (false-truffle) morphologies. Conversely, not all members of the Agaricales, or euagaric clade, possess lamellate hymenophores. Convergent evolution in the bird's nest fungi (*Cyathus* species) and the common puffballs (*Lycoperdon* species) has led to the loss of forcibly discharged spores and the retention of spore-producing tissue within an enclosed sporocarp (Hibbett et al. 1997). This general morphology once was used to unite all such basidiomycetes in the class Gasteromycetes or stomach fungi, a classification rejected based on both morphological and molecular characteristics (Thiers 1984; Hibbett et al. 1997). In fact, of the eight major clades of the homobasidiomycetous Hymenomycetes, five (euagaric, bolete, poroid, russuloid, gomphoid-phalloid) include species that produce some form of gasteroid basidiocarp (Hibbett and Donoghue 1995; Hibbett et al. 1997; Hibbett and Thorn 2001; Humpert et al. 2001; Miller et al. 2001).

Numerous groups of Agaricales *sensu stricto* are recognized and united by morphological traits that molecular analyses support as phylogenetically informative characters (Moncalvo et al. 2000, 2002). However, those traits appear to be most informative at the subordinal and familial levels. Basidiomycete taxonomy is based largely on morphological characteristics of the basidiospore, including color, and sporocarp, including stipe attachment and veil tissue (tissue covering the hymenophore); the microanatomy of unique hyphal elements (e.g., oil-filled hyphae, sterile cells associated with basidia) and staining reactions with certain chemical reagents (e.g., yellowing of the mushroom cap in potassium hydroxide) are also useful traits. Molecular analyses confirm that many of the major families of the Agaricales (e.g., Amanitaceae, Lepiotaceae) are monophyletic, the exceptions being taxa that have long been presumed to be polyphyletic (e.g., Cortinariceae, Tricholomataceae). A detailed discussion of agaricalean systematics is not possible here, but a modern perspective on the current status of the field is available in

Moncalvo and associates (2000, 2002) or on the Basidiomycete Phylogeny website (Appendix III).

KINGDOM STRAMINIPILA (HETEROKONT ZOOSPORIC ORGANISMS)

The eukaryotes (Fig. 1.1) are divided into two large subgroups (Baldauf et al. 2000). As mentioned earlier, one of these groups includes fungi (Fungi), animals (Metazoa), and two groups of slime molds (the mycetozoans or myxomycetes, Dictyostelidae and Myxogastridae). The other large group of eukaryotes is divided into three subgroups, one of which includes the land plants and green algae (Angiospermae and Chlorophyceae), among other taxa. The Straminipila is classified in the second subgroup. It includes certain algae (Phaeophyceae), a group of heterotrophs that previously was classified in the fungi (Oomycota), and a variety of "protozoan" groups not shown in Fig. 1.1. The Phaeophyta, Oomycota, and other straminipiles are characterized by zoospores with anteriorly or laterally biflagellate cells with two flagella, one smooth (whiplash), and one bearing tubular tripartite hairs. In some instances the smooth flagellum (or even both flagella) has been lost evolutionarily. Tubular hairs also may be found as ornamentation on certain zoospore cyst walls. Photosynthetic straminipiles possess chlorophylls *a* and *c* (e.g., brown algae, diatoms, golden-brown algae, chrysophytes). Traditionally, heterotrophic straminipiles that lack the ability to fix carbon (oomycetes, hyphochytrids, labyrinthulids, and thraustochytrids, Fig. 1.7) have been studied by mycologists who for many years have emphasized morphological and biochemical traits (e.g., flagellation, cell-wall composition, pesticide sensitivity) that distinguished straminipiles from true fungi (Bartnicki-Garcia 1970; Fuller 2001). Confirmation based on molecular characters of their phylogenetic position distant from fungi was, therefore, not surprising (e.g., Van de Peer and De Wachter 1997; Hausner et al. 2000).

Various names have been applied to the straminipiles (also known by the variants "stramenopiles" and "straminopiles") and used in conflicting ways (Dick 2001b). Chromista and Heterokonta, for example, are roughly equivalent terms that have been used in place of the pending kingdom name "Straminipila." The kingdom seems not to have been described formally, although it is recognized commonly. Names of taxa within the kingdom have also changed; most notably, Peronosporomycetes has been used as a class name for

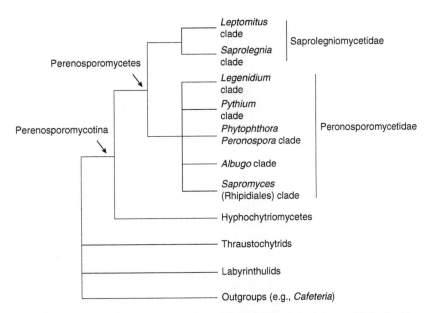

FIGURE 1.7 Relationships in the Kingdom Straminipila based on published analyses of small subunit (SSU) and large subunit (LSU) rDNA. Additional data from the *cox II* gene, physiology, and morphology help to corroborate the basic relationships. (Based on Dick et al. 1999; Hausner et al. 2000; Hudspeth et al. 2000; Petersen and Rosendahl 2000; Dick 2001b; Leander and Porter 2001.)

the monophyletic grouping of Oomycota and Hyphochytriomycota, and Peronosporomycetes has been used as a class name for three groups that are known as Oomycota (Dick 2001b). In addition to the Oomycota and Hyphochytriomycota, the primarily marine labyrinthulids and thraustochytrids have been placed among the straminipiles, although it is not clear whether the heterotrophic straminipiles are monophyletic (see "Labrinthulales and Thraustochytriales," later in this chapter).

OOMYCOTA

Phylogenetic hypotheses for oomycetes (Class Peronosporomycetes; Dick 2001) define two major clades, Saprolegniomycetidae (water molds) and Peronosporomycetidae (plant and animal parasites) (Fig. 1.7). Trees have been produced by analyses based on morphological and biochemical traits, and the genes for SSU rDNA, LSU rDNA, and cytochrome oxidase (*cox II*) (Dick et al. 1999; Riethmüller et al. 1999; Hudspeth et al. 2000; Petersen and Rosendahl 2000). The taxonomic placement of the order Rhipidiales, a group of taxa with unusual metabolism (i.e., obligate fermentative respiration), is uncertain. Analyses of sequences from only one or two species suggest that the Rhipidiales

may represent an independent lineage within the oomycetes of equivalent rank to the Saprolegniomycetidae and Peronosporomycetidae (Dick 2001a) or, alternatively, may be included in the Peronosporomycetidae. A study using the *cox II* gene, morphology, and biochemistry included a single species of *Rhipidium* within the Peronosporomycetidae. Analyses of additional taxa will be required to settle the question (Hudspeth et al. 2000).

Few studies have dealt with taxa at the ordinal level or lower. The Saprolegniales, however, does appear to be monophyletic. Within the order, the large genus *Achlya* is polyphyletic (Riethmüller et al. 1999; Leclerc et al. 2000).

HYPHOCHYTRIOMYCETES

Members of the Hyphochytriomycetes have cells with a single anteriorly inserted flagellum bearing tubular tripartite hairs. The phylogenetic position of hyphochytrids among the straminipiles as the sister group of oomycetes has long been assumed based on ultrastructural comparisons. Analyses of rDNA from three species of hyphochytrids have confirmed that placement (Van der Auwera et al. 1995; Van de Peer and De Wachter 1997; Hausner et al. 2000).

Labyrinthulales and Thraustochytriales (and *Aplanochytrium*)

Several investigators studying rDNA have placed the Labyrinthulales in the straminipiles (Honda et al. 1999). Those organisms inhabit primarily marine and brackish waters and associate with marine plants and bivalves, often as parasites. A broader question concerning the monophyly of all heterotrophic members of the straminipile clade, which includes the Labyrinthulales, has not yet been resolved (Van de Peer and De Wachter 1997; Hausner et al. 2000). The results of a study of SSU and LSU rDNA sequences, however, are consistent with, but do not strongly support, the monophyly of Labyrinthulales and Thraustochytriales with Oomycetes and hyphochytrids (Leander and Porter 2001). A third lineage identified in the same study and composed of two species of *Labyrinthuloides* was basal to Labyrinthulales and Thraustochytriales.

SLIME MOLDS

Plasmodiophorales (Parasitic Slime Molds)

The Plasmodiophorales are plant parasites whose life cycles include multinucleate plasmodia (Braselton 2001). Although they have been studied by mycologists historically, a number of unique characters, such as cruciform nuclear division and mode of penetration during infection, set the group apart from the true fungi. Phylogenetic analyses of rDNA from one species, *Plasmodiophora brassicae*, suggest that the organisms are more closely related to alveolate protozoans (Dinozoa, Ciliophora, and Apicomplexa) than to any of the other slime-mold groups or Fungi (Castlebury and Domier 1998). More recently Cavalier-Smith (2001) included *P. brassicae* as a basal member of the Cercozoa, a protozoan phylum that he erected.

Myxomycetes, Protostelids, and Dictyostelids (Plasmodial and Cellular Slime Molds)

Phylogenetic information on several so-called slime-mold groups indicates that plasmodial and cellular slime molds comprise a monophyletic group (Fig. 1.1; Baldauf and Doolittle 1997; Baldauf et al. 2000). In an analysis of EF-1α amino acid sequences *Physarum polycephalum*,

a myxomycete (as Myxogastridae), was placed as a sister group to several cellular slime molds (dictyostelids); that clade was the sister group to a protostelid (*Planoprotostelium* species) (Baldauf and Doolittle 1997). The monophyly of myxomycetes and dictyostelids is also fairly well supported by data from protein-coding genes but not by data from rDNA analyses.

The history of classification of these intriguing slime molds has been reviewed by Olive (1975), Cavender (1990), Frederick (1990), and Spiegel (1990). Spiegel (1991) used life history and morphological and developmental characters in a phylogenetic analysis of a broader sample of myxomycetes and protostelids. His results indicated that myxomycetes form a monophyletic group with only some protostelids, indicating that the protostelid group as circumscribed by Olive (1975) is paraphyletic.

Acrasid Slime Molds

Acrasid slime molds have been separated from the other groups of slime molds in most studies of the last three decades. Emphasis on the movement and behavior of the amoebae caused a number of workers not only to separate the acrasids but also to suggest a close relationship to organisms with limax-type amoebae such as *Naegleria* species and the vahlkampfiids (Olive 1975; Blanton 1990). The mitotic apparatus of *Acrasis rosea* is very different from those of other slime molds (Roos and Guhl 1996). Furthermore, a phylogenetic analysis of slime molds and amoebae (including acrasids, dictyostelids, and *Naegleria* species) based on genes encoding glyceraldehyde-3-phosphate dehydrogenase supported the grouping of acrasids in the Heterolobosa with Vahlkamfiidae (Fig. 1.1; Roger et al. 1996; Baldauf et al. 2000).

CONCLUSIONS

To date, the lion's share of phylogenetic studies has been based on nucleotide data from nuclear rDNA. The utilization of SSU rDNA in fungal systematics was a watershed event for mycologists and allowed them to answer many long-standing questions in fungal systematics. The continued use of single-gene phylogenies, however, has obvious limitations, and analyses of additional genes and taxa will be required to answer important remaining questions. The refinement of fungal phylogenies will require consideration of the following points:

- *Origin of the Fungi.* Is a choanoflagellate ancestor for fungi well supported? Where is the origin of DAP (diaminopimelate aminotransferase) lysine biosynthesis in the fungal ancestry? Can character (e.g., flagella, hyphae) evolution be traced? Can we infer the general morphology of the "first" fungus?

- *Early diverging events within the Kingdom Fungi.* Limited molecular data do not support the monophyly of either the chytrids or the zygomycetes, suggesting multiple losses of the flagellum. In addition, the association of many zygomycete groups with arthropods suggests the possibility of multiple origins of a terrestrial fungus. We should be able to address the paraphyly of the Chytridiomycota/Zygomycota clades, the origin of non-plant–associated terrestrial fungi (i.e., multiple origins of terrestrial fungi), character evolution (e.g., loss of flagella, modes of sexual reproduction), and realignment of major taxa of early diverging fungi.

- *Phylogenetics of the crown fungi (crown fungi are Glomales, Basidiomycota, and Ascomycota).* These taxa represent the terminal clade within the Kingdom Fungi, and their origin appears to be correlated with the origin and diversification of land plants. However, the relationship of the Glomales with the ascomycete/basidiomycete clade is tenuous, and additional plant-associated zygomycetes must be sampled.

- *Early divergences within the filamentous Ascomycota (Euascomycetes) and macrosporocarp-producing Basidiomycota (Hymenomycetes).* Could symbioses between green algae and fungi have evolved prior to the evolution of land plants and their mycorrhizal symbioses? What is the origin of dikaryotic fungi? When do we find the first organized sporocarps and plectenchyma (fungal tissue)?

ACKNOWLEDGMENTS. We acknowledge support from the Research Coordination Networks program in Biological Sciences of the U.S. National Science Foundation for "A Phylogeny for Kingdom Fungi" (NSF-0090301) to M. Blackwell, J. W. Spatafora, and J. W. Taylor. We also acknowledge the U.S. National Science Foundation for support for "Assembling the Fungal Tree of Life" (DEB-0228725) to J. W. Spatafora, D. Hibbett, F. Lutzoni, D. McLaughlin, and R. Vilgalys. David Geiser, David Hibbett, and John Taylor provided insightful discussions.

2

Preparation, Preservation, and Use of Fungal Specimens in Herbaria

QIUXIN (FLORENCE) WU, BARBARA M. THIERS, AND DONALD H. PFISTER

Preservation of voucher specimens resulting from biodiversity surveys is extremely important for scientific studies. Voucher specimens and their accompanying field notes document the existence of a fungus at a given place and time and provide the raw data from which taxonomic concepts are constructed. The examination of voucher specimens also provides a reliable way to verify or correct the identity of organisms recorded in surveys or used in cytological, ecological, populational, morphological, phylogenetic, and molecular studies. Scientific specimens are also a source of DNA and other compounds for phylogenetic, ecological, and other studies.

Herbaria serve as conservatories of voucher specimens. For the past 200 years, the comprehensive nature of fungal herbaria has increased markedly. Early fungal herbaria (e.g., those of E. Fries and C. H. Persoon—pioneers in fungal systematics) contained one or very few collections of a wide range of taxa; each collection was intended to define its taxon. Herbaria now strive to obtain a variety of specimens that represent the manifestations of a taxon in all its developmental stages and throughout its geographic range, a practice that reflects the incorporation of evolutionary concepts into systematic theory. Methods of preparation and storage of fungal specimens also have changed dramatically. Early collections were dried and preserved according to techniques developed by Linnaeus for flowering plants (i.e., they were pressed and affixed to paper sheets). Such methods are ill suited to fungi. Modern mycologists have recognized the intrinsic differences between fungi and plants and adjusted their preparation and storage techniques accordingly.

Because specimens entrusted to herbaria are to be maintained in perpetuity, herbaria often are seen as static entities. To the contrary, however, herbarium curators and administrators must keep abreast of developments in science and environmental policy. They also must continually seek ways to use herbarium specimens to address current biologic questions, while still fulfilling their primary mission as repositories for documentation of the earth's biodiversity.

Recent accelerated deforestation and general habitat destruction have created a new challenge for herbaria. Because of rapid habitat loss and degradation, many systematists and other members of the scientific community see an urgent need for an inventory of the earth's biota before major portions of it disappear. Systematics Agenda 2000, for example, recommends that the systematics community survey the biota of the world within the next 25 years (Systematics Agenda 2000 1994). Such an undertaking would place unprecedented demands on herbaria (as well as other types of specimen repositories) to house and maintain specimens resulting from those surveys and to meet the needs of users of the specimens and their associated data. It is critical that herbaria and similar museum facilities be fully integrated into the planning of broad inventories and their execution.

COLLECTION ACQUISITION POLICIES

Most large fungus collections that exist today originated as specialist collections and represent the work of generations of scholars. Similarly, new collections reflect new studies by investigators, activities of amateurs, or sometimes fortuitous circumstances. Administrative and curatorial personnel recognize, however, that unplanned growth is neither advisable nor defensible. Most, therefore, have formulated acquisition (accession) policies that reflect the intended scope of their collections and regulate their growth. Such policies focus on a few common

issues that we have formulated as a series of questions as follows.

1. Do the specimens in question adequately document the taxon? Material that is dried improperly, scanty, contaminated by other fungi, or damaged by insects or that represents only a portion of a specimen may be impossible to identify. Such collections are of negligible value to future studies. Given the resources required to process and maintain collections, an herbarium rarely accepts such materials.

2. Do data accompanying a specimen adequately document it? Modern specimens that lack adequate information (e.g., listing only a partial locality, such as the state or province), especially those of common taxa, are also of little value for research and may be discarded or refused.

3. Is the taxon already represented in the collection? Depending on the goals of a particular herbarium, the addition of yet another collection of a common, well-represented taxon from a well-documented locality may be superfluous. Often such collections are used for interinstitutional exchanges that allow other herbaria to enhance their holdings. The number of voucher specimens of a species that should be collected from a locality of limited area (e.g., a small park, a nature trail), if at the discretion of a researcher rather than determined by permit-granting entities, will vary with the nature of the fungi (e.g., soil fungi, plant pathogens, or mushrooms) and the nature of the study (e.g., ecological, morphological, or biodiversity inventory).

4. Has the specimen been identified by a specialist? Such specimens are generally welcome additions to herbaria. They are used to confirm identifications of other collections and as a basis for new studies.

5. Has the specimen been cited in a publication or designated as a voucher? If so, it should be maintained despite other conditions.

6. Do copies of appropriate authorizing permits (e.g., collection, export from a host country, import into the country of deposition) accompany the specimen? Herbaria should not accept specimens unless they are accompanied by copies of such documents, recognizing that collecting in some areas (e.g., national parks, private lands, other countries) may be restricted. Acceptance of specimens newly collected without proper authorization can damage the credibility of the herbarium accepting them, limit its eligibility for federal funds, prevent its scientists from obtaining permits to collect in certain sites or countries, and lead to criminal prosecution.

To summarize, most acquisition policies emphasize selective addition of specimens (1) of poorly represented taxa

(2) from poorly represented geographic areas that are (3) legally collected, (4) authoritatively identified, and (5) cited in the literature and that can be (6) adequately housed and (7) made available for study. Specifics of acquisition policies may differ, according to the goals of the herbarium.

PROCESSING FUNGAL SPECIMENS

Fungal specimens deposited in an herbarium must be accessioned, and their collection data must be recorded. The specimens then are processed, labeled, mounted, and filed. In this section, we discuss methods for the physical preservation of specimens. We discuss associated data and herbarium record keeping later in this chapter (see "Associated Data" and "Record Keeping"). Detailed procedures for specimen preparation also can be found in *The Herbarium Handbook* (Bridson and Forman 1992) and the *Manual for Tropical Herbaria* (Fosberg and Sachet 1965). Because fungal specimens, including lichenized fungi, Myxomycetes, and other protoctistan "fungoid" organisms, are variously shaped and are often very fragile, special care must be taken when they are processed, mounted, and filed.

SPECIMEN PREPARATION

Fungus specimens are preserved by drying, which must be prompt and complete. Fungi commonly are air-dried, desiccated using silica gel, or freeze-dried. All three methods seem to preserve DNA satisfactorily. Air drying or low-heat-accelerated air drying is used most commonly to prepare sporocarpic specimens, although it also can be used to preserve fungi growing on agar plates (see "Maintenance and Preservation of Cultures," Chapter 3, this volume). With silica-gel drying, specimens and field labels are placed in sealed glass tubes or plastic bags with silica crystals. This method can be used only to dry small, easily desiccated specimens. Freeze-drying is a dehydration technique in which a specimen first is frozen, and then the frozen moisture content is vaporized under a vacuum. Equipment for freeze-drying is expensive, but freeze-dried specimens maintain their shapes and therefore are good for displays. They often become brittle and fragile, however, and they also absorb moisture readily. After 5 to 6 years in storage, cells of freeze-dried Ascomycete and Basidiomycete specimens often collapse (H. Burdsall, personal communication). A note indicating preservation method should be included with specimens that are not air-dried. Fungi usually lose their natural color during drying and other preservation

procedures, so detailed color notes should be taken on fresh material before processing.

Sporocarps

Most fleshy fungi can be dried under a constant flow of warm air (40–50°C, preferably, 42°C). Electric fruit dryers with wire mesh shelves work especially well (Singer 1986a). Usually, entire sporocarps are placed loosely on the shelves, although those that are large or watery may be sliced longitudinally to hasten drying. In areas without electricity, shelves or screens are placed above a heat source (e.g., a fire, gas burner, oil lamp, or stove). Special care must be taken to avoid burning specimens that are dried over open flames. Small specimens on dry substrata can be air-dried indoors, in the sun, or under a high-wattage (more than 75 watts) incandescent bulb. They also may be dried directly in paper bags, glassine bags, or small paper packets.

The substratum attached to a specimen also must be dried thoroughly. A small tag with the collection number, placed so as not to block air flows, should accompany each specimen on a dryer. Dry specimens should be kept dry and mold-free. Silica gels can be used to remove moisture from air in a storage container. Completely dry specimens can be wrapped in wax paper or placed in reclosable (zipping) plastic bags for transport back to the herbarium.

Plant Pathogens

Rusts, smuts, and other plant-pathogenic fungi on leaves or delicate stems can be dried as plant specimens (Bridson and Forman 1992). The host-plant tissue with its fungi is spread out on folded drying paper or newspaper along with a collection-number tag. Specimen papers and interspersed padding sheets (blotters) should be sufficient to absorb moisture. Placing corrugated aluminum or cardboard sheets periodically among the blotters will speed drying by enhancing air circulation. The specimens, blotters, and spacers are placed between two wooden frames, and the stack is compressed with tightly drawn straps. Presses can be placed in the sun or over a heat source (35–45°C) for drying. A press should be checked at least twice a day, and the specimen papers and blotters should be changed if they are wet. Completely dried specimens can be placed in paper packets or left in dry, folded newspaper for transport.

Lichens

Specimens should be air-dried on open shelves over gentle heat (42°C) or in the sun. The fungi may be left in paper bags if they are collected dry. Some leaflike or branched lichens can be flattened and pressed as described for plant-pathogenic fungi. Brittle lichens, on soil, should be dried along with their substrata. The lichen thallus is best protected if the soil is glued together and mounted on a card (Ryan and McWhorter 1986). Lichens attached to rocks should be wrapped separately in tissue paper after drying and before packeting.

Myxomycetes

Specimens should be air-dried along with their substrata. Collections should be kept upright during the drying process. Dried specimens are stored separately in small tins or boxes to prevent damage. The substratum to which the specimens are attached is glued to a small piece of herbarium paper (a box insert) cut slightly longer, but narrower, than the bottom of the box. The ends of the paper are folded up at right angles so that the paper insert fits snugly into the box. The investigator uses the folded ends to remove the specimen for closer examination without handling it (W. Sundberg, personal communication).

Fungal Cultures

Here we provide a brief overview of how fungal cultures are preserved. For more complete coverage see "Preservation and Distribution of Fungal Cultures," Chapter 3. Cultures of filamentous fungi can be preserved as voucher specimens by drying (Stevens 1974). The mature fungus colony, in its Petri plate, is placed in a frost-free freezer. The freezer must be frost-free or the preparation will not dry properly. Data are written on the plate bottom or, if they are already written on the lid, the lid can be placed under the open culture plate. After 4 to 6 weeks, the colony should be sufficiently dry so that it is loose in the plate. The colony then is transferred either to an archival-paper envelope or to a thin (e.g., 8-mm) Petri plate and placed in an herbarium packet, and a label with all accompanying data is prepared. The old plate and lid are discarded. Paper envelopes can be used for most pyrenomycetes, many discomycetes, and all loculomycetes. Delicate taxa such as zygomycetes, hyphomycetes, coelomycetes, and some ascomycetes (e.g., Trichocomaceae) are best stored in thin Petri plates (J. Krug, personal communication). The material then is placed in a 100% cotton-rag packet to which the label is attached with acid-free glue.

Liquid-Based Preparations

Phalloid, clathroid, and other delicate fungi can be fixed and stored in liquid preservatives so that the forms and fragile textures of the fungi are not distorted. However, chemical fixation alters the DNA of an organism, so specimens fixed in that manner are not suitable for molecular studies. Consequently, warm-air drying is the preferred method of preserving fungal specimens (Haines and Cooper 1993).

Freshly collected material is submerged in a fixative for 2 to 7 days and then transferred to a storage solution. Fixatives and storage media can be simple solutions such as 70% ethanol, or they can be more complex mixtures of chemicals. Bridson and Forman (1992) recommended the Kew Mixture and the Copenhagen Solution for fixation and storage, respectively (see Appendix II). Both contain glycerol, which keeps the specimen from hardening. Collections stored in ethanol must be monitored closely so that fluid lost through evaporation can be replaced. Generally, specimens are stored in glass jars or vials with neoprene caps or stoppers. Caps may be sealed further with sealing wax, Parafilm, or other materials to limit evaporation. Because chemical preservatives are harmful to human health, safety procedures must be followed when fixative and storage media are handled, both in the field and in the laboratory. Investigators should wear gloves and safety goggles and should avoid breathing fumes by wearing a mask or working in a chemical fume hood.

FIELD LABELS AND FIELD NOTES

To be useful for scientific studies, specimens must be accompanied by certain information. The field label of a specimen should at least include: (1) field identification or tentative name of the specimen; (2) collector's name, collection number, and date; (3) detailed locality information, including coordinates and elevation; and (4) descriptions of the habitat, substratum, and host. A small label, or field tag, with the collector's initials and collection number should be associated with the specimen at all times. In addition, collectors should provide detailed notes on the appearance of the specimen when fresh. Content of field notes varies among fungal groups. Protocols for preparing field notes, including pertinent information about the color, stature, shape, and general ecology of the organism, can be found for various groups in respective chapters in this volume.

Field notes should be deposited with the specimens to facilitate their retrieval and identification, except when specimens are fluid-preserved, in which case they are best stored separately. Field notes should be written on archival-quality paper using permanent ink. Some collectors enter field data directly into a computer. In that case, backup files should be kept separately from the computer, and a hard copy should be printed as soon as possible. Copies generated using a laser printer and archival-quality paper are adequate for long-term preservation.

INCOMING SPECIMENS

Incoming specimens may originate from staff fieldwork or as gifts, exchanges, or loans. All incoming specimens, but especially those coming directly from the field, should be checked on arrival to determine if they are completely dry. Damp specimens must be dried thoroughly (see "Specimen Preparation," earlier). Incoming specimens also must be disinfested (to kill insects) before they are stored in an herbarium. Disinfestation can be achieved by deep freezing. Packages of specimens are wrapped in plastic bags (e.g., heavy-duty garbage bags) and placed in a freezer at or below −18°C for 7 days (Bridson and Forman 1992). Bagging is essential for preventing condensation of moisture on a specimen during the process. Double plastic bags may be used to prevent air leakage. After freezing, specimens should remain in their plastic bags until the whole package regains room temperature.

PACKETING, BOXING, AND MOUNTING SPECIMENS

Specimen containers should be large enough to accommodate fungal specimens of various shapes and sizes to provide appropriate protection. Unpacketed specimens never should be mounted directly on herbarium sheets. Paper packets, small tins, and cardboard boxes usually are used as containers in a fungal herbarium. A packet is analogous to an envelope that is used to hold specimens and accompanying papers.

Paper packets made with 100% cotton-rag paper can be used to preserve leaf parasitic fungi, microfungi, culture plates, lichens, or nonfragile larger fungi. Such packets are mounted on herbarium sheets by gluing, stapling, or pinning or by using double-sided tape. Labels and annotations are affixed to the sheets or packets. Placing only a single specimen on each sheet facilitates annotation and reorganization of specimens, although it uses more space and material than if several specimens were placed on a sheet. Standard size (15.0 × 9.3 cm) packets are made by folding 21.6-cm × 27.9-cm (8 1/2″ × 11″) archival-quality cotton-rag paper. Larger packets can be custom-made. Specimen packets also can be purchased in quantity from herbarium supply companies (see Appendix IV). In many herbaria, packets

are filed vertically in drawers of specially designed cabinets rather than mounted on herbarium sheets. Such arrangement saves space and allows for easy rearrangement of specimens.

Small metal or plastic containers with a cardboard label permanently fixed on top can be used to store small and/or fragile specimens. The containers are placed in a 100% cotton-rag paper packet, which is filed or mounted. Although metal tins may rust in wet or humid areas, they are stable under normal herbarium conditions and inexpensive. In contrast, cardboard boxes are often acidic, and archival-quality boxes are expensive. Specimens of Myxomycetes should be glued to the underside of the lid of small boxes.

Larger specimens are stored in cardboard boxes, preferably of archival quality. Archival boxes are constructed of acid- and lignin-free cardboard held together with metal corners (glue-free). Such boxes can be purchased, but they are expensive and may be hard to locate (but see Appendix IV, University Products). A lower cost, but less satisfactory, alternative is to use regular cardboard boxes but wrap specimens in acid-free tissue paper to prevent direct contact of the specimen with the box. A label and a determination slip can be pasted to the front end of each box. To prevent the lid of a box with a label from being separated from the bottom part of the box containing the specimen, the collector's name and collection number should be written on the outside of the box bottom and on a small tag (e.g., 2 cm^2) placed inside with the collection. Although few Aphyllophorales are fragile, they generally are stored in boxes instead of packets to prevent the specimens from causing the herbarium packets and labels to deteriorate.

Containers should be large enough to hold the specimen and field notes but also as small as possible to save space. Cotton-rag, buffered tissue paper may be used to wrap the specimen and a small collection tag. Field notes and other ancillary material (see "Ancillary Materials," later in this chapter) should be placed in the box but not wrapped in tissue paper. Permanent ink (e.g., India) should be used to write on collection tags and other labels.

Packet and box sizes should be as uniform as possible for easy storage. However, fungi do not compact neatly. Extra-large, oversized specimens that cannot fit in regular herbarium cabinets often are stored separately. They can be placed either in ready-made or custom-made boxes or in polythene bags with reclosable (zipping) tops and stored in custom-made cabinets or on shelves. A duplicate label or a packet containing part of the specimen should be placed in the proper location in the herbarium to serve as a cross-reference. Good cross-referencing greatly facilitates filing and retrieval of specimens.

Associated Information

Specimen Data

The specimen label is an essential part of every specimen. It is attached to the specimen packet or box and provides valuable data. Most institutions use standardized labels of several sizes and formats to accommodate different types of preparations and storage containers (e.g., specimens on sheets, in bottles or boxes, on microscope slides). With computers and printers, generating good-quality labels is an easy task. Handwritten labels can be difficult to read and should be avoided. An herbarium label should include: (1) a title showing the nature of the collection and origin of the specimen (e.g., FUNGI OF CHINA); (2) scientific name of the organism; (3) taxonomic authority and bibliographic information, particularly for type specimens; (4) locality information; (5) substratum/host, vegetation, and habitat information; (6) collector's name, collection number; (7) collection date; and (8) herbarium name and specimen number. The paper used for labels should be of archival quality (i.e., cotton rag and acid-free). Labels are glued to packets, boxes, or mounting sheets using high-quality neutral or buffered glue. Various types of "white glue" (polyvinyl-acetate adhesives) commonly are used. Rubber cement, tape, and most adhesive papers do not provide a long-lasting bond and should be avoided.

Herbaria should develop labels for all in-house-accessioned specimens, even those to be sent out as a gift or an exchange. They also should request that the provider of the specimens include well-made labels with correct and proofread data or at least provide collection data in an automated format.

Annotation Data

Taxonomic (and sometimes other) information about a specimen increases as the specimen is studied over time. Such information enhances the value of the specimen and should be recorded by the investigator on an annotation label provided by the herbarium. An annotation label should include: (1) the identification number of the specimen (i.e., collector and collector and/or herbarium number) in case the label becomes separated from the specimen; (2) annotations for any scientific name newly applied to the organism; (3) authorship of the scientific name; (4) relevant literature citation(s); (5) the name of the annotator; (6) annotation date; and (7) other observations. Annotation labels should be made of archival-quality paper and written with permanent ink. Policies regarding attachment of annotation labels vary among herbaria. We recommend that investigators place annotation labels inside the packet or box or attach them to the packet, box, or mounting sheets with a paper clip or rubber band. An herbarium curator or technician then is notified of their existence and can attach them more securely, if necessary.

Ancillary Materials

Field notes, drawings, and photographs are essential tools for fungus identification. Original field notes should be placed with the specimen; a duplicate set should be kept separately. Spore prints should be placed carefully in packets and stored with the specimens. Transparencies, photographs, prints, drawings, and paintings of specimens should be stored separately and cross-referenced with the species name, collector, collector's number, and the herbarium number. Duplicate transparencies can be placed in small reclosable polythene (zipping) bags stored in the packet or box containing the specimen. This procedure is also appropriate for permanent microscope-slide mounts. Specially constructed cardboard slide holders and polyethylene slide envelopes adequately protect slides. These may be placed in packets and attached to the sheet with the specimen.

Archival-Quality Supplies

Scientific specimens, like other natural history collections, deteriorate under normal conditions. To minimize deterioration, all materials associated with specimens (e.g., paper products, adhesives, ink, and plastic products) should be of archival quality. Unfortunately, the use of such materials often is limited by the financial resources of the herbarium. Updated information on archival products is helpful in making knowledgeable decisions. The Society for the Preservation of Natural History Collections (Appendix III) holds meetings annually to exchange information and discuss conservation issues.

Filing Systems

Herbaria file their specimens according to several different systems (e.g., alphabetically by scientific name, systematic or phylogenetic schemes, numerically according to herbarium accession numbers; Bridson and Forman 1992). It is important to choose an appropriate filing system when the herbarium is organized initially because reorganization of specimens is both labor intensive and time consuming. A good filing system facilitates curation (i.e., verification or updating of label data and repair or rearrangement of specimens in the herbarium but not identification) and retrieval of specimens. The systematics of many groups of fungi are still under investigation; therefore, family delimitation is often difficult. Even so, orders and families can be organized according to a systematic scheme so that closely related families are placed together. Within a family, genera can be arranged alphabetically, and within a genus, species can be arranged alphabetically or possibly geographically. Within species, specimens are arranged by geographic unit and then date.

Ideally, unprocessed or unidentified specimens should be stored in metal cases with sealed doors. They should be arranged in a way that facilitates access (e.g., by collector and collection number or by field trip). Material identified only to genus often is placed with other members of the genus in the herbarium.

Specimen packets mounted on sheets are stored horizontally in compartmentalized, metal cabinets made specifically for that purpose. Packets and boxes also can be oriented vertically in pullout trays or drawers in those cabinets. Drawers must be clearly labeled with family and generic names of their contents. Labeled dividers can be inserted between containers in drawers to indicate taxon boundaries. Packets and boxes should be arranged loosely in the drawers or trays to prevent damage to the specimens and to provide room for additional specimens.

SPECIMEN MAINTENANCE AND PROTECTION

HERBARIUM FACILITIES

An herbarium should provide for safe and permanent storage of valuable scientific collections. The herbarium facility should be fireproof, insect-resistant, and waterproof and equipped with good lighting, heating, ventilation, and air-conditioning systems. Herbarium areas should be isolated from laboratory activities and potential sources of insect infestation. Allowance for the expansion of a collection and the future installation of compactor storage units (which requires a stronger floor) should be included when a new herbarium facility is planned (Touw and Kores 1984). Metal cabinets are more insect-resistant than wooden ones, especially in tropical areas. Cabinet doors should be closely fitted with a strip of neoprene rubber or felt to prevent insect entry.

A temperature between 20° and 23°C at a humidity of 40–60% is ideal for specimen preservation. At higher temperatures and humidities, the probabilities of fungal infections and insect infestations increase. Central-heating and air-conditioning systems are best for controlling temperature. In humid areas, especially tropical ones, fans or dehumidifiers should be used to reduce humidity. Insect-resistant screens (small mesh) should be installed on all herbarium doors and windows that open because adults of many herbarium pests can fly.

PEST CONTROL

By far the most common problem encountered by curators of fungus collections is insect infestation. Drugstore beetles (*Stegobium paniceum*, also known as biscuit beetles or herbarium beetles) and cigarette, or tobacco, beetles (*Lasioderma serricorne*) feed on dried fungus, paper, and adhesives. Fleshy mushrooms and polypores, which are particularly prone to such infestations, are destroyed by these creatures. Insect pests that eat wood can destroy the substratum portion of wood-inhabiting fungus specimens. Such insects include deathwatch beetles (*Xestobium rufovillosum*), drywood termites (*Cryptotermes* species and *Kalotermes* species), and furniture beetles (*Anobium punctatum*). Insect pests that feed on molds or detritus can be merely a nuisance or, at high numbers, can cause serious damage. Such pests include silverfish, booklice, spider beetles, cockroaches, ants, and carpet beetles.

Pest monitoring and control, using a combination of methods including visual inspections and traps with insect hormone attractants, should be carried out as a routine herbarium activity (Hall 1988). Before being incorporated into an herbarium, all specimens, including returned loans, should be treated for insect contamination. Commonly, specimens are frozen to kill eggs, larvae, and adults (see "Incoming Specimens," earlier). Herbaria should be well ventilated and air conditioned. If air conditioning is not available, pesticides or repellents (e.g., naphthalene, paradichlorobenzene) should be placed in each herbarium cabinet. In that case, good ventilation in work areas is needed to reduce the health hazard to staff. Methyl bromide fumigation should be carried out as necessary, if it is permitted by the institution and local environmental laws.

Prompt processing and care of field collections is important. Although freezing or fumigation of specimens often is delayed until field work is completed, it must be done as soon as possible after collection. It is terribly disheartening to find that collections have been destroyed by insects before they could be studied.

In humid climates, molds may grow on herbarium specimens unless proper precautions are taken. Relative humidity should be kept low. Molds damage specimens and introduce foreign DNA that may lead to significant errors in analyses of host DNA.

USE OF HERBARIUM SPECIMENS

Herbaria should promote the use of specimens for scientific purposes by sending specimens to other institutions as gifts, exchanges, or loans and by facilitating working visits by investigators. At the same time, herbarium personnel must prevent specimens from being lost or destroyed. Most herbaria have established centralized policies and procedures for the loan of specimens and for the use of collections. Record keeping is crucial for tracking specimens and is discussed later in this chapter (see "Record Keeping").

OUTGOING SPECIMENS

Herbaria often distribute duplicate specimens to other institutions as gifts (to be kept), gifts for determination (specimens kept, but identifications returned), or exchanges (for other specimens). Distribution of duplicates makes collections more readily available for study. Investigators who collected specimens in other countries often are obligated or at least strongly encouraged to return a set of duplicates to the host country. Herbaria in tropical areas sometimes send duplicates to herbaria in temperate regions where the storage conditions are better.

Specimens are sent on loan in response to a request for study materials or for identification; loans requested by an individual or institution also must be returned once a study is completed. The loan policies of herbaria vary, but most institutions require a borrower to send a formal letter of request indicating who will use the specimens and for what purposes. The loan then must be approved by the person in charge of the herbarium. Loans usually are made only to recognized institutions, not to individuals, to ensure proper handling and storage of the specimens while they are on loan. In addition, specimens are transferred from one institution to another only with written agreement of all parties concerned. When specimens are withdrawn from a collection for a loan, a slip containing information that identifies the loan (e.g., name of the collection, region, borrowing herbarium, date, investigator, and loan number) should be placed in the empty storage area.

Specimens must be packed carefully to avoid damage during transit. Specimens on sheets should be padded, bundled between cardboard separators, packed in a strong box, and sealed and addressed according to postal regulations. When specimens are returned to the loaning institution, the original box and packing materials or materials of equal or better quality should be used. Additional padding may be required inside some containers to hold specimens in place and prevent damage from shifting during transport. Correspondence and packages should be addressed to the director/curator of the herbarium. Information on location, contact persons, holdings, and mailing addresses for herbaria worldwide can be found in the *Index Herbariorum* (Holmgren et al. 1990) or online (Appendix III). Address and customs labels should be attached securely to the outside of the package, and a packing slip identifying both the sender and recipient should be included in the box. Generally, specimens are sent through regular mail, but in certain cases (e.g., especially valuable specimens, shipment to areas with irregular or unreliable postal or custom services) boxes should be registered, certified, or entrusted to special carrier services.

All material shipped from an herbarium should be accompanied by an invoice that lists all specimens shipped and serves as a permanent record. The invoice indicates the nature of the transaction; the contents of the package; and other pertinent information, such as herbarium restrictions and/or requirements regarding the use of loaned material and the duration of the loan. Herbarium personnel returning a loan should check each specimen against the loan invoice and inform the lending institution that the loan (or a portion of it) is being returned. The format and contents of invoices are discussed later in this chapter (see "Record Keeping").

HERBARIUM VISITORS

Requests for loans of large numbers of specimens or of specimens of a wide range of taxa from a broad geographic area (e.g., all fungi from Texas) often are not granted because of the significant investment of resources required to assemble, pack, and ship the material. Instead, investigators are encouraged to visit the herbarium or to provide funds to support preparation of the loan. Investigators requiring material for molecular studies are especially encouraged to visit an herbarium to select their own samples. Herbarium personnel should provide orientation for visitors to ensure efficient use of the herbarium and to prevent damage to specimens during examination. In most herbaria, technicians, rather than visitors, return specimens to cabinets to prevent misfiling.

DESTRUCTIVE SAMPLING

Herbarium specimens frequently are used for morphological and molecular studies in place of fresh material. Use of herbarium specimens enables investigators to include specimens from broad geographic areas and time periods, as well as rare or endangered species in analyses (Mueller 1999). Often, studies require the dissection of specimens and removal or destructive sampling of specimen parts. Herbaria generally regulate and monitor dissection and destructive sampling of specimens to minimize specimen damage and loss.

Herbaria may request that individuals using specimens for morphological studies fully annotate their material and send duplicate glass slides and copies of notes, drawings, photographic prints, and publications resulting from the studies to the herbarium. Users of type specimens are encouraged to publish descriptions of the type material. Investigators should cite the specimens examined, including the herbarium acronym, and should acknowledge the herbarium for its assistance in publications resulting from their studies of herbarium specimens.

Sampling of herbarium specimens for molecular or other chemical studies is often more destructive than sampling for morphological studies. Consequently, policies regarding such activities are generally more complex and may include the following restrictions:

1. No material may be removed from specimens for DNA studies without prior consent of the curatorial staff.
2. Permission is granted on a specimen-by-specimen basis. Sampling from type specimens or from taxa underrepresented in the herbarium is limited or prohibited.
3. Only a portion of any collection may be sampled. The remaining part of the specimen must be annotated with descriptions of the material removed, the nature of the study, the researcher's name and affiliation, and the date.
4. Molecular data must be submitted to a database such as Genbank (Appendix III), and accession numbers must be provided to the herbarium.
5. Storage locations of extracts must be provided to the herbarium whether the data are published or not, because further sampling of the specimen may not be allowed.
6. The procedure used for extraction should be described, as well as its degree of success. Such information is especially important so that researchers can be apprised of the failure of particular procedures.
7. The herbarium that provided specimens used in the study must be acknowledged in any publication, and the specimens themselves should be cited as in a morphological study.

RECORD KEEPING

The importance of an herbarium is judged today largely by the degree to which its specimens are used by the scientific community. As funding for basic scientific research decreases, the need for herbaria to demonstrate their relevance and cost-effectiveness becomes more acute. Record keeping is essential to document herbarium use and to justify the funding needed to maintain it. A computer equipped with database software is indispensable for keeping detailed records. In this section we summarize the types of records that we believe an herbarium should keep to document its holdings, activities, and specimen use.

If an entire herbarium collection is catalogued and the source of every specimen in it is verified, it is possible to document transaction activity solely through the specimen catalogue. That is rarely the case, however, so traditionally, herbarium transactions have been tracked by specimen lot, rather than by individual specimens—a practice likely to continue for the near future at least. Consequently, herbaria will need a system for recording transactions that is separate from, although ideally linked to, the specimen catalogue.

ELEMENTS

The main types of specimen transactions conducted by herbaria are outgoing loans of identified material; outgoing loans for determination (unidentified material); incoming loans for herbarium staff, students, or visitors (determined or undetermined material); incoming or outgoing gifts; incoming or outgoing exchange; purchases; and staff accessions. The following list includes elements of transactions and the information that should be recorded about each. Not all elements listed apply to all types of transactions, but all types of transactions are accommodated in the list.

1. Information about the partner in the transaction, including herbarium name and acronym, address, name of designated correspondent at that institution, and name and status (staff, student, or visitor) of person to whom the transaction is directed.
2. Nature (type) of transaction; transaction number, date, and method of transfer (e.g., shipped, mailed, hand-carried); person(s) at the home institution who approved the transaction and date; material prepared for transaction and date; and creator of the transaction record and date.
3. Contents of transaction, including number of items and number of packing boxes and contents (full specimen record or some part of it such as identification

and barcode number, specimen name, collector, collection number, date).

4. Posttransaction information, including date a full or partial shipment is received or a full or partial loan is returned or canceled; date material is transferred to another institution; revised information about loan contents, as provided by recipient; condition of shipment when received; name of person who updates the transaction and date; dated correspondence related to recall of specimens (loan); dates determinations sent or received (incoming or outgoing gifts or loans for determination); exchange balance with transaction partner; and status of fulfillment of purchase agreement with transaction partner.

FUNCTIONS

The two most commonly needed types of output from a transaction management system are invoices and summary reports. Generally, original invoices are sent separately from a shipment, although a copy usually is included with the specimens. Invoices vary in the amount of detail they contain about a shipment, from just the total number of specimens sent, to an abbreviated version of the specimen record for each specimen. Generally, several copies of an invoice are sent, with the intention that one copy be signed and returned when the shipment is received and that, in the case of loans, another be returned to announce the return shipment of the loan.

Summary reports that include records of every specimen sent or received can serve as an archive for transaction activities. More limited summary reports include only a subset of the data, such as the number of specimens received during a specified period or the number of specimens of a particular taxonomic group that was loaned. Commonly, summary reports list overdue loans. Such reports can be generated periodically and sent as a reminder to borrowing institutions.

HERBARIUM ACTIVITIES

COLLECTION USE

The primary users of herbarium collections are taxonomists who examine specimens as part of their scientific research. Such use can be tracked through loan-transaction records, records of visitors to the herbarium, citations of specimen use or herbarium acknowledgments in published articles, or the specimen database. Records of specimen use can demonstrate the importance of an herbarium and help to limit specimen damage or destruction. If a person requests a loan of specimens for an anatomical study but records indicate that the anatomy of that specimen was studied recently, the herbarium can refer the person to the earlier study and prevent unnecessary additional dissection.

Most herbaria require visitors to sign in on arrival. The information requested from the visitor often includes his or her name, address, and institutional affiliation; date; and duration of and reason for the visit. The U.S. National Science Foundation requires that such visitor information be included in all proposals for funding from its Biological Research Collections (BRC) program (Appendix III). Proposals for the BRC program also must list all publications that cite specimens from the herbarium as tangible evidence that the herbarium collections are important to scientific endeavors. Thus, herbaria must keep track of publications that cite their holdings. Surveying the literature is the only way to keep track of changes in the type status or nomenclature of a specimen and of specimens that have been cited as vouchers in scientific studies. No surefire method of finding all the literature that cites specimens or acknowledges an herbarium exists. Most herbaria request copies of publications that cite their specimens as a condition of specimen use. However, not all users comply with this request. It is possible to identify journals likely to publish the sort of taxonomic papers that cite specimens and to check new issues routinely for articles that cite the herbarium. Nonperiodical publications are difficult to survey, although it is possible to check for names of people who have studied specimens from the herbarium (either as visitors or borrowers) in abstracting journals or online library catalogues. A bibliography of publications that cite herbarium specimens can be maintained as a stand-alone file or, ideally, linked to catalogue records of the specimens cited.

An herbarium with a very detailed specimen catalogue, complete with images of all specimens, may be asked to justify maintenance of the actual specimens. The herbarium must respond that physical examination still is required for the study of internal or microscopic features and that researchers continually are identifying new characters associated with herbarium specimens as well as new methods of study. New characters may involve morphological features or chemical features such as secondary metabolites, isozymes, or gene sequences. No matter how complete we try to make a database, we can never anticipate all the uses for specimens.

Herbarium administrators need to develop strategies to ensure that they can inventory and track images and extracts derived from their specimens, including the procedures used to obtain those images or extracts, the location of results from the examination of the product, and the location of any stored extract or product. Ideally, this information should be in or linked to the specimen database so that both potential users and herbarium curators can determine easily if a specimen has been destructively sampled and whether additional destructive sampling is warranted.

Herbarium specimens have uses other than systematics. For example, they document the existence of a species in a particular area at a particular time (e.g., areas of conservation concern). They also serve as sources of information relevant to ethnological, medical, or pharmaceutical studies. Documentation of nontraditional uses of herbarium data is important for demonstrating the relevance of an herbarium collection outside the taxonomic community. A partial list of such uses can be extracted from specimen catalogues, loan records, scientific literature, and visitor logs; telephone or e-mail requests for information, however, are not always covered by these records. Generally, an additional record is needed. Recording the number of queries made to online specimen catalogues may be relatively easy. To obtain more complete information, an herbarium could require a user to check off a use category or to state briefly the purpose of the query at the point of access to the data.

STAFF ACTIVITIES

It is useful for an herbarium administrator to know approximately how long it takes to complete various herbarium tasks. Such information is needed for grant proposals and is important for planning purposes. Keeping track of staff activities usually involves recording the amount of time spent on a particular task and the number of specimens involved. Herbarium activities for which such records might be kept include pulling loans, checking for damage and mending specimens before they are sent on loan or after a loan is returned (including affixing annotation slips), preparing and filing invoices, recording transaction data, packing and unpacking boxes (including fumigation, if necessary), shipping, processing newly acquired material (including duplicate specimens), preparing new specimen labels, filing returned loans and new accessions (including making and filing new specimen folders or box labels), curation (e.g., rearranging specimens in the herbarium, annotations, updating nomenclature); and assisting visitors.

ANNUAL REPORTS

A report that summarizes the activities supported by an herbarium for a calendar or fiscal year can be prepared from various records kept by the herbarium. The annual report shows the breadth of herbarium accomplishments and demonstrates how herbarium personnel are utilized. The annual report is the usual source of information used to prepare grant proposals and to determine the effects of changes in herbarium staffing and activities over time. In addition to summarizing the records described, an annual report also can include accomplishments that routine record-keeping practices miss, such as curatorial projects begun or finished, public service activities of herbarium staff (e.g., tours or demonstrations for school groups), descriptions of major changes in herbarium policies, or the acquisition of new equipment (e.g., herbarium cabinets, microscopes, or computers).

ELECTRONIC SPECIMEN CATALOGUES

Creating an electronic catalogue of the contents of an herbarium is the only practical way to record exactly what is in the herbarium, although the endeavor requires significant commitments of time and money. The ability to query the catalogue number, however, and select specimens based on attributes other than the scientific name (e.g., locality, habitat, morphological features, collector) dramatically increases the power of an herbarium as a tool for biological studies. A specimen catalogue is also useful in helping managers to identify what sorts of new acquisitions are needed to fill taxonomic or geographic gaps in the collection.

ELEMENTS

Any institution developing or upgrading a specimen catalogue must thoroughly analyze its needs and the functions that it expects the database to serve. It is likely that those needs will differ slightly among institutions, although the basic elements of the database, functions, and the managerial concerns associated with implementing a specimen-cataloguing system likely will be quite similar. Most specimen databases will contain some or all of the following database fields (information):

Name data: Family, genus, species, subspecific rank and name, authorities, synonyms or prior identifications, name(s) of determiner(s), and date(s) of determination(s).

Type information: Category of type (e.g., holotype, isotype), bibliographic citation for original publication, lectotype information, and name of verifier of type.

Collection data: Collector name and number, date of collection, names of other members of the collection team, and project under whose auspices collection was made.

Location data: Name of country and subsidiary political subunits in which the collection was made, the exact locality of the collection expressed in text and/or as coordinates, and elevation.

Habitat data: Vegetation type in which collection was made, substratum, specific characteristics of the collection site (e.g., stream bank, north-facing slope), and host information for parasites (i.e., host name, part of host parasitized).

Organism features: Size, color, texture, reproductive stage, and so forth.

Nature of the specimen: Type of preservation (e.g., dried, fluid), location (e.g., general herbarium or special collection), number and location of duplicates, condition of the specimen, nature and location of associated items (e.g., microscope-slide preparations, photographs, live cultures), and voucher status of specimen.

Transaction history of specimen: Loan status (e.g., to whom, when), how obtained (e.g., staff accession, incoming gift, exchange), and source of specimen (e.g., collector, herbarium).

FUNCTIONS

Electronic catalogues can facilitate a number of routine herbarium activities that contribute significantly to the value and utility of the specimens. Among those activities are the following:

Updating Records

Key fields of specimen-catalogue records, such as elements of the name, may be updated repeatedly. It is important to be able to track changes made to a record and by whom without losing or obscuring the original information. The database record for a specimen must provide the same information about the history of its study that can be obtained from the specimen and its annotation labels.

Reports

Herbarium personnel will need to be able to generate ad-hoc as well as standard reports. A typical ad-hoc report might be a list of specimens from a given locality generated in response to an outside request; a standard report might be a specimen label. Use of relational databases (see "Electronic Information Resources," Chapter 4) will allow users to draw information from different but interlinked databases.

Validation

It is very useful to have authority files (i.e., reference tables) for validating information in critical fields such as those for names of organisms (at least family and genus), people (authors, collectors, determiners), publications, and geographic units. A validation system checks an entry against a standard list of entries for a field (e.g., family names), and if the entry is not found, the user is prompted to verify it and add it to the authority file if it is correct. Using a validation routine may slow data entry, but it also significantly reduces data-entry errors. Authority files that can be used independent of the catalogue are useful references themselves.

Bar Codes

Bar codes, which can be generated from a catalogue database, consist of alternating black and white bars that can be interpreted by a reading device as a string of alphabetic or numeric characters. Bar codes provide rapid and accurate access to the computerized record for each specimen and facilitate tracking specimens (Russell 1999). Bar codes have five elements: (1) the medium, which is the label on which the bar code is printed; (2) symbology, either the Universal Product Code (UPC) from the Uniform Code Council or Code 39 from Bar Code 1 (see "Bar Codes," Appendix III); (3) the coded value, which may be a meaningful value or simply an assigned number; (4) the interpreted value, a number that provides an alternative in case the scan bar cannot be used; and (5) a header, which is a title or an herbarium name.

PROCESS

Herbaria that develop computerized catalogues need to know the effort expended on each part of the process and the conditions that affect that effort to evaluate various aspects of the databasing enterprise such as software and cataloguer efficiency, quality of job training, and amount of time spent on each activity. Such information helps administrators to allocate personnel and funds for the development of the specimen catalogue. If data-entry staff members keep track of time spent, number of records entered, and general type of specimens processed (e.g., types, nontypes, historical or recently collected material), such information is easily obtained.

CONCERNS

Most herbarium administrators are constantly asked to do more with less, and few anticipate significant increases in their near-term operating budgets. Advances in computer technology, however, can provide herbaria with new avenues for addressing some of our most pressing environmental problems—opportunities that should not be ignored. Most herbarium administrators and curators recognize the value of computerized specimen catalogues and are eager to develop them, and granting agencies and environmental organizations are encouraging them to do so. However, the development and maintenance of such databases require large and permanent commitments of financial resources. An electronic specimen-cataloguing system may cut the time required for some herbarium activities (e.g., label production, loan processing), and presumably some retroactive data entry will occur. Nevertheless, time for maintenance and enhancement of the catalogue always will be needed, making it unlikely that the time devoted to it will decline once it is completed. Thus, computerization potentially could reduce the time available for herbarium activities that improve the quality of the collection (e.g., updating nomenclature and filing systems, repairing specimens).

The key to addressing that conflict may be lessening the time required for cataloguing through communication and cooperation among herbaria, many of which currently are grappling with the same problems. By staying in close communication, herbaria can share solutions to common problems and reduce duplication of effort. For example, cooperation among herbaria in the construction and maintenance of authority files will save everyone time and resources. Also, it may be possible to combine curatorial activities with cataloguing. Such an approach would be feasible, however, only if cataloguing were performed by experienced staff (i.e., curators or curatorial assistants) rather than data-entry technicians. Clerical staff without special biological or mycological training would not necessarily recognize nor be able to fix curatorial problems. If staff with curatorial training were to take responsibility for cataloguing, the process likely would take longer and cost more per specimen, but the quality of the data would be higher, to the benefit of the herbarium. The latter approach may be most appropriate for institutions that include only selected parts of the herbarium collection in the database rather than those requiring timely completion of a catalogue including every specimen. It also would produce an accurate and well-documented catalogue.

As specimen catalogues or parts thereof are completed, the data should be made available electronically. Ideally, herbarium catalogues that are put online will be sufficiently compatible in structure to allow data from multiple herbaria to be obtained in response to a single query. Such integration will require the data structures to be the same and database updates to be coordinated. Ultimately, some institution or organization will have to take responsibility for coordinating that effort. No such leadership has emerged from the herbarium community yet, probably because most institutions still are grappling with their own cataloguing issues and because institutions have not achieved consensus on how to move forward together. Meanwhile, herbaria must decide whether to proceed on their own or wait for a joint project to develop. Most herbaria cannot really afford to wait, so they proceed with the knowledge that they eventually may have to make changes in their protocols to be compatible with other institutions.

When a specimen catalogue is ready to go online, herbarium staff must decide which data to make available to whom. Making all data available to anyone is not a wise policy. Obvious exclusions are exact locality information for endangered species or species likely to be sought for nonscientific purposes (e.g., choice edible or hallucinogenic fungi) and determinations for specimens that are part of not-yet-published research projects. Formerly, curators kept specimens that were part of unfinished projects separate from the main herbarium collections and simply locked cases containing *Psilocybe* specimens. Technologically it is equally easy to block certain elements of a database or to withhold entire records. Herbarium curators will need to determine who can see each specimen record, which part they can see, when, for how long, and in what format.

Herbarium administrators also will have to decide whether to make their electronic databases available to commercial users and/or whether to charge them for access. In an ideal world, herbarium cataloguing expenses all would be funded, all data would be available free of charge for noncommercial users, and no one would export and/or use the data from an herbarium without approval from and acknowledgment of the host institution. Herbaria whose data are used without attribution have a hard time receiving due credit for their contributions to science, conservation, and other endeavors and are less likely to attract funding. Unfortunately, some institutions may be forced to charge for data to recoup the expense of providing them. Those and similar topics are the focus of considerable current discussion on listservers such as TAXACOM (Appendix

III), a biological systematics discussion group, and in professional societies such as the Natural Science Collections Alliance (formerly the Association for Systematics Collections; see Appendix III). Herbarium personnel should follow the debates carefully and express opinions based on their own perspectives and experiences.

ACKNOWLEDGMENTS. The authors wish to thank the following people who made valuable suggestions of topics for this paper and commented on the manuscript: Harold H. Burdsall, Jr.; John C. Krug; Gregory M. Mueller; Jack Murphy; and Walter J. Sundberg.

3

Preservation and Distribution of Fungal Cultures

KAREN K. NAKASONE, STEPHEN W. PETERSON, AND SHUNG-CHANG JONG

Maintaining and preserving fungal cultures are essential elements of systematics and biodiversity studies. Because fungi are such a diverse group, several methods of cultivation and preservation are required to ensure the viability and morphological, physiological, and genetic integrity of the cultures over time. The cost and convenience of each method, however, also must be considered. We encourage the reader to investigate the excellent papers on fungal preservation by Fennell (1960), Smith and Onions (1994), Smith (1991), and Simione and Brown (1991).

The primary methods of culture preservation are continuous growth, drying, and freezing. Continuous growth methods, in which cultures are grown on agar, typically are used for short-term storage. Such cultures are stored at temperatures of from 5°–20°C, or they may be frozen to increase the interval between subcultures. The methods are simple and inexpensive because specialized equipment is not required. Drying is the most useful method of preservation for cultures that produce spores or other resting structures. Silica gel, glass beads, and soil are substrata commonly used in drying. Fungi have been stored successfully on silica gel for up to 11 years (Smith and Onions 1983). Drying methods are technically simple and also do not require expensive equipment. Freezing methods, including cryopreservation, are versatile and widely applicable. Most fungi can be preserved, with or without cryoprotectants, in liquid nitrogen or in standard home freezers. With freeze-drying, or lyophilization, the fungal cultures are frozen and subsequently dried under vacuum. The method is highly successful with cultures that produce mitospores. Freeze-drying and freezing below −135°C are excellent methods for permanent preservation, and we highly recommend them. However, both methods require specialized and expensive equipment, as described in the next section (see "Liquid Nitrogen" and "Lyophilization" under "Long-term Preservation," later in this chapter).

The choice of preservation method depends on the species of concern, the resources available, and the goal of the project. Some low-cost methods of preservation, such as storage in distilled water and the silica gel method, are good, but none is considered permanent. The maximum duration of storage varies with each method and with the species being preserved, but it generally is 10 years or less. Whenever possible, fungal strains should be preserved with one of the permanent methods (lyophilization, cryopreservation) described later in this chapter (see "Long-term Preservation"). Permanent preservation is essential for strains with critically important characteristics and for type specimens. Cultures that are permanently preserved in metabolically inactive states now can serve as type specimens, according to Article 8.4 of the International Code of Botanical Nomenclature (Greuter et al. 2000).

MAINTENANCE AND PRESERVATION OF CULTURES

SHORT-TERM PRESERVATION

Short-term preservation involves maintenance of cultures for up to 1 year. Most fungal cultures can be maintained for that period by serial transfer. The method is simple, inexpensive, and widely used. Although time consuming and labor intensive, periodic transfer is a good option for small collections with cultures in constant use for short periods (less than 1 year). The method also has several disadvantages, however. Cultures must be checked frequently for contamination by mites or other microorganisms and for drying. In addition, the morphology and physiology of a cultured fungus may change over time. In particular, the ability to sporulate or to infect a host may be lost after repeated transfers. Because of those disadvantages, the technique is generally inappropriate for long-term (more than 1 year) preservation of cultures.

Inoculum is transferred from an actively growing fungus culture to test tubes (screw cap or plugged with cotton or foam) or Petri dishes (wrapped with Parafilm to reduce drying) containing an agar medium of choice. Alternating nutrient-rich with nutrient-poor media at each transfer helps to maintain healthy cultures. Some fungi, such as endophytic and entomopathogenic species, have specific media requirements (Bacon 1990; Singleton et al. 1992; Humber 1994). After a culture is established, it is kept at room temperature or at 4°C. Cultures must be checked periodically for contamination and desiccation. Fungi such as oomycetes and some basidiomycetes (e.g., *Boletus, Coprinus, Cortinarius,* and *Mycena*) should be transferred monthly if kept at 16°C

(von Arx and Schipper 1978). Most filamentous fungi can survive at least 1–2 years at 4°C. Vigorous, sporulating cultures also can be sealed tightly and stored in a freezer at −20°C (Carmichael 1956, 1962) or stored at −70°C (Pasarell and McGinnis 1992) to enhance survival and increase the interval between required transfers (see "Freezing," later in this chapter).

LONG-TERM PRESERVATION

Sclerotization

Some fungi develop sclerotia or other long-term survival propagules in culture as well as in nature; preserving such structures, usually at 3°–5°C, is a good way to preserve fungal strains. Sclerotia and spherules of various myxomycetes have been germinated successfully after 1–3 years of storage. Many soil fungi, such as *Magnaporthe, Phymatotrichum,* and *Cylindrocladium* species, produce sclerotia or microsclerotia that remain viable for 2–5 years (Singleton et al. 1992).

Instructions for inducing formation of spherules, sclerotia, and microsclerotia are available in Daniel and Baldwin (1964) and Singleton and colleagues (1992). Sometimes rice straw or toothpicks are used as substratum to promote sclerotia production in culture. Jump (1954) described a simple method for inducing sclerotium formation in *Physarum* species. A piece of sterile cellophane cut to the dimensions of a Petri dish is placed over a dish containing 1% water agar (Appendix II). An actively growing plasmodium is then transferred to the cellophane and allowed to grow overnight. The cellophane is removed from the agar; placed in a sterile, dry Petri dish; covered; and allowed to dry for 24 hours. The Petri dish lid is then removed to allow the sclerotia to air-dry until brittle. The cellophane is cut into small pieces, each of which is stored in its own screw-cap vial. Alternatively, the sclerotia are removed from the cellophane and stored in a vial.

Oil Overlay

A low-cost and low-maintenance method for preserving cultures growing on agar slants is oil overlay. Cultures can be kept for several years or, in exceptional cases, up to 32 years at room temperature or 15°–20°C. This method is appropriate for mycelial or nonsporulating cultures that are not amenable to freezing or freeze-drying. As an added benefit, oil also reduces mite infestations. Although many basidiomycetes can be maintained this way, the growth rates of the cultures slow as storage times increase (Johnson and Martin 1992; Burdsall and Dorworth 1994). The major disadvantage of the oil overlay technique is that the fungi continue to grow, and thus, selection for mutants that can grow under adverse conditions may occur.

High-quality mineral oil or liquid paraffin is sterilized by autoclaving at 15-lb (6.8-kg) pressure for 2 hours. Entrapped moisture is removed by heating the liquid in a drying oven at 170°C for 1–2 hours (optional). Fungal cultures grown on agar slants are covered with about 10 mm of oil or paraffin. The entire agar surface and fungal culture should be submerged completely in the oil. The tubes are kept in an upright position at room temperature (15°–20°C; 12°C for *Pythium* species and *Phytophthora* species; G. Adams, personal communication). The oil level in the tubes or vials must be checked periodically, and more oil should be added, if necessary. To retrieve a culture from mineral oil, a small amount of the fungal colony is removed and placed on appropriate media after as much oil as possible has been drained. Lifting the Petri dish on one side to form a slight angle often helps the oil drain. It may be necessary to subculture the colony several times to get a vigorous oil-free culture.

Immersion in Distilled Water

Another inexpensive and low-maintenance method for storing fungal cultures is to immerse them in distilled water. Apparently, the water suppresses morphological changes in most fungi. The method has been used successfully to preserve oomycetes (Clark and Dick 1974; Smith and Onions 1983), basidiomycetes (Ellis 1979; Richter and Bruhn 1989; Burdsall and Dorworth 1994; Croan et al. 1999), ectomycorrhizal fungi (Marx and Daniel 1976), ascomycetes (Johnson and Martin 1992), hyphomycetes (Ellis 1979), plant pathogenic fungi (Boesewinkel 1976), aerobic actinomycetes (van Gelderen de Komaid 1988), and human pathogens and yeasts (McGinnis et al. 1974). Most basidiomycetes survived for at least 2 years at 5°C (Marx and Daniel 1976; Richter and Bruhn 1989); viability decreased after 5–10 years of storage (Burdsall and Dorworth 1994). Although Ellis (1979) reported that most of the basidiomycetes he tested survived for 20 months when stored at 25°C, Johnson and Martin (1992) recovered only 26% of the basidiomycete strains stored at 20°C. Ascomycetes, however, including their mitosporic forms, survived up to 10 years when stored at 20°C (Johnson and Martin 1992).

The procedures used for covering cultures on agar slants with oil also can be used when covering them with sterile distilled water. Alternatively, sterilized straws or Pasteur pipettes (large-diameter end) are used to cut disks from the growing colony edge. The disks are transferred to sterile cotton-plugged or screw-cap test tubes filled with several milliliters of water. To save space, small (1.8 ml), sterile, screw-cap cryovials are filled with several discs and topped with sterile distilled water. Test tubes (loosely capped and wrapped with Parafilm) are stored at room temperature; tightly capped tubes and vials are stored at 4°C. Disks are removed aseptically and transferred to fresh agar medium to retrieve cultures.

An alternative method for sporulating fungi (McGinnis et al. 1974) involves inoculating agar slants of preferred media with fungal cultures and then incubating them at 25°C for several weeks to induce sporulation. Sterile distilled water (6–7 ml) is added aseptically to the culture, and the surface of the culture is scraped gently with a pipette to produce a spore and mycelial slurry. This slurry is removed with the same pipette and placed in a sterile, 2-dram glass vial (or cryovial). The cap is tightened, and the vials are stored at 25°C. To retrieve a culture, 200–300 µl of the suspension is removed from the vial and placed on fresh medium.

Organic Substrata

Over the years, researchers have developed practical, effective, and ingenious methods of preserving fungi on various organic substrata such as wood chips, cereal grains, straw, filter paper, and insect and plant tissues. Many of the techniques were developed for pathogenic or other specific fungi and have not been rigorously tested with a range of fungi.

Wood. Wood-inhabiting fungi can be successfully stored on wood chips or toothpicks as long as the colony is growing vigorously (Nelson and Fay 1985; Delatour 1991; Singleton et al. 1992). Some wood-inhabiting basidiomycetes and ascomycetes can be stored on wood chips for up to 10 years. If the fungi do not vigorously colonize the wood chips, however, the method fails.

Small pieces of untreated beech wood (12-mm diameter × 6-mm thick) are added to 2% malt-extract broth (about 60 pieces of wood per 100 ml broth; Appendix II) and sterilized for 20 minutes at 121°C (Delatour 1991). The mixture is sterilized again 24 hours later. About 15 wood chips are drained and placed on a colony of the fungus that is growing on malt extract agar (Appendix II) in Petri dishes. The Petri dishes are sealed with Parafilm, and the fungus is allowed to colonize the wood chips. After 10–15 days, the inoculated wood chips are transferred to sterile test tubes (18 × 180 mm) containing 6–7 ml of 2%-malt agar. The tubes are plugged with cotton and incubated for about 1 week, after which time the cotton is replaced with sterile Parafilm and aluminum foil. The tubes are stored at 4°C. To retrieve a culture, a piece of wood chip is removed and placed on fresh agar medium. The tube is resealed and returned to the refrigerator.

Cereal Grains. Fungi such as *Sclerotinia*, *Magnaporthe*, *Leptosphaeria*, and *Rhizoctonia* species have been stored for up to 10 years on seeds of oats, barley, wheat, rye, millet, and sorghum (Singleton et al. 1992). To preserve isolates of *Rhizoctonia* species, barley, oat, or wheat grains (Sneh et al. 1991) are soaked overnight in water containing chloramphenicol (250 g/ml). The water is removed, and the grain is autoclaved for 1 hour at 12°C over 2 consecutive days. Screw-cap vials are filled with the grain and autoclaved. The vials are inoculated with transfers from the margins of actively growing cultures and incubated at 23°–27°C for 7–10 days. The cultures then are dried thoroughly in a desiccation chamber. The caps are tightened and wrapped with Parafilm, and the vials are stored at −25°C.

Agar Strips. Nuzum (1989) described a method of vacuum-drying fungal cultures on agar strips. *Pythium*, *Rhizoctonia*, and some basidiomycete species survived 18 months with this method, whereas ascomycetes and their mitosporic forms survived from 3–5 years. Fungal cultures are grown on appropriate media in Petri dishes. Strips 1-cm long are cut from the growing edge of the colony and placed in sterile Petri dishes. After 1 week at room temperature, the pieces of dried agar are transferred to sterile ampoules, vacuum-dried, and sealed. To revive cultures, agar strips are placed on fresh medium of choice.

Insect or Plant Tissue. The host tissue can be used as a substrate on which to maintain and store cultures of some pathogenic fungi. For example, roots of plants infected with *Pyrenochaeta* and *Thielaviopsis* can be dried and then frozen (Singleton et al. 1992). *Neozygites fresenii* cannot be cultured in vitro, but Steinkraus et al. (1993) developed a method of preserving viable conidia on frozen, infected aphid mummies.

Soil or Sand

Some fungi can be preserved easily and successfully for many years in dry, sterile soil or sand. This low-maintenance and cost-effective method is appropriate for fungi such as *Rhizoctonia* (Sneh et al. 1991), *Septoria* (Shearer et al. 1974), and *Pseudocercosporella* (Reinecke and Fokkema 1979). Dormancy caused by dryness can take time to develop, however, and morphological changes in some fungi have been recorded.

Glass bottles (60 ml) are filled to two-thirds capacity with sand or loam soil (water content 20%) and then sterilized by autoclaving for 20 minutes at 120°C. The bottles are allowed to cool and then sterilized again.

Sterile, distilled water is added to a culture, and the colony surface is scraped gently to produce 5 ml of spore or mycelial suspension. One milliliter of the suspension is added to each bottle of soil or sand. After 2–14 days of growth at room temperature, the bottles are capped loosely and stored in the refrigerator at 4°C. To retrieve the fungus, a few grains of soil are sprinkled onto fresh agar medium. Test tubes or vials can be used in place of glass bottles to save space.

Silica Gel

The silica gel method can be used to preserve sporulating fungi if facilities for freeze-drying or for storage in liquid nitrogen are not available. It originally was developed by Perkins (1962) for *Neurospora* species. He found that sporulating fungi protected by skim milk and stored on silica gel remain viable for 4–5 years. Spores and microcysts of dictyostelids can be preserved for up to 11 years on silica gel (Raper 1984). In general, viability after storage on silica gel depends on the strain of fungus and the medium on which it was grown before storage. When cultures are stored in soil, the initial growth period before storage may permit variant vegetative strains to develop and overgrow the wild type, or saprotrophic segregants to overgrow pathogenic ones. The advantage of silica gel is that it prevents all fungal growth and metabolism. Some researchers use glass beads instead of silica gel.

Revival of cultures from silica gel is easy—a few silica gel crystals are scattered on an agar plate. The same storage container can be used for successive sampling. The Fungal Genetic Stock Center (Appendix III) has used this technique successfully since 1962 for preserving genetic stocks of *Aspergillus nidulans* and *Neurospora crassa*. Fungi such as *Pythium* and *Phytophthora* species, however, do not survive this process.

Protocol A. Screw-cap tubes are filled partially with 6- to 22-mesh silica without indicator dye, which has been sterilized with dry heat for 90 minutes at 180°C and stored in tightly sealed containers (Perkins 1962). Spores are suspended in a 10% (v:v) solution of dry powdered skimmed milk in distilled water, previously cooled to 4°C. The silica gel also is chilled to about 4°C and placed in an ice-water bath. The spore suspension is added to the silica gel to wet about three-fourths of the gel (0.5 ml/4 g) and left in the bath for 30 minutes. Tubes are stored with the caps loose at room temperature for 1–2 weeks. Viability is checked by shaking a few crystals onto a suitable medium. If the cultures are viable, caps are tightened, and the tubes are stored in a tightly sealed container at 4°C.

Protocol B. A slight modification of Protocol A also accommodates nonsporulating strains of fungi (Perkins 1977). Cotton-plugged test tubes (13-mm diameter × 100 mm high) partially filled (to 65 mm) with 12- to 20-mesh silica gel without indicator dye are heat sterilized at 180°C for 2 hours and stored at room temperature in tightly sealed containers until needed. Fungal cultures are grown for a week on suitable agar slants. For sporulating strains, about 0.5-ml sterile water is pipetted gently into the tube of a sporulating culture. Spores are suspended using a vortex-type mixer. For nonsporulating strains, aerial hyphae and mycelia are scraped from the agar surface with a sterile blade and transferred to a test tube (10-mm diameter × 75 mm tall) containing 0.5 ml water. The mixture is ground with a pipette or glass rod against the tube wall to obtain a creamy homogenate. About 0.5 ml sterile skim milk is added and stirred gently. The cotton plug covering the silica is removed, and the suspension is pipetted over it, drop by drop. The tube is agitated briefly with a mechanical mixer to distribute inoculum over as many particles of silica gel as possible and then is placed in an ice-water bath for 15 minutes. After 1 day at room temperature, the particles appear dry, and the tube is sealed against moisture with Parafilm. Tubes are stored at 5°C or −20°C in a moisture-proof box. Storage at low temperatures can increase the survival period twofold to threefold over storage at room temperature.

Freezing

Most fungal cultures frozen at −20° to −80°C in mechanical freezers remain viable. Freezing with liquid nitrogen is discussed in the following section. Cultures grown on agar slants in bottles or test tubes with screw caps can be placed directly in the freezer (Carmichael 1956). Overall failure rate for mitosporic ascomycetes, zygomycetes, and yeasts after 5 years in storage at −20°C was 5.1% (Carmichael 1962). The failure rate of medically important fungi, aerobic actinomycetes, and algae stored from 6 months to 13 years at −70°C was 2.3% (Pasarell and McGinnis 1992). Ito (1991) and Ito and Yokoyama (1983) aseptically removed six 6-mm disks from vigorously growing cultures of nonsporulating basidiomycetes and ascomycetes and placed them in sterile cryotubes containing 10% glycerol in water. The cultures were preserved successfully for up to 5 years by mechanical freezing at −80°C. Fungi grown on various organic substrata, such as cereal grains, agar strips, plant parts, and filter paper, and then dried can be frozen (see "Organic Substrata," earlier). In general, vigorously growing and sporulating cultures survive the freezing process better than less vigorous strains. We do not recommend repeated freezing and thawing, which will significantly reduce viability of the cultures.

Liquid Nitrogen

Storage in liquid nitrogen is an effective way to preserve many, if not most, organisms, including those that cannot be lyophilized. It costs somewhat more than lyophilization, however, because liquid nitrogen must be replenished every few days. Liquid-nitrogen storage is recommended for the preservation of dictyostelids (Raper 1984), amoebae (Davis 1956; Evans 1982), zygomycetes including Entomophthorales (Humber 1994), oomycetes (Nishii and Nakagiri 1991), phytopathogenic fungi (Dahmen et al. 1983), and yeasts (Kirsop 1991). Ascomycetes that sporulate poorly in culture, and higher basidiomycetes that generally grow only as mycelia in culture also can be stored in liquid nitrogen.

Because the rates of mutation in cultured fungi are likely to correspond to those of cell division and metabolic activity, storage methods that stop cell division completely and totally arrest metabolism, while still retaining viability, are best. Freezing at or below −139°C, the temperature at which ice crystals do not grow and rates of other biophysical processes are too slow to affect cell survival, accomplishes this. All fungi can be cryopreserved, but this method generally is reserved for fungi that do not sporulate in culture, fungi that have large or delicate spores that will not survive freeze-drying, genetic stocks, and dangerous human pathogens. In addition, many culture collections, such as the American Type Culture Collection (ATCC; Appendix III), store their seed stock in liquid nitrogen so that when distribution stock is depleted, the material used for replenishment will be as genetically close to the original deposit as possible.

Because living cells can be damaged severely by freezing and thawing, chemical cryoprotectants are used in most protocols. Cryoprotectants are of two types: penetrating agents such as glycerol and dimethyl sulfoxide (DMSO), which readily pass through the cell membrane and protect intracellularly and extracellularly, and non-penetrating agents such as sucrose, lactose, glucose, mannitol, sorbitol, dextran, polyvinyl-pyrrolidone, and hydroxyethyl starch, which exert their protective effect external to the cell membrane. Glycerol and DMSO have proved to be most effective for fungi, although polyethylene glycol, another penetrating agent, can be used also (Ohmasa et al. 1992).

The major advantages of liquid nitrogen storage include prevention of increased genetic variability of distributed culture stocks; time-saving, reduced labor requirements compared to handling of living stocks; elimination of the need for repeated pathogenicity tests, prevention of culture loss from contamination; and increased assurance of long-term availability of cultures. The major disadvantages of this technique are the relatively high cost of the apparatus and the liquid nitrogen that must be replaced every 2 days, the space requirement for refrigeration units, and the need for constant surveillance. Culture collections, such as the ATCC, usually maintain a backup off-site liquid-nitrogen storage facility. One disadvantage affecting distribution of strains preserved in liquid nitrogen is that they first must be grown on agar or in liquid medium to avoid the expense of shipping frozen materials.

Storage in the vapor phase of the liquid-nitrogen freezer is an attractive alternative to immersion in the liquid nitrogen. Tubes that are immersed must be sealed carefully to prevent entry of liquid nitrogen. A tube filled with liquid nitrogen that is quick-thawed at 37°C is likely to explode from the pressure created by the expanding nitrogen. Another concern is that cultures in leaky vials may be contaminated with bacteria or spores that may be present in the liquid-nitrogen freezer. Various alternative techniques for liquid-nitrogen storage, such as using plastic straws instead of vials or tubes, have been reported (Stalpers et al. 1987; Kirsop 1991). Other protocols such as the one used by Gulya et al. (1993) to preserve zoosporangia of downy mildew for up to 4 years in liquid nitrogen require neither cryoprotectants nor controlled freezing regimens.

Procedures used to harvest materials for preservation in liquid nitrogen differ depending on whether the fungus sporulates, has mycelia that penetrate below the surface of the agar, or grows only in liquid culture. Samples of human pathogens are scraped from the agar surface, or agar plugs are cut from the cultures. Such samples are never macerated in a mechanical blender because of the hazard of aerosol dispersion of the pathogen.

Protocol A. For fungal cultures that do not sporulate or that produce mycelia that grow deep into the agar, sterilized 2-ml screw-cap polypropylene vials are filled with 0.5–1.0 ml sterile 10% glycerol. Plugs 4 mm in diameter are cut from vigorously growing cultures using a sterilized plastic straw. Several plugs are placed in the vial, the cap is tightened, and the tube is placed directly into the vapor phase (temperature is about −170°C) of a liquid-nitrogen tank. The accession number should be written on each cryovial with a cryoresistant-ink lab marker, printed onto paper, and then taped to the vial, or it should be printed onto a special cryoresistant adhesive label, which is readily available from biotechnology supply companies. The location of storage in the freezer must be indexed for rapid retrieval. Frozen preparations are retrieved by removing the vials from the freezer and rapidly thawing them in a 37°C water bath. The thawed agar plugs are placed on appropriate agar plates. Viability of the cultures should be checked from 2–7 days after storage.

Protocol B. To make suspensions of spores or mycelial fragments from cultures growing on the surface of agar slants or plates, the colony surface is flooded with 10% glycerol or 5% DMSO and gently scraped with a pipette. Preparation continues according to the "General Protocol," described later in this chapter.

Protocol C. The mycelium of a fungus that grows only in liquid culture must be macerated before it can be pipetted into vials. The broth culture is fragmented for a few seconds in a sterile mini-blender and mixed with equal parts of 20% glycerol or 10% DMSO to give a final concentration of 10% glycerol or 5% DMSO, respectively. The mixture is then treated as described in the following section.

General Protocol. The fungal mycelial and spore suspension is pipetted in aliquots of 0.5 ml into sterile 2-ml screw-cap polypropylene vials. Filled vials are placed into prelabeled cans in racks that then are put into the freezing chamber of a programmable freezer and allowed to equilibrate at 4°C for about 10 minutes. They are then cooled from 4° to −40°C at a rate of 1C° per minute and from −40° to −90°C at 10C° per minute. After reaching −90°C, vials are transferred immediately to liquid-nitrogen vapor at −150° to −180°C. Polypropylene vials are not immersed in liquid nitrogen.

Cultures are thawed rapidly by placing vials in a warm-water (37°–55°C) bath until the last trace of ice dissipates. Cultures in glass ampoules and straws thaw in less than a minute; those in polypropylene containers take longer. Culture samples are then transferred aseptically to appropriate growth media.

Lyophilization

Lyophilization, or freeze-drying, a low-cost form of permanent preservation, is not appropriate for all fungi. In fact, the technique is used primarily with species that form numerous, relatively small propagules. However, Croan (2000) demonstrated that mycelial isolates of basidiomycetes can be lyophilized effectively in the presence of trehalose. Lyophilized spores of dictyostelids, with associated bacteria, were maintained successfully for up to 30 years (Raper 1984). This procedure is the preservation method of choice for many spore-forming fungi that produce large numbers of spores 10-µm or less in diameter. Larger spores tend to collapse during the lyophilization process, and the structural damage caused is not reversible by hydration. A significant number of the spores of appropriate size also are physically damaged and killed during the freezing process by the formation of ice crystals. Thus, each ampoule initially must contain many viable spores. Rapid freezing and the addition of a menstruum that dissolves ice crystals minimize growth of ice crystals. The two most common menstrua are nonfat dry milk powder (sterile 5% or 10% solution) and filter-sterilized bovine serum, although other proteinaceous materials also can be used.

Equipment and Supplies. Material required for lyophilization include the following: high-quality mechanical vacuum pump (e.g., Edwards two-stage pump); vacuum gauge; vacuum manifold; cold trap; hoses to connect pump, trap, and manifold; insulated bath; support stand for manifold; oxygen-gas torch; oxygen supply; 10-cm lengths of 6-mm soft glass tubing with one end heat sealed, or lyophilization ampoules; cotton for plugging tubes; Pasteur pipettes; mechanical or electrical pipetting aid; sterile menstruum; and permanent ink suitable for writing on glass.

General Protocol. An agar slant with medium that supports good growth and sporulation is inoculated with the organism, which is allowed to grow until it reaches the resting phase. Lyophilized preparations from cultures much younger or older than resting phase often exhibit very low postlyophilization viability. Five or more lyophilization tubes are sterilized and labeled for immediate use. About 1.5–2.0-ml sterile menstruum is added to an agar slant; spores are suspended in the menstruum by gently scraping the agar surface with a Pasteur pipette. If sporulation has been light, the menstruum-spore mixture is transferred to a second agar slant, whose spores are added to the mixture. Approximately 200 µl of the spore suspension is added to each of several lyophilization tubes. Tubes are plugged loosely with cotton, the open end of the glass is lubricated with castor oil, and the tubes are placed on the vacuum manifold. The manifold is lowered until the lyophilization tubes are immersed in a dry ice and ethylene glycol bath that is maintained between −40°C and −50°C while the contents of each tube freezes (about 5 minutes). A vacuum is applied to the system for about 30 minutes while the bath warms to about 0°C. The manifold then is raised to remove the tubes from the solvent bath. Drying of the lyophilization preparations continues at room temperature until the pressure in the system

reaches about 30 milliTorrs. Evaporative cooling keeps the samples frozen during the drying process. The tubes then are sealed under vacuum using a gas-oxygen torch. Finished lyophilization ampoules are stored in numbered plastic boxes or sealed plastic bags in a 4°C refrigerator. The purity and viability of the preparation in one lyophilization vial should be checked 1–2 weeks after preservation. Viability can be roughly categorized as poor (10–50 colonies/ampoule), moderate (50–100 colonies/ampoule), or good (100–1000 colonies/ampoule).

Bunse and Steigleder (1991) described an alternative method for the lyophilization of cultures grown on agar slants in glass ampoules that preserves the fungus as well as the macromorphology of the cultures. Good results were obtained with *Penicillium, Aspergillus, Cladosporium, Alternaria, Mucor, Candida,* and *Rhodotorula* species but not with any dermatophyte species.

RECORDS AND RECORD KEEPING

For a small research collection (i.e., fewer than 500 cultures) the least costly, yet effective, method of keeping records is through the use of index card files. Records for the Agricultural Research Service Culture Collection (NRRL; Peoria, Illinois) filamentous fungus collection records originally were kept on three sets of 3″ × 5″ index cards. One set of cards was arranged alphabetically by species name; the other two sets were arranged by accession number. In this way one could readily look up species using the alphabetical file and quickly identify cultures labeled with their accession numbers using the numerical file. The cards were the only database kept for many years and included information on more than 10,000 fungal strains. The third set of cards was used as an inventory of vials of lyophilized specimens and specimens stored in liquid nitrogen available for distribution.

Given the common availability of personal computers and the low cost of database programs, it is now easy to enter the records for even small culture collections into electronic databases. The primary considerations when choosing a computer and a database program are universality, or intercompatibility. Most of the common commercial (and some shareware) database programs can export data in a standard format (e.g., comma-delimited fields) or a system data format. If a database can export data in one of those formats, the data can be transported to practically any database platform at will. A second consideration is ease of searching. Some database systems have a user-friendly visual interface that leads the user through all the steps necessary to search fields in the database. Probably the largest single expense in creating the database (after procuring the computer) is data entry. Making a card index or an electronic file, however, would require similar investments of time. Once the data are in electronic form, they can be transferred to other programs and media. It is essential that electronic data files be backed up and stored in different locations to minimize the possibility of their catastrophic loss (e.g., to weather-related factors, fires, mechanical failures of the computer).

Each strain in a culture collection is recorded in the database, which should include separate fields for (1) strain number; (2) genus; (3) species; (4) variety; (5) taxonomic authority; (6) required growth conditions; (7) date of isolation; (8) locale of isolation; (9) isolation substrate or host; (10) person who isolated the culture; (11) provenance of the strain (e.g., isolated by Thom 1917, sent to CBS in 1934 as CBS 225.34, CBS 225.34 sent to IMI as IMI 101557 in 1956, IMI 101557 sent to NRRL as NRRL 28994 in 1995); (12) isolate numbers in other collections that reference the same strain (i.e., CBS 225.34 = IMI 101557 = NRRL 28994 = ATCC 1076); (13) person who identified the strain; (14) person, if anyone, who has reidentified the strain; (15) subjective synonyms; and (16) comments that do not fit well into other fields. Other useful information that might be maintained includes published references to the isolate, metabolites produced, transformations, and so forth. The manner of the preservation of each strain also is recorded, usually in a separate database, along with particulars of the long-term preservation of the strain, number of replicate preservation vials and dates they were prepared, and test viability of the organism following preservation. In addition, if germ plasm is distributed on a regular basis, it is handy to keep a listing of the individuals to whom particular preservation vials have been sent. Such individuals can be contacted to determine (via user feedback) how well a preservation technique is working, whether a strain is pure, and whether a strain conforms to the original description and produces the expected products.

Data-entry audit should be an integral part of any database. In an optimal situation, data are entered into the system by a laboratory technician or a data-entry technician, the entered data are checked for correct spelling using an electronic spell checker, and the checked data are printed out for review by someone who did not participate in the original data entry. In many database systems, data may be entered and then held in a temporary file until audited. Once checked and approved, the data can be added to the database by a batch append command.

DISTRIBUTION AND EXCHANGE

GENERAL PRACTICES

Service culture collections will supply any culture if one is a bona fide scientific investigator associated with a properly equipped laboratory. Some cultures in a service collection, such as plant or animal pathogens, require permits from the U.S. Department of Agriculture (USDA) or the U.S. Public Health Service (USPHS), an export license for shipment abroad, or the acceptance of responsibility statements on certain pathogens. The use of most of the cultures in service collections, including biological materials cited in expired or invalid patents, or in abandoned patent applications is not restricted. Use of cultures cited in valid patents, however, is restricted. Such materials are distributed to the scientific community for research purposes only. In contrast, some patent applicants do not restrict access to their cultures while a patent is pending, but limit use of such cultures to research.

Also available from service collections is biological material that is patented per se (i.e., material that is itself the subject of a patent). Use of such patented material during the life of the patent is limited to research purposes. Other strains to which special conditions apply include safe-deposit strains held on a confidential basis for the depositor and strains isolated by a staff member or currently forming part of a staff member's research. Those strains are not catalogued or generally made available to the public, although they may be supplied under special terms on a case-by-case basis.

PERMITS AND REGULATIONS

Because of the pathogenic or hazardous nature of certain microorganisms, containment and security for domestic and international shipments receive particular attention. Such shipments may require permits or licenses depending on the type of pathogen (human, animal, or plant) and its destination. Permits for shipment serve the following purposes: (1) to ensure that persons receiving the cultures are qualified to use them and that their laboratories are properly equipped to do so; (2) to inform government agencies about the volume of material being shipped and its destination; and (3) to provide information about the shipment in the event of an accident during shipment. For more information on international import and export restrictions, see Rohde and colleagues (1995) and Rohde and Claus (1999).

Regulations are well defined by several U.S. government agencies, including the USPHS, the USDA, and the U.S. Department of Transportation (DOT). International shipments are governed by the rules and regulations of the U.S. Department of Homeland Security (Bureau of Customs and Border Protection) and the U.S. Department of Commerce (DOC). The USPHS, in the Code of Federal Regulations (42 CFR Part 72), defines a hazardous organism or etiological agent as a viable organism or its toxin that causes or may cause human disease. The USPHS requires a permit (CDC0.753) for importation of such organisms into the United States. The regulation is administered by the Centers for Disease Control and Prevention (CDC) through its Foreign Quarantine Program. The CDC also regulates the packaging and shipping of human pathogens for interstate transport (42 CFR 72.3).

The USPHS designated organisms as Class I, II, III, or IV in *Classification of Etiologic Agents on the Basis of Hazard* (USPHS 1974) and described biosafety levels for infectious agents in *Biosafety in Microbiological and Biomedical Laboratories* (Richmond and McKinney 1999).

1. Class I organisms—no recognized hazards, special packaging not required.
2. Class II organisms—ordinary potential hazard, distribution restricted to bona fide laboratories, special packaging required.
3. Class III organisms—special hazards, special packaging and a letter from the receiving institution acknowledging the hazard required.
4. Class IV organisms—potentially dangerous to human health, USPHS and/or USDA permit required for shipping.

Plant pathogens are described in APHIS (Animal and Plant Health Inspection Service of the USDA) 7 CFR 330.100 as organisms that can directly or indirectly injure, cause disease in, or damage any plant or plant part, or any processed, manufactured, or otherwise produced products of plants. The USDA regulates the movement of all plant pathogens and mycotoxin-producing strains across interstate or international boundaries and requires a permit from the Plant Protection and Quarantine (PPQ) Program for doing so. The investigator who receives the culture must complete Form PPQ-526, "Application and Permit to Move Live Plant Pests." The form then must be submitted to the plant regulatory official of the state in which the culture will be used. The state in turn mails the form to PPQ for approval. These safeguards are set in place by USDA and similar authorities in foreign countries to protect the environment from introduced pests and diseases.

The APHIS requires Form VS 16–3 for the importation of extremely virulent pathogens of livestock or poultry or of those for which a national eradication control program exists. The regulations are stated in 9 CFR Part 122.2. An organism that has been genetically engineered via recombinant-DNA techniques from a donor organism, vector, or vector agent that is a plant pest or contains plant pest components requires APHIS Form 2000 (7 CFR Part 340).

The DOT is responsible for establishing and enforcing regulations for safety aspects of transportation (Appendix III), which include, but are not limited to, infectious substances in domestic transport (49 CFR Parts 171–178). The regulations cover (1) classification of materials, (2) packaging, (3) hazard communication, (4) transportation and handling, and (5) incident reporting.

The U.S. Bureau of Customs regulates the importation of materials into the United States. It determines if materials are admissible and if they should be referred to other government agencies for examination, permits, and release. The Customs Bureau judges if import duties should be paid and if the packages being imported do contain the goods that are manifested or declared.

The DOC regulates the export of biological materials through the Bureau of Export Administration. Any organism or toxin that appears under Export Control Classification Number 1C61B requires a validated export license for all foreign destinations except Canada. Materials that do not require a validated license may be shipped under a general license (G-DEST). Institutions apply for this general license once and label each appropriate shipment as such. There is a mandate for the DOC to identify organisms that might be involved in biological warfare and to place export controls on them. Several other countries, such as Australia, New Zealand, and Germany, also require import permits and licenses.

The United Nations (UN) Economic and Social Council staff (2001) provides the international shipping regulations in "Recommendations of the Transport of Dangerous Goods." The current DOT regulations are based on the UN recommendations applicable to hazardous material transport in the United States. Those who export biological material to foreign countries are required to follow the UN regulations. Mandatory requirements for shippers of dangerous goods have been in effect since 1990. A 24-hour emergency number allows for rapid responses to numbers of hazardous materials. These international regulations are explained in detail by Rohde and Claus (1999). In addition, the World Federation for Culture Collections Committee on Postal, Quarantine and Safety Regulations (Appendix III) should be consulted to keep abreast of the ever-changing regulations.

In Canada microorganisms are assigned to one of four risk groups ranging from a low individual and community risk, level 1, to a high individual and community risk, level 4 (Anonymous 1996b). Each risk group has a corresponding containment level. Containment level 1 is found in a basic microbiology laboratory, whereas containment level 4 represents a geographically isolated unit functionally independent of other areas. Agriculture Canada requires an annual import permit for all plant and animal pathogens shipped to Canada. Some pathogens require a single-entry permit; others can be shipped under a multipurpose permit. As of September 1994, all human pathogens (Risk Groups II–IV) require a Health Canada permit. To import a pathogen that is both a human and animal pathogen, permits from Health Canada and Agriculture Canada are required. Shipments of extremely hazardous agents require a Canadian Department of Transport Emergency Response Assistance Plan.

The Working Party on Safety in Biotechnology of the European Federation of Biotechnology (EFB) has classified organisms as harmless (EFB Risk Class 1); low risk (EFB Risk Class 2), which means that they may cause disease and might be a hazard to laboratory workers but are unlikely to spread in the environment; medium risk (EFB Risk Class 3), organisms that severely threaten the health of laboratory workers but represent a comparatively small risk to the population at large; and high risk (EFB Risk Class 4), organisms that cause severe illnesses. Classes 2–4 fall into special containment categories.

Shipping Regulations

Shipping of noninfectious, perishable biological organisms within the United States is generally not a problem once required permits are obtained. Freeze-dried cultures, agar slants, and broth cultures may be shipped via the U. S. Postal Service (USPS) or private freight carrier. Shippers are responsible for the safety of those handling and receiving the culture. However, the USPS, in its Domestic Mail Manuals and International Mail Manuals, and the DOT, in 49 CFR Part 173, all require that etiological agents be packaged in accordance with USPHS guidelines (42 CFR Part 72). Requirements for packages containing infectious substances are described in 49 CFR Part 178.609. Federal regulations and permit requirements are summarized in Richmond and McKinney (1999). Some state governments also may require special permits. For more information on international regulations, see Rohde et al. (1995) and Rohde and Claus (1999).

Etiological agents shipped domestically via the USPS must be sent as first-class mail, priority mail, or express mail. Exceptions are agents listed in 42 CFR 72.3 (f), which must be sent by registered mail or an equivalent

system that notifies the sender when the package has been received. International shipments must be sent via registered mail. For shipping to Canada, any package labeled "infectious substances" must be sent by private carrier following International Air Transport Association (IATA; a trade organization of airlines) regulations. At the present time, only Federal Express will carry such materials.

The USPS will accept only properly packaged cultures. A nonpathogenic culture is placed in a foam insert, which is then put into a fiberboard cylinder. For freight carriers, the cylinder is then placed inside a corrugated box. Both domestic and international shipments of etiological agents or etiological agent preparations are limited to a volume of 50 ml. The primary container (e.g., test tube, vial) must be securely sealed and watertight and must be enclosed in a second, durable, watertight container (secondary container). Several primary containers may be enclosed in one secondary container. Enough absorbent material must be placed between primary and secondary containers to absorb the contents of the primary container(s) if it breaks. For cultures shipped by freight carriers, the set of containers must be enclosed in an outer shipping container constructed of corrugated fiberboard, cardboard, wood, or other material of similar strength.

When volumes of greater than 50 ml are shipped, a layer of shock-absorbent material must be placed between the secondary container and the outer container. Nonpathogenic cultures are limited to 1000 ml per primary container and 4000 ml per outer container.

According to 42 CFR Part 72, if cultures are shipped on dry ice, the ice must be placed outside the secondary container in such a way that the container does not become loose as the ice sublimes. Outer containers must be vented to allow carbon dioxide gas to escape. Dry ice cannot be sent by international mail; cultures in dry ice must be shipped by freight carrier. If wet ice is used, it must be placed between the secondary container and the outer container. The outer container should be designed so that it does not collapse after the ice melts, and the entire package should be leak-proof. Liquid-nitrogen containers must be able to withstand ultra-low temperatures, and DOT regulations in 49 CFR Part 173.316 must be observed.

The Universal Postal Union, the International Civil Aviation Organization (ICAO), and the International Maritime Organization (IMO) regulate the transport of cultures between countries. The United Nations Economic and Social Council Committee of Experts on the Transport of Dangerous Goods (2001) provides recommendations that are enforced by ICAO and IMO. The IATA publishes a manual of air-transport procedures in agreement with the ICAO regulations, which are followed by freight carriers worldwide. Because most cultures are sent by air, IATA regulations must be followed whether the material is sent by the USPS or a freight carrier.

The only requirement when transporting less than 50 ml of an etiologic agent by registered airmail is compliance with packaging regulations. If more than 50 ml of an etiologic agent is shipped, the containers must be specially tested and the material can be sent only by cargo aircraft (DOT 49 CFR 173.387). In addition to the required Biological Substance mailing labels and forms set forth by USPS, private shipping companies require that a telephone number at which a person knowledgeable about the shipment and its contents can be reached 24 hours a day be provided. Growing cultures, which are not acceptable for mailing to certain countries, are sent by freight carrier.

Several agencies require labels for etiological agents or infectious substances, depending on the destination, the state of the shipment (dry or frozen), and the carrier (USPS or freight carrier). All shipments of etiologic agents within the United States must carry the CDC etiologic agent label. Shipments of more than 50 ml of a culture also must carry a "Cargo Aircraft Only" label. Packages sent by air-freight carrier must carry an IATA infectious substance label and a Shipper's Declaration for Dangerous Goods from the IATA and must be marked as a UN 2814, Class 6.2 shipment with the name of the agent, volume, and name and telephone number of the responsible party noted.

All foreign shipments of biological materials, regardless of hazard, must have a Shipper's Export Declaration (DOC Form 7525-V), which verifies the existence of a validated export license or a general license. If sent by the USPS (air mail only), volumes of culture exceeding 50 ml also must carry the green customs label (PS 2976) and a "Cargo Aircraft Only" label. Etiologic agents must carry an IATA infectious substance label and a Shipper's Declaration for Dangerous Goods and must be labeled as a UN 2814 Class 6.2 shipment, specifying the volume, the name of agent, and the name and telephone number of the responsible party.

When a culture is shipped in dry ice, an additional IATA Miscellaneous 9 label and a UN 1845 designation with the amount of dry ice noted in kilograms is required. If the net weight of the dry ice is greater than 5 lb (ca. 2.3 kg), a Shipper's Declaration for Dangerous Goods must be included. When a frozen shipment uses liquid nitrogen, an IATA nonflammable gas label and "Do Not Drop" and "Handle With Care" labels also are required. The words "Keep Upright" with arrows for proper orientation must be placed on each side or at 120° intervals around the package. The outer shipping container is marked "Nitrogen, Refrigerated Liquid, Class 2.2 UN 1977."

ACKNOWLEDGMENTS. We thank Drs. Lynne Sigler and Don Wicklow for helpful comments and suggestions on the manuscript.

4

ELECTRONIC INFORMATION RESOURCES

DAVID F. FARR AND ELLEN R. FARR

In recent years efforts of the world's scientists, industrialists, and government policymakers to improve understanding, use, and conservation of biological diversity have exploded. Increased understanding is based largely on biodiversity studies that generate great amounts of data concerned with, among other things, the systematics and distribution of organisms. Many of the most useful analyses of that information will involve pooled data from many sources. At this time, much of the available information on biodiversity remains with the individual scientists who collected it. Providing a mechanism by which the broader community of biodiversity researchers can consolidate and then easily access those data to answer its questions is a challenge for data managers. Clearly computers will play a major role in the data management.

The needs of the biodiversity community could be met most easily either by having data entered at a central location or by providing scientists with a standard software application for recording their observations. Both are viable options that have been used successfully. In our experience, however, many of the researchers who generate valuable biodiversity information have not been amenable to those approaches, often because they view the programs as irrelevant to their needs and of little value in promoting their individual research goals. Given that situation, it is perhaps more appropriate to stress that centralized collection and management of information is not as important as ensuring that information is available, compatible, and comparable.

Virtually everyone at a modern research facility uses a personal computer for word processing; this is not the case with data management. Until recently, PC database software was not very user friendly, and many scientists used it only for simple applications. As a result, many data that have remained in notebooks or appeared in research publications must be rekeyed to make them available electronically. Over the last few years, however, several manufacturers have introduced new programs that have improved the situation significantly, providing the scientist with new opportunities to manage his or her data using database software.

Few mycologists have advanced skills in manipulating computer databases. Many have only an elementary knowledge of database design and data manipulation tools and lack readily available support. Our goal in this paper is to stimulate scientists to include computer databases in their data management strategies. We provide general guidelines that should help to ensure that the information in the databases they develop will be comparable as well as compatible with that in databases developed by others (see "Database Details," in the Appendix, this chapter). Thus, more information will be integrated into large and readily available datasets necessary to meet the needs of the biodiversity community.

In this chapter we provide a general overview of the way current software packages manage data and how they may be helpful to the museum curator or researcher. We also introduce some general concepts important in building computer databases, using a sample application to develop the concepts of database design. In the Appendix, we provide a detailed list of the entities that form the core of a specimen database—that is, the information that documents the occurrence of an organism in a given place at a given time (see "Core Data Structure for a Specimen Database," in the Appendix, this chapter).

The goals of individual scientists and those of the consumers of biodiversity information are compatible. A well-thought-out personal database management application can serve the needs of the researcher and provide valuable data for comprehensive biodiversity studies. Olivieri and colleagues (1995) have provided an excellent treatment of the vast array of issues involved in managing the information developed in biodiversity studies. Anyone interested in the broader aspects of this subject should consult that publication.

DATABASE DESIGN

FIELDS AND TABLES

Designing a database is best done from the top down. First the purpose of the database is determined, then the subjects that will be covered, and finally the specific bits of data that will be entered. Although a database can have broad coverage (e.g., maintaining all information available about a specimen), we recommend that initial attempts at design be more restricted. For this discussion we will assume that the purpose of the database is to manage a collection of photographs.

Among the best sources of information for the development of a database are the reports or notes that the investigator currently is generating. For example, a notebook entry might contain the following information:

Collection no. 3456, May 10 1996, roll 33-10, *Pholiota terrestris*, 1/30 sec

or

Specimen no. 3333, *Septoria florida*, conidia, koh and phloxine, ×100, exposure reading 0, roll 34-11, tech pan, compound scope

The data in those entries appear to fall into logical groupings. For example, we can assign the information to three categories: name, negative, and roll. We will create a separate table for each of those subjects. We next need to decide which data elements should be assigned to each table (see "Table Design for Specimen Data," in the Appendix, this chapter). For the roll table we will use fields for type of film and type of camera or microscope. The negative table will have fields for exposure, magnification, mounting medium, roll number, negative number, and structure observed. The name table will contain information on the name of the organism—in this case, a fungus. To simplify this discussion, we are putting the full name in one field, although in most cases one likely would place each element of a name (e.g., genus, specific epithet, family) in a separate field. The entry in each table also will need a record number. That can be a single field or a combination of fields that uniquely identify each record. For the name table that would be an arbitrary number, and in the roll table the roll number could be the record number. There are a couple of possibilities for the negative table. One possibility would be to use an arbitrary number; another would be to use the combination of the roll number and negative number. We will use an arbitrary number.

General points important when selecting fields and building tables follow.

For fields:

1. The data in the fields should describe the subject of the table.
2. The data should be broken down into the smallest logical unit.

3. In general it is difficult to parse the data into too many fields.
4. The designated fields should be checked against field notes and reports to ensure that they cover all of the necessary information.

For tables:

1. Tables should only include fields that pertain to the same subject.
2. Tables should not contain fields that are intentionally left blank in many records. That condition often suggests that those fields should be put in a different table.
3. Different tables should not duplicate data except as needed to establish relationships (see the next section).

TABLE RELATIONSHIPS

We now have the basics of our application. The next step is to establish relationships between these tables, so that an application (e.g., production of a report) can use data from all of the tables. Relational databases for personal computers are now widely available and increasingly popular. We assume that readers are familiar with the concept of the relational database. Those who are not should consult one of the many books on this subject in a library or bookstore. The important point for this discussion is that those products have improved greatly the ease with which two or more different tables can be related to produce output containing data from the related tables. What once required significant programming to achieve now can be accomplished with essentially no effort. Different sets of data are stored in separate tables, the relationships among the tables are defined, and then the software uses the relationships to find the data requested.

To be related two tables must share a data element. In addition, data type (e.g., numeric, text) in that element must be the same. The data element (field) common to the negative and roll tables is the roll number.

The negative and name tables do not share a data element and cannot be related. Therefore, we need to add a field to the negative table that will refer back to the name table. That field, SpecimenName, will be a number field to match the ID number field of the name table. With that adjustment, the name table can be related to the negative table. The resulting data display includes the name data.

We now have the ability to generate the types of reports we used as the starting point of this example,

which is one of the goals of a database project. We can do much more with the data, however. We quickly can determine what pictures we have of a particular structure, species, or collection or of a particular structure from a particular species. Reports can be produced in order by roll, by name, by structure, or by collection number.

A primary goal of the table design for the photographs (or for any set of data) is to reduce the keyboarding of redundant information. If the roll and negative fields were in the same table, then the information about the roll would have to be entered for each negative. That is time consuming and increases the possibility of error. The same holds for the name table, but the name table also illustrates another goal of relational database design—reusability. A name table can be used in other applications. Two obvious possibilities are in a specimen label table and in a nomenclature table. A name is entered only once but is used in many places, providing for consistent use of the name through all applications.

RELATIONAL TABLES IN SYSTEMATICS

Now that we have built an application, we can review other areas where the methodology may be helpful. A specimen table should be restricted to information relating to the collecting event: who, what, where, and when. Information dealing with morphological features is best placed in a separate table.

An obvious component of nomenclature information is the original citation. Although we can assume that every name will have a literature citation, we put the citation in a separate table rather than in the name table to facilitate reusability. Literature citations often are maintained in a bibliographic database. By establishing a relationship between the bibliographic table and the name table, it is easy to use a record from the bibliographic database in the name table. Sometimes multiple taxa are described in a single publication. With a bibliographic table, a full citation need only be entered once and then can be linked to all appropriate name tables. A separate table can be used to determine the class and order for the names in the specimen table.

RECURSIVE RELATIONSHIPS

A nomenclature database is often of interest to systematists. A recursive relationship, in which a table is related to itself, can be useful in this regard. To illustrate, consider a simple table that contains only three fields: an ID number ("NameId"), a name, and a synonym. Note that the synonym field is a number, whereas the name field is a text field. The synonym field actually refers to an already-existing record in this same table. To display the actual name in the synonym field you need to relate the table to itself. With this design you can list only the accepted names in the table, the accepted names and synonyms, or only the synonyms. An additional field can be added to the table to control the sorting of the synonyms. Let us call this field "SynonymType." It will have three values, or text designations: *b* for basionym, *o* for obligate synonym, and *t* for taxonomic. Using that field to sort the synonyms will put them in the standard order used in publications. By adding a fourth field to the table, which contains the number that is the link to a literature citation table, it is possible to maintain a nomenclature application. Recalling the theme of reusability of data, that name table now can be used to update the names in the specimen table.

Table design is complex, and, as mentioned previously, it is also somewhat subjective. For more information, consult one of the many books on the subject (see also "Table Design for Specimen Data," in the Appendix, this chapter). In addition, the data models from the Natural Science Collections Alliance (formerly The Association of Systematic Collections) and the Common Data Structure for European Floristic Databases (Biodiversity and Biological Collections Web Server, Appendix III) provide examples based on biological attributes. As with documents from any specialized field, those examples may be rather opaque in the absence of previous experience in information modeling.

STANDARDIZATION OF DATA

Although field structure may be the most important aspect of a database project, failure to standardize both the format and the content of the data entered can greatly reduce the value of the database. In fact, when converting data from one application to another application, a lack of standardization causes the most trouble. The entering of data into a database somehow does not confer a higher level of precision on the data than they actually have.

To maintain the standards of data entered, one routinely can edit the data, or one can enforce standardization at the time of entry. The first requires a minimal understanding of the software being used but suffers from good intentions that almost always fail. The second requires more computer expertise but also delivers the desired level of standardization more reliably. Standardizing data has both an immediate return and long-term benefits. For example, when a database is queried for a particular piece of information, it should produce all of the records with those data. That will not happen, however, if the name of the item has not been spelled the same way in all records. Well-standardized data make feasible the inclusion of data in larger datasets or combinations of datasets. Both format and content should be standardized as data are entered. The format is controlled through the software application, which may check, for example, that the date always is entered in the same way (e.g., YYYY MM DD).

Another major use of related tables is to verify the data being entered into the main table. Tables of that type often are called "lookup tables" or "authority tables." A common example is the use of a table with standardized geographic or political names, such as states, provinces, cantons, districts, and so forth. In the United States, a state table can be used to verify the data added to a field in the specimen table. For example, if a user sets up a relationship between the state field in the specimen database and a table listing all of the states of the United States, most software only will allow the user to enter values for the state field that are in the state table. That is a strong tool for ensuring consistent data entry without the need for post–data entry editing. Many of the fields in a specimen database can benefit from the use of authority tables. Generic names, author names, and place names are obvious candidates.

Authority tables can be obtained from a colleague or institution that already has built such a table available for outside use. The other option is to build the lookup table yourself by entering items from a standard reference into an authority table either all at once or as the data are entered. With the latter, procedures for adding new values to the authority table will have to be set up during the early stages of the project. On a project of any size that involves a continuing long-term effort, authority tables significantly increase the efficiency of data entry. It is difficult to overstate their importance.

GETTING STARTED

APPLICATION SOFTWARE

We have talked about using a single database product to handle many data manipulation needs. In many cases, however, the literature citations may be in bibliographic software, specimen-label data in a database, and morphological data in a spreadsheet. The downside of using different products is the difficulty of reusing the data for different purposes. For example, it can be difficult to integrate literature citations in a specialized bibliographic database with a nomenclatural database. The downside to using a single software package for different applications is that the applications are not likely to be as sophisticated as those developed for a single purpose. One should keep in mind as we move toward electronic publication and real-time display of data over the Internet that those curators and investigators whose data are in a format that makes those activities an extension of existing procedures without the need for any reformatting or rearrangement of the data will be at an advantage.

USER INTERFACE

A user interface controls how data are edited, added to tables, and displayed. This is very much a function of the software and cannot be discussed adequately here. Most of the newer database packages simplify the creation of data-entry screens and reports. As in table design, the interface can be simple or complex. Data managers should start simple and add capabilities as needed.

PROJECT GOALS

Goals for a database project should be realistic, focusing closely on what really is needed for a research project as opposed to what would be "nice" for a project. Many projects get stalled after overzealous beginnings. The biggest culprit is the effort required to collate and enter the information. Even with data entry assistance, time demands for prescreening the information can be high. A small, well-defined database can be integrated into a work routine. Understanding how the database supports data maintenance will help the investigator determine how to expand it to include other activities.

CONCLUSIONS

Modern database software greatly increases the ease with which applications can be developed by end users. If fields are delimited carefully, applications can be developed that meet the researcher's needs yet provide data that are generally useful and easily transferable to the wider community involved in biodiversity-related studies.

Although the software has improved and is easier to use, it will not immediately meet all data management needs. If expectations are reasonable, chances of successful implementation are improved greatly. If designed properly, database tools will be useful throughout a career; thus, they are worth an investment of time and patience.

Relational databases provide logical procedures for building an application table by table over time as needs arise and allow everyone to design an application that best meets the scientists' needs on a daily basis.

APPENDIX

DATABASE DETAILS

In recent years many organizations and institutions have debated, developed, and implemented standards for storing and exchanging specimen data, including conversion of data for use in other databases. Those efforts are ongoing, and although some common principles appear to be emerging, it is difficult to promote a definitive guide for exchange of specimen data. In this appendix we discuss aspects of table design that can be varied to fit particular uses and that may facilitate data exchange. We also suggest a core data structure for mycological specimens. The list of fields is our design, but we have drawn inspiration from the work of many ongoing projects and from our interactions with experts in the field. This core data structure is intended to provide a model framework that can fulfill several functions:

1. Serve as a basic exchange format. We describe a flat file structure with field names, field definitions, and maximum field length. Both the recipient and the donor of data must be aware of the requirements.
2. Serve as a point of departure for communication about the exchange of data. The donor of the data can say, "The data that I am sending differ from the basic exchange format in the following ways ..."
3. Serve as a file structure for collecting and storing specimen information for various purposes. A core specimen-data record includes the "what, who, when, and where" information—the identity of the specimen, the collector, the date, and the place—that enables a specimen to serve as a voucher for observations that support conclusions in biodiversity studies.

TABLE DESIGN FOR SPECIMEN DATA

Field structure is the most important aspect of database design for facilitating conversion to other databases. Table design is more subjective, and different designs may be better for particular purposes. In the text examples we based tables on different subjects. Tables also can be based on different events in an application. With specimen data two events can be distinguished: identification and collecting. The fields associated with such events are assigned to the two right-hand tables. In that layout all of the data involved with the collecting event are treated as one record. The utility of that arrangement when field collecting involves many collections from a few localities is easily imagined. The data for each locality are entered only once, and each specimen record can be referred to that record easily. Consistency in data entry is assured. The identification table includes the scientific name as well as the name of the person who identified the specimen and the date of identification. In fact, that combination of events may not be common enough to justify grouping the fields—that is, the reusability of any given record in the table may be too low to justify the effort required to maintain the table.

Another approach to the identification table would be to replace the "Rec. No." field with the "Coll. No." field from the specimen table. That is little different, however, from putting those fields directly in the specimen table. A third approach is to include both the Rec. No. and Coll. No. fields in that table. By changing the Rec. No. and keeping the Coll. No. the same, you can have multiple records with the same Coll. No. The table then would become the source of information not only for the original determination, but also for all of the subsequent determinations and annotations. The Rec. No. field in this case automatically increases with each new determination, and thus, the records in the table are reusable. If an additional table is inserted between the specimen table and identification table, then both the reusability and historical determination goals are achieved. The additional table would have two fields: the Coll. No. field from the specimen table and the Rec. No. field from the identification table. In that way reusable records in the identification table can be connected to one or more specimen records. In other words, one specimen record can be connected to one or more unique identification records.

Although the theoretic aspects of table design are quite interesting, practical aspects also must be considered. Simply put, the more complicated the table design, the more difficult it will be to construct a usable interface between the user and the tables. Current software is helpful in that area but has limited resolving power. The use of authority tables (see "Standardization of Data," earlier) is incorporated easily into table design. Addition of intermediate tables, however, requires increased knowledge of the more esoteric parts of the software. Another way of handling the name in a specimen table is to have fields for all of the components of the name, each of which is related to a table with names for that component. There, the components of the name can be reused but not the whole name. Although that design includes several tables, which are essentially

authority tables, implementation would be easy. The locality data could be handled in a similar way by putting all of the locality fields in the specimen table and using authority tables for the actual data.

A well-designed database should, at a minimum, achieve the following:

1. Well-designed fields
2. Incorporation of authority tables

CORE DATA STRUCTURE FOR A SPECIMEN DATABASE

Taxonomic Name

Genus (Genus)
  Size: 26   Text   Required Yes
  Format:
  Standard: Farr et al. 1979, 1986; Index of Fungi; Greuter et al. 1993
  Reference: Bisby 1994; Greuter et al. 2000
  Comments:
Species (Species)
  Size: 32   Text   Required No
  Format:
  Standard: Index of Fungi; Saccardo 1882–1972; Reed and Farr 1993
  Reference: Bisby 1994; Greuter et al. 2000
  Comments:
Rank (Rank)
  Size: 6   Text   Required No
  Format: var., f. subsp., subsp., etc.
  Standard:
  Reference: Bisby 1994; Greuter et al. 2000
  Comments:
Subspecific Epithet (SubSpEp)
  Size: 32   Text   Required No
  Format:
  Standard:
  Reference: Bisby 1994; Greuter et al. 2000
  Comments: See Species.
Author (Author)
  Size: 100   Text   Required No
  Format:
  Standard: Brummitt and Powell 1992
  Reference: Bisby 1994; Greuter et al. 2000
  Comments: This is the author at the lowest rank (binomial or trinomial). If there are more than two authors then "et al." is used for all authors after the first. That recommendation comes from the International Code of Botanical Nomenclature (Greuter et al. 2000).

Determiner (Determ)
 Size: 55   Text   Required No
 Format:
 Standard:
 Reference:
 Comments: See Collector for additional details.
Determine Date (DeterDate)
 Size: 11   Text   Required No
 Format:
 Standard:
 Reference:
 Comments: See Collector Date1 for details.

Locality Data

Country (Country)
 Size: 25   Text   Required Yes
 Format:
 Standard:
 Reference: Times Atlas; ISO; Hollis and Brummitt 1992
 Comments: Because field is required, use "unknown" for specimens lacking this information.
Country Subdivision1 (CntrSub1)
 Size: 35   Text   Required No
 Format:
 Standard: For states in U.S., use postal code abbreviations (MD, VA, etc.).
 Reference:
 Comments: First political subdivision for a country. In the United States that would be State. It is not necessary to name the rank of subdivision (State, Province, Canton, District) if it is understood that it is the first major subdivision of a country.
Country Subdivision2 (CntrSub2)
 Size: 35   Text   Required No
 Format:
 Standard:
 Reference:
 Comments: Second political subdivision for a country. In the United States that would be County (or Parish in Louisiana).
Country Place Name (CntrPl)
 Size: 100   Text   Required No
 Format:
 Standard:
 Comments: The lowest identifiable named place not assignable to subdivisions 1 and 2. National Parks, Cities, etc.
Locality (Locality)
 Size: 250   Text   Required No
 Format:
 Standard:

Comments: Specific information about collection site. "At the end of Fungus Lane," or "3 miles from the intersection of Route 1 and Fungus Lane."
Elevation1 (Elev1)
 Size: 5   Numeric   Required No
 Format: Number without commas
 Standard:
 Reference:
 Comments: "Around 100 feet" would be expressed in meters as "around 31 meters" (if rounding is recommended). 0 (zero) implies sea level. Negative values indicate below sea level.
Elevation2 (Elev2)
 Size: 35   Numeric   Required No
 Format:
 Standard:
 Reference:
 Comments: This field is used to express a range of elevations. See Elevation1 for details.
Elevation Source (ElevSr)
 Size: 25   Text   Required No
 Format:
 Standard:
 Reference:
 Comments: Source of elevation data. Map, instrument, estimate, etc.
Elevation Unit (ElevUnit)
 Size: 1   Text
 Format:
 Standard:
 Reference:
 Comments: Enter f for feet or m for meters
Latitude (Latitude)
 Size: ?   Numeric   Required No
 Format: Decimal degrees with north as positive and south as negative. −38.1234
 Standard:
 Reference: Federal Geographic Data Committee 1998.
 Comments: Most mapping/GIS programs appear to use the decimal degree format. The biggest problem with decimal degrees is that converting from a value lacking seconds implies greater precision than was present in the original observation.
Longitude (Longitud)
 Size: ?   Numeric   Required No
 Format: Decimal degrees with east as positive and west as negative.
 Standard:
 Reference: Federal Geographic Data Committee 1998.
 Comments: See latitude for details.
Latitude Longitude Source (LatLonSr)
 Size: 25   Text   Required No

Format:
Standard:
Reference:
Comments: Source of elevation data. GPS, map, estimate, gazetteer, etc.

Collector

Collector (Coll)
  Size: 25   Text   Required Yes
  Format: Last comma First. No spaces between initials.
  Standard:
  Reference:
  Comments: The name associated with the number series.

Team Collectors (TeamColl)
  Size: 55   Text   Required No
  Format: Separate each name by semicolon. If there are more than three then use "et al." for all following the first named collector. Farr, D.F.; Farr, E.R.
  Standard:
  Comments: For additional individuals on collecting expedition. Our MSA Committee is divided about using a separate field for collector name(s) associated with number, but there is precedent in some other projects for doing this.

Collector Number (CollNum)
  Size: 20   Alphanumeric   Required No
  Format:
  Standard:
  Reference:
  Comments:

Submitter (Subm)
  Size: 25   Text   Required No
  Format: Last comma first. No spaces between initials.
  Standard:
  Reference:
  Comments: Name of the person submitting the collection.

Collector Date1 (ColDate1)
  Size: 10   Alphanumeric   Required Yes
  Format: YYYYMMDD or YYYY-MM-DD
  Standard:
  Comments: American National Standards Institute (McCallum 1986) standard.

Collector Date2 (ColDate2)
  Size: 11   Alphanumeric   Required No
  Format:
  Standard:
  Reference:
  Comments: See Collector Date1 for details. Used to express a range of dates.

Season (Season)
  Size: 10   Text   Required No

Format:
Standard:
Comments: Use for nondate time periods. Autumn, summer.

Details of Collection

Substratum (Substr)
  Size: 50   Text   Required Yes
  Format:
  Standard:
  Reference:
  Comments: Leaves. Decayed wood. Small branches. Use "unknown" if data are missing.

Habitat (Habitat)
  Size: 250   Text   Required Yes
  Format:
  Standard:
  Comments: Characterization of the collection spot. In mixed woods. On limestone outcropping. It would be useful for individuals or groups to agree on a standardized set of descriptive phrases for their particular needs. May need more space.

Host Genus (HGenus)
  Size: 26   Text   Required Yes
  Format:
  Standard:
  Reference:
  Comments: See Genus for details.

Host Genus Hybrid (HGenusHy)
  Size: 1   Text   Required No
  Format: Lower case alphabetic × adjacent to the genus name. The alphabetic × substitutes in computers for the multiplication sign specified by the International Code of Botanical Nomenclature. Printouts should use the multiplication sign when possible.
  Comments:

Host Species (HSpecies)
  Size: 32   Text   Required No
  Format:
  Comments:
  Standard:
  Reference: Index Kewensis (Royal Botanic Gardens 1997); Gray Herbarium Card Index.

Host Species Hybrid (HSpeciesHy)
  Size: 1   Text   Required No
  Format: See Host Genus Hybrid
  Comments:
  Standard:
  Reference:

Host Rank (HostRank)
  Size: 8   Text   Required No
  Format: var., f. subsp., subsp., n-subsp.(nothosubspecies) etc.

Standard:
Reference: Bisby 1994; Greuter et al. 2000
Comments:
Host Subspecific Epithet (HostSub)
  Size: 30   Text   Required No
  Format:
  Standard:
  Reference:
  Comments:
Host Common Name (HostCom)
  Size: 50   Text   Required No
  Format:
  Standard:
  Reference:
  Comments: Use to record the common name of the host if that is the only available name.

Housekeeping

Herbarium number (HerbNum)
  Size: 20   Text   Required Yes
  Format: Acronym plus institutional number used to identify the specimen.
  Standard:
  Reference: Holmgren et al. 1990.
  Comments: This should be a unique number for the institution. Use Index Herbariorum codes.

Additional Fields

It has been suggested that a field to indicate the type status of the specimen and a field to indicate if a culture was made from the specimen should be included in the field list. Both of those categories extend the field list beyond the basic; the use of relational tables provides another mechanism for handling that information. If data about the type and the cultures are placed in sepa-rate tables, the specimen table can be related with either one or both, using the collection number to determine if the specimen is a type or has been cultured. If a separate field in the specimen table is used for those data, the information is not directly linked to the information in the culture table and, therefore, it is not automatically updated. For example, if a culture were removed from the culture table, then the data in the specimen table also would have to be changed. Other fields have been suggested for inclusion but are clearly outside the scope of minimum required data. Individuals and institutions, however, should customize their databases to include all information relevant to their needs.

Type (Type)
  Size: 2   Text   Required No
  Format:
  Standard:
  Comments: This is used to indicate that the specimen has some status as a type. A two-letter abbreviation indicates which type (e.g., HO, IS).
Culture (Cult)
  Size: 1   Text   Required No
  Format:
  Standard:
  Comments: Used to indicate that a culture was made from the specimen.

ACKNOWLEDGMENTS. Considerable input regarding the core fields to be used in a specimen database was provided by Nancy Weber, Bob Fogel, Tom Volk, and Jim Ginns, our fellow members of a temporary committee of the Mycological Society of America (MSA) working on this project. The fields listed and associated commentary are a summary of our discussions. Scott Redhead reviewed a draft of the manuscript and also provided many helpful suggestions.

5

FUNGAL BIODIVERSITY PATTERNS

JOHN C. ZAK AND MICHAEL R. WILLIG

No individual, population, or species assemblage exists in ecological isolation, free from interactions with other organisms or populations. Rather, all organisms are integrated within a complex ecological whole, whose higher-order dynamics (i.e., biodiversity) arise from interactions among its abiotic and biotic components (Putman 1994). Consequently, our ability to detect patterns of fungal biodiversity and the mechanisms that shape them is influenced by the approaches we use to evaluate the ecological structure of fungal communities. The param-

eters that define a community and that are important for assessing aspects of its biodiversity include species composition, along with its functional and genetic correlates; the types and intensities of interspecific interactions and how they regulate species densities and species occurrences; and the dynamics of those attributes over time and space or as they change as a consequence of past or present human activities (e.g., agriculture, atmospheric nitrogen deposition). To truly understand patterns of biodiversity and mechanisms contributing to those patterns, one must address all aspects of community organization and dynamics. Most mycologists, however, focus on only one aspect (or at best two) of communities, thereby providing an incomplete assessment of fungal biodiversity.

Our goal in this chapter is to present analytical approaches, ecological guidelines, and appropriate references that will provide mycologists with the tools required to examine many aspects of ecological communities and, thus, permit a more comprehensive treatment of fungal biodiversity. We begin with a discussion of the types of biological information that mycologists usually collect to assess fungal biodiversity, considering as well the potential problems encountered in such work. In the next section we examine traditional approaches to quantifying biodiversity from a taxonomic perspective and provide guidelines for using the various diversity indices and analytical procedures. We conclude our chapter with a discussion of the spatial scale of biodiversity and assess the various methods that can be used to quantify that aspect of fungal biodiversity.

THE FUNGAL UNIT

Most fungi, excluding yeasts and some zoosporic taxa, consist of filaments (hyphae) that increase in length by the deposition of cell-wall material from a growing tip. As these tips expand and produce new growing points, a network termed the mycelium develops. Once established, fungal mycelia are capable of essentially unlimited growth and persistence. This form of indeterminate body structure differs significantly from the determinant body plan of most animal and many plant species. Indeed, as a mycelium continues to expand, it may come to occupy a heterogeneous suite of microenvironments or macroenvironments, but its distant segments remain interconnected, facilitating intercellular communication. Because these heterotrophic organisms can occupy heterogeneous environments, they convert organic material and nutrients into biomass at spatial scales that range from several millimeters to entire landscapes (Smith et al. 1992; Anderson et al. 1994).

The mycelial nature of most fungi affects the definition and interpretations of fungal biodiversity and makes the protocols and assumptions used for estimating fungal biodiversity inherently different from those used for most animals and nonclonal plants. To understand the limitations associated with estimations of fungal biodiversity, one must consider methods for counting fungal units and understand how those units differ from units used for most macroorganisms.

The term individual can be used in a numerical, genetic, or ecological context, depending on the focal organism and question of interest (Andrews 1991). From a numerical perspective, an individual is a countable unit of a particular species (e.g., a tree, a mushroom, or a colony that develops from substratum placed on an agar plate). According to Cooke and Rayner (1984), individuals are discrete and functionally independent units. The level of cellular aggregation that fulfills that criterion for fungi and other microorganisms, however, is not clear (Andrews 1991). For example, isolates of a single species of fungus obtained from decomposing leaves at spatially disjunctive locations on the forest floor may represent the mycelium of a single individual or the mycelia of several individuals. As a result, counting individual fungal colonies on an isolation plate, whether the isolate was obtained from a 1-mm^2 piece of substratum or from a dilution series, does not represent the same

information obtained from counting the insects that occur within a defined area or in a given amount of leaf litter. Usually, one cannot be sure if fungal isolates of a species from a single habitat have been obtained from one individual or from several individuals. Although the latter situation is common, the former circumstance also occurs.

A genetic perspective does not provide an operational resolution to this problem. Genetically, an "individual" can be defined as a single cell or a collection of cells that exhibit the same genotype or multiple genotypes (i.e., a genet *sensu* Kays and Harper 1974). In addition to being an individual from a quantitative perspective, a squirrel is an individual from a genetic perspective. In contrast, the number of discrete, countable units in fungi is not the same as the number of genets (Andrews 1991). For clonal organisms that are capable of asexual growth, the countable units are the ramet (Harper 1977). A ramet, although a member of a specific genet, is capable of essentially independent growth. Thus, sporocarps of one species of basidiomycete on a forest floor can represent multiple ramets of a single genet or the ramets from multiple genets. Only molecular or isozymic analyses of sporocarps can untangle the genetic structure of such a species within a habitat and, thereby, facilitate the quantification of fungal biodiversity in the same manner as for animals and most plants. Unfortunately, that approach is not generally practical. Thus, a clearly stated, operational definition of "individual," relevant to the taxon of interest, must be provided to facilitate unambiguous comparisons among ecosystems.

The presence or absence of species within an area can be used to assess some aspects of biodiversity; other comparisons, however, require the quantification of operational individuals (or units) for each taxon. The operational definition for an individual will differ for different taxonomic or ecological groups of fungi. Those definitions are provided for particular groups in their respective chapters.

CHARACTERIZATION OF COMMUNITIES

COLLECTION EFFORT CURVES

Many, if not most, investigations of populations and communities describe patterns rather than test *a priori* hypotheses. They rely on the collection of presence-absence, biomass, density, or frequency of occurrence data for a suite of species. Although the individual is usually the smallest biological unit for quantitative population and community indices, because of the fungal growth form, the concept of individual as used for animals and plants does not apply (see "The Fungal Unit," earlier). Population and community metrics are designed, however, to accommodate data that are not based solely on the delineation of individuals. For analyses of fungal biodiversity, each investigator must establish a unit by which presence-absence, biomass, density, or frequency of component species can be estimated over space and time. It is critical that biomass, density, or frequency of occurrence be based on the same unit of collection and obtained in the same manner throughout the spatial and temporal domain of the investigation.

Frequencies of occurrence of fungi can be calculated in two ways:

$$\text{Frequency (\%)} = \frac{\text{sample units in (or on) which fungal species occurred}}{\text{total number of sample units examined}} \times 100 \quad (1)$$

For example, if 50 1-mm × 1-mm particles of organic soil matter are plated onto an agar medium, the total number of samples examined is 50. The maximum number of sample units from which any one species can be isolated is also 50, which would yield a frequency of occurrence of 100% for that species. Note that the sum of the frequencies of all species can exceed 100% when calculating frequency of occurrence in this manner. Standardized frequencies, which sum to 100%, may be desirable. Dividing the frequency of each species by the sum of the frequency of all species provides relative frequency, a straightforward standardization.

$$\text{Relative Frequency (\%)} = \frac{\text{number of isolates for each species}}{\text{total number of isolates}} \times 100 \quad (2)$$

Investigators should present average frequency of occurrence and standard error of the mean for each species when sampling has been replicated. Standard error provides an estimate of the amount of variation that is associated with the mean frequency of each species. Often, the variability among species in frequency conveys more insight into mechanisms that are responsible for occurrence patterns in fungal ecological units than do the means alone.

The choice of sampling design for obtaining information on presence-absence, biomass, density, or frequency of occurrence is determined by the type(s) of fungi under investigation. The chapters in Section II of this volume provide recommended operational definitions of an individual for the different ecological groups treated.

Determination of the most effective size and shape for a sampling plot is based on the total time required to erect and characterize individual plots as well as the variability of the resultant data. The investigator should have some *a priori* knowledge of appropriate plot size and shape so that the range of sizes to be evaluated is not excessive. Practicality is also an important concern when determining plot size. A growing body of evidence (Kolasa and Pickett 1991; Waide et al. 1999; Gross et al. 2000) suggests, however, that the patterns and processes evaluated by a study are scale-dependent, so considerable caution must be exercised when comparing the results of studies based on plots of different size. Likewise, the spatial scale at which data are collected should be related to the spatial scales at which causative mechanisms are thought to operate.

The shape of a plot can influence the ease, accuracy, or efficiency of sampling. In grasslands, shrublands, or even forests, circular plots are delineated easily and quickly with a center pole and a freely swinging radius line. Circular plots have the advantage of minimizing edge effects that can strongly bias estimates of density. Krebs (1989), however, suggested that elongated or rectangular quadrats provide more accurate estimates of the species composition of an area than do circular or square plots of the same area. The rectangular plot is especially suited for assessing environmental gradients within a habitat when the long axis is oriented parallel to the underlying gradient of interest (Cox 1996). The reason is that rectangular quadrats are more effective in detecting habitat heterogeneity and accurately estimating the patchy distribution of organisms within a sampling scheme than are square or circular quadrats. A long quadrat potentially covers more patches than does a circular one of the same area.

Determining the number of sampling units required to provide an accurate estimate of species richness or density is always challenging, whether one is sampling macrofungi fruiting on a forest floor or isolating microfungi from decomposing root pieces. The number of plots assessed or particles of material plated onto an agar medium will depend on the characteristics of the fungal assemblage, the objectives of the investigation, and the magnitude of differences considered to be biologically relevant. In reality, sample size often represents the upper limit of effort that can be expended to categorize an ecosystem or the amount of time necessary to culture and identify isolates on an agar medium. Regardless, the investigator must avoid making decisions concerning sample size in a capricious manner and always should expose the criteria on which the decision was based.

If the biological question of interest concerns the total species richness of an area or domain (in practice this could be a leaf, a patch of soil, or a forest), then a random (or stratified random) series of plots should be placed within the inference space (spatial and temporal limits of the domain) to determine the number of samples required for an accurate estimate of richness. In general, the total number of species identified from the area of interest increases as the number of samples increases but eventually attains a plateau (i.e., asymptote). The number of samples required to reach the plateau will depend on the environmental heterogeneity of the area of interest as well as on the dispersion patterns and fruiting phenology of the focal species. Hence, if one is interested in assessing if mean differences in species richness of fungi on leaves (i.e., the richness of a leaf) occur between shade and sun leaves of oak trees, for example, then collector's curves should be generated for each type of leaf. Because even different leaves of the same type can differ in heterogeneity and species composition, collector's curves should be generated for a number of leaves of each type, and, to be conservative, sample sizes should be chosen beyond the minimal value of effort that corresponds to the plateau. Alternatively, one can assume that the relation between richness and effort is logistic and, via statistical techniques, can associate parameters with variables in the logistic equation and graphically determine a conservative value for sample size associated with the plateau (Fig. 5.1 upper graph). A number of alternative methods (e.g., rarefaction analysis; Magurran 1988; Colwell and Coddington 1994) exist for estimation of species richness and should be considered prior to choosing a particular experimental design. Because human and financial resources are often too limited to enable discovery of species richness at a site, quantitative ecologists have developed methods to estimate total richness from a modest amount of sampling. Colwell and Coddington (1994) discussed some of these methods, and a subset of them is treated as part of "Approaches to Sampling Macrofungi" in Chapter 8 of this volume.

QUANTITATIVE INDICES

Richness

Species richness is the most widely used parameter for evaluating aspects of fungal biodiversity. It is deceptively simple to define species richness as an enumeration of the species that are associated with a particular sample, area, habitat, or substratum. In fact, three kinds of species richness can be distinguished: (1) numerical species richness, (2) species density, and (3) total species richness (Hurlbert 1971; Kempton 1979; Brown 1995;

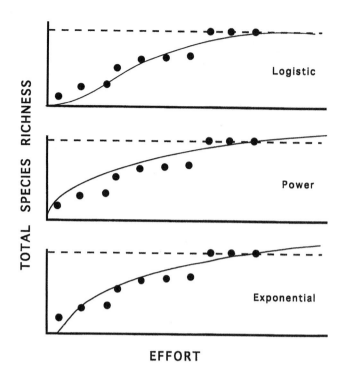

FIGURE 5.1 Graphic representations of the logistic, power, and exponential functions (*solid lines*). Dots represent the way in which total or cumulative species richness increases with effort (cumulative number, area, or volume of samples). The logistic relation approaches an asymptotic value, which is represented by the dashed line, that is equal in value to those of the power and exponential relations. With both the power and exponential curves, richness increases continuously without attaining a plateau.

Rosenzweig 1995). The number of species in a sample in which the biomass or number of individuals has been standardized is numerical species richness. The number of species in a sample in which the area, volume, or weight of the sampling unit has been standardized is species density. Finally, the cumulative number of species based on a series of samples from a habitat or substratum is the total species richness.

Numerical species richness and species density are defined after complete enumeration of the taxa in a sample, whereas total species richness is estimated from a series of samples. The first two are measured without error, assuming that the size of the sample is sufficiently small and that the techniques of isolation and identification are developed sufficiently to allow an investigator to enumerate all taxa. It is important to note that the effects of natural or experimentally induced variation in environmental characteristics on species density and numerical species richness are scale-dependent—that is, the effect of environmental variation may not be the same for samples that differ in area, volume, weight, biomass, or number of individuals (Waide et al. 1999; Gross et al. 2000). In part, the scale-dependence of patterns occurs because the importance of different causal mechanisms depends on the scale at which data are collected. Most ecological studies of fungi cover species densities as a consequence of sampling design, although in most cases that aspect of diversity is not stated explicitly. We strongly recommend that future research always include an explicit definition of scale as well as of the attribute of richness that is being evaluated. Once a particular spatial scale has been selected for an investigation in which species density or numerical species richness is the characteristic of diversity under examination, classical statistical considerations of random sampling and power determine the efficacy of a research design.

It would appear that an unambiguous and straightforward index of total species richness, S, would be the cumulative number of unique species present in a series of samples. The magnitude of S, however, depends on the size, number, and dispersion of samples in a particular habitat, biome, or area. Indeed, three mathematical relations (Power, Exponential, and Logistic) have been championed in the literature to predict the way in which S increases with effort, A (number, area, or volume of samples). In both Power ($S = CA^z$; Arrhenius 1921) and Exponential ($S = C + z\ln A$; Gleason 1922, 1925) models, S increases monotonically as effort increases (Fig. 5.1 middle and lower graphs). Those functions are most appropriate when heterogeneous landscapes are sampled and increases in effort result in increases in the heterogeneity of habitats that are included in samples. All three relations are members of the same family of curves (He and Legendre 1996). Consequently, both Power and Exponential models figure prominently in the theory and practice of island biogeography and conservation biology. In contrast, when the domain of interest is circumscribed geographically and random sampling occurs within the borders of that domain, the logistic relation ($S = B/(C + A)^{-z}$; Archibold 1949) more likely characterizes the manner in which S increases with effort. Unlike the other two species-effort curves, the logistic relation predicts that S eventually will reach a plateau or asymptote (Fig. 5.1 upper graph). The value of S at this asymptote is an accurate estimate of the true species richness of the domain (inference space) of interest. The effort required to attain asymptotic values for particular taxa, however, is likely specific to particular substrata, habitats, or biomes. Consequently, comparisons of S at levels of effort not associated with the asymptote can lead to spurious conclusions.

Given the dependence of S on collection effort and the fact that limited financial resources, personnel, or logistic support often prevent the collection of samples sufficient to attain asymptotic values, S is of limited value as a comparative index (Ludwig and Reynolds 1988). Consequently, a number of indices that are independent of the number of samples taken have been developed to estimate species richness. Those indices are based on the relationship between S and n, the total number of individuals in the collection of samples. Ludwig and Reynolds (1988) also cautioned that two well-known species-richness indices, the Margalef index (1958) and the Menhinick index (1964), make specific assumptions concerning the relationship between S and n ($S = kn^{0.5}$, where k is a constant). In many cases, those assumptions may not hold, and as a consequence, the utility of the indices is limited.

Direct counts of species numbers in samples of equal size (i.e., equal number of individuals) may provide an informative alternative to indices of species richness. Implicit in this approach, however, is the assumption that the collector's curves for the treatments of concern are coincident (or at least nonintersecting). For enumeration of fungi from leaf disks or root surfaces, the approach has merit. However, it is not always possible to ensure that all sample sizes are equal. In situations in which the sample size is not constant, rarefaction, a quantitative method (Magurran 1988), facilitates comparison of species richness among areas or habitats as if they were based on a standardized sample size. The number of species that can be expected in a sample of n individuals is:

$$E(S) = \sum \left\{ 1 - \left[\left(\frac{N - N_i}{n} \right) \Big/ \left(\frac{N}{n} \right) \right] \right\} \qquad (3)$$

where $E(S)$ is the expected number of species in a rarified sample, n is the standardized (rarified) sample size (usually chosen to be equal to the smallest sample available for an area or habitat), N is the total number of isolates (individuals) recorded in the set of samples, and N_i is the number of isolates (individuals) in the species. To calculate the expected number of species in a rarified sample, the abundance of each species is inserted into the quantity defined by braces ({}) in formula (3) and summed to provide the expected number of species. To assist in the computation recall that:

$$\binom{N}{n} = \frac{N!}{n!(N-n)!} \qquad (4)$$

where the exclamation point (!) indicates a mathematical operation termed a factorial. Worked examples can be found in Magurran (1988) and Krebs (1989). Polishook and colleagues (1996) used rarefaction to determine the expected number of species of fungi from decaying leaves from a Puerto Rican rain forest.

Diversity Indices

Diversity is a measure of the complexity of structure in an ecological community. It comprises two distinct

TABLE 5.1
Common Indices of Diversity and Evenness

Index	Equation	Terms and comments
Diversity		p_i is the proportion of individuals (isolates) in the ith species. Proportional abundance is
Simpson	$\lambda = \Sigma p_i^2$	obtained by dividing the density, biomass, or number of plots in which the organism was observed by the total density, biomass, or observations. Used for infinitely large ecological units.
	$D = \sum \dfrac{n_i(n_i - 1)}{N(N-1)}$	n_i is the number of individuals (isolates) in the ith species, and N is the total number of isolates. The reciprocal form of Simpson's index ($1/D$) usually is presented, ensuring that the index increases with increasing diversity. This index is appropriate for sampling from finite ecologic units.
Shannon	$H' = -\Sigma\, p_i \ln p_i$	p_i (n_i/N) is the proportional abundance of the ith (each) species. ln is the natural log (base e).
McIntosh	$U = \sqrt{\left(\sum p_i^2\right)}$	U is the general form of the index; p_i is the proportional abundance of each species.
McIntosh	$D = (N - U)/(N - \sqrt{N})$	D is a dominance measure and is independent of N (total number of isolates).
Berger-Parker	$d = N_{max}/N$	N_{max} is the number of isolates in the most abundant species. The reciprocal of d is used most commonly.
Hill	$N_1 = e^H$	H' is Shannon diversity based on the natural log (base e).
	$N_2 = 1/D$	D is Simpson diversity.
Log-seriesd α	$\alpha = \dfrac{N(1-x)}{x}$	N is the total number of isolates. S is the total number of species. x is estimated from the iterative solution of: $S/N = [(1-x)/x][-\ln(1-x)]$
Evenness		
Shannon	$E = H'/\ln S$	S is species richness.
Hill	$E = N_2/N_1$	N_1 is Shannon Diversity Index; N_2 is Simpson Diversity Index.
McIntosh's	$E = \dfrac{N - U}{N - \left(N/\sqrt{S}\right)}$	N is the total number of isolates. U is McIntosh's diversity index. S is number of species.

attributes: species richness and species evenness. Simpson (1949) proposed the first index of diversity used in ecology (Table 5.1). The index varies from 0 to 1 and is referred to as a dominance measure because it is influenced strongly by the abundance of the most common species. Originally, Simpson's index (λ) was restricted to ecological units in which all members of the unit (in this case community) could be enumerated (Table 5.1). Because we usually sample infinite (in the statistical sense) ecological units for which it is impossible to count all members, Simpson developed an unbiased estimator (D) of diversity based on a sample of N individuals (Table 5.1). Significant differences in Simpson's diversity indices can be tested using parametric analyses (e.g., analysis of variance, regression) or their nonparametric counterparts.

The most widely used measures of diversity are the indices derived from information theory. Those indices are based on the rationale that the diversity of a biological system can be measured in a way similar to that used to estimate the information content of a message. The Shannon Index of Diversity (H') (Shannon and Weaver 1949) is currently the most popular index in community ecology (Table 5.1). The index has sometimes incor-

rectly been referred to as the Shannon-Weaver index (Krebs 1989). H' is a measure of the average degree of uncertainty in predicting the specific identity of an individual chosen at random from a collection of S species and N individuals. Average uncertainty will increase as the numbers of species increases and as the distribution of individuals among species becomes more even. Ludwig and Reynolds (1988) emphasized two properties of the index that make it popular. First, $H' = 0$ if (and only if) the sample includes only a single species. Second, H' reaches its maximum only when all species are equally abundant. The magnitude of H' is usually between 1.5 and 3.5 and is rarely greater than 4.5 (Margalef 1972). May (1975) calculated that to obtain a value of H' of more than 5.0, the ecological unit would need to include 10^5 species. Fortunately, when the Shannon index is determined for replicate samples of the same ecological unit, it exhibits a normal distribution, facilitating the use of parametric statistics to evaluate differences in central tendency (mean) or dispersion (variance) in different habitats (Taylor 1978).

Other useful diversity indices include McIntosh's measure of diversity (U), the Berger-Parker index (d),

TABLE 5.2
Characteristics of Selected Diversity Indices*

Index	Ability to discriminate among sites	Sensitivity to sample size
Shannon	Moderate	Moderate
Simpson	Moderate	Low
McIntosh (U)	Good	Moderate
McIntosh (D)	Poor	Moderate
Berger-Parker	Poor	Low
Fisher	Good	Low

* From Magurran (1988).

and Fisher's log-series alpha (α) (Table 5.1). The McIntosh diversity index is easy to calculate and reflects the Euclidean distance of the sample point (in terms of the abundances of all species) from the origin of an S-dimensional hypervolume. The index, however, is influenced strongly by sample size, limiting its usefulness. Alternatively, the Berger-Parker index reflects the proportional importance of the most abundant species (Table 5.1). This index is independent of S but also is influenced by sample size. Fisher's logarithmic series was the first model to describe mathematically the relationships between the number of species and the number of individuals that compose those species (Fisher et al. 1943). Although the model has been used extensively to examine the distribution patterns of individuals among species within an ecological unit, one parameter of the distribution, alpha (Table 5.1), can be used as an index of diversity (Taylor 1978; Magurran 1988). Taylor (1978) strongly supported the use of the log-series alpha as an index of diversity because of its good discriminant ability. It is not overly sensitive to sample size (Table 5.2), and it is less affected by the abundances of the common species than either the Shannon or Simpson Index. A disadvantage of log-series alpha is that it is unaffected by the actual distribution of individuals (evenness) among the component species in the community or assemblage (it assumes they adhere to a log-series distribution). Thus, two communities that have the same N and S, but different distributions of individuals among the species in a sample, will have the same alpha value.

Lastly, Ludwig and Reynolds (1988) proposed that the series of diversity indices (N_0, N_1, N_2) presented by Hill (1973) are probably the easiest to interpret ecologically (Table 5.1). The units for these indices are numbers of species, and the indices measure the effective number of species present in a sample. N_0 is the number of all species in the sample, regardless of abundance, whereas N_1 and N_2 measure the numbers of abundant and very abundant species, respectively.

The Brillounin Diversity index (HB) was developed to calculate the diversity of an ecological unit when sampling is not random and when the enumeration of the ecological unit is complete (i.e., all discrete individuals are counted) (Magurran 1988). Given the restricted conditions under which the index can be used (cannot be used with biomass or cover data), the complexity of its calculations, and the number of acceptable alternative indices, we do not recommend that it be used to estimate fungal diversity.

A technique termed jackknifing can be used to improve the accuracy of any diversity index. Zahl (1977) first used this technique to obtain confidence limits for the Simpson and Shannon diversity indices. The method makes no assumptions about the underlying distribution but allows one to test hypotheses statistically. Moreover, Zahl (1977) showed that calculated pseudovalues (see later in this chapter) are normally distributed and that random sampling, which might be difficult to achieve, is not required. The technique can be applied only in studies in which a number of samples have been taken to estimate a diversity index (Magurran 1988). The method involves recalculating overall diversity while disregarding data from each of n constituent samples. This creates a number of jackknifed values (VJ_i), equal to the total number of samples. Each of the jackknifed estimates then is converted to a pseudovalue (VP_i):

$$VP_i = (nV) - [(n-1)(VJ_i)] \qquad (5)$$

where n is the number of samples and V is the diversity index based on all samples. The mean of the pseudovalues represents the best estimate of diversity (VP), and the sample influence function is the difference between the jackknifed estimate and the original estimate of diversity. Confidence limits can be calculated as follows:

$$(1-\alpha)\,\text{CI} =$$
$$\text{mean} \pm t_\alpha\ (\text{standard deviation of } VP_i/\sqrt{n}) \qquad (6)$$

where t_α is a value obtained from the critical values of the t distribution for the desired confidence level ($1-\alpha$) with $n-1$ degrees of freedom (for t-table values, see Rohlf and Sokal 1995). No confidence intervals should be calculated for datasets in which n is less than 15 (Adams and McCune 1979). Small datasets also may result in overestimation of diversity, and it may be desirable to set degrees of freedom equal to $S-1$ to provide a more accurate estimate. For large datasets ($n > 100$) and species-rich ecological units (S more than 100), t_{S-1} and t_{n-1} are essentially equal, and use of either $n-1$ or $S-1$ degrees of freedom will provide similar results. Magurran (1988) urged caution in using jackknifed estimates for Shannon and Simpson indices because the technique can produce results that are ecologically absurd.

Little consensus exists on which diversity index gives the best estimate in all situations. Ludwig and Reynolds (1988) favored the use of Hill's (1973) indices, whereas Rosenzweig (1995) recommended the use of Fisher's log-series alpha. Magurran (1988) discussed the sensitivity of common indices to sample size and their ability to discriminate subtle differences between sites; that information is summarized in Table 5.2. Investigators should be aware of the limitations of the diversity measure chosen and of its biological relevance.

Evenness

Diversity indices combine two distinct aspects of a community: the number of species and the degree to which individuals are distributed in an equitable fashion (evenness) among species. Just as species richness can be calculated for any ecological unit, each diversity index also has a corresponding evenness index (Table 5.1). More specifically, indices of evenness measure the degree to which a particular community evinces the maximal diversity possible, given the observed richness—that is,

$$e = \text{Diversity}_{obs} / \text{Diversity}_{max} \qquad (7)$$

where D_{max} is calculated with each taxon assuming $1/S$ of the total abundance.

Species-Abundance Distributions

A limitation of species diversity indices is that they compress the data to a single value that conveys little information about the abundances of the species in the ecological unit. One can examine aspects of the diversity of fungal assemblages, based not only on species composition and richness but also by evaluation of how abundances are partitioned among the component species. Using species-abundance data, one can begin to examine how diversity and the structure and organization of the community are related.

Fisher and colleagues (1943) were the first to recognize that when the number of species and the relative abundances of each species within a community are plotted, a characteristic pattern of species abundances is obtained. These observations led to the development of species-abundance models. Preston (1948) showed that for assemblages of plants or animals, abundance distributions often were described by lognormal functions. Although species-abundance data may adhere to one or more distributions (Pielou 1975), diversity usually is examined in relation to four main species-abundance models: the geometric series, the logarithmic series, the broken-stick model, and the lognormal distribution (Table 5.3). May (1975, 1981) and Southwood (1978) strongly advocated the use of such models as the only sound basis for examining species diversity. Magurran (1988) concurred and emphasized that species-abundance distributions use all available data in quantifying the structure and organization of the ecological unit. Consequently, they provide the most complete assessment of diversity. These distributions can be described mathematically, and perhaps more importantly, each corresponds to specific biological scenarios that account for the form of the distribution (Table 5.3).

Rank-Abundance Plots. Rank-abundance plots are the usual graphic method for presenting species-abundance

TABLE 5.3
Ecological Characteristics of Four Commonly Used Species-Abundance Distributions*

Geometric	Log series	Lognormal	Broken-stick
Abundances proportional to amount of resources utilized	Abundances governed by one or a few factors	Abundances governed by many independent factors	Abundances governed by random division of resources along a continuum
Dominant species preempts largest portion of limiting resource	Dominant species preempts largest portion of limiting resource	Resource utilization characterized as multidimensional	Subdivision of niche space is random and simultaneous
Associated with species-poor habitats	Describes a small assemblage of species	Large and varied communities	Small samples of taxonomically related organisms, with stable populations and long life cycles
Propagules arrive at regular intervals	Propagules arrive at random intervals	—	—
Nonequilibrial assemblages	Nonequilibrial assemblages	Equilibrial assemblages	—

* From May (1975) and Magurran (1988).

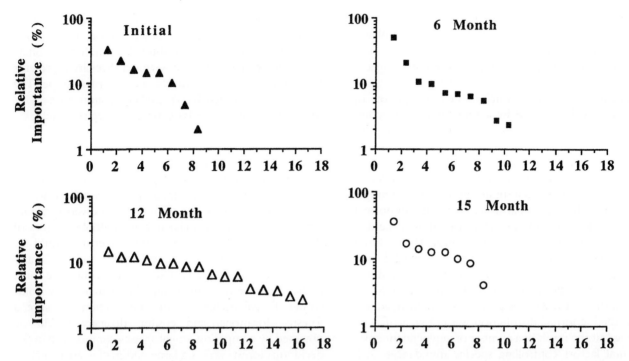

FIGURE 5.2 Abundance distributions for fungal species decomposing oak roots in West Texas at four different times (Zhang 1996). The y-axis (log scale) represents the relative importance of each species in the assemblage as a percent, which was obtained by dividing the number of isolates of each species by the total number of isolates for that particular sample period and multiplying by 100 (see equation 2). The sum of all relative-importance values for each distribution should equal 100. Each species subsequently is ranked from most important to least important along the x-axis, with rank 1 being the most important. The log-series best describes the abundance distributions for all time periods except the 12-month sample, which is best described by the geometric model.

distributions. In these graphs, the abundance or relative importance (expressed as percentages: $P_i = ([n_i/N] \times 100)$ of each species in a collection is plotted on a logarithmic scale against species rank from most abundant to least abundant (Fig. 5.2). Relative importance values are preferred because they allow direct comparison of ecological units with different total numbers of individuals. The four abundance models represent a progression from the geometric series, in which a few species are dominant with the remainder rare, through the lognormal distribution in which most taxa are of intermediate abundance, to the broken-stick model in which species abundances are more equitable.

Geometric Series. The geometric distribution model, sometimes called the niche preemption hypothesis, represents a situation in which the most successful or dominant species preempts some proportion of the total limiting resource. The next most successful species claims the same proportion of the remaining resource, and so on. In general, this model produces the highest dominance and lowest evenness indices compared to other models. The geometric distribution model is typical of species-poor habitats (Zak 1988, 1992) or those that

contain stress-tolerant species. In addition, the geometric series can arise when the arrival of species into an unsaturated habitat is regular and subsequent rates of increase are homogeneous (Magurran 1988).

Log-Series. The log-series model is related closely to the geometric model (May 1975). Like the latter, the log-series occurs when one or a few factors control species dominances in an ecological unit, resulting in only a few taxa becoming dominant. The log-series also can arise if the intervals between arrivals of species into a habitat are random rather than regular and when growth rates are homogeneous (May 1975). Thus, the species-abundance distribution of phylloplane fungi growing on *Lolium perenne* can be described by both the geometric and log-series distributions (Thomas and Shattock 1986). Zak (1988) reported that the geometric and log-series models best described the structure of rhizoplane fungal assemblages on a grass growing on reclaimed mine tailings and that the logarithmic species-abundance distribution may be typical of root-surface fungal assemblages and indicative of a nonequilibrial ecological unit. More recently, Polishook and colleagues (1996) examined the species-abundance distributions of microfungi isolated from decaying leaf litter in a Puerto Rican rain forest. They found that the fungal assemblages were primarily composed of rare species, with only a few moderately abundant and a few highly abundant taxa.

Lognormal. In the lognormal model, occupation of niche space is determined by a number of interacting factors that affect the outcome of interspecific competition. The model is characteristic of ecological units that contain many species (Table 5.3). Lussenhop (1981) reexamined dilution plate data from Wisconsin forest soils as well as from sixteen studies of rhizosphere fungal assemblages. He found that in all cases, the lognormal model best described the fungal species-abundance distributions. Lussenhop (1981) emphasized that if fungi are collected from many microhabitats and mixed during sampling, the resultant distribution is an artifact and will be biased, resembling the lognormal. Nonetheless, he proposed that of the 31 fungal datasets examined, adherence to the lognormal distribution was not an artifact of sampling but the result of a number of independent environmental factors controlling species abundances. Zak (1992) cautioned, however, that in all studies examined by Lussenhop (1981), incomplete fungal abundance data were analyzed. Some rare species were not included in the lists, some taxa were grouped into genera with others identified to the specific level, or all rare taxa were pooled into a single group. When plotting species-abundance distributions, taxonomic units must be applied in the same manner for all ecological units. Therefore, sterile isolates of fungi obtained from litter or soil should be sorted into morphological species groups based on macromorphological and micromorphological characteristics expressed when grown on several agar media. Approaches for fitting each of the four species-abundance distributions are presented in detail along with examples by Magurran (1988). Observed and expected values for each model can be compared using a Chi-square goodness-of-fit test or a G-test (Sokal and Rohlf 1995), thereby determining if a given dataset conforms to model predictions. In addition, these two goodness-of-fit tests can be used to determine whether differences in the proportional abundances of species at different sites are statistically significant.

SPATIAL SCALE OF BIODIVERSITY

Whittaker (1977) was the first to realize that ecological diversity was scale dependent or hierarchical in nature. The primary level of diversity, point diversity, reflects the diversity at a particular location. Alpha diversity, sometimes called within-habitat diversity, is the diversity in a patch (MacArthur 1965). Most studies calculate alpha diversity using the indices described previously. Beta diversity describes the contribution of multiple habitats to overall diversity of a site. Gamma diversity represents the number of species within a region or landscape. The final level envisioned by Whittaker, epsilon diversity, is the diversity of a large biogeographic region, such as a biome.

Such a hierarchical arrangement of diversity could be used to describe the fungi that occur on decaying leaves on the forest floor. For example, point diversity would represent the number of fungal species associated with a leaf of a particular species of tree. Alpha diversity would represent the fungi associated with multiple samples of the same leaf type, whereas beta diversity would be obtained by comparing fungal species richness and composition from different leaf types within the same forest. Gamma diversity would be estimated by comparing fungal richness and composition of several forested locations, and epsilon diversity would be obtained by comparing fungal species composition in the deciduous forests across a large geographic area, such as eastern North America.

DIFFERENTIATION DIVERSITY

Another way to think about beta diversity is to view it as a measure of the degree of similarity or difference in species composition between sites. In other words, beta diversity examines the degree of species turnover as one moves from habitat to habitat, from community to community, or along any ecological gradient (Southwood 1978). The fewer species the various sites or positions along a gradient share, the higher the beta diversity. The term was coined by Whittaker (1977), who devised three levels of differentiation diversity (pattern diversity, beta diversity, and delta diversity). Pattern diversity is the differentiation diversity between samples taken within a homogeneous habitat, and beta diversity is the between-habitat component of diversity. Changes in species composition and abundance between landscape areas are considered delta diversity. Of the three types of differentiation diversity, beta diversity is by far the metric most commonly used to examine the degree to which turnover in species composition characterizes positions along gradients.

The six commonly used metrics of beta diversity (Table 5.4) use presence-absence (binary) data. Each measure has been evaluated regarding sensitivity to community changes, additivity, independence from alpha diversity, independence from excessive sampling, and so forth (Wilson and Shmida 1984). Of the six indices,

TABLE 5.4
Measures of Beta Diversity Using Binary (Presence-Absence) Data

Index	Equation	Terms
Whittaker	$\beta_w = (S/\alpha) - 1$	S is the total number of species (gamma diversity). α is the average sample diversity when each sample is a standardized size, and β diversity is measured as species richness.
Cody	$\beta_c = [g(H) + l(H)]/2$	$g(H)$ is the number of species gained along an entire transect or as one moves from site to site. $l(H)$ is the number of species lost across the same transect.
Routledge	$\beta_r = [S^2/(2r + S)] - 1$	S is the total number of species. r is the number of species pairs with overlapping occurrences. This measure emphasizes species richness and the degree of species overlap.
Routledge	$\beta_I = \ln(T) - [(1/T)\Sigma e_i \ln(e_i)] - [(1/T)\Sigma \alpha_j \ln(\alpha_j)]$	e_i is the number of samples along a transect in which species i is present. α_j is the species richness of sample j. $T = \Sigma e_i = \Sigma \alpha_j$. This measure is based on information theory.
Routledge	$\beta_E = \exp^{(\beta_I)}$	Exponential form of β_I.
Wilson and Shmida	$\beta_T = [g(H) + l(H)]/2\alpha$	$g(H)$ is the number of species gained along an entire transect or as one moves from site to site. $l(H)$ is the number of species lost across the same transect. α is species richness.

β_w (Whittaker's measure) fulfills most of the criteria for an effective index and has fewest restrictions. The index developed by Wilson and Shmida, β_t is an acceptable alternative (Magurran 1988). In addition to binary data, abundances can be used in some equations of turnover; Wilson and Mohler (1983) have provided guidelines for their use.

RESEMBLANCE FUNCTIONS

Similarity, distance, and dissimilarity coefficients represent an alternative analytical approach that can be used to quantify differences in species composition (i.e., beta diversity) among sites. Approximately 20 similarity indices appear in the literature. Those indices differ in the degree to which shared occurrences or shared absences are weighted. We will discuss only the four most widely used indices (Dice, Ochiai, Jaccard, and Sørenson; Table 5.5). Additional information on a variety of indices, including a discussion of their limitations, can be found in Clifford and Stephenson (1975).

Similarity indices are never metric. Hence, they cannot be used to position objects in relation to each other in geometric space as can principal components analysis (Legendre and Legendre 1983). Information obtained from similarity measures is displayed with a similarity matrix (Table 5.6) or after conversion to measures of dissimilarity (Fig. 5.3). Calculation of the four similarity indices requires presence-absence data from a series of sites. A disadvantage of such indices is that all species contribute equally to the index, regardless of abundance. Citing results from an extensive evaluation of similarity measures by Smith (1986), Magurran (1988) reported that indices that use binary data generally give misleading results that indicate a higher degree of similarity than actually exists. Nonetheless, Magurran recommended that Sørenson's Index (Table 5.5) be used for making comparisons if only binary data can be obtained.

Tulloss (1997) reviewed the properties of 15 similarity indices and concluded that all were unsatisfactory. He found that most indices were insensitive to the relative sizes of the two species lists being compared, to the percentage of entries in the longer species list that were common to the two lists, or to the percentage of entries in the shorter list that were common to both. To deal with these weaknesses, Tulloss (1997) proposed a Tripartite Similarity Index (T) that satisfied six requirements for suitability of a resemblance function. The index is tripartite because it includes a cost function for each of the three main, conflicting, requirements for similarity indices. Those requirements are concerned primarily with sensitivity to the number of species in the ecological unit. When the three cost factors are multiplied together, the resultant similarity index is sensitive to the size of each species list that is being compared (Tulloss 1997). The Tripartite Similarity Index is obtained from the following:

$$T = U \times S \times R \qquad (8)$$

TABLE 5.5
Selected Similarity Coefficients

Index	Equation	Terms
Binary		
Dice	$DI = 2j/(2j + a + b)$	j is the number of species in common between two sites; a is the number of species in site A; b is the number of species in site B.
Jaccard	$JI = j/(a + b - j)$	Same as for Dice.
Ochiai	$OI = j/[(\sqrt{j+a})(\sqrt{j+b})]$	Same as for Dice.
Sørenson	$SI = 2j/(a + b)$	Same as for Dice.
Metric		
Bray–Curtis (modification of the Sørensen Index)	$BCI = 2W/(a + b)$	W is the sum of the lower abundances (number of individuals or other measure of density) of species that occur in each site; a is sum of the abundances in site A; b is sum of the abundances in site B. Frequencies of occurrence also can be substituted for abundances, where W is the sum of the lower frequencies of those species in common between the two assemblages; a is the sum of the frequencies of all species in assemblage A; and b is the sum of the frequencies of all species in assemblage B.
Morisita–Horn	$C_{MH} = \dfrac{2\Sigma(an_i - bn_i)}{(da + db)(aN\ bN)}$	aN is the number of individuals in site A. bN is the number of individuals in site B. an_i is the number of individuals in the ith species in site A. bn_i is the number of individuals in the ith species in site B. $da = \Sigma an_i^2 / aN^2$ $db = \Sigma bn_i^2 / bN^2$
Renkonen	$P_{12} = S_{\min}(p_{1i}, p_{2i})$	p_{1i} is the proportional representation of species i in site 1. p_{2i} is the proportional representation of species i in site 2. $S_{\min}$ is the minimum number of species detected from either site.

TABLE 5.6

Hypothetic Matrix of Renkonen Similarity Coefficients Comparing the Compositions and Species Abundances of Fungal Assemblages from Five Sites along an Elevational Gradient

Site	B	C	D	E
A	0.729	0.748	0.748	0.699
B	—	0.733	0.733	0.567
C	—	—	0.965	0.543
D	—	—	—	0.544

where U, S, and R represent cost functions, as follows:

Cost Function 1: $U = \dfrac{\log\left(1 + \dfrac{\min(b,c) + a}{\max(b,c) + a}\right)}{\log 2}$ (9)

where a is the number of taxa common to both areas, b is the number of species in the first area that are not in the second, c is the number of species in the second area that are not in the first, $\min(b, c)$ is the smaller of the two values in parentheses, and $\max(b, c)$ is the larger of the two values in parentheses. The value U is designed to reduce the value of the similarity index if the sizes of the species lists of the two areas being compared differ greatly. If two areas have the same number of species, the function reduces to 1. Note that Tulloss (1997) defined a, b, and c differently from the indices presented in Table 5.5.

Cost Function 2: $S = \dfrac{1}{\sqrt{\dfrac{\log\left(2 + \dfrac{\min(b,c)}{a+1}\right)}{\log 2}}}$ (10)

The value S is designed to reduce the value of the index if the size of the smaller assemblage differs substantially from the size of the list of shared species.

Cost Function 3: $R = \dfrac{\log\left(1 + \dfrac{a}{a+b}\right)\log\left(1 + \dfrac{a}{a+c}\right)}{(\log 2)^2}$ (11)

The value R is designed to increase the similarity between two areas as the percentage of species in the two

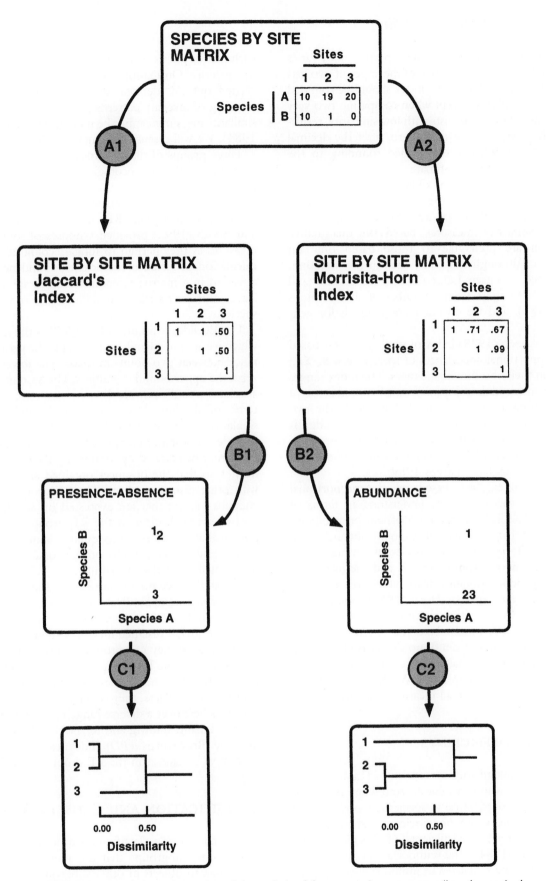

FIGURE 5.3 Schematic representation of the analysis of data on species occurrences (i.e., the species by site matrix) using different resemblance functions. Jaccard's index (A1) based on presence-absence data, and the Morisita-Horn index (A2) based on species' abundances, to give rise to ordinations in species space (B1, B2) and cluster analyses (C1, C2). The species by site matrix was constructed to emphasize the importance of the selected resemblance function to the perceived pattern among sites; sites 1 and 2 are most similar based on presence-absence data, whereas sites 2 and 3 are most similar based on species abundances.

lists that are shared increases. The tripartite index varies roughly as the inverse of the size of the larger of two lists compared (Tulloss 1997). The index should be rounded to two decimal places, except when comparing two very dissimilar lists. For species lists with low similarity, resulting in values with two or more zeros after the decimal point, Tulloss (1997) recommended rounding to the first nonzero digit in the estimate.

If quantitative data can be obtained (i.e., number of isolates, frequencies of occurrence, or other ecologically relevant measure of fungal abundance), they should be used. Similarity measures based on quantitative information are preferable to those based only on binary data. Although frequently used in the literature, the modified Sørensen Index (Table 5.5) of Bray and Curtis (1957) suffers from a number of shortcomings (Wolda 1981), and we do not recommend it for general use.

The Renkonen (1938) index of similarity (P) is based on the proportional abundances of species ($p_i = n_i/\Sigma n_i$) rather than their absolute abundances. The index ranges from 1 (unity; identical proportional abundances of species in both areas) to 0 (no taxa shared in the two areas). Although quite simple, the Renkonen index is affected very little by differences in sample size or diversity between areas (Wolda 1981) and thus is one of the best indices of similarity (Krebs 1989).

Morisita's index, also based on the proportional abundance of species, is best understood in terms of probability. The index is the ratio of two probabilities: the likelihood that an individual randomly obtained from sample 1 will be the same species as an individual obtained at random from sample 2, and the likelihood that two individuals obtained from either area 1 or area 2 will be the same species. Horn (1966) modified the index to allow its use with biomass or cover data as well as numbers of individuals. Like the Renkonen index, Morisita's index ranges from 0 (no similarity) to 1 (total similarity). Except for small samples, the index is essentially independent of sample size. Wolda (1981) recommended it as the best overall measure of similarity.

DISTANCE COEFFICIENTS

Distance or dissimilarity coefficients (Ludwig and Reynolds 1988) assume a value of zero when two areas (sites, habitats) are identical in species composition and a maximum value when no species are shared. Both metric and nonmetric coefficients of dissimilarity are available (Pielou 1984). This is important because metric coefficients may be visualized in multidimensional space (ordination procedures) where the distance between

each pair of points (sites) is equal to their compositional dissimilarity. Ordination and cluster analysis should be carried out on identical data to facilitate the understanding of diversity patterns (Fig. 5.3). Nonmetric dissimilarity measures cannot be used in this manner (Pielou 1984). We will discuss only metric indices.

Three groups of distance measures (Table 5.7) commonly are used in the ecological literature: (1) Euclidian distance coefficients, (2) the Bray-Curtis (BC) dissimilarity index (see "Resemblance Functions," earlier), and (3) relative Euclidean (RE) distance measures. Kenkel and Booth (1992) provided synopses of some of these measures as they pertain to fungal ecological units. For calculation of distance coefficients, abundance data are arranged in a matrix. Column headings are areas (e.g., leaves, tree root systems, quadrats), and row names are species.

The Euclidean distance (Table 5.7) is the metric with which we are most familiar; it is simply the geometric distance between two points in space. The lower bound is zero; the upper bound is limitless. The absolute distance, also known as the city block metric, expresses distance in terms of absolute differences. For the BC measure, a similarity index is calculated and subtracted from 100 to provide the distance component (PD). The dissimilarity index may be rescaled by subtracting the similarity index from 1 so that the distances obtained are more in line with other distance indices. The RE group (Table 5.7) contains metrics that are expressed on standardized or relative scales. Indices in the RE group compare proportional abundances. Consequently, sample units (e.g., root segments) with similar proportional abundances of species will be in close proximity.

Not all distance metrics are equally suitable for assessing the structure of biodiversity in space or time. Ludwig and Reynolds (1988) recommended against using the Euclidean group distance measures (despite their popularity) because they can produce spurious results. They also noted that the RE group indices perform reasonably well, with one metric having little advantage over another. Nonetheless, they concluded that cord distance performs best over a diverse range of ecological circumstances. Finally, Ludwig and Reynolds (1988) suggested that the Bray-Curtis (PD) index can be used as an alternative to indices in the RE group (Table 5.7).

CLASSIFICATION AND CLUSTER ANALYSIS

Cluster analysis is a classification procedure that groups objects into subgroups that are more similar to each other than to objects in other subgroups. The resulting branching diagram is a classification that provides the

TABLE 5.7

Distance Coefficients for Comparing the Structure of Ecologic Units among Locations or Sampling Times*

Index	Equation	Terms
Euclidean distance coefficients Euclidean Distance (ED)	$ED = \sqrt{\Sigma(X_{ij} - X_{ik})^2}$	X_{ij} is the abundance of species i at site or time j_i. X_{ik} is the abundance of species i at site or time k.
Squared Euclidean Distance (SED)	$SED = \Sigma(X_{ij} - X_{ik})^2$	Same as for ED.
Mean Euclidean Distance (MED)	$MED = \sqrt{\dfrac{\Sigma(X_{ij} - X_{ik})^2}{S}}$	Same as for ED. S is number of species.
Mean Absolute Distance (MAD)	$MAD = \Sigma \lvert X_{ij} - X_{ik}\rvert / S$	Same as for MED.
Bray–Curtis Dissimilarity Index	1. Calculate $SI = (2W/a + b)(100)$ 2. Calculate $PD = 100 - SI$ PD may be scaled between 0 and 1 by $PD = 1 - SI$	W is the sum of the lower abundances (number of individuals or other measure of density) of species that occur in each site. a is sum of the species abundances in site A, and b is the sum of the species abundances that occur in site B. SI, similarity index; PD, distance component.
Relative Euclidean Distance coefficients Relative Euclidean Distance (RED)	$RED = \sqrt{\Sigma[(X_{ij}/\Sigma\,X_{ij}) - (X_{ik}/\Sigma\,X_{ik})]^2}$	Same as for ED.
Relative Absolute Distance (RAD)	$RAD = \Sigma\lvert(X_{ij}/\Sigma\,X_{ij}) - (X_{ik}/\Sigma\,X_{ik})\rvert$	Same as for ED. Values range from 0 to 2.
Cord Distance (CRD)	$CRD = \sqrt{2(1 - ccos_{jk})}$	$ccos_{jk}$ = Chord cosine $ccos_{jk} = \Sigma(X_{ij}\,X_{ik}) - \sqrt{\Sigma X_{ij}\,\Sigma X}$
Geodesic Distance (GDD)	$GDD = arccos\,(ccos_{jk})$	This measure is the distance along the arc of a unit circle after projection of the sample units onto a circle of unit radius. Values range from 0 to 1.57.

* Notation follows Ludwig and Reynolds (1988).

sequence of clusters (subgroups) by which a set of objects is subdivided. When a number of ecological units is investigated, cluster analysis is appropriate for representing patterns of species composition among those units (Fig. 5.3). A dissimilarity matrix usually forms the basis of clustering models, although many software packages (e.g., SPSS 1999) allow the input of either similarity or dissimilarity (distance) measures. Essentially, cluster analysis produces a dendrogram or tree whose branches represent each of the ecological units (Fig. 5.3). The data on species composition for these sites determine the branching pattern. The joined branches represent groups or clusters of sites with similar species composition. The length of the branch prior to joining is inversely proportional to the degree of similarity in species composition.

A wide variety of cluster analyses (Sneath and Sokal 1973) are available (e.g., agglomerative and divisive, hierarchic and nonhierarchic, overlapping and nonoverlapping, sequential and simultaneous, local and global, direct and iterative, weighted and unweighted, adaptive and nonadaptive). The choice of a particular method is critical because it determines (in part) the pattern derived from the data on species composition. Nonetheless, most research ecologists and evolutionary biologists follow the unweighted pair group method using arithmetic averages (UPGMA), which provides a good starting point for visualizing the degrees of similarity among sites based on species composition. Like many multivariate statistical analyses, cluster analysis attempts to represent the complex relationships among sites in a simple one-dimensional way. As a result, various amounts of information are lost, and the pattern represented by the dendrogram may not be accurate. The degree to which the observed dendrogram represents the multidimensional relationship among sites based on the similarity or dissimilarity matrix is quantified by the cophenetic correlation coefficient (Sokal and Rohlf 1962). Clustering methods, including considerations of distortion and optimality, are discussed by Sneath and Sokal (1973), Gaugh (1982), and van Tongeren (1995). It is important to remember that the veracity of the pattern detected by cluster analysis is only as reliable as the input data (species presence or absence) and that the pattern is affected strongly by the resemblance function used (Fig. 5.3). The sites to be included in analyses, the suite of species on which to base resemblance functions, and the resemblance function chosen should be considered with care.

POWER ANALYSIS

To understand the effects of abiotic or biotic factors on patterns of biological diversity, differences in species richness among areas must be compared statistically. When the null hypothesis (H_o) is rejected, the likelihood of making an error is the significance level of the test (i.e., P value). Often, researchers believe that "real" biological differences exist between ecological units, although H_o is not rejected. In fact, differences may not be statistically significant for two reasons: (1) no differences exist among the ecological units, or (2) sample sizes are too small to reveal biological differences of a magnitude considered to be important. These problems plague investigators and pose important challenges for the interpretations of patterns in fungal biodiversity. Power analysis can help to distinguish between the two alternatives and should be considered a crucial component of any experimental design and analysis (Taylor and Gerrodette 1993; Thomas and Juanes 1996).

A good discussion of power analysis can be found in Muller and Benignus (1992). The power of a statistical test, such as analysis of variance (ANOVA), is the ability to reject the null hypothesis when an alternative hypothesis is true. Power differs depending on the type of statistical test used to assess differences. For example, parametric tests have more power than nonparametric tests when the assumptions of the former are met. Similarly, a priori tests have more power than a posteriori for detecting differences. Critically, power increases as the number of samples increases (decreasing standard error and increasing the degrees of freedom), as the size of samples increases, if a one-tailed rather than a two-tailed t test is used (if the alternate hypothesis is in the direction of the true difference), or if a higher alpha level (probability of rejecting the null hypothesis) is used. It is important to note that when the sample variance is large, power is low and ability to detect differences among ecological units is small.

Power analysis is most useful when designing and planning a study because it allows the investigator to evaluate the relationships among all five factors, including (1) the range of sample sizes, quadrats, or plots that feasibly can be examined, (2) the magnitude of differences among ecological units considered to be biologically important, (3) the magnitude of variation, (4) desired levels of rejecting the null hypothesis (alpha), and (5) statistical power (Thomas and Krebs 1997). Such preliminary investigations lead to optimal use of financial resources and personnel. Even after a study is completed, power analysis can be useful for interpreting the biological meaning of nonsignificant results.

Although power analysis can be performed using charts and tables provided in several texts (e.g., Cohen 1988; Lipsey 1990; Zar 1996), interpolation between tabled values can lead to errors. Thomas and Krebs (1997) reviewed 29 programs and five statistical packages that perform power analyses (Table 5.8) and are easy to use. In addition, all of the programs cover a wide range of tests and have a number of useful utilities.

TABLE 5.8

Selected, Stand-alone Statistical Software for Performing Power Analysis*

Software	Version	Vendor[†]	Operating System
Nquery Advisor	1.0	Statistical Solutions	Windows
PASS	6.0	NCSS Statistical Software	Windows
SPSS	10	SPSS, Inc.	Windows or Mac

* Information modified from Thomas and Krebs (1997).
† See Appendix IV for contact information.

EPILOGUE

Assessments of fungal biodiversity—whether in tussock grasslands of arctic tundra or in steamy, lowland rain forests of equatorial regions—must be grounded firmly in current methodologies of sampling design that account for effects of scale on patterns of biodiversity. Moreover, research programs should account for both temporal variation and spatial heterogeneity; this is particularly important because of the growing recognition that previous land use and disturbance can have pervasive effects on both. The ability to understand patterns of diversity across landscapes is predicated on a research design that effectively matches the temporal and spatial scales at which processes operate, with a corresponding scale of pattern detection.

Our goal has been to expose mycologists and others examining the diversity of fungi to analytical tools and approaches for understanding how biodiversity is partitioned across space and time. Because diversity is hierarchical, attention to the level of assessment is critical for teasing apart mechanisms that contribute to patterns. Ultimately, we must accept that diversity is a dynamic aspect of ecosystems and landscapes whose importance and contributions to system function are still not well understood. If we simply enumerate long lists of fungal species from various regions without attempting to evaluate and understand the mechanisms that contribute to observed patterns of diversity, we will only have described the shadows of the ecological actors that are participating in a complex and interesting evolutionary play. The challenge is to develop and maintain the taxonomic expertise necessary to identify the fungal taxa that are present in an ecosystem and at the same time to ensure that students and researchers have the strong ecological background necessary to evaluate fungal biodiversity patterns.

6

MOLECULAR METHODS FOR DISCRIMINATING TAXA, MONITORING SPECIES, AND ASSESSING FUNGAL DIVERSITY

RUSSELL J. RODRIGUEZ, DANIEL CULLEN, CLETUS P. KURTZMAN,
GEORGE G. KHACHATOURIANS, AND DWAYNE D. HEGEDUS

Exiting new molecular approaches that allow investigators to assess fungal biodiversity directly are being developed. These approaches are still in their infancies, but they are developing rapidly. The following papers provide an introduction to this burgeoning field of study: Bohannan and Hughes (2003), Cooksey (1997), Jumpponen (2003), Landeweert et al. (2003), Nikolchva et al. (2003), Ranjard et al. (2003), and Tedersoo et al. (2003).—Eds.

INTRODUCTION

Russell J. Rodriguez

The last 30 years have been burdened with a loss of habitats from development, pollution, and poor management of natural resources. As ecosystems are degraded and populations of organisms decrease in size, our need to address a multitude of issues concerning the biological and genetic diversity of remaining individuals increases because of the importance of information on such topics to conservation, management, and restoration efforts. In many species, for example, genetic diversity contributes to reproductive competence, disease resistance, and competitive abilities (Ryman 1970; Ralls and Ballou 1983; Templeton and Read 1983; Allendorf and Leary 1986; Liskauskas and Ferguson 1990). In addition, biological diversity is critical to ecosystem dynamics and health (Naeem et al. 1994). The initial step in dealing with diversity issues is to determine the species present in a particular area and to obtain baseline quantitative population data. To achieve this, accurate, analytical systems for qualitative and quantitative sampling of the diversity must be in place.

Historically, determining the species diversity of fungi in an ecosystem has been complicated by our inability to culture many fungi, the ephemeral nature of sporocarp production, and intraspecific variation in fungal morphology and biochemistry (Barrett 1987). As a result, the number of fungal species and the biochemical functions that can be attributed to them have been difficult to establish. To determine the significance of the species and genetic diversities of fungi adequately and to assign functional importance to specific fungi require increased understanding of several aspects of fungal biology. Some of the impediments to this understanding are as follows:

- Developmental stages of many fungi occur in or on complex substrata such as soil, wood, plants, or animals, making them difficult to monitor; thus the developmental stages of most fungi are unknown.
- Lifestyle habits (i.e., saprotrophic or symbiotic type) of fungi are poorly defined.
- Morphological traits are limited in number, which impedes taxonomic identification.
- Many fungi are difficult to culture, so the number of taxa is unknown.
- The morphological and biochemical variation of fungi *in situ* (natural environments) is undefined, which may obscure systematic analyses.
- Physical and chemical requirements for sporocarp development are poorly defined.
- Genetic and biochemical interactions between fungi *in situ* are poorly defined.
- Expression of primary and secondary metabolic genes *in situ* has not been addressed adequately.
- Anthropogenic factors, such as harvesting of sporocarps, may obscure or eliminate aspects of fungal biology that occur *in situ*.

Some of the greatest challenges confronting the field of mycology involve determining the functional roles of fungi in ecosystems, defining the number of fungal taxa, determining the full range of possible lifestyles for each species, understanding population genetic structure over space and time, and identifying all developmental stages. Advances in these areas require high-resolution methods that allow investigators to study fungi *in situ*, thereby decreasing the need for laboratory culture and *in vitro* analysis.

Although the population genetics of filamentous plant pathogenic fungi have been analyzed extensively (Kelly et al. 1994; Meijer et al. 1994; Anderson and Kohn 1995; Hamelin et al. 1995; Maclean et al. 1995; Theodore et al. 1995; Milgroom 1996; Delmotte et al. 1999; Goodwin et al. 1999; Kumar et al. 1999; Milgroom and Cortesi 1999; Rosewich et al. 1999; Pimentel et al. 2000; Roberts et al. 2000), the genetic structure and diversity of macrofungal populations have received less attention (Jacobson et al. 1993; Smith et al. 1994; Farnet et al. 1999; Högberg and Stenlid 1999; Högberg et al. 1999; Kerrigan et al. 1999; Smith and Sivasithamparam 2000). Traditionally, genetic diversity of fungi has been assessed by measuring morphological and physiological variation among individuals within specific populations. Because such traits depend on gene expression, they are influenced by environmental conditions. As a result, morphological and physiological variation that does not reflect genetic variation at the DNA level may occur among individuals. Likewise, only part of the genetic variation among individuals may be expressed as a variation in morphological and physiological traits.

During the last 20 years, several molecular biological techniques have been developed to assess genetic diversity by screening individuals from many populations for differences at the protein and/or DNA level. The most common protein-based technique used involves screening functional enzymes (allozymes) for differences in electrophoretic mobility as a result of amino acid substitutions, which reflect mutations in the genomic DNA. Because allozyme analysis only assesses functional proteins, however, the number of detectable polymorphisms is limited.

More recently, molecular techniques based on either mitochondrial or nuclear DNA have been developed to assess genetic properties and diversity. Those techniques have increased genetic resolution, allowing for unequivocal taxonomic identification and phylogenetic assessment of fungi. In addition, the genetic markers identified by molecular techniques are not influenced by environmental conditions and may be used to determine the genetic structure of fungal populations. More importantly, development of the polymerase chain reaction (PCR; Saiki et al. 1985; Mullis and Faloona 1987) allows for extensive genetic analysis of very small tissue samples, greatly decreasing the need for lethal sampling or the culturing of organisms. As a result, samples can be collected from individuals *in situ* without having an impact on community structure or balance, regardless of the fastidious nature (i.e., complicated nutritional requirements) of the fungus.

The fungal cell contains two compartmentalized pools of DNA—mitochondrial and nuclear—each of which may reveal valuable information regarding genetic diversity. Depending on the information desired, techniques based on either mitochondrial or nuclear DNA may be more appropriate. Mitochondrial genomes are relatively small (20–80 kb), are transmitted maternally, and function similarly in all fungi (Clark-Walker 1992; see "Mitochondrial Genome," later in this chapter). In general, mitochondrial DNA (mtDNA) represents a single genetic locus; however, low levels of recombination have been observed in some systems (Chung et al. 1996). Analysis of mtDNA is particularly useful for determining phylogenetic differences between closely related fungal species. Nuclear genomes, in contrast, are very large (1.5 $\times 10^7$ bp–820 $\times 10^7$ bp), represent thousands of genetic loci, and contain the genetic information that defines species (see Murrin et al. 1986). As a result, analyses of nuclear genomes are used to identify collections of heritable genetic markers that are correlated with or form the basis for interspecies and intraspecies differences.

An analysis of the genetic diversity of any fungal species addresses the number of differences that exist among individuals within and between populations, the sizes of those differences, and how those differences change over space and time. To determine genetic diversity properly using a molecular genetic system (1) the system must identify differences among individuals and be applicable to other species of fungi; (2) the utility of the system for assessing genetic diversity must be comparable to that of other methods; (3) it must be possible to accommodate large numbers of samples in a short period; (4) the genetic markers identified by the system must behave as normal Mendelian traits; and (5) the results must be reproducible by other laboratories.

Several excellent texts and reviews address applications of molecular biological techniques (Hughes 1998; Taylor et al. 1999; Theron and Cloete 2000). In this chapter we confine our discussion to the application of certain molecular techniques to questions in fungal biology. Specifically, we address ways to discriminate morphologically indistinct taxa, determine relatedness between taxa, quantify genetic diversity in natural populations, monitor species in complex substrata, and quantify specific gene activities in nature.

NUCLEAR GENOME

Russell J. Rodriguez

The nuclear genomes of fungi contain the genetic information that defines species and controls the expression of developmental cycles. The numbers and structure of chromosomes vary among species and may include

single-copy and/or repetitive DNA (Kistler and Miao 1992). Therefore, nuclear DNA is a macromolecule excellent for measuring genetic diversity and deriving molecular fingerprints for taxonomic purposes.

The PCR (Saiki et al. 1985; Mullis and Faloona 1987) has so increased the sensitivity and resolution of genetic analyses that genetic diversity may be assessed using small amounts of starting material. In this section I discuss the potential of using arbitrarily primed PCR (apPCR) (Welsch and McClelland 1990; Williams et al. 1990; Owens and Uyeda 1991), dual-primer PCR (dpPCR), and nested-primer PCR (npPCR) (Haff 1994) to discriminate among taxa, assess genetic diversity, and identify and monitor temporal and/or spatial patterns of fungi *in situ*.

TAXONOMIC DISCRIMINATION

Fungal taxonomy is complicated by intraspecific morphological and physiological variation and the limited number of morphological markers. Including DNA-based characters in taxonomic studies alleviates these problems because genomic DNA is stable under changing environmental conditions. Mutations in DNA do occur, but the frequency is low and the percentage of the genome affected over short periods (100 years) is small.

ApPCR analysis (also known as random amplified polymorphic DNA, RAPD, and single-primer PCR analysis) can be used to discriminate among fungal taxa with few distinguishing morphological characters (Freeman and Rodriguez 1993; Assigbetse et al. 1994; Chiu et al. 1995; Tommerup et al. 1995; Goodwin et al. 1999; Pimentel et al. 2000; Roberts et al. 2000). In this technique, short oligonucleotide primers (10–20 bp) that anneal to complementary DNA (cDNA) sequences are used. When sequences that are complementary to a specific primer are located in opposite orientation on separate strands of DNA, a double-stranded DNA product can be synthesized by DNA polymerase. The number and size of amplified products depend on the frequency and distribution of primer hybridization sites, which cannot be predicted without extensive characterization of the genome. However, the number of apPCR products generated from individual primers ranges from fewer than 5 to more than 20 depending on primer composition and the reaction conditions. Therefore, 10 to 20 oligonucleotide primers may generate more than 100 genetic markers with apPCR. Investigators requiring greater numbers of genetic markers can use additional primers.

Investigators have found that in fungi, nematodes, and insects, 80–100% of the DNA products generated from apPCR amplification are shared among individuals of the

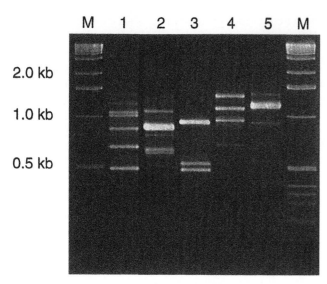

FIGURE 6.1 Arbitrarily primed polymerase-chain-reaction amplifications with lanes 1–5 representing approximately 25 ng of DNA from *Colletotrichum lindemuthianum, C. sublineonela, C. musae, C. coccodes,* and *C. magna,* respectively. The DNA was amplified with the primer (GACA)$_4$. Lane M contains DNA size markers, with the sizes of some markers denoted as kilobase pairs on the left.

same species for any given primer that is 15–16 bp long and composed of simple sequence repeats [e.g., (CAG)$_5$, (GACA)$_4$; Rodriguez and Owen 1992; Freeman et al. 1993; Perring et al. 1993; Van der Knaap et al. 1993; Freeman and Rodriguez 1995]. Alternatively, individuals from different species share from 0–20% of the DNA products generated from these primers. This has led to the generation of species-specific apPCR band patterns that can be used for unequivocal taxonomic identification. For example, five fungal isolates from the genus *Colletotrichum* that could not be identified taxonomically because they had lost the ability to infect plants and to produce conidia, setae, and pigment were easily distinguishable and characterized as five distinct species when analyzed by apPCR (less than 10% band sharing; Fig. 6.1). Comparisons of the apPCR band patterns from those species to apPCR band patterns from known species allowed us to assign species designations (90% or more band sharing; Fig. 6.1). When randomly designed 10-bp primers were used for RAPD analysis, band sharing among conspecifics varied from less than 5% to more than 45%, depending on the fungal system (Kelly et al. 1994; Hamelin et al. 1995; Maclean et al. 1995).

Distantly related species also may differ greatly in degree of intraspecific apPCR band sharing. For example, conspecific isolates from five distantly related basidiomycetes shared from 80% to 100% of their bands, whereas congeners shared only 0–20% (Fig. 6.2).

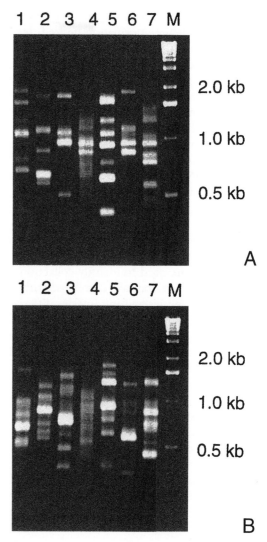

FIGURE 6.2 Arbitrarily primed polymerase-chain-reaction amplifications with lanes 1–7 representing DNA from *Boletus zelleri, Cantharellus formosus, Cortinarious variicolor, Cystoderma cinnabarinum, Stropharia ambigua, Ganoderma oregonense,* and *Inocybe sororia,* respectively. DNAs in A and B were amplified with the primers (GTC)$_5$ and (AGG)$_5$, respectively. Lane M contains DNA size markers with the sizes of some markers denoted as kilobase pairs on the right.

Although apPCR is useful when classifying fungi that lack discriminating morphological features, species designations require comparisons to apPCR patterns from known species. Eventually, it should be possible to generate a library of known apPCR band patterns and make them available on the Internet for comparative analyses.

Several laboratories have been unable to reproduce band patterns revealed by apPCR (Waugh and Powell 1992; Ellsworth et al. 1993; Tommerup et al. 1995). I carried out extensive analyses in an attempt to identify the cause(s) of nonreproducibility. Optimization of apPCR by modification of the reaction buffer, primer length and composition, source of the DNA polymerase enzyme, quality of the template DNA, and programming of thermocyclers have been reported (Yu and Pauls 1992; Ellsworth et al. 1993). Although these modifications may affect apPCR results greatly, my colleagues and I found that a major source of nonreproducibility in apPCR is attributable to the inaccuracy of the thermocyclers. This was determined with 20 thermocyclers from nine companies as described in the "Methods" section (see later in this paragraph). A scanning thermocouple thermometer was connected to 12 temperature probes that were placed in 0.5-ml tubes containing 20-µl PCR reactions. All of the machines tested were precise; however, only three brands were accurate with regard to the programmed denaturing temperatures. In fact, some machines never came within 10°C of the programmed denaturing temperatures. The inability to reach denaturing temperature results in incomplete DNA denaturation, so primer annealing sites may not be available. The ability of thermocyclers to achieve programmed denaturing temperatures was directly correlated with the reproducibility of amplified product patterns from the different reaction tubes. Based on this result, the data presented in this section were generated with thermocyclers manufactured by Barnstead Thermoloyne, MJ Research, and Stratagene (see Appendix IV).

GENETIC DIVERSITY

The intraspecific variation in apPCR products (0–20% difference) generated by some oligonucleotide primers with some but not all individuals of a species allows investigators to assess the genetic diversity among individuals within and between populations. As the number of primers used to analyze a species increases, the probability of generating polymorphic genetic markers also increases. No previous characterization of the genomic DNA is required, and primer sequences may be random or specifically designed. The primers used in the fungal analyses presented here were specifically designed; were composed of three, four, or five base-pair repeat motifs; and were approximately 15–20 base pairs long.

Genetic diversity can be assessed for any species for which a sample of intact DNA is available. A collection and DNA-extraction protocol that assures appropriate-quality DNA, is simple, is inexpensive, and does not involve toxic chemicals is described later in this chapter (see "Methods"). My colleagues and I have analyzed the genetic diversity of several basidiomycete, ascomycete, and anamorphic fungi populations using apPCR. For example, we analyzed clusters of sporocarps representing

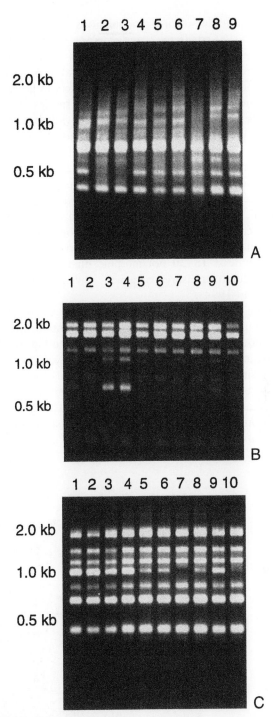

FIGURE 6.3 Arbitrarily primed polymerase-chain-reaction amplifications of fungal isolates. **A.** *Cantharellus formosus*, primer (AGG)₅. **B.** *Laetiporus sulfureus*, primer (GTC)₅. **C.** *Stropharia ambigua*, primer (GTC)₅. All gels contained DNA size markers with the sizes of some markers denoted as kilobase pairs on the left.

geographically distinct populations of *Cantharellus formosus*, *Cystoderma cinnabarium*, and *Laetiporus sulfureus* (Fig. 6.3). The sporocarps of each species occurred in close proximity (1 m or less) to one another. The occurrence of polymorphic markers indicated that the individuals from each of these species were not clones. If we were to screen these individuals with additional primers, we could identify additional polymorphic markers to use in quantifying the genetic diversity of each population. We then could establish a database of the presence or absence of each PCR product in each species isolate. For a measure of genetic diversity to be meaningful, the number of polymorphic markers used must be great enough to ensure that additional markers have minimal effect on that diversity. Experiments suggested that at least 15–20 polymorphic markers are required for accurate measures of genetic diversity (Lynch and Milligan 1994).

The sensitivity of apPCR is such that small amounts of fungal material (0.5 cm³) are sufficient for hundreds to thousands of genetic analyses (e.g., see Fig. 6.4). We took advantage of that property to analyze the genetic diversity of *Cantharellus formosus* populations without disturbing either sporocarps or underground mycelia (R. Rodriguez et al., unpublished data). Using the information we obtained about the genetic structure of *C. formosus* populations, we were able to address several issues relevant to biodiversity studies, including: (1) whether a cluster of sporocarps represents a single individual; (2) how individual genotypes are distributed spatially; and (3) whether adjacent sporocarps are clones. Clusters composed of from 16 to 50 sporocarps were separated by distances ranging from 3 to 30 m. Each cluster represented a genetically distinct population composed of closely related but nonclonal sporocarps. In addition, many adjacent sporocarps were nonclonal.

Many of the genetic markers generated by apPCR are either present or absent, so it is difficult to distinguish heterozygotes, thereby making classical population genetic analyses somewhat complicated. However, theoretical considerations concerning the use of equations and dominant markers for population genetic analyses have been published (Clark and Lanigan 1993; Lynch and Milligan 1994). As a result, apPCR is an excellent tool for rapidly assessing diversity in a population. After a series of polymorphic markers has been identified, the markers can be cloned, sequenced, and used as a basis for designing marker-specific PCR primers for dual-primer PCR (dpPCR) analysis, which permits rapid assessment of populations for levels of genetic diversity.

SPECIES MONITORING

Many questions related to fungal biodiversity are best addressed with *in situ* monitoring of the temporal and spatial occurrence of individual species. Techniques for such monitoring must be sensitive and specific. We have begun to develop such techniques by drawing on the species-specific band patterns of apPCR, the specificity of dpPCR, and the sensitivity of npPCR. Many of the products generated by apPCR represent species-specific DNA sequences, which can be verified by DNA hybridization analysis (Hadrys et al. 1992; Fani et al. 1993). Primers can be generated from the sequences on each end of species-specific bands and used for dpPCR specifically to amplify the band(s) of interest (Fig. 6.5). Primers that amplify products specific to the DNA of a given species are able to detect that species even when it is mixed with four other species. In fact, dpPCR will detect the DNA of a target species even at ratios of 1 : 100 (Fig. 6.6).

The dpPCR process is so sensitive that as little as 25 picograms of DNA amplified with species-specific primers can yield a product (Fig. 6.7). If primers that anneal to sequences 10–20 base pairs from the terminal ends of a species-specific band are constructed, then the sensitivity of PCR amplification can be increased 100- to 1000-fold with a two-step procedure. The first step involves the amplification of genomic DNA with two

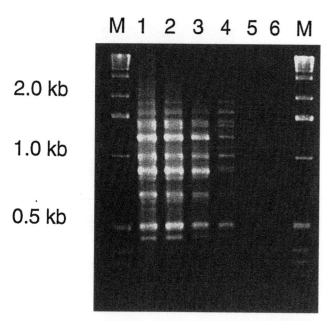

FIGURE 6.4 Arbitrarily primed polymerase-chain-reaction amplifications of a dilution series of DNA of *Colletotrichum magna*. Lanes 1–6 represent 25 ng, 2.5 ng, 250 pg, 25 pg, 2.5 pg, and 250 μg, respectively, of DNA amplified with (GTC)₅. Lane M contains DNA size markers with the sizes of some markers denoted as kilobase pairs on the left.

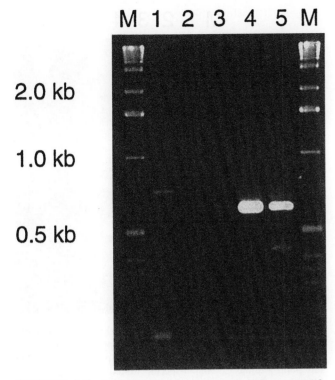

FIGURE 6.5 Dual-primer polymerase-chain-reaction amplifications of DNA from *Colletotrichum* species using *C. magna*-specific primers p365 and p366. Lanes 1–5 represent the amplification of 25 ng of DNA from *C. lindemuthianum*, *C. sublineonela*, *C. musae*, *C. magna*, and a mixture of all four species, respectively. Lane M contains DNA size markers with the sizes of some markers denoted as kilobase pairs on the left.

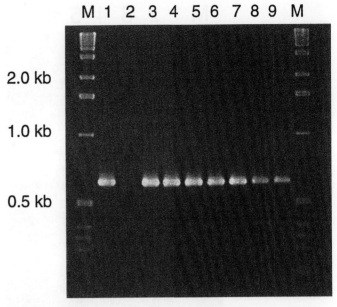

FIGURE 6.6 Dual-primer polymerase-chain-reaction amplifications of varying ratios of DNA from *Colletotrichum magna* and *C. musae* using *C. magna*-specific primers p365 and p366. Lanes 1–9 represent ratios of *C. musae* to *C. magna* of 1 : 0, 0 : 1, 1 : 1, 1 : 2, 1 : 5, 1 : 10, 1 : 20, 1 : 50, and 1 : 100, respectively. In each amplification, the amount of *C. musae* DNA was constant, and the amount of *C. magna* DNA was reduced. Lane M contains DNA size markers with the sizes of some markers denoted as kilobase pairs on the left.

primers that anneal to the ends of a specific sequence. The second step involves the reamplification of a small amount of the product from the first reaction with primers that anneal to sequences 10 base pairs or more from the ends of the amplified product (see "Methods," later in this chapter). This process is designated npPCR (Fig. 6.8; Haff 1994). If greater sensitivity is required, then a second set of "nested" primers may be generated, and a third amplification is performed. This level of sensitivity allows for the detection of a few fungal genomes in any complex sample. In addition, because dpPCR and npPCR require less accuracy and precision from thermocyclers than apPCR, the techniques are easier to use and are faster.

METHODS

The methods described here are used routinely in my laboratory to address the topics discussed in this section. It is important to point out that a wide variety of methods can be used successfully to collect fungal samples, extract DNA, and analyze the genes of the

samples. Each laboratory will adopt, modify, or generate specific methods to perform analyses on specific organisms of interest.

Sample Collection and DNA Extraction

Tissue samples (approx. 0.5 cm³) are collected from sporocarps and placed in 1.5-ml microcentrifuge tubes containing 0.5 ml of a DNA preservation and extraction buffer (150 mM EDTA [ethylene diamine tetraacetic acid], 50 mM Tris pH 8.0, 2.0% N-lauroylsarcosine). Samples can be stored in the buffer at ambient temperatures for at least 1 year without degradation of the DNA. To extract DNA, the samples are pulverized with a small pestle, incubated at 65°C for 30 minutes, and processed as follows (Rodriguez 1993). Cell debris is removed by centrifugation for 10 minutes at 14,000 × g, and the supernatant (S/N) is transferred to a clean 1.5-ml tube. Nucleic acid, protein, and other large macromolecules are precipitated by the addition of 0.7 volumes of PEG/NaCl (20% polyethylene glycol molecular weight 8000 in 2.5 N NaCl), followed by incubation on ice for 5 minutes and centrifugation for 10 minutes at 14,000 × g. The S/N is decanted, and the pellet is resuspended in 0.5 ml of TE buffer (10 mM Tris

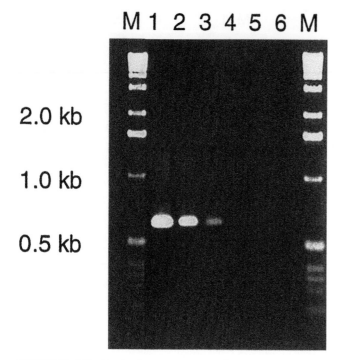

FIGURE 6.7 Dual-primer polymerase-chain-reaction amplifications of a dilution series of DNA from *Colletotrichum magna* using *C. magna*-specific primers p365 and p366. Lanes 1–6 represent the amplification of 2.5 ng, 250 pg, 25 pg, 2.5 pg, 250 μg, and 25 μg DNA, respectively. Lane M contains DNA size markers with the sizes of some markers denoted as kilobase pairs on the left.

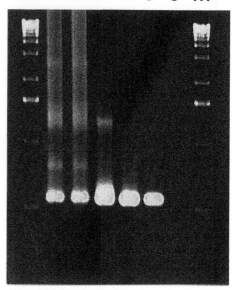

FIGURE 6.8 Nested-primer polymerase-chain-reaction amplifications of the dual-primer amplification products from *Colletotrichum magna* (derived with *C. magna*-specific primers p365 and p366), using *C. magna*-specific nested primers p413 and p415. Lanes 1–6 represent the nested amplification of 2 μl derived from dual-primer products representing 2.5 ng, 250 pg, 25 pg, 2.5 pg, 250 μg, and 25 μg of *C. magna* DNA, respectively. Lane M contains DNA size markers with the sizes of some markers denoted as kilobase pairs on the left.

pH 8.0, 1 mM EDTA). Protein, RNA, and polyphosphates (potent inhibitors of restriction enzymes and DNA-modification enzymes; Rodriguez 1993) are precipitated by the addition of 0.5 volumes of 7.5 N ammonium acetate, incubation on ice for 20 minutes, and centrifugation for 10 minutes at $14,000 \times g$. The S/N is transferred to a clean 1.5-ml tube, and 1 volume of isopropanol is added. The solutions are mixed gently, incubated on ice for 5 minutes, and centrifuged for 10 minutes at $14,000 \times g$ to produce a DNA pellet. The S/N is decanted, and the DNA is air-dried for 30 minutes and resuspended in 0.5 ml of TE buffer. The DNA is again precipitated by the addition of NaCl to 0.1 M and two volumes of 95% ethanol, incubated on ice for 5 minutes, centrifuged for 10 minutes at $14,000 \times g$, and again resuspended in 0.5 ml of TE buffer. Depending on the concentration of the DNA, samples are diluted 1 : 20 with TE buffer prior to PCR amplification.

To obtain DNA from *Colletotrichum* species, mycelia are cultured in liquid media and harvested by filtration (Redman and Rodriguez 1994; Redman et al. 1999). The mycelia are freeze-dried and pulverized in a mortar filled with liquid nitrogen. Two ml of mycelial powder are resuspended with 5 ml of extraction buffer in a

15-ml polypropylene centrifuge tube. The DNA from resuspended mycelia is extracted as described earlier for sporocarp tissues.

PCR Reactions

ApPCR is performed in 20-μl reactions containing 10 mM Tris-HCl (pH 9.0); 50 mM KCl; 2.5 mM MgCl$_2$; 0.2% Triton X-100; 200 M each of dATP, dCTP, dGTP, and dTTP; 0.2 units *Taq* DNA polymerase; 500 ng of an oligonucleotide primer; and 50–200 ng of fungal genomic DNA. Oligonucleotide primers represent repeated DNA-sequence motifs as follows: AGGAG GAGGAGGAGG, GACAGACAGACAGACA (Epplen 1988; Weising et al. 1989; Gupta and Filner 1991), and GTCGTCGTCGTCGTC. In this chapter, those primers are designated as (AGG)$_5$, (GACA)$_4$, and (GTC)$_5$, respectively. Amplification reactions consisted of 35 cycles, with each cycle having a temperature regime of 93°C for 15 seconds, 56°C for 1.5 minutes, and 72°C for 1.5 minutes. Ramp times of 1 minute from denaturing to annealing temperatures and annealing to synthesis temperatures are imposed. Prior to the initiation of the cycles, the reactions are exposed to 93°C for 2 minutes. Electrophoresis of the amplified products is performed for 1.5 hours at 7.0 V/cm in 2% agarose (Sambrook et al. 1989). PCR products are stained with ethidium bromide and visualized with 305 nm ultraviolet (UV) light (Sambrook et al. 1989).

DpPCR is performed with the same reaction buffer components described for apPCR except that two primers are added instead of one. The thermocyclers are programmed the same as in apPCR except that no ramp times are imposed, and the annealing temperature is 58°C.

NpPCR is performed by diluting dpPCR products fivefold with 10 mM Tris-HCl (pH9.0). One to 2 µl of diluted dpPCR product are transferred to a 0.5-ml microcentrifuge tube containing 20 µl of fresh reaction buffer and two nested primers. The thermocyclers are programmed as described for dpPCR reactions.

Testing Thermocyclers

Thermocyclers are tested for accuracy, precision of programmed temperatures, and reproducibility of apPCR products within and between runs. Temperatures are measured with a scanning thermocouple thermometer (Digi-Sense from Cole Parmer Instrument Company, see Appendix IV) connected to 12 temperature probes submerged in 20-µl PCR reactions in 0.5-ml microcentrifuge tubes. The thermocouple may be connected to a computer for data storage and analysis. ApPCR product reproducibility is assessed by aliquoting 20 µl from one large reaction mix containing all components into each of 20 tubes. The reaction tubes along with the 12 thermocouple tubes are placed in a thermocycler programmed for 35 cycles of the temperature regimen. After the cycles are complete, the reaction products are analyzed by agarose-gel electrophoresis.

Cloning Arbitrarily Primed Polymerase-Chain-Reaction Products and Species-Specific Primers

An apPCR band pattern was generated from *Colletotrichum magna* DNA using the primer (TCC)$_5$. A 600–700-bp apPCR product was isolated from agarose gels and purified by the glass bead method (Q-Biogene, see Appendix IV) and cloned into the pT7Blue vector (Novagen, see Appendix IV) for sequence analysis. Sequence analysis was performed using the chain termination method of Sanger et al. (1977). Species-specific dpPCR primers were CGAATCTGTAACTCTTCCTGC (p365) and ACAGACAGGATTCTCAATTTC (p366) and were constructed based on the terminal nucleotide sequences of the cloned apPCR product. Species-specific npPCR primers were AACCGTCTCATGCAAAAGTCA (p413), which was 20 base pairs from the end of p365, and GGTATGTCCCTTCCTGAACAC (p415), which was 10 bp from the end of p366.

MESSENGER RNA TRANSCRIPTS

Dan Cullen

In contrast to DNA, messenger RNA (mRNA) in complex substrata is rarely analyzed, in large part because labile RNA molecules are difficult to purify. Nucleic acid extractions from fungi that colonize soil are particularly difficult and plagued by humic substances that interfere with *Taq* polymerase (Tebbe and Vahjen 1993 and references therein). Magnetic capture techniques have overcome these problems, allowing rapid and efficient purification of microbial DNA and RNA. Magnetic capture involves the use of magnetic beads that are covalently attached to single-stranded DNA, most commonly oligo (dT) chains. The beads are mixed with crude lysates, and hybrid molecules are removed by application of a magnetic field. In the case of beaded oligo (dT) chains, polyadenylated [Poly (A)] RNA, which is a suitable template for reverse-transcription-coupled PCR (RT-PCR), is obtained. When combined with competitive PCR techniques (Gilliland et al. 1990), quantitative assessment of transcript levels is possible. Quantitative transcript analysis is as sensitive and specific as conventional PCR amplification of DNA for identifying fungi. The competitive RT-PCR technique is particularly well suited for differentiating genes within complex gene families such as those encoding the peroxidases, laccases, and cellobiohydrolases of white-rot fungi. In addition, the technique provides a measure of fungal biomass and a glimpse of physiological activities of fungi *in situ*.

The most thoroughly studied gene family includes 10 or more structurally related genes that encode isozymes of lignin peroxidases in *Phanerochaete chrysosporium* (reviewed in Gaskell et al. 1994; Cullen 1997). When the RT-PCR technique was applied to *P. chrysosporium* soil cultures, unusual patterns of peroxidase gene transcription were demonstrated (Lamar et al. 1995). For example, certain lignin peroxidase transcripts, abundant in defined media (Stewart and Cullen 1999), were not expressed in soil during organopollutant degradation. The absence of those peroxidase transcripts argues strongly against a significant role for those genes in the degradation of pentachlorphenol (PCP) and polycyclic aromatic hydrocarbons (Bogan et al. 1996a, 1996b).

IMPORTANT CONSIDERATIONS

Transcripts may not directly reflect protein levels; thus, caution should be exercised in quantitative interpretations of complimentary DNA (cDNA) levels. However, the alternative approach, purification and characterization of fungal proteins, is difficult or impossible in most instances. Even if a protein is abundant and stable, immunological and/or physical characterization seldom permit unambiguous identification. Individual peroxidase isozymes present in soil cultures of *P. chrysosporium* cannot be quantified, but the total extractable manganese peroxidase activity is correlated roughly with transcript level (Bogan et al. 1996a).

Several precautions must be taken if cDNA is to be quantified reliably. First, excessive thermocycling beyond the reaction's plateau phase should be avoided. This is especially important for quantitative comparisons of cDNAs of substantially different lengths that amplify with different primers (Pannetier et al. 1993). Further, comparisons between separate samples may be affected by varying RNA yields and varying efficiencies of reverse transcription reactions. In such cases, simultaneous amplification of constitutively expressed, endogenous genes may provide an internal standard. Such "housekeeping" genes include glyceraldehyde-3-phosphate dehydrogenase (GAPDH; Harmsen et al. 1992), β-actin (Chelly et al. 1988; Horikoshi et al. 1992; Kinoshita et al. 1992), and HPRT (hypoxanthine-guanine phosphoribosyltransferase; Pannetier et al. 1993). Exogenous RNA that differs from the target by the presence of a small intron or restriction site(s) also can be used as an internal standard (Wang et al. 1989; Wang and Mark 1990).

The RT-PCR analysis is applicable to an array of substrata and fungal species. The only prerequisite is nucleotide sequence data of a transcribed region or at least partial coding sequences, flanking intron(s). Although most genes characterized from filamentous fungi contain introns, some do not. In those cases, competitive templates can be constructed with a restriction site not present in the cDNA target. Following PCR amplifications, the products are digested prior to gel fractionation and estimation of the equivalence point. With minor modifications of the soil extraction protocol, we have detected *P. chrysosporium* transcripts in colonized wood chips and in Lambert spawn inoculum, a mixture of sawdust and grain (Lamar et al. 1995; Janse et al. 1998; Vallim et al. 1998; Wymelenberg et al. 2002). All known peroxidase and cellulase transcripts of *P. chrysosporium* in aspen wood chips (Janse et al. 1998; Vallim et al. 1998) have been quantified and shown to differ dramatically from those of *P. chrysosporium* in soil (Bogan et al. 1996a, 1996b) or in defined media (Stewart and Cullen 1999). Laccase transcripts of *Trametes hirsutas* (Kojima et al. 1990) also have been amplified from colonized soil (D. Cullen, unpublished data). In contrast, no peroxidase transcripts were detected in soil samples collected from a hazardous waste site in Brookhaven, Mississippi (Lamar et al. 1994). Peroxidase transcript patterns of *P. chrysosporium*, however, were similar in presterilized and nonsterilized cultures (Bogan et al. 1996b).

Numerous modifications can be envisioned. Magnetic capture need not be confined to polyadenylated regions. Magnetic beads can be attached to virtually any oligonucleotide sequence, including those complementary to abundant ribosomal RNAs or to specific DNA sequences (Jacobsen 1995). Enormous amounts of sequence data are accessible for the small-subunit ribosomal RNA of fungi, so that species-specific primers and probes could be readily designed and tested.

Recent advances in fungal genomics and in high-throughput transcript analysis offer new opportunities. A high-quality draft genome of *P. chrysosporium* has been completed and is freely available from the U.S. Department of Energy's Joint Genome Institute (see Appendix III). Thus, it is now possible to construct microarrays representing all predicted genes (>10,000) and to examine their relative levels under a wide range of conditions. Impediments to this approach include the high costs associated with either DNA- or oligonucleotide-printed arrays. Also, the submicrogram amounts of Poly (A) RNA typically captured from sparsely colonized soil, litter, or wood are inadequate for direct labeling with fluorochromes. However, several mRNA amplification strategies have been devised to overcome limiting quantities of RNA (Iscove et al. 2002; Schoor et al. 2003; Wang et al. 2003). Another increasingly popular method for quantitative transcript analysis is reverse-transcription, real-time PCR (Heid et al. 1996). By continuously monitoring PCR reactions, the key log-linear cycles can be identified precisely in real-time adapted thermocyclers. Competitive and real-time PCR are similar in sensitivity and specificity (Freeman et al. 1999), although equipment and operational costs are considerably higher for real-time PCR approaches. The methodology described below focuses on reverse-transcription, competitive PCR, although the protocols for Poly (A) RNA purification yield mRNA suitable for microarrays or real-time PCR.

METHODS

An array of transcripts has been quantified from *P. chrysosporium* growing in presterilized soil, using magnetic capture of Poly (A) RNA followed by competitive RT-PCR (Lamar et al. 1995; Bogan et al. 1996a, 1996b). The RT-PCR technique used to obtain *P. chrysosporium* GAPDH transcripts from 4-day-old soil culture is illustrated in Figure 6.9. Approximately 10 g of *P. chrysosporium*-colonized soil are wrapped in Miracloth (Calbiochem-Novabiochem Corporation; see Appendix IV), snap frozen in liquid nitrogen, and ground in a mortar and pestle precooled with dry ice. During grinding, 10 ml of extraction buffer (4 M guanidinium thiocyanate, 0.1 M Trizma base, 1% dithiothreitol, and 0.5% Sarkosyl, pH

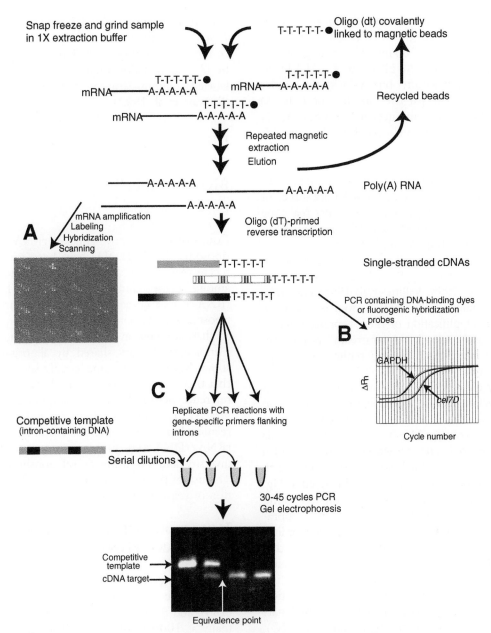

FIGURE 6.9 Protocol used for magnetic capture of Poly (A) RNA from colonized soil or wood, and alternative approaches to quantifying transcripts. Similar capture methods have been used for *Phanerochaete chrysosporium* (Lamar et al. 1995) and yeasts (Tebbe et al. 1995) in soil. (A) Microarray of subset of *P. chrysosporium* peroxidase and cellulase genes. Typically, hybridization involves combining two samples labeled with Cy3 and Cy5 fluorescent dyes (for review see Knudsen 2002). (B) Amplification plots of real time PCR of GAPDH and cellulase gene, *cel7D*. The GAPDH transcripts are more abundant. (C) Competitive PCR of GAPDH. Aliquots of the total complementary DNA (cDNA) pool were subjected to competitive PCR (Gilliland et al. 1990; Stewart et al. 1992). Competitive template, in the form of an intron-containing genomic clone, was added in 10-fold serial dilutions from 10 to 0.1 fg. The far right lane shows the cDNA target only (no competitive template reaction). The approximate equivalence point is 0.05 fg.

8.0) are added in 2-ml aliquots. The slurry is centrifuged at $500 \times g$ for 8 minutes. The supernatant is centrifuged a second time at $500 \times g$ for 8 minutes, and approximately 6 ml of final supernatant is mixed with two volumes of binding buffer (100 mM Trizma base, 400 mM LiCl, and 20 mM EDTA, pH 8.0). Tubes are centrifuged at 8000 $\times$ g for 2 minutes, and each supernatant is added to 1.5 mg of oligo(dT)25 Dynabeads (Dynal Biotech, Appendix IV) that previously have been washed in binding buffer. The mixture hybridizes for 30 minutes on ice.

Bound Poly (A) RNA is then isolated by magnetic concentration with a Dynal Magnetic Particle Concentrator MPC-1, resuspended in 500 ml of washing buffer (10 mM Trizma base, 0.15 M LiCl, and 1 mM EDTA, pH 8.0), and washed twice after retrieval with the smaller MPCE-1 model magnetic particle concentrator. Poly (A) RNA is eluted at 65°C for 2 minutes in 2 mM EDTA, pH 8.0. Dissociated Dynabeads can be recovered by magnetic concentration and regenerated according to the manufacturer's recommendations. RNA is precipitated with 2 volumes of ethanol and stored at –20°C. Typically, yields of Poly (A) RNA from soil or wood samples are sufficient for quantifying dozens of transcripts by either real-time or competitive PCR. (Fluorochrome labeling of mRNA for microarrays would require amplification.)

Oligo(dT)15 may serve as primer in reverse transcription reactions (Lamar et al. 1995; Bogan et al. 1996a, 1996b; Janse et al. 1998; Vallim et al. 1998; Fig. 6.9). The resulting single-stranded cDNA pool is a suitable source for PCR amplification of most expressed genes. Alternatively, a gene-specific "downstream" primer can be used to optimize sensitivity and reduce nonspecific PCR amplifications.

In the example shown in Fig. 6.9, competitive PCR was used to assess relative levels of GAPDH cDNA (Gilliland et al. 1990; Stewart et al. 1992). In brief, replicate PCR reactions, each involving 21 pmol of GAPDH-specific primers, are prepared, and known concentrations of a competitive template are added as a serial dilution to each reaction tube. In the GAPDH example illustrated in Fig. 6.9, the competitive template is an intron-containing genomic clone in plasmid pCRII (Invitrogen Corporation; Appendix IV). However, any vector, cosmid, or liter subclone of GAPDH would be adequate provided that the entire amplification target is included and the region contains at least one intron. A variety of thermocyling regimens can be used. For the GAPDH amplification shown, reactions are subjected to an initial cycle of denaturation (6 minutes, 94°C), annealing (2 minutes, 54°C), and prolonged extension (40 minutes, 72°C), followed by 35 cycles of denaturation (1 minute, 94°C), annealing (2 minutes, 54°C), and extension (5 minutes, 72°C). A final 15-minute extension at 72°C is also included.

Following amplification, the target cDNA and competitive template are size fractionated on agarose gels. The initial concentration of GAPDH cDNA is determined by estimating the dilution point at which the target cDNA and competitive template are equivalent. Typically, 10-fold or 5-fold serial dilutions are used initially to approximate cDNA concentrations. When necessary, increased resolution may be achieved using many smaller dilutions (e.g., 1-fold or 2-fold). Increased resolution of the equivalence point is also attained by scanning the ethidium bromide-stained gels and quanti-

fying band intensity using imaging software such as NIH *Image* (version 1.58 or later; see Appendix IV). To confirm the identity of PCR products, agarose gels can be blotted to nylon membranes and probed with ^{32}P-labeled oligonucleotides specific for cDNA targets (Stewart et al. 1992).

RIBOSOMAL RNA/DNA SEQUENCE COMPARISONS FOR ASSESSING BIODIVERSITY OF YEASTS

Cletus P. Kurtzman

Comparisons of ribosomal RNA (rRNA) and its template ribosomal DNA (rDNA) have been used extensively in recent years to assess both close and distant relationships among many kinds of organisms. The interest in rRNA/rDNA comes from two important factors. First, ribosomes are present in all cellular organisms and appear to share a common evolutionary origin, thus providing a molecular history shared by all organisms. Second, some rRNA/rDNA sequences are sufficiently conserved that they are homologous for all organisms and serve as reference points that enable alignment of the less conserved areas that are used to measure evolutionary relationships. This section examines the impact of rRNA/rDNA comparisons on our understanding of biodiversity and evolutionary relationships among yeasts. The use of this new information for the rapid identification of yeasts is also discussed.

ISOLATION AND CHARACTERIZATION OF rRNAs AND rDNAs

In eukaryotes rRNAs can be assigned to several size classes. The genes coding for large (25S–28S), small (18S), and 5.8S rRNAs occur as tandem repeats with as many as 100 to 200 copies. The separately transcribed 5S rRNA gene often is included in the repeat (Garber et al. 1988) as part of the region termed intergenic spacer (IGS) or nontranscribed spacer (NTS) (Venema and Tollervey 1995; Planta 1997). Each rRNA size class has been examined to determine the amount of phylogenetic information present. The large quantities of rRNAs present in cells make isolation and purification of those molecules relatively easy despite the nearly ubiquitous occurrence of quite stable RNAases. Numerous methods for isolation and purification have been described. The procedures of Chirgwin and colleagues (1979) and their modifications by Kurtzman and Liu (1990) are generally satisfactory. Techniques for the isolation and characterization of rDNA are described in T. J. White and associates (1990) and Vilgalys and Hester (1990).

Methods for sequencing nucleic acids are now commonplace. Techniques for 5S rRNA sequencing are summarized and discussed in Walker (1984, 1985a). The procedures used for sequencing large and small subunit RNAs and DNAs are based on the dideoxy method of Sanger and colleagues (1977). Lane and associates (1985) described the application of this method to rRNA sequencing through use of oligonucleotide primers and reverse transcriptase. Most initial comparisons of yeasts and other microorganisms were based on reverse-transcriptase-mediated sequencing of rRNAs because of the relative simplicity of this method over earlier rDNA sequencing techniques. Complete sequences were determined less often after McCarroll and associates (1983) and Lane and associates (1985) demonstrated that partial sequences of small subunit rRNAs provided essentially the same phylogenies as complete sequences.

T. J. White and colleagues (1990) and Kaltenboeck and colleagues (1992) provided protocols for sequencing rDNA using specific oligonucleotide primers and PCR. Fewer artifacts result from rDNA sequencing than from rRNA sequencing, and both strands of the rDNA are sequenced with their method, which further reduces errors. The introduction of automated sequencers has allowed direct sequencing of PCR-generated amplicons (e.g., see Kurtzman and Robnett 1998).

ESTIMATES OF RELATEDNESS FROM rRNA-rDNA COMPARISONS

rRNA-rDNA Reassociation

Bicknell and Douglas (1970) were the first to compare rRNAs extensively as a basis for yeast systematics, focusing on the degree of reassociation between tritium-labeled 25S rRNA from one species and complementary sites on filter-bound nuclear DNA from another species, and its reciprocal. That and similar methods have been used by other investigators, but because all species pairs must be tested, the comparison of large numbers of taxa is labor intensive. Another aspect of that procedure is that as evolutionary distances increase, a point is reached at which the base sequences are insufficiently similar to allow duplexing of paired molecules. It has been suggested that sequences must exhibit 75–80% or greater similarity before reassociation can occur (Bonner et al. 1973; Britten et al. 1974).

Restriction Fragment Length Polymorphisms of rDNA

Because rDNAs occur in multiple copies, they lend themselves to analysis based on restriction-fragment-length polymorphisms (RFLPs). Magee and colleagues (1987) treated rDNAs from several medically important

Candida species with a variety of restriction endonucleases and concluded that *C. guilliermondii*, *C. tropicalis*, and *C. albicans* produced sufficiently different digestion patterns to allow recognition of each species. Similar results were obtained by Vilgalys and Hester (1990) for several species of the genus *Cryptococcus*. Lachance (1990) used RFLP patterns to map the genetic profiles of 125 isolates of the cactus yeast *Clavispora opuntiae* that had been collected worldwide. Nearly all restriction sites that allowed discrimination of individual strains were located in the hypervariable intergenic spacer region. Montrocher and associates (1998) demonstrated that RFLP analysis of the IGS region also resolved strains of *Saccharomyces cerevisiae* and related species.

Because RFLP patterns allow recognition of individual species as well as individual strains of a species, the method has considerable diagnostic value. Estimates of evolutionary relationships from RFLP patterns have been reported for species assigned to *Candida* (Magee et al. 1987) and to *Cryptococcus* (Vilgalys and Hester 1990). Such estimates are less accurate than estimates derived from sequence comparisons because, as evolutionary distances increase, the degree of pattern similarity becomes less certain.

5S rRNA

Because 5S rRNAs are conservative and small (ca. 120 nucleotides), their nucleotide sequences are determined easily and have been used widely to estimate broad phylogenetic relationships (Hori and Osawa 1979). Walker and Doolittle (1982) compared 5S rRNAs from eight basidiomycetes, including four yeasts, and concluded that sequence similarity among species was correlated with the type of hyphal septum (i.e., simple pores versus dolipores).

Among ascomycetous yeasts, Komiya and associates (1981) determined the 5S sequence of *Schizosaccharomyces pombe*, and Mao and colleagues (1982) noted that it differed sufficiently from that of *Saccharomyces cerevisiae* to suggest that these two organisms are phylogenetically quite divergent. Similar results were obtained by Walker (1985a), who further showed that the ascomycetes fall into three groups: (1) *Schizosaccharomyces* and *Protomyces*, (2) budding yeasts, and (3) filamentous fungi. Despite the importance of those phylogenetic studies, the informationally richer 18S and 26S rRNA molecules, which now can be sequenced easily, have replaced the 5S rRNA molecule for such analyses.

5.8S rRNA

The 5.8S rRNA molecule, with about 160 nucleotides, is not much larger than the 5S rRNA molecule. Unlike the latter molecule, however, 5.8S rRNA contains modified nucleotides as do the 18S and 26S rRNA molecules. Consequently, when enzymatic sequencing is used as was done for 5S rRNA, the modified nucleotides sometimes are difficult to determine. Nevertheless, several 5.8S nucleotide sequences have been analyzed, although only a few phylogenetic relationships have been deduced from the resulting data (Walker 1985b). More 5.8S rDNA data are becoming available as the ITS (internal transcribed spacer) regions of the rDNA tandem repeats are analyzed because the 5.8S rRNA gene lies between the ITS regions.

Small Subunit and Large Subunit rRNA–rDNA Sequences

Close Relationships. The compilation of large subunit RNA sequences by Gutell and Fox (1988) demonstrated that the 5′ end of that molecule is quite variable and of potential use for identifying closely related species. Peterson and Kurtzman (1991) examined sequence divergence in that region for sibling (sister) species pairs from several yeast genera. Their data showed that nucleotide differences in the ca. 300-nucleotide region 25S–635 (domain D2, Guadet et al. 1989) are sufficient to distinguish nearly all sibling species. One exception is the pair *Saccharomyces bayanus/S. pastorianus*. The latter species probably arose as a partial amphidiploid following chance hybridization between *S. cerevisiae* and *S. bayanus*, and it retains the rDNA of *S. bayanus* (Vaughan Martini and Kurtzman 1985; Kurtzman and Robnett

TABLE 6.1
Fungal Gene Probes Used for Mitochondrial DNA Gene Location

Gene	Plasmid	Organism	Reference/source
CO1	pKP600	*Podospora anserina*	Kück et al. 1985
CO1	pTGM-13	*Torulopsis glabrata*	Clark-Walker et al. 1985
CO1,Cytb	pKP402	*P. anserina*	Kück et al. 1985
Cytb/CO1	pTGM-19	*T. glabrata*	Clark-Walker et al. 1985
Cytb	H8	*Phytophthora megasperma*	Forster et al. 1987
CO2	HX-2	*P. megasperma*	Forster et al. 1987
ATP 9	pMW30	*Aspergillus niger*	Ward et al. 1988
ATP 9	KLM6/11	*Kluyveromyces lactis*	Wilson et al. 1989
lrRNA	pPSB2b	*P. anserina*	Kück et al. 1985
lrRNA	H5&H7	*P. megasperma*	Forster et al. 1987
lrRNA	pTGM-12	*T. glabrata*	Clark-Walker et al. 1985
srRNA	pPSB4	*P. anserina*	Kück et al. 1985
srRNA	H5	*P. megasperma*	Forster et al. 1987

1991; Peterson and Kurtzman 1991). *Williopsis saturnus* and its variety *sargentensis*, which show 43% nuclear-DNA relatedness but no nucleotide differences in the D2 domain, have not been studied sufficiently to explain the unanticipated lack of divergence in that domain (Kurtzman 1987; Liu and Kurtzman 1991). Some sibling species pairs show a five-fold difference in substitutions over that of other pairs. That observation cannot be taken as definitive evidence for unequal rates of nucleotide substitutions among species, however, until the genetic processes that initiate species formation are better understood. Many of these differences might be attributed to species formation resulting from amphidiploidy as in the *Saccharomyces* example. With few exceptions, the D2 region of RNA is sufficiently variable to allow recognition of individual ascomycetous and basidiomycetous yeast species, including most sibling pairs. Conspecific strains ordinarily show little (0–1%) divergence in that region, whereas the distantly related species *Pichia bimundalis* and *Schizosaccharomyces japonicus* var. *versatilis* exhibit about 47% substitution (Peterson and Kurtzman 1991). Kurtzman and Robnett (1998) extended this work and sequenced a 600-nucleotide region comprised of both the D1 and the D2 domains for the ca. 500 currently recognized ascomycetous yeast species. Based on that analysis, they demonstrated that ascomycetous strains differing by 1% or greater substitutions in domains D1/D2 are separate species. Consequently, sequencing of that region, which is easily done, provides a rapid means for identifying all but the most closely related species. A similar correlation exists for the basidiomycetous yeasts (Fell et al. 2000).

Distant Relationships Among Ascomycetous Yeasts. The phylogeny of the ascosporogenous yeasts has been debated vigorously since the time of Guilliermond (1912) and even before. Some mycologists view yeasts as primitive fungi, whereas others perceive them to be reduced forms of more evolved taxa. Cain (1972) supported the latter interpretation, arguing that genera such as *Pichia* and *Cephaloascus*, having hat-shaped (galeate) ascospores, are likely to be reduced forms of the perithecial euascomycete genus *Ceratocystis*. Redhead and Malloch (1977) and von Arx and van der Walt (1987) accepted that argument and commingled yeasts and mycelial taxa in their treatments of the *Endomycetales* and *Ophiostomatales*.

Examination of rRNA/rDNA sequence divergence from a limited number of taxa indicated that the ascosporogenous yeasts, with the exception of *Schizosaccharomyces*, form a monophyletic group (clade) distinct from the filamentous species (Walker 1985a; Barns et al. 1991; Bruns et al. 1991; Hausner et al. 1992; Hendriks et al. 1992; Kurtzman 1993, 1994; Nishida and

Sugiyama 1993; Wilmotte et al. 1993). Kurtzman and Robnett (1994a) analyzed rRNA sequence divergence from type species of all cultivatable ascomycetous yeasts as well as yeastlike genera such as *Ascoidea* and *Cephaloascus*. They demonstrated that the yeasts and yeastlike taxa comprise a sister clade to the "filamentous" ascomycetes (euascomycetes). *Schizosaccharomyces*, *Taphrina*, *Protomyces*, and *Saitoella* form a divergent clade basal to the yeast-euascomycete branch. *Eremascus* species, which form asci unenclosed in a fruiting body, are aligned with the euascomycete clade. Those results substantiated the long-held observation that yeasts cannot be recognized solely on the basis of presence or absence of budding. Such members of the yeast clade as *Ascoidea*, *Eremothecium* (and its synonym *Ashbya*) show no typical budding, whereas *Aureobasidium*, *Phialophora*, and certain other genera of euascomycetes are usually dimorphic. Budding is also a common mode of vegetative reproduction among many basidiomycetous genera. Similarly, vegetative reproduction by fission is shared by *Dipodascus* and *Galactomyces*, members of the yeast clade, as well as by members of the distantly related genus *Schizosaccharomyces*. Sexual states of all members of the yeast clade are characterized by asci unenclosed in a fruiting body. This feature is exhibited by only a few taxa outside the yeast clade such as *Eremascus* and *Schizosaccharomyces*.

rRNA/rDNA comparisons are just beginning to have an impact on the taxonomy of ascomycetous yeasts. Major findings to date include the following: (1) yeasts are phylogenetically separate from the euascomycetes; (2) the fission yeast genus *Schizosaccharomyces* is phylogenetically distant from the budding-yeast clade and from the euascomycetes, resulting in the reassignment of the fission yeasts to a separate order, the Schizosaccharomycetales (Eriksson et al. 1993; Kurtzman 1993); and (3) many phenotypic characters, such as ascospore morphology, are poor indicators of phylogenetic relationships (Kurtzman and Robnett 1991, 1994b, 1997, 1998; Liu and Kurtzman 1991). Additional work is required to assign species to phylogenetically circumscribed genera and to evaluate the relationships among genera. Nevertheless, in a number of cases, rDNA-sequence comparisons already have been used to redefine taxa (e.g., Nishida and Sugiyama 1993; Kurtzman 1998; Kurtzman and Robnett 1998). One of the broader redefinitions places the basal ascomycetous lineages represented by *Schizosaccharomyces*, *Protomyces*, *Taphrina*, *Saitoella*, *Pneumocystis*, and perhaps *Neolecta* in a new class provisionally termed the "Archiascomycetes" (Nishida and Sugiyama 1993, 1994).

Distant Relationships Among Basidiomycetous Yeasts. Two morphologically distinct teleomorph states are found among the basidiomycetous yeasts (Boekhout et al. 1993, 1998). In one state, teliospores are formed and germinate to produce a basidium (metabasidium) that bears basidiospores, a sexual cycle that is broadly similar to that of the rust and smut fungi. The other sexual state lacks teliospores. Basidia develop on hyphae or yeast cells and give rise to basidiospores in a manner similar to that of the Tremellales (jelly fungi). Phylogenetic analyses of 18S rDNA (Swann and Taylor 1995a, 1995b, 1995c) and domain D1/D2 of 26S rDNA (Fell et al. 2000) place the basidiomycetous yeasts in three classes of the Basidiomycota: Ustilaginomycetes, Urediniomycetes, and Hymenomycetes. Surprisingly, teliosporic and nonteliosporic taxa are found in each of these classes. Similarly, presence or absence of carotenoids and ballistoconidia were previously considered to be phylogenetically informative, but species having one or both of these characteristics are now found in all three classes of the basidiomycetous yeasts.

At present, the most extensive phylogenetic comparison of basidiomycetous yeasts is that of Fell and colleagues (2000), who sequenced the ca. 600-nucleotide D1/D2 domain of 26S rDNA in 230 species assigned to 42 genera. Examination of that region distinguishes most of the closely related species, but as noted earlier, it has too few phylogenetically informative sites to assess most basal relationships accurately. Nonetheless, rRNA/rDNA sequence comparisons have contributed greatly to our understanding of the phylogeny of basidiomycetous yeasts, although better resolution of taxa will require additional sequencing. The anamorphic genera *Tsuchiyaea* and *Kochovaella*, for example, which were defined from differences in partial sequences, may overlap with some earlier described genera (Fell et al. 2000). Additional comparisons of basidiomycetous taxa can be found elsewhere (Blanz and Unseld 1987; Blanz et al. 1989; Guého et al. 1989, 1990; Laaser et al. 1989; Yamada and Kawasaki 1989; Yamada et al. 1989a, 1989b; 1990a, 1990b; Fell and Kurtzman 1990; Yamada and Nakagawa 1990; Nakase et al. 1991, 1993; De Wachter et al. 1992; Fell et al. 1992; Van de Peer et al. 1992; Berbee and Taylor 1993b; Sugiyama and Suh 1993; Suh and Sugiyama 1993).

RELIABILITY OF PHYLOGENIES INFERRED FROM rRNA GENE TREES

Calculations of phylogenetic distances from changes in the sequences of macromolecules require that substitutions occur at a predictable rate that, preferably, is also constant. Ochman and Wilson (1987) reported that nucleotide substitutions in small (16–18S) subunit rRNAs from both prokaryotes and eukaryotes proceed at the same rate (1%/ per 50 million years). Other workers, however, have presented evidence that rates of nucleotide substitution differ among phylogenetic groups and possibly among species within groups (Britten 1986; Field et al. 1988; Weisburg et al. 1989). Various methods for calculating evolutionary relationships based on macromolecular changes have been described (Lake 1987; Saitou and Nei 1987; Woese 1987; Felsenstein 1988; Holmquist et al. 1988; Avise 1989). Some of the algorithms are designed to compensate for disproportional rates of substitution, but uncertainty concerning their reliability remains (Felsenstein 1988). Evolutionary relationships among the bacteria, for example, are still subject to debate (Lake 1987; Achenbach-Richter et al. 1988; Gouy and Li 1989).

At present, only a few investigators have examined congruence between yeast phylogenies based on rRNA gene sequences and those determined from other molecules. Radford (1993), using sequence analyses of orotidine 5'-monophosphate decarboxylase, demonstrated that budding yeasts and euascomycetes are sister groups, as did comparisons of rRNA/rDNA sequences. One unusual outcome of Radford's work was placement of the Mucorales as a sister group to the basidiomycetes. Tsai and colleagues (1994) showed phylogenetic relationships among species of *Epichloë* (Clavicipitaceae) were the same whether based on rDNA sequences or sequences of the β-tubulin gene. Likewise, relationships among species of *Dekkera* and its anamorph *Brettanomyces* (Boekhout et al. 1994), and among species of *Kluyveromyces* (Belloch et al. 2000), were essentially identical whether determined from nuclear rDNA sequences or from sequences of the mitochondrially encoded cytochrome oxidase subunit II gene.

Another factor that affects gene-tree reliability is the method used for its construction. Most investigators now analyze data using phylogenetic inference programs based on cladistic principles. Those programs often include a statistics package to test the robustness of competing phylogenetic trees. Several researchers have addressed these important issues (Felsenstein 1988; Avise 1989; Saitou and Imanishi 1989; Hillis et al. 1994).

Although evidence is strong that molecular comparisons give an accurate approximation of the evolutionary history of an organism, investigators should be cautious when using phylogenetic trees to define taxonomic hierarchies. An impediment to the appropriate interpretation of those data is the lack of data from "missing taxa." Taxa may be excluded either because they are extinct or because they have not yet been recognized in nature. The presence of long branches on phylogenetic trees often indicates missing taxa if rates of nucleotide substitution are assumed to be reasonably constant for all species compared.

The taxonomy of *Debaryomyces* provides an example of the problems encountered when sampling of species is incomplete. On the basis of nucleotide similarities, Kurtzman and Robnett (1991, 1994b) proposed that *Schwanniomyces occidentalis* and *Wingea robertsii* be transferred to *Debaryomyces*. Likewise, Yamada and associates (1992) proposed the transfer of *Pichia carsonii* and *P. etchellsii* to that genus. A phylogenetic analysis that included the preceding four species as well as previously described *Debaryomyces* species suggested that *Debaryomyces* is monophyletic (Fig. 6.10A). When additional species were included in the analysis (Kurtzman and Robnett 1998), *D. carsonii*, *D. melissophilus*, and *Debaryomyces* sp. n. were no longer closely allied with other species in the genus (Fig. 6.10B). If the circumscription of a genus is to be based on monophyly of its assigned species, then the preceding three *Debaryomyces* species and other members of their clade represent a new genus. Before such taxonomic changes are made, however, all known yeasts need to be included in the comparisons, taking into account taxa located on the long branches of phylogenetic trees and their possible effect on genus circumscription. It is apparent that a large number of species are yet to be found and that their discovery and the analysis of their rRNAs will lead to a continuing reassessment of genera.

MOLECULAR METHODS FOR RAPID YEAST IDENTIFICATION AND DETECTION OF BIODIVERSITY

The specificity of nucleic acid sequences has led to the development of several methods for rapid species identification. Because those techniques often can detect single nucleotide changes, a large variety of genetically defined strains should be tested first so that within-species variation can be understood. Aberrant strains may represent different species.

The RFLP technique was discussed earlier in this chapter (see "Restriction Fragment Length Polymorphisms of rDNA") and has been used extensively in some laboratories. Bruns and colleagues (1991) listed some of the factors that require attention when using RFLPs for species identifications. RAPD is another method whose used (see "Taxonomic Discrimination," earlier). The technique is based on amplification of genomic DNA in the presence of one or more short (ca. 10–15 mers) oligonucleotide primers of random sequence. The amplified products are visualized on an agarose gel, and strains are identified from matching band patterns. Hadrys and colleagues (1992) discussed details of this procedure and points of technical difficulty. In particular, band patterns may reproduce poorly, apparently because of small

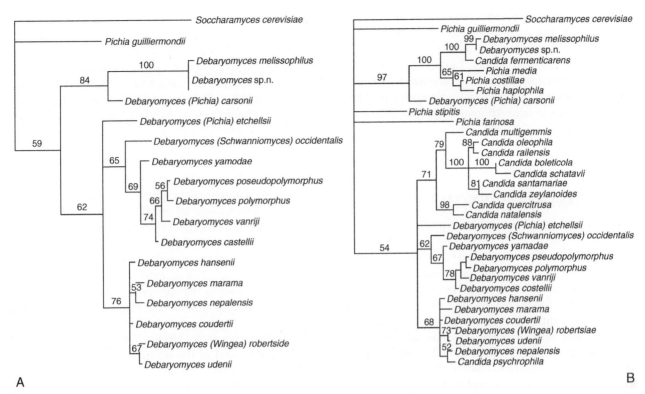

A

B

FIGURE 6.10 Phylogenetic analyses of relationships among species of *Debaryomyces* and reference taxa. The trees are derived from maximum parsimony (PAUP) analysis of ca. 600 nucleotides from the 5′ end of large subunit rDNAs. Branch lengths are proportional to nucleotide differences, and the numbers given on branches are the percentages of frequencies with which a given branch appeared in 100 bootstrap replications. Values less than 50% are not given. **A.** Phylogenetic tree comprised of described *Debaryomyces* species and two reference taxa. **B.** Phylogenetic tree as in *A*, but with eighteen reference taxa. *Debaryomyces carsonii, D. melissophilus,* and *Debaryomyces* sp. n. are separated from other members of *Debaryomyces* in this analysis (C. P. Kurtzman and C. J. Robnett, unpublished data).

differences in reagent concentrations and variations in thermocycler temperatures.

Fell (1993) applied a three-primer, PCR-based technique to yeast identification that appears species-specific. The reaction mix includes genomic DNA, two external primers for the D1/D2 region of large subunit rDNA, and a species-specific internal primer. The external primers allow amplification of the ca. 600-nucleotide D1/D2 region, but in the presence of a species-specific primer (third primer), the amplification product is shorter and easily detected on an agarose gel. Mannarelli and Kurtzman (1998) modified Fell's technique to identify clinically important ascomycetous yeasts. Other technologies include use of labeled species-specific oligonucleotide probes. Arrays of immobilized species-specific oligonucleotides are under development and offer the possibility of rapid, simultaneous identification of multiple species.

The availability of automated DNA sequencers may ensure sequencing as the most effective method for identifying taxa. Databases comprised of the ca. 600 nucleotide D1/D2 region of large subunit rDNAs have been compiled for all known yeast species (Kurtzman and Robnett 1998; Fell et al. 2000), and updates are being added to GenBank. Because all but a few closely related taxa can be discriminated from differences in this region, rapid, reliable species identifications can be realized, and new species are being recognized quickly. For example, in a recent treatment, the genus *Metschnikowia* was comprised of 10 species (Giménez-Jurado 1992), many of which are widely separated from each other on long branches of the phylogenetic tree (Mendonça-Hagler et al. 1993). Additional studies, however, revealed 10 new species (Lachance et al. 1998; C. P. Kurtzman and C. J. Robnett, unpublished data). Those works not only increased the number of *Metschnikowia* species by 100% but also eliminated some of the long branches in the phylogenetic tree that gave the perception of a rapid rate of nucleotide substitution.

As a phylogenetic framework for systematics based on molecular comparisons develops, a better understanding of what constitutes genera, families, and orders will be possible. That, in turn, will allow for assessment of biodiversity at higher levels of taxonomic classification.

The 10 species now assigned to the genus *Saccharomycopsis* based on the analysis of rDNA sequences, for example, formerly were distributed among six genera (Kurtzman and Robnett 1995). In contrast, the genus *Pichia*, which presently includes 91 described species, is clearly polyphyletic, and molecular comparisons likely will divide it into at least 10 genera distributed among several families.

MITOCHONDRIAL GENOME

George G. Khachatourians and Dwayne D. Hegedus

Mitochondria (mt) are small organelles that carry out major functions in cell energetics. Mitochondrial function requires several hundred proteins, only a fraction of which are coded for by the mt genome. In contrast to animal mtDNA, those from fungi show a high degree of diversity in size and organization. The size of fungal mtDNA varies from 17.6 kb in *Schizosaccharomyces pombe* to 175 kb in *Agaricus bitorquis* (Hudspeth 1992). Isolate- and species-specific mtDNAs commonly vary in size, shape, and base sequences, with introns and intergenic sequences accounting for much of the size variation (Gray 1982). Genome-length mutations and nucleotide substitutions are responsible for a 25% mtDNA size variation within *Neurospora crassa* (Taylor 1986). Despite the considerable size variation among fungal mt genomes, the organelles perform identical functions. Mitochondrial DNA offers an excellent system for the study of fungal evolutionary genetics because of small size, ease of purification, presence of RFLPs, and availability of gene sequences for PCR amplification. As a result, mtDNA can be used to study fungal biodiversity, fungal taxonomy, and the role of fungi in ecosystem dynamics.

In this section we discuss the methodology for molecular study of mtDNA with special focus on entomopathogenic fungal species, which we and our colleagues have been studying for the last decade (Khachatourians 1996; Khachatourians et al. 2002).

IMPORTANT CONSIDERATIONS

Mitochondrial Genes

Despite variations in mitochondrial shape and a 10-fold variation in mtDNA-genome size, similar sets of functional genes are found in most fungal mitochondria. Those gene sets control mtDNA replication, transcription, translation, and a limited number of other functions. Among the mtDNA-encoded genes that have been characterized and sequenced are those required for the (1) translation apparatus (tRNA, or transfer RNA, synthesis; large [l]- and small [s]- ribosomal subunit rRNA; or proteins), (2) respiratory chain proteins (cytochrome oxidase subunits, apocytochrome B, NADH [nicotinamide adenine dinucleotide with hydrogen] dehydrogenase subunits), (3) ATP (adenosine triphosphate) synthase complex, (4) several open reading frames (ORFs), and (5) some unidentified reading frames (URFs). The most frequently used gene probes are shown in Table 6.2. The presence of URFs suggests that our understanding of the genetic information contained in the mtDNA is still incomplete. Even after accounting for all the introns, genes, rRNAs, and tRNA in the 94,192 bp genome of the races of *Podospora anserina*, the purpose (if any) of about 23% of the mtDNA is not known (Cummings et al. 1990).

In our studies of the species-level systematics of entomopathogenic fungi, we have examined both conserved and variable regions of mitochondral genomes (Simon 1991; Hegedus et al. 1998). The rRNA genes are the most widely used of the mt genome. Within the rRNA, the separation between the l- and srRNA genes varies. Ribosomal-RNA sequences can be used to determine evolutionary relationships (Förster et al. 1990; see, "Ribosomal RNA/DNA Sequence Comparisons for Assessing Biodiversity among the Yeasts," earlier). The srRNA sequences have been used by Van de Peer and co-workers (1990) to show evolutionary relationships among different life forms and to reveal mitochondrial origins. The sequences of rRNA show both conserved and variable regions, and the structure of rRNA shows secondary and tertiary structures. Although rRNA varies considerably, not all of the short-range mitochondrial rRNA stems are poorly conserved (Simon 1991). The region encoding between the two rRNA genes appears to be highly conserved. In *Beauveria bassiana*, *N. crassa*, *P. anserina*, and *Aspergillus nidulans*, tRNA genes are clustered differently with respect to the two regions, rRNA and the s- and lrRNA (Pfeifer et al. 1993; Hegedus et al. 1998). The rRNA–tRNA gene region in *A. nidulans*, which is 10 kbp, is apparently a condensed version of the *N. crassa*, 17 kbp region. Both species have the same number of tRNA genes, but the lengths of their introns (longer in *N. crassa*) and intergenic and rRNA sequences differ. This pattern is consistent with the evolutionary divergence of rRNA genes (Pace et al. 1986). Likewise, comparisons of the Val-, Ile-, Ser-, Trp-, and Pro-tRNA genes from *B. bassiana* to those of other filamentous fungi (Hegedus et al. 1991) showed a clustering of these genes most like those of *P. anserina* and *A. nidulans*. The tRNA sequence conservation of different functional regions indicates that the most conserved region is the anticodon loop followed by the dihydrouridine stem and the ribothymidine pseudouridine cytosine loop. In sea urchins the mt tRNA genes evolve at a higher rate than their nuclear counterparts. For example, the tRNA leu CUN gene, having lost its function, has been incorporated into the protein subunit of the ND5 gene protein, implying that it has diverged at a higher rate than the ND5 gene (Thomas et al. 1989). The NAD1 and ATP6 genes lie between large ribosomal RNA (lrRNA) and small ribosomal RNA (srRNA) in *B. bassiana* and *A. nidulans* but not in *N. crassa* and *P. anserina*. The gene sequences for C03 and then NAD6 lie between srRNA and lrRNA in *B. bassiana*, *N. crassa*, and *A. nidulans*, but their order in the latter is reversed (i.e., NAD6 and C03; Pfeifer et al. 1993).

By performing extensive physical analysis via cloning, restriction enzyme mapping, and sequencing of the fungal mtDNA, 10 or more genes, including structural and tRNA as well as protein-encoding genes, have been identified. Variations in various aspects of mtDNA within isolates of a species and between many species of a genus and many genera are well known. Comparison of the tRNA genes with those of *A. nidulans*, *B. bassiana*, and *P. aniserina* and the yeasts *Saccharomyces cerevisiae* and *Torulopsis glabrata* revealed that the differences between closely related tRNA clusters are mostly the result of transition-type mutations (Hegedus et al. 1991; Pfeifer et al. 1993). Hybridizations carried out with multiple probes present simultaneously produced unique patterns that can characterize a group and distinguish it from all other fungi. Further, we have been able to assign the mtDNA of the genus *Beauveria* to various mitotypes (Hegedus and Khachatourians 1993b, 2003). Those mitotypes provide a means for testing mitochondrial compatibility and determining natural population structures.

Mapping a Mitochondrial Genome

Using a number of molecular approaches, such as analyses of protein and RFLP data and DNA sequencing, we now can obtain information on the content of the mt genome and its variation. Although earlier studies involving DNA–DNA reassociation and heteroduplex mapping allowed investigators to distinguish species, the use of cloned fragments and analysis of RFLPs provide greater resolution. The mtDNA and mtDNA plasmids (when present) also can be used to determine the extent of species variation and degrees of relatedness. Comparisons of the mitochondral genomes of organisms may provide the best mechanism for measuring intrapopulational and interpopulational genetic variability and for identifying individuals and populations. Mitochondrial DNA restriction-enzyme patterns have been used as markers to monitor cytoplasmic inheritance in interspe-

cific crosses, protoplast fusions, and genetic transformation including molecular cloning. Such maps also can be used to study the evolution and phylogenetic relationship of fungal taxa (Weber et al. 1986; Hintz et al. 1989). In many cases RFLP variability is sufficient to allow use of the resultant banding patterns as taxonomically significant characters. Additionally, mtDNA RFLPs are becoming increasingly important for the identification of genetic loci seemingly unrelated to known mitochondral functions. Particular patterns may be correlated with pathogenicity or mating type. Within-species heterogeneity in mtDNA, restriction-enzyme cutting patterns may or may not be correlated with geographic location (Hintz et al. 1985; Varga et al. 1994; Hegedus and Khachatourians 2003). However, genome variability could occur through intermolecular DNA exchanges, transpositions, and translocations. For certain mitotic fungi the latter process may provide the most plausible explanation for heterogeneity in RFLP patterns. The identification of genetic and physical mitochondrial markers in entomopathogenic fungi has had a significant impact on the study of organelle transmission and reproduction in asexual fungi (see Hegedus and Khachatourians 1993a, 1993b, 2003 and references therein).

Sequencing Mitochondrial DNA

The mtDNA offers an excellent system for the study of fungal evolutionary genetics. Fungal mtDNA possesses unique characteristics, including small size, lack of methylated bases, ease of purification, high copy number, and simple restriction-enzyme patterns. The haploid nature of mtDNA, which eliminates genetic recombination in the mitochondrial genome, is probably responsible for the preservation of allelic forms. Those forms are ideal for building phylogenies using conventional tree-building algorithms and analyses of many taxonomic groupings. Two strategies can be used to estimate fungal diversity at the mitochondria level: RFLP analysis and DNA sequence analysis. Evolutionary trees generated from RFLP data, however, have limited resolution because all sites recognized by one restriction enzyme are considered a single character, and all changes are considered to have an equal likelihood of occurring. DNA-sequence analysis provides greater resolution, producing evolutionary trees that more accurately reflect changes occurring in all regions, silent and active, of the genome. Furthermore, sequencing data allow one to distinguish between conformational mutations (nucleotide substitutions) and changes in genome length.

The costs (time and expense) of RFLP and sequence analyses can be reduced if data for related DNA sequences are available. For example, if the sequence databases of the European Molecular Biology Laboratory (EMBL; see Appendix III) contain DNA sequences of interest, the presence of restriction sites can be determined and used to provide a set of restriction enzymes to map multiple cleavage sites within a population of fungal mtDNA. An alternative technique is known as high-resolution restriction mapping (HRRM). HRRM requires that regions of known sequences that are variable within and between species be identified. Specific PCR primers can then be generated to amplify very small regions of DNA for sequence analysis. Results from the sequence analysis allow one to assess the variation within a population with a minimal amount of DNA sequencing. Finally, if mutational changes occur outside the restriction sites, allele-specific PCR amplification with defined primers or allele-specific primers can be used (Erlich and Bugawan 1992). The resultant DNA can be sequenced and/or tested for single-stranded conformational polymorphism (SSCP; Hegedus and Khachatourians 1996b).

mtDNA Sequences

The first filamentous fungus to have its entire mtDNA (94,192 bp) sequenced was *Podospora anserina* (Cummings et al. 1990). The mitochondrial genome of the s race of *P. anserina* contains 31 group-I introns and two group-II introns. The gene for the ATPase9 subunit (the dicyclohexylcarbodiimide-binding proteolipid) is not located in the mtDNA; it is probably in the nuclear DNA. *Neurospora crassa* and *A. nidulans* each have two copies of the ATPase9 gene, an active one on the chromosomal DNA and an inactive one on the mtDNA. In most organisms, subunit 9 of the ATPase appears to be encoded on the nuclear DNA (Grossman and Hudspeth 1985). Consistent with ancient divergence, mitochondrial genes for cytochrome B; ATPases 6 and 9; cox I and cox II from the yeast, *S. cerevisiae*, and from *Zea maize* did not hybridize with mtDNA from the fungus *Physarum polycephalum*, whereas the rRNA probes did (Jones et al. 1990). Comparisons of mtDNA maps of *Neurospora, Aspergillus, Podospora, Beauveria*, and yeast species showed that the relative positions and orientations of genes are highly variable (Hegedus et al. 1991; Pfeifer et al. 1993). The order of the mt genes in each fungal genus so far examined is unique, although some blocks of genes are conserved among some genera (Wolf and Del Giudice 1988).

The presence of introns such as those in *Neurospora* and *Podospora* species can cause significant variation in the size of mtDNA (Collins and Lambowitz 1983). The apocytochrome B (cob) cytochrome oxidase subunit I (cox I) and lrRNA genes of *A. nidulans* have introns. Examination of other fungi has shown that these specific genes do not always have introns. When they do, however, the intron may occupy the same position in the gene (e.g., in lrRNA) in a number of phylogenetically different fungi (Scazzocchio 1987). Introns also are found in some of the smallest mitochondrial genomes, but their numbers are reduced (Wolf and Del Giudice 1988). Several intron-encoded endonucleases, which assist in intron transposition, are now known (Lambowitz 1989). Some of these may resemble the ORFs of certain *P. anserina* introns (Cummings et al. 1990).

Mitochondrial Genome Codon Usage

Mitochondria represent the first biological system known to deviate from universal codon usage (Heckman et al. 1980). In the universal codon assignment, the codons UGA, AUA, and CUN specify termination, Ile and Leu, respectively. In Ascomycetes (yeasts or filamentous fungi) the same triplets code for Trp, Met, or Ile, and Thr or Leu, respectively (Gray 1982; Hudspeth 1992). In nonascomycetous fungi Trp is coded exclusively by UGG with UGA absent from all reading frames analyzed so far (Hudspeth 1992). Some fungi have an expanded codon recognition pattern in which a single tRNA type recognizes all of the codons as well as the four codon families that specify a single amino acid (CUN = Leu, GUN = Val, UCN = Ser, CCN = Pro, ACN = Thr, GCN = Ala, GGN = Gly, CGN = Arg). For these reasons, the 22 to 26 different tRNAs are fully capable of translating the mt genetic code, without much concern for the influence that the amount of play or "wobble" that the third base in the anticodon would have (Gold et al. 2001). This novel feature of mitochondrial tRNA is to the result of the differing number of base pairs in the anticodon loop (Heckman et al. 1980; Gray 1982; Hegedus et al. 1991; Hudspeth 1992).

Cummings and colleagues (1990) analyzed the codon usage for all genes, introns, and ORFs in *P. anserina*. They found that tRNA codons ending with A or U are preferred, even by tRNA anticodons with a C or G in the wobble position. Furthermore, they showed that differences in polar amino acid residues and hence triplets in intron and exon sequences were 31–42% and 15–24%, respectively. Thus, the mitochondrial genome intronic proteins are not membrane associated (Cummings et al. 1990). In addition, codons UGC, CGC, and CGG are rarely or never present, and codon bias in introns and exons of *P. anserina* mtDNA is virtually the same.

Mitochondrial Genome of Entomopathogenic Fungi

Mitochondrial-DNA probes from entomopathogenic fungi (EPF) can be useful for taxonomy, tracking released fungi, and tracing the progression of a fungal infection (Hegedus and Khachatourians 1993b, 1996a; Khachatourians 1996). Five mtDNA probes used to evaluate the similarity of mtDNAs from EPF genera are described in Table 6.2, and the results of individual DNA probe hybridizations to mtDNA restriction fragments are provided in Table 6.3.

The srRNA and lrRNA genes of *B. bassiana* are highly conserved compared to those of other filamentous fungi (Hegedus et al. 1998; Hegedus and Khachatourians 2003) and are located on the pBbmtE4, pBbmtSE1, and pBbmtXS1 plasmids. Hybridizations (using the method of Southern 1975) of the srRNA with the srRNA gene probes produced identical 1.8 kb DNA bands for all of the *B. bassiana* strains examined. Based on the hybridization patterns, partial restriction enzyme maps of the mitochondrial genomes of *Verticillium lecanii*, *Tolypocladium cylindrosporum*, and *Paecilomyces farinosus* relative to similar maps of the *B. bassiana* mitochondrial genome were derived (Fig. 6.11; photos of the RFLP gels, and probe hybridization data are available in Hegedus and Khachatourians 1993b; Hegedus et al. 1998).

Using a set of mitochondrial probes in hybridization assays, both type A- and type B-mtDNA-containing strains of *Beauveria* species were isolated from aphids from Idaho, Canada, and Russia. No insects were found to contain entomopathogenic *Beauveria* species with both type A and type B mtDNA. All of the *B. bassiana* isolates had three *Eco*RI hybridization bands of 20.0, 5.0 or 4.1, and 1.8 kb, except strains USSR 2274 and 2533, DAOM 210569, and ARSEF 2879 and 2881, which was expected from the individual probing experiments (Table 6.3). Banding patterns of the other fungal species (excluding certain *B. brongniartii*, *B. densa*, and *B. caledonica* strains), as expected, differed markedly from that of *B. bassiana*.

It is interesting that it is was possible to group the *B. brongniartii* isolates into four broad categories (Hegedus and Khachatourians 2003). The type A mtDNAs (e.g., that of *B. brongniartii* isolate F156) had a profile identical to that of *B. bassiana* type A mtDNA. Type B mtDNAs were similar to *B. bassiana* type B (e.g., *B. brongniartii* isolates 32 and 979). The type C mtDNAs were unique in having the srRNA gene located on a 3.6-

kb fragment, which is likely a result of the absence of an internal *Eco*RI site that generates two 1.8-kb fragments in the type A and B strains, an E2 region that binds to a 3.0-kb fragment and overlaps with a fragment of approximately 3.6 kb, and an lrRNA gene that spans 4.5- and 20.0-kb fragments. Type D mtDNAs are intermediate between type A/type B and type C mtDNAs. They are characterized by type A/type B features, including the location of the lrRNA gene solely on a 20.0-kb fragment and the hybridization of the pBbmtE3 probe to 1.8-kb fragment. Conversely, the pBbmtE4 probe, which contains a 5′ portion of the srRNA gene, binds to a 3.6-kb fragment as per a type C mtDNA. The pBbmtE2 probe hybridizes to two fragments of approximately 5.5 and 3.6 kb. Because the srRNA and lrRNA genes are located on two separate probes, the resultant hybridization patterns could be used to derive partial mitochondrial maps for each strain examined (Fig. 6.12). Only about 40% of the mtDNA sequence for *B. bassiana* strain GK2016 is available; thus, the mtDNA map is not complete. However, the type A mtDNA of *B. bassiana* and *B. brongniartii* can be compared directly, and the structures of types B, C, and D mtDNAs can be established clearly.

Further evidence of EPF mitochondrial diversity has been revealed by molecular techniques (Mavridou and Typas 1998; Typas et al. 1998; Kouvelis et al. 1999). Mavridou and Typas (1998) reported intraspecific RFLP polymorphisms in the mtDNA of 25 *Metarhizium anisopliae* var. *anisopliae* isolates from various insect hosts and geographical areas. PCR amplification of the srRNA and the lrRNA, followed by restriction analysis of the PCR products, allowed the classification of all 25 isolates. Other investigators (Kouvelis et al. 1999) examined mitochondrial subgroups in the *Verticillium* species complex (*Verticillium lecanii* and *V. psalliotae* [or *V. lecanii*-like]) from 54 isolates. Those species could be pathogens of plants or insects. Genetic variation among the 54 isolates was considerable, but host and mtDNA patterns and mtDNA patterns and previously described isoenzyme-defined specific groups were not correlated. One important discovery of this study was that genotypes of isolates from tropical and subtropical areas showed considerably more variation than isolates from temperate regions.

Analyses of mtDNA from a larger collection of species should reveal the diversity of mtDNA among Deutromycetous EPF and the relationships of that group to other fungi. Such analyses could determine whether particular mitotypes are confined to particular insects, geographic regions, or abiotic substrata. At this time, correspondence between a *Beauveria* species and a single or fixed number of mitotypes cannot be made. Without such data, the role of mtDNA in the interactions between EPF species and insect hosts remains undefined.

TABLE 6.2
Beauveria bassiana **GK2016 Mitochondrial DNA Probes**

Probe	Genes present*
pBbmtE2	*NAD* 1, *ATP* 6 (5′)
pBbmtE3	*ATP* 6 (3′), *srRNA* (5′)
pBbmtE4	*srRNA* (3′), *C0* 3, *NAD* 6 (5′)
pBbSE1	*NAD* 6 (3′), *tRNA*[Val,Ile,Ser,Trp,Pro], *lrRNA* (5′)
pBbXS1	*lrRNA* (3′)

*NAD, Nicotinamide adenine dinucleotide; ATP, adenosine triphosphate; srRNA, small ribosomal RNA; CO, cytochrome oxidase; tRNA, transfer RNA; lrRNA, large ribosomal RNA.

TABLE 6.3

Hybridization of *Beauveria bassiana* GK2016 Mitochondrial DNA Probes to Other Entomopathogenic Fungal Isolates

Isolate	Source	Hybridization Band (kb) for Probes Shown					
		Mitotype	E2	E3	E4	SE1	XS1
Beauveria bassiana							
GK2016	Laboratory isolate	5.0	1.8	1.8	20.0	20.0	A
ATCC 44860	Soil, Georgia, USA	5.0	1.8	1.8	20.0	20.0	A
DAOM 144746	Soil, Alberta, Canada	5.0	1.8	1.8	20.0	20.0	A
DAOM 195005	Spruce budworm, Quebec, Canada	5.0	1.8	1.8	20.0	20.0	A
ARSEF 2860	*Schizophis graminum*, Idaho, USA	5.0	1.8	1.8	20.0	20.0	A
ARSEF 2880	*S. graminum*, Idaho, USA	5.0	1.8	1.8	20.0	20.0	A
ARSEF 2883	*S. graminum*, Idaho, USA	5.0	1.8	1.8	20.0	20.0	A
ARSEF 2861	*Diuraphis noxia*, Idaho, USA	5.0	1.8	1.8	20.0	20.0	A
ARSEF 2864	*D. noxia*, Idaho, USA	5.0	1.8	1.8	20.0	20.0	A
ARSEF 2882	*D. noxia*, Idaho, USA	5.0	1.8	1.8	20.0	20.0	A
USSR 2274	NA	5.0	1.8	1.8	2.8	20.0, 2.8	A
USSR 2533	NA	4.1	1.8	1.8	20.0	20.0	B
USSR 2533	NA	4.1	1.8	1.8	20.0	20.0	B
DAOM 210569	Beetle, British Columbia, Canada	4.1	1.8	1.8	20.0	20.0	B
ARSEF 2879	*Diuraphis noxia*, Idaho, USA	4.1	1.8	1.8	20.0	20.0	B
ARSEF 2881	*Schizophis graminum*, Idaho, USA	4.1	1.8	1.8	20.0	20.0	B
B. brongniartii							
ATCC 9452	Mumified larvae, Indonesia	3.6, 3.0	3.6	3.6	4.5	20.0	
F5	*Costelytra zealandica*, New Zealand	3.6, 5.5	1.8	3.6, 5.5	20.0	20.0	
F156	*C. zealandica*, New Zealand	5.0	1.8	1.8	20.0	20.0	A
32	Orthoptera, Montana, USA	4.0	1.8	1.8	20.0	20.0	B
656	*Nephotettix cincticeps*, China	3.0, 3.6	3.6	3.6, 4.5	20.0	20.0	
979	*Melolontha melolontha*, France	4.0	1.8	1.8	20.0	20.0	B
B. caledonica							
DAOM 191855	NA	5.0	1.8	1.8	20.0	20.0	A
B. densa							
DAOM 57904	Decaying tree, Canada	5.0	2.2	2.2	20.0	20.0	
Tolypocladium nivea							
ATCC 18981	NA	3.4, 0.51	20.0	—	20.0	20.0	
T. cylindrospora							
ATCC 56519	Soil, Czechoslovakia	7.8, 0.42	6.4	6.4	6.4, 2.7	7.8, 2.7	
Metarhizium anisopliae							
SL 297	NA	—	7.5	19.0	19.0	19.0, 0.47	
SL 549	NA	—	19.0	19.0	19.0	19.0, 0.47	
SL 2165	NA	—	—	19.0	19.0	19.0, 0.47	
Verticillium lecanii							
ATCC 46578	NA	5.0, 10.3	1.8	1.8, 1.7	20.0, 10.3	20.0, 10.3	
Paecilomyces farinosus							
ATCC 1360	NA	5.5	1.8	1.7	4.5, 1.7	4.5	

METHODS

Fungal Growth, Protoplast Formation, and Total DNA Extraction

The spores of entomopathogenic or other filamentous fungi are harvested from a stock plate using a 0.02% Tween solution. One hundred microliters from a stock suspension of 10^6 to 10^8 EPF spores per ml is used to inoculate 10 or 100 ml of yeast-peptone-dextrose (Appendix II) medium in a 50- or 500-ml Erlenmeyer flask. The culture is grown at 27°C for 3–4 days in a gyrotary water bath at 150 rpm. To measure fungal dry-weight biomass, a duplicate 10-ml culture is harvested

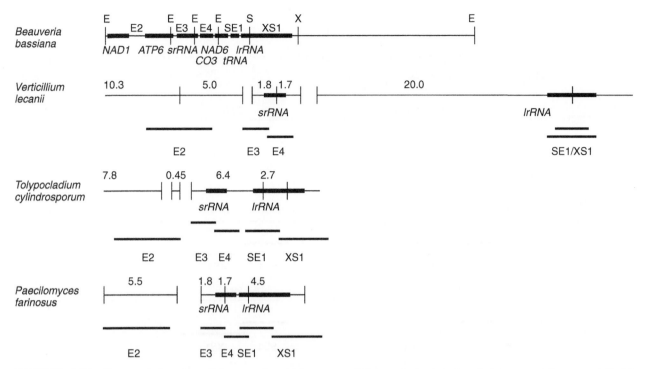

FIGURE 6.11 The restriction map of the mitochondrial genome of *Beauveria bassiana* in relation to partially mapped *Eco*RI regions of other entomopathogenic fungal genomes. For ease of comparison all maps are displayed linearly from a reference point of the E2 fragment, leaving the ends on maps for *Verticillium lecanii* and *Paecilomyces farinosus* contiguous. Restriction enzyme sites: positions of large ribosomal RNA (lrRNA), small ribosomal RNA (srRNA), NADH dehydrogenase subunit 1 (NAD1), NAD6, ATPase subunit 6 (ATP6), cytochrome oxidase subunit 3 (CO3), and transfer RNA (tRNA) cluster of genes. The location of probe hybridizations derived from *Beauveria bassiana* mitochondrial DNA is indicated by heavy lines. Numbers refer to DNA size in kilobases. E = *Eco*RI; S = *Sal*I, and X = *Xho*I. Bar is equal to 2.5 kb.

by centrifugation (3000 × g, 15 minutes), washed once with distilled water, transferred to a preweighed aluminum dish, dried overnight at 65°C, cooled to room temperature, and weighed. Flynn and Niehaus (1997) described an improved method for isolating DNA from slow-growing basidiomycetes. Using an airlift culture system, which decreases the amount of moribund mycelium produced and removes polysaccharides and pigments, they obtained 150–320 µg of DNA per gram of mycelium. The purity of DNA was 83–97%.

The high level of polysaccharides in fungal DNA interferes with restriction enzyme digestion for study of RFLPs and PCRs. Raina and Chandler (1996) used hexadecyltrimethylammonium bromide (CTAB) and sodium dodecyl sulfate (SDS) to recover pure DNA from fungi that have high polysaccharide content. An alternative method involves extracting DNA from protoplasts. Fungal cultures are harvested by centrifugation (20 minutes at 3000 × g) prior to protoplast formation and washed once with 10 mM dithiothreitol in 20 mM Tris-HCl (pH 7.5). Treatment of the cells with buffered dithiothreitol (0.01 M in a 0.02 M Tris-HCl buffer, pH 7.0) enhances protoplast formation (Pfeifer and Khachatourians 1987). The pellet is resuspended in 10

ml of the stabilizer 0.6 M ammonium sulfate, 0.02 M potassium phosphate (ASP) buffer (pH 7.0). The pellet must be thoroughly resuspended before adding 100 mg of *Trichoderma viride* cellulase (BDH Laboratory Supplies, Appendix IV) and 100 mg *Penicillium funiculosum* cellulase (Sigma-Aldrich, Appendix IV) to prevent cell clumping. In some instances protoplast formation is improved by the addition of 10 mg of *Aspergillus niger* hemicellulase. These enzymes also degrade viscous polysaccharides that often are associated with DNA isolated by the grinding procedure. The mixture is incubated at 37°C for 2 hours and then centrifuged as explained earlier. The pellet is washed once in ASP; resuspended in 5 ml 0.1 M Tris (pH 9.5), 2% SDS, 0.15 M EDTA, and depending on the amount of protein, 5–50 mg Proteinase K (Boehringer Mannheim, Appendix IV); and incubated at 55°C overnight or until the solution clears. Incubation for 2 hours with Pronase (Calbiochem-Novabiochem Corporation, Appendix IV) at 37°C can be used instead of Proteinase K at 55°C for the degradation of proteins. Five ml of 10 mM Tris-HCl (pH 8.0), 1 mM EDTA, and 50 mM NaCl is added, and the suspension is extracted once with Tris-saturated phenol (pH 8.0) and with phenol/chloroform (1 : 1, v/v). The

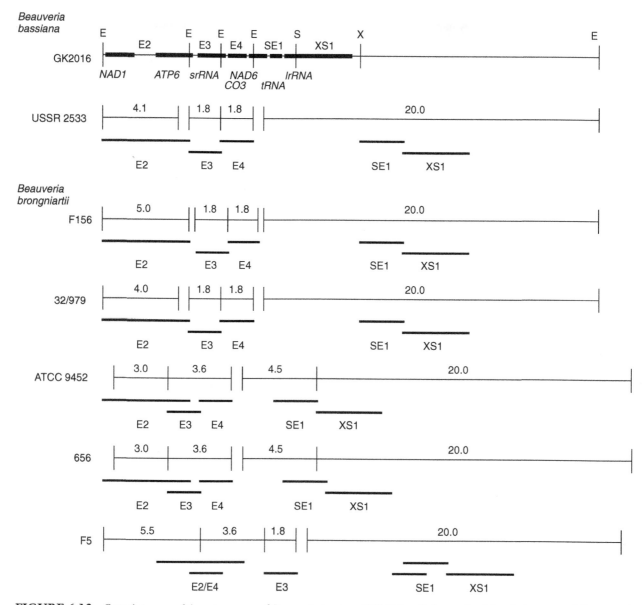

FIGURE 6.12 Complete map of the mt genome of *Beauveria bassiana* GK2016 or *B. brongniartii* in relation to partially mapped *Eco*RI regions of other genomes. For ease of comparison, all maps are displayed linearly using the E2 fragment as the primary point of reference. Restriction enzyme sites, positions of lrRNA, srRNA, NADH dehydrogenase subunit 1 (NAD1), NAD6, ATPase subunit 6 (ATP6), cytochrome oxidase subunit 3 (CO3), and transfer RNA (tRNA) genes, and locations of probe hybridizations derived from *B. bassiana* mtDNA (*heavy lines*) are shown. Numbers refer to DNA size in kilobases. E = *Eco*RI; S = *Sal*I; X = *Xho*I.

aqueous phase is removed and adjusted to 0.3 M sodium acetate; a quantity of isopropanol equivalent to 0.54% of the volume of the adjusted aqueous phase is added, and the solution is mixed to precipitate DNA. After 5 minutes at room temperature, the DNA is centrifuged for 20 minutes at 10,000 rpm. The aqueous phase is removed, and the pellet is completely resuspended in 0.5 M TE buffer (10 mM Tris-HCl [pH 8.0], 1 mM EDTA). The solution is precipitated as before and centrifuged (10 minutes, 10,000 × g) at room temperature to form the DNA pellet. The pellet is washed once with 100 μl of 70%

ethanol, dried under vacuum, and resuspended in 0.5 ml TE buffer containing 20 μg/ml DNAase free RNAase A (Sigma). The DNA is analyzed spectrophotometrically at Å260 and 280 nm to determine its quantity and purity (Glasel 1995). Fungal cellular DNA can be separated into nuclear-, mt-, and plasmid-DNA fractions with methods described in Garber and Yoder (1983).

Pfeifer and Khachatourians (1993) used total DNA extracted from *B. bassiana*, *V. lecanii*, *M. anisopliae*, and *P. farinosus* with the previous protocol, and digested with *Eco*RI, to perform hybridizations for mtDNA.

Purification of mtDNA from Mitochondria and Total Genomic DNA

Purified mitochondria can be obtained from protoplasts (described earlier) either by differential centrifugation or by gradient centrifugation in Percoll or sucrose (Hudspeth et al. 1980; Pfeifer and Khachatourians 1987). After lysing the protoplasts, cell debris is removed by centrifugation (1500 × g, 10–15 minutes). Pellets of mitochondria are then separated from the supernatant by centrifugation at 15,000 × g for 20 minutes. The mitochondrial pellet is suspended in 0.9 M sorbitol, 10 mM Tris HCl (pH 7.5), and 2 mM EDTA and centrifuged (20,000 × g, 20 minutes). To ensure purity of mtDNA, the pellet is resuspended in the previously mentioned buffer, to which 12 mM $MgCl_2$ and 100 µg/ml pancreatic DNAase (to degrade any nuclear DNA) have been added, and incubated on ice for 2 hours. The DNAase is then inactivated by the addition of 20% SDS, and 0.5 M EDTA (0.1 volume/volume of lysate) followed by incubation at 37°C for 1 hour. This solution is extracted once with an equal volume of Phenol/Sevag (1 : 1) and once with an equal volume of Sevag (24 : 1 chloroform : isoamyl alcohol). The mtDNA then is precipitated with ethanol and sodium acetate, pelleted, and resuspended in Tris/EDTA buffer (Sambrook et al. 1989).

Mitochondrial DNA Base-Composition Estimates

Fungal mtDNA is high in A + T (adenine + thymine), although both A + T-rich and G + C-rich (guanine + cytosine) regions are present. DNA-melting temperatures (T_m) can be monophasic or biphasic characteristics with T_m values varying according to A + T content. In this section, we describe the method used by Mandel and Marmur (1968) to estimate the base composition of mtDNA.

DNA samples are dialyzed three times for 12 hours each against 1 × SSC buffer (0.15 M NaCl, 0.015 M sodium citrate [pH 7.0] at 4°C). It is important that the specific conductance of SSC be measured and maintained at 0.016 milliohm/cm. Quartz cuvettes are filled with DNA samples in 0.1 × SSC, sealed with stoppers, and placed into the cuvette holder; temperature is controlled with a waterbath circulator. Electronic temperature probes can be inserted into a third cuvette to monitor the temperature of the housing. The temperature of the housing is changed within 1–2°C, and three readings of the absorbency are recorded. We use a Pye Unicam SP 1800 UV spectrophotometer (Pye Unicam, Appendix IV) with a slit width of 0.4 mm and a bandwidth of 1.2 nm at a 260-nm setting. The T_m then is calculated using the formula of Johnson (1985):

$$T_m = [(A - B) + C]$$

where A is the T_m of a DNA standard in 1 × SSC, B is the same in 0.1 × SSC, and C is the T_m for the unknown sample in 0.1 × SSC. The G + C moles percentage can be calculated according to the DeLey (1970) formula:

$$\%G + C = 2.4(T_m \text{ in } 1 \times SSC - 69.4)$$

Buoyant Density Determination

Determination of buoyant density of the mtDNA is not generally a crucial element in the study of fungal biodiversity, but it does provide information on characteristics of the mtDNA that is important for systematic, taxonomic, and evolutionary studies. The buoyant density of mtDNA is generally equal to or less than that of chromosomal DNA. Furthermore, certain A + T-rich regions within the mtDNA cause buoyant densities to vary. For example, in *N. crassa* long regions of homopurine/homopyrimidine in mtDNA are responsible for illegitimate intramolecular recombination of the mtDNA (Almasan and Mishra 1991). In *Schizophyllum commune* an estimated 22–29 copies of a 49.85-kbp DNA are present per genome (Dons and Wessels 1980). This satellite mtDNA represents 3–4% of the total DNA in the CsCl gradient. Even the genes in the smallest fungal mtDNAs are separated by A + T-rich spacers of a few to 100 bp long (Clark-Walker et al. 1985). These variations in the buoyant densities and the physical organization of large A + T containing regions, which allow DNA dyes such as DAPI and bisbenzamide to bind, facilitate isolation and purification of mtDNA along CsCl density gradients (Garber and Yoder 1984; Pfeifer and Khachatourians 1989).

The buoyant density of fungal mtDNA (5 mg at least) is determined using a CsCl solution with a refractive index of approximately 1.400. It is centrifuged with mtDNA at approximately 30 k rpm for 96 hours at 18°C using a Ti40 rotor (Beckman Instruments, Appendix IV). The gradients are fractionated by puncturing the bottom of the tube and collecting aliquots manually or with a fraction collector. As the fractions are collected, the material may be passed through a spectrophotometer to record Å260 and 280 nm values. The linearity of the gradient is determined with a refractometer. The location of two standard markers—that is, DNA from *Clostridium perfringens* and *Micrococcus lysodeikticus* (1.6915 and 1.7310 g/ml, respectively) at concentrations of 2 to 5 µg/tube—will allow for the calculation of the buoyant density values of mtDNA using the DeLey (1970) equation:

$$\%G + C = 1038.47(r - 1.6616).$$

mtDNA Topology

In most cases mtDNAs have a covalently closed circular (ccc) topology. Although linear configurations of mtDNA also have been reported (Hudspeth 1992), linear molecules may be sheared chromosomal or mtDNA resulting from methods used to isolate it for transmission electron microscopy (TEM) or from poor handling. Alternatively, linear molecules may represent mt plasmids.

For a TEM visualization of DNA topology, the Kleinschmidt (1968) method is best. All solutions must be filtered through a 0.22-μm pore Millipore filter, and all glassware must be cleaned using Nochromix (Godax Laboratories, Appendix IV). Approximately 2–3 ml of 0.2 M ammonium acetate and 0.001 mM EDTA (pH 6.8) are added to the subphase in a 5-cm diameter Petri plate. A clean glass slide is rinsed with 0.2 M ammonium acetate at pH 6.8 and placed in the Petri plate at an angle. The mtDNA sample of approximately 1 mg is added to a solution of 1.0 M ammonium acetate, 0.001 mM EDTA (pH 6.8), and 100 μg/ml of cytochrome C, using a large bore, 1-ml glass pipette. Grids (400 mesh copper grids precoated with colloidon) are placed on the surface of the liquid for 30 seconds to pick up the DNA in the film layer. A quicker method of preparing DNA in a film monolayer involves a placing a hypophase and hyperphase mixture in a watch glass containing plasmid ccc DNAs spread on a monolayer of the basic protein, cytochrome C (Banerjee and Iyer 1995). All three forms of DNA—ccc, replicative intermediate, and linear—can be observed with this method. A grid is then placed in 95% ethanol for 10 seconds, after which it is shadowed with a platinum/carbon Pt/C (60/40) coating. An internal standard such as the plasmid pBR322 (1.39 mm for ccc DNA) can be used (Pfeifer and Khachatourians 1989). Electron micrographs can be projected and traced to measure the contour maps. The contour length measurements of mtDNA range from 6–25 mm and generally agree with size estimates obtained from electrophoresis and restriction enzyme mapping.

Molecular Methods

The *in situ* locations of mitochondrial genes of fungi can be determined by RFLP and hybridization analysis. For RFLP studies, restriction enzyme products are separated by agarose gel electrophoresis, stained with ethidium bromide (EBr), and analyzed with UV light (Sambrook et al. 1989). The identification of mitochondria markers

for the construction of EPF mtDNA maps (Hegedus and Khachatourians 1993b) requires double restriction-enzyme digestion of mtDNA, the cloning of resulting products into standard vectors, and amplification of clones in appropriate host cells (Sambrook et al. 1989). DNA-sequence analysis of cloned mtDNA is performed with the Taq Track sequencing system (Promega, Appendix IV). For mapping, various four-base-pair-recognizing (*Alu*I, *Hae*III, *Sau*3A, and *Taq*I) and six-base-pair-recognizing (*Bam*I, *Bam*HI, *Bgl*II, *Hind*III, *Hinf*I, *Pvu*I) enzymes are used to digest the mtDNA. The enzymes *Msp*I and *Hpa*II are specific for methylated cytosine residues and may be used to determine modified bases. After the digestions are complete, the resultant DNAs are suspended in TBE and kept at 80°C. A sample of 1–2 ml is mixed with dye-loading buffer and run in agarose electrophoresis (Sambrook et al. 1989). The amount of DNA needed per gel will depend on the staining method used. For example, silver staining requires approximately 10–30 pg; EBr staining requires 100–300 ng. Both methods are inexpensive, sensitive, and fast (Tegelstrom 1986; Sambrook et al. 1989). DAPI staining also can be used, but it is more expensive. The gels are documented and analyzed for DNA molecular weight and relative mobility using internal standards (e.g., *Hind*III-digested-phage lambda DNA).

Standard protocols exist for the transfer of DNA from agarose gels to hybridization membranes and the generation of radio-labeled probes (Sambrook et al. 1989). Depending on the heterologous gene probe used, hybridization takes place either at low (50°C) or high (65°C) temperatures.

PCR can be used to separate varieties of fungi for taxonomic purposes and for studies of geographic distributions (see "Taxonomic Discrimination" under "Nuclear Genome" section, earlier in this chapter). Sequences of genes can be used to design specific sets of restriction enzymes for RFLP analysis to determine if mutational changes have occurred in those sequences. If mutations are known to occur in specific sequences, then allele-specific PCR primers can be designed to amplify 200–300 bp regions for genetic analysis (Erlich 1992). The amplified DNA can be used for SSCP analysis (Hegedus and Khachatourians 1996b), RFLP, or sequence analysis.

Recommended Protocols for Sampling Particular Groups of Fungi: Direct Collecting and Isolation Protocols for Macrofungi and Microfungi on Soil, Wood, Leaves, Lichens, and Other Substrata

7

Fungi on Living Plant Substrata, Including Fruits

Brenda E. Callan* and Lori M. Carris†

In this chapter we focus on fungi that produce thalli and reproductive structures on living plants and fruit, exclusive of roots, bark, and dead structures still attached to the plant. Generally, (1) most fungi living on plant substrata are microfungi, with representatives in all major fungal classes and orders; (2) many are pathogenic (cause diseases); (3) many have been well studied by both mycologists and plant pathologists; and (4) the majority are obligate biotrophs (although some are saprotrophs, as discussed in the following paragraphs).

Pathogenic fungi have direct and indirect and overt and subtle effects on their environments. Fungal diseases, for example, have great potential to influence the structure and stand composition of the vegetation in plant communities and ecosystems (Harper 1990). They affect a plant's ability to compete for limited resources, such as light and space, by reducing growth rates and increasing susceptibility to other pests. They can affect a plant's reproductive ability by infecting inflorescences, fruits, and seeds. They affect the genetic diversity of their hosts by selecting for disease resistance. They act as natural thinning and diversifying agents in dense clusters of seedlings and in stands of young plants. Seed and seedling pathogens can select for seed dispersal patterns (Augspurger 1990; Burdon et al. 1990). Augspurger (1983), for example, noted that survival of seeds from the tropical tree *Platypodium elegans* increased with distance from an adult tree. Distantly dispersed seeds were more likely to encounter light gaps where they would be at lower risk from damping-off diseases than they would be in the shaded understory.

The striking manifestations of fungal diseases encountered in the field may in fact be rare events limited to species in which the host and pathogen have not co-evolved. Harper (1990:8) speculated that certain host populations could become extinct after being challenged by a pathogen for the first time, stating, "If a host plant

*Portions of this paper written by Author: Brenda E. Callan (© 2003, Majesty the Queen in right of Canada, Canadian Forest Service)
†Portions of this paper written by Author: Lori M. Carris © 2004 Elsevier Inc.

lacks the genetic variance that allows it to respond to selection it may become extinct or rare (locally extinct) and there may be no hint of why this is so."

Plant pathogenic fungi may be detected on their hosts initially either by the presence of disease symptoms (see illustrations herein) or fungal growth (e.g., mycelia, fruiting bodies) on healthy or diseased hosts. Vascular plants, as well as mosses, liverworts, and hornworts, are potential hosts, although pathogens of the latter three groups have not been studied as widely. With a few exceptions, fungi associated with living roots are discussed elsewhere in this volume (see Chapter 15).

Obligate biotrophs produce either intercellular or extracellular mycelia, often with specialized feeding structures (haustoria) that penetrate and extract nutrients from the host plant. They require a living host, are unable to live saprobically in nature, and rarely are cultured *in vitro* and then only under very specialized conditions. Many fungi are host-specific to certain plant families, often making it necessary to identify the host plant to identify the fungus. Saprotrophs on living plants are, for the purpose of this chapter, primarily epiphytic fungi that fruit on, but do not penetrate, living plant surfaces such as leaves. Many obtain nutrients from extracellular plant exudates or from the excretions of plant-feeding insects, such as aphids.

Facultatively biotrophic fungi also may cause diseases in living plants, but they generally do not fruit on living tissues. Sporulation occurs either after the plant dies or in islands of dead tissue surrounded by living cells (e.g., some types of leaf spots). Facultative biotrophs frequently cause either tissue death, via toxin production, or systemic vascular infections that interfere with water and nutrient translocation. Because they are able to grow and sporulate after the plant is dead (unlike obligate biotrophs), they often are cultured easily *in vitro*. Under field conditions, symptoms produced by facultative biotrophs can be confused with symptoms and signs of obligate biotrophic fungi on living tissues. We discuss techniques for encouraging fruiting of saprotrophic species after substratum collection (to facilitate the separation of these two groups) later in this chapter (see "Standard, Ground-based Techniques").

We have organized the fungi discussed in this chapter into broad groups based on the character that provides the easiest field recognition, such as a symptom (host response) or a sign (i.e., a character of the fungus, such as diagnostic fruiting bodies, that is easily recognizable independent of any symptoms caused). Thus, unrelated fungi may be grouped together because they elicit the same easily recognized symptom, (e.g., a gall) in their host. We begin with three groups of fungi (rusts, smuts, and powdery mildews) whose unique, distinctive appearances in the field make them easily recognizable, regard-

less of the host symptoms associated with infection. We have grouped the other fungi we discuss under five broad symptom/habit headings: foliage, cankers, galls, flowers and fruits, and others. Within each group we have, whenever possible, categorized the fungi according to field recognition characters.

TAXONOMY, DIVERSITY, AND DISTRIBUTION

RUST FUNGI

Rust fungi (Class Urediniomycetes, Phylum Basidiomycota) comprise the largest and most ubiquitous group of obligately biotrophic fungi on vascular plants. The name "rust" is descriptive—often one or more spore stages are rusty orange to brownish because of the color of the spores emerging *en masse* from pustules on the host. Most rust pustules measure a few millimeters or less in diameter (Fig. 7.1), although rusts associated with stem cankers on trees may sporulate in contiguous areas more than a meter long.

Rust fungi are unable to grow saprobically in nature and thus are found only in environments suitable for their hosts. Their relationships with their hosts, on which they are completely dependent and with which they have coevolved, are highly specialized (Laundon 1973; Savile

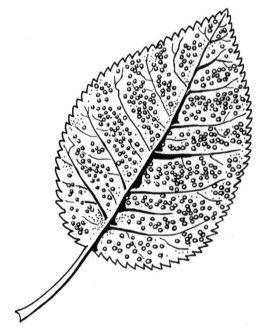

FIGURE 7.1 Leaf rust. Small circles represent rust pustules from which the spores will emerge. (Figure prepared by S. Henrich)

1976). Although rust infections are not generally fatal to plants, they may severely limit growth and fruiting ability. Wheat stem rust (*Puccinia graminis*), coffee rust (*Hemileia vastatrix*), and white pine blister rust (*Cronartium ribicola*) are examples of notoriously damaging, economically important rusts (Agrios 1988).

An estimated 168 rust genera and approximately 7000 species, more than half of which are in the genus *Puccinia*, are currently accepted (Hawksworth et al. 1995). According to Hiratsuka and Sato (1982:2), "the majority of species in temperate regions of the northern hemisphere, Australia, and New Zealand have been well catalogued, but many new genera and species are still expected to be found in tropical and subtropical regions such as South America, Africa, and Southeastern Asia."

Farr and colleagues (1989) listed 53 rust genera from North America and approximately 925 different host-fungus associations (exclusive of subspecies and varieties of rusts). Of those host-fungus records, 47% are for *Puccinia* species, and 18% (168) are from subtropical to tropical localities. South American records (Jørstad 1956, 1959), albeit not as modern or extensive as those from North America, include 23 rust genera and 218 host-fungus associations; again, about half (57%) of the records are for *Puccinia* species. If the subtropical and tropical records from Farr and colleagues are added to this total, including duplicates, rust records from tropical America still number only about one-third of those from North America. Many of the rust genera known to occur in South America, however, are exclusively tropical (e.g., the majority of the *Raveneliaceae*), and according to Oberwinkler (1993), several rusts with unique characteristics and distinct taxonomic features are known from warmer regions. It is probable that the low number of records of rusts from the tropics reflects insufficient study. Buriticá and Hennen (1983) recently collected 100 species of rusts, 20% of which were new taxa, in a 40-hectare tract in the neotropics. About one-third of the plant species in the area had their own rust species. Intensive studies likely will show that the diversity of rust species in the tropics is higher than that in temperate regions.

Potential hosts come from a broad range of vascular plant families, including ferns, gymnosperms, monocots, and dicots. No rusts have been reported from mosses or liverworts, but related small genera of heterobasidiomycetes, such as *Iola*, *Platycarpa*, and *Ptechetelium*, are parasites of mosses and ferns (Oberwinkler and Bandoni 1984; Oberwinkler 1993). The latter two genera are restricted to the tropics and warrant further study.

Rust fungi are highly specialized parasites with several unique features. A single species may produce up to five morphologically and cytologically distinct spore-producing structures (spermagonia, aecia, uredinia, telia, and basidia) in successive stages of reproduction. The various types of spores usually are produced at different times, although they sometimes may be found simultaneously on the same host. Other rust species require several seasons and two unrelated species of host plants, often in different families or orders, to complete their life cycles. In the latter case, a spore stage produced on one host is only capable of infecting the other (alternate) host. A particular spore stage of a rust will often have a narrow range of hosts and may be limited to one or two congeneric plant species. Species with reduced life cycles (microcyclic rusts) are more common in cool temperate to subarctic regions, where a short growing season selects for rusts that can survive perennially on the telial host (Savile 1972, 1976).

Unlike other plant pathogens, rusts usually infect healthy, vigorously growing plants, so if infections are small and limited to certain plant parts, such as foliage, they may be difficult to detect. Perennial, systemic infections may cause deformities such as witches brooms, cankers, or galls. Plants with severe rust infections may appear stunted, chlorotic (yellowed), or otherwise discolored (heavy uredinial sporulation may turn the affected plant part orange). Rusts sporulate on leaves, shoots, fruits, and woody stems.

In temperate regions, aecial stages are generally most prevalent in the spring, with uredinial states developing midseason. Thus, both new and old infected tissues should be collected in case spore stages of early and advanced infections differ. This is particularly important with rusts that produce all five spore stages on a single host. Telia develop in the fall, usually overwintering in dead host tissue. To confuse the issue even further, the telia of microcyclic rusts simulate the habit of the parental macrocyclic forms and occur on the host plants of the aecial stage of the latter (see Cummins and Hiratsuka 1983 for a fuller explanation with examples). In tropical regions, teliospores may develop and immediately germinate at all times of the year. Thick spore walls and wall pigmentation, which often are identifying features for temperate rusts, are far less common. Unless distinctive spermagonia, paraphyses, or urediniospores are present in collected material, a rust may be difficult to identify (Savile 1980).

Identification of a rust species frequently begins with host identification, so healthy, uninfected host foliage and reproductive structures should be collected along with the rust. There are numerous regional indices and taxonomic monographs of rusts. Many of these are listed in Hiratsuka and Sato (1982:35–36). North American species have been treated by Arthur (1934) and Cummins (1962, 1971, 1978); Ziller (1974), who monographed western Canadian tree rusts; and León-Gallegos and Cummins (1981) and Jørstad (1956,

1959), who monographed Mexican and South American rusts, respectively. Doidge (1939), Eboh (1986), Gjaerum (1986, and papers cited therein) surveyed African rusts. Gäumann (1959) and Wilson and Henderson (1966) published treatments of European rusts, and Kuprevich and Tranzschel' (1957) and Savulescu (1953) monographed species occurring in the former Soviet Union and environs. Cummins (1950) and Ito (1950) indexed Chinese and Japanese species, respectively.

SMUT FUNGI

The smut fungi, class Ustilaginomycetes, phylum Basidiomycota (Bauer et al. 2001), are a group of plant parasites occurring on angiosperms, especially grasses and sedges. Modern treatments of smut fungi based on ultrastructural, biochemical, and nucleotide-sequence analyses have shown that Exobasidiales, a morphologically distinct order of phytopathogens, is closely related to *Tilletia* and allied taxa (Bauer et al. 2001). Conversely, smutlike phytopathogens in the genus *Microbotryum* are more closely related to the rust fungi (Urediniomycetes) than to the Ustilaginomycetes (Bauer et al. 2001). For the purposes of this discussion, we include only the teliospore-forming taxa described in the smut fungi. We discuss the Exobasidiales separately (see "Exobasidiales," later in this chapter).

Smut fungi are distributed worldwide and are important pathogens of crop plants, including corn, wheat, barley, oats, rice, and sugarcane. The morphology of smut fungi is relatively simple, consisting of parasitic intercellular mycelia and reproductive structures (Fischer and Holton 1957). The reproductive structures are represented by teliospores and their germination products. In most taxa the teliospores are formed in sori. A sorus consists of a dark mass of powdery teliospores and host tissue. Each sorus contains millions of one-celled, thick-walled, and generally darkly pigmented teliospores ranging in size from 3.5–50 μm in diameter, depending on the species. The teliospore wall may be smooth or ornamented, with variable pigmentation. Ornamentation can be reticulate, cerebriform, echinulate, or tuberculate, and spores of some taxa have narrow, dark equatorial bands. About half of the smut fungi produce teliospores in aggregates called "spore balls" (Durán 1973; Vánky 1998). Spore balls consist of teliospores only or a combination of teliospores and sterile cells. Many species of smut fungi are capable of limited saprobic growth on nonliving substrata, but the predominate stage of the smut fungus life cycle is the mostly intercellular mycelium within the plant host.

Infection by smut fungi is generally recognized only when the sori are formed in the stems, leaves, floral organs, and, occasionally, the roots of the host plant. The taxonomy of the groups is based on the locations and types of sori and spores produced, together with the type of spore germination. Fischer and Holton (1957) grouped the smut fungi by type and location of the sorus into stem smuts, leaf smuts, inflorescence smuts, and gall-forming smuts. The sori of the stem smuts, as the name implies, are typically confined to the stems of the host (Fig. 7.2). The leaf smuts form long, narrow sori in leaves and leaf sheaths. Species of *Erratomyces*, however, produce teliospores that are scattered in intercellular spaces of host mesophyll rather than aggregated in sori (Piepenbring and Bauer 1997). The inflorescence smuts are the most numerous. Some smuts destroy an entire inflorescence, whereas others destroy the spikelets, leaving the rachis intact. The sori of some inflorescence smuts may assume the general shape of the infected host organ—for example, the ovary-infecting species of *Tilletia* (Durán and Fischer 1961). In this group, the sorus may look like a somewhat enlarged seed, and infection can be difficult to detect. Conversely, in a limited number of gall-forming taxa, the sori may be easy to detect because of gross distortion or hypertrophy of the affected host organ. Infection by the corn smut fungus *Ustilago maydis*, for example, causes large, tumorlike growths to form on the ears or vegetative organs of the corn plant (Fig. 7.3).

More than 1300 species of smut fungi in 80 genera are known from 4000 species of host plants, representing more than 75 families of angiosperms worldwide. The number of known species may account for only about half of the extant species (Oberwinkler 1992).

FIGURE 7.2 Sori of *Urocystis trillii* on *Trillium* stem. (Photo by L. Carris)

FIGURE 7.3 Corn smut (*Ustilago maydis*). (Photo by L. Carris)

Each smut species has a limited host range, and host specificity, combined with morphological characters of the sorus and teliospores, traditionally has been used to delimit species. To identify a smut fungus, it is essential to know the identity of the host plant, so collection of the intact host plant is desirable. Once the species of host is known, the morphological characters of the smut fungus can be compared with those of other smuts known to occur on the same host.

The site of sorus formation is generally constant for a smut fungus. In the genus *Tilletia*, for example, the sori of most species are produced in the ovaries of the host. How a sorus is associated with the host tissue is also critical for the delimitation of different genera of smut fungi. For example, sori may remain embedded and intact within the host tissue, or they may become erumpent and gall-like. Durán (1973) and Vánky (1987) have provided excellent descriptions of morphological characters of smut fungi and keys to the genera.

Collection of smut fungi depends on the type of smut (i.e., leaf, stem, root, or inflorescence smut) and the phenology of the host. Presence of mature teliospores on the infected host is important; thus, for the inflorescence smuts, it is necessary to collect hosts that are flowering.

Vánky (1987) listed 18 monotypic genera of smuts; 10 are based on tropical/subtropical species, five on temperate species, and three on species with widespread distributions. Other small genera, such as *Dermatosorus* (four species), *Erratomyces* (five species), *Georgefischeria* (two species), and *Pericladium* (three species) are tropical or subtropical in distribution (Oberwinkler 1993; Piepenbring and Bauer 1997). Zundel (1953), in the only global monograph of the Ustilaginales, provided a basis for estimating the distribution of species in some of the larger genera of smut fungi. Species collections available for *Ustilago*, *Tilletia*, and *Thecaphora* showed

roughly equal distributions of species between tropical and temperate areas. Most species collections of *Entyloma*, *Melanotaenium*, and *Urocystis* were from temperate regions, whereas collection records for species of *Cintractia*, *Farysia*, and *Sphacelotheca* showed a predominately tropical or subtropic distribution. Changes in the taxonomic concepts of the smuts since Zundel's treatment prevents a direct comparison with more recently published treatments (e.g., Vánky and Guo 1986, China; Vánky 1994, Europe). Farr and colleagues (1989) listed 280 species of smut fungi for the United States, whereas 260 species of smuts have been reported from Mexico and Central and South America (Hirschhorn 1986; Durán 1987). Ninety-seven of the latter species, or 37%, also have been reported from the United States (Farr et al. 1989). Durán (1987) reported 126 species of smut fungi from Mexico based on collections made over a 22-year period. He described 22 new species. Of the 126 species, 67, or 56%, also were reported from the United States (Farr et al. 1989).

Mordue and Ainsworth (1984) listed 100 species of smut fungi for the British Isles and noted that only 16 smuts were added to the records from this area between 1950 and 1984. Vánky and associates (1982) reported 119 species of smut fungi from Hungary, 37 of which were collected between 1950 and 1982. The authors speculated that only about half of the smut fungi of Hungary had been reported. Mundkar and Thirumalachar (1952) reported 194 species of smut fungi from India, and Ahmad (1956) reported 86 species from West Pakistan. Additional regional smut mycotas are cited in Hawksworth and colleagues (1995).

EXOBASIDIALES

The order Exobasidiales (class Ustilaginomycetes, phylum Basidiomycota) comprises four families: Exobasidiaceae, Brachybasidiaceae, Cryptobasidiaceae, and Graphiolaceae (Bauer et al. 2001). The largest family and genus are Exobasidiaceae and *Exobasidium*, respectively. Species of *Exobasidium* are parasitic on angiosperms; preferred hosts are temperate Ericales. Infected host tissues on leaves, succulent shoots, flowers, and fruits form misshapen galls or blisters, which often are discolored white or pink. Basidia develop in a dense layer on the outer surfaces of the infected host tissues, often protruding from stomata or from between epidermal cells (Savile 1959; Donk 1966; Blanz 1978). In addition, a yeastlike single-cell stage develops from microconidial propagation. The latter stage is readily cultured (Graafland 1953; Sundström 1964). Overmature *Exobasidium* colonies often are overgrown by hyphomycetes, which may make identification difficult. In temperate regions, symptoms are most visible at midsummer. Other genera of Exobasidiaceae include *Kordyana*, which has a pantropical distribution on Commelinaceae (Gäumann 1922), and *Muribasidiospora*, which causes foliage spots on *Rhus* and *Celtis* (Rajendren 1968).

The Brachybasidiaceae contains a few small or monotypic, predominately tropical genera that cause leaf spots: *Brachybasidium* species occur on Palmae; *Proliferobasidium heliconiae* is a monotypic genus found on *Heliconia*; and *Ceraceosorus bombacis* causes a serious disease of the economically important hardwood *Bombax* (Bombacaceae) in India (Cunningham and Bakshi 1976). The Cryptobasidiaceae is composed of a few small genera associated with hypertrophied host tissue. Preferred hosts are members of the Lauraceae (Bauer et al. 2001).

The Graphiolaceae includes smutlike organisms pathogenic on palms. The family is exclusively tropical and subtropical in distribution (Oberwinkler et al. 1982; Oberwinkler 1993). The fungi occur in the natural range of their palm hosts, although species found on cultivated host species occur wherever the palms are grown under humid conditions. Species on date palms are restricted mainly to coastal areas. The family is based on the genus *Graphiola*, which includes nine species. Early stages of the fungus appear on the palm leaves as dark, 1- to 3-mm diameter, subepidermal pustules surrounded by pale yellow halos. The pustules rupture the epidermis as the fungus matures. Mature fruiting bodies often are opposed to each other on abaxial and adaxial sides of leaves. A mature pustule is composed of a hard, dark outer peridium and an inner, thin hyaline peridium that surrounds the masses of smooth, yellow, thick-walled spores (Fig. 7.4; Cook 1978). *Graphiola* species cause premature shedding of leaves by date palms.

POWDERY MILDEWS

The Erysiphaceae, or powdery mildews, is a family of Ascomycetes whose species are obligate parasites on the above-ground parts of higher plants. They are nourished by haustoria (Fig. 7.5) and develop a characteristic white powdery conidial stage on the surface of the host (Yarwood 1973; Spencer 1978). Although the morphological features of the conidial states are sufficient to separate out four form genera, generally the perithecia must be present to make a species determination using current taxonomic treatments.

Depending on the author consulted, there are from 19 to 22 currently accepted genera and an estimated 400 species, the majority of which are in the genus *Erysiphe* (Braun 1981; Zheng 1985; Hirata 1986). Hirata (1986), in his massive worldwide treatise of powdery mildews, lists 9838 angiosperm host species dispersed among

FIGURE 7.4   Fruiting bodies of *Graphiola* sp. on palm. (Photo by M. Adams)

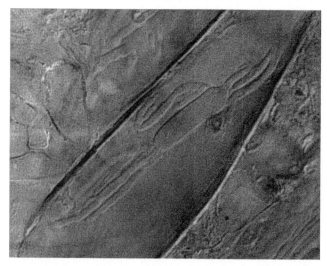

FIGURE 7.5   Digitate haustoria of *Blumeria graminis*. (Photo by L. Carris)

44 orders, 169 families (roughly half of the known angiosperm families), and 1617 genera. Of the hosts, 93% are dicotyledons. Of the 662 known monocotyledonous hosts, 634 are in the Gramineae, and the remainder are scattered among seven other monocot families. No powdery mildews are known to occur on gymnosperms, pteridophytes, or lower plants such as mosses and liverworts. Host ranges vary from large to small. However, because species concepts generally have become narrower and more restricted in recent years,

comparisons of species numbers reported in older regional monographs with those reported in more modern treatments is difficult. Hirata (1986) calculated that the ratio of species of powdery mildews to angiosperm species is 4.5%, based on an estimated total of 220,000 angiosperm species. The areas of greatest abundance and greatest damage resulting from powdery mildews are, according to Yarwood (1973), California and Israel, which are both characterized by warm, rain-free summers and intensive agriculture.

Hirata (1986) divided the world arbitrarily into 10 geographic regions and listed numbers of host species per region. He noted that north-temperate regions had far larger numbers of host species than did tropical, sub-tropical, arctic, or subarctic regions. Future studies on powdery mildews in the tropics will help to determine whether those fungi truly are more species-rich in temperate areas of the world. Hirata also calculated that the number of herbaceous species (8250) hosting powdery mildews was far larger than the number of tree hosts (2645). This difference probably reflects the growth habit of the fungus, which generally is limited to leaves and succulent shoots.

Powdery mildews can be recognized in the field by the white, dusty, conidial state, which coats leaf and small stem surfaces (Fig. 7.6). Infected tissues may be stunted or chlorotic. Conidial states develop first and may be the only stage present on young infections. Perithecia are more likely to occur in older, well-established infections on senescing leaves. On close examination with a hand lens, black, globose, mature perithecia less than a millimeter in diameter may be observed scattered over the mycelium. Immature perithecia, in contrast, are yellowish to orange. If possible, some dark, mature fruiting bodies should be collected. The nonostiolate perithecia (sometimes referred to as cleistothecia) are ornamented with various types of hyphal appendages, which may be simple or ornamented with hooks, branches, or other structures (Fig. 7.7). Genera are delimited based on features of the appendages plus number of asci per perithecium. Because many species of powdery mildews are host-specific, correct host identification is essential. If the diseased host is unidentified, healthy foliage and flowers or fruit should be collected at the same time as the fungus. Hirata (1986) listed close to 4000 publications on powdery mildews, including regional monographs.

FOLIAR FUNGI

Sooty Molds

The sooty molds are saprobic ascomycetes belonging to several different families of Dothideales. They form black

FIGURE 7.6 Powdery mildew on the flower bud, leaves, and stem of a rose. (Figure prepared by S. Henrich)

mats on living leaves and stems and are particularly abundant in the tropics. Sooty molds use the honeydew secreted by plants or sap-sucking insects, such as aphids and scale. They show no host preference, and colonies may consist of mixed populations of eight or more species. Sooty molds may form a thin network of hyphae, a velvety growth, or a dark crust on leaves and smaller twigs (Hughes 1976). On trunks and larger branches, mold growth is usually more robust. Some sooty molds on leaves are found in close association with glandular trichomes (Hughes 1976). Sooty-mold fungi predominate along the continental margins of North and South America, in Central America, and on islands in the Western Hemisphere (Reynolds 1975). They occur in some form in most of the vegetation types of the world but are prevalent around the margins of rain forest communities. Hughes (1976) noted that the trunks and branches of the trees in the *Nothofagus* forests of New Zealand can be covered up to a height of 9 m or more with sooty molds. The growth may be so extensive that only small areas of bark are visible.

Some species of sooty molds produce as many as three different asexual forms (Hughes 1976). The mycelium of a sooty mold is composed of a weft of dark hyphae, which may be several millimeters thick. Conidiogenous

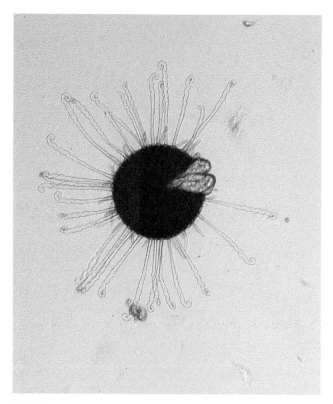

FIGURE 7.8   Using a colloidal leaf peel technique to remove sooty molds from the surface of a *Camillea* leaf. Water-soluble mucilage glue can be used in the same way. (Photo by B. Callan)

FIGURE 7.7   Appendages on perithecia of *Uncinuliella flexuosa*, powdery mildew of horsechestnut (*Aesculus*). (Photo by L. Carris)

cells and various types of conidia are formed within the mycelium. In some species, the weft of hyphae apparently is unorganized, whereas in other species the mycelium is highly organized and has a characteristic appearance (Parbery and Brown 1986). The conidia vary in size, shape, septation, and pigmentation. Species of the Metacapnodiaceae produce distinctive, tapering, moniliform hyphae with rigid smooth-walled cells. The colonies are dense and lustrous (Parbery and Brown 1986). Another sooty mold in the family Seuratiaceae produces dark, gelatinous, lobed thalli, with reproductive structures quite different from those of other sooty molds (Parbery and Brown 1986).

Reynolds (1975) recognized two growth forms for sooty molds: deciduous and permanent. The deciduous growth form is characterized by a thin hyphal layer that peels away from its substratum in drier, cooler months. It is found predominantly in lowland tropical regions but is absent from rain forests. The deciduous form also occurs along the coast of the southeastern United States and commonly is found on citrus trees. *Limacinula* and *Trichomerium* are representative genera (Reynolds 1975). The permanent growth form occurs mainly on twigs and stems where it forms thick, cushionlike black hyphal mats. It occurs predominantly in neotropical

highlands and temperate lowlands, such as those in north coastal regions of North America. Species of *Limacinia* are representative of the permanent growth form.

Sooty molds on leaves are easier to collect and identify than those on tree trunks (Hughes 1976). A drop of 2% collodion/parlodion solution (prepared in a volume/volume mixture of diethyl ether or absolute ethanol) or cellulose acetate is spread gently on the leaf colony using a glass rod or other similar tool. The solution dries in a few minutes to a transparent film in which the fungus is embedded (Stevens 1981). When the film is peeled from the leaf surface, it removes the intact colony. The "leaf peel" is placed on a microscope slide, and the collodion is removed with acetone (Fig. 7.8). The intact sooty mold colony then can be examined under a compound light microscope. Keys for genera can be found in Luttrell (1973), and descriptions of species can be found in Hughes (1976) and Batista and Ciferri (1963). Reynolds (1971, 1978a, 1978b, 1979, 1982, 1983, 1985) reported on collections of sooty molds from the Neotropics.

Black Mildews

Black mildews are obligate plant parasites that are especially abundant in the tropics but also may occur from Chile to the southern United States, from South Africa to central Europe and Scotland, and from Japan to Tasmania (Hansford 1961). The name "black mildew" is derived from the scattered, dark, superficial, circular colonies produced by the fungi on the surface of their plant hosts. Black mildews belong to the order Meliolales in the Ascomycota. The order contains more than 1580 species (Hawksworth et al. 1995), most of which (1400) are in the genus *Meliola* (Parbery and Brown 1986). Most (>90%) plant hosts are dicotyledonous Angiosperms, but black mildews also occur on monocotyledons, Gymnosperms, and Pteridophytes. Reports of black mildew infections on species in some of the large plant families common in rain forests, including Orchidaceae, Gesneriaceae, and Araceae, are limited.

Black mildew colonies can be thin or dense and velvety with setae. Each patch normally represents the colony of a single species, but colonies can become confluent with heavy infections. Colonies are found mostly on mature leaves but also can occur on leaf petioles and young twigs. Fruit rarely are infected. Colonies are characterized by very dark, superficial, hyphopodiate hyphae radiating from a central mass of globose ascomata (Parbery and Brown 1986). The hyphopodia can be either mucronate or capitate. The ascomata are either cleistothecia or perithecia, are darkly pigmented, and can have appendages. The ascospores are dark brown, usually composed of four or five cells, and rarely more than 10 μm long. The black mildews lack asexual states. Unlike the sooty molds, most species of black mildews produce hyphae of a uniform 6- to 10-μm diameter and have a characteristic branching pattern. In fact, the most distinctive feature of black mildew hyphae is the presence of short, two-celled lateral branches (capitate hyphopodia), which are distributed in opposite, alternate, or mixed arrangements at the ends of hyphal cells. A fine hyphal filament beneath the hyphopodium penetrates the plant host cell wall to produce an haustorium within the host cell. Species of *Meliola* are characterized by mycelial setae, which arise laterally from hyphal cells and immediately bend to assume a perpendicular position. Setose colonies tend to collect debris such as pollen and insect frass that absorbs water (Parbery and Brown 1986).

Black mildews are most diverse in the tropics, where they are most common in scrub or open parkland; they are relatively rare on undergrowth and shorter trees in dense tropical rain forest, but may be abundant in the upper canopy (Hansford 1961). The greatest infections of black mildew in rain forests occur in naturally or anthropogenically disturbed areas (Parbery and Brown 1986). Black mildews are absent from the arid regions of the subtropics, although they are common in some areas of South Africa, with annual dry seasons of up to 6 months duration (Hansford 1961). Techniques for collecting black mildews are similar to those for the sooty molds. Keys to the genera of black mildews can be found in Müller and von Arx (1973); species descriptions and illustrations are in monographic treatments by Hansford (1961, 1963).

Fly-Speck Fungi

The name "fly-speck" fungi refers to genera of the Microthyriaceae (Microthyriales *sensu* Batista 1959) of the Ascomycotina. The term name was derived from the dark, flattened, dimidiate-scutate (shield-shaped) ascomata, which often appear as small, dark, superficial dots on a leaf surface. Microthyriaceae fruit primarily on living leaves but also on young stems and occasionally on fruits; they occur mostly in tropical and subtropical regions. In the sole monograph of this group, Batista (1959) recognized 447 species in 45 genera. The largest genus, *Micropeltis*, includes more than 100 species. The ascospores are uniseptate to multiseptate, and the mycelium is usually entirely superficial with haustoria forming in the host epidermal cells. The mycelium can be hyaline and inconspicuous, form dark networks of anastomosing hyphae, or form plates of radiating hyphae. Some members of the Microthyriaceae cause necrosis and produce ascomata on dead host tissue (Luttrell 1973). For keys to families and genera, see Luttrell (1973, as Hemisphaeriales). For descriptions and illustrations of species, see Batista (1959).

Downy Mildews

The downy mildews are a group of more than 300 species of plant-parasitic Straminipiles belonging to the Peronosporales and Sclerosporales in the Oomycota. They are characterized by microscopic, erect, treelike, determinate sporangiophores that protrude singly or in tufts through stomata on the leaves of the flowering plant host, giving the macroscopic appearance of a white to gray "down" on the affected tissues (Moore-Landecker 1990). Many downy mildews are associated with serious plant diseases.

Traditionally, downy mildews have been studied by mycologists and thus are included in this chapter. Fifty-four plant families of Dicotyledoneae and two families of the Monocotyledoneae serve as hosts for downy mildews. Several genera of Sclerosporales occur on monocots (e.g., *Verrucalvus*, *Sclerophthora*, *Sclerospora*, *Peronosclerospora*), whereas genera of Peronosporales (e.g., *Peronospora*, *Bremia*, *Basidiophora*, *Plasmopara*, *Pseudoperonospora*, *Bremiella*) occur on dicots. *Peronospora* is the largest genus, with 260 species.

Downy mildews are obligate parasites that produce intercellular hyphae with intracellular haustoria. Sporangia or conidia are produced singly at the branch tips. Sporangia form zoospores when they germinate, whereas conidia germinate by the formation of germ tubes. The genera are delimited based on the characteristic branching of the sporangiophores and sporangia. Sexual reproduction by the downy mildews, represented by the formation of resistant oospores, is infrequent (Moore-Landecker 1990).

The downlike appearance of downy mildews on the leaves of infected hosts frequently is accompanied by chlorosis or necrosis of host tissue. All species require free moisture for sporulation, which thus often is restricted to periods of high relative humidity and darkness. With the exception of *Verrucalvus* species, the downy mildews cannot be grown in pure culture, so identifications are based primarily on the sporangiophores produced on host tissue. Downy mildews are able to sporulate repeatedly for several successive nights from the same lesions, and depending on the species and host, the sporulation period for systemic infections can extend for several months or more under favorable conditions. Keys for the identification of genera of downy mildews can be found in Waterhouse (1973).

Taphrinales and Protomycetales

These two closely related orders of Ascomycota produce similar host symptoms (Kramer 1973). The Taphrinales are represented by one genus, *Taphrina*, with about 95 recognized species. *Taphrina* species are biotrophic microfungi that occur on ferns or higher plants, often causing hyperplasia and formation of galls or witches' brooms (Mix 1949, 1954). Some species are associated with deformed fruit, leaf spot, or leaf curl, and discolor infected tissues yellow or pink, often with a whitish bloom, which indicates sporulation. Asci are produced in a palisade on or from within the epidermis of the infected host tissue; there are no specialized fruiting bodies. The common leaf curl of peach is caused by *Taphrina deformans*. In temperate regions, young leaves and flowers are infected each spring; asci and ascospores are produced in late spring or early summer. The spores bud during warm, moist weather, giving rise to a yeast phase that can survive saprobically on plant surfaces. Cultures of *Taphrina* species isolated from ascospores or conidia can be maintained readily but do not produce a sexual state. They behave *in vitro* like slow-growing, pale pink yeasts.

Species of *Taphrina* appear to be fairly host-specific in that no species occurs on more than one family of hosts. Although Mix (1949) reported a species of *Taphrina* from the family Zingiberaceae in the Philippines, *Taphrina*, as currently understood, appears to occur almost exclusively in temperate regions. Hypertrophic symptoms (galls, witches' brooms) caused by *Taphrina* are observed easily, but leaf spots may be inconspicuous or confused with eriophyid mite damage. Foliar damage in temperate regions is most visible in the spring. If collections are not made at this time, *Taphrina* infections may be overlooked.

The Protomycetales includes one family, Protomycetaceae, which contains five genera. The largest genus, with about 10 recognized species, is *Protomyces*, a predominately temperate-region fungus that causes galls to form in leaf and stem tissues of Apiaceae and Asteraceae. The galls contain large, thick-walled, overwintering chlamydospores. Many species are readily cultured, with colonies resembling those described above for *Taphrina*. Little has been published on tropical species, although one species, *P. andinus*, has been reported from Asteraceae in Ecuador and Brazil (Reddy and Kramer 1975).

Other Groups

Living plant leaves are colonized by a large and diverse group of fungi. Those fungi are distributed widely, occurring on a wide range of plants in different climates, although Dix and Webster (1995) distinguished between fungi that are found on angiosperm leaves and those that occur on the leaves of gymnosperms. Some of the early colonizers are saprobic or weakly parasitic fungi that are restricted to the leaf surface until the leaf becomes senescent, at which time they colonize other leaf tissues as well. Among the earliest colonizers of angiosperm leaves are yeasts; red yeasts in the genera *Sporobolomyces* and *Rhodotorula*, and white yeasts, including *Cryptococcus* species, are common (Dix and Webster 1995). *Aureobasidium pullulans* is a ubiquitous and cosmopolitan filamentous fungus that occurs on angiosperm leaf surfaces in early summer and autumn in temperate regions. In culture, this fungus produces copious quantities of slimy conidial masses ranging in color from creamy white or pink to dark brown or black. *Sclerophoma pythiophila*, the anamorph of *Sydowia polyspora*, resembles *Aureobasidium* and is common on conifers (Dix and Webster 1995). Other common fungi on angiosperm leaves include species of *Alternaria*, *Cladosporium*, *Botrytis*, *Epicoccum*, and *Stemphylium*. The hyphae and reproductive structures of many leaf-colonizing fungi such as these are pigmented. Leaf-colonizing fungi in temperate regions tend to be more numerous on lower leaves and on more peripheral (further from the trunk) leaves in the canopy (Dix and Webster 1995). Fungal growth often is better on the abaxial (lower) than on the axial (upper) leaf surface because the former surface is more sheltered and has a higher number of stomata than the latter.

Common leaf-spotting microfungi in tropical, subtropical, and temperate regions include species of *Alternaria*, *Ascochyta*, *Cercospora*, *Cladosporium*, *Corynespora*, *Phyllosticta*, *Pestalotia*, and *Pestalotiopsis* (Cook 1975, 1978; Holliday 1980). Many produce lesions or other symptoms of infection on stems, flowers, and fruits as well as leaves. In temperate forests, common leaf-spotting fungi on broad-leaf trees and shrubs belong to the genera *Coccomyces, Discula, Gnomonia, Mycosphaerella, Rhytisma,* and *Venturia* (Funk 1985). The life cycle of a typical foliage parasite of temperate trees involves the production of ascomata on infected leaves either during the fall or early spring. Ascospores are discharged about the time that the new foliage appears in the spring. The fungus overwinters as a saprobic mycelium or as ascomata in infected leaves or twigs to complete the cycle (Manion 1981).

The presence of foliar pathogens is recognized macroscopically by water-soaked areas; distortion; or more commonly, chlorotic and/or necrotic lesions on the leaves. Occasionally, the fungi will cause total necrosis or shriveling of the leaves. To identify a fungus, it is usually necessary to examine its reproductive structures under a compound light microscope. Fungi causing foliar diseases on plants commonly produce reproductive structures such as acervuli, pycnidia, perithecia, or apothecia in association with the necrosis. If fungal reproductive structures are not present on the leaves or the reproductive structures are not mature, it may be necessary to incubate the leaves in a moist chamber (see "Standard Ground-Based Techniques," later in this chapter). Keys and descriptions for genera and species of fungi occurring on leaves of temperate trees, shrubs, and woody climbers can be found in Ellis and Ellis (1985), and the same for foliar fungi of trees of western North America can be found in Funk (1985).

Parasitic fungi occurring on grasses in temperate regions include the ascomycetes *Gaeumannomyces, Gibberella, Monographella, Phyllachora,* and *Pleospora (Stemphylium* anamorphs) as well as the specialized groups of pathogens previously noted. The common hyphomycetous fungi on grasses include *Alternaria, Cladosporium, Drechslera, Fusarium, Pseudocercosporella, Rhynchosporium,* and *Ulocladium* species. Common coelomycetes on grasses include *Ascochyta* and *Colletotrichum.* Few parasitic fungi occurring on living rushes and sedges in temperate areas have been reported other than rusts, smuts, and members of the Clavicipitaceae. Included among those fungi are the ascomycete *Didymella,* the hyphomycete *Pseudocercosporella,* and the coelomycetes *Ascochyta* and *Colletotrichum.* For keys, additional literature references, and descriptions of taxa see Ellis and Ellis (1985).

Common foliar parasites of grasses in tropical regions include species of *Cercospora, Drechslera, Magnaporthe* (anamorph *Pyricularia*), *Rhynchosporium,* and *Sphaerulina. Pyricularia* species are particularly important in the tropics as parasites of rice, other cereals, and grasses. *Pyricularia grisea* (synonym, *P. oryzae;* teleomorph, *Magnaporthe grisea*), the causal agent of rice blast, is an important pathogen in many rice-growing countries. The fungus also parasitizes more than 50 other

species of grasses and sedges (Ou 1985). Symptoms consist of lesions on leaves, nodes, panicles, and grain. The leaf spots are elliptical with more or less pointed ends and grayish centers with dark margins. They begin as small, water-soaked areas and develop into lesions of up to 1.5×0.5 cm. Heavy spotting or infection at the tillering stage can kill the host (Holliday 1980).

Needle-cast fungi occur in temperate regions where conifers (gymnosperms) are common. Many needle-cast diseases are caused by facultative parasites in the Rhytismataceae (Ascomycota). This family comprises 43 genera and 344 species, some of which fruit on foliage and stems of angiosperms (Cannon and Minter 1986). Species commonly causing needle blights and casts of conifers are members of genera such as *Lophodermium, Lophodermella, Elytroderma, Hypodermella,* and *Lirula* (Hunt and Ziller 1978; Sinclair et al. 1987 and references therein); they often produce symptoms and initiate fruiting bodies in the first growing season. However, those fruiting bodies often do not sporulate until the foliage is a year old or older. Hence, symptoms such as chlorotic, discolored foliage and repeated defoliation (only current-year needles remain on the tree) may be readily observed, but fruiting bodies collected on current-year needles could be immature. It may be necessary to collect older foliage remaining on the tree or to augment the collection with the previous year's cast foliage to make species determinations.

FUNGI ASSOCIATED WITH CANKERS

A canker is a sunken necrotic lesion on a stem or branch, caused by localized death of the cambium. Cankers on the main stem may result in death of the tree by expanding to the point of girdling the tree. They are also the point at which wood-decay fungi enter the heartwood of trees. The size and shape of cankers vary with environmental conditions (e.g., temperature, aspect), species of fungus, and host response, as well as size and age of the host and time of year that it was infected. Many types of cankers are caused by facultatively parasitic ascomycetes or their anamorphs, and they become established at the site of a wound or other host stress (Sinclair et al. 1987). Cankers also are caused by obligate parasites such as rusts, as discussed earlier (see "Rust Fungi").

Cankers can be grouped into three general categories: woody stem perennial cankers, woody stem diffuse cankers, and twig and shoot blights, all of which are described in the following paragraphs. References for major taxa associated with cankers are cited in the descriptions; the citations provide an introduction to relevant literature rather than exhaustive coverage. General references for canker diagnosis include Ellis and

Ellis (1985), Funk (1981), and Sinclair and associates (1987).

Woody Stem Perennial Cankers

Perennial or target cankers are generally circular to lens-shaped with sunken centers. These cankers persist for years and slowly expand at the same rate as the radial growth of the tree. They often are surrounded by concentric ridges of host tissues formed during each growing season in response to the infection, resulting in a pattern reminiscent of a target. Pyrenomycetes, such as *Nectria* and related Hypocreales (Lohman and Watson 1943; Samuels et al. 1990), and *Ceratocystis* and related Ophiostomatales (Wingfield et al. 1993) are associated with target cankers, particularly on broadleaf trees.

Woody Stem Diffuse Cankers

Diffuse cankers expand more rapidly than a host tree grows radially and so include little or no callus. Infected tissues often are discolored, and the margin between healthy and diseased tissues may be raised slightly. Fruiting bodies of the fungus causing the canker often are found within a few centimeters of this margin. Rapid growth of diffuse cankers can kill the tree, sapling, or shrub in a few years. Diffuse cankers often are seen in shade-stressed trees in the understory.

Many members of the 98 genera of Diaporthales (Barr 1978) are associated with diffuse cankers. Examples in the Valsaceae are *Valsa* (*Cytospora* anamorph), *Diaporthe* (*Phomopsis* anamorph), *Gnomonia*, *Cryptosporella*, *Cryphonectria*, and *Endothia* (Spielman 1985; Micales and Stipes 1987). Members of the Diatrypales, such as *Eutypa* and *Leucostoma*, are frequently associated with similar symptoms. The Leotiales are a family of inoperculate discomycetes, many of which, particularly the Leotiaceae, are associated with diffuse cankers (Dennis 1978, as Helotiales). Examples are *Cenangium*, *Tympanis*, and *Encoelia* (Zhuang 1988a). Members of the genus *Lachnellula* (Hyaloscyphaceae) are associated with cankers on conifers (Baral 1984). *Hypoxylon* (Xylariaceae) species are also associated with diffuse cankers (Sinclair et al. 1987, Ju and Rogers 1996).

Twig and Shoot Blights

Twig and shoot blights contain little or no callus, usually develop rapidly, and persist for only one season. They may be detected by discolored or slightly sunken tissues in infected areas, which may take on a pimpled or roughened appearance as a result of the presence of fruiting bodies. Blights frequently occur on young stems or twigs that have been predisposed to opportunistic fungal

infection by abiotic factors such as frost, drought, sun scalding, or wounding. The dead bark may be overgrown by callus in the following growing season, resulting in complete healing if the external stresses are no longer present. Examples of blight fungi are *Melanconis* (Melanconiaceae, Diaporthales; Wehmeyer 1941) and pycnidial mitosporic fungi (Coelomycetes) such as *Sclerophoma*, *Cytospora*, *Phoma*, *Gelatinosporium*, and *Diplodia* (Sutton 1980). Twig blights also commonly are caused by anthracnose fungi (see "Other Groups" earlier).

Cankers are observed best on deciduous trees in temperate regions in late winter and early spring when leaves are absent. Fruiting bodies generally are present at this time. Often the bark of diffuse cankers will be discolored and sunken, with a raised ridge around the leading edge. Excessive pitch production often is associated with cankers on conifers.

FUNGI ASSOCIATED WITH GALLS

Certain rust fungi, *Endocronartium harknessii* on *Pinus contorta* (Fig. 7.9), and *Gymnosporangium* species on

FIGURE 7.9 Galls on lodgepole pine (*Pinus contorta*) caused by the rust fungus *Endocronartium harknessii*. (Figure prepared by S. Henrich)

Juniperus, for example, are associated with conifer galls (Ziller 1974). *Apiosporina morbosa* is associated with *Prunus* galls in temperate climates. *Phomopsis* species cause galls on a number of woody temperate angiosperms, as do *Nectriella pironii* and *Sphaeropsis tumefaciens* on many woody and herbaceous plants in the subtropics (Sinclair et al. 1987). *Cyttaria*, a parasitic discomycete, is known only from galled branches and smaller twigs of 12 south-temperate species of *Nothofagus*. Korf (1983) recognized three groups of *Cyttaria* based on geographic position, host range, and other features.

Synchytrium (Chytridiales) is a large genus of obligately biotrophic microfungi commonly associated with galls on host species of algae, mosses, ferns, and flowering plants. The roughly 120 *Synchytrium* species occur worldwide in tropical, temperate, and arctic regions; 106 species have been reported from North America, 24 from South America, five from Central America, eight from Australia and New Zealand, 16 from Africa, 56 from Europe, and 88 from Asia (Karling 1964). More than 1350 plant species in more than 800 genera and 165 families have been reported as hosts. Some species occur on only one host genus; others are limited to a single host family; and one species, *S. macrosporum*, has been shown to infect more than 1300 host species under greenhouse conditions.

Synchytrium species infecting flowering plants cause the formation of galls on leaves, petioles, flower buds, and stems. Two species, *S. fragariae* and *S. endobioticum*, infect roots and underground stems. The galls range from scattered, small, and inconspicuous to large and clustered or compound; some galls are brightly colored. The galls may be formed by the enlargement of a single, infected host cell or by the proliferation of cells around the infected cell. The infected host organ may become noticeably hypertrophied and deformed. Species that infect algae cause enlargement and elongation of the infected host cell.

Identification of *Synchytrium* species is based primarily on host determination. Keys and descriptions can be found in Karling (1964). Voucher specimens should consist of dried plant material, including galls or other deformed host tissues containing resting sporangia. *Synchytrium* species cannot be grown *in vitro*.

The Plasmodiophoromycetes (Protoctista) comprises 15 genera of obligate endoparasites of plants, algae, aquatic protoctists, and fungi that frequently evoke gall-like symptoms. These fungi are found in soil and fresh water. Their distributions follow those of their hosts.

Macroscopic above-ground symptoms of infection include stunting, wilting, chlorosis, and discoloration of leaves of infected plants. Species of *Plasmodiophora*, *Spongospora*, *Sorosphaera*, *Sorodiscus*, and *Tetramyxa*

cause enlargement of infected host tissues. Underground symptoms can be striking and may include galls, lesions, cankers, warts, and dry rot. Species of *Ligniera* and *Polymyxa*, found in roots and root hairs, do not cause host cells to enlarge and, therefore, are difficult to detect on underground plant parts, although above-ground symptoms may be present. Members of the Plasmodiophoromycetes cannot be grown in culture away from their hosts.

Identification of the Plasmodiophoromycetes requires microscopic examination of infected cells (Karling 1968). Those fungi are characterized by multinucleate, unwalled protoplasts that develop into sporangia or cystosori within the host. Sporangia produce motile cells called zoospores, whereas cystosori produce aggregates of thick-walled cells (cysts). Cysts are liberated into the soil with the death and breakdown of host tissue. After a resting period, a cyst germinates to produce a single zoospore.

The genera are distinguished by the organization of the cystosori in the host cells (Dylewski 1990). The cysts of *Plasmodiophora* are irregularly arranged and lie free within the host cell. *Sorosphaera* produces cysts arranged in hollow spheres; *Spongospora* produces cysts in loosely arranged, spongelike clusters; *Woronina* produces cysts arranged in a single layer; and *Sorodiscus* produces cysts arranged in bilayered disks. The cysts of *Octomyxa* are united in groups of eight; those of *Tetramyxa* are united in groups of two or four. *Membranosorus* produces cysts of irregular shape and size in a single layer conforming to the size and shape of the host cell. Voucher specimens should include permanent slides of infected host cells containing the cysts.

FUNGI ON FRUITS, NUTS, AND FLOWERS

Many of the fungi that infect leaves infect flowers and fruits on the same host. Other fungal parasites are encountered more frequently on flowers and fruits. Symptoms of infection include discoloration, shriveling, wilting, and premature drop of flowers. These symptoms can be described as a blossom blight. Some of the predominant fungi associated with blossom blights in both tropical and temperate areas include species of Sclerotiniaceae, such as those in the anamorphic genus *Botrytis* (Coley-Smith et al. 1980) and in *Monilinia* (Batra 1991). Those fungi also are associated with fruit rots. Fruits infected by *Monilinia* species frequently become "mummies," dried shriveled fruits that may remain attached to the plant. *Monilinia* species also are associated with brown rots of fruits (as well as twig cankers and gummosis). Other symptoms of infected fruits and nuts include discolored lesions or streaks, shriveling, wrinkling, hardening, and cracking. Scabs—roughened, crustlike diseased areas on fruits and nuts—are caused in the tropics predominantly by species of *Elsinoë* and their associated *Sphaceloma* anamorphs (Dothideales; von Arx 1963, as Myriangiales). In temperate regions, *Venturia* species commonly cause scabs on fruits (Sivanesan 1977). Scabs occur on a wide range of agriculturally important hosts, including avocado, cashew, castor, groundnut, papaw, poinsettia, and soybean in the tropics and Rosaceae in temperate regions. Other fungi associated with fruit rots include species of *Aspergillus*, *Colletotrichum*, *Dothiorella*, *Guignardia*, *Phytophthora*, and *Rhizopus*. For descriptions and illustrations of genera and some species, see Cook (1975), Ellis and Ellis (1985), and Holliday (1980).

Anthracnose

Species of *Glomerella* (Phyllachoraceae, Ascomycota) and its anamorph *Colletotrichum* are among the most important causes of preharvest and postharvest diseases in both temperate and tropical regions (Sutton 1992). Thirty-nine species and numerous subspecies of *Colletotrichum* are recognized (Sutton 1992). These taxa cause diseases generally referred to as anthracnoses. *Glomerella cingulata* (anamorph: *Colletotrichum gloeosporioides*) is the most common pathogen in the tropics and causes anthracnose of avocado, citrus, mango, papaya, passion fruit, guava, cocoa, rubber, and other wild and cultivated plants (Holliday 1980; Dodd et al. 1992). Anthracnose symptoms are fairly conspicuous as typically sunken, irregular-shaped necrotic lesions on fruits, leaves, and stems (Holliday 1980). Lesions on fruits may be raised and cankerous, scablike, or wartlike in appearance. Branch die-back also may occur. The fungus can survive long periods on plant debris in the soil.

Colletotrichum species can be recognized macroscopically by the mucilaginous masses of often pink- or salmon-colored conidia oozing from acervuli associated with anthracnose lesions under humid conditions. Microscopically, *Colletotrichum* species are characterized by acervuli that frequently produce dark brown setae and hyaline, aseptate conidia that germinate to produce appressoria. *Colletotrichum* species are grown readily in culture, but the production of the *Glomerella* stage is less consistent. Production of appressoria can be induced by placing conidia in a drop of water on a microscope slide in a moist chamber overnight (see Appendix I). For descriptions and keys to *Colletotrichum* species, see von Arx (1957), Baxter and van der Westhuizen (1984), and Sutton (1980). Holliday (1980) and Sutton (1992) have provided descriptions of species.

Clavicipitaceae

The Clavicipitaceae (Hypocreales, Ascomycota) contains 27 genera, most of which are systemic endophytes or insect parasites. A notable exception is the familiar genus *Claviceps*, which contains approximately 35 species worldwide. In grasses, *Claviceps* species cause local ovarian infection known as ergot, in which the seed is replaced by a (usually) black sclerotium, approximately the same size as an uninfected seed (Fig. 7.10). The notorious *C. purpurea* causes ergot of rye and other grasses. Ergot is known for its toxic effect (ergotism, caused by fungus alkaloids) on humans and other mammals who ingest sclerotia mixed with grain.

Field collection and subsequent identification of *Claviceps* species can be difficult for a number of reasons. In warm, humid climates the hyperparasitic fungi *Cerebella* and *Fusarium* infect the honeydew or conidial exudation stage of *Claviceps* (Rykard et al. 1984) and inhibit development of sclerotia (Brady 1962). However, the prominent black *Cerebella* fruiting bodies also can be used as indicators of the cryptic early stages of *Claviceps* infections, and searches of plants nearby may yield uncontaminated *Claviceps* infections. Germination of sclerotia to produce teleomorphic stromata has some practical difficulties. Some species have ripening periods of up to 9 months. Sclerotia, provided they are ripe, may be

FIGURE 7.10 Ergot of rye, caused by *Claviceps purpurea*. Infected ovaries develop black sclerotia instead of seeds. (Figure prepared by S. Henrich)

stimulated to germinate after they are placed on moist filter paper, although cold pretreatment may be required. If the species is unknown, ripening periods of several different lengths may have to be tried before germination is achieved (Langdon 1954). Several small Clavicipitaceous genera (e.g., *Aciculisporium, Konradia, Mycomalus, Ascopolyporus*) are associated with bamboo and warrant further study (Rogerson 1970).

ESTIMATING THE NUMBERS OF FUNGI ON LIVING PLANTS

Estimating of the number of fungi on living plants in a specified area requires two types of information: (1) the numbers and types of plant hosts present in the area and (2) estimates of the number of fungal species associated with those plants. Good estimates of the numbers of plant species in tropical and temperate regions are available as are estimates of the proportions of trees, shrubs, and herbs. Unfortunately, the sites that have been studied generally cannot be compared directly because of differences in plot size and sampling techniques. Gentry (1992) estimated that less than 25% of the species at most neotropical sites are trees. Epiphytes account for from 12% to 16% of the species at moist forest sites, and for about 25% in wet forests. Up to 50% of the plant species at individual sites are herbs and shrubs. Hubbell (1979) recorded a total of 135 species of woody plants with stems greater than 2 cm in diameter at breast height (dbh) in a mapped area of approximately 13.4 ha in Guanacaste Province, Costa Rica. The species recorded included 87 overstory and understory trees, 38 shrubs, and 10 vines. The site was in a "dry forest life zone," and Hubbell noted that epiphytes were rare. Some areas in the neotropics are considered "tree-rich." For example, two 1-ha-sample sites from lowland forests near Iquitos, Peru, each had nearly 300 species of woody plants greater than 10 cm dbh (Gentry 1992). In paleotropical forests, plants other than trees seem to predominate. At a forest site in Gabon, 34% of the species recorded were trees, 25% were climbers, and 37% were herbs or shrubs (Gentry 1992).

Plant-species diversity in temperate forests generally is measured as species richness, which is defined as the number of species per 0.1 ha (Glen-Lewin 1977). Whittaker (1965) used species richness to compare diversity among temperate communities. He noted that low tree-species diversities can be associated with highly productive climax forest, such as the redwood forests of coastal California and Oregon, in which only four tree species (20% of total species), one shrub species (5% of total), and 15 herb species (75% of total) were recorded. The highest tree diversity noted by Whittaker was at an intermediate moisture gradient in the Great Smoky Mountains (18 tree species, 10 shrubs, 16 herbs; 41%, 23%, and 36% of the total plant species, respectively).

Plant-species diversities are important measures for estimating species richness of fungi parasitic on plants as a result of the preponderance of host-specific taxa. Strong and Levin (1979) determined that species richness of plant-parasitic fungi in temperate zones was highly correlated with the growth form of the hosts: the more complex the growth form, the greater the number of associated fungal parasites. Hence, the mean number of fungal parasites reported per host species was 5.4 for herbs, 7.1 for shrubs, and 15.4 for trees. The authors did not specify whether the fungi were host-specific, and some parasitic species may occur on more than one host. It is uncertain whether these figures are representative of tropical regions. Large margins of error exist in estimates of extant fungal species, in part a reflection of the degree of host specificity of fungi on tropical plants (Cannon and Hawksworth 1995).

The highest known alpha diversity of trees is in the Neotropics (Gentry 1992). Groups of host-specific parasites, such as the black mildews and other foliicolous fungi and parasitic leaf Ascomycetes, that are especially abundant in the tropics (Hansford 1961; Nishida 1989) occurring on trees and other woody hosts, might be expected to contribute to greater species diversity. Studies by A. C. Batista in Brazil showed that a single leaf could support 10–12 species of fungi, including up to five undescribed species (Cannon and Hawksworth 1995).

Estimates of fungal species richness certainly have been influenced by the economic importance of the host. At least twice as many species of fungi have been reported from species of Gramineae as from species of Cyperaceae (Cannon and Hawksworth 1995). It is interesting that in a table of ratios of fungal species to host species, the size of the host seemed to be the major determinant of species richness, with a ratio of 35:1 for *Zea* and 21.2:1 for *Phragmites* as compared to 8.1:1 for *Triticum* (Cannon and Hawksworth 1995). The same study reported relatively low ratios for *Saccharum* (4.3:1), but the authors suggested that host associations in this genus were probably under-recorded. Parasitic fungi associated with epiphytic plants are poorly known and may be another potentially diverse group in the tropics based on the species richness of the potential hosts. Sillett and colleagues (1995) recorded 127 bryophyte species in the inner crowns of six *Ficus* trees in Costa Rica.

Hirata (1986) estimated that 4.5% of vascular plant species host powdery mildews. This figure is in line with the percent of plant species that host smut fungi in the British Isles (4.8%) and the United States (3.2%). In Latin America (Mexico, Central and South America), in contrast, only 0.3% of the estimated 88,000+ species of vascular plants (Raven 1988) are known hosts for smut fungi (based on numbers from Zundel 1953 and Durán 1987). This low percentage probably reflects our ignorance of the Latin American smut mycota. Of the collections made by Durán (1987) in Mexico over a 22-year period, 17% were of species new to science. Vánky (in Oberwinkler 1992) estimated that only half of the extant smut species have been described. Moreover, powdery mildew fungi and smuts may not be representative of plant parasitic fungal diversity in tropical regions, particularly in tropical forests, given that neither group is abundant on trees and shrubs. Powdery mildews reach their greatest diversity in temperate regions (Hirata 1986), and smut fungi are most abundant on grasses and sedges. A more representative group may be the rusts; Nishida (1989) commented on a study by Buriticá and Hennen (1980) in the Neotropics reporting that one in three of the plant species in the study area was parasitized by a unique rust species.

The highest levels of endemism among the plant parasitic fungi should occur in areas with the highest levels of host-species endemism. Gentry (1992) noted four such areas in the Neotropics, most of which represent biomes isolated as islands, intermountain regions, mountain ridges, and other unusual habitats. Determining the degree of endemism in plant-parasitic fungi, as in their hosts, requires much more complete information on parasite and host distributions than currently exists. A survey of additional areas may show that many apparently endemic species have a wider distribution than presently believed. Sutton and Hodges (1976) erected the genus *Pseudotracylla* based on *P. dentata*, which was found on *Eucalyptus* leaves in Brazil. Carris (1992) described a second species, *P. falcata*, from *Vaccinium macrocarpon* leaves in a cultivated cranberry bog in Massachusetts, and later (Carris 1994) reported *P. dentata* from the same bog. That was the first time the fungus had been reported since its description in 1976. Conversely, many plant-parasitic fungi have yet to be described, particularly those from areas of the tropics that have not been adequately explored.

All estimates of fungal diversity are based on ratios of vascular plants to fungi for a specific area (Hawksworth 1991). Those figures are not based solely on plant-parasitic fungi but also include a wide range of saprotrophic species as well as symbiotic species such as those that form mycorrhizae. The ratios are based primarily on data from temperate regions, and ratios in tropical areas may be higher (Hawksworth 1991), although it is not known whether the host specificity of fungi on vascular plants occurs to the same extent in the tropics as in temperate regions.

As noted earlier in this section, Hubbell (1979) recorded 135 species of woody plants with stem diameters greater than 2-cm dbh in a 13.4-ha forest tract in Costa Rica. The species included 87 overstory and understory trees, 38 shrubs, and 10 vines. Epiphytes were described as infrequent at the site. Using Gentry's (1992) estimates of 25% tree species at most neotropical sites, an estimated total of 348 plant species should occur at this site, including 48 shrubs and vines and 213 herbs and epiphytes. Taking those numbers and the ratios of Strong and Levin (1979), the estimated number of plant-parasitic species of fungi present in that small forest tract is 2833 [(87 × 15.41) + (48 × 7.05) + (213 × 5.42)].

At a larger scale, approximately 10,000 plant species have been recorded in a 110,000-ha plot in Costa Rica. If Gentry's (1992) estimates are used, 2500 of the species (25%) are trees, 1600 species (16%) are epiphytes, 1400 (14%) are shrubs and vines, and the remaining 4500 (45%) are herbs. The estimated total number of species of plant-parasitic fungi present in the plot then would be 81,457, more than the number of all currently described fungal species!

Similar calculations were used to estimate numbers of fungi on living plants in an equal-size tract in the Great Smoky Mountains in the eastern United States. Whittaker (1956) described the region as characterized by forested areas with both conifers (hemlock, pine, spruce) and hardwoods (beech, oak and chestnut), interspersed with heath balds, and other vegetation types. It is interesting that the hardwood stand composition has been altered dramatically during the past century, largely because of the introduction of the chestnut blight fungus, *Cryphonectria parasitica*. Prior to the introduction, *Castanea dentata* (American chestnut) formed 30–60% of canopy stems of some forests; large trees often dominated the forest strata. Now, 70% of the tree stems in those forests are composed of other, smaller tree species (Whittaker 1956).

Approximately 1300 species of flowering plants, including 130 species of native trees, have been found in the Great Smoky Mountains (Whittaker 1956). The variety of habitats, which are largely influenced by moisture regimens, gives the impression of greater plant species diversity than actually is present. If we assume that 10% (130) of the plant species are trees, 25% (325) are shrubs (common on submesic sites in this region), and 65% (845) are herbs, including epiphytes (which are rare, and nearly all lichens, bryophytes, or liverworts, according to Sharp 1957), then using Strong and Levin's (1979) ratios, an estimated 8873 fungus species should be present. This is slightly more than one-tenth of our prediction for numbers of fungi at the Costa Rican site.

These are very rough estimates, which may prove inaccurate for a number of reasons. Kranz (1990) noted that when a plant ecosystem is explored for plant pathogens, the majority of plant species encountered are likely to be healthy. Kranz based this statement on a 3-year study of six distinct plant communities at three sites in Germany. He found that from 28% to 52% of the plants were hosts for plant pathogens. Conversely, the leaf-surface mycota is usually richly developed, often consisting of 12 or more species of mostly saprobic sooty molds (Hirsch and Braun 1992). The phylloplane fungi are especially conspicuous in the tropics, and under humid conditions the leaves of many plants are covered with a thick mat of dark fungal mycelium (Hirsch and Braun 1992).

A number of factors in addition to host-species composition should be considered in determining diversity of plant-parasitic fungi. For example, diversity of plant-parasitic fungi in a study plot depends also on the density and distribution of host plants. Hirsch and Braun (1992) cited studies showing an increasing diversity of plant-parasitic fungi with increasing moisture levels of the plant community and a decreasing level of diversity with increasing elevation. They also noted that phytoparasitic fungi are often especially diverse in border-plant communities. Hence, the diversities of plant-parasitic fungi are not uniform across a study plot, and sampling strategies must take this uneven distribution into account.

ASSESSING SPECIES RICHNESS AND ABUNDANCE

Many of the fungi fruiting on living plants and fruits are obligate or facultative parasites and are often host-specific. Therefore, extensive knowledge of the taxonomy and diversity of plant species in an ecosystem under study is essential. The mycologist investigating host-fungal associations must first determine the identity and heterogeneity of plant populations. That information, gained by working closely with botanists and plant ecologists, is important in determining the number and size of sample plots required to sample the fungi occurring on these hosts adequately.

Standard techniques designed by plant pathologists to monitor the introduction and spread of a single pathogen in a genetically uniform crop field (monoculture) are not appropriate for measuring populations of parasitic fungi in natural ecosystems. We cannot expect to encounter similar levels of pathogen populations in a natural ecosystem. In addition, a host population in the wild is likely to have a higher degree of genetic variability and may be composed of different-age individuals, clustered or scattered throughout the terrain and interspersed and interacting with nonhost species and individuals.

To survey the biodiversity of fungi on living plants and fruits comprehensively, an investigator(s) should first determine plant species richness and then analyze the incidence and levels of infection that occur in a given ecosystem (abundance). According to Williams and Humphries (1994:285), "Empirically, in the absence of all other information, unweighted species richness provides a good first approximation to more demanding taxonomic diversity measures." Endemism, however, is not a measure of biodiversity but of range size. Assessment of endemism is, therefore, not a substitute for information on richness and abundance, and its exclusive use can reduce the effectiveness of procedures for conservation of biodiversity (Williams and Humphries 1994).

It would seem logical, considering that the fungi in this functional group are fruiting on living plants, to survey for parasitic fungi in conjunction with botanical surveys. Species accumulation curves for plant parasitic fungi should be related to, but not necessarily parallel, those of their host plants.

TESTING METHODS FOR SAMPLING SPECIES ABUNDANCE

In most ecosystems, plants and animals show characteristic patterns of species abundances that approximate a log normal distribution, with a small group of dominant species, a larger middle class of relatively important species, and a smaller number of rare species (Whittaker 1965). Patterns of species' abundances among fungi associated with plants may be somewhat different from those for plants and animals. Recent studies indicate that within an ecosystem a few species of those fungi are dominant, but most species are rare (Dix 1964; Swift 1976; Bills and Polishook 1991, 1994).

The frequency of a specific plant parasite could be determined using modified sampling schemes for determining incidence and severity of plant diseases. The choice of sampling scheme would depend on the vegetation type; for example, sampling in a liana forest would be very different from sampling in a savanna dominated by grasses. The type of sampling also would depend on the objective of the sampling, the type of fungus being sampled, and its interaction with the host plant. Systemic infections, such as those of many *Tilletia* species, can be detected from the presence of sori in host reproductive organs. The presence of several sori on a plant would be counted as a single infection. When fungi such as the *Colletotrichum* species associated with anthracnose diseases cause localized infections, however, each lesion represents a unique infection.

A surveyor must decide whether to assess incidence (percent of diseased plants or plant parts per sample) of infection or severity (percent of the relevant host tissue or organ covered by symptoms). Disease-severity measurements frequently are used for such fungi as rusts, powdery mildews, and fungi associated with foliar spots. Determining incidence of parasitic fungi requires fewer samples than measuring disease intensity and is useful for early stages of epidemics and for organisms that become systemic in their hosts after a single infection. To determine incidence, the simple counting of diseased versus healthy sampling units is all that is required. Sample sizes can be roughly estimated from pilot sampling. The mean and standard deviation for the number of fruiting bodies (i.e., pustules) of a particular fungus on a sampling unit (leaf, stem, entire plant) are computed for a number of different size samples. The means and standard deviations are plotted against sample size; when variations in the curves for both mean and standard deviation become more or less steady, the investigator can choose a sample size according to the desired level of precision (Kranz 1988). A similar method can be used to determine the number of samples required from a particular type of vegetation to estimate the number of fungal species present. In this case, the number of different fungal colonies is used instead of the number of fruiting bodies or lesions of a single fungus. Frequency of sampling is another consideration in determining both incidence and intensity of disease. A general rule is that at least five points of measurement are necessary for establishing a reliable curve (Kranz 1988).

Valentine and Hilton (1977) developed a sampling procedure for estimating foliage characteristics of *Quercus* species. It is based on a randomized branch sampling technique designed by Jessen (1955) for estimating numbers of fruit on orchard trees. Valentine and Hilton (1977:298) indicated that, using their method, "two experienced people could easily sample three large trees in 8 h." This two-phase method involves following a given major branch until it reaches a fork, at which point a selection probability is assigned for each branch emanating from the fork. A fork is randomly selected, and the process is repeated at each fork until a single cluster of leaves is reached. At this point a second measurement (e.g., leaf surface area or, potentially, number of spots caused by a foliar fungus) is taken.

COLLECTING AND EXAMINING SPECIMENS

Sampling techniques for all types of fungi on living plants are discussed later in this chapter (see "Studying Fungi on Living Plants"). Descriptions and references to additional collection techniques can be found in Deighton

(1960) and Hawksworth (1974). To obtain a full species list of fungi in a given forest, either temperate or tropical, an investigator will have to collect samples many meters above the ground. The Canadian Forest Service, which is conducting forest canopy studies on Vancouver Island, British Columbia, is using semi-permanent platforms reached by ropes and pulleys (L. Humble, personal communication) to access collecting sites (Fig. 7.11). Campbell (1989) listed other techniques that have been used, such as employing human climbers who use spiked foot griffes or a foot strap to scale trees. However, the former apparatus punctures the tree, and the latter may be used only on trees with small-diameter trunks. Shotguns and rifles often are used to obtain samples for grafts or voucher specimens from the upper canopy (Fig. 7.12). According to Campbell (1989), E. J. H. Corner trained monkeys in Malaysia to climb trees and break off

FIGURE 7.11 Platform used for tree canopy research in British Columbia. (Photo by P. Kingham)

FIGURE 7.12 Rifles are used to sample branches from tree canopies. (Photo by W. J. Beese)

branches. Wolf (1993), who studied epiphytes in a tropical rain forest, collected 10 samples each from habitats within trees (e.g., tree base, trunk, inner crown, middle crown, outer branches, branchlets in the outer tree crown) by shooting an arrow with a nylon rope attached through the tree crown and then climbing the tree using jumars and mountaineering rope. Any of these techniques also could be used to search for fungi in the upper canopy.

Denison and colleagues (1972:148) provided excellent, illustrated descriptions of tree-climbing techniques and equipment, noting, however, that "climbing old-growth trees is dangerous and a fall would probably be fatal." Falling branches and equipment also may endanger the support crew on the ground. A team of three experienced climbers should be able to rig a tall tree in 1.5 days. Once anchors and climbing ropes are rigged, a collector can ascend the tree in approximately 20 minutes. Denison and associates (1972) described how to select tree-branch systems (branch attached to main trunk, including all of its branchlets and foliage) for sampling based on the size and branching complexity of the

tree. They mapped each branch system on a tree and assigned an "importance value" to it, based on its size, number of branchlets, and various other features. They selected at random two branch systems from small, uniform trees; three from normal, full-sized trees; and four or five from large trees with more than one leader (main axis) for extensive sampling. This procedure was designed for and has been used with temperate trees; tropical trees, which are often huge and branched throughout, would be considerably more difficult and time-consuming to sample.

Using the climbing techniques described above, Pike and colleagues (1975) sampled epiphytes from the trunk of a 450-year-old Douglas fir (*Pseudotsuga menziesii*). They separated the tree into six zones: base, moist side of the trunk, dry side of the trunk, upper trunk, axes of branch systems greater than 4 cm in diameter, and branchlets less than 4 cm in diameter. The last category included small branches attached directly to the trunk as well as branch tips less than 4 cm in diameter. They identified 74 species of lichens and 32 species of bryophytes. The techniques they developed for systematically collecting samples from the tree also could be used for fungi associated with a tree or its epiphytes.

Few tree trunks are perfectly vertical, and the upper side of an angled trunk or branch receives more moisture than and has a very different morphology and epiphyte flora from the lower side. In the upper canopy, light intensities and fluctuations of ambient temperatures and moisture levels are more extreme than in the more sheltered lower canopy. Representative areas of each tree, therefore, must be sampled to create comprehensive lists of host-fungus associations. Pike and colleagues (1975, 1977) marked tree trunks with tape at 5-m intervals and mapped the location of each branch system. They then sampled two 10- × 25-cm quadrats at each 5-m mark, at 45 degrees and 135 degrees from the trunk, on either side of the climbing path, which was the designated vertical axis. All epiphytes were harvested from the quadrat and were bagged for laboratory examination. Large diameter branches were sampled at 0.5-m intervals, the sampling unit being a cylindrat—the cylindrical surface of a 1-dm-long axis section. The diameters of branchlets were measured, and the total area per branch was calculated; 20% of the branchlets were removed and bagged for laboratory examination. Cutting down trees to study epiphytic organisms is counterproductive because bark and branches usually are shattered and scattered on impact (Pike et al. 1975).

Lesica and associates (1991) described sampling regimens for the study of lichen and bryophyte communities on old-growth and managed second-growth grand-fir forests in Montana. Their techniques are also appropriate for sampling fungi associated with living plants. They used 0.081-ha circular plots, established a 32-m reference line through the plot center, and then ran three 30-m long sampling lines perpendicular to the reference line. Samples were taken in three strata: lower canopy, trunk, and ground. In the lower canopy stratum, a total of 24 branches were collected by sampling the branch closest to each line at eight equidistant points on each line. In the trunk stratum, epiphytes were sampled from within 0.5×0.2 m flexible quadrats 1.5 m from the ground, on both north and south sides of trunks larger than 13 cm dbh. Ground samples were taken along three 30-m-long and 1-m-wide belt transects, on both sides of the sampling line, for a total plot size of 180 m².

STANDARD, GROUND-BASED TECHNIQUES

Small herbaceous plants may be collected in their entirety and pressed between sheets of newspaper. Small, woody branches with leaves attached can be removed using pruning shears or pruning poles if they are within reach of the ground; they also can be pressed between sheets of newspaper. Under wet field conditions or in rough terrain, samples may have to be stored in plastic bags in the field and then pressed at the end of each day on return to the field camp. Branches that are not easily pressed and conifers or other dense spiny plants can be stored in large paper bags in dry weather. In humid climates or during wet weather, it is easier to tie a label to the branch itself and leave it unwrapped. Soft fruit may be stored temporarily in perforated plastic vegetable storage bags and carried in a rigid container until desiccated or cultured. Large fruits may have to be cut into sections to ensure that they dry properly. Erumpent fruiting bodies or cankers can be removed from the trunk or a main branch of a large tree by excising a piece of bark or underlying wood. If the tissue is too hard to cut with a knife, it can be removed with a chisel and mallet.

If fleshy plant tissue or woody plant sections in need of culturing have to be stored or transported for a few days prior to delivery to a laboratory, they should be wrapped in newspaper or rolled in a paper bag to allow for some air exchange but prevent desiccation. They should not be sealed in air-tight plastic bags, which encourages the growth of molds or bacteria. Conversely, under extremely humid conditions, a small amount of desiccant can be added to the bag to prevent molding. In drier climates, zipper-locking perforated plastic bags designed for vegetable storage work well for temporary storage. As a general rule, however, slight desiccation is better than excess humidity. If material cannot be cultured or observed microscopically within a few days, an alternative method of preservation sometimes works if the diseased tissue contains immature fruiting bodies

close to sporulation. In that case, most of the collection is dried with warm air, but a small amount is dried with ambient air. Attempts can be made to obtain cultures from the ambient-air-dried material at a later time.

The previously described sampling techniques sometimes will have to be modified to deal with special circumstances, as follows:

1. Obligate parasites usually infect healthy, vigorously growing plants, so the plant symptoms usually used by plant pathologists to detect the presence of those fungi (e.g., stunting, chlorosis) may not be evident. That is especially true of low-level infections on the foliage of trees, which are overlooked easily. Thus, when it is not possible to survey every part of a plant visually, it may be necessary to plot second, independent species-accumulation curves for parasitic fungi on the foliage, on the fruits, on the shoots, and on the stems of individual large hosts. For example, a set number of living leaves (e.g., 50) on a branch, starting at the youngest and working toward the stem, should be examined for the presence of fungi. This procedure could be repeated for a randomly selected branch within each of the four compass quadrants, in the upper, middle, and lower crown of the tree.

2. Host tissues parasitized by fungi may be overgrown rapidly by secondary saprobes that obscure symptoms and/or signs of the primary cause of the disease. Saprobes also greatly reduce the likelihood that a pure culture of the parasite will be obtained, unless highly specialized selective media are used. When fungi are not sporulating on obviously diseased tissues (i.e., canker or nonsporulating lesion on a fruit), cultures should be initiated as soon as possible from material excised from the zone between the healthy and diseased tissue, where the pathogen is most active. If the tissue is thick, such as in a stem or fruit, a clean tissue sample may be obtained by cutting away layers of tissue and flame-sterilizing the knife between cuts. That precaution will help to prevent contamination from surface fungi and bacteria. If the infected plant part is too small to section, a portion may be surface-sterilized with a 10% solution of household bleach and then rinsed several times with sterile, distilled water. Acidified malt often is used to suppress bacteria and encourage growth of pathogenic fungi from host tissue.

3. Techniques that permit obligate parasites to grow outside their hosts have not yet been developed for many species. Consequently, it is essential to dry or otherwise preserve collections of obligate parasites quickly after collection, especially if a fungus is fruiting on fleshy fruits or succulent stems. Because they

are unlikely to be cultured, those parasites should be heat-killed and dried after their fresh appearance is described adequately. Such treatment should prevent ingress of saprobes.

4. Fruiting bodies associated with cankers, blights, or leaf spots are often immature when collected. Many of those fungi are facultative parasites that do not fruit until the host tissue is dead or nearly so. The causal agent may be cultured using the techniques described earlier in item 2. If fruiting bodies are very close to sporulation, however, the following inexpensive shortcut can be used. The investigator places the leaf, fruit, or stem segment in a plastic bag, inflates the bag (humidity from one's breath is usually enough for incubation; addition of water or damp paper towels is generally excessive), and lets the fungus incubate from 1 to 3 days, checking for sporulation daily (Fig. 7.13). The fruiting bodies may be overcome by aggressive saprobes, but if the sample is not contaminated by soil or damaged by insects, this method often works well.

5. Not every host plant will have symptoms or signs of parasitic fungal infections when collected. Some plants will have escaped or resisted infection. Thus, a sample area adequate to determine the species richness of the host plants could be several times too small to determine the species richness of parasitic fungi on those hosts. Kranz (1990), based on a 3-successive-years study of six distinct, multispecific plant communities located at three sites in Germany, showed that only 28–52% of the plant species were hosts for plant pathogens. The host-species richness ranged between 80 and 155 in the communities. In five of the six communities, the majority of the plant species remained healthy during the 3 years of study.

6. Certain groups of plant parasites, such as rusts, have alternate hosts and may not be reliably identified to species from spore stages produced on only one of the hosts. Care should be taken to ensure that as many spore stages as possible are collected by collecting tissues of all ages and degrees of infection.

7. In temperate regions, fruiting bodies of certain groups of fungi, in particular those causing foliar diseases, must overwinter before the sexual stage will mature. Symptoms and fruiting bodies will be clearly evident on diseased plants collected during the current growing season, but there will be no sporulation. To find the sporulating stage, the investigator may have to collect fallen, overwintered foliage in the spring, which then can be compared with newly infected foliage later in the growing season.

STUDYING FUNGI ON LIVING PLANTS

The colonization of living plant leaves by fungi has been investigated using a variety of techniques described in Dix and Webster (1995). With the exception of obligate biotrophs, many of the leaf-colonizing fungi, such as the downy mildews, rusts, black mildews, and powdery mildews, can be grown in culture for identification purposes. Culture methods are similar to those used for the study of endophytic fungi (see "Methods" in Chapter 12), except that the parasites generally are isolated from the margins of lesions. Care must be taken to include living plant tissue in the sample for isolation because isolation from necrotic tissue usually yields a range of secondary colonizers. Direct observational methods for fungi that sporulate readily include incubating the leaf tissue in a moist chamber (e.g., sealed Petri dish with moistened filter paper or sealed plastic bag; Appendix I) for a few days. The incubated leaves then can be examined under a dissecting microscope for the presence of fungal fruiting structures. This is especially useful for *Colletotrichum* species. Alternatively, leaf surface impressions or peels are useful for examining fungal structures on leaf surfaces. The surface of a leaf is coated with nail varnish or similar material that is allowed to dry and then stripped off, taking many of the superficial fungal structures with it. The film is mounted on a microscope slide and examined with a compound light microscope. An easy method for making leaf peels using double-sided adhesive tape is described in Langvad (1980). Water-based mucilage also may be applied as a thin, uniform coat to cover slips, allowed to dry until tacky, then lightly touched to the leaf surface (Fig. 7.14). The cover slips are then mounted in water on a microscope slide for

FIGURE 7.13 A plastic bag can be used to create a moist chamber for the incubation of foliar fungi. (Photo by B. Callan)

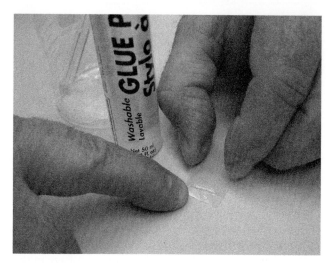

FIGURE 7.14 Coating a cover slip with water-soluble mucilage glue. The coated side is then touched to plant surfaces covered with molds and mildews while the glue is still tacky. The glue holds the spores in place but remains transparent after the cover slip is placed on a drop of water on a microscope slide. (Photo by B. Callan)

examination. An alternative method for examining sooty molds is discussed earlier (see "Sooty Molds").

Pike and colleagues (1977) described an interesting technique that can be used to sample twig microepiphytes. Edges of cylindrats from twigs were scored by turning the twig against two razor blades mounted 3.3 mm apart in a wooden block. The bark surface then was moistened with water, and the microepiphytes were scraped off using a flattened probe, dispersed on a slide in a drop of mounting medium (1:1 aqueous KOH [potassium hydroxide] and glycerin), and covered with a cover slip. All fungal cells found in a 40-μm field were counted and identified. The cover slip was divided into transects running vertically, horizontally, and diagonally. Spore counts from each field were multiplied by a factor directly related to the distance of each field from the center of the cover slip to allow for uneven distribution under the cover slip. Bernstein and colleagues (1973) described methods for estimating foliar coverage of Douglas fir needle microepiphytes using fluorescence microscopy.

ACKNOWLEDGMENTS. We thank Soren Henrich and John Weins for their help in preparing the figures.

8

TERRESTRIAL AND LIGNICOLOUS MACROFUNGI

D. JEAN LODGE, JOSEPH F. AMMIRATI, THOMAS E. O'DELL, GREGORY M. MUELLER, SABINE M. HUHNDORF, CHUN-JUAN WANG, JOGEIR N. STOKLAND, JOHN PAUL SCHMIT, LEIF RYVARDEN, PATRICK R. LEACOCK, MILAGRO MATA, LOENGRIN UMAÑA, QIUXIN (FLORENCE) WU, AND DANIEL L. CZEDERPILTZ

This chapter explores methods of measuring the diversity of macrofungi that are found in soil, in litter, and on woody substrata. We emphasize project planning and techniques for collecting and documenting specimens and data. We also provide some ideas for monitoring populations of macrofungi, but such activities have been attempted so rarely that little information is available.

Macrofungi are distinguished by having fruiting structures (sporocarps) visible to the naked eye. Most macrofungi are Ascomycota or Basidiomycota, but a few are Zygomycota. Most terrestrial macrofungi are saprobes or mycorrhizal symbionts, but some are pathogens of plants, animals, or fungi. Fungi fruiting on woody substrata are usually either saprobes or plant pathogens. The primary feature that distinguishes the ascomycetes covered in this chapter from the ascomycetes treated as microfungi later in the book is their visibility on the substratum surface. All of the ascomycetes referred to here have fruiting structures that are visible with the naked eye, with a hand lens, or under a stereomicroscope. For the most part, a moist chamber is not used to observe or induce fruiting of these species. Much of the information on microfungi, covered in Chapters 7, 11, and 12, such as limitations on their study and their taxonomic and biotic statuses, is equally relevant to the ascomycetes treated in this section.

We do not specifically consider lichenized fungi (see Chapter 9), plant or animal pathogens (see Chapters 7 and 19), Myxomycota, or hypogeous fungi (see Chapter 10).

Although macrofungi have perhaps the longest history of diversity studies of any mycota, they nevertheless are understudied over most of the world. More data are available from Europe than from any other region, and many of those data have been reviewed in a book edited by Winterhoff (1992). Even for Europe, however, knowledge of macrofungal diversity is incomplete. Taxonomic obstacles and the absence of long-term studies prevent us from conclusively answering even basic questions about the number of species at a specific location or whether diversity is greater in one type of forest than in another. Having struggled with our own attempts to inventory macrofungi, we offer this overview in hopes that it will facilitate the work of others.

COLLECTING AND DESCRIBING MACROFUNGI

D. Jean Lodge, Joseph F. Ammirati, Thomas E. O'Dell, and Gregory M. Mueller

Before initiating a survey or a monitoring program of any group of organisms in an area, an investigator should carry out some preliminary background research. Essential materials for the research include maps of the area and descriptions of its climate, geology, and vegetation. Learning to recognize the woody plant species and major plant associations likely to be encountered is particularly important. Knowledge of the anticipated species diversity and distributions of the macrofungi is also very helpful. Such background information increases an investigator's understanding of the habitat requirements of the fungi encountered, aids in delimiting plots that accurately represent the habitats present, and helps in the

design of an effective sampling regimen. Color photographs of the vegetation and landscape also are useful.

In any study, the investigator must choose where, when, and how to collect data. Field studies are particularly sensitive to timing and location of observations. Macrofungi exhibit patterns of diversity that are related largely to substratum and host availability, and their fruiting (and, hence, our opportunity to observe them) is climate driven. The productive seasons for sampling macrofungi and the frequency and duration of sampling are dictated largely by the local climate. Fungi fruit when temperatures are above freezing and moisture is available. Different species, however, exhibit different fruiting phenologies, which vary from year to year and at different elevations and latitudes. Maximum richness of fruiting species occurs only during brief periods and differs among years. Environmental variables and ecological processes that affect the likelihood of recording a species during a study must be considered irrespective of the study's goals.

FACTORS INFLUENCING SPECIES RICHNESS

SEASONALITY AND YEAR-TO-YEAR ANNUAL VARIATION

Within a geographic region fruiting is influenced by elevation and latitude and their effects on temperature and precipitation (Ohenoja 1993). Thus, a particular species may fruit at different seasons across wide geographic distances or along strong elevational gradients. In addition to different seasonal peaks of abundance for individual species, annual variation in presence of sporocarps can be enormous. For example, only 5% to 20% of the ectomycorrhizal species encountered at each of eight sites in Olympic National Park fruited in 2 successive years (Fig. 8.1). Combining the data from all sites, about 30% of the species were detected in both years (O'Dell and Ammirati 1994; O'Dell et al. 1999). Similarly, Winterhoff (cited in Arnolds 1992) studied five sites for 5 years and found that from 8% to 88% of the total reported species from a single site were detected in any given year. Lodge (1996) found that several species in the Entolomataceae fruited every second or third year in a wet subtropical forest in Puerto Rico, whereas a few other species were found only during 1 year of a 13-year survey. Schmit et al. (1999) reported a significant annual difference in species composition during a 3-year study of macrofungal diversity in oak-dominated forests of the Chicago region. In a 21-year study of fungal fruiting

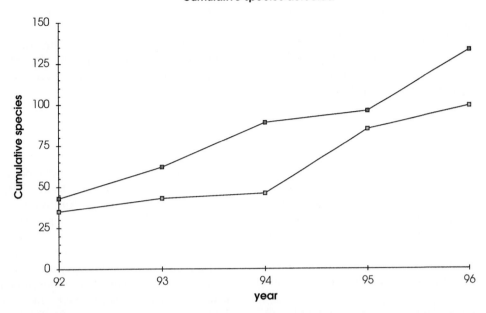

FIGURE 8.1 This cumulative species-accumulation graph shows the increase in the number of recorded species with additional samples over time (Data from T. O'Dell). If the community is relatively homogeneous, the curve will level off (reach an asymptote) more quickly than if several communities are mixed together. The asymptote indicates the species richness for the area. For macrofungi, up to 8–12 years of sampling may be required to approach an asymptote. In general, methods used to estimate the total species richness for an area will not work well if no 'shoulder' is present in the species accumulation curve.

phenology in Switzerland, Straatsma and colleagues (2001) found that some species fruited only during the 1 year that was particularly rich in species and that fruiting abundance and species richness were correlated. Species richness estimators did not stabilize during the 21 years.

The challenge of measuring species diversity of macrofungi increases in years when many species fruit simultaneously at a given site. Unless a large, efficient workforce is available, specimens may decay before they can be adequately documented, resulting in significant loss of data. That situation, together with the fact that some species may fruit only 1 year out of 4 or more, underscores the need for long-term (at least 5-year, preferably 10-year) studies. The time scale of fungal succession is not well known, but the first observations of some species on a site may reflect their recent migration into the community. Annual variation in species recorded because of variable fruiting patterns is difficult to distinguish from succession.

In temperate regions with summer drought, spring and fall are the main fruiting seasons. The "fall" season is progressively later at lower latitudes, in some regions occurring after the winter solstice. In temperate regions with summer rain and at high elevations and latitudes,

summer may be the most important season for fruiting. Unusual weather events can trigger unusual patterns of fruiting, however; in regions with dry summers, for example, fruiting of species rarely observed in the region may follow a summer shower. Although the fruiting of macrofungi most often is limited by lack of precipitation, excess moisture also can prevent fruiting in some species.

Temperature also has a major impact on macrofungi fruiting, an effect that may not be limited to the fruiting season. Ohenoja (1993) documented the effects of year-round temperatures on fall fruiting of macrofungi. She found that temperature interacted with the habitat and ecological guild of the fungus in stimulating or retarding fruiting. For example, warm summers increased the production of mycorrhizal sporocarps in spruce forests but had no effect on sporocarp production in pine forests.

If a measure of species richness is the goal of a study, then baseline data for a site may involve intense collecting over a period of several days when sporocarp production is high. Ideally, however, one samples fleshy macrofungi in an area every 1–2 weeks throughout the fruiting season to maximize the number of species observed. When such a schedule is not possible, it is useful to do a pilot study (see discussion of complementarity tests under "Determining Adequate Sampling," later in this chapter) before deciding when and how often to sample. Studies of annual variation and seasonality of macrofungal fruiting have been limited largely to Europe. Few studies have examined multiple taxa or lasted long enough to have predictive value. Some data on seasonality, however, may be gleaned from herbarium records.

VEGETATION

The type of vegetation in an area affects the species richness and composition of macrofungi at that site. Grasslands, deserts, forests, tundra, and other habitats all have characteristic species. Plant species composition also influences the number and species of macrofungi present because plants constitute the habitat and energy source for most fungi, and all fungi show some degree of host or substratum specificity.

Vegetation zones or plant associations are useful criteria to use when dividing a landscape for sampling. Many fungi occur only in association with particular families or genera of plants. Villenueve and colleagues (1989), Bills and associates (1986), and Nantel and Neuman (1992) found the distribution of ectomycorrhizal fungus species to be correlated with forest type (forests dominated by deciduous trees versus conifers). O'Dell and others

(1999) found precipitation to be better than vegetation type as a predictor of species richness and community structure of ectomycorrhizal fungi at local scales. The large variation found among sites in a single forest zone in their study points to the need for sampling multiple sites with similar host tree species (Figs. 8.2 to 8.5).

GEOGRAPHY

Mycologists with broad geographic experience were surveyed to determine what factors they believed most strongly affect diversity in the groups that they study (Lodge et al. 1995). For the fungi treated in this chapter, diversity of habitats rather than geography was believed to have the strongest influence on fungal species richness. In much of Europe, tree diversity is low as a result of local extinctions of species caused by Pleistocene glaciation, and it is probably not coincidental that Europe was ranked lowest in fungal diversity when compared with other regions, although it has been collected more intensively (Dennis 1986). Some ectomycorrhizal fungi (e.g., Cortinariaceae) and discomycetes were thought to be most diverse at middle and high latitudes, whereas Agaricales in general (especially saprobic Tricholomataceae) were considered most diverse at low latitudes.

FIGURE 8.2 Ectomycorrhizal fungi include those with fleshy indented caps and thick wrinkles instead of gills, such as *Gomphus* (shown here) and species of *Cantharellus* (Chantarelles). (Photo by T. O'Dell)

FIGURE 8.3 Ectomycorrhizal fungi include all species of *Lactarius* (milkcaps), including *L. indigo* shown here, and species of the related genus, *Russula*. Latex presence, color and color changes, taste, odor, and spore-print color, and color reactions of pileus, lamellae, and stipe to FeSO$_4$ should be recorded to identify species in this family, Russulaceae. (Photo by G. Mueller)

FIGURE 8.4 Ectomycorrhizal fungi include some of the typical gilled agaric fungi, such as this *Laccaria laccata*. (Photo by G. Mueller)

FIGURE 8.5   Most species of boletes, such as this *Suillus spragueii,* form ectomycorrhizal associations with tree roots. Boletes can be recognized by their texture, tubes that peel easily from the cap flesh, and their typical habit of fruiting on the ground. (Photo by G. Mueller)

SUCCESSION

Succession of macrofungi must be considered from several perspectives. First, there are successions of sporocarp production on particular substrata, although all species may be present in the substrata from the beginning. Herbivore dung, for example, has characteristic fungi that fruit consecutively over time (Chapter 21). Succession involving changes in community composition often are related to changes in the quality of the substratum. Hedger (1985) found, for example, that some species of *Lepiota* only grow well on leaf litter that previously has been decomposed by other fungi, such as certain *Marasmius* species. Similarly, trunks of large fallen trees host a cadre of fungi that fruit early in log decomposition and others that fruit only later (Lange 1992; Renvall 1995; Heilmann-Clausen 2001). One approach for determining changes in fungal community composition over long periods is to study a chronosequence. To document changes in the composition of the macrofungal community on decomposing beech logs, Heilmann-Clausen (2001) studied a 30-year record of aerial photographs of a Danish forest to determine when each tree more than 60 cm dbh had died. He used that information to divide the 130 logs into seven age cohorts and determined the macrofungi present on the logs in each one.

Secondly, successional changes occur in the vegetation at a site. Such changes may have a direct impact on fungi through the establishment of new host taxa and changes in the amount and quality of available organic matter.

Fungi are also likely to be affected by changes in understory microclimate. Lodge and Cantrell (1995b) found that the opening of the forest canopy in the Caribbean National Forest in Puerto Rico (El Verde field station) by hurricane Hugo caused the extirpation of several mycelia of *Collybia johnstonii,* especially on ridges where the canopy has been slowest to close. That leaf-decomposing fungus has superficial mycelia and hyphal strands that are very sensitive to drying.

Apparently, ectomycorrhizal fungi on host root systems also exhibit succession. Some species fruit in very young stands of trees, some are restricted to older stands, and others occur across many age classes (Mason et al. 1983; Dighton et al. 1986; Luoma et al. 1991; Deacon and Fleming 1992; Last et al. 1992; O'Dell et al. 1992a, 1992b). Ectomycorrhizal succession is poorly documented, having been studied largely in plantations of trees outside of their native ranges. Successional phenomena illustrate the need for including diverse microhabitats and a range of stand ages among the sampling sites when studying the macrofungal diversity of a landscape.

COLLECTION, CULTURE, AND PRESERVATION

After sampling design, the core of any fungal diversity study is the collection and documentation of specimens. That includes collecting sporocarps at the field site, labeling them, taking photographs, taking tissue cultures, setting up spore prints, writing descriptions, and preserving the specimens for later identification. Adequate time must be budgeted for all of those activities. One approach is to spend part of the day in the field and the remainder at the laboratory or field station working up specimens. Alternatively, if field sites are remote and collecting is good, 2 or 3 full days of collecting can be followed by up to a week of laboratory work. In that case, the specimens must be prioritized in order of preparation, with the most fragile and important specimens processed first and the remaining specimens stored in as cool a place as possible.

Identification of macrofungi relies heavily on the attributes of fresh specimens, and it is generally better to prioritize documenting these attributes over identifying specimens to species. If time permits, identifying a specimen is beneficial but not at the expense of documenting its morphological characters or letting other specimens spoil.

COLLECTING

Data management starts before collecting begins because specimens should be labeled as they are collected. Investigators must decide in advance which ecological or habitat data to record, prepare field labels, choose a cataloging system, and so on.

Not every specimen encountered needs to be collected. Deciding whether to collect a particular mushroom is not a trivial problem and depends on the condition and quality of the specimen as well as the goals of the study. General collecting in an area for inventory purposes or for a generic monograph allows one to be selective among specimens, although representatives of all species encounters in a plot-based study should be collected. Larger amounts of material are required for biomass or chemical studies.

For many taxa it is important to have data on all stages of sporocarp development. Thus, collectors should obtain sporocarps exhibiting a range of developmental stages for each taxon. Mature sporocarps for some groups, such as some Pezizales, may appear decrepit. Often a careful search of an area of several meters radius around the initial collection site will yield a full range of developmental stages. In general, however, it is usually best to leave behind sporocarps that are too immature or too old to provide useful data (Figs. 8.6 to 8.8).

Making a collection requires care and patience. Specimens are removed from the substratum by excavating around the base of the stipe to reveal any volva, rooting base, bulb, or attachment to a sclerotium or buried substrata. Such substrata include wood, cones or fruits, other fungi, and animals. If the specimen is on wood or litter, including some of the substratum with the collection will facilitate its identification. Unidentified plants near the specimen also can be included with the

FIGURE 8.6 It is best to collect all of the developmental stages of macrofungi to identify them and to document morphological variation; that is true for the species of *Crucibulum* shown here. (Photo by G. Mueller)

FIGURE 8.7 Fruit bodies in all stages of development are required for confident identification of almost all species of puffballs and earthballs, including the *Geastrum* species shown here. If only immature specimens are collected, it is sometimes possible to observe the color changes through maturation of the spores in the interior (the gleba) for several days after they have been collected. (Photo by D. J. Lodge)

FIGURE 8.8 Some agaric fungi, such as the *Coprinellus* (*Coprinus*) *mexicanus* shown here, change rapidly as they mature, so all stages should be collected. Most species of *Coprinus* (in the broad sense) digest themselves as the spores mature thus destroying the structures on the cap surface and the gill edges that are critical for identification. (Photo by D. J. Lodge)

collection for later identification. After the site is examined carefully, it can be described as wet, boggy, dry, or whatever is appropriate. Collectors also should note if digging into the substratum was required to remove the sporocarp, if rhizomorphs were associated with it, or if it was attached to a moss. Those observations should be recorded on field labels, although if a specimen is collected properly, the substratum and/or rhizomorphs still should be apparent when the specimen is examined during processing (Fig. 8.9).

Sporocarp color is often important in the identification of macrofungi. Collectors should note color changes caused by bruising or handling of specimens. In many agaric genera, loss of moisture from the pileus occurs rapidly after collecting (such specimens are termed hygrophanous), so it is helpful to note the general color in the field. For example, color changes in basidiomes of some *Cortinarius, Ramaria,* and *Mycena* species can be rapid and, if undetected, can lead to incorrect descriptions and identifications. Small or fragile basidiomes of species, such as some *Lepiota* and *Psathyrella,* may have delicate veil material that can be damaged or lost on handling. Such features should be noted in the field.

EQUIPMENT

Every mycologist has his or her preferred collecting paraphernalia, and to a degree preferences depend on the taxa being collected. At least four items are required for collecting macrofungi: (1) a tool for cutting and digging, (2) a container or wrapping material for each specimen, (3) a larger container for transporting specimens in the field and back to the lab, and (4) a label for each specimen.

A thick-bladed, moderately sharp knife can be used to cut woody substrata or dig in soil. Some collectors carry both a knife and a trowel for collecting sporocarps from soil. Different types of fungi occurring on wood require different types of collecting equipment. An ax or hatchet often is needed to extract wood to a depth sufficient to enable identification of the host if it is unknown. However, a mallet and wood chisel, a heavy sheath knife,

FIGURE 8.9 When collecting species of *Cordyceps,* which are parasites of insects or other fungi, it is important to collect and identify the host as well. (Photo of *Cordyceps* cf. *amazonica* parasiting an insect in Belize by D. J. Lodge)

FIGURE 8.10 A sheath knife or a folding knife with a locking blade is one of the most useful pieces of collecting equipment for macrofungi. Fungi fruiting on the ground, such as the species of *Macrocybe* shown here, should be pried up from the base to ensure than none of the basal parts are lost. (Photo in Costa Rica by G. Mueller)

or a folding knife with a locking blade are usually sufficient for removing the fungus. A pair of pruning shears and a folding pruning saw are also helpful for cutting smaller diameter twigs and branches to a uniform length. Care must be used to avoid undue damage to the plant if collecting from a living tree (Figs. 8.10 and 8.11).

The type of container or wrapping material used for individual specimens depends in part on the type of fungi being collected. Aluminum foil, waxed paper bags, or sheets of waxed paper rolled into packets and twisted in opposite directions at the ends often are used for fleshy fungi. Fragile specimens are best wrapped in aluminum foil or placed in small boxes. Foil keeps the specimens from dehydrating, affords good protection, remains

intact when wet, and easily can be reused. Small plastic boxes, vials, fishing tackle or toolboxes, or other containers also work well (Fig 8.12). Boxes prevent fungi from being crushed or bruised, keep them moist, afford protection from rain, and allow for easy viewing and comparison of specimens. Paper bags of various sizes work well for "woody" basidiomycetes (e.g., bracket fungi) and small, durable ascomycetes.

Appropriately wrapped specimens are placed in a basket, bucket, pack basket, backpack, or bag (plastic or otherwise) chosen to protect the specimens from being crushed and for ease of carrying in the field (Fig. 8.13). Basidiomes of *Amanita* and certain other fungi will reorient their pilei after collection (the pileus is positively

FIGURE 8.11 Macrofungi that fruit on wood, such as this tooth fungus in the genus *Hericium*, can present a challenge for collectors. If the wood is soft, a sheath or buck knife or a knife with a locking blade can be used to remove the specimen with part of the substratum. Some collectors prefer to use a hatchet or a chisel and rubber mallet to remove specimens from wood. (Photo by G. Mueller)

FIGURE 8.12 Plastic boxes are useful for holding fungi, especially small and medium-size macromycetes. Collecting boxes of the type used for fishing tackle or hardware supplies allow for rapid viewing of the specimens and also protect them from being crushed. Placing leaves or pieces of moss or lichens in with the specimens will prevent them from rolling and becoming bruised and will also help small specimens to retain moisture. (Photo by D. J. Lodge)

FIGURE 8.13 Many collectors, especially those working with large fleshy agarics and boletes, wrap their specimens in waxed paper or aluminum foil and carry them in a basket. Both waxed paper and foil provide physical support and help prevent moisture loss without suffocating the fruit bodies. (Photo by D. J. Lodge)

geotropic) and should be carried standing straight up to prevent the twisting and distortion of their stipes.

Fleshy fungi dehydrate and decompose very quickly. Specimens transported in a cooler often hold up better and should remain in the cooler or a refrigerator until they are described and cultured. Small coolers with an ice pack that can be frozen and attached to the lid work well for small specimens, especially in tropical and

subtropical climates (Fig. 8.14). If a collector working in a hot or sunny site places the ice pack below rather than above the specimens, a strong temperature gradient develops inside the cooler. That gradient causes the specimens in the upper part of the container to desiccate; condensation that forms in the lower part accelerates decomposition of specimens there. Securing very small specimens inside their containers with a fresh, green leaf

FIGURE 8.14 Coolers are a good alternative to baskets for collecting small specimens in hot tropical climates. Ideal models come with an ice pack that can be attached to the lid of the cooler to prevent drying of specimens at the top and moisture condensation at the bottom. (Photo of D. J. Lodge by S. Schmieding)

or moss can reduce desiccation and physical damage. Larger specimens are less likely to desiccate, and they can be placed on the bottom of a large cooler lined with frozen ice packs. Separating the ice packs and the specimens with a layer of paper bags or newspaper keeps specimens from freezing.

DOCUMENTATION

Fleshy fungi are particularly difficult to work with because sporocarps dry or decompose readily; the characteristics needed for identification are often ephemeral; and many macromorphological features such as size, shape, and color are lost with preservation. Specimen

documentation includes obtaining cultures, data on macrochemical tests, photographs, spore prints, and written descriptions. Not all taxa or specimens require equal documentation. When fruiting is at its peak, one must maximize field time. Specimens that do require documentation—that is, those that are fragile, decay rapidly, or otherwise require immediate attention—must be processed quickly. It is, therefore, necessary to organize specimens before working through them. Specimens are put into priority order by taxon; within each taxon they are processed in order according to condition of the specimen (best to poorest), which provides the most accurate and efficient documentation of diversity.

Collectors of biological materials have long recognized the need to catalogue their specimens. Typically, that includes a label for each specimen and a catalogue that contains the data for all collections. In addition, some biologists keep a field journal wherein they record their travel routes, daily activities, and observations. A field journal can be valuable in relocating a collection site or verifying a location or other information when a question arises later on. General methods and justification for keeping a field journal are described by Herman (1986). Whereas keeping a field journal is an elective activity, consistent labeling and cataloging of collections are mandatory.

Field labels are the first step in tracking collection data. The data should be recorded in pencil or permanent black ink (e.g., India ink) on waterproof paper at the time each collection is made. Minimum data for each label are collection number (if only an interim one), collector's name, date, and location. One also can include a tentative identification; a plot number; and notes on the associated vegetation, substratum, microhabitat, and ephemeral characters of the sporocarp.

Preprinting labels ensures that important data are not omitted, saves time while collecting, and facilitates efficient recording of ecological and host data. Some collectors use prenumbered labels; this has advantages and disadvantages. If a specimen is discarded, the collection number goes with it, so one must keep track of "discarded" numbers. An alternative to prenumbering the field labels is to use labels that are numbered sequentially beginning at one on each date and later assigning "final" collection numbers to specimens. If that method is used, a separate entry of the daily collection numbers should be included in the catalogue to facilitate tracking of specimens when they are sorted or identified.

A catalogue is a list of collections in numerical order that includes the date, collector, and location. It also can contain site and weather data, photographic data, and other relevant information. Every collector should maintain a catalogue with a unique number for each specimen. Various numbering systems can be used to construct a collection number; the critical factor is that one be consistent in the system used. Some mycologists' collection numbering systems incorporate the collector's last name or initials and a unique number (e.g., O'Dell 2401, JLL 1014, or JLL 1995-17); other systems include an institutional acronym or an abbreviation for the country of origin followed by a unique number. Because collection numbers constitute a historical record that others may use someday, the consistent use of a chosen numbering system is imperative.

PRESERVATION

Maintaining voucher specimens is a critical aspect of measuring or monitoring diversity (Ammirati 1979).

Most agarics and other fleshy fungi require a heat source for drying. For most polypores, some Corticiaceae, and some ascomycetes, air-drying is preferable because the fungus is not killed but merely goes dormant. The specimens subsequently can be used for isolating cultures. To facilitate drying, the specimens should include minimal adhering wood, especially if the wood is particularly wet. Submerging small specimens in repeated changes of silica gel until the sporocarps are dry is also an option, but specimens should not come into direct contact with the silica crystals because adhering crystals can cause problems when microscopic mounts are made.

Dryers with wire screen shelves suspended on a rack over a heat source of 38–55°C are most effective. The source of the heat is not critical; incandescent light bulbs, hot plates, and space heaters all can be used, but a fan to increase air movement is helpful. Commercially made food dehydrators are popular because they are self-contained, reasonably compact, complete units, with no assembly required. An alcohol burner with something to disperse the heat (e.g., aluminum foil or a metal steamer basket) placed between the flame and the specimens is effective and will burn for 8–12 hours without refilling. Small kerosene stoves or white gas catalytic heaters also can be used in the absence of electricity. The advantage of the latter is the lack of flame. Specimens should not be placed too close to any heat source; they can char or catch fire easily.

Large agarics and boletes should be split in half or quartered from top to bottom before being placed in a dryer. That practice speeds drying, prevents the context (the interior of the sporocarp) from decaying, and stops the feeding activities of insect larvae. If drying racks are stacked over the heat source, the largest fleshy fungi should be placed on the lower shelves, and the smaller fungi should be placed on the upper shelves. When many collections of small specimens are placed together on the same drying tray the risk of losing part of some collections or incorrectly matching the specimens to the labels is high. Very small specimens can be placed inside paper packets or small, thin, cloth bags before drying.

Fixing Material for DNA Analysis

Subsamples of material collected as part of an inventory can serve as sources of DNA for laboratory-based studies and should be preserved accordingly. Pieces of sporocarp are excised carefully and submerged in a fixative solution in cryovials or microcentrifuge tubes. Several fixative solutions prevent degradation of sporocarp DNA during long-term storage. The two most commonly used solutions are 2X CTAB buffer and a supersaturated solution of DMSO (Wu et al. 2000; Appendix II).

Specimen Storage

Temporary storage of preserved macrofungi can be difficult, especially when the ambient relative humidity is high, because dried specimens are hygroscopic. Self-closing (e.g., zipper), airtight, transparent bags provide excellent temporary storage for dried macrofungi. The bags come in different sizes and accommodate sporocarps of various sizes. Wax paper and plastic sandwich bags are less expensive alternatives. Bags protect the specimens from moisture and insects while allowing for visual inspection or sorting of specimens; spore deposits also may be included. Specimens to be shipped in the mail first should be wrapped in blank newsprint and then put in a plastic bag. The paper will afford the specimens more protection. A label always is included with each specimen.

In tropical regions special care must be taken to prevent rehydration of preserved specimens. Filling and closing the plastic bags over a drier will prevent moisture from entering the bags. Cracker and cookie or biscuit tins are airtight, crush-proof containers suitable for transporting specimens in a completely dry state. When air is extremely humid, tins should be packed and sealed over a heat source, or the packed tins should be exposed to heat for 30 minutes before the lids are sealed with duct tape. Cloth or paper packets of silica gel, preferably mixed with indicator dye, also can be dried and packed with specimens.

For long-term storage, specimens should be maintained in herbarium boxes or packets according to the methods outlined in Chapter 2 (see "Packeting, Boxing, and Mounting Specimens").

CULTURING MACROFUNGI

Cultures are sometimes helpful in distinguishing similar taxa and identifying them to species. Cultures are especially important for some groups of ascomycetes (e.g., Xylariaceae; Hypocreaceae; other pyrenomycetes, such as *Chaetosphaeria*) in which the anamorph (asexual stage) is required for accurate species identification. Because obtaining, maintaining, and studying fungal cultures is time consuming and requires specialized materials and training, such preparations will be beyond the scope of many biodiversity studies. In those cases, only generic identifications will be possible for some ascomycetes.

Terrestrial macrofungi have been cultivated from tissue or from germinated basidiospores or ascospores with varying success. Saprobic fungi generally are cultured more easily than ectomycorrhizal species, but exceptions exist. For example, certain *Lepiota* have been grown on leaf litter that has been decomposed by *Marasmius* and then autoclaved (Hedger 1985). Tissue cultures of ectomycorrhizal fungi such as boletes, *Laccaria*, *Tricholoma*, *Hebeloma*, and *Amanita*, can be obtained using the proper procedure and media; other genera, such as *Lactarius*, *Russula*, *Entoloma*, *Cortinarius*, and *Hygrophorus* are difficult or impossible to grow. For general information on the cultivation of macrofungi, see Watling (1980) or Stamets and Chilton (1983).

Many wood-inhabiting fungi grow well on malt extract agar, but potato dextrose agar and oatmeal agar also are used in some laboratories (Appendix II). Special agar containing organic matter from a particular substratum can be used to grow some fungi difficult to culture on standard media; some wood-inhabiting basidiomycetes, for example, will fruit on wood-containing media with sawdust used as a base (Etter 1929). Special media also have been tried with varying success with some hard-to-germinate wood-inhabiting ascomycetes (S. M. Huhndorf, personal communication) and basidiomycetes such as species of *Laetiporous*, H. H. Burdsall, personal communication) and *Pluteus* (O. K. Miller, personal communication).

Culturing from Basidiospores and Ascospores

The most common way to obtain macrofungi in culture is to germinate their spores. An easy method for obtaining spore cultures is to let a fungus drop or shoot spores directly onto an agar surface. Ascomata or portions of the hymenial surface of basidiomycetes are suspended near the edge on the lid of a Petri plate of agar (petroleum jelly, small chips of agar, or water drops can be used to attach the fungus to the lid). The wrapped plate can be tilted at a steep angle with the hymenium of the fungus located at the upper end (gills of agarics or pores of polyporaceae are aligned vertically in the direction the plate is to be tilted). Such positioning produces a cascade with spore densities high at the upper end and decreasing toward the lower end and along the side. Both polyspore and single-spore isolates can be obtained from the same spore print. The source specimen should be removed as soon as spores are detected. Alternatively, the lid of a horizontal Petri dish can be rotated periodically (e.g., every 30 seconds to 10 minutes, depending on the specimen) to obtain distantly spaced spores for single-spore isolates. Individual spores can be identified and then removed using a fine sterilized needle while under a dissecting microscope. For single-spore isolations, only germinated spores that are well separated from other germinated and ungerminated spores should be transferred to individual culture plates. The microscope should be equipped with an adjustable mirror below the stage for the light source. The mirror should be adjusted so that the spores appear bright against a dark background; spore deposits and germination can be observed by turning a plate upside down and focusing through the agar. Usually 8–12 hours (sometimes much less, especially with dark spores) is sufficient time for a spore print to be observable, especially when using a dissecting microscope. For tropical fungi, the hymenium or ascomata should be removed within 12 to 24 hours. Additional incubation will not result in additional spore production but likely will lead to contamination of the plates with spores from opportunistic fungi such as *Fusarium* and *Trichoderma*. In other cases (e.g., temperate and boreal fungi) several days to a week may be required to obtain a spore drop culture. Changes of barometric pressure that are associated with changes in elevation of more than several hundred meters are probably responsible for the failure of some tropical basidiomycetes to drop spores (Dennis 1970; D. J. Lodge, personal observation; R. Singer, personal communication); it may be necessary, therefore, to set up spore drops in the field if the investigator's laboratory is located at a very different elevation. In addition, tropical basidiomycetes that were placed in a refrigerator for more than an hour before culturing failed to drop spores, and their cultures often died when held at 5–10°C (D. J. Lodge, personal observation). The elevational effect has not been observed among temperate species (H. H.

Burdsall, personal observation), nor has the cold-temperature effect been observed among temperate basidiomycetes (H. H. Burdsall, personal observation) or tropical ascomycetes (Fig. 8.15; D. J. Lodge, personal observation).

Spore prints for agarics, most aphyllophorales, and some ascomycetes can be obtained in the field on sterile 60-mm-diameter Petri plates from which spores are transferred to tubes or agar. For some fungi, spore prints can be obtained on paper or aluminum foil, air-dried, and used later. Spore prints of some temperate fungi (e.g., morels) can be taken on aluminum foil and kept frozen for a year or more without affecting viability (T. Volk, personal communication). For others, such as *Armillaria*, air-dried spores more than several days old are not usually viable, and air-dried spores of *Marasmius* remain viable for only a couple of hours (D. Desjardin, personal observation). Likewise, the spores of many tropical rain forest fungi are sensitive to desiccation and lose their viability after 1–2 hours. When the period of spore viability is in doubt, it is best to drop the spores directly onto an agar surface.

Spore prints of basidiomycetes obtained in the field can be used to obtain single-spore isolates by suspension plating or streaking. Streaking is less complicated and

FIGURE 8.15 Spores can be dropped directly onto agar plates under field conditions. Many fungi have fragile spores that will not germinate after being dried. (Photo of Dr. Karen Nakasone by D. J. Lodge)

involves transferring a drop of spores suspended in sterile water to an agar plate and then streaking the suspension over the agar surface using a piece of agar or a flame-sterilized loop or rod.

To obtain a spore print from air-dried basidiomycetes, the basidiome is refrigerated for 2–3 days in an agar-filled Petri dish (usually this procedure works only with certain temperate-zone aphyllophorales). When returned to room temperature, the basidiomes will develop new basidia and produce a spore print. Resupinate basidiomycetes are more likely to produce spores if the hymenium is taken from wood rather than from bark. Many basidiomycetes, especially fleshy fungi like mushrooms, fail to drop spores if they have been refrigerated. Tropical fungi usually drop spores within the first 2 hours following setup; suspension of an hymenium over a plate for more than 12 hours leads to contaminated plates rather than spore maturation (D. J. Lodge, personal observation). Immature specimens of tropical fungi are better placed in a moist chamber with the original substratum intact or tagged and placed back in an easily accessible part of the forest until the basidiomes or ascomata mature.

Mature ascomata (moist and turgid, either fresh from the field or air-dried and rehydrated) are used for obtaining ascospores. Mature air-dried ascomata of Xylariaceae usually can be rehydrated by soaking them in water for 30 minutes to 2 hours prior to attempting the spore shoot or drop. For Xylariaceae and other ascomycetes that forcibly discharge their spores, contamination by hyphomycetes (and other fungi that do not forcibly discharge their spores) can be avoided by inverting the Petri dish, placing the ascomata on the lid below the agar surface, and allowing the ascomata to shoot the spores upward (J. D. Rogers, personal communication). It may be necessary to elevate the perithecia with a small piece of clay, small Petri dish lid, or something similar so that the discharged spores can reach the agar surface. The ostioles must not be clogged with spore deposits, which can be picked off with a sterile needle and streaked onto the agar surface after adding a drop of water from the lid. If ascomata fail to discharge spores, the ascal contents can be extracted with a sterile scalpel, flamed needle, dental tool, or insect pin and streaked onto the agar surface. Single spores usually can be obtained, but spore germination within the ascus is common in many species. We do not recommend attempting to culture other tissues, such as the ascomal wall or internal stromatal cells, of small ascomycetes, but that procedure may be used as a last resort for some larger Xylariaceae. However, cultures of *Hypoxylon* have been obtained from the internal stromatal tissues of *Xylaria* species (D. J. Lodge, personal observation), which illustrates that it is possible to isolate a contaminate rather than the intended

fungus. Consequently, one cannot assume that a culture obtained from stromatal tissue necessarily belongs to the species that produced the ascocarp from which it was isolated.

Isolating Tissue from Basidiomes or Vegetative Structures

As mentioned earlier, we do not recommend isolating hyphae or spores from stromatal or ascomal wall tissues of ascomycetes. In contrast, cultures of basidiomycetes can be obtained by isolating sterile hyphae from the tissue of the basidiome. For mushrooms, the basidiome is split down the middle, and a small piece of tissue is removed from the upper stipe or pileus with a sterile scalpel or forceps. Tearing the basidiome apart instead of cutting it reduces the likelihood of contamination and reduces tissue damage. Reflexed or ungulate species that have tough basidiomes also can be isolated from the tissue or context. In that case, we recommend cutting part way through the basidiome and tearing the part from which the tissue will be removed. The extracted tissue then is placed directly on small test tube slants for growth. Media recipes and other procedural details can be found in Molina and Palmer (1982), Watling (1980), and Stamets and Chilton (1983) (also see Appendix II).

Some thin, resupinate basidiomycetes that are difficult to culture from basidiospores or tissue can be isolated by means of "hyphal bridges" (D. J. Lodge, personal observation). The specimen on its substratum is placed inside a moist chamber, and aerial hyphae or mycelial strands that attempt to bridge air gaps by growing toward the paper toweling or lid can be transferred to agar plates. Vegetative structures of species that produce hyphal strands, rhizomorphs, or cordons can be fastened to the lid of an agar plate. Clean hyphae will grow from the broken ends and make contact with the agar in a day or two. Hyphal tips then may be transferred to fresh agar plates. Some species are difficult to culture from hyphal tips but will grow from larger masses of hyphae, in which case the agar plate with the rhizomorph should be inverted to reduce the risk of contamination once contact has been made. In such cases, large masses of hyphae are transferred on agar chips, some of which are placed on new agar with the hyphal side up (Fig. 8.16).

Direct Isolation from the Host or Substratum

Some fungi present in wood produce fruiting bodies infrequently, posing problems for diversity surveys. Fungi can be isolated from reasonably solid wood by placing chips or cores taken from the source on agar media. Recording the location from which a sample was

FIGURE 8.16 Some rhizomorph-forming corticiod species, such as *Phanaerochete flava* shown here, can be cultured by inverting an agar plate and placing the fruit body on the lid. Rhizomorphs growing up to the agar surface are generally free of contaminants. (Photo by D. J. Lodge)

taken relative to the location of disease symptoms is especially important for many pathogens that can be isolated only from the disease front. Samples can be obtained using a flame-sterilized increment borer. The surface of the substratum to be sampled is cleaned thoroughly with a 70%-ethanol or a 10%-sodium hypochlorite wash, or a clean surface is exposed with a flame-sterilized knife. The increment borer is screwed into the wood to form the core. When the core is removed, it is placed in a sterile container and returned to the laboratory. Plastic drinking straws frequently are used to hold the cores. It is not possible, however, to place a sample from a friable woody substratum into the straw without destroying it. In such cases, the core can be placed in a piece of aluminum foil that then is rolled around it. Cores can be taken from sections of a substratum appropriate to the information desired. For example, if determining the internal distribution of the fungi in a log is important, then the position of the sections in the core must be recorded. Chips are handled in a similar fashion. Before plating, chips or cores are dipped in 95% ethanol and flamed for surface sterilization. Alternatively, the surface can be sterilized by washing for several minutes under a strong stream of distilled water. Surfaces of large pieces of wood or roots can be sterilized by washing with and then submerging in a 0.5% solution of sodium hypochlorite for 10–15 minutes. Chips then are cut using a flamed knife, and fungi are isolated as they grow from the core or chip.

Another method used to extract internal hyphae is to drill into a log and sample a few of the resulting shavings at given depths. The shavings are more difficult to handle than the cores, but they can be collected directly into culture tubes or plates and allowed to grow. The intermediate steps required by the coring method then are avoided.

Cultures of polypores and some other fungi can be identified using the keys and descriptions provided in Nobles (1965) and Stalpers (1978). Appropriate monographs must be used for other taxa.

DESCRIBING MACROMORPHOLOGICAL AND MICROMORPHOLOGICAL CHARACTERS

MACROMORPHOLOGICAL FEATURES

Descriptions of macromorphological characters (e.g., color, shape, size, odor) of sporocarps are among the most critical data for the identification of Agaricales and other macrofungi. In many species of fleshy macrofungi, those characters are lost when the sporocarps are dried, and so they must be documented carefully before drying. Conversely, sporocarps of many macrofungi that grow on woody substrata are nonputrescent, and they usually retain their macromorphological features when preserved. Thus, data requirements vary among taxa, and the detail required will depend on the scope of the

FIGURE 8.17 Describing agarics and boletes requires good light so that all of the characters can be observed and the colors can be matched against a standardized color guide. Dr. Orson K. Miller is shown matching colors against a Methuen Color Guide, Hope Miller is describing specimens, and Dr. Clark Ovrebo, in the back, is taking photos against a neutral gray photo card in Belize. (Photo by B. Ortiz Santana)

project and the resources available. If two or more workers are available for descriptive work, it is usually most efficient to assign each worker responsibility for some groups or genera of fungi each day. When several workers are on a project for an extended period, assigning taxa leads to more consistent descriptions. If a lot of culturing is needed, however, one individual might do that instead of descriptive work (Fig. 8.17).

The procedure for describing macrofungi depends on the type of fungus under consideration and also on the investigator. Forms, worksheets, or computer formats should be organized so that a description is recorded in logical sequence (Figs. 8.18, 8.19, and 8.20). Most investigators begin with the pileus and move downward through the lamellae or tubes, stipe, and finally to any veils that are present. Recording information about the context is also important (e.g., color, color changes, consistency, presence or absence of latex), as is the inclusion of data on all available developmental stages. Procedures for studying macrofungi that lack lamellae and/or a stipe (e.g., coral fungi, puffballs) vary slightly, and monographs should be consulted for details. The goal in all cases is to describe the sporocarp well enough for someone else to picture it (Figs. 8.21 to 8.23).

Good lighting is critical for documenting color, vesti-ture, and other features. The type of light used is very important and should be noted (e.g., natural, incandescent, fluorescent). Color notes can be taken in natural daylight, but an appropriate artificial light system is more reliable and consistent. A light system with two Vita-Lite fluorescent tubes works extremely well because it provides illumination that is close to natural light. Portable daylight-corrected light sources, such as Ott-Light True Color Lamps (Appendix IV), also provide high-quality light. Those lights and fixtures are smaller and more easily transported than large Vita-Lite fluorescent tubes. Normal fluorescent light tubes and incandescent light sources, however, distort colors, especially those in the red to purple range.

Experienced mycologists often record descriptions on cards, paper, or directly into a computer file. However, preprinted or computerized forms ensure inclusion of all diagnostic features, and we recommend their use, especially if a team is describing the specimens. Equipment required for describing specimens include a pencil or pen with permanent ink; a pocket knife with a sharp, thin blade and single-edge razor blades for sectioning sporocarps; a hand lens or dissecting microscope for small

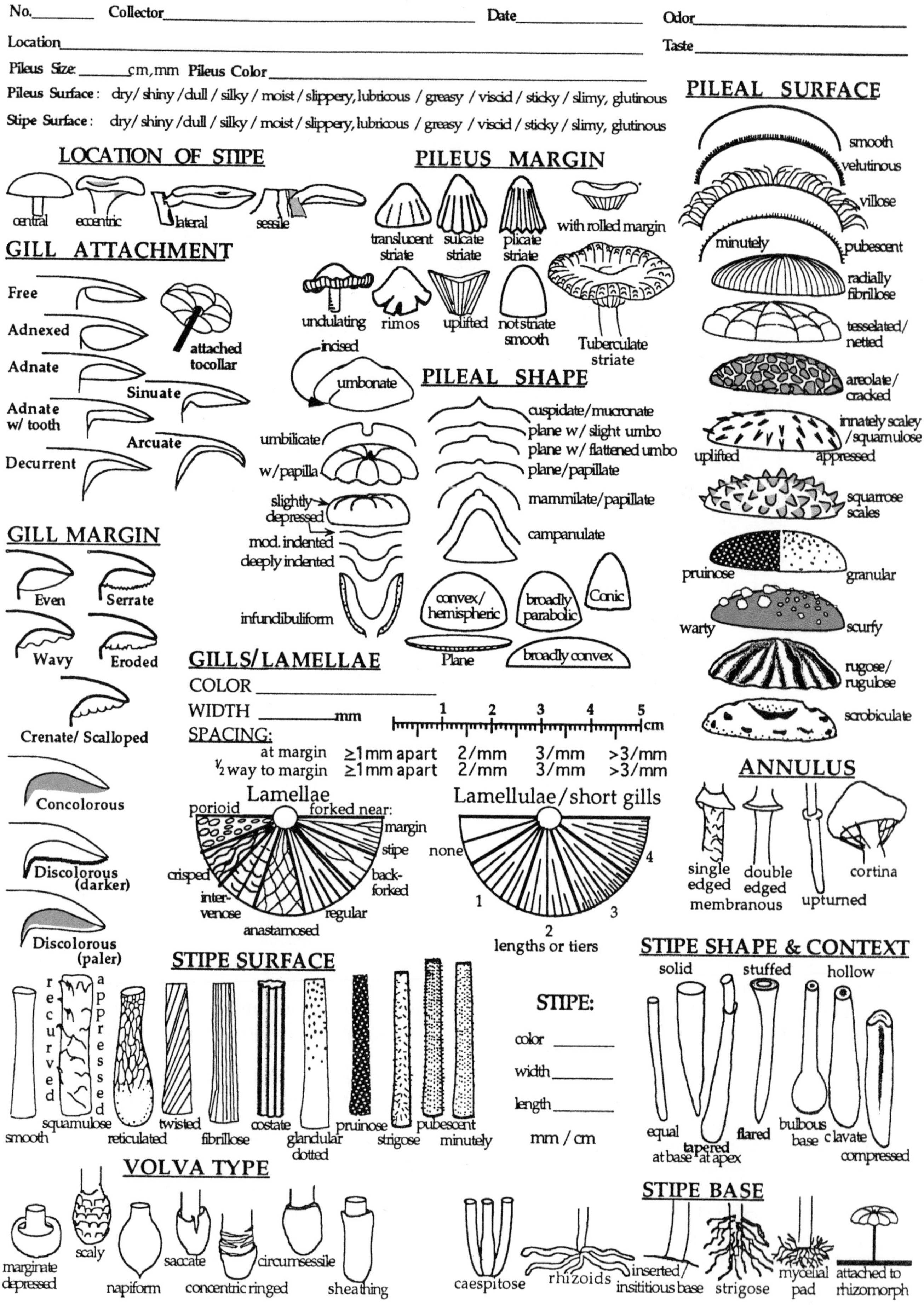

No._____ Collector_____ Date_____ Odor_____

Location_____ Taste_____

Pileus Size:_____cm,mm Pileus Color_____

Pileus Surface: dry/ shiny /dull / silky / moist / slippery,lubricous / greasy / viscid / sticky / slimy, glutinous

Stipe Surface: dry/ shiny /dull / silky / moist / slippery,lubricous / greasy / viscid / sticky / slimy, glutinous

LOCATION OF STIPE

central eccentric lateral sessile

GILL ATTACHMENT

Free
Adnexed
Adnate
Adnate w/ tooth
Sinuate
Arcuate
Decurrent
attached to collar

GILL MARGIN

Even Serrate
Wavy Eroded
Crenate/ Scalloped

Concolorous
Discolorous (darker)
Discolorous (paler)

PILEUS MARGIN

translucent striate sulcate striate plicate striate with rolled margin
undulating rimos uplifted not striate smooth Tuberculate striate
incised
umbonate

PILEAL SHAPE

umbilicate
w/ papilla
slightly depressed
mod. indented
deeply indented
infundibuliform

cuspidate/mucronate
plane w/ slight umbo
plane w/ flattened umbo
plane/papillate
mammilate/papillate
campanulate

convex/ hemispheric broadly parabolic Conic
Plane broadly convex

GILLS/LAMELLAE

COLOR_____
WIDTH_____mm
SPACING:

	at margin	½ way to margin
≥1mm apart	≥1mm apart	
2/mm	2/mm	
3/mm	3/mm	
>3/mm	>3/mm	

1 2 3 4 5 cm

Lamellae
porioid forked near:
margin
stipe
back-forked
crisped
inter-venose
anastamosed
regular

Lamellulae/short gills
none
1 2 3 4
lengths or tiers

PILEAL SURFACE

smooth
velutinous
villose
minutely pubescent
radially fibrillose
tesselated/ netted
areolate/ cracked
innately scaley /squamulose
uplifted appressed
squarrose scales
pruinose granular
warty scurfy
rugose/ rugulose
scrobiculate

ANNULUS

single edged membranous
double edged upturned
cortina

STIPE SHAPE & CONTEXT

solid stuffed hollow

STIPE:
color_____
width_____
length_____
mm / cm

equal flared
tapered at base at apex
bulbous base clavate
compressed

STIPE SURFACE

recurved appressed
squamulose
smooth
reticulated
twisted
fibrillose
costate
glandular dotted
pruinose
strigose
pubescent minutely

VOLVA TYPE

marginate depressed
scaly
napiform
saccate
concentric ringed
circumsessile
sheathing

STIPE BASE

caespitose
rhizoids
inserted/ insititious base
strigose
mycelial pad
attached to rhizomorph

FIGURE 8.18 Agaric annotation sheet, in English. These are copied, and one is filled for each collection. (D. J. Lodge)

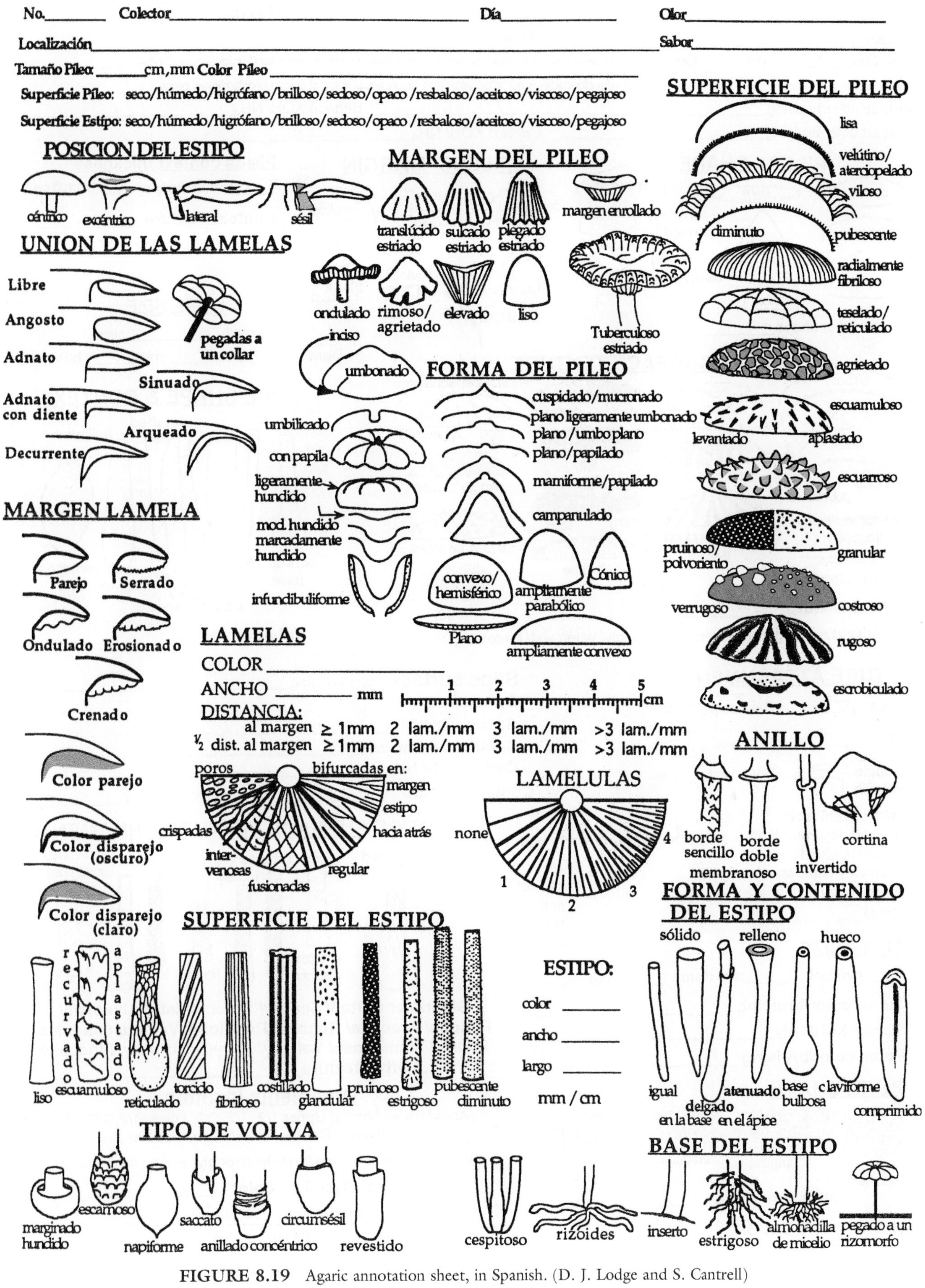

No._____ Colector_____ Día_____ Color_____

Localización_____ Sabor_____

Tamaño Píleo: _____ cm, mm Color Píleo _____

Superficie Píleo: seco/húmedo/higrófano/brilloso/sedoso/opaco/resbaloso/aceitoso/viscoso/pegajoso

Superficie Estípe: seco/húmedo/higrófano/brilloso/sedoso/opaco/resbaloso/aceitoso/viscoso/pegajoso

SUPERFICIE DEL PILEO

lisa
velútino/aterciopelado
viloso
diminuto
pubescente
radialmente fibriloso
teselado/reticulado
agrietado
escuamuloso
aplastado
escuarroso
pruinoso/polvoriento
granular
verrugoso
costroso
rugoso
escrobiculado

POSICION DEL ESTIPO

céntrico   excéntrico   lateral   sésil

MARGEN DEL PILEO

translúcido estriado   sulcado estriado   plegado estriado   margen enrollado

ondulado   rimoso/agrietado   elevado   liso

inciso

umbonado

Tuberculoso estriado

UNION DE LAS LAMELAS

Libre
Angosto
Adnato
pegadas a un collar
Adnato con diente
Sinuado
Arqueado
Decurrente

FORMA DEL PILEO

cuspidado/mucronado
plano ligeramente umbonado
plano/umbo plano
plano/papilado
mamiforme/papilado
campanulado

umbilicado
con papila
ligeramente hundido
mod. hundido
marcadamente hundido

convexo/hemisférico   ampliamente parabólico   Cónico

infundibuliforme

Plano   ampliamente convexo

levantado
aplastado

MARGEN LAMELA

Parejo   Serrado

Ondulado   Erosionado

Crenado

Color parejo

Color disparejo (oscuro)

Color disparejo (claro)

LAMELAS

COLOR _____

ANCHO _____ mm

DISTANCIA:

al margen ≥ 1mm   2 lam./mm   3 lam./mm   >3 lam./mm
½ dist. al margen ≥ 1mm   2 lam./mm   3 lam./mm   >3 lam./mm

1   2   3   4   5 cm

poros   bifurcadas en:
margen
estipo
hacia atrás
crispadas
inter-venosas
regular
fusionadas

LAMELULAS

none   4
1   3
2

ANILLO

borde sencillo membranoso   borde doble   cortina   invertido

FORMA Y CONTENIDO DEL ESTIPO

sólido   relleno   hueco

ESTIPO:

color _____
ancho _____
largo _____

mm / cm

igual   delgado en la base   atenuado en el ápice   base bulbosa   claviforme   comprimido

SUPERFICIE DEL ESTIPO

recurvado   aplastado

liso   escuamuloso   reticulado   torcido   fibriloso   costillado   glandular   pruinoso   estrigoso   pubescente diminuto

TIPO DE VOLVA

marginado hundido   escamoso   napiforme   saccato   anillado concéntrico   circumsésil   revestido

BASE DEL ESTIPO

cespitoso   rizóides   inserto   estrigoso   almohadilla de micelio   pegado a un rizomorfo

FIGURE 8.19 Agaric annotation sheet, in Spanish. (D. J. Lodge and S. Cantrell)

No._____ Date_____ Collector_____ Fungus_____

Location_____ **Host** .

Pileus Size_____mm   **Odor:** none/ not distinctive/ mild/ sweet/ nutty/ quickly acrid/ spicy/ fungoid/ unpleasant/ not determined/ other_____

Taste: none/ not distinctive/ mild/ sweet/ nutty/ acidic/ slowly acid/ quickly acrid/ bitter/ very bitter/ farinaceous/ spicy/ sl. spicy/ not determined/ other_____

Surface color/bruising_____/_____   Surface KOH/ NH$_4$_____/_____

Context color/bruising_____/_____   Context KOH/ NH$_4$_____/_____

PILEAL SHAPE

convex plane umbonate broadly convex conical slightly depressed mod. indented deeply indented hemispheric broadly conical

Other shape_____

PILEUS MARGIN

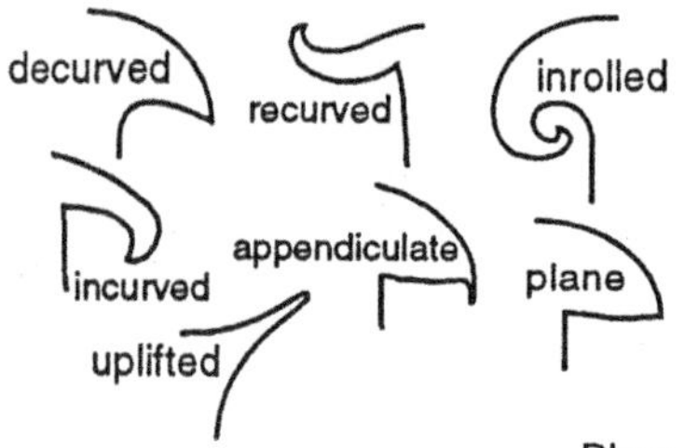

decurved recurved inrolled incurved appendiculate plane uplifted

Pileus context thickness

.
_____mm @ margin _____mm @ center

Context Texture_____

Worm Hole Color_____

STIPE SIZE

Length_____mm

Diam. _____apex _____mid _____base mm

PILEAL SURFACE TEXTURE

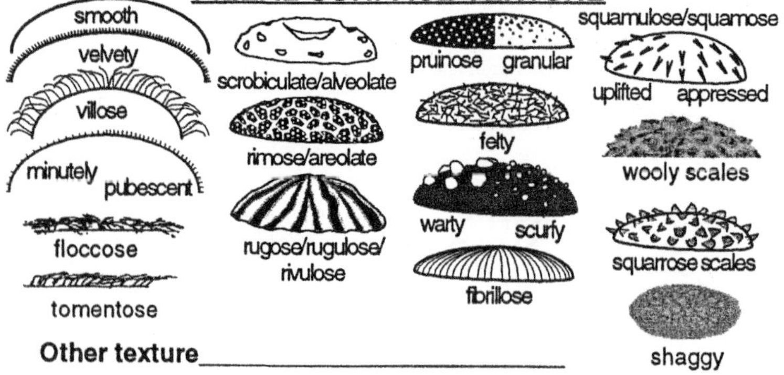

smooth velvety villose minutely pubescent floccose tomentose

scrobiculate/alveolate rimose/areolate rugose/rugulose/rivulose fibrillose

pruinose granular felty warty scurfy

squamulose/squamose uplifted appressed wooly scales squarrose scales shaggy

Other texture_____

Viscid not /slightly/ moderately/ strongly/ when wet/ becoming

STIPE SHAPE & CONTEXT

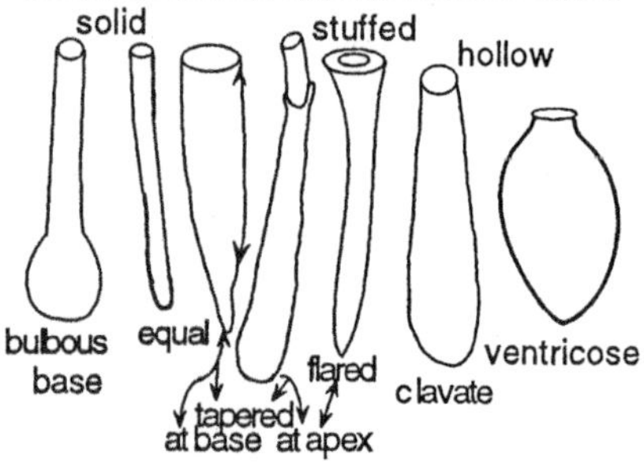

solid stuffed hollow

bulbous base equal tapered at base tapered at apex flared at apex clavate ventricose

Other shape_____

TUBE ATTACHMENT

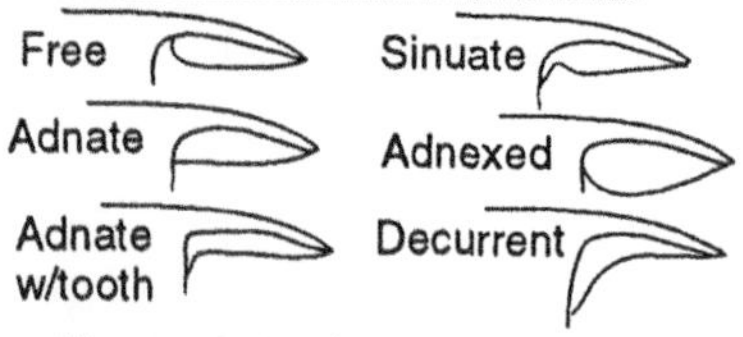

Free Sinuate
Adnate Adnexed
Adnate w/tooth Decurrent

w/decurrent tooth
w/ long decurrent tooth
Depressed/ca. stipe
Shallowly depressed/ca. stipe
Deeply depressed/ca. stipe
Other_____

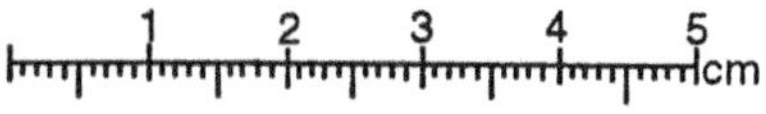

1 2 3 4 5 cm

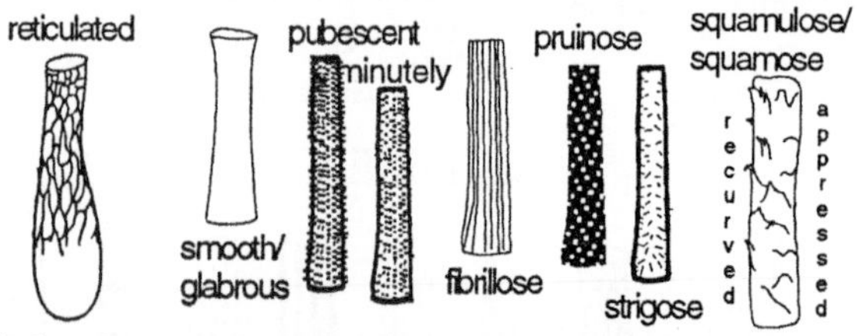

Surface Reticulated:
not/ finely/
moderately/
strongly/ lacerate

TUBE SIZE

_____mm long _____mm diam _____/mm

Tubes color/bruising_____/_____

Staining KOH/ NH$_4$_____/_____

Pores color/bruising_____/_____

PORE SHAPE

round subrounded angular irregular radially elongated

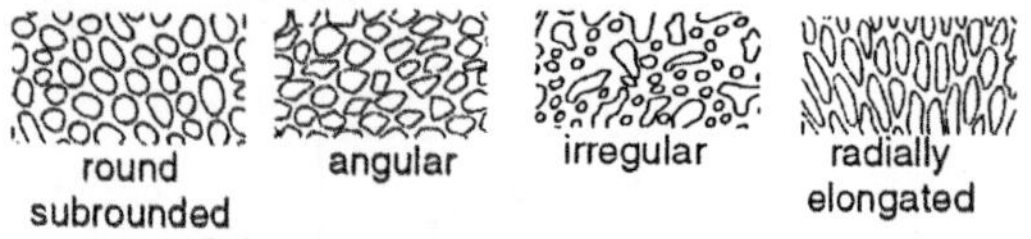

Other pore shape_____

SPORE PRINT COLOR_____

Stipe surface color/bruising_____/_____
Ground color_____
Surface KOH/ NH$_4$_____/_____
Context color/bruising_____/_____
Context KOH/ NH$_4$_____/_____
Context Texture_____   **Worm Hole Color**_____

STIPE SURFACE TEXTURE/ORNAMENTS

reticulated pubescent minutely pruinose squamulose/ squamose
smooth/ glabrous fibrillose strigose recurved appressed

Glandular dotted/ Scaber/ Scrobiculate/ Alveolate/ Rimose/ Areolate/ Rugose/ Rugulose/ Veined/ Granular/ Scurfy/ Tomentose/ Velvety/ Felty/ Scabrous/ Floccose
Other texture_____

Ornaments location

Apex/ Base/ Middle/ Upper 1/2, 1/3, 2/3/ Lower 1/2, 1/3, 2/3/ Overall/ Denser below/ Denser above

Viscid not /slightly/ moderately/ strongly/ when wet/ becoming

Basal Mycelium color_____

VEIL/DESCRIPTION_____

ANNULUS/DESCRIPTION_____

FIGURE 8.20 Bolete annotation sheet, in English. (B. Ortiz-Santana)

FIGURE 8.21 Nonagaric macrofungi, such as this *Ramaria cyanocephala*, are difficult to annotate, because the terms used to describe patterns of forking and the shape of the ultimate branches are different from those used to describe other groups. (Photo from Puerto Rico by D. J. Lodge)

FIGURE 8.22   Describing stink-horn fungi (Phallales), such as this *Dictyophora* species, is challenging for individuals who are not experts in this group. Consult Dring (1980) for appropriate descriptors and terminology. (Photo by G. Mueller)

FIGURE 8.23 Preparing thorough descriptions of some of the ascomycetes that produce complex macroscopic fruit bodies, such as this species of *Morchella*, requires a significant amount of time. Consult Weber (1995b) for appropriate terminology. (Photo by G. Mueller)

fungi and fine detail; forceps for handling specimens and cleaning particles of dirt and debris from specimens; a millimeter rule; and depending on the goals of the study, a spot plate and chemicals for macrochemical tests (see "Testing for Macrochemical Color Reactions," later in this chapter). References such as the *Dictionary of Fungi* (Kirk et al. 2001), *How to Identify Mushrooms to Genus: Macroscopic Features* (Largent 1986), *Modern Genera* (Largent and Baroni 1988), and other taxonomic references (see "Taxonomic Resources for Identification," later in this chapter) are helpful, but not essential, books to have in the field laboratory as sources of names, descriptors, and definitions.

Accurate and consistent notation of sporocarp color, including color changes of mature sporocarps and colors of different developmental stages, is important when describing macrofungi, so we recommend use of standardized color names. Commonly used color guides are those by Ridgway (1912), Kornerup and Wanscher (1978), Munsell (1966), Rayner (1970), Maerz and Paul (1950), and Kelly (1965). Unfortunately, most of those books are out of print, but libraries often have copies of at least some of them. The Munsell system is the most comprehensive; it is still in print, but it is expensive.

COLOR PHOTOGRAPHS

Color photographs (slides, prints, or digital images taken with a megapixel digital camera) should be taken whenever possible because they are extremely valuable for documenting macromorphological features and can be used in a variety of ways. Photographs taken in the field are excellent for publications and lectures. Because field photography is time consuming, however, it is not always practical. The alternative is to photograph specimens in the laboratory or studio. Shaw (1987) gives a wealth of valuable information on field macrophotography.

The most important considerations in photographing mushrooms are adequate close-up capability and appropriate film for the light source. Magnification of 1:2 is the minimum, although for very small taxa 1:1 or better is necessary and can be obtained with a bellows or extension tube. The best color generally is obtained with daylight type film in combination with either sunlight or a flash. A ring flash, multiple flashes, or a bounce flash setup eliminates the "shot-at-night" look that often comes from a single directed light source. Some full-feature megapixel digital cameras give excellent results and can serve as an alternative to traditional film-based cameras (macro capability and control over depth of field through aperture priority settings are important features to look for when choosing a digital camera).

Specimens should be photographed on a black or neutral gray background in a way that illustrates important diagnostic features of the sporocarps and allows easy comparison of features. All stages of development available, a sporocarp cut lengthwise to reveal the interior, a top view of the pileus, and a view of the underside of the pileus showing the edges of the lamellae should be included. If the sporocarp is large, or if the specimen consists of only one sporocarp, pie-shaped sections of the top and underside are sufficient (Figs. 8.24, 8.25, and 8.26).

Testing for Macrochemical Color Reactions

Macrochemical color reactions, or spot tests, have been used for decades in the identification and classification of fungi. Such tests can be helpful in identifying genera or groups of species, especially in the genera *Ramaria* and *Russula* (Romagnesi 1967; Marr et al. 1986; Singer 1986b; but see Thiers 1997 for an opposing view). The

tests are performed by placing reagents on pieces of the sporocarp and noting any color changes. Although reactions are not routinely checked in large biodiversity sampling programs because the tests are time consuming, they may prove useful when identification of key fungal taxa require such information. Marr and colleagues (1986) and Singer (1986b) provided recipes for and discussed the application of the most commonly used macrochemicals. Chemicals are applied to detached portions of sporocarps to prevent degradation of the specimen and interference with other chemical tests or extraction of DNA. No more than one reagent or solution should be applied to a piece of sporocarp. Some of the test chemicals are dangerous to handle so appropriate safety measures should be used. Both negative and positive color-change results should be recorded.

Obtaining Spore Deposits for Identification

The main reason for taking spore deposits is to determine spore color. Spore deposits should be observed on white paper, clean glass slides, or foil. Using spores from spore deposits during micromorphological analyses eliminates

FIGURE 8.24 Obtaining the proper lighting (from the side but without hard shadows) for taking photographs of fungi in the field requires significant time. Such photographs are often the most esthetically pleasing, however, and they are best for delicate specimens that are easily damaged by handling. (Photo of *Laccaria amethystina* by G. Mueller)

FIGURE 8.25   Field photographs are often taken using a fill-flash to provide more light, thereby decreasing the exposure time required for hand-held shots, as well as softening hard shadows. (Photo of *Mycena* aff. *leana* in Costa Rica by T. O'Dell)

FIGURE 8.26   Taking photographs against a neutral gray card or another neutral background is generally faster than taking good field photographs, since the same lighting and background can be used for multiple specimens. Such photographs are often good for recording the critical characters of a fungus, and they are easily color-corrected if using the standard background. (Photo of *Tylopilus balouii* in the Dominican Republic by D. J. Lodge)

questions about the maturity of the spores because only mature spores are dropped. Spore deposits can be obtained from some macrofungi (e.g., coral fungi) by placing a piece of white paper directly under the branches of the sporocarp before it is wrapped or put into a container. With agarics, the cap is removed from the stipe and placed on a piece of white paper. Alternatively, a round hole (or a large "X") is cut in a piece of paper through which the stipe is inserted until the lamellae make contact with the paper. Only a portion of the pileus of a large basidiome is needed to obtain a spore deposit.

Mature sporocarps are placed in a relatively humid environment at 10–15°C. A large box (e.g., a cigar box) can be used to raise the humidity and can accommodate several caps. A plastic container also works well, as does wrapping the setup in aluminum foil or wax paper. Providing a moist environment may increase success if it is particularly dry. That can be done by placing a moistened piece of paper towel in the container, by stretching a piece of moistened paper towel tightly over the container before closing, or by taking the spore print in a closed plastic bag. The sporocarp should not be in water, and

the collection number should be included with each spore deposit.

MICROMORPHOLOGICAL FEATURES

A compound microscope sometimes is used in the field laboratory to study spore characteristics, pileus anatomy, or other features that can help in initially assigning a specimen to a taxonomic group. It also can be used to assess the condition and maturity of small ascomycetes to help determine if a specimen should be kept. A quick look under the microscope can provide information that improves the overall description of the fresh material and allows one to check for features that are lost on drying. Some workers like to do a complete description of micromorphological features in the fresh condition. That is time consuming, however, and often impractical. Well-preserved specimens work as well as fresh material as a source of micromorphological data, so those analyses normally are done between collecting trips and off-season. Watling (1968), Largent and associates (1977), and Singer (1986b) have discussed micromorphological characters and the techniques and reagents needed to describe them (Fig. 8.27).

TAXONOMIC RESOURCES FOR IDENTIFICATION

Assignment of a fungus to family is an important initial step in obtaining an accurate species determination because it directs the investigator toward appropriate, more-specialized monographs. Publications with general keys or guides include *Fungus Flora of Venezuela and Adjacent Countries* (Dennis 1970), *The Fungi* (Ainsworth et al. 1973a, 1973b) volumes IVA and IVB, and *The Fungi of Switzerland* (Breitenbach and Kränzlin

1984, 1986, 1991, 1995, 2000). *Ainsworth and Bisby's Dictionary of the Fungi* (Kirk et al. 2001) is a good place to look for references to taxa of various ranks.

BASIDIOMYCETES

Dependable publications for identification of nonagaric basidiomycetes from woody substrata include Eriksson and Ryvarden (1973, 1975, 1976), Eriksson and colleagues (1978, 1984), Gilbertson (1974), Jülich and Stalpers (1980), and Gilbertson and Ryvarden (1986). Burt (1914–1926, reprinted as a book in 1966 with an index to new names and synonyms) can be used to identify many resupinate and erect nonagaric basidiomycetes from North America and the Caribbean, but his generic

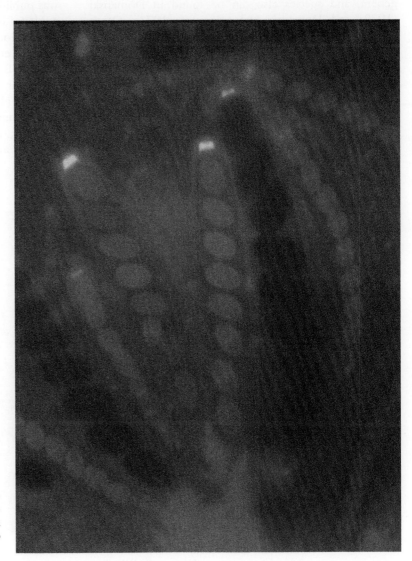

FIGURE 8.27 Some micromorphological characters in ascomycetes, such as the presence of a fluorescing ascus ring in *Valsaria rubicosa* treated with Calcafluor, are ephemeral and are lost or difficult to see after drying. (Photo by S. Huhndorf)

concepts are outdated and the species descriptions are not always dependable. The keys to genera of temperate fungi do not always work for tropical species, but annotated checklists of tropical and subtropical species facilitate identification once a genus has been determined (e.g., Hjortstam and Larsson 1994). The conspicuous Tremellales can be identified in North America using McNabb (1964a, 1964b, 1965a, 1965b, 1965c, 1965d, 1965e, 1966a, 1966b, 1969, 1973); in Europe using Breitenbach and Kränzlin (1984); and in the Neotropics, at least to genus, using Dennis (1970), Lowy (1952, 1971) or Ryvarden (Keys to Neotropical Polypores, unpublished manuscript) (Fig. 8.28).

Keys to North American and some neotropical polypore species are found in Gilbertson and Ryvarden (1986). A preliminary polypore mycota for East Africa was published by Ryvarden and Johansen (1980). Keys and descriptions for poroid basidiomycetes families, genera, and species also can be found in Domanski (1972) and Domanski and associates (1973). Other useful references for neotropical polypores include Murrill (1915, reprinted in 1973), Fidalgo (1968), Fidalgo and Fidalgo (1968), Furtado (1981), Ryvarden (2000), and Lodge and colleagues (2001). Pleurotoid, lamellate polypores (i.e., *Lentinus*) and agarics (i.e., *Pleurotus* and *Panus* sensu Corner) can be identified using the monographs by Pegler (1983b) and Corner (1981), respectively.

Keys to orders, families, and genera of gasteromycetes can be found in Breitenbach and Kränzlin (1986), Coker and Couch (1928), Miller and Miller (1988), Pegler and colleagues (1995), and Smith and Smith (1973). Keys to the genera and species of Clathraceae are available in Dring (1980), and keys to the genera and species of West Indian Nidulariaceae can be found in Brodie and Dennis (1954).

Watling and Watling (1980) provided an annotated list to the taxonomic literature for Agaricales. However, considerable literature on Agarics has appeared since 1980. Anyone working with those fungi, or any group of fungi for that matter, should visit a major library and conduct computer searches by genus and/or author to access that literature. Older taxonomic literature (1753–1821) is reviewed in Pfister and associates (1990). There are many field guides to temperate agaric fungi, especially for North America and Europe. A key to families and genera was published by Largent and Baroni (1988) as part of a Mad River Press series that also includes guides to morphological features of agarics (Largent 1977; Largent et al. 1977). Useful guides to European fungi include Bas and colleagues (1988–1995), Breitenbach and Kränzlin (1991, 1995, 2000), Courtecuisse and Duhem (1995), and Moser (1978). There is nothing approaching a complete agaric mycota for North America, but popular field guides by Arora (1987), Bessette and associates (1997, 2000), Lincoff (1981),

FIGURE 8.28 Many identification guides are available for polypore fungi such as this *Bridgidoporus* species. (Photo by T. O'Dell)

McKnight and McKnight (1987), Miller (1972), Phillips (1991), and Pomerleau (1980) are useful. Horak (1979) published an agaric mycota for Tierra del Fuego in temperate South America. Teng (1996) made a major contribution to the fungi of China. References for basidiomycetes of the Neotropics can be found in Nishida (1989). Artificial keys to agaric genera may or may not work for tropical agarics. It is usually most expedient to try artificial keys (Dennis 1970; Largent and Baroni 1988) and to compare the fungus with the generic description first and to resort to the natural keys in the *Agaricales in Modern Taxonomy* (Singer 1986b) only if the results are unsatisfactory. References that are useful for identifying tropical agaric species are Corner (1994), Dennis (1970), Pegler (1977, 1983a, 1986), and Singer (1970, 1973, 1974, 1977, 1978, 1982, 1989).

ASCOMYCETES

Ainsworth and colleagues (1973a), Breitenbach and Kränzlin (1984), and Dennis (1970, 1978) are useful references for identifying both North American and European ascomycetes. Useful references with keys to loculoascomycetes and pyrenomycetes include von Arx and Müller (1954, 1975), M. E. Barr (1987, 1990), Hanlin (1990), Luttrell (1973), Müller and von Arx (1962, 1973), and Sivanesan (1984).

Some ascomycetes occurring on wood can be identified using keys to groups at various taxonomic levels, including Aptroot (1991, 1995) and M. E. Barr (1972, 1978, 1990b, 1990c). Keys to genera of xylariaceous ascomycetes can be found in Dennis (1956, 1957, 1958, 1970), Ju and Rogers (1996), Laessøe and colleagues (1989, unpublished data), Laessøe and Spooner (1994), Petrini (1992), Rappaz (1995), Rogers and colleagues (1987, 1988), and San Martín-González and Rogers (1989, 1993). North American, European, and some tropical discomycetes can be identified using Breitenbach and Kränzlin (1984), Dennis (1970, 1978), Gamundi (1979), Le Gal (1953, 1959), Korf (1973), Rifai (1968), and Seaver (1942, 1951) (Figs. 8.29, 8.30, and 8.31).

FIGURE 8.29 Many of the 'macro' Ascomycetes on wood, such as *Striatosphaeria codinaeaphora* shown here, are rather small. They must be magnified before they can be described and photographed. (Photo by S. Huhndorf)

FIGURE 8.30   Discomycetes, such as the *Sarcoscypha* shown here, are members of the Ascomycota. (Photo by G. Mueller)

FIGURE 8.31   Species, such as dead man's fingers (*Xylaria*); resupinate or pulvinate ones, such as most species of *Hypoxylon* (shown here); and related genera in the Xylariaceae are included in the Ascomycota. (Photo of *H. haematostroma* from St. John, U.S. Virgin Islands, by D. J. Lodge)

MACROFUNGI ON WOODY SUBSTRATA

Sabine M. Huhndorf, D. Jean Lodge, Chun-Juan Wang, and Jogeir N. Stokland

GENERAL CONSIDERATIONS

Wood-inhabiting macrofungi inhabit substrata that differ in size, state of decay, and moisture content. Thus, the ecologies of fungi growing on wood differ. Wood-inhabiting species include fungi that were in or on the wood when a tree or tree part fell to the ground (e.g., pathogens and nonpathogenic endophytes of living tissue, heart rots), as well as the succession of fungi that follow in an organized fashion as the wood deteriorates.

In addition, many putative and known ectomycorrhizal species "climb up" on wood to fruit and simultaneously may obtain nutrients from the wood by degrading organic compounds (Read et al. 1989). Although we do not know how most wood-inhabiting ascomycetes functions, some have been shown to be pathogens, endophytes, or saprobes. Thus, surveys of wood-inhabiting fungi will not be limited strictly to wood-decaying species (Figs. 8.32 and 8.33).

All of the decay fungi, to some degree, recycle lignocellulosic and mineral nutrients back into the ecosystem. In addition, their decay activities soften the woody tissues, making them more amenable to bird and small-mammal habitation and use by arthropods, nematodes, and other invertebrates as well as other fungi. It also has been shown that wood in advanced stages of decay on the ground is important for establishment of mycorrhizal associations with seedlings, and that decayed woody debris acts as a moisture sink for the maintenance of mycorrhizal fungi in seasonally dry forests (Harvey et al. 1978; Larsen et al. 1982; Jurgensen et al. 1986). In forests, brown-rot residues can be a major soil component. In standing trees, decay fungi can be root-, butt-, or heart-rot organisms, which are generally mutually exclusive. Those fungi cause decay of the roots and/or stems and thereby predispose the tree to wind throw or trunk breakage.

Different groups of wood-inhabiting fungi use quite different volumes of substratum (resource bases) for the production of sporocarps. Consequently, fungal sub-

FIGURE 8.32 Some of the macrofungi, such as *Armillaria tabescens* shown here, are plant pathogens. (Photo by G. Mueller)

FIGURE 8.33 Many of the large macrofungi, such as this species of *Volvariella*, are wood decomposers. (Photo by G. Mueller)

strata from tiny twigs to large trunks must be sampled if a majority of fungal species within the study area are to be found. The optimum size of employed plots or subplots will vary from one to several thousand square meters depending on the density of downed wood at a sampling site. In addition, live woody-stem and trunk substrata support conks or other fruit bodies arising from decaying parts of the tree or bark.

In general, gymnosperm hosts/substrata support a more restricted mycota than do angiosperm hosts/substrata (S. M. Huhndorf and H. H. Burdsall, personal observation). The numbers of species of wood-inhabiting fungi in a given area probably depends more on the range of decay classes of substrata than the number of species of phanerogams present because the majority of those fungi probably are not particularly host-genus or host-species specific.

Although numerous standardized surveys have been conducted on terrestrial macrofungi, few have been conducted on the inhabitants of wood. Examples of surveys of wood-inhabiting fungi include the studies of Scandinavian pyrenomycetes (Mathiassen 1993); wood-rotting basidiomycetes (Renvall 1995); and basidiomycetes, ascomycetes, and slime molds (Heilmann-Claussen 2001). Ryvarden and Nuñez (1992) and Lindblad (2000, 2001) studied polypores in the tropics, and Heinrich and Wojewoda (1976) studied the effects of air pollution on basidiomycetes. Additional references are available in Vogt and associates (1992).

Mycodiversity, as with all other subsets of biodiversity, exhibits distinct patterns in both space and time. Such fungal patterns are to a large extent unexplored, as evidenced by the almost complete lack of mycological examples in the increasing body of biodiversity literature (see Rosenzweig 1995 for a comprehensive and taxonomically broad treatment of species richness patterns). A number of ecological and evolutionary questions related to wood-inhabiting fungi need investigation. A sample of such questions follows:

- What are the shapes of the species-area curves of wood-inhabiting fungi in various environments or on woody substrata of different size and decay class?
- How do habitat variety and woody-plant diversity affect mycodiversity?
- How do forest productivity and substratum abundance influence mycodiversity?
- Does the local species richness of wood-inhabiting fungi exhibit mainland-island patterns?
- Does the species richness of wood-inhabiting fungi follow a latitudinal gradient, as is found in many other taxa (low latitudes being richer)?
- How does sporocarp production (sexual reproduction) vary within and between years?
- How do species composition and richness change through the course of decay?
- How do natural disturbances, such as hurricanes and fires, perturb species composition and richness patterns?
- How do anthropogenic disturbances, such as logging and wood removal, affect diversity of wood-inhabiting fungi?

ISSUES OF IMPORTANCE WHEN DESIGNING SAMPLING PROTOCOLS

In the following section we cover concepts specific to studies concerned with sampling macrofungi found on woody substrata. The material supplements the general information presented in the next section ("Approaches to Sampling Macrofungi").

As with other groups of macrofungi, the number of species of fungi on wood increases with the size of the area sampled, or more specifically, the amount of substratum sampled. The distribution and amount of substratum present therefore will dictate the size of the area to be sampled at a particular site. Small-size woody debris generally is encountered much more frequently than large-diameter wood. Thus, the area that must be searched to obtain an adequate sample of macrofungi will depend on the distribution and frequency of the different diameter classes of downed wood, as well as the frequency of the fungi under study. Discovery of cryptic fungi is often aided by more intensive searching of designated sampling areas or subplots. Fungi fruiting on large-diameter wood (logs and standing-live and standing-dead trees) are easily undersampled if collecting is confined to small subplots because of the low density of large pieces of wood in those subplots and because few of those large pieces will be in the right stage of decay to support fungal fruiting. For a complete survey of macrofungi on all classes of woody debris, we recommend augmenting small subplot-based samples of small debris with another method for surveying many large pieces of wood. Even when many logs were surveyed for macrofungi in Scandinavia, the species-substrata curve did not reach an asymptote, indicating that many more species would be discovered if additional logs were sampled (Fig. 8.34; Lindblad 1998). Had those samples been stratified by diameter class, decay class, or tree species, however, they might have produced flattened species-substrata curves within strata. Aggregating samples from different habitats tends to produce steadily increasing species-effort curves. A gradient analysis of communities of macrofungi and slime molds on beech logs of different ages and decay classes in Denmark (Heilmann-Clausen 2001) showed that the correlated factors of age, decay stage, and contact with soil, together with microenvironmental variables such as exposure to sun, moss cover, and soil moisture were the most strongly correlated with community structure. Studies in Puerto Rico have indicated that diameter class and microclimate are among the most important factors influencing wood-inhabiting macrofungal community

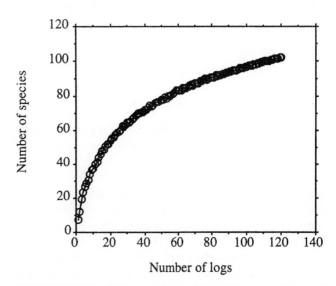

FIGURE 8.34 This graph shows the positive correlation of cumulative species diversity with substratum quantity. The number of macrofungal species (Polypores, Corticiums, Agarics, Heterobasidiomycetes, Hymenochaetaceae, and a few other groups) are plotted as a function of substratum quantity (dead logs of *Alnus*). Each point represents an average of 100 random samples drawn from a pool of 150 logs inspected twice in a Norwegian study area. (Adapted with permission from Kauserud 1995.)

composition in tropical climates (Lodge and Cantrell 1995b; Lodge 1996; S. M. Huhndorf and D. J. Lodge, unpublished. data). Inspection of higher reaches of trees may add new species, but most of the canopy species probably also will be found on fallen branches or in gaps (Ryvarden and Nuñez 1992; Lodge and Cantrell 1995b).

INCONSPICUOUS AND INDISTINGUISHABLE SPECIES

Two main problems arise when sampling fungi on wood. First, many resupinates, inconspicuous basidiomycetes, and small ascomycetes either are not detectable or not recognizable in the field. That problem can be resolved by removing all woody substrata to the laboratory for examination. Although doing so precludes resampling of the same subplot, that problem can be solved, as discussed later. Second, it frequently is difficult to determine whether or not a species has been collected already, either within a subplot during one sampling or in previous samplings. More than one individual of a species may be taken from one piece of substratum when only one is intended because different stages of maturity of some species are difficult to recognize. However, some species look so similar in the field that species may be missed. No assumptions should be made regarding the identity of the small ascomycetes and other inconspicuous fungi, and all separate patches of those fungi should be collected.

PLOT SIZE

Small to Medium-Size Woody Debris

Collecting macrofungi in randomly or regularly spaced subplots is an effective way to sample fungi on small to medium-size (1.0–15.0-cm diameter) woody debris. The investigator examines each piece of debris or stem for macrofungi that produce recognizable fruiting structures and collects one specimen of every species from each piece of substratum within the plot. The data outlined later in this chapter (see "Data Collection") are recorded for each specimen. Portions of hosts or substrata that protrude into a subplot also are sampled. Samples of each species present on each class of substratum should be sufficient to enable identification. After sampling, a substratum should be returned to its original position for future sampling. Substrata from which specimens are taken may be tagged for future recognition. For fungi that cannot be seen easily or distinguished to morphospecies with the naked eye, all woody debris should be collected from a subplot for examination under a dissecting microscope. Small ascomycetes and resupinate heterobasidiomycetes fruiting on small woody debris are examples of taxa that require such destructive sampling. If the subplots are to be sampled repeatedly, they should be divided into halves or quadrants so that debris is collected from each sector only once. Subplots should be sampled throughout the growing season.

A few forests have been surveyed and gridded in such a way that grid cells may be selected as sample plots, but new plots will need to be established for most studies. Circular plots are usually the most efficient to set up and sample. The precisely oriented, 90-degree corners of square plots are unnecessary, and the only required equipment is a string premeasured to the desired radius extending from a center pole. The diameter of the plots depends on the density of fallen wood in the diameter classes of interest and the type of fungi under study. For example, although detecting small ascomycetes on branches and twigs is difficult, those fungi have been sampled successfully using 0.5-m^2 circular plots positioned at the central core of larger plots (S. M. Huhndorf, personal observation). Plots 10 m in diameter can be used to sample medium-size pieces of wood, although larger plots may be needed if fallen branches are rare. The practical upper limit for the diameter of a circle plots is between 10 and 20 m (78.5–314 m^2).

Large Woody Debris, Standing Trees, and Recent Treefalls

It is often advantageous to use large woody substrata or dead trees instead of plots as the sample units for fungi that fruit on those types of substrata. A fallen tree or log can be sampled for visible fruiting structures from base to top, and the location of each specimen and the type and decay class of the tree or log noted. Specimen collection sites can be tagged or mapped and resampled periodically.

If plots or subplots for sampling fungi on small debris are set up along transect lines, sample units can be selected and density of large fallen or standing-dead trees in the area estimated using the point quarter method (Cottam and Curtis 1956). In that technique, points are identified along the transect line at distances selected using a random numbers table. Then, the distance from each point to the base of the nearest fallen log or dead tree in the northwest, northeast, southeast, and southwest quarters is measured. Those distances are used to calculate density of standing or fallen trees. Recording the diameters and heights or lengths of the trees and logs and the compass orientation on which they fell is useful for relocating them. Alternatively, if the forest is gridded, or if parallel transect lines are established at regular intervals, then the locations of large woody debris, tree falls, and snags can be mapped. If the sampling area is large, global positioning systems (GPS) can be used to identify the locations of logs and dead trees, and locations can be mapped using latitude and longitude in a geographic information system (GIS). Logs to be sampled are selected from the mapped ones. One or more of the recently developed adaptive sampling protocols (see "Adaptive Sampling," later in this chapter) may prove useful when sampling those relatively rare substrata (Thompson 1992; Thompson and Seber 1996).

FREQUENCY OF SAMPLING

How often a plot should be sampled depends on the amount of species turnover between sampling dates (see discussion of complementarity under "Determining Adequate Sampling," later in this chapter, and in Colwell and Coddington 1994). For Xylariaceae and small ascomycetes in the tropics, 3-month sampling intervals appear to provide optimal balance between discovery of previously unrecorded species and sampling effort. Turnover in fruiting agaric species is high in both temperate and tropical forests such that these species should be sampled at weekly or biweekly intervals. Corticioid and polypore fungi, however, may be sampled at approximately 2-month intervals (H. H. Burdsall and D. J. Lodge, personal observation).

DATA COLLECTION

In addition to date of sampling, macrohabitat information (e.g., forest type, plant association, climatic zone, elevation), host, plot number, and other standard collection data, the following information should be collected when sampling from woody substrata:

1. Substratum/host class: information on substratum type (e.g., trunk, branch, twig, stump, or root), condition (e.g., living, dead, or dead on living plant), and position (e.g., vertical, >60°–90°; prostrate; on ground; or suspended (<1 m or >1 m above ground).
2. Substratum preference(s) (from Mathiassen 1993): where the fungus is fruiting, for example, on wood only (W), mainly on wood but some on bark (Wb), equally frequent on wood and bark (wb), mainly on bark but some on wood (Bw), or on bark only (B).
3. Substratum size class: diameter of the substratum, using the following size classes: less than 1.0 cm, 1.0–2.5 cm, 2.6–5.0 cm, 5.1–10 cm, 10.1–20 cm, or more than 20 cm.

4. Decay class: condition of the wood; for example, as hard and not decayed; hard but decayed and either discolored or not; softened and punky (light weight, fibrous, spongy; this condition is termed white rot); well decayed (humus-like; dark; friable or, if the remaining wood is primarily of lignin, breaking into small hard blocks; this condition is termed brown rot).
5. Condition of bark: tight, loose, or off.

APPROACHES TO SAMPLING MACROFUNGI

Thomas E. O'Dell, D. Jean Lodge, and Gregory M. Mueller

INVENTORYING VERSUS MONITORING

At present most inventory work is conducted to increase our knowledge of fungal diversity and to learn about habitat preferences and geographic distributions of different taxa. A recent development in that respect is the inventory of target taxa for conservation purposes legally mandated as part of the federally approved management plan for old growth forests in the Pacific Northwest (Forest Ecosystem Management Assessment Team 1993; O'Dell et al. 1996; Molina et al. 2001). Although no nonlichenized fungus is protected yet by that plan, 225 species of old-growth-associated macrofungi in the Pacific Northwest are legally required to be surveyed and managed (USDA Forest Service and USDI Bureau of Land Management 2000; O'Dell et al. 2003). Reasons for undertaking inventories include prioritization of potential sites for protection and assessment of the impacts of different land management practices on biodiversity (e.g., O'Dell et al. 1992a; Luoma et al. 1996b, 1996c, 1996d). Monitoring is undertaken to assess effects of commercial harvest on edible fungal sporocarp production and on population size (Pilz and Molina 1996), to determine long-term trends in populations of target species, and to determine effects of pollutants on different fungi (Gulden et al. 1992). Baseline data on the diversity and distributions of the taxa to be monitored (i.e., inventories) of the region under study are normally necessary before monitoring can begin.

Several differences exist between sampling for biodiversity, as discussed throughout this chapter, and monitoring. Monitoring programs are designed for determining population trends (rather than overall richness), extend for longer periods, generally focus on fewer taxa (often indicator species), and include many more sites. Because monitoring efforts need to be carried out over long periods, archiving of data and careful documentation of sites and methods are priorities. One should consider the documentation required to facilitate the repetition of a study at a site 20 years in the future and plan accordingly.

The use of fungal "indicator species" to monitor environmental change or assess ecosystem health should be considered carefully. Pearson (1994) proposed several criteria for the selection of indicator taxa, concluding that they should: (1) be taxonomically stable and well known; (2) have a well-known natural history; (3) have readily surveyed and manipulated populations; (4) be sensitive to some disturbances; (5) be a component of a widely distributed higher taxon; (6) include species exhibiting habitat specificity; (7) include species with economic value; and (8) have distributions or patterns of species richness that are correlated with those of other, unrelated groups.

Compared with vertebrates and plants, macrofungi are poorly known taxonomically, especially at the species level (although some taxa are well known); they are difficult to survey and manipulate, especially in natural ecosystems; and their natural histories, beyond a cursory level, are poorly known. However, microfungi can be sensitive to human-generated disturbances, such as air pollution (Gulden et al. 1992; Fellner 1993; Carreiro et al. 2000) and logging (O'Dell et al. 1992a), as well as to natural disturbances such as hurricanes, volcanic eruptions, and landslides (Lodge and Cantrell 1995b; Lodge 1996; Miller and Lodge 1997; Heilmann-Clausen 2001). They have globally distributed genera as well as locally endemic species; include economically valuable taxa (Molina et al. 1993); and influence the distribution of other organisms such as fungivorous insects (Worthen and McGuire 1990), rodents (North 1993), and plants (e.g., Bidartondo et al. 2000; Blackwell 2000; Kretzer et

al. 2000). Thus, macrofungi appear to be of intermediate value (five or eight attributes) as indicator organisms.

Drawbacks to monitoring macrofungi include limited taxonomic expertise, annual fluctuations in sporocarp occurrence and abundance, and lack of well-tested protocols for sampling. The first problem can be overcome through careful selection of taxa and species documentation with voucher specimens, which are essential for addressing taxonomic problems and authenticating the taxa being monitored. Annual variation in occurrence of species remains troublesome. The fact that a particular species may not fruit on a site for a number of consecutive years but then be abundant in other years detracts from the utility of macrofungi for monitoring efforts. Species once present at a site that fail to fruit at that site in subsequent years may be present and physiologically active but not fruiting, present as dormant propagules, or extirpated. The causes of such variation are poorly understood, but efforts to improve detection of, for example, ectomycorrhizal species, have met with limited success. Several studies comparing fungi that fruit versus those colonizing roots (as detected by polymerase chain reaction–amplified DNA) have reported a low correlation between ectomycorrhizal fungal diversity and species composition estimates based on sporocarp data and analysis of mycorrhizal roots (e.g., Gardes and Bruns 1996; Peter et al. 2001). Therefore, each phase of monitoring macrofungal species—that is, detection of the species and assessment of its status—requires many years of observations. Macrofungi are most useful for monitoring in areas for which long-term records of their occurrences are available (Arnolds 1991; Ohenoja 1993) or in which sites being compared follow a known environmental or disturbance gradient (Gulden et al. 1992; O'Dell et al. 1992a; North 1993).

INVENTORYING MACROFUNGI

As with most fungi, it is difficult to quantify populations of macrofungi. Individuals rarely can be distinguished in the field because an individual mycelium may produce from one to many sporocarps. The sizes, numbers, and dry weights of sporocarps vary greatly among taxa, and sporocarps represent an unknown fraction of total biomass, which also includes the vegetative mycelium and sclerotia.

Mycologists traditionally have sampled sites by walking through the area collecting conspicuous specimens of select taxa. That opportunistic approach, however, does not allow for rigorous comparisons of different sites, which requires that sampling intensity be standardized at each site. Collection of all fungi within a series of plots or transects ensures that all taxa fruiting at the time are scrutinized and reduces the likelihood that cryptic species will be overlooked. Often with such an approach many specimens are identifiable initially only to genus. If the same plots or transects are sampled repeatedly for several years, however, most taxa eventually will be identified.

Currently, the best measure of species abundance is plot frequency (the number of subplots in which a taxon occurs) because it reflects the minimum area occupied by that species in the study site. Because the sporocarps of a species distributed over several sampling units (subplots) may have originated from one large or several small mycelia (cf. Jacobson et al. 1993; Dahlberg and Stenlid 1994), however, plot frequency provides only a rough estimate of abundance and importance (Schmit et al. 1999). Consequently, size, shape, and spacing of sampling units (plots, transects) are important aspects of sampling design.

TYPES OF SAMPLING UNITS

Studies should be designed to encompass the spatial and temporal scales appropriate to the taxa of interest. A fundamental distinction must be made between fungi that occur on substrata that form readily observable, discrete, natural sampling units and those that occur on more continuous or concealed substrata that require arbitrary plot boundaries. Natural sampling units are appropriate for species fruiting on leaves, logs, or cones, whereas arbitrary sampling units are required for decomposers of litter or humus and for mycorrhizal fungi.

Studies in which terrestrial macrofungi are surveyed usually use arbitrary sampling units, or plots. Plots range in size from 1 m² to 1000 m² and can be square, rectangular, or circular. The same plots often are scrutinized for several years. When studies involve removal of most sporocarps (e.g., to determine sporocarp productivity), however, some investigators move plots on each sampling occasion to avoid effects of disturbance (Luoma 1991; O'Dell et al. 1999). An important drawback of changing plot location, deriving from the fact that macrofungi have patchy distributions, is that annual variation in the occurrence of fruiting species becomes confounded with variation in plot location, which can greatly influence the results. Moving the plot location results in a single sampling event over the area, and the data collected are not comparable to data from permanent plots sampled over time. Permanent plots provide a good estimate of diversity for a defined area, as well as information on annual variation in fruiting phenology. They also are easy to relocate and measure. Although plots may underestimate the diversity in a larger area beyond the plot borders, such problems are associated with any sampling method. The effects of long-term removal of

sporocarps and perturbation of a site by repeated sampling are not well known, but preliminary studies have provided little evidence of negative effects (Egli et al. 1990; Norvell and Roger 1998). Moving plots may be appropriate, however, when they are raked for truffles (see Chapter 10) in addition to mushroom sampling.

SAMPLING VEGETATIVE MYCELIA

Most of our discussion has focused on the sampling of sporocarps because these are the most readily identifiable structures of macrofungi. Other structures that can be sampled include sclerotia, mycorrhizae, mycelia, mycelial strands, cords, and rhizomorphs. Those vegetative structures can be useful in determining the proportion of taxa that are present but not fruiting. Molecular analyses of ectomycorrhizae have shown that substantial numbers of species present on host roots at some sites rarely fruit (Gardes and Bruns 1996). Standard methods for sampling ectomycorrhizae are given in Johnson and associates (2000).

Although we rely largely on field sampling of sporocarps, encouraging fruiting of some ascomycetes and small agarics by placing substrata in humid chambers may increase detection of litter inhabiting taxa. See Appendix I.

CONVENIENCE VERSUS PLOT SAMPLING

Methods of sampling must be chosen to fit the goals of a particular study. If the goal is to get a preliminary estimate of richness at a site, then some sort of survey, or convenience sampling, may be most efficient. In that type of sampling, experienced individuals visit a location and document the taxa that they find. The approach is a traditional one among taxonomists and can be an efficient way to obtain a snapshot of the diversity present. It also may be useful for rapid assessment of potential sites for reserves. For example, convenience sampling was used to survey the macrofungi of Barlow Pass, Washington (Ammirati et al. 1994). Volunteers from the local amateur mycological society made weekly visits to a site for 2 years. During each visit, they walked through the site collecting samples of each taxon encountered. Although more of the taxa present, at least of the large and showy species, may have been detected with this approach than would have been detected with plot sampling, that hypothesis remains untested.

Relying solely on convenience sampling, however, precludes comparisons of diversity and species composition of macrofungi between sites, or at a single site over time, because convenience sampling does not provide for the necessary standardization of sampling effort. Additionally, collector bias during convenience sampling may cause some habitats or taxa to be overlooked. Sampling in plots provides quantitative data because sampling effort is standardized, either through the sampling of an explicitly defined area or by sampling for an explicitly defined amount of time. Removal and examination of all sporocarps in plots also ensures that species difficult to distinguish in the field are collected, and the imposed scrutiny of a defined area results in the detection of more inconspicuous taxa.

DISPERSED SUBSAMPLES

Ecologists have been sampling macrofungi for several decades but have made little effort to optimize or standardize sampling approaches (Cooke 1955; Arnolds 1992). For example, in many studies, large rectangular plots, divided into subplots, are observed over several years. Permanent plots have the advantage of being monitored easily over time. Because of the clumped distribution of sporocarps of many fruiting macrofungi, however, subplots adjacent to one another are more likely to contain the same species than those at a distance (Murakami 1987). Therefore, the use of contiguous subplots may underestimate species richness (O'Dell et al. 1999). That spatial autocorrelation can be addressed by dispersing plots in various ways (Luoma et al. 1991; Lodge and Cantrell 1995a; O'Dell et al. 1999; Mueller and Mata 2000), which are discussed under "Recommended Protocols for Sampling Macrofungi," at the end of this chapter.

More spatially explicit studies for various groups of macrofungi are needed. For example, saprobes may exhibit distributions different from those of ectomycorrhizal species. G. Walker and J. Ammirati (unpublished data) found that species-area curves for saprobic fungi leveled off sooner than those for ectomycorrhizal fungi in four Sitka spruce (*Picea sitchensis*) stands on the Olympic Peninsula in Washington State. Schmit and colleagues (1999), however, did not observe a difference in the shape of species-effort curves among the different ecological guilds of macrofungi sampled in a northwest Indiana oak-dominated forest over a 3-year period.

ADAPTIVE SAMPLING

Studies of macrofungi typically use sampling designs of fixed area. In other words, investigators decide prior to sampling how many plots of what size to examine at each site. A recent innovation with possible value to mycologists is adaptive sampling, which refers to any case in

which sampling effort is modified in response to observations made in the course of sampling. Usually some threshold value for the number of samples of the target organisms is set. When the number in a plot exceeds that value, another area is sampled in some standard predetermined way. Adaptive sampling is useful for increasing the number of interesting observations (e.g., macrofungal sporocarps), particularly for organisms that are rare and have patchy distributions. Thompson (1992) has provided statistical methods for dealing with bias resulting from unequal sampling effort that must be accounted for when comparing values from different sites sampled using adaptive sampling methods (Figs. 8.35A and B).

DETERMINING ADEQUATE SAMPLING

Deciding how much area to sample; the size, number, and distribution of plots; and the frequency of sampling is not a trivial undertaking and, with regard to fungi, deserves much more attention than it has received. When possible, investigators carrying out biodiversity studies should apply methods that allow for spatially and temporally explicit analyses (e.g., see Colwell and Coddington 1994; Lodge and Cantrell 1995a; Schmit et al. 1999).

One rule regarding the amount of area to sample that sometimes is applied to diversity studies states that sampling is adequate when every taxon occurs in at least two sampling units. Our experience indicates that this criterion is difficult, or probably impossible, to meet when sampling macrofungi. Ecologists have developed species-effort curves for other organisms to address this issue, and such curves occasionally have been applied to fungi (Arnolds 1992). For a given landscape, stand, or habitat, the number of species observed tends to increase steeply with increasing sample area and then level off as fewer new taxa are encountered.

Applying species-effort techniques to macrofungi has demonstrated the need to sample an area larger than that required for plants. Arnolds (1992) recommended plots of 1000 m² in forest or 500 m² in grasslands, with a minimum of five plots per community type being sampled. One also can plot cumulative species richness versus the number of years of sampling to estimate adequate duration. J. Ammirati and T. E. O'Dell (unpublished data), however, observed a continuous increase in mycorrhizal species richness over 5 years of sampling. Similarly, Schmit and colleagues (1999) documented an

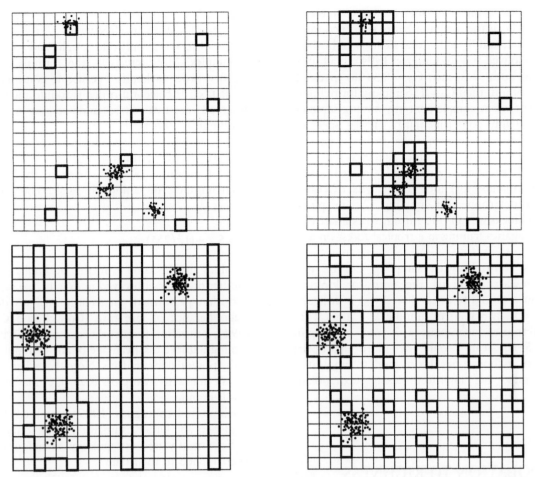

FIGURE 8.35 These figures show how an adaptive sampling is carried out. The upper panels illustrate adaptive cluster sampling to estimate the number of point objects in a study region of 400 units. An initial random sample of 10 units is shown on the left. Adjacent neighboring units are added to the sample whenever one or more of the objects of the population is observed in a selected unit. The resulting sample is shown on the right. (Both upper panels from S.K. Thompson, "Adaptive cluster sampling," 1990, *Journal of the American Statistical Association* 85: 1050–1059. With permission from the American Statistical Association.) The lower left panel illustrates adaptive cluster sampling with initial random selection of five strip plots. The final sample obtained is indicated with the heavy line. The lower right panel illustrates adaptive cluster sampling with initial random selection of two systematic samples. The final sample obtained is indicated with the heavy line. (Both lower panels from S.K. Thompson, "Adaptive cluster sampling: Designs with primary and secondary units," 1991, *Biometrics* 47: 1103–1115. With permission from the Biometric Society.)

increase in species richness for saprobic and mycorrhizal taxa over a 3-year period, and Straatsma and associates (2001) continued to detect similar numbers of new species during each year of a 21-year study. We are unaware of any recommendation for the minimum duration of an inventory, but based on our experience, 10 years would be desirable.

Sampling design should be considered an iterative process. Data from early efforts can be used to make decisions about ways to modify later sampling. For example, the occurrence in multiple plots of most taxa observed in a study (an unlikely circumstance) would indicate that sample area could be reduced. Conversely, if most samples included unique taxa, sampling intensity should be increased.

Complementarity (dissimilarity) is an empirical measure of the degree of species uniqueness between different samples. It is expressed as the proportion of total species encountered that is unique to one sample or the other. An empirical test for complementarity is as follows:

$$C_{jk} = U_{jk} / S_{jk}$$

where C_{jk} is the complementarity between samples (or sites) j and k, U_{jk} is the total number of taxa encountered at only one site, and S_{jk} is the total richness of both sites combined (Colwell and Coddington 1994). If all species are found at both sites, then $C_{jk} = 0$; if no species are found in common, then $C_{jk} = 1$. Sampling for richness is most efficient when complementarity among samples is about 0.5. High complementarity (fewer species in common) indicates that many taxa probably are being overlooked, and a lower value suggests that some samples are redundant. Lodge and Cantrell (1995a; unpublished manuscript) used an analysis of complementarity to determine that sampling 12 1-m² plots in each of two blocks was an efficient way to estimate diversity of litter agarics fruiting in a tropical forest on a particular date.

Resources constrain what sampling can be done. Although research indicates the need for long-term studies of macrofungi to assess diversity, it is difficult to obtain more than 3 years of funding at a time. Therefore, most studies incorporate 1–3 years of sampling, although 5–10 years would be more appropriate. Estimates of richness based on modest amounts of sampling may be improved using extrapolation techniques.

EXTRAPOLATION TO RICHNESS

Because large numbers of species inhabit small areas, studies of fungal diversity tend to focus on fairly narrow taxonomic or ecological groups. That is obviously inappropriate when estimates of total species richness are sought. Proposals are being made to assess total biotic diversity. Although that is a worthwhile goal, available resources are obviously insufficient to support all taxa inventories at more than a few sites. More often, total diversity for specific taxa or ecological guilds will have to be estimated from some modest amount of sampling.

Several methods for estimating total diversity from a limited number of samples have been developed by mathematical ecologists (see discussion in Colwell and Coddington 1994). A full discussion of extrapolation procedures is beyond the scope of this chapter, but techniques generally are based on the proportion of rare species, typically those encountered in one or two samples. Simply put, the greater the proportion of rarely encountered taxa, the more a diversity estimate should be revised upward. The jackknife is one such procedure and is carried out according to the formula:

$$S^* = S_{obs} + L[(n-1)/n]$$

where S^* is estimated total richness, S_{obs} is the total number of species in all samples, L is the number of species observed in only one sample, and n is the number of samples. More precisely, the formula is known as the first-order jackknife because it considers only the rarest taxa; the second-order jackknife includes species observed at two sites, and so on (Colwell and Coddington 1994).

Another measure for estimating richness from samples is Chao's second estimator, "Chao 2." It is one of the simplest measures as well as one of the most reliable when numbers of samples is small. The formula is as follows:

$$S^* = S_{obs} + (L^2/2M)$$

Variables are the same as in the jackknife, and M is the number of species observed in exactly two samples. This algorithm performed particularly well on data with a preponderance of rare taxa (Colwell and Coddington 1994), a situation frequently encountered with fungi. It also highlights a disadvantage of any extrapolation—that is, the potential for artifacts. Note that when the number of taxa in exactly two plots is zero, estimated total richness is infinite.

Schmit and colleagues (1999) examined the utility of these two and other estimators. They found that none of the currently used extrapolation techniques was robust when applied to their data.

RECOMMENDED PROTOCOLS FOR SAMPLING MACROFUNGI

Gregory M. Mueller, John Paul Schmit, Sabine M. Huhndorf, Leif Ryvarden, Thomas E. O'Dell, D. Jean Lodge, Patrick R. Leacock, Milagro Mata, Loengrin Umaña, Qiuxin (Florence) Wu, and Daniel L. Czederpiltz

Many factors influence the diversity of macrofungi at a particular site (see "Factors Influencing Species Richness," earlier in this chapter). Those factors include the nature of the habitat (e.g., grassland or forest, natural or planted, early successional or mature), the diversity of plant species (particularly host plants), the diversity of substrata, geographic location (latitude and elevation), soils types, and climate. Site factors are compounded by the amount of natural and manmade disturbance, site-management practices, and exposure to pollution. The protocols used in a study also have an impact on the observed diversity of macrofungi.

Schmit and associates (in review) undertook a meta–data analysis of plot-based studies that measured the diversity of both macrofungi and trees. They performed two analyses on 25 studies involving a total of 184 plots located in North America, Europe, China, and Costa Rica. The first analysis determined that, although plots contained more macrofungi than trees, the macrofungi were neither more nor less widely distributed than tree species. The second analysis looked at the effects of sampling effort (plot size, number of years sampled, number of visits), tree diversity (number of tree species present), number of ectomycorrhizal tree species present, and the type of tree present (conifers versus hardwoods), and latitude on the number of macrofungal species present in the plot. Although sampling effort had a major impact on the diversity of macrofungi discovered at a particular site (in general, more taxa are found with an increased effort), habitat type and tree diversity played a larger role in explaining differences between studies than did sampling effort. However, differences in the sampling protocols used in the 19 studies interfered with direct comparisons of their results.

Fortunately, protocols used to sample terrestrial macrofungal diversity have begun to converge in recent years (e.g., Richardson 1970; Jansen 1984; Tyler 1985, 1989; Bills et al. 1986; Villeneuve et al. 1989; Gulden et al. 1992; Salo 1993; Leacock 1997; Rossman et al. 1998; O'Dell et al. 1999; Schmit et al. 1999; Straatsma et al. 2001; P. R. Leacock and D. J. McLaughlin, unpublished manuscript). Many of the investigators used variations of a common sampling theme—that is, sampling all of the macrofungi occurring in relatively small circular plots set out evenly along transects of defined lengths, with a total area of 0.1 hectares sampled per site. The most commonly used protocols differed in the sizes of the circular subplots (4 m^2 versus 5 m^2) and, consequently, the numbers of subplots and the lengths of the transects. Even if the total area surveyed is the same, using subplots of different sizes could influence frequency data because species frequency is calculated as the proportion of all subplots in which it is present. A 0.1-hectare area contains 250 4-m^2 subplots but only 200 5-m^2 subplots. We recommend the use of 5-m^2 circular subplots for two reasons: (1) the remaining forest fragments in many of the areas where we are working (e.g., Central America, China, Midwestern North America) are small, making it difficult to establish long transects; and (2) we can make use of grid markers (usually based on 10-m × 10-m grids) established as part of plant inventory projects as the same study sites. Using established grids facilitates setting up transects and, more importantly, facilitates comparison of the plant and macrofungal diversity data. Setting up the 200 5-m^2 subplots is straightforward, and sampling them is efficient; in addition, our collecting team of students, interns, volunteers, and parataxonomists has expressed satisfaction with the layout. Sampling frequency and number of plots per site given here (see "Recommended Protocols") are minimal values. If resources are available, we suggest increasing the number of plots per site and/or sampling intensity.

Protocols for sampling macrofungi occurring on large woody substrata also are becoming more similar (Lindblad 1998, 2000, 2001; D. L. Czederpiltz, unpublished data). The primary difference between our recommended protocols and some other currently used protocols (e.g., protocol used by Czederpiltz et al. 1999; D. L. Czederpiltz, unpublished data) is that the latter protocols require destructive sampling (e.g., rolling logs, removing bark), whereas our protocols minimize disturbance to the site. Destructive sampling enables one to record all of the fungi fruiting on a substratum at the time of sampling but precludes resampling of that substratum. Our protocol allows for resampling, thereby allowing one to record macrofungi fruiting on the surface of the same logs over time. Because of temporal changes in the species fruiting, we prefer to sample over time, although we may miss fungi that potentially are fruiting on the underside of logs or under their bark.

We recommend using an integrated set of sampling protocols for macrofungi, including opportunistic sampling, sampling of fixed-size plots, and sampling of a fixed number of downed logs. That practice will optimize the number of macrofungal species recorded at a site by including all the conspicuous fungi while also providing quantitative data that are comparable with data from other sites. Repeated sampling of relatively small fixed-size plots and large trunks ensures that inconspicuous fungi and fungi on relatively scarce substrata are included in the study. These protocols are being used in the Costa Rican National Fungal Inventory (Mueller and Mata 2000).

RECOMMENDED PROTOCOLS

OPPORTUNISTIC SAMPLING OF MACROFUNGI

By opportunistic sampling we mean carefully walking through a study site and collecting conspicuous sporocarps. Collectors should sample as many habitats in the site as possible. This technique does not yield quantitative data. Nevertheless, it is an important adjunct to the plot-based quantitative methods discussed next because additional species of macrofungi will be seen "off plot" as a result of the patchy distribution of sporocarps. Thus, a combination of opportunistic and plot-based sampling is necessary to maximize the macrofungal diversity documented at a site.

SAMPLING CONSPICUOUS MACROFUNGI USING FIXED-SIZE PLOTS

The protocol for sampling conspicuous macrofungi using fixed-size plots includes the following steps:

1. The investigator selects an area within the site for the permanent plot. The area should be representative of an important forest or grassland type at the site and should be chosen to optimize the diversity of habitat types sampled by the entire study (i.e., if the study covers grasslands, open woodlands, and dense forest, at least one plot should be set up in each habitat type). The plot should be established in as homogeneous an area as possible and as easily accessible from the road or trail as possible, without being susceptible to "edge effects" because it will be visited repeatedly.
2. Each plot consists of 10 transects that are 100 m long. Typically, transects are laid out parallel to one another at 10-m intervals. If the shape of the area to be sampled does not allow that, however, some transects can be laid out end to end. Transects are marked every 5 m with a flag or stake. Each transect is assigned a unique letter, and each flag should be numbered sequentially within a transect (i.e., A1–A20; B1–B20; . . . ; J1–J20).
3. Each person sampling carries a plastic pipe or wooden pole and a rope that is 1.262 m long; the rope is used to circumscribe 5-m^2 circular subplots around each flag in a transect, giving rise to 20 5-m^2 subplots per transect, for a total of 200 subplots, or a sampling area of 1000 m^2 (0.1 ha) per plot. Care should be used to not walk in or unnecessarily disturb the subplots.

4. All macrofungi occurring in a subplot are collected, labeled with the transect letter and subplot number, and placed in an appropriate bag or container. At the end of the collecting day, specimens are transported back to the field station for sorting, describing, photographing, and drying. The substratum (soil, leaf litter, wood) is noted for each specimen.

Ideally, subplots should be sampled every 2 weeks during the fruiting season. Often, however, sampling intensity must be reduced because of limited resources. Generally, it is possible to accomplish that without compromising the quality of the diversity information being collected. G. M. Mueller and his colleagues (unpublished data) intensively sampled plots in the Chicago area for 1 year. Based on the data from that year and on herbarium records, they determined peak fruiting times for the region and adjusted their sampling schedule accordingly so as to optimize their sampling effort.

SAMPLING SMALL ASCOMYCETES USING MICROPLOTS

The goal when sampling small ascomycetes is not to cover a lot of area but to obtain a quantitative sample of the microfungi growing on small substrata, especially those species that are infrequently or rarely collected. The sampling protocol for microfungi is similar to that for macrofungi and involves the following steps:

1. Investigators establish a microplot for sampling microfungi adjacent to a macrofungal subplot along the transects. Because substrata in the plots are collected, different plots are laid out each time a sample is taken.
2. Each person sampling carries a 0.56-m-long (or 1.128-m-long) plastic pipe or wood pole, which is used to circumscribe 1-m^2 circular subplots. The number of subplots per site to be sampled depends on the team's resources. Sampling the plots is time consuming and labor intensive, involving careful examination of all small substrata present. The time necessary to complete a sample varies with the diversity of substrata in the plot and the number of people sampling them. In Costa Rica, team members were instructed to sample as many plots as they could within 1 week. Most individuals sampled only two or three 1-m^2 subplots in a week because of the time involved in scanning the substrata to observe the microfungi.

3. All microfungi on twigs, branches, and leaves occurring within a plot are collected, labeled with the transect letter and subplot number, and placed in appropriate bags for transport back to the field station at the end of the collecting day. The samples then are sorted, examined under a stereomicroscope to determine if they are fertile, described, divided, and dried. The substratum (branch, twig, leaf) is noted for each specimen.

4. At a minimum, microfungal plots should be sampled twice a year—in the middle and at the end of the fruiting season.

SAMPLING A FIXED NUMBER OF DOWNED LOGS

Restricting quantitative sampling to the 0.1-ha plots and the microplots would exclude most of the fungi found on large pieces of wood because the frequency of those larger substrata within small subplots is generally very low. Therefore, fungi occurring on larger pieces of wood must be sampled separately. The following protocols should capture a good percentage of the fungal diversity on those substrata as well as provide quantitative data on abundance and host and size-class specificity that can be compared with such data from other areas and studies.

1. Logs to be sampled should measure more than 20 cm in diameter and more than 2 m in length and should be lying on the ground.

2. At each site, the investigator selects 30 logs in each of the following decay classes:
   Class 1: Relatively newly fallen, usually retaining its bark
   Class 2: Medium rotten; bark fallen off; knife can penetrate—2 cm into the wood without undue pressure
   Class 3: Thoroughly rotten; knife can penetrate into the wood without much pressure; the wood can be partly destroyed with the fingers.

3. Each log is marked with a colored plastic band and given a number, and its position is mapped. If possible, the tree is identified to genus or species. The length and diameter of each log is measured and recorded. If the diameter of the log varies greatly from one end to the other, the investigator measures the diameter near both ends and at the middle. If the log has major branches, each of those also is measured.

4. Generally, only two to three sporocarps of each common species are collected from each log; multi-

ple sporocarps of species not or only infrequently encountered previously are collected. If many sporocarps of a particular species are present, a pin or other marker can be used as a reminder not to collect the same species again on the next trip.

5. Small specimens, such as those of corticoid fungi and ascomycetes, are collected whole, whereas large polypores are sectioned radially at a width of approximately 1.0–1.5 cm to enhance drying and discourage mold. Specimens are labeled with the log number and placed in appropriate containers. At the end of the collecting day, specimens are transported back to the field station for sorting, describing, photographing, and drying. When possible, the genus or species of the host tree is noted.

As mentioned earlier (see "Determining Adequate Sampling" and "Sampling Conspicuous Macrofungi Using Fixed-Size Plots"), sampling should be carried out every 2 weeks to ensure that all fleshy agaric species are collected. If that sampling intensity is not possible because of limited human and/or financial resources, good diversity information on lignicolous macrofungi still can be obtained with less frequent sampling. Sampling logs four times per year (in the dry season and at the beginning, middle, and end of the rainy season), for example, has worked well for the Costa Rican National Fungal Inventory.

SUMMARY

The material in this chapter is intended to provide investigators with some direction in planning and conducting inventories of macrofungi. We hope that the recent growth of such studies will continue and be nurtured by the information that we have provided. In particular, we emphasize the need for well-planned research with clearly stated goals. Monitoring of fungi is a more recent undertaking that may be useful in detecting anthropogenic disturbances, such as air pollution, and quantifying their impacts.

Project planning should include background literature research on the vegetation and geology of the study area, as well as on the taxa of fungi likely to be encountered. Pilot studies or preliminary sampling of the fungi are useful in determining the intensity of sampling required to achieve the goals of the survey. Such studies also will provide insight into the numbers of specimens likely to be acquired and the taxonomic difficulty of the project.

Executing the project involves careful collection of specimens; documentation of the resulting specimens with written descriptions, photographs, spore prints, cultures, and macrochemical tests; preservation of specimens; and archiving of voucher specimens and data in recognized herbaria. The processing of large numbers of specimens involves some prioritization because not all specimens are equally ephemeral or useful. Data analyses can include extrapolation of site richness from samples and complementarity tests to evaluate sampling efficiency. Use of the sampling protocols recommended in this chapter will yield standardized data on the diversity of macrofungi found fruiting on soil, leaf litter, and woody substrata comparable to data from equivalent studies at other times or sites. Nevertheless, research on sampling design remains a priority for improving sampling efficiency at all stages of the study and the quality of the data obtained.

At present, the greatest constraints on studies of macrofungal diversity are the paucity of fungal taxonomists and identification resources. No region of the world as yet has a complete mycota equivalent to a vascular-plant flora, a condition likely to persist for some time. We must not let that obstacle prevent us from carrying out inventories, but it is a limitation that we should strive to overcome.

9

LICHENIZED FUNGI

SUSAN WILL-WOLF, DAVID L. HAWKSWORTH, BRUCE MCCUNE,
ROGER ROSENTRETER, AND HARRIE J. M. SIPMAN

lichenized fungi pursue a lifestyle different from that of most other fungi but similar to that of plants, in that they operate as autotrophic, photosynthetic units. These units are more physically compact, spatially circumscribed, and distinguishable from the substratum than are units of other fungi. As a consequence, lichenized fungi have been subjects of far more studies involving quantitative sampling than have other fungi.

In this chapter we briefly review the ecology of lichenized fungi, summarize current knowledge of their taxonomy and patterns of diversity, and discuss important variables and criteria for designing inventories of their biodiversity. Nonquantitative biotic surveys can make important contributions to our knowledge of biodiversity, especially in poorly collected areas. This chapter will focus, however, on organized biodiversity surveys with at least minimum levels of quantification and repeatability. Information on the biodiversity of lichenized fungi can also contribute to the development of management strategies for their conservation in nature reserves and to efforts to retard loss of biodiversity and ecosystem function on lands managed for human use and resource consumption (Hawksworth 1991; Galloway 1992a; Rose 1992; Rosentreter 1995).

Lichens are composite organisms consisting of a fungal partner (the mycobiont) and a photosynthetic partner (the photobiont) that grow symbiotically with one another to form a coherent structure (the thallus). Through their participation in the lichen symbiosis,

LICHEN CHARACTERISTICS

The relationship between the fungal partner and the photosynthetic partner of lichens has been variously described

(Nash 1996). It sometimes is considered a mutualism in which both partners benefit; the fungus gains carbohydrates, and the photosynthetic partner is protected from desiccation. Alternatively, the association is considered to be a controlled parasitism in which the fungal partner benefits and the photobiont loses by growing more slowly than when alone. From an ecological perspective, the relationship is usually mutualistic because the integrated lichen thallus survives in more habitats and is more abundant than is either partner when alone.

An individual thallus usually is composed of one species of fungus and one species of either alga or cyanobacterium. Many thousands of fungal species, but only about 150–200 species of photobiont, have been identified as participants in lichen symbioses. Only the fungal partner reproduces sexually while in the association.

In contrast to most other fungi, the thallus of a lichen usually has a definable boundary; a distinctly layered structure; and a characteristic, often species-specific morphology. Most lichen thalli fall into one of five general morphological categories (Fig. 9.1): leprose, composed of mealy particles of intertwined fungal hyphae and algal cells; crustose, embedded on or in the surface of the substratum; squamulose, composed of small flakes of thallus; foliose, leaflike, with distinct upper and lower surfaces; and fruticose, shrubby or stringlike (Fig. 9.2). Although the fungal partner produces the characteristic structure of the lichen thallus, the morphological form of the composite thallus is a result of the interaction; isolated mycobionts on agar plates form colonies similar to nonlichenized fungi (and without typical "lichen" structures), although they grow more slowly than most fungi and are very compact. Scientific names given to the lichen thallus are based on and refer to the fungal partner (see "Identification," under "Taxonomy, Diversity, and Distribution," later in this chapter).

A number of functional characteristics are common to most lichens. These include autotrophic nutrition via photosynthesis, (mostly) slow growth, small size, long life, long-lasting (nonseasonal) vegetative morphology and reproductive structures, mineral nutrition mostly from airborne sources, and greater tolerance of desiccation than most other photosynthetic organisms in the same habitat.

Clearly, the partners have independent phylogenies, with the integration represented by the lichen thallus best described as a product of coevolution. In addition, the fungus and/or the photobiont cells of a single lichen thallus are not necessarily products of a single gene line, even when they are from a single species (see "Basis for Taxonomic Distinctions," under "Taxonomy," later in this chapter). The concepts of "organism" and "individual" for the lichen thallus thus do not necessarily have the same implications of genetic coherence as they do

when applied to humans or to vascular plants (Allen and Hoekstra 1992).

The lichen habit is remarkably widespread among fungi (>13,500 species, or about 20% of all described fungal species; Sipman and Aptroot 2001; Hawksworth et al. 1995). Lichenized fungi belong to several distantly related orders and families and are not a cohesive taxonomic group (see "Taxonomy," later). Rather, they represent a biological strategy that has evolved on separate occasions in different groups of fungi. Their cohesiveness as a group thus relates to their similar ecological and functional roles and the common constraints of the lichen habit. For more information on lichen biology, see Brodo and colleagues (2001), Galun (1988), Hale (1983), Hawksworth and Hill (1984), Lawrey (1984), and Nash (1996).

The roles of lichenized fungi in communities and ecosystems are best understood by studying the lichen thallus as a photosynthetic unit that in many ways is equivalent to plants. For this reason, methods for assessing the diversity of lichenized fungi can be more similar to those for plants than for other fungi. To study the biogeographic and taxonomic relationships of lichenized fungi, however, one needs to investigate the phylogenetic relationships of the fungi and, to a lesser extent, of their photobiont partners.

LICHEN ECOLOGY

Lichenized fungi are found in every terrestrial habitat capable of supporting photosynthesis, and a few lichens occur in aquatic habitats as well. The photosynthesizing lichen units compete with plants for light, space, and possibly other resources. Because of their small stature and extremely slow growth, lichens are poor competitors, perhaps best described when compared to vascular plants as extreme examples of Grime's (1977) stress-tolerator lifestyle. The only plants with which lichens sometimes appear to compete on an equal footing are bryophytes, with which they often are compared in community studies (Nash and Egan 1988). Major factors affecting the presence and abundance of lichenized fungi are the following: (1) substratum chemistry, stability, and longevity; (2) light availability, with effects often mediated through competition with faster growing, larger plants and other lichens; and (3) moisture availability. Common substrata that support lichen growth include rock surfaces, woody plant bark and wood, soil and dead organic matter in low productivity environments and microhabitats, and broad evergreen leaves in the humid tropics. Most species have at least some substratum and habitat preferences, although individual species vary widely in substratum and habitat specificity.

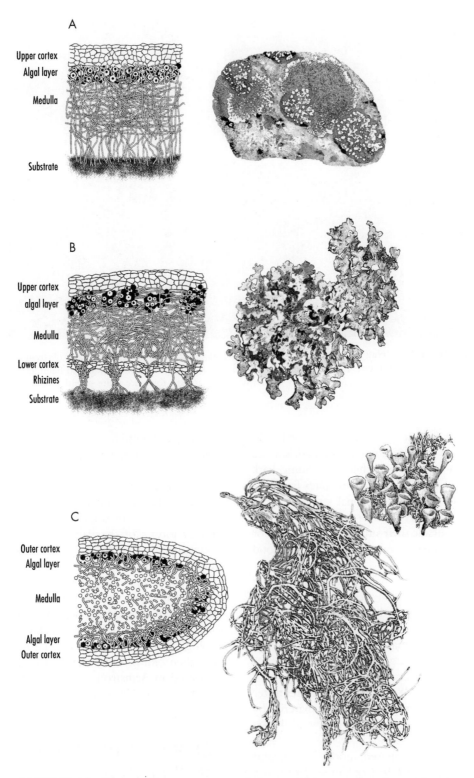

FIGURE 9.1 Three of five common growth forms of lichens: surface views and cross-sections. **A.** Crustose lichens usually have several layers, with the medullary fungal hyphae attached directly to the substratum. **B.** Foliose lichens have distinct upper and lower cortex layers and usually are attached to the substratum by special groups of rootlike fungal hyphae called rhizines. **C.** Fruticose lichens are stalked, tufted, or pendulous and usually have a uniform outer layer. (Redrawn with permission from fig. 2 of Ahmadjian 1993 by Kandis Elliot.)

FIGURE 9.2 Hanging fruticose lichens such as *Ramalina menziesii,* growing on a *Quercus garryana* in the Willamette Valley of Oregon, United States, are often most abundant in habitats with much moisture available. Many lichen species with this growth form are quite sensitive to air pollution. (Photo by Bruce McCune)

In general, lichens interact minimally with their substrata, although some species penetrate the bark surface or cause changes in rock surface chemistry. Some bark lichens may obtain at least some carbohydrates from the surface of the host tree, but additional studies of this phenomenon are needed (Hawksworth 1988a). Lichens with thalli above or outside their substrata usually harbor minicommunities of invertebrates and serve as food for some.

Terrestrial and arboreal lichens are important regular winter foods for large mammal herbivores, including reindeer, caribou, and deer, in boreal forest and arctic tundra (Stevenson and Rochelle 1984; Seaward 1988) and for some small mammals (Maser et al. 1986). Lichens are an important emergency food for many other large mammalian herbivores during particularly hard winters (Fox and Smith 1988; Thomas and Rosentreter 1992). They also are used regularly as nest material by birds and mammals in some habitats (Seaward 1988; Hayward and Rosentreter 1994).

Industrial melanism in peppered moths in England (Kettlewell 1973; Majerus 1989; Cook et al. 1990) is a famous example of indirect impacts of lichens on animals. Moth populations whose color patterns mimicked patterns of lichens on trees ("cryptic coloration") were affected first by loss of lichens because of air pollution, and second by the lichens' subsequent return with improving air quality. Relationships of lichens on trees or rocks to other instances of cryptic coloration in insects,

reptiles, and amphibians are largely unstudied, and possible indirect effects of loss of lichens on those animals are unknown.

In some desert and tundra ecosystems, lichens constitute a significant proportion of the biomass of autotrophs (Kappen 1988). In some low-nutrient habitats, such as conifer forests of the Pacific Northwest of the United States, lichens may fix a significant proportion of nitrogen for the system (Nash 1996). In temperate and boreal forests, they may alter the availability of nutrients and buffer (by sequestration) heavy metals entering the system from atmospheric deposition via canopy throughfall and stemflow (Seaward 1988). Summaries of the ecological roles of lichens can be found in most books on lichen biology; more extensive literature reviews can be found in Armstrong (1988), Galun (1988), Gilbert (2000), Nash (1996), and Slack (1988).

TAXONOMY, DIVERSITY, AND DISTRIBUTION

TAXONOMY

Lichenization appears to have evolved independently several times during the history of fungi (Gargas et al. 1995; Nash 1996). A few Basidiomycota form lichens, particularly some genera in Agaricales (*Omphalina*),

Cantharellales (*Multiclavula*), and Stereales (*Dictyonema*), but mostly fungi that form lichen associations[16] belong to the Ascomycota. Of the 46 orders of that phylum (Kirk et al. 2001), 14 include lichen-forming representatives, and five of those comprise lichenized species exclusively. Molecular studies have shown that many species in the former order Caliciales actually fall in the Lecanorales (Wedin and Tibell 1997; Wedin et al. 1999). Some lichenologists place the Peltigerales, Pertusariales, and Teloschistales within the Lecanorales; we list them separately. No consensus exists yet on assignment of 15 additional lichen-forming families to an order, although Harris (1995) suggested placing the Aspidotheliaceae (as part of Thelenellaceae) and Strigulaceae in the order Melanotommatales. Most of the unassigned families are exclusively lichenized, generally with green algae. In Table 9.1, we list the 14 orders and their families that include lichen-forming fungi, the unassigned families, and estimates of the number of lichenized species in each.

Proposed changes in classification of Ascomycota were summarized twice per year in *Systema Ascomycetum* (1982–1998), and a revised "Outline of the Ascomycetes" now appears regularly on Myconet (see Appendix III). The latest revision is by Eriksson and colleagues (2002).

Many lichenized Ascomycota also form conidial structures. In contrast to most other fungi, however, the conidial forms (anamorphs) of lichenized species are not given independent scientific names. In some cases conidia act as spermatia; in others they act as dispersal propagules. Some lichen-forming genera are either completely sterile or mitosporic; those forms are given holomorphic names (Vobis and Hawksworth 1981; Hawksworth and Poelt 1986). Most are probably states of Ascomycota, but in some the sexual phase now may have been completely lost (e.g., *Blarneya, Cheiromycina, Flakea, Lepraria, Siphula, Thamnolia*).

Basis for Taxonomic Distinctions

Currently orders, families, and usually genera of lichen-forming fungi are delimited primarily by characters of the fruiting bodies (Henssen and Jahns 1974; Hafellner 1984, 1988). As a result, lichens whose thalli are quite

TABLE 9.1

Orders of Fungi in the Phylum Ascomycota and Their Families That Include Lichen-forming Species

AGYRIALES. 3 families: Agyriaceae, Anamylosporaceae, Elixiaceae; 100 species; mostly lichenized but some saprobic on wood.

ARTHONIALES (including OPEGRAPHALES). 4 families: Arthoniaceae, Chrysothricaceae, Melaspileaceae, Roccellaceae (= Opegraphaceae); 1000 species* (of 1200); mostly lichenized with the green algae *Trentepohlia* species, but some lichenicolous or saprobic.

DOTHIDEALES *sensu lato*.[†] 3 families* (of 58): Arthopyreniaceae, Microtheliopsidaceae, Pyrenothricaceae, 120 species (of 4770); the biological status of many members of Arthopyreniaceae is unclear; the order also includes some families primarily with lichenicolous species (e.g., Dacampiaceae).

GYALECTALES. 1 family: Gyalectaceae; 100 species; all lichen-forming, especially with the green algae, *Trentepohlia* species.

LECANORALES. 34 families (of 40): the larger families include Acarosporaceae, Bacidiaceae, Catillariaceae, Cladoniaceae, Collemataceae, Lecanoraceae, Lecideaceae, Pannariaceae, Parmeliaceae, Physciaceae, Ramalinaceae, and Stereocaulaceae. Most former CALICIALES belong here, but family assignments are uncertain; 7150 species (of 7250); the majority of lichen-forming fungi belong here; those that are lichenicolous or saprobic probably evolved from lichenized species; mainly forming lichens with green algae, especially Chlorococcales and Pleurastrales (e.g., *Trebouxia* species).

LICHINALES. 3 families: Gloeoheppiaceae, Lichinaceae, Peltulaceae; 280 species all lichen-forming with cyanobacteria.

OSTROPALES (including GRAPHIDALES). 6 families (of 7): Asterothyriaceae, Graphidaceae, Heppiaceae, Solorinellaceae, Stictidaceae Thelotremataceae; 1600 species (of 1800); mainly lichen-forming with *Trentepohlia* species; some lichenicolous species in Odontotremaceae in this order.

PATELLARIALES. 1 family (of 2): Arthrorhaphidaceae; 5 species (of 50); lichen-forming with green algae, or lichenicolous.

PELTIGERALES.[‡] 4 families: Lobariaceae, Nephromataceae, Peltigeraceae, Placynthiaceae; 510 species; all lichenized with either green algae or cyanobacteria, sometimes with more than one photobiont in the same thallus, or forming morphologically different thalli according to the photobiont present.

PERTUSARIALES.[‡] 1 family: Pertusariaceae; 300 species; all lichenized with green algae.

PYRENULALES. 4 families (of 5): Megasporaceae, Monoblastiaceae, Pyrenulaceae, Trypetheliaceae; 500 species; mainly lichenized with the green algae *Trentepohlia* species.

TELOSCHISTALES.[‡] 3 families: Fuscideaceae, Letrouitiaceae, Teloschistaceae; 570 species; almost all species lichenized with green algae, commonly *Trebouxia* species, but some lichenicolous species in a few genera.

TRICHOTHELIALES. 1 family: Trichotheliaceae; 240 species; all lichenized with *Phycopeltis* species or other trentepohlioid green algae.

VERRUCARIALES. 1 family (of 2): Verrucariaceae (= Dermatocarpaceae), 700 species; mainly lichenized with green algae; some genera with all or a few species lichenicolous; the other family of the order, Adelococcaceae, exclusively lichenicolous.

FAMILIES OF UNCERTAIN POSITION. 15 families: Aphanopsidaceae, Aspidotheliaceae, Baeomycetaceae, Epigloeaceae, Gomphillaceae, Icemadophillaceae, Mastodiaceae, Pachyascaceae, Phlyctidaceae, Protothelenellaceae, Strigulaceae, Thelenellaceae, Thelocarpaceae, Thrombiaceae, and Umbilicariaceae.

* Estimated number of lichenized species or families in the order.
[†] Now used in a restricted sense; the three families of uncertain ordinal placement.
[‡] Often included in LECANORALES by recent authors.

different morphologically often are united in the same genus, family, or order. For example, the family Teloschistaceae includes lichens with thalli that are endolithic (completely embedded within a rock surface), leprose (granular), crustose, placodioid (crustose with a lobed margin), squamulose, foliose, and fruticose. Recent taxonomic revisions based on fungal reproductive structures have led to major changes in generic delimitations (e.g., in *Lecidea, sensu lato*; Purvis et al. 1992).

Genera (e.g., *Arthonia, Arthrorhaphis, Buellia, Caloplaca, Catillaria, Mycomicrothelia, Rhizocarpon, Toninia*), as well as some orders and families, can include both lichenized and nonlichenized fungal species. The nonlichenized species are most commonly lichenicolous or more rarely saprobic on wood. In some cases the biology of a single species may vary during its life cycle; for example, several species of *Diploschistes* and *Rhizocarpon* are initially lichenicolous, parasitizing another lichen and then taking over its photobiont to form an independent lichen thallus.

Molecular approaches bring significant modifications to classical morphological and chemical species concepts in lichenized fungi (Grube and Kroken 2000). Secondary chemistry is used extensively for identification (see "Identification," later in this chapter); morphologically identical species differing only in their secondary chemistry have been recognized as separate species by some workers, generating considerable debate (Lumbsch 1998). Studies of morphological, chemical, and genetic variation within a species used to be uncommon (Culberson et al. 1988; DePriest 1993, 1994) but are moving ahead rapidly with the advent of molecular approaches. Morphologically defined species may include two or more cryptic genetic species (Kroken and Taylor 2001; Crespo et al. 2002). Lichens that are identical in most features (including secondary chemistry) but differ in that one reproduces by ascospores and the other by vegetative means (usually isidia or soredia) have been considered "primary" (sexually reproducing) and "secondary" (vegetatively reproducing) species pairs (Mattsson and Lumbsch 1989). However, molecular data suggest that some "species pairs" may be better treated as single species (Articus et al. 2002).

Occasionally different propagules, perhaps genotypically distinct, grow together to form a single lichen thallus. Although the propagules are generally of the same species, that is not always the case, and both interspecific and intergeneric "mechanical" hybrids have been postulated (Hawksworth 1994), the latter being similar to the presence of parasymbionts in lichens. Such hybrid thalli can exhibit features of both fungal partners as their hyphae intertwine in the lichen tissues and around the shared photobiont cells, and they can confound the unwary. Luckily they are uncommon.

Identification

Scientific names given to lichens refer to the fungal partner (mycobiont); the photobiont(s) has an independent scientific name. Officially, the lichen association itself has no name, and under the current International Code of Botanical Nomenclature (Greuter et al. 2000), it is not possible to "identify" or "name" a lichen, only the lichen-forming fungus (and/or the photobiont, Hawksworth 1998). In practice the name assigned to a specimen is the name of the lichen-forming fungus. A consequence of this restriction is that when more than one species of photobiont occurs with the same fungus (either in the same or in separate thalli), the same fungal name is used for all resulting lichen thalli even if they have different morphologies.

Keys to families are included in the *Dictionary of the Fungi* (Kirk et al. 2001). No world key to lichenized genera has appeared since that of Clements and Shear (1931), but several major regional treatments include keys to genera—for example, those of Awasthi (1988, 1991), Brodo and colleagues (2001), Clauzade and Roux (1985), Foucard (1990), Nash and colleagues (2002), Poelt (1969), Poelt and Vězda (1977, 1981), Purvis and colleagues (1992), Swinscow and Krog (1988), Thomson (1984, 1997), and Wirth (1995a, b).

Catalogues of names of lichen-forming fungi are provided by Zahlbruckner (1921–1940); Lamb (1963); Hawksworth (1972); since 1970, in the twice yearly *Index of Fungi* (begun 1940), and the free Index Fungoram web database (see Append III) with some 340,000 species names of fungi, including lichens. The *Bibliography of Systematic Mycology* (also twice yearly, since 1946) indexes all fungal systematic literature, including that on lichens, down to the rank of genus, and data since 1980 are available as a searchable CD-ROM. The "Recent Literature on Lichens" series begun by W. L. Culberson in 1951 and currently compiled by T. L. Esslinger in the journal *The Bryologist* (including a searchable reference database on the web page of University of Oslo, Lichen Herbarium) notes new and principal species treated in each listed paper.

Monographs and key regional or national treatments, which often include keys to species, are listed under family and generic entries in the *Dictionary of the Fungi* (Hawksworth et al. 1995), and major modern national compilations are listed by country. Another recent bibliographic guide organized by geography (Hawksworth and Ahti 1990) includes about 1390 literature citations. Many Internet sources for keys and information about lichens are now available, including, for example, the LIAS web site and H. J. M. Sipman's *Lichen Determination Keys* web site (Appendix III). In addition, the web pages (see Appendix III) of the American Bryological and Lichenological Society and the British Lichen Society include numerous links to Internet resources for lichens.

Lichen chemistry is a valuable tool for identifying and separating species, particularly for sterile material. Spot tests or color reactions can be used on the thallus surface, on internal tissues exposed with a razor or scalpel blade, or on compound microscope sections of fruiting bodies or thallus tissue. The chemicals most commonly used are known to lichenologists as K (10% solution of potassium hydroxide in water); C (undiluted commercial bleach); P or PD (saturated alcohol [95% ethanol] solution of p-phenylenediamine, or Steiner's Stable Solution); I (mainly Lugol's Iodine; sometimes Melzer's Iodine is used); N (10–50% solution of nitric acid), and the staining compound LCB (lactophenol–cotton blue). Most lichen identification guides (e.g., Hale 1979; Purvis et al. 1992; Malcolm and Galloway 1997; Brodo et al. 2001) include summaries of how to make and use these chemicals. Wright (1996) has provided hints on interpreting test results.

The identification of secondary metabolites in lichen thalli, including those responsible for color changes with spot tests, generally is performed by thin-layer chromatography using standardized solvent and spray systems, and such equipment routinely is present in lichenology laboratories (Arup et al. 1993; Orange et al. 2001). Culberson and others have compiled the available data on the chemistry of each species (Culberson 1969, 1970; Culberson et al. 1977); Hunech and Yoshimura (1996) have compiled structural information. Computer programs to aid in the use of chemical contents for species identification are available also (Mietzsch et al. 1993).

Discussion of identification of photobionts in lichens is beyond the scope of this chapter. Identification to species often requires isolation of the organism into unialgal cultures. For information on identification procedures, see Ahmadjian (1993) and Tschermak-Woess (1988).

Major Reference Collections

Lichens have been collected extensively for hundreds of years. Most major botanical collections (herbaria) have old lichen material and also serve as depositories for voucher specimens from fresh inventories. Air-dried samples of lichens can be maintained as reference collections for extended periods when they are kept in a stable, dry atmosphere and thus provide a rich source of background material for inventory studies. They also add a historical dimension to a survey and provide comparative material for identifications. In addition, metabolites in appropriately maintained material (some dating as far back as the 1690s) still can be detected with thin-layer chromatography for identification of specimens.

The largest dried-specimen collections (Appendix III) are those of the University of Uppsala, Sweden (UPS), with about 500,000 lichen specimens and the Natural History Museum in London, United Kingdom, with about 400,000 specimens. The Smithsonian Institution in Washington, DC, with about 250,000 specimens is the largest collection in the United States of America. Hall and Minter (1994) and Holmgren and colleagues (1990) listed institutions maintaining dried collections of lichens, and lichen-related Internet sites often have lists of and links to such herbaria (e.g., New York Botanical Garden; Appendix III).

Living cultures of lichen-forming fungi (without photobionts) are difficult to maintain for extended periods, mainly because of their slow growth and susceptibility to contamination. Therefore, few of these fungi are available from the main collections of microbial cultures. The American Type Culture Collection (Appendix III) has the largest number publicly available. About one-third of those lichenized fungal species tested have been cultured successfully on the first attempt (Crittenden et al. 1995). Growth rates are very slow on solid media, so shaken liquid cultures are preferred. Sophisticated technology now allows individualized freezing and thawing protocols for preserving cultures of lichen-forming fungi in liquid nitrogen (Smith and Onions 1994).

DIVERSITY AND DISTRIBUTION

The count of lichenized fungal species accepted by Hawksworth and colleagues (1995) and Sipman and Aptroot (2001) is just less than 13,500, a number unchanged from the 1930s. This somewhat surprising result indicates that the numbers of newly described lichenized fungi have been roughly balanced by recognition of synonymy. Larger figures have been claimed during this period by some workers (e.g., 17,000 species by Nash and Egan 1988 using rough estimation rather than a precise count). The complete world inventory is expected to total about 18,000 species (Sipman and Aptroot 2001). Thus, lichenized fungi are much better known than most other fungal groups, with perhaps 60–80% of the species already described, as opposed to the now generally accepted 5% for fungi overall (Hawksworth 1991; Heywood 1995).

Taxa of lichenized fungi tend to be found over larger geographic areas than do equivalent taxa of vascular plants. For example, a significant proportion of lichen genera occur on all continents, and several individual species have worldwide distributions (e.g., *Chrysothrix candelaris*, *Parmelia sulcata*, *Placopsis gelida*, *Thamnolia vermicularis*, *Xanthoria elegans*). Others are bipolar (e.g., *Pseudephebe minuscula*: Galloway and Aptroot

1995) or pantropical (e.g., *Strigula smaragdula*, Santesson 1952, as *S. elegans*). Regional distribution patterns also have been and continue to be strongly influenced by human activity, especially sulfur dioxide air pollution, habitat destruction, loss of ecological continuity, the introduction of new substrata such as buildings, and the spread of exotic lichen species.

In contrast to diversity patterns for vascular plants and many other fungi, the world's richest areas for lichen-forming fungal species currently appear not to be the humid tropics, but rather the southern temperate rain forests (Galloway 1992a, 1992b, 1995), northern temperate forests, and high-latitude zones. This may be an artifact of limited knowledge; the richest lichen biota known for a single tree is reported from an Asian tropical forest (Aptroot 1997). When whole countries, including all available habitats and biomes, are compared, the boreal zone currently appears to be the richer of the two high-latitude zones. For Sweden and Norway

combined, 2271 lichen species are listed, whereas for New Zealand, a southern hemisphere area of comparable latitude, size, and climatic range, only 1378 species are known (Table 9.2). This difference may exist partly because the lichens of New Zealand are currently less completely known than those of Sweden and Norway.

Distribution patterns may differ when biodiversity is measured at the level of family or order rather than species because several exclusively lichenized families are primarily tropical (e.g., Gomphillaceae). Sipman (1995) reported a significant contribution of Ostropales (including Graphidales) and Pyrenulales to the lowland lichen mycobiota of Colombia, whereas in the high Andes more than 90% of all species belong to the Lecanorales, a percentage similar to that in the temperate zones of the world. Aptroot and Sipman (1997) reported that the diversity centers of three of the principal orders (Ostropales, Pyrenulales, Arthoniales; 3100 species combined) of lichenized fungi are in the tropics. Accurate

TABLE 9.2

Numbers of Species of Lichens Reported from Large (≥5–10 Million ha) Geographic Areas

Location	Number of species*	Citation
Africa		
East Africa	628 (macrolichens)	Swinscow and Krog 1988
Antartica and South Georgia	427	Øvstedal and Smith 2001
Asia		
China	1766 (macrolichens)	Wei 1991
India, Nepal, and Sri Lanka	1850	Awasthi 1988, 1991
Japan	1557	Kurokawa 2003
Papua New Guinea and Irian Jaya	1500+	Aptroot and Sipman 1997[†]; Streimann 1986
Thailand	554	Wolseley et al. 2002
Australasia		
Australia	2500+	Filson 1996
New Zealand	1378	Malcolm and Galloway 1997
Europe		
Austria	2101	Hafellner and Türk 2001
Belgium and Luxembourg	930	Diederich and Sérusiaux 2000
Germany	1835	Scholz 2000
Great Britain and Ireland	1660	Coppins 2002
Italy	2145	Nimis 1993
Netherlands	706	Aptroot et al. 1999
Spain and Portugal	2426	Llimona and Hladun 2001
Sweden and Norway	2271	Santesson 1993
North America		
British Columbia	1100	Goward et al. 1994[†]
California	1000	Hale and Cole 1988[†]
North of Mexico	3580	Esslinger and Egan 1995
American Arctic	996	Thomson 1984, 1997
Pacific Islands		
Hawaiian Islands	723	Smith 1991[†]
South America		
The Guyanas	800+	Aptroot and Sipman 1997[†]
Colombia	1000–1500	Sipman 1995[†]

* Species numbers represent the combined efforts of many collectors over many years, rather than the efforts of any single person or project.
[†] Authors provide number but do not list all species in the publication.

estimation of the diversity of lichens in most tropical areas is currently impossible because of inadequate collections and taxonomic knowledge (Galloway 1991).

Estimation of diversity by substratum also provides different patterns. For instance, foliicolous (living on leaves) lichens (Fig. 9.3) are almost exclusively tropical (Lücking 1995). Rocky substrata (Fig. 9.4), however, which support much of the diversity in the temperate and cold zones of the world, are rare in the lowland tropics. The wide variety of habitats, substrata, and climates available in temperate areas of the world probably explain why they are so rich in lichens (Galloway 1992a). In general, the number of lichenized fungi present in an area of a particular size depends heavily on the range of substrata available.

Despite the differences in broad patterns of distribution, the proportion of common to rare genera and species among lichenized fungi in a particular area appears to be generally similar to that of vascular plants. Patterns relating to the number of species per area have long been discussed in ecology, but the comparability of particular numbers is often in question (see Chapter 5). For ecologists, species recorded as present in fixed areas

provide the basis for calculating frequency values (see "Designing Sampling Protocols for Biodiversity Inventory," later in this chapter), which have long been recognized as scale-dependent (Grieg-Smith 1983). Palmer and White (1994) have shown that species-area relationships at small scales (<1 km^2) depend on the grain (size of the smallest sample unit) and other aspects of field study design. We present species/area figures for lichenized fungi in Tables 9.2 and 9.3. The data come either from knowledge accumulated by many investigators over many years or from studies with many different sample designs, so we present them with minimal discussion of general relationships.

Both the regional species pool and the diversity of available microhabitats are important contributors to the diversity of lichenized fungi in a particular area. Rose (1992) compiled extensive surveys of lichenized fungi in deciduous temperate woodlands in the British Isles and found that up to 227 species can be expected per square km (100 ha) on bark and wood alone, depending on the degree of air pollution and other human influences. The epiphytic (growing on plants) lichen community in the tropics, however, can be more diverse than in temperate

FIGURE 9.3 There may be 10 to 30 species of foliicolous lichens on each leaf and, perhaps, 40 to 60 species on the whole twig of this rain forest plant (Bignoniaceae) in Costa Rica. (Photo by Robert Lücking)

areas. Lücking (1995) reported 177 foliicolous (growing on leaves) lichen species from a single forest plot in Costa Rica, a number similar to that reported for corticolous (growing on bark) species in a forest studied by Mont-foort and Ek (1990) in French Guyana (Table 9.3). Together, these two studies suggest that a local forested area (<200 ha) in old-growth tropical lowland forest may have more than 400 species of epiphytic lichens, more than in a local area of old-growth temperate forest in the United Kingdom (Table 9.3).

Cumulative lichen species numbers from published checklists and regional treatments (Table 9.2) provide

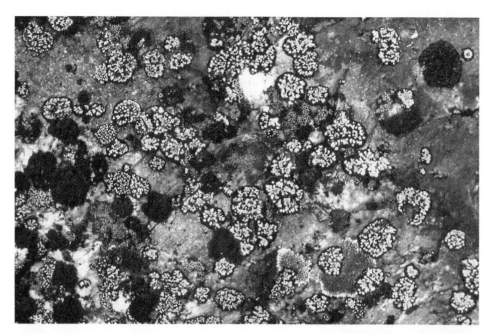

FIGURE 9.4 This section of a granite boulder in the northern Rocky Mountains of Alberta, Canada, supports 10 to 15 species of crustose saxicolous lichens. (Photo by John Wolf)

TABLE 9.3
Numbers of Species of Lichens Reported from Local and Regional Areas of Various Sizes[a]

Location	Biome	Area (ha)	Number of species	Topography	Reference
Regional geographic areas[b]					
NE Unites States (US)	Conifer-deciduous forest	6272	126	Hilly	Wetmore 1995
N Central US	Conifer-deciduous forest	2560	190	Rolling	Wetmore 1993
Western US	Alpine-tundra forest	31,200	137	Mountainous	Hale 1982
	Conifer forest-savanna	1.3 million	404	Mountainous	Wetmore 1967
	Savanna-grassland	98,550	171	Hilly	Will-Wolf 1998
SE Alaska, US	Conifer forest	6.8 million	381	Mountainous	Geiser et al. 1994a
Venezuela	Tropical forest, scrub	40,000	216[c]	Mountainous	Sipman 1992
Greek Islands	Mediterranean scrub, rocks	20,000	295	Hilly	Sipman and Raus 2002
Hong Kong	Tropical forest	107,000	261	Hilly	Aptroot and Seaward 1999
Local geographic areas[d]					
Great Britain	Deciduous forest	211	323 (of 2500 fungi)	Rolling	D. L. Hawksworth, unpublished data[e]
French Guyana	Tropical rain forest	<100	209[f]	Flat	Montfoort and Ek 1990

[a] Based on comprehensive surveys of all possible substrata, except as noted.
[b] Each report represents the results of a single survey project with moderate survey effort.
[c] Total lichen mycobiota estimated to be twice as many species (H. J. M. Sipman, personal communication).
[d] Results of exceptionally intensive survey efforts.
[e] D. L. Hawksworth's Slapton NNR study is discussed elsewhere in this volume (see Chapter 11).
[f] Does not include leaf epiphyte lichens, of which >100 species may occur in the area.

some idea of the available species pool for large geographic areas. Species found per area for smaller areas (Table 9.3) in single surveys give rough estimates of the variation in species per area to be expected at smaller scales. From comparison of numbers of species in Tables 9.2 and 9.3, one could suggest that in a single biodiversity survey of a 50,000-hectare area, one might find between 5% and 20%, depending on survey intensity, of the species in the large regional species pool for a 5- to 10-million-hectare (or larger) area.

DESIGNING SAMPLING PROTOCOLS FOR BIODIVERSITY INVENTORY

In this section we address the design of biodiversity studies with at least minimal levels of quantification and repeatability. We do not address the design of strictly qualitative studies. For a broad review of monitoring lichens and using lichens as monitors see Nimis and colleagues (2003).

RESOURCES REQUIRED

Personnel

It is important that trained lichenologists participate in the design of sampling protocols for lichen surveys, and it is essential that they identify or confirm identifications for all voucher specimens. It is desirable to have lichenologists conduct the field work; a team of specialists in different lichen groups will maximize species capture in high-intensity inventories. Nevertheless, the use of nonspecialist field personnel trained to distinguish between species and collect them without necessarily identifying them often allows much larger areas to be inventoried for the same costs. Species capture, however, especially for crustose lichens, will be much less complete in the latter case.

For low-intensity inventories, the ability of a properly trained observer to distinguish species in the field maximizes species capture with limited field time. Nonspecialist field personnel can be appropriate for moderate-intensity inventories of large areas, for example, regionwide status and trend inventories that form part of a national program (e.g., for the United States, see McCune 2000), or as team members in high-intensity inventories.

Training for nonspecialist field personnel should be designed and conducted by a lichenologist familiar with the region of the proposed survey. Both habitat selection criteria and morphological variation of lichens differ from the norms for vascular plants. Nonspecialist field technicians trained by an expert(s) can, with continued experience, become quite proficient, especially with macrolichens, and those having fruticose, foliose, and squamulose growth forms (McCune et al. 1997b).

Collecting Equipment

Field equipment needed to collect lichen voucher specimens is minimal, including the following items: hand lens; rock hammer and chisel for hard mineral substrata; stout knife or wood chisel for woody and soft substrata; hand pruning shears for twigs, paper packets, or bags; and small card boxes for fragile soil-surface specimens. Individuals surveying special habitats, such as vertical cliff faces or tree canopies, may require special equipment and training (e.g., Pike et al. 1975; ter Steege and Cornelissen 1988).

FIELD PROCEDURES

Lichens should not be placed in watertight or water-resistant containers because molds can overgrow specimens confined under warm, humid conditions in as little as a few hours. Specimens should be stored in the field one to a paper bag or packet, with substratum, location, and date recorded on the bag in pencil or waterproof ink. Abundance data should be linked to individual field specimens when they are collected. Wet or damp specimens must be air-dried thoroughly, ideally within a few hours of collection, and should never be stored damp.

Both for conservation reasons and for efficient processing, the field-sampling protocol should require collection of the minimum numbers of specimens needed to ensure accurate identifications, while achieving an appropriate level of quantification. The number of specimens to be collected should be decided partly based on the expertise of the collector. A minimum of one specimen for each species encountered in the whole study area should be collected as a voucher. Nonspecialist technicians can be trained to distinguish some species accurately in the field; an expert familiar with the study region should decide which species can be treated this way. Other species may be identifiable, or at least unambiguously distinguishable, in the field by an expert. Presence and abundance of species distinguished in the field, along with pertinent information, can be recorded on data sheets or in an electronic data recorder. For species distinguished in the field, two to four specimens from sites widely separated within the study area and linked to specific field data sheets should be collected to provide multiple vouchers and to confirm correct field characterization. At least one specimen per site should be collected of species for which field distinction is relatively difficult. Most crustose lichens fall into the latter category for most field personnel. We recommend that one specimen per species per site be collected in poorly surveyed regions.

Sampling protocols also must take into account conservation issues. Protocols requiring massive collection of specimens from small areas may result in overcollection of uncommon species and thus be inappropriate for many species in some areas. Field etiquette for lichenologists dictates that if only one specimen is seen during a search, only half of the specimen is collected. This practice should be adopted for all biodiversity sampling protocols. Overcollection is usually not a problem for most species in general surveys on large plots and when several years elapse between repeat surveys (McCune et al. 1997a). Collecting epiphytic lichens from recent litterfall has no impact on current or future abundance because these thalli will die even if not collected.

Field personnel should learn to identify any threatened or endangered species suspected to occur in an area by sight and should record but not collect them. When collecting is not allowed or is limited severely, inventory goals and sampling design must be planned carefully, and field surveys should be carried out only by lichen experts. For material on historical buildings, monuments, or similar structures, microscope preparations made in the field directly from specimens can facilitate identifications. Finally, it is important to recognize that many lichens simply cannot be identified to species with certainty unless they are collected and subsequently studied in the laboratory.

INFORMATION PROCESSING AND ARCHIVING

Most lichen taxa must be identified to species for an inventory to be meaningful. A minimum of one voucher specimen per species reported should be housed in an accessible permanent collection. At minimum the scientific name, latitude and longitude of the collection locality, specimen substratum, date of collection, name(s) of the collector and identifier, and a unique collection number should accompany each voucher specimen. Voucher specimens are critical for assessing the validity and reliability of data. Species reports not supported by voucher specimens generally are considered unreliable. Arrangements with a lichenologist to identify voucher specimens, provision for time and expense to curate specimens, and agreement with a herbarium to house voucher specimens permanently should be completed as part of the planning for a biodiversity survey.

When field collecting and preliminary sorting are done by nonspecialist personnel, several specimens of each "species" (sorted group) should be identified by expert lichenologists. Resorting of "species" groups then also involves reassigning associated abundance values.

Voucher specimens are archived dry in acid-free paper packets. Most lichen field guides include instructions for proper curation of lichen specimens (e.g., Hale 1979; Purvis et al. 1992; Wirth 1995a; Brodo et al. 2001; Chapter 2 of this volume). To reduce fragmentation, the substratum of soil specimens (not the lichen specimen itself) can be dipped in a solution of water-soluble glue as soon as is practical, dried, and then glued to a stiff card (Rosentreter et al. 1988).

Reports of inventory and monitoring studies should include descriptions of sampling protocols that are described in sufficient detail to allow other professionals to duplicate the study methods. Lists of species should report the names of experts consulted for identifications, the taxonomic authorities followed (e.g., published checklists, keys), and the location(s) of voucher specimens. Summary statements and conclusions should be supported by data included in the report or cited from published sources.

Both hard copy and electronic versions of raw data and data summaries from biodiversity inventories should be archived in some standard format (ASCII is a commonly accepted minimal standard format for electronic data) with the institute or organization responsible for managing the area surveyed (or funding the survey) as well as at other appropriate locations. Comparison with future surveys is usually an important goal for biodiversity inventory and monitoring projects. Electronic technology can be expected to continue to change at a fast pace in the foreseeable future, so translatability should be a major criterion in choice of electronic formats for archiving data.

Estimated ratios of field time (including travel) to laboratory and office time for lichen biodiversity studies range from approximately 1:4 to 1:10. Some time estimates for biodiversity studies are given in Table 9.4.

CRITERIA FOR DESIGN OF SAMPLING PROTOCOLS

Three aspects of designing sampling protocols are critical: (1) matching the sampling protocol to the goals of the survey, (2) selecting sites for sampling, and (3) selecting a within-site sampling protocol. As we use the terms, a sample site is an independently chosen geographic location. The number of these independently chosen sites is the sample size for the inventory as a whole and the basis for all between-site comparisons. The within-site sampling protocol designates the size of the area to be searched at a site and specifies how within-site subsampling will be carried out. It has no effect on the sample

TABLE 9.4

Characteristics of Biodiversity Survey Protocols with Different Intensities of Effort

Size of sample site (ha)	No. of sites sampled	Area covered (ha)	Site placement	Subplot placement	Lichen groups	Abundance measure
Species presence inventory						
~0.8, varied	27 (21)*	2,560 (31,000)*	Stratified by habitat, vegetation	None (permanent on rocks)*	All lichens	None
Rapid, semiquantitative inventory						
0.4, fixed	79	9 million (estimated)	Regular grid, forest only	None	Macrolichens, standing woody substrata	4 abundance classes
Semiquantitative inventory						
0.05, fixed	188 (+77 partial)*	6.8 million	Stratified by habitat, vegetation	None	All macrolichen, some crustose	5 density classes
Intensive site inventory, proposed[†]						
0.1, fixed	At least 2/ site-class	unknown	Stratified by habitat, vegetation	Stratified by tree species, forest layer, microhabitat	All lichens	Frequency from presence in samples

Field personnel	Time field	Time lab, reports	Data Products		References
			Within-site	Between-site	
[Species presence inventory]					
Specialist	14 person—days	6 person—months	Species richness, notes on abundance, substrata	190 species (137 spp.); species frequency by site class	Wetmore 1993, Wisconsin USA (Hale 1982, Colorado, USA)
[Semi-quantitative inventory, rapid]					
Nonspecialist	40 person—days	7 person—months	Abundance by species; diversity indices	155 species; species response to climate/air-quality gradients; correlations with vegetation variables; critique of methods	McCune et al. 1997a, 1997b, SE United States
[Semi-quantitative inventory]					
Nonspecialist	20 person—months (much travel time)	35 person—months	Abundance by species; diversity indices	381 species; species abundance by site class; multivariate community analyses; correlations with habitat/ vegetation variables; critique of methods	Geiser et al. 1994a, 1994b, SE Alaska, USA
[Intensive site inventory, proposed]					
Teams of specialists and nonspecialists	4–10 person—days/site	3–5 person—months/site	Species frequency and diversity indices by subplot class	Proposed, no species numbers available; species abundance by site and subplot class; correlations with vegetation/ habitat variables	Rossman et al. 1998, Costa Rica; Wolseley et al. 1995, Thailand; for methodology, Lücking and Lücking 1996, Sipman 1996b

* Refer to study by Hale (1982).
[†] Several reports pertain to this proposed inventory design.

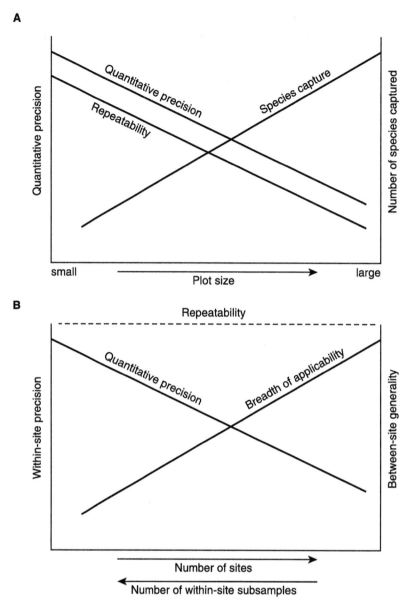

FIGURE 9.5 Trade-offs involved in allocation of effort for different biodiversity sampling designs. **A**. Allocation of a fixed level of survey resources per plot for a range of sample plot sizes involves a trade-off between species capture and quantitative accuracy. **B**. Allocation of a fixed level of resources for an entire survey involves trade-offs between inclusion of many independent sites for generality and elaboration of within-site sample protocol.

size for the inventory as a whole (Hurlbert 1984; Will-Wolf 1988; Will-Wolf et al. 2002). Variations in within-site sampling protocols affect the precision (closeness of repeated measurements of the same entity) of descriptions of the lichen community at that site. When the available resources for a project are fixed, design of the protocol involves trade-offs such as (1) sampling many sites with low within-site quantitative precision for assessment at broad geographic scales versus fewer sites but with greater within-site quantitative precision for assess-

ment of smaller, less diverse areas or (2) using many small sample subplots for quantitative precision versus fewer large subplots for greater species capture at a site (Fig. 9.5).

Matching Survey Goals with Sampling Protocols

The first step in designing a biodiversity survey is to state clearly the overall purpose and specific objectives of the survey. If maximizing species capture for small areas is

A

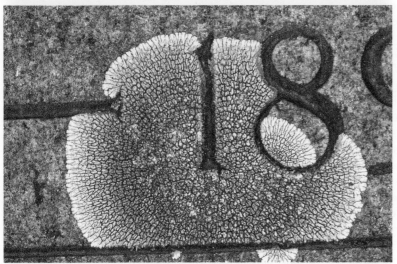

B

FIGURE 9.6 Human activities can introduce unusual substrata into regions. **A**. Five species of lichens and two species of nonlichenized fungi colonized these discarded cotton briefs, which were exposed for perhaps 15 to 20 years in the arid environment of Badlands National Park, South Dakota, United States. Such small patches of unusual substratum have little impact on regional lichen biodiversity. (Photo by Claudia Lipke) **B**. In other situation, such as buildings and grave markers erected in regions otherwise poor in rocklike substrata, the impact on lichen diversity could be very large. *Dimelaena oreina* has colonized this granite tombstone in a rural cemetery in central Wisconsin, United States. (Photo by John Wolf)

the main goal of the survey, one can focus on completeness of coverage within sites with a minimum of replicate sample sites to optimize use of time and resources. If monitoring for trends in change across time or space is a major goal, adequate replication of appropriate sample sites at the expense of some completeness of coverage within sites is preferred if time and resources are limiting.

Next, to focus the design of a sampling protocol, three factors should be considered: (1) size and overall diversity (topographic, climatic, vegetation, habitat, degree and diversity of human influence (Fig. 9.6) or other disturbance) of the area to be characterized; (2) time and resources available for the project; and (3) the need to select multiple independent sample sites per class so that results can be generalized across that class of sites (e.g., a type of vegetation or habitat).

One further practical constraint on the design of sampling protocols is the need to search slowly, intensively, and at very close range (often with a hand lens) in all potential microhabitats of an area to maximize the number of lichenized fungi found. This means that a careful survey of a discrete site of fixed size will give more repeatable results than a superficial inventory of a large, diffuse area. Many species are likely to be overlooked unless each microhabitat is searched carefully.

The two major attributes of sampling protocols that can vary with resource availability and required sampling intensity are (1) the number of independent sites to be sampled in the area being inventoried and (2) the degree to which within-site search strategy is quantified and subdivided. To assess variation between site classes, or changes in a site class over time, one must have a valid estimate of variation between sites within each class (at each time), which requires replicate sites within each class (at each time). The number of sites needed to achieve the desired precision in estimating differences or changes for the survey as a whole should be decided first, followed by a decision on how detailed the within-site sample protocol can be and still allow completion of the whole survey within the allotted time. Increased precision (reduction in variance of measurements) at the scale of within-site sample protocol does not automatically improve the accuracy (closeness of an estimate to the "true" value) of comparisons at the scale of between-site comparisons (Sokal and Rohlf 1995).

An extensive literature, including hundreds of citations going back more than 100 years, exists on monitoring air quality by surveying lichen communities (see Ferry et al. 1973; Nash and Wirth 1988; Stolte et al. 1993; Nimis et al. 2002; and the recurrent feature "Literature on Air Pollution and Lichens" in the journal *The Lichenologist*). Much of the discussion of sample design in that literature is pertinent to biodiversity inventory. However, one goal of biodiversity inventory is as complete a representation of the lichenized fungi of an area as possible at the chosen degree of effort, so all major substrata and habitats should be sampled. In contrast, air-quality monitoring (as well as other studies of lichen response to specific factors) often targets a relatively homogeneous subset of lichen habitats for convenience and comparability. The goal of choosing the subset is to reduce or constrain variation over time and space as a result of all factors other than the air-quality gradient or other target factor. Literature on air-quality monitoring should be consulted with this difference in mind. Lichen-community composition can be thought of as a composite response to constraints operating at many different spatial and temporal scales (Allen and Hoekstra 1990). Seasonal variation in community composition usually is negligible. Variation in species response and community composition along successional and disturbance gradients, in contrast, is important. The greatest range of variation in species composition among lichenized fungi can be related to microhabitat and habitat differences, which vary at spatial scales from centimeters to hundreds of kilometers. Variation as a result of habitat differences and macrovegetation successional and disturbance gradients is addressed with location of sample sites. Variation as a result of microhabitat differences is addressed through within-site sampling protocol.

Many topics pertaining to the design of biodiversity studies in general are discussed in greater detail in Chapter 5. Chapters in Stolte and colleagues (1993) provide extensive reviews of lichen survey methods, including all aspects of planning and implementing the sampling protocol. Although the authors focus on use of lichens for air-quality monitoring, their coverage is comprehensive and pertinent to biodiversity studies.

Sample Site Location

In many cases in which an organized biodiversity survey of lichenized fungi is planned, classifications of the landscape, habitats, and/or vegetation of the area either already exist or are developed concurrently. Stratification of samples by classification units (selecting sample sites to represent classification units, rather than locating them at random within a survey area; Grieg-Smith 1983) is useful for relating lichenized fungi to other organisms and to the environment (Fig. 9.7) and will encourage the use of the information for management of the area (McCune and Antos 1981a, b; Wolseley and Aguirre-Hudson 1991; Wolseley et al. 1995). Another advantage to stratifying by classification units is that one sometimes can choose readily accessible sites that adequately represent more remote areas.

Spatial autocorrelation (the tendency for geographically closer samples from a single habitat to be more similar to one another than to more distant samples from the same habitat) has been shown to be an important aspect of ecological pattern in nature at many scales (Legendre and Fortin 1989). To represent the biodiversity of an area fairly, sample units should be dispersed over the area they are to represent, whether it be sample sites within the whole survey area or subsample plots within a site (Hurlbert 1984). Whatever spatial autocorrelation exists then is represented in the samples, allowing characteristics of the samples to be extrapolated to the whole study area from which they are drawn.

Investigators should not assume that variation among lichen communities will be represented adequately by classification units of an area when the latter are based on macrovegetation composition or on environmental and habitat attributes important to macrovegetation. Just as worldwide distribution and diversity patterns of lichenized fungi do not match those of vascular plants (see "Diversity and Distribution," earlier), local diversity patterns (scale of 0.1–1.0 ha) of lichenized fungi also often diverge from those of vascular plants (McCune and

Many survey designs call for the sampling of lichens and macrovegetation at the same sites to provide a strong basis for relating lichen and macrovegetation variables. Surveying for lichenized fungi in preexisting, permanently marked, macrovegetation sample sites offers many of the same advantages. If placement of individual sample sites and subplots for macrovegetation does not match needs for lichen surveys, then additional sites for sampling lichens alone can be selected to supplement the joint survey. Habitats that contribute much to the diversity of lichenized fungi, such as rock outcrops, talus piles, coarse woody debris, and desert pavements, are often less important for vascular plant communities. The advantages of being able to relate patterns of lichens to data from a large sample of other organisms outweigh the problems presented by data analysis for such a mixed-strategy survey.

In practice, statistical aspects of sample site selection often are subordinated to the particular conservation goals of an inventory. For example, an agency may want to monitor an area that is being considered for development (e.g., logging, urban, mining), so sample sites are not selected randomly. Inventories and monitoring programs also often are requested for protected areas that are expected to remain intact for long periods and can serve as benchmarks for comparison with human-dominated areas.

Within-Site Sampling Protocols

A within-site sampling protocol designates the size of a site, whether and how to subsample that site, and what information to record. Within-site sampling protocols can vary from time-constrained qualitative surveys of the whole site to intensive, quantitative surveys of subsamples within the site. Field time can range from 1 to 2 hours to several days. Choices within this range should be firmly constrained to meet goals of the whole project with available resources. The kinds of information obtained with different sampling protocols differ significantly (Fig. 9.5b; Smith et al. 1993). Contrast, for example, the results of a rapid survey of many sites representing widely differing habitats with the results of intensive surveys of a few sites that replicate a narrow range of habitats. There is no one best way to inventory lichens, and even with the most intensive survey protocol discussed in this chapter, it is unlikely that all species in an entire site will be found in a single survey. Repeated sampling by different observers and with different methods can be used as a basis for estimating "species capture" rates of different sampling designs (McCune and Lesica 1992). Smith et al. (1993) discussed in detail many aspects of within-site sampling design, including few-and-large versus many-and-small subsample units;

FIGURE 9.7 The 6 to 8 species of lichens on this section of an *Acer saccharum* trunk grew in the well-illuminated mid-canopy of a large tree in an old-growth forest (Sylvania Wilderness Area, Michigan, United States). Biodiversity of epiphytic lichens is strongly affected by forest management and other human land-use practices. (Photo by Susan Will-Wolf)

Antos 1981a, 1981b). There is no particular reason to expect that diversity patterns of lichenized fungi always should match those of vascular plants at spatial scales between those extremes. For that reason, it is likely that some macrovegetation class distinctions are not important for lichens and vice versa. Macrovegetation classes with similar species compositions but different ages (e.g., old-growth versus young forest) may or may not be classified separately, but the communities of lichenized fungi differ (Lesica et al. 1991; Wolseley and Aguirre-Hudson 1991; Tibell 1992; Selva 1994; Wolseley et al. 1995). The well-known sensitivity of lichens to air pollution also means that diversity patterns of lichenized fungi may differ markedly from those of less pollution-sensitive organisms in an area affected by local air pollution.

they also compared various abundance estimates and types of data analysis.

Selecting the Size of the Sample Plot. Lichens occupy surfaces at scales corresponding to microhabitat variation. Microhabitats are distributed discontinuously so that in most cases at spatial scales larger than the $0.01–0.25$ m^2 range, the distribution of lichenized fungi will be heterogeneous. Above this very small scale, heterogeneity of lichen habitats at many different scales can contribute to the lichen diversity and community composition of an area; there may be no "best" scale (plot or microplot size) for measuring lichen diversity. Microhabitat variables known to be important to lichens include light regime; moisture status; and type, hardness, chemistry (especially pH and calcium content), and age of substratum.

We strongly recommend surveying a fixed-area plot at each site to facilitate comparability of site data, especially when diversity indices or species frequency are used. Sites corresponding to habitat or macrovegetation classes typically occur in the 0.1–5 hectare (0.25–12 acres) range, with sites larger than 0.5 hectares usually subsampled in some way. Sizes of plots for surveys of lichenized fungi at independently chosen sites have been at the lower end of that range (0.05–1 ha). One hectare probably represents a practical upper limit for a circumscribed search plot for lichenized fungi; the largest plot used in a study we cite was about 0.8 hectares (Wetmore 1993, 1995). Larger sample sites (>1 ha) can be inventoried by locating more than one plot at a site, but these plots are not independent of one another and do not increase the sample size for between-site comparisons. Most published lists of lichen species have been generated from strictly qualitative surveys of areas of no fixed size.

The relative advantages and disadvantages of sample plots of different shape, and the effects of shape on species capture, also should be considered. Compact shapes, such as squares or circles, are more likely to be homogeneous within, whereas extended shapes like rectangles are more likely to capture heterogeneity within (Grieg-Smith 1983). Compact shapes often are faster to lay out accurately in field conditions, and circular quadrants can be relocated precisely from a single permanent center marker, so they can be recommended for field efficiency. Also, because compact shapes are more likely to be homogeneous within, correlation between lichen community composition and other plot-level ecological factors can be estimated more efficiently for them. However, because capturing and representing heterogeneity is a goal for a biodiversity study, extended shapes may be preferred.

Within-Site (Within-Plot) Sampling Strategies. An investigator can search an entire plot to determine an average microhabitat heterogeneity or can search subsamples of the plot to represent this heterogeneity (stratified sample design). McCune and Lesica (1992) explicitly considered the trade-offs involved in these alternative strategies for estimating lichen diversity in three forest layers (ground, tree trunks, and branches), as well as the more general problem of subsample size, for 0.08-hectare plots in a conifer forest. They found that whole-plot sampling was more accurate at measuring species richness and less accurate at estimating cover (see "Estimating Abundance," in the next section) than quantitative subsampling strategies. Whole-plot sampling was much quicker for branches but took the same time or longer than quantitative subsampling of tree trunks and the ground. Because repeatable estimates of species richness for whole site samples are desirable for comparing sites with one another and for comparing diversity of lichenized fungi with other site variables, a whole-plot qualitative search at each site visited should be a part of any sampling protocol. More quantitative subsampling strategies, if included, will provide improved assessment of abundance for the more common species (Fig. 9.5; Will-Wolf 1988).

The degree to which lichen sampling is stratified by microhabitat again will depend on the level of intensity of the survey. Lichen species often have strong substratum/habitat preferences, but relatively few species are completely substratum/habitat specific. Major differences in species composition of lichens on rocks or soil are related to differences in the calcareous composition of the substratum (Hale 1982). Occurrences of lichen species on trees vary with bark pH and surface texture and less often with tree species (Oksanen 1988; Schmitt and Slack 1990). Community composition, which reflects abundance as well as presence of lichen species, may differ strongly among tree species (Schmitt and Slack 1990). The number of subplots, stratified by microhabitat within site, that are needed to represent differences between microhabitats adequately depends on within-microhabitat variability and the repeatability desired (Mueller-Dombois and Ellenberg 1974). Some estimates from studies of temperate-zone lichens are as follows: for tree trunks, 10–25 trees per tree species (e.g., Schmitt and Slack 1990; McCune and Lesica 1992); for tree branches, 25–60 branches for abundance and 100 or more branches for species capture (e.g., McCune and Lesica 1992; Geiser et al. 1994a; both constructed species-area curves to estimate number of sample units needed); and for ground and rock surfaces, 40–60 subplots (Lawrey 1991; McCune and Lesica 1992). Sipman (1997) observed in Guyana that two adjacent *Licania densiflora* trees shared only about 50% of their foliicolous lichen species, suggesting that several trees per species should be sampled in tropical forests also.

Shape and size of the subsample unit should be tailored to the specific habitat or microhabitat being sampled (Goldsmith et al. 1986; Longman and Jenik 1987; Wolf 1993; Lücking and Lücking 1996; Sipman 1996b; Nimis et al. 2002). Rock or ground surfaces, shrub branches, tree trunks, and tree canopies all have particular characteristics that influence the choice of an appropriate sampling unit. For trees, trunk plots from 0.1 to 0.5 m² in area and branch lengths of 0.25 to 1 m are recommended (Johansson 1974; Will-Wolf 1980, 1988; Cornelissen and ter Steege 1989; McCune 1990; McCune and Lesica 1992). For rock and ground surfaces, plots from 0.2 to 1 m² in area are recommended (Anderson et al. 1982; Rosentreter 1986; James and Wolseley 1992; McCune and Lesica 1992); ground plots often are located along transects.

Much field time and effort can be spent sampling microhabitats that are difficult to reach (e.g., upper tree trunks, tree canopy branches, cliff faces), so they should be included only in very intensive surveys. Investigators should take advantage of low-effort opportunities to collect data on such microhabitats. For example, samples from recently fallen ("recent" means lichens still look healthy) or felled trees and branches can represent the compositions of tree-canopy lichen communities, yielding important information even if the data acquired must be analyzed separately from the rest of the inventory data. Tree canopies are known to harbor lichen species different from (and with abundances different from) those of the more accessible lower trunks (Yarranton 1972; Pike et al. 1975; Lang et al. 1980; Oksanen 1988; McCune and Lesica 1992; McCune et al. 2000). This technique may be less successful in moist tropical forests, where thalli decay rapidly when they die.

Estimating Abundance. Abundance of lichens can be estimated or measured in a variety of ways to ascertain "how much" or "how many." Field assessment of abundance of lichens is constrained in practice by limits on the ability of even the best lichenologists to identify and distinguish species of some lichenized fungi in the field. If one cannot distinguish a species group in the field, one must make complete collections and estimate abundance in the laboratory from proportions of specimens of different species in the collections. The latter practice quickly generates massive numbers of specimens when subplots within sites are grouped by microhabitat. A much higher proportion of macrolichens (squamulose, foliose, and fruticose growth forms) than microlichens can be readily distinguished in the field. Usually, lichenologists only attempt to assess abundance of macrolichen species in the field.

Another factor affecting abundance estimates is the surveyor's ability to delineate individuals, which is some-times easy and sometimes impossible. For general community inventory purposes, the investigator usually records abundance in the field based on easily countable units, whether they be individuals, clones, or multi-individual complexes. At a minimum, a surveyor can record qualitative notes about the apparent abundance of species or species complexes distinguishable in the field. A further step in increased effort would be to assign species to abundance classes. For example, McCune and colleagues (1997b) used two density classes combined with two abundance classes, whereas Geiser and colleagues (1994a) used five density classes (Scenarios 2 and 3 in Table 9.4). Mueller-Dombois and Ellenberg (1974) compared several abundance class schemes for ecological sampling. Alternatively, an investigator can record species presence in subplots, then use frequency across subplots as the estimate of abundance. The known problems of using frequency as an estimator of abundance (see "Data Analysis," later) must be weighed against the ease and efficiency of collecting presence data in the field.

Yet another approach to assessing abundance is to assign abundance classes to morphological groups of lichens (e.g., all small foliose rock lichens, or all long, hanging fruticose bark lichens; McCune 1993); the usefulness of data for particular groups is likely to vary by habitat (Rosentreter 1986, 1995; Eldridge and Rosentreter 1999). For example, vagrant (unattached) macrolichens are an important group in grasslands and steppes (Rosentreter 1993), whereas they are unimportant in forests. Cover classes or percent cover of morphological groups often are recorded in subplots for studies of terricolous lichens in grassland and steppe habitats, where even macrolichens are difficult to distinguish in the field (Kaltenecker et al. 1999). In some cases, those morphological groups may be surrogates for functional groups (Pike 1978; Rosentreter 1995). Gelatinous ground lichens in steppe communities, for example, include *Collema* species, *Leptogium* species, and *Polychidium* species, all of which fix nitrogen and protect the soil surface (Anderson et al. 1982; Brotherson et al. 1983; Nash 1996).

For completely quantitative estimates of abundance of species in replicated fixed-area subplots stratified by microhabitats within a site, cover classes or percent cover of lichen species provide simpler and more accurate measures of abundance than counts of individuals (Will-Wolf 1988; James and Wolseley 1992; Marcelli 1992). Cover estimation protocols are probably too labor-intensive both in the field and for data analysis to be recommended for general biodiversity inventories and monitoring programs (McCune and Lesica 1992). Measurements of lichen biomass require similarly labor-intensive, rigorous subsampling and weighing of lichens

(Lang et al. 1980; McCune 1993; Hayward and Rosen-treter 1994; Rosso and Rosentreter 1999). Biomass of canopy macrolichens has been estimated efficiently by analysis of litterfall (McCune 1994) and lichens on lower canopy branches (Esseen et al. 1996) in conifer forests.

Because many lichen thalli, unlike most other fungi, are reliably visible, an alternative to assessing abundance in the field is to establish permanent photoplots (sizes as for small subplots) to monitor trends in abundance over time (e.g., Will-Wolf 1980; Hale 1982; Bureau of Land Management 1996). Species may or may not be recognizable in photographs, but morphological types are, and identifications of voucher specimens as well as cover estimates for recognizable lichen groups can be made in the laboratory. Use of photoplots for monitoring lichen abundance has been simplified by recent advances in digital camera technology and development of more sophisticated computer programs for spatial analysis.

Biodiversity Indicators. The use of individual species as indicators for the condition of a whole guild or community of species presupposes detailed knowledge of what the species indicate and how reliably they do it. Such detailed knowledge of ecological and physiological relationships for many lichenized fungi in an area currently exists only in Europe (Seaward 1988), and interpretation of relationships there is still subject to discussion. Also, even common lichen species seldom are present at all sites one might wish to monitor. We, therefore, do not recommend the indicator-species approach for baseline biodiversity studies. We also do not recommend exclusive use of single species as indicators of air pollution (or other specific factors affecting lichen biodiversity), based on the extensive literature review of Smith and colleagues (1993).

Use of groups of species to indicate the condition of whole communities has much more promise, although again, some knowledge of relationships among species must precede acceptance of groups as indicators. The identification of an easily quantifiable subset of lichen species (including both rare and common species) to serve as monitors of biodiversity for repeat sampling is an appropriate goal for a baseline biodiversity survey. Communities of macrolichens on tree bark, for example, are now widely accepted as good indicators of the response of all lichens to sulfurous air pollutants (Ferry et al. 1973; Nash and Wirth 1988; Bates and Farmer 1992; Smith et al. 1993) and as indicators of forest ecosystem biodiversity and function (McCune et al. 1997b; Will-Wolf et al. 2002). Nitrogen-fixing lichens (lichenized fungi associated with cyanobacteria photo-

bionts) have been shown in several regions to be more common and more diverse in older, less disturbed forests (Fig. 9.8) and grasslands than in more disturbed areas and also in areas with cleaner air (Eldridge and Rosentreter 1999; Gauslaa 1995; Kondratyuk and Coppins 1998; Sillett et al. 2000). "Pin-lichens," species in the former order Caliciales, have been recommended for use as indicators of long, continuous occupation of a site by mature ("old-growth") forest in northern temperate and boreal regions (Tibell 1992; Selva 1994). Selva (1994) developed a group of forest continuity (old-growth) indicator species, including pin-lichens, for northeastern North America. Ground lichens have been used as indicators of ecosystem function for grassland and steppe habitats (Anderson et al. 1982; Brotherson et al. 1983; Rosentreter 1986; Eldridge and Tozer 1996).

Habitat Information

Minimum habitat information required for each sample site includes location; general vegetation description/classification, including life forms and structure; brief vegetation history (e.g., old, young, disturbed); landform; topography; and rock or soil type(s). As much as possible, this information should be acquired from the literature and previous studies rather than as part of the sampling protocol. If no background information about a specific site is available, then investigators need to collect information in each of the previously mentioned categories as part of the sampling protocol (Stolte et al. 1993; Nimis et al. 2002) rather than relying solely on generalized map information or general landscape unit, habitat type, or vegetation class descriptions.

Detailed habitat information should be recorded at a site as needed to characterize subplots at the spatial scale used for collecting lichen data. For instance, instead of compiling a list of common or dominant plant species, an investigator might estimate the relative importance of those species. Likewise, quantitative estimates of structure, such as shrub or ground cover, percent standing dead stems, cover of dead wood on the ground, and canopy cover in several forest layers, may provide more useful data than just a brief characterization of vegetation structure (shrubland, old forest). Notes on rock type and ground habitats can be augmented with more detailed information such as estimates of cover of different types of rock and ground surface, patch sizes, percent of aspect exposures available, and percent in sun or shade. Information should be collected by categories that match lichen microhabitat subdivisions. Lichenologists with little ecological training should consult an ecologist familiar with the region for help in designing appropriate sampling protocols.

FIGURE 9.8 *Lobaria pulmonaria*, the lung lichen, has both green alga and cyanobacteria photobionts and thus fixes atmospheric nitrogen. It is found in northern hemisphere temperate zone moist forests, where it is considered to be a bioindicator of old-growth forests. It also has been shown in several regional studies to be sensitive to air pollution. (Photo by Bruce McCune)

DATA ANALYSIS

An investigator must analyze his or her data in a way appropriate to the survey design protocol. Species richness will be available for each site from all survey protocols we discussed, as will at least qualitative estimates of commonness or rarity and habitat and microhabitat affinity. More complex within-site sampling protocols will support more complex data analyses. Relative abundance for species or groups can be estimated with varying degrees of precision from cover-class or abundance-class data. Relative abundance estimates, although often only semiquantitative, can provide repeatable measures of community composition useful for monitoring change over time in one area and for comparing areas, if methods of data collection are the same.

BASIC SUMMARY STATISTICS

Species frequency, the proportion of fixed-area sample units in which a species is found, is probably the most commonly used estimator of species abundance because the data are so easy to collect, but it is somewhat difficult to interpret and compare among areas. Frequency values for the same site vary with the size of the sample unit (Mueller-Dombois and Ellenberg 1974; Grieg-Smith 1983; see "Collection Effort Curves," Chapter 5), and frequency estimates have nonlinear relationships with quantitative abundance estimates such as density or cover.

Absolute measures of density (counts of individuals per unit area) and dominance (e.g., cover or biomass) for whole sites are difficult to calculate for lichens. They are not readily attainable using the sampling designs recommended here.

For any single within-site sampling protocol, data aggregated from single sites into site classes and from site classes into all sites can be subjected to increasingly detailed data analyses. For example, if the chosen within-site protocol is species presence in fixed-area sites with abundance classes for macrolichens, one could summarize the data at different spatial scales as follows:

1. Within-site: presence list for all species; abundance and relative abundance for macrolichens
2. Within-site class (e.g., habitat or vegetation class, climate zone): frequency and relative frequency for all species; average abundance and variation in abundance for macrolichens
3. All sites combined: average frequency and variation in frequency among site classes for all species; comparisons of average abundances and variation among site classes for macrolichens; frequency estimates for all species and abundance comparisons for macrolichens for the whole survey area, including any nonreplicated sites.

DIVERSITY INDICES

Common diversity indices can be calculated from any semiquantitative data, but such numbers are strongly scale-dependent (see "Spatial Scale of Biodiversity," in Chapter 5) and therefore difficult to compare using numbers from studies with different sample designs. Species richness (the total number of species in a sample unit) is a simple, effective, and easily communicated measure of alpha diversity (Whittaker 1972). Diversity indices for equal-area sample sites within the survey can readily be compared with one another and can be compared over time. McCune and associates (1997b) used the total species pool across sites in a region or subregion as an estimator of gamma, or landscape diversity, and average alpha diversity (number of species/site) divided by gamma diversity as an estimator of beta diversity, or turnover rate across environmental gradients or between habitats (Whittaker 1972). Measures of beta and gamma diversity are not as standardized as are the familiar indices for alpha diversity (e.g., Stoms and Estes 1993).

SPECIES-AREA CURVES

Species-area curves can be constructed, but they have complex interpretations and can differ notably for the same area depending on the method of data aggregation used to produce the curves (Palmer and White 1994) and the size of the smallest sample unit used. They should be extrapolated only cautiously to predict diversity for areas larger than, or other than, the area represented in the survey. Differences between species-area curves constructed with data on lichenized fungi obtained using the same methods in similar habitats that differ in air quality have been interpreted to indicate alteration of community function (Lawrey 1991, 1992). Comparisons of such curves for lichenized fungi of similar microhabitats in different parts of a survey area or in the same areas over time deserve investigation as a potential tool for monitoring biodiversity changes.

PARTITIONING DATA BY MORPHOLOGICAL OR FUNCTIONAL GROUPS

During analysis, partitioning data *a posteriori* by important groups of lichens that differ in function or morphology may help to clarify the nature of lichen community response to habitat and other environmental variables. Correlations of such groups of similar lichen species with explanatory variables often differ. For example, cyanolichens, alectorioid lichens, and green-alga foliose lichens in Pacific Northwest forests of the United States respond differently to such variables as position in canopy, forest age, tree density, and habitat heterogeneity (McCune 1993; Sillett and Neitlich 1996; Peck and McCune 1997; McCune et al. 2000).

CORRELATION WITH ENVIRONMENTAL VARIABLES

All recommended sampling protocols support investigation of relationships between patterns of lichen data and habitat and macrovegetation variables. Nonstatistical, but quantitative, presentations of patterns of species diversity often are used to interpret survey results (Geiser et al. 1994a, 1994b; Vitt and Belland 1997). Single-factor statistical analysis can include correlation, regression, and contingency table analyses. Multivariate quantitative analyses of species data, including ordination (Will-Wolf 1980; Rosentreter 1986; Oksanen 1988; Marcelli 1992; Geiser et al. 1994a; Wolseley et al. 1995; McCune et al. 1997a, 1997b), classification (Tibell 1992; Geiser et al. 1994a), and gradient analysis (Oksanen 1988; McCune et al. 1997b), can facilitate description of differences between communities, highlight habitats and microhabitats with high diversity, and help define relationships between communities and habitats.

IMPLEMENTATION OF BIODIVERSITY INVENTORIES

In this chapter we have summarized the current status of classification of lichenized fungi and have discussed in detail the important design elements for surveys of their biodiversity with examples (Table 9.4) to show how these elements are combined in practical situations.

Within-site survey protocols can be grouped into four categories: (1) Species presence at sites; all species, sites grouped into classes (Table 9.4: example 1). (2) Abundance at sites, for an easily surveyed subset of species (indicator species); sites grouped into classes or placed along gradients (Table 9.4: example 2). (3) Species presence at sites, all species; plus abundance at sites for a subset of species; sites grouped into classes (Table 9.4: example 3). (4) Species presence or abundance in subplots within site, all species; sites grouped into classes (Table 9.4, example 4).

Each category of within-site protocol is useful for biodiversity surveys with particular sets of goals:

1. Rapid (1-year) inventory of large regions. Protocols 1 and 2 are appropriate; 1 results in greater species capture; 2 provides better quantification and repeatability for monitoring trends.
2. Medium-term (3–5-year) inventory of regions. Protocol 3 results in good species capture plus more precise-abundance estimates than protocol 2; variation in number of sites affects accuracy and time to completion.
3. Intensive inventory of selected sites. Variations on protocol 4 give very accurate single-site inventories but require considerable time. For a large region, an inventory may take many years to complete.

Each study used as an example in Table 9.4 relied primarily on one within-site sampling protocol. This need not be the case, however, for an effective biodiversity survey. For example, a two-phase inventory might first use a protocol with a low-intensity effort per site to obtain an areawide survey that can be completed within a year or two. That would be followed by high-intensity sampling of a selected subset of sites to be completed over a longer period. Results from the first-phase inventory would form the basis for selecting the subset of sites to be emphasized in the high-intensity inventory and at the same time would provide a completed inventory to be used for making management decisions before the more intensive surveys were finished. Another strategy to make information available for reference before the completion of a several-year survey is to distribute sites across a survey area or between site classes in an area for each field season. Then preliminary areawide summaries can be made available before the completion of the entire inventory.

ACKNOWLEDGMENTS. All coauthors contributed substantially to the text; after the first author, order is alphabetical and does not necessarily reflect relative contribution. The authors thank B. Aguirre-Hudson for extensive comments on the content of the manuscript and J. Wolf for editing and proofreading the text. L. H. Geiser, A. Y. Rossman, and C. M. Wetmore provided valuable information on their studies. K. Elliot drafted the figures. G. M. Mueller and anonymous reviewers provided helpful critiques. All opinions expressed are the responsibility of the authors.

10

SEQUESTRATE FUNGI

MICHAEL A. CASTELLANO, JAMES M. TRAPPE, AND DANIEL L. LUOMA

Sequestrate fungi (Fig. 10.1) are fleshy, leathery, rubbery, or cartilaginous species that sequester, or seclude, their spore-bearing tissues, preventing the spores from discharging into their surroundings (Kendrick 1992). Usually, the spore-bearing tissues are enclosed within a persistent peridium. Sequestrate fungi, with the exception of the "secotioid" species, are hypogeous, without a stipe or with only a rudimentary one (true and false truffles, respectively). The secotioid species are partly to almost entirely exposed at maturity and may have a well-developed stipe. About 150 genera and 1200 valid species of sequestrate fungi have been described in 38 families, representing 11 orders in the Ascomycetes, Basidiomycetes, and Zygomycetes. Much of the world has yet to be explored for sequestrate fungi, however, including most of Asia, Africa, and South America. New species are being discovered regularly in North America and Australia, which have been searched extensively for members of this group. Only Europe has been covered thoroughly by collectors of sequestrate fungi, and even there, new species continue to be discovered. The total number of species existing in the world is likely more than double that currently known.

In the Ascomycetes and Basidiomycetes, most sequestrate fungi appear to be ectomycorrhizal mycobionts (Molina et al. 1992). However, saprobic taxa occur in both groups, especially in New Zealand (e.g., *Paurocotylis* species and *Weraroa* species) and in tropical rain forests in Australia (e.g., *Stephensia* species). Sequestrate forms of Zygomycetes are found in the Endogonales and Glomales. Their fruiting bodies range from loose clusters of spores to dense, somewhat organized masses of spores. Most sequestrate species in the Endogonales

FIGURE 10.1 Sequestrate fungi.

appear to form ectomycorrhizae (Gerdemann and Trappe 1974), but *Endogone pisiformis* and probably other closely related taxa are saprobic (Berch and Castellano 1986). We do not include arbuscular mycorrhizal (AM) Zygomycetes, which fruit as individual spores in soil or roots, in our definition of "sequestrate fungi." The Glomales, in general, form arbuscular mycorrhizae with assorted host plants, but sporocarpic sequestrate forms are associated primarily with woody perennials (Gerdemann and Trappe 1974).

ECOLOGY AND EVOLUTION

DISTRIBUTION AND HABITATS

Sequestrate fungi usually associate with trees or shrubs and hence are most commonly found in forests, woodlands, and shrublands. Ectomycorrhizal forests once were thought to occur primarily in temperate to boreal zones (Meyer 1973), but now ectomycorrhizae are known to abound in many tropical forests as well (Allen et al. 1995). Usually, wherever ectomycorrhizal hosts occur, ectomycorrhizal sequestrate fungi also occur. Similarly, sequestrate AM fungi may be found among roots of tropical to boreal AM trees and shrubs. Saprobic

sequestrate species may occur similarly within this range of habitats. Some mycorrhizal sequestrate fungi occur in deserts (Gilkey 1939; Alsheikh and Trappe 1983a, 1983b; Alsheikh 1994), where they may form special types of mycorrhizae with annual or perennial hosts (Awamah et al. 1979). Some saprobic species (Zeller 1943; Singer and Smith 1958; McKnight 1985) also are found in deserts, although in general few fungal species and specimens of any type are found in this harsh habitat.

Data on species diversity of sequestrate fungi are scant, but our observations from extensive collecting of sporocarps suggest some generalities (J. M. Trappe and M. A. Castellano, unpublished data). Species diversity of all sequestrate groups can be relatively high in the tropics, but in general it seems highest in subtropical to temperate forests (Allen et al. 1995). Diversity, based on sporocarp collections, tends to decline in austral or boreal habitats with increasing proximity to the polar regions. Similarly, species diversity in mountains tends to be greater in low-elevation forests than in subalpine to treeline habitats. In alpine and arctic habitats, diversity of sequestrate fungi is low, even when suitable hosts such as krummholz conifers, willows, or birches are present (Laursen and Miller 1978; Cázares and Trappe 1990; Graf and Horak 1993; J. M. Trappe, unpublished data). Hosts suitable for sequestrate fungi are rare or lacking in Antarctica.

ORIGIN OF LIFE FORMS

Sequestrate fungi in the Ascomycetes and Basidiomycetes share a common ancestry with or were derived from epigeous, nonsequestrate forms of cup fungi and mushrooms (Trappe 1979; Miller 1983; Thiers 1984; Bruns et al. 1989; O'Donnell et al. 1996). They have evolved along several phylogenetic lines into fungi ranging from mushroomlike to trufflelike. The Basidiomycete order Boletales exemplifies these forms particularly well (Miller 1983; Thiers 1989; Cázares and Trappe 1991). In species of the mushroom genus *Suillus,* for example, spores are produced in hymenium-lined tubes from which they are discharged forcibly. That genus is related closely to *Gastrosuillus,* which, although retaining a modified mushroom form, has hymenium-lined tubes that are contorted and plugged so that the spores are sequestered—they cannot escape to the air (Bruns et al. 1989; Thiers 1989). The stipes of *Gastrosuillus* species are much reduced, so that the spore-bearing tubes are not raised out of the ground.

Truncocolumella species represent a further progression toward reduced reproductive structures (Smith and Singer 1959). In this genus, the spore-bearing tubes found in *Suillus* and *Gastrosuillus* have been replaced by hymenium-lined chambers, and the entire basidioma is enclosed in a persistent peridium. *Truncocolumella* lacks a stipe or, at most, a vestigial ancestral stipe occasionally projects from the base of the columella. Further reduction of the sequestrate basidioma is found in *Rhizopogon* species, which resemble *Truncocolumella* in form but lack a columella (Miller 1986; Bruns et al. 1989). Finally, genera such as *Alpova* and *Melanogaster* lack the chambers and hymenia; the gleba is a gelatinous-solid mass of pockets packed with gelatinized basidia and spores (Trappe 1975).

Similar morphological progressions occur in several other families of Basidiomycetes. In the Cortinariaceae, the mushroom genus *Cortinarius* is analogous to the stipitate sequestrate genus *Thaxterogaster.* The basidiome is reduced further in the genera *Cortinomyces* and *Hymenogaster,* with some species having a columella and others losing all trace of it (Bougher and Castellano 1993). Two major progressions similarly occur in the Russulaceae: from *Russula* species to the stipitate, sequestrate *Macowanites* species, with a contorted-lamellate to loculate gleba, followed by the astipitate, loculate *Gymnomyces* species, and from *Lactarius* species to *Arcangeliella* species, followed by *Zelleromyces* species (Singer and Smith 1960). In the Tricholomataceae, the reduction from mushroom to simple sequestrate forms occurs from *Laccaria* species, through *Podohydnangium* species, to *Hydnangium* species (Bougher et al. 1993; Mueller and Ammirati 1993).

Two puffball genera, *Radiigera* and *Pyrenogaster,* are sequestrate. The sequestrate forms are hypogeous and resemble unopened specimens of their epigeous analogues such as *Geaster* species. They have lost the sutures that permit the peridium to open to produce the earth star effect of mature *Geaster* species (Toledo and Castellano 1996). The genus *Scleroderma* consists mostly of epigeous species with a peridium that opens apically to expose the powdery spore mass to the air. Some *Scleroderma* species, however, consistently remain below ground and never open (Zeller 1922, 1947; Beaton and Weste 1982). Their spore mass still may be powdery, but in some species the spores adhere to each other to produce a more or less solid mass. Even when these species happen to emerge from the soil, their peridia remain closed (J. M. Trappe, unpublished data).

Similar phylogenetic links from epigeous to sequestrate forms are seen in the Ascomycete order Pezizales. Epigeous cup fungi, such as *Humaria,* are related to the genus *Geopora,* which has sequestrate species that fruit hypogeously and have large chambers lined with hymenia that are enclosed in a peridium (Burdsall 1968; Trappe 1979). *Stephensia* is similar to *Geopora* except that its hymenia do not line open chambers but are embedded in solid tissue (Trappe 1979). *Phaeangium* represents a further simplification: the asci are packed in pockets rather than disposed in a hymenial layer (Alsheikh and Trappe 1983a). Similarly, the epigeous cup-fungus genus *Peziza* is related to the genus *Hydnoplicata,* in which the spore-bearing tissue is folded, producing sequestrate chambers lined with hymenia (Gilkey 1954). Reproductive structures are reduced further in *Hydnotryopsis* species, in which a solid gleba encloses veins of hymenia or pockets stuffed with asci. O'Donnell et al. (1996) provided molecular evidence linking other sequestrate Ascomycetes to epigeous genera and families.

Sequestrate groups in the Zygomycetes may represent evolutionary progression in the opposite direction. The fossil record of the arbuscular mycorrhizal Glomales suggests extremely early occurrence of species that form individual spores among roots in soil (Taylor et al. 1995). Reproductive structures of present-day sequestrate species range from simple, tight masses of spores to complex fruiting bodies (Gerdemann and Trappe 1974), which in this case could be regarded as an evolutionary advance. The origin of the Endogonales is obscure, but reproductive structures of sequestrate *Endogone* species also range from simple masses of spores to organized arrangements of spore-bearing tissues (Gerdemann and Trappe 1974).

These "evolutionary experiments" involve simplification of structure (except, perhaps, in the Zygomycetes) and increased protection from climatic stress accompanied by increased energy efficiency and specialized adaptation to spore dispersal. Selection favoring protection of sporocarps from heat, drought, or freezing leads to partial or complete immersion of sporocarps in the protective, insulating soil. Consequently, the mushroom stem no longer is needed to raise spore-bearing tissues into the air, nor is the cap needed to spread those tissues for forcible spore discharge. If spores are not forcibly discharged, no orderly hymenial palisade is required to expose ascus or basidium tips to the air. Host-fungal mycorrhizal associations in sequestrate species no longer produce the nonfertile structures and thereby save resources. Although reduction from a form with an elaborate cap, stem, and hymenial tissue ultimately to a simple, truffle form with less differentiated tissue seems morphologically radical, molecular evidence indicates that such changes do not require massive numbers of mutations (Bruns et al. 1989).

Loss of forcible discharge of spores to air must be accompanied by mutations that adapt sequestrate fungi to other spore dispersal tactics. In most cases, the dispersal agents are animals, from arthropods and gastropods to mammals and birds. The fungi have developed a means to attract animals when spores are mature and cause them to ingest the spores. The attractants are aromatic compounds, including pheromones (Claus et al. 1981). Each sequestrate species produces its own array of aromas, usually a combination of several to many compounds (Marin et al. 1984). Immature sporocarps have little or no distinctive odor. As spores begin to mature, the attractant compounds are produced, and as additional spores mature, the aroma increases in intensity and pungency (Trappe and Maser 1977).

Some species visually attract birds. In New Zealand, which lacks native mammals, birds appear to be important vectors of sequestrate fungi. *Paurocotylis pila* has a scarlet peridium. As its ascomata expand, they lift themselves to the surface of the humus, sometimes detaching completely and lying loose on the surface. Their size and color mimic fruits of nearby *Podocarpus* species and other plants that are eaten by birds; *P. pila* also matures at the same time as the *Podocarpus* fruits. *Paurocotylis* ascomata lying on the forest floor among podocarp fruits almost surely are eaten by birds. Several other New Zealand sequestrate fungi also emerge from the forest floor and are brightly colored (e.g., species of *Chamonixia, Thaxterogaster,* and *Weraroa*). *Phaeangium lefebvrei* in deserts of the Arabian Peninsula and North Africa produces different visual signals. Several to a dozen or more small ascomata cluster to form a pronounced hump on an otherwise flat desert floor. Birds spot these humps, scratch away the overlying soil, and eat the ascomata (Alsheikh and Trappe 1983a).

Animals disperse spores in response to nutritional inducements; most mammals, including carnivores, excavate and eat sequestrate fungi (Fogel and Trappe 1978; Maser et al. 1978). Sequestrate fungi contain carbohydrates, nonprotein amino acids, proteins, and elevated concentrations of some minerals (Al-Delaimy 1977); their nutritional value is evidenced by those animals that eat little but sequestrate fungi (Maser et al. 1978; Maser et al. 1985; Claridge and May 1994). All sterile tissues are digested, but the spores pass through the digestive tract unharmed. As spore-containing feces weather, the spores are washed into the soil and potentially contact receptive mycorrhizal host feeder roots (Trappe and Maser 1977).

Some sequestrate fungi such as *Gummiglobus, Mesophellia, Castoreum, Radiigera,* and *Scleroderma* species in the Basidiomycetes and *Elaphomyces* species in the Ascomycetes have a powdery gleba. The first two of these genera have a central columella or core that mammals eat; to get at it, they peel off the outer, leathery, or carbonaceous peridium. As this is done, the powdery spores are released to the air, cling to fur, or fall to the soil (Claridge and May 1994). The other genera all have thick peridia that the animals consume. They dig up the sporocarps, often taking them to a feeding spot, and then eat the peridium. They discard the powdery spore mass inside, releasing the spores to the wind (Trappe and Maser 1977).

Most desert-dwelling sequestrate fungi are adapted for passive spore dispersal by wind. Species of *Carbomyces, Terfezia,* and *Tirmania* fruit only in years when adequate rain falls at appropriate times. The fungi produce succulent ascomata composed of large, thin-walled cells ill-adapted to withstand drying. As the relatively small *Carbomyces* ascomata expand and mature, they tend to lift themselves out of the soil. When fresh, they may be eaten by animals (Zak and Whitford 1986). If not eaten, they dry and the inflated, thin-walled cells of the gleba disintegrate into a spore-bearing powder. The ascomata of *Carbomyces* species are blown about the desert floor by the wind, dispersing the spores as the peridium abrades. *Terfezia* and *Tirmania* species tend to have relatively large ascomata. As they expand, they push the overlying soil upward. The wind blows away the soil as it dries to expose the sequestrate ascomata, which quickly dry. The initially large, succulent cells of the gleba disintegrate into a spore-bearing powder, which wind-blown sand abrades away.

Development of sequestrate sporocarps in the Zygomycetes could enhance spore dispersal because, without animal mycophagy, the Endogonales and Glomales mostly are dispersed by movement of soil through the action of wind, water, landslide, or animals feet. The individual spores produced in soil by most Glomales, although large in comparison to spores of other fungi, are probably too small to attract even the smallest mammals or birds. Large masses of spores clustered in compact sporocarps, however, are readily found and consumed by small mammals (Maser et al. 1978). Two species show adaptations that may promote dispersal by birds. *Endogone pisiformis* and *Glomus convolutus* are both bright yellow and fruit in the spring on the soil surface at the edge of melting snow banks (Gerdemann and Trappe 1974). They are spotted easily, and birds may be attracted to them or mistake them for large seeds or small fruits.

The wide distribution, considerable diversity, and local abundance of sequestrate fungi imply that this morphology confers broad evolutionary fitness. The sequestrate habit has arisen independently in parallel evolutionary lines in different places. Australia and North America provide a striking comparison. The *Suillus–Gastrosuillus–Truncocolumella–Rhizopogon* line is host-specific to members of the Pinaceae, a family that does not occur naturally in Australia. The *Descolea–Setchelliogaster–Descomyces* line appears likely to be host-specific to eucalyptus, which does not occur naturally in North America. The *Russula–Macowanites–Gymnomyces*, the *Lactarius–Arcangeliella–Zelleromyces*, and the *Cortinarius–Thaxterogaster–Protoglossum* lines are found on both continents, although no species occur on both continents.

So far as is known, no sequestrate fungi are toxic, which reflects their dependence on mycophagy for spore dispersal. Killing or discouraging one's dispersal agents hardly would confer reproductive advantage. No phylogenetic lines from toxic epigeous fungi to edible sequestrate fungi are known. The genus *Cortinarius* contains both edible and toxic species. We hypothesize that its sequestrate relatives, *Thaxterogaster* and *Protoglossum*, will prove to be related to the edible *Cortinarius* species.

It generally is believed that epigeous fruiting taxa are more numerous than sequestrate fungi, but presently no evidence suggests that one life form is more successful than the other. In all likelihood, one may be favored over the other in certain niches or habitats or in certain climates. Why the sequestrate habit occurs overwhelmingly among mycorrhizal, rather than saprobic, fungi is a matter of speculation.

TROPHIC RELATIONSHIPS AND FUNCTIONAL ROLES

The mycorrhizal or saprobic habits of the epigeous relatives of most sequestrate fungi carry over into the sequestrate fungi themselves, often to the same degree of host specificity (Molina and Trappe 1982; Molina et al. 1992). Mycorrhizal fungus–host interactions have been covered extensively elsewhere (e.g., Harley and

Smith 1983; Miller and Allen 1992) and need not be repeated here. In summary, the fungus absorbs water and nutrients from the soil and transfers them to the host and, in return, receives photosynthates in the form of carbohydrates from the host. The saprobes, in contrast, produce the enzymes needed to extract energy and nutrients from dead organic matter without recourse to symbiosis with plant rootlets.

The adaptations found in organisms associated with sequestrate fungi are a tribute to interdependence and coevolution. Perhaps the ultimate example is provided by old-growth forests of Pinaceae in the Pacific Northwestern United States. The trees require ectomycorrhizal fungi for adequate mineral nutrition, and many of those fungi are sequestrate. The mycorrhizal fungi need energy from the trees to fruit, and the sequestrate fruiting bodies are a primary food of the northern flying squirrel, *Glaucomys sabrinus* (Maser et al. 1985). The squirrel, in turn, disperses the spores and is the primary food of the northern spotted owl, which nests in and hunts from the trees (Carey 1991). Once an owl captures a squirrel, it may carry it a considerable distance before gutting it; the entrails are likely to contain spores of sequestrate fungi. In summary, the trees need the fungi for their nutrition. The fungi need trees for energy, and they need squirrels for spore dispersal. The squirrels need trees for habitat, and they need fungi for their nutrition. Owls need squirrels for their nutrition, and they need trees for nesting and hunting. Owls also serve as long-distance vectors of spores.

Little is known about the trophic characteristics of the relatively few sequestrate saprobes. *Nivatogastrium nubigenum* fruits on or near brown-cubical-rotted wood in western North American mountains, as do its epigeous *Pholiota* relatives (Singer and Smith 1959). *Endoptychum depressum*, *Paurocotylis pila*, *Stephanospora* species, and *Weraroa* species fruit on forest or other humus, as do some species of their related, epigeous genera (e.g., *Agaricus*, *Aleuria*, *Lindtneria*, and *Stropharia*). *Endogone pisiformis* fruits on a variety of organic substrata, including humus, rotten wood, and dead basidiomata of various species of the Polyporaceae (Trappe and Gerdemann 1979).

Some sequestrate fungi also have special capabilities. *Hysterangium* species and *Gautieria* species, for example, form dense mycelial mats in forest soil. The mats have specific biogeochemical properties (Griffiths et al. 1991). *Rhizopogon* species typically induce prolific branching of host rootlets through production and release of auxin (Ho and Trappe 1987). It is interesting that the ancestral *Suillus* species do this as well. *Rhizopogon* species also have been shown to enhance the survival and growth of seedlings of *Pseudotsuga menziesii* in forest plantations (Castellano and Trappe 1985). *Tuber melanosporum* produces herbicidal compounds that reduce weed competition with the host tree (Montacchini and Lomagno 1977).

LIMITATIONS TO THE STUDY OF SEQUESTRATE FUNGI

Sequestrate fungi are difficult to study because most fruit hypogeously. When they fruit in compacted soil with little humus, however, they may emerge or at least raise a hump in the soil because lateral and downward expansion is difficult. Campgrounds, picnic grounds, roadsides, and old logging skid trails or landings thus are places where sequestrate fungi can be found easily when they fruit. Forest soil, in contrast, is generally not compacted and has a layer of humus and forest litter. Sequestrate fungi can expand in all directions without providing visual clues to their presence.

The holes left by animals that have excavated sequestrate fungi can indicate the location of a colony because these fungi commonly produce multiple sporocarps that often mature at different times. Raking in the vicinity of an animal dig and to about the same depth often will reveal additional specimens (Fig. 10.2). During the peak of the fruiting season, this approach can be particularly productive because production of sequestrate sporocarps often exceeds the food demands of resident animals. At the beginning and end of the fruiting season, however, when demand exceeds supply, animals remove a signifi-

cant portion of the mature specimens (North and Trappe 1994).

In Europe, dogs are trained to find commercially valuable sequestrate fungi such as truffles. A collector lacking a trained dog must rely on animal digs, knowledge of the kinds of habitats that are productive, and chance to find sequestrate fungi in forests. This poses special problems for inventory and monitoring of fungal diversity or of rare species (see "Surveying Rare Species," later in this chapter).

Fruiting of sequestrate fungi varies by season and geographic area. In general, however, sequestrate species fruit at about the same time as epigeous fungi in any particular area. Fruiting seasons of sequestrate species tend to be extended because of their mostly hypogeous habit, which protects them from frost, heat, and drought.

In areas where sequestrate fungi have pronounced spring and autumn fruiting seasons, the species fruiting generally differ between the two seasons (Luoma 1988; Luoma et al. 1991). At low elevations in the Pacific Northwest United States, for example, fruiting sequestrate Ascomycetes are more abundant in spring than in autumn, and fruiting sequestrate Basidiomycetes are more abundant in autumn than spring. Those fruiting patterns mirror those of their epigeous relatives. At higher elevations, in contrast, the season is confined to a few summer months, and the seasonal separation of fruiting species is much less distinct. That is also true of areas with a single rainy season, such as much of Asia and Australasia.

FIGURE 10.2　Raking often will reveal specimens.

TAXONOMY, DIVERSITY, AND DISTRIBUTION

SYNOPSIS OF TAXA

Sequestrate fungi occur in two phyla of the Kingdom Eumycota: the Zygomycota (in the class Zygomycetes) and the Dikaryomycota (in classes Ascomycetes and Holobasidiomycetes). Presently recognized genera are listed in Table 10.1 by order and family. Assignments of genera to families differ in many cases from previous placements (e.g., Trappe 1979; Jülich 1981; Castellano et al. 1989; Trappe and Castellano 1991) because more recent data from additional collections or DNA studies have provided new insights into phylogeny. In many cases, the family to which a genus belongs is uncertain based on morphological data alone. Studies of ultrastructure and DNA are providing new evidence that is helping to resolve many of the uncertainties, but that

work has only begun (O'Donnell et al. 1996). All taxonomic decisions are hypotheses that are tested repeatedly and sometimes refuted. Some genera do not fit into any described family and are listed under the heading "uncertain status." We are aware of many undescribed genera that are not included in our compilation.

HERBARIUM COLLECTIONS

The study of sequestrate fungi began in Italy, where several pre-Linnaean botanists had a special interest in truffles. Carlo Vittadini (1831, 1842) laid the foundations for our modern generic concepts. His collections, which are housed at the Botanical Institute of the University of Torino (Torino [or Turin], Italy), comprise a seminal collection of type specimens (Mattirolo 1907). Oreste Mattirolo's types and other extensive collections also are housed there.

TABLE 10.1

Genera and Families of Sequestrate Fungi in the Kingdom Eumycota, by Class and Order

Phylum Zygomycota
Class Zycomycetes
  Order ENDOGONALES: Endogonaceae: *Endogone*
  Order GLOMALES: Glomaceae: *Glomus, Sclerocystis*
  Order MUCORALES: Mortierellaceae: *Modicella*

Phylum Dikaryomycota
Class Ascomycetes
  Order ELAPHOMYCETALES: Elaphomycetaceae: *Elaphomyces*
  Order GLAZIELLALES: Glaziellaceae: *Glaziella*
  Order PEZIZALES: Ascobolaceae: *Muciturbo*; Carbomycetaceae: *Carbomyces*; Helvellaceae: *Balsamia, Barssia*; Humariaceae: *Geopora, Hydnocystis, Phaeangium, Stephensia*; Discinaceae: *Hydnotrya*; Otidiaceae: *Genea, Genabea, Paurocotylis, Picoa, Sphaerosoma*; Pezizaceae: *Amylascus, Cazia, Hydnoplicata, Hydnotryopsis, Mycoclelandia, Pachyphloeus, Ruhlandiella, Sphaerozone, Tirmania*; Tuberaceae: *Choiromyces, Delastria, Dingleya, Hydnobolites, Labyrinthomyces, Loculotuber, Paradoxa, Reddellomyces, Terfezia, Tuber*
  UNCERTAIN STATUS: *Fischerula: Leucangium: Petchiomyces*
Class Homobasidiomycetes
  Order AGARICALES: Agaricaceae: *Endoptychum*; Amanitaceae: Torrendia; Bolbitiaceae: *Agrogaster*; Coprinaceae: *Gasteroagaricoides, Podaxis*; Cortinariaceae: *Cortinarius, Cortinomyces, Descomyces, Destuntzia, Hymenogaster, Kjeldsenia, Quadrispora, Setchelliogaster, Thaxterogaster*; Cribbiaceae: *Cribbia*; Entolomataceae: *Rhodogaster, Richoniella*; Lepiotaceae: *Neosecotium, Notholepiota*; Octavianinaceae: *Octavianina*; Russulaceae: *Arcangeliella, Cystangium, Gymnomyces, Macowanites, Zelleromyces*; Strophariaceae: *Nivatogastrium, Tympanella, Weraroa*; Tricholomataceae: *Hydnangium, Podohydnangium, Gigasperma*
  Order BOLETALES: Boletaceae: *Alpova, Amogaster, Gastroboletus, Gastroleccinum, Gastrosuillus, Hallingia, Melanogaster, Rhizopogon, Royoungia, Truncocolumella*; Gomphidiaceae: *Brauniellula, Gomphogaster*; Paxillaceae: *Austrogaster, Gymnopaxillus, Paxillogaster, Singeromyces*; Strobilomycetaceae: *Austrogautieria, Chamonixia, Gautieria, Mycoamaranthus, Rhodactina, Timgrovea, Wakefieldia*
  Order GASTROSPORIALES: Gastrosporiaceae: *Gastrosporium*
  Order LEUCOGASTRALES: Leucogastraceae: *Leucogaster, Leucophleps*
  Order LINDTNERIALES: Stephanosporaceae: *Stephanospora*
  Order LYCOPERDALES: Geastraceae: *Pyrenogaster, Radiigera* Mesophelliaceae: *Andebbia, Castoreum, Gummiglobus, Malajczukia, Mesophellia, Nothocastoreum*
  Order PHALLALES: Clathraceae: *Protubera*; Gelopellaceae: *Gelopellis, Phallobata, Phallogaster*; Hysterangiaceae: *Chondrogaster, Claustula, Gallacea, Hysterangium, Phlebogaster, Rhopalogaster, Trappea*
  Order SCLERODERMATALES: Sclerodermataceae: *Horakiella, Scleroderma*; Sedeculaceae: *Sedecula*
  Order TREMELLOGASTRALES: Tremellogastraceae: *Clathrogaster, Tremellogaster*
  UNCERTAIN STATUS: *Brauniella, Hysterogaster, Mycolevis, Protogautieria, Sclerogaster, Smithiogaster*

Other important nineteenth and early twentieth century pioneers in the study of sequestrate fungi and the herbaria (Appendix III) that presently contain their collections include the brothers L. R. and C. Tulasne, Cryptogamic Museum of Paris (Paris, France); M. J. Berkeley, C. E. Broome, M. C. Cooke, and G. Massee, Royal Botanic Gardens (Kew, United Kingdom); N. Patouillard and F. Bucholtz, Farlow Herbarium, Harvard University (Cambridge, Massachusetts, United States); R. Hesse, University of Marburg (Marburg, Germany); L. Hollós, Hungarian Natural History Museum (Budapest, Hungary); H. W. Harkness and C. G. Lloyd, National Fungus Collections (Beltsville, Maryland, United States); and L. Rodway, Tasmanian Museum (Hobart, Australia). Unfortunately, the African and west Asian types and other collections of desert truffles of A. Chatin have been lost (R. Heim, personal communication).

Many mycological herbaria have a few collections of sequestrate fungi, but the most important mid- to late-twentieth century collectors and the locations of their collections are as follows (see Holmgren et al. 1990; Appendix III):

Europe—M. Soehner, Botanisches Staatssammlung, Munich (M); A. Knapp and E. Horak, Eidgenössische Technische Hochschule Zürich, Zürich (ZT); C. and L. R. Tulasne, Laboratoire de Cryptogamie, Paris (PC); M. G. Malençon and L. Riousset, Institut de Botanique, Montpellier, France (MPU); M. Lange, Botanical Museum, University of Copenhagen (C); L. Kers, Bergius Foundation, Stockholm (SBT)

North America—S. M. Zeller, New York Botanical Garden, The Bronx (NY); R. Thaxter, Farlow Herbarium, Harvard University, Cambridge (FH); H. E. Parks, University of California at Berkeley (UC); A. H. Smith and R. D. Fogel, University of Michigan (MICH); M. A. Castellano, J. W. Gerdemann, H. Gilkey, D. Luoma, and J. M. Trappe, Oregon State University (OSC); H. Saylor and H. D. Thiers, San Francisco State University (SFSU); J. S. States, Deaver Herbarium, Northern Arizona University, Flagstaff (ASC); D. R. Hosford, Central Washington University, Ellensburg (ELRG)

Australasia—J. W. Cribb, Department of Plant Industries, Indooroopilly, Queensland (BRIP); J. Cleland and J. Warcup, State Herbarium of South Australia, Adelaide (AD); G. Beaton and T. Lebel, Royal Botanic Gardens, Melbourne, Australia (MEL); N. Bougher and N. Malajczuk, C.S.I.R.O. Division of Forestry, Wembley, Perth; L. Rodway, Tasmanian Herbarium, University of Tasmania, Hobart (HO); G. H. Cunningham and R. Beever, Landcare Research, Auckland (PDD)

LIVE CULTURE COLLECTIONS

Some sequestrate fungi (e.g., *Rhizopogon* species) grow reasonably well in culture and are easy to isolate. Most such fungi, however, grow poorly or not at all in culture media tried so far. This is particularly true of the Ascomycetes, Zygomycetes, and those Basidiomycetes in the Russulaceae and Cortinariaceae. The saprobic species probably can be cultured with greater success than mycorrhizal species, judging from the few attempts with which we are familiar. We are aware of only two culture collections that include several to many taxa of sequestrate fungi: that of the U.S. Forest Service, Pacific Northwest Research Station and that of C.S.I.R.O., Division of Forestry (Appendix III). Identity of any cultures not represented by voucher collections in herbaria should be regarded as questionable because misidentification of species is common due to lack of good keys for most taxa.

PATTERNS OF SPECIES RICHNESS AND ENDEMISM

Much of temperate Europe, western North America, New Zealand, temperate and tropical Australia, and the deserts of North Africa and western Asia have been examined extensively for sequestrate fungi, although large areas and many habitat types in all these places are, as yet, largely unexplored. The rest of the world remains unexplored or only scantily explored.

Species richness is generally greatest at subtropical to middle latitudes in ectomycorrhizal forests. In the Northern Hemisphere, richness declines from middle latitudes northward into the boreal forest and farther north. The only equivalent forests in the Southern Hemisphere, in South America and New Zealand, are not well enough explored to draw conclusions concerning the existence of similar gradients. Sequestrate species richness appears to be greater in North America and Australia than in Europe, the same as for ectomycorrhizal hosts. In North America and Australia, the predominance of north–south-oriented mountain ranges permitted fungal and host-species migrations in response to climate changes during the Ice Ages. In Europe the predominantly east–west orientation of major mountain ranges impeded species migrations, a factor of particular importance to fungi depending on animals for spore dispersal.

Detailed data on species richness by habitat are not available, but our studies in western North America and Australia, together with studies of collections from Europe, suggest some generalities. Presence of ectomycorrhizal host plants is associated strongly with relative richness. Because many sequestrate fungi are specific to individual host genera, a diversity of host genera in a region also contributes to richness. Moist forests appear to support greater richness than either dry or very wet forests. The *Pseudotsuga* forests of the Pacific slope of western North America support a very rich sequestrate mycota, as do the *Eucalyptus* forests of eastern Australia. Richness is reduced considerably in rain forests or strongly xeric forests in both regions.

Stand age and history probably influence sequestrate species richness. Several studies of ectomycorrhizal fungi currently in progress in the Pacific Northwest include stand age as a variable, but, so far, no general conclusions can be drawn. Many more than 100 sequestrate species are associated with mature to old-growth forests in the Pacific Northwestern United States (USDA Forest Service and USDI Bureau of Land Management 1994).

Endemism is common among the sequestrate fungi, in part because of host specificity. Pines, Douglas fir, spruce, true firs, oaks, Eucalyptus, and other ectomycorrhizal hosts harbor numerous sequestrate fungi that do not fruit in association with other host genera. In addition, the dependence of sequestrate fungi on animals for spore dispersal restricts their ability to cross major geographic barriers, such as oceans, deserts, or major mountain ranges. The combination of host specificity, narrow habitat requirements, and limited dispersal may confine successful sequestrate mutations to their region of origin for extended periods.

Sequestrate taxa known only from the type or from a few collections in a limited area are inferred to be locally endemic. Because finding ephemeral fruiting bodies is a matter of chance, at least for hypogeous species, some taxa now inferred to be locally endemic subsequently may prove otherwise. For example, when the Ascomycete *Cazia flexiascus* was described (Trappe 1989), it was known only from the type collection in southwest Oregon. It later was found 80 km to the north, and recently it has been collected in southern California, 1500 km to the south. Although still regarded as rare, this species no longer can be regarded as locally endemic.

SURVEY AND INVENTORY

GLOBAL ASSESSMENT

Sequestrate fungus communities have not been assessed adequately in any large geographic region. It is only in the last 20 years that stand-level assessments have been conducted in restricted areas of Oregon, Washington,

and northern California (studies by R. Fogel, G. Hunt, D. L. Luoma, J. Smith, W. Colgan III, M. North, and J. Waters). Trappe and Castellano (personal communication) in collaboration with N. Malajczuk, N. Bougher, T. Lebel, P. Reddell, and A. Clardige recently have finished a series of 20 collecting trips over 13 years to ectomycorrhizal forests of Australia for assessment of sequestrate fungi. The volume of material collected in fewer than 400 person-days of collecting was astonishing: 7500 collections representing approximately 600 species, 90% of which are undescribed. Sampling was neither systematic nor intensive; rather, broad-based surveys were conducted across large tracts of land in just a few fruiting periods.

MINIMUM RESOURCES REQUIRED

Not enough systematic baseline information has been collected to allow numbers of species to be found in "typical" temperate or tropical forests to be estimated accurately. In general, however, it appears that the diversity of sequestrate fungi increases with the diversity of ectomycorrhizal hosts and the variability of soils and topography.

A crew of six trained personnel can survey about 10 hectares in a day. Each area surveyed should be visited every 2–3 weeks during the fruiting season(s). Surveys should be conducted for a minimum of 3 years, and preferably 5 years, to increase the detection of rare and infrequently fruiting species. Three to four days of laboratory work are required for each day of fieldwork.

SAMPLING ACROSS SPACE AND TIME

Most forests contain diverse microhabitats. Even in "uniform" plantations, the microtopography varies with localized wet or dry soil conditions. Distribution of woody debris is also variable and can be patchy, buried, or exposed. Some sequestrate fungi commonly are associated with or found in rotten wood (e.g., *Radiigera* species, *Hydnotrya variiformis*). The patchiness of ground cover and shrub and herb layers can dramatically affect the microclimate in restricted areas. Sites with heavy ground cover are more difficult to search for fungal specimens because laying out plots is difficult and the view is obstructed. Slope and aspect have an important effect on water relations and temperature. In the Pacific Northwest United States, steep, south-facing slopes tend to be the driest, and north-facing, gentle slopes are the wettest. All these variables must be accounted for when designing sampling procedures for each sampling objective.

VOUCHER SPECIMENS, COLLECTING PROTOCOLS, AND PERMITS

Some species are specially protected at the local, state, or federal level and can be collected only if authorized under scientific collecting permits. Local authorities usually can help investigators to secure permits needed. Collecting of some species (e.g., nonsequestrate polypore *Oxyporus nobilissimus*) is not permitted, and photographs, along with detailed locality and microhabitat information, should be used as a record of occurrence.

When permitted, voucher specimens must be collected to document species occurrence. The specimens should be annotated with appropriate information (see next paragraph) and then sent to a recognized herbarium for long-term storage. Except in the case of multiple collections of extremely common species from the same locality in a narrow time frame, all specimens should be kept. Large collections of common species do not provide any additional information, particularly for a location that has been collected previously. A single representative of each of the common species per collecting period is adequate to document presence over time. Most, if not all, specimens of rare or uncommon species should be harvested carefully and sent to an herbarium; such specimens may provide additional morphological information after careful study or may represent incompletely known taxa.

The date, specific location, collector, habitat, notes on the plant community, and transient attributes that may be useful for identification should be noted on the specimen label. Habitat characteristics, particularly the presence of the large woody plants, are important for determining the ecology of the fungi. In addition, identification of at least some sequestrate fungi to species requires information on fresh colors and odors, subsequent color after exposure and handling (after 10–20 minutes and again after 2–3 hours or the next day after storage in a refrigerator), color after drying, and exudation of latex from or color changes of the cut surface of a specimen. Note whether the specimens were found on the soil surface, emergent, or completely hypogeous. Specimens are best kept in waxed-paper sandwich bags under cool conditions until processed. Plastic wrap and closed "airtight" containers should never be used because they lead to anaerobic conditions that stimulate growth of resident bacteria and other microorganisms that quickly can degrade the condition of the sporocarp(s).

Each sporocarp should be cut at least in half to promote drying; large specimens (>3 cm in diameter) should be cut in several vertical slabs ±5 mm thick. Many species have hard, somewhat impermeable peridia that do not allow the specimens to dry appropriately. In other species, the sporocarp dries to the hardness of bone and disintegrates when broken open for access to the contained spores. A cross-section cut can readily be rehydrated with water or KOH (potassium hydroxide) and sectioned with a razor blade. Many species resemble one another on the surface but differ strikingly in the interior. Examining the interior of the sporocarp minimizes the chance of including more than one species in a single collection. Well-dried specimens are much easier to work with than those preserved in liquid. Specimens should be described and then dried as soon as possible, preferably within 1 day. Specimens of some species, if collected in prime condition, handled properly, and stored correctly, can be kept for up to 5 days before drying. Once begun, deterioration proceeds rapidly, and then much of a specimen's value for later study is lost.

Rapid drying at relatively low temperatures is the most successful method for preserving sequestrate fungi. A food dryer set at approximately 30°C works well. Good air circulation is critical. Specimens can deteriorate quickly when heat alone is used. When electricity is not available, specimens should be thinly sliced (±2 mm thick) and placed in a sealed, air-tight container with predried silica gel. Care should be taken to pack the specimens closely in the silica gel. Specimens should not touch each other within the container. Air space within the container should be kept to a minimum to ensure the effectiveness of this method. No more than one collection should be put in a container because, when dried, species often can be difficult to separate on the basis of macroscopic characters. One to two days will dry specimens sufficiently if the volume of silica gel is adequate. Specimens dried by silica gel should be transferred to a more conventional dryer at the first opportunity to ensure that they dry completely.

If silica gel is unavailable or impractical, specimens can be strung together with waxed dental floss and a large needle and suspended over a campfire. Alternatively, lightweight frames covered with a fine aluminum mesh screen can be used. The screens can be suspended over the campfire or exposed to a steady, but not forceful, breeze. In either case, the thin slices should be spaced to allow air to move around them, and their height above the heat should be adjusted to prevent cooking while drying.

LIVE CULTURES

We direct the reader to Molina and Palmer (1982), Smith and colleagues (1994), and Chapter 3 for a comprehensive protocol for isolation and maintenance of live fungal cultures of both mushrooms and sequestrate fungi. Isolation of young sporocarps is most successful. The specimens from which isolates are obtained plus mature specimens from the same collection should serve as vouchers for confirmation of the species identity at a later time.

ABUNDANCE AND DIVERSITY MEASURES

Relative Importance and Abundance

Sporocarp abundance often is quantified to provide a measure of the relative importance of a species in an ecosystem. Sporocarp biomass is a direct measure of importance when considering sporocarps as a food source for animals (Fogel and Trappe 1978) or the variation in individual species' presence on mycorrhizae in different stands (Luoma et al. 1997), and it has been shown to vary within species across moisture gradients (Luoma 1988). The value of sporocarp biomass as an index to levels of ecological and/or physiological functioning of the mycelial "body" of a fungus, however, only can be inferred or assumed. The influence of sporocarp size on such variables as efficiency of nutrient uptake and translocation or competitive ability in interactions for root space are unknown. Furthermore, community structure as determined by sporocarp biomass may not reflect community structure determined by examination of ectomycorrhizae in the same stand (Gardes and Bruns 1996).

The use of numbers of sporocarps as a measure of importance is less desirable because across all species, the number of sporocarps for an individual species in a plot is poorly correlated with total sporocarp dry weight in that plot (Luoma et al. 1991). Species frequency, or presence in plots or stands, provides another mechanism for interpreting biomass values and importance and can serve as a measure of "commonness." For example, high frequency offers a different perspective on the potential "importance" of a species that has relatively less "important" biomass values for a particular season or habitat (Luoma et al. 1991). In studies of vascular plant communities, frequency traditionally evaluates the regularity of a species' distribution throughout a community (homogeneity) and has been interpreted with caution because variations in plot size, number, and vegetation structure strongly affect results (Cain and de Castro Oliveira 1959; Grieg-Smith 1983). Frequency may be used as a measure of abundance to convey importance by presence or absence at the stand level or by plot within stands. A constraint on the use of frequency in studies

of fungal sporocarps is that species that fruit predominantly in spring or fall are restricted to fewer potential samples (on a habitat basis) than species fruiting in both seasons.

Methodologies used in vegetation surveys are not adequate for fungal surveys because of the need for repeated sampling of often cryptic populations. Sporocarps of sequestrate fungi vary in abundance and size. A major difficulty with using sporocarps to determine presence or importance of such fungi is the lack of data on the correlations (if any) between the size and extent of the thallus and the number or biomass of sporocarps. Some fungal species may have a relatively large thallus but produce few sporocarps; other species have a relatively small thallus but produce many sporocarps. Some species produce sporocarps irregularly or infrequently, regardless of thallus size. Some sequestrate species (e.g., *Hysterangium setchellii*, *H. coriaceum*, and *H. crassirhachis*) form clusters of numerous, small sporocarps; others (e.g., *Gautieria monticola*) form clusters of large sporocarps; and still other species (e.g., some *Tuber* species and many sequestrate Russulaceae) produce solitary, dispersed sporocarps. Using sporocarp biomass as the sole indicator of dominance is clearly tenuous. Variation in sporocarp production can serve, however, as an indicator of community response to environmental conditions. Luoma (1988) found that sequestrate community structure, as assessed by sporocarp biomass, changed in accordance with a vegetation/moisture gradient.

If sporocarp biomass estimates are combined with species-frequency data, comparisons between species and communities then are appropriate. Luoma and colleagues (1997) found a positive correlation between sporocarp biomass and mycorrhiza frequency for some epigeous (e.g., *Lactarius rubrilacteus*) and sequestrate (e.g., *Hysterangium coriaceum*, *Leucogaster rubescens*, and *Truncocolumella citrina*) mycorrhizal species. T. Bruns (personal communication), however, found little correlation between biomass and mycorrhiza frequency for several other fungal species.

Several investigators in the Pacific Northwestern United States have assessed the composition of sequestrate-fungus communities quantitatively (Fogel 1976; Fogel and Hunt 1979; Hunt and Trappe 1987; Luoma 1988, 1991; Luoma et al. 1991, 1996a, 1996b; Smith et al. 1996; M. North, personal communication; D. L. Luoma, unpublished data). Claridge and associates (1993) did likewise for some habitats in southeastern Australia. Fogel (1976) was the first to assess the community structure of sequestrate fungi in 50 1-m^2 quadrats, sampled every month for 3 years. Fogel (1981) reviewed techniques for quantification of sequestrate fungi and correlated sporocarp production with temperature and moisture.

Vogt and associates (1981) reported on hypogeous sporocarp biomass production in two *Abies amabilis* stands. Each stand was sampled once a month for 6 months using 12 randomly chosen 4-m^2 quadrants to yield a total sampled area of 288 m^2. Luoma and colleagues (1991) compared sporocarp biomass values from a range of total sample areas: 5900, 2800, 1400, 800, and 500 m^2 and noted that small sample area increases the likelihood of overestimating sporocarp biomass at the stand level. They concluded that when using dispersed 4-m^2 plots, the sample is not adequate until the sample area totals between 800 and 1400 m^2. Vogt and associates et al. (1992) reviewed sporocarp production studies, including those of sequestrate species. Adequate replication and sampling intensity are keys for statistically sound quantitative comparisons (Stafford 1985).

Fungal sporocarps are relatively clustered (Fogel 1976, 1981; States 1985). Therefore, when quadrants fall in localized biomass concentrations, stand level biomass may be overestimated. Use of a relatively small number (with respect to the selected stand area) of random plots may exaggerate this overestimation because of the effects of sample clustering. An alternative to the impracticably large number of randomly distributed plots necessary for adequate sampling is the systematic placement of fewer plots, which will reduce the coincidence of plots with localized concentrations of biomass and reduce the tendency to overestimate production (Luoma et al. 1991).

Nonrandom sequestrate fungus distribution at the plot level is determined by the gregarious to caespitose sporocarp formation that many species exhibit (Fogel 1976; Arora 1987). The mycelial colony will occupy roots and soil in a discrete area of forest floor, and sporocarp formation is restricted to that area. With the growth of adequate mycelial biomass, a single colony can produce several to many sporocarps (Dahlberg and Stenlid 1994). If hierarchic scales of clustering were to be detected, then explanatory hypotheses could be developed and tested. One hypothesis to be tested is that older stands may exhibit a greater incidence of clustered sporocarp production as compared to younger stands. An older stand has had more time for colonies of different cluster-forming species to become established and may contain a greater number of host tree species. Greater host species richness will increase the number of host-specific sequestrate species present(Molina et al. 1994).

Species Richness

Richness is the total number of species counted in the unit of assessment, be it a subsample, plot, experimental unit, stand, or higher level landscape segment. A species-area curve (Cain and de Castro Oliveira 1959) commonly is used to examine adequacy of sampling for species richness. Determination of species richness is particularly problematic with fungi because their presence is not obvious. Hunt and Trappe (1987) pointed out that documenting all species of sequestrate ectomycorrhizal fungi in a forest stand requires collecting over several years. After 32 months of collecting over a total sample area of 1536 m^2, their species-area curve still had not stabilized.

Even when sporocarps are sampled over several years, mycorrhizal species richness may be greatly underestimated. During three spring and two autumn seasons, Luoma and colleagues (1996a) found 43 sequestrate species and about 100 species of ectomycorrhizal mushrooms from sporocarp production plots that covered a total area of 27,000 m^2. In marked contrast, 192 ectomycorrhizal morphological types (from 189 soil cores in the same study area) were found under a total soil surface area of only 0.45 m^2 (Eberhart et al. 1996; Luoma et al. 1997). Ectomycorrhizal morphological typing is somewhat subjective and is quite labor intensive, which prohibits its widespread use for rapid assessment of community composition. Until voucher collections of morphological types are analyzed with molecular tools to confirm species or generic identities, morphological typing remains inconclusive and cannot be compared easily between communities. Until molecular techniques become incorporated routinely into field studies, sporocarps will continue to provide the most common index of sequestrate species richness in a given area.

INVENTORY, SURVEY, AND MONITORING

Definitions

We define inventory as cataloging or listing of individuals on hand, such as extant specimens in herbaria. We define survey as the determination or delineation of the extent of individual species at any one location. Monitoring is defined as the observation of extant individual populations through time. Adaptive monitoring/surveying is used following the discovery of a target species. Intensive sampling serves to delineate population extent, using intensive sampling procedures, such as strip plots or plots in concentric circles centered around a known location.

Sampling Protocols

Protocol implementation should be supervised by personnel trained in the use of the protocols and in fungal identification. Prior to sampling, personnel should familiarize themselves with the general biology, ecology, habitat associations, and the specific morphological features of target species. That will improve accuracy of fungal identifications in the field and increase efficiency of field searches.

For analysis of alpha diversity, it is extremely important to restrict sample sites to relatively homogeneous habitat because of the variation associated with habitat diversity. Sampling habitat characteristics such as vegetation, stand age, stem density, elevation, aspect, and topographic position at multiple sites within an area will help to determine habitat homogeneity.

Periodicity

Sequestrate fungi can fruit any time of the year, depending on weather and substratum. Some species, for example, fruit in the middle of the dry season in buried rotten wood or near streams or standing water. For the most part, however, sequestrate fungi should be sampled during the rainy season(s). Periods of sporocarp formation in some sequestrate fungi are restricted (Fogel 1976; Hunt and Trappe 1987; Luoma et al. 1991). *Rhizopogon parksii*, for example, fruits only in late summer or autumn in western Oregon, whereas *R. vinicolor* in the same stands fruits mostly in spring and early summer. In some Ascomycetes sporocarps form in autumn but do not mature until the following winter or early spring. Sporocarps of other species found out of season may have been produced in season and then frozen (e.g., some collections of *R. parksii*) or dried *in situ* (e.g., some collections of *R. vinicolor*) rather than decaying. Temperature as well as rain can affect the fruiting season. Freezing weather truncates or delays the maturation of sporocarps, and high temperatures may accelerate drying of the substrata, thus curtailing fruiting.

Because timing of fruiting is uncertain and strongly seasonal (Luoma 1988), each site should be sampled repeatedly during 2 consecutive months of each fruiting season. Sampling within the 2-month window can be carried out at biweekly intervals because sequestrate sporocarps usually persist for about 2 weeks. When sampling across an elevational gradient, one should visit low-elevation, south-facing slopes first in spring, but last in autumn, and high elevation, north-facing slopes last in spring and first in autumn (Luoma 1988). Despite the restricted fruiting seasons of most sequestrate fungi, some sporocarps are likely to appear at any time during the year. Thus, sporocarps must be sampled throughout the year, rather than just during the peak fruiting period, to encompass diversity. Year-round sampling is particularly important in surveys for community structure.

Three or more years are needed to assess community structure or rarity of large fleshy fungi (Richardson 1970;

Lange 1978; Arnolds 1981; Fogel 1981; Luoma 1991, 1996b, 1996c; Luoma et al. 1991; Vogt et al. 1992; O'Dell et al. 1999). The cryptic sequestrate fungus sporocarps, nearly all of which fruit below the litter, some within the mineral soil layer, are more difficult to detect than epigeous sporocarps. Therefore, one should not conclude that a species is absent from an area or that the area has been reasonably well surveyed until data from 2 to 4 years are available.

SURVEYING COMMUNITY STRUCTURE

Community structure and richness can be assessed in several ways. The three preferred methods are plotless transects, line transects, or randomized plots, which can be permanent or temporary (moving). Once a clear objective is identified and the resources available for sampling have been assessed, the best method for meeting objectives with the available resources can be selected.

Line Transects

In this method sample plots are located along a line, which may or may not be straight, through the area of interest. The transects should be well dispersed in the sample area so as to intercept a wider variety of microsites than would be present in a single circular plot of the same area (Ruhling et al. 1984; Mehus 1986; Ohenoja and Metsänheimo 1982; Luoma et al. 1996b). If transects are established repeatedly in the same area, new transects should be at least 20 m from previously used transects. Plots are placed every 6 m along a 50-m transect (Luoma et al. 1996b). Usually, 25 4-m^2 sample plots are used. On slopes, transects of 8, 9, and 8 plots, respectively, are established on the upper-, mid-, and lower-slope strata. A "collection" is defined as those sporocarps of the same species from a particular 4-m^2 plot. A total area of 100 m^2 per 5- to 15-hectare stand in 25 4-m^2 circular plots gives a reasonable pooled estimate of major species diversity for that stand sample. Plots are marked with a flag or stake to avoid resampling. Another approach is to space plots 25 m apart on transects in the horizontal direction (along contours) and to space transects 75–150 m apart in the vertical direction (across contour). A statistician should be consulted about the sample design prior to sampling.

Randomized Plots

Although statistically sound, the randomized-plot method is logistically cumbersome because it is difficult to lay out a truly random array of plot samples. That logistical difficulty with the placement of plots should

FIGURE 10.3   Personnel performing a time-constrained search.

not be confused with the need, in most cases, to assign "treatments" to "experimental units" in a random fashion. To meet the requirements of a random placement of sample plots, every potential plot location in a study area must have an equal chance of being selected. To achieve that, each potential location must be identified and enumerated, often by use of a fine grid. A random numbers table then is used to select the sample points from the total pool. O'Dell and associates (1999) successfully used a stratified random method to locate permanent strip plots for mushroom sampling. They needed to locate only three plots per experimental unit, however. The location of many small plots, especially repeatedly over an extended period, is more readily accomplished by systematically dispersing them throughout the study area. Systematic placement has the additional advantage of providing better coverage of the range of variation encountered in a stand because randomly placed plots tend to cluster in some areas (Mueller-Dombois and Ellenberg 1974).

Plotless Transects (Time-Constrained Search)

A time-constrained search is one in which a survey area is searched for a set amount of time. Claridge and colleagues (2000) have outlined the procedure, which is more effective than a series of small plots of fixed size for determining the species diversity of an area or finding particular species. The personnel doing the search are free to sample any spots in the search area that they wish

(Fig. 10.3), although they are instructed to include as many microhabitats (e.g., near to and far from trees, under thick and thin litter layer, under shrubs and in the open) as possible.

The time needed for a search will depend on the objectives of the survey and available resources. The number of person-minutes devoted to searching a given site is predetermined, preferably by preliminary trials of habitats representative of those to be searched. For example, a timer is set for 5 minutes for two people to search a site for the target fungi. At the end of the period, all specimens found are labeled; they represent the species and number of sporocarps of each found in 10 person-minutes. This procedure is repeated until the species curve levels off, thereby indicating how many person-minutes were needed to find all or nearly all taxa on the site. The procedure can be replicated several times for statistical analysis to produce a probability level for finding all taxa at a site. Claridge and colleagues (2000) determined that the probability of finding all sequestrate taxa in 1000-m^2 plots in the forest types they studied with 100 person-minutes of searching (four persons for 25 minutes) was very high.

To evaluate the taxonomic diversity of a site adequately by time-constrained sampling, the site must be sampled repeatedly over the fruiting season for 2–4 or more years. A truly complete assay of species diversity may require far more sampling per year for many years (Egli et al. 1997), but in the case of sequestrate fungi this is not feasible because the surface of an entire study area likely could be disturbed by raking with that intensity and duration of sampling.

SURVEYING RARE SPECIES

Surveys of rare species may focus on assessment of existing populations or detection of new ones. Two techniques, both tied to habitat assessment, are used in searching for new populations. Many sequestrate taxa have specific habitat requirements (e.g., soil type or associated host species). Rare sequestrate fungi first should be sought in habitats similar to those where extant specimens have been found. The most likely habitats are identified and located on the landscape. They then are searched during the most likely fruiting period, using plotless transects guided by ecological indicator or systematic placement of standard plots.

Plotless transects are useful for finding rare species because large numbers of target microsites can be searched by each individual collector. Indicators of fruiting sequestrate fungi include fresh digs by small animals and raised compacted humus material. Productive microhabitats generally include soils adjacent to well-decayed logs and protected areas (and areas adjacent to them) within small areas heavily disturbed by animals such as wild pigs. Small animals such as squirrels, mice, and voles commonly unearth sporocarps of sequestrate fungi one at a time as they mature, leaving a small pit 2–8 cm deep and 5–8 cm in diameter. Such digs sometimes are hard to distinguish from digs for seeds or insects or from hoof prints. Sometimes an animal will eat only the upper portion of a sporocarp, and the remaining portion can be observed at the bottom of the pit. Many sequestrate fungi fruit in clusters, so further exploration within a radius of 30–60 cm around a suspected fruiting spot often reveals additional specimens. It is best to rake into the soil to the depth of the nearby small animal dig. Needles, leaf fragments, or other debris in or a spider web across a small animal dig indicates that it is not fresh. Further exploration may reveal specimens, however, particularly if fresh digs are scattered about in the habitat. Plotless transects also can be useful in habitat with compacted soil or where the humus layer is thin. Under such circumstances even small specimens form small humps at the soil surface that look suspicious to a trained observer. Often, large specimens emerge from the small humps. Campgrounds, abandoned roads, road banks, and used or abandoned walking trails sometimes are productive.

If a particular rare species is sought, then plots subjectively placed in a microhabitat that is most likely to harbor the species have a much higher probability of including the target species than do randomly placed plots. Some species, for example, commonly fruit in association with well-decayed logs. When searching for these species, plots should be located along well-decayed wood in the appropriate habitat. As with all sampling procedures for sequestrate fungi, timing is critical, and repeated sampling over a number of years is required to assess rare fungi adequately.

Existing populations whose locations are fairly well known may be revisited to (1) assess population viability, (2) further delimit the population, or (3) detect nearby populations. A previously sampled circular plot can be resampled if enough time has elapsed to allow recovery or a concentric zone that extends 0.47 m in radius beyond the original plot can be sampled. The latter approach will provide additional information concerning the potential of the original thallus. To explore for other nearby populations, additional plots that extend away from the site of the original thallus can be installed in a starburst pattern, in which eight 0.5-m^2 plots surround the original plot.

Because the sampling procedure for sequestrate fungi is disruptive, particularly for rare species, it should be used cautiously. Disturbance of the microhabitat may have an adverse impact on the habitat and render it uninhabitable by the fungus that once was present. This is particularly evident in habitat such as coarse woody debris that is dismantled in sampling. Woody debris thus sampled does not return to its former structure rapidly, if at all. In our experience, soil substrata and their concomitant herbs and forbs return to predisturbance levels 1–2 years after sampling.

SAMPLE SIZE

Total area to be sampled depends on the objective of the study. It takes less sampling to validate the occurrence of a species at a known location than to discover new locations. Likewise, less area is required when searching for new locations of rare species than when assessing community diversity of a stand or watershed. In general, a minimum total area of 800 to 1500 m^2 (0.5–2.0% of a stand area) must be sampled to assess species richness and community composition in a 5- to 15-hectare stand (Hunt and Trappe 1987; Luoma et al. 1991). That total is best achieved over 3–4 years of sampling.

DATA MANAGEMENT

Field data sheets should be reviewed several times during the sampling process. A data sheet is checked for completeness when the plot to which it pertains is finished. At the end of each transect, data sheets are checked to ensure that all plots were completed. When an individual stand is finished, data sheets should be collated and proofed. Questions should be resolved before moving to the next stand. In the evening, collections must be reconciled with data sheets. Unresolved questions may necessitate return to the stand or plot of concern. Castellano and colleagues (1999) provide excellent examples of data sheets for a variety of uses including documenting the field site, describing various life forms of fungi, and making field tags for collections.

Computer database design should be considered before commencement of data collection in the field to ensure that all necessary information is collected at the most appropriate and cost-effective time and recorded with the most useful notation. Consultation with a computer database specialist will ensure that the end product meets objectives.

LABORATORY PROCEDURES

In the laboratory, sporocarps should be described further and identified (see "Description of Taxa," later in this chapter). If biomass quantification is required, then after description, sporocarps are oven-dried and weighed to the nearest 0.01 g.

CHARACTERISTICS OF SEQUESTRATE FUNGI

DESCRIPTION OF TAXA

The evolution of sequestrate fungi has tended toward morphological simplification coupled with specialization of spore-dispersal tactics. Because of their reduced morphology, sporocarps of sequestrate fungi generally are easier to describe than their epigeous relatives; spores, however, are ornamented and pigmented and tend to be more developed than in their epigeous relatives. Spore characters are, thus, important for distinguishing genera and species. Spore ornamentation also may serve as protection against chemical or physical hazards of passing through animal digestive tracts.

Careful notes on fresh material are important for complete descriptions of taxa, although some fresh characters vary so much among specimens from different habitats or collected at different seasons that they are only marginally useful in taxonomy. Developmental stages should be described to the degree possible. Sporocarps of some species, such as *Genabea cerebriformis,* are always small (never >1 cm in diameter). However, species

that reach much larger dimensions, such as *Rhizopogon villosulus* (up to 10 cm in diameter), also can be less than 1 cm under some conditions. Color is also highly variable for most species. *Rhizopogon vinicolor* is white in youth and stains pink where bruised. As it matures, it develops patches of yellow and still stains pink. Later the yellow becomes mottled with reddish brown, and the pink staining becomes obscure. At full maturity, the peridium may be reddish brown to a large extent and may not stain where bruised. Standard and generalized color terms, such as those used by Kelly and Judd (1955), are accordingly more useful in describing sequestrate fungi than long lists of very specific terms from other color dictionaries. For example, Thiers and Smith (1969:532) described the pileus color of the hypogeous *Cortinarius velatus* as "near 'pale vinaceous drab' to 'pale purple drab' (lilac to pale lavender) to occasionally as dark as 'light purple drab' ..." The color terms they used are from Ridgway (1912) and mean little unless one has a copy of Ridgway in hand. In the Kelly and Judd terminology, these terms translate into the more meaningful grayish pink, grayish purplish pink, and grayish purple, respectively. Because grayish purplish pink is intermediate between the other two colors, the pileus color of *C. velatus* can be described conveniently as "grayish pink to grayish purple." Responses of tissues to reagents such as KOH can be striking in a few genera of Basidiomycetes (particularly *Rhizopogon*), but they, too, vary so much with the developmental stage of the sporocarps that their use in species determinations is limited.

Other important characters of fresh sporocarps include ranges of vertical and lateral dimensions; shape (e.g., even and globose to ellipsoidal, flattened, irregular or furrowed, and lobed); presence of surface cavities or holes to interior; nature of surface (e.g., glabrous, scabrous, tomentose, verrucose, and, if verrucose, then size and configuration of the warts); color of peridium, including variations on single specimens (e.g., streaked, mottled) and differences between epicutis and subcutis; thickness of peridium; color changes where bruised or exposed in cross section; nature of gleba or interior (e.g., solid, chambered, enfolded, hollow, gelatinous, pulverulent); presence of columella or stipe; color of the various tissues of the interior and of spore deposits that line chambers; odor; taste of peridium and gleba (determine separately when possible); and response of peridium and gleba to a few drops of a 5% solution of KOH. Terms for the various types of ascomata have been proposed by Weber and associates (1997).

Because relatively few sequestrate fungi are macroscopically distinctive, micromorphological characters are especially important in delineating taxa. Fresh material can be mounted on microscope slides using water, 3%

KOH, or Melzer's reagent (Appendix II). The Melzer's reagent should be refrigerated and less than 1 month old for best results. Spores and tissue cells should be measured using water mounts because KOH and other reagents can cause swelling or shrinkage of cells, particularly ascospores. Some spore ornaments swell or dissolve in KOH (Trappe 1979; N. S. Weber, personal communication). Dried tissues can be mounted in KOH for rapid rehydration, but ascospores always should be measured from mounts rehydrated in water only. The color reactions of spores, asci, or other tissues to Melzer's reagent should be recorded directly (orange, red, purple, blue, black) rather than as "amyloid" or "dextrinoid." The cells or tissues that appear to react to Melzer's reagent should be compared to the analogous cells or tissues in KOH or water mounts to ensure that the color is indeed the result of Melzer's reagent and not a natural color.

Whenever possible, all micromorphological features of spores and nonfertile tissue from both young and well-matured specimens should be noted. Several "new" species have been described from specimens later demonstrated to be immature stages of an earlier-described species. Some characters, such as spore ornamentation, are absent or smaller on young spores than on mature spores. Other structures, such as cell walls, change progressively as sporocarps mature. In some cases the walls thicken; in others, they are thick and mucilaginous in youth, but thinner and firm in age. Some characters, such as the dextrinoid reaction of spores, may occur in young spores but not in mature spores. Zygomata can be judged as immature if all spores are thin walled. They are considered fully mature if all spores are thick walled and the spore attachments are occluded by wall material or a well-developed septum. Basidiomata can be judged as immature if spores are scattered over the hymenium and mature if the spores are crowded on the hymenium or in spore-bearing tissues and are of similar size, shape, wall thickness, pigmentation, and ornamentation (when present). Ascomata are judged to be immature if only a small proportion of the asci contain spores and fully mature if all asci contain spores of similar size, shape, wall thickness, pigmentation, and ornamentation (when present).

Characters to be recorded for each collection include ranges in size and shape of spores (means and standard deviations based on measurements of 50 randomly selected spores sometimes are useful); size, shape, and nature of spore ornamentation; spore-wall thickness; reaction of young and mature spore parts to Melzer's reagent (compared with spores in water or KOH); range of sizes in basidia or asci; reaction of asci to Melzer's reagent (intensity and location of color); range of sizes and shapes of other hymenial structures (e.g., paraphyses, brachybasidioles, cystidia, setae); structure, cell size and shape, cell-wall thickening, and other distinctive features of subhymenium, trama, stipe-columella, and each layer of peridium; presence, location (intracellular or extracellular), nature, and color of pigment deposits in KOH and Melzer's reagent; and additional distinctive reactions of cells or tissues to Melzer's reagent.

HABITAT ASSESSMENT

Detailed information on habitat is potentially useful for understanding the ecology of fungi. The collector must decide how to allocate field time between collecting fungi and collecting habitat data for those fungi. Habitat in the broad sense includes a descriptive geographic location as well as latitude, longitude (degree, minutes, and seconds), and elevation (in meters). At a given location, the general habitat or forest type (complete overstory and major understory species) and estimated stand age, slope, aspect, and dominant ground-cover plants are useful. The specific habitat description for a given collection should include substratum (e.g., terrestrial, decayed wood), and fruiting habit (gregarious versus solitary, single versus ceaspitose).

ACKNOWLEDGMENTS. Partial funding to prepare this manuscript was provided by National Science Foundation Grant BRS 9201421 to J. M. Trappe. We appreciate the helpful editorial comments of James B. Eblin.

RECOMMENDED PROTOCOLS FOR SAMPLING PARTICULAR GROUPS OF FUNGI: ISOLATION PROTOCOLS FOR READILY CULTURABLE MICROFUNGI ASSOCIATED WITH PLANTS AND FUNGI

11

MICROFUNGI ON WOOD AND PLANT DEBRIS

PAUL F. CANNON AND BRIAN C. SUTTON

Microfungi are defined as fungi with microscopic spore-producing structures (Hawksworth et al. 1995). They comprise an unnatural group of organisms that cuts across the classification schemes for Fungi and for the fungal-like organisms in the Protista, Chromista, and Myxomycota, some members of which previously were placed in the Fungi. The single criterion for inclusion in the group is the small size of the spore-producing structures, in contrast to the large macroscopic fruiting bodies of the macrofungi. Despite their lack of a common phylogenetic history, the microfungi are grouped together because they, or their asexual states, are collected using the same techniques. In this chapter, we focus on procedures and protocols used to collect microfungi that live on or in wood and plant debris. Microfungi also can be found in other habitats, but techniques for collecting them from those environments are covered in other chapters (see "Culture Techniques," later in this chapter).

Although the criterion for defining the microfungi is arbitrary, at least some species from most groups of fungi qualify for membership. We include the following groups:

Ascomycota. The Ascomycota comprise the largest division of fungi including approximately 3250 genera containing 32,250 recognized species (Hawksworth et al. 1995). Most ascomycetes are regarded as microfungi, and so members of this phylum account for the bulk of the group. Nevertheless, the Ascomycota includes some large-size species with macroscopic fruiting bodies, especially among the Pezizales and the Xylariales, that are collected and observed with techniques used for macrofungi.

Anamorphic Fungi (deuteromycetes, asexual fungi, anamorphs, conidial fungi). This unnatural group comprises around 2600 genera and 15,000 species, characterized by mitotic rather than meiotic production of spores (conidia). Clear cytological evidence of the mode of spore production is available for only a small proportion of those species. Likewise, the number of meiosporic states that have been correlated with teleomorphs is limited. Most of those meiospores have teleomorphs in the Ascomycota; a lesser number have teleomorphs in the Basidiomycota. The majority of species in the group have no known link; those forms are presumed to be ascomycete-related.

Microscopic Basidiomycota. The basidiomycetes comprises about 1400 genera and 22,250 species. Although most macrofungi are basidiomycetes (a few are ascomycetes), not all basidiomycetes are macrofungi. Some species—for example, the Tulasnellales (tiny resupinate saprotrophs)—are only microscopically visible and thus are sufficiently small to be grouped, for convenience, with the microfungi. Basidiomycetous anamorphs frequently have minute fruit bodies or lack them completely, and many are observed directly and cultured using microfungus protocols. Microscopic basidiomycetes are found especially among microfungi colonizing woody, rather than soft, plant debris.

Zygomycota. The zygomycetes that coexist with other plant-inhabiting microfungi are species in the order *Mucorales.* Although we include them in our treatment, almost all species in the order can be sampled readily using techniques described in the chapters on fungi in soil (see Chapter 13) or on animal excrement (see Chapter 21). Fifty-six genera and around 300 species currently are known from the order.

Other groups of microfungi associated with nonplant substrata are considered elsewhere. They include the following:

1. Lichenized ascomycetes encountered on plant surfaces (see Chapter 9).
2. Lichenicolous and fungicolous fungi treated together as fungicolous fungi (see Chapter 17). Many fungi gaining nutrition from other microfungi will be sampled using the protocols described later in this chapter.
3. Chytridiomycota. Some are encountered on aquatic plants and plant detritus (see Chapter 23); others are found in soil (see Chapter 13) or associated with living terrestrial plants (see Chapter 7).
4. Oomycota. Almost all are found either associated with living plant tissue (see Chapter 7) or in soil or sediment samples (see Chapters 13 and 23).
5. Myxomycota (treated in Chapter 25).

EXTENT OF HABITATS

The most important factor influencing the distribution and lifestyles of fungi is their heterotrophic nutrition (Cooke and Rayner 1984). Fungal growth therefore is confined to environments in which suitable supplies of organic compounds (primarily organic polymers) are more or less immediately available for use, although dispersal propagules and resting spores independent of their mycelium may be sampled from a very wide range of habitats. Fungi vary widely in their capacity to gain access to and exploit carbon compounds, which differ in availability, location, and configuration. As a result, microfungi grow on plant debris in every conceivable habitat, including those subjected to the extremes of climate that characterize the Arctic and Antarctic and deserts and rain forests, as well as the milder subtropics and temperate environments. The nature and quantities of nutrients together with physical features of the substratum, water availability, humidity, and temperature determine the success of colonizations, subsequent survival of individuals, and species composition of a community.

Large numbers of fungal species colonize plant debris, a rather general term, which at its simplest can be defined as nonliving plant fragments. More precise characterizations may be required, depending on the objectives of the study, to ensure use of an appropriate sampling strategy. Such characterizations involve consideration of additional factors such as the following:

1. Origin of the material: roots, pneumatophores, stems, trunks, bark, wood, branches, twigs, petioles, leaves, phyllodes, lamina, veins, hairs, bracts, flowers, fruits, seeds
2. Condition of the material: age and degree of senescence (recently fallen or otherwise), fragmented mechanically or by a biological agent, such as insects or mammals; digested by insects; rotted by other fungi; tunneled; secondarily utilized; degree of decomposition; chemical composition
3. Location of material in the habitat: superficial, buried, immersed, litter layer, canopy, on or in other organisms

Microfungi, because of their small size, can occupy minute sites. The importance of precise definitions of such microhabitats cannot be overemphasized. Attempts to quantify and define microhabitats within particular environments are few, although this should be an important first step in designing an efficient sampling program. What information is available generally has been gathered as part of studies on successional patterns (e.g., for colonization of woody tissue; Dix and Webster 1995). Hawksworth and colleagues (1998) noted the importance of this issue for the design of sampling programs in tropical ecosystems, although the list of niches included was not exhaustive and the authors did not define them specifically.

TROPHIC RELATIONSHIPS

Three nutritional strategies, which are not necessarily mutually exclusive, have been recognized for fungi (Cooke and Rayner 1984). Individual species can, in different circumstances, manifest all three, and fungi frequently shift their nutritional strategies in the course of their life cycles. Saprotrophy, which is the most frequent nutritional habit of microfungi colonizing plant debris, occurs when a fungus uses nonliving organic material from a plant that it did not kill. In necrotrophy, which frequently involves processes similar to saprotrophy, the fungus kills the living tissues of a plant and then uses them for nourishment. For some necrotrophs, the completion of a full or abridged life cycle depends on a switch of modes from necrotrophy to saprotrophy; often the transition is gradual. Biotrophy, in which only living host cells are exploited as nutrient sources, is much less common among microfungi colonizing plant debris, and biotrophic species largely are excluded by definition. Some biotrophs later become necrotrophs, however, and a small proportion subsequently can function saprotrophically. The timing and rate at which the transition occurs clearly is important in the context of plant debris colonization.

Because the processes by which dead plant material is altered and returned to the environment are many and varied, it is inevitable that niches and trophic strategies of microfungi overlap. For example, we include leaf lesions in this chapter because the dead or dying host tissues resulting from action by a necrotroph or biotroph almost always are colonized by secondary saprotrophs. Bark of living trees, which is dead, often supports growth of microfungi (bark fungi are treated with the endophytic fungi; see Chapter 12). Consequently, temporal, spatial, and physical transitions from species of microfungi that colonize living tree tissue and to those that colonize dead tissues occur at the interface between the bark and the living tree cambium.

A large proportion of endophytic fungal species become saprobic on the death of their host, an advantageous strategy because their intimate association with the plant allows immediate utilization of the dead tissue. Furthermore, many endophytic species enter an intermediate necrotrophic stage as the host tissue becomes moribund, allowing even earlier colonization and utilization of energy resources. Nutritional metamorphoses have not been studied adequately, although observations such as that of Bills and Peláez (1996) integrating information from endophytes and saprobes are beginning to reveal what are frequently very complex ecological systems. Culturing techniques for examining the diversity of microfungi associated with dead plant tissue are largely the same as those for endophytes, although initial surface sterilization is not necessary. Endophytic fungi are treated in detail in Chapter 12.

It may be years before plant debris decays fully and its residues are incorporated into the soil (Dix and Webster 1995), although in humid tropical ecosystems the process may be much more rapid (Anderson and Swift 1983). Therefore, a temporal, as well as a physical, interface between plant debris and soil needs to be taken into account in microfungal community analyses. Microfungi that sporulate can be readily observed directly even on heavily decayed plant debris (often after incubation in moist chambers), which means that fruit bodies, spores, and mycelium may be removed and isolated if required. Although the latter technique may substantially undersample species that are present in plant debris compared to culture methods such as particle washing (Polishook et al. 1996), identification problems caused by nonsporulating cultures are largely avoided.

LIFE STRATEGIES

Concepts of life strategies in fungi have been derived largely from the work of Grime (1977, 1979) on higher plants. First brought to the attention of mycologists by Pugh (1980), they subsequently were accepted, modified, and extended by Cooke and Rayner (1984) and Andrews (1992) and summarized by Dix and Webster (1995). Pugh and Boddy (1988) emphasized that trophic nutritional types, which may change at different phases of growth, should not be used to classify fungi. Rather, they should be used to define behaviors shown at particular stages of fungal life cycles because their behaviors may change according to different phases of growth.

The three primary life strategies are explained in the following paragraphs.

RUDERAL

Short life spans and high reproductive potential characterize fungi found in disturbed but productive environments. Species are ephemeral, but growth can be rapid and devoted almost entirely to reproductive structures. Many *Mucorales* are ruderals, and they are abundant in soil. Ruderal fungi appear to be particularly sensitive to transient factors such as water stress (Pasricha et al. 1994) and play an insignificant role in litter decomposition (Webster 1957). A modified ruderal strategy is shown by coprophilous fungi, which include zygomycetes, basidiomycetes, and ascomycetes.

COMPETITIVE

High competitive ability is dependent on characteristics that maximize growth in productive, relatively undisturbed conditions, by active, antagonistic means, rather than through invasion of a niche vacated by a previous occupant. They are persistent and long-living, are capable of defending captured resources, have rapid or slow growth and spore germination, exhibit slow or intermittent reproduction, and have good enzymatic competence.

STRESS-TOLERANT

Endurance of conditions of environmental stress or resource depletion select for other communities of microfungi. Most known stress-tolerant fungi, which are not symbiotic, can be grouped in well-defined ecological niches, each of which is adapted to a particular type of stress. They are persistent as long as stress conditions prevail and are replaced if stress is alleviated. They lack noticeably rapid growth, spore germination, or reproduction, but are enzymatically competent.

LIMITATIONS TO THE STUDY OF MICROFUNGI

SIZE

Large numbers of fungal species are so small as to defy sampling except by the most meticulous and time-consuming means. Their lack of conspicuousness (often even when fruiting) is a serious barrier to understanding their diversity and the roles they play in an ecosystem. Fruiting bodies are frequently as small as 50 μm in diameter, and some hyphomycetes have conidiophores only about 10 μm long. Spores may be as small as 2–3 μm long and are often less than 1 μm wide, although a few species have spores greater than 100 μm long.

The minute nature of microfungi makes their direct observation in the field difficult; consequently, collecting focuses on adequate sampling of entire substrata and niches rather than on the collection of individual species. The success of this approach varies according to collector experience, but even an expert, who may be able to achieve targeted collecting for some groups, often has little idea of what actually has been collected until the material is examined microscopically unless the habitat is already well known. Almost all plant debris supports more than a single species. Painstaking mounting of vegetative and spore-forming structures and their examination using a compound microscope is the only way to find out what is actually present in mixed populations of microfungi without culturing from the material using techniques borrowed from soil analysis, such as particle washing or particle filtration (see "Principal Isolation Methods" in text and "Appendix" of Chapter 13). Some systems are overwhelmingly complex. Individual leaves in tropical litterfall can harbor up to 30 different species of hyphomycetes (Sutton 1986), only some of which grow in colonies well defined by zone lines. In temperate regions it is common to find similar substrata—for example, cupules of *Fagus* and *Quercus* (the seeds are colonized by another group of microfungi) or leaves of *Laurus* and *Ilex*, colonized by similar numbers of species. The spines of fallen cupules of *Castanea sativa* in the United Kingdom are found regularly with 5–10 species of hyphomycetes and ascomycetes (Sutton 1973b, 1975); other fungi colonize the fleshier tissues.

ACCESSORY VEGETATIVE STRUCTURES

Most microfungi are relatively long-lived, but sporulate (and thus are directly identifiable) only for short periods. They may produce melanized fruiting structures, which serve as protection for the enclosed spores during adverse conditions of drought, low temperature, high ultraviolet rays, and so forth. Melanized hyphae also can withstand adverse conditions. Other vegetative structures such as chlamydospores, sclerotia, bulbils, appressoria, and setae also may be produced partially or entirely for dispersal or survival of cold or dry periods. Unless the morphology of such structures is very distinctive, species are difficult to identify in this condition and must be cultured to induce formation of their sporulating states.

TELEOMORPHS AND ANAMORPHS

Many microfungi have more than one morphologically distinct spore-bearing stage in their life cycles and can produce not only a teleomorph but also one or more anamorphs (Kendrick and Di Cosmo 1979). Anamorphs may be formed without teleomorphs and vice versa, and additional conidial states (synanamorphs) may be formed. Large numbers of species appear only to produce mitotic spores, but because the production of the meiotic morph is frequently cryptic in comparison with the formation of conidium-bearing structures, the proportion of species that actually reproduce sexually is unknown. Even if both (or all) morphs are produced, the timing of events is often unknown even in relative terms. The anamorph may or may not precede the teleomorph, which may or may not be produced at the same location on the mycelium. Two or three synanamorphs may be formed (sometimes even in the same conidioma), and one may be formed earlier in the season than the other. This is the situation in some species of *Phomopsis* (B. C. Sutton, personal observation). For ascomycetes such as *Diplocarpon rosae* or *Rhytisma acerinum*, the anamorph (*Marssonina* or *Melasmia*, respectively) is formed as a

necrotroph on leaves. After leaves fall, the anamorph ceases to sporulate, and a teleomorph stroma is formed. The fungus overwinters in this state, and at leaf emergence, the ascospores reinfect new leaves to produce the anamorph again (Knight and Wheeler 1977). The need to recognize and link these different reproductive and vegetative states is a significant restraint on biodiversity sampling strategies.

ECOLOGY

Detailed ecological information is lacking for most microfungi, except the more well-known and widespread plant pathogens, which have been studied mainly in agricultural systems. The only systems that have been explored systematically are some temperate successional systems in litter and wood (Hudson 1968; Cooke and Rayner 1984; Dix and Webster 1995). Microfungi overcome the constraints on colonization, especially in extreme environments of high or low temperatures and limited water, by exploiting narrowly defined microhabitats. Hudson and Webster (1958) discovered that fungal succession varies even between upper and lower internodes of grass stems, and Webster and Dix (1960) found differences between standing and uprooted culms. A. Rambelli (in litt., March 1996) observed that primary colonizers of newly fallen *Pistacia lentiscus* leaves in Sardinia are different on the upper and lower surfaces, even of individual leaves.

Knowledge of microfungal ecology, and of the range of species colonizing particular niches and microhabitats, is restricted, especially in regions that are subject to climatic extremes. To take two opposite examples, in *Agave* and similar desert plants healthy leaves and stems seemingly are devoid of epiphytic fungi (although they may be colonized by endophytes). However, microfungi do develop in sites where moisture is available (i.e., in the rosettes and leaf axils). Thus, the bases of older leaves are often colonized. The novelty and composition of the mycota of this niche do not appear to depend on the plant species (B. C. Sutton, personal observation). Further evidence for this apparent lack of host specialization comes from examination of the restricted diversity of microfungi associated with cacti (Cannon and Hawksworth 1995). At the other end of the water-availability gradient, similarly distinctive guilds of species occur in association with the water traps within bromeliad rosettes in wet tropical forests (M. M. Dreyfuss, personal communication) and among the wide range of aeroaquatic fungi that grow in conditions of periodic inundation (Park 1974; Ando 1992).

Emerging evidence suggests that many saprobic fungi are specialized to colonize substrata with particular physical characteristics rather than to particular host taxa (Polishook et al. 1996). Features such as substratum longevity, topography (e.g., spininess, venation patterns), rigidity, and water retention capability must have encouraged the evolution of species guilds, and climatic features such as extent and pattern of rainfall also play a part in promoting specialization.

Currently no easily available source of information on distributions and ecological preferences of fungal species exists. This is yet another major barrier to inventory work, to identification and characterization of taxa, and to knowledge of conservation status. Such a publication would allow surveyors of ecosystems to reduce dramatically the numbers of species with which their taxa should be compared when attempting identifications. Such data would have to be amassed on a systematic basis with well-developed habitat definitions such as those used by phytosociologists (Barkman et al. 1986) and supplemented by experimental work detailing temperature ranges, nutrient levels, water levels, and other environmental parameters necessary for growth of individual species in culture. Such an undertaking would be a mammoth exercise, but agreement on a logical framework for the data would allow a gradual buildup of information.

SPECIES CONCEPTS

Most species of microfungi associated with dead plant tissue continue to be identified based on morphological characters observed from collections from host material rather than cultures or molecular characters. Identifications are sometimes unreliable because of difficulties of definition and delimitation of genera and species. Furthermore, key morphological features are often subject to environmental modification over time, often to unquantifiable degrees. Therefore, many species are probably distinguished unsatisfactorily from their relatives on morphology alone. Lack of information on morphology of fungal structures in culture for many species makes identification of these species particularly difficult, not only because sexual structures are produced more rarely in culture but also because other structures frequently used for taxonomic analysis in field specimens, such as fruit bodies and conidiophores, may be absent or poorly developed in culture.

Nutritional status clearly plays a large role in the evolution of fungal species, a fact that has not been sufficiently taken into account for the vast majority of taxa (Brasier 1987; Cannon and Hawksworth 1995). For many groups of biotrophic fungi, current opinion is that species are restricted to relatively small taxonomic units

of their associates (often genera, sections or groups of related genera), although most of the evidence presented is based on observations of morphological similarity rather than by experimental techniques (e.g., Cannon 1991). Many necrotrophic fungi have traditionally been regarded as strongly host-specific, resulting in a plethora of species names, in which the species are defined more in terms of their host identity than by features of the fungi themselves. The accepted species concepts for some necrotrophic genera with traditional taxa largely based on host identity, such as *Colletotrichum* and *Phomopsis,* are in the process of revision (Sutton 1992; Rehner and Uecker 1994). DNA analysis (particularly ribosomal DNA) is showing that such genera are systematically diverse, but the species recognized by this new evidence do not always correlate significantly with host-based classifications. Aproot arrived at similar conclusions while studying distribution of the common species of *Didymosphaeria* on decaying dicotyledonous plants (Aproot 1995). Such work increases the need to develop appropriate and, as far as possible, comprehensive species concepts that conform to modern evolutionary theory.

Saprobic fungi do not suffer the same nutritional constraints as necrotrophs because there is no need to evolve ways to penetrate living plant defenses. That does not mean that saprobes never prefer particular substrata because many are adapted to specific microhabitats or hosts (e.g., Læssøe and Lodge 1994). Such adaptations may involve the ability to metabolize, or at least to tolerate, noxious plant compounds, especially the wide range of defensive metabolic products produced as protection against pathogenic heterotrophs of all kinds (Bharat et al. 1988). Species that are apparently saprobic and that show clear host/substratum preferences may be endophytes during their initial growth stages; many species of Xylariaceae fall into this category (Petrini and Petrini 1985). Fungal taxonomic groups often include species varied in their nutritional mode; Laessøe and Lodge (1994) found that the strictly saprobic *Xylaria axifera* was strongly host-specific to dead petioles of species of Araliaceae and is not found as an endophyte. Similarly, Whalley (1987) found that *X. carpophila* does not develop stromata from freshly fallen fruits but only from partially degraded fruits in the litter layer. Some substratum-specific saprobes, particularly species of the Xylariaceae, may be present as endophytes in unrelated hosts but do not develop after the host material dies. *Verticicladium trifidum*, typically observed on conifer debris, is widespread and a non–host-specific endophyte (G. F. Bills, personal communication). Variations in nutritional mode, type of host specificity, and transient states of the life cycles present particular problems in estimating diversity of endophytic fungi and in understanding their interaction with the host plant.

Species definitions for many microfungal groups are problematic at present as a result of the apparently frequent lack of sexual recombination, absence of information about gene exchange, and arguments as to the relative merits of biological and phylogenetic species concepts (Cannon and Hawksworth 1995; Colwell et al. 1995). Lack of information on anamorph phylogeny and anamorph/teleomorph connections (Kendrick and Di Cosmo 1979; Sutton and Hennebert 1994; Sutton 1996) are further barriers to proper characterization of microfungal species. Molecular data already are improving our knowledge in these areas, but current techniques of nucleic acid analysis must be adapted for groups that cannot be cultured. Unfortunately, DNA extracted from herbarium specimens is often too degraded for full analysis (Haines and Cooper 1993).

NUMBERS OF SPECIES

Described species of fungi of all groups number about 72,000 (Hawksworth et al. 1995), which represents less than 5% of the conservatively estimated total number of fungi (Hawksworth 1991). The probability of encountering a large proportion of unknowns in niches, habitats, and geographic areas previously unexplored for fungal diversity is high. This issue is considered in detail by Cannon and Hawksworth (1995).

AVAILABILITY OF DATA

Diversity studies of fungi, especially of microfungi, are currently seriously hindered by the lack of information in easily accessible and usable formats. Comprehensive databases of fungal species names are available only for limited geographic areas and for limited taxonomic or ecological fungus groups (e.g., Cannon et al. 1985; Farr et al. 1989). Few data on names accepted by modern experts or placed into synonymy are easily available. The lack of modern monographs and other accounts of fungi that can be used by nonspecialists is even greater. The work by Ellis and Ellis (1997) on microfungi in northern Europe is a shining example of what can be achieved. The United Kingdom's Natural Environment Research Council has provided funding for important descriptive manuals of the ascomycetes of Great Britain and Ireland that are currently in preparation, but, for tropical regions little is available that does not require extensive training to use. Problems caused by the high proportion of undescribed taxa are even more acute.

TAXONOMIC AND BIOTIC STATUS

PATTERNS OF DISTRIBUTION

With a few notable exceptions, knowledge of fungal biogeography is very limited, in large part because of lack of data. Except for some economically important groups (mainly plant pathogens) and a few relatively conspicuous species, the number of collections of microfungus species is inadequate for portraying the extent of their distributions and likely biogeographic origins. Difficulties in accessing available data (most collection records are not computerized) compound the problems faced even in assessing current knowledge. General treatments of this subject are provided by Pirozynski and Weresub (1979) and Galloway (1994).

Distribution patterns of fungi are linked intimately to their modes of nutrition. Clearly, fungi that are confined to particular host groups (i.e., social groups; not necessarily implying parasitism or pathogenicity) have distributions that are, at most, coextensive with those of their plant associates, although other factors may limit their range further. Some fungi, especially ruderal species, are distributed widely, and many of these are observed abundantly in ecosystem analyses. This has led to a traditional view that fungi tend to be universal in their occurrence. Some taxa, such as members of the Caliciales (Tibell 1984), for example, are presumed to be cosmopolitan as a result of their ancient evolutionary history. Many dung fungi and perhaps soil and litter fungi, in contrast, are distributed widely (e.g., Gams 1992) at least partly because of extensive human introduction. The most prominent (or, at least, most commonly encountered) members of the microfungal component of ecosystems will be those that sporulate profusely, are easy to isolate into pure culture, and have short generation times. Such species are inherently more likely to spread widely than those with restricted growth and sporulation and are recorded much more reliably in samples. Research on fungal diversity in genuinely undisturbed habitats is badly needed as a basis for proper assessment of natural distribution patterns of saprobic microfungi. Recently, many fungal species that previously were considered to be cosmopolitan have been subdivided into geographically well-defined segregates (e.g., Brasier 1987; Otrosina et al. 1993; Mueller 1992). This trend may prove widespread as additional taxa are analyzed.

It is difficult to prove that individual species are rare or have limited ranges, given that evidence supporting such hypotheses is always negative. Nevertheless, it is likely that distributions of a significant proportion of saprobic microfungi are restricted. Questions of distribution and rarity of many microfungi will be answered authoritatively only after extensive inventory of a series of sites.

CURRENT KNOWLEDGE OF SPECIES DIVERSITY

The number of species of microfungi currently known that are associated with dead plants is difficult to estimate, even roughly, for two reasons. First, although a few reasonably complete databases (although much editing is required) for fungal names now exist, there are no comprehensive registers of accepted species in which synonyms are indicated and linked with correct names. Such information is available only for restricted groups (rarely higher than family in the taxonomic hierarchy) and is confounded by disagreements over species concepts and the status of poorly known taxa. Second, taxonomic and ecological divisions rarely coincide, and data on substratum preferences of fungi are not available in an easily accessible form. The value of organizing distribution data both taxonomically and ecologically is frequently significant for distributional, as well as systematic and nomenclatural, reasons. Unless species designations have changed significantly over time, historical occurrence records should be valid and can provide valuable information on changes in distribution patterns, conservation status, and similar responses.

Some basic estimates of species diversity are possible. Hawksworth and colleagues (Hawksworth et al. 1995) gave a conservative estimate of 72,000 described species, although other experts (e.g., Rossman 1994) have suggested that the number may be much more than 100,000. The large discrepancy is probably at least partly a result of differences in the ways poorly known taxa were included in the estimates. Hawksworth's group included only species that were established as "good" by modern experts. It is not unreasonable, perhaps, to speculate that we have at least basic knowledge of about 100,000 fungal species, although large numbers of those have not been characterized adequately.

A crude estimate of the proportion of fungal species that are microfungi on dead plant tissue can be made by examining the numbers of fungi known that belong to the major taxonomic group that inhabits such substrata (see also Dick and Hawksworth 1985). According to the *Dictionary of the Fungi* (Hawksworth et al. 1995), about 13,000 of a total of 32,250 species of Ascomycota, or about 40%, are likely to be microfungi found on dead plant parts. The remaining species are lichenized, confined to living plant tissue, or primarily associated with soil. About 45% of the Ascomycota are lichenized, so about 73% of nonlichenized Ascomycota are likely to be sampled using the techniques described in this chapter.

About 15,000 species of mitosporic fungi are accepted in the *Dictionary of the Fungi*. Some of those have established teleomorph connections, although the methodology for establishing connections needs further development (Samuels and Seifert 1995) and inevitably some species have been counted twice. A large proportion of mitosporic fungi on dead plant tissue are categorized as microfungi because few lichenized mitosporic fungi have been described so far, although the number is increasing rapidly. Counterbalancing this to some extent are the many necrotrophic fungi that produce conidia from living plant tissue and form teleomorphic structures once the plant has died. Given our poor knowledge of mitosporic fungi, it is likely that a larger proportion of fungi currently classified as mitosporic than of fungi of Ascomycota will be sampled using living plant protocols. Because information is not available to assess these data accurately, the ratio of 73% (for nonlichenized taxa) therefore is reduced to a somewhat arbitrary 65% of mitosporic fungi likely to be sampled using dead plant techniques. A known species estimate of 10,000 is established, therefore, for mitosporic fungi on dead plant parts.

The numbers of species of microfungi in the Basidiomycota and Zygomycota are relatively small compared with those of the Ascomycota, and most species of the former two groups will be sampled from dung and soils. We include a token 500 species to represent the Basidiomycota and Zygomycota on dead plant tissue that would not be discovered using the techniques described in other chapters.

We estimate, therefore, that the total known diversity of fungi associated with dead plant tissues is 23,500 species, or 32% of the 72,000 described species estimated in the *Dictionary of the Fungi* (Hawksworth et al. 1995). Estimations of total fungal diversity are currently of considerable interest (Rossman 1994; Cannon and Hawksworth 1995). The figure of 1.5 million species proposed by Hawksworth (1991) remains the best-researched estimate to date, although that number includes organisms from other kingdoms traditionally treated as Fungi. About 1500 known species fall into that category, with perhaps a proportion of species detectably associated with dead plant tissues similar to the proportion estimated for the true fungi. If we use the proportion of known fungal species associated with dead plant tissues with the 72,000 estimated total species of microfungi, then the total species associated with dead plant material can be assessed. Those calculations suggests that about 480,000 fungal species could be collected using the protocols described in this chapter. Some investigators have suggested even higher estimates for total fungal diversity, whereas others believe that the very high local diversity of fungi will not translate into such large numbers on a global scale (Hawksworth and Kalin-Arroyo 1995).

Assumptions that known microfungi on dead plant material represent the same proportion of the total of known fungi as the actual figures may not be justified. Hammond (1995) demonstrated that the proportion of organisms in a given group that is known to science varies according to size of the organisms concerned. Hawksworth (1993) provided some limited evidence of such a relationship within the fungi. Those arguments suggest that the figure of 480,000 may underestimate the number of fungi associated with dead plant materials. The sheer number of species involved (more than 95% of which have not even been described) presents special problems for designing and carrying out fungal inventories.

STUDIES OF LITTER DIVERSITY

A number of studies of microfungi associated with dead plant parts have been carried out in recent years,

although formal diversity assessment has been the primary aim of only a few. The studies may be grouped roughly into those that focus on fungi associated with particular species of substrata and those that concentrate on habitats.

HOST-SPECIFIC DIVERSITY ESTIMATION

The large number of fungi that are host- or substratum-specific, combined with the fact that plants frequently occupy well-defined ecological niches, suggests that estimates based on the ratios of plant to fungal species may provide the best working hypotheses of total diversity for planning inventories. Even so, such estimates must be used cautiously, especially when incomplete surveys are being considered. The diversities associated with selected plant species are listed in Table 11.1. Many of the earlier studies were not intended to be comprehensive, focusing primarily on species succession rather than diversity. The number of fungi associated with an individual plant species may be quite high, even with incomplete sampling.

TABLE 11.1

Fungal Diversity Associated with Dead Tissues of Selected Species of Plants

Plant species	Plant part	No. fungal species	Reference
Abies alba	Leaf	95	Aoki et al. (1992)
Abies firma	Leaf	104	Aoki et al. (1990)
Acanthus ilicifolius	Decayed tissue	130	Vrijmoed et al. (1995)
Agropyron pungens	Debris	98	Apinis and Chesters (1964)
Alchornea triplinervia	Submerged leaves	81	Schoenlein-Crusius and Milanez (1995)
Atlantia monophylla	Leaf	73	Subramanian and Vittal (1979a, 1979b)
Carpinus caroliniana	Living and dead bark	155	Bills and Polishook (1991)
Carex paniculata	Leaf litter	60	Pugh (1958)
Castanea sativa	Cupule	>27*	Sutton (1975)
Eucalyptus regnans	Leaf litter	>24	Macauley and Thrower (1966)
Fagus sylvatica	Leaf	29	Hogg and Hudson (1966)
Gymnosporia emarginata	Leaf	52	Subramanian and Vittal (1980)
Helianthus annuus	Achene	98	Roberts et al. (1986)
Heliconia mariae	Decaying leaves	56–98	Bills and Polishook (1994)[†]
Heracleum sphondylium	Stem	51	Yadav (1966)
Juncus roemerianus	Dead leaves	86	J. Kohlmeyer (personal communication)
Laurus nobilis	Leaf	137	Kirk (1983, personal communication)
Nypa fruticans	Leaf	63	Hyde and Alias (2000)
Pinus sylvestris	Leaf	120	Hayes (1965)
Pinus sylvestris	Leaf	70	Kendrick and Burges (1962)
Pinus sp.	Leaf	73	Tokumasu et al. (1994)
Pinus sylvestris	Decaying leaves	127	Tokumasu et al. (1997)
Pteridium aquilinum	Petiole	114	Frankland (1966)
Shorea robusta	Leaf litter	38	Bettucci and Roquebert (1995)
Quercus germana, Q. sartorrii, Liquidambar styraciflua	Leaf	46	Heredia (1993)

* Incomplete survey.

† Four individual plants sampled.

Frankland's (1966) study of the succession of fungi colonizing dead *Pteridium aquilinum* rachides remains one of the most influential. She found a total of 114 fungal species. Some recent studies have charted similar or greater levels of diversity. Cornejo and colleagues (1994) reported the recovery of about 500 species of fungi from leaf litter of only six tree species in Panama, although their methodology might not have accurately discriminated litter fungi from soil fungi. Bills and Polishook (1994), using a particle filtration method, recorded between 56 and 98 species of microfungi for samples derived from individual leaves of *Heliconia mariae* from Costa Rica. Roberts and associates (1986) isolated 98 fungal species from achenes ("seeds") of cultivated sunflowers (*Helianthus annuus*). Aoki and co-workers associated more than 90 fungal species with litter from each of two species of *Abies* in Germany and Japan (Aoki et al. 1990, 1992). Other recent studies have reported relatively low levels of diversity. Bettucci and Roquebert (1995), for example, identified only 38 taxa from leaf litter of *Shorea robusta* in Malaysia. That number may in part reflect deficiencies in experimental technique; their incubation of dilution plates at 25°C or 35°C and use of high-nutrient media probably resulted in overgrowth by ruderal saprobes. The high proportion of species detected from genera such as *Trichoderma*, *Gliocladium*, and *Penicillium* suggests that this was the case.

There is little evidence that even the more detailed studies of fungal diversity associated with litter of particular plant species have produced a near-complete inventory. Perhaps the most comprehensive survey available is one made by P. M. Kirk who carried out a series of direct observations of fungi on leaf litter of *Laurus nobilis* (Lauraceae; Kirk 1981, 1982, 1984). He identified 137 species of microfungi from 44 collections of leaf litter made over a period of 12 years in the southern United Kingdom. Because data were obtained by one person, inconsistencies of observations and their interpretations associated with multiple collectors and identifiers are reduced. We produced a rarefaction curve (Fig. 11.1A) for his data by plotting cumulative species number against cumulative number of collections (P. F. Cannon and B. C. Sutton, unpublished data). It suggests that, at least for this host in the southern United Kingdom, the total number of species of associated microfungi is likely to be about 150, although use of a different sampling technique might reveal another suite of fungi. The data also show (Fig. 11.1B) that no species was found in more than 75% of the 44 samples, that 35% of the species were found only once, and that a majority of species (64%) was found in less than 9% of the samples. Those data strongly suggest that many species of microfungi are rare, although lack of

knowledge of host specificity and the fact that *Laurus nobilis* is not native to Britain make such interpretations less reliable.

The estimates provided earlier in this section and those in Table 11.1 suggest that the number of microfungi associated with the litter for a plant species is likely to range well over 100. In Kirk's study, and in those listed in Table 11.1, a large proportion of the species recovered were wide-spectrum saprobes. Some crude estimates of the diversity of host-specific species might be attained by combining those disparate character sets, but variations in species concepts and sampling techniques and the large number of incompletely identified taxa included in each study would render any estimates obtained of very uncertain value. Studies by Polishook and associates (1996) on the complementarity of fungal species isolated from litter of two plant species growing together in a forest in Puerto Rico addressed the question of specificity and ubiquity of microfungi on different host plants. Additional studies of the variation in complementarity between different host species and the effects that are genuinely host-related versus those that primarily are the result of physical differences in the host tissue are essential. Such an experimental program would be of considerable value in planning future sampling protocols, especially if it were carried out in a speciose site such as a tropical rain forest.

NON–HOST-DIRECTED DIVERSITY ESTIMATION

A small number of surveys have focused on the diversity of microfungi associated with litter in general (Table 11.2). Such an approach has advantages and disadvantages. Identification of species recovered can be more difficult because host identity cannot be used as a limiting character. Sampling is faster, however, and can be carried out by someone untrained in plant fragment identification.

Two of the most extensive general litter studies are those of Rambelli and colleagues (1983) and Bills and Polishook (1994). Rambelli and his co-workers studied leaf litter samples from four tropical forest plots in the Ivory Coast, comparing undisturbed vegetation with sites that had been cleared using traditional agricultural practices. They periodically examined litter samples incubated in moist chambers. They identified between 129 and 165 species from undisturbed and disturbed plots, respectively. The disturbed sites showed greater diversity than the undisturbed ones, presumably because native species persisted in competition with taxa introduced or encouraged by the agricultural process. The similarity of the fungal species lists from the sites was high, suggest-

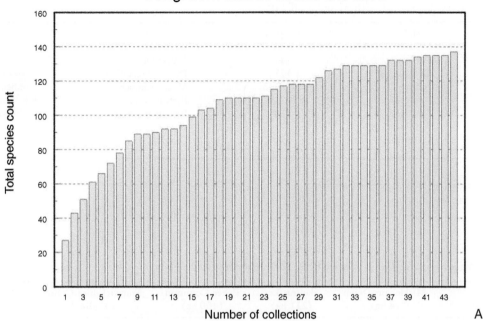

Fungi from *Laurus nobilis* leaf litter.

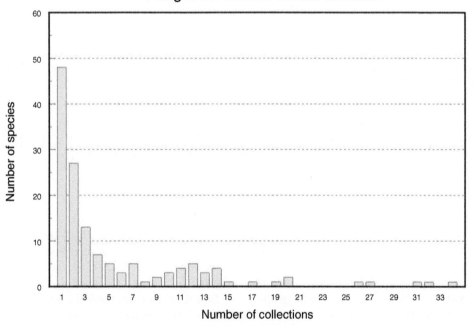

Fungi from *Laurus nobilis* leaf litter

Figure 11.1 Fungi from *Laurus* leaf litter in the United Kingdom. **A.** Rarefaction curve showing cumulative species of fungi against cumulative number of collections. **B.** Frequency diagram of incidence of fungal species.

ing that many species were distributed widely and evenly. The plots constituted two pairs on somewhat different soil types, which appeared to have very little effect on the microfungal litter diversity, although how the mycota varied between plot pairs was not stated. In continuing studies, Rambelli and colleagues (1984) found that the fungal biota in the cultivated plots appeared to be gradually returning to the natural, precultivated state, although in a later paper (Rambelli et al. 1991) they noted that continuing cultivation had the potential of

TABLE 11.2

Microfungal Diversity Associated with Mixed Leaf Litter

Locality	Habitat	No. species	Methods	Reference
Great Britain	Salt marsh	141	Unspecified	Apinis and Chesters (1964)*
Austria	Alpine sedge	128	Direct observation	Nograsek (1990)[†]
Ivory Coast	Forest	129–165	Moist chamber	Rambelli et al. (1983)
Costa Rica	Wet forest	78–134[‡]	Particle washing	Bills and Polishook (1994)
Puerto Rico	Wet forest	338	Particle washing	Polishook et al. (1996)
Puerto Rico	Wet forest	24	Direct observation	Polishook et al. (1996)
Panama	6 tree species	500	Dilution plating	Cornejo et al. (1994)

* Ascomycota only.

[†] Ascomycota only, excluding inoperculate discomycetes.

[‡] Number of species per sample.

TABLE 11.3

Proportions of Major Taxa of Fungi Recorded at Different Sites in England

Taxon	Mickleham Down (4 ha)	Slapton Ley (211 ha)	Esher Common (400 ha)	Warwickshire County (250,000 ha)	Southeastern England (1,050,000 ha)
Mycetozoa	5%	5%	3.5%	6%	No data
Oomycota, Zygomycota	1%	3%	1.5%	6%	No data
Ascomycota	31%	43%	33.5%	40%	28%
Basidiomycota	46%	32%	40.5%	44%	42%
Mitosporic fungi	16%	17%	21%	8%	24%
Total species recorded	1380	2500	2700	2700	No data

Data sources: Mickleham: A. Henrici (in litt. 2000); Slapton: D. L. Hawksworth (in litt. 1997); Esher: B. Spooner and P. Kirk (in litt. 2000); Warwickshire: Clark (1980); Southeastern England: (Dennis 1995).

reducing diversity as a result of the climatic changes induced by cultivation.

Bills and Polishook (1994) carried out an intensive study of litter microfungi from four sites in lowland Costa Rica by culturing fungi from washed litter particles, in part with the aim of developing methodologies for diversity assessment. They detected from 78 to 134 species in each collection and found that rarefaction curves comparing total species number with total culture number showed little sign of leveling off despite identification of an average of 424 isolates from each sample. That finding suggests that they did not achieve anywhere near a complete sampling of the fungal diversity. The undersampling of the species diversity was underscored by Lodge and Cantrell (1995b), who reanalyzed Bills and Polishook's data and found only a 15–28% overlap of named species between samples.

The diversity of fungi in leaf litter is significantly greater than that in soil (Bills and Polishook 1994; Bettucci and Roquebert 1995). In a study in São Paulo State, Brazil, only six of 87 taxa identified from soil and litter samples were found only in the soil (Attili and Grandi 1995; Schoenlein-Crusius and Milanez 1995).

SPECIES DIVERSITY OF MICROFUNGI OVER LARGE AREAS

STUDY SITES

The data available on which to base estimates of fungal species diversity in large study areas, wherever they are located, are minimal. The most complete (or, perhaps more accurately, the least incomplete) information on fungal species numbers for specific areas is from three sites in southern England (Table 11.3): Slapton Ley National Nature Reserve (211 ha, of which nearly half is open water), Esher Common (ca. 400 ha), and Mickleham Down (4 ha). Each has been surveyed over a long period by a succession of expert collectors. Slapton Ley has been studied for more than 25 years by a large group of mycologists led by D. L. Hawksworth (International Mycological Institute [IMI]); Esher Common is largely a joint effort between B. M. Spooner (Royal Botanic Gardens, Kew) and P. M. Kirk (IMI); and Mickleham Down is primarily by an expert amateur mycologist, A. Henrici. Esher and Mickleham are rather drier than Slapton, and the diversity of habitats is somewhat greater at Slapton. Plant species overlap considerably among the sites, however, which makes limited comparisons valid. More than 2500 fungal species are known from both Slapton Ley and Esher Common, but only about 30% of those are common to both areas. That observation and the fact that large numbers of microhabitats at both places have not yet been examined in detail suggest that the inventories are nowhere near complete for either site. An estimate of 3000 species for the total fungal diversity for either site does not seem unreasonable. Because Slapton Ley contains around 500 plant species, the figure corresponds well with Hawksworth's (1991) working hypothesis that fungal diversity outstrips plant diversity by a factor of around six. Of the species currently known from each site, the proportion associated with dead plant parts is somewhat higher than the overall estimates calculated earlier, but the research interests of the collectors may explain the high numbers. More than 1300 species of fungi currently are known from Mickleham Down, despite its small area, uniform vegetation, and lack of open water.

Extrapolation of these incomplete data to a large (50,000-ha) sample area, even in southern England, is more or less impossible because information on fungal distribution patterns is so inadequate and comparable data for species numbers over larger areas do not exist. The best data for a large study area in Britain are for Warwickshire County (Clark 1980), which covers approximately 250,000 hectares. Around 2700 fungal species have been recorded, although some speciose groups (especially the mitosporic fungi) are substantially underrecorded (Table 11.3). The relatively small increase in number of species encountered in this large area in comparison with the figures for Slapton, Esher, and Mickleham suggests that extrapolating from data associated with such a large area may not be a particularly good way of estimating fungal diversity. The large variation among the sites in proportions of species from major taxonomic assemblages (Table 11.3) provides further evidence that historical sampling procedures have been biased and that the proportion of species remaining to be sampled is significant.

TROPICAL VERSUS TEMPERATE AREAS

Many circumstantial data, but no direct evidence, support the commonly held view that species diversity of the fungi as a whole is greater in tropical regions than it is in temperate regions (Lodge et al. 1995). Dennis (1970) compared the mycotas of Venezuela and the British Isles in fair detail, noting particularly that leaf epiphytes and inhabitants of living leaves in general are significantly more diverse in tropical, than in temperate,

zones. Tubaki and Yokoyama (1973) observed that the diversity of fungi associated with litter of *Castanopsis cuspidata* and *Quercus phillyraeoides* was greater in southern Japan than in central Japan. Batista and co-workers frequently observed 10 or more microfungus species during direct examinations of single leaves in Brazil (Cannon and Hawksworth 1995; da Silva and Minter 1995), emphasizing the enormous diversity associated with neotropical plants. The same workers recorded as many as 26 fungal species from a foliicolous lichen community (da Silva and Minter 1995). However, large numbers of microfungi also can be found on plant parts in temperate (e.g., Sutton 1975) and Arctic regions. H. Knudsen (in litt.) estimated that the number of fungal species in Greenland was about 3000, conforming closely to the 6:1 ratio of fungal-to-plant species suggested by Hawksworth (1991).

Direct evidence that species diversity is greater in tropical than temperate regions does exist for individual fungal groups. For example, Cannon (1997a) found that the number of species of the Phyllachoraceae (Ascomycota) known between latitudes 40 and 60 N and 40 and 60 S is only 38% of that between latitudes 20 N and 20 S, and the corresponding figure for latitudes above 60 N (there is no vegetation below 60 S) is only 6% of the figure for the tropics. The true proportions are likely to be even more extreme given the under-recording of tropical species compared with temperate zones. Similar data might be gathered for a range of other fungal groups, providing a more balanced estimate of relative diversity in the two regions. Inventories are the only way to obtain such data.

FACTORS INFLUENCING SITE DIVERSITY

Many of the factors influencing site diversity affect all fungal groups similarly, but in some circumstances effects on diversity of microfungi associated with dead plant material are specific to that group. Factors affecting plant distribution and diversity, for example, will have proportionately greater influence on such microfungi than on fungi from some other groups (e.g., lichenized fungi or animal-associated taxa). The greater within-site habitat variation in plants and microhabitats, the larger is the expected microfungal diversity.

TEMPERATURE, WATER RELATIONS, SEASONALITY, AND PHYSICAL FACTORS

There is little reliable information charting the effects of temperature on fungal diversity (in contrast to a plethora of information for individual species), but physiologists appear to agree that the majority of fungi grow and sporulate best at temperatures greater than 15–20°C (e.g., Cooke and Whipps 1993). Although supporting data are based on a species assemblage that is not representative of the world mycota, Cooke and Whipps (1993) suggested that most fungi are adapted to grow in warmer climates. That then might lead to the conclusion that diversity is greatest in warm regions. A small proportion of fungi is known to be psychrophilic or thermophilic (e.g., Petrini et al. 1992b; Mouchacca 1995; see Chapter 14). It is likely that microfungi on dead plant parts conform to this rule. It is also likely that humidity plays a much more prominent role in influencing fungal diversity than does temperature alone. Elevation effects also may be relevant.

Again, data on the requirements of individual species for particular levels of moisture in the environment are abundant, but statistical analyses of overall species diversity under different water regimens are rare. Field mycologists know, at least, that spore production in a wide range of fungi depends on adequate rainfall. The considerable diversity of fungi in aquatic environments (e.g., Kohlmeyer and Kohlmeyer 1991; Bärlocher 1992; Shearer 1993) is indicative of the importance of water as a factor limiting fungal existence. Conversely, some species are adapted to xeric environments, both natural and manmade (e.g., Hocking 1991; Zak 1993; Cannon and Hawksworth 1995; see "Xerotolerant and Xerophilic," in Chapter 14). The diversity of microfungi closely associated with plants is small in arid areas compared with that in mesic areas, reflecting reduced host-plant diversity and restriction of fungi to very specific microniches where water is more available (Zak 1993; Fisher et al. 1994). In contrast, soil fungus diversity seems similar regardless of moisture level, if sufficient nutrients are available (see Chapter 13), although individual species may have specific water requirements.

Water availability directly affects rates of decomposition (Dix and Webster 1995). Abundance of water promotes fungal growth in general. It also may affect diversity, because fungi vary in their tolerances to water stress, and interspecific competition (Pasricha et al. 1994). Relatively low water availability may result in (at least in the sampling of) greater diversity in culture because slower-growing taxa are less likely to be crowded out by weedy species. For microfungi on dead plant tissue, water relations are important, although the substratum itself may act as a reservoir, allowing continued growth in dry conditions. Variation (seasonal fluctuation) in water supply can have dramatic effects on the microbial biomass of soil and litter (Lodge 1993; Lodge et al. 1994, 1996a), although not necessarily on microbial diversity.

The effects of seasonality on microfungal diversity are connected intimately with the effects of temperature and water. Significant seasonal variation in either temperature or water availability will result in a fungal biota different from that developing when conditions are constant (Lodge and Cantrell 1995b). Factors such as the aspect of the site, topographic variation, and soil type will affect microfungal diversity but primarily through their influences on the flora of the study area.

PLANT DIVERSITY AND DISTRIBUTION

The taxonomic diversity and structural variety of plant taxa exert the greatest influence on microfungal diversity at a site because such a large proportion of the microfungi has at least some degree of host and substratum preferences. Fungal diversity also is affected by variations in host life form; lignified tissues, such as wood, bark, and bamboo stems, decay much more slowly than non-lignified remains and therefore exhibit more extensive succession of fungal species. Distinct guilds of fungi are associated with dead bryophytes, algae, grasses, and other such groups, and although direct evidence is not available, it seems likely that the diversity of the fungi will be strongly positively correlated with the numbers of these plant groups.

DIVERSITY OF OTHER ORGANISMS

The diversity of organisms other than plants should have little effect on the diversity of plant-associated microfungi but will be relevant in multitrophic relationships involving animal dispersal of fungal propagules. In addition, the rate of decomposition of litter material may be affected substantially by animal species, especially termites and bark-boring beetles (Anderson and Swift 1983), and competition for nutrients among fungi and other organisms may influence fungal diversity considerably. Again, the data available are too limited for proper analysis.

HUMAN INFLUENCE

The effects of disturbance on microfungus diversity rarely have been quantified, but any physical disturbance, such as logging, trampling, and chemical pollution, should have a dramatic effect on microfungus diversity. Fungi may be affected indirectly through the influence of the disturbance on plant diversity, but increasing evidence (primarily from studies of lichenized and endophyte taxa) indicates that fungi also may be affected directly.

A comprehensive study of the effects of human disturbance on microfungus diversity in litter was carried out in seasonally dry forests of the Ivory Coast by Rambelli and associates (1983). They compared two sample plots from each of two locations that differed in soil type and plant-species composition. One plot was left undisturbed. The other was cleared, planted with rice using the traditional agricultural techniques of the region, and then left to revert to natural forest after a single planting cycle. Sampling was initiated weeks after forest clearing and continued for a 3-year period. Fungi were sampled using moist chamber studies and direct observation of developing sporulating structures. Species identified from the litter of each sample plot ranged from 129 to 165, with species diversities at the disturbed sites exceeding those in the undisturbed ones. That increased diversity presumably reflects the persistence during cultivation of most fungi associated with the undisturbed habitat and the influx ruderal species and species associated with the introduced crop. However, the widespread practice of clearing forest and growing crops continually until the soil fertility deteriorates has a marked deleterious effect on diversity of litter microfungi (Rambelli et al. 1991).

A major survey of soil fungi is also of interest because fungal diversity in soil and litter may be closely correlated. Using isolation techniques, Miller and colleagues (1957) also showed higher diversity of fungi in cultivated soil in Georgia than in undisturbed forest soil. They found that of 165 species isolated, 115 were only found in forest soil, 141 only in cultivated soil, and 91 in both habitats. Although the differences in species profiles may not be numerically significant, it is likely that disturbance favored increased diversity of ruderal and rapidly colonizing species, which are sampled more efficiently using standard isolation techniques such as suspension plating. Other smaller-scale soil studies, however, have shown that persistently disturbed ecosystems exhibit lower microfungus diversity than natural, undisturbed sites (e.g., Joshi et al. 1994).

UNDERTAKING A MICROFUNGAL INVENTORY

SAMPLING STRATEGIES

The resources necessary for a complete microfungal survey of even a small area are enormous, and most species are so small that it is impossible to be certain of their absence even after extensive sampling. Although the case for conducting a limited number of comprehensive inventories in distinct types of ecosystems throughout the world is scientifically compelling, realities of current funding levels argue for surveys that emphasize reproducibility rather than completeness (Cannon 1997b). Partly because of the lack of widely promulgated protocols, fungal surveys historically have been poorly defined, making it difficult to compare them.

Sampling strategies will differ somewhat depending on whether a total inventory or an assessment of diversity is required. For either goal, standardization of sampling strategies is crucial, both for comparative purposes and for the establishment of stop rules. Methods can involve direct observation (either at the time of collection or after incubation in moist chambers) and protocols for cultures.

DIRECT OBSERVATION

Moist chamber (Appendix I) and isolation techniques have been compared by several research groups. Watson and colleagues (1974) and Visser and Parkinson (1975) concluded that each method sampled distinct species assemblages, with a significant level of complementarity (Colwell and Coddington 1994). Polishook and associates (1996) found that at least 79% of the microfungal species identified by direct observation of decaying leaves also were obtained by particle filtration and subsequent culturing. Complete surveys require that both direct observation and culture techniques be used. Fungal cultures provide major benefits; they offer additional systematic characters for use in identification and determining relationships, living cells for genetic analyses, and the opportunity to screen for technological applications.

Advantages of direct observation methods over cultural techniques include the following: (1) lower cost because agar media and incubators are not required; (2) less airborne contamination; and (3) more complete identification because a higher proportion of observed species produce spores or other distinctive structures. Disadvantages of direct observation include the following: (1) greater difficulty in observing many taxa simultaneously; (2) vegetative structures immersed in host tissues; (3) greater difficulties in recognizing and separating species in mixed communities; (4) suppression of growth of some taxa as a result of competition among fungal and bacterial species in host tissue; (5) conditions in moist chambers suboptimal for sporulation of some species; and (6) absence of cultures for subsequent studies.

To ensure comparable results among samples, incubation conditions of natural substrata must be monitored closely, and the host tissue used must be clearly defined. Similar samples usually are achieved by cutting areas (circles or squares) of designated size from broad-leaf or bark tissue or using given lengths of petiole, needle, or stem. Samples should be as uniform as possible, for example, including a given proportion of midrib for leaf tissue. Samples should be placed in closed containers of appropriate size on damp filter paper, paper towel, moistened sterile vermiculite, or moistened sterile sand. Heredia (1993) attached microfungi from leaf surfaces to microscope slides with double-sided adhesive tape to facilitate repeated observation and to make voucher slides. Insecticides and/or molluscicides can be used to prevent degradation of specimens by small invertebrates, but some fungi are probably sensitive to these chemicals. Emerging pests also can be removed by hand. Samples must be inspected frequently for newly emerging fungal species and to ensure that moisture levels are correct. Variation in moisture content through the experiment may alter the diversity being recorded (J. Zak, personal communication). For additional information, see Appendix I on preparation of moist chambers.

CULTURE TECHNIQUES

The methods of analysis of litter microfungi in culture are similar to those described for soil fungi (see Chapter 13), and thus we only briefly summarize them here. The most effective method described to date is that of Bills and Polishook (1994), in which plant material is mixed with sterile water and pulverized in a sterilized blender. The resulting slurry is washed through a series of mesh filters using distilled water. Particles from 100 to 200 μm in diameter are suspended in a small volume of distilled water and washed again. The resulting suspension is pipetted onto agar plates of two selective media, a malt agar medium amended with cyclosporin A and dichloran-rose bengal medium (Appendix II). New emerging colonies are subcultured onto a range of standard media, and sporulation is encouraged, where necessary, by additional culturing on cornmeal agar media containing pieces of sterilized banana leaves. The method was developed for tropical rain forest leaf litter but is equally appropriate for temperate litter and for analysis of dead twigs, wood fragments, and other substrata.

Various factors influence the diversity of fungi isolated from litter samples. Those include the following:

Particle size. If litter fragments are too large, multiple colonies may arise from the same fragment and overgrow and obscure one another. If the particles are too small, then the fungal cells may be disrupted. Bååth (1988) recommended fragmentation resulting in a colony to a particle ratio of 1:1 or less. A technique in which plant tissues are cut with a sterile scalpel into small, equal-size pieces before culture has been used for endophytes (see Chapter 12). The spatial order of the plated fragments is recorded, and the number and kinds of fungi are referenced to each fragment. That procedure allows for the investigation of the spatial relationships of the various fungal mycelia or propagules (Wilson and Carroll 1994; Lodge et al. 1996a). The technique is labor-intensive compared with fragmentation in a blender and cannot be used effectively with well-degraded samples.

Growth medium. Nutritionally selective media can also play a major role in maximizing the diversity of taxa recovered, although their effects are not always obvious and the extent of their selectivity has rarely been tested adequately (Cannon 1996; G. F. Bills, personal communication). Many of the most effective media are nutrient-poor (e.g., soil extract agar and carboxymethylcellulose medium, CMC; Appendix II). These media restrict hyphal growth for most species and are inexpensive to prepare. Many of these media are transparent, which facilitates observation of minute colonies.

Growth inhibitors. If growth, especially of the common ruderal species, is inhibited, slower-growing fungi are more likely to be recognized and subcultured before they are overwhelmed by invasive molds. Growth can be inhibited either by incubation of cultures at relatively low temperatures or by the inclusion in the medium of chemical inhibitors such as cyclosporin (Dreyfuss 1986). Cyclosporin-containing media permit germination of a wide range of fungal propagules, but subsequent hyphal extension is strongly restricted. The media, therefore, are not strictly selective but allow easy separation of colonies on agar plates from materials with high fungal densities. See Appendix II for media containing growth inhibitors.

Inventories that rely heavily on culture protocols can produce large numbers of nonsporulating cultures, which are difficult to identify. Whereas specific identification is not essential for diversity studies, separation of taxa is necessary. Bills and Polishook (1994) adopted a wide taxon concept and aggregated individual nonsporulating cultures into a general category. At least 25% of species detected belonged to the category of once-observed unidentifiable species. The separation of nonsporulating cultures into taxa is inevitably more difficult than if spores are produced, but careful study of characters such as growth rates, colony morphology, pigmentation, and hyphal structure usually allows for an acceptable estimate of species diversity.

Molecular techniques such as random amplified polymorphic DNA analysis (see Chapter 6) can be used to confirm the taxonomic integrity of groups of nonsporulating cultures, although their cost may preclude widespread use. Basic biochemical tests and molecular fingerprinting are also of value for characterization, although these methods are expensive compared with traditional classification techniques. The lack of reference data for other than a very small proportion of fungal taxa means that direct identification of isolates from diversity investigations using sequencing techniques is impractical for the foreseeable future.

For large surveys, the division of taxa into both nonsporulating cultures and sporulating taxa of uncertain identity is problematic because of the large numbers both of individuals and species. We recommend, therefore, that investigators set up a detailed database of character expressions to facilitate comparison among taxa.

OBSERVATION AND ISOLATION FROM WASHED LITTERS

Fungi whose dormant propagules land on or become mixed with a substratum can be excluded from a sample of fungi nutritionally associated with the dead plant material by litter washing (Harley and Waid 1955; Parkinson and Williams 1961). Tokumasu and associates (1997) successfully adopted such protocols in studies of fungi associated with pine needles in Thailand. They washed the needles in detergent and rinsed them in distilled water using a shaker. After washing, material either can be placed in moist chambers for observation, or it can be cultured.

COMPLETE INVENTORIES

There are two primary ways to carry out a total inventory of microfungi on dead plant tissue. The first uses a sampling strategy based entirely on host-plant identity and will cover most microfungi, excluding only those that occur on unidentifiable, rotted wood. The second approach involves sampling substrata, especially those in a highly decayed and unrecognizable state, uniformly across a defined geographic area. With a double-pronged protocol, both host-specific and plurivorous taxa are sampled adequately.

The two approaches must be balanced to prevent unnecessary duplication of sampling and to avoid overlap with other sampling strategies (primarily for living plants/endophytes and soil). Preliminary research is needed to identify that balance because few data on this subject are available and site characteristics (especially patterns of host diversity) may affect the outcome significantly. Minimum resources necessary to carry out the inventory will depend on characteristics of the sampling site (e.g., ease of access, climate, seasonal variation) and familiarity of the investigators with the mycota.

SAMPLING ON A HOST-SPECIES BASIS

Sampling using host-plant identity as an organizational tool depends first on an adequate knowledge of the plant species at the site, although common, familiar host species can be surveyed while rarer and difficult taxa are located and identified. If possible, an investigator should sample several widely separated populations of each species at a large site, taking into account differences in local community structure caused by varying physical features, soil types, aspect, and so forth. Fungi associated with living plant tissue (biotrophs and endophytes) should be sampled at the same time. It will be necessary to have detailed knowledge of the locations of rare and targeted host species (preferably using geographic information system, or GIS, technology) or to sample alongside plant collectors. Ideally, suitably trained plant experts should be consulted during the planning stages. In some cases, it may be feasible for plant collectors to collect dead plant materials for examination by fungal experts.

SAMPLING ON A GEOGRAPHIC BASIS

A small proportion of the fungal species associated with plant parts in late decay stages will probably be missed with host-species–directed sampling because plant parts will no longer be identifiable to species. The soil specialists will sample some of this material among the litter and humus, but microfungi on rotten wood and bark will need special attention. Such material is best sampled on a geographic basis, using a series of transects designed to sample the widest possible environmental diversity at the site by crossing valleys, streams, rocky outcrops, and other physiographic features. Similar transects may be used by macrofungus specialists, and sampling can be carried out in conjunction with that group. Samples with recognizable fungal presence should be taken from rotten wood and bark at specified points along each transect. A small number of apparently asymptomatic specimens should be taken in addition for close examination in the laboratory and for study with moist chambers. Samples should be taken at intervals during seasons of active growth and at least once during dormant spells. Care should be taken to minimize the effects of site disturbance and repeated and destructive removal of samples on the study sites. A formal monitoring process to ensure that species are not being lost from a habitat as a result of the sampling may be appropriate.

General litter study (i.e., well-decayed, fragmentary plant parts) is carried out more effectively with culture techniques than with direct observation because of the difficulties in manipulating small pieces of substratum for microscopic observation.

RAPID ASSESSMENT TECHNIQUES

A primary aim of rapid estimation techniques is to minimize reliance on experienced fungal systematists, who are too few in number to provide extensive widespread diversity assessment (Cannon 1997b, 1999). Assessment of the diversity of poorly known and speciose groups will require radical methods if meaningful results are to be obtained in an acceptable timeframe. There is considerable interest in developing methods for biodiversity estimation that do not rely on identification of all the species sampled. The use of parataxonomists (highly trained collectors without formal academic credentials; Janzen et al. 1993) has been discussed widely, and similar workers are potentially able to process collections in laboratories. Untrained staff have assessed the diversity of speciose groups of spiders, ants, polychaetes, and bryophytes successfully (Oliver and Beattie 1993). Some of the most extensive fungal surveys have been performed almost exclusively by dedicated nonprofessionals (e.g., Clark 1980).

Industrial screening programs represent a practical extension of those concepts; such programs use methods developed to maximize diversity of target organisms extracted from natural samples so that their metabolic, biochemical, or genetic attributes can be assessed. In some cases, the work may be carried out by technicians without taxonomic training; formal identifications are required only if a strain producing potentially valuable metabolites is discovered (Bills 1995).

METHODS OF ESTIMATION

Inventories can be carried out using direct or indirect methods and with or without reference to plant associates. Because of the cryptic nature of microfungi, direct observation methods require extensive field trips by highly experienced collectors. Even then, identifications will rely heavily on laboratory studies of materials collected for confirmations. Direct observation protocols may be valid for well-known ecosystems where suitably qualified staff members are available.

One of the simplest methods to reduce the sampling required for a diversity assessment is to restrict collection of material to substrata derived from a small range of plant species. Ideally, a range of types of substrata from a number of plant species that represent the range of botanical diversity of the sample site should be chosen, but even this can result in an unacceptable level of work. Of the various substrata available, leaves are probably the most satisfactory to work with because they are abundant; have a range of structural features such as laminae, midribs, and petioles; and decay relatively quickly so that successional studies in moist chambers can be completed rapidly. They can be identified relatively easily, even when detached, compared with bark or wood and are present in the vast majority of vascular plants.

PROTOCOLS FOR SAMPLING PROGRAMS

We suggest protocols here for analysis of microfungi associated with dead tissues of a single (woody) plant and for a complete survey of a small wooded area. We stress that even restricted surveys involve considerable work and that their value for comparative purposes can be increased immeasurably if protocols are rigidly defined and followed. Nevertheless, environmental variation both within and among sites is complex and difficult to define, and local conditions sometimes may preclude adhering to the quoted methods precisely. In those circumstances, investigators should record and explain departures from the planned protocols.

SITE SELECTION

Sites may be chosen for practical reasons, such as ease of access, to complement surveys of other organisms that have been carried out or are planned or because of environmental threats such as proposed development or encroaching offsite pollution. Because of the lack of baseline data on fungal diversity and distribution patterns, it is often useful to select sites that are typical of well-defined vegetational or ecological categories such as old-growth northern coniferous forest, tundra or tropical deciduous forest. Ideally, development of international consortia of biologists will promote integrated surveys of particularly valued environments with benefits for all.

SITE ANALYSIS

Once a study site is chosen, a comprehensive analysis of within-site environmental variation to identify the full range of microclimates and niches is critical both for single-species surveys and complete inventories. Such information allows for an efficient and complete survey of the area and can ensure the comparability of results between sites. The site should be surveyed fully, with variation in physical features, such as elevation and aspect, accurately mapped. Temperature, precipitation, and insolation should be measured regularly, ideally over a number of years. Geologic features such as underlying rock type, soil acidity, composition, and water content must be recorded for the entire site. Special features such as rivers, caves, and rock outcrops should be noted as providing distinct microclimates that potentially may harbor different fungi, even if the host plants present are the same.

Even if a single-species survey is envisaged, information on the plant species present and their distributions will provide accurate characterization of the vegetation at the site. For a complete inventory, investigators must sample fungi associated with each plant species at the center of its within-site distribution as well as in isolated spots.

SAMPLING FREQUENCY

To determine sampling frequency, the investigator must know the climatic regime, including the timing of dormant periods during hot, cold, or dry spells, as well as which perennial host species are deciduous. Sampling can be less frequent (but not suspended) during dormant periods. Many fungal species in both temperate and tropical ecosystems sporulate seasonally. In temperate regions, that seasonality often is related to nutritional demands and requirements for spore dissemination. Most larger basidiomycetes fruit in the autumn, when the nutrient supply available from leaf fall is at its maximum. Many pathogenic microfungi, in contrast, produce spores in the spring when host tissues emerge from dormancy and have not yet developed extensive physical and chemical defenses against infection. In the tropics, sporulation can be associated with dry or wet seasons (Cornejo et al. 1994; Lodge and Cantrell 1995b).

Frequency of sampling may be reduced by extensive use of moist chambers, which allow samples collected at perhaps 2-month intervals to be observed continuously. Such an approach requires adequate laboratory space but is advantageous in reducing collection time and site dis-

turbance and in allowing for more extensive and objective successional studies. Succession is extended in many systems, so moist chamber observations must be continued over a considerable period. Sealing the ends of cut branches slows desiccation so that a wider variety of fungi, particularly endophytes adapted to colonize senescent bark, may be sampled (see Chapter 12).

FIELD OBSERVATIONS

Detailed field notes are an essential part of objective and reproducible sampling. Information to be recorded includes identity of the associated plant species, the precise plant part colonized, date of collection, location within the sample site, and microclimate (e.g., sun or shade, aspect). Prevailing weather conditions also should be recorded. Samples should be defined precisely and separated or cut into appropriate portions if needed to achieve internal homogeneity.

Notes on the population structure of the host and its position within the ecosystem also may be valuable; some fungi may be present only where there is a critical mass of the associated plant. An assessment of the general health of the host also may be important, although less so than with sampling protocols for biotrophic fungi inhabiting living plant tissue.

FIELD COLLECTIONS

Samples should consist, where practical, of several individual, defined plant parts with visible signs of fungal growth. Asymptomatic dead tissue probably can be ignored because additional species presumably will be sampled using endophyte techniques, although no studies have been carried out to confirm that. Multiple collections should be taken of tissues in various states of decay, especially when the dead tissues differ in appearance as a result of the presence of different fungi. Tissues affected by other organisms (e.g., beetles) should be treated as separate samples. Samples collected from fallen or from dead, attached plant parts should be separated.

Voucher material of the host should be gathered along with the fungal specimens if the identity of the host plant is not obvious. Unrecognizable hosts can be identified based on wood anatomy and by comparison with herbarium specimens, but such methods should be used only in the absence of an alternative.

An investigator sampling a given plant population needs to specify the part of the plant surveyed (e.g., leaf, petiole, stem, branch, trunk, wood, bark, root) and record its dimensions. In addition, its size should be taken into account. Fallen flowers, fruits, and seeds will have specific mycotas. Specialized plant structures, such as aerial roots, pneumatophores, and cladodes must be sampled separately using protocols adapted from those used for more standard plant parts. Some standardization of sample sizes may be appropriate, although this will depend on the fungal structures present and the heterogeneity of the plant material. Collectors should cut samples into suitable sizes for specimen packets or moist chambers on-site to avoid carrying extra material to the laboratory, but samples should be large enough to allow analysis and division into multiple specimens. The size of the organ collected also will influence decisions of sample size. Standardization of terminology for the degree of host-plant decay also may be valuable for sampling purposes, but generalization can be difficult because decay rates and fungal agents of decay vary among plant species.

Collectors should be trained to recognize mature fruiting material of at least the medium-size pyrenomycetous Ascomycota and coelomycetes to prevent the uncontrolled gathering of senescent specimens. Samples apparently containing only senescent and dead fungi still may be valuable, however, because small, secondarily saprobic species, difficult to observe in the field, can be found by microscopic inspection in the laboratory.

Collections are placed in paper bags or glassine packets (plastic bags result in an uncontrollable buildup of humidity). Tiny or fragile specimens are put or pinned into small boxes (e.g., those sold for fishing tackle) to protect them from physical damage. Collecting equipment includes a sharp knife, clippers (secateurs), small axe, wood chisel, folding saw, and a good-quality hand lens. *In situ* photography may be appropriate for larger specimens using a camera with a ring-flash attachment, but for microfungi in general this is best carried out in the laboratory with dedicated equipment.

LABORATORY ANALYSIS AND SPECIMEN PROCESSING

As soon as practical after returning from the field, the investigator should examine the material collected using a dissecting microscope for preliminary identification and culturing. Slides of fungal structures made with a range of mounting media also will be needed (Appendix II), along with materials for standard tests such as those for reactions to iodine. The choice of a mountant is part personal preference and part taxon dependent, but observations in aqueous preparations as well as the preparation of permanent mounts is essential. Baral (1992) vividly illustrated the benefits of water mounts, in which cells are turgid and gelatinous sheaths and appendages are more obvious. Material that is overmature should be dis-

carded, but immature collections should be placed in moist chambers to encourage further development. Once collections have been processed, representative samples should be dried and fumigated or frozen to kill animals such as mites and beetles, placed in labeled packets, and deposited in an herbarium. Specimen data (i.e., field notes, identification, laboratory observations) should be entered into a database from which packet labels can be produced. Air-drying is appropriate for most microfungi, but large and fragile specimens may benefit from freeze-drying. Some physical protection may be need to prevent crushing, especially of large hyphomycetes and long-necked ascomycetes but also for some small specimens in herbarium packets. Noting the appearance and precise position of fruit bodies on a specimen (e.g., in pale patches on both surfaces of the leaf laminae, along the abaxial surface of the midrib) and any apparent interactions with other organisms (e.g., saprobic on another fungus) is valuable, particularly for substrata such as large slowly decaying leaves, which may harbor a number of different fungal species. Such data are critical for type specimens to prevent unnecessary depletion of the original material by other mycologists.

We recommend that investigators photograph both whole fungi and relevant details of microscopic structures, especially for unknown or critical taxa. Line drawings made with a microscope and camera lucida are also valuable and ideally should be made in addition to photographs. The value of both photographs and drawings in characterization and subsequent recognition of species is obvious from the work of Matsushima (e.g., Matsushima 1995).

CULTURES

It may be necessary to culture many microfungi to characterize and elucidate their life histories properly and to confirm their identities when the morphologies of field collections are not optimal. Investigators should transfer spores to suitable media in culture plates using a sterile needle or, for particularly small fungi or sparse samples, a micromanipulator. With care, hyphal fragments can be treated similarly. Some pyrenomycetes with forcible spore discharge can be attached to the lids of the Petri dishes so that they eject their ascospores onto the media surfaces. Tap water agar is an effective culture medium for initial isolation and often for stimulating sporulation. Low-strength cornmeal-dextrose or potato-carrot agars are used widely (Appendix II). We recommend use of low-nutrient media and moderate incubation temperatures to discourage weedy mold contaminants from overgrowing new colonies before they can be subcultured. However, optimal temperatures, media, and germination times will depend on the kinds of fungi; in the absence of prior knowledge, investigators should approximate conditions at collection sites. Antibiotics discourage bacterial contamination, but anecdotal evidence suggests that some fungi are sensitive to such compounds, so they should not be used universally. Sterilized fragments of a substratum from which propagules have been isolated may be placed on a culture plate to provide both a physical structure on which colonies can develop and basic nutrients to which the species are adapted.

HANDLING DATA

The most intractable problem with laboratory analysis of collections, especially in tropical regions, is how to collect data on unknown taxa so that further collections can be grouped correctly. A sophisticated, well-planned, computerized database with well-defined fields that contain homogeneous information is essential. Drawings and photographs are useful for rapid confirmation of similarity and can be included within a database or stored in file folders. Two linked databases are required. A specimen database gives details of places of collection, associated organisms, field and laboratory notes, and (where known) an identification. A species database organizes information from the specimen database into standardized species accounts, which then provide descriptions. The two databases may be part of a formal relational structure (see "Table Relationships" in Chapter 4), or they may be linked using a descriptor field (typically the accepted name of the species) rather than a numeric value. The interface between the databases is particularly important, and automated transfer of data from one to the other presents a considerable challenge. Collection data also may be combined for statistical analysis of sampling, design, and data. Other databases may be appropriate also, covering such details as nomenclature and bibliography. Minter (1996) provided a comprehensive description of one computerized data-gathering system for fungal collections (see also Chapter 4).

SAMPLING MICROFUNGI ASSOCIATED WITH DEAD WOODY PLANT MATERIAL

CHOICE OF HOST SPECIES AND SUBSTRATA

The size and life form of the host species and the selected substratum components will affect significantly the number and profile of associated fungal species recov-

ered. Woody plants offer a much wider range of microhabitats to colonizing fungi than do herbaceous species. In addition, the long life span of a woody plant and its parts and its extended decay period promote the diversity of associated organisms. The ultimate choice of species may be largely personal or project driven, but ease of identification (both in living and decayed states), ecological role (dominance in the community), and conservation status all may play a role in the decision.

LIFE-FORM ANALYSIS

The variety of microhabitats provided by a woody plant must be analyzed prior to the sampling program. Dead plant parts are likely to include the following:

Leaves: fallen; buried; dead attached; suspended; dead parts of leaf laminae affected by necrotrophic fungi, action of other organisms, or physical damage; petioles and rachides; upper and lower lamina surfaces; leaf hairs, spines and glands; extrafloral nectaries

Twigs, stems, branches and trunks: fallen; buried; dead attached; suspended; killed by die-back or disease; bark surfaces; decorticated sections; cut or fractured surfaces; areas damaged by rodents, bark beetles, or other animals; natural cavities and bark fissures; spines or other special morphological features

Roots: cortex; exposed or covered by soil; physically damaged sections with decay caused by fungi or other organisms

Flowers: fallen, either complete or as component organs; attached but damaged as a result of other organisms; nectar; pollen

Fruits and seeds: fallen; buried; dead attached; decaying fleshy tissues; damage as a result of animal action; spines, wings; seeds decaying as a result of physical damage or action from other fungi; unfertilized ovules

The substratum descriptors are not strictly comparable, and the list is not exhaustive.

TEMPORAL AND SUCCESSIONAL ISSUES

The age at death and state of decomposition of the plants sampled must be taken into account. For some woody plants such as conifers, leaves and young stems are formed in a precise sequence, which allows their ages to be ascertained whether attached to a living tree or to fallen branches. For other trees, age of lignified parts can be estimated by counting annual rings or can be estimated roughly from diameter. Ideally, all substrata should be sampled from trees representing a range of ages at death and various degrees of decomposition. Speed of decay varies considerably depending on the plant part, environmental conditions, and the spectrum of accompanying organisms. In ideal circumstances, local rates of decay are determined before the sampling begins, so that fungi can be collected at regular intervals in the decay process under completely natural conditions. This is rarely possible, for practical reasons, but long-term observation of material in moist chambers offers a reasonably effective alternative.

SAMPLING RECOMMENDATIONS

We already have emphasized the importance of comparability of biodiversity sampling. Because many of the substrata described previously (see "Life-Form Analysis," earlier in this chapter) may be encountered infrequently, we recommend that a minimum set of substrata that is widely reproducible be sampled, even if those substrata do not yield maximal measurement of diversity. The minimum set will vary to some extent among plant life forms and major taxonomic groups—for example, between deciduous and evergreen trees, herbs and shrubs, and angiosperms and gymnosperms. For a deciduous oak species (*Quercus*), the minimum set might include:

Leaves: newly fallen, partly decayed (with the lamina having lost most of its strength), fully decayed (reduced to petiole and midrib)
Twigs (<20 mm in diameter): dead attached; recently fallen, with bark intact; well rotted with bark lost
Branches (>50 mm in diameter): dead attached; recently fallen, with bark intact; well rotted with bark lost
Fruits: recently fallen cupules, well-rotted cupules, recently fallen seeds, well-rotted seeds

A minimum of 20 collections of each of those categories of plant parts should be obtained from each of 10 trees that are widely spaced within the sample site, including young, old, isolated, and clustered individuals. The collections should be spread evenly throughout the available collecting seasons; for the oak species of the example, leaves and fruit would be sampled throughout autumn, winter, and early spring, whereas woody tissues could be collected over the entire year. Some compromise may be appropriate to restrict the total physical volume of the collections. Individual parts should be studied using a hand lens as they are collected and selected to maximize the range of physical appearances included, with special attention to evidence of external fungal growth.

In the laboratory, each individual plant part should be examined under a dissecting microscope. Samples of

apparently sporulating fungal material should be removed with a scalpel or wetted needle and mounted on glass slides in water and lactofuchsin before examination using a compound microscope with oil immersion and, ideally, Nomarski optics. If recognizable sporulating structures are present, identification can proceed. Multiple voucher collections of each species should be preserved along with permanent microscope mounts. Ideally, cultures should be prepared by transferring spores or hyphal fragments to culture plates. We recommend use of tap water agar with sterilized pieces of the plant being studied and half-strength cornmeal or potato-carrot agar (Appendix II).

A portion of at least 10 collections of each component of the minimum sample set should be placed in a moist chamber (Appendix I) and observed at regular intervals (weekly) until the plant tissue has decayed completely or no additional fungal species are recorded. That process may take 6–9 months, even with leaves. It will provide information on succession and the occurrence of additional species that sporulate at intermediate stages of the decay cycle.

An additional portion of at least five collections of each component of the minimum sample set should be cultured using the particle filtration technique described by Bills and Polishook (1994) and Bills and colleagues in Chapter 13 (see "Particle Filtration," in the Appendix) after being washed thoroughly to remove surface contaminants (Tokumasu et al. 1997). Investigators should excise the rapidly growing colonies from all but one or two isolation plates with a sterile scalpel or kill rapidly germinating colonies with a soldering iron so that more slowly growing species can develop.

In areas where the fungal biota is poorly known (e.g., in tropical regions), each individual strain should be described and recorded and features of each recognized species (whether named or not) should be drawn and photographed. Such records are helpful for subsequent identifications and document the sampling process. A computerized database (see Chapter 4) will facilitate use of the information.

COMPLETE SURVEY OF A SMALL WOOD

A complete survey of a small area largely involves repetition of the process described for a single plant species, although for practical reasons, the numbers of samples studied in culture rather than by moist chamber observation is likely to be substantially reduced. The culturing process (if carried out carefully) is used to isolate species from one another, thereby eliminating the competitive interactions operating among species in moist chamber environments. As a result, dormant propagules dispersed onto a host plant from surrounding vegetation that might not be expected to develop in the presence of species well adapted to the host material will grow. Once the initial site survey is complete, fungi on dead tissues of at least one representative of each plant genus present should be studied, taking special notice of overall dominant taxa and those occupying restricted environmental niches. That process inevitably will be protracted and result in considerable duplication of species collected as a result of the absence of detailed information on host specificity for most fungal groups. In the absence of information from well-documented inventories of similar ecosystems, it is valuable to plot rarefaction curves (see "Richness" under "Quantitative Indices," Chapter 5) in the course of the sampling process to assess its efficiency.

ACKNOWLEDGMENTS. We gratefully acknowledge valuable comments on and contributions to various versions of this manuscript made by Gerald Bills, David Hawksworth, Alick Henrici, Sabine Huhndorf, Paul Kirk, Henning Knudsen, Jan Kohlmeyer, Jean Lodge, and Brian Spooner.

12

ENDOPHYTIC FUNGI

JEFFREY K. STONE, JON D. POLISHOOK, AND JAMES F. WHITE, JR.

Higher plants furnish complex, multilayered, spatially and temporally diverse habitats that support species-rich assemblages of microorganisms. Microfungi are dominant components of those assemblages, colonizing foliar and twig surfaces (epiphytes), internal tissues of foliage (foliar endophytes), young and old bark (bark endophytes), and wood (xylem endophytes and wood decomposers). Increasing interest in cryptic occupation of internal tissues of healthy plants by endophytic microfungi has led to a growing awareness that higher plants likely harbor a reservoir of undiscovered fungi.

DEFINITION AND CIRCUMSCRIPTION

During the past 30 years the terms *endophyte* and *endophytic fungi* have appeared frequently in the mycological literature to describe the internal mycota of living plants. Although the origin of the terms can be traced back to the nineteenth century, their contemporary meaning is different from the original one (Large 1940; Carroll 1986). The terms often are combined with modifiers to refer to a specific host type, a taxonomic group of hosts, or the type of tissue occupied (e.g., systemic grass endophytes, bark endophytes). Contemporary applications of the terms are not always consistent nor are they accepted by all investigators (Petrini 1991; Wennström 1994; Wilson 1995b; Saikkonen et al. 1998; Stone et al. 2000). In general, however, the terms apply to fungi capable of symptomless occupation of apparently healthy plant tissue. In the broadest sense, endophytic fungi are fungi that colonize living plant tissue without causing any immediate, overt negative effects (Hirsch and Braun 1992). This definition includes virtually the entire spectrum of symbiotic interactions in which fungi and plants participate: parasitism, commensalism, and mutualism.

For grass hosts (primarily Poaceae), the word *endophyte* has been used to denote a particular type of systemic, nonpathogenic symbiosis. Grass endophytes provide their hosts with a number of benefits, such as protection against herbivory and pathogens, that increase their fitness (reviewed by Clay 1988, 1990, 1994; Saikkonen et al. 1998). Taxonomically these fungi are primarily *Neotyphodium* anamorphs of Balansiae (Clavicipitaceae); they colonize leaf, culm, and root tissues of species of cool-season grasses extensively and are transmitted in their hosts' seeds. Sporulation on the host is suppressed completely, and host and fungus function together essentially as a single organism. These symptomless endophytes of *Lolium, Festuca,* and other genera of pooid grasses are interspecific hybrid strains derived from *Epichloë* species that cause partial or complete host sterility (choke disease; Schardl et al. 1994; Tsai et al. 1994; Moon et al. 2000).

BIOLOGY AND ECOLOGY

Many of the fungi commonly reported as endophytes are regarded as minor or secondary pathogens by forest pathologists. Their common occurrences in both healthy and diseased tissues underscore the uncertainty of boundaries separating endophytes, facultative pathogens, and latent pathogens. Indeed, the behavioral differences between many fungi considered as "endophytic" and those considered to be "latent pathogens" are slight and simply may reflect differences in the duration of the latent or quiescent phase and the degree of injury sustained by the host during active growth of the fungus. Pathogenic fungi capable of symptomless occupation of their hosts during a portion of the infection cycle, "quiescent infections" (Williamson 1994), and strains with impaired virulence can be considered endophytes (Schardl et al. 1991, 1994; Fisher et al. 1992; Fisher and Petrini 1992; Freeman and Rodriguez 1993), as can a variety of commensal saprobic and mutualistic fungi that have cryptic, nonapparent patterns of host colonization. Fungi described as "endophytic" characteristically exhibit a prolonged, inconspicuous period in which growth and colonization cease temporarily, resuming after a physical, or maturational, change in the host. This episodic growth is a defining feature of endophytes, whether they ultimately are considered commensal saprobes, latent pathogens, or protective mutualists. Although such a definition may seem too broad, most fungal biologists agree that the species composition of the internal mycobiota is distinct for various hosts, organs, and tissues although some species of endophytic infections also may be found in the epiphytic or rhizosphere mycobiota.

DISTRIBUTION

Fungal surveys of various hosts during the past 20 years have demonstrated that endophytic colonization of land plants by fungi is ubiquitous. Endophytes are known from plants growing in tropical, temperate, and boreal forests; from herbaceous plants from various habitats, including extreme arctic, alpine (Petrini 1987; Fisher et al. 1995), and xeric environments (Mushin and Booth 1987; Mushin et al. 1989); and from mesic temperate and tropical forests. Endophytic fungi occur in mosses and hepatics (Döbbler 1979; Pocock and Duckett 1985a; Ligrone et al. 1993), ferns and fern allies (Fisher et al. 1992; Schmid and Oberwinkler 1993), numerous angiosperms and gymnosperms, including tropical palms (Rodrigues and Samuels 1992; Fröhlich and Hyde 2000; Hyde et al. 2000), broad-leaved trees (Arrhenius and Langenheim 1986; Lodge et al. 1995), the estuarine plants *Salicornia perennis* (Petrini and Fisher 1986), *Spartina alterniflora* (Gessner 1977), and *Suada fruticosa* (Fisher and Petrini 1987), diverse herbaceous annuals, and many deciduous and evergreen perennials (Table 12.1). Larger woody perennials also may support parasites such as mistletoes and dodders and complex assemblages of epiphytic plants, which in turn may harbor endophytic fungi (Dreyfuss and Petrini 1984; Petrini et al. 1990; Richardson and Currah 1995; Suryanarayanan et al. 2000). Detailed investigations of the internal mycobiota of plants frequently uncover novel taxa and reveal new distributions of known species. Because endophytic infections are inconspicuous, the species diversity of the internal mycobiota is relatively high (both within and among individual host species), and a relatively small proportion of potential hosts have been examined, endophytes may represent a substantial number of undiscovered fungi (Stone et al. 1996; Arnold et al. 2000). Investigation of even well-characterized, economically important plants for endophytic fungi frequently yields novel taxa. Studies of endophytic fungi are needed to provide information fundamental for evaluating global fungal diversity and distribution.

Endophytic microfungi may be diverse at an exceedingly small scale; a single conifer needle may harbor several dozen species. Endophytic microfungi typically are present as internal, unseen, microscopic hyphae; their presence is revealed externally only when they sporulate, usually a seasonal and ephemeral event. Many endophytes are highly host- or tissue-specific. Conventional methods for sampling fungi are inadequate for accurately enumerating microfungi, and the details of distributions of even the most familiar taxa remain sketchy. Detection and quantification generally require selective isolation procedures.

Identification usually involves microscopic examination of host tissue and often requires a high degree of taxonomic expertise. That is especially true for isolates in pure culture that fail to produce spores or identifiable structures; determination of growth conditions that induce sporulation is very important. Fungi that neither grow nor sporulate in culture must be detected and identified by other means, such as comparisons of ribosomal DNA (rDNA) gene sequences, which also can elucidate phylogenetic position (Guo et al. 2000). The absence or deficiency of basic taxonomic information is a major obstacle to ecological studies of endophytic fungi. The problem can be overcome partially by integrating existing databases (host indices, nomenclature), but fundamental biological survey work also is needed.

ECOLOGICAL ROLES

The ecological roles played by endophytic fungi are diverse and varied (Saikkonen et al. 1998). Endophytes have been described as mutualists that protect both grasses (Clay 1990) and conifers (Carroll 1991) against insect herbivory, and many of those fungi produce biologically active secondary metabolites (Fisher et al. 1984a; Polishook et al. 1993; Peláez et al. 1998). Fisher and colleagues (1984b) reported antibacterial or antifungal activity for more than 30% of the endophytic isolates from ericaceous plants, and Dreyfuss (1986) reported antibiotic activity from isolates of the endophytic *Pleurophomopsis* species and *Cryptosporiopsis* species, as well as from a sterile endophyte from *Abies alba*. Strains of the endophytic *Pezicula* species (and its anamorph *Cryptosporiopsis*) from several deciduous and coniferous tree hosts produce an ensemble of bioactive secondary metabolites in culture (Fisher et al. 1984a; Noble et al. 1991; Schulz et al. 1995). Endophytic species of the Xylariaceae frequently produce compounds with high biological activity, including cytochalasins (Dreyfuss 1986; Brunner and Petrini 1992) and indole diterpenes (Hensens et al. 1999). Although diverse endophytes produce toxins in culture, such compounds have been difficult to detect in plant host tissue.

Nongrass endophytes produce antifungal (Peláez et al. 2000) or antibacterial substances, as well as insecticidal compounds (Johnson and Whitney 1994; Hensens et al. 1999), *in vitro*. We do not know, however, whether these metabolites are produced (1) in plants during the period of quiescent occupation of host tissue by the endophytes or (2) in sufficient concentrations to benefit the host in a protective mutualism (e.g., by deterring insect herbivory). *In vitro,* many of those compounds are intracellular and so, although the substances may have survival value for the endophyte (e.g., through interference competition), their general role (if any) in protection of living hosts has not yet been determined (Saikkonen et al. 1998).

PROTECTIVE MUTUALISTS OR SAPROBIC COMMENSALS?

Although the systemic, clavicipitaceous grass endophytes and the nonsystemic fungi of grasses and other hosts both are considered endophytes, they differ in important ways and should not be regarded as biologically or ecologically homologous (Table 12.2). Much has been published on the highly specific nature of grass-endophyte symbiosis, the effects of fungal alkaloids in infected hosts

TABLE 12.1
Examples of Endophytic Mycobiota in Various Host Plants Worldwide

Host	Tissue or organ	No. of species	Notes	Location	Reference
Abies alba	Branch bases	44	17 common, 2 endemic	Germany, Poland	Kowalski and Kehr 1992
A. alba	Twigs	50		Switzerland	Sieber 1989
A. alba	Needles	120	13 common	Switzerland	Sieber-Canavesi and Sieber 1993
Acer macrophyllum	Leaves, twigs	9		British Columbia	Sieber and Dorworth 1994
A. pseudoplatanus	Branch bases	28	16 common, 5 endemic	Germany, Poland	Kowalski and Kehr 1992
A. pseudoplatanus	Leaves	22		Germany	Pehl and Butin 1994
A. spicatum	Roots	7	Aquatic hyphomycetes	Nova Scotia	Sridhar and Bärlocher 1992a, 1992b
Alnus glutinosa	Branch bases	24	17 common, 3 endemic, 2 new species	Germany, Poland	Kowalski and Kehr 1992
A. glutinosa	Aquatic roots	46	14 common, 12 aquatic hyphomycetes	United Kingdom	Fisher et al. 1991
A. rubra	Leaves	25	12 common	British Columbia	Sieber et al. 1991
A. rubra	Twigs	27	13 common	British Columbia	Sieber et al. 1991
Arctostaphylos uva-ursi	Leaves	176	23 common	Switzerland	Widler and Müller 1984
A. uva-ursi	Twigs	35	29 common	Switzerland	Widler and Müller 1984
A. uva-ursi	Roots	14	8 common	Switzerland	Widler and Müller 1984
Betula pendula	Branch bases	23	14 common	Germany, Poland	Kowalski and Kehr 1992
Carpinus caroliniana	Bark	155	11–12 species/tree, 5 Basidiomycetes	New Jersey, West Virginia	Bills and Polishook 1991a
Calocedrus decurrens	Foliage	15		Oregon	Petrini and Carroll 1981
Chaemacyparis lawsoniana	Foliage	18	1 Basidiomycete	Oregon	Petrini and Carroll 1981
C. thyoides	Leaves, twigs	88	8–12 species/tree	New Jersey	Bills and Polishook 1992
Cuscuta reflexa	Stems	45		India	Suryanarayanan et al. 2000
Dryas octopelata	Leaves	4		Spitsbergen	Fisher et al. 1995
D. octopelata	Leaves	23		Switzerland	Fisher et al. 1995
Eucalyptus globulus	Stems	41	9 Basidiomycetes	Uruguay	Bettucci and Saravay 1993
Euterpe oleracea	Leaves	62	21 common	Brazil	Rodrigues 1994
Fagus sylvatica	Branches	18		United Kingdom	Chapela and Boddy 1988b
Gaultheria shallon	Leaves	13		Oregon	Petrini et al. 1982
Heisteria concinna	Leaves	242		Panama	Arnold et al. 2000
Hordeum vulgare	Leaves	14		New Zealand	Riesen and Close 1987
Juncus bufonius	Leaves	14		Oregon	Cabral et al. 1993
Juniperus communis	Leaves	114		Switzerland	Petrini and Müller 1979
Licuala ramsayi	Leaves	11		Australia	Rodrigues and Samuels 1992
Livistona chinensis	Fronds	45		Hong Kong	Guo et al. 2000
Manilkara bidentata	Leaves	23		Puerto Rico	Lodge et al. 1996a
Musa acuminata	Leaves	24		Hong Kong, Australia	Brown et al. 1998
Opuntia stricta	Stems	23		Australia	Fisher et al. 1994
Oryza sativa	Leaves, roots	30		Italy	Fisher and Petrini 1992
Ouratea lucens	Leaves	259		Panama	Arnold et al. 2000
Picea abies	Twigs	85		Sweden	Barklund and Kowalski 1996
P. mariana	Roots	97		Ontario	Summerbell 1989
Pinus densiflora	Needles	9		Japan	Hata and Futai 1995
Pteridium aquilinum	Roots, stems, leaves	61	6 common	United Kingdom	Petrini et al. 1992a
Quercus ilex	Twigs, leaves	149	10 dominant species	Spain	Collado et al. 2000
Salicornia perennis	Stems	31		United Kingdom	Petrini and Fisher 1986
Sequoia sempervirens	Leaves	26		California	Espinosa-Garcia and Langenheim 1990
Tilia cordata	Leaves	17		Germany	Pehl and Butin 1994
Vitis vinifera	Leaves, stems	46		South Africa	Moustert et al. 2000
Zea mays	leaves, stems	23		United Kingdom	Fisher et al. 1992

TABLE 12.2
Comparison of Characteristics of Endophytes Occurring in Grass and Nongrass Hosts

Endophytes of grass hosts	Endophytes of nongrass hosts
Few species, Clavicipitaceae	Many species, taxonomically diverse
Extensive internal colonization	Restricted internal colonization
Occurring in several host species	Most species with limited host species
Systemic, seed transmitted	Nonsystemic, spore transmitted
Host colonized by only one species	Hosts infected by several species concurrently

on vertebrate and invertebrate herbivores, and on drought tolerance, and on the apparently greater vigor of endophyte-infected grasses compared with noninfected ones. Pervasive systemic colonization of host tissue with endophyte hyphae ensures that herbivores, whether large mammals or small arthropods, will encounter fungal metabolites in their meal. Host colonization by foliar endophytes of nongrass hosts, however, is generally nonsystemic, limited, and disjunct. The latter fungi are apparently physiologically quiescent during the lives of both deciduous and evergreen host tissues and generally are found in greater abundance in older tissues. Young foliage is generally less heavily colonized. Level of consumption of fungal metabolites by herbivores of endophyte-infected nongrass hosts, therefore, may be relatively unpredictable. Endophytes of nongrass hosts also represent a broader range of taxa, mainly from several orders and families of Ascomycetes or anamorph genera but also from some Basidiomycete families. Several species and/or genera often infect the same host tissue concurrently.

The occupation of host tissue prior to either natural senescence or induced necrosis gives endophytes an advantage over saprobes normally excluded from healthy tissue. Endophytes that are quiescent during the normal lifespan of deciduous host organs immediately can intercept and use host metabolites mobilized during early senescence (Chapela and Boddy 1988b; Griffith and Boddy 1988; Boddy and Griffith 1989). Competitive interactions (especially interference competition or denial of access to the resource) with later-invading saprobic fungi may account for the apparently widespread production of antagonistic metabolites by endophytic fungi. If so, such compounds would be of competitive value primarily to the endophytes and of minimal value to the host as a basis for a protective mutualism. Other metabolites produced by some endophytes modulate host growth responses, accelerate or delay senescence (Petrini et al. 1992a; Saikkonen et al. 1998), or act as pathogens (Desjardins and Hohn 1997). Future

investigations might include studies aimed at detecting the production of antibiotics and pest deterrents in plants as a first step toward evaluating the ecological significance of secondary metabolite production by endophytes and the potential use of endophytes in biological control. Identification of specific physiological cues that promote or modulate synthesis of antagonistic substances in endophytic fungi, which often involve coordination of biosynthetic pathways, is also of fundamental interest (Desjardins and Hohn 1997).

Species of endophytes inhabiting leaf and stem tissue in the canopy of coniferous forests also can be isolated during the early stages of litter decomposition (Kendrick and Burgess 1962; Mitchell et al. 1978; Minter 1981; Stone 1987; Aoki et al. 1990; Sieber-Canavesi and Sieber 1993; Tokumasu et al. 1994). The behavior of endophytes from initial infection of young foliage through their decomposition in forest litter has been examined in successional studies of deciduous (Wildman and Parkinson 1979; Pehl and Butin 1994) and evergreen (Ruscoe 1971; Sieber-Canavesi and Sieber 1993) hosts. Many common endophytic fungi represent the earliest fungi to colonize tissue as latent invaders. They grow and sporulate rapidly in response to senescence and can be isolated from litter in the early stages of decomposition, but they gradually are replaced by saprobic fungi more typical of decomposer assemblages (Stone 1987; Tokumasu et al. 1994) in the forest litter.

LATENT, QUIESCENT PATHOGENS

Fungal pathogens of particular hosts also commonly are isolated as endophytes. Such fungi are not usually among the most abundant isolates from apparently healthy tissue of a given host; they are, however, consistent and recurrent components of a characteristic host mycobiota. Typical pathogens include anthracnoses, such as *Apiognomonia venita* on *Platanus* species, *Apiognomonia errabunda* on *Fagus* species, and *Colletotrichum* species on numerous hosts. The causal agent of "Dutch Elm" disease, apparently the normal (virulent) strain of *Cryphonectria parasitica*, was isolated from a small proportion of *Castanea sativa* coppice shoots in Switzerland (Bissegger and Sieber 1994). The canker pathogens *Melanconis alni* and *Diplodina acerina* were minor components of the twig mycobiota of their respective hosts, *Alnus rubra* and *Acer grandifolia*, in British Columbia, and several leaf-spot pathogens (e.g., *Septoria alni*) were present in foliage (Sieber et al. 1991). Conifer needle pathogens, such as *Cyclaneusma minus*, *Lophodermium seditiosum*, and *Rhizosphaera kalkoffii*, recurrently are found in asymptomatic foliage of coniferous hosts in Europe and North America (Carroll and Carroll 1978;

Sieber 1989; Franz et al. 1993). *Fusarium* species, many of which are associated with wilt diseases, cankers, or root diseases, are frequent, but seldom dominant, components of the endophyte biota. The presence of weakly phytopathogenic fungi in healthy tissues emphasizes the heterogeneous ecology of endophyte associations and the evolutionary continuum between latent pathogens and symptomless endophytes (Saikkonen et al. 1998).

ENDOPHYTIC, EPIPHYTIC, CAULOPLANE, AND RHIZOSPHERE FUNGI

Most studies of endophytes have dealt with infections occurring in natural host populations. Although higher plants have evolved a variety of general resistance mechanisms that prevent infection by most opportunistic fungi, endemic symbiotic fungi, including endophytes, have coevolved with their hosts and adapted to them. Adaptations include methods for host recognition, means of overcoming the complement of host defenses, mechanisms for host-specific attachment, host-induced spore germination, and diversification of infection structures (Stone et al. 1994). The fungi are largely unaffected by anthropogenic selection.

The frequent occurrence of species typical of plant surfaces as internal fungi (Cabral 1985; Fisher and Petrini 1987; Legault et al. 1989; Cabral et al. 1993) suggests that host barriers are not completely effective; however, the interface between the external surface and internal tissue of a plant is not always clearly delimited. Epiphytic fungi are generally much less common in internal tissue than in external tissue. Conversely, those endophytes represented by high proportions of isolates apparently are adapted in varying degrees to overcome general host barriers to infection and establish internal symbioses with their hosts but are absent or infrequent epiphytes. Guilds of endophytic colonists can contain species that are shared in common with epiphytic or rhizosphere assemblages, but these tend to be comparatively infrequent. A few fungus species, infrequent or absent from the epiphyte or rhizosphere assemblages, tend to be the dominant endophytic colonists for a given host.

Many of the fungi most commonly isolated as endophytes are considered typical epiphytic saprobes (Fisher and Petrini 1992; Cabral et al. 1993). *Hormonema dematioides*, for example, is a dominant epiphytic colonist of foliar surfaces but is regularly isolated as an internal colonist as well (Legault et al. 1989). Similarly, *Alternaria alternata* and *Cladosporium cladosporioides* are ubiquitous epiphytes but also are capable of internal colonization of healthy tissue (O'Donnell and Dickinson 1980; Cabral et al. 1993). Soil fungi rarely are found in foliage but are common colonists of the cauloplane

(Cotter and Blanchard 1982; Bills and Polishook 1991) and are among the most common fungi isolated from roots (Fisher et al. 1991; Holdenrieder and Sieber 1992). Ascomycetous coprophilous fungi, mainly Sordariaceae, are isolated consistently, but with low frequency, from leaves and stems of woody plants (Petrini 1986). Those fungi often possess ascospores with thickened cell walls and gelatinous sheaths or appendages; the ascomata are adapted to launch their spores onto the cauloplane or phylloplane. Zygomycetes and Basidiomycetes tend to be poorly represented in endophyte inventories. The generally low proportion of Basidiomycetes may reflect sampling bias. Endophytic Basidiomycetes have been reported from tree bark and sapwood (Chapela and Boddy 1988a; Griffith and Boddy 1988; Bills and Polishook 1991) and from foliage samples.

METHODS

OBJECTIVES OF ENDOPHYTE RESEARCH

Methods for studying patterns of infection and colonization by endophytic fungi are essentially the same as those used in the study of fungal plant pathogens (Stone et al. 1994). Investigations of endophytic fungi, however, emphasize the autecology, synecology, and biodiversity of fungi infecting hosts in natural environments (Hirsch and Braun 1992; Carroll 1995). In mycobiotic surveys, host tissues are sampled methodically, and the spatial and temporal distributions of the fungal colonists encountered are described using methods adapted from vegetation ecology (Hirsch and Braun 1992).

Ecological studies emphasize patterns of the mycobiota, of host genera and families, or of specific habitat types (Petrini and Carroll 1981; Petrini et al. 1982; Petrini 1985) or the distribution of fungal taxa as endophytes (Petrini and Petrini 1985; Petrini 1986). Infection frequencies for specific hosts have been related to foliage age (Stone 1987; Espinosa-Garcia and Langenheim 1990), host distribution (Petrini 1991; Bills and Polishook 1992; Rollinger and Langenheim 1993), and temporal and spatial variation in patterns of endophyte infections (Wilson and Carroll 1994). Several investigators have studied the role of endophytic fungi in complex symbioses involving hosts, fungi, and insects (Todd 1988; Butin 1992; Wilson and Carroll 1997; Bultman and Conrad 1998; Raps and Vidal 1998; Omacini et al. 2001; Wilson and Faeth 2001). Others have studied ecological factors affecting distribution patterns among endophytes (Petrini 1991; Rodrigues 1994; Sieber and Dorworth 1994; Hata and Futai

1996; Schulthess and Faeth 1998; Elamo et al. 1999; Sahashi et al. 1999) and the influence of anthropogenic factors on endophytic assemblages (Petrini 1991; Helander et al. 1994; Ranta et al. 1995).

ISOLATION AND CULTURE

The method most commonly used to detect and quantify endophytic fungi is isolation from surface-sterilized host tissue. For inventories of species occurrences and diversity, that is presently the most practical approach, although fungal biologists recognize that certain groups (e.g., obligate biotrophs) may be undetected or under-represented and that isolates failing to sporulate in culture may need to be characterized by other means. Detection of organisms from natural substrata and their identification are influenced by the sampling procedures, isolation methods, composition of the culture media, and physiological adaptations of the fungi. In some cases, such problems can be resolved by comparing cultures obtained from tissue isolations with those from sporulating states on the host (e.g., Bills and Peláez 1996). Another method for identification is molecular taxonomy (see "Molecular Sequence Approaches," later in this chapter).

Investigations of colonization patterns and "fungal community structure" based solely on isolation data must be interpreted carefully. Dimensions of sampling units are critical given the microscopic scale of the fungal distributions. For analyses of species dominance and diversity, an investigator must know the relative proportions of individuals present. Isolation methods may provide an approximation of this relationship, but direct microscopic examination of endophyte infections often reveals that many more individual infections are present than can be detected using manageable tissue segment sizes (e.g., Stone 1987). Similarly, serial dissection and plating of host material gives only approximate information about host colonization patterns; it may be impossible to differentiate between systemic colonization of contiguous tissue by a single infection or multiple infections by the same species where the domain of infection is very small in relation to the size of the sample unit (Stone 1987; Carroll 1995). Direct microscopy also may show that internal host tissue is not always colonized by all fungi isolated from surface sterilized tissue (Viret and Petrini 1994; J. K. Stone, unpublished data). Ideally, observations from direct examination of infected tissue should be used to confirm patterns detected by surface sterilization and pure culture (Cabral et al. 1993). Detection and enumeration methods based on biochemical approaches offer promise, but currently no such methods are practical for large-scale surveys.

SAMPLING CONSIDERATIONS

Host species, host-endophyte interactions, interspecific and intraspecific interactions of endophytes, tissue types and ages, geographic and habitat distributions, types of fungal colonization, culture conditions, surface sterilants, and selective media all influence the efficiency of a sampling strategy for detection and enumeration of endophytic fungi. Bacon (1990) and Bacon and White (1994) have reviewed the techniques and materials used for isolation, maintenance, identification, and preservation of grass endophytes. Petrini (1986) and Schulz and colleagues (1993) have compared the efficacy of several surface-sterilization procedures on various host plants and organs. Much practical information on methods for isolation of filamentous fungi from natural substrata, including techniques, selective agents, and common media, can be found in Bacon and White (1994), Bills (1996), C. Booth (1971), and Seifert (1990).

HOST COLONIZATION PATTERNS: SYSTEMIC VERSUS LIMITED DOMAINS

The infection domain of endophytes has a profound effect on sampling efficiency for species diversity. Clavicipitaceous endophytes of grasses form systemic associations with their hosts; their fungal hyphae colonize virtually all plant tissues and are found both in the seed coat and in close association with the embryo in certain species. Nonsystemic infections of "P-endophytes" of grasses, mainly *Phialophora* species and *Gliocladium* species, are more limited but also can be seed-borne (An et al. 1993). There also are scattered reports of systemic, seed-borne endophytes in nongrass hosts. Bose (1947) reported that hyphae of *Phomopsis casuarinae* permeated the tissues, including the seed coat of every *Casuarina equisetifolia* plant he examined. Boursnell (1950) documented an unidentified systemic fungus in *Helianthemum chamaecistus*, and Rayner (1915, 1929) found unidentified fungi infecting Ericaceae. Histological studies detailing endophyte infection patterns of endophytes that colonize mostly non-grass hosts are available for only a few host species (Stone 1987; Suske and Acker 1987; Cabral et al. 1993; Viret and Petrini 1994). In those cases, however, the domain of the endophyte colonization in healthy tissue often is restricted, usually limited to no more than a few cells (Figs. 12.1 to 12.4, *Rhabdocline parkeri* infections and *Phyllosticta* infections). The differences between systemic infections and those of limited domain dictate that sampling strategies take patterns of host colonization into account if recovery of greater diversity of species or if

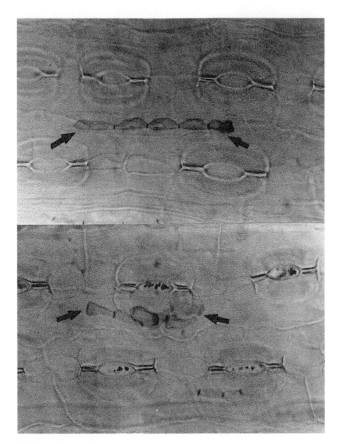

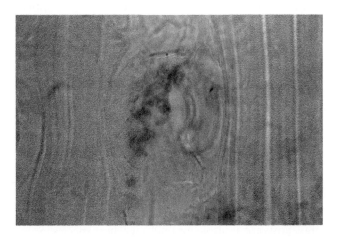

FIGURE 12.1 Intracellular *Rhabdocline parkeri* hyphae (*arrows*) in Douglas fir (*Pseudotsuga taxifolia*) needles (×500).

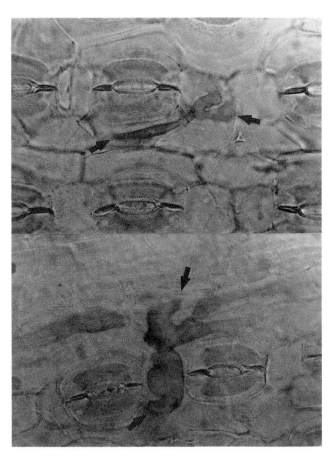

FIGURE 12.2 Intracellular *Phyllosticta abietis* hyphae (*arrows*) in Giant fir (*Abies grandis*) needles (×500).

FIGURE 12.3 Hypha of an unidentified endophyte in epidermal cells of *Picea pungens*. Needles were cleared in 10% KOH and stained with 0.05% trypan blue in lactoglycerol.

FIGURE 12.4 Hypha of *Stagonospora innumerosa* in an epidermal cell of *Juncus effuses* var *pacificus*. The epidermis was excised with a razor blade, cleared by boiling in lactophenol-ethanol (1:2 v/v), and stained in acid fuchsin-malachite green (Cabral et al. 1993).

precise estimation of relative species importance in specific tissues or organs is the objective. Where sample units are not appropriate to the microscopic scale of infections, undue bias will be introduced. Unfortunately, in the majority of published studies selection of sample units

was apparently arbitrary and is highly variable (Carroll 1995); inferences regarding species dominance and diversity drawn from those may be suspect as a consequence.

MICRODISSECTION

Distribution patterns of fungi in host tissue also can be investigated using microdissection and culture. Recovery of species from microdissected tissue of a few hosts also has revealed disjunct, discrete patterns of fungal occupation occurring on a minute scale. Bissegger and Sieber (1994) divided 1-cm × 1.5-cm segments of phloem tissue into 25 2-mm × 3-mm units and recorded the pattern of fungal growth from each. A mosaic of occupation patterns of eight endophyte species was obtained that revealed discontinuous distribution of the individuals within the sample. Multiple infection of some segments suggests that even smaller segments would have revealed greater heterogeneity. Lodge and colleagues (1996a) used a similar procedure to isolate endophytes from leaves of a tropical broad-leaved tree, *Manilkara bidentata*. Patterns of colonization of 5-mm × 20-mm leaf panels cut into 1-mm × 2-mm fragments were highly heterogeneous, with apparently noncontiguous distributions of the 28 taxa recovered. Of the 28 taxa, 21 were found on two of the three leaves sampled, and some panels contained up to 15 taxa. G. C. Carroll and colleagues (unpublished data) dissected and cultured needles of Douglas fir (*Pseudotsuga menziesii*) on an even finer scale and compared the effects of sample unit dimensions on infection frequency. Not surprisingly, incidence of infection decreases precipitously with smaller sample units (Carroll 1995). Results of such microdissection experiments agree with the histological measurements of infection density of *Rhabdocline parkeri* in Douglas fir, which varied from 0.2 to more than 30 infections per mm^2 (Stone 1987). Experiments such as that of Bissegger and Sieber (1994) and G. C. Carroll and associates (unpublished data) suggest that maceration of host tissue and serial dilution plating, as described by Bills and Polishook (1994) to process leaf litter samples, may yield more accurate estimates of fungal infection frequencies.

GENERAL GUIDELINES

Some general guidelines regarding protocols for sampling endophytes are as follows:

- The smaller the sampling unit, the greater the recovery of diverse species/genotypes. Also, conversely, the larger the sampling unit, the greater the potential to miss rare or slow-growing species and to recover mixed genotypes of the same species.
- Older foliage is likely to harbor greater species diversity than younger foliage. Perennial species thus can be expected to harbor greater diversity than annuals, and plants with evergreen foliage are likely to harbor more diversity than deciduous or annual plants.
- The relative constancy of the mycobiota of a host species over its geographic range and across age classes suggests that sampling many different hosts species in one area is a more time- and cost-effective way to survey endophytes than extensively sampling one host species throughout its range. Similarly, sampling older host foliage (i.e., foliage with the greatest endophyte species diversity) will result in the greatest recovery of species.
- The greatest diversity of fungi probably can be recovered with intensive selective sampling of a limited amount of host tissue from individuals growing on ecologically varied sites and in different community associations. Varying the culture conditions; segment size used, including size of the host tissue; and composition of the medium also will enhance the variety of fungal groups isolated and enumerated from a limited sample.
- Frequently a host will harbor one to several endophyte species that are unique to that particular host. Thus, biodiversity of endophytic fungi also can be a function of the number of different hosts species sampled.

SAMPLE COLLECTION AND STORAGE

Rapid changes in endophyte colonization probably do not occur immediately following collection. Nevertheless, it is important that samples be handled carefully and processed as quickly as possible following collection, usually within 48 hours. Samples should be air-dried to remove any surface moisture before transport or storage. During transport, samples should be kept cool and dry. Cotton, Tyvek, or paper collecting bags or paper envelopes are preferred for holding samples. We discourage the use of plastic bags for holding samples, but if plastic bags are used, they should be left open for air circulation to prevent condensation and the growth of superficial molds.

SURFACE STERILIZATION AND CULTURE PROTOCOLS

Size of the sampling unit and surface sterilization procedures vary according to the preferences of the investigator, the species of host plant, and host tissue type sampled. Some investigators have compared carefully the effects of surface-sterilization procedures (Petrini 1992; Schulz et al. 1993; Bissegger and Sieber 1994), isolation medium (Bills and Polishook 1991), and sample-unit size (Carroll 1995) on isolation frequencies. We recommend that investigators experiment with those factors prior to initiating detailed investigations so that protocols optimal for recovery of endophytes from particular host species or specific organs and tissues can be developed. For root tissues, serial washing may be preferable to surface sterilization to obtain representative frequencies of fungal colonists (Summerbell 1989; Holdenrieder and Sieber 1992).

Surface sterilization of plant material usually entails treating the plant material with a strong oxidant or general disinfectant for a brief period, followed by a sterile rinse to remove residual sterilant (Table 12.3). Household chlorine bleach (NaOCl), usually diluted in water to concentrations of 2–10%, is the most commonly used surface sterilant. Because commercial hypochlorite solutions vary in concentration, the percentage hypochlorite or available chlorine, as well as the duration of exposure, should be specified. Similar oxidant treatments include 3% H_2O_2 and 2% $KMnO_5$ or 0.03% peracetic acid (M. M. Dreyfuss, personal communication). Efficacy of surface sterilants often is improved by combining them with a wetting agent, particularly for hydrophobic or densely pubescent leaves. Ethanol (70–95%) is the most commonly used wetting agent; it has limited antibiotic activity and should not be used alone as a surface disinfectant (Schulz et al. 1993). Sometimes surfactants, such as Tween 80, are combined with the sterilant. Tissue is rinsed in sterile water or 70–95% ethanol after treatment for 1 minute to remove the sterilant.

Other sterilants, not commonly used in endophyte studies, include silver nitrate, mercuric chloride, formalin, and ethylene or propylene oxide. C. Booth (1971) described methods and apparatus for surface sterilization of plant material using several of these substances. Silver nitrate (1%) commonly is used for surface sterilization of roots and stems of grasses for isolation of *Gauemannomyces graminis*. The silver nitrate can be precipitated following treatment by rinsing in 5% NaCl (Cunningham 1981). Mercuric chloride (0.01% for 1 min) was used for surface sterilization of *Acer* leaves (Pugh and Buckley 1971), *Eucalyptus* leaves (Cabral 1985), and *Picea* roots (Summerbell 1989) in studies comparing internal and external fungal assemblages on the respective hosts. It seldom is used now because of its residual toxicity and hazardous nature. Equally effective substances are available.

Formalin, at concentrations of 30–50%, is also an effective surface sterilant (C. Booth 1971; Schulz et al. 1993). Propylene oxide and ethylene oxide, because of their slow rates of penetration, are useful for sterilization of natural media, for field sterilization of equipment, and for surface disinfection of woody plant tissue. Both materials are explosive and toxic and should be handled with extreme care. Volumes of sterilant, size of sterilization vessel, thickness and type of tissue, and temperature all should be noted for reproducible gas surface sterilization. Generally, absorbent cotton is soaked in propylene

oxide or ethylene oxide and placed in the sterilization vessel (e.g., a screw-cap jar) with the sample and left for a time sufficient for penetration to occur. Optimal conditions should be determined for particular host species and tissues by experimentation.

Serial washing often is used to remove soil from root tissues, to remove incidental spores from leaf surfaces, and to remove surface contamination in cases where a nontoxic method is desired. This is best accomplished using a large vessel so that the inflowing water vigorously agitates the sample (C. Booth 1971). The serial washing method of Harley and Waid (1955) is relatively simple and can be used for study of fungi colonizing roots, shoots, and leaves (Mushin and Booth 1987; Holdenreider and Sieber 1992). An ultrasonic cleaning apparatus removes surface contamination most completely (Holdenreider and Sieber 1992).

MEDIA AND INCUBATION

Routine mycological media are suitable for primary isolation of endophytic fungi and for subculturing for identification. Malt extract agar (1–2%) is used most commonly, sometimes in combination with yeast extract (0.1–0.2%; see Appendix II). Colony-limiting agents and antibiotics also are often used for primary isolations (see "Selective Isolation Agents," later in this chapter). Some workers prefer to use water agar for isolations to reduce contamination, although many fungi produce more diffuse, spreading, and less recognizable colonies on weak media. Effects of isolation medium on species richness were investigated by Bills and Polishook (1992), who found that a mixture of 1% malt extract and 0.2% yeast extract with 50 ppm each of streptomycin and chlortetracycline gave the highest species richness for isolations from twigs and leaves of *Chamaecyparis thyoides*. Greater species richness was obtained in isolations from bark of *Carpinus caroliniana* after fungal growth inhibitors were added to the media (Bills and Polishook 1991). Fungi growing on selective media should be subcultured as quickly as possible onto media without inhibitors to enhance normal sporulation for better identification.

Optimal incubation conditions vary according to the provenance of the host tissue. Because endophytic fungi are slow to emerge, prolonged incubation sometimes is required, and media may dry out. Sealing plates with Parafilm helps to prevent desiccation of the medium, but it also can inhibit sporulation; slow desiccation often promotes sporulation, particularly of coelomycetes. Incubation of plates in a growth chamber with a humidity control or in plastic boxes also can help prevent rapid desiccation. The effects of incubation temperature and light cycles on emergence of endophytes are unknown,

TABLE 12.3
Surface Sterilization Materials and Protocols

Disinfectant, concentration and duration	Host/tissue	Reference
Formaldehyde 37–40%, 1–5 min	Various hosts leaves	Schulz et al. 1993
NaOCl, 10% available Cl, 5 min	*Festuca* leaves and culms, *Anemone,* *Crataegus, Glechoma, Potentilla, Salix,* *Sorbus, Teucrium, Vaccinium* leaves	Schulz et al. 1993
Ethanol 96%, 1 min; NaOCl, 10 % available Cl, 5 min; ethanol 96%, 30 s	*Crataegus, Glechoma, Potentilla, Salix* leaves	Schulz et al. 1993
Ethanol 96%, 1 min; NaOCl, 2% available Cl, (1:2 bleach), 7 min; ethanol 96%, 30 s	Conifer twigs	Petrini and Müller 1979
Ethanol 99%, 1 min; NaOCl 8.7% available Cl, 5–120 min; ethanol 99%, 30 s	*Castanaea* shoots	Bissegger and Sieber 1994
Ethanol 96%, 1 min; NaOCl 3% available Cl, 10 min; ethanol 70%, 30 s	*Sequoia* leaves	Espinosa-Garcia and Langenheim 1990
Ethanol 96%, 30 s; NaOCl 2.5% available Cl, 1–3 min; ethanol 96%, 30 s	Lichens, mosses, ferns	Petrini 1986
	Arctostaphylos leaves	Widler and Muller 1984
	Rhododendron, Vaccinium leaves	Petrini 1985
Ethanol 96%, 30 s; sterile water, 30 s; NaOCl 5% available Cl, 5 min; ethanol, 3 s; sterile water, 30 s	*Crataegus, Glechoma, Salix, Sorbus* *Teucrium, Vaccinium* leaves	Schulz et al. 1993
Ethanol 95%, 1 min; NaOCl 20% available Cl, 3 min; ethanol 95%, 30 s	*Pteridium* rhizomes, rachis, pinnules	Petrini et al. 1992a
Ethanol 75–96%, 1 min; NaOCl 2–4% available Cl, 3–5 min	Conifer needles	Carroll and Carroll 1978
Ethanol 75–96%, 30 s, rinse with sterile water	*Quercus* leaves and twigs	Halmshlager et al. 1993
	Ulex twigs	Fisher et al. 1986
	Pinus, Fagus twigs	Petrini and Fisher 1988
	Salix, Quercus twigs	Petrini and Fisher 1990
	Quercus leaves, twigs	Fisher et al. 1994
	Abies, Picea twigs	Sieber 1989
	Acer, Betula, Picea roots	Sridhar and Bärlocher 1992a, 1992b
	Fagus leaves, twigs	Sieber and Hugentobler 1987
	Alnus leaves, twigs	Sieber et al. 1991
	Fagus buds, twigs	Toti et al. 1993
	Chamaecyparis leaves, twigs	Bills and Polishook 1992
	Pinus needles	Helander et al. 1994
	Abies, Larix, Picea, Pinus, Acer	Kowalski and Kehr 1992
	Alnus, Betula, Carpinus, Fagus, Fraxinus, *Quercus* branch bases, *Acer, Quercus,* *Tilia* leaves	Pehl and Butin 1994
Ethanol 99%, 1 min; H₂O₂ 35% available Cl, 5–120 min	*Castanea* shoots	Bissegger and Sieber 1994
Ethanol 99%, 30 s	*Abies, Fagus, Picea, Pinus* roots	Ahlich and Sieber 1996
Ethanol 70%, 1 min; H₂O₂ 15% available Cl, 15 min; ethanol 70%, 1 min; sterile water, 2 rinses	*Pinus* needles	Hata and Futai 1995
Ethanol 96%, 1 min; peracetic acid 0.35%, 3–5 min; ethanol 96%, 30 s	*Alnus* stems	Fisher and Petrini 1990
HgCl₂ 0.01%, 3 min	*Picea* roots	Summerbell 1989
HgCl₂ 0.1%, 1 min; ethanol 5%, 1 min	*Eucalyptus* leaves	Cabral 1985
	Acer leaves	Pugh and Buckley 1971
HgCl₂ 0.001%, 1–5 min; ethanol 70%, 1 min; sterile water, 1 min	plant material	C. Booth 1971

but they can influence sporulation and characters used to differentiate species. Incubation temperatures should reflect natural conditions; the range typically used is from 18°C to 25°C.

Pieces of host tissue usually are placed on the surface of agar medium in a serial order so that positional and distribution effects can be determined. The use of multiwell plates instead of Petri dishes may help to prevent cross contamination of segments by fast-growing or sporulating fungi. Isolates from each well can be noted separately and, thus, aid the reconstruction of the spatial distribution patterns of fungi in the host. In addition, the presence of host tissue often promotes sporulation. Wells can be filled with molten media rapidly and reproducibly by means of a repeating pipette.

Cutting tissue into many small pieces can be laborious, but some simple devices can be used to speed the process. Glass microscope slides can be used to "sandwich" leaves and provide a straight edge guide for slicing thin strips with a razor blade or scalpel. Some workers use a kitchen pasta cutter to obtain thin leaf strips (G. F. Bills, personal communication; M. M. Dreyfuss, personal communication). If positional effects are not a concern, an alternative to cutting tissue into consecutive segments is tissue maceration, particle filtration, and dilution plating. Ordinary kitchen blenders or more specialized laboratory blenders, such as the "Stomacher Blender" (Tekmar-Dohrmann, Appendix IV), macerate tissue efficiently into small fragments suitable for direct dilution plating. Donegan and colleagues (1996) used this method to examine fungal diversity in potato leaves, and Bills and Polishook (1994) used a similar procedure combined with particle filtration to investigate fungal diversity from Costa Rican leaf litter. The use of particle filtration markedly improves the recovery of rare species over simple dilution plating. Plant material is surface-washed and disinfected, macerated in a laboratory blender, and filtered through a wire mesh prescreen. The fine particles then are forced between polypropylene mesh filters with a stream of (sterile) distilled water. The filters used by Bills and Polishook (1994) trapped 105–210-μm particles, which led to one colony or none per particle. The trapped particles are resuspended in water (or 0.2% agar to slow sedimentation) and plated on standard isolation media. Plates require daily attention so that newly appearing colonies can be isolated and overgrowth by fast-growing species prevented (see Chapter 13, "Particle Filtration").

SELECTIVE ISOLATION AGENTS

Enrichment (enhancement) of media with different carbon or nitrogen substrata and use of selective and general growth inhibitors (as commonly used in soil microbiology) may be of value for isolation of certain groups of endophytic fungi from plant tissues. Suppression of bacteria with antibiotics may be necessary for some host tissues. More often, rapidly growing fungi obscure the presence of more slowly growing species. Weak media (those with low nutrient levels) often are used for initial isolations to prevent overgrowth. Selective growth inhibitors and antibiotics (Table 12.4) also can be used to retard growth of particular groups, suppress bacteria, and enable detection of less aggressive fungi. Cyclosporin A used at 2–10 ppm causes a general growth inhibition of fast-growing filamentous fungi (Dreyfuss 1986; Bills and Polishook 1994; Bills 1996). Surfactants (benzyltrimethylammonium hydroxide, sodium dodecyl sulfate) and organic acids (tannic acid, lactic acid) also sometimes are included as differentially selective agents in culture media. As with surface sterilization procedures, we recommend initial experimentation with several media and incubation conditions to determine optimal combinations for recovery of endophytic fungi from a specific host.

MOLECULAR SEQUENCE APPROACHES

Nucleic acid sequencing makes it possible to determine the approximate phylogenetic position of any sterile isolate. Construction of partial phylogenies of ascomycetous and basidiomycetous fungi has been achieved by sequence analysis of polymerase chain reaction (PCR)–based amplification of DNA copies of several regions of ribosomal RNA (rRNA) genes from an array of representative taxa (Bruns et al. 1991; Berbee and Taylor 1992a, 1992b; Carbone and Kohn 1993; Zambino and Szabo 1993; Swann and Taylor 1993, 1995a, 1995b, 1995c; Monreal et al. 1999). Through alignment and cladistic analysis with homologous nucleotide sequences of known fungi, phylogenetic relationships can be inferred and the unknown sterile strain can be assigned to a taxonomic category (order, family, and sometimes genus), even without assignment of names. In this way, an approximation of endophyte diversity can be obtained without sporulation of individual isolates. Knowledge of the approximate phylogenetic placement of an unknown isolate may allow an investigator to select conditions that will control growth and promote sporulation or to seek a sporulating fruit body from the natural substratum that may correspond to the unknown endophyte.

The use of molecular analyses to establish connections between anamorphs and teleomorphs (LoBuglio et al. 1993; Rehner and Samuels 1994), as well as phylogenetic relationships of autonomous anamorphs and closely

TABLE 12.4
Selective Agents Useful for Isolation of Endophytic Fungi

Agent	Activity against	Concentration	Comments
Amphotericin B	Filamentous fungi	0.5–10 mg/l	Sterol synthesis inhibitor
Ampicillin	Bacteria	100–300 mg/l	
Dichloran (Botran)	Mucorales, *Penicillium*	2–100 mg/l	Less hazardous substitute for PCNB
2-6-dichloro-4-nitroaniline PCNB (pentachloronitro-benzene)	*Aspergillus*, filamentous fungi	100 mg–1.0 g/l	Carcinogen
Benzimidazole fungicides	Fungi	50–500 mg/l	Substitute thiophanate thiabendazole fungicides for benomyl
Chloramphenicol	Bacteria	50–200 mg/l	Autoclavable
Cycloheximide	Filamentous fungi	100–200 mg/l	Autoclavable
Cylcosporin A	Filamentous fungi	10 mg/l	Heat labile
LiCl	*Trichoderma, Mortierella*	1–6 g/l	
Natamycin (pimaricin)	Filamentous fungi	2–30 mg/l	Autoclavable, photosensitive
Nystatin	Filamentous fungi	2–10 mg/l	Photosensitive
OPP (orthophenylphenol)	*Trichoderma*	5–50 mg/l	Na salt is water soluble
Oxgall (bovine bile)	Bacteria, Mucorales, Oomycetes	0.5–1 g/l	
Penicillins	Gram-positive bacteria	30–100 IU/ml	Heat labile, pH sensitive
Rifampicin	Bacteria	5–25 mg/l	Photosensitive
Rose bengal	Bacteria, filamentous fungi	50–500 mg/l	Photosensitive
Streptomycin	Gram-negative bacteria	50–500 mg/l	Heat labile
Tetracycline	Bacteria	25–100 mg/l	Heat labile
Vancomycin	Bacteria	50–200 mg/l	Heat labile

related teleomorph genera, has become routine. Such a connection between the asexual fungus *Trichoderma reesei* and the teleomorphic fungus *Hypocrea jecorina* was established using a combination of RAPD (random amplified polymorphic DNA) fingerprinting and ITS (internal transcribed spacer) sequences (Kuhls et al. 1996). ITS sequences also were used to demonstrate the connection between *Meria laricis,* an autonomous anamorphic fungus, and the teleomorph genus *Rhabdocline* (Gernandt et al. 1997), including the common endophyte of Douglas fir, *R. parkeri*. Approaches such as these also can be extended to the analysis of plant-associated symbiotic fungi, even those that cannot be cultured or for which no reference cultures exist (Egger 1995).

Because new techniques develop rapidly, we are not recommending any specific methods for identification of endophytes based on amplified sequences. Endophytes comprise a large and diverse group of fungi, so no identification methods will apply to endophytes in general. Direct amplification of DNA for detection and quantification of endophytes from infected hosts may be of more general interest. Primers that differ in sequence composition, length, restriction sites, presence of intron sequences, and similar characters can be exploited for selective PCR amplification of fungal DNA directly from infected plant host tissue. In particular, the large difference in size of the ITS region of rDNA between conifers

and fungi has been used to selectively amplify ITS DNA from conifer endophytes and pathogens (Liston et al. 1996; Camacho et al. 1997; Gernandt et al. 1997). The discovery of nonorthologous ITS-2 types in *Fusarium* (O'Donnell and Cigelnik 1997), however, demonstrates the need for cautious interpretation of results from PCR amplification of genes whose structure is not fully understood.

Camacho and colleagues (1997) and Liston and Alvarez-Buylla (1995) used "conserved motifs" in ITS-1 and small subunit rDNA for provisional characterization of fungal sequences that had been accidentally amplified from spruce foliage (Klein and Smith 1996). Fungal endophytes were suspected sources of the contaminating DNA, and identity of the putative endophytes was sought. ITS sequences were determined, aligned, subjected to phylogenetic analysis with PAUP (Phylogenetic Analysis Using Parsimony) (Swofford 1989), and compared with ITS sequences for filamentous fungal species in the GenBank database. Conserved sequence motifs were consistent enough to enable disposition of the sequences at least at the family level and in some cases at the genus level. One group of unidentified fungal sequences was grouped with inoperculate Discomycetes, one with the Hypocreales, and one with the Dothidiales. By constructing a complementary probe to the ITS-1 sequence, more than 60 different endophytic isolates were tested by southern blotting for complementarity with one of the groups of unidentified fungal sequences. Southern blotting was positive for one isolate of *Hormonema dematioides,* which proved to have a greater than 98% homology to the unidentified sequence (Camacho et al. 1997). Such procedures have proved valuable in assigning taxonomic rank to sterile isolates and have the potential for use in preliminary screening of endophyte isolates for unique sequence attributes (Monreal et al. 1999; Guo et al. 2000).

HISTOLOGICAL METHODS

Relatively few investigators to date have used direct microscopy to demonstrate endophytic colonization of host tissue or histological techniques to document correlated infections based on isolations. Most suitable for such investigations are hosts in which only one or a few species, as determined by isolation methods, are responsible for a high proportion of the recorded infections. In many cases, endophytic fungi can be visualized easily in cleared whole mounts with light microscopy. A simple procedure is to clear leaves for several days in a solution of KOH (potassium hydroxide) at 40–60°C. Leaves then are rinsed in water, bleached lightly in 3% H_2O_2, and rinsed in two changes of 2% hydrochloride followed by staining in either 0.05% trypan blue, 0.05% acid fuchsin in lactic acid, or 0.05–0.1% Calcofluor white M2R (ACS Chemical Index Number 40622) in 0.2 M tris buffer, pH 8.0. Stained leaves then are dehydrated in an ethanol series to absolute ethanol, two changes in xylol, and mounted in a permanent medium such as Permount, which works well for a variety of host plants and tissue types. Useful concentrations of KOH vary from 2–10%; the most effective concentration for a particular host should be determined by experimentation. Leaf tissue initially turns dark brown in the KOH; the solution should be changed daily until it remains colorless. The tissue eventually will become uniformly straw-colored (5–10 days, depending on the material). Clearing unfixed material is usually preferable because fixing makes it difficult to remove all cell material. This clearing method is useful for visualizing endophyte hyphae in conifer foliage and roots. Stone (1987) used this technique to obtain infection densities of *Rhabdocline parkeri* in Douglas fir by means of direct counts; he then compared his results with frequency data obtained by isolation methods. Treatment of the tissue with a protease, such as papain (1–2 g/100 ml 0.1 M phosphate buffer, pH 7.2) for 24–72 hours prior to clearing in KOH can improve removal of cellular residue that is resistant to clearing in KOH. Soaking tissue for 12–24 h in a saturated solution of chloral hydrate (250 g/100 ml)

after the KOH treatment improves transparency; however, chloral hydrate is a closely regulated substance in most countries, and special permits must be obtained before it can be purchased.

Generally, 0.05% trypan blue is a suitable stain for transmitted light microscopy; 0.05% Calcofluor in 0.2 M tris buffer, pH 8.0 is excellent for epifluorescence microscopy. Other stains, such as 0.1% Chlorazol black E in lactoglycerol, have been used for examination of vesicular-arbuscular mycorrhizae in roots (Brundrett et al. 1984) and may be useful for examination of endophytes in other tissues. Cabral and colleagues (1993) examined several endophyte species in *Juncus* leaves cleared by boiling in lactophenol-ethanol (1:2 v/v) for 5–10 minutes. The material was stored overnight in this solution, then stained with either 0.05% trypan blue or malachite green-acid fuchsin (Appendix II).

The clearing-staining method of Wolf and Frič (1981) also can be applied to the study of endophytic fungi. Tissue is cleared for 10–60 minutes in a mixture of ethanol-chloroform (3:1 v/v) with 0.15% trichloro acetic acid, using several solution changes. It is stained with Coomassie brilliant blue R-250 (Appendix II), which is protein specific and useful for cytoplasm-rich structures such as germ hyphae and haustoria.

Periodic acid–Schiff (PAS) stain also has been reported as a satisfactory stain for fungi in plant tissue (Dring 1955; Farris 1966; Nair 1976). PAS stain works well with tissue that has been fixed in FAA (50% ethanol, 5% acetic acid, 3.7% formaldehyde) and is used as follows: Tissue is immersed in 1% aqueous periodic acid for 5 minutes; rinsed in tap water for 10 minutes; immersed in Schiff's reagent for 5 minutes; washed again in tap water for 10 minutes; and then immersed in a solution containing 5 ml 10% aqueous $K_2O_5S_2$, 5 ml 1 M HCl, and 90 ml distilled water for 5 minutes. The solution is changed; the tissue is immersed for another 5 minutes; and then it is washed in tap water for 10 minutes, completely dehydrated through an absolute ethanol to xylol series, and mounted in Permount.

Pearce (1984) recommended a rhodamine B/methyl green method for staining fungal hyphae in wood because that stain, unlike trypan blue and aniline blue, is not taken up by cytoplasm. It is also suitable for differential staining of hyphae in foliage of KOH-cleared specimens. The rhodamine B stains lignified cells; the methyl green stains fungal hyphae. Pearce (1984) stained this material for 20 minutes in 1% aqueous rhodamine B; rinsed it in distilled water; stained it again in freshly prepared 15% methyl green in phosphate buffer (0.2 M, pH 8.0) for 5 minutes; rinsed it for 10 seconds in 50% 1,4-dioxan; transferred it to 70% 1,4-dioxan for approximately 20 seconds; submerged it in two 3-minute changes of 100% 1,4-dioxan; cleared it in xylene; and mounted it in Permount.

DISTRIBUTION PATTERNS AND SAMPLING CONSIDERATIONS

Species composition of endophyte assemblages and infection frequencies vary according to host species; site characteristics, such as elevation, exposure, and associated vegetation; tissue type; and tissue age. For large woody hosts, growth stage and position in the canopy also may affect distribution (Johnson and Whitney 1989). Generally, assemblages of foliar endophytes for a given host comprise a relatively consistent, cohesive group of species characterized by a few dominant species (Carroll 1995). In *Sequioa sempervirens,* for example, Rollinger and Langenheim (1993) found the endophyte composition to be relatively constant over the north-to-south distribution of the host.

In addition to the core group of species consistently isolated as endophytes from any given host, surveys of plant hosts for endophytes invariably generate long lists of incidental species that are not known to sporulate on the host. Each incidental species often is represented only once or twice in several hundred samples. In general, the number of rare and incidental species isolated is proportional to the intensity of sampling; distribution of rare species is influenced more by site than by host (Petrini 1986). The high diversity of endophytic fungi that has been demonstrated repeatedly for a variety of host species and the bewildering numbers of species often found on individual hosts contribute to the appeal of endophytic fungi for ecological studies.

Variation in species assemblages on the same host at different sites usually is attributable to recovery of incidental species with more disjunct distributions. Bills and Polishook (1991, 1992) noted differences in species richness among sites, but a core group of taxa was recovered in relatively constant proportions from all sites for each of two host species. Species richness (i.e., number of species per host) increases at a constant rate, eventually becoming asymptotic. The number of species recovered per isolate generally is comparable to that observed for soil habitats (Christensen 1981a; Lussenhop 1981; Bills and Polishook 1992, 1994). In tropical forests, host specificity has been more difficult to demonstrate, requiring more intensive sampling, and species richness may be considerably higher than in temperate forests (Arnold et al. 2000).

SPATIAL AND TEMPORAL DISTRIBUTION

Several investigators have documented different distribution patterns of endophyte species within individual leaves. Carroll and Carroll (1978) noted consistent differences in species composition in leaf blade versus petiole segments among several conifer species. Hata and Futai (1995) also noted position-specific differences in distributions of *Phialocephala* species and *Leptostroma* species in pine needles. Halmschlager and colleagues (1993) showed that isolation frequencies of foliar endophyte species on *Quercus petrea* varied spatially in leaves and also exhibited a temporal periodicity. Frequency of infection by *Aureobasidium apocryptum* tended to increase over the entire leaf from May through September, whereas *Discula quercina* tended to decrease. Wilson and Carroll (1994) also found that leaf midveins of *Quercus garryana* were colonized more heavily by *D. quercina* than leaf blades. Overall infection frequencies increased sharply between May and June but then declined slightly through August. Pronounced seasonal differences in colonization frequencies might be predicted for climates with a distinct wet/dry seasonal cycle. Rodrigues (1994), however, found relatively small seasonal differences in overall infection frequencies of *Euterpe oleracea* in Amazonian Brazil, although distribution of certain species was strongly seasonal. Species composition and relative abundances in endophyte assemblages may reflect spatial distributions as well as sampling times.

EFFECT OF TISSUE AGE

A consistent trend repeatedly confirmed for foliar and stem endophytes is that overall infection frequencies increase with the age of host organs or tissues. This is best observed from evergreen plants or plants with long-lived foliage but is also apparent to a lesser degree in deciduous trees and annuals. Infection densities (infections/mm^2) of *Rhabdocline parkeri* in epidermis increase at a constant rate with age of Douglas fir needles. Because infections of *R. parkeri* are limited to single cells and each infected cell represents a discrete infection event, the increased densities are caused by repeated infection of a needle by fungal propagules, not by extended colonization from *a priori* infection sites (Stone 1987). Age of foliage of the tropical palm *Euterpe oleracea* strongly influences fungal colonization frequencies (Rodrigues 1994), as does the age of gorse (*Ulex* species) stems (Fisher et al. 1986).

Endophyte species diversity, as well as total infection frequency, also increases with age of foliage for *Sequioa sempervirens* (Espinosa-Garcia and Langenheim 1990). Species evenness among age classes, however, was high, indicating a well-spread dominance rather than an age-related species succession. Barklund and Kowalski (1996) noted a sequential pattern of bark endophyte species composition in Norway spruce internodes. *Tryblidiopsis pinastri* occurred in greater abundance in the upper crowns of trees and was the dominant endophyte in younger internodes, whereas *Phialocephala scopi-formis*, *Mollisia cinerea*, *Tapesia livido-fusca*, and *Geniculosporium serpens* were more prominent in the lower branches and also in older internodes.

TISSUE SPECIFICITY

Specificity of endophytes for particular host tissues or organs can be assessed through careful dissection and separate culturing and analysis of samples from those tissues or organs (Carroll 1991, 1995). Differences in the assemblages of endophytic species in leaves and twigs of *Acer macrophyllum* (Sieber and Dorworth 1994), *Alnus rubra* (Sieber et al. 1991), and *Quercus petraea* (Halmschlager et al. 1993) have been documented. Similarly, assemblages of endophytic species in the outer bark differ from those in xylem for species of *Alnus* (Fisher and Petrini 1990), *Castanea sativa* (Bissegger and Sieber 1994), *Quercus ilex* (Fisher et al. 1994), *Salix fragilis*, and *Quercus robur* (Petrini and Fisher 1990). Petrini and Fisher (1988) found that several of the most common endophytic fungi in a mixed stand of *Pinus sylvestris* and *Fagus sylvatica* occurred on only one host, even when host species were growing adjacent to one another. However, *Verticicladium trifida*, a species generally associated with conifers and dominant on *P. sylvestris*, also occasionally was isolated from *F. sylvatica*.

Generally, the diversity of fungal species that colonize inner bark is less than that of species colonizing outer bark. Fungal communities of outer bark include many species with general host distributions (i.e., that lack host specificity) in contrast to species that colonize inner bark, which tend to exhibit greater host specificity. Colonists of inner bark, such as *Tryblidiopsis pinastri* and *Phialocephala scopiformis*, are termed "phellophytes" (Kowalski and Kehr 1992). Xylotropic endophytes (Chapela 1989), quiescent colonists of sapwood, have been demonstrated to occur deep within sound 55- to 60-year-old *Picea abies* stems (Roll-Hansen and Roll-Hansen 1979, 1980a, 1980b; Huse 1981).

SCREENING GRASSES FOR ASYMPTOMATIC CLAVICIPITACEOUS ENDOPHYTES

Examination of diversity among grass endophytes requires that grass hosts be screened for the presence of endophytic mycelia. The most rapid method for assessing distribution of endophytes in grasses is to screen herbarium collections. Following preliminary screening of dried collections, fresh collections can be made and the endophytes can be isolated for examination in culture. For identification purposes, endophytes frequently must be isolated and grown in culture to assess morphological features or subject them to molecular methods of identification.

SCREENING HERBARIUM SPECIMENS

Herbaria collections often contain large numbers of specimens from diverse localities. Although the sample is neither random nor systematic, it can provide a good indication of the geographic distribution of an endophyte (White 1987). Either culm or seed tissues are examined to assess the presence of an endophytic mycelium within an individual plant. The tissue is removed from the herbarium specimen, stained, and examined under a microscope. Stains required for this procedure include either nonacidified aniline blue (Appendix II), which can be diluted with water to improve visibility of mycelia in thick slide preparations, or rose bengal (Appendix II), which can replace aniline blue for viewing endophytic mycelia. Contrast is enhanced by use of a green interference filter over the light source (Saha et al. 1988).

Examination of herbarium specimens for presence of endophytes must be done carefully to minimize damage to the plant specimen. The least destructive procedure is a seed examination. One to 10 seeds are softened by being placed in a test tube containing 10 ml of concentrated nitric acid maintained at 60°C in a hot water bath. After 40–60 seconds of continuous agitation, the seeds and acid are poured into 1 liter of cold water to stop the digestion process. After 15–30 minutes, a seed may be removed from the water and placed on a slide in a drop of nonacidified aniline blue stain. Seeds can be squashed under the coverslip and examined microscopically for the blue-stained mycelium associated with aleurone cells around the periphery of each squashed seed (Figs. 12.5 and 12.6). This method of seed preparation is rapid but can result in the overdigestion of the seeds, so altering the structure of the aleurone cells and associated mycelium that determination of endophyte presence is impossible. If aleurone cells seem abnormally swollen, overdigestion likely has occurred, and time of seed digestion should be reduced. In general, smaller seeds require a shorter digestion time than larger seeds. To eliminate problems of overdigestion, seeds can be softened in 5% sodium hydroxide for 8 hours at room temperature and then rinsed for 20–30 minutes in continuously running tap water. Seeds then are examined as described earlier.

Probably the simplest method for assessing dried specimens for the presence of endophytes is to examine culm tissue for evidence of endophytic hyphae. A short segment (1–2 cm) of the culm is split longitudinally using a scalpel blade. The upper half of the split segment

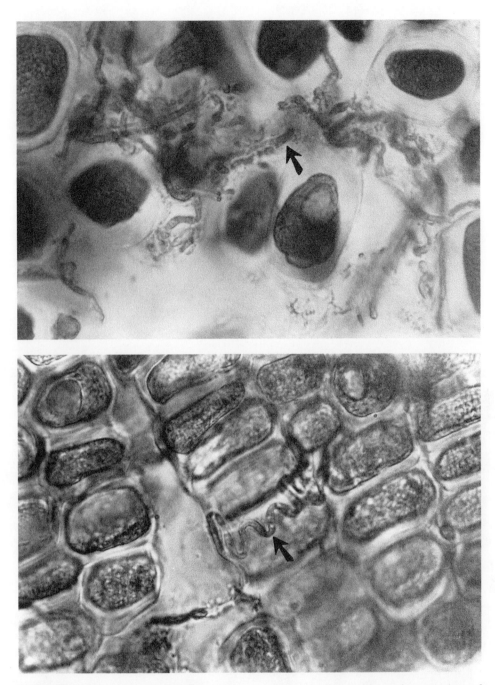

FIGURES 12.5 Convoluted hyphae (*arrows*) on aleurone cells in seed-squash preparations of *Neotyphodium*-infected fescue (×2000).

is removed using forceps. The parenchyma tissue within the culm then is moistened using nonacidified aniline blue or rose bengal stain. After approximately 1 minute, the moistened tissues within the culm are scraped onto a clean glass slide with the scalpel blade. This tissue then is moistened using distilled water, macerated with a scalpel blade, covered with a coverslip, and examined under the 40× objective of a compound light microscope for the presence of typical nonbranching endophytic

mycelia (Figs. 12.7 to 12.9) in close association with external walls of parenchyma cells.

After examining tissue from a herbarium specimen, the investigator should affix a label to the herbarium sheet indicating the tissue examined, infection status, any notable characteristics of the endophytic mycelium, date, and investigator. The label facilitates the relocation of specimens for later reexamination and the use of endophyte data by other scientists.

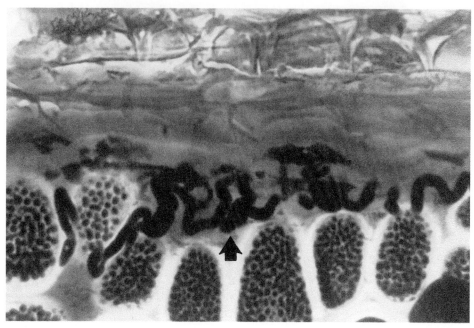

FIGURE 12.6 Cross section showing the convoluted hypha layer (*arrow*) of *Neotyphodium coenophialum* between an aleurone layer and a seed coat of tall-fescue seed (×2500).

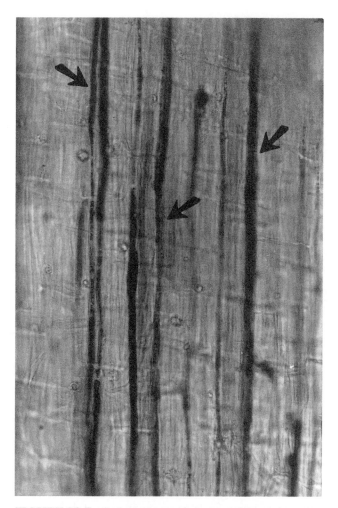

FIGURE 12.7 Endophytic mycelium (*arrows*) in a culm-scraping preparation from *Achnatherum robustum* (×2000).

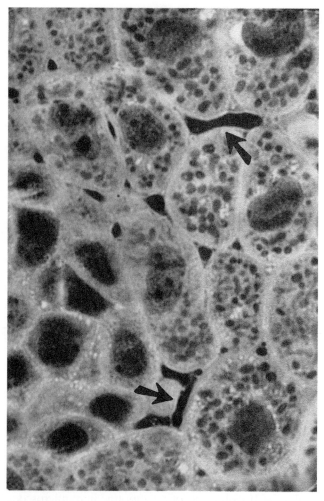

FIGURE 12.8 Endophytic mycelium (*arrows*) in an embryo or *Festuca versuta* (×2000).

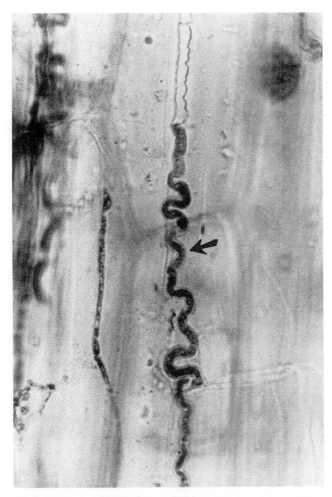

FIGURE 12.9 Endophytic mycelium (*arrow*) in culm scraping from *Festuca* species (×2000).

We have used this procedure to determine the presence of endophytes in plant specimens that were more than 100 years old. Perhaps as many as 20% of grass specimens, regardless of when they were collected, are in poor condition because of saprotrophic activities of other fungi, consumption by mites, or other factors. The presence of frequently branching hyphae, hyphae that are not oriented longitudinally, hyphae that are closely appressed to parenchyma cells, or poorly preserved parenchyma cells indicates that a herbarium specimen is too degraded to assess for presence of endophytes.

SCREENING FIELD POPULATIONS

After a preliminary assessment of endophyte distributions using herbarium material, living plants can be obtained from areas corresponding to collection sites identified from herbarium labels. Several plants or plant samples from the site can be collected randomly and transported into the laboratory for microscopic examination. Plant samples can be kept on ice during transport, frozen, and later thawed for examination (Clark et al. 1983). Alternatively, infected individuals can be identified at the site using a field microscope. Living culm, rhizome, or leaf sheath tissues are all suitable for examination.

Living tissues tend to resist penetration of nonacidified aniline blue stain; acidified aniline blue is more effective (Bacon and White 1994). The latter stain is prepared by adding 0.1 g of aniline blue powder to 100 ml of sterile distilled water, mixing vigorously until the powder is dissolved, and then adding 50 ml of lactic acid (85%) and mixing again. The stain can be stored for months at room temperature without losing its effectiveness. Rose bengal stain prepared as previously described (see "Screening Herbarium Specimens," earlier) also may be used when examining living tissues (Saha et al. 1988).

Culms and rhizomes should be split longitudinally, and the moist inner tissue should be scraped out and placed in a drop of acidified aniline blue stain. The tissue then is macerated and heated for a few seconds to facilitate penetration of the stain. Excess stain then is blotted off. The tissue is remoistened with distilled water and examined using the 40× objective.

If plants are not in flower, leaf sheaths that are close to the crown of the plant where very little pigmentation is evident can be examined (Bacon and White 1994). The upper epidermal layer is cut laterally across the sheath with a sharp scalpel blade. The epidermis is peeled back to expose a 5-mm-long area of mesophyll. That region of the leaf sheath then is placed on a slide with the mesophyll portion facing up, stained as described earlier for culms and rhizomes, and examined for mycelia.

ISOLATION PROCEDURES

After plants are screened for presence of endophytic mycelia, isolations should be made to confirm the clavicipitaceous identity of the endophytes. To make isolations from leaf or stem, young tissues of culms or leaf sheaths are cut into segments 3–5 mm long and then agitated continuously for 15 minutes in a solution of 50% bleach. After 5 minutes, two to three pieces of tissue are removed every 2–3 minutes and vigorously rinsed in sterile distilled water. These pieces then are pressed into potato dextrose agar media, and the plates are sealed with Parafilm and incubated at room temperature for 3–4 weeks. Rapidly growing fungi that appear within the first 2 weeks should be discarded. After 2–4 weeks, the white to off-white colonies of the endophytes will be visible (Fig. 12.10).

Before fungi can be isolated from seeds, the seeds must be deglumed to remove contaminants associated with

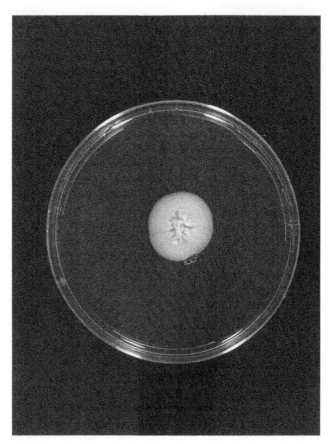

FIGURE 12.10 Three-week-old culture of *Neotyphodium starrii* grown on potato-dextrose agar at room temperature (×0.5).

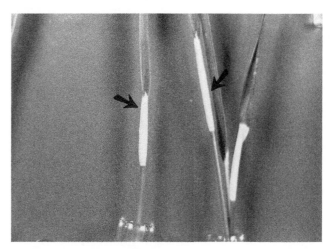

FIGURE 12.11 Stroma of *Balansia epichloë* on leaves of *Sporobolus* species (×3).

the dry glumes. This can be done by rubbing seeds vigorously between the hands for several minutes and periodically collecting the seeds that are freed from the glumes. After 30–40 seeds have been collected, they are placed in a 250-ml beaker and covered with a 50% bleach solution. Seeds should be agitated in the sterilizing solution for 15–20 minutes. The bleach solution is decanted and replaced with 100 ml of sterile distilled water. After the seeds are agitated for 5 minutes, they are removed using sterile forceps and pressed into potato dextrose agar. We recommend using about 20 plates with three seeds per plate. The plates are sealed with Parafilm to reduce drying and contamination. Rapidly growing fungi appearing during the first 2 weeks are discarded.

TAXONOMIC STATUS, DIVERSITY, AND DISTRIBUTION

Endophytic fungi comprise a highly diverse ecological and taxonomic group. We consider some of these major taxa and ecological categories considered in this section, including methods for detection and enumeration, and resources for identification.

ENDOPHYTIC BALANSIEAE

Two genera of Balansieae (Clavicipitaceae, Ascomycetes) contain endophytes, *Epichloë* and *Balansia* (Diehl 1950; White 1993, 1994a). Several species of *Balansia* are endophytic. Stromata bearing reproductive structures, conidia, and perithecia in this genus may form on host inflorescences, as in *Balansia claviceps* and *B. obtecta;* on culms at nodes, as in *B. aristidae, B. nigricans, B. strangulans,* and *B. gaduae;* or on leaves, as in *B. epichloë* and *B. henningsiana* (Diehl 1950). The typical conidial stroma of *Balansia* is white, purple, or brown (Fig. 12.11). Ascomata on the stromata are black, and stipitate or flattened (White 1994a). The conidia are filamentous. The conidial stages (anamorphs) of *Balansia* are classified in the genus *Ephelis.* Asci are cylindrical with thick refractive tips and filamentous, multiseptate ascospores that disarticulate to form 1-septate cylindrical units. Endophytic mycelium has been found in leaf and culm tissue but does not appear to enter ovaries and seeds (White and Owens 1992).

In all species of *Epichloë,* white conidial stromata, within which perithecia develop, form on meristem of the host inflorescence but also surround part of a host leaf that emerges from the apex of the stroma (Fig. 12.12). This stromatic structure is consistent throughout *Epichloë* (Leuchtmann et al. 1994; White 1994b). As perithecia develop, the stomata become yellow to orange. Asci at this stage are cylindrical, with a thick refractive tip, and ascospores are filamentous and hyaline (White 1994b). Mycelia of all species of *Epichloë* are

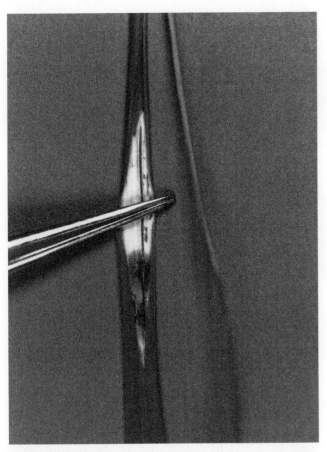

FIGURE 12.12 Conidial stroma (*arrows*) of *Epichloë amarillans* on culms of *Agrostis hiemalis* (×2).

endophytic in leaves, culms, and rhizomes; in many grasses; and in seeds.

Most endophytes that infect grasses elicit no external symptoms. Those endophytes have been classified in the genus *Acremonium* sect. *Albo-lanosa* (Morgan-Jones and Gams 1982). Based on their unique biology and phylogenetic affinities to *Epichloë* in the Clavicipitaceae, these fungi were reclassified in a more natural anamorph genus *Neotyphodium* (Glenn et al. 1996). *Neotyphodium* endophytes consistently show a close relationship to the genus *Epichloë* (White 1987; Schardl et al. 1991; Moon et al. 2000). Many of these endophytes appear to have developed from *Epichloë* species through loss of the ability to form the *Epichloë* stage and by interspecific hybridizations (Schardl and Leuchtmann 1999; Moon et al. 2000; Schardl and Wilkinson 2000). *Neotyphodium* endophytes commonly are encountered in cool-season grasses (White 1987).

Most colonies of *Neotyphodium* endophytes are white and have a cotton or feltlike texture (White and Morgan-Jones 1987). Conidiogenous cells project laterally from hyphae forming a mycelium (Fig. 12.13). Conidia, which are produced apically on conidiogenous cells, typically are reniform to subulate (Fig. 12.14). Under a dissecting microscope, the conidium lies crosswise at the apex of the conidiogenous cell, forming a characteristic T-shape, with a conidium lying (Fig. 12.15; J. F. White, unpublished data).

Several species of endophytes can be readily identified on the basis of host association and characteristics in

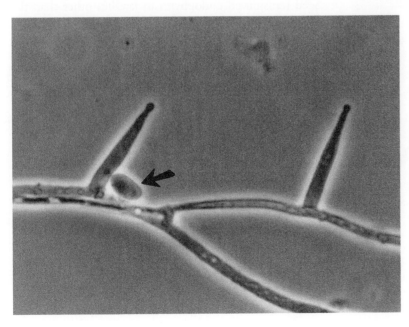

FIGURE 12.13 Conidiogenous cells and conidium (*arrow*) of *Neotyphodium typhinum* from culture (×3000).

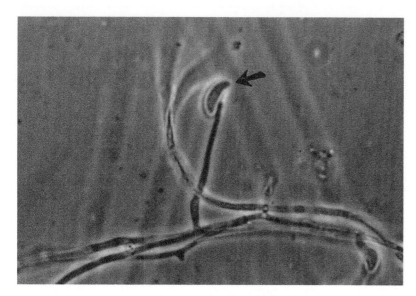

FIGURE 12.14 Conidium (*arrow*) and conidiogenous cell of *Neotyphodium coenophialum* (×3000).

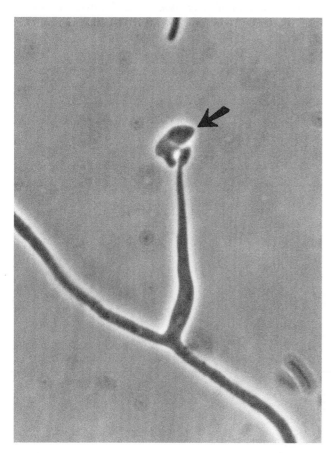

FIGURE 12.15 Germinating conidium (*arrow*) at apex of conidiogenous cell of *Neotyphodium starrii* (×3000).

culture. These include *Neotyphodium coenophialum*, an endophyte of tall fescue (Morgan-Jones and Gams 1982), and *E. uncinatum*, an endophyte of *Festuca pratensis*. Several other endophytes have been described

but are not commonly encountered (White et al. 1987; Schardl and Leuchtmann 1999). *Neotyphodium typhinum* encompasses the conidial state of several distinct species of *Epichloë* (White 1992). Recent work using molecular sequence analyses has provided evidence that the asymptomatic endophytes have evolved through hybridization of *Epichloë* species. It is becoming increasingly clear that classification of asymptomatic endophytes must be linked to classification of the *Epichloë* states (Moon et al. 2000; Schardl and Wilkinson 2000).

NONCLAVICIPITACEOUS SEED-TRANSMITTED GRASS ENDOPHYTES

Seed-transmitted endophytes in families other than the Clavicipitaceae have been encountered in grasses. An endophyte identified as *Pseudocercosporella trichachnicola* is widespread in the warm-season grass *Trichachne insularis* (J. F. White et al. 1990). The histological features of the endophytic mycelium of *P. trichachnicola* are similar to those of the clavicipitaceous endophytes: it is intercellular; longitudinally oriented; unbranched; and present in leaf sheaths, culms, and seeds. No one knows whether *P. trichachnicola* ever produces an external stage. The impact, if any, that this endophyte has on its host is also unknown, although *T. insularis* has been reported to be toxic under some circumstances (White and Halisky 1992). Procedures for detecting this endophyte are the same as those used for detecting clavicipitaceous endophytes.

Gliocladium-like and *Phialophora*-like endophytes can be isolated from stems and seeds of numerous festucoid grasses (Latch et al. 1984). Those endophytes are referred to as "P-endophytes" ("P" for *Phialophora*) by some fungal biologists (An et al. 1993; Siegel et al. 1995). Procedures for visualizing mycelium in leaf sheaths and culms are the same as for clavicipitaceous endophytes. When grown on potato dextrose agar, the mycelium remains sterile, and growth rate and colony appearance are similar to that of the endophytic Clavicipitaceae. Thus, it is frequently difficult to distinguish the "P-endophytes" from clavicipitaceous endophytes. When "P-endophytes" are grown on starch-milk agar, however, colonies produce a bright-yellow pigment that is not seen in clavicipitaceous endophytes (Bacon and White 1994). *Gliocladium*-like and *Phialophora*-like endophytes can be induced to sporulate by storing cultures in a refrigerator at 4–5°C for 6–10 weeks. Conidiogenous cells (Fig. 12.16) are clavate, with a single, apical conidiogenous locus, borne singly or in clusters of two to three on short, lateral, hyaline conidiophores. The conidia are ovate to ellipsoidal in shape, and hyaline. Systematic studies on these endophytes have not been done, and taxonomic information on these fungi is scant.

NONSYSTEMIC GRASS ENDOPHYTES

Few graminaceous hosts have been examined for nonsystemic, non–seed-borne fungi. Dominant nonsystemic endophytes of these hosts are generally familiar epiphytes, such as *Alternaria alternata*, *Cladosporium* species, and *Epicoccum purpurascens*, or pathogens typical of grass hosts. Barley (*Hordeum vulgare*) leaves in New Zealand are infected primarily by the pathogen *Didymella phleina*, *Alternaria* species, and *Stemphyllium botryosum* (Riesen and Close 1987). *Phaeosphaeria* (*Stagonospora*) *nodorum*, a common leaf and culm

blotch, is the most common of 196 endophytic colonists of winter wheat (*Triticum aestivum*) in Switzerland (Riesen and Sieber 1985). *Alternaria alternata*, *Arthrinium* species, *Cladosporium tenuissimum*, and *Epicoccum purpurascens* are the dominant endophytes of rice (*Oryza sativa*) and maize (*Zea mays*) (Fisher and Petrini 1992; Fisher et al. 1992). They occur with latent pathogens such as *Phoma sorghina*, *Fusarium equiseti*, *F. oxysporum*, *F. graminearum*, and *Ustilago* species. Leslie and associates (1990) found universal infection of maize, and near universal infection of sorghum (*Sorghum bicolor*), by at least one species of *Fusarium* section *Liseola*, primarily *F. moniliforme*. Both rice and maize frequently were colonized concurrently by several *Fusarium* species. Species inhabiting live tissue were different from those in plant debris and soil. In symptomless maize plants infected by *F. moniliforme*, intercellular hyphae occur throughout the host plant (Bacon and Hinton 1996). Cabral and colleagues (1993) investigated endophytes of *Juncus* species in Oregon and used both culture methods and direct microscopy to document unique patterns of internal colonization of leaves and culms by *Alternaria alternata*, *Cladosporium cladosporioides*, *Drechslera* species, and *Stagonospora innumerosa*.

ENDOPHYTES OF WOODY PERENNIALS AND OTHER HOSTS

An endophytic habit apparently has evolved independently numerous times and is represented by fungi in various orders of the Ascomycetes (Tables 12.5 and 12.6). A large proportion of the genera frequently encountered as endophytes of woody perennials are inoperculate Discomycetes. The endophytic habit similar to that of *Rhabdocline parkeri* may be widespread in the Rhytismatales (Livsey and Minter 1994) and in the Phacidiaceae and Hemiphacidiaceae (Leotiales). *Fabrella tsugae* commonly fruits in late winter on the oldest needles of several species of *Tsuga*, where its appearance coincides with normal senescence. *Lophodermium* species, conspicuous on senescent and fallen conifer needles and recognizable in culture by their anamorphs and culture morphology, are among the most common endophytic isolates from *Abies*, *Picea*, and *Pinus*. Species of *Rhytisma*, such as *R. punctata* on *Acer grandifolia*, and of *Coccomyces* may have similar endophytic niches in broad-leaved trees and shrubs. Their fruiting bodies usually appear on leaves coincident with leaf senescence, but maturation and release of ascospores coincides with bud opening and leaf emergence, a pattern typical of "latent pathogens." Anamorphs of *Lophodermiun* and *Coccomyces* frequently are isolated from healthy leaves of *Mahonia nervosa* (Petrini et al. 1982). Genera of

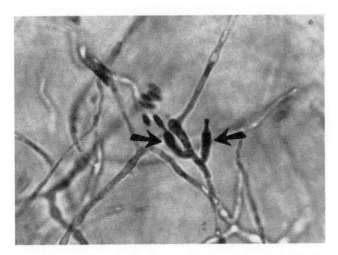

FIGURE 12.16 Conidiogenous cells (*arrows*) and conidia of a *Phialophora*-like endophyte from *Festuca igantea* (×3000).

TABLE 12.5

Genera of Endophytes Commonly Isolated from the Foliage of Woody Perennials

Holomorph order	Endophyte genera
Leotiales	Pezicula, Cryptosporiopsis, Phlytema, Chloroscypha Sirodothis, Gremmeniella, Brunchorstia, Phragmopycnis, *Rhabdocline*
Dothidiales	*Hormonema, Stagonospora, Phyllosticta*
Pleosporales	*Pleospora, Alternaria, Curvularia, Sporormia, Sporormiella, Stemphyllium*
Diaporthales	*Diaporthe, Phomopsis, Apiognomonia, Discula, Cytospora, Gnomonia, Ophiognomonia*
Diatrypales	*Libertella, Diatrypella, Diatrype, Eutypa*
Rhytismatales	*Ceuthospora, Lophiodermium, Tryblidiopsis, Cyclaneusma*
Xylariales	*Coniochaeta, Hypoxylon, Biscogniauxia, Camillea, Geniculosporium, Nodulisporium, Virgariella, Periconiella, Xylaria*
Sordariales	*Chaetomium, Sordaria, Gelasinospora*
Hypocreales	*Clonostachys, Cylindrocarpon, Dendrodochium, Fusarium, Gibberella, Gliocladium, Nectria, Trichoderma, Stilbella, Volutella*
Amphisphaeriales	*Pestalotiopsis, Seiridium, Pestalotia, Seimatosporium*
Polystigmatales	*Glomerella, Colletotrichum*
Uncertain	*Phialocephala, Cryptocline, Gelatinosporium, Acremonium, IdriellaFoeostoma, Kabatina, Sirococcus*

TABLE 12.6

Genera of Endophytes Commonly Isolated from Bark and Shoots

Holomorph order	Endophyte genera
Leotiales	*Mollisia, Pezicula, Cryptosporiopsis, Tympanis, Sirodonthis, Durandiella, Godronia, Brunchorstia, Xylogramma Cystotricha, Phleosporella*
Dothidiales	*Sphaeropsis, Hormonema, Sclerophoma, Botryosphaeria, Tripospermum, Ramularia, Cladosporium, Didymosphaeria, Diplodia*
Diaporthales	*Amphiporthe, Coryneum, Diaporthe, Cytospora, Fusicoccum, Diplodina, Melanconis, Gnomonia, Phomopsis, Phragmoporth, Pseudovalsa*
Pleosporales	*Alternaria, Pleospora, Sporormia, Sporormiella*
Hypocreales	*Albonectria, Beauveria, Bionectria, Cosmospora, Cylindrocarpon, Didymostilbe, Gliocladium, Fusarium, Haematonectria, Nectria, Trichoderma, Tubercularia*
Xylariales	*Anthostomella, Biscogniauxia, Camillea, Coniochaeta, Creosphaeria, Daldinia, Hypoxylon, Geniculisporium, Nodulisporium, Rosellinia, Rhinocladiella, Periconiella, Virgariella, Xylaria*
Rhytismatales	*Colpoma, Tryblidiopsis*
Pezizales	*Chromelosporium, Oedocephalum, Verticicladium*
Diatrypales	*Libertella, Cryptosphaeria, Diatrypella*
Basidiomycetes	*Coniophora, Coprinus, Peniophora, Rhizoctonia, Sistotrema, Sporotrichum, Stereum*
Uncertain	*Melanconium, Coniella, Gelatinosporium, Phialocephala, Acremonium, Phialophora, Microsphaeropsis, Leptodontidium, Acrodontium, Rhinocladiella, Nigrospora, Phaeococcus*
Mucorales	*Mucor, Mortieriella*
Amphisphaeriales	*Pestalotiopsis, Seiridium*
Sordariales	*Chaetomium, Podospora, Sordaria, Gelasinospora, Spadicoides*

Leotiales repeatedly isolated as endophytes include *Tiarosporella* and *Ceuthospora* in the Phacideaceae; *Pezicula, Dermea,* and *Mollisia* in the Dermateaceae; and *Chloroschypha* in the Leotiaceae. Dothidiales, such as *Phyllosticta* anamorphs of *Guignardia* species and *Hormonema* anamorphs of *Dothiora* and *Pringsheimia*; Diaporthales, such as *Phomopsis* species; and various Hypocreales are also ubiquitous endophytes.

Species of the Xylariaceae are a ubiquitous and exceptionally speciose group of endophytes (Petrini and Petrini 1985; Petrini 1986; Petrini et al. 1995), especially in the tropics, where the family is most diverse (Whalley 1993; Petrini et al. 1995). Endophytic Xylariaceae infecting temperate zone hosts are also quite diverse (Petrini and Petrini 1985; Brunner and Petrini 1992). Endophytic isolates usually produce anamorphic states in culture. The genera are differentiated easily, and many of the more common temperate species can be identified after careful comparisons with cultures derived from identified teleomorphs (Petrini and Petrini 1985). Identification of many other species, however, is either challenging or impossible. Anamorphic Xylariaceae were the predominant endophytes recovered in two detailed investigations of tropical hosts (Rodrigues 1994; Lodge et al. 1996a). Anamorphs of *Hypoxylon* species

and related genera (e.g., *Biscogniauxia, Camillea,* and *Nemania*) are ubiquitous in virtually all temperate hosts but are less frequent compared to *Xylaria* species in tropical hosts.

Although anamorphic Xylariaceae frequently are encountered as endophytes and saprobes from diverse substrata, the teleomorphs are more restricted in occurrence. In fact, the distribution of teleomorphs from field collections might lead one to conclude that many xylariaceous species are host specific; the relatively common recovery of anamorphic states in culture from diverse substrata suggests otherwise (Petrini et al. 1995; Rogers 2000). Only certain hosts or substrata evidently meet the specific requirements for formation of teleomorphic states of Xylariaceae. The broad distribution of Xylariaceae, both as endophytes and as saprobes, together with their well-documented ability to produce a variety of bioactive metabolites, point to a significant, but as yet unelucidated, role in the ecosystem (Petrini et al. 1995; Whalley 1993; Rogers 2000).

TERRESTRIAL AQUATIC HYPHOMYCETES

Ingoldian hyphomycetes are fungi whose conidia are tetraradiate, are sigmoid, or have appendages and are specialized for aquatic dispersal. Staurosporous tetraradiate conidia are characteristic of species typically associated with senescent and decaying leaf litter from trees growing near rapidly flowing streams (Webster 1981). Conidia characteristic of those fungi, however, also have been recovered from rainfall collected beneath canopies of mature forests in upland sites far from streams (G. C. Carroll, personal communication) and from roofs of buildings (Czeczuga and Orlowska 1997). Sigmoid, helicoid, tetraradiate, and branched conidia representing several anamorph genera have been collected from rain-washed trunks of several tree species in the Pacific Northwest (Bandoni 1981). Many of the conidia thus collected can be readily assigned to existing genera and species (e.g., *Gyoerffella biappendiculata*), but apparently undescribed taxa also commonly are found in rain. Different species seem to be associated with different host trees, but until recently the source of the conidia remained enigmatic. Bandoni (1981) suggested that some aquatic hyphomycetes may be endophytelike early leaf colonists or actually may be parasitic on land plants.

Ando (1992) demonstrated the origin of tetraradiate-spored fungi resembling Ingoldian aquatic hyphomycetes, which he termed "terrestrial aquatic fungi," from living leaves of intact plants. Those fungi produce typically staurosporous conidia from minute conidiophores of apparently endophytic origin in droplets of fog, dew, and rain on living leaves. Webster and Descals (1981) reasoned that the tetraradiate spore shape allows a more stable attachment of conidia to substrata in the flowing current of streams. Ando (1992), in contrast, proposed that the tetraradiate shape is an adaptation to retain water about the conidium. To obtain conidia of these fungi, water droplets from leaf or stem surfaces are collected in plastic bags. The liquid is centrifuged gently, and the resultant sediment is either fixed for microscopic observation or spread onto isolation media for selection of single conidia isolates (Ando and Tubaki 1984a, 1984b). The hyphomycete genera *Geminoarcus*, *Kodonospora*, *Tetraspermum*, *Trifurcospora*, and *Trinacrium*, as well as several hyphomycete species, have been described from material collected and isolated using these methods (Ando and Tubaki 1984a, 1984b; Ando et al. 1987; Ando 1993a, 1993b).

LICHENS

Cryptic endophytic microfungi, which have been isolated at high frequencies from lichen thalli, require somewhat specialized methods for maximum recovery (Petrini et al. 1990; Girlanda et al. 1997). Seventeen fruticose lichen samples yielded 506 fungal taxa, the majority of which (306) were isolated only once (Petrini et al. 1990). A more intensive study of two lichen species from a common site revealed differences in their fungal assemblages but similar levels of biodiversity (Girlanda et al. 1997). Most isolates were not representative of lichenicolous fungi but represent genera and species known from various other substrata. The high level of fungal diversity may have been the result of the highly porous and heterogeneous nature of the lichen thalli.

MOSSES, HEPATICS, LIVERWORTS, AND PTERIDOPHYTES

The association of endophytic fungi with terrestrial and epiphytic moss and hepatic hosts is intriguing. Although comprehensive studies of the endophyte assemblages occurring on these hosts are lacking, numerous reports document occurrences of individual fungal species on such hosts. *Selenospora guernisacii*, an inconspicuous Discomycete, is associated with mosses in northwestern North America (Weber 1995b). Döbbler (1979) reported pyrenocarpous and pezizalean parasites of mosses in Europe. Intracellular associations between achlorophyllous gametophytes of hepatics and Pteridophytes and various fungi are apparently widespread (Pocock and Duckett 1984; Ligrone and Lopes 1989; Ligrone et al. 1993).

Similar associations between endophytic fungi and nonvascular plants, such as the Anthocerote *Phanoceros laevis* (Ligrone 1988), are known primarily from histological studies. Unidentified endophytic Ascomycetes, Basidiomycetes, and Zygomycetes have been reported to form associations with a variety of nonvascular hosts in a range of cytological specializations ranging from the simple to the complex (Pocock and Duckett 1984, 1985a, 1985b). Symbioses between primitive vascular plants and fungi have been described as mycorrhizalike, although Schmid and Oberwinkler (1993) coined the term *lycopodioid mycothallus interaction* to recognize the distinct nature of the association between fungal endophytes and the achlorophyllous gametophytes of *Lycopodium clavatum*. *Endomycothalli* is a general term for the fungal colonization of hepatics (Ligrone et al. 1993).

Colonization of roots of Pteridophyte sporophytes is well known (Boullard 1957, 1979). Most terrestrial Pteridophytes are considered to be endomycorrhizal, although reports of septate hyphae in Pteridophyte roots are also numerous (Boullard 1957; Schmid et al. 1995). Few comprehensive surveys of fungi-colonizing Pteridophyte roots exist. Roots of *Pteridium aquilinum* are colonized by a variety of fungi, including Zygomycetes (*Absidia cylidrospora*, *Mortierella* species), several anamorphic Ascomycetes, and a sterile Basidiomycete (Petrini et al. 1992a). An undetermined Ascomycete was found to have colonized the roots of several species of tropical, arboreal, epiphytic ferns. The fungus invaded epidermal and cortical cells in the manner of ericoid mycorrhizae, with hyphal coils occupying the epidermal and outer cortical cells (Schmid et al. 1995).

BARK ENDOPHYTES

Many species of fungi inconspicuously colonize living bark on twigs and small branches of coniferous and broad-leaved trees, but almost nothing is known of their biology. The resinous young bark of conifers, such as Douglas fir, as well as the smooth-bark of several deciduous trees such as *Alnus*, frequently is colonized by nonlichenized members of the Arthopyreniaceae, including *Arthopyrenia plumbaria*, *Mycoglaena subcoerulescens* (*Winteria coerulea*), and *Mycoglaena* species ("*Pseudoplea*"). In eastern North America, *Arthonia impolita*, another nonlichenized member of a normally lichenized genus, is ubiquitous on young bark of *Pinus strobus*. *Vestigium felicis*, an unusual coelomycete with "cat's paw"–shaped conidia, is known only from young living twigs of *Thuja plicata* in the Pacific Northwest (Pirozynski and Shoemaker 1972). Other bark endophytes, primarily Ascomycetes, that fruit on recently dead twigs still attached to otherwise healthy trees—notably members of the Rhytismataceae but also including *Lachnellula* species (Hyaloscyphaceae), *Pezicula*, and *Mollisia* (Dermateaceae)—are relatively common. *Tryblidiopsis pinastri*, a common circumboreal species, which occurs on *Picea* species, and *Discocainia treleasei*, which occurs on *P. sitchensis*, are representative. Both species fruit in abundance in the spring on twigs that have been dead for less than a year and thus must be suspected of routinely colonizing bark of living twigs. Bark-colonizing endophytes may behave like some inconspicuous foliar endophytes that colonize healthy young tissue and fruit only on necrotic tissue (Table 12.1). Other species that appear to follow a similar strategy are *Therrya pini* and *T. fülii* on *Pinus* species, *Coccomyces strobi* on *P. strobus*, *Coccomyces*

heterophyllae on *Tsuga heterophylla*, and *Lachnellula ciliata* and *L. agasizii* on *P. menziesii* and *Abies* species. *Colpoma* species and *Tryblidiopsis pinastri* frequently fruit on recently killed twigs of oaks and spruces, respectively, but also commonly are isolated as endophytes from healthy inner bark. Fungi that are normally insect parasites, such as *Beauveria bassiana*, *Verticillium lecanii*, and *Paecilomyces farinosus*, have been isolated from living bark (Bills and Polishook 1991) and are not uncommon as endophytes of foliage. The endophytic occurrence of insect parasites has prompted the suggestion that bark may provide an interim substratum for saprobic growth (Carroll 1991; Elliot et al. 2000).

XYLOTROPIC ENDOPHYTES

Xylotropic endophytes are a distinct guild of xylem-colonizing species that are ecologically similar but apparently encompass a wide range of taxa (Table 12.1). The group is composed mainly of xylariaceous species, such as those of the genus *Hypoxylon* and related genera; Diaporthales (e.g., *Phragmoporthe*, *Amphiporthe*, *Phomopsis*); Hypocreales (*Nectria* species); a few Basidiomycetes (e.g., *Coniophora*); and a few other species more typical of the periderm mycobiota (Bassett and Fenn 1984; Boddy et al. 1987; Chapela and Boddy 1988a). In general, species diversity and abundance are low in this group compared to bark, shoot, and foliar endophytes. Some species are host specific. *Hypoxylon* species, for example, have specialized mechanisms for recognizing and attaching to a host (Chapela et al. 1990, 1991, 1993). In *H. fragiforme* germination is triggered only by host-specific monolignol glucosides.

An endophytic mycobiota peculiar to each host colonizes healthy, attached branches of alder (Fisher and Petrini 1990) and conifers (Sieber 1989; Kowalski and Kehr 1992) and oak and beech (Chapela and Boddy 1988b; Boddy 1992) in Europe and beech and aspen in North America (Chapela 1989). The fungi colonize the host tissue initially as disjunct infections that remain quiescent in healthy wood. The high water content of functional sapwood prevents active invasion and/or colonization, but when host stress, injury, or death causes water content to drop, active colonization resumes (Chapela and Boddy 1988b). Active growth and eventual sporulation occur in response to drying of the substratum (Chapela 1989; Boddy 1992). Xylotropic endophytes have life-history strategies analogous to those of foliar endophytes that infect healthy tissue early on; interrupt their growth for a prolonged period; and then grow rapidly again, engaging in saprobic exploitation of the substratum at the onset of physiological stress or senescence. Facultative pathogens, such as *Entoleuca mammata* on aspen (Manion and Griffin 1986), as well as many wood-decaying fungi, apparently have adopted this strategy of early endophytic occupation.

ROOT ENDOPHYTES

Mycorrhizal fungi are also endophytes. However, because they are primarily macrofungi that form well-characterized, specialized symbioses with their hosts, they are considered separately (see Chapter 15) from more generalized root endophytes. Root endophytes as we describe them refer to nonmycorrhizal microfungi that infect roots or associate with mycorrhizae. Roots of forest trees are colonized by a variety of nonmycorrhizal endophytes, although detailed investigations of healthy roots exist only for a few hosts. Soil fungi, saprobic rhizosphere fungi, fungal root pathogens, and endophytes overlap considerably, although certain taxa appear to be isolated repeatedly and preferentially as symbionts from living roots. Nonmycorrhizal microfungi isolated from serially washed mycorrhizal roots of *Picea mariana* (Summerbell 1989) were primarily sterile strains of *Mycelium radicis atrovirens* and *Penicillium* species. Holdenrieder and Sieber (1992) similarly used serial washing to compare populations of endophytic fungi colonizing roots of *Picea abies* in relation to site and soil characteristics. Of the 120 taxa recovered, *Mycelium radicis atrovirens*, *Penicillium* species, *Cylindrocarpon destructans*, and *Cryptosporiopsis* species were isolated most frequently.

Phialocephala fortinii, *P. dimorphospora*, *P. finlandia*, *Oidiodendron* species, *Geomyces* species, and *Scytalidium vaccinii* (Dalpe et al. 1989) are common components of a guild of endophytes forming root associations with alpine ericoid and other perennial hosts. *Mycelium radicis atrovirens*, generally regarded as a heterogeneous taxon, is the name commonly applied to sterile dematiaceous isolates. Roots colonized by these fungi have a unique morphology, particularly when associated with ericoid hosts; consequently, they sometimes are termed ericoid mycorrhizae, although the fungi apparently have a much broader host range (Stoyke and Currah 1991; Stoyke et al. 1992). Dark, septate endophytes dominated the mycobiota isolated from the fine roots of several species of forest trees and shrubs in Europe and western Canada. A large proportion of those isolates proved to be *Phialocephala fortinii*, a root-inhabiting fungus with a very broad host distribution and geographic range (Ahlick and Sieber 1996).

Hyphae in roots appear rhizoctonialike with "monilioid hyphae" and frequently produce a loose weft on the outer root surface. Root colonization is relatively extensive, but intracellular colonization of outer cortical cells is limited. Coiled or branched hyphae and intracellular microsclerotia may be present. In contrast, hyphae associated with conifer hosts form ectomycorrhizalike structures in which the intercellular colony resembles a Hartig net (Wilcox and Wang 1987; O'Dell et al. 1993). In culture, the fungi characteristically have thick-walled, dark-pigmented, septate hyphae and are usually sterile or very slow to sporulate.

Although root endophytes are apparently quite common, with wide geographic and host distributions, the ecological role of most species is unknown, although some may form mycorrhizae or be root pathogens. In addition to the roots of their ericoid hosts, *Phialocephala fortinii* and *Mycelium radicis atrovirens* commonly are isolated from roots of hardwoods (*Fagus sylvatica*), conifers (*Abies alba*, *Picea abies*, *Pinus sylvestris*, *P. resinosa*, *P. contorta*), and various alpine perennials (Wang and Wilcox 1985; Holdenrieder and Sieber 1992; Stoyke et al. 1992; O'Dell et al. 1993; Ahlick and Sieber 1996). Root morphology, depending on the extent of fungal infection, is described as ectomycorrhizal, ectendomycorrhizal, pseudomycorrhizal, nonmycorrhizal, or possibly pathogenic (Wilcox and Wang 1987). Species designations are based on morphotypes, which are not very informative. A current trend, therefore, is to use biochemical or genetic markers to distinguish host- or site-specific strains. This approach is exemplified by the restriction-fragment-length polymorphism analyses of sterile *P. fortinii* isolates from various alpine hosts (Stoyke et al. 1992) and of E-strain mycorrhizal fungi, a relatively uniform morpho-group that produces chlamydospores on and within infected roots (Egger and Fortin 1990; Egger et al. 1991).

The number and identities of species are uncertain. Repeatedly reported taxa in the group are *Rhizoctonia* species, *Phialocephala* species, *Phialophora* species, and *Chloridium* species. *Scytalidium vaccinii*, *Gymnascella dankaliensis*, *Myxotrichum setosum*, and *Pseudogymnoascus roseus* also may form ericoid root associations (Stoyke and Currah 1991). Species of *Exophiala*, *Hormonema*, *Monodictys*, and *Phaeoramularia* have been reported from the roots of several forest trees (Ahlick and Sieber 1996). Although most appear to have affinities among the orders of Discomycetes (Monreal et al. 1999), too little is known to generalize about the possible involvement of basidiomycetes. Sterile, basidiomycetous root endophytes have been reported (e.g., Ahlick and Sieber 1996); typically they are not melanized. The paucity of morphological characters and difficulty of inducing sporulation in root fungi contribute to the difficulty of identification. Inoculation experiments on host responses to infections or pathogenicity of *Mycelium radicis atrovirens* have led to contradictory and inconclusive results ranging from beneficial to pathogenic reactions. Host range is apparently broad, based on inoculation studies (Wilcox and Wang 1987). Roots and other tissues of various tropical epiphytes have been examined by Dreyfuss and Petrini (1984), Petrini and Dreyfuss (1981), and Richardson and Currah (1995).

INGOLDIAN HYPHOMYCETES IN ROOTS

Aquatic hyphomycetes, or Ingoldian fungi, are another common component of the root mycobiota, having been isolated from the living xylem and bark of submerged roots of various hosts and also periodically from terrestrial roots (see Chapter 23). *Heliscus lugdunensis*, *Tricladium splendens*, *Lunulospora curvula*, and *Varicosporium elodeae* were found as endophytes of terrestrial roots of *Alnus* species in Europe (Fisher and Petrini 1990; Fisher et al. 1991). *Campylospora parvula*, *Filosporella fistucella*, and 11 other Ingoldian hyphomycetes were recovered from submerged aquatic roots (Fisher et al. 1991; Marvanová and Fisher 1991). Eleven aquatic hyphomycete species were isolated as endophytes in submerged roots of *Picea glauca;* fewer were found in roots of *Acer spicatum* and *Betula papyrifera* (Sridhar and Bärlocher 1992a, 1992b). Among stem endophytes, aquatic hyphomycetes are more common in the outer bark than in the xylem (Fisher et al. 1991; Sridhar and Bärlocher 1992a, 1992b).

ENDOPHYTES IN ABNORMAL HOST TISSUES

Galls in plant tissues may harbor fungal populations distinct from those of normal tissues. Gall midges (Lasiopterini and Asphondyliidi), for example, introduce a variety of coelomycetous fungi into the galls, which serve as a source of food for the developing larvae (Bissett and Borkent 1988). Cecidiomyid midge galls on Douglas fir needles often support the heavily fruiting *Meria* anamorph of *Rhabdocline parkeri*, giving the appearance of a fungal disease (Stone 1988). Other fungi invading galls may be saprobes, insect parasites, or inquilines (organisms that inhabit insect galls and feed on gall tissue but do not parasitize the gall maker; Wilson 1995a). Several investigators have focused on the associations of endophytic fungi with foliar insect galls and cysts of root-infecting nematodes. *Phialocephala* species and *Leptostoma* species were the most common endophytic fungi in both healthy needles and needle galls of *Pinus densiflora*, but *Phomopsis*, *Pestalotiopsis*, *Alternaria alternata*, and an unidentified coelomycete preferentially colonized galls made in needles by *Thecodiplosis japo-*

nensis (Hata and Futai 1995). Wilson (1995a) compared fungal populations of leaves and galls of three host-insect pairs and found that the fungus species colonizing cynipid wasp galls on *Quercus garryana* and *Q. agrifolia* were typical of the endophyte species on those hosts—that is, the galls were invaded secondarily by foliar endophytes.

A possible role of endophytic fungi as antagonists of insect herbivores frequently is proposed as is the exploitation of such a relationship for biological control. Secondary invasion of leaf galls by foliar endophytes has been reported repeatedly in connection with larval mortality (Carroll 1988; Butin 1992; Halmschlager et al. 1993; Pehl and Butin 1994). Fungi isolated from galls of the aphid *Pemphigus betae* on cottonwood (*Populus angustifolia*) leaves, however, were not found as endophytes of cottonwood. The fungi included probably saprobic *Penicillium* species, *Cladosporium cladosporioides*, and *Verticillium lecanii*, which may act either as an insect parasite or as a mycoparasite. An unusual *Lophodermium* species confined to galls of the midge *Hormomyia juniperina* on *Juniperus* foliage, but distinct from the endophytic *L. juniperinum*, is mentioned by Cannon and Hawksworth (1995). Multi-level interactions between host plants, insects and other invertebrates, and internal fungi should be considered in studies of fungal diversity. A relationship between endophytic fungal infection, leaf miner injury, and premature leaf abscission in the emory oak (*Quercus emoryi*) has been demonstrated (Saikkonen et al. 1998). When fungal endophytes were excluded, leaves with high levels of miner injury were not dropped, but endophyte infection together with high minor injury caused premature leaf abscission.

ENDOPHYTIC PENICILLIA

Penicillium nodositatum forms specialized root associations with *Alnus incana* and *A. glutinosa*. These structures, or "myconodules," resemble actinorhizae, but like the ericoid mycorrhizae, are confined to the outer cortical layer of the root. The fungus invades and eventually kills the cortical cells as its highly branched and convoluted hyphal mass expands to occupy the entire cell (Cappellano et al. 1987). *Penicillium janczewskii* reportedly forms similar myconodules on *A. glutinosa* (Valla et al. 1989). Although *P. nodositatum* first was described from root nodules of *A. incana* in France (Valla et al. 1989), it since has been recovered together with *Aspergillus tardus* as a foliar endophyte from *Linnea borealis* in Oregon (G. C. Carroll and J. Frisvad, personal communication), suggesting the existence of a broader ecological and geographical range for this fungus. Other *Penicillium* and *Aspergillus* species have been recovered as endophytes from various hosts in Oregon and from *Sorbus* species in Germany, including the more common *P. expansum*, *P. westlingii*, *P. pinophilum*, *P. citreonigrum*, *A. sydowii*, and *A. terreus* (J. Frisvad, personal communication). Summerbell (1989) reported 20 species of *Penicillium* from roots of *Picea mariana*, of which *Penicillium spinulosum* and *P. montanense* were relatively frequent. Endophytic Penicillia and Aspergilli are apparently widespread and polyphyletic.

ENDOPHYTES AND GLOBAL SPECIES DIVERSITY

It is unlikely that anyone following the very broad circumscription of endophytic fungi that we have used in this chapter will attempt to census all endophytic fungi in a landscape level study. A more realistic approach will be to characterize the endophyte species from a single host or group of hosts. Sampling needs will depend largely on host abundance and distribution (e.g., dominant, rare, disjunct) of the host plant and the tissue types (foliage, stem, bark, xylem, root) to be sampled. Studies to date suggest that more intensive sampling increases the recovery of rare species, which are likely also to occur on many hosts, but the most common species on a specific host will be widely distributed on that host. Sampling levels include within tissue (e.g., leaf) on plant, within plant (i.e., several leaves), and within site (i.e., leaves of several host individuals). Estimates of numbers of species present can be made through effort/recovery trajectories, diversity indices, and/or rarefaction (Magurran 1988; Bills and Polishook 1994; Lodge et al. 1996b; Arnold et al. 2000), all of which can be developed for all levels to evaluate adequacy of sampling effort.

The magnitude of sampling required can be illustrated by considering the leaf area included in a 50,000-hectare site. An appropriate scale for consideration of the domain of endophyte colonies is on the order of square millimeters (Stone 1987; Carroll 1995). A sampling area of 50,000 hectares converts to of 5.0×10^{14} mm^2. Leaf area indices (a ratio of plant cover to ground surface area) range from 4.0 (savanna, shrubland) to 12.0 (temperate evergreen forest, tropical rain forest), so the total leaf area from which to sample in 50,000 hectares ranges from 2 to 6×10^{15} mm^2, depending on vegetation type (e.g., Barbour et al. 1987). Leaf area indices give values for one surface only, so for studies in which both upper and lower leaf surfaces are examined (e.g., histological studies), the area doubles. The density of infections of a single endophyte species, *Rhabdocline parkeri*, can vary from 0.2 to more than 20 per mm^2 (Stone 1987), and the total number of infections in a single old-growth Douglas fir tree was estimated to be on the order of 1×10^{11} (McCutcheon et al. 1993). Numbers of foliar endophyte infections in 50,000 hectares of tropical or boreal forests thus could be estimated to range between 1×10^{14} to 2×10^{16} or greater.

Estimating the number of endophytic species occurring in a given biome is, at present, guesswork. If the total number of fungal species on earth approaches the 1.5 million proposed by Hawksworth (1991), much more than 1 million species remain to be discovered (perhaps 15 times the number of fungi already described). It is almost certain that a substantial proportion of these undiscovered species will be what we consider here to be "endophytes." It follows that more accurate estimates of numbers of endophytic species will lead to more accurate estimates of global fungal species diversity. It generally is accepted that there exist at least as many species of fungi globally as phanerogams, which number about 250,000. Endophytic fungi are ubiquitous in phanerogams, and intensive surveys of hosts invariably yield new species. It is possible that the number of species of endophytic fungi alone equals that of phanerogams and could exceed it. The exact range of this multiple, however, is a matter of conjecture.

Table 12.1 gives the number of endophytic species endemic on certain hosts and lists some taxa described recently as a result of endophyte surveys. Data from the table suggest that the ratio of endemic endophyte species to hosts, at least in temperate plants, is greater than 1, but at present the information does not permit more precise estimates. Whether this ratio holds true in tropical forests remains to be investigated (Arnold et al. 2000). Investigators who conducted the studies listed in Table 12.1 used various methods to sample and enumerate fungi, so comparisons are only approximate. Numbers of species recovered clearly reflect the intensiveness of sampling effort. An endophyte-to-host ratio of 4.0, which could account for the bulk of the "missing" 1.4 million species, is not improbable. Genera such as *Phyllosticta*, which are widespread, highly speciose, and nearly exclusively endophytic, may be vastly underrepresented by our current species lists. The magnitude of the ratio of host-endemic species to generalists very likely is influenced by regional differences in patterns of host-species diversity and distributions (e.g., between temperate and tropical forests), as pointed out by May (1991). Ratios of fungi to phanerogams in temperate regions range from 6:1 in Britain, which has a well-characterized mycobiota, to 1:1 in the United States, where the fungi are less well studied.

May (1991) argued that there should be fewer endemic species in the tropics, where host distributions often are disjunct. Investigations of tropical endophytes, however, have produced data that support and data that refute that hypothesis (Arnold et al. 2000). A study of endophytes in the tropical palm *Euterpe oleracea* yielded three new species of *Idriella* (Rodrigues and Samuels 1992) and a new genus of loculoascomycete, *Letendraeopsis*, suggesting that the diversities of tropical and temperate endophytes may be similar. In contrast, the endophyte mycobiota reported by Lodge and colleagues (1996a) for *Manilkara bidentata*, a broad-leaved tropical tree, included 23 species, most of which were known from a broad range of hosts. They estimated that the sample of one leaf from each of three trees yielded more than 80% of the endophyte mycobiota. Genetic variation among endophytic isolates of *Xylaria cubensis*, a species widespread on tropical and subtropical hosts, is high, presumably as a result of a sexual recombination (Rodrigues et al. 1995). If this pattern is representative for endophytes of tropical hosts, then the magnitude of the "endophyte multiple" necessarily would be diminished.

Clearly, a more accurate estimate of the ratio of endophytic fungi to phanerogams will affect significantly estimates of global fungal diversity, and so comparative surveys aimed at evaluating this proportion should be accorded a high priority. Equally important is the examination of variation and speciation within particular genera of fungi, such as *Phyllosticta*, that appear to have evolved a specialized endophytic habit. Symbioses may be a fundamental factor in speciation within certain endophytic genera. Intensive examination of such genera may reveal a large number of new species.

13

SAPROBIC SOIL FUNGI

GERALD F. BILLS, MARTHA CHRISTENSEN, MARTHA POWELL,
AND GREG THORN

Soil is a highly complex medium, an ecosystem with multiple abiotic and biotic components. It is not one habitat—it consists of a myriad of minute and microscopic habitats and microenvironments. Consequently, microbes, fungi, and soil fauna in soils fill multiple niches. That biotic complexity and the fact that we cannot look very far into soil without disturbing it make soil one of the most difficult ecosystems to study *in situ*. Human-caused disturbances, especially cultivation, erosion, and contamination, affect soil habitats and, thus, have an impact on the diversity of the soil biota. Knowledge of the constituent organisms in soil is of practical importance, therefore, for the conservation of that nonrenewable resource.

Among microorganisms inhabiting soils, fungi commonly rank as the most abundant in terms of biomass and physiological activity (Kjøller and Struwe 1982; Schnürer et al. 1985). Fungi comprise an estimated 78–90% of the total decomposer biomass in grassland soils, and in a British deciduous forest, fungal mycelia contributed approximately 89% of the total living microbial biomass (Frankland 1982). Soil fungi are of interest to ecologists and to applied researchers because of their importance in decomposition, carbon and nitrogen storage, biogeochemical cycles, soil stabilization, and plant parasitism. Their ubiquity and well-known capacity for formation of mycorrhizae, complex biosyntheses, biodegradation, and internecine warfare account for their interest to researchers studying mycotoxins, natural products chemistry, biodeterioration, mycoses and allergies, bioremediation, enzymology, and biological control. Fermentation products of soil fungi—for example, penicillin, cephalosporin, cyclosporin (Sandimmune, Norvartis AG), and lovastatin (Mevacor; Merck and Co.)—are important in world medicine. All potential users of fungi can benefit from a broad base of information on the natural history of soil fungi and require usable taxonomic guides. Users of fungi may require on-demand access to certain fungal species and may need to know how to plan and execute an inventory of soil fungi (Bills 1995).

For the purposes of our discussion, we use a broad definition of soil—an unconsolidated material developed through the interaction of plants and intact or pulverized rock. The three abiotic constituents of soil—mineral particles (sand, silt, clay), water, and air—are intermingled to varying degrees with the dead and broken remains of plants, animals, and microorganisms that constitute an energy-rich organic matter. Together, the biotic and abiotic components support a teeming community of soil-dwelling fungi, bacteria, and microfauna and mesofauna. Grassland soils are enriched by organic accretion throughout the soil profile, especially from grass roots that are annual, finely divided, and contribute a biomass that far exceeds above-ground biomass. Forest soils, in contrast, are enriched by debris from a massive above-ground biomass plus substantial input from mycorrhizae (Fogel 1980; Vogt et al. 1982; Vogt et al. 1983). In deserts, desert-grasslands, and some forest communities, the above-ground debris often consists of relatively intact organic matter (litter) only. In mesic or wet grasslands and in most shrublands, heathlands, and forests, the superficial debris is underlain by an amorphous, aromatic layer of organic material (humus) that rests on mineral soil impregnated with humus. In nearly all soils, organic matter decreases downward through the soil profile. In the deep-lying horizons of forest soils especially, it may be primarily in the form of highly stable molecules—polyphenolic compounds, for example, that are hundreds or thousands of years old. Technically, the soil system is all horizons, from litter or standing water to bedrock; as will become evident, however, investigators of soil fungal communities generally have limited

their studies to forest soils in the mineral horizon immediately adjacent to humus in forest soils or to grassland soils in the upper 2–10 cm of the soil profile. Fungi of litter and unconsolidated plant debris are treated separately in Chapter 11.

The trophic relationships of saprobic, mutualistic, and parasitic soil fungi explain most of what soil fungi do as components in ecosystems. In addition to functioning as primary, secondary, and tertiary decomposers in the often complex process of degrading organic matter (Hudson 1968; Hayes 1979; Swift 1984), they are significant elements in many food webs involving soil fauna and microbes. Diverse soil fungi parasitize living members of those webs and attack remains of virtually all organisms. In turn, they are consumed by mycophagous nematodes; protozoans; collembolans; oribatida; enchytraeids; dipteran larvae; and to a lesser extent by gnats, earthworms, certain water bears (Tardigrada), and mites in the suborders Prostigmata, Astigmata, and Mesostigmata (Krantz 1978; Newell 1984a, 1984b; Moore et al. 1988; Ingham 1992; Shaw 1992). Fungivore-fungus symbioses, wherein the fungivore facilitates growth of the fungus, have been reported for wood-boring beetles, attine ants, and certain termites (M. M. Martin 1987; Shaw 1992). In many instances, soil animals feeding on fungi are either species-specific or species-selective. From their perspective, all fungi are not equal, and biodiversity matters.

All major taxonomic groups of fungi are encountered in soils. Especially prevalent among saprobic filamentous forms are the Saprolegniales, Pythiales, Mucorales, Eurotiales, Microascales, Hypocreales, Sordariales, Onygenales, Leotiales, Pezizales, and a plethora of mitosporic fungi (which are almost exclusively anamorphs of ascomycete or basidiomycete species). Many soil fungi are saprobic on organic material in various stages of decay and, apart from several notable exceptions, are readily culturable. Chesters (1949:206) referred to basidiomycetes as "the missing link in soil mycology," and, unfortunately, his comment is still accurate more than 50 years later. In the total litter and humus of a British deciduous forest, including woody debris and dead roots, Frankland (1982) reported that basidiomycetes accounted for approximately 60% of the total living microbial biomass; under-representation of basidiomycetes or their absence in general surveys is evidence of a technique-based bias (Warcup 1959; Warcup and Talbot 1962; Thorn et al. 1996). Biotrophic mutualists and parasites of plants, animals, and fungi, as well as many zoosporic fungi, also are underestimated by most soil plating techniques. Many of the species regularly isolated from soils are adapted for colonization of specific root- or animal-derived organic materials. Others appear to be widely distributed, plurivorous fungal taxa that colonize a variety of organic substrata. A smaller fraction may be incidental or transient invaders of the soil; their spores or other propagules are introduced accidentally and then remain dormant until they perish. Chytridiomycetes (chytrids), typically considered aquatic fungi because their asexual reproductive propagule is a flagellated zoospore dependent on water for dispersal, also are recognized as common soil-inhabiting fungi (Willoughby 1956; Gilman 1957; Sparrow 1960; Tribe 1960; Booth 1969). They are not readily isolated with techniques applicable for hyphal fungi and, consequently, often are overlooked in biodiversity studies.

Direct observation of fungi in soil is possible, but usually only indistinct and featureless mycelia are visible (Barron 1971; Frankland et al. 1990). Thus, soil fungi *in situ* generally cannot be identified without a heavy preinvestment in the development of species-specific probes based, for example, on nucleic acids (see Chapter 6) or immunofluorescent labelings. At present, taxonomic surveys of soil fungi nearly always are based on indirect isolation techniques in which fungal propagules (zoospores, spores, mycelia, sclerotia) in soil are dispersed on an artificial medium or trapped on organic baits. Following growth and the reculturing of isolated mycelia, the species represented in the soil suspension are identified or characterized on the basis of a combination of morphological, physiological, and biochemical features. Indirect surveys may have very little to do with the spatial and temporal patterns of fungal occurrence, and slight variations in technique can result in different perceptions of fungal activity (Griffin 1972).

The understanding of genetics and mating systems and delimitations of population structure have progressed enormously in recent years, especially for soil-borne phytopathogens. Even so, researchers still are plagued by their inability to observe and identify taxonomic entities directly in the soil. On-site descriptions are possible only for large, identifiable, but often ephemeral, structures such as fruiting bodies, sclerotia, rhizomorphs, and mycelial cords (Warcup and Talbot 1962; Frankland et al. 1990). Community analyses and comparative community and biogeographic studies of soil fungi continue to be based on qualitative and quantitative data obtained indirectly through soil plating. A key issue in the use of any soil-plating technique is the degree to which presence of isolated taxa and their relative abundances are related to the distribution and physiological activity of soil mycelia *in situ,* especially because multiple propagules do not necessarily represent separate mycelia or fungal individuals (Zak and Visser 1996). Although the indirect, cultural approach is fraught with technical limitations and results are difficult to interpret, that approach remains the preferred method for inventorying species composition of the soil mycota.

Methods for the study of soil fungi and the limitations of those methods have been reviewed extensively during the past four decades (Parkinson and Waid 1960; Garrett 1963; Barron 1971; Griffin 1972; Frankland et al. 1990; Gams 1992; Singleton et al. 1992; Parkinson 1994; Watanabe 1994). This chapter does not review all imaginable methods for investigation of soil fungal communities and their historical development; its goal is to outline and recommend the authors' preferred methods for capturing living soil fungi in the field and in the laboratory, for describing community structure, and for documenting diversity. At the same time, we intend in this chapter to inform readers about the complexities and applications of certain methods by reference to literature and to advise them against certain mycological practices that have led to erroneous perceptions of the soil fungal community and enormous loss of biodiversity information. Varied new approaches to the evaluation of soil and rhizosphere communities have demonstrated that inventories using a single-isolation method detect only a portion of the organisms present (Punsola and Guarro 1984; Carreiro and Koske 1992; Petrini et al. 1992b; Mouchacca 1995; Okuda et al. 1995; Thorn et al. 1996).

Several groups of filamentous fungi and fungus-like organisms, such as thermophilic, osmophilic, and keratinophilic fungi; yeasts; yeastlike fungi; dictyostelids; protostelids; and acrasids are abundant in soils. Those groups are covered separately in Chapters 14, 16, 20, and 25 because they have specialized life cycles and require unique isolation methods. Mycorrhizal and endophytic fungi, intimately associated with plant roots, are treated in Chapters 15 and 12, respectively, and fungal parasites of the soil microfauna are discussed in Chapter 19.

TAXONOMY, DIVERSITY, AND DISTRIBUTION

TAXONOMIC LITERATURE

To our knowledge, no complete inventory of soil fungi has been undertaken for any single geographic region, but numerous intensive and a few extensive surveys have been carried out in Europe, North America, the Middle East, India, Japan, Taiwan, Australia, New Zealand, and the Arctic. Tropical soils; soils from austral South America, southern Asia, and much of Australia and Africa; and soils in other geographically and ecologically remote areas have been examined less thoroughly. Conclusions drawn from studies of tropical soils are discussed in the section on "Species Distributions and

Assemblages," later in this chapter. Accounts of historically important studies of soil mycotas can be found in other reviews (Burges 1958; Griffin 1972; Christensen 1981a; Christensen and Tuthill 1985; Watanabe 1994).

Several modern guides to the major genera and species of soil fungi have been compiled by researchers in temperate regions (Barron 1968; O'Donnell 1979; Domsch et al. 1980; Watanabe 1994); those authors have included many common tropical species as well. In soils of temperate or boreal regions, probably 60–80% of the principal, sporulating species can be identified to species with a moderate degree of reliability. A higher percentage of unknown forms will be encountered in tropical regions and on nontropical sites that are geographically remote or ecologically unusual. The development of comprehensive guides to soil fungi, especially tropical soil fungi, should be a high priority in fungal systematics. Identification of many soil fungi remains difficult, and trained systematists often are needed to identify isolates obtained in surveys and in the course of work in biodegradation, phytopathology, and natural products discovery. More importantly, ecologists and applied researchers rely on systematists to produce and refine basic monographic and floristic works that will make accurate identification possible.

The number of species of soil-inhabiting fungi is unknown. Even within the universe of known species, the number of taxa is formidable. Domsch and colleagues (1980) listed 153 common genera in their compendium, and Barron (1968) treated 202 genera of soil-dwelling mitosporic fungi. New species of soil fungi, even in common genera, are being described at a rapid rate (Udagawa et al. 1994; Cannon et al. 1995; Matsushima 1995). Anyone venturing into an inventory of soil fungi, even on a restricted scale of a few hundred isolates from a single soil sample, should expect to identify dozens of genera and from 50–100 species. Species belonging to taxonomically complex, species-rich genera often are isolated, and specialized taxonomic references are required for their identification. Monographs and recent revisions of older works must be consulted to confirm identifications in taxa such as ascomycetes (Lundqvist 1972; Currah 1985; von Arx et al. 1988; Hanlin 1990, 1998b); dark-spored hyphomycetous mitosporic fungi (Ellis 1971, 1976); pycnidial and acervular mitosporic fungi (Sutton 1980; Nag Raj 1993); *Acremonium* (Gams 1971, 1997); *Aspergillus* and its teleomorphs (Raper and Fennell 1965; Christensen 1981b, 1982; Christensen and States 1982; Klitch and Pitt 1988; Kozakiewicz 1989); *Chaetomium* (Hawksworth and Wells 1973; von Arx et al. 1986); *Chrysosporium* (Van Oorschot 1980); *Cylindrocarpon* (Booth 1966; Brayford 1992); *Cylindrocladium* (Crous and Wingfield 1994); *Fusarium* (C. Booth 1971;

Gerlach and Nirenberg 1982; Nelson et al. 1983); *Mortierella* (Linnemann 1941; Gams 1977); *Mucor* (Schipper 1978); *Myrothecium* (Tulloch 1972); *Oidiodendron* (Barron 1962); *Paecilomyces* (Stolk and Samson 1972; Samson 1974); *Penicillium* and its teleomorphs (Stolk and Samson 1972; Pitt 1979; Ramirez 1982; Gams 1993); *Phoma* (Dorenbosch 1970; Gruyter and Noordeloos 1992; Boerema 1993; Gruyter et al. 1993; Boerema et al. 1994); *Phytophthora* (Waterhouse 1970; Ho 1981); *Pythium* (Waterhouse 1968; Van der Plaats-Niterink 1981; Martin 1992); *Rhizoctonia* (Sneh et al. 1991); *Trichoderma* (Rifai 1969; Bissett 1991a, 1991b; Samuels et al. 1998); and *Verticillium* (Gams and van Zaayen 1982; Gams 1988b, 1997). Computer-assisted diagnostic aids are available for *Penicillium* (Pitt 1990; Bridge et al. 1992) and *Fusarium* (K. Seifert, FusKey, see Appendix III). Additional taxonomic works referring to species groups or minor genera are available (Zycha et al. 1969; Carmichael et al. 1980; Domsch et al. 1980; Hanlin 1990, 1998; Rossman et al. 1987, 1999; von Arx 1981, 1987; Gams 1993; and Hawksworth et al. 1995).

Ensuring accuracy of identifications and descriptions of new taxa may require comparisons of unknown with authentic specimens. The most valuable specimens for comparative taxonomic studies are live cultures, although dried culture mats and semipermanent slides of sporulating structures sometimes are useful (see Chapters 2 and 3). Many of the world's major service culture collections (ATCC, CBS, DAOM, IFO, IMI, MUCL, NRRL; see Appendix III explanation of abbreviations) have large holdings of soil fungi. Specialized collections may reside in laboratories of expert research groups (Hall and Minter 1994). Mycotoxin-producing strains of *Penicillium, Aspergillus,* and *Fusarium,* for example, are maintained at the Technical University of Denmark, Lyngby, Denmark (Singh et al. 1991), and *Fusarium* species are maintained at the *Fusarium* Research Center, Pennsylvania State University, College Station, Pennsylvania (Nelson et al. 1983).

SPECIES DIVERSITY

The specific biotic and abiotic factors influencing species occurrence and species diversity in soil fungal communities are numerous and interactive (Christensen 1989; Beare et al. 1995). Rigorous statistical analyses including canonical correlation analysis and multiple regression have confirmed the existence of many independent variables that presumably influence and are influenced by compositional dynamics in the saprobic soil fungal community (Bissett and Parkinson 1979a, 1979b, 1979c; Lussenhop 1981; Widden 1986b). In fact, biotic interactions may equal or exceed abiotic factors as effectors of composition and diversity in soil fungal communities (Swift 1984; Christensen 1989; Ingham 1992).

Species richness in soil fungal communities apparently increases through both primary and secondary succession in plant communities (Swift 1976; Christensen 1981a; Zak 1992). As has been pointed out in Chapter 5 and elsewhere, however, structural change in the community (an increase in equitability, for example, as a result of diminishing densities among principal species) is a parameter different from species richness and unless specifically analyzed confuses the matter of total species present (i.e., species richness; Christensen 1981a). Structural and compositional responses of soil fungal communities to disturbance have been addressed in various reviews (Gochenaur 1981; Zak 1992; Zak and Visser 1996).

High numbers of species may be characteristic of fungal communities occurring in complex mixtures of resources (Swift 1976; Christensen 1989). High species richness commonly is attributed to extensive niche differentiation, specifically the packing of more species with tighter niches along resource axes. In soil fungal communities, environmental variation through time, and fungal capacities for survival undoubtedly are additional contributing factors. The primary factors promoting high species diversity in soil fungal communities are probably the following: (1) microhabitat heterogeneity; (2) intermingling of primary, secondary, and tertiary decomposers; (3) faunal grazing; (4) persistence of organic matter that promotes niche partitioning; (5) temporal changes in climate and on-site vegetation; and (6) the remarkable capacity of the cells and mycelia of fungi to behave individualistically (Christensen 1989; Rayner 1994; Zolan 1995).

Estimates of species richness for soil fungi vary widely with the extent and intensity of the survey (Tables 13.1 and 13.2). Slight variations in the isolation methods used can have dramatic effects on quantitative and qualitative results, but primarily, the bias, skill, and tenacity of the investigator greatly influence the outcome of an inventory, especially with respect to the length of the species list. Numbers of isolates examined obviously affect numbers of species reported. From examination of selected surveys (Tables 13.1 and 13.2), one can gain a

TABLE 13.1

Species Richness and Sampling Efforts for Saprobic Soil Fungi in Alpine, Temperate, and Tropical Soils*

Location	Habitat(s)	Number of isolates	Number of taxa	Reference(s)
Alberta	3 alpine plant communities	4643	128 species	Bissett and Parkinson 1979a
Germany	Wheat fields	23,500	220 species	Gams and Domsch 1960, 1969
Sweden	Clear-cut conifer forest	3891	96 species	Bååth 1980
New York	*Quercus-Pinus* forest	15,770	267 species	Gochenaur and Woodall 1974
New York	*Quercus-Betula* forest	10,684	89 species	Gochenaur 1978
South Dakota	Grassland	250	62 species	Clarke and Christensen 1981
N. Arizona, S. Utah	Desert	19,000	228 species, 87 genera	States 1978
Veracruz, Mexico	Rhizospheres of *Coffea arabica*, 2 plantations	3584 and 2349	114 and 116 species	Persiani and Maggi 1988
Tai National Park, Ivory Coast	Native; burned, cultivated, subhygrophilous forests	6246	215 identified species, 120 genera; 164–194 species/study site	Rambelli et al. 1983; Maggi and Persiani 1992
Galapagos Islands, Ecuador	Xerophilic soils and litter	Unknown	250 species, 50% identifiable	Mahoney 1972
Taiwan	Various soils and organic materials	Unknown	47 *Penicillium* species + teleomorphs; 42 *Aspergillus* species + teleomorphs	Tzean 1994
India	Various soils	Unknown	>60 *Aspergillus* species	Rai et al. 1974
Australia	Transect from north coast to south coast	4484 *Fusarium* isolates	22 named *Fusarium* species + 3 undescribed	Sangalang et al. 1995

*Compare numbers of isolates and taxa with those listed in Table 13.2.

TABLE 13.2

Species Richness of Fungi in Major Vegetation Types in Wisconsin*

Vegetation type	Stands	Isolates	Taxa	References
Prairie	25	5700	>111	Orpurt and Curtis 1957; Wicklow 1973
Southern upland hardwood forests	13	1080	53	Tresner et al. 1954
Southern wet-mesic hardwood forests	5	3000	>210	Christensen et al. 1962; Novak and Whittingham 1968
Southern wet forests	5	2500	154	Gochenaur and Whittingham 1967
Southern wet shrublands	5	2500	115	Gochenaur and Backus 1967
Northern upland conifer-hardwood forests	36	8061	>476	Christensen 1969; Wicklow and Whittingham 1974, 1978
Northern conifer swamps and bogs	15	3000	>133	Christensen and Whittingham 1965
Sand barrens	3	1000–2000	>18	Phelps 1973

*Compare numbers of isolates and taxa with those listed in Table 13.1.

sense of the numerical range of species extractable from particular soils and the effort necessary for a minimal inventory. Intuitively, one might expect a specific volume of tropical soil to be more species rich than an equal volume of temperate soil, but recent evidence indicates that species diversity in tropical soils is similar to that in temperate soils (Pfenning 1993; Bettucci and Roquebert 1995) and that the species richness of fungi in humid, tropical soils is comparable to that of fungi in temperate forests (Table 13.1).

Two graphical methods of representing species-abundance data—that is, species-isolate curves and rank-abundance plots (see Chapter 5; Fig. 13.1)—are useful for examining species diversity of fungi isolated from different soils or by different methods. Once species abundance is plotted, the nonlinear relationship of species richness to number of isolates examined is readily apparent. For this reason, simple species-to-isolate ratios cannot be used to compare species richness among different studies (Tables 13.1 and 13.2; also see Chapter 5). Rarefaction (see Chapter 5) is a mathematical technique that can be useful for projecting levels of species richness within a small number of isolates based on ranked abundance of the isolated population. Rarefaction also permits random resampling of subsets of isolates from larger pools, so that species richness can be compared directly among unequal sample sizes (see "Selecting Colonies," later in this chapter).

THE WISCONSIN SURVEYS AND THEIR RELEVANCE

The Wisconsin studies were a series of surveys of soil fungi in different vegetation types in Wisconsin for the purpose of relating soil fungi community structure to that of the overlying forests. The variety and pattern of vegetation types in Wisconsin are landscape features that reflect the occurrence in the state of three biomes and four major floristic centers (Curtis 1959). Prairies interspersed with southern hardwood forests and savannas (Prairie, Ozarkia, and Alleghenian floristic elements) are prevalent communities in the southwestern half of Wisconsin. In the northeastern half of the state, Alleghenian and Boreal floristic elements blend in hardwood forests, conifer-hardwood forests, and conifer bogs. That diversity of plant communities and the availability of detailed, quantitative ecological descriptions for specific sites (stands) provided an ideal basis for a series of quantitative surveys of soil microfungal studies in the state over a span of more than 20 years. Investigators ranked the microfungal species by commonness in each of the state's vegetation types (Table 13.2). In all, they surveyed 107 prairies or forests in 18 of the 21 major community types recognized by Curtis (1959) and prepared more than 25,000 fungal isolates from soil immediately underlying humus or litter.

The Wisconsin survey program evolved under the guidance and encouragement of two well-known botanists at the University of Wisconsin, professors M. P. Backus and J. T. Curtis. The enthusiasm of Professor Backus for mycological detail was complemented by the broader view and stimulating thoughts about fungal ecology of Professor Curtis. The guiding hypothesis in the early studies was that soil fungi, like plants and animals, are responsive to the totality of their environment. If true, then soil fungal surveys using the developing quantitative methods of plant ecology would (1) reveal the existence of differing assemblages of principal species in the various vegetation types of Wisconsin and (2) provide a means for discovering specific correlations between fungal and plant communities by multivariate analysis.

As reported elsewhere, the general hypothesis of concomitant species in the Wisconsin plant and soil fungal communities was supported (Griffin 1972; Christensen 1981a)—that is, fungal communities varied widely

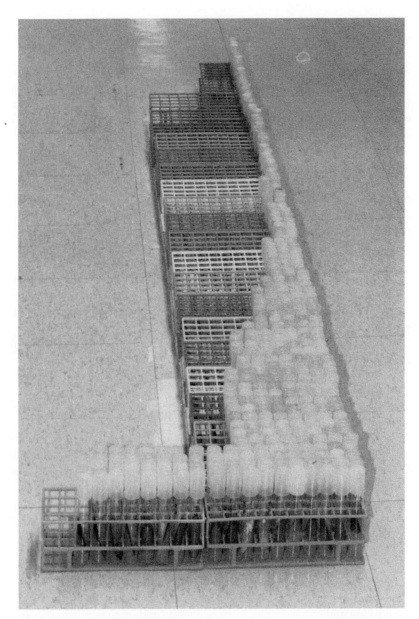

FIGURE 13.1. Isolates of soil fungi recovered from a single soil sample on a single medium arranged in descending order of abundance to represent a species-abundance curve. The most abundant species (72 isolates) is in the first two racks in the foreground. Each of the next 14 racks contains one species. The species represented by single isolates are the last racks in the background.

among biomes, within single communities or community types. However, the imprint of vegetation could be overridden by a more proximate factor. In the conifer-hardwood forests, calcium content of the litter, irrespective of tree species, correlated with the first axis of an ordination of the fungal communities based on compositional similarity; dry to mesic sequences of deciduous and coniferous forests, respectively, correlated with the second and third axes (Christensen 1969). In the three-dimensional (3D) ordination of the soil fungal commu-

nities, spheres of occurrence of fungal assemblages were apparent for all major tree species.

In several studies in Wisconsin and elsewhere, effects of disturbance and treatment (e.g., burning, fumigation, and logging) on the community of soil fungi have been discerned and quantified using the Wisconsin protocol or a similar technique (Wicklow 1973; Bissett and Parkinson 1980; Bååth et al. 1984; Banerjee and Anderson 1992). Such effects, however, were usually quantitative rather than qualitative.

At the biome level, certain arrays of taxa from among the 30 or more principal microfungal species in the soil are unmistakable indicator assemblages for a biome. Christensen (1981a) reviewed the patterns of occurrence of species commonly encountered in the soils of deserts, grasslands, forests, heathlands, and tundra on six continents. The presence of such forms as *Fusarium* species, *Paecilomyces lilacinus*, *Penicillium restrictum*, *Mortierella alpina*, and *Pyrenochaeta* species in temperate grassland soils and *Mortierella vinacea*, *M. isabellina*, *M. nana*, *Paecilomyces carneus*, *Penicillium daleae*, *P. janczewskii*, *Torulomyces lagena*, and *Oidiodendron* species in conifer forest soils is predictable, irrespective of geographic location, because of the nature of the biome in its broadest sense (Gochenaur and Woodall 1974; Christensen 1981a).

The pioneer studies in Wisconsin and elsewhere (Gochenaur and Woodall 1974; Morrall 1974; Gochenaur 1978; Bissett and Parkinson 1979a, 1979b, 1979c) could be extended in two exciting and potentially productive ways:

1. First, the quantitative survey could be expanded, using the same (or comparable) techniques used in the Wisconsin studies. Increasing the numbers of samples would increase detection of infrequent and rare species (e.g., 500 isolates per soil sample and 10 or more samples per relatively homogeneous site, such as a permanent plot in an All Taxa Biodiversity Inventory study area). The resulting sample of the soil microfungal community at the given site would consist of 5000 or more isolates and an unknown number of species, but perhaps 50–500 (Gochenaur 1978; Christensen 1989; Bills and Polishook 1994). With tabulation of both frequency and density for each apparent taxon, a variety of qualitative and quantitative data either would be immediately available or calculable, including the following:

   a. Total number of species represented among the 5000 or more isolates;

   b. Structure of the community in terms of equitability, species diversity, and species richness in this limited but defined portion of the total spectrum of species present;

   c. Species composition of the community under study, with the species ranked by relative abundance.

   Such studies would (1) contribute to our knowledge of the geographic distribution, occurrence and prominence by habitat, fungal-faunal associates, and other autoecological features of individual species (Domsch et al. 1980; Arnebrant et al. 1990); (2) provide the basis for improved taxonomic treatments; (3) facilitate the discovery of concomitant patterns of distribution of species in the microfungal community

and their correlation with biotic and abiotic factors and processes, through multivariate analysis (Bissett and Parkinson 1979a, 1979b, 1979c; Widden 1986a, 1986b, 1986c; Arnebrant et al. 1990); and (4) stimulate the generation of hypotheses in such areas as plant–animal–fungal interactions, including food-web and mutualistic interactions, biological control, and nutrient accumulation or transformation in fungi (Christensen 1989; Cromack and Caldwell 1992; Wainwright 1992).

Several aspects of the research plan just outlined have been carried out recently by Bissett and Parkinson (1979a, 1979b, 1979c), Bååth and colleagues (1984), Pfender and Wootke (1988), Arnebrant and associates (1990), Maggi and associates (1990), and Fritze and Bååth (1993). All of those investigators have assessed the fungal communities quantitatively and have analyzed their results using a variety of multivariate techniques; their work follows in the spirit of the early Wisconsin studies.

2. Second, the species inventoried could be expanded through the use of rigorously selective techniques. As has been pointed out here and elsewhere (Christensen 1981a; Nordgren et al. 1985; Arnebrant et al. 1990), the top-ranking 20–40 species obtained with soil plating (Warcup 1950), suspension plating (Tresner et al. 1954; Christensen 1969), and particle filtration (Gams and Domsch 1967; Bissett and Widden 1972; Söderström and Bååth 1978) are similar when a nutrient-poor medium is used and the plates are incubated at 17–24°C or at *in situ* temperatures first and then at room temperature. To add to species lists when inventory is the primary objective, the investigator can expand the number of isolates beyond the 180–500 per stand used in the Wisconsin studies or can undertake isolation of species not readily obtainable by application of the standard protocol. Ingenuity and patience are the only factors limiting the development of rigorous, selective techniques. Four procedures among several used by Wisconsin workers or their students are ethanol pasteurization (Mahoney et al. 1969; Huang 1971; Mahoney 1972; Huang and Schmitt 1975), soil steaming (Gochenaur 1964), addition of botran (dichloran) to the medium (Griffin and Garren 1974; Tuthill 1985), and high-temperature incubation (Gochenaur 1975; Huang and Schmitt 1975).

SPECIES DISTRIBUTIONS AND ASSEMBLAGES

During the past 40 years, researchers describing soil fungal communities consistently observed correlations between fungal community composition and environmental parameters and historical patterns of land use. The correlations are especially pronounced and predictable among the most common fungi of the humus–mineral soil interface. An adjunct observation, at least in temperate regions, is that a strong positive correlation often exists between vascular plant diversity and soil-fungal diversity. Probably the more important generalizations are that the compositions of soil fungal communities are regulated strongly by climatic and edaphic factors and ecological similarity among habitats supersedes geography. An experienced microfungal taxonomist or ecologist reading the list of the 10–20 most abundant fungi in a soil sample can guess with great accuracy the prevailing vegetational and climatic regimens of the sample's area of origin. Soil fungal similarity is an outstanding indicator of environmental similarity (Bissett and Parkinson 1979a, 1979b; Christensen 1989).

The influence of environment on occurrence of soil fungi usually is manifested in inventory lists when certain genera or species are encountered more commonly in one type of biome than in others or are restricted to certain biomes. Comparisons of lists of the most common soil fungal taxa in arid, temperate, or tropical zones have revealed that certain vegetation types are characterized by distinctive assemblages of soil fungi (Christensen 1981a, 1989; Sangalang et al. 1995). For example, *Cylindrocarpon*, *Chrysosporium*, *Mortierella*, *Oidiodendron*, certain *Penicillium* species, and sterile fungal strains are common in Antarctic, arctic, alpine, and boreal regions (Dowding and Widden 1974; Bissett and Parkinson 1979a; Bailey and Wynn-Williams 1982; Pugh and Allsopp 1982; Fletcher et al. 1985). In cold bogs and coniferous swamps, yeasts, yeastlike forms, *Tolypocladium*, *Trichoderma* species, sterile dematiaceous forms, and certain penicillia and eurotialean fungi prevail (Christensen and Whittingham 1965; Dowding and Widden 1974; Bissett and Parkinson 1979a). Soils of beech (*Fagus* species) forests of North America and Europe show striking similarities in their principal saprobic fungal species (Christensen 1989).

Aspergillus species are isolated more often from desert, subtropical, and tropical soils than from soils of other regions; in some temperate deserts and grasslands and in some tropical forest soils, they may account for up to 20% of the total species isolated (Christensen and Tuthill 1985). Desert soils generally are characterized by low propagule densities but high species diversity (Christensen 1981a; Mouchacca 1995). Mitosporic fungi with melanized, often thick-walled conidia (e.g., *Alternaria*, *Ulocladium*, *Curvularia*, *Periconia*, *Papulaspora* species); fungi with melanized pycnidia; dark-spored ascomycetes; and fungi producing chlamydospores, bulbils, or sclerotia are well represented in desert soils (Nicot 1960; Gochenaur 1970). Thick cell walls and melanized spores and cells are considered adap-

tations to extreme desiccation and insolation (Durrell and Shields 1960; Bell and Wheeler 1986). In grasslands and agricultural soils, *Acremonium*, *Arthrinium*, *Cladosporium*, *Drechslera*, *Fusarium*, *Idriella*, *Periconia* species, *Microdochium bolleyi*, and *Phoma eupyrena* frequently are recovered, whereas the number of species of Mucorales is reduced. Clarke and Christensen (1981) listed a suite of 12 species of *Aspergillus*, *Fusarium*, and *Penicillium* that are characteristic of soils in temperate American grasslands; nine species of *Fusarium* may occur consistently in grasslands worldwide.

Less is known about characteristic assemblages of taxa in soils of humid tropical forests, except that *Aspergillus*, *Penicillium*, and *Trichoderma* species usually are abundant (Goos 1960, 1963; Maggi and Persiani 1992; Pfenning 1993; Bettucci and Roquebert 1995); and among *Aspergillus* species, a large group of typically tropical species exists (Christensen and Tuthill 1985; Maggi and Persiani 1992). Tropical soils also contain many minor or rare genera that appear to be primarily tropical in distribution and that commonly are associated with overlying litter; their rarity in mineral soils could indicate incidental introduction (Matsushima 1971, 1975; Rambelli et al. 1983; Maggi and Persiani 1992).

Various investigators have isolated filamentous fungi from soils, sands, and sediments of estuaries, mangroves, and marine and freshwater habitats (Tubaki et al. 1975; Cooke 1979; Kohlmeyer and Kohlmeyer 1979). Almost all taxa isolated from those habitats, with the exception of a few facultative marine mitosporic fungi (e.g., *Asteromyces cruciatus*, *Dendryphiella arenaria*, and *D. salina*) appear to be terrestrial soil fungi. Terrestrial fungal propagules are thought to accumulate, and to form an inactive spore bank in marine sediments, because of the mycostatic effect of seawater (Kohlmeyer and Kohlmeyer 1979).

At present, we have no data on fungal communities of "aerial soils" that result from humus formation and accumulation on large limbs in forest canopy trees. Fungal species composition in those specialized soils may be quite different from those in mineral soils because of the lack of mineral input, extreme fluctuations in water content, distinctive microfauna, and the generally exposed environment in the canopy.

Chytridiomycetes are distributed widely in soils from the Arctic to the tropics and appear abundantly in soils when specialized enrichment methods are used to detect them (Gaertner 1954; Sparrow 1960; Booth and Barrett 1976; Powell 1993). As with other organisms, the species diversity of chytrids increases along a gradient from higher latitudes toward equatorial regions (Gaertner 1954). The distributions of chytrids in soils are sensitive to moisture, ion content, pH, and temperature (Willoughby 1961; T. Booth 1971; Booth and Barrett 1976) and therefore argue that changes in soil conditions caused by soil-collection methods are likely to decrease species richness in a sample.

EVALUATING THE DIVERSITY OF SOIL FUNGI

PRELIMINARY PLANNING, SCALE, AND DISTRIBUTION PATTERNS

Preliminary phases prior to the main inventory should focus on clear definition of habitats, development of an efficient sampling design, exploration of the magnitude of species diversity, and determination of minimum sample sizes (see Chapter 5). Scale of sampling is an important issue, and the goals of the inventory will determine what is appropriate. If the goal is comprehensive evaluation of a park or a reserve that spans a complex landscape, then a network of sampling sites encompassing the major habitats is essential (see "The Wisconsin Surveys and Their Relevance," earlier). An inventory that focuses on the impacts of land management systems, agricultural practices, or other disturbances calls for limited sampling regimens coinciding with introduced variables. If no previous data on the soil mycota are available, preliminary sampling will be needed to determine a profitable level of effort in exploring fine-scale patterns, particularly in relation to soil horizons, soil types, topography, and disturbance.

In general, we recommend that samples be collected from the upper surface of the mineral soil (A1 horizon) or its equivalent. In areas of moderate to abundant precipitation, water and organic matter contents and fungal abundance and diversity are highest in that soil layer. In arid and semiarid areas, a subsurface moist horizon, characterized by high evapotranspiration, can harbor an assemblage of species substantially different from that in the much drier surface soil; two or three levels should be sampled in inventory studies when there is evidence of midprofile soil-water storage and a consequent lenticular layer of organic matter accumulation.

The litter and humus layers that overlie mineral horizons harbor distinct mycotas. The litter is a transition area between the living plant community and the soil; thus the composition of litter mycota is tied closely to that of the plant community and contains many plant-inhabiting species as well as a distinctive assemblage of genera and species that largely are restricted to decaying plant material (see Chapter 11). The assemblage of plant-associated species usually intergrades with the assemblages of soil species at the humus–soil interface. As a result, the litter–humus profile is exceptionally species

rich. In humid tropical forests, however, the litter–soil interface is often abrupt, and the mycotas of the two layers are distinct. The overlying litter is species rich, whereas the underlying soil is depauperate by comparison (Pfenning 1993; Bills and Polishook 1994).

Because fungi exhibit succession downward through the litter and to a lesser extent through soil horizons (Wicklow and Whittingham 1974; Kjøller and Struwe 1982; Aoki et al. 1990, 1992), vertical dissection of soil strata via pit studies or dissection of soil cores probably should be included in all preliminary studies. Species composition and propagule density of individual species clearly are affected by the sequence of changes through fragmentation, decomposition, and humidification of plant detritus. Stable organic compounds, such as waxes, lignins, and phenolics, may persist for hundreds of years, are incorporated into mineral soil horizons, and are colonized by characteristic groups of fungal species (Gochenaur 1978; Bååth 1980). In the humid tropics, species richness declines rapidly with increasing depth in mineral soil (Stotzky et al. 1962; Pfenning 1993). Beiswenger and Christensen (1989) found soil fungi characteristic of present-day tundra communities at 1 m below native short-grass vegetation in an intermountain basin in Wyoming. They considered the fungi to be relics of the mycota that existed under the periglacial conditions known to have been present in the basin in the late Pleistocene.

Relative abundances of the common species of fungi in forest and grassland soils vary seasonally (Christensen 1969; Gochenaur 1978; Clarke and Christensen 1981; Widden 1986b); such variation apparently is less conspicuous in agricultural soils and those of some boreal conifer forests (Gams and Domsch 1969; Söderström and Bååth 1978; Bååth 1980; Gams 1992). Soil mycobiota also change over successional time. Thus, intensive and extensive sampling are necessary if the aim of the study is to inventory the principal soil fungi both spatially and temporally. Sampling from a representative mosaic of vegetation types in the study area in a single season logically should precede sampling designed to examine compositional variation related to season, microsite, or disturbance events (Gochenaur 1978).

Studies also indicate the number of chytrid species detected in a habitat increases as the numbers of samples increase (Booth and Barrett 1976). Chytrids typically are extracted from the top 5 cm of soils; vertical zonation of chytrids in soils has not been established. Different types of soils, although sharing common species, harbor totally different assemblages of taxa (Willoughby 1961). Moreover, some seasonal variation in species composition occurs at a given soil location, but typically, within seasons, compositions are very similar from year to year (Willoughby 1956; Willoughby 1961).

When planning a soil fungal inventory, the numbers of soil samples and isolates taken for examination need to be considered carefully. Even small numbers of soil samples, examined intensively, yield hundreds of taxa. Time constraints and costs of materials must be weighed against the priority of experimental questions or inventory goals. As a practical guide, the maximum number of isolates that can be managed adequately by two full-time, experienced researchers at a single time in a well-equipped laboratory is about 2000–3000. This estimate is based on the assumption that the investigators are already familiar with a high proportion of the genera and species. During the design phase, the number of soils, number of horizons, and number of isolation media or isolation protocols must be balanced against the time and resources available to investigators for media preparation, vigilance of isolation plates, identification, and culture preservation. Species-abundance curves and species-isolate curves are the mycological analogues of species-area curves in plant ecology. By conservatively sorting isolates into morphological types and ranking them by decreasing abundance (Fig. 13.1), one can foresee rapidly the complexity of a full-scale survey.

If suspension plating or particle filtration is used to isolate fungi in small composite samples of temperate or tropical soils, asymptotes for species-isolate curves range from about 60 to 150 species for 600–1000 isolates. As has been demonstrated repeatedly, however, new species continue to be encountered indefinitely (Christensen 1981a; 1989; Gams 1992). Determination of all taxa present in a given soil or litter type seems impossible (Christensen 1989; Gams 1992; Bills and Polishook 1994). We believe that the common taxa of a homogeneous vegetation type can be determined with a random sampling of 200–250 strains isolated from 6 to 10 separate samples or a composite soil sample.

ISOLATION TECHNIQUES FOR FILAMENTOUS FUNGI

Background

The isolation techniques described in the following section involve the separation of fungal propagules from the soil and from each other and initiation of growth on an isolation medium. Placement of selective nutritional baits in soil *in situ* is not usually practical in surveys of saprobic soil fungi but may be considered for certain zoosporic fungi (see "Isolation Techniques for Soil Chytridiomycetes," later in this chapter), root-pathogenic fungi, wood-associated fungi (Chapter 8), and animal-associated fungi (Chapters 18, 19, 20, 21, and 22). A key to representing diversity is adequate separation of propagules and maintenance of separation during colony maturation. Propagules are disrupted physically and separated prior to application and must be dispersed adequately on the medium. On agar media, physical, nutritional, and chemical means are used to limit colony expansion and interaction and to bias the selection against some of the germinated propagules. Although mechanical disruption of substrata has been criticized because it obscures 3D spatial patterns, local patterns of fungal colonization still can be discerned if small, homogeneous samples are compared (Frankland et al. 1990).

To assure comparability in both qualitative and quantitative inventories and ecological surveys, we recommend use of at least one of the three soil plating methods—Warcup soil plates, suspension plating, or particle filtration (see "Principal Isolation Methods," later in this chapter, and the "Appendix," this chapter)—and a predetermined sampling protocol that dictates a standard number of isolates. Application of the species-isolate curve concept indicates that a minimum of 600–1000 isolates per site should be examined if the goal is to list the majority of species obtainable using suspension plating or particle filtration. The number of isolates is less important in quantitative comparative studies so long as the taxon-site data matrix contains sufficient isolates to allow identification of the top-ranking 30 or more species (Christensen 1981a). In several surveys, 180–250 isolates obtained by soil suspension or particle filtration for each site have provided quantitative data for each species that are repeatable and hence reliable for revealing the principal soil fungal species at any relatively homogeneous forest or grassland site. This general isolation approach can be complemented by application of a variety of more selective or exploratory techniques that target specific taxonomic, functional, or physiological groups of fungi and thus extend the inventory list (see "Qualitative Inventories," later in this chapter).

Collection and Storage of Soil

To collect soil from the A1 horizon, one carefully removes litter and humus from an area of about 20 cm^2. The digging instrument (trowel, knife, spatula, spoon, coring tube) is sterilized in 70% ethanol and dried before 6–10 plugs (10–30 g) are collected from the exposed soil surface and placed in a labeled, sterile bag. Several such samples are taken for each homogeneous site and analyzed separately or pooled to produce a sample representative of a wider area. If field sterilization is not possible, contaminants on hands, tools, and collection bags can be reduced to a negligible amount by repeated immersion in the soil prior to collection. In warm weather, a cooler (0–10°C) should be used to transport samples from the field.

Isolates should be prepared as soon as possible (within hours or a few days) after collection. If immediate processing is not possible, cold storage or freezing of samples may be necessary. For reconnaissance of fungi, soils can be sprinkled, swabbed, or stamped directly onto isolation media in the field (see "Principal Isolation Methods," later in this chapter, and the "Appendix," this chapter). We recommend that soil samples be air-dried and stored in breathable cotton or perforated Tyvek, in soil collection bags (Hubco sand sample bags, Hubco, Inc.; see Appendix IV), or in strong paper envelopes for long-distance transport. Plastic bags may be used if soils are processed or frozen within a day or two.

If immediate processing is not feasible, soil samples can be air-dried, preferably in a clean room or under a laminar flow hood, and maintained at a cool temperature (2°–5°C). Degradation of soil during the course of analysis as well as isolations from soil samples collected at different seasons may confound comparisons of different soils or methods for isolation of soil fungi. In cases in which change in the soil mycota needs to be eliminated as an experimental variable, temperate soil samples can be frozen (–10° to –20°C) to stop temporal changes, with minimal effects on species composition and abundance, for at least 9 months (Christensen et al. 1962; Christensen 1969). The effects of physical manipulation (e.g., drying, cold storage, or freezing) on soil fungi have not been studied adequately, and investigators should be cautious in drawing conclusions about natural populations of fungi if samples have been manipulated heavily.

Media

Some media used in studies of soil mycotas are considered to be "nonselective" media (e.g., Czapek agar, malt agar, or potato-dextrose agar). In reality, such media are highly selective. For example, media with high sugar contents can create conditions that favor recovery of species with rapid germination and robust and rapid radial growth. In the ideal fungal inventory, one should use a truly nonselective medium or a standard array of selective media that permits equitable growth of all species active in the substratum, but that medium has not yet been developed. Thus, recovery of a complete mycota requires media and isolation procedures that recover a broad range of fungi; also, a series of isolation conditions is required to unveil the entire spectrum of fungi present, insofar as they are able to grow at all in artificial culture.

In some cases, nutritionally rich media may be needed to support slow-growing fungi. To prevent inundation by rapid-growing fungi, interfering species must be eliminated or constrained by physical or chemical means.

However, such destructive chemical and physical procedures must not be applied to the degree that slow-growing species also are eliminated. Sublethal doses of fungitoxic agents that restrict radial extension of hyphae (e.g., rose bengal, cyclosporin A, cycloheximide, dichloran), either singly or in combination, effectively can suppress rapidly growing fungi. Phytopathologists have formulated specialized media, often based on considerable experimentation, to recover targeted pathogens from various substrata (Tsao 1970). These media are useful for an autoecological study of a particular fungal species or, in the case of inventory, to extend the species list by acquisition of fungi that are otherwise difficult to isolate.

Considerable attention has been paid to media for isolation of soil fungi, and many anecdotes are in circulation about the effects of different isolation media. Unfortunately, only a few investigators have directly compared the number of taxa recovered from soils using different isolation media or soil treatments (Mahoney 1972; Andrews and Pitt 1986; Wildman 1991). New antifungal agents that interfere with distinct cellular functions increase the possibilities for selectively fractionating the fungal community. However, the labor-intensive nature of handling and evaluating statistically relevant numbers of isolates no doubt has discouraged investigators from evaluating the ability of different media to separate and characterize fungal communities. Nevertheless, it is widely understood that even with use of a range of nutrient media and a range of incubation regimens, the description of a community of soil fungi will be incomplete (Griffin 1972).

For general inventories and comparative analyses of the principal saprobic soil species using suspension plating or particle filtration techniques, we recommend use of a medium very low in nutrients (e.g., soil extract-rose bengal medium, carboxymethyl cellulose medium; see Appendix II) that simulates the conditions of oligotrophic soils. Expansion of inventory lists and identification of physiologically specialized groups can be achieved by incubation of replicate plates at low (0–10°C) or high temperatures (40–55°C) (see Chapter 14). To augment the inventory further, we recommend use of a nutrient-rich medium containing one or more colony-restricting agents (e.g., dichloran-rose bengal-chloramphenicol agar [DRBC] or malt cyclosporin medium) in conjunction with early weeding and/or low-temperature incubation to increase the number of slow-growing or rare species detected (see "Selecting Colonies," later in this chapter).

Use of some of the special media designed to select for specific difficult-to-detect taxa of pathogenic fungi will be required to isolate and culture the full range of saprobic soil fungi. Media used for selective isolation of filamentous oomycetes, principally but not exclusively the genera *Aphanomyces, Phytophthora,* and *Pythium,* contain the antifungal polyene antibiotic pimaricin (natamycin), which is toxic to most true fungi. A common formulation used by phytopathologists for oomycete isolations is P10VP, in combination with suspension plating, direct plating of washed or disinfected root fragments, or direct plating of plant baits (see "Appendix," this chapter). Additional formulations of selective media designed to isolate *Aphanomyces, Phytophthora,* and *Pythium* species in agricultural situations can be found in Singleton and colleagues (1992) and Tsao (1970). Cycloheximide (actidione) is another antibiotic that has been used to isolate anamorphs of Ophiostomatales selectively from wood and the rhizosphere of their conifer hosts (Swai and Hindal 1981; Harrington 1992). Cycloheximide is extremely toxic to most filamentous fungi and at recommended levels (0.25–0.5 g/liter) will eliminate most filamentous fungi from soil suspension plates (G. F. Bills, unpublished observation.). Another useful antifungal agent for extending the inventory is benomyl (see "Selective Isolation of Saprobic Basidiomycetes," later in this chapter, and in "Appendix," this chapter), which is toxic toward most ascomycetous fungi but less so toward basidiomycetes, Pleosporales, and Mucorales (Bollen and Fuchs 1970; Edgington et al. 1971).

Incubation

Incubation temperature affects growth and thus can affect the kinds of fungi developing on isolation plates. Below the soil surface, temperatures are buffered by the cooling effects of soil water during evaporation and the poor heat conductivity of soil. Isolation schemes for soil fungi should take into account prevailing soil temperatures and their seasonal amplitudes and localized patterns of shading and insolation. In forest openings, grasslands, and deserts, insolation may elevate surface soil temperatures. Isolation protocols often call for incubation at room temperature or 25°C, and at these temperatures most colonies develop within 2–10 days. Prevailing temperatures in temperate forest soils, however, are generally much lower (Carreiro and Koske 1992). Similar cool-mesic temperatures (15–20°C) have been observed during the rainy season in mid-elevation forest soils in southern Mexico (G. Heredia and G. F. Bills, unpublished data). In a coastal rain forest east of São Paulo, Brazil, soil temperatures at a depth of 10 cm ranged from 15–20°C in the winter and from 20–25°C in the summer (Attili et al. 1996). For species from such habitats, we recommend that isolation plates be maintained at 15–20°C in a cooling incubator. In this temperature range, colonies of different species may appear sequen-

tially on isolation plates for up to 3 weeks. Incubation at cool temperatures (less than room temperature) reduces the rate of colony expansion of mesophilic species, thereby allowing more time for separation of species (Petrini et al. 1990). Likewise, fungi with lower temperature optima, such as terrestrial aquatic hyphomycetes (Bandoni 1981) and psychrophilic fungi (Carreiro and Koske 1992; Petrini et al. 1992b), may grow readily only at cool temperatures. Conversely, incubation at high temperatures (>40°C) can reveal thermotolerant and thermophilic fungi (see Chapter 14).

The effects of light on isolation of soil fungi has not been investigated systematically, although light can affect significantly the growth and sporulation of fungi (Leach 1971). Investigators have incubated isolation plates in darkness and under ambient and controlled light regimens. We have obtained satisfactory results with both darkness and controlled light regimens. However, some media components may photodegrade or their properties may change when illuminated. A notable example is the increased toxicity of rose bengal during incubation in light.

Selecting Colonies

Removal of colonies from Petri plates may follow different strategies depending on the goals of the survey. Overall, we discourage the practice of taking only "different-looking" colonies arbitrarily from isolation plates, identifying them, and representing the list as an inventory. Instead, we strongly advocate that all isolates be gathered according to a predetermined pattern and ranked by their abundance (Fig. 13.1). Collection of quantifiable data in a systematic way is valuable for several reasons. First, it forces investigators to consider and evaluate all fungi, not only those that interest them or seem easy to identify. Often, organisms thought to be uncommon turn out to be common, and eventually the investigator is forced to cope with taxonomically difficult species. Secondly, less experienced investigators naturally tend to overselect strongly pigmented and heavily sporulating colonies. *Epicoccum nigrum, Stachybotrys chartarum,* and *Aspergillus terreus,* for example, are more likely to be selected as "representative," especially by beginners, than are dingy or unremarkable-looking *Acremonium* and *Fusarium* species and featureless basidiomycetes. Third, if all isolates are considered, recurring patterns can be detected and measured and the biological significance and underlying cause of the pattern can be investigated. A list that does not indicate the relative commonness of each species rarely establishes a pattern that can be revisited and retested. Finally, lists of species identified at a site are valuable for establishing biogeographic patterns. However, the quantity of material at a site that cannot be identified is also important and should be recorded. The inability of an expert to name species may indicate a high degree of taxonomic uniqueness or endemism.

We present two main options for obtaining abundance data from isolation plates and a third option, a variation of the second, that is useful in exploring for rare or specific groups of soil fungi. Another strategy, plating of individual washed soil particles (Gams and Domsch 1967; Bissett and Parkinson 1979a), yields relative quantitative data and indicates soil microhabitat preference but may be too tedious for general surveys. A dissecting microscope that has both epistage and substage lighting and variable magnification of up to at least 45×, but preferably up to 80 or 120×, is essential for dissection of fungi from isolation plates.

Option 1. For a quantitative survey involving relatively few isolates per site (25–35 isolates/soil sample; 6–10 samples/site), we recommend that a predetermined number of isolates be selected from plates with low colony densities. After the plates are incubated for 3–5 days, colonies are isolated in sequence beginning at the plate trademark or serial number. Because numbers of isolates will be approximately equal among sites, perhaps 300 ± 10 compositional comparisons can be made based either on the 30–50 most abundant species or all the taxa in the sample population, according to the investigator's preference.

Option 2. To obtain thorough community characterizations, including ranked abundance data for comparisons of levels of diversity, we recommend use of a predetermined number of isolation plates per sample per site. Under this option, all isolates are removed from a limited number of isolation plates and saved. Isolation of all colonies at first may seem numerically overwhelming, but with practice and some preliminary testing of the soils, an investigator can judge the number of plates needed to obtain the desired number of colonies. For example, 5–10 plates that yield 30–40 colonies each should adequately represent the principal saprobic species as well as provide many rare isolates. Colony selection should start within a day or two after plating because propagule germination and elongation can start within hours after contact with the medium. Selection and elimination of fast-growing colonies from plates at early stages of incubation permit slower-growing forms less often seen in species lists to develop (Dreyfuss 1986; Dreyfuss and Chapela 1994). Colonies to be eliminated can be transferred onto separate agar slants for preservation until total counts of fungi are obtained. Different sites or treatments inevitably yield unequal numbers of colonies. To make comparisons among samples with different total number of isolates, the number of isolates of each species in a sample can be divided by the total isolates in the sample to calculate a relative frequency or percent frequency. Rarefaction indices permit direct comparison of fungal species richness between samples with different numbers of total isolates and between independent studies (see Chapter 5; Bills and Polishook 1994; Bills 1995).

Option 3. For exploratory surveys whose goal is to increase the recovery of specific taxonomic groups, rarely seen species, or slow-growing species, we recommend a variation of Option 2 that eliminates most of the rapidly growing species from the isolation plates. That approach should be applied after the principal species have been identified and quantified and investigators are no longer interested in reexamining those species. The technique exploits the fact that colonies usually appear sequentially on the plates as a result of differences in germination times and growth rates. Colonies originating from propagules embedded in soil particles may emerge later than those originating from propagules directly on the agar. During the first few days of incubation, colonies of fast-growing species (e.g., *Trichoderma* and *Mucor* species) are removed by dissection or burned off with a soldering gun (Dreyfuss and Chapela 1994). Late-emerging colonies are dissected from the plates and saved in tubes. Because quantitative data are not sought, multiple colonies of the same species, if they can be recognized, are eliminated or left on the plates. Plates are examined until no new colonies appear. Colony separation and control of colony expansion can be enhanced by intermittently lowering incubation temperatures. The morphology of colonies on low-nutrient media or on media with antifungal antibiotics may be ambiguous or aberrant. We recommend overselecting different colony types because even colonies that appear similar on the isolation plate may in fact represent different species.

PRINCIPAL ISOLATION METHODS

SUSPENSION PLATING

The most often used method for isolation of soil fungi is suspension plating. This method is usually and incorrectly called "dilution plating." Dilution refers to the fact that soils must be "diluted" in increasingly dilute series of aqueous suspensions before application to media. The mixture actually applied to plates is a suspension of soil and propagules. The method is simple and rapid, gives reasonably repeatable results, and yields excellent comparative data. Soils are suspended in sterile water or solutions of agar (0.05–0.2%), dextrin, or carboxymethyl cellulose (0.1–0.2%) to retard sedimentation and then pipetted onto a Petri plate.

Because soils vary in propagule density, we recommend trial analyses to determine a soil/water ratio that will yield 5–30 colonies in a 100-mm Petri plate. Low colony densities that minimize intercolony interactions are essential. Higher colony densities can yield excellent results when used in conjunction with the colony-restricting measures previously outlined in the section on "Media." A generalized suspension plating protocol is presented in the Appendix at the end of this chapter.

WARCUP SOIL PLATES AND OTHER DIRECT INOCULATION TECHNIQUES

Warcup soil plates are prepared by dispersing minute quantities (0.2–5.0 mg) of finely pulverized soil across the bottom of a sterile Petri dish. Cooled agar is poured over the soils, and the plates are agitated gently to disperse the particles (see "Appendix," this chapter). Warcup soils plates yield a spectrum of taxa similar to that obtained by suspension plating (Warcup 1950) and also may yield basidiomycetes not recovered by that procedure. Because of its simplicity, the method is good for preliminary or rapid assessments of soil species. However, young colonies embedded in agar are more difficult to remove.

Soils with low propagule densities can be sprinkled or swabbed directly on the surface (see "Appendix," this chapter). Drechsler's method (Drechsler 1941b; Barron 1977), in which minute quantities of fresh soil or humus are sprinkled directly on water agar or very dilute cornmeal agar, has been used to create microcosms of soil fungi and soil invertebrates and an environment that favors infection and sporulation of parasitic fungi on senescent fungal hyphae and soil invertebrates (Chapter 19). Drechsler's method also is used for isolation of fast-growing filamentous oomycetes from fresh soil, litter, and root fragments (Watanabe 1994) and of cellular slime molds (Chapter 25). Some basidiomycetes whose mycelia appear to be sensitive to the disturbance of suspension plating or particle filtration may appear on Drechsler soil plates.

Variations of the Warcup soil plate and the suspension plate methods can be used to isolate rhizosphere and rhizoplane species (Harley and Waid 1955; Persiani and Maggi 1988). Tiny root fragments are removed aseptically from soil and placed in vials with 5 ml of sterile water or 0.1% sodium pyrophosphate and agitated for 1 minute. Washed root fragments or 0.1–1.0-ml aliquots of the washing water can be placed in Petri dishes and mixed with a cooled agar medium or spread on the agar surface.

PARTICLE FILTRATION

Particle filtration techniques, also known as "soil washing," have been developed in laboratories in Canada, Germany, and Sweden as an alternative to suspension plating for characterization of soil fungal communities (Harley and Waid 1955; Parkinson and Williams 1961; Parkinson and Thomas 1965; Gams and Domsch 1967; Söderström and Bååth 1978; Bååth 1988; Gams 1992; Hestbjerg et al. 1999). Particle-filtration techniques favor isolation of fungi from mycelial fragments immersed in the substratum while reducing recovery of fungal colonies initiated from spores. Isolation of species with chlamydospores embedded in organic particles may be favored because their propagules probably are not entirely removed by washing (Gams and Domsch 1969).

Particle filtration has not been used as widely as suspension plating, partly because good results are achieved only after considerable practice and partly because preparation of materials can be more time-consuming. Often, however, the method reveals numbers of rarely seen species beyond those acquired by conventional techniques (see "Incubation" and "Selecting Colonies," earlier). Small particle sizes that achieve a colony/particle ratiio of less than 1 prevent intraparticle competition and assure accurate frequency data. Choice of particle size depends on soil or substratum texture (Bååth 1988; Kirby et al. 1990). Bååth (1988) recommended particles in the range of 50–80 μm in diameter; Bills and Polishook (1994) used 105–210-μm particles to study fungi in forest litter.

Particle filtration provides two kinds of quantitative data, the simplest being relative density (percentage of isolates/species). Usually, a constant volume of washed particles is spread onto isolation medium, and resultant colonies are counted and identified. A second type of data is frequency of colonization of particle types. Separation of soils into microcomponents (e.g., organic versus mineral fragments) and plating of those components permits calculation of ratios of colonization frequencies of different soil constituents, providing information about microhabitat preferences and active colonization of mineral components of soil (Gams and Domsch 1969; Bissett and Parkinson 1979a). Predetermined numbers of particles within a specified size range are plated on isolation media, and the number of particles colonized by each species is determined. If the soil or substratum is heterogeneous (e.g., a mixture of sand and organic matter), microhabitat preferences can be detected. Likewise, frequency of particle colonization in small homogeneous samples may approximate the density of principal species (Kirby et al. 1990). Plating of individual particles can reveal preferences of species for different types of organic matter in soil horizons (Hestbjerg et al. 1999).

Washing devices range from simple vials, filtered syringes, and nested sieves to complex self-flushing and refilling devices (Gams and Domsch 1960; Parkinson and Williams 1961; Parkinson and Thomas 1965; Bissett and Widden 1972; Söderström and Bååth 1978; Bååth 1988; Kirby and Rayner 1988; Kirby et al. 1990; Gams 1992). The effectiveness of these devices is a measure of their ability to produce well-washed particles, clearly separated into descending size categories. Choice of device seems to be a matter of personal preference, cost, and availability of materials. A simple device, made entirely from materials available from a standard scientific supply catalog, is illustrated in the Appendix (this chapter; Fig.

13.5) along with a protocol for particle filtration. An alternative method is described in "Selective Isolation of Saprobic Basidiomycetes," later in this chapter).

Particle filtration is readily adaptable to many kinds of organic substrata and theoretically can be used to partition anything that can be broken into smaller units. The method has been applied successfully to isolate fungi from terrestrial and aquatic litter and decaying wood and to separate fungicolous, lichenicolous, and coprophilous fungi (Fig. 13.2).

FRACTIONATING THE SOIL FUNGAL COMMUNITY

The previously mentioned methods rapidly produce large numbers of isolates of a wide range of species that grow readily in culture. High propagule densities,

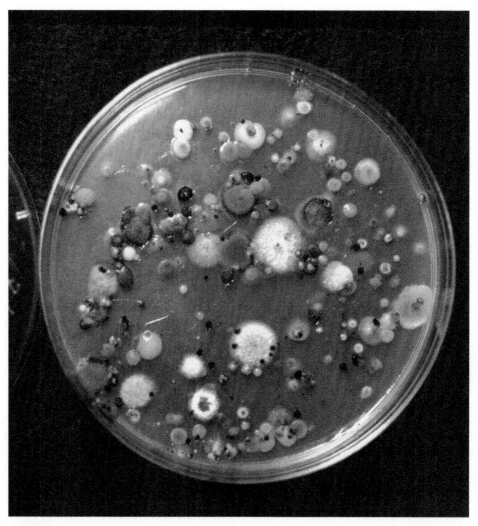

FIGURE 13.2.   Three-week-old isolation plate with colonies of fungi isolated from deciduous-forest litter. Filtered and washed litter particles were plated onto malt-cyclosporin A agar.

however, can interfere with isolation of rare or slow-growing species. For certain purposes, such as extension of the inventory list, assays for pathogens, or natural products screening (Dreyfuss and Chapela 1994; Bills 1995; Okuda et al. 1995), rare taxa are of equal or greater interest than common ones. Special techniques, described in the following paragraphs, can increase encounters with less commonly seen species.

One technique involves preseparation of propagules from soil by density-gradient centrifugation before they are applied to isolation media. An early study indicated that large ascospores might be isolated preferentially from soils by centrifugation in sucrose density gradients (Paden 1967); although some ascomycetes were isolated, no comparative data were presented to confirm the observation. More recently, a Percoll density-gradient method that separates different-sized spores prior to plating and standard suspension plating were compared side-by-side; a higher number of taxa was isolated from a given soil with the former method than with the latter (Okuda et al. 1995). However, groups of fungi not ordinarily obtained by suspension plating were not isolated. A similar method using a sucrose density gradient was used to separate, concentrate, and directly enumerate basidiospores and sclerotia from soil in *Pinus contorta* forests (Miller et al. 1994). Use of differential centrifugation to separate large-spored predatory nematophagous fungi and to enhance isolation of small-spored endoparasites of soil microfauna is described in Chapter 19.

Partial killing of a fungal population with dry heat, steam, or solvents has been used for preferential isolation of species that have resistant propagules. Such soil treatments shift the population of isolates toward ascomycetes (Fig. 13.3) and basidiomycetes that have ascospores, thick-walled chlamydospores, or sclerotia. Some investigators have combined ethanol treatment (also called "ethanol pasteurization") with mild heat treatment, which is thought to stimulate ascospore germination (Warcup and Baker 1963). Other researchers have examined ethanol-treated soils for taxonomically interesting fungi (Novak and Backus 1963; Nelson et al. 1964; Mahoney et al. 1969; Huang 1971, 1973). Extensive side-by-side comparisons of suspension plating on glucose-ammonium nitrate agar, ethanol pretreatment, ethanol-heat pretreatment, suspension plating on a strongly saline medium, and high temperature incubations were carried out on soils from the Galapagos Islands (Mahoney 1972). An important conclusion of that study is that isolates resulting from ethanol pasteurization alone do not differ significantly from those of the sequential ethanol and heat treatment. Ethanol pasteurization was effective at extending the species list. Of 250 taxonomic entities recovered by all methods, 44 were

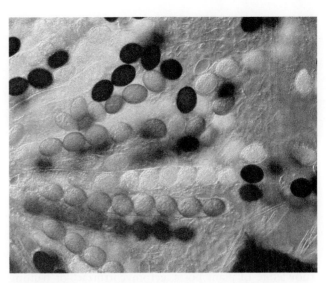

FIGURE 13.3. Asci and ascospores of *Gelasinspora retispora*, isolated from ethanol-pasteurized soil from Veracruz, Mexico.

isolated only after ethanol pasteurization. Ethanol pasteurization also dramatically lowered the total number of isolates of some yeast species as well as many species with thin-walled conidia, thus increasing isolation of ascosporic Eurotiales, *Chaetomium* species, *Xylaria* species, numerous genera of coprophilous fungi (Fig. 13.3), and possibly some ascosporic yeasts. A protocol for ethanol pasteurization of soil is given in the "Appendix" (this chapter). The use of heat treatments for separating soil fungi is discussed elsewhere in this volume (see "Thermotolerant and Thermophilic" under "Sampling and Isolation Methods for Groups of Fungi" in Chapter 14).

Phenol (Furuya and Naito 1980; Udagawa and Uchiyama 1998) and formalin pasteurizations (Warcup 1951b), which are variations of ethanol pasteurization, have been used effectively to increase the frequency of ascomycete isolations from soils.

BAITING

Baiting is a technique in which an organic substratum is enriched nutritionally and then placed in contact or mixed with soil thought to contain a target organism(s) in hopes that the desired organism will preferentially colonize and reproduce on the substratum. The need to assay for difficult-to-isolate destructive plant pathogens has driven the development of baiting techniques, as it has the development of selective media (Tuite 1969; Singleton et al. 1992). With the exceptions of cellophane and other cellulosic materials for isolation of cellulolytic fungi (Gams 1960), hair for keratinolytic fungi (Chapter

20), and insects for isolation of entomopathogenic fungi (Chapter 18), baits generally have not been used in broad soil fungal surveys. Rather, they have been used most widely to survey wild populations of Chytridiomycetes and other zoosporic fungi not usually recovered with soil-plating methods (see "Isolation Techniques for Soil Chytridiomycetes," later, and "Appendix," this chapter).

SCLEROTIA

Populations of sclerotium-forming fungi in soil have been assessed primarily during autoecological studies of root-associated or plant pathogenic fungi (Coley-Smith and Cooke 1971; Singleton et al. 1992). In those cases, knowledge of plant hosts, ectomycorrhizae, or fruiting bodies was used to locate sclerotia, which were separated by wet-sieving (Trappe 1969; Adams 1979), elutriation (Porter and Steele 1983), or density-gradient centrifugation (Miller et al. 1994) or observed directly on the soil surface (Cotter and Bills 1985). Density-gradient centrifugation was used to survey soils in burned and unburned *Pinus contorta* stands for two sclerotium-forming species, *Cenococcum geophilum* and *Morchella* species in Wyoming (Miller et al. 1994). Colonization of soil-borne fungal sclerotia by other soil-dwelling fungi is discussed in Chapter 17.

SELECTIVE ISOLATION OF SAPROBIC BASIDIOMYCETES

As mentioned earlier, surveys of soil fungi based on isolation into culture markedly underestimate the presence and diversity of saprobic basidiomycetes (Warcup 1959; Warcup and Talbot 1962; Thorn et al. 1996); see Chapters 8 and 10 for techniques for study of ectomycorrhizal basidiomycetes. Saprobic basidiomycetes are isolated from soil by suspension plating only rarely because of the following: (1) their propagules are greatly outnumbered by the abundant asexual spores of many Zygomycetes and mitosporic fungi; and (b) they typically are overgrown on culture media (Bisby et al. 1935; Warcup 1955). Warcup isolated numerous basidiomycetes from soil, first with soil plates (Warcup 1950, 1959; Stenton 1953) and then by plating individual hyphal fragments plucked under a dissecting microscope from minute quantities of soil dispersed in water (Warcup 1955, 1957, 1959; Warcup and Talbot 1962). Unfortunately, the skill and effort demanded by the latter technique have deterred others from using it (Frankland et al. 1990).

Thorn and associates (1996) used the particle filtration technique to remove spores of mitosporic fungi and Zygomycetes from soil and plated the washed particles on a medium made selective for basidiomycetes by the incorporation of benomyl. The medium also contained alkali lignin to encourage growth of ligninolytic fungi and guaiacol, which acts as a colorimetric indicator of the lignin-modifying enzymes laccase and peroxidase (see "Appendix," this chapter). The agar adjacent to a few of the 5–25 colonies that developed on each plate turned dark brownish-red as guaiacol was oxidized to a quinone. Those colonies could be quickly scanned under low magnification (40–100×) and putative basidiomycetes could be selected for transfer, whereas undesired mitosporic fungi (e.g. *Alternaria* and *Trichoderma* species) were avoided. Thorn and colleagues (1996) recovered 67 basidiomycete isolates from 64 soil samples from 35 sample plots. Investigators using the particle filtration technique in surveys should consider plating some of the washed particles onto lignin–guaiacol–benomyl agar (see "Appendix," this chapter) for the recovery of saprobic basidiomycetes that might be missed otherwise. Additional basidiomycete species may be recovered from very fresh soils using Drechsler or Warcup soil plates. Basidiomycetes, sometimes recognizable by their lax, white, cottony colonies, may appear among soil crumbs but fail to colonize the isolation medium, in which case their hyphae must be picked from the soil with a needle or fine forceps and transferred to other nutrient media.

ISOLATION TECHNIQUES FOR SOIL CHYTRIDIOMYCETES

COLLECTION OF SOILS

When isolating Chytridiomycetes, both fresh and air-dried soils should be studied. Generally, chytrids produce a thin-walled sporangial stage, a sexual or asexual thick-walled resting spore stage, or both (Johnson 1977). However, some chytrid species do not produce resting spores. Because the sporangial stage often dies after prolonged desiccation, the more ephemeral taxa may not be recovered from dried soils. Yet, because drying of soils reduces the numbers of many competing organisms, some chytrids with resting spores are best extracted from dried soils (Whisler 1987). Multiple soil samples (at least 14) should be collected at each of a range of sites at a given location. A site also should be sampled at least once during each major season (see "Selective Baiting and Enrichment Techniques for Isolating Soil Chytrids" in "Appendix," this chapter).

IDENTIFICATION

Many taxa of Chytridiomycetes have distinctive and stable morphologies that make them easy to recognize. Other taxa, particularly those with simple morphologies, such as a spherical sporangium with an unornamented wall and rhizoids, are difficult to identify to genus, let alone to species. To further complicate matters, some chytrids exhibit a range of morphologies that cross generic boundaries (Powell and Koch 1977a, 1977b). Unlike the majority of fungi, most Chytridiomycetes have simple thalli and do not grow by hyphal tip elongation. Consequently only a few chytrids can be isolated using strategies applicable to saprobic hyphal fungi (Willoughby 1984). Because chytrids reproduce by zoospores, motile spores requiring water for motility, isolation techniques exploit their motile abilities. Chytrids break down relatively refractive substrates, such as chitin, cellulose, keratin, and plant pollen, which can be used to attract and bait chytrid zoospores from inundated soils.

For ecological studies, problems of identification are exacerbated by the fact that descriptions of Chytridiomycete genera now are based on the ultrastructural features of zoospores, which are stabile, reliable, and conservative (Barr 1980). Although this approach reflects a natural classification system based on phylogenetic relationships rather than morphological similarities, which may be the result of convergence, ready identification in field studies is difficult. Modern revision of generic characters has resulted in two parallel systems of classification and nomenclature. For example, the morphological concept of the genus *Phlytochytrium* was defined on the basis of morphological characters of the monocentric thallus, which consists of an epibiotic sporangium bearing an endobiotic rhizoidal system. The rhizoids emerge from the sporangium at a single point, near a subsporangial swelling, the apophysis. Zoospores are released from one to several inoperculate discharge areas that are either flush with the surface of the sporangium or raised into papillae or tubules. When zoospores of this morphological genus were examined by electron microscopy, however, four distinct forms that could be ascribed to four different genera were revealed based on zoosporic ultrastructure (Barr 1980). New species, however, were described for only one of those four genera. The morphologically defined genus, *Phlyctochytrium*, remained as a taxonomic relic for taxa not transferred into the four new genera. Thus, in diversity studies dealing with large numbers of taxa, the dilemma becomes what system to use. The ultrastructure-based system is still rudimentary, and if the chytrids cannot be brought into single-fungus cultures or if access to or expertise in electron microscopy is lacking, ultrastructure-based classification becomes impractical.

The most recent monograph of the Chytridiomycetes is by Sparrow (1960), although a few genera (e.g., *Septosperma;* Blackwell and Powell 1991) have been monographed since that publication. Karling (1977) revised several genera. The primary literature must be used to identify more recently described species.

Work with a large number of chytrid taxa necessitates identification based on the classical morphological criteria. The descriptions of many chytrid species and the majority of classical genera were based on morphology of thalli growing on natural substrata rather than studies of pure cultures. However, investigators should use this approach cautiously because thalli are morphologically plastic, which can confound identifications (Powell and Koch 1977a, 1977b).

When a large number of isolates is to be identified, bringing them all into pure culture may be impractical. They can, however, be removed as single thalli (see "Appendix," this chapter) and transferred to enriched water-bait cultures (Johnson 1973, 1975) rather than being purified on nutrient agar. Although bacteria may accompany these cultures, cultures will be unifungal and allow for observation of different developmental stages and evaluation of the range of morphological plasticity, at least under the given culture conditions. Newly encountered genera will have to be brought into enriched or pure culture so that zoospore architecture can be determined.

ENUMERATION OF SOIL CHYTRIDIOMYCETES

Because indirect means are used to isolate chytrids from soils, measuring the numbers of propagules per unit weight of soil is difficult. Willoughby (1984) estimated the numbers of propagules per unit weight of soil for a filamentous chytrid by mixing released zoospores with a polycell wallpaper paste solution and counting the number of thalli produced. Typically chytrid taxa in soil are recorded as present or absent. Abundance at a site is expressed quantitatively in terms of percent frequency: number of collections in which a taxon is located/total number of collections (×100) (Willoughby 1961). This method does not distinguish between species that rarely are seen and those that are abundant. As Willoughby (1961) noted, however, abundant species also have higher frequency across multiple samples in a landscape.

Investigators of aquatic chytrids have adapted the bacteriological method of most probable number of propagules (or individuals) per unit water tested based on dilution series of water samples and pollen baits (Ulken and Sparrow 1968); the method is unreliable for soil chytrids (Booth and Barrett 1976). Another approach for quantifying diversity of soil chytrids is the distribution intensity index (DII) calculated by multiplying the percent occurrence by the square root of the percent frequency (Booth and Barrett 1976). Typically the number of chytrid species observed within a given habitat ranges from about 2 to 35 (Willoughby 1961; Booth and Barrett 1976). Biodiversity of soil chytrids can be quantified most easily by total species per site. To estimate the ecological prominence of taxa, the DII measurement (Booth and Barrett 1976) is used most often.

CHARACTERIZATION OF HABITATS AND ORGANISMS

CHARACTERIZATION OF SAMPLING SITES AND SOILS

Specific questions regarding vegetation, climate, soil types, and patterns of disturbance or land use often determine the objectives of a soil-fungus survey. As in any ecological or biotic study, the site characteristics emphasized depend partly on the objectives. The account of a research area should include at least a description of the overlying vegetation and the history of its development; climate data; the physical and chemical characteristics of the litter and soil; and, if appropriate, some information on the components and activities of the soil fauna. Soil organic matter, temperature, and water potential strongly affect soil fungal activity. For guidelines and methods for analysis and description of the soil microhabitat, see Carter (1993). Site perturbations, manmade or otherwise, may need to be measured to explore their effects on the soil fungal community. We highly recommend use of sites recently characterized by other biologists for inventory.

DOCUMENTATION AND PRESERVATION OF ISOLATES

Records must be kept on all isolates, and all isolates must be tracked. We recommend the following: (1) labeling each agar tube or culture with a simple code that indicates the sample origin (e.g., site, date, soil horizon) and other treatments (e.g., pretreatments, medium, incubation temperature); (2) permanently maintaining at least one representative isolate as a voucher of each taxon after sorting and enumeration (see next section); (3) assigning culture collection accession numbers; and (4) linking of the collection accession numbers and working identification numbers to a careful tabulation of occurrence data, qualitative and quantitative, in both hard copy and in an electronic spreadsheet. We urge lyophilization or ultra-cold freezing of at least one representative of each taxon and reference to the culture collection accession numbers in all publications. Dried culture mats and semi-permanent slides of sporulating structures can serve as additional vouchers (see Chapter 2).

ENUMERATION AND PRELIMINARY TAXONOMIC ANALYSIS

For quantitative studies, we recommend that nascent colonies be transferred from isolation plates to agar slants for growth, which will promote maximum colony differentiation (Christensen 1969; Bills and Polishook 1994). Preparation of agar slants is tedious and observation of microscopic structures through tube walls is difficult, but slants offer several advantages over Petri dishes or multiple well cell-culture plates. Agar slants resist desiccation and contamination better than Petri plates and are easier to move and sort for tentative comparisons and classification. We prefer large glass tubes, 15–20 mm diameter (×125 or 150 mm long. When slants are inoculated uniformly at a single point at the same level in the tube, growth is easily compared among isolates. We emphasize that this sorting of morphological types is only a preliminary step. Microscopic examination or molecular genetic analysis is essential to confirm that isolates are equivalent, especially in genera where more than one of a group of very similar species may inhabit the same soil (e.g., *Penicillium, Trichoderma, Mortierella, Fusarium*).

Use of wide-diameter tubes for slants permits dissection of sporulating structures during preliminary sorting and facilitates transfer of isolates. Furthermore, colony characteristics such as surface color, texture, margin, exudates, and soluble pigments that are used to sort isolates into morphological groups easily are observed. Preparation of uniform agar slants can be expedited by use of a peristaltic pump to fill tubes; inclined test tube racks (Kim-racks); and reusable, fluted polypropylene or permeable membrane caps to permit air exchange. Slants must be prepared in tubes of uniform size, and each tube must contain medium of the same composition, volume, and age to ensure uniform appearance among isolates of a given species. Although isolations may be carried out at different or variable temperatures, all colonies from the sample need to mature at the same constant temperature. The isolates should be incubated under a variable light regimen (e.g., periodic fluorescent light or indirect daylight) to promote colony zonation.

After incubation for 1–2 weeks, the process of sorting isolates into morphological types can begin. The experienced investigator often can identify some genera and a few species from slant cultures. Representatives of all morphological types should be transferred to Petri dishes for characterization and identification.

GENERAL MACROSCOPIC AND MICROSCOPIC CULTURAL CHARACTERS

Methods for the characterization of soil isolates vary widely in accordance with the genera encountered. When a genus is known (e.g., *Fusarium*, *Penicillium*, *Aspergillus*), representative isolates can be transferred immediately to the medium recommended for their identification (see "Taxonomic Literature," earlier). Unknown or sterile, nonsporulating fungi should be transferred to two or three different mycological media, such as alpha-cellulose agar; cornmeal agar, with or without a piece of autoclaved banana leaf (Matsushima 1971, 1975); oatmeal agar; V8-juice agar; and Synthetic Nutrient Agar (SNA), likely to support sporulation. If sporulating structures are induced on subculture, then diagnostic keys from general guides to soil fungi can be applied.

ENCOURAGING SPORULATION IN STERILE ISOLATES

The frustrating problem of sterile (nonsporulating) isolates occurs whenever large numbers of fungi are isolated indirectly from vegetative hyphae immersed in a substratum (see also Chapters 11 and 12). Sterile isolates can be nonmitosporic anamorphs of filamentous ascomycetes and basidiomycetes (i.e., those that produce only sexual reproductive structures that require mating of compatible haploid-monokaryotic mycelia). Other sterile isolates may be anamorphs that have particular requirements for the production of mitospores, or they may be truly sterile forms. Induction of fruiting in the former groups may require considerable experimentation with media and environmental conditions. Time is an especially important variable, and prolonged incubation for weeks or months may be necessary. With persistence, many isolates can be induced to sporulate eventually, but such efforts probably are impractical in large-scale surveys of soil fungi. Nonsporulating isolates can be grouped on the basis of shared unique vegetative and physiological characters, a standard practice in bacteriology and yeast systematics.

MICROSCOPIC CULTURAL AND VEGETATIVE CHARACTERS

Elaborate systems have been developed to identify basidiomycetes that usually do not produce a teleomorph or distinctive anamorph *in vitro* (Nobles 1965; Stalpers 1978). Suites of characters in culture are especially helpful for describing and identifying cultures of basidiomycetes, particularly the temperate wood-decaying Aphyllophorales and Agaricales. Many of those characters can be used to describe and group the nonsporulating isolates of basidiomycetes that arise from soils or other substrata. In addition, hyphae of nonsporulating isolates can be evaluated with acriflavine or DAPI (4′,6-diamidino-2-phenylindole dihydrochloride) staining and fluorescence microscopy (Williamson and Fennell 1975; Raju 1986) or with other staining techniques and brightfield microscopy (Raju and Newmeyer 1977). The nuclear number of hyphal cells can be used as a character for grouping and to indicate taxonomic affinities. Haploid monokaryotic basidiomycetes will often have uninucleate cells. Consistently binucleate cells are characteristic of dikaryotic basidiomycetes. Multinucleate cells indicate the cells are those of ascomycetes (Sneh et al. 1991). Staining large-diameter, thin-walled hyphae with trypan blue (Sneh et al. 1991) or ammoniacal Congo red (Nawawi et al. 1977) may enable microscopic observation of a septal apparatus characteristic of ascomycetes (simple septal pores) or of holobasidiomycetes (dolipore septa).

PHYSIOLOGICAL CHARACTERIZATION OF ISOLATES

Ability to grow on or use certain complex carbon or nitrogen compounds; sensitivity to or tolerance to defined concentrations of selected fungicides or mineral salts, including sodium chloride (NaCl), benomyl, and cycloheximide; and staining of mycelia with diazonium blue B (DBB) (Summerbell 1985; Hutchison and Summerbell 1990) can be used to sort unknown isolates into groups or to assign them to named species if reference isolates are available. We recommend a simplified suite of media and incubation conditions (see "Appendix," this chapter) designed to stimulate sporulation of recalcitrant isolates and provide physiological characters for grouping of persistently nonsporlulating isolates. The latter are left unidentified with an arbitrary designation (e.g., nonsporulating white #1) or characterized and identified by analysis of ribosomal DNA sequences (Chapter 6).

SUBSTRATE-UTILIZATION TESTS AND SPOT TESTS FOR HYDROLYTIC ENZYMES

Many tests for utilization of complex carbon and nitrogen compounds for growth and spot tests for hydrolytic enzymes are available for characterization of fungal cultures (Hutchison 1990a, 1990b; Paterson and Bridge 1994). We recommend use of tests for the utilization of cellulose (Smith 1977), lipids, (Hankin and Anagnostakis 1975), creatine (Frisvad 1981), milk solids (Fischer and Kane 1971), and urea (Kane and Fischer 1971). The latter tests use a pH indicator that allows monitoring of pH shifts during substrate catalysis and utilization. Colorimetric spot tests for the ligninolytic enzymes laccase, tyrosinase, and peroxidases also can be helpful in characterization of isolates, although results should be interpreted with caution because of variable expression of these enzymes in different media and similarities in results with different bates. We recommend syringaldazine for the assay of laccase and p-cresol for tyrosinase; if laccase is absent, peroxidase(s) can be assayed by adding dilute H_2O_2 (hydrogen peroxide) to the syringaldazine spot. Formulae and protocols for these tests are in the "Appendix" at the end of this chapter.

Several simple physiological tests can be used to characterize nonsporulating isolates rapidly; among those, we recommend tests for sensitivity to or tolerance of defined concentrations of benomyl, cycloheximide, and NaCl (Hutchison 1990c). Those tests are particularly valuable in determining whether a nonsporulating isolate is ascomycetous or basidiomycetous. Basidiomycetes, for the most part, are tolerant of benomyl (10 mg/liter) but sensitive to cycloheximide (2 mg/liter) and NaCl (10 g/liter), and their mycelia stain positively with DBB following pretreatment with 1 M KOH (potassium hydroxide). Cultures of ascomycetes usually have opposite reactions to each of these tests.

RECOMMENDATIONS FOR INVENTORIES OF SAPROBIC SOIL FUNGI

QUANTITATIVE INVENTORIES

We have used the series of fungal studies carried out at the University of Wisconsin as models in developing procedures for recognizing principal species in and comparing patterns of species diversity among soil fungal communities (Fig. 13.4). Site selection should be guided by prior analysis of vegetation and environmental gradients. This more extensive, less isolate-intensive approach is especially appropriate if one is interested primarily in the initial segment of the species-abundance curve (Fig. 13.1; see Chapter 5) and relative abundance of the 10–40 most common species. Those species are recognized most easily, and their responses to environmental or successional variables can be monitored most easily.

Generally, quantitative inventories using suspension plating or particle filtration techniques involve the following:

1. Use of 6, 8, 10, or 20 soil samples per site so that the prominence of a species in the fungal community can be assessed quantitatively both as frequency (F)—that is, the percentage of sites at which the species occurred—where

$$F = \frac{\text{no. sites of occurrence} \times 100}{\text{total no. of sites sampled}}$$

and as relative density (RD), where

$$RD = \frac{\text{no. of isolates of the ith species} \times 100}{\text{total no. of isolates}}$$

Relative density is the percentage of total isolates contributed by the ith species (Chapter 5). Each sample of 5–20 g of soil is mixed thoroughly prior to preparation.

2. Isolation of a standard number of colonies per soil sample, at minimum 25–30 from temperate forest soils and 30–40 from temperate grasslands soils, where densities for top-ranking species are often lower. Sampling is repeated during several seasons to measure seasonal variation in ranking of the principal species.

3. Use of a standard, low nutrient medium (see "Media," earlier) such as soil-extract agar (SEA) made with soil from the area under study, Gochenaur's glucose ammonium nitrate agar (GAN), carboxymethyl cellulose agar (CMC), or dilute malt extract agar with antibiotics (MEA) (see Appendix II). After a primary isolation medium is selected, experimental trials should be run to determine the approximate propagule densities (i.e., soil/water ratios) that will yield 5–30 colonies per plate and to determine the medium and antibiotic combination that will eliminate bacteria yet yield the greatest number of fungal taxa.

4. Incubation of the surface-inoculated plates at 15–20°C or at *in situ* soil temperature, or, if a cooling incubator is unavailable, at room temperature for 4–14 days or longer.

5. Isolation by hyphal tip dissection of all colonies encountered in sequence until the predetermined

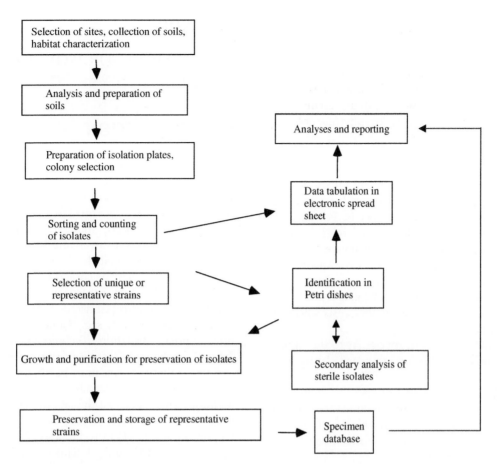

FIGURE 13.4. Organization of work flow for indirect isolations of soil fungi or fungi from complex organic substrata.

"representative sample" of the fungal community has been obtained (Option 1, under "Selecting Colonies," earlier). Thus, if the representative sample is 40 isolates per site, 42–44 tiny blocks of hyphal tips (to allow for unsuccessful transfers) are cut from the agar with a sterilized microscalpel or needle. Each is placed on an identical slant of nutrient agar (e.g., malt-yeast extract agar or potato-dextrose agar).

6. Sorting of the equal-aged, identically grown cultures to reduce redundancy in identification. Depending on the intensity of sampling, 200-, 500-, or 1000-member sample population of colonies will result. After each population of isolates has grown for 5–10 days, the population of tube cultures are sorted into apparent taxonomic entities based on appearance (Fig. 13.1). Permanent culture numbers are assigned to representatives of each taxon, and occurrences for all entities are tabulated in an electronic spreadsheet. After fungi are identified, the numbers of occurrences of each taxon may be combined or separated. Finally,

frequencies and relative densities are calculated and recorded for all taxa on the final list of identified or characterized isolates.

A key component of the Wisconsin studies was quantitative data in the form of species frequencies and relative densities. Such data allow for compilations of species ordered by relative commonness (Christensen 1969; Gochenaur and Woodall 1974). Comparisons among surveys then are possible despite differences in numbers of sites sampled and numbers of isolates examined (Christensen 1981a). Additionally, precise data are available for depictions of equitability, diversity, quantitative prominence of single species in relation to other variables; for computations of coefficients-of-similarity among communities using either simple presence or quantitative data for the separate species (Christensen 1969, 1981a; Chapter 5); and for multivariate analyses that attempt to correlate differences in fungal composition analyzed quantitatively with biotic and abiotic variables among sites or sample treatments (Chapter 5).

QUALITATIVE INVENTORIES

In retrospective surveys of soil fungi, whose goals are mainly taxonomic, specialists describe and comment on fungal strains, prepare diagnostic keys, and revise taxa. Such inventories often are based on strains accumulated over several years of ecological or phytopathological studies of fungi within a given region and on strains deposited in service culture collections.

Several techniques are appropriate for the compilation of species lists for samples or sites, the mycological equivalent of a plant systematist's floristic analysis, and such compilations can serve a variety of needs. For example, acquisition and selection of fungal isolates for natural products screening is, in essence, a qualitative inventory; qualitative analysis is more reliable than quantitative analysis when samples are shipped from sites distant from the investigator's laboratory (Gochenaur 1970). The qualitative approach also is used by specialists who want to survey for specific taxa.

To discover the identities of species far to the right in the species-abundance curve for a standard plating-isolation procedure, we recommend the following: (1) preparation of 500–1000 or more isolates; (2) intentional, subjective bypassing of apparently redundant taxa in an effort to isolate as many taxa as possible (see Options 2 and 3, under "Selecting Colonies," earlier); and (3) use of rigorous, selective techniques.

To capture taxa rarely obtained with a single, standard plating-isolation procedure, the investigator can vary the medium and incubation conditions and use an array of selective techniques. For example, different soil subsamples could be processed with or without ethanol and phenol pasteurization, or heat pretreatments, and each treatment could be incubated at low and high temperatures. Washed soil particles can be plated on Thorn's basidiomycete medium, P10VP, yeast isolation medium (Chapter 16 and Appendix II), or other selective media. The organisms targeted by these procedures may appear sporadically and at very low frequencies. Selective plating techniques would follow procedures outlined in Option 3 (under "Selecting Colonies," earlier) to emphasize selection of rarely seen isolates. Baiting for pathogenic root fungi, keratinophilic fungi, oomycetes, and Chytridiomycetes would expand further the species lists.

APPENDIX

TECHNIQUES FOR ISOLATION OF SOIL FUNGI

SOIL COLLECTION

Ten to 30 g of soil should be collected per sample, and 6–20 samples should be collected per site, using a soil-coring tube or other tool sterilized in 70% ethanol and dried just prior to collection. The sample should be taken from a soil profile or mineral soil surface, carefully exposed by removal of overlying litter and humus. The samples are placed in labeled, sterile bags or vials; transported in an ice chest or cooler; refrigerated; and either plated out within 4 days or frozen for examination within the next 6–9 months (Christensen 1969). For additional details on procedures, see "Collection and Storage of Soil" under "Isolation Techniques for Filamentous Fungi," in the main section of this chapter.

Phytosanitary permits must be obtained before foreign soils can be imported legally into the United States. In addition, those soils and the laboratory ware exposed to them must be properly stored, cleaned, and disposed of to prevent accidental introductions of exotic organisms. Specific regulations and guidelines pertaining to research with foreign soils can be obtained from the U.S. Department of Agriculture, Animal Plant Health Inspection Service, National Center for Import and Export (Appendix III) and from individual state Departments of Agriculture.

TOOLS, EQUIPMENT, AND WORK AREA

Procedures for indirect isolation of fungi from soils or other organic substrata require a minimum of tools and equipment generally found in most microbiology laboratories. First, a clean work area is essential. In humid or dusty climates, use of a closed air-conditioned room will help to prevent entry of air-borne contaminants. A laminar flow hood may be needed for pouring media, seeding isolation plates, and transferring cultures. If working with foreign soils, a class-2 biosafety containment hood is recommended. Incubation at cool temperatures in a refrigerated incubator or an air-conditioned room will improve results. Media preparation, a process very similar to cooking, requires measuring and weighing equipment, clean water, a sink and wash area, an autoclave or pressure cooker to sterilize media, and appropriate glassware. Media are dispensed into Petri dishes and tubes with manual or automatic pipettes, self-refilling syringes, or peristaltic pumps. Media, reagents, antibiotics, and substrata are refrigerated or stored in a cold incubator.

Both a bright-field microscope and a high-quality dissecting microscope able to attain 50–120× magnification are required, as is having both epistage and substage lighting. Long hours at the microscope are the rule, and comfort greatly enhances operator productivity and attention. An ideal lab has low benches or tables with fully adjustable seats and good lighting. Primary identification guides should be at hand in the work area.

A variety of tools (e.g., insect pins, small-bore hypodermic needles, or sharpened sewing needles mounted in capillary tubes, wooden handles, or pin vises) are necessary for precise dissection of minute colonies. Needles must be sterilized between colony transfers by flaming. However, flames rapidly destroy fine dissecting tools. Therefore, we have used minute insect pins for precision dissection of newly germinated spores or washed soil particles. Pins are disinfected prior to use by immersion in 70–95% ethanol followed by immersion in clean agar. Fine gauge (26–28 gauge) disposable hypodermic needles mounted on a handle are also excellent tools for colony and spore dissections. Equipment required for preservation and long-term maintenance of isolates and techniques are described in Chapter 3.

DIRECT INOCULATION PROTOCOLS

DRECHSLER'S METHOD

Sprinkle a few milligrams or less of fresh, finely divided soil or humus onto water agar, cornmeal agar, or carboxymethyl cellulose medium, and incubate at 15–20°C; antibacterial antibiotics can be added to the medium. Review plates daily, and transfer emerging hyphal tips to agar slants. Within a few days mitosporic or zygomycetous species may start to sporulate. Conidia or sporangiospores of moist-spored species can be touched with a fine needle and transferred to agar slants. Conidia of dry-spored species are transferred with a needle moistened with sterile glycerol, sterile dilute agar (0.2%), or clean agar obtained by plunging the needle into the medium near the colony edge.

SOIL SWABBING AND STAMPING

The swabbing and stamping method (Hunter-Cevera and Belt 1992) works well with soils in which propagule densities are low, as in deserts. With care and a light touch, swabbing or stamping can be applied to any soil. Weather permitting, plates can be inoculated in the field.

In the laboratory, spread pulverized or sifted soil in a sterile Petri plate lid or plastic weighing pan. With a sterile cotton swab, touch the soil surface and streak particles, following a bacteriological purification pattern, onto the surface of DRBC, carboxymethyl cellulose, water, or malt-cyclosporin agar (see Appendix II). Cotton balls or felt replicate pads also can be used to disperse soil particles accoss agar surfaces.

Place finely divided soil from each sample in a Petri dish lid. Prepare a row of 6–10 plates with the desired isolation medium, and place them to one side. Stamp a cotton ball or a felt pad lightly onto the soil. Then, starting at the first plate, stamp the entire surface of the agar in each plate in succession in a clockwise pattern. Usually, colonies overwhelm the first two or three plates, but the number of colonies on later plates is manageable. Incubate plates at 15–20°C, and review daily, transferring emerging hyphal tips to agar slants.

WARCUP SOIL PLATES

Prepare molten soil extract–rose bengal medium, carboxymethylcellulose medium, glucose ammonium nitrate agar, or other medium (Appendix II) and maintain it at about 45°C in a hot water bath. Place 0.2–5.0 mg of finely pulverized soil on the bottom of a sterile Petri dish. The optimal quantity of soil will vary widely with propagule density and soil texture. The soil can be dispersed by agitation in 0.5 ml of sterile water or with a few milligrams of sterile sand grains. Overlay the soil particles with 8–10 ml of medium while gently agitating the plates to disperse particles. A thin layer of agar facilitates dissection of colonies from beneath the agar. Young submerged colonies need to be cut from beneath agar. Incubate the plates at 15–20°C and review daily; transfer emerging hyphal tips to agar slants (Warcup 1950; Parkinson 1994).

SUSPENSION PLATING

Specific accounts of suspension plating can be found in the main section of this chapter. We outline a simplified, basic protocol as follows:

1. Collect and prepare soils (see "Collection and Storage of Soil" under "Isolation Techniques for Fil-

amentous Fungi," in the main section of this chapter, and "Soil Collection," earlier).

2. Determine the approximate fungal propagule densities in 1 ml of a soil suspension by a preliminary plating of a series of soil/water ratios of 1:100, 1:5000, and 1:200,000.

3. Prepare a dilution series that in as few steps as possible will yield a final concentration of 30 or fewer fungal propagules per ml from a 5–20-g subsample of thoroughly mixed soil. One possibility is 5 g soil in 125 ml sterile water in a 500-ml flask (suspension A, 1:25); 1 ml suspension A in 39 ml sterile water in a 250-ml flask (suspension B, 1:1000); 2 ml suspension B in 18 ml sterile water in a 50-ml flask (suspension C, 1:10,000).

4. Prepare and sterilize the medium; cool, add acid or antibiotics for bacterial inhibition, and dispense into Petri plates.

5. Prepare and sterilize water, 0.15% agar, or 0.1–0.2% carboxymethyl cellulose in cotton-stoppered Erlenmeyer flasks for hand shaking or agitation in an omnimixer or on a gyrotary shaker. Replicate sterile quantities of water or aqueous solutions are called "blanks." Sterilize a supply of wide-mouth pipettes (tip internal diameter 1–2 mm).

6. Weigh the appropriate amount of thoroughly mixed soil using a sterilized weigh pan and immediately add to first water blank, repeating for up to 10 soil samples; continue to shake the suspension vigorously to maintain a uniform suspension; prepare remaining suspensions, withdrawing each subsample quickly from a rapidly moving suspension; place 1 ml of final suspension onto poured, cooled agar in three to six Petri plates; tilt plates to distribute the suspension evenly or spread it with a flamed bent-glass rod. Place water, agar, or carboxymethyl cellulose blanks on several plates of isolation medium as controls for sterility.

7. Cover plates to ensure dark incubation (absolutely essential if the medium contains rose bengal). Incubate at standardized temperature for a predetermined period (see "Incubation," under "Isolation Techniques for Filamentous Fungi," in main section of this chapter).

8. Isolate all colonies present on a plate or a standardized number of colonies following a predetermined sequence, thereby assuring a random sample, after a standardized incubation temperature and interval (see "Selecting Colonies," under "Isolation Techniques for Filamentous Fungi," in the main section of this chapter).

9. Prepare the required number of identical tube slants of medium from a single large batch of medium. The medium should be relatively rich (e.g., potato-

dextrose agar [PDA], malt-extract agar, or Czapek yeast extract agar) to facilitate sorting of week-old isolates by growth rate and colony surface area and to reverse pigmentation. Label the tubes to indicate site and sample.

10. Sort isolates by macroscopic appearance after 7–10 days of growth; assign permanent culture numbers to all apparent taxonomic entities; for each entity, record the number of isolates by sample and site for use in calculating frequencies and relative densities.

11. Lyophilize or ultra-cold freeze one or more representatives of each apparent taxonomic entity before beginning characterizations and identifications.

PARTICLE FILTRATION

Apparatus

References describing various particle filtration devices for fungal isolations are listed in "Particle Filtration," in the main section of this chapter. The filtration system illustrated in Fig. 13.5A is a micro-sieve set (Bel-Art Products) that consists of a stack of four or five interchangeable screen-holding cylinders of molded polypropylene (inner diameter 51 mm). The cylinders unscrew easily for changing mesh inserts, cleaning, and disinfecting (Fig. 13.5C). Three meshes are placed in the device in descending order of mesh size. The first insert, a brass mesh with 1-mm openings, excludes large fragments; the second insert is made of polypropylene (Spectra/Mesh), with mesh openings of 210 μm, and a third mesh insert, also of polypropylene, has openings of 105 μm. The polypropylene inserts, which are deformed easily by water pressure, are supported inside the locking cylinders by brass inserts with 2-mm openings underneath the polypropylene inserts (Fig. 13.5C). The polypropylene inserts sit directly on the supporting brass meshes, which are locked into the sections. Pressure from the interlocking cylinders secures the polypropylene meshes during washing. The assembled device is held under the faucet by a burette clamp mounted on a ring-stand (Fig. 13.5B).

Protocol

1. Collect and prepare soils (see "Soil Collection," earlier). Prepare isolation media and tube slants as described for suspension plating (see "Suspension Plating," earlier).

2. Disinfect the assembled filtration device by soaking in ethanol (70–95%), sodium hypochlorite solution (two parts bleach/one part water), or dilute peracetic acid (0.03%). Autoclave or heat sterilize

FIGURE 13.5. Particle filtration device. **A.** Assembled device in sink with distilled water source and vacuum flask for trapping effluent. **B.** Close-up of interlocking cylinders. **C.** Single disassembled cylinder, from left to right: polypropylene filter with 105 µm pores; upper locking ring that supports brass mesh with 1-mm pore size; bottom section of cylinder.

Waring blender jars or mortars and pestles and allow to cool. Assemble filtration device near a water source such as a sink (Fig. 13.5B).

3. Pulverize soils with the sterilized mortar and pestle or in the Waring blender.

4. Load soil on the uppermost brass prescreen. With a fine spray of water, gradually wet the soil and wash it for 10 minutes (4–6 liters of water), forcing the soil particles downward through the 1-mm brass mesh and the two polypropylene filters. Clean tap water can be used for washing if appropriate controls are used to check sterility of the water source.

5. Remove the cylindrical chamber of the filtration device from the burette clamp, and separate cylinder sections. The lowest filter should trap a thin layer of washed particles. Remove the 105-µm filter with the trapped soil particles from the lowest cylinder section and place it in a sterile 50-ml polystyrene centrifuge tube. The flexible filter is folded for insertion into the centrifuge tube. Add 50 ml of sterile, distilled water, and agitate the tube vigorously to suspend and wash the particles; remove washed filter with sterile forceps.

6. Allow particles to settle, wash again with sterile water, and again allow particles to settle to the bottom of the conical centrifuge tube.

7. Dilute particles with sterile water to obtain a 20:1 (v/v) ratio of water/particles so that the density of particles from each sample will be uniform. The suspension can be made with 0.2% agar or carboxymethyl cellulose to impede sedimentation.

8. Agitate and resuspend the particles. Pipette 0.1-ml portions of suspended particles onto 5–10, 100-mm Petri plates of isolation medium. Spread the suspension over the surface of the agar with a flamed bent-glass rod.

9. Prepare control plates with samples of the water source on the same fungal isolation medium to assay for fungal contamination.

10. Incubate the plates at the desired temperature and continue the isolation strategy as outlined earlier. Isolate colonies as outlined in "Selecting Colonies," under "Isolation Techniques for Filamentous Fungi," in the main section of this chapter.

Overloaded meshes tend to clog, resulting in autofiltration and a population of particles skewed toward smaller sizes; occasional stirring of trapped particles helps to prevent clogging. A no. 11 one-hole stopper with a pipette attached to the hose of a vacuum pump can be inserted into the bottom section of the filtration device to pull the water through the sieves and speed water flow (Fig. 13.5A). The vacuum hose must be attached to a side-armed flask (2–4 liters) or equivalent to trap water and prevent it from entering the pump. The trap also prevents discharge of the washed effluent from foreign soils or organic substrata into the environment. The trapped effluent is autoclaved before discarding.

Incomplete washing is apparent when stray colonies, not associated with particles, appear on plates. However, some very fine textured soils, composed of extremely minute particles of silt or clay, do not wash well because most of the soil particles are washed through the sieves.

ETHANOL AND PHENOL PASTEURIZATION

Overview

Several partial sterilization methods for selective recovery of ascomycetes from soil habitats use steam, dry heat, or chemical treatments, either singly or in combination (Warcup and Baker 1963; Mahoney 1972; Furuya and Naito 1980; Bills and Polishook 1993). Here is presented an ethanol and phenol pasteurization method, which also has been adapted for isolation of fungi from dung of herbivorous animals (Bills and Polishook 1993). Substitution of 2% phenol or 5% acetic acid for ethanol is a variation that has been used to isolate ascomycetes selectively from soils (Furuya and Naito 1979, 1980). When soil is pasteurized properly, coprophilous ascomycetes (e.g., *Chaetomium, Gelasinospora, Thelebolus, Sordaria, Ascobolus, Saccobolus, Sporormiella* species), as well as ascosporic and sclerotial Eurotiales and other cleistothecial and perithecial ascomycetes, are isolated at higher frequencies. Occasionally, common soil fungi also will survive ethanol pasteurization, probably because propagules deep within particles are protected.

Our experiences using ethanol pasteurization are based on soils from Antarctica and many arid and humid tropical regions (G. F. Bills and J. D. Polishook, unpublished data). Soils vary widely in numbers of ethanol-resistant propagules, so quantities of soil needed will vary accordingly. In addition, soil texture and organic matter content can significantly affect the results of ethanol pasteurization. Fine-textured soils are sterilized more effectively than those with coarse textures or large quantities of organic matter, such as plant fibers and roots, which can interfere with and prevent complete pulverization. Some preliminary adjusting of soil volumes or pasteurization times may be needed to compensate.

Protocol

1. Collect and prepare soils (see "Soil Collection," earlier).

2. Prepare media (GAN, DRBC, CMC, or MEC [Malt Extract Agar plus Copper]; see Appendix II) in 500-ml aliquots in 1-liter Erlenmeyer flasks. Immediately

after autoclaving, place flasks in a 45°C water bath and allow them to stand until the medium temperature equals the water temperature.

3. Pulverize 5–10 g of air-dried soil with a sterile mortar and pestle for each sample. Place powdered soil in a sterilized 125-ml or 250-ml Erlenmeyer flask and label appropriately. A series of trial runs may be needed to determine the optimal amount of soil.

4. Add enough (20–30 ml) 60% ethanol (or 2% phenol) to each flask to wet the soil and cover it completely. Agitate flasks on shaker for 5 minutes. Allow the flasks to stand and soil to settle for an additional 5 minutes. At the end of the treatment period, decant the ethanol (or 2% phenol). Rinse the remaining soil twice with sterile water. Some soil will be lost, but more than sufficient quantities will remain in the flask.

5. Place five Petri dishes in a row for each pasteurized soil sample. Pipette 0.5–1.0 ml of treated soil suspension into the bottom of each Petri dish with a wide-tipped pipette. Dispense 10–15 ml of medium over the soil suspension in each of the Petri dishes. Slowly agitate each plate to mix the molten medium and treated soil suspension to disperse soil particles evenly.

6. After the agar solidifies, Petri dishes are incubated at the desired temperature. Control plates are made by treating the soil with water instead of ethanol or phenol.

7. Remove of colonies from plates following the procedures described in "Warcup Soil Plates," earlier, and in "Selecting Colonies," under "Isolation Techniques for Filamentous Fungi," in the main section of this chapter.

Selective Isolation of Saprobic Soil Basidiomycetes

The method of selective isolation combines a modified particle filtration procedure with plating of washed particles on a selective, indicator medium containing lignin, guaiacol, and benomyl (LGB; see Appendix II; Thorn et al. 1996). Steps include the following:

1. Collect and prepare soil samples (see "Soil Collection," earlier). Prepare plates of LGB isolation medium and tube slants (see "Suspension Plating," earlier).

2. Following aseptic procedures, weigh 5 g of mixed soil and add it to a 1-liter jar containing 500 ml of sterile 0.1% sodium pyrophosphate. Shake the jar on a platform shaker at low speed for 1 hour at 4°C, and pour the resulting suspension through two standard, 20-cm-diameter soil sieves of 250-μm and 53-μm mesh. Between soil samples, rinse sieves with tap water and disinfect them with 70% ethanol. Wash particles retained on the 53-μm mesh sieve for 5 minutes with a spray of tap water. Tilt the sieve to separate organic from inorganic particles, and pick up 1 ml of dense suspension of washed organic particles in a sterile, wide-mouth pipette.

3. Disperse particles in 9 ml of sterile water, and plate 0.2–0.4 ml on each of 5–10 plates of LGB agar, spreading them with a flamed bent-glass rod. This rapid particle filtration technique may be suitable for general isolations on a routine medium, such as SEA, CMC, or DRBC, following a second dispersion and keeping in mind the possible consequences of using large-size particles. To test for possible fungal contamination of the wash water and cross-contamination by use of inadequately sterilized soil sieves, replicates of 0.2–0.4 ml of wash water should be pipetted from the 53-μm mesh sieve following ethanol sterilization after the last soil sample is processed and plated on LGB agar.

4. Incubate LGB plates for up to 4 weeks. Place plates on the movable stage of a bright-field microscope, and at 40–100× examine colonies that have turned the agar bright red to reddish-brown; avoid colonies with conidia of undesired mitosporic fungi (e.g., *Alternaria* species, *Trichoderma* species). Transfer putative basidiomycetes to slant tubes.

When dealing with foreign soils, all wash water and all suspensions not plated should be autoclaved prior to disposal.

SELECTIVE BAITING AND ENRICHMENT TECHNIQUES FOR ISOLATING SOIL CHYTRIDS

Method

In one of the most widely used methods for isolation of soil chytrids, approximately a teaspoon of soil (3–5 g) is placed in a Petri dish and covered with double-distilled water, charcoal water (Couch 1939), sterilized pond water, or SEA (Hanson 1945). Characteristic baits that correspond to substrata utilized by chytrids are floated on the water's surface. Chytrid populations selectively colonize and multiply on those baits. Empiric studies of propagule numbers for aquatic chytrids (Ulken and Sparrow 1968) have shown that as the number of chytrids in a sample increases, the amount of soil per plate should decrease to assure that the amount of substratum is adequate.

Typically, baits are floated at the water surface, separated from the soil below. The lack of contact with the soil discourages hyphal growth directly into the substratum and forces zoosporic colonization of the substratum. Larger baits sometimes are submerged so that air-borne spores do not colonize the substratum. The quantity of bait per culture should be minimized to assure that the culture does not become overly enriched (most substrata leach soluble nutrients into water). In nutrient-rich cultures, bacteria and hyphal fungi can outcompete the chytrids.

BAIT PREPARATION

Some baits may be used directly, but most must be pretreated, as outlined as follows.

Cellulose Baits

Cellophane. Boil dialysis tubing or other cellophane in water for 5 minutes. Open the tubing, split the side wall, and cut the sheet into 5–10-mm squares. Transfer the squares to fresh water and autoclave.

Filter Paper or Lens Paper. Cut the paper into 10-mm squares and use directly. Paper is less transparent than cellophane, but the loose fibers are readily accessible to cellulolytic chytrids. However, filter paper is not degraded as rapidly as other natural baits (e.g., young grass leaves), and hence, the interval for examination of colonized baits before their total decomposition is longer (Haskins 1939).

Grass. Cut young grass (such as *Paspalum* species) leaves into trapezoids on the diagonal to expose the ends of the vascular bundles. Clear the leaf pieces in 100% ethanol, rinse in water, and autoclave in water.

Onion Epidermis. Strip the epidermis from storage leaves of bulbs, boil in water, and autoclave in water. The epidermis should be floated on water with the disrupted inner surface down.

Seeds. Seeds are complex stores of nutrients and also contain cellulose. Sterilized *Cannabis* seeds are used most commonly to isolate chytrids that are parasitic on oomycetes that also may colonize the seeds. Seeds of other species also can be used (e.g., sesame, buckwheat, mung bean). Cut or boil seeds to open the seed coat and expose the cotyledons and embryo.

Keratin Baits

Hair. Cut hair into 15-mm lengths, and remove natural oils with ether. There is no consensus about the types of hair that make the best substrata. According to Karling (1977), human baby hair is among the best keratin substrata for chytrids.

Horn, Hoof, and Nails. Thinly slice cow horn and cow or sheep hoofs, or collect nails. These materials are used with no additional treatment.

Skin. Molted snake skin and human skin (e.g., peeled from a sunburn) can be used directly.

Chitin Baits

Arthropod Exoskeleton. Cut shrimp exoskeletons into squares, selecting the clearest regions. Treat the squares with dilute hydrochloride (HCl) followed by 1% KOH (Karling 1945) and store in 100% ethanol. Wash squares in boiling water and then autoclave in water before use.

Fish Scales and Insect Exuviae. Fish scales and insect exuviae are favored for their transparency. Pretreat with dilute HCl and autoclave in water.

Insect Wings. Cut termite, fly, and roach wings into pieces of manageable size and autoclave in water.

Pollen Baits

Plant pollen is a particularly successful bait because it is widely available in nature and because it is refractory and decay-resistant. Other than some bacteria, few microbes decay pollen. Pollen of *Pinus* species and *Liquidambar styraciflua* (sweet gum) often are used as baits. Sweet gum pollen lacks the wings of *Pinus* pollen, which increases visibility of internal structures. Male cones or flowers releasing pollen are placed on the surface of stacked graduated-mesh sieves. A collection pan is placed under the 1-μm sieve where debris-free pollen accumulates. The pollen is air-dried or heat-dried at about 60°C. Pollens of some species (e.g., *Pinus* species) can be used after drying. Sweet gum pollen should be sterilized with propylene oxide. Propylene oxide is a potential carcinogen and should be used only under a fume hood. Alternatively, dried pollen can be autoclaved briefly (121°C for 5 minutes). The pollen is sprinkled on the water's surface with a sterilized cotton plug. Pollen is removed easily from cultures for examination by dragging one edge of a coverslip over the surface of the water. The coverslip with pollen then is lowered gently onto a glass slide. Once chytrids are isolated from soil (2 days to 6 weeks), typically on bait, chytrid populations on substrata are enriched by removing the substratum from the gross culture, washing its surface, and transferring it to a new culture dish containing water and fresh bait.

PROTOCOLS FOR NONSPORULATING ISOLATES

PROMOTING SPORULATION

Nonsporulating isolates are transferred to a small selection of specialized media designed to enhance sporulation. We recommend cornmeal agar, with or without a piece of autoclaved banana leaf; oatmeal agar; V8-juice agar; synthetic nutrient agar; Alphacel agar; or Weitzman and Silva-Hutner agar (Appendix II). Incubation for up to 1 year at low temperatures (5–10°C) may be required to induce sporulation in some species (Wang and Wilcox 1985; Petrini 1986). More labor-intensive and media-intensive physiological tests are used only if sporulation does not occur.

CHARACTERIZATION OF NONSPORULATING ISOLATES

1. Persistently nonsporulating isolates are transferred to a medium that promotes vegetative growth (e.g., PDA) so that fresh, uniform inocula of all strains will be available for transfer to all test media. Sterile polypropylene transfer tubes (Spectrum Laboratories) or a flamed-cork borer (4–5 mm diameter) are convenient for cutting inoculum plugs for transfer, mycelium-side down, onto test media.
2. Plates or tubes of the media listed as follows under specific tests are prepared.
3. At least three replicates of all tests for each isolate are carried out and, ideally, repeated once.

Metabolic Tests

Cellulose Azure-Dye Test. Cellulose azure-dyed cellulose is layered on top of a colorless basal agar medium. Release of the blue dye, initially bound to the cellulose substrate, indicates cellulolysis; disappearance of the dye is correlated with ligninolysis (Thorn 1993). Incubation time required will vary with the growth rate of the test isolates but can range from 2–8 weeks. This test may be done with concentrated broth from liquid-grown cultures or with partially purified enzymes. Dye released is quantified spectrophotometrically at an absorbance of 595 nm (Smith 1977).

Lipase-Activity Test. Lipase activity is assayed conveniently using Tween-20; enzymatic hydrolysis of the ester bond releases a fatty acid that combines with calcium ions in the medium to form a visible precipitate (Hankin and Anagnostakis 1975). With some fungal strains, precipitates may form within 24 hours of inoculation, but incubation times of 1–2 weeks are routine. Be aware that some fungi subsequently clear the crystalline precipitate.

Identification of Some *Penicillium* Species. Some *Penicillium* species can be identified by shifts in the pH of the medium during growth, when the medium contains creatine as a nitrogen source and sucrose as a carbon source (Frisvad 1981). Isolates first may acidify the medium causing it to turn yellow and then, after 2–3 weeks, cause the pH to increase until the agar turns purple.

Differentiation of Certain *Trichophyton* Species. Certain *Trichophyton* species (Fischer and Kane 1971; Summerbell et al. 1988) can be differentiated when grown in a medium containing milk solids as a nitrogen source and glucose as a carbon source. The test is also useful for identification of ectomycorrhizal and wood-decay basidiomycetes (Hutchison 1990a; Thorn 1991). Clearing by some strains is apparent 24 hours after inoculation; in other strains 1–2 weeks may be required. Strains that acidify the medium as they slowly clear the milk solids turn it yellow. Strains that rapidly clear the milk solids tend to increase pH, staining the agar purple.

pH shifts in the medium during growth also have been used to differentiate certain *Trichophyton* species (Kane and Fischer 1971) when a liquid medium containing urea as a nitrogen source and glucose as carbon source is used. The test is also useful for identification of ectomycorrhizal and wood-decay basidiomycetes (Hutchison 1991; Thorn 1991). Incubation time is typically 2 weeks.

Ligninolytic Enzymes. Colorimetric spot tests for the ligninolytic enzymes laccase, tyrosinase, and peroxidases also can be helpful in characterization of isolates, although the results should be interpreted with caution because (1) the substrates of the enzymes may overlap and the action of one enzyme may influence the actions of another, and (2) the expression of the enzymes varies in different media (Käärik 1965; Harkin and Obst 1973; Marr 1979). We recommend syringaldazine (0.1% in 95% ethanol) for the assay of laccase and p-cresol (0.1 M in 95% ethanol) for tyrosinase (Käärik 1965). If laccase is absent, peroxidase(s) can be assayed by adding one drop of freshly diluted H_2O_2 (0.4%) to the syringaldazine spot (Harkin and Obst 1973). Peroxidases cannot be assayed by spot tests if laccase is present. Broth from liquid-grown fungi can be assayed spectrophotometrically (particularly if enzymes have been separated chromatographically) using ABTS-$(NH_4)^2$ [2,2′-azino-bis (3-ethylbenzothiazoline-6-sulfonic acid) diammonium salt] as a substrate for laccase (Niku-Paavola et al. 1990), D-L-DOPA (DL-3,4-Dihydroxyphenylalanine) for tyrosinase (Horowitz et al. 1960), veratryl alcohol with H_2O_2 for lignin peroxidases (Tien and Kirk 1984), and phenol red or vanillylacetone with H_2O_2 and Mn(II)

Sensitivity Tests. Sensitivity to or tolerance of defined levels of benomyl, cycloheximide, and NaCl have been used to distinguish *Ophiostoma* species from *Ceratocystis* species (Harrington 1992; Seifert et al. 1993) and to characterize ectomycorrhizal and wood-inhabiting basidiomycetes (Hutchison 1990c; Thorn 1991). Results are assessed after 1–2 weeks and expressed as growth on an amended medium as a percent of growth on a non-amended control.

Distinguishing Filamentous Ascomycetes from Basidiomycetes

Diazonium blue B (DBB) can be used to distinguish filamentous ascomycetes from basidiomycetes. Basidiomycete mycelia stain positively with DBB following pretreatment with 1 M KOH; cultures of ascomycetes usually do not. Isolates are grown statically in a standard liquid medium in Petri dishes. Mycelia are harvested while actively growing; rinsed three times with distilled water; incubated for 12 hours in 1 M KOH at 4°C; rinsed thoroughly (three or more times) with distilled water; transferred to cold (4°C) 0.1 M Tris-HCl buffer

at pH 7.0; and maintained on ice while fresh DBB (1 mg/ml) is prepared in the same, cold buffer. Disposable capped centrifuge tube (17-mm tall × 100-mm diameter) are convenient for rinsing, KOH-hydrolysis, and testing procedures. Excess buffer is removed from a mycelium, a few drops of DBB solution are applied, and the tubes are returned to ice for 1–2 minutes. Basidiomycete mycelia will begin to turn red to purple within 1–2 minutes; longer incubation, elevated temperatures, or pH levels of more than 7.0, however, will cause rust-red false positives. Heavily pigmented mycelia make assessment difficult, and a few ascomycetes tested do produce red reactions. DBB is extremely carcinogenic and is highly sensitive to oxidation at temperatures above 0°C; it should be stored desiccated at −20°C.

Distinguishing Ascomycetes and Holobasidiomycetes

Examination of septal structure in nonsporulating isolates with large-diameter, thin-walled hyphae is possible with a good light microscope after staining with trypan blue or ammoniacal Congo red. Simple septal pores with Woronin bodies, characteristic of ascomycetes, or dolipore septa and sometimes parenthesomes, characteristic of holobasidiomycetes, can be seen under optimal conditions. Examination of mycelia grown in microscope slide culture improves the ability to discern details of individual hyphae.

14

FUNGI IN STRESSFUL ENVIRONMENTS

JOHN C. ZAK AND HOWARD G. WILDMAN

The physical environment in which most organisms exist is not constant, with both temporal and spatial fluctuations in abiotic conditions the rule rather than the exception. Organisms have evolved to survive such conditions, however, and they show adaptations to the environments from which they originate. Extreme environments are conventionally those in which temperature, pH, or nutrient concentrations are very low or high, availability of water or oxygen is very low, or concentrations of salts or other solutes are very high. Realistically, an extreme environment is one that differs significantly from the range of culture conditions that we believe are normal, either in natural settings or in the laboratory. Therefore, extreme conditions are relative, not absolute, and represent an anthropocentric view. Furthermore, organisms that have become adapted to a highly specialized physical environment should be distinguished from those that can cope with considerable variations in environmental factors.

Perhaps a better term for the types of habitats we examine in this chapter is "stressful," which implies comparisons with other environments that may support the growth of a fungus of interest (Jennings 1993). Stressful environments are those with abiotic conditions that restrict or prevent the growth and reproduction of an organism. Even that definition is of limited use, however, because the environments of the communities compared often have been studied at very different spatial and temporal scales. Even at the scale of the microorganism, environmental stress and stressors cannot be defined clearly, except with respect to a particular organism. For example, all nonhalophilic and most terrestrial organisms deal with osmotic stress, whereas halophiles whose internal environment is fairly isosmotic with the external one do not (Kushner 1978).

Life-history traits of organisms have evolved in response to two types of environments: one in which organisms live near the carrying capacity of their environment and one in which abundant resources allow population expansion. The life-history characteristics associated with each type of environment have been termed K- and r-strategies, respectively (MacArthur and Wilson 1967). K-strategists often are characterized by long, slow-growth periods and low reproductive effort; r-strategists are characterized by short, fast-growth periods accompanied by a high reproductive effort. Within the spectrum of r- to K-strategies, organisms are influenced by environmental factors such as competition, disturbance, and stress. For fungi in extreme environments, stress in the form of some "extreme" environmental parameter(s) is of great significance. Fungi are able to occupy niches, even in stressful environments, by adopting different ecological strategies.

The concepts of ecological strategies, developed for higher plants (Grime 1977, 1979), have been modified and extended to fungi (e.g., Pugh 1980; Cooke and Rayner 1984; Rayner and Boddy 1988; Andrews 1992). Three primary life strategies now are recognized—competitive (C-selected), stress-tolerant (S-selected), and ruderal (R-selected)—with many fungal species displaying combinations of those strategies. The strategies define an organism's behavior at a particular time only, and different strategies may be adopted under different environmental conditions or during different stages in a fungus life cycle with different modes of growth (Boddy and Wimpenny 1992). The relative absence of stress and abundance of soluble carbohydrate-rich resources favor R-selected fungi, which have fast growth rates, high reproductive capacity, and short life spans. Stress, imposed either by poor access to nutrients in the substratum or by abiotic factors, tends to limit interspecific fungal competition and favors long-term exploitation of scarce resources. Under such conditions, S-selected fungi that exhibit characteristics that include slow increase in biomass, specialized physiological adaptations and high enzymatic competence for resource exploitation, temporal persistence, and slow germination are favored. Some wood-decomposing basidiomycetes, for example, fall into this category. S-selected fungi also exhibit low competitive ability in the absence of environmental stress, but they may compete among themselves. Under a situation of either reduced stress or reduced availability of unexploited resources, C-selected species may predominate. These fungi are persistent, long-lived, and capable of defending captured resources by interspecific antago-

nism, and they exhibit good enzymatic capabilities. Again, certain wood-decay fungi provide good examples of fungi with this strategy. In some stressful environments (e.g., polar and alpine regions, exposed soil surfaces), species richness is low, with relatively few species adapted to the abiotic conditions of these environments. Other stressful environments, however, may contain diverse communities of fungi that are active for short periods (e.g., soil or dung in desert ecosystems).

In this chapter, we discuss ecological groups of fungi that are considered stress tolerant. We also cover various methods for isolation and sampling that are necessary for effective enumeration of species and for understanding patterns of biodiversity. We present the ecological groups as separate, nonoverlapping entities, although they may indeed overlap, with fungi expressing adaptations to several forms of stress. In temperate deserts, for example, fungi deal with low moisture availability as well as extreme diurnal and seasonal variations in temperature.

From a practical standpoint, stress-tolerant fungi hold much promise for the development of novel industrial applications, biocontrol or bioremediation agents, or other utilitarian functions. Their stress-adapted enzymes, as well as the genes that code for those enzymes, may be especially important. Selection of a fungus for ecological research or industrial development from the range of species isolated from a stressful habitat should be guided by physiological information obtained through research that explains the distribution of fungi along a continuum of specific environmental factors (Jennings 1993).

ECOLOGICAL GROUPS OF FUNGI

THERMOTOLERANT AND THERMOPHILIC

Temperature is one of the most important abiotic factors affecting the distribution and activities of fungi. Most fungi are mesophilic and grow at temperatures ranging from 5°–35°C, with optima for growth occurring between 20°C and 30°C. Fungi can be found, however, in habitats where temperatures exceed those commonly experienced by most fungal species. Fungi that grow only at temperatures ranging from above 20°C to above 50°C are defined as thermophilic (thermophiles) (Cooney and Emerson 1964; Crisan 1973). Fungal species that can grow when temperatures are below 20°C but also can grow at 40°–50°C or more are considered thermotolerant. Thermotolerant and thermophilic fungi generally can be isolated from large, damp, well-insulated piles of organic matter where the internal temperatures increase as a result of microbial respiration (thermogenesis). Examples of such sites include compost piles (Chang and

Hudson 1967), mushroom beds (Fergus 1964), municipal waste and compost deposits (Kane and Mullins 1973), and home and industrial wood chip piles (Tansey 1971). Thermophilic and thermotolerant fungi also can be isolated from bird nests (Apinis and Pugh 1967), alligator nests (Tansey 1973), coal tips (Evans 1971), and volcanic hot springs (Brock 1978).

It is interesting that thermophilic and thermotolerant fungi are ubiquitous in soils wherever the sun can heat soil to temperatures that are suitable for germination and growth (Tansey and Jack 1976). They can be a significant component of the soil-fungal community in arid and tropical regions (Moustafa et al. 1976; Abdel-Hafez 1982a; Kuthubutheen 1982; Bokhary et al. 1984; Singh and Sandhu 1986; Abdullah and Al-Bader 1990; Mouchacca 1995) but are also found in temperate grasslands, forests, and agricultural fields (Apinis 1963, 1972; Ward and Cowley 1972; Taber and Pettit 1975). Moreover, a reservoir of propagules of those fungi exists in most soils and persists throughout the year (Tansey and Jack 1976). Tansey and Jack (1976) reported that thermophilic and thermotolerant fungi are ubiquitous. The key to detecting these fungi in any ecoystem is using appropriate incubation temperatures.

In the first comprehensive treatment of this ecological group of fungi, Cooney and Emerson (1964) reported that thermophilism was strongly developed in the phycomycetes, ascomycetes, and mitosporic ascomycetes (hyphomycetes), but it was absent from the basidiomycetes. A key for identifying 13 species of thermophilic fungi is included in their book. Tansey and Jack (1976) and Tansey and Brock (1978) reported that 12 or so species commonly are isolated from soils, whereas an additional 12 or so less common species can be isolated with sufficient effort. More than 20 thermophilic taxa are considered to have cosmopolitan distribution (Table 14.1). Recent estimates put the number of thermotolerant and thermophilic fungi at approximately 70 species (Abdullah and Al Bader 1990; Mouchacca 1995). Undescribed species frequently are encountered (Tansey and Jack 1976). Thus the number of thermophilic and thermotolerant fungi will increase as additional habitats are evaluated. Closer examination of the ecological role of fungi in decomposition processes, especially cellulose and lignin degradation, is warranted, particularly in arid and semi-arid ecosystems and in habitats in which organic matter accumulates and becomes thermogenic.

PSYCHROTOLERANT AND PSYCHROPHILIC

Psychrotropic microorganisms are those that are capable of growth below 10°C. They are widespread in natural environments as well as in foods (Smith 1993), which is not surprising given that 80% of the biosphere exists at −3 to −7°C and that 90% of the volume of marine waters is below 5°C. Gounot (1986) noted that 0°C is not in itself an extreme condition and that cold environments may be considered extreme only if another factor creates adverse conditions (e.g., low water activity in arid Antarctic soil, low nutrient availability and high pressure in the deep sea).

True psychrophilic microorganisms are restricted to permanently cold habitats, such as oceans, polar areas, alpine soils and lakes, snow and ice fields, and caves. They have optimum growth temperatures that range up to 16°C and a maximum growth temperature of 20°C, but they also can grow at 0°C. Psychrotolerant microorganisms, in contrast, have maximum growth temperatures above 20°C, although they can grow at 10°C. Well-known psychrophilic and psychrotolerant fungi are found in genera such as *Alternaria*, *Cladosporium*, *Keratinomyces*, *Leptomitus*, *Mucor*, *Penicillium*, *Rhizopus*, and *Typhula* (Gounot 1986). The composition of the fungal community isolated from forest soils in Rhode Island at incubation temperatures of 0°C differed completely from that of the community obtained when soil and litter samples were incubated at 25°C (Carreiro and Koske 1992). In the former study, the psychrophiles were predominately *Mortierella* species and *Mucor* species, whereas at 25°C, the community was dominated by mitosporic ascomycetes. Eighteen isolates representing five species of mitosporic ascomycetes (*Heteroconium chaetospira*, *Leptodontidium obscurum*, *Microsphaeropsis* species, *Myrioconium* species 1 and species 2) and five sterile strains were isolated from soil and plant litter samples collected from glacier cones, snow fields, and a pond at a glacier front in Switzerland (Petrini et al. 1992b). Those taxa were considered psychrophilic. (Yeasts and yeastlike fungi were not cultured as part of the study.)

In polar regions, the occurrence of fungi is coupled to water availability, as it is in warm desert ecosystems. Yeasts tend to predominate in the undisturbed areas of the dry interior valleys of Antarctica. *Cryptococcus albidus* is one of the most prevalent psychrophilic species (Cameron et al. 1976). However, as research activity has increased at the poles, so has the detection of previously unknown fungi (see also "Species Distributions and Assemblages" in Chapter 13,). Cameron and colleagues (1976) listed two species of phycomycetes; two species of ascomycetes; 27 species of mitosporic ascomycetes, of which *Chrysposporium*, *Penicillium*, and *Phialophora* represented the majority of the taxa; and 10 species of yeasts from Antarctic soils. Species diversity of fungi isolated from dry Arctic soils was similar to that from dry regions of Antarctica (Bergero et al. 1999).

TABLE 14.1
Some Widespread Thermophilic Fungi and Their Known Habitats

Taxon	Habitat	Reference
Acremonium strictum	Soil	Bokhary et al. 1984; Mouchacca 1995
Chaetomium thermophilum var. *coprophile*	Desert soil	Cooney and Emerson 1964; Mouchacca 1995
C. thermophilum var. *dissitum*	Desert soil	Cooney and Emerson 1964; Mouchacca 1995
C. thermophilum var. *thermophilum*	Compost	Cooney and Emerson 1964
Corynascus thermophilus (anamorph *Myceliophthora fergusii*)	Soils, compost	Mouchacca 1995
Malbranchea cinnamomea	Compost	Cooney and Emerson 1964
Melanocarpus albomyces	Compost	Cooney and Emerson 1964; Guarro et al. 1996
Myriococcum thermophilum	Soils	Mouchacca 1995
Ochroconis gallopava	Soils, hot springs, coal spoils, animals	Tansey and Brock 1973; Horré et al. 1999
Rhizomucor miehei	Soils, compost	Cooney and Emerson 1964; Schipper 1978
R. pusillus	Soils, compost	Cooney and Emerson 1964; Schipper 1978; Domsch et al. 1980
R. microsporus	Desert soil	Bokhary et al. 1984
Scytalidium thermophilum	Soils, compost	Straatsma and Samson 1993; Mouchacca 1995
Talaromyces byssochlamydoides	Soils	Stolk and Samson 1972; Mouchacca 1995
T. emersonii	Soils, compost	Stolk and Samson 1972; Domsch et al. 1980; Bokhary et al. 1984
T. thermophilus	Soils, compost	Cooney and Emerson 1964; Stolk and Samson 1972
Thermoascus aegyptiacus	Desert soil	Ueda and Udagawa. 1983; Mouchacca 1995
T. aurantiacus	Soils, compost	Cooney and Emerson 1964; Domsch et al. 1980
T. crustaceus	Soils, compost	Bokhary et al. 1984; Chen and Chen 1996
Thermomyces languinosus	Soils, compost	Cooney and Emerson 1964; Domsch et al. 1980
T. stellatus	Soils, compost	Ellis 1971; Mouchacca 1995
Thermophymatospora fibuligera	Desert soil	Udagawa et al. 1986; Mouchacca 1995
Thielavia heterothallica	Soils, compost	Domsch et al. 1980; Mouchacca 1995
T. terrestris (anamorph *Acremonium alabamense*)	Soils	Samson et al. 1977; Mouchacca 1995

Snow molds are low-temperature–tolerant soil-borne fungi that can damage and kill grasses, cereals, and other plants (Lebeau and Logsdon 1958; Nelson and Sturges 1982). The name "snow mold" is based on the fact that these fungi can grow at the bases of snow-covered plants, and often their extensive mycelial mats are located in association with damaged plants soon after snow melts. Snow molds also may damage fruits and crops stored at low temperatures. The most common species of "snow molds" are *Coprinus psychromorbidus* (often called low-temperature basidiomycete; Traquair et al. 1987), *Microdochium nivale*, *Myriosclerotina borealis* (Saito 1998), and *Typhula ishikariensis* and other *Typhula* species (Hsiang et al. 1999).

HEAVY-METAL TOLERANT

Although copper, nickel, and zinc are trace elements essential for fungal growth, an overabundance of those micronutrients or the presence of class B, borderline heavy metals (cadmium, copper, lead, nickel, mercury, silver, zinc) in soil can be toxic, leading to a general decline of the soil mycota and subsequent deterioration of the entire ecosystem (Brown and Hall 1989). Fungal responses to naturally occurring heavy and human-generated metals range widely. In most cases, species richness declines and community structure changes. Nevertheless, some heavy-metal–tolerant fungi can be isolated from most metal-contaminated sites (Brown and Hall 1989). Although a higher percentage of metal-tolerant fungi can be isolated from metal-contaminated sites (as would be expected), copper-tolerant fungi also have been isolated from locations without high concentrations of the metal (Arnebrant et al. 1987). Likewise, lead sensitivities of *Aureobasidium pullulans* isolates from contaminated and noncontaminated sites did not differ (Mowll and Gadd 1985). Yamamoto and associates (1985) reported similar findings for fungi from copper-contaminated soil.

OSMOTOLERANT AND OSMOPHILIC

An important characteristic of water in any environment is not the amount present but its availability for biological activity. Water in the form of ice or in complex with a mineral, for example, is not accessible to most living organisms. The water-activity value (a_w), which ranges from 0 to 1, is a physiochemical index of water availability. Life exists over water-activity values that range from 0.60 to 1.0 a_w, which is defined as the water-availability range for biological activity. Animal life is confined almost entirely to 0.99 to 1 a_w, whereas the permanent wilting point of plants is around 0.98 a_w. Most microorganisms are restricted to substrata with a_w values of 0.95 and higher (Pitt and Hocking 1997). In this chapter, we distinguish between osmophilic fungi, which can grow on substrata whose low water activities result from osmotic effects, and xerophilic fungi, which can grow on substrata that lack water. The effects of low water activity on growth may be similar for both groups of fungi, but the causes of the low water activities differ.

High osmotic concentrations outside a cell cause it to lose water and can kill a fungus or prevent spore germination. The ability to preserve foods with highly concentrated sugar or salt solutions, for example, is because of the low water activity and high osmotic pressure that result from the high concentration of ions and solutes in the liquid environment. In environments in which salt concentrations are naturally high, such as salt brines or some soils, a fungus must cope with both high ionic strengths and little available water. The environment indeed may be aquatic, but the high salt concentrations limit microbial growth.

Whereas fungi and other microbes are common in moderately saline environments, such as salt marshes, saline soils, and seawater, only a few halophilic fungi, algae, and bacteria can grow in a saturated NaCl (sodium chloride) solution (6 mol/kg with a 0.75 a_w). Seawater has an average NaCl concentration of 0.5 mol/kg, an osmolality of 1.0, and a water activity of 0.981 a_w. Those microorganisms tolerant of a_w values between 0.75 and 0.981 that result from the a high concentration of NaCl usually are considered to be halophilic or halotolerant (Trüper and Galinski 1986) and are discussed later in this chapter (see "Halotolerant and Halophilic").

Foods with low water activities as a result of high sugar or salt concentrations that are subject to spoilage by fungi include salted fish, jams, jellies, and candy. Although a wide variety of mitosporic fungi can be isolated from spoiled or deteriorating materials with low water activities, the important species are confined to the genera *Aspergillus*, *Aureobasidium*, *Chrysosporium*, *Cladosporium*, *Fusarium*, *Geotrichum*, *Myrothecium*, *Oidiodendron*, *Paecilomyces*, *Penicillium*, *Rhizopus*, *Scopulariopsis*, *Trichoderma*, *Wallemia*, and *Xeromyces* (Pitt 1981; Pitt and Hocking 1997). Important osmophilic yeasts include *Debaryomyces hansenii* and *Saccharomyces rouxii* (Jennings 1984; Pitt and Hocking

1997). It is interesting that the number of colonies of osmophilic fungi isolated per gram of desert soil in Saudi Arabia was low, but species richness was high (Abdel-Hafez 1982b). One hundred species belonging to 38 genera were isolated on 40–60% sucrose Czapek's agar. *Aspergillus* was the most common genus isolated.

XEROTOLERANT AND XEROPHILIC

Xerotolerant and xerophilic fungi are designated as such for their abilities to grow on substrata in which the low matrix potential, rather than osmotic stress from sugars and salts, accounts for low substratum water activity. Pitt (1975) defined a xerophilic fungus as one that would grow at a water activity value below 0.85 under at least one set of environmental conditions. Most fungi able to tolerate water activities below 0.90 a_w are ascomycetes or their anamorphs (Pitt and Hocking 1997). Fungi in this ecological group most frequently are isolated from arid and semi-arid ecosystems and most often are associated with stored cereal grains and dried foods (Magan and Lacey 1984a, 1984b). Species of *Aspergillus* are considered the most xerotolerant taxa in this ecological grouping, although some *Penicillium* species are important xerotolerant species (Dix and Webster 1995). In an extensive review of xerophilic fungi, Pitt (1975) reported 44 species that spoil foodstuffs as belonging to this ecological group. Three of the species were yeasts; the remainder were in the genera *Aspergillus*, *Chrysosporium*, *Eremascus*, *Paecilomyces*, *Penicillium*, *Wallemia*, and *Xeromyces*. *Penicillium* species tend to be more common in temperate environments, whereas *Aspergillus* species predominate in warmer climates (see "Species Distributions and Assemblages" in Chapter 13). Magan and Lacey (1984b) reported that *Aspergillus* species are better competitors than *Penicillium* species at high temperatures and low water activities.

Information on the distributions and species compositions of fungal communities in arid ecosystems is meager; most investigations having focused on soil mycota (Mouchacca 1995; Sangalang et al. 1995). At this point, instead of listing the fungal species present in arid soils, mycologists should focus on predicting and understanding how arid environments may regulate the composition and biodiversity of fungal communities within those ecosystems. Furthermore, it will be crucial to determine whether fungi from arid environments are more xerophilic or xerotolerant than fungi from more mesic environments. One of the first hints that fungal communities and biodiversity patterns in xeric environments might differ from those patterns characteristic of mesic environments came from Gochenaur's study (1975) of the species composition of soil fungal communities at mesic and xeric sites in the Bahamas. Fungal taxonomic diversity at the mesic forest site was lower than that at the xeric site. Desert ecosystems also have a higher number of species than would be predicted based solely on abiotic conditions (Wicklow 1981). Zak (1993) attributed the greater species richness to the greater spatial and temporal variability in abiotic conditions and resources that is typical of desert ecosystems. Recently, Zak and colleagues (1995) noted that functional diversity (estimated from the numbers and types of substrata that can be used by the species comprising a fungal community and range of rates at which they are used) in deserts also should be greater than predicted because of the high spatial and temporal heterogeneity in moisture patterns at various scales of resolution. Moreover, they suggested that the scale at which interactions between a fungal mycelium and its environment are perceived biases our understanding. At the scale of the mycelium, fungi in arid systems colonize spatially and temporally heterogeneous mesic habitats, which are embedded within the xeric matrix that we perceive. Fungal communities that develop on litter under shrubs should be fundamentally different in composition and functional diversity from those that occur on buried litter and those growing on exposed litter. They concluded that the species diversity of xerotolerant or xerophilic fungi in arid ecosystems is no greater than one would find in mesic ecosystems. The key point is that in arid ecosystems investigators must be keenly aware of the type and scale of the habitat that they are sampling (Zak et al. 1995).

ACIDOPHILIC AND ALKALOPHILIC

Acidity and alkalinity are stress factors often observed in nature. pH values for acid mine or volcanic spring effluents can range from 1 to 2; for fruit juices or acid soils, the pH values range from 3 to 4. In contrast, alkaline soils have a pH around 9; alkaline lakes can reach a pH of 10, and concrete surfaces have a pH of up to 11 (Bachofen 1986). Because the pH of most environments is between 4 and 9, the pH optima of most microorganisms lie within this range. Fungi generally germinate and grow well in weakly acidic to neutral pH ranges. Microorganisms able to grow or survive exposures at pH values outside these limits are considered acid- or alkali(ne)-tolerant.

Mildly acidic natural habitats with pH values of 3 to 4 are relatively common, including many lakes (increasing in number with acid rain deposition), some pine soils, and acidic bogs. More extreme acid environments with pH values of 3 or less are quite rare. They commonly are associated with hot springs, coal mine refuse piles, drainage waters, and mining effluents. Those habitats often are characterized by high concentrations of dissolved sulfates and iron as well as hydrogen ions.

Microorganisms whose pH optima for growth exceed pH 8, usually falling between 9 and 10, are defined as alkalophiles. Although the majority of the organisms described as growing under very alkaline conditions are prokaryotes (Grant et al. 1990), alkalophilic fungi have been isolated. The early work of Johnson (1923) indicated that *Fusarium bullatum*, *F. oxysporum*, and *Penicillium variables* could grow at pH values of 11. Those fungi neutralize the alkaline culture medium, however, to ca. pH 7, so their characterization as alkalophiles may be inappropriate. An alkalophilic strain of *Aspergillus oryzae* isolated from a soil sample from Japan was able to grow well at pH values of 9 and 9.5 (Horikoshi 1991). Goto and associates (1981) identified *Exophiala alcalophila*, a pleomorphic species of a black yeastlike member of the Herpotrichiellaceae (Haase et al. 1999). It was growing in soil at pH 10.4 with its accompanying teleomorph stage, *Phaeococcomyces alcalophilus*. Approximately 30% of the fungi isolated from slightly alkaline forest soils or from limestone caves are alkaline-tolerant or alkalophilic (Nagai et al. 1995, 1998). Species of *Acremonium* and *Fusarium* are the fungi most frequently isolated from these soils.

Alkaline soda lakes and soda deserts, low Ca environments with pH values greater than 10, are distributed widely in tropical and subtropical regions. They are characterized by high concentrations of $NaCO_3$ resulting from evaporation (Grant et al. 1990). Deposition of animal excreta in normally nonalkaline soils also may result in transient patches of high alkalinity (Horikoshi 1991). Commercial processes such as cement manufacture and casting, electroplating, paper manufacturing, and the lye treatment of potatoes and animal hides generate alkaline conditions in soils.

HALOTOLERANT AND HALOPHILIC

Microbes, including fungi, can be categorized as halophilic based on their requirements for NaCl in the growth medium (Kushner 1978). High sodium concentrations not only exert an osmotic affect but adversely affect protein structure and enzyme function (Cooke and Whipps 1993). Extreme halophiles grow best in media that contain 2.5–5.2 M NaCl. Borderline extreme halophiles grow best in media that contain 1.5–4.0 M NaCl. Halotolerant microbes can tolerate salt to the point of saturation of the medium. Nonhalophiles grow best in media containing less than 0.2 M NaCl. No extreme halophiles have been found among the eukaryotes.

Halotolerant fungi, including species in the genera *Alternaria*, *Aspergillus*, *Penicillium*, *Myrothecium*,

Stachybotrys, and *Trichoderma*, are commonly isolated from arid soils (Malik et al. 1982). Halotolerant fungi, such as *Basipetospora halophila*, *Hortaea werneckii*, and *Aspergillus penicilloides*, often are isolated from salted or dried fish. Those species have been found to grow optimally at salinities of 2.5 M NaCl (Blomberg and Adler 1993). However, these fungi also will grow on more dilute media. *Basipetospora halophila* and *H. werneckii* are considered halophilic because they grow at higher rates on NaCl media than on media with other osmoticants (e.g., glucose/fructose, glycerol) (Andrews and Pitt 1987). *Basipetospora halophila* may be restricted to salted foods because it has not been isolated yet from hypersaline lakes, although in the lab it will grow on media containing saturated levels of NaCl (Blomberg and Adler 1993). Presumably, other abiotic factors, such as pH or magnesium concentrations, prevent its successful colonization of those habitats.

Naturally hypersaline environments, such as salt flats, saltpans, and brine pools, appear to be the domain of the prokaryotes and algae (Post 1977). Although only a few fungi are considered to be obligate halophiles, some fungi grow better in the presence of salt than when it is absent, and thus have a competitive advantage over less salt-tolerant competitors (Malik et al. 1982; Hocking 1993). Cellulose decomposition by some *Aspergillus* species, for example, increases with salt concentrations from 3–10% (Malik et al. 1982). *Aspergillus halophilicus* and *Scopulariopsis halophilica* are halophilic species that have been isolated from various saline and arid-region soils (Abdel-Hafez et al. 1989). Guiraud and colleagues (1995) recently reported that no strictly halophilic fungi (only halotolerant fungi) were isolated from 56 localities along the Dead Sea valley. *Hortaea werneckii*, *Phaeotheca triangularis*, and *Aureobasidium pullulans* of the Dothideales and *Trimmatostroma salinum* of the Chaetothyriales are black, yeastlike fungi, which are associated with hypersaline saltpans of 15–30% salinity (Zalar et al. 1999a, 1999b; Gunde-Cimerman et al. 2000). *Phaeotheca triangularis* is an obligate halophile, but *H. wernckii*, *T. salinum*, and *A. pullulans* are facultatively halotolerant. Typical marine fungi have not been detected in saltpan environments.

OLIGOTROPHIC

Laboratory culture usually provides fungi with an excess of nutrients, but in nature, fungi often confront acute shortages of some nutrients. Oligotrophy is defined as the ability of microbes to grow when concentrations of nutrients are very low or absent. Although oligotrophy generally has been considered the domain of prokaryotes, sufficient evidence is available to support the idea that fungi also can grow oligotrophically (Wainwright 1993). That ability is important for survival in some low-nutrient environments, which mycologists generally consider to be stressful.

Carbon plays an important role in fungal community dynamics and taxonomic diversity, particularly in soil. In fact, fungal growth generally is considered to be limited by the amount of exogenous organic carbon available, with continuous growth occurring only in specialized habitats such as the plant rhizosphere. Wood-decomposing fungi, such as *Phanerochaete chrysosporium*, and several mycorrhizal species cannot grow on a carbon-free substratum.

The ability to grow in habitats where carbon is either limited or absent is called oligocarbotrophy (Poindexter 1981). Although there is evidence that obligate oligocarbotrophy exists in bacteria, no evidence to date suggests that fungi possess that characteristic (Wainwright 1993). It is interesting that numerous zygomycetes and mitosporic ascomycetes are able to grow oligocarbotrophically on nutrient-free silica gel (Wainwright 1993). Possible explanations for that ability are that the fungi fix atmospheric CO_2, that the silica gel adsorbs nutrients from the atmosphere, or that silica gel itself stimulates spore germination and hyphal growth.

Soil microbes generally are divided into two groups: those that are considered true inhabitants of the soil and are capable of slow but continuous growth and those that respond quickly to carbon inputs but are dormant most of the time in the absence of sufficient nutrients. Fungi were found to grow readily from soil particles and roots onto nutrient-free agar by Parkinson and associates (1989), suggesting that soils may contain concentrations of nutrients and carbon that, although low, are sufficient to support some fungal growth. Calculations of the amount of carbon necessary to support fungal growth in soil have not taken into account the ability of fungi to scavenge carbon from the soil solution and atmosphere (Wainwright 1993). Under low carbon conditions, colonies do not readily form on agar as they would on more nutrient-rich media, and fungal growth on carbon-free media is not obvious and should be verified with a microscope (Wainwright 1993). Although not usually detected using traditional isolation approaches, oligotrophic fungi may be prevalent in deep spoils or in habitats where carbon availability is extremely low. These environments represent new frontiers for fungal ecologists.

ROCK-INHABITING

Rocks that are exposed to high solar radiation, high temperatures, low nutrient availability, high electrolyte concentration, and low relative humidity, where moisture is available only sporadically, can be considered extreme environments and are colonized by unique species of stress-tolerant fungi. Black yeasts and dematiaceous-mycelial fungi have been isolated from various rock surfaces and from crevices in rock surfaces and stone monuments (Urzì et al. 1995; Sterflinger and Krumbein 1997; Sterflinger et al. 1999). Those fungi are phylogenetically diverse, are nonlichenized, and grow on bare surfaces of diverse rocks types in different climates. Such fungi share high melanin production; resistance to high temperatures and ultraviolet (UV) radiation; and a restricted, meristematic growth morphology. The latter feature has led to the name of "microcolonial fungi" for this ecological group. Many new species of fungi have been characterized as their unique habitats have been investigated (Simmons 1981; Wollenzein et al. 1997; Sterflinger et al. 1997, 1999; De Leo et al. 1999).

Rock surfaces in arid regions sometimes are coated by "desert varnish," which is a brown, black, or orange coating, often rich in oxides of manganese and iron and in clay minerals (Staley et al. 1992). The activities of lichenized epilithic and endolithic cyanobacteria and chemoorganotrophic bacteria and fungi have been implicated in the formation of desert varnish (Krumbien and Jens 1981; Staley et al. 1992). Some filamentous fungi that occur on rock surfaces are associated with yeasts and bacteria that produce a layer of extracellular slime that protects the hyphae from desiccation. The fungi and bacteria that grow within desert varnishes are capable of enriching the substrata with iron and manganese, resulting in precipitation and accumulation of the dense black layers on the rock surfaces. For both the microcolonial and the rock varnish fungi, production of a dark pigment or a dark layer on rock surfaces appears to be an adaptation to reduce exposure to extended periods of high UV irradiation (Urzì et al. 1995).

PHOENICOID

Phoenicoid fungi are a specialized group of primarily ascomycetes and basidiomycetes that fruit on soil; plant material; burned wood following a forest fire, prairie fire, or fireplace fire; and on steam-sterilized soil in greenhouses (Wicklow 1973; Carpenter and Trappe 1985). The terms pyrophilous, anthracophilous, and carbonicolous also have been used to refer to fungi that fruit on heat-treated substrata.

Although a habitat that has burned is not a stressed environment in the classical sense, forest fires usually lead to a 3–5 unit increase in pH because of ash deposition (Petersen 1970). The plant community in which a fire has occurred and the depths of burn play major roles in determining the types of fungi that will appear following the fire. Heterothallic *Neurospora* species, for example,

are common on charred vegetation in the moist tropics and subtropics (Shaw 1990, 1998; Pandit and Maheshwari 1996). In temperate forests, operculate discomycetes (Pezizales) and a few agarics dominate the phoenicoid community (Dix and Webster 1995), whereas in prairie ecosystems, it is the pyrenomycetes (Wicklow 1973, 1975).

Volcanic eruptions and subsequent depositions of hot ash, as occurred on Mount St. Helens in the state of Washington (United States), also can lead to abundant fruiting of phoenicoid fungi. Six weeks after the eruption of Mount St. Helens, apothecia of *Anthracobia melaloma* and perithecia of *Gelasinospora reticulospora* appeared within the tephra (Carpenter et al. 1987). One year after the eruption, large discomycetes (e.g., *Peziza* species, *Rhizinia undulata*, *Trichophaea hemisphaerioides*) and agarics (e.g., *Coprinus plicatilis*, *Pholiota carbonaria*, *Psathyrella carbonicola*, *Schizophyllum commune*) were common.

When designing sampling schedules for phoenicoid fungi, it is important to know how long after a burn different species fruit and how long they will continue to produce sporocarps. Dix and Webster (1995) assigned phoenicoid fungi from forested sites to four groups based on the timing of fruiting following fire (Table 14.2). The fruiting times are relative given that the initiation and duration of fruiting vary depending on the time of year in which a burn occurs and the duration of favorable conditions for fruiting.

Although phoenicoid fungi occupy similar habitats, the abiotic factors that account for their presence in a particular location may vary. Soil heating may stimulate some species to fruit, whereas other fungi may respond to changes in soil pH that follows a fire. Sporocarp production by several phoenicoid fungi in undisturbed forest soils, for example, can be stimulated by the application of lime (Hora 1958; Petersen 1970).

SAMPLING AND ISOLATION METHODS FOR GROUPS OF FUNGI

GENERAL CONSIDERATIONS

Species compositions of fungal communities from soils and litter of stressful habitats in most cases should be evaluated using modifications of the particle-washing technique (see "Particle Filtration" under "Principal Isolation Methods" in Chapter 13 and under "Direct Inoculation Protocols" in the Appendix in Chapter 13). This method allows for isolation of fungi that actively are growing in various substrata at the time of sampling.

TABLE 14.2
Fruiting Periods of Phoenicoid Fungi after Fire in Forest Ecosystems

Category	Representative taxa
Group 1: Species that appear as early as 7 weeks but no later than 80 weeks after a burn.	*Pyronema* spp., *Trichophacea abundans*
Group 2: Heterogeneous assemblage of species that begin to fruit 10–15 weeks after a burn and may continue to fruit for up to 190 weeks after the fire.	*Ascobolous carbonarius, Peziza trachycarpa, P. praetervisa, Rhizinia undulata, Tephrocybe carbonaria*
Group 3: Species that appear 20–50 weeks after a fire and that continue to fruit for 130–200 weeks.	*Peziza endocarpoides, Trichophaea hemisphaerioides*
Group 4: Heterogeneous taxa that do not generally appear until 50 weeks after a burn and that may continue to fruit for 150 weeks. Apothecial forms that can be associated with the development of a moss carpet are included in this group.	*Myxomphalia maura, Neotiella hetieri, Ripartites tricholoma*

From Dix and Webster (1995).

Suspension plating (see "Suspension Plating" under "Principal Isolation Methods" and under "Direct Inoculation Protocols" in Chapter 13), a technique used for assessing changes in fungal species compositions of various community types, should be used only in conjunction with particle-washing techniques because it preferentially selects for fungi that are present in the form of spores.

Most media used for fungal community isolations favor those fungi that can grow on nutrient-rich substrata and may not accurately assess the species composition of fungal communities from soil (Wainwright 1993). Although from a laboratory perspective low-nutrient conditions have been considered to be stressful, Foster (1949) argued that secondary metabolites produced by fungi when grown under the high-nutrient conditions of most isolation media are indicative of stressful conditions. Clearly, additional work is needed to evaluate the implications of using traditional media to isolate fungi from all environments, including those considered "stressful." Moreover, development of additional media attuned to the real conditions fungi encounter (e.g., nutrient-poor media such as those developed by Wainwright [1993]) will be necessary if fungal community compositions and biodiversity patterns are to be evaluated realistically. The use of more "natural" media that more closely mimic the conditions from which fungi are isolated should be less stressful to the extremeophiles and improve isolation success. After all, the extreme conditions from which fungi are to be isolated are not extreme to an extremeophile.

We do not provide an extensive survey of media or methods for isolating fungi from extreme environments in this section. Rather, we describe some general approaches that will enable an investigator to carry out preliminary sampling and to identify or develop the best approaches for enumerating fungi from environments of interest. Most current methods for isolating fungi from extreme environments are merely modifications of existing practices, such as incubation at high temperatures or modification of the pH of existing media.

THERMOTOLERANT AND THERMOPHILIC

Thermophilic fungi can be isolated from plant material through removal and subculturing of their spores (Cooney and Emerson 1964), by incubation in moist chambers, or by selective washing and enrichment procedures (Tansey and Jack 1976). Appropriate safety measures must be used when working with thermotolerant and thermophilic fungi growing at or near 37°C because fungi isolated from high-temperature sites can be serious human pathogens (Tansey and Brock 1973).

Successful isolation of thermophilic and thermotolerant fungi from soil and plant materials, regardless of the method used, requires incubation of material at temperatures between 45°C and 50°C. The primary isolation cultures never should be grown below 40°C. Primary transfer plates should be incubated initially at 50°C. Once good growth is observed, cultures can be moved to 40°C to prevent the agar from drying (Cooney and Emerson 1964). Moisture loss from cultures incubated at 50°C is a problem. Evaporation can be reduced by placing the plates in sealed containers lined with moist paper towels, if the air in the incubator cannot be humidified directly. Sealing plates with Parafilm helps, but moisture condensation then can be a problem.

Preferred isolation and culture media vary among investigators. Media most commonly used are listed in Appendix II. Potato-dextrose agar, a medium commonly used to culture fungi, however, is unsatisfactory because it tends to shrivel along the edges when kept at 50°C (Cooney and Emerson 1964). Tansey and Jack (1976) used a series of enrichment cultures to isolate taxa that were present in low numbers at exposed sites. They added one of seven different carbon sources to agar on a 1% w/w basis: cellulose powder (Whatman CF110), carnauba wax (No. 1 yellow), unbleached chitin, a mixture of keratin from various animal sources, soluble starch, lactose +100 mg/g soil gentamicin sulfate, and lignin. Bottles with soil and an enriched-carbon source were incubated from 12 to 25 days at 50°C and kept moist with the periodic addition of sterile water. After incubation, pour-plates were prepared from each incubated bottle of soil, and plates were incubated at 50°C.

The addition of recalcitrant materials (i.e., those very resistant to decomposition) to agar plates for selective isolation of thermophiles and thermotolerant fungi can be an effective means of isolating taxa that can degrade cellulose, lignin, keratin, and other complex polymers. In addition, by choosing the types of carbon sources added to the isolation medium, functional guilds of thermophilic fungi can be enumerated. *Rhizomucor pusillus,* for example, is a primary sugar fungus on wheat-straw compost, whereas *Humicola insolans* can hydrolyze pure cellulose (Chang 1967). *Phanerochaete chrysosporium*, a fast-growing thermotolerant fungus that causes white rot, degrades lignin (Rosenberg 1978). By using selective isolation procedures, the taxonomic and functional diversity of thermophilic and thermotolerant fungi on any substratum can be obtained.

As with all fungal community analyses, the isolation procedure, along with the isolation medium, will determine what component of a community will be isolated. Tansey and Jack (1976) used pour-plates to enumerate thermophilic and thermotolerant fungi in temperate soils that were present primarily as spores. To isolate rare species, they used two washing procedures to kill spores and hyphae of the more abundant fungi. In one procedure they soaked 15 g of soil in 70% ethanol for 10 minutes, and then washed the soil with sterile water prior to pour-plating (see "Ethanol and Phenol Pasteurization" in the Appendix in Chapter 13). The second procedure involved soaking 1 g of soil in 5.25% sodium hypochlorite for 5 minutes, and then washing the soil by repeated centrifugation and resuspension in sterile water. Apinis (1963) isolated fungi actively growing as hyphae at the time of sampling by plating washed particles of soil and plant and organic debris.

PSYCHROTOLERANT AND PSYCHROPHILIC

A number of investigators have isolated undescribed species of fungi or rediscovered rare ones by incubating soil, dung, and plant remains at low temperatures (Carreiro and Koske 1992). Such species may be "rare" or may have been overlooked only because they do not grow or survive at higher incubation temperatures. Carreiro and Koske (1992) plated forest soil and litter directly onto chilled MYP (Malt-Yeast-Peptone) agar (Appendix II) or diluted the sample and made pour-plates using agar cooled to 43°C. Plates were incubated at 0°C and examined twice weekly for up to 4 weeks. In the study by Petrini and associates (1992b) samples were incubated for 10 weeks at 4°C, subsequently transferred to 2% malt-extract–agar slants, and incubated again at 4°C until substantial growth was observed (3–10 weeks).

Undoubtedly, additional studies of fungi from permanently or seasonally cold environments, by incubation of samples at low temperatures, will expand species inventories and provide insights into the relationships between fungal community activity and decomposition dynamics. Psychrophilic fungi should coexist with mesophiles in soils and plant litter, given that their spores can survive warm periods (Petrini et al. 1992b). Subsequent incubation of isolates at temperatures above 20°C will indicate whether they are psychrophilic or psychrotolerant.

HEAVY-METAL TOLERANT

Changes in the composition of soil-fungal communities in response to presence of heavy metals can be assessed along transects that follow a decreasing concentration gradient away from the source. Plots are established at specified distances determined by the metal concentration in the soil profile, litter layer, or phylloplane. Plot size should be determined by the number of samples that will be taken over a specified period. More than one plot should be established for each position sampled along the gradient to avoid the problem of pseudoreplication (Hulbert 1984).

The concentrations of the heavy metal used for isolation purposes should be determined from a preliminary assessment of the fungal community response to a range of concentrations that mirrors amounts found at the study sites. If cultures are to be tested for metal tolerance after isolation, the storage medium must contain the metal of interest to maintain tolerance (Ashida 1965).

OSMOPHILIC

The a_w values of the media used for the isolation of osmophiles usually are adjusted by adding NaCl, sugars, glycerol, or polyethylene glycols of various weights (Table 14.3). Glycerol often is preferred for lowering the

TABLE 14.3

Molalities of Common Solutes and their Corresponding Specific Water Activities*

Glycerol	Sodium chloride	Sucrose	a_w
0.277	0.150	0.272	0.995
0.554	0.300	0.534	0.990
1.11	0.607	1.03	0.980
2.21	1.20	1.92	0.960
3.32	1.77	2.72	0.940
4.44	2.31	3.48	0.920
5.57	2.83	4.11	0.900
8.47	4.03	5.98	0.850
11.5	5.15	—	0.800
14.8	—	—	0.750
18.3	—	—	0.700
22.0	—	—	0.650

Adapted from Corry (1987).
*Molality = moles/liter water.

TABLE 14.4

Relationships Between Water Activity and Water Potential at 25°C

Water activity (a_w)	Water potential (Mpa)	Water activity (a_w)	Water potential (Mpa)
0.999	−0.138	0.70	−49.1
0.995	−0.69	0.65	−59.3
0.99	−1.38	0.60	−70.3
0.98	−2.78	0.50	−95.4
0.97	−4.19	0.40	−126
0.96	−5.62	0.30	−165
0.95	−7.06	0.20	−221
0.90	−14.5	0.10	−317
0.85	−22.4	0.01	−634
0.80	−30.7	0.001	−951
0.75	−38.6		

Data adapted from Griffin (1981).

a_w values of media because NaCl at high concentrations may be toxic to some fungi (Cooke and Whipps 1993; Pitt and Hocking 1997). The preferred solute will depend on the growth conditions and the fungal species. Moreover, the growth rates may change at selected values of a_w, depending on incubation temperature, pH, and nutrient availability (Wheeler et al. 1988; Hocking 1993). If a selective medium is not used, osmophiles may not be isolated. Furthermore, if dilutions are the source of the inoculum, the diluent should contain from 20% to 30% glucose or fructose in peptone water to prevent osmotic shock (Corry 1987).

XEROPHILIC

The study of fungal community dynamics and biodiversity patterns in xeric environments requires careful attention to the distribution of organic matter within those ecosystems and its seasonal variation. For example, Zak and colleagues (1995) reported that the community of fungi associated with wood located under shrubs in the Chihuahuan desert did not change substantially over a 9-year period but that composition of fungal communities associated with wood in woodrat middens changed considerably within a few months. Isolation media and incubation temperatures are also important. Hocking and Pitt (1980) recommended using dichloran-glycerol-18 medium over other standard media (Appendix II) for isolating and enumerating xerophilic fungi from low-moisture foods and from arid environments with low water availabilities.

TABLE 14.5

Amounts of Glycerol Required for Solutions with Selected Water Potentials at 25°C

Water potential (Mpa)	Glycerol (g/l)
−0.5	8.36
−1.0	19.02
−1.5	27.32
−2.0	35.59
−2.5	46.14
−3.0	54.28
−4.0	72.68
−5.0	91.08

Adapted from Dallyn and Fox (1980).

Water availability in soils often is measured as "water potential," which is defined as the sum of matric, osmotic, and gravitational potentials. Water potential of soil can be related to water activity by the following equation:

$$\Psi = 8.314T \ln a_w/V_w$$

where Ψ is water potential (MPa), T is absolute temperature (°K), V_w is the partial molar volume of water, and 8.314 is a constant. Relationships between a_w values and water potential at 25°C are provided in Table 14.4. The amounts of glycerol needed to create water potentials ranging from −0.05 to −5.0 MPa are provided in Table 14.5. The concentrations of glycerol, NaCl, and sucrose needed to adjust the water potential or water activity are provided in Table 14.3.

ACIDOPHILIC AND ALKALOPHILIC

Sterile acid or base solutions, but not concentrated acids or bases, should be used to prepare media with pH values outside the normal range of pH 4.5–7.5. The pH should be adjusted after the medium has been prepared and autoclaved. Autoclaving at extreme pH values can destroy the agar matrix (i.e., the medium may not gel) and change the color and texture of medium. In some cases, it may be necessary to increase the agar concentration to obtain satisfactory solidification on the medium.

For most studies, potato-dextrose, malt, and cornmeal agars can be acidified with tartaric, lactic, citric, phosphoric, or hydrochloric acids (Jarvis and Williams 1987). An alkaline medium containing sodium carbonate, sodium bicarbonate, potassium carbonate, or sometimes sodium borate can be used for the isolation of alkalophilic fungi (Horikoshi 1991; Horikoshi I and II media; Appendix II). The recommended salt concentrations range from 0.5–2%, with the pH of the medium held between 8.5 and 11. Horikoshi-I and II media can be used for isolating fungi using methods described for soil fungi (Chapter 13) or for fungi from extreme environments. Alkaline cornmeal agar (ACMA) also can be used for general isolation of alkalophilic and alkali-tolerant fungi (Appendix II; Nagai et al. 1995, 1998).

HALOPHILIC

Standard media for the isolation of halophiles can be modified by the addition of products such as Instant Ocean (used to create artificial sea water for tropical fish, Carolina Biological Supply; Appendix IV) or sea salts (Sigma Chemical Company; Appendix IV). The ion concentration (predominantly NaCl) then can be modified by making the salt solution as concentrated or diluted as required. The salt solution is added to the halophilic isolation medium until ion concentrations reflect those in the environment (Table 14.3). The average NaCl concentration of seawater, for example, is 0.5 mol/kg (approximately 3.0%) and has the osmolality of about 1.0, which corresponds to a water activity of 0.981 (Trüper and Galinski 1986). The Na^+ and Cl^- ion concentrations of the Great Salt Lake are 105.4 and 181 g/liter, respectively (Post 1977). To isolate halophilic fungi from saline soils in Egypt, Abdel-Hafez and associates (1989) used a glucose-substituted Czapek-Dox medium with 15–20% NaCl added per liter (Appendix II).

Methods for physiological characterization of dematiaceous halophilic and halotolerant fungi are described in de Hoog (1999).

OLIGOTROPHIC

Minimal carbon media or a complete silica gel medium (Appendix II) that lack a carbon source (Parkinson et al. 1989) can be used to isolate and evaluate the species composition of oligotrophic fungi within any fungal community and from any chosen ecosystem. To ensure that oligotrophic conditions are established, media must be prepared to exclude trace contamination. In addition glass, rather than plastic, Petri dishes should be used, and all glassware should be washed in 5% hydrochloric acid before use. The glassware and silica gel must be heated in a muffle furnace to eliminate all traces of organic compounds (Wainwright 1993).

ROCK-INHABITING

For enumeration of fungi associated with rocks, samples either from the surface or from the interior are dissected and placed onto either dichloran-rose bengal-chloramphenicol (DRBC) medium (Appendix II) or on diluted (1:50 suspension) Czapek-Dox agar (Appendix II). The inhibitors in DRBC and the oligotrophic conditions of diluted Czapek-Dox are necessary for isolation of slow-growing taxa. Samples also can be washed according to procedures used for soil washing and particle filtration (see Chapters 12 and 13) to remove spores prior to isolation. The books *Dematiaceous Hyphomycetes* and *More Dematiaceous Hyphomycetes* (Ellis 1971, 1976) are useful for identification of isolates. Methods for physiologically based and ribosomal DNA-based phylogenetic analyses of dematiaceous rock-inhabiting fungi are described in de Hoog (1999).

PHOENICOID

The presence of phoenicoid fungi following a fire in a forested ecosystem is indicated by the presence of sporocarps. Collection of sporocarps of macrofungi should be conducted as described in Chapter 8, and collection of the soil microfungi should be done as described in Chapter 13.

The effects of fire on the fruiting of macrofungi can persist for a significant period (Table 14.2), so investigators should maintain collecting schedules for up to 4 years following a forest burn. Genera and species of fungi associated with heat-affected forest sites are listed in Ramsbottom (1953), Petersen (1970, 1971), Carpenter and colleagues (1987), and Ellis and Ellis (1988). In addition, Chapter 8 (see "Taxonomic Resources for Identification") cites literature useful for the identification of basidiomycetes from heat-affected substrata.

Detection of microscopic phoenicoid fungi following prairie fires and in soils or litter from burned forested areas is difficult because the sporocarps, primarily asco-carps, are small and usually inconspicuous. Their presence is best detected by incubating soil samples and burned plant litter in Petri dishes in the laboratory. Sterile pieces of straw or filter paper can be placed on soil to induce fruiting of fungi. Material should be moist but not saturated. Fruiting can begin as soon as 1 week after incubation begins and can continue for up to 50 days. The majority of phoenicoid ascomycetes appearing after prairie fires tend to be coprophilous. Those taxa can be identified using Richardson and Watling (1969) and Bell (1983). See also Dennis (1978) for a more in-depth analysis of the temperate ascomycetes.

Steaming of soil simulates the effects of fire. Steaming soils at 100°C for 4 minutes markedly increases the numbers of ascomycetes isolated from soil isolation plates (Warcup 1951a). Likewise, an aliquot of soil can be diluted 1:100 in a water blank and heated in a water bath at 60°C for 30 minutes, further diluted, and then used to prepare isolation plates as needed. Zak and Wicklow (1978) obtained the greatest number of ascomycetes after steaming prairie soils for 60 seconds at 55°C (rather than at higher or lower temperatures) and then placing them in moist chambers. They determined species richness by identifying ascomycete sporocarps that appeared on the soil surface over a 19-week period, which eliminated the need for fungal cultures.

Conidia of *Neurospora* can be collected from burned vegetation in the humid tropics and subtropics (Perkins et al. 1976). Charred sugarcane stems can be used to bait *Neurospora* species from sugar cane soils (Pandit and Maheshwari 1996). Burned or unburned canefield soil is placed in pans, and sections of burned cane stems are placed on the soil surface. Furfural (2-furaldehyde), formed in the heated cane stems, diffuses into the soil and stimulates ascospore germination, and the stems are colonized. Selective isolation of *Neurospora* species from heat-treated soils is described in Appendix II (see Vogel's Medium N).

has been difficult because of their varied habitats and life cycles and the different techniques that have been used for their study. What we have suggested, except for specific isolation conditions, should apply to most of the groups of fungi discussed elsewhere in this volume—that is, investigators should obtain prior knowledge of habitat factors that influence the fungi to be isolated before sampling those fungi. Investigators looking for acidophiles, for example, should use a medium adjusted to the pH of the habitat being examined. Using a medium adjusted to pH 4 to look for acidophilic fungi in a habitat with a pH of 2.5 is useless. This may seem obvious, but some investigators may use pH 4 because of practical difficulties in preparing media at low pH values (viz, gelling). Moreover, as Carreiro and Koske (1992) pointed out, many investigators studying soil microfungi in temperate forests do not report soil temperatures or even seem to consider them. Rather they continue to incubate samples at 25°C, although soil temperatures may be considerably lower for most of the year.

Learning about the habitat to be studied (viz, abiotic information such as physical and chemical parameters) and incorporating that information into isolation regimens will enable investigators to add additional unusual fungi to species inventories, particularly in permanently "extreme" environments. Moreover, use of the same isolation methods in more "normal" habitats likely will uncover other new species with affinities for or tolerance of extreme conditions.

Probably many species of fungi that could be considered "-tolerant" or perhaps even "-philic" survive and reproduce in so-called extreme microenvironments within apparently normal macroenvironments. Parallel use of multiple isolation regimens will provide a more complete assessment of fungal community composition and biodiversity patterns, reflecting the resource heterogeneity that exists in natural habitats. Only then can we begin to understand the mechanisms linking fungal community dynamics and biodiversity patterns to ecosystem processes.

CONCLUSIONS

Proposing standard sampling techniques and approaches for the different fungal groups discussed in this chapter

15

Mutualistic Arbuscular Endomycorrhizal Fungi

JOSEPH B. MORTON, RICHARD E. KOSKE, SIDNEY L. STÜRMER, AND
STEPHEN P. BENTIVENGA

A "mycorrhiza" is a two-way mutualistic symbiosis between a fungus and a plant host. Of seven types of mycorrhizal associations described thus far (Brundrett 1991; Smith and Read 1997), the most widespread association involves fungi in the order Glomales, phylum Zygomycota (Morton and Benny 1990), formerly Endogonales *sensu* Gerdemann and Trappe (1974). They are termed arbuscular mycorrhizae (AM) here because all of the fungi classified in the group colonize plant roots and form arbuscule structures that interface with the plasmalemma of root cells. Botanists, ecologists, and agriculturalists have focused considerable attention on AM associations because they enhance plant growth, reproduction, and survival of the host plant (Safir 1987; Smith and Read 1997) and because they play important roles in succession and in maintenance of plant-community diversity (Allen 1991; Brundrett 1991; Allen et al. 1995). Arbuscular mycorrhizal associations are widespread among agricultural and horticultural crop plants (Abbott and Gazey 1994), although their impacts on those systems under field conditions can be highly variable and, thus, not always definitive (Jeffries and Dodd 1991).

Glomalean fungi reside in soil, although they also have been found in aerial epiphytes (Gemma and Koske 1995) and in vegetative fragments of plant debris (Koske and Gemma 1990). The mycelium is coenocytic; septa form only with age or wounding, much like other zygomycetous fungi. Tommerup and Sivasithamparam (1990) claimed that one fungus, *Gigaspora decipiens,* has a sexual reproductive phase, but most evidence, including genetic analysis (Rosendahl and Taylor 1997), indicates that AM fungi reproduce clonally via mitospores formed in soil and/or in roots. The morphology of glomalean fungi and organism life-history traits are discussed more fully later in this chapter (see "Organism Characterization").

Research during the past two decades has focused on plant–nutrient interactions in the AM symbiosis. Greenhouse experiments have shown repeatedly that the growth-promoting activities of different fungal isolates vary and that those properties are reasonably stable under disparate environmental conditions provided that soil phosphorus levels are low (Sylvia et al. 1993). Much less work has been done to measure the effects of introduced or indigenous fungal organisms on community dynamics under natural conditions. At the plant level, that type of investigation involves monitoring plant-community dynamics in mycorrhizal and non-mycorrhizal plots, the latter obtained either through fumigation (Shumway and Koide 1994) or applications of benomyl to the soil (Carey et al. 1992). On the fungus side, complicated and time-consuming work is required to define the fungal community taxonomically, to isolate and culture organisms from each community, and then to screen and compare individual organisms for their functional diversity under a range of natural conditions. Although surveys have been conducted to ascertain the presence and composition of arbuscular fungal communities in some parts of the world, many areas remain unsampled. In addition, little effort has been directed toward characterization of the fungal symbionts, their life histories, and their community dynamics.

RELATIONSHIPS WITH HOST PLANTS

Perhaps the largest obstacle to understanding the biology and ecology of AM fungi is our inability to culture them apart from their plant hosts. The plant provides carbon to the fungus largely via an arbuscule-plant cell plasmalemma interface. It also provides a protected site in root cells where the fungus can live. The external fungal hyphae improve phosphorus acquisition by the plant in soils with low levels of phosphorus (Safir 1987; Smith and Gianinazzi-Pearson 1988; Smith and Read 1997). In soils in which phosphorus levels exceed requirements of the host, however, the AM symbiosis often is inhibited. Under those conditions for certain host–soil interactions, mycorrhizal development can reduce plant growth and thus become pathogenic (Modjo and Hendrix 1986).

A mycorrhiza is initiated by hyphae that emanate from another mycorrhizal root or from germinating spores. As colonization spreads within the root cortex via hyphal extension and branching, other structures, such as "vesicles" and "auxiliary cells," develop (see "Organism Characterization"). Both structures contain lipids and therefore are thought to have a function in carbon storage. Auxiliary cells are much more transitory than vesicles, so their storage function may be tied closely to developmental events in fungal life cycles. Infectivity of different fungal propagules appears to vary among fungi at the subordinal level. Mycorrhizal roots, detached hyphae, and spores of fungi in the suborder Glomineae (Fig. 15.1) generally are highly infective, whereas auxiliary cells and detached hyphae of fungi in the suborder Gigasporineae are not (Biermann and Linderman 1983). Generally, new sporulation commences between 4 and 8 weeks after onset of colonization and plateaus in 12–16 weeks in greenhouse pot cultures (Morton et al. 1993). Under field conditions, however, sporulation patterns are more variable and fluctuate with season or host phenology (Gemma et al. 1989; An et al. 1993). Sporulation does not appear to occur in response to nutrient deprivation or environmental stress as in other fungi. When fungi are cultured on plant hosts in a pot containing sand, soil, or some other growth medium (henceforth called pot cultures), new sporulation generally commences several weeks after onset of colonization and plateaus in 12–16 weeks (Morton et al. 1993). Research currently suggests that spores begin to form after a threshold level of mycorrhizal biomass is achieved (Gazey et al. 1992; Franke and Morton 1994), and sporulation increases dramatically after roots cease to grow but continue to function as a

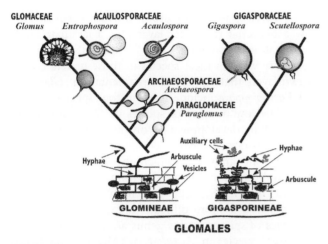

FIGURE 15.1 Phylogeny of arbuscular endomycorrhizal fungi based on morphological characters of the mycelium and reproductive spores. (From Morton 1990a; Morton and Benny 1990; Morton and Redecker 2001.)

sink for carbon from photosynthesizing shoot tissue of the plant.

In nature, fungal communities are taxonomically complex and rarely, if ever, consist of only one species (Morton 1999). When individual fungi are cultured on plants, host specificity appears to be minimal or absent (Smith and Gianinazzi-Pearson 1988; Brundrett 1991; Smith and Read 1997). Investigators at the International Culture Collection of (Vesicular) Arbuscular Mycorrhizal Fungi (INVAM) in West Virginia showed that more than 1000 isolates of 98 species of fungi in all genera are able to grow and sporulate on one plant host, *Sorghum sudanense*, or Sudangrass (Morton et al. 1993). Roots of Sudangrass accommodate colonization by as many as 10 species of fungi at one time in pot cultures using field soil as inoculum (referred to henceforth as trap pot cultures), and those fungi can be members of any genus (Morton et al. 1995; Morton 1999). Lack of host specialization may be the result of mutualistic coevolution of the plants and their fungal partners over the 400 million years since their origin (Simon et al. 1993; Taylor et al. 1995).

A combination of host and environmental factors can differentially influence rates and degrees of colonization and/or sporulation by different AM fungi in a community (Johnson et al. 1992; Bever et al. 1995), which are manifested as changes in species richness and relative abundance in sporulation. In general, those responses represent compatibility adjustments rather than specificity, although McGonigle and Fitter (1990) and Brundrett (1991) explained them as "ecological specificity" and "edaphic specificity," respectively. Divergence in physiological and life-history traits within a fungus

species and between species is expected in these asexual organisms as a natural response to local and regional selection pressures (Morton 1990b, 1999).

Members of a small number of plant orders do not form AM associations (Tester et al. 1987). Many other orders, however, include both mycorrhizal and non-mycorrhizal families or genera. Nonmycorrhizal taxa are assumed to have evolved away from the symbiosis, based on evidence from their distributions, relative to those of their mycorrhizal relatives (Trappe 1987), and from the presence of activated defense mechanisms in chemically induced mycorrhiza-resistant mutants (Peterson and Bradbury 1995). A few plant genera are able to support both arbuscular fungi and ectomycorrhizal fungi, one of the most widely studied being *Eucalyptus* (Lapeyrie and Chilvers 1985).

RELATIONSHIPS WITH ASSOCIATED RHIZOSPHERE ORGANISMS

Complex interactions take place in the volume of soil around roots, which traditionally has been termed the "rhizosphere." More appropriately, that soil volume constitutes a "mycorrhizosphere" (Rambelli 1973) because of the dramatic influence an abundance of fungal external hyphae has on root- and soil-associated microorganisms, as well as the effects of those microorganisms on the establishment and spread of mycorrhizae (Bagyaraj 1984). Mycorrhizal roots also have altered root exudation patterns (Marschner 1998). The *Rhizobium-Bradyrhizobium* association with legumes most notably is affected by mycorrhizal fungi, largely as a result of increased availability of phosphorus in host roots, which drives nitrogenase activity in nodules (Azcón et al. 1991). Additional interactions between nitrogen-fixing bacteria and plant growth–promoting rhizobacteria have been inferred from controlled greenhouse experiments (Azcón 1993). Other indirect interactions affect both pathogens and beneficial organisms, either through effects of mycorrhiza formation on root exudates or through competition (Linderman 1988; Garbaye 1991).

The influence of fungal hyphae in the mycorrhizosphere is much greater than previously thought (Tisdall et al. 1997) with the discovery of "glomalin," a heat-stable glycoprotein that coats hyphae and spore surfaces and accumulates in soil (Wright et al. 1996, 1998). Evidence indicates a strong involvement of glomalin in soil aggregate stability; researchers found a highly significant correlation between glomalin concentration and soil aggregate stability (Wright and Upadhyaya 1998).

TAXONOMIC STATUS

TAXONOMIC DIVERSITY

All AM fungi are classified in the order Glomales, which is divided into two suborders based on characteristics of the mycorrhizae (arbuscules/vesicles in Glomineae and arbuscules/auxiliary cells in Gigasporineae; Fig. 15.1). Six genera were recognized at the time Glomales was erected (Morton and Benny 1990). Almeida and Schenck (1990) then transferred most *Sclerocystis* species to *Glomus* based on shared morphological traits and characteristics of sporocarp development. The degree of shared 18S ribosomal DNA (rDNA) sequences also placed the remaining species, *S. coremioides*, in *Glomus* (Redecker et al. 2000a). Some species of *Acaulospora* and *Glomus,* including two species that produce *Acaulospora*-like and *Glomus*-like spores from the same thallus, were discovered to be genetically more closely related to *Geosiphon pyriforme* than to other AM fungi (Redecker et al. 2000b). *Geosiphon pyriforme* does not form mycorrhizal associations, but it is involved in a mutualistic symbiosis and serves as the physical host for photosynthetic *Nostoc* cyanobacteria (Schüssler et al. 1996). Those phylogenetically ancient species were placed in two new families, Archaeosporaceae and Paraglomaceae, with one genus in each, *Archaeospora* and *Paraglomus,* respectively (Morton and Redecker 2001).

The number of species of Glomales described so far totals 163. That number is low, especially when compared with the more than 3500 species of ectomycorrhizal fungal (Law and Lewis 1983). Such low morphological diversity is predictable, however, given the relatively simple designs of taxonomically important structures in arbuscular mycorrhizae and spores and the lack of complexity in spore or sporocarp organization (Morton 1999). Unique characters may be discovered as new species are found, but we do not expect the number to be high. If the range of possible species types has not decreased very much as a result of species extinctions through geologic time, as hypothesized by Morton (1990b), then newly discovered species likely will fill ecological gaps between existing species. A taxonomic consequence of such discoveries will be greater difficulty in defining species boundaries from morphological traits alone.

The numbers of species in each of the seven genera differ considerably. Developmental evidence indicates that properties of the spore wall resolve species primarily, other traits grouping species into higher taxa. *Glomus,* with 91 species, is disproportionately specious relative to other genera, but species in this genus also produce spores with the greatest variety of spore wall layers. In contrast, *Acaulospora* and *Scutellospora* have only a bilayered (an outer layer plus a laminate layer) spore wall and contain 35 and 28 species, respectively. The spore walls of most species in those genera consist of two layers that co-vary in ornamentation. *Gigaspora* and *Entrophospora* include five and four species, respectively. The low number of species in *Gigaspora* can be attributed to developmental constraints on spore wall variation associated with germination; that genus is the only one known in which the terminal event in spore wall differentiation is germination (Bentivenga and Morton 1995). The low number of species in *Entrophospora* is more difficult to explain. Spore morphologies of species in this genus are identical to those of "sister" species in *Acaulospora* (Morton and Benny 1990), with the exception of the spore relative to a "sporiferous saccule." We hypothesize, therefore, that the position of spore formation may have shifted repeatedly and that *Entrophospora* likely is a polyphyletic genus (each species having evolved from a different *Acaulospora* ancestor). *Archaeospora* and *Paraglomus* contain only two species each, but those genera have been delimited only recently, based on differences in selective rDNA primer sequences in polymerase chain reactions (Morton and Redecker 2001); it is likely that additional species, presently included in other genera, will be transferred to *Archaeospora* and *Paraglomus* as new molecular analyses are conducted.

The low number of species in Glomales is offset by the considerable adaptation of spatially separated conspecifics to conditions in their local environments. Adaptations are expressed in terms of fungal tolerance to soil variables such as pH, heavy metal ion concentrations, and salinity levels, as well as growth responses to different soil–fertility regimens and to changes in habitat or growing conditions (Brundrett 1991; Weissenhorn et al. 1994). Adaptation to particular host species is more difficult to ascertain, but indirect evidence suggests that it occurs (McGonigle and Fitter 1990; Johnson et al. 1992; Bever et al. 1995). Undoubtedly, therefore, the physiological and ecological diversities among populations of all globally distributed species far exceeds taxonomic diversity.

SPECIMEN REPOSITORIES

Living germ plasm now can be deposited in established and well-maintained collections supervised by individuals experienced in taxonomy and in dealing with cultures. The oldest and largest centralized collection is INVAM in West Virginia (Appendix III). In 2001, the collection consisted of almost 1100 accessions of more than 86 species. An extensive Website is associated with this collection. The Bank of European Glomales (BEG; Appendix III), a smaller collection started in 1994, is organized as a network with a central coordinator (Dodd et al. 1994). That collection differs from others in that accessions are registered at one location, but they are cultured and maintained by individual researchers at their home institutions. Names of Coordinating Committee members, the culture database and gene archives, and other resources and information are available on the BEG website. Regional collections of fungi exist as well, in established, centralized research facilities such as those of Lynette Abbott, University of Western Australia, Perth; Ricardo Herrera, Instituto de Ecología y Sistemática, Havana, Cuba; Chi Guang Wu, Taiwan Agricultural Research Center, Wufeng, Taichung, Taiwan; and Centro Internacional de Agricultura Tropical (CIAT) in Cali, Colombia.

Nonliving specimens (e.g., vouchers, types) can be obtained from individual researchers who have followed recommended procedures for characterizing their experimental organisms. Larger repositories conserving valuable type specimens are located in the Oregon State University Mycological Collections (Corvallis, Oregon), the Farlow Herbarium of Harvard University (Cambridge, Massachusetts), and the Royal Botanic Gardens Kew Herbarium (Surrey, England) (see Appendix III).

DIVERSITY AND DISTRIBUTION

RANGE OF HABITATS

Fungi in the Order Glomales now have been found in almost every habitat in which plants are able to grow (Brundrett 1991). The occurrence of a species and its abundance are likely to be determined largely by the presence and abundance of compatible host plants that allow the fungus to grow and reproduce. Arbuscular fungi are least abundant and heterogeneous in coniferous, *Quercus*, *Fagus*, *Thuja*, *Sequoia*, and *Eucalyptus* forests, which are largely ectomycorrhizal (Allen et al. 1995), and in the high arctic (Bledsoe et al. 1990). They are also rare in tundra regions, where the dominant ericaceous plants have their own type of mycorrhizae (ericoid). Arctic habitats show little evidence of AM, although endemic plants are capable of forming mycorrhizal associations when inoculated with an introduced fungus (DeMars and Boerner 1995). The highest numbers and diversity of fungal species are found in maritime and lacustrine sand dunes (Koske 1987), prairie grasslands (Bentivenga and Hetrick 1992), forests (Louis and Lim 1987), deserts (Bethlenfalvay et al. 1984), salt marshes (Hoefnagels et al. 1993), agricultural fields (Hetrick and Bloom 1983), and orchards (Miller et al. 1985) of temperate and tropical latitudes.

PANDEMISM OR ENDEMISM

Endemism has not been established conclusively for any species, although some species appear to have more limited distributions than others. Temperature appears to be important (Koske 1987) in limiting distributions of some species to certain latitudes. In contrast, a number of species hypothesized to be ancient as well as highly derived either morphologically (Morton 1990a) or genically (Redecker et al. 2000a; Morton and Redecker 2001) are distributed globally. For example, *Glomus intraradices* is found on all continents except Antarctica, at tropical and temperate latitudes, in arid and mesic climates, and in a wide range of habitats. Fungal species from other genera, such as *Paraglomus occultum*, *Acaulospora scrobiculata*, *Archaeospora leptoticha*, *Gigaspora gigantea*, and *Scutellospora calospora* are distributed widely on most continents.

At the generic level, no clear-cut habitat preferences are discernible. For example, similar ratios of species per genus are obtained from deserts, grasslands, tropical forests, and agricultural communities (INVAM Biogeography Database). The number of *Glomus* species often exceeds those of other genera by an average of 3 to 1, but the number of described species in that genus also exceeds those in others by almost 3 to 1. Thus, the apparent dominance of this genus may be a function of number of species on the planet rather than geography or ecology. Fungal communities in arid sites are made up of small-spores species, mostly in the genera *Glomus*, *Acaulospora*, and *Entrophospora* (Stutz and Morton 1996; Stutz et al. 2000). In tropical forests, *Acaulospora* appears to dominate communities, but that is an erroneous conclusion resulting from the great abundance of spores produced by *Acaulospora* species in field settings.

The distributions of many species are difficult to define because of sampling and logistic obstacles and difficulties in making definitive identifications (Walker 1992; Morton 1993). Few spatially or temporally separated locations that share common habitat characteristics have been studied. When such comparative studies are carried out, then some selectivity in taxonomic structure of fungi in plant communities begins to appear. For example, arid sites from the Namib Desert of Africa and the Sonoran Desert of North America contain from four to eight species in the suborder Glomineae (*Glomus*, *Acaulospora*, *Entrophospora*). Among the 21 species recovered from the two continents, 61% of them occurred on both (Stutz et al. 2000).

PATTERNS OF SPECIES RICHNESS

Species richness of arbuscular fungi is measured as the number of species whose spores are extracted from soil samples, with a few exceptions for fungal species whose mycorrhizal roots are sufficiently well characterized to be used for identifications (McGee 1989; Abbott and Gazey 1994). Gazey and colleagues (1992) demonstrated that a minimum level of fungal biomass (mycorrhizal colonization) is needed before reproduction commences. Many fungi, therefore, may colonize roots at low levels and grow through hyphal spread among plants. That is evident in pot cultures trapping native fungi from arid sites in the southern United States and Namibia, where sporulating species increased from zero to one in the first trap-culture cycle and from four to eight by the third cycle (Stutz and Morton 1996). The level of species saturation in the root system of a single host plant has yet to be determined, but trap-culture yields indicate that it can be as high as eight species (Morton et al. 1995). In addition, species composition can vary considerably among communities from different habitats, especially those in which moisture availability is limited (Table 15.1). In the absence of such constraints, most genera can be represented in the same root system.

In the field, species richness varies considerably among samples from a given site (e.g., 3–11, Walker et al. 1982), largely because of all the variables having an impact on sporulation potential. Richness also varies among sites in temperate (e.g., 5–10, Benjamin et al. 1988; 12–22, Johnson et al. 1991) and tropical (e.g., 10–18, Johnson and Wedin 1997) regions. Richness appears to be low in arid habitats, often from zero to two species (Rose 1981; Bethlenfalvay et al. 1984), but higher at more mesic sites

(Allen et al. 1995). In sites lacking any sporulation, most plants are still mycorrhizal, except in habitats such as the arctic (discussed earlier).

The widespread trend among living organisms of increasing species richness along a gradient from high to low latitudes (Rohde 1992) is not so clear-cut among arbuscular fungi (Allen et al. 1995). Koske (1987) measured from two to eight species per site in the root zones of *Ammophila* and *Solidago* and from six to 14 per site in root zones of *Uniola* in dunes along a 300-km transect of the Atlantic coast of the United States, with some evidence of a latitudinal temperature gradient. However, at the landscape level, species richness (N = 23 species) in 75 m^2 of an old field in North Carolina at latitude 36° N (Bever et al. 1995) was nearly equivalent to that (N = 25) in 20 km^2 of grassland in a dry forest in Costa Rica at latitude 10° N (Johnson and Wedin 1997).

Fungi at southern latitudes have been sampled less intensively, but we predict that patterns of species richness will be similar to those north of the equator because of similar dependencies of plant-community compositions on broad climatological patterns (Allen et al. 1995). Species diversity on the Antarctic Peninsula appears to be limited. Cabello and colleagues (1994) described at least one arbuscular species associated with the grass *Deschampsia antarctica* along the Danko coast, whereas none was associated with the same plant species on Anvers Island (DeMars and Boerner 1995; J. B. Morton, unpublished data). Greenland and the Northwest Territories of North America are virtually unexplored, although the few surveys carried out thus far indicate that many plant species are nonmycorrhizal (Bledsoe et al. 1990; Kohn and Stasovski 1990). Plant communities along a glacial front in Alaska appear to be

TABLE 15.1

Generic and Species Compositions of Second-cycle Trap Cultures* of Indigenous Fungi from Different Habitats[†]

Genus	Numbers of species recorded for each genus from each sample						
	KE104[‡]	VZ106	NB106	AZ225	MN416	VA105	WV109
Acaulospora	2	1	2	1	1	1	1
Archaeospora	0	1	1	0	0	0	1
Entrophospora	0	0	0	1	0	1	1
Gigaspora	0	2	0	0	1	1	0
Glomus	3	1	2	2	2	1	1
Paraglomus	1	1	1	1	1	1	0
Scutellospora	1	2	0	0	1	2	2

* Started on Sudangrass in a sand–soil mix (pH 6.2) and maintained in a growth room (photon flux density 150 mol m^{-2} s^{-1}, day length 15.5 hours, temperature and relative humidity ranging from 24–31°C and 30–70%, respectively) for 4 months.

[†] Habitats = tropical forest in Kenya, grassland in Venezuela, desert scrub sites in Namibia and Arizona, prairie in Minnesota, sand dune in Virginia, and acid mine soil in West Virginia.

[‡] Locality and International Culture Collection of (Vesicular) Arbuscular Mycorrhizal Fungi (INVAM) accession number. KE, Kenya; VZ, Venezuela; NB, Namibia; AZ, Arizona; MN, Minnesota; VA, Virginia; WV, West Virginia.

mycorrhizal, based on sporulation of *Glomus etunicatum* and *Archaeospora trappei* in trap cultures from two sites (J. B. Morton, unpublished data).

A few studies have suggested that fungal species richness is correlated with plant species richness in some temperate and tropical plant communities (Sieverding 1990; Johnson et al. 1991); other studies show no relationship between the two (Wilson et al. 1992; Johnson and Wedin 1997). Regardless, richness of fungal species can remain relatively high (3–5) when plant communities consist of only one species (Morton 1986).

Allen and associates (1995) hypothesized that the AM association is an important regulator of plant diversity at community and ecosystem scales. Plants differ considerably in their dependency on symbiosis, thus affecting successional patterns (Janos 1980) and perhaps even levels of species saturation. As plant-community composition changes, the structure of the fungal community also appears to change, at least with respect to the species sporulating at the times of sampling (see Brundrett 1991 and references therein; Johnson et al. 1992).

Differences in species richness and composition among sites within a study area or between study areas of similar scale are likely the result of several factors:

- Stochastic dispersal events over time and space
- Host or edaphic characteristics that influence colonization and sporulation
- Spatial and temporal differences in intensity of sampling
- Quality and quantity of fungal material collected for identification
- Differences in the taxonomic expertise or species concepts of investigators

LOCAL VERSUS REGIONAL DIVERSITY IN TEMPERATE AND TROPICAL CLIMATES

Most measures of species richness have been based on samples from small areas. Only a few studies have covered areas of a size sufficient for estimating regional diversity. A combination of field sampling (Koske 1987) and trap cultures (Morton et al. 1995) from a 300-km latitudinal gradient along the dunes of the U.S. Atlantic coast yielded 27 species. Beta diversity (calculated as species richness of the region divided by the species richness in each local sample) ranged from 1.5 to 2, with a mean of 1.6. Bever and colleagues (1995) obtained a species richness of 23 in a 75 m^2 plot of an old field in Durham, North Carolina, based on field collections and various trap culture methods (e.g., seeding inoculum with Sudangrass or cuttings of field host; a field-collected host transplanted into sterile soil). That number

increased to 32 when the same methods were used to cover a larger (400 m^2) area. When the mean beta diversity value derived from the dune transect is applied to a similar-size region around the field plot, the predicted species richness is 51. Based on those calculations, we estimate that 25–50 species are likely to be recovered from a 50,000-hectare region at temperate latitudes.

Sampling in tropical climates has been limited to individual localities, and even then the number of surveys is low. Species richness in field-collected samples was 28 in a tropical dry forest of Costa Rica (Johnson and Wedin 1997) and 41 and 17, respectively in *Terminalia* plantations of Côte d'Ivoire (Wilson et al. 1992) and Cameroon (Mason et al. 1992). Insufficient data were available to calculate beta diversity between *Terminalia* plantations. We would predict, however, that it would be close to the value obtained in temperate regions because of similar levels of species diversity (in at least two of the study areas). Although plant communities generally are more diverse in tropical than temperate regions, a similar pattern is not predicted for glomalean fungal communities. Correlations between fungal and plant species richness in a given region are neither strong nor consistent. Moreover, trap-pot cultures of soil samples from tropical regions (Costa Rica, Kenya, Mexico, Nicaragua) often produce fewer sporulating species than those from temperate regions (J. B. Morton, unpublished data; Table 15.1). Those results suggest that the complexities of fungal communities in temperate and tropical climates are likely to be similar.

STATE OF THE WORLD DIVERSITY INVENTORY

Results of sampling for arbuscular fungi have been published for less than 25% of all politically defined countries in the world. For those countries from which fungal diversities have been reported, the geographic area of sampling often is small. Obviously, the most thorough inventories have been carried out in regions with active researchers, especially researchers with taxonomic expertise. Species inventories from five of the 13 countries in South America are available, for example, but only in Brazil and Colombia have more than two regions been sampled. A relatively large number of Brazilian scientists with mycorrhizal research programs are widely dispersed through their country, judging from attendance at biannual conferences. In Colombia, workers from CIAT (Appendix III; Sieverding 1991) have carried out intensive sampling of agricultural areas.

Some geopolitical regions, such as Russia and the former Soviet Republics, Asia, many Middle East countries, and many Atlantic and Pacific islands, are virtually unsampled (INVAM Biogeography Database). Fewer than half the provinces of Canada have been sampled and then mostly in agricultural areas. The Yukon and the Northwest Territories remain relatively untouched. Even within the continental United States, nine states remain to be sampled.

RESOURCES FOR DIVERSITY INVENTORY

SAMPLING REGIMENS AND SPATIAL, SEASONAL, AND SUCCESSIONAL VARIABLES

A variety of diversity indices have been formulated in plant and animal ecology, based on different assumptions of population dynamics (Magurran 1988). Many of those indices can accommodate spore count data and have been used to measure the taxonomic structure of communities of arbuscular fungi (Koske 1987; Johnson et al. 1991, 1992); realistically, however, they assess only resource allocation to sporulation by each member of the population. The Shannon-Weiner Index and the Dominance Index, to cite two of the most commonly used indices, assume that each unit of measure represents an individual (see Southwood 1978). That assumption is not met with arbuscular (or most other) fungi because the units being sampled and counted are spores or sporocarps. The functionally important parts of the fungus (i.e., the mycelium, arbuscules, and vesicles or auxiliary cells) cannot be distinguished among species coinhabiting a mycorrhizal root system. The exception is *Glomus tenue* (McGonigle and Fitter 1990), although there is some doubt about whether this species is a true mycorrhizal fungus or even a single species. Other fungi may be monitored within stained roots, using mycorrhizal colonization patterns when morphological traits of each fungal species can be characterized and distinguished (Abbott and Gazey 1994). That usually occurs only under uncommon conditions in which single species of different genera are present.

Estimates of richness, abundance, importance, and evenness of species colonizing one or more plant root systems are possible only when sporulation is sufficient to allow for identification and recovery of spores from at least 50% of the samples collected (Tews and Koske 1986). A preliminary survey to determine the correct depth of sampling and to obtain rough estimates of species richness and spore abundance as well as some notion of the amount of variation between samples should precede any serious sampling program. Populations of spores vary considerably over both horizontal

and vertical distances of only a few centimeters (Anderson et al. 1983; Friese and Koske 1991) because the spores are highly aggregated (Walker et al. 1982; Sylvia 1986; St. John and Koske 1988). The appropriate sampling depth is determined by location of the active root system (Abe and Katsuya 1995). Roots of crop and pasture plants are most abundant in the top 15 cm of soil, and it is here that most spores are likely to be found (An et al. 1990). In some sand dunes, spore abundance can be high even at depths of 100 cm or more (Abe and Katsuya 1995). Although the choice of sampling depth is seemingly obvious, many collectors sample at and just below the soil surface where organic matter may be high but roots (and sporulation) may be rare or absent.

Once the correct sampling depth has been ascertained, then the horizontal pattern of sampling can be planned. One method, called stratified random sampling (Mueller-Dombois and Ellenberg 1974), involves collecting root and soil samples at randomly selected points along one or more transects in "representative" areas of interest to the investigator. For sites with sparse plant cover, it is useful to sample the plant nearest the random point on the transect (if no plants occur directly under the transect line). In habitats such as sand dunes and some forested areas, characterized by plants with rhizomes and well-developed root systems, however, truly random samples are appropriate (St. John and Koske 1988).

The number of samples to collect from each habitat (general surveys may include several habitats) has not been determined for most plant communities. The appropriate number of samples can be estimated with the species-increment plot method if measuring species richness; it can be measured with the spore-abundance plot method if measuring populations in the fungal community (Tews and Koske 1986). Thirty separate samples were sufficient for quantifying the abundance of the dominant and subdominant species and estimating species richness in some New England sand dunes (Fig. 15.2). Similar calculations have not been made for other sites. Pooling or "bulking" of samples often results in the loss of valuable information on the frequency of occurrence and variation in abundance, so we do not recommend it.

Seasonal fluctuations in the abundance of live spores have been documented for a number of fungal species (Gemma et al. 1989; An et al. 1993), but the longevity of the propagules is unknown for most species (Tommerup 1992). The numbers of live spores in soil increase as a result of sporulation by colonizing fungi; they are reduced by germination, mortality, parasitism, and ingestion by soil animals (Rabatin and Stinner 1988; Lee and Koske 1994). Such losses may be especially acute in wet tropical environments (Fischer et al. 1994). In some

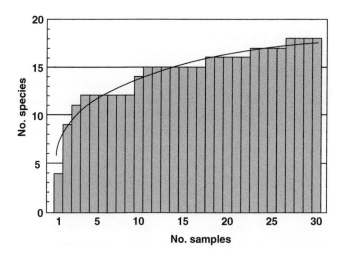

FIGURE 15.2 Species-increment curve based on sporulation by arbuscular fungi in a sand-dune habitat. The likelihood of recovering additional species from a site as more samples are examined decreased after the eleventh sample. (Data from Tews and Koske 1986.)

other environments, especially arid and other habitats in which mycorrhizal development is low (Dhillion and Zak 1993), sporulation may not be detected because carbon appears to be invested only in vegetative growth in roots and soil (Bethlenfalvay et al. 1984; McGee 1989). Sporulation frequently has been linked to host phenology in the field (e.g., maximum spore production occurs near the middle to end of a growing season), but it clearly is affected by abiotic factors, especially temperature and moisture. Under field conditions, different fungal species can have unique seasonal patterns, so samples from a single date may not assess the composition of the fungal community accurately. Whereas monthly sampling over a period of 1 or 2 years is most desirable, quarterly sampling may be more practical. Given the lack of information from many areas of the world (see "State of the World Diversity Inventory," earlier), even sampling an area once can provide valuable information, especially if it is coupled with trap cultures to recover species not sporulating in the field.

Spores of arbuscular fungi come in many sizes (20–800 μm) and colors (white to red-black); the spore wall and the one or more flexible inner walls have complex subcellular characters (Morton 1988; Morton et al. 1995). A dissecting microscope is required for observation of macroscopic properties and a compound microscope with excellent optics for resolution of subcellular layers. A microscope equipped with differential interference contrast optics is preferable so that thin, and often colorless, flexible layers less than 1-μm thick can be clearly differentiated.

Whether a spore collected from soil can be identified to species will depend in part on its condition. In soils with high organic matter and concomitantly high microbial activity, especially those in tropical climates (Fischer et al. 1994), most spores are likely to be too degraded or parasitized to be unrecognizable even at the generic level. For example, *Gigaspora tuberculata* was described from parasitized spores (Neer et al. 1993), but a detailed comparison with other species in Gigasporaceae revealed that the species was synonymous with *Scutellospora persica* (Bentivenga and Morton 1995).

The relationship between aboveground succession and the composition of the arbuscular fungal community has been studied in a number of habitats (e.g., Allen and Allen 1984; Boerner 1988; Rose 1988; Janos 1992; Johnson et al. 1991; Miller and Jastrow 1994), but most of the studies have emphasized plant- rather than fungal-community dynamics and have considered successional changes in fungal composition only with respect to those species able to sporulate at the time of sampling. Interpretation of field-collected data at minimum requires multiple sampling over the seasons. For those observations to be most informative, characteristics of the soil (e.g., pH; organic matter; cation exchange capacity; N, P, K, Al, and other cation concentrations) also should be measured.

In sites of low species richness, a great number of samples must be collected to validate the preliminary observations of few species. Because of the common correlation between spore abundance and frequency of recovery of fungal species, seasonal decreases in spore numbers may suggest a seasonal decrease in species richness. Nonspeciose sites thus also should be sampled with trap cultures to search for nonsporulating species that may colonize host roots sparsely, especially in many arid habitats or saturated soils (see "Culture Protocols," later in this chapter).

In some studies, the goal is to determine the presence or absence of mycorrhizal plants rather than to determine fungal community structure. The general occurrence of AM associations in plants usually cannot be determined from direct examination of untreated root samples, although a few plant species, such as *Allium* species, *Zea mays*, *Hordeum vulgare*, and *Triticum aestivum*, may develop noticeable yellow pigments in root areas that are colonized (Klingner et al. 1995). Root cells must be cleared of cytoplasm and acidified, and the fungal cell walls must be differentially stained. Trypan blue in a lactic acid and phenol mixture (Phillips and Hayman 1970) was used widely for many years, but the hazards of phenol precipitated development of a safer, but equally effective, procedure (Koske and Gemma 1989). Acid fuschin (Kormanik et al. 1980) and chlorazol black E (Brundrett et al. 1984) are alternative stains, the latter producing the highest contrast among fungal structures. Sylvia (1994) summarized, in detail, procedures for clearing and staining roots and estimating percentage of a root that is mycorrhizal.

COLLECTION AND ISOLATION PROTOCOLS

Procedures used to extract spores from soil or roots have changed considerably over the years. A majority of the species first described were sporocarpic because the fungi were detected easily in soil sievings or they fruited at the soil surface (Gerdemann and Trappe 1974). As practices changed to wet-sieving of soil followed by sucrose-density gradient centrifugation to remove soil particles and organic matter (Daniels and Skipper 1982), many more species that produce spores singly or in loose aggregates or clusters were discovered. Currently more than 80% of described species are nonsporocarpic, and the noticeable absence of new sporocarpic species may be the result of lack of attention to whole soil sievings. Also a number of undescribed *Glomus* species that produce spores less than 60 μm in diameter may exist undetected, despite being the dominant species in some plant communities (Johnson et al. 1991; Allen et al. 1995). Investigators must pay close attention to the sizes of the openings in the sieves or filters they use in separation procedures.

McKenney and Lindsey (1987) described a rigorous procedure for capturing spores of any size or arrangement. A somewhat less laborious version of that procedure includes the following steps: (1) visual inspection of soil for epigeous sporocarps at the site of collection; (2) collection of soil to a predetermined depth, usually to 15–20 cm, ensuring that the sample contains all roots available; (3) mixing the sample with water in a blender for 30 seconds; (4) passing the slurry through nested sieves with 500-, 250-, and 38- (or 45-) μm openings using a water spray; (5) examining (separately) material from the 500- and 250-μm sieves for small sporocarps and large spores; and (6) centrifuging the material from the 38- (or 45-) μm sieve in sucrose gradients of 20% and 60% using 50-ml tubes. A swinging bucket rotor provides optimal separation of soil particulates and spores. The supernatant in each tube is poured into a 38- (or 45-) μm sieve and washed for 1–2 minutes, rinsed in distilled water, and placed in a glass Petri dish for storage. Various modifications of this technique have been discussed by Daniels and Skipper (1982), Brundrett and associates (1994), Pacione (1994), Sylvia (1994), and Tommerup (1994).

Some spores of different species, sizes, and colors can be separated with nested sieves with openings of specific sizes. Because volumes at this stage are small and many different sizes may be needed, we cut 2.5-cm diameter PVC pipe into 2-cm sections and glue a circular nylon mesh on one end of each section. Sieves of almost any size mesh are available from Tetko, Inc. (Appendix IV). An alternative separation method is to wash the sievings through prewetted filter paper in a Buchner funnel under vacuum (Brundrett et al. 1994). The filter paper can be stored in Petri dishes for 5–7 days at 4°C without obvious damage to spores or much apparent parasitism. Filter paper also can be dried and stored for longer periods and rewetted 24 hours before making spore counts.

Spores stored in water in a plastic Petri dish with intersecting grid lines can be counted directly in all fields under a stereomicroscope when numbers of spores are low. When numbers are high, spores from 30 to 40 randomly chosen cells can be counted, averaged, and multiplied by the number of grid cells in the Petri dish. With either method, one must agitate the Petri dish initially to distribute the spores relatively evenly but then must minimize further disturbance. Placing spores from water on gridded filter paper eliminates movement, but differentiation of similar spore morphotypes is more difficult for an inexperienced observer. The use of gridded filter paper in Buchner funnel extractions eliminates spore transfer steps.

CULTURE PROTOCOLS

Pot Cultures

Pot cultures refer to the growth of fungi on host plants grown in pots under artificial conditions (e.g., in a greenhouse). Pot cultures can serve as experimental microcosms (Grime et al. 1987) for examining changes in fungal-species composition and patterns of sporulation in untreated simulated plant communities or in those subjected to various treatments. They also can be used to trap and induce sporulation of fungi-colonizing roots at successive stages of plant root (or community) development.

Arbuscular fungi must be cultured in greenhouse pots to obtain inoculi of single isolates of the full range of taxonomic groups. Various culture methods have been used successfully to grow arbuscular fungi in association with host root tissue. These include growth of (1) plants in a soil medium (Ferguson and Woodhead 1982; Morton et al. 1993) or sand (Millner and Kitt 1992); (2) plants with roots bathed in enclosed mist chambers (Jarfster and Sylvia 1993); and (3) root explants on a defined agar medium (Bécard and Piché 1989, 1994). The more culture conditions diverge from natural conditions of a field setting, however, the smaller the number of taxa that seem to be cultured. We focus on pot-culturing methods because the pots provide a compatible environment for at least 95% of all species of fungi and for organisms from a wide range of habitats, and they are the only means to "trap" or bait sporulating and non-sporulating fungi from field-collected roots (Morton et al. 1993).

Other culture protocols have been sought primarily because of the effects of soil conditions, rhizosphere activity, and open-pot cultures on inoculum yields and purity. The amounts of available nitrogen and phosphorus affect mycorrhizal development and hence sporulation. We recommend use of a balanced fertilizer low in phosphorus (Sylvia 1994) because high phosphorus levels generally inhibit mycorrhizal development (Smith and Gianinazzi-Pearson 1988; Bolan 1991). Temperature, soil pH, and the amount of available water must closely approximate conditions to which the fungi are adapted. Optimum light conditions also are needed to maximize carbon sequestration to roots.

Contamination by other arbuscular fungi and antagonistic organisms can be controlled with a vigilant quality-control program. Insect problems especially must be avoided because infestations by fungal gnats or *Collembola* can build up rapidly in pots and consume most, if not all, fungal propagules. It is essential to do the following: (1) prohibit field soils or other potted plants in the culture rooms (they often are the source of insect problems), (2) remove surface dust and disinfect all surfaces at regular intervals (to reduce movement of airborne propagules), (3) handle pots carefully and avoid touching contents (to prevent transfer of propagules), (4) use watering methods that prevent splashing and free movement of water between pots, and (5) have in place an immediate chemical control response plan should insects appear. To monitor for contamination, several pots that contain sterile soil and the culture host should be placed randomly on culture tables and examined for mycorrhizal colonization at the end of a culture period. We find that the most persistent source of contamination is dust from dry soils or cultures at the time of harvest. Solutions are to keep "raw soils" out of the culture room, to carry out all setup and harvest procedures in a separate room, and to mist the air each time a culture is handled to eliminate suspended particles in air.

Choosing a host plant is a major decision that often requires comparisons of the sporulation productivity of fungi cultivated on different plant hosts using the same inoculum source. Sudangrass is used routinely in INVAM because it supports colonization of so many different species (Morton et al. 1993). However, other hosts may be better suited at other locations. At the University of Minnesota, for example, big bluestem grass (*Andropogon gerardii*) was superior to Sudangrass (F.

Pfleger, personal communication), and in semitropical Florida, bahiagrass (*Paspalum notatum*) was used to culture many fungal species (Schenck and Pérez 1990). It is interesting that the most compatible and productive hosts are C4 grasses, possibly because their root system is abundant and because they produce and partition photosynthetic carbon efficiently.

Seeding rate depends on light quality and quantity, temperature, space availability, and other greenhouse conditions. At INVAM, where Sudangrass is used routinely, plants are overseeded to accomplish three important objectives: (1) produce shorter plants at a more uniform height; (2) ensure widespread infiltration of roots throughout the pot (as opposed to only along the sides and bottom); and (3) ensure that no soil at the pot surface is exposed, to prevent splashing and contact with any abrupt infestation of foliage-feeding insects.

Another major decision is choice of growth medium. Sand is inert and therefore neutral in its effects on plant–fungus combinations, but it also requires daily inputs of fertilizer that must be monitored closely (Millner and Kitt 1992). Sand in combination with a low phosphorus loamy soil (Morton et al. 1993) or vermiculite (Schenck and Pérez 1990) requires fewer applications of fertilizer and has greater buffering capacity. Initial properties of the medium must be defined, however, and adjusted as desired. At INVAM, a soil–sand mix adjusted to pH 6.2 provides a compatible growing environment for a majority of fungal isolates.

Pot size also is important. Sporulation patterns appear to be regulated by amount and duration of root (and concomitant shoot) growth. In our experience, trap cultures with volumes of less than 300 cm^3 rarely induce sporulation of more than one to two species. Volumes of from 1000 to 1500 cm^3 appear to generate the most diverse sporulation within a 4- to 5-month period. Larger volumes can achieve similar results, but duration of the culture period increases proportional to increase in pot size.

Soils with high sporulation, low organic matter, and low microfloral activity are likely to contain spores that can be extracted and used immediately as inoculum to set up monospecific cultures. Most field soils do not meet those criteria, and many spores are not healthy or infective. A safer approach is to propagate fungi from a sample of roots and soil in pot culture. Not only are healthy spores obtained, but spore-abundance measurements may indicate which fungi are most culturable under a particular set of host-environment conditions.

Establishing Trap-Pot Cultures

Trap-pot cultures are established with the following steps:

1. At least 250 cm^3 of field soil containing abundant roots is collected. The roots are chopped into fragments small enough for uniform mixing with the soil. The soil-root material is mixed thoroughly with coarse sterilized sand, 1:1 (v/v), to fill a pot of at least 600-cm^3 volume.

2. Seeds of the host are sown and plants are grown for a minimum of 4 months, fertilizer being applied only when plants show symptoms of nitrogen (yellow leaves) or phosphorus (red leaves/stems) deficiency. Ideal harvest time (peak sporulation) appears to be 30–40 days after shoots (and hence roots) have ceased to grow. Sporulation patterns in numerous pot cultures suggest that photosynthetic carbon is reallocated at this time to spores rather than to putatively dormant fungal vegetative structures.

3. At harvest, some root samples are taken for estimates of mycorrhizal development (percent colonization or mycorrhizal root length), and spores are extracted to obtain a measure of abundance.

4. If colonization and spore abundance are high, then watering of pots is terminated. Pots are dried slowly without disturbance, preferably in a cooler, shaded part of the greenhouse or growth room. If the intent is to store the material long term, then pot contents should be dried sufficiently to ensure that when it is refrigerated in bags, condensation will be low or absent. In contrast, if the intent is to extract spores, separate species and start monospecific cultures, then at least half the pot contents should be placed in storage bags when it is still slightly moist to the touch. Just before bagging, shoots are removed by slicing stems 0.5–1 cm below the soil line with a sharp utensil. The root mass often is dense enough to allow the entire pot contents to be removed intact. That mass should be disturbed only enough to fit it into bags. Before extracting spores for inoculation, bags should be sealed and stored at 4°C (temperate isolates) or at room temperature (tropical isolates) for a minimum of 30 days to accommodate any spore dormancy factors.

5. If spore numbers are low or only one or two species are sporulating, a second propagation cycle often will increase spore abundance and species richness. Again, shoots are removed with a scalpel 0.5–1 cm below the soil line, but pot contents are not removed. Instead, they are loosened with a scalpel or other cutting instrument and reseeded with the same host as used previously. These pots are maintained for another 3–4 months, and harvest time is again determined by signs of cessation of shoot growth.

6. In rare cases, a third propagation cycle may be needed. At that time, pot contents need to be mixed vigorously before reseeding. We transfer the pot contents to a zipper-closing plastic bag, gently break apart the root masses, and then return it to the pot.

Pot cultures also can be used to trap and induce sporulation of fungi-colonizing roots at successive stages of plant root (or community) development. In that case, only lightly washed roots are used as the starter inoculum so as to exclude fungal propagules in soil. One note of caution: use of only roots as inoculum may enrich AM fungi with a high frequency of intraradical sporulation (e.g. *Glomus clarum*, *G. diaphanum*, *G. intraradices*) and result in abnormal dominance at the end of a culture cycle (J. B. Morton, unpublished data).

Culturing Species from Trap Cultures

The procedure followed at INVAM (Morton et al. 1993) to establish a culture of each fungal species obtained in a trap culture is as follows:

1. Spores are extracted from stored cultures (see previous section), and those of similar morphology (color, size, other macromorphological features) are grouped together. A minimum of 10 spores from that collection should be mounted on glass slides to verify that spore morphologies are identical; those mounts then can serve as permanent vouchers.

2. Spores (usually 50–100) of each morphotype are placed in a watchglass, all extraneous material (hyphae, root fragments, organic matter) is removed with a pipette, and the "pure" spores then are placed in a microtiter plate well or some other container. They are stored at 4°C for at least 24 hours and examined again to remove any unhealthy or visibly hyperparasitized spores.

3. A pot with 10- to 12-day-old host seedlings (we routinely use sorghum because roots are robust and transplantation shock is minimal) growing in sterile medium is placed in a sterilized glass bowl filled with water, and the roots are teased apart gently.

4. Three seedlings gently are removed from the water; a stream of water is applied to the roots, gently washing away any remaining soil or sand, and the roots become entwined.

5. Spores in the microtiter well are pipetted directly along the length of entwined roots (excess water drips from the roots without dislodging any spores).

6. The seedlings are planted into 3.8 × 21-cm cone-tainers (Stuwe and Sons; Appendix IV) that include a sterile growth medium. Wolf and Skipper (1994) discuss methods of soil sterilization and their drawbacks. The seedlings grow for at least 90 days and then are examined for mycorrhizae and sporulation.

For single-spore cultures, only one spore is pipetted onto seedling roots, preferably at a location midway between the root crowns and tips. Usually a minimum of ten replicates of single spore cultures must be prepared to ensure at least one successful culture. That varies, however, with species dormancy requirements and the condition of spores.

An alternative method (Brundrett et al. 1994) for culturing species from trap cultures is to collect spores on filter paper, separate those of each morphotype onto individual filter paper triangles using forceps or some other fine-tipped instrument, place a triangle in a premade hole in a pot of growth medium, fill the hole with sterile sand or growth medium, and then add seed or pregerminated seedlings of the host plant to the hole. For single spore cultures, individual spores are transferred from filter paper to 10×10 mm membrane filters with 0.45-μm openings. The filter squares are placed on nylon mesh (50-μm openings) atop a layer of steamed soil and are watered to field capacity in a Petri dish. The Petri dish is incubated in the dark at $20°C$ until a spore germinates. Each filter square with a germinating spore is placed in a premade hole (as described earlier) with a young (age varying with host species) transplanted seedling (Brundrett and Juniper 1995).

Once a culture is known to contain only one species, the volume of inoculum can be multiplied by transferring whole inoculum (spores, root fragments, hyphae, and growth medium) from containers to larger pots. Colonization from whole inoculum is generally rapid and more predictable than spore inoculum. The occurrence of only one morphotype in the first cycle of a monospecific culture does not mean, however, that undetected hyphae or spores of associated fungi within spores in the parent culture have not become established. Those "internal" contaminants usually do not appear until the third or fourth propagation cycle, when biomass sufficient to induce sporulation has developed. One way to avoid that contamination problem is to collect spores and again inoculate seedlings when beginning the second propagation cycle (few if any contaminant hyphae are likely to be present at this time) and to increase with whole inoculum in the third cycle. The potential of "internal" (or external) contamination makes it critical to check the composition of a culture at the end of each propagation cycle and to prepare vouchers.

MEASURING RELATIVE IMPORTANCE AND ABUNDANCE

The presence and number of spores of a particular species in a sample and the biological importance of that species in the sampling area often are assumed to be correlated. Soil mycologists have struggled with the validity of that assumption for years but without resolution. At present, only arbuscular fungal species sporulating in the root zone can be counted and identified. The number (or volume) of spores produced in the field by a species then is assumed to indicate the relative biological importance of that species. "Importance" does not necessarily mean that a species is most beneficial to its mycorrhizal host. Rather, it represents a value of the capacity of that species to acquire a major portion of plant carbon resources available to the fungal community colonizing roots at that site. No other routine, practical methods are available to characterize species composition at a field site, although some progress is being made using selective enrichment of polymerase chain reaction–amplified DNA to probe for both sporulating and nonsporulating fungal colonizers of roots (Clapp et al. 1995; Vandenkoornhuyse et al. 2002).

The question of whether nonsporulating species in a fungal community can be the dominant colonizers has yet to be resolved. However, evidence based on mycorrhizal morphology (McGee 1989), trap cultures (Morton 1993), and molecular probes (Clapp et al. 1995) that nonsporulating fungi may be present at some level of colonization is accumulating. The ratio of sporulating and nonsporulating fungi is likely to vary considerably with site and habitat.

Another assumption has been that arbuscular fungi resemble small plants, possibly to accommodate terminology and concepts borrowed from the plant ecology literature. The assumption has some merit because both plants and fungi are modular organisms (Andrews 1995). An important distinction, however, is that the functional component of the fungal organism is the mycelium, which is not the part measured when determining abundance. Various infectivity assays have been used to estimate all arbuscular fungal propagules in a community, such as the "most probable number method" and the "mean infection percentage" method (see Sylvia 1994), but these measurements do not partition the contributions of individual species.

SAMPLING PROCEDURES

At present, there is no standard sampling size. In areas where fungal populations are unknown, typically large volumes of soil are collected initially (generally $250–1000$ cm^3). Different habitats and plant communities vary considerably in the number of spores that such a sample may yield. For example, prairie soils with high organic matter in the midwestern United States contain low numbers of spores but a relatively high species richness (Hetrick and Bloom 1983). Generally, in soils such as those, sample volumes greater than 200 cm³ are required for recovery of a majority of the sporulating species. Conversely, the low-organic-matter sand dunes of the Atlantic coast yield high numbers of spores (Koske 1987), and only 50–100 cm³ samples are needed. It is important to recognize that the aggregated distribution of spores in a field of soil persists in soil samples. Thus, one sample or a number of smaller subsamples removed from a sample should be mixed thoroughly. For example, if 10 subsamples of approximately 10 cm³ each are removed from soil at one site, they are combined in a 100-cm³ beaker and then mixed. For some soil types, experimentation is necessary to determine the optimum sampling regimen.

QUANTITATIVE ESTIMATION OF SPECIES ABUNDANCES, RICHNESS, AND RARITY

The composition of arbuscular fungi at a site can be determined by direct analysis of spores recovered from the soil, from trap cultures, or from a combination of both methods. Data from both techniques can be used to generate a species list for a sample location. Because of the effects of age, parasitism, chemical action in soil, and other factors, field-collected spores may intergrade so much morphologically that some attempt should be made to trap the fungi to induce production of healthy spores for accurate identification (Morton 1993). Once a researcher is somewhat familiar with the taxonomic composition of the fungal community, then field-collected specimens are much easier to identify and count. Clearly, the relative importance of species in the field generally is best obtained by direct observation of field samples. Trap cultures provide only an indirect measure of abundance, and measures of sporulation in them are fraught with potential error. For example, the following may occur: (1) the dominant fungal colonizers may have the highest inoculum potential and, therefore, will colonize more root and sporulate; (2) a less dominant fungus may be a more aggressive colonizer in the confines of the pot space and sporulate more than it would under field conditions; and (3) host, soil, and greenhouse conditions influence interactions among co-colonizing fungi.

The relative importance of individual species in an arbuscular fungal community can be estimated by measuring two variables: spore abundance (spores/100 cm³ of soil) and frequency of occurrence. Both variables can be used to estimate the importance of a species directly or as a basis for calculating other indices of importance, such as relative abundance, relative frequency, spore biovolume (volume of spores/100 cm³ of soil), and dominance (based on either abundance or frequency). Additional measures of importance can be calculated by combining those variables. For example, an importance value (IV) for each species can be calculated by summing frequency and abundance or summing those variables and spore volume (Koske 1987). Plotting abundance (or relative abundance) versus rank is less informative because of the great variation typically found in spore count data.

Spore volume (Dickman et al. 1984; Koske 1987) is the most easily measured approximation of the amount of host carbon transferred to a colonizing fungus, which converts it to cytoplasm, lipids, and cell-wall components. It is calculated by multiplying mean spore abundance by the mean spore volume in a sample. Volume is calculated by the following equations:

$$4/3 \, r^3 \tag{1}$$

and

$$(lw^2)/6 \tag{2}$$

These equations are for spores that are spherical (most species) or oblong (e.g., *Scutellospora erythropa*, *S. pellucida*), respectively, where r = spore radius, l = spore length, and w = spore width. A spore of *Gigaspora gigantea* that is 300 μm in diameter has a volume of 1.4×10^7 μm³, whereas a spore of *Glomus caledonium* that is 150 μm in diameter has a volume of 1.8×10^6 μm³. Spore volume is not often correlated with the amount of root colonized (niche occupation) nor is it consistent with season or fungal species (Gazey et al. 1992). In addition, volume estimates are based on mean diameters, and the range of diameters of spores of a species and their abundance distribution can vary considerably. Despite the drawbacks of spore volume measurements, they seem to provide the most biologically appropriate estimates of the carbon resources being transferred to coinhabiting fungal species.

Rank-IV plots appear to be useful for visualizing the contribution of each species in the fungal community to reproduction (Koske and Gemma 1992). As with many of the methods to assess species significance, such plots are borrowed from plant ecology (Whittaker 1975). They are constructed by calculating an IV for each species in the fungal community and then plotting each IV on a graph with a logarithmic scale on the Y-axis. "Rank" values are placed on the X-axis. The species with the highest IV would rank "1," the species with second-highest IV would rank "2," and so on. The initial slope of the plot is of special interest. A steep slope appears to be associated with habitats dominated by severe stress at the level of the plant community (drought, cold, reduced photosynthetic activity), whereas flatter slopes appear indicative of more equitable habitats in which competition is an important component in determining community structure (Whittaker 1975; Koske 1987).

Obviously, the method by which IV is calculated affects slope and the interpretation of community structure.

In some soils and in some seasons, dead intact spores can outnumber live spores (Lee and Koske 1994). For most studies, only the number of live spores is informative. Viability of spores is determined on the basis of healthy appearance (intact spore wall and internal flexible walls in the described color range, clear lipid globules in spore lumen). Vital stains have been used more recently to determine viability, but results are not reliable (Meier and Charvat 1993; Walley and Germida 1995).

Rare species make up large proportions of communities in both temperate (Koske 1987) and tropical regions (Johnson and Wedin 1997). They present two main problems. First, specimens may be too few to allow for accurate identification, unless they have clearly diagnostic characters that show little variation in response to biotic and abiotic agents. Many species in the genera *Glomus* and *Gigaspora* cannot be identified with fewer than 40–50 spores and then only if all diagnostic characters are intact and in good condition. Second, the large variation in spore counts tends to preclude any statistical comparisons among associated species. For example, 11 soil samples were collected as part of a study of fungi in sand dunes (R. E. Koske, unpublished data). Spores of *S. pellucida* were found in only one sample but were abundant. Those of *S. calospora* occurred in much lower numbers in five samples (Table 15.2). Based on average spore abundance, *S. pellucida* would be overwhelmingly dominant. Yet statistical tests revealed no significant difference in spore abundance between the two species. Frequency of occurrence may be the more informative parameter when dealing with rare species. The example in Table 15.2 illustrates the importance of combining abundance and frequency values when calculating an index of importance. In quantitative analysis of AM fungal communities, as with those of most other organisms, only the more abundant species can be measured accurately (Mueller-Dombois and Ellenberg 1974).

The number of described species is far less than the number of species predicted to exist globally (Morton et al. 1995). Therefore, exploration of any new site (and even old ones in different years) is likely to yield one or more undescribed species. The inclusion in a number of surveys of long lists of rare species, which are difficult to identify even with abundant specimens, suggests that many of the identifications are doubtful. It is far more valuable for investigators who are unsure of an identification to prepare a voucher; identify to genus; identify the isolate with a code number; and if space is available, provide a short summary description with photographs or illustrations.

QUALITATIVE FIELD SAMPLING

Qualitative sampling is appropriate when time is insufficient for the more extensive quantitative sampling procedures. Such surveys are useful because they provide information on local, regional, and worldwide distributions of arbuscular species. Nevertheless, frequency of occurrence is one of the most easily determined variables for arbuscular fungi, so that, with relatively little additional record keeping, a purely qualitative study can become quantitative. The principles of sampling and associated problems in dealing with microscopic spores with highly aggregated distributions remain the same with qualitative sampling, and the same methods described in "Sampling Procedures," earlier, are appropriate for these types of studies.

It is in the arena of qualitative measurements that greenhouse trap cultures can offer considerable insight into species composition as long as one or more members of the species assemblage (whether dominant or not) are not so aggressive that they colonize most roots and inhibit sporulation by associated species. If that occurs, then trap cultures significantly underestimate natural diversity. Fortunately, in more than 600 trap cultures established from a multitude of habitats, species diversity usually exceeds that found in one-time field samples (J. B. Morton, unpublished data). In arid or wet habitats, mycorrhizal colonization is often low for most, if not all, of the year and the inoculum potential of col-

TABLE 15.2

Abundance and Frequency of Occurrence of Two *Scutellospora* Species from a Sand Dune

Species	Spores/100 cm³ sample*	Mean spore abundance[†]	Frequency of occurrence (%)
S. pellucida	727, 0, 0, 0, 0, 0, 0, 0, 0, 0, 0	66	9
S. calospora	57, 23, 16, 8, 4, 0, 0, 0, 0, 0, 0	9.8	45

Data from R. E. Koske, unpublished data.

* N = 11 samples.

[†] Mean spore abundances do not differ between species as transformed (log [x + 1]) or untransformed data using a Student's *t*-test.

onizing arbuscular fungi is presumed to be low. Sporulation may not occur in trap cultures set up for the first time, although mycorrhizae are present, or it may occur in only one or two species. A second trap culture cycle, however, set up by reseeding the host plant into the first pot culture after host-plant shoots have been removed, can double or triple the number of species sporulating. A third cycle can lead to even higher fungal-species richness, although results vary considerably from site to site (Bever et al. 1995; Morton et al. 1995).

ORGANISM CHARACTERIZATION

SPECIES DESCRIPTIONS

Procedures

The first task in describing a species is to obtain healthy, fresh spores with all pertinent morphological characters present and intact, a rarity in field soils with a highly active soil microbiota. Spores showing any evidence of parasitism or degradation should be discarded. Because species differences in some genera, such as *Glomus*, are small, misidentification of degraded specimens is likely. Two approaches can be used to obtain enough spores for an accurate description: (1) samples can be collected from a site at different times or seasons but especially just after a period of sustained root growth in the plant community, and (2) trap cultures using field-collected soil, roots, or both can be established (see "Culture Protocols," earlier). The latter approach often is critical for providing experimental verification that the fungi sporulating form mycorrhizae, for obtaining spores with all diagnostic characters intact, and for producing inoculum for establishment of monospecific cultures.

After spores are extracted from the soil and cleaned (see "Collection and Isolation Protocols," earlier), all observable macromorphological and microphological features must be described accurately. Spore size and color can be determined from spores stored on filter paper or, preferably, in water where variation as a result of age, degradation, or parasitism is more discernible. Size is measured with an ocular micrometer, and reflected color of the specimen should be measured by comparison with a color chart at the same illumination, which is best accomplished with a dual-beam fiber-optic unit. INVAM has generated such a chart specifically for this purpose.

A minimum of 75 spores should be mounted in water and a permanent medium, such as PVLG (polyvinyl-alcohol-lactic acid-glycerol) or PVLG + Melzer's reagent (see "Methods for Archiving Specimens and Cultures,"

later in this chapter). One-third of the spores should be mounted whole, another third should be mounted and broken in PVLG and water, and the final third should be mounted and broken in a mixture of PVLG + Melzer's reagent (1:1 v/v). The number of slides generated depends on the number of spores placed together under one coverslip. Generally, no more than 5–10 spores should be mounted together when spores exceed 200 μm in diameter; as many as 20–30 spores can be mounted together when spores are less than 150 μm in diameter. We recommend fewer spores on more slides for the novice. Slides of whole spores should be examined the day they are made, but those of broken spores should be stored for at least 72 hours to clear spore contents and allow delineation of subcellular structures.

Most layers of spore cell walls are less than 10 μm thick, and layers of flexible inner walls can be less than 1 μm thick. Inner walls fold on each other, stick together or separate completely from the spore, or sometimes expand with pressure (Morton 1988). Layers of the spore wall vary in their continuation from the subtending hyphal wall (some are in both, others are unique to the spore wall). Description of hyphal wall structure and the transition region between hypha and spore is important. Recognition and description of these characters requires a microscope with excellent optics. Standard light microscopes may be adequate with excellent optics, but those equipped with a differential interference, contrast light system enhance the contrast between layer boundaries, which makes thin transparent structures easy to see.

Even if all of the previously mentioned procedures are followed scrupulously, many hurdles still must be overcome to arrive at an accurate species description. The problem is that 90% or more of the published descriptions are vague or incomplete to varying degrees, including many redescribed taxa. Type specimens often are in poor condition with missing, parasitized, or degraded characters; sometimes their morphology has changed as a result of effects of the preservative (Morton 1986). Schenck and Pérez (1990) published a synoptic key to mycorrhizal fungi, but their book is no longer in print. In the absence of more suitable keys, the best solution at this time is for the investigator to do the following:

1. Establish as close a match as possible between a specimen and a species description.
2. Note all shared characters and any characters that conflict with the species description.
3. Assign a species name equivocally to the specimen, and prepare a permanent voucher consisting of whole spores preserved in 0.05% sodium azide or other preservative of choice, broken spores mounted in PVLG

alone and PVLG mixed with Melzer's reagent (1:1 v/v), and black and white or color photographs and/or digital files of important characters *from fresh material.*

4. Contact an experienced taxonomist to obtain a verification, especially if the identification is important to an experiment. The taxonomist is likely to respond positively if the investigator has prepared vouchers of healthy material and made some attempt at diagnosis.

5. Seek specimens from living reference cultures of diagnosed species from repositories such as INVAM or BEG for comparison and verification. Photographs of reference accessions of cultured species are available on the INVAM Website.

6. When a specimen cannot be matched with a described species, designate it as undescribed and define it by an accession number.

7. If the organism has been established in culture, register it with INVAM or BEG. Because INVAM is a central repository, inoculum can be mailed to the collection and accessed for maintenance or increase.

Characters

Mycorrhizal structures differ only at the family level and higher, so they are not particularly informative in describing species. They are important, however, in providing experimental evidence of membership in Glomales and in defining variation within a suborder or a family. They should not be ignored, therefore, if a fungal isolate can be cultured.

Arbuscules, which clearly have been established as the pivotal site of communication between the host and fungus, provide important taxonomic characters (Smith and Gianinazzi-Pearson 1988). Each arbuscule is composed of fine, dichotomously branched hyphae that develop from a trunk hypha following its penetration of a host cortical cell wall. A unique interface forms between the arbuscule cell wall and the host plasmalemma (Gianinazzi 1991). In species of the suborder Glomineae (see Fig. 15.1), arbuscule branches appear to be incrementally thinner than the arbuscule trunk. Those of some *Glomus*, most *Acaulospora*, and *Entrophospora* species stain only faintly in roots and thus may be hard to detect. In species of the suborder Gigasporineae (Fig. 15.1), the first-order branches decrease abruptly in width from the much thicker arbuscule trunk and typically stain dark in cleared roots. Staining of those structures varies considerably among families as follows (darkest to faintest): Gigasporaceae, Glomaceae, and Acaulosporaceae. The mycorrhizal structures staining very faintly or not at all are in Archaeosporaceae and Paraglomaceae (Morton and Redecker 2001).

Intraradical hyphae vary considerably in morphology and architecture within families of Glomales (Abbott and Gazey 1994). In *Glomus* species, relatively straight hyphae ramify in parallel along the long axis of the root cortex (if root structure permits), often producing "H" branches that result in simultaneous growth in two directions (Abbott 1982). In *Acaulospora* species, the hyphae are more irregularly branched and looped; they are thin-walled, and their intensities of staining are highly variable. *Gigaspora* and *Scutellospora* species have wide hyphae with inflated and constricted regions that produce knobs and projections; they also form dense loops or coils in entry cells and in many cortical cells.

Intraradical vesicles are produced only by taxa in the suborder Glomineae (Fig. 15.1). These thin-walled saclike intercalary or terminal structures vary in shape by genus: globose to ellipsoidal in *Glomus* and irregularly shaped to lobed in many species of *Acaulospora* and *Entrophospora*. Vesicles can develop concurrently with arbuscule formation, but they often increase in abundance as arbuscular colonization declines later in mycorrhizal development. As colonization progresses in some species of *Glomus*, the vesicles become thick-walled and cannot be distinguished morphologically from extraradical spores. The extent to which thin-walled vesicles are infectious remains unclear. Biermann and Linderman (1983) reported that intraradical vesicles are infective, but their experimental organism was *Glomus intraradices*, a fungus that typically transforms most vesicles into intraradical spores (J. B. Morton, unpublished data).

Auxiliary cells are formed only by taxa in Gigasporineae (Fig. 15.1). The cells are thin-walled and form as clusters on the ends of branching external hyphae. They have noticeable spiny ornamentations in the genus *Gigaspora* and a knobby to almost smooth surface in the genus *Scutellospora*. Auxiliary cells often form on germination tubes from spores prior to mycorrhiza formation and then develop concurrently with early mycorrhizal development. They appear to decline in abundance as sporulation progresses, at least in pot cultures (Franke and Morton 1994). Evidence to date suggests that auxiliary cells are not infective propagules (Biermann and Linderman 1983).

Extraradical hyphae vary in form and function, with the former appearing to correlate to some extent with the latter (Friese and Allen 1991). Fine "absorptive" hyphae grow beyond the root zone for uptake of nutrients and as a pipeline for translocation of nutrients back to host roots; coarser "runner" hyphae extend along roots and initiate new sites of colonization to coincide with additional root growth. The extent to which external mycelium is infective varies with taxonomic group. Biermann and Linderman (1983) were unable to obtain new mycorrhizae from hyphae or mycorrhizal roots of fungi in *Gigaspora*, a result corroborated in numerous pot cultures that lose all infectivity when mycorrhizal roots are present and all spores have died (J. B. Morton, unpublished data). However, hyphal fragments of *Glomus* are highly infective. Length and architecture (branch patterns) of hyphae can be highly variable among fungal isolates (Abbot and Gazey 1994) and possibly even within a fungus on one plant. The length of the external mycelium appears to be inversely correlated with host root length (Miller et al. 1995).

Mode of formation of spores differs among families and genera. Most species-level diagnostic characters are found in phenotypic properties of spore wall layers (including size and shape, which are manifestations of spore wall growth) (Morton et al. 1995). In general, spores are produced asexually, occurring singly on a sporogenous hypha, in aggregates, or within more highly organized sporocarps. Spores are multinucleate, although the number of nuclei varies considerably among taxa. Direct counts range from 2600 to 3850 nuclei per spore in species of *Gigaspora* (Cooke et al. 1987). Spores germinate in various ways to infect and colonize new roots (Morton 1988; Morton and Benny 1990).

MICROHABITAT DESCRIPTIONS

All fungal hyphae are in direct contact with the soil, so soil factors have a strong impact on germination of spores or growth of other infective propagules (hyphal fragments or hyphae from colonized roots) during the mycorrhizal establishment phase; on growth of runner hyphae, which are important in mycorrhizal spread; and on growth and architecture of absorptive hyphae, which provide for nutrient uptake and other plant benefits. As many soil parameters as possible should be measured or described because any one or all may help to explain mycorrhizal form or function. Chemical and physical properties of soil, especially pH and major cation concentrations, are most important (see Black et al. 1965 for general methods of soil characterization).

Host species is also important for a number of reasons. First, knowledge of the host can provide relevant information about the fungal–host association, such as its potential duration (annual versus perennial plant), relative degree of mycorrhizal dependency, tolerance of the host to soil conditions and climate, and root architecture. Second, anatomical features of the host root have a significant influence on the appearance of a mycorrhiza (Brundrett and Kendrick 1990). Third, knowledge of the sample host and its associated plant species (i.e., of plant-community structure at the site) and their densi-

ties around the sample host provides clues to the extent, if any, to which colonizing fungal species may be sporulating or may be affecting density-dependent interactions in the community (Koide et al. 1994). For example, dense pastures or grasslands typically do not contain the abundance of spores or species diversity found in more isolated sand dune habitats.

Climatological data also are important. Temperature, light, elevation, rainfall, and seasonal patterns all have an impact on mycorrhizal development and life-history characters of fungal symbionts because they influence soil conditions, host physiology and phenology, spore germination, and hyphal extension. If possible, microbiological assays for general microbial activity should be carried out (Sylvia 1994), and the taxonomic composition of the arbuscular fungal community associated with organisms of interest should be described.

METHODS FOR ARCHIVING SPECIMENS AND CULTURES

Currently, species diagnoses are not decisive for many taxa of arbuscular fungi because spores from living pot cultures do not correspond well with many published species descriptions (Walker 1992; Morton 1993). Many early taxonomic investigations were carried out on field-collected (and sometimes obviously parasitized) specimens bearing only vague resemblance to healthy spores. Type specimens carefully prepared from healthy, intact specimens help to resolve some of these problems. Unfortunately, many older type specimens were preserved in lactophenol, which we now know causes irreversible loss or degradation of characters and inhibition of histochemical reactions with long-term storage (Morton 1993). Therefore, more than one type of specimen is needed for vouchers or herbarium accessions.

LIVE SPECIMENS

Viable spores can be extracted from soil, washed three to five times, and then placed on filter paper and dried slowly. The filter paper then can be folded and stored in a Petri dish or glassine envelope. Spores often collapse with desiccation, but they return to their original shape with added water. Length of viability of spores stored in this way has not been established. Sporocarps can be collected individually or in soil and placed in plastic vials and dried slowly. They retain their morphological integrity, but viability often is destroyed with desiccation. Both spores and sporocarps may be stored in dry soil at

4°C or at room temperature for at least 1 or 2 years, although viability differs considerably among species (Jarfster and Sylvia 1993). Fungal propagules (spores, sporocarps, roots) established in pot cultures also can be stored in liquid nitrogen (Douds and Schenck 1990), but success is highly variable (J. B. Morton, unpublished data). It is critical that pot contents be dried slowly and without disturbance before being transferred to cryovials. Propagules exposed to any liquid (e.g., from extraction or storage in water) do not survive freezing.

PRESERVED SPECIMENS

Whole spores can be stored in plastic or glass vials (usually 2-ml volume) containing 5% glutaraldehyde, 3.5% formalin, or 0.05% sodium azide. Sodium azide causes the fewest morphological changes over time, and 5% glutaraldehyde causes the greatest changes, but all of those preservatives fail to prevent deterioration of some morphological features (J. B. Morton, unpublished data). Other common preservatives (Appendix II) such as FAA (formalin + acetic acid + alcohol) and lactophenol (lactic acid + phenol) have been used extensively in the past, but they cause major changes in the subcellular structure of spores and should be avoided (Morton 1993). Whole spores also can be mounted permanently in PVLG-based (a mix of polyvinyl alcohol, lactic acid, glycerol, and water) media on glass slides (Appendix II). Spore contents become plasmolyzed, however, and spore size and shape can be distorted. Spore color also changes, most noticeably in colorless spores that become yellow or brown.

Some spores should be broken in a mountant on glass slides to obtain a physical record of subcellular structure and organization. Some mountants are temporary, such as water or glycerin, and provide a view of spore characters as they exist in nature. Clearing of spore contents requires a 5-day (or longer) incubation period. Even then, details often are obscured by overlapping flexible inner walls. Slides prepared with PVLG mountant can be stored for 5 years or longer without desiccation or decomposition. Histochemical reactions in spores contribute significantly to species diagnosis, the most informative being the staining response to Melzer's reagent (Appendix II). For permanent mounts and better clearing of spore contents, Melzer's reagent can be mixed 1:1 (v/v) with PVLG. Spores in the latter mixture are slightly easier to break and are less prone to "slide out" from under the coverslip when pressure is applied.

Broken spores should be mounted permanently in both PVLG and the PVLG-Melzer's reagent mixture after being subjected to different degrees of pressure on the coverslip (to expose all subcellular structures) using a blunt-tipped instrument. Both mountants can be placed on the same slide, with spores placed in each. The mountants are viscous, so some care must be taken to prevent spores (especially small ones) from sliding outside the coverslip edge. After the mountant is applied to the slide with dropper bottle or pipette, spores are added (with tweezers, capillary tube, or pipette) and allowed to set for 5–10 minutes before the coverslip is added. The surface layer of the mountant dries and creates enough surface tension to retard flow of the mountant when pressure is applied to the coverslip. Individual spores should be broken to different degrees to expose inner flexible walls when present. That is best accomplished by applying localized pressure on the coverslip over a spore with a narrow- but blunt-tipped instrument such as a pencil.

Roots stained with 0.05% trypan blue, chlorazol black E, or acid fuchsin can be mounted permanently in PVLG-based media (Appendix II) to make a voucher of mycorrhizae. Roots should be added to the slide first, drops of PVLG should be added at one end, and the coverslip then should be positioned at the edge of the drop and lowered slowly. That will allow the PVLG to spread without causing excessive migration of root fragments. After 30 minutes, pressure is applied to individual roots to "spread" cortical tissue and allow visualization of colonization in different focal planes. This process is most important if the goal is to quantify the relative proportions of arbuscules, vesicles, and intraradical mycelium (Toth and Toth 1982).

For critical specimens, such as type material whose morphological changes in preservative or mountants must be traceable, whole spores, broken spores, and mycorrhizal structure should be photographed in color or black and white and either developed as slides or prints or archived as digital images on permanent media such as compact discs. Such vouchers are best prepared within 2 weeks of specimen preservation.

FUTURE NEEDS

Much about the biology (ecology, physiology, molecular biology, etc.) of AM fungi has yet to be discovered; the main obstacle to our obtaining such information is our inability to culture the fungi independent of a host plant. Sampling techniques need to be refined to account for the patchy nature of spore distributions. Problems such as autocorrelation of samples must be solved. Trap-culture methodologies should be adopted as much as possible to ensure that nonsporulating species in the field are recovered. Even when both field extractions and trap-culturing techniques are combined, the choice of method can influence results greatly (Bever et al. 1995). With standardization of techniques, cross-site comparisons can be made more easily.

Realistic interpretation of data is crucial to elucidating mechanisms causing different patterns of diversity. As mentioned earlier, species identifications are made exclusively from spores. Even with their limitations, spore counts can provide valuable information given the appropriate questions. Particular attention needs to be given to issues such as qualitative versus quantitative sampling and scale (Janos 1991; O'Neill et al. 1991). Basic biological questions need to be answered to fully understand the implications of spore data. Sporulation appears to be a result of the fungus thriving (achieving a threshold level of biomass) together with the degree of partitioning of carbon between vegetative structures and spore biomass. The response differs from that of other large fungal groups such as the Ascomycota, in which sporulation is more commonly induced by stress. One wonders whether those fungi that are most fecund (however measured) obtain the greatest amount of resources from the host and confer the greatest benefit to the host plant (i.e., are the most effective mutualists). Those questions remain unanswered for the most part, and their resolution will require more than one investigative approach.

The relationship between taxonomic diversity and functional diversity has received little attention. It is well established that glomalean fungi can be grouped consistently into distinct species and higher taxa using morphological and some biochemical characters, despite clonal reproduction. However, do they also share genealogically conserved functional characters that can provide additional grouping criteria? It is not enough to show that variation exists among isolates of a species in their ability to promote plant growth (Sylvia et al. 1993) or that functional diversity exists in external hyphae (Friese and Allen 1991) and other fungal parts. To date, no data have been collected to determine the extent to which different taxa share these properties or are affected by a cascade of other effects such as soil aggregation and biogeochemical cycling (Miller and Jastrow 1994). Molecular techniques used to detect individual species within roots (Clapp et al. 1995; Di Bonito et al. 1995) and perhaps even in bulk soil have the potential to provide valuable new tools for defining variation at the genetic level. Efficacy of these procedures will depend on intraorganismal heterogeneity of gene sequences within the multitude of mobile nuclei present in single cells, as well as the rate of fixation and frequency distribution of each genotype. Equally important are conserved products of gene expression, which can be especially valuable if they are linked to symbiotic functions.

Researchers interested in environmental problems should strive to identify indicator taxa that are sensitive to changes in edaphic and environmental conditions. Those fungi may assist researchers in monitoring ecosystem health and at the same time provide additional tools for minimizing negative impacts of human actions.

ACKNOWLEDGMENTS. Aspects of this work were supported by funds from National Science Foundation grant DBI-9600699 to J. B. Morton.

16

YEASTS

CLETUS P. KURTZMAN AND JACK W. FELL

Yeasts are a polyphyletic group of ascomycetous and basidiomycetous fungi characterized by having a unicellular growth phase and sexual stages that are not enclosed in fruiting bodies. Approximately 100 genera of yeasts comprising more than 700 species have been described (Kurtzman and Fell 1998). Current evidence suggests, however, that those species represent less than 1% of the species that occur in nature.

Yeasts are ubiquitous, and we will discuss their occurrence in and isolation from terrestrial, aquatic, and marine environments. Most yeasts are saprotrophs that assimilate plant- or animal-derived organic compounds. Although they are usually decomposers, some species are pathogens of plants and animals. Because yeasts are metabolically diverse, they are used in many industrial processes, such as the production of ethanol (beverage, industrial, and fuel), vitamins, organic acids, carotenoids, and enzymes. A large number of yeasts catabolize benzene compounds and may prove useful for cleaning up spills of industrial chemicals as well as for biosynthesizing new compounds useful to humans (Middelhoven 1993). In particular, recent advances in molecular techniques make yeasts highly attractive for synthesis of pharmacologically active proteins and other compounds.

TAXONOMIC STATUS

One of the major impediments to determining the distributions of yeasts in nature and their functions in ecosystems is the poor taxonomic resolution provided by currently used methods of identification. Often, species and genera are defined on the basis of phenotypic characters such as cell morphology and responses on various growth tests. In recent years, however, research with genetic crosses, as well as molecular comparisons, have shown that many of the phenotypic characters considered to be taxonomically definitive vary among strains of the same species. As a consequence, divergent strains of known species may be mistaken for novel taxa, whereas new species may go unrecognized. Currently, in several laboratories, DNA nucleotide sequences of unknown strains of yeasts are being compared with those of authentic reference strains for species identification. It is clear that molecular genetic methods must be used for the identification of yeasts if we are to obtain an accurate appraisal of their biodiversity. However, more rapid molecular methods of identification will be needed if large numbers of isolates are to be examined.

The definition of *yeast* has changed markedly in the last few decades. *Saccharomyces cerevisiae* and related ascomycetous species appeared to represent the "typical" yeast, until Banno (1967) demonstrated a basidiomycetous life cycle for a species of *Rhodotorula*. That discovery was followed quickly by the demonstration of basidiomycetous life cycles in numerous other species (Fell et al. 1969, 1973; Kwon-Chung 1975). Studies of molecular evolution have further broadened and refined the definition of the yeasts. Analyses of ribosomal RNA (rRNA) and ribosomal DNA (rDNA) have shown that the ascomycetous yeasts, which now include most of the hemiascomycetes such as *Cephaloascus*, *Ascoidea*, and *Dipodascus*, form a sister group to the euascomycetes rather than a more primitive group (Kurtzman 1994; Kurtzman and Robnett 1998, and references therein). Furthermore, *Schizosaccharomyces*, *Taphrina*, *Protomyces*, *Pneumocystis*, and *Saitoella* comprise a divergent group that is phylogenetically basal to both the ascomycetous yeasts and the euascomycetes (Kurtzman 1993; Nishida and Sugiyama 1993; Sugiyama 1998 and references therein). According to rDNA-based phylogenies, the basidiomycetous yeasts are a distinct group that includes such common genera as *Rhodosporidium* and *Filobasidiella*, as well as the morphologically more complex *Tremella*. Molecular systematics of the yeasts and related filamentous fungi (Fell et al. 1992, 2000; Suh and Sugiyama 1993; Swann and Taylor 1993) indicate a phylogenetic distribution of yeasts, as shown in Table 16.1.

TABLE 16.1

Taxonomic Orders for Classification of Teleomorph Genera of Yeasts*

Ascomycetous yeasts
Saccharomycetales
Schizosaccharomycetales
Taphrinales
Pneumocystidiales
Protomycetales

Basidiomycetous yeasts
Hymenomycetes
  Tremellales
  Trichosporonales
  Filobasidiales
  Cystofilobasidiales
Urediniomycetes[†]
  Microbotryales
  Sporidiales
  Erythrobasidiales
  Agaricostibiales
Ustilaginomycete
  Ustilaginales
  Microstromatales
  Malasseziales

---

* Primary references include Kurtzman and Fell (1998) and Barnett et al. (2000).
[†] Orders in the Urediniomycetes require redefinition for inclusion of yeasts.

DISTRIBUTION

As noted earlier, yeasts commonly are distributed throughout the biosphere (Phaff et al. 1966). Many yeasts are isolated from soils. Phaff and Starmer (1987), for example, noted that occasional soil populations of 10^5–10^6 colony-forming units (cells) per gram of soil suggest the presence of actively growing cells. Some species, including *Debaryomyces* (*Schwanniomyces*) *occidentalis*, *Lipomyces* species, *Schizoblastosporion starkeyi-henricii*, and certain *Cryptococcus* species, are isolated exclusively from soils. Population levels of aquatic yeasts are usually highest in fresh waters and decrease in marine waters with increased depth and increased distance from land (Hagler and Ahearn 1987). Populations in the open ocean occur at densities often as low as 10 cells/g of sample but may increase to 10^3 cells/g in waters associated with plankton blooms, current boundaries, surface slicks, thermoclines, or pollutants (Fell 1976).

Despite these observations, the majority of yeast species are collected from fallen plant materials and other organic matter (Phaff and Starmer 1987). The highest densities of yeasts are usually associated with concentrations of assimilatable sugars and other carbon sources. Leaf surface tissues and plant exudates commonly sustain large numbers of yeasts. di Menna (1959) reported 10^5–10^7 viable cells/g of fresh foliage. Flowers and decaying fruits and other plant material also support a wide spectrum of species in high numbers; densities as great as 10^6 cells/g of tissue were reported from decaying cladodes of some cacti (Phaff and Starmer 1987). Insects represent rich sources of yeasts, especially wood-borers and species of *Drosophila*. Phaff and Starmer (1987) summarized the yeasts found with different types of insects.

Reference cultures are essential for correct identification of yeast species. The primary collections containing reference strains for yeasts are listed in Table 16.2.

PROTOCOLS FOR INVENTORIES

COLLECTING SAMPLES

The majority of yeasts are obligate aerobes and therefore require oxygen for growth and reproduction; consequentially, yeasts usually are found in the upper layers of soil and sediment. Exceptions are yeasts preserved in deep layers of ice. Soil and sediment samples can be collected manually from terrestrial and shallow-water environments using sterile jars, vials, reclosable plastic bags with zippers, and coring devices (Parkinson et al. 1971). Deep-water sediments can be collected remotely using

TABLE 16.2
Major Yeast Culture Collections*

Italy: Industrial Yeasts Collection (DBVPG), Dipartimento di
Biologia Vegetale, Universita di Perugia
Japan: (i) Institute for Fermentation (IFO); (ii) Japan Collection of
Microorganisms (JCM)
The Netherlands: Centraalbureau voor Schimmelcultures (CBS), Yeast
Division
Peoples Republic of China: Center for Collection of General
Microbiological Cultures (CCGMC), Institute of Microbiology,
Academia Sinica
Portugal: Portuguese Yeast Culture Collection (PYCC)
Republic of Germany: Deutsche Sammlung von Mikroorganismen
und Zellkulturen GmbH (DSMZ)
Russia: Department of Type Cultures of Microorganisms (VKM),
Institute of Biochemistry and Physiology of Microorganisms,
Russian Academy of Sciences
Slovakia: (i) Slovak Collection of Yeasts (CCY), Institute of
Chemistry, Slovak Academy of Sciences; (ii) Research Institute for
Viticulture and Enology (RIVE)
United Kingdom: National Collection of Yeast Cultures (NCYC),
AFRC Food Research Institute
United States: (i) Agricultural Research Service Culture Collection
(NRRL), National Center for Agricultural Utilization Research;
(ii) American Type Culture Collection (ATCC); (iii) Culture
Collection, Department of Food Science and Technology (FST-
UCD), University of California—Davis

---

* See Appendix III for contact information.

---

grabs, gravity and piston cores, or submersibles. Deep-sea water samples can be collected through the water column with samplers such as the Sterile Bag Sampler (General Oceanics; Appendix IV; Nisken 1962). Temperature and salinity must be obtained simultaneously with the yeast collection to determine the specific water mass characteristics and depths of the sample collection. A 50-ml sample is usually adequate, and multiple samples can be taken from a single location or grab. Minimal water should be included with the sample because water often promotes an anaerobic condition deleterious to cell survival. Cores can be sectioned, and samples can be taken from recorded depths along the core.

Samples should be processed immediately, if possible, or held on ice until processing can be initiated. Maintenance at low temperature is particularly critical for samples from low-temperature habitats. Sample containers should not be air tight to avoid formation of anaerobic conditions, although the containers should be closed adequately to eliminate contamination.

Samples should be processed in a contaminant-free environment, preferably a culture room or hood, although this is not always feasible in the field. Air drafts are the primary source of contaminants, and inexpensive, temporary hoods can be fabricated or purchased for the field. Utensils can be sterilized by flaming with alcohol,

and bench workspace should be kept clean with a disinfectant.

ISOLATION STRATEGIES

Direct Plating

Isolation procedures include direct plating, dilution, and enrichment. In direct sampling, soil particles are placed directly on an agar medium and allowed to incubate for at least 24 hours. Particles are then examined microscopically *in situ* under low power magnification to locate yeast growth at the particle/medium interface. Suspected yeast colonies are transferred to a microscope slide for closer inspection. Confirmed yeast growths are then transferred from the isolation plates or slides to a growth medium. Streak cultures should be made to purify the culture. Colonies resulting from single cells can be transferred to slant culture tubes for further study.

Water collections can be plated directly, or they can be passed through 0.45-μm nitrocellulose filters. The filter is placed face-up on solid nutrient agar containing bacterial inhibitors. The quantity of water filtered depends on population density and may range from a few milliliters to 2–3 liters. Culture plates are incubated, and after appropriate periods of development, the colonies are enumerated and isolated (Fell 1976).

Soil Dilution

Soil dilution procedures involve thoroughly mixing a known quantity of soil with a given volume of sterile water (sea water in the case of marine samples) in a test tube and then making a dilution series. At each successive step of the series, the soil is maintained in an increasingly dilute aqueous suspension. The suspensions are spread onto isolation media, and colonies are recorded and enumerated as described in the section on "Suspension Plating" in the Appendix "Techniques for Isolation of Soil Fungi," at the end of Chapter 13.

Enrichment

Enrichment procedures consist of placing soil or other samples in flasks with a nutrient medium tailored to the needs or preferences of the specific yeast or group of yeasts being sought and reflecting the types of substrata on which the target yeast grow. The medium is intended to favor the growth of the target species over other species. The composition of the growth medium is customized by adding specific carbon and nitrogen compounds, vitamins, or other growth factors or by manipulating substratum levels (particularly for osmophilic species), salinity (for marine species), and so forth. For example, *Cyniclomyces guttulatus* requires a medium containing certain amino acids as well as an incubation temperature of 35–40°C and an elevated level of atmospheric CO_2. *Rhodosporidium lusitaniae* can be isolated from forest soil with media whose sole carbon source consists of lignin-related phenolic compounds.

The preparations are incubated for 12–24 hours (or more) while being shaken continuously. The liquid is sampled by streaking or spreading the enrichment suspension onto isolation plates. Shake culture usually causes the filamentous fungi to grow in pellets, whereas the yeasts remain free cells in the liquid. Such techniques are highly selective and therefore cannot be used to enumerate natural populations.

Isolation and incubation procedures should be designed to mimic ambient environmental conditions. For example, all samples from polar and other cold regions should be kept cold (ca. 5°C) during incubation and when being examined. In contrast, when sampling for *Cyniclomyces guttulatus*, a yeast common in rabbit feces, an incubation temperature of 35–40°C and an isolation atmosphere high in CO_2 are required.

Problems

Overgrowth of yeast cultures by filamentous fungi is a constant problem. Shake culture (see "Enrichment," earlier in this chapter) provides a partial solution by causing most filamentous fungi to grow in pellets. Another partial solution is to incubate the samples at temperatures lower than ambient temperatures of the natural environmental. For samples from temperate and tropical areas, a temperature of 12°C usually slows the growth rate of filamentous fungi. Culture plates must be examined daily so that yeast strains can be isolated as soon as they appear and before they are overgrown by filamentous fungi.

Bacterial contamination can be reduced by lowering the pH of the isolation medium. Most yeasts are tolerant of acidity in the 3.5–3.8 pH range, but some bacteria also grow well in that range. Adding broad-spectrum antibiotics that do not affect yeasts to the growth medium is usually more successful than pH manipulation at inhibiting bacterial growth. Chloramphenicol or a combination of penicillin and streptomycin are particularly effective (see "Antibacterial Antibiotics," in Appendix II).

ISOLATION MEDIA

A variety of media for the isolation of yeasts has been described (Yarrow 1998). Yeast Isolation Medium (PYG;

Appendix II), for example, commonly is used. An alternative is Wickerham's YM medium (Appendix II). Various carbon sources can be substituted for glucose. Fresh water or filtered sea water can be used with either medium. For oceanic sampling, sea water at 37 ‰ is used, whereas brackish water at 15 ‰ is used for samples from inshore and estuarine waters, whose salinity is reduced. Media are solidified by the addition of 2–3% agar. Antibiotics can be added to the medium (see "Antibiotics," Appendix II). Chloramphenicol can be added at a concentration of 200 mg/liter of medium (0.02%) prior to autoclaving. An alternative combination of penicillin G and streptomycin sulfate, each at a concentration of 150–500 mg/liter, can be added dry to autoclaved, cooled medium.

ABUNDANCE AND RELATIVE IMPORTANCE

When selecting collection methods, investigators must take into account the usually patchy distribution of organisms in nature. Single-celled organisms, such as yeasts, are no exception; they aggregate in both soils and sediments. In low-nutrient environments, cells may lie dormant in a low metabolic state and then grow and reproduce rapidly when organic particles or solutes are encountered or introduced into the habitat. Introduced species may or may not be able to survive in soils. Therefore, results from a sample taken at one time and place can be extrapolated to other times and locations (even a few centimeters away) only with caution.

Investigators also must be concerned with sampling all the species that may be present at a particular site. For one thing, all isolation media are selective and artificial. Strains that are prevalent in soils may not grow, let alone reproduce, when removed from their natural substrata. In addition, certain dominant "weed" species rapidly may overgrow other strains and species on artificial media. Another limitation arises from the maximum number of cells (approximately 300) that can be accommodated on a single plate. If rare species represent 1% or less of a yeast community, they may not be sampled or they likely will be overgrown by weed species and go undetected. Finally, cell numbers represent the standing crop of a given yeast population and do not reflect factors such as turnover rates, hyphal fragmentation, spore release, or rates of consumption by various invertebrates. As a result, analyses of community structure may not portray natural population accurately.

Dealing with the preceding concerns requires careful consideration of environmental conditions and substrata. A variety of media and incubation conditions should be used, followed by close and repeated inspection of the culture plates to ensure that representatives of all species have been identified. Multiple collections at each site are required to delineate representative spatial and temporal distribution patterns and community structure. Collection protocols must be designed in conjunction with investigators from other biological disciplines so that an optimal understanding of interactions with other organisms can be achieved. Environmental data collected should be sufficient to provide an understanding of the ecology of the site and for future comparisons in light of current and often rapid human-generated environmental degradation. A global positioning system should be used to locate a site accurately, and climate, soil, vegetation, and other environmental conditions should be described in detail.

CHARACTERIZATION OF ORGANISMS

The standard description of a yeast consists of the results of tests that measure fermentation rates on seven to eight different sugars; growth rates on approximately 40 different carbon sources and one or two nitrogen compounds, such as nitrate and lysine; and ability to grow at various temperatures (Kurtzman and Fell 1998; Barnett et al. 2000). In addition, descriptions of the vegetative growth and any sexual state located after a search are included. The diazonium-blue-b (DBB) test usually discriminates ascomycetes from basidiomycetes (see "Distinguishing Filamentous Ascomycetes from Basidiomycetes" under "Characterization of Nonsporulating Isolates," in the Appendix at the end of Chapter 13). Despite all of this work, isolates often are misidentified. Because of that, many yeast systematists have begun to rely on molecular characterizations for accurate species identification. Molecular methods are discussed elsewhere in this volume (see Chapter 6), but two commonly used methods are nuclear-DNA reassociation and comparison of nucleotide sequences from the 5′-end of large (26S) rRNA and rDNA (Kurtzman and Robnett 1998; Fell et al. 2000).

Molecular comparisons provide accurate species identifications as well as a means for predicting phylogenetic relationships among species. A database of 26S 5′-end base sequences is adequate for the identification of most species. Additional sequences are needed to predict more distant relationships. Sequence determinations can be made rapidly on an automated sequencer, and the methods are quite straightforward. Even so, new technologies are likely to expedite present methods. For example, the length of short (approximately 600–1000) nucleotide sequences can be determined with great precision on a gene scanner. If the lengths and/or nucleotide sequences of a particular region were to vary in a large numbers of isolates, the determination could serve an exclusionary function. Species- or genus-specific probes might prove useful for assessing diversity directly on microscope slides. Considering the inadequacy of conventional taxonomic characters, molecular traits must be used.

Key organisms, particularly undescribed species, should be preserved following a biodiversity study. If identifications are accurate, fewer isolates of known species need to be maintained. If conventional taxonomic methods are used for identification, however, more strains must be preserved because of likely misidentifications. Such misidentification may prevent recognition of new taxa or, alternately, lead to interpretation of divergent strains as new taxa, both of which can result in inaccurate estimates of the biodiversity of a particular habitat. Isolates that are suspected to represent undescribed species may be deposited, with the permission of the curator, in culture collections such as those listed in Table 16.2. Accurate identifications during initial studies reduce the workload that such studies impose on culture collections.

INTEGRATION OF INFORMATION

Investigators should construct a computer database that includes general phenotypic characters; photographs of organisms (and collecting sites); collection data; molecular sequences; and information on observed sexual states and habitat. The database needs to be constructed so that authorized investigators can update files with new information, while allowing the general scientific public access to all data currently available. With the current availability of the GenBank (Appendix III) yeast 26S D1/D2 database, yeasts can be identified by sequence analysis and a BLAST search. That is an effective method and a necessary step for the discovery of new species. Despite the accuracy, sequence analysis is too slow and expensive for the identification of large numbers of isolates. Large numbers of isolates can be screened more quickly with species-specific PCR (polymerase chain reaction) primers (Fell 1995; Haynes et al. 1995; Mannarelli and Kurtzman 1998). Through the use of

GenBank data, primers can be designed for any known species. However, the need to coordinate PCR stringency conditions for a variety of primers often places unrealistic requirements on the design of a multiple assay for simultaneous identification of several species or strains. Hybridization assays, with high-density placement of oligonucleotide probes, provide a more versatile approach.

17

FUNGICOLOUS FUNGI

WALTER GAMS, PAUL DIEDERICH, AND KADRI PÕLDMAA

The term *fungicolous fungi* refers to species of fungi that consistently are associated with other fungi, including the lichenicolous fungi that consistently grow on lichens. The term is used even when the biological nature of the association and its trophic relationship are obscure (Jeffries 1995). Hawksworth (1981b) and others have spoken, for example, of "fungi growing on other fungi as parasites ('mycoparasites'), commensals, or saprobes. . . . The fungi may be 'necrotrophic' (destructive), or 'biotrophic' (forming balanced relationships)" (Hawksworth et al. 1995:172). Gilman and Tiffany (1952), Barnett (1963, 1964), and many others used the term for the totality of fungi associated with other fungi. However, fungicolous fungi (or mycophilic fungi) also have been restricted to those growing on sporocarps of other fungi or slime molds. Here we use the term sporocarp-inhabiting fungi (SCIF) to refer to the latter group.

Fungicolous fungi have been known for centuries (Ainsworth 1976). Various aspects of fungicolous associations and mycoparasitism have been reviewed many times in both the older and more recent literature (e.g., Buller 1924; Weindling 1938; Barnett 1963, 1964; Boosalis 1964; Madelin 1968; Barnett and Binder 1973;

Cooke 1977; Lumsden 1981; Baker 1987; Jeffries and Young 1994; Jeffries 1995). Jeffries and Young (1994) provided a highly readable survey of the general basis of mycoparasitism, including numerous examples of fungicolous associations categorized mainly according to the type of host–parasite interface, its physiological and ecological characteristics, and its applications in biocontrol. Since the 1980s, the taxonomy of fungicolous fungi has progressed considerably, particularly with many discoveries of anamorph–teleomorph connections in the Hypocreales, as well as discoveries of new mycoparasitic heterobasidiomycetes. Regional inventories mostly have been confined to taxa of SCIF (e.g., Helfer 1991). Besides published information, this chapter also refers to unpublished data available in the files of the Centraalbureau voor Schimmelcultures in Utrecht, The Netherlands (Anonymous 2001) with the designation "CBS, unpublished data."

Fungicolous and lichenicolous fungi are widespread in nature. Rudakov (1978), in his survey of fungicolous fungi, counted about 1700 (nonlichenicolous) species. He also inventoried fungicolous fungi (Rudakov 1981) occurring in the former Soviet Union, but many of his identifications need to be revised. In an exhaustive revision of the conidial fungicolous fungi, Hawksworth (1979a, 1981a) reported 1100 species on approximately 2500 species of host fungi (including lichens). Some 550 species of lichenicolous fungi were recognized in 1979, and their number now probably is more than 1500.

In this chapter, we do not cover the mere decomposers of fungi, and we ignore most of the unspecific colonizers of decaying substrata. Therefore, our admittedly incomplete estimates of numbers of fungicolous species are in some instances lower than those reported by Hawksworth (1981a) and Rudakov (1978). However, we incorporate new data that have become available only during the last two decades.

Wherever fungi are found, fungicolous fungi also are found. Fungicolous fungi are represented in many ecological categories, and the fungicolous mode of life has numerous variations. Those fungi play an important role in ecosystems through their detoxification of substrata not otherwise accessible to decomposition. In addition, many fungicolous fungi are economically important as biocontrol agents of plant-pathogenic fungi or as parasites that damage crops of edible fungi.

TYPES OF FUNGICOLOUS ASSOCIATIONS

Interspecific relationships between fungicolous fungi and their hosts can be characterized broadly as neutralistic,

mutualistic, or antagonistic/competitive (Cooke and Rayner 1984). Different kinds of fungicolous associations are difficult to distinguish in the field, but mycoparasitism can be inferred when a colonized fungus shows signs of deformation. The term *mycoparasite* (Butler 1957) applies strictly to relationships in which one living fungus serves as a nutrient source for another, although this kind of nutrient flow rarely has been demonstrated (Jeffries 1995). Within that group, necrotrophic and biotrophic mycoparasites generally are distinguished from each other (Barnett 1963; Boosalis 1964; Barnett and Binder 1973; and many others). Biotrophs live in a balanced relationship with a living host, without causing it immediate, overt harm (reviewed by Manocha 1987, 1991), whereas necrotrophs kill the cells of the host. Some biotrophic parasites become necrotrophic in the later stages of a relationship and may cause at least some degeneration of the host when they begin to sporulate. Necrotrophic mycoparasites are relatively unspecialized and tend to have a broader range of fungal hosts than biotrophic mycoparasites, which also often form specialized infection structures (Jeffries 1995).

Biotrophic and necrotrophic mycoparasites are not sharply delimited, and many intermediate forms exist. Rudakov (1978) assigned fungicolous species to six groups: biotrophs, facultative biotrophs, necrotrophs, facultative necrotrophs, "semisaprophytic" mycophiles, and "saprophytic associates" (associated with mycophilic fungi without having antibiotic activity). Some mycoparasitic species change their behavior toward a host during the course of their development (Rudakov 1978, 1981), and other species (e.g., *Pythium oligandrum*) develop different types of contact structures on different hosts. Trophic relationships also may be host-dependent; many species grow as biotrophs on certain hosts but as necrotrophs on others. Species causing polypore and mushroom rot develop biotrophically in the hyphae of certain other filamentous fungi. *Hypomyces chrysospermus* (generally its anamorph, *Sepedonium chrysospermum*), for example, always causes necrosis of the cells of its mushroom hosts. It, however, can grow biotrophically inside the cells of fungi such as *Botrytis cinerea* and *Trichothecium roseum*, when the latter two species colonize a mushroom that it has already parasitized, or when it is grown with one of those fungi *in vitro* (Rudakov 1981).

A phenomenon termed *hyphal interference* (Ikediugwu and Webster 1970a; Dennis and Webster 1971c) occurs when the mycelium of one fungus, growing either close to (within 50 μm) or in contact with that of another species, reduces the growth rate and causes cytoplasmic disruption of the second fungus. Electron-microscopy studies have revealed vacuolation, or an abundance of lipid droplets and invaginations of the host plasmalemma at the point of contact with the parasite—for example, when *Coprinus heptemerus* touches *Ascobolus crenulatus* (Ikediugwu 1976a), or *Phlebiopsis* (*Peniophora*) *gigantea* encounters *Heterobasidion annosum* (Ikediugwu 1976b). Similar vacuolation induced by *Fusarium oxysporum*, *Trichoderma viride*, and *Penicillium expansum* in hyphae of *Aspergillus niger* has been described (Park and Robinson 1964; Robinson and Park 1965). This very localized phenomenon usually has been observed only in dual culture *in vitro*.

The regular association between a fungicolous fungus and its host that has little visible effect on the host is termed commensalism (e.g., Jeffries and Young 1994; for lichenicolous fungi, Hawksworth 1988b). Mycoparasites growing on other fungi that are parasitic on either plants or animals have been called hyperparasites, but we confine this term to those fungi that parasitize fungicolous fungi.

Many arguments support the view that a semi-biotrophic mode of parasitism (biotrophic organisms later becoming necrotrophic or saprotrophic) is more primitive in plant parasites than are holobiotrophy and saprotrophy (Cooke and Whipps 1980); that view may also be true of mycoparasitic fungi. Obligate parasitism may not be a belated evolutionary development, but a fundamental attribute of the primitive groups, from which saprotrophic subgroups have arisen repeatedly. Necrotrophic parasites that are facultatively saprotrophic, however, may in some cases have been derived from saprotrophs (Jeffries and Young 1994). It is evident that the specific associations between parasites and hosts are the products of a long coevolutionary process that is likely to be as old as the fungi involved (Hass et al. 1994).

MYCOPARASITE–HOST INTERFACE

The hyphae of a mycoparasite generally contact a host by hyphal apposition, by coiling around the hyphae, or by growth of short hyphal pegs. Appressoria and haustoria may or may not form, and the host hypha may or may not be penetrated. Coiling, which does not occur in all host–antagonist interactions, is not necessarily evidence for mycoparasitism. Penetration may occur immediately after hyphal contact in the absence of any coiling. Coiling occurs particularly in those cases in which some resistance to entry is encountered (Deacon 1976; Jeffries and Young 1994). Depending on the circumstances and the condition of the host, different structures may be produced by the same parasitic fungus (e.g., van den Boogert et al. 1989). Swart (1975) described a defense reaction, in which callosities, or wall thickenings, form around penetrating haustorial branches of *Verticillium dahliae* or *Fusarium solani* in the walls of sporangiophores of Mucorales and conidiophores of *Aspergillus*. The term *callosity* originally was applied to similar structures, also called lignitubers, that are produced by plants as a defense reaction against fungi (Young 1926). In a few cases, transmission-electron-microscopy studies have shown a micropore connecting the haustorium of the parasite and the host cell. A specific organelle of the parasite that facilitates nutrient exchange between the host and certain parasites is the colacosome or lenticular body (see "Urediniomycetes, Platygloeomycetidae" in the section on "Taxonomic Groups of Fungicolous Fungi and Fungus-like Microorganisms," later in this chapter).

Jeffries and Young (1994) and Jeffries (1995) distinguished five principal types of mycoparasite–host interfaces. The first two types are necrotrophic interfaces, and the other three types are biotrophic interfaces.

With *necrotrophic interfaces* the cytoplasm of the host degenerates, and hyphal lysis often occurs:

1. Contact necrotrophic. Neither hyphae nor haustoria of the parasite penetrate the host mycelium; it is damaged by hyphal interference. Examples include *Arthrobotrys superba* and *A. oligospora*, predaceous, nematophagous fungi that can also function as contact mycoparasites; *Tilletiopsis* species, which contact and kill cells of the powdery mildew *Sphaerotheca fuliginea*.
2. Invasive necrotrophic. Hyphae of the parasite penetrate the cell wall and enter the host cell; they show considerable growth within the host hyphae (Jeffries and Young 1994). Examples include *Talaromyces flavus*, *Schizophyllum commune*, *Trichoderma* species, and many others.

With *biotrophic interfaces* the host cytoplasm remains healthy (at least initially).

3. Haustorial. A short haustorial branch from the hypha of the parasite penetrates the hypha of a host. Examples include Mycoparasites among the Zygomycota and the Tremellales; *Sporidesmium sclerotivorum* parasitizing *Sclerotinia sclerotiorum*.
4. Fusion. Micropores develop in the walls of the host and parasite hyphae that are in contact or in the walls of a short penetrating hyphal branch of the parasite, allowing cytoplasmatic contact. This is an unusual type of interface. Examples include *Gonatobotrys simplex*, *Hansfordia parasitica*, *Melanospora zamiae*, and related fungi. *Tetragoniomyces uliginosus* on *Rhizoctonia solani* and *Syzygospora pallida* on *Phanerochaete* produce micropores in their haustoria.

Anchoring cells of *Parasitella parasitica* have a wide cytoplasmic connection with the host *Absidia glauca*.

5. Intracellular. The complete thallus of the mycoparasite enters a hypha of the host. Examples include many species of Chytrids and some Oomycetes growing inside other fungi.

Additional categories may be needed to accommodate commensal and gall-forming fungicolous associations. Lichenicolous fungi growing on a lichenized fungus exploit and often kill the associated alga, generally penetrating it with haustoria.

DETERMINING THE MODE OF INTERACTION

The significance of mycoparasitism in nature has been underrated, although interfungal parasitic relationships likely play an important role in the development of fungal community structure. Mycoparasitism is difficult to observe in the field but can be assumed to occur in nature when growth abnormalities are seen in a host. In contrast, many examples of mycoparasitism have been described from laboratory cultures. Even the unequivocal demonstration of a parasitic association under laboratory conditions, however, does not prove that such a relationship occurs in nature (Jeffries 1995).

Infection structures can be observed in various ways. Preparations in water, lactic acid (better and safer than lactophenol) with cotton blue, or other standard mountants are appropriate for microscopic observation of fungi. Other stains, however, may do a better job of differentiating living from dead cells. For example, when agar blocks taken from paired slide cultures are stained with a drop of dilute phloxine, dead cells absorb the dye immediately, whereas living cells remain unstained for several minutes (Griffith and Barnett 1967). Specific organisms have been located by means of immunolabeled dyes (Mendgen and Casper 1980). Fluorescence methods also have been used to locate areas of great enzymatic activity or the locations of lectins (see "Mycoparasites of Mycelia, Ectomycorrhizae, Sclerotia, and Spores in Soil," later in this chapter). Infrared photomicrography also has been used to detect sites of strong enzymatic activity in the interaction between *Trichoderma harzianum* and *Rhizoctonia solani*; bright regions were visible in the coiling cells of the parasite in slide cultures (Elad et al. 1983b). Electron-microscopic investigations of infected host structures are also numerous.

In a few cases, the flow of nutrients from host to parasite has been proved definitively by means of radiola-beled compounds. ^{32}P was translocated from *Rhizoctonia solani* to the parasite *Arthrobotrys oligospora* (Olsson and Persson 1994). Foley and Deacon (1986) demonstrated that mycoparasitic *Pythium* species probably obtain carbon and nitrogen, as well as thiamine and sterols, from their host, with detrimental consequences for the host. Many mycoparasites can grow in culture without a host, but a number of contact biotrophs require at least host extracts for growth in culture. If axenic growth is possible, nutritional requirements for mycelial development of the parasite can be determined. Requirements for *in vitro* growth, however, are likely to differ from those of the fungus growing on a host. Dual cultures may represent a more natural system for physiological studies, but then the nutrient requirements of the host also must be determined, and the parasitized host may modify the nutrients in the growth medium before they are absorbed by the parasite. Moreover, the parasitized host may have nutrient requirements different from those of the unparasitized host (Jeffries and Young 1994).

The ability of some fungicolous fungi to reduce the growth rate of some test-host fungi before making contact has been interpreted as evidence of production of antibiotic metabolites, an effect that is strongly medium-dependent (e.g., Lerner and Sidorova 1978; Whipps 1987). Some efficient mycoparasites, such as *Acremonium strictum*, *Clonostachys rosea*, *Trichothecium roseum*, and *Sistotrema brinkmannii*, are able to overgrow other fungi without being inhibited themselves. After such an infection, the parasitized fungus usually cannot be recovered.

The effects of antibiotics and volatile substances on antagonist–pathogen interactions in dual cultures should be distinguished from the effects of temperature, water potential, and individual isolates (Whipps 1987; Jeffries and Young 1994). *Trichoderma* species exhibit various modes of interspecific interaction that are mediated by volatile and nonvolatile metabolites (Dennis and Webster 1971a, 1971b; Whipps 1987). Besides studying interactions in dual culture, those investigators grew potential antagonists of a particular fungus on cellophane disks placed on potato-dextrose agar and other media for a few days. After removal of the cellophane, the medium was inoculated with test fungus in the same spot, and the degree to which it was inhibited (if at all) was assessed. The action of volatile metabolites was assessed in Petri dishes inoculated with a host organism and then fastened opposite plates inoculated with the antagonist. Jackson and colleagues (1991) used this methodology to screen for potential biocontrol capacities of numerous soil isolates against *Sclerotium cepivorum*.

Lutchmeah and Cooke (1984) adopted a slide-culture technique to observe the parasitic action of *Pythium oligandrum* microscopically. They coated large cover slips with water agar and inoculated opposite sides with blocks of potato-dextrose agar—cultures of the fungi to be tested. Interactions were observed usually for 1–1.5 hours at the margins of the sparse developing colonies by bright-field microscopy, using a 70 × oil-immersion objective. Laing and Deacon (1991), Berry and Deacon (1992), and Berry and associates (1993) thus beautifully illustrated the aggressiveness of *P. oligandrum* and *P. nunn* toward a range of hosts using electronically enhanced video-micrography.

The physiological bases of mycoparasitism are diverse. Necrotrophic mycoparasites often release toxins and lytic enzymes (particularly chitinase) into the environment, whereas no biotrophic types have been demonstrated to do so (Jeffries 1995). Rudakov (1978) grew each of nearly 200 species of fungicolous fungi in paired culture with a test host, looking for production of antibiotics (*in vitro* antagonism). He found that production of inhibitory metabolites was most pronounced in necrotrophs but also common in many fungicolous fungi that grow on macromycetes (e.g., Turner and Aldridge 1983; Bastos et al. 1986; Hoßfeld 1990). The production of inhibitory metabolites and wall-degrading hydrolytic enzymes is probably less important for fungi that can make physical contact with a host than it is for fungi whose hyphae do not directly infect a host (Jloba et al. 1980). Production of inhibitors may be positively correlated with intensity of pigmentation in *Hypomyces aurantius*, *H. odoratus*, and *H. rosellus* because strains with only weak pigmentation or none had no inhibitory effect (Lerner and Sidorova 1978). When a typically pigmented strain of *H. odoratus* was transferred to a medium that caused it to lose color, inhibitor production also decreased and interactions with test fungi were modified. Hypodoratoxide, a sesquiterpene eremophilane ether with phytotoxic properties, is a major component of the volatile metabolites of that species (Kühne et al. 1991). Toxins occurring in *Hypomyces aurantius*, *H. orthosporus* (anamorph *Cladobotryum orthosporum*), *H. semitranslucens* (*C. fungicola*), *Eudarluca caricis*, and *Sesquicillium microsporum* are also effective against various test fungi (Hoßfeld 1990).

Lawrey and associates (1994) examined selected hypocrealean fungi for their sensitivity to lichen metabolites. Subsequently, he (Lawrey 1995) reviewed data on the tolerance of lichenicolous fungi to secondary metabolites of lichens. Tolerance is a prerequisite for that mode of life. Some nonlichenized host fungi that contain acrid metabolites also are colonized by specific mycoparasites; it is likely that those metabolites are the selective agent determining which mycoparasites are present (see "On Fleshy Sporocarps of Boletales and Agaricales" under "Major Groups of Fungicolous Fungi, Arranged

by Host Group," later in this chapter). Diverse mycoparasites can tolerate cyanide to levels of 0.01% in a synthetic medium, although this compound is not generally present in their natural substrata (Singh and Plunkett 1967).

Most mycoparasites are unspecialized and can infect a wide range of host fungi across diverse systematic groups. Not only can certain fungi parasitize other fungi belonging to diverse orders, but some are also entomogenous or nematophagous (see Chapters 18 and 19). More specific associations, however, also exist (see "Factors Determining Host Specificity," later in this chapter). Lichenicolous fungi, for example, generally are specialized and rarely have been observed to grow on nonlichenized host fungi. Several fungicolous fungi are hyperparasites of other fungicolous (see "Fungi on Sporocarps of Fungi or Myxogastrea," "Lichenicolous Fungi," and Aquatic Fungi and Fungus-like Organisms," later in this chapter).

The tremelloid *Aporpium caryae* has features of both Tremellales (cruciate basidia) and polypores (consistency of sporocarp). Setliff (1984) speculated that such an unusual morphology may have resulted from the horizontal transfer of genetic material from a host fungus to its mycoparasite. A similar hypothesis could be applied to many other tremelloid fungi. Bandoni (1984) argued, however, that Auriculariales–Aporpiaceae are closer to the Aphyllophorales than to the Tremellales, which may account for the morphological similarities among the groups. However, gene transfer from parasite to host has been demonstrated experimentally in the mucoralean association between *Absidia glauca* and its parasite *Parasitella parasitica* (Kellner et al. 1993; Wöstemeyer et al. 1995). In those species a plasmatic connection is established between a sikyotic ("sucker" or "anchor") cell of the parasite and the host (Burgeff 1924). The prototrophic parasite was found to compensate for artificially induced deficiencies in the host, compensation that became established in 0.4% of the uninucleate sporangiospores of the offspring. Transfer of a neomycin-resistance–conferring plasmid also was observed. Although the "pararecombinants" were unstable, such a natural gene transfer mechanism could have important evolutionary implications (Jeffries 1995).

TAXONOMIC GROUPS OF FUNGICOLOUS FUNGI AND FUNGUS-LIKE MICROORGANISMS

Although the fungicolous habit is widespread throughout the fungi, it is particularly common in certain taxonomic groups. We characterize those groups here. We provide more detailed treatments of the fungicolous fungi in the section dealing with "Major Groups of Fungicolous Fungi, Arranged by Host," later. Table 17.1 gives estimates of numbers of species in these taxonomic groups, distributed over the five major ecological groups.

For some fungi, we include often-used generic names in parentheses after the presently correct genus, but we do not mean to imply that the previous genus as a whole is a synonym of the correct name given. We do not capitalize names suggesting class or ordinal rank, but consisting of very heterogeneous elements, such as heterobasidiomycetes and aphyllophorales. We also adopt the convention from Seifert and colleagues (2000) in regarding generic names of hyphomycetes in connection with "-like" as morphological terms rather than as formal generic designations and, therefore, neither italicize nor capitalize them.

OOMYCOTA

One-celled members of the Olpidiopsidales (*Olpidiopsis*) and Rozellopsidales (*Rozellopsis* and *Dictyomorpha*) can form biotrophic associations with a diversity of other Oomycota (particularly Saprolegniales and Pythiales), some Chytridiomycetes, and many algae. Sparrow (1960) and Karling (1981) listed 18 and 22 fungicolous species, respectively, of *Olpidiopsis*, four of *Rozellopsis*, one of *Skirgiellopsis*, two of *Pythiella*, two and three of *Lagenidium*, one of *Myzocytium*, and one and three of *Petersenia*. With the exception of the endobionts of *Rozellopsis*, the endobiotic thalli of the taxa are surrounded by walls. The walled endoparasites usually form several zoosporangia in one compartment of the host, which is hypertrophied considerably. Only a single thallus of the wall-less endoparasites forms in a host compartment, which then is delimited by a septum induced in the host. The species of *Rozellopsis* known to have naked thalli fall into several groups analogous to the subgroups composing the genus *Rozella* (Held 1981).

In the soil several specialized species of *Pythium* parasitize some congeneric plant pathogens, as well as other fungi. In particular, the four *Pythium* species with spiny oogonia (i.e., *P. oligandrum*, *P. acanthicum*, *P. periplocum*, and *P. acanthophoron*) are mycoparasitic, as are two species with smooth oogonia (*P. nunn* and *P. mycoparasiticum*). Those species differ from the plant-pathogenic *Pythium* species in that they require organic nitrogen and thiamine for growth; they are noncellulolytic and thus can play the role of "secondary sugar fungi,"—that is, they profit from a surplus of reducing sugars liberated by cellulolytic fungi from cellulose (Foley and Deacon 1986; Jones and Deacon 1995;

Ribeiro and Butler 1995). In addition, zoospores of the mycoparasitic species tend to encyst on chitinous substrata, whereas those of plant-pathogenic species preferentially encyst on cellulose; the form of cyst attachment is a particularly sensitive indicator of taxonomic affinity and determines host–parasite specificity (Deacon 1988a; Jeffries and Young 1994)

HYPHOCHYTRIDIOMYCOTA

Two genera of Hyphochytridiales, *Hyphochytridium* and *Rhizidiomyces*, include one and two species, respectively, that parasitize oospores of *Phytophthora* species.

PLASMODIOPHOROMYCOTA

A few species of *Woronina*, *Octomyxa*, and *Sorodiscus*, in the Plasmodiophorales parasitize species of the Oomycota (genera of the Saprolegniales and *Pythium*). They form intracellular naked plasmodia and later transform into cystosori and sporangiosori (Batko 1975; Held 1981).

CHYTRIDIOMYCOTA

One-celled members of the Chytridiomycota live in either intracellular or epibiotic parasitic associations, mostly with algae, occasionally with other Chytridiomycota and Oomycota, and rarely with Zygomycota. They cause some host fungi to hypertrophy. Epibiotic taxa of *Caulochytrium* and *Sparrowia* grow on a diversity of hosts (Karling 1977). Endobiotic taxa form an intracellular thallus that either is surrounded by a wall ensheathed in the endoplasmatic reticulum of the host (e.g., *Catenaria allomycis* in *Allomyces*, Sykes and Porter 1980; Powell 1982), or is naked (e.g., *Rozella allomycis* in *Allomyces*, Karling 1942; Held 1981; and *Rozella polyphagi* in *Polyphagus euglenae*, an ectoparasite on flagellates, Powell 1984). The naked endobionts consume part of the host cytoplasm by phagocytosis. Their thallus is surrounded only by the plasmalemma, which disintegrates when sporogenesis of the parasite begins, leaving the mature endoparasitic thallus in direct contact with the host cytoplasm. The developing thallus induces a septum to form in its vicinity so that each compartment of the host contains only one zoosporangium. The segmented host of *R. allomycis* then appears to have multiple infections (Held 1980). When *R. allomycis* first invades an *Allomyces* hypha, it elicits a host reaction at the site of penetration. The host forms lomasomes to which wall material is apposed giving rise to an internal papilla. The parasite enters through the center of the

TABLE 17.1

Approximate Numbers of Species Known from Orders Containing Fungicolous Fungi*

Taxon	Sporocarp-inhabiting	Lichenicolous	Biotrophic plant parasites	Hyphal parasites	Aquatic	Total[†]
Oomycota				11	37	43
Hyphochytriomycota				2	3	3
Plasmodiophoromycota					3	3
Labyrinthulomycota				1		1
Chytridiomycota				3	50	50
Zygomycota						
Mucorales	10					10
Zoopagales				50		50
Dimargaritales				13		13
Ascomycota (and associated anamorphs)						
Saccharomycetales	12					12
Leotiales	40	67				107
Ostropales		22				22
Patellariales		3				3
Pezizales	2	1				2
Eurotiales	5			1		6
Chaetothyriales	5	10				15
Pyxidiophorales	15			2		17
Arthoniales		150				150
Caliciales‡		44				44
Verrucariales		191				191
Dothideales	10	224	30			264
Lecanorales		112				112
Pyrenulales		1				1
Ophiostomatales	6					6
Trichosphaeriales		1				1
Xylariales		1				1
Sordariales	58	37				95
Hypocreales (excl. Clavicipitaceae)	185	59	3			247
Clavicipitaceae	18	4	4	5		22
Phyllachorales		26				26
Ascomycetes incertae sedis		84				84
Ascomycetous conidial fungi	30	198	27		1	256
Basidiomycota						
Urediniomycetes	92				1	93
Ustilaginomycetes		4	8			12
Tremellales	120	50				170
Aphyllophorales	15	2		3		19
Boletales	3	1				4
Agaricales	15	3				18
Basidiomycetous condial fungi	2					2
TOTAL	643	1295	72	91	95	2175

* Numbers of species/genus taken from Hawksworth et al. (1995) and Lawrey and Diederich (2003), plus some additional observations; lichenicolous fungi do not include lichenized species.

[†] Because some fungi occur on multiple substrata, totals may be less than the sum of the columns in a particular row.

‡ The Caliciales are no longer accepted, but are considered to be a synonym of the Lecanorales; some species previously included in the Caliciales are now placed in the Mycocaliciales or in the Microcaliciaceae family incertae sedis.

papilla with minimal cell disruption. The reaction does not occur in hyphae that already are infected, so renewed penetration by the parasite cannot occur (Held 1972). Held (1981) assigned the 25 known mycoparasitic species of *Rozella* to five groups according to morphol-ogy and host affinities. The groups consisted of fungi parasitic on (1) monocentric Chytrids, (2) *Blastocladia* and Leptomitales, (3) *Monoblepharis* and Pythiaceae, (4) *Allomyces* and Saprolegniaceae, and (5) miscellaneous hosts. Sparrow (1960), Karling (1960, 1977), and Batko

(1975) described and illustrated about 45 potentially mycoparasitic taxa (Table 17.2).

ZYGOMYCOTA

Obligate, biotrophic, contact parasites are found in the orders Zoopagales and Dimargaritales. These haustorial mycoparasites grow on other members of the Mucorales or, exceptionally, on nonmucoralean fungi (Benjamin 1979). The mycoparasites and some unrelated saprotrophic groups sometimes are combined in an unnatural group of "merosporangiferous Mucorales." The biotrophic zygomycete associations have been reviewed by Jeffries (1985) and Jeffries and Young (1994).

On contact, the mycoparasite forms an haustorium that induces gall formation in the host. Burgeff (1924) noted that mating type factors often determine the compatibility of a host and its parasite. He also observed plasmatic continuity between a sikyotic cell (contact cell) of the parasites *Parasitella parasitica* and *Chaetocladium brefeldianum* and their hosts.

The order Zoopagales includes numerous endoparasites, as well as predators of amoebae, nematodes, and

TABLE 17.2
Numbers of Mycoparasitic Chytridiomycota Taxa*

Chytridiales
    Chytridiaceae
        Chytridium—3, 1, 3[†]
        Chytriomyces—1, 4, 5
        Rhizophydium (incl. *Phlyctidium*)—9, 3, 2
        Phlyctochytrium—3, 1, 0 (parasites in Glomales recognized later,
            see Sylvia and Schenck 1983)
        Blyttiomyces—2, 2, 1
        Rhizidium—0, 2, 1
        Septosperma—2, 2, 2
        Solutoparies—1, 1, 1
        Sparrowia—0, 1, 1
Spizellomycetales
    Olpidiaceae
        Olpidium (incl. *Pleotrachelus?*)—5, 5, 4
    Spizellomycetaceae
        Rhizophlyctis—1, 1, 0
    Caulochytriaceae
        Caulochytrium—1, 1, 1
Blastocladiales
    Catenariaceae
        Catenaria—1, 2, 1
        Rozella[‡]—19, 3, 16

* Classification modified according to Barr (1980) and Hawksworth et al. (1995).

[†] The three numbers given represent taxon estimates from Sparrow (1977), Karling (1977), and Batko (1975), respectively.

[‡] The placement of *Rozella* in this order is debated; three additional species of *Skirgiellia* are close to this genus (Batko 1975).

other small animals. Members of the family Piptocephalidaceae are obligate haustorial mycoparasites (Fig. 17.1A,B). *Piptocephalis* and *Kuzuhaea* species are purely biotrophic, forming small haustoria in the host cell. *Piptocephalis xenophila* can grow on *Penicillium* species and *Chaetomium* species (Dobbs and English 1954; Shigo 1960b), but all other species in the family grow only on Mucorales *sensu lato*. When inoculated on suitable host fungi (e.g., *Umbelopsis* [*Mortierella*] *isabellina*), they develop best on media containing sources of organic nitrogen and a low concentration of sugar (Berry 1959). The haustoria are enucleate and delimited from the host cytoplasm by a very thin layer of electron-transparent wall material (Jeffries and Young 1976).

Host morphology hardly is affected by species of *Piptocephalis*, apart from an increase in branching and a decrease in marginal density of the hyphae (Evans and Cooke 1981), but overall growth of the host can be inhibited. Other parasitic species stimulate growth of certain host species (Curtis et al. 1978); such cases are assumed to be mutualistic. Axenic growth of *P. virginiana* on a malt extract–yeast extract medium was very limited, and the culture could not be maintained (Manocha and Deven 1975). Apparently *P. virginiana* can only parasitize fungi in which the concentration of linolenic acid exceeds a certain threshold, although other nutrients also are required (Manocha and Deven 1975). Trehalase activity in sporelings of the parasite increases during the interaction, indicating that trehalose reserves of the host may be exploited by the parasite (Evans et al. 1981).

Syncephalis species are more necrotrophic than *Piptocephalis* species (Hunter et al. 1977). They form a highly branched haustorial system in the hypha of the host and impair its growth and sporulation (Fig. 17.1B). Axenic cultivation of *S. californica* succeeded after transfer from a monoxenic culture onto a liver medium (Emerson 1958; Ellis 1966).

The order Dimargaritales comprises the mycoparasitic genera *Dimargaris* (Fig. 17.1C), *Dispira*, and *Tieghemiomyces*. *Dispira cornuta* (Ayers 1935; Brunk and Barnett 1966) is a biotrophic parasite that forms nucleate haustoria from gall-like swellings and is found on rat dung. It grows on *Mucor* species and other Mucorales but not on *Mortierella*. It can be grown axenically on media rich in proteins, with the addition of growth factors (Kurtzman 1968; Barnett 1970; Barker and Barnett 1973; Singh 1975). *Dispira simplex* and *D. parvispora* grow on *Chaetomium*; the latter also grows on *Monascus* species. As far as is known, the Dimargaritales cannot utilize glucose; axenic growth requires glycerol, organic nitrogen, and some vitamins (Brunk and Barnett 1966; Kurtzman 1968; Barnett 1970). Singh (1975) found an exceptional strain of *D. cornuta* for which starch was the best carbon source.

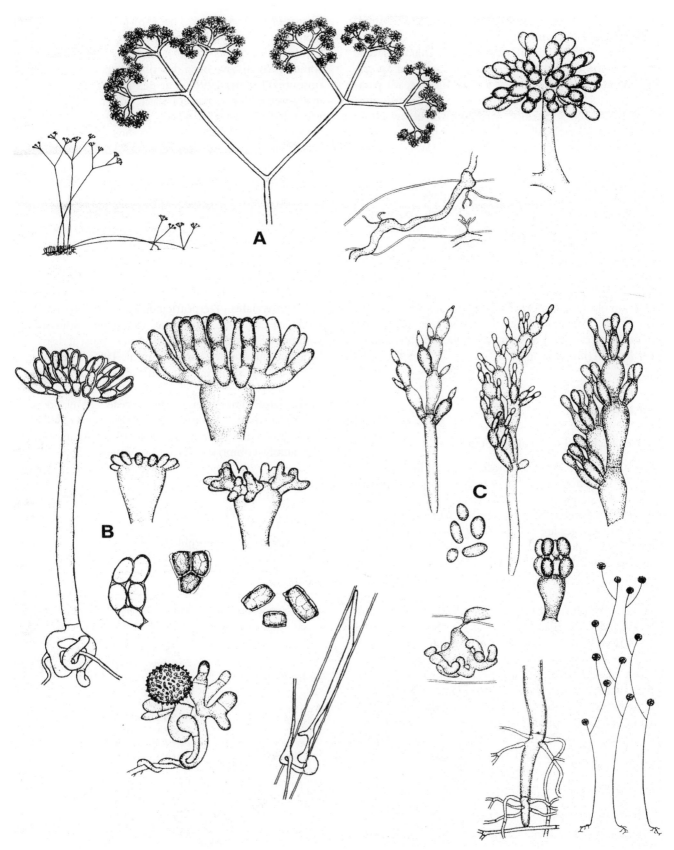

FIGURE 17.1 Biotrophic mycoparasites from the Piptocephalidaceae and Dimargaritaceae. **A.** *Piptocephalis lepidula*, habit sketch of colony, sporangiophore with merosporangia; hypha of the parasite overgrowing the host hypha forming minute branched haustoria. **B.** *Syncephalis nodosa*, sporangiophores with merosporangia, a zygospore, and hypha in contact with a *Mucor* host, showing appressorium and extended haustorium. **C.** *Dimargaris cristalligena*, habit sketch of sporangiophores, details of sporangiophore head with sporiferous branchlets, complex haustorium inside host hypha. (All reprinted with permission, from Benjamin 1959.)

Nonspecific mycoparasites growing on other Mucorales also are found among the Mucoraceae (*Parasitella parasitica* and *Absidia parricida*), Thamnidiaceae (*Chaetocladium*), and Mortierellaceae. Species of *Spinellus, Syzygites,* and *Dicranophora* (all Mucoraceae) parasitize agarics. Axenic cultivation of *Spinellus* requires organic nitrogen and temperatures near or below 20°C; its spores require chemical stimulants from host extracts or ascorbic or gluconic acid to germinate (Watson 1962, 1965; Jeffries 1985; Jeffries and Young 1994). *Syzygites,* however, grows easily on any culture medium (Hesseltine 1957), as does the weakly parasitic *Chaetocladium* (Benny and Benjamin 1976). *Parasitella parasitica* is a weak parasite of various Mucoraceae, Thamnidiaceae, and Choanephoraceae, which it contacts by means of a kind of capturing hypha (Burgeff 1924; Zycha et al. 1969). The genus *Mortierella* comprises highly chitinolytic fungi that often appear indiscriminately on old fungal sporocarps; exceptions are the more specific *M. bainieri,* which commonly is found on *Amanita* species, and *M. armillariicola,* which is known only from *Armillaria* species.

ASCOMYCOTA

Orders Rich in Fungicolous Fungi

Laboulbeniales. Pyxidiophoraceae. In contrast to the majority of Laboulbeniales, which is associated with insects, the genus *Pyxidiophora* (anamorphs in *Gabarnaudia* or *Chalara*) is heteroxenous, which means that some of its species are able to change hosts during their life cycles (Malloch 1995). Most species are biotrophic mycoparasites that parasitize their hosts by means of noninvasive contact cells. Their ascospores generally are translocated by mites, and the fungus can penetrate its vector with haustoria (Lundqvist 1980; Malloch and Blackwell 1993a; Blackwell 1994; see "Spore-Dispersal Interactions" in Chapter 18). *Gabarnaudia tholispora* is a common fungicolous anamorph species, the teleomorph of which has not yet been established. *Pleurocatena acicularis,* which can be found in association with *Hypocreopsis,* probably also belongs to this family (G. Arnold, personal communication). An unnamed *Pyxidiophora* (CBS 665.93 and 626.94) can be grown on *Cylindrocarpon destructans* but usually not axenically (G. Fischer and W. Gams, unpublished data).

Leotiales. Species of *Orbilia,* in the family Orbiliaceae (mainly *O. inflatula*), are found on stromata of ascomycetes and even more frequently on decaying basidiomata of aphyllophorales. *Orbilia* species have anamorphs in the genera *Arthrobotrys, Dactylella,*

Dicranidion, and others. The *Arthrobotrys* anamorphs are mainly nematophagous but also mycoparasitic, whereas *Dactylella* includes only nonnematophagous species, some of which have been described as parasites of fungal oospores (Rubner 1996). The genus *Helicogonium,* originally erected for two mycoparasites of Stereales, has been expanded to cover 15 species growing on inoperculate discomycetes, a situation in which their asci easily are mistaken for a deviating form of the host asci (Baral 1999). The genus is classified in the Leotiales, although the species lack discrete ascomata. The genus *Unguiculariopsis,* in the family Helotiaceae, includes at least 24 species, all of which grow on fungi; 10 of those species also grow on lichenized ascomycetes. The species of *Llimoniella* (six species), *Rhymbocarpus* (nine species) and *Skyttea* (17 species) are all lichenicolous (Diederich and Etayo 2000).

Chaetothyriales. Herpotrichiellaceae. *Capronia* (Untereiner et al. 1995) comprises four fungicolous species commonly found on pyrenomycetes, plus at least five lichenicolous species (Aptroot et al. 1997). Anamorphs of these fungi mainly belong in the genus *Exophiala.*

Sordariales. The Ceratostomataceae (Melanosporaceae), as revised by Cannon and Hawksworth (1982), includes many mycoparasites (Table 17.3). Anamorphs of this family comprise *Harzia* (synonym *Acremoniella*),

TABLE 17.3

Mycoparasitic Teleomorphic Ceratostomataceae and Their Hosts

Parasites	Hosts
Melanospora species	Various discomycetes, *Fusarium, Botrytis, Paecilomyces, Beauveria, Tubercularia* species, and several aphyllophorales
Persiciospora species	*Fusarium* species
Sphaerodes species*	*Labyrinthomyces* and *Sphaerozone* species (truffle genera)
	Hypomyces armeniacus
	Sclerotinia sclerotiorum and *Trametes* species
Syspastospora parasitica (= *Melanospora parasitica*)	*Paecilomyces, Beauveria, Hirsutella,* and *Verticillium* species
Sphaeronaemella helvellae	*Gyromitra* species
Viennotide fimicola[†]	*Ascobolus* species

* *Microthecium* in part, see also Hawksworth and Udagawa (1977).
[†] Other species of *Viennotidea* (formerly merged with *Sphaeronaemella,* anamorph *Gabarnaudia*) apparently are not fungicolous.

Gonatobotrys, Nematogonum, Gonatobotryum, Olpitrichum, and *Papulaspora,* which are mainly mycoparasitic and commonly associated with an aspergillus-like hyaline, phialidic synanamorph (*Proteophiala*). The best-studied anamorphic biotrophic contact mycoparasites are in that group: *Gonatobotrys simplex* (teleomorph *Melanospora damnosa*), heterothallic and fruiting *in vitro* (Vakili 1989); *Nematogonum ferrugineum* (synonym *Gonatorrhodiella highlei*); and *Gonatobotryum fuscum.* Species of *Gonatobotrys, Gonatobotryum,* and *Nematogonum* were revised by Walker and Minter (1981). Another well-studied species, *Hansfordia* (*Calcarisporium*) *parasitica,* may be related to the Xylariales; it is unusual in that it develops 0.2–1.0-μm-wide pores at the point of contact with its host fungi (Hoch 1977a). Some of the ecologically obligate mycoparasites can be grown axenically on media with thiamine, biotin, and fungal extracts that contain mycotrophein (Gain and Barnett 1970; Calderone and Barnett 1972; Barnett and Binder 1973; Hwang et al. 1985). Mycotrophein, found in extracts of various fungal cultures, is a mixture of tetraethyleneglycol and pentaethyleneglycol mono-(nonylphenyl) ethers. It may act as carrier of a biologically active factor (Hwang et al. 1985).

Gonatobotrys simplex normally grows on *Alternaria* and *Cladosporium* species (Whaley and Barnett 1963; Hoch 1977b). It contacts a host by means of fingerlike branches (Hoch 1977b). Its nutritional requirements have been analyzed by Whaley and Barnett (1963). *Gonatobotryum fuscum* can grow on *Polyporus, Poria, Ganoderma, Tremella, Leptographium, Hypocrea,* and *Hypomyces* species (Shigo 1960a, 1960b; Calderone and Barnett 1972; Walker and Minter 1981). Jordan and Barnett (1978) found similar biotrophic relationships in *Melanospora zamiae* and also showed that the parasite can withdraw all required nutrients from washed host mycelium; mycotrophein is not required. The biotrophic contact parasite *Hansfordia parasitica* has similar requirements. It shows a unique mutualistic relationship with *Graphium fuscum.* Under certain circumstances, the parasitic relationship of *H. parasitica* with its hosts can be reversed, and old aerial mycelium and conidiophores are overgrown and attacked by the *Graphium.* The parasite produces pyridoxine and the host biotin; each partner requires the compound produced by the other (Shigo 1960a; Barnett 1968). *Stephanoma phaeosporum* is a biotrophic contact mycoparasite of *Fusarium* species and 14 other fungi (Butler and McCain 1968; Rakvidhyasastra and Butler 1973; Hoch 1978; Hawksworth 1981a). It also can be grown axenically with fungal extract, but its dependence on mycotrophein is uncertain. *Olpitrichum tenellum* is a contact mycoparasite of *Fusarium* species and various dematiaceous hyphomycetes. It first was discovered on *Fusarium*

moniliforme on maize ears in Illinois (Kuykendall et al. 1983); it requires thiamine and biotin, but not mycotrophein, for axenic growth.

In several genera of Sordariales, including *Globosphaeria, Lasiosphaeriopsis, Reconditella, Rhagadostoma, Roselliniella, Roselliniomyces,* and *Roselliniopsis,* all species are lichenicolous.

Hypocreales. The order Hypocreales consists of the families Hypocreaceae, Nectriaceae, Bionectriaceae, Clavicipitaceae, and Niessliaceae. The former three families were partly revised by Rossman and colleagues (1999). The positions of the Clavicipitaceae and Niessliaceae relative to the other families are not yet settled. A large portion of hypocrealean fungi is mycoparasitic or mycosaprotrophic, a fact that often is unrecognized (Rossman et al. 1999). They are extremely versatile in their abilities to exploit fungal substrata (Rossman 1996). A key to 24 genera of the Hypocreales known to contain one or more fungicolous, lichenicolous, or myxomyceticolous species can be found in Samuels (1988), although some of the poorly known genera may not actually belong to the order.

Hypocreaceae. Within the family Hypocreaceae, *Hypocrea* is reported less frequently as a mycoparasite than are its anamorphs in *Trichoderma.* However, the genus includes fungicolous species that are either host-specific or occur on a variety of aphyllophorales in addition to woody substrata. Several *Trichoderma* species are polyphagous mycoparasites. The species are difficult to distinguish from each other on morphological grounds, even using the most refined criteria (Bissett 1991b; Gams and Bissett 1999). Strains of *T. harzianum* that are beneficial biocontrol agents of plant pathogens can hardly be distinguished microscopically from genetically distinct strains that are aggressive competitors of the cultivated mushroom (Muthumeenakshi et al. 1994). A related species, *T. virens,* was distinguished clearly from other *Trichoderma* species by Webster and Lomas (1964) and redescribed as a member of *Gliocladium,* but it is unrelated to other species of that genus and is best classified in *Trichoderma* (Bissett 1991b; Rehner and Samuels 1994) as *T. aggressivum.* In view of the poor differentiation of these anamorphs, accurate reference to the strain used in experiments with *Trichoderma* is crucial.

Generally, *Trichoderma* species produce a great diversity of secondary metabolites (Ghisalberti and Sivasithamparam 1991) that, in addition to their enzymatic actions, antagonize host fungi over some distance, causing cell vacuolization, collapse, and disintegration of the cytoplasm. A volatile fungal inhibitor from *T. harzianum* has been identified as an alkyl pyrone

(Claydon et al. 1987). Peptaibols are a class of potent polypeptide antibiotics, commonly found in species of *Trichoderma, Gliocladium,* and *Clonostachys,* and the often-observed bursting of hyphal tips induced by *Trichoderma* has been ascribed to such compounds (Brückner and Przybylski 1984; Brückner et al. 1989). The mycoparasitic interaction between *T. harzianum* and *Lentinula edodes* on agar has been characterized as (1) parasitic contact, (2) direct penetration of host cells, and (3) production of diffusible toxic agents prior to physical contact (Hashioka et al. 1961; Hashioka and Komatsu 1964; Komatsu and Hashioka 1964; Hashioka and Fukita 1969; Komatsu 1976; Tsuneda and Thorn 1995). In contrast with other species of the genus, *T. virens* produces gliotoxin, which has a very specific fungitoxic effect (Weindling 1941; Webster and Lomas 1964).

Several species of *Hypocreopsis* are known to be fungicolous, including *H. lichenoides,* which occurs with species of *Hymenochaete* on tree bark (Cauchon and Ouellette 1964), and *H. xylariicola,* found on old perithecia of *Xylaria* species (Samuels 1988). The *Hypocreopsis* species cannot be cultured, and their physiology is unknown.

The genus *Hypomyces* comprises the most characteristic mycoparasites on macromycetes (Fig. 17.2). The genus is heterogeneous (Rogerson and Samuels 1994; Põldmaa 2000). Some species colonize host sporocarps systemically, are not known to have an anamorph, and produce ascospores that do not germinate *in vitro* (Fig. 17.3). Other species cause decay of sporocarps in various stages of development, form copious anamorphs, and grow easily in culture (Fig. 17.4). Ecological and morphological differences may justify the recognition of several genera (Põldmaa 2000). Species of *Hypomyces* can be grouped according to host type (viz., discomycetes, boletes, aphyllophorales, or agarics) but members of the last two groups overlap. The largest number of *Hypomyces* (23 species) parasitizes aphyllophorales and are not usually host-specific (Rogerson and Samuels 1993, 1994; Põldmaa 1999, 2000). Anamorphs, which originally were distributed among 12 genera, are confirmed for most of these species. *Hypomyces* species that parasitize boletes or agarics tend to be specific to a host family or genus (Rogerson and Samuels 1989; Sahr et al. 1999).

Rogerson and Samuels (1993) combined most anamorphs of agaricicolous and aphyllophoricolous *Hypomyces* species, including species of *Arnoldiomyces, Eurasina, Helminthophora, Pseudohansfordia, Sibirina, Sympodiophora,* and part of *Trichothecium,* in the expanded anamorph genus *Cladobotryum* (Figs. 17.2 to 17.4). The phialidic anamorphs of other species are mostly verticillium-like. Chlamydosporic or aleurial or bulbillar synanamorphs often are formed and classified in the genera *Sepedonium, Mycogone, Blastotrichum,* *Stephanoma,* or *Papulaspora.* The dormant aleurioconidia can be activated by basidiome tissue or extracts from several basidiomycete species (Holland and Cooke 1990).

Several smaller genera that share characters with species of *Hypocrea* and/or *Hypomyces* also contain fungicolous species. Most of the members of *Sphaerostilbella* (anamorphs *Gliocladium* sensu stricto) grow mainly on *Stereum* species (Seifert 1985). *Sporophagomyces* (anamorph acremonium-like) (Fig. 17.5) is characterized by a subiculum that traps hosts basidiospores, which obviously serve as source of nutrients. The hosts are mostly members of the Ganodermataceae (Põldmaa et al. 1999). *Arachnocrea, Protocrea,* and *Podostroma* also include fungicolous species, but whether these genera are distinct from *Hypocrea* is dubious, and their taxonomy currently is being revised.

Nectriaceae. *Nectria* was, until recently, a vast and heterogeneous aggregate that now is subdivided into several distinct genera (Rossman et al. 1999). Most of the approximately 60 species of *Cosmospora* (formerly *Nectria* subgenus *Dialonectria* = *N. episphaeria*-group) generally are fungicolous (Rossman et al. 1999). The most conspicuous species occur on carbonized perithecia of pyrenomycetes, most notably members of the Xylariales and Diatrypales, and less frequently on loculoascomycetes; one species is found on a discomycete. A few species that have been collected repeatedly are known to be restricted to certain host fungi. Cosmospora leptosphaeriae, for example, occurs mainly on *Leptosphaeria* and *C. coccinea* (synonym *Nectria cosmariospora*) on polypores (Samuels et al. 1991).

Bionectriaceae. The family is centered around *Bionectria,* the former *Nectria ochroleuca* group (Booth 1959; Schroers et al. 1999; Schroers 2000, 2001). The genus is comprised of a series of destructive fungicolous species with anamorphs in *Clonostachys* (*Gliocladium* sensu lato, particularly *C. rosea* and related species). The *Nectria* species with *Sesquicillium* anamorphs now also are classified in *Bionectria* (Schroers 2001). Some of them are fungicolous (Samuels 1989). A group of morphologically similar mycoparasitic pyrenomycetes with yellow perithecia (five fungicolous and seven lichenicolous species, from the genera *Hypocreopsis, Nectriopsis,* "*Nectria*", *Ijuhya* [synonym *Peristomialis*]), and *Trichonectria*) was revised by Samuels (1988); the *Nectria* species of that study are now classified in *Hydropisphaera* and *Bionectria. Pronectria* (14 species) is mostly lichenicolous (one algicolous species known). The three species of *Paranectria* are all lichenicolous.

Nectriopsis includes 39 species, although they are somewhat heterogeneous. Most of them grow on

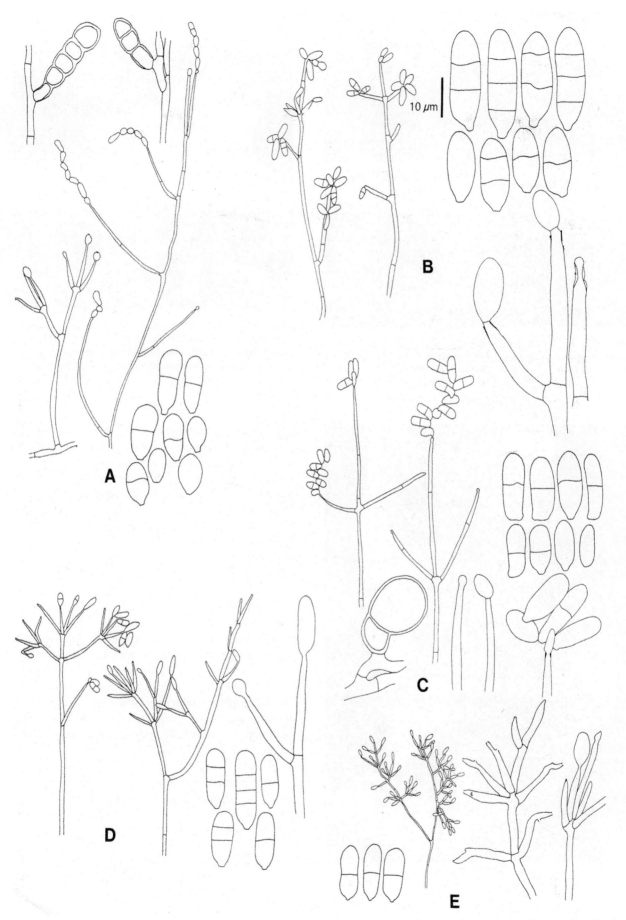

FIGURE 17.2 Morphologic variation in *Cladobotryum* anamorphs of *Hypomyces*. **A.** *H. aurantius*. **B.** *H. rosellus*. **C.** *H. subiculosus*. **D.** *H. semitranslucens*. **E.** *H. sympodiophorus*. Scale bar applies to detailed drawings of conidiophores and conidia. (Reprinted with permission, from Rogerson and Samuels 1993.)

10 μm

FIGURE 17.3 Mummified fruitbody of an agaric (probably *Lactarius* species) parasitized by *Hypomyces lactifluorum*. (Photo by K. Põldmaa)

FIGURE 17.5 Fanlike colony of *Sporophagomyces chrysostomus* hanging under a basidioma of *Ganoderma applanatum*. The brownish color of the subiculum is due to the basidiospores of the host trapped in the subiculum. (Photo by K. Põldmaa)

FIGURE 17.4 Subiculum and perithecia of *Hypomyces rosellus* on *Basidioradulum radula*. (Photo by U. Kõljalg)

myxomycetes, pyrenomycetes, loculoascomycetes, or discomycetes, and at least six grow on lichens. Three species of *Trichonectria* are known from nonlichenized ascomycetes, and six, *T. hirta* and *T. rubefaciens,* are from lichens (Samuels 1988; Sérusiaux et al. 1999). White to yellow perithecia commonly are observed in fungicolous taxa (Samuels 1988). That feature also is found in the superficially similar Tubeufiaceae (Dothideales), which contains several fungicolous taxa (Rossman 1987).

Clavicipitaceae. Whereas numerous species of the genus *Cordyceps* are entomogenous, three species commonly grow on *Elaphomyces* species. Among numerous anamorph genera associated with *Cordyceps, Verticillium* sect. *Prostrata,* now *Lecanicillium* (Gams 1971; Zare et al. 2000; Zare and Gams 2001), is comprised of species, some of which exhibit high levels of chitinase activity and have great potential as entomogenous, nematophagous (mainly parasitizing eggs), and fungicolous biocontrol agents.

Niessliaceae. Although the majority of species in this family are saprotrophic, some appear to be fungicolous, particularly the *Monocillium* anamorphs of *Niesslia.* Whereas species of *Circinoniesslia* and *Valetoniella* have been found only on perithecia of various *Nectria* species, *Trichosphaeriella decipiens* mostly is reported on sporocarps of aphyllophorales (Samuels and Barr 1997). Three species of *Niesslia* are lichenicolous.

Dothideales (Pleosporales). The Tubeufiaceae comprise a large number of fungicolous taxa found on melioaceous (18 species) or phyllachoraceous (nine species) leaf-inhabiting ascomycetes, as well as rusts (three species). Other taxa grow on decaying wood (20 species) (Rossman 1987). Associated anamorphs often have helicoid conidia. The Dacampiaceae include several genera with exclusively lichenicolous species: *Clypeococcum* (six species), *Dacampia* (three species), *Kalaallia* (one species), *Polycoccum* (30 species), *Pyrenidium* (15 species, mostly undescribed), and *Weddellomyces* (eight species). Most of the species are specialized on a single host genus or species. The genus *Lichenopeltella* (Microthyriaceae) comprises a large number of lichenicolous species, most of which parasitize macrolichens (Aptroot et al. 1997). Other genera of the Dothideales that include only lichenicolous species are *Buelliella*, *Cercidospora*, *Didymellopsis*, *Endococcus*, *Lichenostigma*, *Sagediopsis*, *Sphaerellothecium*, and *Zwackhiomyces*.

Orders Relatively Rich in Lichenicolous Fungi

Arthoniales. Most species of the genera *Arthonia*, *Lecanographa*, and *Opegrapha* are lichenized. Approximately 45 of the species of *Arthonia* are lichenicolous (Fig. 17.6) (Clauzade et al. 1989; Grube et al. 1995), however, as are three of the *Lecanographa* species (Egea and Torrente 1994), and 35 of the *Opegrapha* species (Fig. 17.7) (Clauzade et al. 1989; Hafellner 1994). In addition, at least 23 species of *Plectocarpon* are lichenicolous (Diederich and Etayo 1994).

Caliciales. Species of the Caliciales order can be lichenized, lichenicolous, fungicolous, or saprotrophic on

FIGURE 17.7 *Opegrapha cladoniae*, an ascomycete confined to *Cladonia* species. (Photo by P. Diederich, width of picture 1.7 mm)

bark or wood. Several genera (e.g., *Chaenothecopsis*, *Microcalicium*, *Sphinctrina*) comprise a high percentage of lichenicolous species, many of which grow on other lichenized species of Caliciales. *Microcalicium arenarium* develops either on the saxicolous lichens *Psilolechia lucida* and *P. clavulifera* (lichens with a *Stichococcus* photobiont) or on saxicolous *Stichococcus* species.

Lecanorales. *Toninia* comprises numerous lichens, including several species parasitic on other lichens, and some nonlichenized, lichenicolous fungi. The genera *Buellia*, *Carbonea*, *Catillaria*, *Rhizocarpon*, *Tephromela*, and *Thelocarpon* include mostly lichenized species as well, some of which develop on other lichens; a few species are nonlichenized but lichenicolous. *Phacopsis* (14 species) is exclusively lichenicolous; most species grow on members of the Parmeliaceae.

Ostropales. In the Odontotremataceae, *Odontotrema* (including *Lethariicola*) comprises 15 lichenicolous and some nonlichenicolous species (Lumbsch and Hawksworth 1990; J. Etayo and P. Diederich, personal

FIGURE 17.6 *Arthonia molendoi*, an ascomycete growing on *Xanthoria elegans*. (Photo by P. Diederich, width of picture 2.2 mm)

communication) and *Nanostictis* (3 + 2 undescribed; J. Etayo and P. Diederich, personal communication).

Verrucariales. Although most species of the Verrucariales are lichenized, a number of species of *Verrucaria* are parasitic lichens. Several related genera include only non-lichenized, lichenicolous species: *Adelococcus* (three species; Matzer and Hafellner 1990), *Merismatium* (nine species; Triebel 1989), *Norrlinia* (two species; Santesson 1989), and *Stigmidium* (approximately 70 species; Clauzade et al. 1989; Roux and Triebel 1994). Species of *Muellerella* (12 species; Triebel 1989) are lichenicolous or hepaticolous.

Genera Incertae Sedis. The genus *Abrothallus* includes at least 20 species that grow on macrolichens (mainly belonging to the Parmeliaceae). At least 30 species of *Dactylospora* are lichenicolous, whereas other members of the genus are muscicolous or saprotrophic on wood.

Anamorphic Conidial Ascomycota

Species of the heterogeneous genera *Acremonium* (mostly anamorphs of Hypocreales), *Verticillium* and *Lecanicillium* (anamorphs of Hypocreales and Clavicipitaceae), and *Chalara* (anamorphs of Leotiales) commonly are found in fungicolous associations. In particular, the species *A. strictum*, *A. berkeleyanum* (synonym *A. butyri*), *A. domschii*, *A. crotocinigenum*, *A. psammosporum*, *V. luteo-album* (synonym *V. tenerum*, often cited erroneously as anamorph of "*Nectria*" *inventa*), *V. fungicola*, *Lecanicillium lecanii*, *L. psalliotae*, *Simplicillium lamellicola*, and *V. leptobactrum* grow on a great diversity of substrata. A new section, *Lichenoidea* of *Acremonium*, was described for nine mostly common lichenicolous species with unbranched conidiophores, some of which are anamorphs of *Pronectria* (a segregate of *Nectria*) (Lowen 1995). More than 100 lichenicolous conidial fungi are known currently (see Hawksworth 1979a, 1981b). In addition, the genera *Acrodontium*, *Calcarisporium*, *Gabarnaudia*, and *Rhinotrichella*, related to various other ascomycete orders, contain common fungicolous species that will be dealt with in the following sections.

BASIDIOMYCOTA

Mycoparasitism is particularly widespread in various groups of Heterobasidiomycetes (Urediniomycetes, Ustilaginomycetes, and phragmobasidial Hymenomycetes).

Urediniomycetes, Platygloeomycetidae

Two major lineages are distinguished within the class Urediniomycetes (Swann and Taylor 1995a, 1995b, 1995c), the Urediniomycetidae and Platygloeomycetidae. The subclass Platygloeomycetidae (or Sporidiales clade) comprises the majority of the species formerly included in the Auriculariales sensu lato, for the most part parasites of fungi, mosses, ferns, and flowering plants (Bandoni 1984). Besides a series of saprotrophic yeastlike genera and the dicot-inhabiting smuts (*Microbotryum*), this subclass includes four orders with mycoparasitic genera.

The parasite and host are generally in contact through haustoria. In a few cases, the parasites have large numbers of a specialized organelle in the periplasmic region. These organelles first were described by Kreger-van Rij and Veenhuis (1971) for *Sporidiobolus* species and called lenticular bodies. When described from *Cryptomycocolax* species, and later from *Colacogloea*, the structures were referred to as colacosomes (Oberwinkler and Bauer 1990; Bauer and Oberwinkler 1991). The organelles are vesicular structures with an electron-opaque core and an electron-transparent sheath (Bauer and Oberwinkler 1991); the vesicular content projects through the cell wall of the parasite and makes contact with the plasmalemma of the host. Lenticular bodies also have been found in *Bensingtonia* and *Rhodosporidium*, which suggests that those yeasts also may have mycoparasitic capabilities (Boekhout et al. 1992).

Atractiellales. Although not originally described as such, species of *Chionosphaera* and *Stilbum*, in the Atractiellaceae, may be mycoparasites (Bandoni 1995; Roberts 1997). Two species of *Chionosphaera* grow on lichens (Diederich 1996).

Platygloeales. Cystobasidiaceae. The genera *Platygloea*, *Cystobasidium*, *Mycogloea*, and *Colacogloea* contain mycoparasitic species. *Platygloea fimetaria* and *Colacogloea* (*Platygloea*) *peniophorae* parasitize corticiaceous hosts, but also *Dacrymyces* and *Poria* (Bauer and Oberwinkler 1991; Bandoni 1995). *Cystobasidium lasioboli* is found on *Lasiobolus*; two other species in the genus elicit formation of conspicuous galls on lichens (Diederich 1996). *Naohidea* forms pustular basidiomes on old pyrenomycetes, and the intrahymenial mycoparasites *Occultifur* and *Kryptastrina* are found in basidiomata of *Dacrymyces* and on corticioid fungi, respectively (Oberwinkler 1990).

Cryptomycocolacales. Cryptomycocolacaceae. *Cryptomycocolax*, possibly one of the most primitive basidiomycetes, is unusual in having simple septal pores and apparently some ascomycete features. It was found parasitizing an unidentified ascomycete in *Cirsium* culms in Costa Rica. The host fungus produces botryose outgrowths that are engulfed by the parasite (Oberwinkler and Bauer 1990).

Heterogastridiales. Heterogastridiaceae. *Heterogastridium pycnidioideum* (Oberwinkler et al. 1990a, b), with the anamorph *Hyalopycnis blepharistoma*, parasitizes various fungi on rotting plant material, including decaying walnuts (*Juglans ailanthifolia*) in Japan (CBS, unpublished data), and various fungal sporocarps (Bandoni and Oberwinkler 1981). It can be grown axenically, but some strains require a host, such as *Plectosphaerella cucumerina*, for pycnidium formation *in vitro* (G. Fischer and W. Gams, unpublished data). T. Boekhout (unpublished data) noted the presence of lenticular bodies in *Heterogastridium*.

Ustilaginomycetes

Tilletiopsis (revised by Boekhout 1991) is an anamorph genus that contains mycoparasites of powdery mildews. The redefined genus *Pseudozyma* (Boekhout 1995), which is characterized by dimorphic growth, variously shaped blastoconidia borne on sympodially proliferating conidiophores, and the absence of ballistoconidia, comprises anamorphs of *Ustilago* and also includes the mycoparasites of powdery mildews formerly described as *Stephanoascus* species (Traquair et al. 1988).

Hymenomycetes, Tremellomycetidae

The order Tremellales includes species with a dimorphic life cycle, a septal pore apparatus with sacculate caps, and a tremelloid basidium. That most or all species of the Tremellales are mycoparasites has been reported repeatedly. The dikaryotic, somatic phase shows restricted growth, which may reflect the mycoparasitic nature of many, if not all, of the species. Many species have monokaryotic "haustorial branches" subtended by a clamp that consists of a bulbous base and a tubular appendage. The tubular appendage has been shown to attach to and/or penetrate the hyphal wall of a host fungus (Bauer and Oberwinkler 1990a, 1990b; Oberwinkler and Bandoni 1981, 1982; Oberwinkler et al. 1984). The haustorium may have an absorptive role, leading eventually to the degeneration of the host cytoplasm (Oberwinkler et al. 1984). Micropore openings have been demonstrated in the haustoria of some species (*Syzygospora, Phragmoxenidium*; Bauer and Oberwinkler 1990b; Oberwinkler et al. 1990b). The pore membrane appears to be continuous with the plasmalemma of both cells (Zugmaier et al. 1994). Somewhat larger pores are present in *Tetragoniomyces uliginosus* (Oberwinkler and Bandoni 1981; Bauer and Oberwinkler 1990a). Haus-

toriumlike cells also can be observed in pure cultures without a host fungus.

Explicit information on fungicolous relationships is available for six of the 10 families included in the order (Kirk et al. 2001).

Tremellaceae. *Tremella* (more than 170 species known), *Holtermannia, Sirotrema, Trimorphomyces, Xenolachne, Bulleromyces,* and probably *Itersonilia* (anamorphic) contain numerous mycoparasites. *Tremella* is the largest and most heterogeneous genus in the Tremellaceae. It includes mycoparasitic species growing on hymenia of aphyllophorales, in basidiomes of Dacrymycetales, on ascomata and/or stromata of pyrenomycetes, and on hymenomycetes and lichens (Bandoni 1995; Diederich 1996). *Trimorphomyces* and *Sirotrema* species have been found on conidial masses of *Arthrinium* or on hymenia of Hypodermataceae, respectively (Oberwinkler and Bandoni 1983), and *Holtermannia* has been found on a polypore (Bandoni 1995). *Tremella*-type haustoria also have been found in the yeastlike *Bulleromyces albus*, suggesting that it has mycoparasitic capacities (Boekhout et al. 1991). *Itersonilia* species have been found as mycoparasites on powdery mildews. *Xenolachne* is found on inoperculate discomycetes. The lichenicolous *Biatoropsis usnearum* may also belong here.

Syzygosporaceae. (synonym Carcinomycetaceae). *Syzygospora* (fide Ginns 1986; originally *Syzygospora, Christiansenia,* and *Carcinomyces*; see Oberwinkler and Bandoni 1982). The Syzygosporaceae comprise parasites on *Phanerochaete* species, *Gymnopus* species, and *Marasmius* species, often causing gall-like deformations. They have holobasidia of various shapes, and in some species, "zygoconidia" arise from the fusion of blastoconidia thrust forward by adjacent cells; their septa lack parenthesomes (Oberwinkler and Bandoni 1982). Ginns (1986) synonymized *Syzygospora, Christiansenia, Carcinomyces* in *Syzygospora*, recognizing nine species in three subgenera, distinguished according to basidial and conidial shape. A key for the species found in Norway is given in Torkelsen (1996). The species *C. mycetophila, C. effibulata,* and *C. tumefaciens* grow on *Gymnopus dryophilus,* inducing gall-like fructifications (Fig. 17.14) (Ginns and Sunhede 1978; Oberwinkler and Bandoni 1982; Oberwinkler et al. 1984; Ginns 1986). *Syzygospora alba* produces tremelloid sporocarps on an unknown host (Oberwinkler and Lowy 1981). The common and widespread lichenicolous species *Syzygospora bachmannii* (Fig. 17.8) and *S. physciacearum* grow on *Cladonia* species and on Physciaceae (Diederich 1996).

Filobasidiaceae (raised by some authors to Filobasidiales). *Filobasidium, Filobasidiella,* and *Cystofilobasidium* are more or less mycoparasitic. The otherwise

FIGURE 17.8 *Syzygospora bachmannii*, a heterobasidiomycete inducing the formation of galls on *Cladonia* species. (Photo by P. Diederich, width of picture 10 mm)

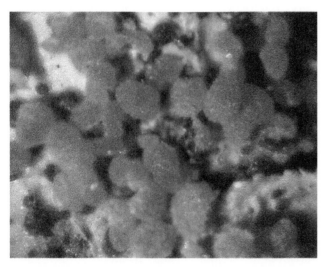

FIGURE 17.9 *Marchandiomyces corallinus*, a sclerotial basiciomycete anamorph (Corticiaceae?) parasitizing thalli of *Parmelia*. (Photo by P. Diederich, width of picture 1.4 mm)

yeastlike genera *Filobasidium* (Olive 1968) and *Cystofilobasidium* have hyphal structures with haustoria only in the dikaryotic phase; species of *Filobasidium* and the vertebrate parasites in *Filobasidiella* (best known as the *Cryptococcus neoformans* anamorph) also have haustoria, suggesting mycoparasitic capacities (Bandoni 1995). *Cystofilobasidium capitatum* was found in the gleba of Phallaceae (Oberwinkler et al. 1983).

Rhynchogastremaceae. Mycoparasitic species of *Rhynchogastrema* contact their various hosts in the soil with haustoria, but axenic growth is possible (Metzler et al. 1989).

Phragmoxenidiaceae (doubtful classification). *Phragmoxenidium mycophilum* (Oberwinkler et al. 1990b) grows on *Uthatobasidium fusisporum*, contacting its host through micropores; its septa lack a parenthesome.

Tetragoniomycetaceae. *Tetragoniomyces uliginosus* (Oberwinkler and Bandoni 1981) grows on *Ceratobasidium*.

Hymenomycetes, Hymenomycetidae

Aphyllophorales. The aphyllophorales is an artificial assemblage of divergent elements of homobasidiomycetes. Destructive necrotrophic mycoparasites of some ascomycetes (e.g., *Ceratocystis*) have been found *in vitro* among diverse genera such as *Bjerkandera*, *Lenzites*, *Trichaptum*, *Trametes*, *Schizophyllum*, and *Pleurotus* (Griffith and Barnett 1967; Traquair and McKeen 1978). The conidia and hyphae of the hosts are affected by hyphal interference. Mutual displacement of several aphyllophorales seems to be an important phenomenon in wood decay (see "On Polypores" under "Fungi on Basidiomycota," later in this chapter). Tzean and Estey (1978b) discovered that the wood-decaying *Schizophyllum commune* is a potent mycoparasite of many plant-pathogenic fungi. It coils around and penetrates the host hyphae, which it then destroys. A suitable host is the nematophagous and fungicolous *Arthrobotrys oligospora*.

Tzean and Estey (1991, 1992) also described an arthroconidial basidiomycete, *Geotrichopsis mycoparasitica*, with characteristic spindle-shaped chlamydospores. The parasite attacks *Arthrobotrys oligospora*, *Monacrosporium cionopagum*, and other nematophagous fungi; some Mucoraceae; *Mortierella* species; *Pythium* species; *Chalara elegans*; and *Rhizoctonia solani* but not species of *Aspergillus*, *Penicillium*, *Geotrichum*, *Drechslera*, and *Nematoctonus*. *Athelia* (*Corticium*) *rolfsii* and *Heterobasidion annosum* are resistant. However, *Athelia arachnoidea* is mainly lichenicolous. The lichenicolous sclerotial genus *Marchandiomyces* (Fig. 17.9) (teleomorph *Marchandiobasidium*) probably belongs to the family Corticiaceae.

Boletales and Agaricales. A few fungicolous boletes and agarics from diverse groups are known. Host-specific associations are found in the genera *Asterophora* and *Squamanita*.

MAJOR GROUPS OF FUNGICOLOUS FUNGI, ARRANGED BY HOST GROUP

For the present survey, we divide the fungicolous fungi into five major groups according to the substrata on which they grow and their biotopes: (1) fungal sporocarps; (2) lichens; (3) biotrophic plant parasites; (4) mycelia, sclerotia, and spores in soil; and (5) aquatic fungi and fungus-like microorganisms. The five groups are not sharply delimited, however, and they also merge with saprotrophs that grow incidentally on other fungi. For example, certain fungi normally obtained as soil isolates also commonly are associated with macromycetes or lichens in decay. Similarly, the presence of microscopic endoparasitic *Tremella*-related species in *Asterophora* chlamydospores (Laaser et al. 1988; Prillinger et al. 1993) is consistent with the definitions of both groups 1 and 4; and zoosporic fungi can parasitize oomycete hosts both in water (group 5) and in soil (group 3).

Different methods are used to study each of these major groups. We discuss those methods and identification sources, as well as the species richness and geographic distribution of each group. The numbers of fungicolous species in various fungal orders that pertain to each major group are summarized in Table 17.1.

FUNGI ON SPOROCARPS OF FUNGI OR MYXOGASTREA

It is easy to single out a group of fungi that live on fresh or decaying sporocarps of other, nonlichenized fungi. The substrata of such SCIF usually consist of dry or soft macromycete sporocarps (which may show more or less conspicuous deformations as a result), but the SCIF often colonize their anamorphs, conidiomata, conidiophores, and conidia as well. We also include under this heading fungi growing on myxomycete sporangia and a few that grow on above-ground conidiophores or conidiomata of anamorphic fungi. The SCIF are mainly ascomycetes or their anamorphs, but they also include some Zygomycota and many Basidiomycota; most of the latter are Tremellales, but a few others are Agaricales (Buller 1924; Redhead et al. 1994). The species of SCIF are relatively well-known.

The SCIF on a decaying sporocarp gradually are replaced by a wide array of saprotrophic fungi (many species of *Penicillium* and *Cladosporium*) that are in no way specialized to the substratum; we do not consider them here. Certain species of *Penicillium*, however, specifically are associated with the decay of sporocarps of particular fungi, as in *Chroogomphus* (*Gomphidius*) *rutilus* and *Boletus parasiticus*. *Hygrocybe virginea* (synonym *Camarophyllus niveus*) is colonized regularly by the common soil fungus *Paecilomyces marquandii* (Ellis and Ellis 1988; CBS, unpublished data).

The fungal substrata available to SCIF are by no means homogeneous. Here we group the SCIF into a polyphagous group and major-host groups.

Polyphagous Fungicolous Fungi

Many kinds of fungal substrata can function as hosts of polyphagous fungicolous fungi, including *Clonostachys* (*Gliocladium*) *rosea*, *Trichothecium roseum*, *Calcarisporium arbuscula*, *Trichoderma* species, *Acremonium strictum*, *Lecanicillium* [*Verticillium*] *lecanii*, and *V. luteo-album* (Domsch et al. 1980).

Clonostachys rosea is a ubiquitous mycoparasite whose hyphae penetrate and destroy those of many host fungi, including some Mucorales, *Ceratocystis fimbriata*, and even the potent toxin producer *Trichothecium roseum* (Barnett and Lilly 1962; Barnett and Binder 1973; Berry and Deacon 1992; Jeffries and Young 1994). In natural infections of *Botrytis aclada* on onions, *C. rosea* hyphae grow along the host hyphae, which they contact with appressorial branches, and then penetrate the hyphal wall (Walker and Maude 1975). Hyphae of *C. rosea* also can attack the sclerotia of *Botrytis* species.

Calcarisporium arbuscula grows equally well on many kinds of Ascomycota and Basidiomycota (Barnett and Lilly 1958; Nicot 1968), which usually do not survive the infection. *Acremonium strictum* has been found on many saprotrophic fungi, several plant-pathogenic fungi (Gams 1971), and on the mycoparasite *Mycogone perniciosa* (Gandy 1979), whose growth it inhibited.

Verticillium luteo-album has a wide range of hosts, but host susceptibility seems to vary among taxa, according to the type and age of the host structure attacked, and with the culture medium used. *In vitro*, the species has been observed to overgrow and destroy colonies of *Alternaria brassicae*, *A. raphani*, *Phoma lingam*, *Trichothecium roseum*, *Ulocladium atrum*, and *Rhizoctonia solani*. *Trichoderma harzianum*, in turn, destroys *V. luteo-album* (Tsuneda et al. 1976; Tsuneda and Skoropad 1980).

Sistotrema brinkmannii, found mainly on decaying sporocarps of aphyllophorales, is a common air-borne contaminant that can overgrow and destroy any *in vitro* fungal culture. *Athelia arachnoidea* is a common corticiaceous species that overgrows and kills epiphytic lichens (and algae) and can destroy whole lichen communities, especially in areas with significant air pollution (Arvidsson 1976, 1978; Parmasto 1998). The anamorph of this species, *Fibulorhizoctonia carotae*, is a cold-storage pathogen of carrot in Europe, North America, and India. *Athelia arachnoidea* is the only known lichenicolous fungus whose anamorph is nonlichenicolous (Adams and Kropp 1996).

Fungi on Slime Molds

Sporocarps (but not the plasmodia) of slime molds (Myxogastrea) often are colonized by more or less specialized fungi. Most obligate myxomyceticolous conidial fungi smother the sporangia, usually turning them white. Whether most of those should be viewed as parasites is doubtful because they colonize an already terminal phase of the sporocarp and flourish after a great many spores already have been liberated (Hawksworth 1981b). However, the relationship between myxomyceticolous fungi and their hosts could be regarded as parasitic because the mycelium usually grows over the entire surface of the host fructification (Rogerson and Stephenson 1993). Hyphae penetrate the spore mass of the host, killing the spores and preventing their release. In the vast majority of the collections examined, only a single colonizing species was present, and usually most of the fructifications of the host were colonized.

Rogerson and Stephenson (1993) distinguished two groups of parasites: those that parasitize the calcium-rich fructifications of the Physarales (*Gliocladium album, Nectriopsis violacea, Sesquicillium microsporum*) and those that attack only noncalcareous myxomycetes (*Aphanocladium album* sensu stricto; *Byssostilbe stilbigera,* usually observed as the anamorph, *Blistum tomentosum*). More specifically, *N. violacea* has been found only on *Fuligo* species, and *B. tomentosum* had been found only on members of the Trichiales. Other myxomyceticolous ascomycetes occur on different host species or are too poorly collected to allow for conclusions about their host specificity. Most myxomyceticolous conidial fungi occur on various host species. *Nectriopsis exigua* (anamorph *Verticillium rexianum*), one of the most ubiquitous species, has been recorded from all major groups of myxomycetes and so is *Stilbella byssiseda* (Fig. 17.10). All the species mentioned here are found exclusively on sporocarps of myxomycetes.

Some less specific colonizers are also known. The hemiascomycete *Dipodascus macrosporus* can live in the slime trail of *Badhamia utricularis* cultured on cornmeal agar. If ingested by the plasmodium, the fungus resists digestion and seems to play the role of a facultative parasite (Madelin and Feest 1982). The yeast cells of *D. macrosporus* multiply within the plasmodium as well as on the supporting medium. Also, various fungi other than ascomycetes and their conidial relatives (e.g., the Zygomycetes *Umbelopsis* [*Mortierella*] *ramanniana* and *Mucor hiemalis;* Helfer 1991) have been encountered on sporocarps of myxomycetes. The occasional presence of hyphae of unidentified saprotrophic or wood-decaying basidiomycetes on myxomycete fructifications is considered to be coincidental co-occurrence (Rogerson and Stephenson 1993).

Fungi on Ascomycota

On Sporocarps of Discomycetes (Cup Fungi) and on Fleshy Epigeous Sporocarps of Discomycetes

FIGURE 17.10   Synnemata of *Stilbella byssiseda* growing on a slime mold. (Photo by U. Kõljalg)

TABLE 17.4

Characteristic Ascomycetes and Their Anamorphic Relatives Parasitic on Nonlichenized Discomycete Sporocarps

Fungicolous species	Host
Hypomyces cervinigenus (anamorph *Mycogone cervina*)	*Mycolachnea* (*Humaria*) *hemisphaerica*
Hypomyces stephanomatis (anamorph *Stephanoma strigosum*)	*Helvella* species
Stephanoma tetracoccum	Geoglossaceae
Hypomyces papulasporae (anamorph *Papulaspora candida*)	Geoglossaceae
Hypomyces mycogones	Geoglossaceae
Sphaeronaemella helvellae	*Gyromitra* species
Hormiactis species	*Morchella* and *Peziza* species
Hypomyces leotiicola (anamorph *Sepedonium leotiarum*)	*Leotia lubrica*
Nectriopsis discophila	*Lachnum* species
Nectria sepultariae	Inoperculate discomycetes
Nectria discicola	Inoperculate discomycetes
Trichonectria albidopilosa	Inoperculate discomycetes
Exochalara longissima	*Bulgaria inquinans* (CBS)

Data from Rogerson and Samuels (1985), Helfer (1991), and CBS (unpublished data). Additional species that grow on discomycetes are listed by Hawksworth (1981a).

(Pezizales, Leotiales). The most characteristic colonizers of ascomata of discomycetes are listed in Table 17.4. Most of them occur only on sporocarps of this group of hosts.

The heterobasidiomycete *Xenolachne longicornis* can grow on ascomata of *Cudoniella* species and *Disciniella* species where the parasite replaces the hymenium (Hauerslev 1977; Jülich 1983). A stronger displacement of ascal hymenia by a parasite occurs in *Helicogonium* (Baral 1999; see "Leotiales" under "Ascomycota," in the section "Taxonomic Groups of Fungicolous Fungi and Fungus-like Microorganisms," earlier). *Cystobasidium lasioboli* was found in Europe growing on *Lasiobolus pilosus*, and *Exidiopsis fungicola* was found on *Mollisia cinerea* (Jülich 1983). *Clitocybe sclerotoidea* can parasitize and deform *Helvella lacunosa* in the western United States (Trappe 1972). Many fungal interactions can be observed on dung; the growth of *Ascobolus* species, *Pilobolus crystallinus*, and other coprophilous fungi, for example, is strongly inhibited by *Coprinus heptemerus*, a case of lethal hyphal interference (Ikediugwu and Webster 1970a, 1970b; Ikediugwu 1976a).

Penicillium glabrum has been found on twigs affected by *Monilinia laxa* peach twig blight. Its antifungal substances seem to render it suitable as a biocontrol agent (de Cal et al. 1988, 1990).

On Sequestrate Ascomycetes. Species of *Tuber* and other tuberoid hypogeous Pezizales often are colonized by species of *Acremonium*, *Verticillium* sect. *Prostrata*, and *Mortierella* (CBS, unpublished data). *Nodulisporium* (*Sporothrix*) *tuberum* was described from *Tuber maculatum* (Fontana and Bonfante 1971). A few species of *Sphaerodes* (synonym *Microthecium*) are specialized on hypogeous fungal substrata (Hawksworth and Udagawa 1977). *Battarrina inclusa* (Bionectriaceae) also is found inside ascomata of *Tuber puberulum* (Jeffries and Young 1994). The subterranean sporocarps of *Elaphomyces* often can be detected when the conspicuous epigeous ascomata of their parasites, *Cordyceps ophioglossoides*, *C. capitata*, and *C. canadensis*, appear. What impact they have on their hosts has not been investigated. The ascomata of *Elaphomyces* species also often are colonized by *Gabarnaudia tholispora* and other conidial fungi (CBS, unpublished data).

On Coriaceous Discomycetes (Phacidiales, Rhytismatales). Hawksworth (1981b) listed a few rather unspecific species of mycoparasites occurring on members of these orders. Species of *Sirotrema* (*Pseudostipella*) have been found growing on apothecia of *Lophodermium*, *Hypoderma*, and *Hypodermella* species (Ellis and Ellis 1988). A rather specific competitive mycoparasitic relationship seems to exist between *Hendersonia pinicola* or *H. acicola* and *Lophodermium concolor* or *Lophodermella sulcigena*. *Hendersonia pinicola* causes needle cast in *Pinus contorta* (Stahl et al. 1988); as secondary invaders, it and *H. acicola* can prevent ascoma formation in *L. sulcigena* and thus naturally control epidemics (Funk 1985; Jalkanen and Laakso

1986). Also *Hemiphacidium* species regularly follow *Bifusella* or *Lophodermella* species and other fungi on *Pinus* species, preventing them from fruiting (Funk 1985). *Cosmospora* (*Nectria*) *ganymede* has been reported from the rhytismataceous *Zeus olympius* (Minter et al. 1987).

Species of *Platygloea* (*Achroomyces*), *Mycogloea*, and *Sirobasidium* occur on species of *Colpoma* and *Lophodermium*. *Tremella juniperina* is found on sporocarps of *Colpoma juniperi*, and *T. translucens* is found on *Lophodermium* (Jülich 1983).

On Stromata and Ascomata of Pyrenomycetes and on Dry, Carbonized, Mostly Stromatic Fructifications (Especially Xylariaceae and Diatrypaceae). Common colonizers of old carbonized stromata include *Polydesmia pruinosa* and species of *Cosmospora*—for example, *C.* (*Nectria*) *episphaeria* and more rarely *C. purtonii* (e.g., on *Diatrype stigma*, *Hypoxylon fragiforme*, and other stromatic ascomycetes; Samuels 1976; Domsch et al. 1980; Samuels et al. 1991; Helfer 1991). The associated anamorphs, classified as varieties of *Fusarium aquaeductuum*, are equally common on these substrata. Old stromata of *Hypoxylon* species, as well as polypores, are commonly colonized by *Cosmospora vilior* (synonym *Nectria viridescens*) and its *Acremonium berkeleyanum* (synonym *A. butyri*) anamorph and by *Monocillium tenue* (Gams 1971). A unique fungus growing on carbonized perithecia of ascomycetes was described as *Hypomyces triseptatus* (Rossman and Rogerson 1981). Many Xylariaceae and *Physalospora obtusa* are among the hosts of *Calcarisporium arbuscula* (Hoch 1977b). *Capronia parasitica* grows frequently on old stromata of *Diatrype* and *Hypoxylon* but is easily overlooked because it is small and the same color as its substratum (Helfer 1991). The ascomata of *Capronia* (*Berlesiella*) *nigerrima* are found on *Eutypa* (Jeffries and Young 1994). Less common discomycetes found on such stromata are species of *Phaeohelotium*, *Bisporella*, and *Cistella* (Ellis and Ellis 1988). *Orbilia inflatula* (*O. auricolor*) is rather common on *Hypoxylon fragiforme* (Baral and Krieglsteiner 1985; Helfer 1991).

Tubeufia cerea (often with its *Helicosporium* anamorph), the most common north-temperate representative of the genus, occurs on old stromata of *Diatrype stigma*; *Graphostoma platystoma*; and species of *Eutypa*, *Eutypella*, and *Hypoxylon*, as well on other fungi (Barr 1980). Hawksworth (1981b) listed numerous potential colonizers of pyrenomycetes. Many more hypocrealean species growing on this and other groups of fungi were revised by Samuels (1988).

Certain perithecial Ascomycota, such as species of *Didymosphaeria* and *Phaeodothis* (*Didymosphaeria*) *winteri*, live inside the ascomata of other pyrenomycetes (see Munk 1957 and a revision by Aptroot 1995).

Ascodichaena rugosa is parasitized in southern Germany by *Keissleriella bavarica* (Lophiostomataceae), whereas *Tripospermum myrti* and *Eriomyces* species grow on both fungi (Butin 1981). *Neobarya* and some undescribed *Acremonium* species are obligate parasites on *Bertia moriformis* and some other pyrenomycetes; their physiology is not yet understood (Munk 1954, 1957; Helfer 1991). *Chaetosphaerella fusca* and *C. phaeostroma* (with *Oedemium* anamorphs) frequently are found on stromata of diatrypaceous fungi (Ellis and Ellis 1988). *Epicoccum nigrum* and *Coniothyrium olivaceum*, which can suppress canker development, have been considered as promising control agents of the anamorphic *Cytospora cincta*, the causal agent of canker in peach twigs (Royse and Ries 1978).

Catulus aquilonius (Mycosphaerellaceae) has been found on *Seuratia millardetii* (Myriangiales) growing on needles of *Abies balsamica* (Malloch and Rogerson 1978).

On Soft Perithecia (Particularly Hypocrealean Fungi). *Nectria cinnabarina* and its *Tubercularia vulgaris* anamorph are commonly overgrown by the hyphomycete *Tympanosporium parasiticum*, which can be grown in axenic culture (Gams 1974), as well as by many nonspecific colonizers. *Gabarnaudia tholispora* and an obligate parasitic species of *Acremonium* also have been observed on *N. cinnabarina* (W. Gams, personal observation). *Nitschkia parasitans* grows only on *Nectria cinnabarina* (Ellis and Ellis 1988). *Circinoniesslia nectriae*, *Valetoniella crucipila*, and *V. pauciornata* have been reported on *Nectria* species. *Nematogonum ferrugineum* (synonym *Gonatorhodiella highlei*) commonly colonizes *Neonectria* (*Nectria*) *coccinea* and related species; *in vitro*, it also grows on *Cladosporium*, *Penicillium*, and *Sporobolomyces* species (Walker et al. 1982). *Sphaerodes episphaeria* (synonym *Microthecium epimyces*) also has been found on *Hypomyces armeniacus* (Cannon and Hawksworth 1982).

On Anamorphic Ascomycota. A number of condial Ascomycota are hosts to biotrophic parasites. The entomogenous *Hirsutella citriformis*, a destructive parasite of brown planthoppers on the Solomon Islands, was colonized by *Calcarisporium* (*Cladobotryum*) *ovalisporum*; because this fungus requires sterilized host hyphae to grow on agar media, it also may be dependent on the fungal metabolite mycotrophein (Rombach and Roberts 1987). The biotrophic *Debaryomyces hansenii* can control penetration of citrus fruit by *Penicillium digitatum* (Droby et al. 1989); in high concentrations, the parasite inhibits the conidial germination and hyphal growth of its host, apparently through competition for nutrients. This yeast also acts similarly against *Rhizopus stolonifer*,

Botrytis cinerea, and *Alternaria alternata* on tomatoes and grapes.

Anamorphic Ascomycota also host necrotrophic parasites. *Hansfordia* (*Dicyma*) *pulvinata* is a destructive parasite on many dematiaceous conidial fungi (Hepperly 1986), particularly *Cercospora* (Hawksworth 1981b), but also *Passalora* (*Mycovellosiella*, *Cladosporium*, *Fulvia*) *fulva* on tomato (Peresse and Le Picard 1980; Le Picard and Trique 1987), and *P.* (*Cercosporidium*, *Phaeoisariopsis*) *personata* on peanut leaves (Mitchell et al. 1986, 1987). The fungus coils over hyphae and conidia of its host and kills them (Peresse and Le Picard 1980) with a fungistatic sesquiterpene metabolite, deoxyphomenone (Tirilly et al. 1991). A conidial suspension of this fungus with carboxymethyl cellulose sprayed on *P. personata* prevented secondary spread of the pathogen under moist conditions (Mitchell et al. 1986, 1987). *Passalora personata* also can be parasitized by *Cladosporiella cercosporicola* (Esquivel-R. 1984).

Biocontrol of the plant-pathogenic *Verticillium dahliae* has been attempted repeatedly, as reviewed under "Mycoparasites of Mycelia, Ectomycorrhizae, Sclerotia, and Spores in Soil." *Alternaria brassicae*, *Pleospora* species, and *Trichothecium roseum* are highly susceptible to hyphal interference by *Verticillium luteo-album* (Tsuneda et al. 1976; Tsuneda and Skoropad 1978, 1980), which forms coils and appressoria and sometimes penetrates the host hyphae. Turhan (1993) found a number of additional, interesting antagonists of *Alternaria alternata*, particularly *Dicyma olivacea*, *Stachybotrys elegans*, a species of *Sesquicillium*, five species of *Myrothecium*, and *Coniothyrium sporulosum*.

Penicillium funiculosum can smother various fungi in culture and thus prevent pineapple fruit diseases (Lim and Rohrbach 1980). *Aspergillus luchuensis* and a diversity of other fungi are parasitized in India by *Fusarium udum* (Upadhyay et al. 1981), which produces a diffusible toxin and induces vesicular deformations in its host.

The ascomycete *Letendraea helminthicola* parasitizes the hyphomycete *Helminthosporium velutinum* (Ellis and Ellis 1988). The discomycete *Bisporella pallescens* (synonym *Calycella monilifera*) fruits densely on the conspicuous black conidial patches of *Bispora antennata* on tree stumps (Jahn 1968).

In a search for parasites of *Macrophomina phaseolina* (synonym *Rhizoctonia bataticola*), *Arachniotus* species and *Aspergillus aculeatus* were found to be promising control agents (Dhingra and Khare 1973). In Varanasi, India, *Colletotrichum dematium* is parasitized by *Acremonium sordidulum*, which smothers the host and reduces its sporulation on the plant (Singh et al. 1978). Hawksworth (1981a) gives a long list of additional mycoparasites growing on conidial fungi.

Sphaeronaemella helvellae, *Trichoderma viride*, *Exobasidiellum* species, and others (Vakili 1985) are efficient antagonists of pathogens, such as *Fusarium verticillioides* (synonym *F. moniliforme* sensu stricto), *Cochliobolus* (*Helminthosporium*) *carbonum*, and *Colletotrichum graminicola* involved in stalk rot of corn.

Fungi on Basidiomycota

On Aphyllophorales. Most fungicolous colonizers of sporocarps of aphyllophorales are Ascomycota or related anamorphic fungi. Usually, they colonize the less protected hymenophore.

On Crustose, Mainly Resupinate, Aphyllophorales. *Sphaerostilbella aureonitens* (anamorph *Gliocladium penicillioides*), *S. lutea* (anamorph *G. aurifilum*), and *S. novaezelandiae* grow on basidiomata of several wood-decaying aphyllophorales. The first species and *Gliocladium polyporicola* (Fig. 17.11) are particularly common on *Stereum* species (Samuels 1976; Seifert 1985). *Sphaerostilbella berkeleyana* also grows on polypores (Samuels 1976; Helfer 1991; Candoussau and Magni 1995). Some species of *Hypomyces* (*H. sympodiophorus*, *H. albidus*, *H. corticiicola*) are found most frequently on the hymenophore of resupinate apyllophorales. *Cladobotryum stereicola* grows preferentially on *Chondrostereum purpureum* (K. Põldmaa, personal observation).

Hypocreopsis lichenoides has been found in central and northern Europe on species of *Hymenochaete*, especially *H. tabacina* (Cauchon and Ouellette 1964; Niemelä and

FIGURE 17.11 Synnemata of *Gliocladium polyporicola* growing on *Stereum* species. (Photo by K. Põldmaa)

Nordin 1985; Læssøe 1989; Kristiansen and Torkelsen 1994). In addition to species in the Hypocreales, several ascomycetes with dark perithecia grow on the hymenium of species in the aphyllophorales; *Helminthosphaeria odontiae* and *H. hyphodermatis* are found on *Hyphoderma tenue* and *Odontia cristulata*, and *Helminthosphaeria corticiorum* is found on *Peniophora* species (Doll 1973; Samuels et al. 1997).

Only a few ascomycetes have been found in the basidiocarps of corticioid fungi. *Helicogonium* (*Myriogonium*) *odontiae* occurs on *Dacryobolus sudans* (Cain 1948) and *H. jacksonii* occurs on *Ceraceomyces* (*Athelia*) *sublaevis* (White 1942; Parmasto 1974). Vegetative mycelium of *H. jacksonii* grows in the subiculum and subhymenium of the host, and scattered asci emerge among the basidia of the normally developed carpophore. It apparently does little harm to its host. The other species of the genus parasitize members of the Leotiales (see "Leotiales" under "Ascomycota," in the section "Taxonomic Groups of Fungicolous Fungi and Fungus-like Microorganisms," earlier).

Phaeospora corae and several undescribed species have been recorded from the lichenized corticiaceous *Cora glabrata* (Santesson 1989). *Phaeocalicium polyporaeum* (Caliciales) grows on basidiocarps of *Trametes versicolor* and *Trichaptum biforme* (Tibell 1984).

Acremonium psammosporum and other species of the genus commonly are found on *Peniophora* species (Helfer 1991; W. Gams, personal observation). *Denticularia limoniformis* is a host-specific parasite of *Hyphodontia* (*Grandinia*) *breviseta* (de Hoog 1978). Mycoparasitism of nematophagous species of *Arthrobotrys* that form mats on corticiaceous fungi is described in the section "On Sclerotia" under "Mycoparasites of Mycelia, Ectomycorrhizae, Sclerotia, and Spores in Soil."

Stilbella flavipes, *S. sebacea*, and *S. stereicola* grow on sporocarps of aphyllophorales (Seifert 1985). *Zakatoshia hirschiopori* is so far known from decayed basidiomes of the polypore *Trichaptum* (synonym *Hirschioporus*) species in Canada, whereas *Z. erikssonii* is known from *Sistotrema brinkmannii* in Europe (Sutton 1973a; Gams 1986).

In Denmark, galls on *Coniophora puteana* are caused specifically by a biotrophic parasite, *Nodulisporium cecidiogenes* (Koch 1994), a fungus that cannot be grown axenically.

Several intrahymenial heterobasidiomycetes occur also on the sporocarps of aphyllophorales. *Peniophora* species can host *Colacogloea* (*Platygloea*) *peniophorae* (Burdsall and Gilbertson 1974; Bauer and Oberwinkler 1991). The firm nucleus that regularly is seen in the basidiomes of the common *Tremella encephala* is nothing other than its host fungus, *Stereum sanguinolentum* (Bandoni 1984; Helfer 1991). One species of *Tulasnella* and three of

Exidiopsis have been found growing inside corticioid fungi, whereas two species of *Platygloea* (*Achroomyces*) and *Syzygospora* and four of *Tremella* have been found growing on their hymenia (Jülich 1983). Gelatinous and hyaline basidiocarps of *Syzygospora* species grow on basidiocarps of *Leucogyrophana* and *Phanerochaete* species (Hauerslev 1969; Ginns and Sunhede 1978). *Tremella mycophaga* and *T. simplex* parasitize *Aleurodiscus amorphus* (Ginns and Sunhede 1978).

Clavariaceous fungi often are colonized by *Mycogone calospora* (Gams 1983) and *Helminthosphaeria clavariarum* (anamorph *Spadicoides clavariarum*; Samuels et al. 1997).

On Jelly Fungi. While jelly fungi (mainly Tremellales) are often mycoparasitic, few reports are available about their function as hosts for other parasites. *Acremonium psammosporum*, *Verticillium*, and *Chalara* species; *Hypomyces aurantius*; and *Ophiostoma epigloeum* have been found as parasites on *Tremella* species, and *Dactylaria lanosa* has been isolated from *Pseudohydnum gelatinosum* (CBS, unpublished data). *Tremella danica* can grow on or in *Myxarium nucleatum* (Hauerslev 1979). Fungi parasitizing *Auricularia* species include *Acremonium* and *Verticillium* species, *Hypomyces semitranslucens*, and three species of *Cladobotryum* (CBS, unpublished data). *Hypomyces mycophilus* (anamorph *Cladobotryum polypori*, synonym *Pseudohansfordia* [*Dactylaria*] *mycophila*) also was reported from *Hirneola* (*Auricularia*) *polytricha* (Tubaki 1955). *Hypocrea sulphurea* is always found on remnants of *Exidia* species (B. Overton, personal communication).

An epibiotic chytrid, *Rhizophlyctis* species, was found on *Dacrymyces stillatus* (Canter and Ingold 1984). *Tremella obscura* (Reid 1970), *Sebacina penetrans* (Hauerslev 1979; Oberwinkler and Bandoni 1982), and *Itersonilia perplexans* (CBS, unpublished data) also can attack *Dacrymyces* species. Two species of *Platygloea* (*Achroomyces*) and two of *Tremella* were found on or in the hymenium of Dacrymycetaceae (mainly *Dacrymyces* species). While *Achroomyces peniophorae* has a thin, gelatinous, pustulate basidiocarp, which later turns confluent and resupinate; the other species of *Platygloea* and *Tremella* lack basidiocarps (Jülich 1983).

Several ascomycetes and mitosporic fungi that are normally lichenicolous grow preferentially on basidiomata of lichenicolous heterobasidiomycetes (Diederich and Christiansen 1994; Diederich 1996).

On Polypores (Bracket Fungi). A considerable range of fungicolous species can grow on sporocarps of polypores; the most commonly recorded species are listed in Table 17.5. Those fungicolous fungi often cover the surface of the perennial carpophores without harming the host (Rudakov 1981). *Hypomyces aurantius* disrupts the cytoplasm and causes irreversible changes in the host cells (Kellock and Dix 1984). Besides *H. aurantius* and several other *Hypomyces* species (Table 17.5), a number of closely related anamorphic species (*Cladobotryum* species) also inhabit the sporocarps of aphyllophorales. Several anamorph–teleomorph connections have been established in this complex (Helfer 1991; Rogerson and Samuels 1993; Põldmaa 1996; Põldmaa et al. 1997; Põldmaa and Samuels 1999).

Fungi occupying decaying wood, particularly polypores, are frequently strong antagonists of a wide range of other fungi, at least in laboratory culture (Cooke 1977). Antagonism may facilitate colonization of freshly cut stumps or logs (i.e., primary resource capture); it also may suppress predecessors from older, precolonized substrata (secondary resource capture) (Griffith and Barnett 1967; Rayner et al. 1987). Antagonistic activities of any kind are influenced by the available carbon and nitrogen sources (Griffith and Barnett 1967; Barron 1992). Displacement reactions between several wood-decaying aphyllophorales, such as *Trametes versicolor* by *Lenzites betulina*, *Bjerkandera adusta* by *Pseudotrametes gibbosa*, various fungi by *Schizophyllum commune*, and *Datronia mollis* by *Phanerochaete magnoliae*, have been examined in experiments using agar cultures or wood blocks and interpreted as cases of hyphal interference (Ainsworth and Rayner 1991; Jeffries and Young 1994). It is not clear whether laboratory-based observations of the replacement reactions will be supported by field observations because a succession need not be followed by sporocarp formation in nature (Niemelä et al. 1995).

Polypores commonly are colonized by some *Cosmospora* species. *Hydropisphaera* (*Nectria*) *peziza* is also frequent on *Polyporus*, mostly *P. squamosus*. *Pseudonectria tilachlidii* and the synnemata of its anamorph, *Tilachlidium brachiatum*, common on agarics, also can be found on the hymenium of polypores, especially of *Tyromyces* species (Helfer 1991; K. Põldmaa, personal observation).

The ascomata of *Ophiostoma polyporicola*, often hidden inside pores of bracket fungi, have been discovered only recently (Constantinescu and Ryman 1989; Helfer 1991). The species frequently is found on sporocarps of *Fomitopsis pinicola* and *Piptoporus betulinus* and almost always occurs together with *Hypocrea pulvinata*, sometimes growing out from its stromata (K. Põldmaa, personal observation). It has been recorded from sporocarps of other aphyllophorales (*Antrodia*, *Tyromyces*) as well (Helfer 1991). *Melanospora lagenaria* (Fig. 17.12) also grows on *F. pinicola*, often together with *O. polyporicola* and *H. pulvinata* (Fig. 17.13). It also has been recorded on species of *Bjerkandera*, *Trametes*, *Polyporus*, *Pipto-*

TABLE 17.5
Common Fungi That Grow on Polypores

Species (anamorph)	Host	Reference
Hypomyces aurantius (anamorph *Cladobotryum varium*), *H. rosellus* (*C. dendroides*), *H. semitranslucens* (*C. fungicola*), *H. orthosporus* (*C. orthosporum*), *H. odoratus* (*C. mycophilum*)	Various aphyllophorales	Kellock and Dix 1984; Rogerson and Samuels 1993; Pöldmaa and Samuels 1999
H. polyporinus (*C. clavisporum*)	Mostly *Trametes versicolor*, occasionally other aphyllophorales	Kellock and Dix 1984
H. subiculosus (*Cladobotryum* species)	Various aphyllophorales, often *Trametes* species	Kellock and Dix 1984
H. mycophilus (*C. polypori*)	Various aphyllophorales, often *Polyporus* species	Kellock and Dix 1984
Cosmospora vilior (*Acremonium berkeleyanum*)	Various polypores	Gams 1971
C. coccinea (syn. *Nectria cosmariospora*; anamorph *Verticillium olivaceum*)	*Inonotus*	Gams 1971; Helfer 1991; Samuels et al. 1991
Cosmospora purtonii (*Fusarium aquaeductum* var. *aquaeductuum*)	Polypores	Wollenweber and Reinking 1935; Domsch et al. 1980
Hydropisphaera (*Nectria*) *peziza* (*Acremonium* species)	*Polyporus* species, mostly *P. squamosus*	Samuels 1976
Pseudonectria tilachlidii (*Tilachlidium brachiatum*)	Agarics, polypores, especially *Tyromyces*	Helfer 1991
Sphaerostilbella (*Hypomyces*) *broomeana* (*Gliocladium microspermum*)	*Heterobasidion annosum*	Pöldmaa 1999; Pöldmaa et al. 1999; Gams and van Zaayen 1982
Hypocrea pulvinata (syn. *H. fungicola*)	*Fomitopsis pinicola, Piptoporus betulinus*	Doi 1972; Helfer 1991
Hypocrea pallida	Various aphyllophorales, often *Tyromyces* species	Doi 1972; Helfer 1991
Trichoderma polysporum	Polypores	Helfer 1991
Sporophagomyces chrysostomus	Ganodermataceae	Rogerson and Samuels 1993; Pöldmaa et al. 1999
Ophiostoma polyporicola	*Fomitopsis pinicola, Tyromyces* species, *Piptoporus betulinus*	Constantinescu and Ryman 1989; Helfer 1991
Melanospora lagenaria	*Fomitopsis, Bjerkandera, Trametes, Polyporus, Piptoporus, Stereum* species	Cannon and Hawksworth 1982
Albertiniella polyporicola (syn. *Cephalotheca splendens*)	*Ganoderma* species	Helfer 1991
Orbilia inflatula	*Fomes fomentarius*	Helfer 1991
Cistella hymeniophila (*Phialophora rhodogena*)	*Antrodia, Piptoporus* species	Helfer 1991
Bisporella citrina	*Daedaleopsis confragosa*	Helfer 1991
Eleutheromyces subulatus	*Trametes* species	Helfer 1991
Endomyces polyporicola	*Piptoporus betulinus*	Schumacher and Ryvarden 1981; de Hoog et al. 1986
Rhinotrichella globulifera	Polypores	Helfer 1991

porus, and *Stereum* (Cannon and Hawksworth 1982), whereas the related *M. caprina* is found primarily on *Tomentella* species and other tomentelloid fungi.

Some discomycetes occur on carpophores of polypores and other aphyllophorales (Helfer 1991). The apothecia of *Hyphodiscas hymenophilus* rarely are found on sporocarps of *Antrodia* and *Piptoporus*, whereas its anamorph, *Catenulifera rhodogena*, commonly stains the hymenium of many aphyllophorales red (Helfer 1991). *Hyaloscypha epiporia* has been found only on polypores growing on softwood, usually covering the pore surface of old, partly decayed sporocarps of *Amylocystis lapponica* (Huhtinen 1989). It is one of the few *Hyaloscypha* species that fruits in culture. A few other discomycetes (e.g., *Bisporella*

citrina) have been reported on sporocarps of polypores. However, the apothecia usually also are found on the wood nearby.

Coelomycetes are not frequent on sporocarps of aphyllophorales. A characteristic and common representative is *Eleutheromyces subulatus* with an almost yeastlike synanamorph; it is not easily found because it may be hidden among the pores of the host. The tiny black pycnidia of the rare *Eleutheromyces* (*Eleutheromycella*) *mycophilus*, which only occur on *Trametes* species, are even harder to find.

Endomyces (*Dipodascus*) *polyporicola* is a parasitic hemiascomycete found on *Piptoporus betulinus* (Schumacher and Ryvarden 1981; de Hoog et al. 1986).

FIGURE 17.13   Stromata of *Hypocrea pulvinata* on a decaying basidioma of *Piptoporus betulinus*. (Photo by U. Kõljalg)

FIGURE 17.12   Perithecia of *Melanospora lagenaria* growing on the hymenophore of *Ganoderma* species. (Photo by K. Põldmaa)

Mycelia of different mycoparasites, particularly species of *Acremonium*, *Monocillium*, *Fusarium*, *Gliocladium*, *Scopulariopsis*, *Sporotrichum*, and *Rhinotrichella* often are found growing inside perennial carpophores. *Calcarisporium arbuscula*, *Rhinotrichella globulifera*, *Clonostachys rosea*, *C. catenulata*, *Gliocladium polyporicola*, and *G. viride* are found on decaying polypore sporocarps.

Antagonists of the mycelium of the root rot fungus, *Heterobasidion annosum*, have been studied intensively. *Phlebiopsis* (*Peniophora*) *gigantea* is its most successful competitor (Rishbeth 1963). Among the antibiotically most active antagonists are species of *Scytalidium*; polyphenol oxidases that induce brown discoloration also seem to be involved in their competitive activity (Klingström and Johansson 1973).

Two species of heterobasidiomycetes associated with poroid fungi are known. *Tremella polyporina* can replace the hymenium on the basidiocarps of *Postia* (*Tyromyces*) *caesia* and *P. lactea* (Reid 1970), and *Exidiopsis opalea* is found on other old polypores (Jülich 1983). A yeast isolated from a sporocarp of *Ganoderma applanatum* was identified as *Ustilago maydis* when examined with molecular methods (Prillinger et al. 1989). Hawksworth (1981a) lists many additional, often rather nonspecific, colonizers of polypores.

Sporocarps of wood-rotting aphyllophorales often appear on those of other polypores. A special group of wood-rotting fungi (called successors) inhabits woody material that previously has been decayed by certain other species (predecessors) (Niemelä et al. 1995). *Fomes fomentarius*, *Fomitopsis pinicola*; and species of *Inonotus*, *Trichaptum*, and *Phellinus* are common predecessors. The great majority of successors are white-rot fungi, with *Antrodiella* species being the best known. The sporocarps of the successors often develop on those of the predecessors. For example, *Antrodiella citrinella* grows on sporocarps of *Fomitopsis pinicola*, and *A.* (*Trametes*) *hoehnelii* grows mainly on those of *Inonotus radiatus* (Jahn 1963). Possibly the repeatedly observed association of the rare *Boletus* (*Buchwaldoboletus*, *Pulveroboletus*) *lignicola* and *Phaeolus schweinitzii* on the same conifer stump is such a phenomenon (Szczepka and Sokól 1984; Lipka 1985). Sporocarps of successors also can develop away from those of predecessors on the same trunk. Because an interaction between the mycelia may take place within the decaying wood, the fruiting of one fungus on another could signal a competitive, parasitic, or dominance relationship (Niemelä et al. 1995). However, channels left inside the wood by mycelial strands of dead fungi may act as canals for hyphae of succeeding fungi, which may result in their fruiting on the dead basidiocarps of the first colonizer. Thus, an observed succession in fruiting does not indicate whether a successor is behaving as a selective parasite or whether it is just profiting from the way opened by its predecessor (Niemelä et al. 1995). Many observations of this kind simply may represent the accidental co-occurrence of different wood-rotting fungi. Tomasi (1977), building on the work of Bourdot and Galzin (1928), and Besl et al. (1989) presented overviews of aphyllophorales growing on different polypores. The common occurrence of *Sistotrema brinkmannii* on old basidiomes is obviously a case of succession of a rapid and competitive colonizer onto a suitable substratum (Eriksson et al. 1984).

On Fleshy Sporocarps of Boletales and Agaricales.

Zygomycetes that grow on agarics and boletes include species from common genera such as *Mortierella* and *Umbelopsis* (*M. bainieri*, particularly on *Amanita* species, and the less specific *U. ramanniana*) and nonspecific species of *Mucor* and *Rhizopus* found in soil and other substrata. The genera *Spinellus*, *Syzygites*, and *Dicranophora* comprise only mycoparasitic species. *Spinellus fusiger* and four congeners occur mainly on *Mycena* species but also have been recorded from other agarics (*Amanita*, *Gymnopus* [*Collybia*], *Hygrophorus*). *Syzygites megalocarpus* grows on decaying agarics and boletes (Hesseltine 1957). *Dicranophora fulva* is a rare species found on *Paxillus*, *Gomphidius*, and *Leccinum* species (Jeffries and Young 1994; Voglmayr and Krisai-Greilhuber 1996).

The most common and representative fungicolous ascomycete genus is *Hypomyces*. Many species of the genus and their anamorphs attack fleshy sporocarps (Rogerson and Samuels 1989, 1994). Boleticolous species of *Hypomyces* more or less malform host basidiocarps, although the tube layer usually is still recognizable. Such partially altered hosts still may be identifiable to genus and species (Rogerson and Samuels 1989). Most boleticolous *Hypomyces* species can use a wide range of hosts, and only a few appear to be specialized to a host genus or species (Sahr et al. 1999). *Hypomyces chrysospermus* (anamorph *Sepedonium chrysospermum*) attacks all kinds of Boletales with tubular and lamellar hymenophores, including the gasteroid *Scleroderma* and *Melanogaster*. The species also has been reported as a mycoparasite of *Sclerotinia sclerotiorum* (F. Marziano, personal communication; CBS, unpublished data) and powdery mildews (Hijwegen 1992a). *Hypomyces completus* (anamorph *S. brunneum*) and *H. transformans* colonize species of *Suillus* (Rogerson and Samuels 1989; Sahr et al. 1999). *Sepedonium chalcipori* specifically parasitizes the acrid *Chalciporus piperatus* (Helfer 1991), and *H. melanocarpus* is found only on the bitter *Tylopilus* species (Rogerson and Samuels 1989).

Eight of the agaricicolous species of *Hypomyces* occur only on members of the Russulaceae. *Hypomyces luteovirens* is restricted to *Russula* species; *H. lithuanicus* (*H. torminosus*) is restricted to the *Lactarius torminosus* group; and *H. lateritius*, *H. lactifluorum*, *H. macrosporus*, and *H. banningiae* also are restricted to species of *Lactarius*. *Hypomyces hyalinus* grows only on *Amanita* species, Three other species are found on brown- (or pink-) spored agarics: *H. succineus* on *Pholiota* species, *H. porphyreus* on *Entoloma* (*Leptonia*) *strigosissimum*, and *H. tremellicola* on *Crepidotus* species (Rogerson and Samuels 1994). In addition, *Nectriopsis tubariicola* (probably also a member of *Hypomyces*) is found on *Tubaria* species (Gams and van Zaayen 1982).

Some *Hypomyces* species transform the gill surface of the host's pileus into an ascomycetous hymenium. In fact, *H. lactifluorum* and *H. hyalinus* render their hosts unrecognizable so that only healthy basidiocarps in the neighborhood allow identification of the host (Fig. 17.3). One of the most ubiquitous parasites on species of *Russula* and *Lactarius* is *H. armeniacus* (synonym *H. ochraceus*, anamorph *Cladobotryum verticillatum*, synonym *Monosporium agaricinum*), which occurs mainly as an anamorph and completely destroys its host. After the host basidiome has decayed, *H. armeniacus* grows away from it onto soil, mosses, or other nearby substrata, where the perithecia are formed. The anamorphs of the common aphyllophoricolous *H. aurantius*, *H. odoratus*, and *H. rosellus* also often are found on various agarics but form perithecia only on aphyllophorales, wood, or bark (Põldmaa 1999). Some anamorphic species related to *Hypomyces* also colonize the sporocarps of agarics (de Hoog 1978; Helfer 1991). *Mycogone rosea*, for example, has been recorded on agarics belonging to nine genera, but particularly *Amanita* (Arnold 1976), whereas *Cladobotryum apiculatum* is known to grow mainly on sporocarps of *Lactarius* and *Russula* species (Gams and Hoozemans 1970; Helfer 1991). Other anamorphic species have so far been collected only once or twice on agarics. *Hypocrea avellanea* and its *Verticillium* anamorph have been found only on *Marasmius subnudus* (Carey and Rogerson 1976).

Melanosporopsis subulata and *Dendrostilbella mycophila* rarely are found on dark, discolored agarics. Their coexistence, identical nuclear DNA content, and other evidence indicate that the latter is the anamorph of the former (Helfer 1991).

The *Geotrichum* anamorph of *Dipodascus armillariae* (Gams 1983) is observed regularly and specifically on the gills of old sporocarps of *Armillaria* species, but its asci could not be obtained on agar media. The redefined genus *Endomyces*, which is characterized by hat-shaped ascospores and slimy blastoconidia (Redhead and Malloch 1977), comprises a group of SCIF found on agarics. Redhead and Malloch also found a related taxon, *Phialoascus borealis*, on a decaying *Cortinarius*. An additional species of *Endomyces* occurs on *Tricholoma* species (Helfer 1991).

Putrefying agarics and boletes provide suitable substrata for a number of yeasts. Ramírez Gómez (1957) obtained more than a hundred yeast species from about the same number of macromycetes. Several anamorphic yeasts, including *Candida anomala* (*Debaryomyces hansenii*), *C.* (*Rhodotorula*) *buffonii* (on *Boletus edulis*), *C. obtusa* var. *arabinosa* (*Pichia mississipiensis*, on *Clitopilus prunulus*), and *C.* (*Torulopsis*) *kruisii* (on *Boletus purpureus*) are known only from agarics. The ubiquitous *Sporobolomyces roseus* has been isolated from *Leccinum aurantiacum* (Hawksworth 1981a).

Among the fungicolous hyphomycetes, one of the most common species is *Calcarisporium arbuscula*, found in the growing sporocarps of many agarics, including species of *Russula* and *Lactarius* (Watson 1955). It either sporulates quickly and profusely or develops as a symptomless "endophyte," revealing its presence only when it begins to sporulate on old host sporophores or when small pieces of the sporophores are placed on agar (Barnett and Lilly 1958; Nicot 1968). It generally colonizes the basidiomes of *Cantharellus cibarius* so that isolation of that species from tissue is impossible (Schouten and Waandrager 1979). Damaged hyphae of many Russulaceae produce isovelleral, a defense metabolite, which *C. arbuscula* detoxifies by reducing it to isovellerol; other SCIF tolerate isovelleral without transforming it (Anke and Sterner 1988).

Verticillium fungicola and its varieties occur frequently on agarics in nature (see also "On Fleshy Sporocarps and Mycelium of Cultivated Agaricales," later in this chapter). An undescribed species of *Verticillium* regularly occurs on *Gymnopus* (*Collybia*) *peronatus* in the Netherlands (W. Gams, personal observation). The synnemata of *Tilachlidium brachiatum* are rather common on the stipes of old agarics, particularly *Hypholoma* and *Mycena* species (Gams 1971); its rare teleomorph, *Pseudonectria tilachlidii*, also has been found on an old agaric (Gams 1975a). The species has not been recorded outside fungal sporocarps, but it is difficult to distinguish *in vitro* from *Acremonium berkeleyanum*.

Various other anamorphic fungi occur on sporocarps of agarics and boletes. *Gabarnaudia tholispora* has been found on *Russula nigricans* (Hawksworth 1983). Rarer mycoparasites include *Amblyosporium* on *Peziza*, *Paxillus*, and *Lactarius* species (Nicot and Durand 1965, citing a discomycete teleomorph; Pirozynski 1969; CBS, unpublished data) and *Harziella capitata* on *Lepista nuda* (Fontana 1960).

Only a few species of coelomycetes are known to grow on agarics, particularly *Eleutheromyces subulatus* (Seeler 1943; Helfer 1991) and the basidiomycete anamorph *Hyalopycnis blepharistoma* (Bandoni and Oberwinkler 1981).

Among the fungicolous basidiomycetes, species of *Syzygospora* are causal agents of gall formation. This is a rare phenomenon confined to particular agaric hosts and a few associated parasites (Fig. 17.14) (Hauerslev 1969; Ginns and Sunhede 1978; Oberwinkler et al. 1984; Ginns 1986). Their hymenia cover galls, which appear on the lamellae, the surface of the pilei, and the stipes of *Gymnopus* and *Marasmius* species, sometimes almost enveloping the mushroom. The inner part of the gall is formed of the hyphae of the host.

Only in two agaric genera are all species fungicolous. Two species of *Asterophora* (= *Nyctalis*; Tricholomataceae) attack species of Russulaceae and rarely *Gymnopus* (*Collybia*) *fusipes* and form abundant chlamydospores in their basidiomes. Both *Asterophora lycoperdoides* (synonym *N. asterophora*) and *A. parasitica* complete their entire development on an agaric in about 3 weeks (Buller 1924). These species grow and sporulate well in pure culture when the substratum contains a high ratio of organic nitrogen to carbohydrate (Jeffries and Young 1994).

The irregular swellings always present at the base of *Squamanita* species (Tricholomataceae) have been described as protocarpic tubers. *Squamanita* is recognized as an obligately parasitic genus whose species grow on sporophores of other basidiomycetes. Systemic infection of host sporophore primordia causes gall formation (Redhead et al. 1994). The galls vary from pileate or stipitiform to amorphous, are composed of host basidiome tissue infused with *Squamanita* hyphae, and bear one to several *Squamanita* basidiomes. The 10 known species have hosts in the genera *Cystoderma*, *Phaeolepiota*, *Tubaria*, *Galerina*, and others.

Isolated fungicolous species occur in some larger genera. *Volvariella surrecta* (synonym *V. loveiana*) (Fig. 17.15), for example, specifically colonizes intact or deformed caps of *Lepista nebularis* and *Clitocybe*

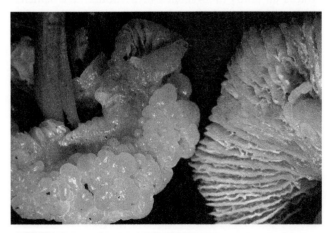

FIGURE 17.14 Galls of *Syzygospora tumefaciens* on *Gymnopus* (*Collybia*) *dryophilus*. (Photo by W. Gams)

FIGURE 17.15 *Volvariella surrecta* fruiting on old *Lepista nebularis*. (Photo by W. Gams)

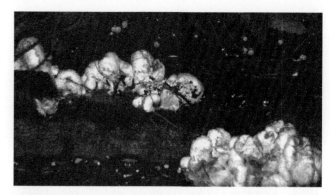

FIGURE 17.16 Abortive basidiomata of *Entoloma abortivum* parasitized by *Armillaria mellea*. (Photo by K. Põldmaa)

clavipes. Psathyrella (*Stropharia*) *epimyces* fruits on deteriorated *Coprinus atramentarius* and *C. comatus* in North America. *Entoloma* (*Claudopus*) *parasiticum* has been reported from *Cantharellus cibarius*, *Craterellus cornucopioides*, and *Coltricia* (*Polyporus*) *perennis* (Noordeloos 1992; Helfer 1991; Jeffries and Young 1994); *Entoloma pseudoparasiticum*, with paler basidiocarps, also grows on *Cantharellus cibarius* and *Craterellus lutescens* (Noordeloos 1992), and the related *E. byssisedum* grows on hypogeous sporocarps of truffles and gasteromycetes (Malençon 1942). Conversely, the phenomenon of abortive basidiomes in *E. abortivum* (Fig. 17.16) is ascribed to mycoparasitism by *Armillaria mellea* (Watling 1974; Jeffries and Young 1994).

Four species of *Collybia* sect. *Collybia* have been found to attack a wide range of hosts from several orders of basidiomycetes. *Collybia cookei*, *C. racemosa*, and *C. tuberosa* form sclerotia in mummified host basidiomes and may develop their own basidiomes directly from their sclerotia in soil, often after all visible remains of the hosts have vanished; *C. cirrhata* is similar, but lacks sclerotia. Mycophagy (not necessarily mycoparasitism) by two species of *Lyophyllum* on rotting basidiocarps of *Russula* and *Lactarius* species and unidentified rotting basidiomycetes also has been documented (Redhead et al. 1994).

A phenomenon termed sporophagy was observed when the large pigmented basidiospores of *Leccinum aurantiacum* and other basidiomycetes (including *Paxillus, Suillus, Gomphidius, Pluteus,* and *Thelephora*) were offered as bait to other basidiomycetes. Hyphae of *Coprinus comatus, Pluteus cervinus,* and numerous aphyllophorales, but not ectomycorrhizal fungi, penetrated

the spores (Fries and Swedjemark 1985), but the hyphae of the same fungi were not attacked. The effect was influenced by the nutrient content of the supporting medium.

Examples of hyperparasitism on SCIF include species of *Asterophora* that can host *Pyxidiophora asterophorae* (Lundqvist 1980) and an intracellular *Tremella*-like yeast (Laaser et al. 1988). *Eleutheromyces subulatus*, which also grows on old sporocarps of *Russula* species, supports growth of *Nematogonum mycophilum* (Gams 1975b). *Boletus parasiticus*, growing on *Scleroderma citrinum*, is attacked by a particular species of *Penicillium* (W. Gams, personal observation).

On Thalli of Lichenized Agarics. *Norrlinia peltigericola, Lichenopeltella minuta,* and *Stigmidium joergensenii* parasitize the thallus of the lichenized *Lichenomphalia* (*Omphalina*) *foliacea* (Santesson 1989). An undescribed species of *Cercidospora* has been collected twice on the lichenized thallus of *Lichenomphalia umbellifera* (J. Hafellner, personal communication). *Merismatium nigritellum* parasitizes *Lichenomphalia hudsoniana* and various lichenized ascomycetes (Triebel 1989). The congeners of all of the parasitic species grow on lichenized ascomycetes.

On Fleshy Sporocarps and Mycelium of Cultivated Agaricales. Fungal infections may cause great losses in mushroom farms. Infections are of two types: weed molds that contaminate the mushroom compost and compete with the mycelium of the commercial mushrooms and parasites of the sporocarps. Weed molds include *Myceliophthora* (*Chrysosporium*) *lutea, Peziza ostracoderma* (anamorph *Chromelosporium fulvum*), and *Trichoderma* species (Jeffries and Young 1994; Samuels et al. 2002). Generally, the cobweb disease damaging the cultivated *Agaricus* has been ascribed to the anamorph of *Hypomyces rosellus* (*Cladobotryum dendroides*). McKay and colleagues (1999), however, showed by the analysis of molecular data that most of the isolates responsible for the disease represent the anamorph of *H. odoratus* (*C. mycophilum*). The cobweb disease can be controlled relatively easily by hygienic measures. The false truffle, *Diehliomyces microsporus*, is mainly a competitor that prevents colonization of the substratum by the mushroom. It is found primarily in warmer countries and always is associated with cultivated mushrooms (van Zaayen and van der Pol-Luiten 1977).

Hypocrea and *Trichoderma* species adversely affect cultivated wood-decaying fungi (Hashioka et al. 1961; Hashioka and Komatsu 1964; Komatsu and Hashioka 1964; Komatsu 1976; Tsuneda and Thorn 1995). Tsuneda and Thorn (1995) found that frequency of occurrence and strength of mycoparasitic activity of *Trichoderma harzianum* on the wood decay fungi *Lentinula edodes* and *Pleurotus ostreatus* were influenced by the degree of wood decay. Tsuneda and associates (1997) found that the *Hyphozyma* synanamorph of *Eleutheromyces subulatus* is highly pathogenic to and causes black spot symptoms in the shiitake mushroom (*Lentinula edodes*) grown outdoors on *Quercus* bedlogs. Reper and Pennincky (1987) reported that a diffusible toxin from *T. hamatum* damaged *P. ostreatus*. The presence of cellulose and chitin microfibrils seemed to enhance the mycoparasitic activity. In the last decade, highly aggressive strains of a taxon initially identified as *T. harzianum* (genetically different from the biocontrol taxon and then named *T. aggressivum*) are causing heavy losses in cultivated mushrooms in the United Kingdom, Canada, the United States, and Australia (Seaby 1987; Muthumeenakshi et al. 1994; Samuels et al. 2002). Recently, aggressive strains of *Trichoderma virens* also have been found to attack cultures of *Pleurotus ostreatus* (CBS, unpublished data).

However, *Agaricus bisporus* benefits from a preceding colonization of the compost by *Scytalidium thermophilum*, which prepares the substratum during a thermophilic phase and then is inactivated by the *Agaricus* (Straatsma et al. 1989, 1994). The regular and intimate association of *Xylaria* and *Termitomyces* species in the fungus gardens of termites may be a comparable case, in which the *Xylaria* seems to prepare the substratum for colonization by the agaric, but the relationship is not fully understood (Batra and Batra 1977).

One of the best-known sporocarp diseases of *Agaricus bisporus* is the wet bubble caused by *Mycogone perniciosa*. Symptoms range from external infection of the normal sporophore to its total distortion into a spherical sclerodermoid form (Smith 1924). Up to 30% of apparently healthy sporophores from an affected crop may be infected at the base of the stipes (Fletcher and Ganney 1968). Other hosts of this species include wild species of *Agaricus* and the cultivated *Volvariella esculenta* (CBS, unpublished data). The top cell of the

Mycogone aleurioconidia survives a long time but requires stimulation (e.g., mushroom extract or a period of low temperature) for germination (Holland et al. 1985). In paired cultures of some anamorphs of *Hypomyces* species or of *Verticillium* species with *A. bisporus*, the parasite usually overgrew the host and caused necrotization of its hyphae, an indication of intense mycoparasitism (Gray and Morgan-Jones 1981).

Verticillium fungicola var. *fungicola* (synonym *V. malthousei*), the common causal agent of dry bubble, also causes considerable damage to crops of *A. bisporus*. After *A. bitorquis*, which is immune to virus disease and cultivated at higher temperatures than *A. bisporus*, was introduced as a commercially grown mushroom, another parasitic form, *V. fungicola* var. *aleophilum*, appeared. It produces brown spots on the cap resulting in inferior quality mushrooms (van Zaayen 1981; van Zaayen and Gams 1982). *Lecanicillium psalliotae* is also pathogenic to *A. bitorquis* (van Zaayen and Gams 1982), whereas *Simplicillium lamellicola* leads to "gill mildew" at high cropping temperatures or development of dark brown spots in large, open mushrooms. The polyphagous *Lecanicillium aphlanocladii* (= *Aphanocladium album sensu lato*) also has been reported to infect mushroom sporocarps and to reduce yield in Australia (van Zaayen and Gams 1982). Apparently, *Verticillium fungicola* can be controlled by spraying with a mix of *Lecanicillium* species conidia (de Trogoff and Ricard 1976).

Lecanicillium psalliotae also has been found to parasitize conidiophores and spores of the nematophagous *Rhopalomyces elegans* and other Mucorales (Dayal and Barron 1970). *Mycogone perniciosa* has the same capacity, and the infected spores of a host may contribute to the spread of this parasite (Barron and Fletcher 1972). *Verticillium dahliae* and *V. albo-atrum* can parasitize the nematophagous *Rhopalomyces* (Barron and Fletcher 1970). The nonspecialized mycoparasite *Pythium oligandrum* causes a disorder in *Agaricus bisporus* that results in black patches on the cap (Fletcher et al. 1990). *Mortierella bainieri* occasionally attacks mushrooms causing "shaggy stipe" (Jeffries and Young 1994).

On Epigeous and Hypogeous Sporocarps of Gasteromycetes. Relatively few taxa have been studied for parasites, and few specific colonizers have been found. Most observations are of species of *Scleroderma*, which are colonized by many nonspecific fungicolous fungi, including *Verticillium luteo-album* and other *Verticillium* species, *Trichothecium roseum*, and *Sepedonium chrysospermum*. The latter species also has been found on *Melanogaster* and *Rhizopogon*, genera supposedly related to the Boletales. *Boletus parasiticus* (also classified in *Pseudoboletus* or *Xerocomus*) consistently fruits on sporocarps of *Scleroderma* species, particularly *S. citrinum*, whereas *Xerocomus astraeicola* is found on *Astraeus hygrometricus* (Sclerodermataceae). *Xerocomus astraeicolopsis* has been reported from the same host in China (Redhead et al. 1994). The relationship between *B. parasiticus* and *Scleroderma* may not actually be parasitic but just coincidental fruiting stimulated by the host (Rayner et al. 1985). Agerer (1991) observed a probable association of *B. parasiticus* with the mycorrhiza of *Scleroderma* species in *Picea abies*.

A discomycete, *Gelatinipulvinella astraeicola* (yeast-like anamorph named *Aureohyphozyma astraeicola*), was found on the decaying peridium of *Astraeus hygrometricus;* the glebal portion of the host clearly was dead, suggesting that *G. astraeicola* does not need a living host for fruiting (Hosoya and Otani 1995). The stromata of *Hypocrea latizonata* form bands on the cuplike sporocarps of *Cyathus striatus* (Lohman 1938). *Podostroma solmsii* is another host-specialized fungus that grows on unexpanded sporocarps of *Phallus impudicus* (Doi 1978). *Hypomyces odoratus* also has been recorded from *Lycoperdon piriforme* (CBS, unpublished data). Additional colonizers are listed in Hawksworth (1981a). *Cystofilobasidium capitatum* is a rare case of a mycoparasite found in the gleba of Phallaceae (Oberwinkler et al. 1983), but this fungus also is found as a saprotroph on waxberries (*Symphoricarpus albus*; R.C. Summerbell, personal communication).

Methods of Study

Methods used to detect SCIF are similar to those used to collect macromycetes (Arnolds 1992; see Chapter 8). Parasites of slime molds usually are studied only by experts in the myxomycetes who have their own methods of collecting (see Chapter 25). Parasites of macromycetes also usually are studied by specialized investigators. Collection and quantification of SCIF are subjected to the same constraints as that of their hosts, with the additional difficulty that the occurrence of the parasites is often even more erratic than the fructification of the hosts (Arnolds 1992). The short-lived existence of macromycete sporocarps can be shortened even further by the attack of fungicolous fungi. Some of the latter outlive their hosts and fruit only on significantly decayed sporocarps; others sporulate abundantly with their anamorph on the basidiome, whereas perithecia appear only on the almost vanished host substratum or other substrata nearby.

SCIF usually are easy to detect in nature owing to macroscopically visible deformations and color changes of the host fungi. Still, mycelia with conidia and tiny perithecia (e.g., those of the species of Hypocreales) are scarce and easily overlooked on the sporocarps of Aphyllophorales. To study the fungi inside the hymenium of perennial polypores, the sporocarps must be cut open.

Specimens are collected and stored separately in boxes or paper bags to keep them clean for culture work. We recommend that observations be made and fungi be isolated as soon as possible after collecting because soft sporocarps, especially those of agarics and boletes, always are inhabited by insects, which very quickly destroy the whole fungus under moist conditions. Also bacteria accumulate rapidly, reducing the chance of obtaining pure cultures. Alternatively, the specimens can be kept in a cool place to slow down their degradation. Preservation of voucher specimens following the procedures outlined in Chapter 2 is crucial (Agerer et al. 2000).

SCIF often sporulate abundantly on a host substratum and can be identified easily. Sometimes the invader can be named in the field, especially if found on a selective host. However, even species of agarics, in which host specificity seems to be the most pronounced, can be colonized by different fungicolous fungi with similar appearance, in which case microscopic examination is needed. In the case of *Hypomyces*, anamorphs often occupy large parts of the host sporocarp, whereas ascomata are scarce and appear only in a late stage of host decay. The anamorphs are often more strongly differentiated than the correlated teleomorphs, facilitating identification (Põldmaa and Samuels 1999). No general guidelines can be given for recognizing the extremely diverse fungicolous Sporidiales and Tremellales in the field (F. Oberwinkler, personal communication). When it is impossible to prepare pure cultures during field expeditions, parts of the infected fungi can be air-dried and used for isolations, even after several months. Conidial material should be kept in a cool place until isolation; dried conidia seem to lose their germination capability rapidly.

In many genera (mainly anamorphic ones), species determination is not possible unless different developmental stages of the fungus are examined in culture and growth rate, texture, pigmentation, and odor are observed. Small amounts of conidia or ascospores transferred with the tip of a fine (glass) needle to media containing antibiotics usually yield pure cultures directly or after a few transfers. However, old decaying sporocarps of polypores often host several fungicolous fungi so that mixed cultures are obtained on transfer. We recommend, therefore, that a dilute spore suspension (taken from the surface of the hymenium) be streaked on the agar medium to allow transfer of pure cultures. Fractions of a deformed host fungus not showing a visible parasite can be incubated in a moist chamber (Appendix I). Such incubation also may serve to promote ripening of ascomata and ascospore production. Transfer of intact remnants of a host onto agar as a whole, before purifying the cultures, also can activate some biotrophic mycoparasites (Rudakov 1981).

The majority of the SCIF (and fungicolous fungi in general), even true biotrophs and necrotrophs, are able to grow and produce their conidial apparatus and sometimes also the sexual fructifications on ordinary culture media. Some investigators prefer cornmeal agar (CMA; a nutritionally poor, transparent medium), sometimes with 2% dextrose or oatmeal agar; others use malt extract agar (MEA), MEA with peptone, or potato-dextrose agar (PDA). Rudakov (1981) used Czapek agar with yeast and malt extracts. A mixture of PDA and MEA, called *Hypomyces* fruiting agar, enhanced perithecial production of *Hypomyces polyporinus* (Carey and Rogerson 1981). The sporulation of *Cladobotryum* anamorphs of *Hypomyces* tends to deteriorate rapidly after maintenance of a culture on rich media; we recommend use of dilute media such as potato-carrot agar (Gams et al. 1998) or CMA, particularly for long-term preservation. Formulae for the media are available in Appendix II.

Improved sporulation can be obtained by using various mushroom agars containing either mushroom extract (Tubaki 1955, 1975) or homogenized and sterilized carpophores of various host fungi (e.g., *Russula* agar, de Hoog 1978). A polypore extract agar (200 g polypore flour infused with 1 liter of distilled water) or 1% glucose agar with gas-sterilized pieces of polypore blocks has been proposed for the same purpose (Udagawa and Horie 1971).

Reliable isolation from germinated ascospores (preferably single spores) is required to establish the connection of a particular anamorph with a presumed teleomorph. To induce teleomorph formation of heterothallic species in culture, compatible isolates have to be mated. This is often successful in species of *Hypomyces* (Gams and Hoozemans 1970; Rogerson and Samuels 1993, 1994), as well as in fresh ascosporic isolates of other ascomycetes (e.g., *Albertiniella polyporicola*, *Hypocrea pulvinata*, *Hydropisphaera [Nectria] peziza*, *Sphaerostilbella [Nectriopsis* or *Hypomyces] broomeana*, and *Ophiostoma polyporicola*).

Identification

The myxomyceticolous fungi are treated and keys are provided in Gams (1971) and Ing (1974) and in a major contribution by Rogerson and Stephenson (1993), who treated nine species of ascomycetes and 26 species of hyphomycetes.

Helfer (1991) prepared one of the most comprehensive treatments of fungi that grow on fungal sporocarps. Other keys for identifying common SCIF include Arnold (1969) and Ellis and Ellis (1988). Species of SCIF are among the best-studied fungicolous species; nevertheless, many teleomorph–anamorph connections as well as new species, particularly in tropical areas, still await dis-

covery. Tubaki (1955, 1975) surveyed the SCIF in Japan. By far the most commonly observed mycophilic ascomycete genus is *Hypomyces* (Hypocreales) (approximately 55 species described) with (syn-)anamorphs in *Cladobotryum* and in several aleurioconidial genera. Comprehensive descriptions of species growing on discomycetes (five species), boletes (10 species), aphyllophorales (23 species), and agarics (13 species) are available (Gams and Hoozemans 1970; de Hoog 1978; Rogerson and Samuels 1985, 1989, 1993, 1994; Helfer 1991; Põldmaa and Samuels 1999). Several conidial species of *Cladobotryum* and related anamorphic genera have been described from Cuba (Arnold 1986, 1987, 1988; Castañeda 1986; Arnold and Castañeda 1987). The only key for identifying these as well as the other anamorphs associated with *Hypomyces* is provided by Helfer (1991). A review of the literature published on the *Hypomyces* complex was compiled by Arnold (1976).

Wells (1994) surveyed the modern classification of the Tremellales, but other than a key to 35 species of fungicolous heterobasidiomycetes known from Europe (Jülich 1983), no comprehensive literature for the identification of nonlichenicolous, fungicolous heterobasidiomycetes exists.

Major repositories (Appendix III) of authenticated specimens of SCIF include herbaria in Beltsville (BPI) and New York (NY) (collections from C. T. Rogerson and G. J. Samuels), CABI Bioscience in Egham (IMI, for species on tropical foliicolous fungi), Regensburg (REG, collection of W. Helfer), Tartu (TAA, collection of K. Põldmaa), and Weimar (now University of Jena, JE, collection of G. Arnold). The richest collection of living cultures is preserved in the Centraalbureau voor Schimmelcultures, Utrecht.

Geographic Distribution and Diversity

In contrast to the mostly cosmopolitan soil fungi (see Chapter 13), the distributions of SCIF are determined largely by the presence of their substrata. Macromycete species themselves are more strongly localized as a consequence of continental barriers and differences in climate, soil, and vegetation than are other saprotrophic fungi. Therefore, some of the SCIF are localized considerably over continents and units of vegetation.

Although many fungicolous Hypocreales, particularly their teleomorphs, are found mainly in the tropics, other species are distributed mainly in temperate regions. *Cosmospora (Nectria) epishaeria*, for example, is common at temperate latitudes but less so in the tropics. Of the other species in *Cosmospora*, however, about 25 species have tropical distributions and 21 species have warm-temperate distributions (Samuels et al. 1991; Rossman et al. 1999). Although species diversity for the genus

Hypomyces is higher in temperate regions, several species, including anamorphic *Cladobotryum* species, have been found only from the tropics. The impression that polyporicolous species of this genus are more numerous in the tropics, but that the agaricicolous are not, may be explained by the distributions of their hosts and the short lives of the host sporocarps. *Hypomyces aurantius* is very common in north- and south-temperate regions but rare in tropical regions. The similar *H. subiculosus* is a species of warm regions, being one of the most common tropical and subtropical species in the genus (Rogerson and Samuels 1993). Many of the species of *Hypomyces* (e.g., *H. aurantius, H. semitranslucens, H. rosellus)* are probably cosmopolitan. The "cause" of the common lobster mushroom, *H. lactifluorum*, and two other agaricicolous species (*H. banningiae* and *H. macrosporus)* occur only in North America. While those three species obviously have restricted distributions, *H. polyporinus* is frequent on *Trametes* species in North America but also has been found occasionally in Europe, where it has a much wider host range (Põldmaa and Samuels 1999). Several species of *Hypomyces* are known as single records from different regions. They cannot be considered as endemics, however, because of the scarcity of data. However, species reported only from New Zealand are probably endemics. Approximately a dozen species from *Cladobotryum* and closely related anamorphic genera have been described from the tropics, mainly from Cuba (Castañeda 1986; Arnold 1987; Arnold and Castañeda 1987, Bastos et al. 1982), and the same number from Europe (Arnold 1969, 1970; de Hoog, 1978; Helfer 1991; Põldmaa 1996, 1999). According to present knowledge, the distributions of those two groups of species do not seem to overlap.

Oberwinkler (1993) listed several tropical genera among the heterobasidiomycete mycoparasites, particularly *Cryptomycocolax* and *Kryptastrina* in the Platygloeomycetidae. The center of diversification of the Atractiellales is probably in the tropics; the genera *Atractiella* and *Helicogloea*, in particular, are mainly tropical. Similarly, a considerable proportion of the Tremellales are tropical, but their geographic distributions are still poorly known. *Holtermannia* has four tropical species and two in Japan; some species of *Syzygospora* are also tropical.

Species richness of SCIF is high, both in temperate and tropical countries, although the tropics are less well sampled. Apparently the Americas are richer in species than Europe, corresponding to higher numbers of host fungi. Rudakov (1981) found that regions of the former Soviet Union (Moldova, Ukraine, Northern Caucasus, Georgia, Central Russia, Kirgizia, some parts of Siberia, and the Far East) with a humid climate were much richer in mycoparasites growing on either macromycetes or biotrophic plant parasites than regions with a dry climate. Alexopoulos (1970) argued that the colonization of myxomycete sporocarps by filamentous fungi, which is promoted by constant high humidity, is responsible for a limited development of myxomycetes in tropical forests. In accordance with that argument, Rogerson and Stephenson (1993) observed that a very high proportion of myxomycete collections from northwest India during the wet monsoon season were colonized. Although studies of isolated species suggest a locally confined distribution of SCIF, that conclusion should be viewed with caution because sampling in most parts of the world is fragmentary, especially compared with that of macromycetes and other ecological groups of fungi. So far, known distribution patterns probably reflect the distribution of collectors.

Although seasonal variation in the presence and abundance of SCIF depends strongly on that of their hosts, their occurrence is even less regular, making them difficult to locate. Many pleomorphic ascomycetes sporulate abundantly with their anamorph on the host sporocarps, whereas the ascomata (in most cases perithecia) appear only on the strongly decayed sporocarps or on adjacent substrata (bark, mosses). Only the long-lived carpophores of polypores can serve as a substratum throughout the year.

The optimum time for the development of agaricicolous and boleticolous species of *Hypomyces* in temperate zones is usually in the autumn when fleshy fungi are abundant. For example, both *H. luteovirens* and its *Russula* hosts are most common in Norway in August (Eckblad and Torkelsen 1974). In the Moscow region, Rudakov (1981) recorded highest species richness and frequencies of SCIF in the second part of the summer. He ascribed this to the abundance of the hosts, the deposition of dew, and the presence of the inoculum in the air. He also followed the succession of SCIF on agarics, which begins with the development of anamorphs of various species of *Hypomyces*; those forms then are followed by *Calcarisporium arbuscula, Acremonium species*, and others. On almost decayed sporocarps, species of *Mucor, Mortierella, Penicillium, Trichoderma, Alternaria*, and *Cladosporium* generally supersede the more specific SCIF. Undamaged overwintering of the mycoparasites was observed only after a dry autumn (Rudakov 1981).

Quantification and Relative Importance

A quantitative census of SCIF ideally would involve regular observation of permanent plots. To our knowledge, however, only qualitative inventories of the species present in particular areas, with only rough quantitative estimates, have been carried out.

With some parasites, such as *Hypomyces chrysospermus*, an investigator can tell at a glance whether the host population (many diverse boletes and *Paxillus* species) in an area is infested heavily. The occurrences of other fungicolous taxa, such as species of *Asterophora*, *Squamanita*, and *Volvariella surrecta*, are geographically and seasonally so erratic that their detection requires an extended search over large areas. *Asterophora*, moreover, seems to be affected strongly by environmental pollution and is declining significantly in the Netherlands (Nauta and Vellinga 1995).

Microscopic SCIF, although probably not rare, are usually hardly visible and impossible to identify in the field. Assessment of such fungi would require extensive culturing. To date, however, such time-consuming quantitative methods have not been applied as widely to SCIF as they have been to soil fungi. Of the polypore carpophores studied by Rudakov (1981), 5% contained other fungi, mainly symbiotic so-called semisaprotrophs.

For 5 years Rogerson and Stephenson (1993) collected infected and normal sporocarps of myxomycetes in the Smith Mountain Lake area of southwestern Virginia and then assessed the relative abundances of myxomyceticolous fungi among different host orders. Members of the *Ceratiomyxales*, *Liceales*, and *Trichiales* turned out to be less susceptible to colonization than members of the *Physarales* and *Stemonitales*.

LICHENICOLOUS FUNGI

We use the term *lichenicolous* to refer only to those fungi that are obligate residents of lichens, excluding forms that have been found only incidental to culturing. Lichenicolous fungi live on the thalli or ascomata of lichenized fungi. Certain lichens (e.g., species of *Cladonia*, *Lobaria*, *Parmelia sensu lato*, *Peltigera*, *Pertusaria*, *Pseudocyphellaria*, *Usnea*) provide extremely rich habitats for parasites and merit increased study, whereas other lichen groups (e.g., *Pyrenula*, *Thelotrema*) are rarely hosts. As a rule, macrolichens are richer hosts than crustose lichens. In general, however, lichen thalli are resistant to colonization by most saprotrophic fungi, forming a very selective substratum. The firmer lichen ascomata support parasites that are quite different from the SCIF that parasitize nonlichenized discomycetes.

It is not always clear which of the two lichen symbionts—the mycobiont (an ascomycete or rarely a basidiomycete) or the photobiont(s) (an alga or/and a cyanobacterium)—is parasitized. Many species of lichenicolous fungi grow on only a single monophyletic group of lichenized fungi; one can assume that those species, at least, have developed a close relation with the mycobiont. In some species (e.g., *Tremella* species), special-ized haustoria have been observed attacking the mycobiont, whereas haustoria of other species attack and kill the algal cells. Very specialized parasites, such as *Blarneya hibernica*, kill the mycobiont and go on to form an independent lichen with the surviving algal cells (Hawksworth et al. 1979).

A number of lichenicolous fungi (e.g., *Arthonia* species) (Fig. 17.6) are closely related to the lichenized fungi and are considered to represent lichens that have lost their ability to form their own thallus. In contrast, species in another group of parasitic, lichenicolous fungi have their own, sometimes reduced, thalli, which appear at the onset of development. Other groups of lichenicolous fungi (e.g., species of *Hypocreales*) are related to nonlichenized taxa.

Lichenicolous fungi that damage their host lichens are considered to be parasitic, whereas those that cause no visible damage are assumed to be commensals (or parasymbionts) (Hawksworth 1982, 1988b). Some lichenicolous fungi induce galls or gall-like growths to form on the lichen thallus but do not damage the lichen. Such fungi sometimes form multiple symbioses with a lichen involving one or more photobionts and two or more mycobionts.

Some lichenicolous fungi have hyperparasites. Pycnidia of the lichenicolous *Lichenoconium usneae*, for example, have been observed parasitizing ascomata of *Abrothallus usneae*, which in turn were growing on the basidiomata of *Biatoropsis usnearum* developing on the lichen; simultaneously, the basidiomata of *B. usnearum* were developing on the lichen *Usnea rigida* (Bricaud et al. 1992; P. Diederich, personal observation).

Historically, few scientists have studied lichenicolous fungi, and the group largely has been neglected. Knowledge of the European and American forms is incomplete; knowledge of such fungi in the rest of the world is nearly nonexistent. Lichenicolous fungi are collected almost exclusively by lichenologists who generally have little experience with nonlichenized fungi; other mycologists rarely collect fungi on this unusual substratum.

Methods of Study

The best way to find and collect lichenicolous fungi is to examine many thalli of different lichen species in the field. Concentrating on abnormal-looking, damaged, or dying lichens may increase the numbers of fungi located, although numerous species grow on healthy thalli without damaging them. By using a hand lens lichenicolous species can often be recognized in the field, although some intrahymenial hyphomycetes or heterobasidiomycetes are detected only after the lichen is studied under a microscope.

Techniques for studying lichenicolous fungi are normally the same as those used for studying lichens—that is, examination under a light microscope of material mounted in water, KOH (potassium hydroxide), lactophenol-cotton blue, or other mountants (see Appendix II, "Mounting Media"). Although culturing is useful for the study of many groups, lichenologists normally do not culture material. Extensive studies of cultures of lichenized and lichenicolous fungi are, however, now in progress (Hawksworth and Jones 1981; Crittenden et al. 1995). Considerable numbers of fungi were isolated from fruticose *Cladonia* species and *Stereocaulon* species in Germany, including soil or litter fungi, symbionts or pathogens of higher plants, and true lichenicolous fungi (Petrini et al. 1990).

Most descriptions of lichenicolous fungal species are, at present, based exclusively on macromorphological and micromorphological characters, and many studies of those species have been carried out on herbarium specimens. Lichenologists, who regularly encounter and collect such fungi, but do not study them, should keep them separated and send them to experts. Major host groups that have been surveyed for lichenicolous fungi are listed in Table 17.6. References for recent monographs on genera of lichenicolous fungi are listed in Table 17.7.

Identification

Major publications useful for the identification of lichenicolous fungi include Vouaux (1912–1914), Keissler (1930), Hawksworth (1983), Clauzade and colleagues (1989), Triebel (1989), and Alstrup and Hawksworth (1990). Hawksworth (1979a) reviewed 44 species of lichenicolous hyphomycetes, among which 10 genera are

TABLE 17.6
Lichenized Fungi That Have Been Inventoried for Lichenicolous Fungi

Host group	Reference
Arthrorhaphis	Hafellner and Obermayer 1995
Baeomyces, Dibaeis, Icmadophila	Ihlen 1998
Brigantiaea	Hafellner 1985
Foliicolous lichens	Matzer 1996; Etayo 1998; Lücking et al. 1999
Haematomma	Kalb et al. 1995
Lecideoid lichens	Triebel 1989
Lepraria neglecta-group	Kümmerling et al. 1993
Omphalina foliacea	Santesson 1989
Peltigerales (*Lobaria, Peltigera, Solorina, Sticta*)	Hawksworth 1980; Etayo and Diederich 1996; Hawksworth and Miadlikowska 1997; Martínez and Hafellner 1998
Thamnolia	Ihlen 1995

TABLE 17.7
Recent Monographs Dealing with Genera of Lichenicolous Fungi

Genus	Reference
Arthonia (in part)	Grube et al. 1995; Wedin and Hafellner 1998
Biatoropsis	Diederich and Christiansen 1994
Corticifraga	Hawksworth and Santesson 1990
Dacampia	Henssen 1995
Didymellopsis, Zwackhiomyces	Grube and Hafellner 1990
Endococcus	Hawksworth 1979b
Gelatinopsis, Geltingia, Phaeopyxis	Rambold and Triebel 1990
Hemigrapha	Diederich and Wedin 2000
Hobsonia	Lowen et al. 1986
Karschia and similar genera	Hafellner 1979
Lichenochora	Hafellner 1989; Navarro-Rosinés et al. 1998
Lichenoconium	Hawksworth 1977
Lichenopeltella	Aptroot et al. 1997
Llimoniella	Hafellner and Navarro-Rosinés 1993; Diederich and Etayo 2000
Minutoexcipula	Atienza and Hawksworth 1994
Odontotrema	Diederich et al. 2002
Phacopsis	Triebel et al. 1995
Plectocarpon	Diederich and Etayo 1994
Polycoccum	Hawksworth and Diederich 1988
Pronectria	Rossman et al. 1999; Etayo 1998
Refractohilum	Roux et al. 1997
Rhagadostoma	Navarro-Rosinés and Hladun 1994
Rhymbocarpus	Diederich and Etayo 2000
Rosellinia	Matzer and Hafellner 1990
Sagediopsis	Triebel 1993
Sarcopyrenia	Navarro-Rosinés and Hladun 1990
Skyttea	Diederich and Etayo 2000
Sphaerellothecium, Stigmidium	Roux and Triebel 1994
Tephromela	Rambold 1993
Weddellomyces	Navarro-Rosinés and Roux 1995
Wentiomyces	Roux et al. 1994

exclusively lichenicolous. He (Hawksworth 1981b) also recognized 44 species of lichenicolous coelomycetes, including 16 genera that grow obligately on lichens. Diederich (1996) monographed 54 species of lichenicolous heterobasidiomycetes.

Lichenicolous fungi also can be identified by comparison with voucher specimens in herbaria. The herbaria (Appendix III) in Graz (GZU), Egham (IMI), München (M), Upsala (UPS), and Luxembourg have the richest collections of lichenicolous fungi.

Since 1980, interest in the study of lichenicolous fungi has grown among lichenologists, and the most recent checklists of lichens include lichenicolous fungi. Checklists of lichenized and lichenicolous fungi have been produced for North America (Esslinger and Egan 1995, listing 219 species); Britain (Hawksworth et al. 1980,

183 species); the Netherlands (Aptroot et al. 1999, 70 species); Belgium, Luxembourg, and northern France (Diederich and Sérusiaux 2000, 201 species), Germany (Wirth 1994, 161 species), and Norway and Sweden (Santesson 1993, 314 species), and other sites. Those publications can facilitate taxon identifications.

Geographic Distribution and Diversity

More than 1500 species of lichenicolous fungi are known. Based on recent studies, we believe that the actual number of species is probably much larger, possibly exceeding 3000 species. New taxa are being discovered and described at an increasing rate (at present, about 30 species/year; Sipman 1996a), for several reasons: (1) numerous taxa still remain to be discovered and described, even in well-explored areas; (2) outside Europe and North America, lichenicolous fungi are still poorly known, despite their extraordinary richness and diversity in some areas, including the montane tropics, especially in the Southern Hemisphere; and (3) revisions of genera based on modern techniques often lead to the division of single heterogeneous, poorly differentiated, and apparently nonspecialized species into several highly specialized species with slight morphological differences (e.g., Roux and Triebel 1994).

Many well-known lichenicolous species appear to be cosmopolitan, growing wherever the host lichens occur. A few lichenicolous fungi have much narrower distributions. *Plectocarpon macaronesiae*, for example, is specialized on widespread species of *Lobaria* but, to date, has been found only in Macaronesia (Diederich and Etayo 1994). *Stromatopogon baldwinii*, which grows on various species of *Usnea*, is known only from Tasmania and the South Sandwich Islands (Diederich 1992). So far, no well-documented species are known endemics of smaller areas. Because of directed collecting, some 50 lichenicolous heterobasidiomycetes now are known, about one-third of which come from montane forests in Papua New Guinea (Diederich 1996).

Quantification and Relative Importance

Most lichenicolous fungi are rare or rarely have been recorded, and only a few species are considered to be really abundant. *Athelia arachnoidea* can invade the epiphytic vegetation (mainly lichens and algae) of whole trees. In polluted areas of central Europe, this fungus can destroy the lichen vegetation of entire forests. Because of its intermittent appearance, particularly in autumn, only lichen species (e.g., *Lecanora conizaeoides*, *L. expallens*, *Lepraria incana*, *Soliciosporum chlorococcum*) that can recolonize the trees within a few months survive (see Arvidsson 1976, 1978). The little-known *Trichonectria*

hirta and its *Cylindrocarpon* anamorph are frequent on bark in polluted areas and often are present on every tree in a forest. Lichens that are killed by *Athelia arachnoidea* or other agents frequently are scavenged by the coelomycete *Lichenoconium erodens*. Lichen communities of the Xanthorion in central Europe often are attacked by numerous parasites; species of *Marchandiomyces* (Fig. 17.9), especially *M. aurantiacus*, as well as *Hobsonia christiansenii*, can completely destroy thalli of *Xanthoria* species and members of the family Physciaceae. Some lichenicolous heterobasidiomycetes, such as *Tremella lichenicola* (growing on *Mycoblastus fucatus*) and *T. phaeophysciae* (growing on *Phaeophyscia orbicularis*) are present on 20% of herbarium specimens of the host and can be found in most localities where the hosts are abundant.

MYCOPARASITES ON BIOTROPHIC PLANT PARASITES

Biotrophic plant pathogens are attacked frequently by mycoparasites, many of which can penetrate the spores (or conidia) of their host fungi. Some of the mycoparasites attack specific groups of plant pathogens and are, thus, of interest as potential biocontrol agents (see "Fungicolous Fungi as Biocontrol Agents of Plant Pathogens" later in this chapter). Many of the species are obligate parasites, unknown in nature outside their hosts, although they can be grown in axenic culture. The most prominent hosts include powdery mildews, rusts, and smuts.

On Peronosporales (Downy Mildews)

The downy mildews on above-ground parts of green plants rarely are attacked, and then it is mostly by generalist mycoparasites. *Fusarium incarnatum* (synonym *F. semitectum*) has been found as a destructive mycoparasite of oospores of *Sclerospora graminicola* on *Pennisetum typhoides* in India (Rao and Pavgi 1976).

On Erysiphales (Powdery Mildews)

By far the best-known mycoparasite of the Erysiphales (powdery mildews) is *Ampelomyces quisqualis* (synonym *Cicinnobolus cesatii*), which can eradicate whole populations of its host (Emmons 1930; Sundheim 1982; Falk et al. 1995a, 1995b). Initially, it grows biotrophically, suppressing sporulation of the host; later it becomes necrotrophic (Philipp 1985). The parasite contacts the host hyphae with appressoria and forms its pycnidia inside the host cells within 5 days (Sundheim and Krekling 1982). It is grown easily *in vitro* and produces several cell-wall-dissolving enzymes (Philipp 1985). It can penetrate developing ascomata of the host and overwinter in them on bark, but mature ascomata of the host are immune (Falk et al. 1995b). Although the species is found on many species of *Erysiphales*, it more frequently parasitizes those with well-developed and persistent superficial mycelia (e.g., *Erysiphe, Sphaerotheca*) than those with poorly developed surface mycelia (e.g., *Podosphaera, Microsphaera*); the latter sometimes are colonized inside the ascomata (Põldmaa 1966).

The nonspecific mycoparasite *Lecanicillium muscarium* (formerly identified as *Verticillium lecanii*) also commonly inhabits powdery mildews and kills the conidia (Heintz and Blaich 1990). In Crete, *Acremonium alternatum* was found commonly as a parasite when mildewed cucurbit leaves were incubated in moist chambers for 4 days; after conidia of *A. alternatum* were sprayed on the leaves, the mildew *Sphaerotheca fuliginea* was parasitized completely within 3 days (Malathrakis 1985). *Acrodontium crateriforme, Lecanicillium aphanocladii* (formerly identified as *Aphanocladium album*), *Isaria (Paecilomyces) farinosa, Ramichloridium apiculatum, Tilletiopsis minor, Trichothecium roseum,* and *Dissoconium apiculatum* also appear to be active mycoparasites on powdery mildews (Hijwegen and Buchenauer 1984). The first two species also commonly are found on rust fungi. *Isaria (Paecilomyces) farinosa,* like *Lecanicillium* species, is otherwise entomogenous and highly chitinolytic. Two species of *Dissoconium* probably also parasitize Erysiphales and other fungi in the phyllosphere (de Hoog et al. 1991).

Some species of *Tilletiopsis* and *Pseudozyma* of the Ustilaginales are efficient biocontrol agents, particularly *T. pallescens* and *T. minor*. The destructive interaction seems to be a case of hyphal interference (Hoch and Provvidenti 1979; Klecan et al. 1990). Urquhart and colleagues (1994) demonstrated that *Tilletiopsis* species produce β-1,3-glucanase, an enzyme associated with mycoparasitism in a wide array of fungi (Hijwegen 1992a). Host hyphae collapse on contact with a *Tilletiopsis* parasite (hyphal interference), without appressorium formation, leading to a drastic reduction in hyphal expansion and sporulation throughout the colony.

On Black Mildews, Sooty Molds, and Similar Leaf-Inhabiting Fungi

Numerous mycoparasites on taxonomically unrelated leaf-inhabiting fungi have been described (Hansford 1946; Deighton 1969; Deighton and Pirozynski 1972; Pirozynski 1974, 1976). Pirozynski (1976) revised many genera of parasites growing on Meliolales, including the pale-colored bitunicate ascomycetes *Melioliphila, Paranectriella, Puttemansia, Tubeufia,* and *Hyalocrea* and the unitunicate genera *Schweinitziella, Hyaloderma,*

Rizalia, and *Nematothecium*. All species of *Byssocallis, Malacaria,* and *Melioliphila* and some species of *Paranectriella, Hyalocrea, Hyalosphaera,* and *Nematothecium* live on meliolaceous fungi. Species of *Dimerosporiella* (the *Nectriopsis leucorrhodina* group, Bionectriaceae) and the discomycete *Calloriopsis gelatinosa* also parasitize hyphae on living leaves (Rossman 1987). *Dimerosporiella (Calonectria, Nectriopsis) cephalosporii* with an acremoniumlike anamorph, is a widely distributed parasite of *Meliola* species in the tropics (Gams 1971; Samuels 1988).

Deighton (1969) described 10 species and Deighton and Pirozynski (1972) described 46 species of fungicolous conidial fungi from leaf-inhabiting fungi. Species of *Spiropes* are generally mycoparasitic (Ellis 1968; Katumoto 1983). Additional species commonly found on such substrata include *Pleurodesmospora coccorum* (synonym *Oospora meliolae*) and species of *Acremonium, Lecanicillium, Eriomycopsis, Titaea, Trichoconis, Sympodiophora, Chionomyces, Cylindrocarpon, Hansfordia, Cercospora, Arthrobotryum, Triposporium, Trinacrium,* and *Tripospermum*. A key to the fungi in those genera that have elongate, septate, or nonseptate blastoconidia is provided in Braun (1995, 1998).

The biotrophic Phyllachoraceae, along with the Meliolaceae and Erysiphaceae, are among the most heavily parasitized fungus families (Parbery 1978; Hawksworth 1981a, 1981b; Cannon 1991). Parbery (1978) emphasized the risk of confusing parasitized stromata with conidial anamorphs of *Phyllachora*, which are exceedingly rare, apart from the *Linochora*-type spermatia with filiform conidia; he listed particularly the parasites *Phaeodothis winteri* and species of *Mycosphaerella, Shanoria, Discomycopsella, Seimatosporium, Cercospora,* and some other dematiaceous fungi. A parasitized *Phyllachora* can be recognized by its dull surface and the appearance of necrotized host tissue around the colony.

The sooty molds (sensu Hughes 1976) comprise at least six orders of saprotrophic leaf-inhabiting bitunicate ascomycetes. Hughes (1993) noted two parasitic ascomycetes, three hyphomycetes, and a few coelomycetes of the genus *Cicinobella* that grow on *Meliolina*. Several species of *Paranectriella, Puttemansia,* and *Hyalocrea* occur on carbonous stromata of various ascomycetes on living leaves. Long lists of conidial parasites for all these groups have been compiled by Hawksworth (1981b).

On Uredinales (Rust Fungi)

Parasites of rust fungi often have been investigated as possible biological control agents. By far the most common mycoparasite of rust fungi is *Eudarluca caricis* (for taxonomy and nomenclature, see Eriksson 1966;

anamorph the pycnidial *Sphaerellopsis filum*, synonym *Darluca filum*); it is especially destructive, growing in the sori (mostly uredinia). Its ascomata develop after the pycnidia and eventually occupy the whole sorus (Hawksworth 1981b). Unspecialized hyphae penetrate the urediniospores (Carling et al. 1976). The species has a wide spectrum of hosts, including more than 370 species of rust, worldwide (Eriksson 1966; Kranz 1974; Kranz and Brandenburger 1981). The parasite penetrates the urediniospores with an enzyme (Carling et al. 1976) and strongly inhibits their germination on wheat seedlings and the branching of their germ tubes (Stähle and Kranz 1984). *Eudarluca* is not regarded as a practical biocontrol agent of *Cronartium strobilinum* (a fungal pathogen of wheat), however, because of the short and irregular cycle of that host (Kuhlman et al. 1978).

The sporodochial genus *Tuberculina* (teleomorph *Helicobasidium*) comprises 10 species, most of which are exclusively uredinicolous (Hubert 1935; Wicker and Woo 1973; Kuhlman and Miller 1976; Hawksworth 1981b; Wicker 1981). The species *T. persicina* is known from at least 26 rust species, particularly affecting their aecial stage. Natural populations of *T. maxima* also seem to have a controlling effect on rusts, particularly *Cronartium* species. The parasite exploits the rust gall and, by destroying its nutrient source, displaces the rust (Wicker and Woo 1973; Wicker 1981). Hubert (1935) envisioned the spread of the mycoparasite by insects.

Scytalidium uredinicola (Kuhlman et al. 1976) was first found on aecia of *Cronartium fusiforme* on *Pinus* species in the southeastern United States, but later also was recorded from galls of *Endocronartium harknessii* in western Canada (Tsuneda et al. 1980). It penetrates the woody tissue of the galls and destroys the rust in the wood. It inhibits spore germination in *E. harknessii* with the metabolite maltol, a substance known to occur in some plants to which it may confer a natural resistance to some fungal pathogens (Cunningham and Pickard 1985). This parasitic fungus is distributed by Nitidulid beetles (Currie 1995).

Lecanicillium dimorphum (formerly identified as *Aphanocladium album*, a taxon different from the myxomyceticolous one with this name) is a necrotrophic parasite that efficiently invades the aecidiospores of *Puccinia graminis*. It inhibits urediniospores formation and induces precocious teleutospore formation, a phenomenon regarded as a defense reaction on the part of the fungal host (Koç et al. 1981; Koç and Défago 1983). It produces large amounts of endochitinases or exochitinases, depending on the inducing substratum (Srivastava et al. 1985a; Studer et al. 1992). Like *Verticillium muscarium*, it also produces a β-1,3-D-endo-mannanase, which lyses the germ pore plug in urediniospores of its host (Langen et al. 1992).

Other efficient mycoparasites are *Lecanicillium muscarium* (Mendgen and Casper 1980; Spencer 1980; Mendgen 1981) and *L. psalliotae* (Lim and Wan 1983; Saksirirat and Hoppe 1990b). The same isolates of *L. muscarium* can parasitize both rusts and aphids (Hall 1980); they penetrate the urediniospore wall directly (Spencer and Atkey 1981) and destroy the germ tubes of the rust (Mendgen and Casper 1980; Uma and Taylor 1987). Enzyme activities of *L. psalliotae* allow for the penetration of the urediniospores of *Phakopsora pachyrhizi* in soy beans (Saksirirat and Hoppe 1990a, 1990b). This fungus also can control coffee rust in Malaysia. Srivastava and associates (1985b) and Leinhos and Buchenauer (1992) screened other fungi for their parasitic properties and found that *L. muscarium* and *L. psalliotae* were much more efficient than their congeners and also induced early teliospore formation.

Cladosporium uredinicola and *C. gallicola* were described as specific rust parasites, but whether they are specifically distinct from saprotrophic taxa has not yet been demonstrated. *Cladosporium uredinicola* penetrates the urediniospores of *Puccinia violae* (Traquair et al. 1984), and *C. gallicola*, which originally was described from *P. violae* (Sutton 1973a), efficiently parasitizes aecial galls of *Endocronartium harknessii* (Tsuneda and Hiratsuka 1979). The otherwise saprotrophic *C. tenuissimum* (Ellis 1976) is an efficient antibiotic-producing parasite of *Melampsora larici-populina* in Australia. It also can grow epiphytically on poplar leaves, but its growth is promoted by rust pustules (Sharma and Heather 1978, 1987). Species of *Cladosporiella*, *Cercospora*, and other genera also parasitize rusts (Deighton 1969).

Monocillium nordinii is known to kill rust spores of *Cronartium coleosporioides* and *Endocronartium harknessii* with the toxic metabolites monorden (radicicol) and monocillin I, which also damage *Ophiostoma ulmi* and *Alternaria alternata* (Ayers et al. 1980; Tsuneda and Hiratsuka 1980). Additional rust parasites include *Ramichloridium schulzeri* (Uma and Taylor 1987), *Acrodontium crateriforme*, *Fusarium bactridioides* on blister rusts (Wollenweber and Reinking 1935), the ascomycete *Scopinella gallicola* in galls of *Endocronartium harknessii* (Tsuneda and Hiratsuka 1981), species of *Paranectriella* and *Uredinophila* (Tubeufiaceae; Rossman 1987), and *Colletoconis aecidiophila* on *Puccinia* in warmer regions (CBS, unpublished data). Many other potential parasites are listed by Hawksworth (1981a).

On Ustilaginales (Smut and Bunt Fungi)

Reports of mycoparasites on smuts and bunts are relatively scarce (Hawksworth 1981a). An oomycete parasitized *Ustilago bullata* on the grass *Bromus mollis* in the field. *Pythium vexans* was isolated from the same host and was mycoparasitic in culture but not when the host was growing in a plant (Roberson et al. 1990). Several species of *Fusarium* have been reported from *Ustilago* hosts (Wollenweber and Reinking 1935). Although the galls caused by *Ustilago maydis* and other species are used as food in some places, infections by toxinogenic *Fusarium* species can render them poisonous (CBS, unpublished data). Species of *Itersonilia* and *Tilletiopsis* have been found growing on lesions caused by *Entyloma* (Brady 1960). *Itersonilia perplexans* and its congeners in turn can be parasitized by the chytrid *Rozella itersoniliae* (Barr and Bandoni 1979).

Methods of Study

Attacks by mycoparasites are normally inconspicuous, except when *Ampelomyces quisqualis* or other species cause large patches of powdery mildew to disintegrate. Whitish flakes of mycoparasites on rust fungi are rarely conspicuous. Mycoparasites usually are located by a directed search in which voluminous collections from the field are screened under a dissecting microscope, often after incubation in a moist chamber. *Eudarluca caricis* and other mycoparasites grow readily on ordinary agar media *in vitro* (Calpouzos et al. 1957).

To selectively isolate *Tilletiopsis* species from field material of powdery mildews, Urquhart and colleagues (1994) used the active spores shed from mildewed mats onto a dichloran-containing CMA. Spore fall is also successful for the isolation of *Dissoconium* species, which shoot off their conidia (de Hoog et al. 1991).

Identification

There is no comprehensive treatment of biotrophic plant pathogens and their parasites. Most papers deal with individual groups of host fungi (see the respective sections under "Taxonomic Groups of Fungicolous Fungi and Fungus-like Organisms," earlier). Those data taken together suggest that about 70 parasite species are known, plus an additional ca. 75 from black mildews and others (Deighton and Pirozynski 1972; Rossman 1987). In tropical countries, mycoparasites of black mildews and other leaf-inhabiting fungi are particularly abundant.

Major publications by Hansford (1946), Deighton (1969), and Deighton and Pirozynski (1972) and keys to hyphomycete genera with holoblastic conidiogenesis by Braun (1995:46, 1998:7) are helpful for identifying leaf-inhabiting mycoparasitic fungi.

Geographic Distribution, Biodiversity, and Ecology

The best-studied representatives of this group (e.g., *Eudarluca caricis*) are cosmopolitan. Such mycoparasites develop best under moist tropical conditions. Not surprisingly, then, these parasites of rusts and other leaf-inhabiting ascomycetes are particularly common in tropical countries. In temperate regions, parasites of rusts and powdery mildews most frequently are found in habitats characterized by a higher-than-average relative humidity, such as those in the vicinity of water bodies and in bogs (Põldmaa 1966). Similarly, species of rusts on cultivated and wild grasses growing on raised places usually lack parasites, whereas those near water bodies or in moist places frequently are colonized by several species of parasites (Rudakov 1981). Black mildews and sooty molds that are highly susceptible to mycoparasites are distributed most widely in the tropics (Jeffries and Young 1994:37). Field studies generally have been confined to a qualitative assessment of the parasitic association.

Abundance of *Ampelomyces quisqualis* on powdery mildews varied among mildew species found on different plants in the Eurasiatic countries (Rudakov 1981). In material collected from Northern Estonia, 60% of 370 specimens belonging to 24 species of *Erysiphe* and 30% of 140 specimens of eight species of *Sphaerotheca* appeared to be infected (Põldmaa 1966). *Ampelomyces quisqualis* appeared at the beginning of summer and spread widely later in summer. Infections on colonies of powdery mildews that had overwintered declined considerably in the spring as a result of overgrowth by saprotrophs (Rudakov 1981).

MYCOPARASITES ON MYCELIA, ECTOMYCORRHIZAE, SCLEROTIA, AND SPORES IN SOIL

Fungi of divergent taxonomic groups that parasitize or otherwise antagonize the hyphae, vegetative propagules, or survival structures of nonlichenized fungi in the relatively moist soil environment are included under this heading. Both biotrophic and necrotrophic relationships are observed in the soil. Some necrotrophs invade any host structure that they encounter, but most are specialized and colonize only certain structures such as vegetative hyphae, sclerotia, or spores of particular host fungi (Whipps 1991; Lumsden 1992; Jeffries 1995). The highly specialized biotrophic mycoparasites of zygomycetes treated earlier (see "Zygomycota" under "Taxonomic Groups of Fungicolous Fungi and Fungus-like Microorganisms") belong in this section, as do the mycoparasitic species of *Pythium*. *Trichoderma* and *Clonostachys* (*Gliocladium*) species are particularly well-known mycoparasites that are destructive, little-specialized, and highly successful in biocontrol (see "Fungicolous Fungi as Biocontrol Agents of Plant Pathogens," later in this chapter).

On Oomycetes

Species of *Pythium* often are antagonized by congeners, particularly *P. oligandrum*, *P. periplocum*, and *P. acanthicum*, which have spiny oogonia (Deacon 1976; Vesely 1977; Deacon and Henry 1978; Lutchmeah and Cooke 1984; Martin and Hancock 1987; Berry et al. 1993), and the more distant *P. nunn* and *P. mycoparasiticum* (Deacon et al. 1991). *Pythium oligandrum* is the most common (Ribeiro and Butler 1992). The density of *P. nunn*, which suppresses *P. ultimum*, can be increased by adding crumbled dried bean leaves to the soil (Lifshitz et al. 1984; Paulitz and Baker 1988); *P. nunn* is less aggressive than *P. mycoparasiticum* and *P. oligandrum* (Laing and Deacon 1990) but is better at competing for nutrients (Elad et al. 1985). These parasites cause hyphal interference (see "Types of Fungicolous Associations," earlier) and induce a rapid hyphal lysis in many hosts (Elad et al. 1985). *Pythium oligandrum*, *P. mycoparasiticum*, and *P. nunn* needed an average of 4.5, 4.8, and 13.3 minutes, respectively, after a first contact to stop a host hypha; penetration occurred after about 50 minutes, disrupting the hypha up to 1.2 mm ahead of the point of contact. However, *Stachybotrys chartarum* parasitizes *P. oligandrum* (Deacon and Henry 1978), and *Olpidiopsis gracilis* also attacks certain species of *Pythium* and *Phytophthora* (Pemberton et al. 1990).

Soil-borne species of *Pythium* and *Phytophthora* are affected negatively by numerous other antagonistic fungi. The toxinogenic *Trichoderma virens* efficiently suppressed *Pythium ultimum* in cotton fields (Howell 1991). *Geomyces pannorum* var. *pannorum* is a very common soil-borne fungus that also has been recorded from *Ramaria* and *Trametes* (Helfer 1991); its varieties *asperulatus* and *vinaceus* were found to antagonize *Pythium ultimum* in cucumber cultivation. A peat substratum strongly colonized by this fungus suppressed *P. ultimum* (Danielsen and Wolffhechel 1991).

On Oospores

The oospores of certain Oomycetes, particularly *Pythium* and *Phytophthora* are, because of their persistence in soil, a special substratum for a diversity of mycoparasites. Conidial fungi of the genera *Dactylella* (some formerly in *Trichothecium*) and *Trinacrium* have been described from *Pythium* oospores (Drechsler 1938, 1943, 1952, 1962, 1963). *Microdochium fusarioides* was found in oospores of *Phytophthora* syringae (Harris 1985). The hyphochytridiomycete *Hyphochytrium catenoides* (Ayers

and Lumsden 1977) and the Chytridiomycete *Catenaria anguillulae* are common endobiotic colonizers of *Phytophthora* oospores (Humble and Lockwood 1981; Daft and Tsao 1984). The epibiotic *Rhizidiomyces japonicus*, and species of *Fusarium*, *Acremonium*, *Verticillium*, *Humicola*, and *Clonostachys* colonized buried oospores of several genera (Sneh et al. 1977; Wynn and Epton 1979). *Hyphochytrium catenoides* is one of the most active parasites of *P. cinnamomi* and *P. parasitica*, as well as of *Humicola fuscoatra* and *Anguillospora pseudolongissima* (Daft and Tsao 1983). Oospores of *P. ultimum* also are attacked by *Fusarium merismoides* (Hoch and Abawi 1979a).

On Zygomycetes

Besides some nonspecific zygomycete parasites of *Chaetocladium*, *Parasitella*, and some other *Mucorales* (mainly *Rhizopus*), *Curvularia* species, *Bipolaris spicifera*, *Alternaria alternata*, and *Exserohilum* (*Drechslera*) *rostratum* have been found on the sporangiophores of Mucoraceae (El Shafie and Webster 1979; Gupta et al. 1983). Also, species of *Aspergillus*, *Penicillium*, *Myrothecium*, *Trichoderma*, and *Fusarium* may penetrate the sporangiophores, but whether these are mycoparasitic associations remains doubtful (Hawksworth 1981b). For biotrophic parasitism, see under "Taxonomic Groups, Zygamycota."

On Spores of Glomales

The resting spores of the arbuscular mycorrhizal fungi (order Glomales) are particularly susceptible to necrotrophic parasites (reviewed in Paulitz and Linderman 1991b). Large proportions of spores extracted from field soils usually are parasitized and nonviable. Their walls are perforated by fine radial canals, often with internal projections (Bhattacharjee et al. 1982; Lee and Koske 1994). Both amoebae and fungi can make radial pores in the thick spore walls (Boyetchko and Tewari 1991). Such pores first were described, apparently, from spores of *Glomus microcarpum* and were associated with an acremonium-like fungus (Malençon 1942). Necrotrophic mycoparasitism by chytrids and other zoosporic fungi that sporulate either inside or on the surface of spores is probably widespread and may limit populations of mycorrhizal fungi in wet soils (Sylvia and Schenck 1983). A heavy infestation with *Phlyctochytrium* can reduce soil populations of *Glomus macrocarpum* and *Gigaspora gigantea* (Ross and Ruttencutter 1977). In contrast, the colonization of dead spores of *Gigaspora* species by *Spizellomyces* (*Phlyctochytrium*) *punctatus* is regarded as a saprotrophic relationship and does not reduce the mycorrhizal population (Paulitz and Menge

1984). A species of *Labyrinthula* also parasitized the arbuscular fungus *Gigaspora gigantea* in sand dune soils (Koske 1981). Many mycoparasites, easily isolated from the spores of arbuscular mycorrhizal fungi, appear to be facultative parasites that are to some degree saprotrophic and not dependent on the presence of the spores for survival (Paulitz and Menge 1986). *Trichoderma virens* does little harm to Glomales (Paulitz and Linderman 1991a), but *T. harzianum* does (Rousseau et al. 1996). Species of Glomales with pigmented spores are less susceptible to parasitism than those with hyaline spores (Ross and Ruttencutter 1977; Bhattacharjee et al. 1982). *Humicola fuscoatra* and *Anguillospora pseudolongissima* were found to be highly parasitic on *Glomus epigaeum* and *G. fasciculatum* in California (Paulitz and Menge 1980, 1986; Paulitz and Linderman 1991b). *Stachybotrys chartarum* is also an efficient parasite (Siqueira et al. 1984). In a maritime sand dune soil, 44 species of higher fungi, most belonging to the genera *Acremonium*, *Verticillium*, *Chrysosporium*, *Exophiala*, and *Trichoderma*, were found on *Gigaspora gigantea* (Lee and Koske 1994). An experiment with healthy, surface-disinfected spores of *G. gigantea* showed that species of *Acremonium* and *Verticillium* were the most pathogenic. The mycorrhizal colonization of citrus trees can be diminished considerably by mycoparasites (Daniels 1981; Jeffries 1995). Nonsporulating fungi also may be common inside the spores but may not be recognized without culturing. Contrasting with the Glomales, the subterranean sporocarps of Endogonales often are colonized by species of *Mortierella* (W. Gams, personal observation).

On Sclerotia

Several soil-borne plant pathogens of ascomycetes and basidiomycetes form sclerotia in soil. Fungal sclerotia provide a rich source of nutrients for the fungi that are capable of attacking them. Most sclerotial parasites are necrotrophs, although an initial biotrophic phase has been observed in some species.

On Ascomycete Sclerotia. Species of *Sclerotinia* frequently have been examined for their parasites, many of which also attack the related white-rot fungus (*Sclerotium cepivorum*) of onion (Jackson et al. 1991). The most common and best-studied parasite of *Sclerotinia* species is the pycnidial *Coniothyrium minitans*, which is distributed worldwide (Turner and Tribe 1976; Whipps and Gerlagh 1992; Sandys-Winsch et al. 1993). It does not form appressoria (Huang and Kokko 1988). *Coniothyrium minitans* significantly inhibits *Sclerotinia sclerotiorum in vitro* on nutrient-poor media (Whipps 1987). *Microsphaeropsis centaureae*, a related fungus from British Columbia, causes necrotic lesions on *Centaurea diffusa*

and attacks the sclerotia of *S. sclerotiorum* (Watson and Miltimore 1975). These fungi are unable to parasitize sclerotial basidiomycetes, which appear to produce toxic metabolites (Whipps et al. 1991).

Three biotrophic hyphomycetes, *Sporidesmium* (or *Teratosperma*) *sclerotivorum* (Uecker et al. 1978; Ayers and Adams 1979), *Teratosperma oligocladum* (Uecker et al. 1980), and *Laterispora breviramosa* (Uecker et al. 1982), are more specialized. *Sporidesmium sclerotivorum* penetrates host cells by means of branched haustoria that, nevertheless, do not penetrate the plasmalemma of infected cells (Bullock et al. 1986). None of these parasites infects sclerotial fungi in other families (Ayers and Adams 1979, 1981).

A few strains of *Trichoderma koningii*, *T. harzianum*, and *T. pseudokoningii* parasitize and have been isolated from the sclerotia of *Sclerotinia sclerotiorum* (Dos Santos and Dhingra 1982). *Clonostachys* (*Gliocladium*) *rosea* f. *catenulata*, isolated from sunflower field soil, is an efficient destructive parasite of *S. sclerotiorum* and several *Fusarium* species, which it contacts with pseudoappressoria (Huang 1978). *Trichoderma virens* also destroys sclerotia of *S. sclerotiorum* and related fungi (Tu 1980; Phillips 1986). *Trichothecium roseum* attacks the sclerotia on bean plants in the field with a slightly lower frequency than *Coniothyrium minitans*; because it is potentially harmful to plants, however, it has not been considered as a control agent (Huang and Kokko 1993). *Dictyosporium elegans* is one of the most active parasites on *S. sclerotiorum* in West Australia (McCredie and Sivasithamparam 1985). In a screening of 10 known mycoparasites, *Trichoderma virens* and *C. minitans* singly and in combination were the most efficient control agents (Whipps and Budge 1990).

Talaromyces flavus is an aggressive parasite of many kinds of sclerotia and one of the most successful biocontrol agents against *Verticillium dahliae* (Boosalis 1956; Dutta 1981; McLaren et al. 1986, 1989; Madi et al. 1992; Fahima and Henis 1995; Nagtzaam 1998; Nagtzaam et al. 1998). Additional efficient parasites of *V. dahliae* are *Clonostachys rosea* and *C. rosea* f. *catenulata* (Keinath et al. 1991). *Talaromyces flavus* produces antibiotic metabolites, including glucose oxidase, which has been identified as the main antifungal agent (Kim et al. 1990). That enzyme releases hydrogen peroxide from glucose, which is highly toxic to *V. dahliae*. The mycoparasite also produces β-1,3 glucanase and chitinase. When *C. minitans* was applied in combination with *T. flavus* against *Sclerotinia*, it had a nearly equivalent effect (McLaren et al. 1994).

Paecilomyces lilacinus, a pathogen of nematodes, also colonizes sclerotia of *Aspergillus flavus* and *A. parasiticus* in soil and thus shortens their survival. It also has been reported to parasitize *Sclerotinia* species (Wicklow and Wilson 1990). Sclerotia of *Phymatotrichopsis*

(*Phymatotrichum*) *omnivora* buried 50-cm deep in soil were attacked by *Clonostachys rosea*, *C. rosea* f. *catenulata*, and *Trichoderma* species (Kenerley and Stack 1987).

Fusarium heterosporum can parasitize the sclerotia of the ergot fungus *Claviceps purpurea* and also attacks its mycelium in culture (Wollenweber and Reinking 1935; Hornok and Walcz 1983). Mower and colleagues (1975) proposed that it be used as a biological control agent against ergot. *Neobarya* (*Barya*) *aurantiaca* also grows on *Claviceps purpurea* (Ellis and Ellis 1988; Eriksson 1992).

On Basidiomycete Sclerotia. A different spectrum of parasites colonizes sclerotia of the important plant pathogens *Rhizoctonia solani* (teleomorph *Thanatephorus cucumeris*) and *Sclerotium rolfsii* (teleomorph *Athelia rolfsii*). Some 30 species of mycoparasites have been recorded for *R. solani* (Butler 1957; Chand and Logan 1984; Jeffries 1995). *Trichoderma* species are both common and efficient mycoparasites (Elad et al. 1980, 1984; Lewis and Papavizas 1985; Howell and Stipanovic 1995). Hyphae of *Trichoderma* species exhibit directed growth towards *R. solani*, mostly coiling around the host, sometimes forming appressoria, and penetrating the host hyphae; antibiotic action has been detected only in *T. virens* (Chet et al. 1981; Elad et al. 1982). *Trichoderma harzianum* has a complex method of action, involving chitinase and β-1,3-glucanase (Elad et al. 1984), whereas *T. virens* affects host fungi mainly with its toxic metabolites gliotoxin and gliovirin (Tu 1980; Howell and Stipanovic 1983; Howell 1991). *Trichoderma virens* also forms appressoria, penetrates host hyphae, and forms intracellular hyphae in *R. solani*, which eventually lead to its collapse and prevent it from forming sclerotia (Tu and Vaartaja 1981). Antibiotic effects also prevail in *Gliocladium viride* (synonym *G. deliquescens*), which disorganizes the cell walls and organelles of *R. solani* (Hashioka and Fukita 1969). When *G. viride* or *T. harzianum* are cultured together with *R. solani*, high β-1,3-glucanase and chitinase activities are observed. The localization of these enzymes at points of contact has been observed by means of fluorescein-conjugated lectins (Elad et al. 1983c).

Verticillium biguttatum is a very efficient biocontrol agent of *R. solani*. It is an obligate parasite and has many biotrophic traits (van den Boogert and Deacon 1994). It is found only on *Thanatephorus cucumeris* (*Rhizoctonia solani*) and related fungi (van den Boogert et al. 1989; Morris et al. 1995b) and strictly is associated with host fungi under natural conditions. It does, however, grow easily and axenically *in vitro* (van den Boogert et al. 1990). This fungus requires biotin and grows best with mannitol or galactose as carbon sources and glutamine or ammonium salts as nitrogen sources (van den

Boogert 1989). Various other species of *Verticillium* and *Lecanicillium* (in fact, anamorphs of three ascomycete families) interact with the hyphae of *R. solani* by appressed growth, coiling, and sometimes penetration (Kuter 1984). *Pythium oligandrum* also can attack *Rhizoctonia solani* (Hoch and Fuller 1977), suppressing its cellulolytic activity and sclerotium formation (Al-Hamdani and Cooke 1983).

The sclerotial, strand-forming basidiomycete *Laetisaria arvalis* (Burdsall et al. 1980) is a potent biocontrol agent against *Rhizoctonia solani* and *Pythium ultimum* (Hoch and Abawi 1979b). The former fungus seems to be closely related to a still incompletely identified "sterile red fungus" (Dewan and Sivasithamparam 1988, 1989), which has similar capacities. Another unidentified basidiomycete first recognized as a mycoparasite of *Macrophomina phaseolina* on pine seedlings was found to parasitize several other fungi by coiling, invasion, and lysis (Cerrato et al. 1976). *Rhizoctonia solani* also is controlled by nonpathogenic conspecific isolates and by binucleate isolates of *Rhizoctonia* species (now *Ceratorhiza*; e.g., in sugar beet) even better than it is by *Laetisaria arvalis*. The mechanism is not well understood, although it may involve competition (Ichielevich-Auster et al. 1985; Herr 1988; Lewis and Papavizas 1992).

The nematophagous *Arthrobotrys superba*, *A. oligospora*, and *A. cladodes* are also contact mycoparasites. *Arthrobotrys oligospora* acts on *Rhizoctonia solani* by hyphal interference (Udagawa and Horie 1971; Tzean and Estey 1978a, 1987b), which involves an accumulation of membranous vesicles and greatly increased proteolytic activity in the coiling cells (Persson et al. 1985; Persson and Friman 1993). That species also can attack *Aphanomyces euteuches*, *Mucor silvaticus*, *Penicillium spinulosum*; and its own relatives, *A. superba* and *Dactylellina* (*Monacrosporium*) *haptotyla* (synonym *Dactylella candida*). The mycoparasitic effects of *A. oligospora in vitro* are strongest on a dilute CMA or on water agar with 0.2 g/liter glucose (Persson and Bååth 1992). The fungus excretes chitinase in liquid culture with cell walls of *R. solani* but not in culture with colloidal chitin (Persson 1991). Twigs are often densely covered with *A. superba*, which apparently overgrows corticiaceous fungi as a mycoparasite (Corda 1839).

A strain of *Stachybotrys elegans*, isolated from soil in Turkey (Turhan 1993), efficiently destroyed *R. solani*, but only the anastomosis groups 1–6 were susceptible (Benyagoub et al. 1994, 1996). Although the host cells were killed on contact, no inhibition zone was observed in dual culture. Mortality seemed to be caused by chitinases and β-1,3-glucanases, which are induced by the substratum (Tweddell et al. 1994).

Additional parasites of *R. solani* and *P. ultimum* include *Fusarium oxysporum* (Gupta et al. 1979), *Neo-*

cosmospora vasinfecta var. *africana*, *Acrophialophora levis* (Turhan and Turhan 1989), *Stachybotrys chartarum*, *Trichothecium roseum*, and *Verticillium luteo-album* (Turhan 1990). Conversely, some isolates of *R. solani* also parasitize other fungi, such as species of *Rhizopus*, *Mucor*, *Pythium*, *and Amblyosporium*, by hyphal penetration (Butler 1957). Sclerotia of *Athelia* (*Sclerotium*) *rolfsii* usually are parasitized by similar fungi. Some isolates of *Trichoderma harzianum* can attack this pathogen; others cannot (Wells et al. 1972). That difference has been ascribed to differences in enzyme production; both β-1,3-glucanase and chitinase are required for successful parasitism (Elad et al. 1984). *Aspergillus terreus* also efficiently parasitizes and penetrates sclerotia of *A. rolfsii* (Shigemitsu et al. 1978). Parasites have not been found on sclerotia of *Typhula* species, apart from one instance of the widely distributed *Cylindrobasidium parasiticum* (Woodbridge et al. 1988).

On Conidiophores and Hyphae of Conidal Fungi (Biotrophic Parasites)

Cladosporium cladosporioides can be parasitized by an epibiotic chytrid, *Caulochytrium protostelioides* (Powell 1981). *Gonatobotrys simplex* sometimes is found on its *Alternaria* hosts during analyses of soil and litter. Biotrophic contact mycoparasites of hyphomycetes commonly are found among the Ceratostomataceae and associated *Papulaspora*, *Harzia*, *Gonatobotrys*, and *Olpitrichum* anamorphs. Host specialization varies, even among closely related species (Jordan and Barnett 1978; Cannon and Hawksworth 1982). *Harzia acremonioides* grows easily without a host fungus, but it parasitizes *Stemphylium botryosum* with lobed, sometimes branched contact cells that function as appressoria, causing little damage (Urbasch 1986). In contrast, *H. velata* can grow on species of *Verticillium*, *Fusarium*, and *Cylindrocarpon* and is strongly dependent on a host fungus (G. Fischer and W. Gams, unpublished data). Similar observations apply to several species and strains of *Melanospora* and *Papulaspora*. The possibly related hyphomycete *Gliocephalis hyalina* also is biotrophic and can grow *in vitro* only with a host such as *Cylindrocarpon destructans* (G. Fischer and W. Gams, unpublished data).

On Ectomycorrhiza

In contrast to sporocarps, ectomycorrhizal mantles are usually little affected by mycoparasites, and the array of secondary invaders isolated from this substratum is similar to that of mycorrhiza-free root tips (Summerbell 1989). Only occasionally do species of *Trichoderma* inhibit the establishment of *Laccaria bicolor* mycorrhiza

on *Picea mariana* (Summerbell 1987). *Pochonia* (*Verticillium*) *bulbillosum*, which commonly is isolated from forest soils and tree roots, also can parasitize the hyphae of *Laccaria laccata* (Girlanda et al. 1995). The frequently observed adjacent fructification of *Gomphidius roseus* and *Suillus bovinus* has been traced to the mycorrhizal mantle where *G. roseus* grows partially within the *Suillus* mycorrhiza (Agerer 1991).

Methods of Study

Methods used to isolate soil mycoparasites are often the same as those used to isolate soil fungi (e.g., Gams 1992; Chapter 13). The methodological problems are also similar, particularly the difficulty of extrapolating *in vitro* results to the situation in the soil. In suspension plating, the soil material usually becomes so thoroughly dispersed that the associations between biotrophic parasites and their hosts are broken. The likelihood of keeping them together is higher when washed soil particles, or even fragments of unwashed soil or root material, are plated. When the latter method was used, *Gliocephalis* species, *Harzia velata*, and *Gabarnaudia* species were found in soil in parasitic associations with *Cylindrocarpon destructans*, and *Heterogastridium pycnidioideum* was found in association with *Plectosphaerella cucumerina* (G. Fischer and W. Gams, unpublished data). *Heterogastridium pycnidioideum* hitherto had been found only on species of the *Russulaceae* and associated *Hypomyces* and on various litter-inhabiting fungi (Seeler 1943; Bandoni and Oberwinkler 1981).

"Drechsler's soil plate" method used to study nematophagous fungi (see "Drechsler's Technique" in Chapter 19 and "Drechsler's Method" in the Appendix of Chapter 13) is also useful for the observation of mycoparasitic associations (particularly of Mucorales); on water agar or other transparent media inoculated with soil, dung, or other organic material, a dense mosaic of sparsely growing fungi develops. The opportunity for parasites to encounter their hosts increases in such a mat.

The incubation of dung samples in moist chambers (Appendix I) is usually an efficient way to obtain biotrophic members of the Zoopagales and Dimargaritales for observation. The spores they produce are transferred onto *Cokeromyces recurvatus*, which is a convenient host for their maintenance (Benjamin 1959); *Mycotypha microspora*, *Umbelopsis (Mortierella) longicollis*, and *M. isabellina* are also suitable for certain species (Richardson and Leadbeater 1972). Choosing a substratum on which the parasite can overgrow the host efficiently is important. *Piptocephalis* species and their host fungi were detected from soil crumbs spread on potato-carrot agar with streptomycin (Richardson and Leadbeater 1972). Jeffries and Kirk (1976) improved the technique by inducing the formation of yeast-phase cells of *Cokeromyces recurvatus* in peptone-glucose liquid medium at 25°C. When the yeast-phase cells are spread on malt extract-yeast extract agar, vegetative hyphae grow within a few hours, providing excellent sites for infection by parasites originating from a soil sample.

It is generally impossible to assess the effects of mycoparasites in the soil because of the difficulty of direct observation (Jeffries 1995). Only in special cases is such an assessment possible; for example, hyphal swellings induced in *Rhizopus oryzae* by *Syncephalis californica* indicate levels of parasite activity (Hunter et al. 1977). Lumsden (1981) emphasized the need for developing imaginative techniques to study mycoparasitism in natural systems, such as enumeration on improved selective media, trapping methods for examining mycohosts, and direct observation by scanning electron microscopy.

Because of the roles that mycoparasitic fungi may play in natural, or biological, control of plant pathogens (see "Fungicolous Fungi as Biocontrol Agents of Plant Pathogens," later in this chapter), it is important to quantify them in the soil. Use of a target organism as bait makes a directed search for specific mycoparasites possible in many cases. Sclerotia or spores of the bait organism can be recovered easily either by physical extraction from field soil or by trapping them in a membrane filter or nylon gauze sandwich in which the bait organism has been buried in the soil. The colonization by mycoparasites of oospores of *Pythium* or *Phytophthora* (Ayers and Lumsden 1977; Sneh et al. 1977; Hoch and Abawi 1979a; Wynn and Epton 1979; Daft and Tsao 1984), microsclerotia of *Verticillium dahliae* (Keinath et al. 1991), or the large spores of *Glomus* or *Gigaspora* (Lee and Koske 1994) has been studied with the latter technique. Sneh and colleagues (1977) observed parasitism by *Rhizidiomycopsis*, *Canteriomyces*, *Leptolegnia*, and *Hyphochytrium* species as well as some hyphomycetes on oospores of *Phytophthora*, *Pythium*, and *Aphanomyces* species buried in soil. Harris (1985) buried apple leaves colonized with oospores of *Phytophthora syringae* and later recovered them from soil. Oospores enclosed between membrane filters can be examined microscopically after embedding the membrane in water agar (Sneh 1977). Colonization of *P. megasperma* by *Hyphochytrium* species was observed in that way.

Sclerotia often have been used for the selective isolation of mycoparasites after recovery from soil and superficial disinfection (Chand and Logan 1984; McCredie and Sivasithamparam 1985; Woodbridge et al. 1988). A mycoparasite population also can be enriched by the repeated addition of the host fungus to the soil—for example, *Rhizoctonia solani* for *Verticillium biguttatum* (van den Boogert and Jager 1983) and *Sclerotinia scle-*

rotiorum for *Coniothyrium minitans* (Gerlagh and Vos 1991). Ayers and Adams (1979) and Adams and associates (1984, 1985) quantified *Sporidesmium sclerotivorum* and other sclerotial parasites in relation to the number of host sclerotia retrieved from soil (naturally infected or buried artificially as bait) after incubation on moist filter paper.

The sclerotial parasite *Verticillium biguttatum* can be quantified in soil particles that are spread over Petri dishes colonized with *Rhizoctonia solani* (van den Boogert and Gams 1988). Morris and colleagues (1995a) refined the method using suspension plating on *R. solani* plates with potato-dextrose agar pH 4 (APDA, Appendix II). Mulligan and Deacon (1992) extended the method using PDA plates overgrown with various other host fungi and recovered *P. oligandrum* on *Fusarium culmorum*, *Trichoderma* species on *R. solani*, and *Papulaspora* species on *Botrytis cinerea*; *Clonostachys* (*Gliocladium*) *rosea* appeared on all three hosts. For the selective isolation of *Talaromyces flavus* from soil, a PDA medium was amended with 0.1% lactic acid, antibacterial antibiotics, 4 mg pimaricin, 30 mg nystatin, and 0.5 g oxgall per liter (Marois et al. 1984).

Selective techniques for the isolation of *Pythium* species are described in Chapter 13 (see "Media" under "Isolation Techniques for Filamentous Fungi"). *Pythium oligandrum* and related species are more sensitive than other congeners to the inhibitors commonly used in isolating *Pythium* species (particularly pentachloronitrobenzene). Placing soil crumbs on a Petri-dish culture of a potential host, such as *Phialophora* species, as a bait is an alternative method for recovering their mycoparasites (Deacon and Henry 1978; Foley and Deacon 1985). Sclerotia of *Sclerotinia sclerotiorum* can serve as baits for mycoparasitic species of *Pythium* (Ribeiro and Butler 1992). Conversely, Petri dishes colonized with a *Pythium* have been used to isolate parasites of their oospores (Drechsler 1943).

The isolation of fast-growing and heavily sporulating colonies of *Trichoderma* is usually easy, but semi-selective techniques can be used to quantify the species of this genus in soil (Elad et al. 1981; Elad and Chet 1983; Chet 1987; Askew and Laing 1993). Elad and Chet (1983) modified a previously published formula for the isolation medium (see *Trichoderma* Isolating Medium, Appendix II) by adding 20 mg/liter captan after autoclaving. Askew and Laing (1993) replaced the fenaminosulf with propamocarb or metalaxyl, which suppressed oomycetes even more efficiently. To isolate *Clonostachys rosea* and *Trichoderma virens* selectively, Park and associates (1992) used a medium with benomyl, sodium propionate, rose bengal, and antibacterial antibiotics, supplemented with either 1 mg gliotoxin and 60 mg acriflavine or 20 mg gliotoxin (per liter), respectively.

Laetisaria arvalis was isolated selectively by placing soil pellets, or table beet seeds that had been buried in soil, on a medium containing an ethanol solution of phenol, dichloran, benomyl, and thiabendazol with antibacterial antibiotics (Papavizas et al. 1983). Concentrations of down to seven propagules per gram of soil thus could be detected.

The large spores of the Glomales are extracted from soil by a wet-sieving procedure (see "Particle Filtration" under "Principal Isolation Methods" and in the Appendix in Chapter 13). Plating of the surface-disinfected spores allows the isolation of many mycoparasites, as noted earlier (see "Methods of Study" under "Mycoparasites of Mycelia, Ectomycorrhizae, Sclerotia, and Spores in Soil").

Identification

For the identification of Zygomycota, see Zycha and colleagues (1969). The genera of the biotrophic parasites have never been monographed exhaustively, but Benjamin (1959) provided much information on species of those genera. Benjamin (1965) also published a key to the species of *Dimargaris*. For other soil fungi, readers should refer to the literature compiled in Chapter 13.

Geographic Distribution and Diversity

Soil-inhabiting mycoparasites form a very heterogeneous group that has never been surveyed as a whole. About 50 species of biotrophic zygomycetous mycoparasites are known, as are about 50 parasites of Glomales, besides the other groups outlined earlier. The true numbers are undoubtedly much higher.

Most fungi discussed in this section are cosmopolitan in distribution, although most records are from temperate regions. The biotrophic parasites of Mucorales include seven tropical and three cosmopolitan species of the *Dimargaritales*. Two of 20 species of *Piptocephalis* are strictly tropical, at least two are temperate, and two are cosmopolitan; little information is available for the others (Kirk 1993). *Coniothyrium minitans* is distributed worldwide (Whipps and Gerlagh 1992; Sandys-Winsch et al. 1993). *Sporidesmium sclerotivorum* has been found so far in Australia, Canada, the United States, Finland, Hungary, Japan, and Norway but not in other European countries (Adams and Ayers 1985). *Laterispora brevirama* is known only from Australia, Canada, Finland, and Japan. The distribution of *Verticillium biguttatum* apparently follows the worldwide distribution of its host fungus, *Rhizoctonia solani*, when associated with potatoes (van den Boogert and Saat 1991).

Distribution patterns of species of *Hypocrea* and *Trichoderma* are correlated with temperature (e.g., Domsch et al. 1980; Samuels et al. 1994).

Quantification and Relative Importance

Biotrophic parasites of Mucorales commonly are found on dung that is rich in mucoralean host fungi. Richardson and Leadbeater (1972) reported that *Piptocephalis* also could be isolated frequently from litter and humus horizons, particularly in woodland and pasture habitats, if the material were examined systematically.

The majority of the spores of Glomalean arbuscular mycorrhizal fungi retrieved from soil are parasitized and nonviable. For example, large numbers of empty spores (ghosts) were observed during an assessment of the abundance of arbuscular mycorrhizal fungi on various crops in different agricultural soils of Northern Greece. Ghosts usually outnumbered intact spores up to fivefold (Jeffries et al. 1988). Few other calculations of the relative proportions of parasitized spores in field soils have been published (Jeffries 1995).

Mycoparasitic species of *Pythium* have been recovered from most soil samples taken in California, regardless of vegetation cover, soil pH, or soil texture (Ribeiro and Butler 1992), exemplifying their wide distribution in temperate zones. *Coniothyrium minitans* colonized host sclerotia on and in roots more successfully than it reached those inside the stems of sunflower (Huang 1978). The population dynamics of *Verticillium biguttatum* tracks the population fluctuations of its host after a lag period (van den Boogert et al. 1990; van den Boogert and Velvis 1992).

AQUATIC FUNGI AND FUNGUS-LIKE MICROORGANISMS

Fungicolous organisms in aquatic environments include representatives of the zoosporic fungi in the Oomycota, Plasmodiophoromycota, and Chytridiomycota, as treated in "Taxonomic Groups of Fungicolous Fungi and Fungus-like Microorganisms," earlier. They often parasitize hosts from the same orders. The intracellular, often wall-less, thalli of certain Chytridiomycota and Oomycota in particular commonly grow inside other zoosporic fungi. Most records of associations between mycoparasites and aquatic fungi are based on single observations, and we cannot generalize about their distributions.

The hyphae of several Saprolegniaceae and algal cells serve as hosts to *Phlyctochytrium planicorne*, a chytrid with an epibiotic thallus, and endobiotic apophysis and rhizoids (Milanez 1967). In aquatic environments, *Woronina* species (Plasmodiophorales) parasitize *Saprolegniaceae*; the zoospores of the parasite encyst on the hypha of the host, and the contents of the cyst pass into the hypha via a penetration tube (Jeffries and Young 1994). Zoospores of *Olpidiopsis incrassata* encyst on and penetrate the hyphae of *Saprolegnia* (Slifkin 1961,

1963); those of *Rozellopsis inflata* attack *Pythium* species (Prowse 1954). Surprisingly, the chitinolytic (not cellulolytic) *Mortierella alpina* can penetrate and disintegrate the cellulosic hyphae and oogonia of *Saprolegnia* species (Willoughby 1988) as can certain stemphylium-like and acremonium-like parasites (Moreau 1939). Chytrids themselves can act as hyperparasites. *Chytridium parasiticum* grows on *C. suburceolatum*, which itself parasitizes *Rhizidium richmondense* (Willoughby 1956). The rotifer-capturing *Zoophagus insidians* can be parasitized by naked thalli of *Rozellopsis inflata*, which also can grow in species of *Pythium* (Prowse 1954). We found no recent literature about galls induced in *Pilobolus* by the chytrid *Pleotrachelus fulgens*.

The mostly aquatic Ingoldian hyphomycetes, the conidia of which float in water (see "Summary of Existing Knowledge," under "Mitosporic Fungi," in Chapter 23), can be parasitized by a few specialists. *Sphaerulomyces coralloides* parasitizes conidia of the aquatic hyphomycetes *Anguillospora crassa* and *Trichocladium splendens* (Marvanová 1977); it is characterized as a biotrophic contact parasite, cannot be grown in culture, and apparently causes little harm to its host. The tetraradiate anamorph fungus *Crucella subtilis* (teleomorph *Camptobasidium hydrophilum*, Atractiellales) cannot grow on leaf material without a host fungus, for which several species of Ingoldian hyphomycetes are suitable; *in vitro* it coils around the hyphae of the host and behaves like a biotrophic mycoparasite. Growth and sporulation of the host were reduced considerably when host and parasite were coinoculated on leaf material (Marvanová and Suberkropp 1990; Howe and Suberkropp 1993). *Naiadella fluitans* (probably also a heterobasidiomycete anamorph) produces tremelloid haustorial branches on agar media, suggesting mycoparasitic activity (Marvanová and Bandoni 1987). *Nectriopsis indigens* regularly is associated with the small pyrenomycete *Naetrocymbe saxicola* on mats of algae and cyanobacteria in slightly polluted streaming water (Molitor and Diederich 1997).

Zoosporic mycoparasites usually are discovered in specially devised studies. Baiting with substrata suitable for the host in combination with samples of water (see "Collection, Identification, and Deposition of Specimens" under "Chytridiomycetes and Hyphochytridiamycetes," in Chapter 23 and "Isolation Methods" in Chapter 24) or soil (see "Baiting" and "Selective Baiting and Enrichment Techniques for Isolating Soil Chytrids" in Chapter 13) from nature often also yield the parasites. Fuller and Jaworski (1987) have provided detailed descriptions of methods that can be used to observe the host-parasite combinations of *Olpidiopsis varians-Achlya*, *Woronina pythii-Pythium*, and *Rozella allomycis-Allomyces arbuscula*.

FACTORS DETERMINING HOST SPECIFICITY

Host specificity requires that a mycoparasite recognize its host (see the reviews by Barak and colleagues [1985]; Baker [1987, 1991]; and Manocha and Sahai [1993]). Recognition appears to depend on interactions of surface sugars of the parasite and lectins or agglutinins (often two specific glycoproteins) of the host that control attachment and appressorium formation in biotrophic zygomycetes (Manocha 1991) as well as necrotrophic mycoparasites such as *Trichoderma* (Elad et al. 1983a). Lectin probes have been developed to study those interactions (Manocha and Sahai 1991). Nylon fibers impregnated with concanavalin A or *Athelia rolfsii* agglutinin induced coiling in the appropriate strain of *Trichoderma* (Manocha and Sahai 1993). When a mycoparasite is paired with a resistant host, the parasite elicits increased wall deposition in the host, a defense reaction that prevents a nutritional relationship from developing between the two. A nonhost species does not even induce attachment by the parasite (Manocha 1981; Manocha and Golesorkhi 1981; Manocha and Graham 1982; Manocha 1987; Manocha and Sahai 1993). *Phascolomyces articulosus* is a resistant host (Manocha 1981, 1987; Manocha and Graham 1982), which, when infected by *Piptocephalis unispora*, prevents further development of the parasite (Jeffries and Young 1978).

Manocha and Golesorkhi (1979, 1981) studied the ultrastructure of compatible and incompatible interactions. Growth of the germ tubes of parasites toward suitable hosts such as *Umbelopsis* (*Mortierella*) *vinacea* and *Mycotypha microspora*, but not the equally suitable host *Circinella mucoroides*, is directed along diffusible factors (proteinaceous compounds) from the host (Evans et al. 1981; Evans and Cooke 1982). Germ tubes of the parasite attach to protoplasts of host, but not to those of nonhost, fungi (Sundari and Manocha 1991). In the case of the combination *Parasitella-Absidia*, which functions only between complementary sexual mating types, the sexual hormone trisporic acid is likely involved in recognition (Wöstemeyer et al. 1995). Isolates of *Trichoderma harzianum* antagonize different plant pathogens by producing different enzymes (Elad et al. 1984). An interaction with *Rhizoctonia solani* led to increased production of an extracellular fibrillar polysaccharide (Elad et al. 1987). Different strains of *Trichoderma virens* produce either gliotoxin (Q strains), which acts against *R. solani*, or gliovirin (P strains), which acts against *Pythium* species (Howell 1999).

Only a few decomposer fungi are able to attack lichens, presumably because of the presence of noxious lichen metabolites. Nonlichenicolous species of *Nectria* and

Pronectria thus are unable to grow on various lichens, whereas their lichenicolous relatives can (Lawrey et al. 1994).

FUNGICOLOUS FUNGI AS BIOCONTROL AGENTS OF PLANT PATHOGENS

Pesticides needed to control plant pathogens or pests easily can upset the natural equilibrium. Many of the side effects of pesticides are quite undesirable, particularly in soil (Bollen 1979); others, in contrast, may enhance a control action (Hofman and Bollen 1987; Bollen 1993). Regardless, with growing consciousness of the negative environmental effects of these substances, the application of pesticides is being increasingly restricted. As a result, the importance of biological control of plant pathogens is growing. Fungicolous mycoparasitic fungi often make suitable biocontrol agents, their many antagonistic effects reducing the populations of pathogens in artificial (agricultural) systems as they do in natural systems. Biocontrol with fungi, which was started by Weindling (1932), has been reviewed many times (e.g., Baker and Cook 1974; Burge 1988; Deacon 1988b; Philipp 1988; Lumsden and Lewis 1989; Whipps and Lumsden 1989; Adams 1990; Hornby 1990; Harman 1991; Lumsden 1992). Organisms used for this purpose have been tabulated by Jeffries and Young (1994).

SOIL-BORNE PLANT PATHOGENS

Biological control is particularly appropriate in soil; the most successful cases concern mycoparasites of fungal mycelia and sclerotia. Sclerotia normally survive in the soil for years and are difficult to control with other methods. Normally, it is hardly possible to establish a microorganism in a soil where it was not present previously or only reached low densities. If a host organism is present, however, introduced parasites are at an advantage. Biocontrol agents usually are applied at high densities, the inundative approach, to reach a sufficient proportion of the target propagules. Another important criterion for the selection of control agents is their capacity to colonize the rhizosphere. *Talaromyces flavus*, for example, is rhizosphere-competent (Marois et al. 1984; Nagtzaam 1998). In *Trichoderma*, in contrast, rhizosphere competence varies among strains (Ahmad and Baker 1987). Strains that are good colonizers have been combined successfully with those that are highly antagonistic by protoplast fusion (Sivan and Harman 1991).

Indirect control may be achieved when a pesticide suppresses certain organisms and favors others that are able to control a disease. For example, when pimaricin or thiram was used instead of the banned mercury products to control *Fusarium oxysporum* basal rot in cultivated *Narcissus* in the Netherlands, populations of *Penicillium janthinellum* and *Cylindrocarpon destructans* increased; the latter two species then controlled the disease (Langerak 1977). *Trichoderma harzianum* and *T. virens* most often are used to control fungal plant pathogens (Elad et al. 1982; Papavizas 1985; Chet 1987). By means of nonvolatile and volatile toxic metabolites, enzymes, and direct mycoparasitism, they limit many soil-borne plant pathogens. *Trichoderma* species are particularly effective because they are excellent at recolonizing fumigated or otherwise decontaminated soils and are highly antagonistic.

Pythium oligandrum can be used against damping-off caused by *P. ultimum* (Deacon 1976; Vesely 1977); its oospores are mass-produced in liquid culture and air-dried (McQuilken et al. 1990) and then are applied as a seed coating (Lutchmeah and Cooke 1985; Martin and Hancock 1987). *Harpophora* (*Phialophora*) *radicicola* is the host most susceptible to *P. oligandrum*; other fungi are less affected by it or not affected at all (Deacon 1976; Laing and Deacon 1990, 1991; Berry et al. 1993).

CONTROL OF PATHOGENIC SCLEROTIAL ASCOMYCETES

Various potential mycoparasites of *S. sclerotiorum* sclerotia have been tested to determine their efficacy as biocontrol agents of that species. *Coniothyrium minitans* was tested because it had been observed that when it was abundant in the phyllosphere of oilseed rape, *S. sclerotiorum* was suppressed (Whipps et al. 1993a). In fact, among several potential antagonists, *C. minitans* and *Trichoderma virens* were the most active (Whipps and Budge 1990). *C. minitans* is the most successful agent against *Sclerotinia* species (Whipps and Gerlagh 1992; Lewis et al. 1995), however, because it attacks the sclerotia and also can grow inside the hyphae and thus follow the host into plant tissue (Huang and Hoes 1976; Huang 1978; Trutmann et al. 1982; Phillips and Price 1983; Tu 1984; Huang and Kokko 1988; Whipps and Gerlagh 1992; Whipps et al. 1993a, 1993b). A successful preparation of this agent is now on the German market (Lüth 1998). *Trichoderma virens*, which is abundant in soil, is most efficient in preventing carpogenous germination (Mueller et al. 1985).

Talaromyces flavus is also effective against *S. sclerotiorum*. Once introduced into the soil, its effect, as with that of *C. minitans*, lasts for more than 2 years, conferring suppressive properties on the soil (Whipps et al. 1993a, 1993b).

The more specialized biotrophic *Sporidesmium sclerotivorum* (Uecker et al. 1978; Ayers and Adams 1979) is a highly successful biotrophic mycoparasite, despite its poor growth on synthetic media and its reproductive difficulties (Adams et al. 1984; Adams 1990). It deprives sclerotia of food and eventually kills them. Five conidia per gram of soil can be sufficient to infect the host sclerotia present; the fungus then multiplies enormously, producing thousands of conidia from one infected sclerotium (Adams et al. 1984).

The aggressive parasite *Talaromyces flavus* has been applied successfully against the microsclerotia of *Verticillium dahliae* (see "On Sclerotia," earlier). Applications in potato fields led to a concentration-dependent reduction of microsclerotium formation on potato stems. The effect lasted for 2 years after the introduction into the soil of ascospores incorporated in alginate wheat-bran pellets (Nagtzaam 1998). *Clonostachys rosea* has been applied with some success against *Phomopsis sclerotioides* cucumber black root rot (Moody and Gindrat 1977).

CONTROL OF PATHOGENIC BASIDIOMYCETES

An isolate of *Trichoderma harzianum* that efficiently controls *Athelia rolfsii* (anamorph *Sclerotium rolfsii*) produces large amounts of protease and lipase (Elad et al. 1982).

Verticillium biguttatum can attack both hyphae and sclerotia of *Rhizoctonia solani*. It controls its host by draining away nutrients and preventing sclerotium formation. Its use on a large scale has been limited because of the relatively high minimum temperature (13°C) required for its activity in the soil and its dependence on a host fungus for multiplication; it has been used successfully, however (Jager and Velvis 1985; van den Boogert and Velvis 1992). Recently a new harvesting technique ("green-crop harvesting") has considerably improved its potential as a biocontrol agent. Conidia are sprayed onto freshly harvested potatoes, which then are covered again with soil and left on the field for a couple of weeks to harden (Mulder et al. 1992). The fungus also can be applied in combination with certain fungicides (van den Boogert et al. 1990; Jager et al. 1991).

Rhizomorphs of *Armillaria* species provide a suitable substratum for many fungicolous fungi (CBS, unpublished data), including *Pseudographiella rhizomorpharum* (Helfer 1991). So far, however, no efficient direct biological control agent has been found for those pathogens. Nevertheless, various chemicals stimulate the mycoparasitic *Trichoderma* species to attack *Armillaria*

species, leading to indirect control (Bliss 1951; Ohr and Munnecke 1974; Fox et al. 1991).

Antagonists of *Heterobasidion annosum* have been studied intensively (e.g., Lundborg and Unestam 1980). Freshly cut stumps can be inoculated with one of its most successful competitors, *Phlebiopsis (Peniophora) gigantea* (Rishbeth 1963), which outcompetes it (primary resource capture) or displaces it, killing the host hyphae by hyphal interference (secondary resource capture) (Ikediugwu 1976b).

ABOVE-GROUND PLANT PATHOGENS

Eudarluca caricis has been applied with some success against certain rust fungi, particularly *Cronartium strobilinum*, and less against *C. fusiforme* (Kuhlman and Matthews 1976; Kuhlman et al. 1978); the natural infestation already provides considerable control.

Control of powdery mildews with *Ampelomyces quisqualis* has been attempted in many places (Sundheim 1986; Philipp 1988). The fungus requires high moisture for germination and penetration of its host; adding paraffin compounds to the conidial suspension protects the germinating conidia (Philipp et al. 1990). Falk and colleagues (1995a) used inoculated cotton wicks suspended over grapevines as a lasting source of inoculum, which was successful during wet weather only. *In vitro*, the fungus sporulated well in submerged, shaken cultures on potato-dextrose broth at 20–25°C; sporulation was best if glucose was omitted from the culture medium. Sztejnberg and associates (1990) recommended this method for mass production of the control agent.

Other suitable agents for the control of powdery mildews are found among species of *Tilletiopsis* and *Pseudozyma*. *Tilletiopsis pallescens* and *T. washingtonensis* efficiently controlled *Sphaerotheca fuliginea* in greenhouse tests with three weekly applications of a conidial spray (Urquhart et al. 1994). Sprays with *Tilletiopsis* also eradicated the host population both *in vitro* on cucumber leaves (Hoch and Provvidenti 1979) and in the field on barley (Klecan et al. 1990). *Pseudozyma (Stephanoascus) flocculosa* and *P. rugulosa* kill mildew conidia without penetration (Jarvis et al. 1989); *P. flocculosa* acts on *Sphaerotheca* mainly by antibiosis and less by chitinase (Hajlaoui et al. 1992). *Tilletiopsis* species and *Pseudozyma* species are easier to handle, demand less moisture, and control the host more efficiently than *Ampelomyces* (Jarvis et al. 1989; Klecan et al. 1990; Urquhart et al. 1994).

The success of many fungal biocontrol agents against leaf pathogens is hampered because of lack of sufficient moisture. Exceptions may be the successful application of *Trichoderma harzianum* against *Botrytis cinerea* infections (eye rot) in apple (Tronsmo and Ystaas 1980) and foliar application of *Hansfordia pulvinata* against *Passalora (Mycovellosiella, Cladosporium) fulvo (Mycovellosiella, Cladosporium) fulva* (Dubos 1987).

In the phyllosphere, some yeastlike fungi and a few hyphomycetes normally compete with plant pathogens (Fokkema 1978; Williamson and Fokkema 1985; Roberts 1990). Foliar application of fungicides can damage those yeast populations, which control several plant pathogens in the phyllosphere, thus causing undesired side effects above ground (Andrews and Kenerley 1978; Cullen and Andrews 1984; Dik and van Pelt 1992).

Cladobotryum amazonense, which produces high levels of antibiotic, was used experimentally to control witches' broom (*Crinipellis perniciosa*) in cacao (Bastos et al. 1982, 1986). Witches' broom of cacao also can be controlled by strains of *Trichoderma stromaticum* (Samuels et al. 2000) that grow as a direct parasite on the mycelium of the pathogen (Bastos 1996). The *Trichoderma* prevents the formation of new inoculum through the suppression of basidioma formation.

MODES OF APPLICATION OF MYCOPARASITES

Mycoparasites can be applied to the plant or to the soil. For soil applications, material propagated in wheat bran (e.g., Elad et al. 1980) or in spore suspensions immobilized in alginate pellets (e.g., Magan and Whipps 1988) are suitable. When seeds are coated with fungal material, the control agent can become active in the soil during germination of the plant. *Trichoderma* species are particularly used for seed-coating (e.g., Chet 1987; Howell 1991). *Pythium oligandrum* (reviewed by K. Lewis et al. 1990) now is used mainly to coat seeds of cress and sugar beet against *Pythium ultimum* and *Aphanomyces*.

Solid matrix priming is another successful procedure in which germination of both the seed and the control agent is induced before seeding (Harman 1991). *Coniothyrium minitans* also is applied as a solid substratum preparation (wheat-grain inoculum) for celery, lettuce, and other greenhouse crops (Whipps et al. 1993b; Gerlagh et al. 1995; Lüth 1998). *Sporidesmium sclerotivorum* (Adams 1990) can be spread using infected sclerotia of *Sclerotinia minor*, or low concentrations of conidia can be sprayed on infected lettuce fields and then plowed into the soil. For the production of biomass, a formulation on a vermiculite base has been developed by Ayers and Adams (1983).

Successful application of *Tilletiopsis* and other control agents to above-ground plant parts requires relatively high moisture levels in greenhouses. Lipophilic additives to the suspension can compensate for lack of moisture (Hijwegen 1992b).

GENERAL CONSIDERATIONS

Biocontrol has many constraints that so far have limited its large-scale use. The effects are usually less reliable and less complete than those with chemical treatments. Environmental factors have a greater effect on the success, and only under certain conditions can the application of particular strains assure the desired control effects (Lumsden and Lewis 1989). Biocontrol programs require a much more refined system of monitoring of target organisms and selecting the control agent than do conventional chemical applications. The fact that *T. harzianum* also can be recommended for improving plant growth (Baker 1988) probably makes it more attractive in practice than its role as a control agent. At the moment, about 10 *Trichoderma*-based products are on the market or on the verge of being introduced worldwide.

Biological agents have the advantage of building up an antagonist population that may survive for some years. Antagonists usually can cope with only a limited density of pathogens and are inefficient at high degrees of infection. Nevertheless, the integration with low concentration of chemicals often can give sufficient results. Overall, studies of the application of biocontrol agents are progressing rapidly and such agents rapidly are gaining importance. A strategy to promote the research required to identify appropriate biocontrol agents and understand their functions has been proposed by Whipps (1992).

RECOMMENDATIONS FOR THE INVENTORY OF FUNGICOLOUS FUNGI

Each individual group of terrestrial and aquatic fungicolous fungi requires special techniques for detection, many of which have been described in preceding sections. No general technique will recover a good number of them in one step. It is important to look for mycoparasites under ecological conditions suitable for each group of host organisms. We confine ourselves here to give some suggestions for the SCIF and lichenicolous groups. Mycoparasites of soil-borne plant pathogens normally are investigated with methods devised for each host individually as described under Major Groups of

Fungicolous Fungi: Mycoparasites of Mycelia, Ectomycorrhizae, Sclerotia, and Spores in Soil.

For fungi growing on sporocarps, an inventory is possible. It must be carried out in the right season, particularly when fungi fruiting on soft and caducous hosts are involved. Because of the often rather unspecific host relationships, it may suffice to examine sporocarps of abundant species, which are likely to show all SCIF of the area and their relative frequencies. With the more persistent Aphyllophorales, a detailed investigation, also in deeper layers of the sporocarp, is worthwhile to detect a wider spectrum of microscopic species. To study SCIF, cooperation with mycologists collecting macromycetes is profitable and helps in the determination of host fungi. More SCIF can be collected when experienced agaricologists point to probable infections so that a first sorting of infected material can be done in the field.

For the lichenicolous group, an inventory is carried out by lichenological fieldwork (study of numerous lichen thalli and apothecia of numerous species) and by the study of lichen herbaria as described under Major Groups of Fungicolous Fungi: Lichenicolous Fungi.

ACKNOWLEDGMENTS. We are greatly indebted to Prof. D. L. Hawksworth and Drs. P. Jeffrics and T. W. K. Young, who provided the background for this chapter with their voluminous reviews, which we have exploited thoroughly. Prof. Hawksworth; Drs. P. Jeffries and G. J. Samuels; and our colleagues D. van der Mei, T. Boekhout, R. C. Summerbell, and G. J. Bollen have kindly read and commented on parts of previous versions of this text.

RECOMMENDED PROTOCOLS FOR SAMPLING PARTICULAR GROUPS OF FUNGI: COLLECTING AND ISOLATION PROTOCOLS FOR FUNGI ASSOCIATED WITH ANIMALS

18

Insect- and Other Arthropod-Associated Fungi

RICHARD K. BENJAMIN,[†] MEREDITH BLACKWELL,[*] IGNACIO H. CHAPELA,
RICHARD A. HUMBER, KEVIN G. JONES, KIER D. KLEPZIG, ROBERT W. LICHTWARDT,
DAVID MALLOCH, HIROAKI NODA, RICHARD A. ROEPER, JOSEPH W. SPATAFORA,
AND ALEXANDER WEIR

[†] Deceased
[*] Corresponding author

Vast numbers of fungi are associated with a variety of insects and other arthropods to form symbioses of various types. In some cases these associations are obvious; at other times only thorough observations throughout the life cycles of the organisms involved and careful dissection and microscopic examination of insects reveal a fungal presence. The fungi of these associations include necrotrophic (killing host cells and using them as a nutrient source) and biotrophic (requiring living host cells) parasites, which may be dispersed by their hosts. In other interactions insects use fungi directly as food or as sources of enzymes. Symbioses of this type allow the insects to use refractory nutrient resources. A few of these fungi are merely dispersed by arthropods in their environments.

The number of species of these fungi is so large and their morphologies are so specialized that only a few specialists study them. It is not surprising, therefore, that our knowledge of many species is so poor. This situation can be improved, however, only when these fungi are better collected and better known, making their inclusion in biodiversity studies compelling. In this chapter we discuss their biology and techniques for their discovery as well as their collection and transportation to the laboratory, storage until study, preparation for examination, culture when applicable, deposition as vouchers, and identification

COMMON SAMPLING METHODOLOGIES

Although many groups of insect-associated fungi can be studied only with specialized techniques, there are, nevertheless, some generalizations that can be made about the study of all groups. In some circumstances it is possible to observe and collect insect-associated fungi directly in the field. Often, however, these fungi cannot be seen with the unaided eye, and it becomes necessary to bring promising substrata into the laboratory for further processing. In either case it is essential to have at least a basic understanding of the associated insects (Borror et al. 1992). For example, some fungi may be associated only will the larval stage of some insects, which may appear for a very short time during a single season each year. Furthermore, the fungi may be ephemeral in both somatic and sporulation phases of the life cycle. Investigators should consult general references dealing with a variety of insect–fungus associations, including Batra 1979, Wheeler and Blackwell 1984, Pirozynski and Hawksworth 1988, Wilding et al. 1989, Carroll and Wicklow 1992, Humber 1992, and Alexopoulos et al. 1996.

COLLECTING SPECIMENS

Few protocols are available for quantitative sampling of insect fungi. Problems involve the unpredictable timing of the appearance of fruiting structures of some species, the microscopic nature of others and the need to culture them, and, most importantly, the patchy distribution of the associated arthropods. In most cases sampling methods for the fungi must be targeted at the host insects. The protocols used also depend on the condition (i.e., dead, dying, or alive) and the stage (i.e., larval or adult) of the host likely to support infection. For fungi found on living, adult hosts, quantitatively gathered samples of potential hosts can be used to assess host utilization patterns and species richness of the fungi (Weir and Hammond 1997a, 1997b). For these groups of fungi mass collections of hosts can be obtained by several trapping methods.

Flight-Interception Traps

Large-area flight-interception traps are about 3 m across and consist of black, fine-mesh netting suspended vertically above trays of preservative. The nets provide an effective means of obtaining quantitative samples rich in some groups of flying insects, such as Coleoptera and Diptera, and are not especially influenced by incidental variables. Samples obtained in this way have been used to compare insect species richness in tropical and temperate forests (e.g., Hammond 1990) and to assess infection patterns of laboulbenialean fungi (Weir and Hammond 1997b). Catches usually contain large numbers of predaceous beetles, including Staphylinidae, and also may prove useful for bark- and fungus-feeding species belonging to Scolytinae and Platypodinae. In lowland tropical moist forest in Indonesia average weekly catches of beetles ranged from 100 to 250 species (Hammond 1990).

Light Traps

Ultraviolet and mercury vapor lamps, usually used to attract lepidopteria, also attract a wide range of other insects including Coleoptera and Diptera. Light traps frequently attract beetles associated with water such as Gyrinidae, Dytiscidae, and Hydrophilidae, and these insects can be a rich source of Laboulbeniales. Between site and within site samples can be quantified crudely by trapping effort, and information on faunal (and fungal) differences with elevation can be assessed using transects.

Pitfall Traps

Pitfall traps consist of plastic or metal containers sunk into the ground so the lip is level with the soil surface.

They provide a convenient means of trapping active ground-dwelling invertebrates such as carabid beetles. The catch is determined by both population size and activity and is a measure of the "effective abundance" of the host (den Boer 1977). Traps usually are placed in a grid system for predefined time periods, and the catches lend themselves well to mathematical manipulation (Luff et al. 1989; Corn 1993).

Canopy Fogging

The study of samples of insects obtained by insecticidal fogging has provided an enormous insight into the structure and richness of the fauna of tropical and temperate forests. Individual trees are fogged from the ground using a synthetic pyrethroid, Reslin E, a nonresidual insecticide with high knockdown and low kill components. Arthropod samples are collected on both suspended trays (for statistical analysis) and on plastic sheets on the ground. Large numbers of beetles, especially Chrysomelidae and Corylophidae, hosts of species of *Corylophomyces*, *Dimeromyces*, *Laboulbenia*, and *Rickia* (Laboulbeniales) can be obtained in this way (Weir and Hammond 1997b).

Litter Samples

Known volumes of litter can be sampled for ground- and litter-dwelling invertebrates. The litter must be gathered quickly to avoid loss of insect hosts and can be deposited into large bags and processed through Berlese funnels on return to the laboratory. Litter samples can provide large numbers of beetles (principally Carabidae, Staphylinidae including Pselaphiniae, and Ptiliidae), ants, millipedes (Diplopoda), and mites (Acarina).

Other Samples

Malaise traps provide another rich source of arthropod material, but comparison of catches can be difficult because the contents of the traps will vary depending on the precise placement of the traps. For Trichomycetes associated with aquatic insects, it may prove possible to quantify catches using techniques already developed by river authorities for water-quality assessment.

Samples of dead and/or dying arthropods and their fungal associates are much more difficult to quantify because the hosts are usually located by direct searching techniques. Nevertheless, sampling effort may be amenable to calibration, and the results could be analyzed statistically.

SITES

The basic spatial unit of study for biodiversity investigations is the "site," and if sites are to be compared directly, the sampling effort undertaken at each site must be precisely equivalent to eliminate or minimize bias. Ideally, surveys of insect fungi at a site should be carried out throughout the year, although little may be collected in temperate climates during the winter.

HANDLING AND STORAGE

Once collected, fungi of interest or their insect hosts can be placed in a variety of containers in the field; each collector usually has her or his own preferences. Substratum samples can be placed in paper bags, vials, or any other suitable containers. Generally, specimens should not be allowed to become too moist; consequently, most collectors shun the use of plastic bags, although in some circumstances they may be preferred (e.g., when collecting aquatic samples). If many small containers are used, it is convenient to have a single larger container in which to keep them together. Baskets, fishing tackle boxes, and mesh bags such as those sold in diving shops are useful for this purpose. Individual samples should be kept separate from others collected at the same time to avoid contamination. Containers must be sterilized or at least new to prevent contamination from previous collections. The collector writes and includes field labels with the specimen at the time of collection.

Material collected from the field should be maintained in as unchanged a condition as possible until study, especially when it is to be cultured. Although many fungi tolerate drying and can be cultivated from herbarium specimens many years old, others die within hours of drying. When no information on drought tolerance is available and obtaining a culture is important, material in fresh condition should be used. In some cases it may be necessary to attempt cultivation in the field.

Refrigeration is a good way to keep some material in the laboratory for a few days; humidity in refrigerators often is very high, however, and can lead to contamination by other fungi on the substratum. Walk-in refrigerators, in particular, can unintentionally become incubators containing a rich diversity of fungi. Freezing fresh specimens may be practical, even with living insects. Care must be used, however, if the insects and fungi are to be kept alive, because many organisms, even those occurring in cold climates, seem to be relatively frost-intolerant. Cooling or freezing may be a good way to slow insects and mites for examination or to kill them for dissection.

Some groups of insect-associated fungi have never been cultured. Nevertheless, those fungi may be of interest and certainly should be collected. Most commonly, fungi that are not to be cultivated are dried or preserved in various solutions before examination. Large fleshy specimens are best dried on a dryer with a source of heat. Smaller specimens can be dried in silica gel or simply air dried. Oven drying is never recommended because specimens tend to cook. Freeze drying is also possible by lyophilization or in a frost-free freezer. When drying is undesirable, specimens can be preserved in solutions such as alcohol or formalin. Most solutions are not recommended, however, because nucleic acids or secondary products may be extracted over time.

PREPARATION OF SPECIMENS FOR STUDY

Specimen preparation is highly specialized for certain groups of organisms. It usually requires mounting diagnostic parts of fungi from nature or cultures on microscope slides. Part of the arthropod may be mounted as well. Although certain stains usually are used for a mounting medium, we almost always recommend that a water mount be made as well, if living material is available. In several cases when this was done, spores germinated in unexpected ways and provided more information on life histories of the fungi in question. Also, all features of the mounted specimen are more natural, and slight pigmentation can be seen easily. Many mycologists insist that measurements be made on water mounts, but whatever mounting medium is used for specimen measurements should always be stated.

Several general mounting media contain lactophenol-cotton blue because it has a high affinity for fungal cells. However, precautions should be taken when using phenol, because it is a carcinogen. Glycerine jelly can be used as a mounting medium with water-soluble stains, including cotton blue, and offers an alternative to lactophenol with the added value that it gels on cooling. Hardening mountants are available commercially, but the refractive properties may obscure some fungi. Slides, even those prepared with aqueous mounting media, can be made to last more than a hundred years if carefully prepared and can serve as voucher specimens. Sandwiching a specimen between two glass cover slips is a particularly effective technique because it ensures that the sealant will fill the gap between the cover glasses and the environment. These topics are discussed in more detail in sections to follow.

Some of the fungi associated with arthropods are minute and have few diagnostic characters. Electron microscopic preparations may be necessary to observe certain structures, such as septal pores, that can provide phylogenetic information (Blackwell and Kimbrough 1976a, 1976b). Transmission electron microscopy has been of greatest value, but scanning electron microscopy can provide some information as well (Weir and Beakes 1996).

More recently, molecular techniques have been applied to arthropod-associated fungi to solve previously intractable taxonomic problems. All collections may be valuable for such studies. For that reason it is best not to store specimens in liquid and not to dry them at high temperatures for long periods. However, DNA can degrade when moist specimens lie about.

CULTIVATION AND DEPOSITION OF CULTURES

Many fungi are identified more readily from cultures. Even when that is not so, it is advantageous to have cultures for physiological and genetic studies and for the production of secondary metabolites for analysis. Dried cultures or those in a physiologically inactive state also may serve as type specimens for new taxa. Media for culturing specific kinds of fungi are given later in this chapter. A number of culture collections accept cultures of arthropod-associated fungi. These include many of the major general collections and more specialized collections such as the USDA Agricultural Research Service Collection of Entomopathogenic Fungi (ARSEF) (Appendix III, Humber 1992). It is best to inquire ahead about the willingness of the collection curator to accept them. Unfortunately, not all groups of fungi associated with arthropods have been cultured, and sometimes colonies of infected insects must be maintained as sources of fungi that cannot be cultured.

PREPARATION AND DEPOSITION OF VOUCHERS

Vouchers should always be prepared and maintained for all specimens of interest. Many of the fungi to be discussed are minute and are best preserved when mounted on permanent slides for deposition in a collection. Dried cultures and usual herbarium specimens in packets or boxes may be prepared for other fungal groups. Publications citing specimens that are not available to subsequent workers are of greatly diminished value compared to those with vouchered specimens. Vouchers of insect hosts are equally important for obtaining identifications and documenting associations. In both cases, voucher specimens should exhibit the features that are used for identification. Some care should be taken to deposit them in a collection that is likely to be well maintained for many years (Blackwell and Chapman 1993). It is convenient for subsequent workers if specimens of a certain type are deposited together in collections with similar specimens. However, there are good arguments for not putting all specimens of rare organisms in a single collection, which could be damaged or destroyed in a natural or manmade disaster. Methods for the preparation of voucher specimens vary greatly from group to group, as we discuss in sections that follow.

IDENTIFICATION

Even when fairly up-to-date manuals are available, identification of fungi is difficult. The task is even more difficult, however, when the literature is scattered as it so often is with this assemblage of organisms. Identification by specialists is the most reliable way to name an organism, and nonspecialists must be prepared to pay for that essential service. Identification services are offered by most major culture collections as well as by many individual specialists. If such a service is to be used, the investigator should select one with a specialist in the particular group of interest. Home-institution mycologists should be able to help locate an appropriate specialist. Investigators wishing to identify an unknown fungus on their own can find appropriate manuals for some groups of fungi in the literature. Hawksworth and colleagues (1995) provided keys to families of the four fungal phyla and oomycetes (Hawksworth et al. 1995). However, other groups mentioned here in passing, such as Myxomycetes and acrasid and dictyostelid slime molds, although studied by mycologists, are not included in the keys. Identification of the arthropod associate at some taxonomic level at the very least will simplify identification of the fungus. In fact it is much easier to have an arthropod specialist involved in a project from its inception. Another way to discover diversity of some poorly known fungal groups (such as Laboulbeniales) and avoid the problem of arthropod identification is to collect fungi from specimens in well-curated arthropod collections where the identifications already have been made.

TECHNIQUES FOR SAMPLING NECROTROPHIC PARASITES

Plant pathologists use the term *necrotroph* to describe the development of parasites on plant tissues; the parasite kills living plant cells before it consumes them. Necrotrophy is more broadly applied with animal hosts; we use it for parasites that kill cells but also for parasites that ultimately kill their hosts, even though they may not kill

individual cells. For example, a fungus may fill the haemocoel of an insect and ultimately cause its death by interfering with normal physiological processes. Necrotrophs may be host-specific, and many are potential biocontrol agents. Most of these fungi are terrestrial and filamentous, although the arthropod-parasitic chytrids and oomycetes are aquatic. Most species do not sporulate until the host is dead or nearly so. In some cases the host actually moves around and disperses spores actively; in other cases, however, the body of the dead host is a platform for forcible spore discharge.

The fungi that kill insects are a very diverse group (Table 18.1). As an example of the group, we will discuss the life histories of the Clavicipitaceae, which contains several genera of fungi that are necrotrophs of a diversity of arthropods *Cordyceps*, with more than 200 described species, is the largest of those genera (Kobayasi 1982). Species of *Cordyceps* most frequently attack species of Lepidoptera, Hymenoptera, Coleoptera, and Orthoptera. Several stages of the life cycle of a particular host may be infected but not necessarily by the same species of fungus. Consequently, host identity and stage of life cycle may be important criteria in

TABLE 18.1

Major Orders and Genera of Necrotrophic Fungal Parasites That Attack Arthropods

Chytridiomycota
  Chytridiomycetes
    Blastocladiales—*Coelomomyces*
Zygomycota
  Zygomycetes
    Entomophthorales
      Entomophthoraceae—*Entomophaga, Entomophthora, Erynia, Furia, Massospora, Pandora, Zoophthora*
      Neozygitaceae—*Neozygites*
      Ancylistaceae—*Conidiobolus*
Ascomycota
  Pyrenomycetes
    Hypocreales
      Clavicipitaceae—*Cordyceps, Cordycepioideus, Ophiocordyceps, Torrubiella, Gibellula, Pseudogibellula, Akanthomyces, Nomurea, Hymenostilbe, Hirsutella, Paraisaria*
      Hypocrealean anamorphs—*Aschersonia, Beauveria, Fusarium, Hirsutella, Metarhizium, Nomuraea, Paecilomyces, Tolypocladium, Verticillium*
  Loculoascomycetes
    Myriangiales
      Myriangiaceae—*Myriangium*
    Pleosporales
      Tubeufiaceae—*Podonectria*
    Unclassified anamorph—*Entoderma*
Oomycota
  Oomycetes
    Lagenidiales
      Lagenidiaceae—*Lagenidium*

differentiating particular taxa. In addition to the necrotrophs of arthropods, several species of *Cordyceps* are parasites of fungi in the genus *Elaphomyces;* we do not consider them here.

Species of *Cordyceps* infect their insect host via part-spores (ascospores that have fragmented) or conidia (asexual spores). The fungus fills the hemocoel, thereby killing the host, and produces an endosclerotium that may fill the inside of the arthropod corpse. The endosclerotium produces one to several stromata, depending on the species of *Cordyceps*. The stromata rupture the exoskeleton of the host, typically at joints, or protrude through orifices; the stromata bear the perithecia. The stromata and perithecia are fleshy and often brightly colored. One of the most commonly encountered species, *C. militaris*, is a vibrant red to reddish orange. Other species, however, are a dusky olive-brown to black (e.g., *C. gunnii*). The ascii and ascospores of *Cordyceps* species are unique and easily identified. The cylindrical ascal apex has a pronounced thickening surrounding the apical pore that may or may not function in ascospore dispersal. The ascospores are long and filiform and break into partspores in most species.

COLLECTION

Ground-dwelling adult insects and larvae are often hosts of necrotrophic fungi, especially in years with high levels of precipitation at an appropriate season. Many terrestrial necrotrophs of arthropods can be collected directly in the field by looking for dead or dying insects. The insects may crawl up on living plants to which they become bound by hyphae of the parasite, or they may die hidden away in soil or under wood and stones (Keller and Zimmermann 1989). In some cases healthy insects are associated with spores of necrotrophic fungi in their environment but become infected only under certain environmental conditions. For that reason placing insects in a moist chamber may lead to infection in the laboratory. Subterranean termites, for example, often become infected with *Conidiobolus coronatus* in flooded moist chambers.

Species of the chytrid genus *Coelomomyces* are noted pathogens of mosquito or chironomid larvae; a second life-cycle stage of each species parasitizes a copepod host. These fungi can be found by collecting potential hosts. For this purpose one may use a turkey baster (a very large plastic or glass syringe with a rubber bulb, used in cooking) as a syringe to draw water and larvae up for transfer into a plastic bag or other container. The arthropod hosts should be held in the laboratory in shallow water in enameled pans or some other suitable container and examined over several weeks for signs of infection.

The fungal incidence may be low in nature but generally is higher in laboratory situations.

Entomopathogenic species of *Cordyceps* are not so frequently collected as other macrofungi (e.g., mushrooms), although they are abundant in particular habitats. Species of *Cordyceps* are found in the habitats of their hosts or in the specific habitats of the particular phases of the host life cycle that they parasitize. The genus displays its highest species diversity in the tropics but occurs in other regions of high insect diversity as well. The majority of species fruits during hot and humid seasons, but phenology varies, and exceptions to this generalization exist. When searching for any fungus, including *Cordyceps* species, an investigator must maintain a particular search image and focus on appropriate microhabitats. Species of *Cordyceps* that parasitize subterranean larvae or pupae, for example, protrude from the soil. Similarly, many species of *Paecilomyces* (an asexual state of *Cordyceps*) parasitize lepidopteran pupae that are found in the leaf litter or within decaying wood. Those fungi are best located by focusing just above the leaf litter. The stroma appears to project from leaves on the forest floor, but actually it originates from a pupa on the underside of the leaf. Species that parasitize adults or other non–soil-dwelling stages of the host life cycle occur on leaves, stems, or other parts of living plants, which comprise different microhabitats to be searched. Finally, the plant community can be an important indicator of where to search, because the hosts of *Cordyceps* are pollinators or pests of particular flowering plants. More than 30 species of *Cordyceps*, for example, seem to occur more frequently in rhododendron communities in the southern Appalachian mountains of the United States than in any other plant community (J. W. Spatafora, unpublished data).

STORAGE

The most important step in preserving newly collected fungal pathogens of insects is to dry them quickly (e.g., by air-drying, using desiccants, or very gentle heating). Drying will suppress the saprobic bacteria and fungi or fungivorous invertebrates collected with the specimens that can quickly overwhelm or destroy the desired pathogens. The spores of entomopathogens can remain viable for weeks or months on properly dried specimens and can be cultured after returning to the laboratory. Entomophthoralean fungi are isolated most readily from very fresh specimens, but if dried before or during active production and discharge of spores, they may revive and continue sporulating when rehydrated later. Because many insect necrotrophs can be cultured, we do not recommend preservation in alcohol except under unusual circumstances.

Fresh specimens of insect fungi must not be shipped in air- and water-tight containers unless the specimens are already quite dry. Whenever it is reasonable to do so, the specimens should be shipped in paper and cardboard containers that allow any remaining water vapor to escape; if a desiccant is included, it must be packed so that it will not damage a specimen with jostling during shipment.

PREPARATION OF SPECIMENS FOR STUDY

Specimens of most insect pathogens can be handled like other filamentous fungi, and squash mounts usually suffice for study. However, some hard parts, such as sclerotia, may have to be sectioned either with a freezing microtome or after routine fixation and embedding. Damage to an insect can be determined from sections, especially those that are plastic embedded. The mounting medium used for slides is rarely critical except to ensure consistent measurements. Some mycologists recommend measuring water-mounted material which does not shrink. Aceto-orcein is an outstanding routine mounting medium for many insect pathogens because its high acidity serves to fix the fungus and wet the taxonomically critical structures of common entomopathogens, such as *Beauveria bassiana*, more easily than lactic acid-based mounting media.

Aceto-orcein or other nuclear stains (e.g., Bismarck brown, methyl green, and aceto-carmine; see Appendix II) may be required to identify fungi in Entomophthorales (Humber 1989). Family-specific differences in nuclear cytology are seen readily in unfixed fresh or preserved specimens. Fungi in the Entomophthoraceae, many of which are necrotrophic parasites of insects, have large, readily stained nuclei with highly granular contents (Fig. 18.1). Fungi in the Ancylistaceae (*Conidiobolus* species), few of which are entomogenous, have small nuclei that generally fail to stain in aceto-orcein (Fig. 18.2) because of the absence of condensed heterochromatin which is so prominent in nuclei of Entomophthoraceae.

Slides, even those in which the fungi are in aqueous mounting media, can be made to last for many years by following one of the variant techniques of the double coverslip method (see "Preparation of Specimens for Study" in the Laboulbeniales section later in this chapter for a detailed protocol).

CULTIVATION AND DEPOSITION OF CULTURES

The great majority of necrotrophic parasites can be cultured from conidium or ascospore inoculum on simple

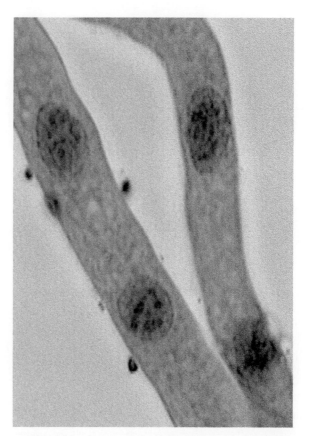

FIGURE 18.1   Fresh unfixed hyphae of the insect-inhabiting fungus *Pandora nevaphidis* (Entomorph thoraceae) stained in aceto-orcein. Four large, readily stained nuclei, which have highly granular contents, are visible.

media. Sabouraud dextrose agar + 1% yeast extract (Appendix II) is very commonly used for these fungi. The most fastidious fungi, however, especially some entomophthoralean species, may not grow *in vitro* from conidial inoculum; instead, somatic inoculum obtained from surface-sterilized (and preferably living) infected hosts should be used for cultures. In some entomophthoralean fungi, especially species of Entomophthoraceae, the somatic phase consists of protoplasts. Those fungi must be grown in more complex liquid culture media (e.g., Grace's insect tissue culture medium + 5–15% fetal bovine serum; Appendix II) but usually will not sporulate in such a medium. Humber (1994) discussed some of the problems of culturing and maintaining strictly obligate insect pathogenic fungi.

Coelomomyces species present bigger problems, but they can be reared with their mosquito and copepod hosts in the laboratory. This process of course, requires a constant source of uninfected laboratory-reared hosts.

The United States Department of Agriculture (USDA) ARSEF collection (Appendix III), comprising about 5,000 isolates of more than 300 fungal taxa, is the

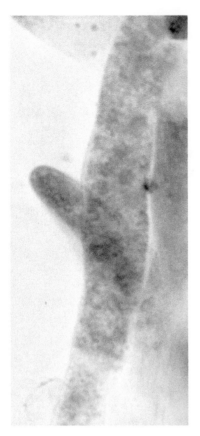

FIGURE 18.2 Fresh unfixed hyphae of the insect-inhabiting fungus *Conidiobolus thromboides* (Ancylistaceae) stained in aceto-orcein. The small nuclei fail to stain in aceto-orcein because they lack condensed heterochromatin.

soil as well as mites or other mycophagous organisms from specimens before storage is essential. We do not recommend that these fungi be preserved in alcohol unless drying is impossible.

Major herbaria (see Appendix III) containing insect necrotrophic fungi include Kew and Commonwealth Agricultural Bureau International (both with Tom Petch's extensive collections), Farlow Herbarium (with Thaxter's rich collections of Entomophthorales and Laboulbeniales), and the University of Michigan and University of Tennessee (with extensive collections of *Cordyceps* and related fungi made by E. B. Mains and K. Kobayasi, respectively).

IDENTIFICATION

Identification aids for the majority of entomopathogenic fungi are spread through the literature and often are distressingly out of date, even for major taxa. The most current and convenient general guide to insect fungi is by Samson and colleagues (1988). Several extensive manuals for Entomophthorales (e.g., Keller 1987, 1991; Balazy 1993) and for *Cordyceps* (Kobayasi 1982) are available. The indices in the ARSEF culture-collection catalog (Humber 1992) offer the most comprehensive listings of fungal pathogens by their species, hosts, and collection localities.

world's largest and most comprehensive repository for cultures of insect fungi (Humber 1992). Among the major general service culture collections (see Appendix III), Centraalbureau voor Schimmelcultures (CBS) maintains many diverse insect fungi; the American Type Culture Collection (ATCC) also has some holdings of necrotrophic insect fungi. Both collections have online databases that can be searched by taxon or substratum (including host).

PREPARATION AND DEPOSITION OF VOUCHERS

It is best to deposit necrotrophic insect fungi in herbaria as dried specimens enclosed in packets or small boxes that prevent or minimize compression or mechanical damage of specimens. Specimens of infected hosts should be included with any stems, twigs, leaves, or other substrata to which they were attached when collected. This is, obviously, not possible with specimens scraped from large, hard substrata (e.g., sound wood, stones) or dug from soil, plant detritus, or rotting logs. Removing loose

TECHNIQUES FOR SAMPLING BIOTROPHIC PARASITES

Many highly specialized members of Ascomycota, Zygomycota, and Basidiomycota are biotrophic parasites that attack insects and other anthropods. Because they are highly specialized, biotrophic fungi often have unique characteristics that make individualized techniques for their collection and study necessary.

TRICHOMYCETES

Trichomycetes exist throughout the world, and in many regions they are common and abundant. Because they are obligate gut parasites, their geographic and habitat distributions depend on those of their hosts. All species are associated with mandibulate arthropods that are detritivores, algivores, or ominivores, but apparently not with those that are predators or carnivores or that consume tissues of living vascular plants. In almost all instances the fungi are hidden within the host's gut and not discernible until the animal is dissected in the laboratory and examined microscopically. Special techniques are needed to find, study, and culture them.

Because of several convergent similarities including host preferences, Amoebidiales traditionally have been studied by investigators of Trichomycetes, but evidence suggests that they are not closely related to the other three orders (Lichtwardt et al. 2001). Nonetheless, we include Amoebidiales in this treatment because collectors of Trichomycetes often encounter them. Currently, about 225 species and 55 genera of Trichomycetes (including Amoebidiales) are known (Table 18.2). Six genera and more than 60 species have been discovered on various continents since the work of Lichtwardt's monograph (1986), attesting to the fact that the species richness of Trichomycetes is far greater than formerly realized. Information on the biology and taxonomy has been updated in several publications, including a monograph published on the Internet (Misra and Lichtwardt 2000; Lichtwardt et al. 2001).

Special adaptations that have evolved in trichomycetes have led to the success of the group as obligate inhabitants of the gut of particular kinds of arthropods. Although the fungi vary in the degree to which they infect populations of their hosts, in some cases infection of individuals may approach 100%. This level of incidence is remarkable, considering that the fungi are shed at each molting event along with the linings of the guts to which they are attached and that in most trichomycete species the number of reinfection propagules per thallus is significantly less than the number of spores produced by most other fungi, whether free-living or parasitic.

Most species of Trichomycetes appear to be commensals and are innocuous, obtaining their nutrients from ingested substances passing through the gut (Fig. 18.3). Some evidence suggests that certain species of *Smittium* (Harpellales) may provide sterols and B vitamins to mosquito larvae that are deprived of those essential nutrients. Detecting such subtle nutritional relationships

FIGURE 18.3 Thalli of *Asellaria armadillidii* (Asellariales) attached to the hindgut cuticle of a terrestrial isopod, *Armadillidium vulgare* (Crustacea). Asellariales reproduce by developing arthrospores from the hyphal branches. Scale bar = 50 μm.

TABLE 18.2
Taxa, Hosts, and Habitats of Trichomycetes

Order	Family	Number of genera	Number of species	Hosts	Habitats
Harpellales	Harpellaceae*	33	141[†]	Dipteran larvae (mayflies, stoneflies)	Freshwater (streams, ponds, pools)
Asellariales	Asellariaceae	3	11	Isopods, springtails	Freshwater, terrestrial, marine
Eccrinales[‡]	Eccrinaceae	14	52	Millipedes, crabs, anomurids, isopods, amphipods, insects	Freshwater, terrestrial, marine
	Palavasciaceae	1	3	Isopods	Marine
	Parataeniellaceae	2	6	Isopods	Terrestrial
Amoebidiales[‡]	Amoebidiaceae	2	12	Insect or crustacean larvae	Freshwater

* Includes Legeriomycetaceae and Harpellaceae.
[†] Includes new species not yet formally described.
[‡] Not phylogenetically related to Trichomycetes but traditionally included in the class.

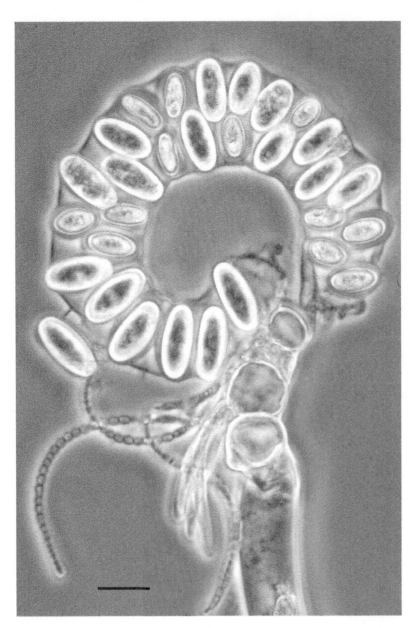

FIGURE 18.4 Coiled, sporangiospore-producing tip of a *Palavascia sphaeromae* (Eccrinales) thallus from an intertidal marine isopod, *Sphaeroma quadridentatum* (Crustacea). Scale bar = 20 μm.

between fungi and arthropods is hampered, however, by our inability to culture most trichomycete species and the need for experimental hosts to be free of other gut microorganisms. At least one very widespread but apparently uncommon species, *Smittium morbosum*, kills mosquito larvae by inhibiting ecdysis. It also has been demonstrated that some species of Harpellales in blackfly larvae occasionally grow from the gut into the developing ovaries, resulting in adult females that are sterile but disseminate the fungus by flying to new sites and ovipositing the ovarian fungal cysts in place of eggs. Whether that type of dissemination generally occurs in other Harpellales, all species of which grow and reproduce in nonflying larval forms, is unknown.

Although Asellariales (Fig. 18.3) and Eccrinales (Fig. 18.4) can infect immature host stages, full development of gut fungi occurs primarily in the sexually mature arthropod.

Collection and Storage

In this section we provide a brief overview of techniques that can be used to obtain suitable hosts for Trichomycetes. These fungi parasitize a wide variety of arthropod (Table 18.2), however, so collectors should seek advice from specialists on the host groups or their habitats, such as entomologists, benthologists, invertebrate zoologists, and marine biologists, especially those familiar with local faunas. Lists of fungal species, their known hosts, and host habitats have been published by Lichtwardt (1986).

All Harpellales (Fig. 18.5) and Amoebidiales are aquatic, as are most Asellariales (see Figure 18.3) and some Eccrinales. Harpellales and species of *Paramoebidium* (Amoebidiales) occur in the guts of insects that live mostly in lotic (flowing) waters. The usual habitats are actively flowing streams, but streams can include edges of waterfalls and seeping cliffs. Collecting in smaller streams is easier that collecting in large ones, and small streams often contain a greater diversity of larvae. Harpellales are common in nymphs of particular genera of mayflies and stoneflies and in larvae of a number of lower dipteran families, such as nonbiting midges (Chironomidae), blackflies (Simuliidae), mosquitoes (Culicidae), and to a lesser extent in certain genera of craneflies (Tipulidae), biting midges (Ceratopogonidae), moth flies (Psychodidae), and solitary midges (Thaumaleidae).

Lentic (still water) insects (e.g., mosquitoes, midges such as bloodworms, and a few other kinds of arthropods) in ponds, pools, lakes, and swamps may host Harpellales. A number of Asellariales and Eccrinales live in either lotic or lentic isopods and amphipods, as well as in some kinds of crayfish, freshwater crabs, hydrophilid beetles, and springtails (Collembola). Species of *Amoebidium* (Amoebidiales) can be found on the exoskeletons of water fleas (Cladocera), mosquitoes, bloodworms, and crayfish.

The most useful collecting instrument is an aquatic D-shaped net with a small size mesh. Stream substrata consisting of rocks and gravel can be kicked with the feet, releasing insects that drift downstream and collect in the net. Lifting larger rocks or scraping them with the hand often releases a variety of insects that can also be caught in the net. Many lotic insects prefer riffles and other agitated stretches of streams that are well aerated, although some seek zones where sediment collects. Good sources of insects are vegetation, sticks, and small rocks. These substrata can be lifted from the water, and the attached insects can be harvested with blunt forceps. A woven-metal food strainer or sieve (12-cm diameter) with a handle and with the support prongs bent downward is useful in waters with abundant vegetation (e.g., borders of streams, marshes). Such strainers also can be used to sample muddy stream bottoms.

Large, white plastic trays (about 40 × 30 cm), such as those used in photographic darkrooms, are useful repositories for the contents of nets and strainers or for materials plucked from the water. The animals can be picked from trays with forceps or plastic droppers and placed into wide-mouth collecting jars containing shallow layers of water. Preferably, all arthropods will be kept alive for dissection; consequently specimens should be kept cold in the field in an ice chest. Some lotic insects may die soon after collection if they are not kept cold; others are hardier and will survive longer, provided the container with insects is kept in the shade.

It almost always takes much longer to dissect, study, and process specimens than it does to collect them. Depending on the species, aquatic insects can be kept alive for from one day to several weeks after collection. They should be refrigerated in shallow layers of water at low densities in containers such as Petri dishes or collecting jars. Mosquito larvae, lentic bloodworms, and similar insects can be kept at room temperature. It usually is best to separate the living specimens by type. Predaceous insects that may have been collected should be discarded.

Several genera of Eccrinales and one species of Asellariales (*Asellaria ligiae*) are marine (Hibbits 1978). The fungi live in crustaceans that are mostly intertidal or in the splash or high-tide zone. Hosts include isopods, amphipods, crabs, and anomurids such as hermit crabs and mud shrimps, which can be collected by hand or with nets at low tide. Mud flats are home to several kinds of infected crabs (e.g., fiddler crabs) and anomurids; they sometimes appear on the surface but may have to be dug from the substratum with a trowel or shovel. Some Eccrinales in galatheid anomurids are found below the low-tide zone, even at abyssal depths (*Arundinula abyssicola*, around hydrothermal vents). Crustaceans in deeper zones obviously require special equipment for collecting, but those that are intertidal or live along shorelines.

Marine specimens can be placed in pails or other suitable containers with shallow layers of seawater or damp seaweed for transport back to the laboratory. They must be kept from overheating but should not be refrigerated. If the hosts are to be kept for awhile before dissection and circulating seawater is not available in the laboratory, then small amounts of freshwater may be added to the seawater from time to time so that the seawater does not become too concentrated through evaporation.

Terrestrial hosts of Asellariales and Eccrinales include millipedes, isopods, and a few kinds of beetles. These invertebrates can be collected by hand and should be placed in a container that is not tightly sealed, preferably with some of their natural substratum. Many terrestrial arthropods can be kept alive for long periods in a terrarium. They need to be kept moist but not wet.

Preparation of Specimens for Study

Dissection techniques are often a matter of individual preference. In this section we present only basic methods for groups of arthropods. Most dissections must be done under a dissecting microscope. The most useful tools include two pairs of fine jeweler's forceps; a sharp,

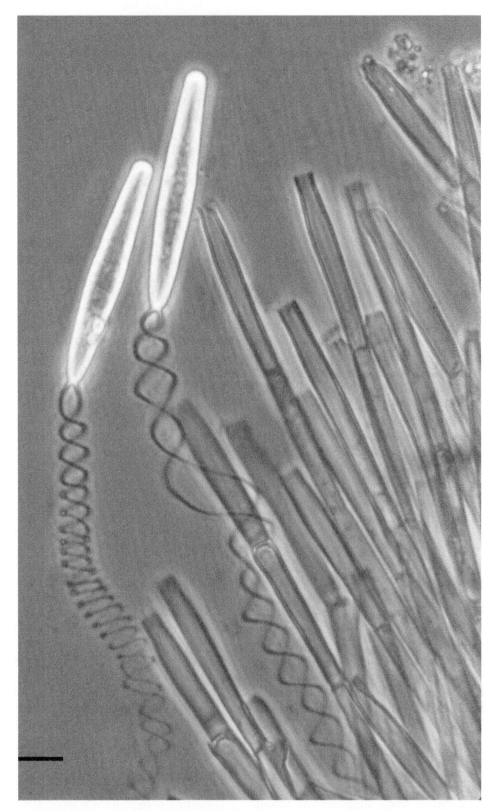

FIGURE 18.5 Two trichospores of *Genistelloides helicoides* (Harpellales) growing in a stonefly nymph (Plecoptera); in this genus trichospores have two basal appendages. Scale bar = 20 μm.

single-edged razor blade; two very fine dissecting needles; and dissecting scissors, including a pair of fine iris scissors.

Nymphs of mayflies and stoneflies are grasped with jeweller's forceps at the posterior abdominal segment; when the segment is pulled from the body, it brings the hindgut with it. The epithelial layer is removed with fine forceps in a drop of water, and the gut is opened, if necessary, to reveal any Harpellales. With dipteran larvae, such as those of midges and blackflies, Trichomycetes may be present in either the hindgut and/or the midgut. The posterior end and the head of such larvae can be detached with a razor blade, and the gut removed. Fungi in the midgut are always attached to the peritrophic membrane, a loose and transparent lining that is attached only at its anterior end. The peritrophic membrane can be cleared of algae and debris by grasping one end and lifting it several times through the surface layer of the dissection water. The hindgut epithelium should be removed to reveal Harpellales (or *Paramoebidium*) attached to the chitinous lining.

The hindgut of amphipods also can be removed by pulling away the posterior segment of the body. With isopods, the anal structures under the telson are grasped with forceps and pulled. This procedure can also be used to remove the hindgut of some larger beetles and minute springtails (a special challenge!). Part of the exoskeleton of crustaceans such as crabs and anomurids usually must be cut away with scissors to reach the hindgut and stomach (foregut) where some eccrinids live. The abdomen of crabs is folded under the animal; it can be detached and dissected to obtain the entire hindgut. The simplest method for removing the hindgut of millipedes is to cut off the posterior end with a razor blade as well as about one-fourth of the anterior body.

Host identification to the lowest taxonomic level possible is desirable, so voucher specimens should be preserved if the host is not already known. Ethanol (70%) is generally a good preservative for most arthropods. Undissected specimens are preferable for identification, but are not always available if few specimens were collected. It is especially important to preserve the head of midge larvae and the male genitalia of millipedes and some crustaceans, which may be necessary for identification. Some aquatic dipteran larvae may be difficult, if not impossible, to identify to species. For this reason, if pupae or adults are available in the field or emerge in the laboratory, they also should be made available to the specialist.

Preserved arthropods can be examined for fungi under a microscope, but this is not the method of choice in most instances because dissection usually is more difficult and artifacts almost always are introduced. Recently dead specimens can be dissected, but trichomycete thalli tend to deteriorate very soon after the host dies.

Water mounts on slides should be used for microscopic examination of most Harpellales because if thalli are placed directly in fixative, fine details such as trichospore appendages may be difficult, if not impossible, to discern properly. In most cases hindguts can be torn open with fine needles or forceps, and the thalli of harpellids can be spread. The transparent peritrophic membranes need not be opened. After structures mounted in water have been identified, studied, or photographed, a partial drop of lactophenol with cotton blue (Appendix II) can be placed on one edge of the coverslip and allowed to infiltrate as the water evaporates. After sealing three sides of the coverslip with clear fingernail polish, the slide can be washed with water and dried, and then the fourth side can be sealed.

Larger arthropods sometimes present a problem because the chitinous lining with attached fungi often cannot be stripped to reveal the thalli clearly under higher magnifications. It may be possible to remove individual fungi or small pieces of the lining, in some cases, and to mount them in water. Eccrinales have unbranched, nonseptate thalli that are easily damaged. The opened and washed gut of millipedes and some large crustaceans can be placed in dilute (10%) lactophenol for a few hours or overnight to loosen the chitinous lining. Mounting larger pieces of the cuticle with many attached thalli often reveals several stages of development on one slide.

Cultivation and Deposition of Cultures

Only some Harpellales (>190 isolates) and *Amoebidium parasiticum* (three or more isolates) currently exist in axenic culture. The genera of Harpellales include *Capniomyces*, *Furculomyces*, *Genistelloides*, *Simuliomyces*, *Smittium*, and *Trichozygospora*. Only *Smittium*, the largest genus, is represented by more than one species in culture; these currently consist of 13 named species plus many yet to be described.

Some species of Harpellales culture rather easily; others require a good deal of persistence. The preferred isolation medium is a dilute brain–heart infusion (1/10 BHIv) amended with thiamine and biotin (Appendix II).

The two vitamins may not be essential, but some evidence suggests that thiamine is stimulatory to some species of *Smittium*. This same medium can be used for storage of cultures in the refrigerator and generally produces good sporulation.

The following technique has proved successful in many isolations: An insect larva is dissected, and the hindgut is removed. It is not necessary to remove the outer layers of the gut, provided microscopic examination of a water mount indicates that a fungus is present.

The gut is washed at least twice in an antibiotic solution consisting of a stock solution of 40,000 units of penicillin G and 80,000 units of streptomycin sulfate per milliliter of distilled water. This solution can be filter-sterilized directly into serum bottles and dispensed with a syringe. Three to five drops of the concentrate are added to the wash water in a plastic Petri dish 35-mm in diameter × 10-mm high. It is convenient to use a small loop (approximately 4 mm in diameter) to handle the specimen.

After washing, the gut is transferred to a Petri dish (60-mm in diameter × 15-mm high) with a thin layer of medium that has been overlayered with sterile, glass-distilled water, to which three to five drops of stock antibiotic solution have been added. Most trichomycete species grow well at room temperature, but some, such as those from winter stoneflies (Capniidae), may have an optimum closer to 18°C. The culture must be monitored daily, using the low-power objective of a compound microscope. If contamination appears, the specimen can be rewashed and replated. Successful cultures usually will show growth in from 2 days to 2 weeks. When growth is evident, the fungus can be transferred to medium without antibiotics and later to a test-tube slant containing a small amount of sterile distilled water (ca. 20-mm deep at the bottom of an upright tube) added after the agar has gelled. Until growth is well established, the tube should be rotated daily to allow some of the water to flow over the slant until eventually some of the fungus adheres to (but does not usually grow into) the agar. In Petri dishes, most harpellids produce many scattered colonies within the water overlayer, and colonies of some species may release trichospores in great abundance.

Culturing Harpellales is different from culturing most other fungi because (1) water is needed as an overlayer on agar medium, and (2) trichospores of most isolates do not extrude (germinate) *in vitro*, consequently requiring all transfers to be made by breaking up pieces of colonies with a loop. Most harpellid isolates grow and sporulate well in shaken liquid culture, and the mycelium can be chopped with a sterile blender, as with most other fungal cultures.

Cultures of *Amoebidium parasiticum* can be started from thalli attached externally to pieces of the host. Once growing, the small, unbranched thalli and sporangiospores are best transferred with a Pasteur pipette rather than with a loop.

Long-term storage in liquid nitrogen is preferred. Trichomycetes do not withstand lyophilization. Cultures can be maintained at refrigerator temperatures and, when thus stored, should be transferred every 2–4 months, depending on the hardiness of the particular isolate.

Identification

Accounts of morphological and other characters used to identify taxa of Trichomycetes can be obtained from descriptions, illustrations, and keys. Knowing the type of host quickly narrows the possibilities of trichomycete identity, and in a few cases the genus of the host identifies a known species. Both sporulating and non-sporulating features are important in trichomycete identifications. For all groups of Trichomycetes, as with most other fungi, it may be necessary to measure reproductive structures and determine their shapes before identification is assured. Excellent identification keys and biological information for all species of Trichomycetes are available (Lichtwardt 1986; Misra and Lichtwardt 2000; Lichtwardt et al. 2001).

Taxa of Harpellales are identified primarily by thallus type (whether branched or not, amount and form of branching), basal (holdfast) structures, number of trichospore appendages, the presence or absence of a trichospore collar, and zygospore type. Thalli often are immature or devoid of zygospores; consequently finding trichospores or zygospores may require preparations from several to many individuals. Appendages appear in trichospores only after their release from generative cells. If a water-mounted specimen with sporulating branchlets with maturing trichospores is kept in a moist chamber for several hours to overnight, some trichospores may be released. Occasionally, zygospores may mature under the same conditions.

In Asellariales, the holdfast structure is especially important in the identification of species. The holdfast of Eccrinales is a useful character, but the shape and size of the thallus and the various types of sporangiospores produced are more important. Many genera and species of these orders are easily identified. Other taxa require many preparations before identification is assured.

LABOULBENIALES

Laboulbeniales is a distinctive group of obligately biotrophic parasitic ascomycetes that lack a mycelium. They live on a diverse group of arthropods. Most species grow on true insects (Hexapoda) and are known from the following insect orders: Coleoptera, Dermaptera, Diptera, Heteroptera, Hymenoptera, Isoptera, Mallophaga, Orthoptera, and Thysanoptera. Relatively few (54) species infest mites (Class Arachnoidea; Order Acarina) and millipedes (Class Diplopoda, Order Juliformia). None has completed its life cycle (i.e., produced ascospores) in axenic culture. In this section we discuss species classified in Ceratomycetaceae, Herpomycetaceae, Euceratomycetaceae, and Laboulbeniaceae.

Currently, Tavares's (1985) treatise, *Laboulbeniales* (*Fungi, Ascomycetes*), is the most comprehensive source of general information available on families and genera, development, morphology, sexuality, and distributions of Laboulbeniales. This work is a necessity for anyone interested in the systematics of these fungi. Equally important is the classic, beautifully illustrated, monograph of Thaxter, *Contribution Towards a Monograph of the Laboulbeniaceae* (Thaxter 1896, 1908, 1924, 1926, 1931). A supplement to Thaxter's work (Benjamin 1971) is also an essential aid to the study of the group. Regional studies that offer much useful information on ecology, general biology, collection, and preparation of specimens, as well as taxonomy, include those of Huldén (1983), Majewski (1994), and Santamaría (1989).

Because the thalli of the group are so different from those of other fungi, it is important to consider their morphology. All Laboulbeniales are relatively small, ranging in length from about 50 μm to 1 mm. The fungal thallus develops directly from a germinating ascospore, which may undergo a precise sequence of cellular divisions, at least during the early stages of growth. The main part of the body of the young thallus is termed the receptacle and consists of few to many cells often arranged in a particular order.

The receptacle is attached to the host by its modified basal cell, or foot, from which a simple or sometimes branched haustorium develops. Haustoria usually penetrate the host no farther than the living cells of the epidermis; in some species, however, haustoria are less localized and may penetrate and even ramify some distance into the body cavity (Thaxter 1896, 1908, 1924, 1926, 1931; Tavares 1985). Laboulbeniales appear not to be pathogenic, and evidence suggests that they cause little, if any, damage to their hosts.

The cells of the receptacle may bear simple or branched, often several-celled, appendages. The appendages may be sterile or fertile. The latter produce minute, uninucleate, nonmotile cells, the spermatia, which are assumed to have a sexual function. In the Ceratomycetaceae, tiny branchlike spermatia appear to develop directly from the cells of an appendage. In the Herpomycetaceae, Euceratomycetaceae, and Laboulbeniaceae, spermatia develop inside distinctive structures termed antheridia. Antheridia may be intercalary cells of an appendage; simple, free phialides that discharge spermatia directly to the outside; or more or less complex assemblages of closely associated fertile cells that discharge spermatia into a common chamber from which they escape to the outside via a single opening. As the thallus develops, it gives rise to one or more perithecia. Each immature perithecium gives rise to a female receptive structure, the trichogyne. Spermatia appear to be transferred passively from antheridium to trichogyne. Actual fusion of sexual nuclei never has been observed, but it is presumed to occur. In any event, a centrum (the perithecial contents), develops within the maturing perithecium and forms one or more ascogenous cells that produce a succession of asci containing usually four ascospores (Tavares 1985). Ascospores of all known Laboulbeniales are more or less acicular and two-celled.

Of the 137 genera of Laboulbeniales currently recognized, 120 appear to be monoecious; the other 17 are exclusively dioecious or include dioecious as well as monoecious species. Two species of *Triceromyces*, which have both monoecious and dioecious morphs, represent the only known examples of apparent trioecism in the fungi (Benjamin 1986). Dioecism apparently has arisen several times in the order. In some genera (e.g., *Dimeromyces*, *Dimorphomyces*, *Trenomyces*, and *Laboulbenia*, which includes only a few dioecious species out of many hundreds), males and females are morphologically similar except for the production of sexual organs. In other genera (e.g., *Amorphomyces*, *Dioicomyces*, *Aporomyces*, *Corylophomyces*, and *Rhizopodomyces*), the male may be reduced to a single series of two or more cells bearing a terminal antheridium. In *Aporomyces* and *Dioicomyces* the ascospores giving rise to the males are often greatly reduced in size compared to those giving rise of the females (Benjamin 1989).

Collection

Field collection of Laboulbeniales depends on collection of the hosts. Few of the thalli can be seen well in the field, and the success of a field trip truly can be judged only after microscopic examination of the insects. The Laboulbeniales parasitize many groups of true insects. Beetles (Coleoptera) (Fig. 18.6) and flies (Diptera) are hosts of a number of cosmopolitan species and are relatively easy to collect. Staphylinidae harbor species of many genera (e.g., *Corethromyces*, *Monoicomyces*, *Rhachomyces*, *Teratomyces*); carabid beetles are hosts of many species of *Laboulbenia*; and flies are hosts to more than 100 species of *Stigmatomyces*.

Likely hosts for Laboulbeniales live in a wide variety of habitats: water; soil; decomposing plant and animal remains of all kinds; flowers, stems, and foliage of living plants; and on the bodies of living animals such as bats and birds. Collecting insects from many such habitats may call for specialized techniques, which can be found in the entomological literature pertaining to given groups.

In tropical or subtropical regions where insects may be active throughout the year, collecting Laboulbeniales may be profitable at any time, being influenced primarily by whether the season is wet or dry. In northern or

FIGURE 18.6 Darkly pigmented thalli of *Laboulbenia* species on the dorsal surface of a carabid beetle. The mature thalli range in size from 500 to 800 μm. Laboulbeniomycetes are restricted to the ectoskeleton of arthropods. (Photomicrograph by Alex Wier)

southern climes where winters can be severe and insects are more or less inactive until the return of clement weather, collecting should be best in spring or early fall when the degree of infection of host populations may peak. Often, only a few individuals in a given population of an insect may be infected with fungi. Thus, the goal always should be to make mass collections of a variety of hosts to assure success in obtaining a variety of parasites.

Appropriate equipment can make collecting Laboulbeniales much easier and more profitable. Such items include heavy gloves; forceps with both broad and fine points for handling living or dead insects; a knife, trowel, or other strong tool for stripping bark, breaking open rotten logs and stumps, and digging in soil or detritus of all kinds; assorted plastic vials with screw caps with tight seals; a small funnel; a hand lens; a plastic bottle of preservative (70% alcohol); pencils; paper for labels; and a notebook. A pump-spray can of insect repellent (30% deet or more) not only makes collecting in the haunts of insects more comfortable but also is especially useful for killing flies or other insects collected in a net. Specialized equipment needed for certain insects includes a deep, flat-bottomed insect net constructed of lightweight nylon cloth for capturing flies and other insects on the wing, one or two small nets for capturing aquatic insects (those sold by dealers in tropical fish are excellent), and an aspirator for capturing small terrestrial insects. An aspirator easily is made as follows from a glass or, preferably, plastic bottle (a 50-cc centrifuge tube is excellent; Walter Rossi, personal communication) with a screw cap: (1) Two holes with diameters of ca. 7 mm and 10 mm are drilled through the lid a few millimeters apart. (2) Ends of flexible vinyl tubing of appropriate diameter and about 30-cm long are inserted through the holes to a depth of about 1–1.5 cm. (3) A bit of fine-mesh cloth affixed to the short end of the small-diameter tube pre-

vents insects or detritus from moving upward into the tube. (4) The lid is screwed to the container. The collector sucks on the long end of the small-diameter tube while holding the long end of the large-diameter tube close to but not quite touching a desired insect. The insect will be drawn into the bottle along with the inrushing air.

A sifter to aid in separating insects from ground litter, flood debris, and detritus of all kinds can be made of a square or rectangular piece of hardware cloth having 8, 10, or 12 squares to the inch. Material to be checked for insects is placed on the sifter and shaken over oilcloth or a plastic pan with more or less vertical sides. Insects falling through the screen are captured using the aspirator.

A Berlese funnel set up in the laboratory is ideal for recovering large numbers of arthropods from materials returned from the field. The apparatus consists of a large funnel, about 10–12 inches (25–30 cm) in diameter at the top, with a circle of hardware cloth similar to that used for the sifter secured approximately one-third of the way down inside of the funnel. The funnel is suspended in a ring stand or other support with the small end inserted into a container of preservative such as 70% ethyl alcohol (added after the funnel has been charged with material to be examined) for catching insects and other arthropods. Debris is placed on the screen, and the top of the funnel is covered with a cloth or plastic sheet held in place by a strong rubber band. Heat, not so intense that it kills the insects, can be applied above the retaining cloth (an electric light with suitable reflector). As the debris in the funnel dries out top to bottom, insects migrate downward and fall through the screen into the preservative, from which they can be recovered.

Beetles (e.g., Hydrophilidae, Dytiscidae, Haliplidae, and Gyrinidae) living in water are captured best using small fish nets or tea strainers. Other insects (e.g., Carabidae and Staphylinidae) that live under stones, logs, wood fragments, and piles of flood debris are collected most easily using a sifter and aspirator. Many insects live on or in mud or sand; these materials as well as other debris can be immersed in near-shore water; insects floating to the surface are netted and removed with tweezers. Members of several families of Diptera that are hosts of Laboulbeniales, especially Ephydridae and Sphaeroceridae, frequent mud flats at the margins of streams, lakes, and ponds; these insects are best captured with an aspirator or sweep net. A very light application of insect repellent to the net will stun or kill the flies (and other insects), which can be removed with forceps and transferred to alcohol. Rich communities of infected beetles can be found under algal drifts on the coast or under general debris around the margins of reservoirs. In those environments beetles are more or less confined to linear habitat strips, and levels of infestation by Laboulbeniales and other fungi can approach 100% (P. M. Hammond and A. Weir, unpublished data).

Leaf mold on the forest floor; accumulations of rotting vegetation, such as grass piles and garden debris; and litter in hollow trees and stumps are inhabited by many kinds of insects and other arthropods. The best way to collect insects living in such materials (e.g., Anthicidae, Carabidae, Staphylinidae) is by sifting debris in the field or by processing samples in a Berlese funnel in the laboratory. Other forest insects can be caught in flight-interception, light, or pitfall traps, or, for the more specialized collector, by canopy fogging. Fruiting bodies of larger fungi also support a wide range of invertebrates that can be collected using an aspirator or tweezers.

Material obtained by sweeping emergent vegetation with an insect net or beating and shaking branches and flowers is emptied into a large pan, a cloth spread on the ground, or a beating sheet (a square of cloth stretched taut between the ends of a pair of crossed sticks or thin boards). The dislodged insects are caught with an aspirator and transferred to alcohol.

Insects collected in the field are preserved in 70% ethyl alcohol for transport to the laboratory. Transfers from aspirator bottles to vials should be made frequently to reduce the likelihood of mechanical damage to fungi and hosts by debris that inevitably is drawn into the aspirator along with the insects. Single insects can be transferred to vials by means of forceps. Those captured in quantity with the aspirator can be transferred to a vial of alcohol with the aid of a small funnel.

Preparation of Specimens for Study

Removal of the thalli of laboulbenialean fungi to a slide mount is a challenge eased somewhat by the proper equipment and supplies. Tools and other equipment include: (1) High-quality, stainless steel, watchmaker's tweezers, whose points can be sharpened with a fine-grained grinding stone. These tweezers are essential for sorting hosts and manipulating insects when preparing slide mounts. (2) Several sizes of flat-bottomed porcelain imbedding dishes or other suitable containers for sorting and examining insects in alcohol. (3) Large-capacity Maximov depression slides having a concavity about 35 mm in diameter and 5-mm deep in which to manipulate insects in glycerol when removing parasites. (4) Ordinary depression microscope slides for temporary storage in glycerol of parasites removed from a host. (5) Stainless steel Minuten insect pins (available from entomology supply companies such as BioQuip; Appendix IV) mounted in the end of a matchstick or held in a pin vise, for detaching parasites from insects. The pins can be kept sharply pointed using the grinding stone. (6) Microscope slides (2.5 × 7.5 cm) and both 22-mm and 18-mm cover glasses (#0 or #1), either square or round.

Mounting media used include the following: (1) Aqueous glycerol, with or without a trace of a dye such as cotton blue or acid fuchsin, is the preferred medium for mounting Laboulbeniales for general morphological studies. Its refractive properties are such that the relationships of cells comprising the thallus are easily observed. (2) Amann's solution, which can be used with or without a trace of acid fuchsin or cotton blue, and the solution can be substituted for aqueous glycerol. (3) Hoyer's medium, which is used sparingly only as a medium in which to position fungi on slides or cover glasses, before adding mountant.

As already stressed, insects collected in the field or by means of a Berlese funnel in the laboratory should be stored in 70% ethyl alcohol. Gross collections usually will be contaminated with dirt or other debris. The investigator segregates the insects from such material with the aid of a dissecting microscope and fine tweezers and transfers them to fresh alcohol before examining them for fungi. All parts of the body of each insect are examined carefully under medium or high magnification. In some cases it is helpful to have the insects in a glass container with light directed from below to show the thalli in profile. Parasites are sometimes relatively numerous, more or less generally distributed on the surface of the host, and easily found. Other parasites may be limited in number and/or distribution and more difficult to detect. The color of some fungi may contrast sharply with that of the host, whereas other fungi may be nearly colorless. Fungi may be more or less erect or closely appressed to the host integument. Insects often are covered with setae that tend to obscure or be confused with the parasites. Most Laboubeniales have a blackened foot that may contrast in color with that of the rest of the ascoma or insect and can reveal the presence of a parasite on a host's integument.

Parasitized insects should be stored in alcohol along with a label giving complete collection data.

Collectors examine each parasitized host carefully under a dissecting microscope, recording the exact locations of parasites on the host's body, either scattered over the surface or concentrated in clumps of few to many individuals. Some species of Laboulbeniales can grow anywhere on the host body, whereas others are highly specific in this regard. Thus, groups of closely associated thalli could represent several distinct species. Specimens from different groups should be mounted on separate slides, preferably several slides for each group. Each group is sampled separately. As an additional precaution against mixing species, the insect and the micropin should be rinsed with alcohol after preparing slides from one group and before proceeding to the next group. Some species of Laboulbeniales are dioecious, and the males may be extremely small, often little larger than an ascospore. Their detection requires special effort.

Because someone may wish to study development of a given species, immature as well as mature specimens should always be mounted if possible. Usually, development involves several kinds of transitory stages, many of which may have great value in assessing relationships among species and genera. Examples include antheridia, which often do not persist in mature ascomata, and trichogynes, which typically degenerate soon after perithecium development begins.

The best procedure for preparing slide mounts is a slight modification of the double-cover glass method adopted some years ago by Benjamin (1986: 247–248; see also Chupp 1940 and Kohlmeyer and Kohlmeyer 1972):

1. An infected insect is placed on the stage of a dissecting microscope immersed in plain glycerol solution in a large concavity slide. The viscosity of the solution facilitates manipulating the insect and the fungi to be removed from it.
2. While the insect is grasped with fine forceps held in the nondexterous hand, the parasites are detached carefully with a Minuten needle. Great care must be used to avoid damaging the insect and fungi.
3. The fungi are immediately mounted or stored temporarily in a small drop of glycerol in a depression slide.
4. A minute amount of Hoyer's medium is dropped on the center of a 22-mm cover glass on a microscope slide (squares cut from gummed labels affixed to the slide to the right and left of the cover glass help to maintain its position).
5. The depression slide carrying the fungus and the slide carrying the cover glass are placed side by side on the stage of a dissecting microscope and, with the nondexterous hand, are moved backward and forward as needed while the specimens are transferred with the dexterous hand.
6. The fungal specimens are positioned with the receptacle directed upward so that the thallus will appear erect when viewed with the compound microscope.
7. A few fibers of lens tissue placed adjacent to the fungal specimens will help to prevent their being flattened or distorted when the second cover glass is added.
8. A small drop of mountant is centered on an 18-mm cover glass, which is carefully inverted over the specimens on the 22-mm cover glass. Mountant is best dispensed using a squeeze-type plastic dropper bottle having a tiny orifice. When placed on the cover glass, the drop should spread to a diameter of no more than about 3–4 mm. Only enough mountant to fill the space between the two cover glasses is needed.
9. Another small drop of clear glycerine is placed in the center of the 18-mm cover glass.
10. Then, without reversing, the two-cover glass combination is inverted onto a clean microscope slide in a position sufficiently offset to allow a suitable label to be attached to the slide. The glycerol should reach the edge of the small cover glass, which is now in direct contact with the microscope slide and will help prevent inwelling of the sealing compound.
11. The large cover glass, now uppermost, is sealed to the microscope slide by spreading (on opposite edges of one side of the cover glass) an excess of a medium such as Canada balsam or Fisher Permount 7, which hardens in time. The sealant will dry more quickly if the slides are placed on a slide warmer set at approximately 40°C for several days. The heat also hastens diffusion of the Hoyer's medium surrounding the fungi into the glycerol and the plumping of the specimens. During the first day or two, while the sealant hardens around the edges of the large cover glass, it is advisable to place a small weight on the cover glass. Small metal nuts weighing about 0.5 g each and available at a hardware store are ideal. When the sealant has hardened sufficiently, sealant extending beyond the edge of the cover glass can be carefully removed, if desired, using a razor blade.
12. As the sealant hardens, some inwelling may occur at the margins of the cover glass; airspaces thus formed can be filled with sealant or with nail lacquer.
13. The slides are labeled and stored flat.

Cultivation

As was mentioned earlier, no member of the Laboulbeniales has been cultivated from spore to spore despite several attempts. There is, however, one report of cultivation of *Stigmatomyces ceratophorus* (Whisler 1968). Autoclaved wings of the host, the lesser house fly, *Fannia canicularis,* placed on brain–heart infusion agar (Appendix II) fortified with typtose, a mixture of enzymatically digested protein, and overlain with horse serum were used to grow thalli up to a 20-celled antheridium-producing stage.

Preparation and Deposition of Vouchers

Data should be written legibly in permanent ink on a label securely attached to the slide. Host information should be included so that voucher material of the host can be located years later if necessary. Vouchers of host specimens should be kept in 70% alcohol, preferably with the slide collection or linked to it by a common collection number. Addition of some glycerol to the alcohol will retard evaporation of the solution; it is imperative, however, to have containers that are well sealed. Plastic vials with caps having o-rings are especially suitable for long-term storage.

Well-sealed slide mounts should be deposited in permanet collections. Herbaria that already have extensive fungal collections include the Farlow Herbarium at Harvard University, which is the repository for the extensive R. Thaxter collection of Laboulbeniales. Large collections made by R. K. Benjamin and I. I. Tavares are in the herbaria of the Rancho Santa Ana Botanic Garden and the University of California, Berkeley, respectively (Appendix III).

Identification

Only very few specialists attempt to identify members of the Laboulbeniales to species, and it is difficult to know when one has collected a new species. Many keys to the group are based on the system of Tavares (1985), and host identification is an important aid in the identification process. Tavares (1985) recognized 132 genera of Laboulbeniales: one in the Herpomycetaceae, 12 in the Ceratomycetaceae, five in the Euceratomycetaceae, and the other 114 in the Laboulbeniaceae. Since the publication of Tavares's book in 1985, eight additional genera, including *Majewskia* (Lee and Sugiyama 1986), *Sugiyamaemyces* (Tavares and Balazuc 1989), *Cupulomyces* (Benjamin 1992a), *Phalacrichomyces* (Benjamin 1992b), *Corylophomyces* (Benjamin 1995), and *Parvomyces* (Santamaría 1996), *Triainomyces* (Rossi and Weir 1998), and *Monandremyces* (Benjamin 1999), all Laboulbeniaceae, have been described, and one genus, *Fanniomyces,* has been reduced to synonymy under *Stigmatomyces* (Weir and Rossi 1995) for a total of 139. Good evidence from ongoing collection studies (Rossi and Weir 1998) and from quantitative assessment of well-inventoried tropical

beetle faunas (Weir and Hammond 1997a, 1997b) indicates that large numbers of species remain undescribed.

Tavares (1985) recognized two suborders of Laboulbeniales based on the origin of the centrum in relation to the development of the perithecial wall. In the Herpomycetineae, with only one family, Herpomycetaceae, the perithecial wall begins to develop prior to the appearance of the carpogonial upgrowth, whereas in the Laboulbeniineae, which includes the other three families, Ceratomycetaceae, Euceratomycetaceae, and Laboulbeniaceae, the carpogonial upgrowth is formed prior to the appearance of the first perithecial wall cells. The latter three families are distinguished from each other by the relationship of their perithecial stalk cells to the cells of the receptacle during early stages of development of the perithecium. Genera are distinguished by differences in the morphologies of their receptacles, appendages, antheridia, and perithecia, as well as the positions of their appendages, antheridia, and perithecia on the receptacle. The literature for identification begins with the monographic studies of Thaxter (1896, 1908, 1924, 1926, 1931) in which many species are illustrated beautifully. The most recent complete key to genera can be found in Tavares (1985), who discussed the genera and provided numerous references to primary literature.

FUNGI OF UNKNOWN AFFINITIES

A number of fungi found on the surfaces of arthropods do not appear to harm them. The fungi are observed only by specialists interested in them specifically or by those studying Laboulbeniales. In fact many of these species have been seen by few mycologists since R. Thaxter and C. Spegazzini described them more than 50 years ago (see Blackwell and Rossi 1986; Blackwell et al. 1986; Blackwell 1994).

Several genera in this group (e.g., *Termitaria, Mattiroella, Laboulbeniopsis, Coreomycetopsis, Antennopsis, Hormiscioideus*) appear to be specialized on termites; other genera (*Termitariopsis*) specialize on attine ants or (*Muaiaria, Muriogone,* and *Chantransiopsis*) on other insect hosts. Little is known of their biology. Some, but not all, have haustoria that penetrate the host cuticle. The few characters available for these morphologically reduced forms indicate that they do not form a monophyletic group. A few of them may have alternate states or growth forms that should be sought in the habitats of the host insects (J. W. Kimbrough, personal communication). That is the situation with *Thaxteriola*, an ascospore-derived anamorph that we discuss later in the chapter (see "Spore Dispersal Interactions").

Because these species have much the same habit as the laboulbenialean species discussed earlier, the techniques for their study are essentially similar. However, in many cases the fungi are much smaller than Laboulbeniales. A high-quality dissecting microscope capable of magnifications of at least 200× and equipped with fiber optics makes discovery and mounting of these forms easier.

SEPTOBASIDIALES

A number of ascomycetes are serious pathogens of scale insects (Homoptera: Coccidae); species of *Septobasidium*, however, appear to have a mutualistic relationship with their insect associates and to cause them little harm. A colony of scale insects can be covered entirely by a fungal colony and thus derive partial protection from predators. In return a few individuals of the colony are invaded by helical haustoria of the fungus. Infected individuals do not die; in fact they may outlive uninfected neighbors. The only apparent disadvantage is that they cannot reproduce.

Collection

Species of *Ordonia, Septobasidium* (Fig. 18.7), and associated anamorphs (*Janetia* and *Johncouchia* species) form perennial colonies on the surfaces of plant structures with colonies of scale insects. The basidiocarps are usually brown to black, rarely more brightly colored, normally resupinate, and felty in texture. Their surfaces may be smooth, warty, or spiny. Inexperienced collectors may mistake them for corticioid basidiomycetes or even lichens. *Septobasidium* occurs on living leaves, stems, and branches of a great variety of perennial plants, including gymnosperms, monocots, and dicots. As is the case with most basidiomycetes, basidia are produced so that they project toward the ground. Thus the resupinate basidiocarps often are found on the lower sides of branches. Their occurrence on living rather than dead plant parts, and away from the extreme tips of branches distinguishes them from some resupinate species of Aphyllophorales. *Coccidiodictyon inconspicuum* differs from other members of Septobasidiaceae in producing an inconspicuous colony that is not easily seen in the field and probably will be found more readily by examining collections of scale insects with a dissecting microscope.

Specimens of Septobasidiaceae can be collected using methods suitable for collecting Aphyllophorales or lichens—that is, they are normally removed from the substratum with a knife or ax—and should include a generous portion of substratum and associated scale insects. Because the identity of the substratum plant is essential for identifying the associated scale insect and the fungus, a voucher specimen of the plant should be collected and include a sample of leaves, flowers, fruit, and other diagnostic structures. The materials should be placed in paper bags or in well-aerated containers, not plastic bags.

Septobasidiales are perennial and thus exhibit distinct seasonal responses. Growth occurs during the wet season and ceases or slows at the onset of the dry or cold season.

FIGURE 18.7 *Septobasidium burtii* growing on a scale insect attached to the lower side of a small tree limb. (Photo by Daniel Henk)

Probasidia (telia) are produced at the end of the growing season or during the dormant period and germinate at the onset of the new growing season. Because basidia are taxonomically important structures, the most useful material for identification will be collected toward the end of the dormant period or at the beginning of the growing season.

Storage

The perennial colonies of Septobasidiales are adapted to periodic drying, and probasidia function as dormant perennating structures. Nevertheless, drying over extreme heat will kill a probasidium and prevent subsequent development. Air-drying is preferable in most cases, although in very humid climates it may be necessary to use desiccants or even low heat. The drying air should be moving, not stationary as it is in an oven. The top rack of a large mushroom dryer is usually sufficient for drying in any climate.

Identification of the fungus may require identification of the host insect as well as examination of haustoria, probasidia, basidia, and basidiospores. Separating a colony from insect requires delicacy, and the material should be soaked in water first to wet it thoroughly. Couch (1938) recommended pulling a colony away from its insect with the specimen still submerged in water. In that way only the connections between the colony and the parasitized insect individual will remain. Finding parasitized individuals, which may be greatly outnumbered by healthy ones, is important if one is to observe the taxonomically diagnostic haustoria. Parasitized individuals may be smaller than healthy ones but not necessarily so.

Healthy scale insects should be saved for identification, although only adult females can be identified reliably. The insects are dried with a bit of the plant substratum or placed in alcohol.

Preparation of Specimens for Study

The critical structures used for identification of these fungi can be observed in mounted specimens. In most cases whole infected individual insects can be mounted on a microscope slide in water; they are transparent enough to allow internal structures to be viewed. If the insects are dry, they can be mounted in 7% potassium hydroxide (KOH); if they are very old and dry, they even can be boiled in KOH. In some cases it may be necessary to crush the insect or even dissect the exoskeleton away to reveal haustoria or other important structures. The port of entry (or exit) of the fungus into the insect also may be important and should be noted.

Probasidia, basidia, and basidiospores are also critical in identifying Septobasidiales. Basidia may be one-, two-,

three- or four-celled, and the probasidium may be persistent or not. Basidia may be straight, bent, or coiled. Unless material has been collected at precisely the right moment, however, it will not exhibit basidia. Nevertheless, many live specimens can be induced to produce basidia, with mature basidiospores being discharged actively from the sterigmata. The specimens are soaked in water until they are wet and then placed in a covered container with wet paper towels. Usually, the probasidia develop within 24–48 hours and produce basidia and basidiospores, which can be mounted in water or 7% KOH for examination. Haustoria are also important taxonomically and must be observed carefully. Couch (1938) recognized six types of haustoria.

Cultivation and Deposition of Cultures

Cultures can be established by moistening a colony producing probasidia and suspending it over low-nutrient or water agar. This is done easily by attaching a small fertile portion of the basidiocarp to the top of a Petri dish using white glue, but a piece of agar cut from the edge of the plate often adheres to the lid and can be used to hold the basidiocarp. This procedure requires less care than one might imagine, but crumbs from the specimen that carry faster-growing fungi must not be allowed to fall on the agar surface where the ejected basidiospores are expected to germinate. Common laboratory media such as malt–yeast–peptone used to culture many "jelly fungi" appear to be adequate. As with any group of fungi, cultures should be deposited in an established culture collection. In view of the ease with which at least some species can be cultivated it is rather surprising that few of these fungi are available in culture collections.

Preparation and Deposition of Vouchers

Vouchers should consist of all or part of a colony, including the associated insects and some of the substratum. The material can be placed in boxes or packets of the kind used for lichens and Aphyllophorales. Labels should include the names of the fungus, the associated insect, and the plant substratum, as well as the date and usual geographic and habitat data. The J. N. Couch collection is at the University of North Carolina (Appendix III); however, until there is an active mycologist at that institution, we cannot recommend sending vouchers there. A more appropriate collection that contains a large collection of *Septobasidium* is the Farlow Herbarium (Appendix III).

Identification

Unfortunately, there is remarkably little literature on Septobasidiales. Couch's 1938 monograph is still the most authoritative source, and few species have been described since that work was completed. Couch recorded the largest number of species (36) from the United States, although these fungi probably are predominantly tropical. Collectors in tropical regions likely will discover a large number of new taxa, a point Couch (1938:50) recognized when he stated that "no discussion of geographic distribution will be of much value." Other useful references to Septobasidiales are those of Couch (1935) and Azema (1975).

TECHNIQUES FOR SAMPLING FUNGI INVOLVED IN GARDENING SYMBIOSES

Mutualistic associations between insects and the fungi on which they feed or from which they acquire enzymes for digestion often are referred to as gardening symbioses (M. M. Martin 1987). Not all of the fungi in these associations are obligate members of such symbioses; rather there is a continuum of associations ranging from those in which the fungus is only dispersed by the insect to those that are true gardening associations. Some of the associations are of interest because they may provide systems for evolutionary studies of a spectrum of interactions. In other cases the interactions are of economic importance because they involve dispersal of serious fungal pathogens or sapstain fungi that damage trees, crop plants, and forest products. Some of the fungi rely on the insect for survival because they are poor competitors with saprobes in their habitats. The fungi that form associations include ascomycetes (yeasts, *Ophiostoma* [Figs. 18.8 and 18.9], *Ceratocystis* and related conidial forms, and aphyllophoralean basidiomycetes) that are symbionts of various groups of beetles, intracel-

lular yeastlike forms (Fig. 18.10 [*Symbiotaphrina* and undescribed taxa]) that inhabit specialized host cells and coelomic cavities, and the basidiomycetes and ascomycetes associated with siricid wood wasps, ants, and termites (Table 18.3).

BARK BEETLES AND FUNGI

Fungi are often associated with beetles that inhabit bark and wood of living or recently dead trees. Those fungi fall into two categories. One category encompasses

TABLE 18.3
Fungi Involved in Gardening Symbioses with Arthropods

Ascomycota
  Saccharomycetes
    Saccharomycetales (*Ascoidea, Dipodascus, Pichia, Candida*)
  Pyrenomycetes
    Hypocreales (undescribed yeastlike forms associated with
      planthoppers)
    Xylariales
      Xylariaceae (*Xylaria*)
    Microascales
      Ceratocystiaceae (*Ceratocystis; Chalara; Ambrosiella,* in part)
    Ophiostomatales
      Ophiostomataceae (*Ophiostoma; Leptographium; Ambrosiella,*
        in part; *Sporothrix; Raffaelea*)
  Loculoascomycetes-Discomycetes
  Unknown affinites—*Symbiotaphrina*
Basidiomycota
  Hymenomycetes
    Aphyllophorales
      Corticiaceae (*Entomocorticium* and others)
    Agaricales
      Lepiotaceae (*Chlorophyllum, Leucoagaricus, Termitomyces,* and
        undescribed forms)

FIGURE 18.8 Ascospores oozing from the neck of an *Ophiostoma* species perithecium. (Photo by Kier Klepzig, USDA Forest Service)

FIGURE 18.9 Resinous lesions on the bark of a pine tree caused by mass inoculation with *Ophiostoma minus*, a fungal associate of the southern pine beetle. (Photo by Erich Vallery, USDA Forest Service)

species that commonly occur with phloem-feeding beetles, usually in living trees (grouped here as bark beetles). The other category includes species associated with beetles that require fungi as a primary nutrient resource in all life history stages (ambrosia beetles). The distinction is artificial, and the fungi often are closely related. However, because techniques used to study the phloem-feeding and ambrosial associations differ somewhat, we discuss them separately (see "Ambrosia Beetles and Fungi" later in this chapter). We include our discussion of phloem-feeding weevils (Curculionidae) with the bark beetles.

Bark beetles colonize both hardwood and conifer trees, and although we will emphasize those that colonize conifers and their fungal associates, many of the methods we describe are applicable to colonizers of hardwoods as well. Numerous fungi occupy almost all parts of the body surface and gut of a beetle, as well as the tree tissue the beetle infests. Among the fungi found on the beetle surface and within the digestive tract are yeasts (Callaham and Shifrine 1960; Bridges et al. 1984; Leufven and Nehls 1986), various saprobes (Bridges et al. 1984), and ophiostomatoid fungi (see "Identification," later in this chapter), especially in Ophiostomatales (Upadhyay 1993). Species of *Ophiostoma* and related conidial fungi associated with beetles include many of the stain fungi known to discolor wood (Fig. 18.11) (Harrington 1988 and references therein). Beetle-associated ophiostomatalean fungi also have been implicated as conifer pathogens (Harrington and Cobb 1988;

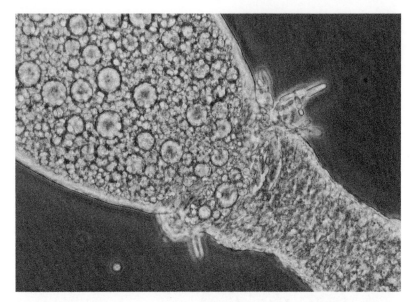

FIGURE 18.10 Yeastlike symbiotes gathered at the epitherial plug (the junction of the ovariole and pedicel) of a planthopper. The symbiotes of planthoppers are transmitted to the next generation through the ovary, entering the terminal oocyte in the ovarioles posteriorly. The symbiotes move from the insect fat body to the epitherial plug when the ovary matures.

FIGURE 18.11 Cross-section of a red pine infected with blue-stain fungi. (Photo by Kier Klepzig, USDA Forest Service)

Harrington 1993), and certain members of this group are capable of killing trees (Brasier 1988; Harrington 1993; Solheim et al. 1993). More often, however, they are associated with resinous lesions that may cause the occlusion of sapwood (Harrington 1993). Some of these fungi also are antagonists of beetles, reducing reproductive success and larval development (Barras 1970). Although the exact ecological roles played by the various ophiostomatoid fungi have yet to be determined, they are undoubtedly closely associated with bark beetles and weevils and their tree hosts. Efforts to examine the diversity of beetle-associated microorganisms center around these fungi.

Certain beetles of the curculionid subfamilies Scolytinae and Platypodinae have evolved specialized structures known as mycangia, the purpose of which appears to be the storage, culture, and transport of fungi (mycangia occur also in ambrosia beetles). The mycangia of a few bark-beetle species are complex and include secretory cells (Harrington 1993). More commonly, beetle mycangia are less developed, simple pits in the exoskeleton of the head, pronotum, or elytra. These simple structures may contain yeasts, ophiostomatalean fungi, and other fungi, including corticioid basidiomycetes (Harrington 1993; Lewinsohn et al. 1994). Mycangial fungi are thought to be mutualists of their beetle hosts, possibly by receiving nutrients from the host (Bridges 1983; Bridges and Perry 1985; Goldhammer et al. 1990). Often fungi have a yeastlike morphology while they are in a mycangium, rather than the hyphal form outside the mycangium and in the environment of the wood. The taxonomy and ecology of many of these fungi are not fully known (Moser et al. 1995).

Collection

Bark-beetle fungi can be found in or on insects, other than the beetles that they colonize. True hosts of these fungi are found in only a few coleopteran families, including the Curculionidae, especially in the subfamilies Scolytinae and Platypodinae (Harrington 1988; Malloch and Blackwell 1993a). Knowledge of the host insect's biology facilitates effective collection of specimens for isolation of fungi. These insects colonize the lateral roots, the root collar, the main stem, the branches, developing shoots, and even fruits of a variety of trees (S. L. Wood 1982; Drooz 1985); the best-studied insect–fungal complexes are found in conifers. Many bark beetles and weevils use tree- and insect-produced compounds to locate suitable hosts as well as mates (D. L. Wood 1982). Using host material, host compounds, and/or beetle pheromones as attractants during times of seasonal insect abundance, it is possible to collect large quantities of beetles from which fungal associates can be isolated. Adult beetles actively seeking host substratum and/or mates respond to indicators of attacking beetles and/or susceptible trees. Aggregation pheromones, either alone or in combination with host compounds, serve as potent attractants for such beetles (D. L. Wood 1982). Later-arriving species of beetles and weevils (secondary invaders), ordinarily associated with stressed or dead trees, may respond strongly to volatile substances produced by the host (Tunset et al. 1993). A variety of sampling devices has been designed to capitalize on these key aspects of bark beetle and weevil biology.

Pitfall traps have been used to capture root-infesting weevils and beetles within forests and plantations (Harrington et al. 1985; Witcosky et al. 1987; Hunt and Raffa 1989). One such trap (Hunt and Raffa 1989) can be constructed from capped sections of plastic pipe, drilled with small entrance holes and placed so that the holes are even with the soil surface. Vials of ethanol and turpentine are hung inside the trap, and a section of pine stem is placed in the bottom of the trap. Weevils and beetles crawl through the soil and into the entrance holes in response to the volatile materials (Hunt and Raffa 1989) and are unable to escape. Stem sections (billets) also have been used to collect root-feeding beetles and weevils (Lewis and Alexander 1986; Tunset et al. 1988). The stem section usually is placed in contact with the soil

where it is left for several days. The host material is checked daily for insects moving from the soil onto the stem-section surface. In addition, some root weevils may be collected from the lower stem as they ascend at night to feed on branches (Klepzig et al. 1991). Walking weevils are forced into a collection jar atop a screen funnel that is wrapped around the main stem.

Bark beetles that attack the lower stem of trees can be collected in various types of flight traps. A lower-stem flight trap consisting of an inverted, plastic jug modified by having a collection jar attached can be baited with turpentine and ethanol and used to collect turpentine beetles and some root insects (Klepzig et al. 1991). Turpentine beetles also can be captured in bounce traps in which the flying beetle strikes a black pipe baited with ethanol and turpentine, and falls into a water-filled pool below, from which it is collected (Fatzinger 1985; Phillips et al. 1988). Lindgren multiple-funnel traps (Lindgren 1983) can be hung near the ground for collection of lower flower-stem insects flying toward attractants (Phillips et al. 1988).

Most of the aggressive, tree-killing bark beetles attack the central and upper portions of the stem (S. L. Wood 1982; Drooz 1985). Flight traps hung in the mid to upper canopy and baited with species-specific pheromones often are used to sample those insects. The proper choice and use of pheromones is, however, a complicated matter. Species may respond to different compounds or to the enantiomers of those compounds (D. L. Wood 1982; Payne et al. 1984; Raffa and Klepzig 1989). In addition, methods used to sample populations of the beetles often are at odds with obtaining viable cultures of associated-fungi. For example, collection jars of Lindgren multiple-funnel traps can be filled with soapy water (Klepzig et al. 1991) or an insecticidal strip (Hayes and Strom 1994) to kill trapped beetles. Both the soap and the insecticide are likely to influence the fungal flora of the beetles. In addition, dead beetles soon become overwhelmed by saprobic fungi that may interfere with isolation of beetle-associated fungi. An alternative is to leave the collection jar empty so that beetles are not killed on falling into the trap. However, bark beetle pheromones may serve as kairomonal attractants for predaceous insects, in which case the collection jars become a feeding ground for predators (D. L. Wood 1982). The result is a pile of bark beetle parts with little value as a source of fungi. If traps are sampled the same day they are deployed, however, the method can be used successfully. Another approach is daily collection from a collection cup contained within an electric cooler to slow predators (B. S. Lindgren, personal communication).

Many bark beetle-associated fungi can be isolated from infested host tree material. Phloem and xylem tissue from areas around insect feeding sites, entrance holes, and adult or larval galleries can be collected aseptically for fungal isolation as can frass from larval galleries (Bridges et al. 1984; Harrington 1992; Solheim 1995).

Various phoretic tarsonemid mites have been implicated in the transmission of bark beetle–associated fungi, and these arthropods also can be a source of fungi for cultures (Bridges and Moser 1983; Moser et al. 1995). Mites typically are removed from the beetle exoskeleton (often underneath the beetle elytra) with fine needles. They can be cultured directly or mounted on slides for later microscopic examination of spores.

Storage

Fungi should be isolated as soon after the insects are collected as possible. When this is not feasible, insects, mites, and tree tissue should be placed in sterile vials, transported to the laboratory in ice-filled coolers, and refrigerated until they are processed. Placing small pine twigs and/or moist paper in collection vials may increase survival of insects during periods of extended refrigeration. When possible, insects should be kept in separate vials or containers to minimize the likelihood of cross contamination between hosts. Tree-tissue samples can be refrigerated or stored at room temperature in moist chambers, which helps to keep the fungi viable and actually may promote sporulation of the fungi on the host substratum (Seifert et al. 1993). Specimens can be dried, but subsequent isolation into culture often is not successful.

Preparation of Specimens for Study

Structures associated with sexual or asexual sporulation in the tree wood or bark provide diagnostic characters for identification of the fungi. It is important to make good slides of material when it is available, even if the fungus is to be cultured, because some of the species will not produce sexual reproductive structures in culture. To serve as vouchers, specimens must be mounted on microscope slides using the double-cover slip method discussed earlier (see "Preparation of Specimens for Study" in the Laboulbeniales section for detailed protocol) and made permanent. In addition, spores of fungi found in pits or mycangia on beetle and mite exoskeletons can sometimes be identified using light and transmission, and scanning-electron microscopy (Happ et al. 1971; Lewinsohn et al. 1994).

Cultivation and Deposition of Cultures

Mycangial fungi may be isolated from beetles by dissecting the mycangium from the beetle, surface sterilizing the structure, and subdividing and placing it on selective media (Barras and Perry 1972). For example, for the southern pine beetle, the pronotum is removed from the head and abdomen of an adult female, and the legs are removed from the pronotum using a pair of sterile, fine forceps. The entire, intact pronotum is placed in sterile distilled water for 1 minute; placed in modified White's solution (Appendix II; Barras 1972) for 4 minutes; and then passed through two successive rinses with sterile distilled water. The pronotum then is quartered aseptically with fine forceps, and the four sections are placed on malt extract agar (MEA) or MEA amended with 2 mg/ml benomyl to facilitate the semiselective isolation of certain species such as *Ceraotocystiopsis ranaculosus*, an isolate known as *Entomocortocium* species A, and perhaps other mycangial fungi (Ross et al. 1992).

Beetles can be crushed on an agar medium to isolate phoretic fungi (Harrington 1992). Alternately, beetles may be ground in sterile distilled water in a glass tissue homogenizer. Aliquots of the homogenate are plated directly or diluted a number of times and then plated on an agar medium (Klepzig et al. 1991). This technique has the added advantage of allowing for quantitative estimates of the number of propagules of each fungal species being carried by each insect. Samples of beetle-infested tree tissue may be collected aseptically and placed directly on water agar (WA). In many cases, fungal fruiting structures can be found lining beetle galleries; masses of spores can be transferred directly to media with a fine needle (Seifert et al. 1993).

A major confounding factor in all of the isolation techniques that we have described is the presence of saprobic contaminants either on the beetle exoskeleton or within beetle-infested tree tissue. Although these fungi may be significant components of the beetle fungal flora, many of them grow so quickly that they overwhelm other fungi of interest (Gibbs 1993; Seifert et al. 1993). Diluting samples before plating often results in separation of fungal colonies of interest that can be subcultured (Klepzig et al. 1991). Another technique is useful for isolating members of the Ophiostomatales, which includes many of the well-studied beetle associates. A characteristic of this group of fungi is a high degree of tolerance of the antifungal compound cycloheximide (Seifert et al. 1993). Accordingly, MEA, WA, or potato dextrose agar (PDA) can be amended with 200 ppm cycloheximide, to inhibit growth of nonophiostomatalean fungi, and 100 ppm streptomycin sulfate, to inhibit bacterial growth (Harrington 1992; Seifert et al. 1993). Some yeasts, however, and species of filamentous fungi, including *Penicillium*, may also grow on these media (Harrington 1992). Single colonies on plates may be transferred (via hyphal tip transfer and/or transfer of spore masses) to unamended PDA or MEA for identification of pure cultures.

Plates normally are incubated at from 20–25°C, although some bark beetle associates may grow better at cooler temperatures (Harrington 1992). Cultures can be grown on a weak agar medium in vials and stored at −20°C, or they can be lyophilized for preservation. Serial transfer and storage on rich media generally are not recommended (Seifert et al. 1993). All of these fungi should be of interest to most general culture collections because of their interesting biological associations and economic importance.

Preparation and Deposition of Vouchers

Vouchers can be prepared by placing fungal structures on plant material or cultures that have been dried, usually in moving air at room temperature, in packets or small boxes. The specimens should have mature sexual-reproductive structures that include diagnostic characters. Permanent slide mounts are also desirable as vouchers. In some cases parts of beetles and associated mites can be mounted or prepared for scanning electron microscopy to show phoretic spores of the fungi. Arthropod material can be maintained in 70% alcohol to which a little glycerine has been added.

Identification

The fungi that occur in bark-beetle associations are a diverse lot. We mentioned ophiostomatoid fungi, which have distinct morphological features such as evanescent asci and long-necked ophiostomatoid perithecia through which the sticky ascospores are passively discharged for arthropod dispersal. These species belong to several orders, primarily Ophiostomatales (*Ophiostoma*) but also Microascales (*Ceratocystis*) and *Pyxidiophora*. In addition, these orders include a number of derived asexual forms such as *Sporothrix*, *Leptographium*, and *Chalara*. There are yeasts, mycangial basidiomycetes, and other filamentous saprobic fungi as well. Several helpful references to these fungi are available (Barras and Perry 1975; Harrington and Cobb 1988; Perry 1991; Schowalter and Filip 1993; Upadhyay 1981; Wingfield et al. 1993; S. L. Wood 1982).

AMBROSIA BEETLES AND FUNGI

As we mentioned earlier, the ambrosia and bark beetles differ primarily in their nutrient sources, with adult, pupal, and larval stages of the former relying on fungi, rather than phloem used by the latter. Because the separation of these two groups is artificial, many of the fungal genera that we discussed earlier in the section "Bark Beetles and Fungi" are discussed again here in association with the same beetle families. However, the interactions are different and less flexible. Roeper (1995) described two feeding categories of ambrosia beetle

larvae: those that consume only fungi (mycetophagous) and those that enlarge their gallery or larval cradles in the xylem as they develop, thus consuming both wood and fungal material (xylomycetophagous). The latter habit approaches the behavior of some mycangial bark beetles. As adult beetles make brood galleries by tunneling into the new woody host material, they transmit species-specific obligatory fungal symbionts in ectodermal mycangia. The damage done by ambrosia beetles is caused by their boring activity and subsequent staining of the wood (McLean 1985).

The primary symbionts of ambrosia beetle species are those that are consistently isolated from the mycangia of adult beetles collected during flight, from adults excavating their new brood galleries, or from brood galleries in the presence of actively feeding larvae (Batra 1967, 1985; Roeper et al. 1980). Those fungi include *Ambrosiella* and *Raffaelea*, asexual genera related to *Ophiostoma* and *Ceratocystis*. In warmer regions *Fusarium* may be an ambrosial associate. Other fungi, referred to as auxiliary or secondary ambrosial fungi, are not usually isolated from mycangia but are regularly present in the brood tunnels after beetle pupation. Many of those fungi have mucilaginous spores that may be transmitted by phoretic mites and beetles. Their presence in tunnels parallels that of the saprobic fungi associated with bark beetles and interferes with the isolation of the slower-growing primary symbionts in culture.

Wood-boring scolytine and platypodine ambrosia beetles usually inhabit dead or dying trees (see reviews by Batra [1967], Francke-Grossmann [1967], Norris [1979], Beaver [1989], and Roeper [1995]). S. L. Wood (1982) described the North and Central American Scolytinae and listed their woody hosts and geographic distributions. In addition, Wood and Bright (1987, 1992) catalogued all Scolytinae and Platypodinae and their plant hosts, distribution records, and literature references.

Collection

Ambrosia fungi are collected with their beetle associates. The fungal form is dimorphic; in the well-developed mycangia the ambrosia fungus is yeastlike, and within the galleries it is filamentous but usually cropped by the feeding beetles.

Beetles during Dispersal Flight. In northern temperate regions, adult beetles fly only during a short period (generally a month) in spring when temperatures reach 18°C (Roling and Kearby 1975; Turnbow and Franklin 1980; Weber and McPherson 1991). Most species have only a single annual flight, but some species (e.g., *Monarthrum* species) have two generations and two flights each year. In semitropical and tropical regions, many of the species have multiple generations but seldom fly during dry seasons. The beetles tend to fly in the late afternoon and early evening.

Beetles can be caught live in fine-mesh nets, in mechanical rotary traps (Rudinsky and Daterman 1964), or with hand nets. Unprocessed timber at sawmills and logging operations, and wind-thrown, wind-damaged, standing suppressed, and/or diseased trees attract flights of ambrosia beetles. The beetles are attracted to ethanol produced by the fermentation of host timber. Thus ethanol or beer can be used in addition to timber to attract beetles to a collection site. Beetles collected should be placed individually into sterile vials or stoppered tubes with damp sterile filter paper and then cooled during transport to the laboratory.

Beetles Infesting Wood. Most ambrosia beetles in temperate regions infest cut timber, wind-thrown trees, wind-broken limbs or boles of trees, and/or suppressed or diseased trees. Woody timber dead for more than a year is seldom infested. However, temperate-region ambrosia beetles of the genus *Corthylus* and warm temperate-, semitropical-, and tropical-region members of the scolytine beetle tribe Xyleborini (*Xyleborus* and *Xylosandrus*) are capable of attacking apparently healthy and undamaged woody hosts. Ambrosia beetles infest the bole of the tree, boring directly through the bark into the xylem. The entrance hole is seldom more than 1 mm in diameter; the boring frass is light in color initially, but darkens as the primary symbionts begin to grow or as it is contaminated with larval fecal pellets. By comparison, scolytine bark beetles typically produce brownish boring frass because they mine the inner bark as they construct galleries.

Storage

Once infested timber is located, it should be cut into manageable lengths, generally about a meter, and returned to the laboratory. The surface of the infested timber is cleaned of dirt and biota and disinfected by lightly wiping several times with ethanol. Once the beetles have bored into the xylem, they generally will not reemerge unless the wood begins to dry. Painting cut ends with melted paraffin wax slows log dehydration. If the wood is stored out of direct sunlight, the beetles will continue gallery construction, produce broods, and complete a generation of their life cycles.

Preparation of Specimens for Study

These methods generally are the same ones used for bark-beetle associates; however, ambrosia beetles may lack diagnostic characters such as sexual stages and even conidia may be lacking. Consequently, cultures are important for identification of these species because they provide additional diagnostic characters such as growth rate and pigment production.

Isolation

The primary fungi of an ambrosia beetle are abundant in a gallery only when larval stages are present (Kajimura and Hijli 1992). Thus, the best isolates of primary fungal symbionts can be made a month or two after initial infestation. Galleries are exposed by sawing thin sections from the infested bole. It is important to work as quickly and as aseptically as possible, using alcohol-flamed saws, wood chisels, and/or pruning shears. Adult insects can be removed, and visible fungal growth within the several-millimeter-diameter gallery can be isolated using sterile fine forceps. Thin slices or chips of galleries should be preserved, dried, and mounted, or mounted directly on slides with fixative mounting medium, such as lactophenol-aniline blue, for later study.

Ambrosia fungi in the genus *Corthylus* and most *Xyleborus* species generally form a thick, whitish palisade layer on the walls of galleries if eggs and/or larvae are present. That fungal growth can be isolated easily by streaking or spot plating on isolation media (see next section on "Culture").

Fungal growth usually is not so evident on the gallery walls or larval cradles of xylomycetophagous insects; thus, small slices and chips of wood should be removed aseptically for plating. Slices or fragments of galleries can be placed aseptically in a sterile moist chamber (Appendix I) to encourage fungal growth in the absence of actively feeding larvae, so that primary ambrosia fungi can be isolated, often within a few days, before contamination from saprobic fungi.

Live beetles trapped in flight or taken from galleries are difficult to handle because of their small size and smooth cylindrical shape. A simple vacuum apparatus consisting of a sterile micropipette tip with a small aperture attached to a rubber hose fixed to a vacuum pump or vacuum line allows one to pick up individual beetles and transfer them easily from dish to dish or to sterile glass slides for dissection.

Beetles can be surface disinfected to reduce the presence of nonmycangial microbes by washing in sterile 0.1% $HgCl_2$ solution or dilute sterile bleach ($NaHCl_2$) for 2–4 minutes, followed by several rinses in sterile water. Investigators can also free adult beetles of external nonmycangial microbes by placing them alternately in plates of sterile wet filter paper for 18 hours and then on dry sterile filter paper for 6 hours. Several transfers typically remove most external microbes. Individual beetles can be stored on sterile moist filter plates for months at refrigerator temperature until needed for dissection and isolation. Prevention of dehydration appears to be the critical factor for keeping them alive during long-term storage.

Culture

Primary ambrosia fungi are abundant in mycangia at the time of flight and/or early stages of gallery development (Roeper 1988; Kajimura and Hijli 1992). For that reason timing is important for isolation of the true primary symbiotic fungi from a beetle. The sex of the beetle also is important because mycangia usually develop only in the sex that initiates the brood gallery system. The beetles usually have a single pair of mycangia, whose positions can vary even between very closely related beetles. Mycangia of scolytine beetles can be oral, pronotal, mesonotal, prothoracic pleural, promesonotal, or elytral; in platypodines the mycangia are pitlike. If the investigator does not know where mycangia are located, he or she must examine all the beetle's body parts carefully under a microscope.

Mycangia are dissected from the beetle, and the contents should be plated onto agar medium for isolation. The beetle should be dissected under a dissecting microscope on sterile alcohol-flamed glass slides in three separate drops of sterile saline or bovine serum using alcohol-flamed fine watchmaker forceps, fine needles, and sterile micropipettes. The body part of the beetle containing the mycangium(a) should be separated from the rest of the body in the first of the three drops. In the second drop, the mycangium(a) is separated. The mycangium(a) is broken apart in the third drop and plated. The presence of fungal cells can be verified using the low power of a compound microscope. Mycangial fungal cells sink to the bottom of the drop, whereas insect fat droplets, with which they may be confused, stay in suspension. The mycangial fungi tend to be yeastlike budding forms or, if the fungi are proliferating actively in the mycangium, monilioid chains. All parts of the mycangia are spot plated and/or streaked on isolation media (see following paragraphs) and incubated at 22–25°C. Sterile micropipettes can be used to pick up small masses of fungal material to be plated.

Precoxal mycangia (e.g., in species of *Monarthrum* and *Gnathotrichus*) can be dissected to remove their fungi, or the mycangia can be sampled directly. The adult beetle is killed and fixed ventral side up on a sterile glass slide with a drop of molten paraffin. Under a dissecting microscope, the forelegs are removed, and then a sharp needle is used to remove the contents of the enlarged coxal mycangium; that material is plated directly onto isolation medium.

Plates should be examined daily after isolations have been made from a gallery or mycangium. Hyphal-tip

subisolation of filamentous fungal growth and restreaking of yeastlike colonies usually is necessary for purification. Some *Ambrosiella* species (*A. hartigii, A. ferruginea, A. xylebori, A. sulphurea*) grow rapidly in filamentous form and often produce melanin pigments. *Ambrosiella brunnea, A. gnathotrichi,* and most *Raffaelea* species form yeastlike mycelial colonies initially and should be subcultured by streaking and hyphal tipping. Yeasts commonly are encountered and should be subcultured by streaking.

Several culture media, including PDA, malt extract–yeast extract agar (MEYE), and dilute malt extract–yeast extract-glucose agar (see Appendix II), can be used to culture primary mycangial symbionts and isolate fungi from galleries and mycangia of a particular beetle. As many isolations as possible should be attempted from available collected material. Frequency of occurrence of a particular microbe should establish the presence of associated symbiont microbes. Prokaryotic microbes are seldom encountered, so antibiotics generally are not used in isolation media. Once axenic cultures have been made, they can be stored on slants of dilute malt extract–yeast extract-glucose agar for future study.

Preparation and Deposition of Vouchers

Procedures for preparation and deposition of vouchers are similar to those for bark-beetle associates; however, these fungi do not produce sexual states. It is important to prepare vouchers from early cultures because ambrosia fungi may stop producing conidia after a few transfers.

Identification

Primary fungi (*Ambrosiella* and *Raffaelea* species) can be identified using the works of Batra (1967) and Roeper and colleagues (1980). Identification of the beetle associate is important because the fungi are usually host specific. Many of the filamentous ambrosia fungi fail to sporulate in culture or after repeated subculture. However, increasing nitrogen content (L-proline) of the medium, buffering the agar to a pH near neutral, and elevating CO_2 levels in the culture container may induce sporulation in the fungi. Molecular studies have been used to characterize species of *Ambrosiella* and *Raffaelea* and have shown that *Ambrosiella* is polyphyletic with some species being related to *Ceratocystis* and others allied with *Ophiostoma* (Cassar and Blackwell 1996). A similar study showed that *Raffaelea* species are related to *Ophiostoma* (Jones and Blackwell 1998). Reference cultures of most known primary ambrosia fungi are available from ATCC and CBS.

YEASTLIKE ENDOSYMBIONTS AND EXTRACELLULAR SYMBIONTS

The term endosymbiotic is used to denote the situation in which one organism, here an insect, harbors an intracellular symbiont, here a fungus, usually within cells (mycetocytes) of specialized structures (mycetomes); in some cases mycetomes are modified fat bodies. The yeastlike fungi are single-celled and reproduce only asexually. Yeastlike symbionts are associated with anobiid (death-watch) and cerambycid (long-horn) beetles (Jurzitza 1979; Nardon and Grenier 1989), planthoppers, and some species of scale insects. The symbionts of cerambycid beetles and of some anobiid beetles have been cultured. Extracellular hemocoel-inhabiting symbionts have been observed in a group of gall-forming social aphids (Fukatsu and Ishikawa 1992) and in a wasp (*Comperia merceti*) that parasitizes cockroaches (Lebeck 1989; Table 18.4). Although these fungi have garnered little attention from mycologists, their biology is fascinating. The associations apparently have arisen independently in several fungal lineages.

The location of the symbiont within each host group varies. In anobiid beetles, symbionts are located in the ceca of the larval midgut at its junction with the foregut. During metamorphosis, the ceca disintegrate, and the adult mycetome is formed within smaller ceca. In adults, the symbionts are released into the intestinal lumen from the ceca and eventually reach the vaginal pockets. Cells from the vaginal pockets are smeared on the eggs when they are oviposited. Newly hatched larvae ingest the symbionts, which infect the ceca of the gut. Cerambycid beetles also harbor symbionts in ceca located around the midgut. Transmission from adult to larvae is similar to that of the symbionts of anobiid beetles. The larval ceca, however, disappear shortly before pupation, and in the female the symbionts multiply in the hindgut.

The yeastlike symbionts of planthoppers are located within inner cells of the fat body (see Fig. 18.10). Soon after males emerge, symbiont cells decrease in number, and they are gradually lost. In adult females some of the symbionts infect the ovarian epithelial plug (the part that connects the ovariole and pedicel); they enter the terminal oocytes situated at the most posterior part of the ovariole from the posterior pole. Yeastlike symbionts are found in the egg by the time of shell formation. Within the egg, they remain in a mass called the symbiont ball from which they infect the fat-body cells of the next generation of planthoppers at the later stages of embryogenesis. The fungi are never found outside the host insects.

Extracellular symbionts usually are found in the hemolymph of their hosts rather than within cells. Scale insects, however, harbor symbionts in the hemolymph or in the fat-body cells and transmit them transovarially. Symbionts of cerataphidine aphids occur in the hemocoel and also are transmitted transovarially. The symbionts of the wasp, *Comperia merceti*, occur in the hemolymph, digestive tract, and poison gland. All of these fungi are known poorly, and some have not yet been formally described.

Collection

Hosts of the yeastlike symbionts can be found under the following conditions. Most anobiid and cerambycid beetles live in dry plant material, logs, or under the bark of dead trees. Anobiids such as *Stegobium paniceum* (the drugstore beetle) and *Lasioderma serricorne* (the cigarette beetle) occur in dried tobacco or various dry foodstuffs. Other anobiid beetles can be found in dead logs and twigs. The larvae of cerambycid beetles are wood-boring and live in dead trees or freshly cut logs. They sometimes attack living trees.

Rice planthoppers, *Nilaparvata lugens* (brown planthopper), *Sogatella furcifera* (white-backed planthopper), and *Laodelphax striatellus* (small brown planthopper), are found in rice paddies. *Nilaparvata lugens* is distributed throughout the warmer regions of Asia and can be found easily in autumn. It lives on the lower part of the rice plant and thus can be distinguished from the other two rice planthoppers that occupy the upper part of the plant. *Sogatella furcifera* is more widely distributed in Asia than *N. lugens* and is abundant in summer. These two species do not undergo diapause and are often collected from summer to autumn. *Laodelphax striatellus* is distributed widely in Eurasia, where it attacks wheat, rice, and other monocotyledonous plants; it enters diapause in the cooler parts of its range, and diapausing nymphs can be collected in grasses near already-harvested paddy fields in late autumn. Adults can be collected in late spring from wheat plants.

Insect nets and aspirators are useful for collecting leafhoppers. Small test tubes or other small containers are appropriate for holding the insects until laboratory study. Plates spread with a sticky glue can be used to catch insects knocked from rice plants.

TABLE 18.4

Primary taxonomic groups of insects that harbor yeastlike symbionts

Intracellular symbiosis (endosymbionts)
Coleoptera—Anobiidae, Cerambycidae
Homoptera—Delphacidae, Flatidae, Ricaniidae
Extracelluar symbiosis (ectosymbionts)
Homoptera—Cerataphidini (Aphids)
Hymenoptera—Encyrtidae

Storage

The following procedures are useful for storage and culture of host insects. Anobiid beetles (e.g., *Stegobium paniceum* and *Lasioderma serricorne*) can be reared on wheat bran or other dry food. Without food, they survive best at low temperatures, but we do not recommend that they be kept for a long time in a refrigerator. The beetles do not live more than several hours at 4°C, and after they die, their symbionts degrade while undesired microorganisms proliferate quickly. Some planthoppers, such as *Nilaparvata lugens*, weaken at refrigerator temperatures, but *Laodelphax striatellus* survives several days of refrigeration at 4°C. For nucleic acid extraction (but not culture), the symbionts from planthoppers can be frozen at −70°C.

Preparation of Specimens for Study

It is important to remember that insects have many yeasts and fungi on the integument and within the alimentary canal, which makes contamination a real problem for culturing, observing, and extracting nucleic acids of the fungal species of interest.

To observe *Symbiotaphrina* (yeastlike symbionts of anobiid beetles) in the mycetome (in this case the ceca) of the midgut, the gut must be dissected. First, the beetles are surface disinfected by being dipped in 70% ethanol. Special care must be taken when handling larvae because their soft integument is easily broken with fine forceps. The larvae are rich in fat bodies, which can be removed carefully into a physiological saline or 0.85% NaCl (sodium chloride), along with the gut. The several swollen lobes of the cecum, which are located between the foregut and midgut, are easily recognized. The yeast-like symbionts are several micrometers in diameter and can be observed under the light microscope by smearing the mycetome on a microscope slide.

Numerous cells of the symbiont are scattered in the fat bodies of planthoppers. The cells can be observed by pressing the planthopper nymphs, female adults, or eggs in a droplet of physiological saline on a glass slide. The fine structure of the mycetocytes (the cells in which the symbionts occur) can be observed with a transmission electron microscope. The insect body is cut into pieces and fixed in glutaraldehyde and osmium tetroxide or in MnO_2, which may give better results. The yeastlike symbionts of planthoppers have not been cultured. They can be isolated from the planthoppers for study, however, by buoyant density gradient centrifugation using Percoll (Amersham Biotechnology; Pharmacia LKB; Appendix IV). A drawback of this method is that it requires a minimum of several grams of planthoppers (more than 1,000 mature nymphs or adults), although when too large a volume of planthoppers is treated at once, host tissue may contaminate the symbiont fraction. Fresh or frozen insects are dipped into 70% ethanol, washed with distilled water or physiological saline, and then homogenized in 0.85% NaCl solution using a blender or homogenizer. The homogenate is filtered through cotton cloth and centrifuged for 5 minutes at 100 g. The pellet is resuspended, and centrifuged twice. The resuspension of the final pellet is mixed with Percoll and centrifuged for 30–40 minutes at 85,000× g. The symbionts are recovered with a pipette or syringe. This method is useful for planthoppers, which possess a large number of symbionts in the fat bodies. It may not work for other fungal symbionts that form fewer cells within their host insects. Freeze-dried samples of the symbionts of *N. lugens*, *S. furcifera*, and *L. striatellus* are available from H. Noda (National Institute of Sericultural and Entomological Science, Japan; Appendix III). Both nucleic acids and proteins can be extracted from these samples.

Air-dried cultures or, especially in cases when the fungi cannot be cultured, permanent slides can serve as voucher specimens of symbionts. Cell samples and insect parts stored at −70°C or lyophilized can also serve as voucher specimens.

Cultivation and Deposition of Cultures

The intestines of anobiid beetles should be dissected, as described in the previous section, in sterile physiological saline. The cecum is cut from the intestine and homogenized with a small amount of the saline. The homogenate is inoculated on yeast–peptone–dextrose (YPD) agar, potato–peptone–glucose agar (PPGA), or similar media (Appendix II). Antibiotics usually are added to the medium to suppress bacteria. The plates are incubated at 26–28°C. Colonies of the anobiid symbionts can be observed within about 10 days. The medium then is centrifuged at 600× g for 5 minutes to concentrate the cells. As we mentioned earlier, the yeast-like symbionts of planthoppers have never been cultured. They must be isolated directly from planthoppers, which can be frozen or lyophilized after collection, for future study.

Identification

Identification of organisms with few morphological characters, especially when they cannot be cultured, is a real problem. Some of the endosymbionts and extracellular symbionts discussed in this section have not even been formally described, and investigators rely on host identity as the criterion for identification. However, strict host specificity never has been demonstrated. DNA analysis has revealed a broad taxonomic diversity of symbionts across the major host groups, and development of molecular identification tools is a distinct possibility both for higher and lower taxonomic levels once closely related hosts have been surveyed more thoroughly.

Some endosymbionts of anobiid beetles have been shown, based on ribosomal DNA (rDNA) sequence analysis, to be among the radiation of discomycetes and loculoascomycetes (*Symbiotaphrina*) (Noda and Kodama 1996; Jones and Blackwell 1996). True yeasts (asexual forms related to *Pichia*) also are associated with anobiids. Symbionts of cerambycid beetles are members of the genus *Candida*, likely *Pichia* relatives as well. The symbionts of planthoppers (Delphacidae) appear to be hypocrealean pyrenomycetes (Noda et al. 1995) specifically related to *Cordyceps* (Suh et al. 2001). Other planthoppers may be infected with yeastlike symbionts as well.

ANTS AND FUNGI

Two close associations between insects and fungi fascinate lay and scientific minds alike. These associations involve termite and ant cultivation of insect-edible fungi. Fungus-culturing ants and termites have been considered as biogeographic counterparts of each other, with the ants taking up the ecological niche left vacant in the Neotropics by the absence of fungus-growing termites like those that inhabit Africa and Asia. Relatively recently, however, the basidiomycete *Termitomyces* has been reported from a termite nest in Costa Rica (see "Termites and Fungi," later in this chapter). If verified, the report will support the view that neotropical ants and termites obtain nutrients for their fungi from separate resource pools (Hölldobler and Wilson 1990:596).

The ant–fungus association alone is responsible for an estimated 20% of the photosynthate turnover in the Neotropics, and some specific associations rank among the most important causes of crop loss in Latin America (Cherrett et al. 1989). The broadest and most complete treatment of the ants can be found in Hölldobler and Wilson (1990: chapter 17), although most of their statements about the fungi have been superseded by new information. M. M. Martin (1987) reviewed some biochemical aspects of the symbiosis, but Möller's (1893) initial work remains the basis for all research on this system.

The fungus-culturing habit probably was acquired by ants only once in evolutionary time (Chapela et al. 1994), and all ants able to culture fungi belong to a single tribe (Attini) in the family Myrmicidae (Schultz and Meier 1995). Cultural and molecular studies of the fungi suggest that associations between attine ant species and their fungi differ, ranging from relatively loose to very intimate (Chapela et al. 1994). Most derived ant lineages appear to have domesticated a single species of fungus that is asexually (clonally) transferred from generation to generation, whereas more primitive ant lineages seem to be able to capture free-living fungi from a somewhat broader selection in the field. Nevertheless, most of these fungi clearly belong in the family Lepiotaceae; only a single shift to a separate (still to be identified) fungal lineage, in the ant genus *Apterostigma*, is known (Chapela et al. 1994).

Collection

If the goal of a survey is to obtain as diverse a sample as possible of fungi associated with ants, an investigator should emphasize the inconspicuous primitive symbioses of the "lower" attine genera *Mycocepurus*, *Mycetophylax*, *Mycetarotes*, *Mycetosoritis*, and *Sericomyrmex* and the nonlepiotoid associate, *Apterostigma*. A closer association of ants and fungi is suspected in the genera *Cyphomyrmex* and *Trachymyrmex*, in which inhabitants of different nests in one locality may cultivate the same fungal strain (clone). This trend is fixed in the "higher" attines *Atta* and *Acromyrmex* as well as in *Apterostigma*. Because fungi are transferred asexually from one generation to the next in this last group of genera, sampling in a given location and even a given country or region may produce a very low diversity of ant-associated fungi. This generalization should not be construed to give a greater importance of one ant group over another because regulatory mechanisms that fine-tune the relationships between ants and fungi may be most diverse in the otherwise taxonomically least diverse fungi of the higher attines.

Fungi can be isolated from nests or from queens soon after their nuptial flights. Sampling the large nests of higher attines is relatively easy; sampling the nests of lower attines requires more knowledge, time, and patience. Nests are usually located by following worker ants to the entrance. Workers can be attracted by baiting with starchy substratum, such as raw oat flakes, cream of wheat, or bits of pasta. The fungus garden can be excavated or simply uncovered (e.g., from under a log), and nest material, brood, and worker ants can be placed in a container. Attempts to isolate a cultivated fungus directly from the nest in the field or immediately on arrival at the laboratory usually fail because cultures are highly likely to be contaminated with faster-growing fungi. However, if ant workers are kept together with the nest in a small, moist container (e.g., a 250-ml, plaster-bottom glass jar) for about 3 days, they soon will reconstitute the quasi-axenic conditions of the nest, greatly reducing the risk of contamination. Fungi can then be isolated in a clean-air environment with almost assured success. Tracking lower attines and discovering their small nests can be difficult and is clearly the most critical stumbling block to obtaining informative collections. Many attine ants are more active at night, and searches with head lamps tend to be more successful than daytime searches. Even for the more conspicuous "higher" attines, tracking and excavating a nest can take a couple of days or longer.

Another method for obtaining attine-associated fungi that is more suitable for use in an established laboratory or permanent field station involves the trapping (e.g., light trapping) of inseminated females (queens) as they return from their nuptial flights. Females of several genera carry a pellet of mycelium with which to start the garden in their new nests. Trapped queens can be maintained in a plaster-bottom jar containing moist (not saturated) sandy soil 15 cm deep, in which they can dig a gallery and start cultivating the fungus. The time required for a fungus culture to mature enough for isolation varies among the different ant species but can be as short as 4 days or as long as 7 weeks. Isolation again is performed under clean-air conditions. Use of disposable plastic drinking cups instead of glass jars allows for easy sectioning of the soil column to expose the gallery with its fungus culture. Nuptial flights occur only at certain periods during the year, sometimes only on a single night, limiting the use of this method. Another limitation is that the queens of some species probably do not carry mycelia on their nuptial flights (Chapela et al. 1994).

Storage

Basidiocarps and nests containing fungal mycelia should be kept dry and free of insects. The rather extensive nest material is fragile, and investigators should use extreme care to prevent its disintegration. The basidiocarps can be dried and placed in packets as is done with other agarics.

Preparation of Specimens for Study

Basidiocarps can be studied by examination of the hymenophore structure and spore and hymenial characteristics. In addition we recommend that DNA be extracted from hyphae or basidiocarps whenever possible so that species can be characterized genetically in the absence of other characters.

Cultivation and Deposition of Cultures

The fungi cultivated by attine ants are saprobes that grow well on most general media such as PDA and MEA. Spores, tissue explants from basidiocarps, and hyphae can be used for isolation. If isolations are made from well-tended nests, neither antibiotics nor any other medium supplement is necessary. Much more important are the conditions of incubation and storage of the cultures. In our experience, the clonally derived fungi cultivated by the higher attines and *Apterostigma* species cannot withstand temperatures below 4°C, and no cryopreservation regimen has been developed. Our results contradict the report by Cazin and colleagues (1989) on preservation methods. In addition, cultures of these forms tend to loose viability relatively quickly and are best subcultured every 3 months. Although the fungi cultured by the lower attines do well at low temperatures during long-term storage, care should be taken to ensure their viability.

Particular care must be taken when isolating pellets of the yeast cultured by some *Cyphomyrmex* species. In this case, the 1–4-mm diameter pellets develop short (1 mm) hyphal tips but stop growing within 48 hours of plating. If left untouched, yeasts in these pellets will not grow or divide. They can, however, be broken into many smaller pellets that grow when they are strewn over the isolation plate. After repeated subculturing, these yeasts return to mycelial growth. It is difficult to determine when isolation of a fungus into culture has been successful. The only attine fungi that form diagnostic clamp connections are those associated with *Apterostigma* species. With some practice, however, researchers can distinguish the basidiomycetes cultivated by the lower attines from those of the higher attines based on characteristic features of the culture.

Fungi other than the host symbionts are often associated with attine fungi. This observation has been used to suggest that attine nests in fact harbor mixed cultures of a great variety of organisms, including yeasts and bacteria (e.g., Goetsch and Stoppel 1940; Hölldobler and Wilson 1990). In our opinion, the observation is misleading. Attine nests usually are maintained under axenic or quasiaxenic conditions, which means that a nest is formed mostly of the mycelium of a single basidiomycete strain. Indeed, fungi other than the basidiomycete component do seem to be truly active in the attine nest, but they are uncommon and of only two types. First, endophytic fungi, such as xylariaceous species, can be found, which probably are brought into the nest in the leaves used as substrata. Second, a small group of presumably highly specialized fungi are parasites of the attine fungi. *Escovopsis*, a rather recently described genus, although restricted to that substratum, is rarely isolated (Seifert et al. 1995).

Preparation and Deposition of Vouchers

Given the difficulty of maintaining ant-associated fungi in artificial culture, collections of viable cultures are rare and easily lost. A relatively large culture collection obtained by N. A. Weber in the 1960s and 1970s has now all but disappeared from the New York Botanical Garden (Hervey et al. 1977). A more recent collection of live nests encompassing most of the phylogenetic and biogeographic breadth of the attine symbiosis has been used to establish multiple culture collections of these fungi. At present, duplicates of the collection are maintained at Cornell University and at the National Fungus Collection (Systematic Botany and Mycology Laboratory) (see Appendix III). In the future the cultures will be maintained in a dried, nonviable state. Living cultures from these collections will be maintained in the Department of Entomology, United States Museum of Natural History, Smithsonian Institution (c/o Dr. T. R. Schultz) and the Department of Environmental Science, Policy, and Management at the University of California, Berkeley (c/o Dr. I. H. Chapela). Additional live cultures can be deposited at either of these places. With a growing need to perform molecular determinations and systematic studies from cultures, dried colonies and extracted DNA samples of each specimen should be sent in addition to live cultures.

Identification

The nomenclature of attine-associated fungi is still in a very fluid state. The only applicable binomial is *Leucoagaricus gongylophorus*, which refers to the fungus cultured by higher attines. That fungus is most probably a single species in which all representatives may be clonal descendants of a common ancestor (Chapela et al. 1994; Fisher et al. 1994). Fungi associated with lower attines eventually produce basidiocarps in culture, which are consistent with those of *Leucoagaricus* species or *Leucocoprinus* species, although specific associations are still unknown. Each basidiocarp reported appears distinct at the species level based on traditional morphological characters. Until all the fungi are better known, a conservative approach is to refer to each isolate by the name of the ant species that cultured it, together with an exact record of the location and date.

TERMITES AND FUNGI

Termite gardening symbioses are restricted to old-world higher termites, primarily of the tribe Macrotermitinae of the derived family Termitidae. A single report of a basidiomycete from the nest of another group of higher termites in Costa Rica, if verified, will be an important host and geographic record (Gómez 1994). The Macrotermitinae tribe occupies large regions of tropical Africa, and its range extends eastward to southeastern Asia. The nests built by these termites can be large structures that extend above ground to dominate local landscapes, and the termites often interfere with agriculture by attacking crops.

Within the nests the termites construct a "comb" of fecal pellets on which the fungus grows (Sands 1969); it is the combs that are referred to as the fungus garden. The fungal associates are species of *Termitomyces*, a genus of the same basidiomycete family, Lepiotaceae, associated with most attine ants. Most species of termite-associated fungi form basidiocarps annually, at the onset of the rainy season. The fungal inoculum of some, but not all, species is dispersed in the guts of alates (winged workers that establish new nests and become reproductives). Workers of some species also have been observed foraging on basidiocarps, but the role of this behavior in dispersal is unknown. Because the methods used in the study of these fungi are similar to those used with attine-ant associates, we will discuss them only briefly here. The topic is also reviewed elsewhere (Sands 1969; Batra and Batra 1979; Wood and Thomas 1989).

Collection

Collection of basidiocarps depends on being at the termite nest at the right time; however, local people may compete with researchers for the mushrooms, which are considered by many to be choice edibles (Sands 1969). The fungus also can be collected from the combs themselves, where spherules made up of masses of conidiophores and conidia develop. However, excavation of the nest to reveal the combs may take several days. Insects are preserved in 70% alcohol. Combs can be carried to the laboratory in plastic bags to avoid desiccation, but they should be studied immediately on arrival.

Cultivation

Cultures can be established from basidiospores or from the combs themselves. Although Batra and Batra (1979) were unable to germinate conidia from spherules, yeast-like cells associated with them did germinate in culture. A number of commonly used media will support growth of most species (e.g., oatmeal flakes and Sabouraud glucose, malt extract, and yeast extract agars).

Culture of *Termitomyces* species from older combs in nests or from combs that have been removed from a nest is difficult because the combs become overgrown by saprobic fungi. Apparently termite secretions and nest conditions such as high CO_2 levels deter growth of the saprobic fungi (Wood and Thomas 1989). Older combs are the source, therefore, of a number of interesting species that do not appear to have a specialized relationship with termites, except possibly to be dispersed by them or associated termitophiles. One saprobic genus,

Xylaria, is regularly associated in combs with *Termitomyces*, but the relationship between these fungi has not been determined. Both fungi can grow in the same combs even in the absence of termites if the nest remains intact; however, *Xylaria* quickly overgrows *Termitomyces* when the comb is removed from a nest (Batra and Batra 1979).

Identification

Species of *Termitomyces* and termite-nest-inhabiting *Xylaria* are listed and discussed by Batra and Batra (1979).

WOOD WASPS AND FUNGI

Several genera of resupinate aphyllophoralean wood-decaying basidiomycetes have specific interactions with wood wasps of the family Siricidae, but some of the fungi may be free-living. Apparently, the wasps require the fungi for their nutrition, but the fungi do not depend on the wasps, even for dispersal; they are air-dispersed when not associated with the insects. Although the wasps usually invade conifers, one species is associated with hardwoods in North America. The wasps generally invade dead wood or dying trees; an exception is the wasp *Sirex noctilio*, which invades living trees and is associated with the fungus *Amylosterium areolatum*. Both species have been introduced into New Zealand and Australia, where the fungus has become a serious pathogen of *Pinus radiata*, yet another introduced species. Wasp-associated fungi may cause die back when they are introduced into trees already infected with certain stain fungi associated with bark beetles (Redfern 1989).

The symbiont produces yeastlike cells called oidia in mycangia (hypopleural organs) usually located in a fold at the end of the first abdominal segment of adult female wasps. The mycangia pits are deep and filled with fungal hyphae, oidia, and an oily fluid produced by the wasp. By the time of egg-laying the fluid has solidified to hold the oidia in platelets that are moved by contractions of the ovipositor to intersegmental pouches located at the base of the ovipositor. The platelets dissolve in the environment of the intersegmental pouches, and the oidia are freed to produce new cells that fill the pouches. Oidia are deposited with the eggs (Madden and Coutts 1979; Gilbertson 1984).

Collection and Cultivation

The basidiocarps can be collected directly from wood in the manner of many wood-decaying species (see Chapters 8 and 11 in this volume). In addition to collection and incubation of decayed wood associated with wasp borings, is it possible to obtain cultures by allowing basidiospores to be ejected to an agar surface from a suspended hymenium or, in some species, from conidia. The fungi also can be isolated directly from the intersegmental pouches of female wasps. The Center for Forest Mycology, USDA Forest Service (Appendix III) maintains a large number of cultures of wood-decaying fungi and the voucher specimens from which each culture was derived.

Identification

Identification of these fungi usually is based on the identification of their wasp associates. Only certain species of *Amylosterium* (Stereaceae) and *Cerrena unicolor* (Polyporaceae) have been found in association with wood wasps. *Cerrena* is associated with *Tremex* species in *Fagus* species. All other wood wasps that harbor fungi (*Sirex* and *Urocerus* species) are associated with conifers. *Xeris*, which also occurs in conifers, does not have fungal associates. Degrees of interfertility have been used as criteria to establish that some wood wasp associates are conspecific with free-living species. These fungi are difficult to identify; Gilbertson and Ryvarden (1986, 1987) and Ginns and Lefebvre (1993) have provided information helpful for identification and included a wealth of references. Some wood-decaying fungi also can be identified from cultures (Nakasone 1990).

TECHNIQUES FOR SAMPLING FUNGI INVOLVED IN SPORE-DISPERSAL INTERACTIONS

In addition to the fungi already discussed, many free-living fungi and lichens that are associated with arthropods in their usual terrestrial habitats may be dispersed by arthropods. Some of the associations are obligate; others occur when arthropods pick up the fungi and lichens on which they feed, when arthropods actively camouflage themselves with fungi and lichens, or when arthropods casually encounter fungal propagules in their environments. Dispersal may be on the surface of the arthropod or via gut passage.

In more casual dispersal associations spores may be caught and carried among the setae of arthropods, or they may adhere to smooth surfaces. Because many spores have gelatinous outer parts or simply may be damp, they occasionally adhere to the outer surfaces of mites and insects and coincidentally are dispersed. In fact, spores from genera such as *Scolecobasidium*, *Sordaria*, *Penicillium*, and *Coprinus* can be observed on arthropods from nature; such observations may provide insight into the functions of ephemeral sporulation structures that are present in some fungi.

Arthropods that routinely use fungi as a nutritional resource, grazing on hyphae or spore-containing structures, may disperse the food source as they themselves disperse to new habitats (Hammond and Lawrence 1989). Although many of the insects are not confined to specific fungi, evidence indicates that they avoid particular fungi.

Bees and other insects that usually visit flowers actively collect fungal spores on occasion. Such behavior is precipitated by the fungus. The crucifer rusts, such as *Puccinia thlaspeos* and *P. monoica*, inhibit flowering of their hosts and cause "pseudoflowers" to develop from host leaves. The physiologically and morphologically transformed leaves have the color and nectar production of flowers (Roy 1993, 2001; Roy et al. 1998) and are highly successful in attracting various types of pollinating insects (Fig. 18.12), which may affect the reproductive success of neighboring plants. Bees also collect spores of *Neurospora*, rusts, and smuts (Alexopoulos et al. 1996).

Although a number of insects rely on free-living yeasts for food, the phenomenon is poorly studied. Many insects, for example, are drawn to fermenting substrata such as fruit or slime fluxes that are rich in a variety of yeasts. The yeasts may not strictly require specific fungal species, and the associations are not gardening symbioses; they may, however, approach that condition. Interactions between cactophilic and other plant yeasts and *Drosophila* have been studied extensively in this regard (Phaff and Starmer 1987).

Myxomycete sporangia often harbor spore- or plasmodium-eating beetles, flies, and mites. In fact, beetles in three or more families are referred to as slime-mold beetles. The surfaces of beetles in some species resemble an egg carton with spore-size depressions. Slime-mold dispersers may also carry yeasts that serve as food for somatic stages of the slime molds (Blackwell 1984; Madelin 1984; Wheeler 1984). Some yeast cells remain viable after gut passage in insects and mites. The effects of these dispersal relationships on a group with spores that also are wind-dispersed has not been determined.

Some fungi are entirely dependent on arthropods for dispersal of diaspores, often to ephemeral substrata. In some cases mites are essential in the dispersal association because some of the fungi sporulate deep within beetle galleries, and the beetles may undergo their last molt in the outer bark of the tree losing previously accumulated spores (see "Bark Beetles and Fungi," earlier). In addition to *Ophiostoma* and *Ceratocystis*, other filamentous ascomycete genera such as *Melanospora*, *Kathistes*, *Subbaromyces*, and *Sphaeronemella* may depend completely on arthropod dispersers, as suggested by their long-necked ophiostomatoid perithecia. Also included in this group is *Pyxidiophora*, which sometimes is assigned to the Laboulbeniales and which has a well-developed ascospore attachment region that may penetrate the

FIGURE 18.12 Tachid fly on a rust-infected *Arabis denissa*, W.S.C., Gunnison, Colorado. (Copyrighted and photo by B.A. Roy)

arthropod with an haustorium (Eriksson and Hawksworth 1993; Blackwell 1994). The ascospore develops into a few-celled, conidium-producing anamorph, *Thaxteriola* (Blackwell et al. 1986; Blackwell and Malloch 1989; Blackwell 1994). Within the substratum, *Pyxidiophora* often is a contact mycoparasite with some host specificity (Malloch and Blackwell 1993a). It is interesting that several of the other ascomycete genera with ophiostomatoid morphology and arthropod associations also are known or suspected mycoparasites.

In addition to forcibly dispersed spores, *Basidiobolus* and several members of Entomophthorales produce another diaspore, the capilliconidium, which is specialized for dispersal by arthropods or other invertebrates. At the time of dispersal an attached spore produces a sticky droplet at its distal end by which it adheres to passing objects, breaking from the capilliconidium at a weak place. In *Basidiobolus* the attached spore, which was independently described as *Amphoromorpha*, can cleave internally into a number of ovoid segments; these segments have not been observed to germinate.

COLLECTION

To discover fungi on arthropod surfaces the investigator observes potential substrata from the animal's environment (most easily in the laboratory in moist chambers) or examines arthropod surfaces using a high-powered dissecting microscope or a compound microscope. Some spores can be found by examining insects and mites in collections. This is easiest with mites, which usually are mounted on microscope slides. Such observations of mites led to the discovery of *Pyxidiophora* species that live in beached seaweeds (Malloch and Blackwell 1993a). Other habitats of fungi found on arthropod exoskeletons include dung, bark beetle galleries, mushrooms, and decaying vegetation. Insects, especially beetles and flies, can be caught in traps as they are attracted to or leave substrata. Fungal life cycles usually are synchronized with those of the dispersers, so that mature diaspores are available when the arthropods are ready to disperse.

STORAGE

Many of the spores described here cannot withstand desiccation and do not germinate if potential substrata have been allowed to dry. They should be processed immediately if moist chamber techniques are to be productive. Some of the species in this group sporulate early in the cycle of decomposition and are ephemeral and, thus, are difficult to find without careful daily observation. It is important to judge the time of arrival of dispersers and to collect substrata for moist chambers after that point.

PREPARATION OF SPECIMENS FOR STUDY

Sporulating fungi can be mounted on microscope slides for further study. It is best to do this as soon as the perithecia are observed because they do not persist and are overgrown by other fungi within a day or two. Conidia may be produced for several days before the appearance of perithecia; however, it often is possible to find anamorphs associated with the perithecia, and these should be included in the slide mount. Arthropods collected from nature or moist chambers can be mounted on microscope slides in glycerine jelly or other mountant. Because large numbers of arthropods may need to be examined, permanent slides usually are made only after suitable spore-bearing material is discovered.

CULTIVATION AND DEPOSITION OF CULTURES

Culture of some of these species is routine, requiring only transfer to common media such as cornmeal agar or MEA. However, because many of these fungi grow only (or best) with their arthropod host, the host may need to be included. That is difficult when the host is not known, but sometimes transfer of material growing in the vicinity of the fungus of interest can result in the discovery of the host and result in growth and sporulation of the fungus.

Transfer of sticky spores is easy with the coverslip shard technique. This method consists of breaking a cover glass and picking out pieces in the shape of isosceles triangles using fine forceps that have been heat sterilized and cooled. The base of the shard is held tightly (but not so that it will crack) in the forceps and flamed briefly. When it is cool, the shard can be used to pick up a spore by touching the spore mass held at the tip of the ophiostomatoid perithecium; the spore-containing shard is placed spore surface down on an agar plate. The advantage of this method is that it allows the investigator to observe the spores through the cover glass shard with a compound microscope to check for germination and contamination. The initial stages of mycoparasitic relationships also can be observed if a spore or hypha of a host is transferred. Some of these species are uncommon in culture collections because they may not grow vigorously without a host and may not sporulate in pure culture. Such fungi can be deposited as pure or two-membered cultures in general collections. Lyophilization has not been tried widely, but it might be effective for sporulating cultures.

PREPARATION AND DEPOSITION OF VOUCHERS

Permanent microscope-slide mounts are required for voucher specimens. As we mentioned earlier, it is useful to include anamorphs that often are present several days before sexual states develop. It also is helpful to include other material from the substratum that might provide a hint to the fungal host of the mycoparasites and, for some species, to ensure that the specimen is complete. *Kathistes* species, for example, have a distinctive structure, the sporidioma, at the base of the perithecium that may be missed with a clean specimen. Type material is an exception and should not contain other species. Finally, the dispersal stage, often a spore found attached to an arthropod, should be mounted as well. These fungi should be of interest to curators of many herbaria; some arthropod-associated stages are at the Farlow Herbarium.

IDENTIFICATION

Ophiostomatoid ascomycetes are usually identified on the basis of sexual structures, whose descriptions are scattered through the literature. A key to genera and literature references are available (Malloch and Blackwell 1992; 1993b).

MOLECULAR METHODS

The need for nucleic-acid–derived taxonomic characters is especially great for arthropod-associated fungi in which morphological convergence and lack of known teleomorphs frequently obscure phylogenetic relationships. Moreover, because of the sensitivity of the polymerase chain reaction (PCR), molecular characters potentially may be gleaned from minute quantities of and/or nonculturable types of entomogenous fungi (Blackwell 1994).

Fungal molecular systematics generally has relied on PCR amplification and characterization (restriction digestion, sequencing) of discrete genomic DNA targets. Those methods are beyond the scope of the present review, and several excellent technical manuals cover such techniques (Hillis and Moritz 1990; Innis et al. 1990; Weising et al. 1995). Our aim is to provide a cursory overview of how a collector of arthropod-

associated fungi should process material for molecular studies.

Specimens that are amenable to culture present the fewest problems for molecular analysis, and long-term storage of cultures in itself safeguards material for DNA extraction. Blackwell and Chapman (1993) provided an overview of methods available for storage of cultures, and culture techniques for particular groups of entomogenous fungi can be found elsewhere in this chapter. Generally, for molecular studies, isolates are grown under conditions optimal for rapidly generating biomass. For mycoparasitic, insect-dispersed fungi, such as some *Pyxidiophora* species, which must be cultured with a fungal host, culture conditions are manipulated to favor sexual reproduction; ascospores extruded from the long-necked perithecia are collected on glass shards and used as a pure source of DNA (K. G. Jones and M. Blackwell unpublished data).

Mycological herbaria are also a valuable resource for material for molecular studies. Preparation methods for freshly collected specimens that will best preserve DNA have been reviewed by Haines and Cooper (1993). Such methods are applicable to genera such as *Cordyceps* that routinely are collected as macroscopic fruiting bodies.

Extraction of DNA from nonculturable fungi (e.g., Laboulbeniales, many Trichomycetes, some endosymbionts) is difficult because of the minute sizes of the organisms and the possibility of contamination by DNA of the host or other organisms from the environment. The following observations, based on our experience with extraction and amplification of 18S rDNA from fungi with minute thalli (K. G. Jones and M. Blackwell, unpublished data), are useful for DNA extraction from ectosymbiotic fungi: (1) Infested insects should be scrutinized under a microscope to assess gross contamination by other fungal material. (2) Either whole insects or excised, thallus-bearing body parts can be stored for extended periods at −20°C prior to extraction. (3) For DNA extraction, three to five thalli are crushed dry between silanized microscope slides; the efficacy of that homogenization is monitored microscopically. (4) Homogenate is collected in sterile water and utilized directly in PCR reactions with primers that discriminate between insect and ascomycete DNA templates. Occasionally, coamplification of DNA from, for example, contaminating insect-borne yeasts, occurs. One strategy used to alleviate this problem is to clone the PCR products and then screen the clones for heterologous inserts. As more recalcitrant taxa are sequenced, it will become feasible to design amplification primers specific to particular entomogenous fungal groups.

Isopycnic density centrifugation has been used to purify yeastlike endosymbionts of leafhoppers, but this approach requires gram-quantities of host material

(Noda and Omura 1992). Techniques of *in situ* PCR, in which DNA amplification is performed on 5–10 µm sections of fixed, embedded material (Wright and Manos 1990; Sarkar et al. 1993), appear to have great potential for molecular analysis of endosymbiotic fungi. The development of fixation schedules that both preserve fine-structure and retain nucleic-acid accessibility (e.g., see McFadden et al. 1988) suggests that material ulti-mately may be prepared for both structural and *in situ* PCR studies. A recently published volume devoted to *in situ* PCR techniques should prove a useful starting point for the exploration of this methodology (Gosden 1997).

ACKNOWLEDGMENTS. M. Blackwell acknowledges the support of NSF (DEB-0072741).

19

FUNGAL PARASITES AND PREDATORS OF ROTIFERS, NEMATODES, AND OTHER INVERTEBRATES

GEORGE L. BARRON

A diversity of fungi make up a guild of fungal species that obtain part or all of their nutrition by attacking and consuming living microscopic invertebrates such as nematodes and rotifers. The existence of fungi using that nutritional mode first was revealed in the detailed studies of Charles Drechsler (1933, 1941a, 1941b, 1941c). Drechsler described fungi that attack nematodes and other microfauna by means of trapping devices. More than 300 recognized species of nematode- and rotifer-destroying fungi, belonging to the Ascomycota, Basidiomycota, Chytridiomycota, and Zygomycota, as well as the fungus-like Phylum Oomycota are now recognized.

Fungal parasites and predators of microfauna have three fundamental modes of existence: predation, parasitism of adult invertebrates, and parasitism of invertebrate eggs and cysts. Predatory species form extensive mycelial networks and capture many animal prey per mycelial individual. Organs of capture may include adhesive knobs or nets and constricting or nonconstricting rings (Barron 1977; Gray 1987). Once an animal is captured, the capture organ produces one or more penetration pegs that pierce the animal's cuticle and then germinate, forming digestive hyphae within the body. Those species generally produce noninfective conidia on their external mycelium.

Species that are known parasites (including endoparasites) form, at most, extremely limited mycelia external to their animal hosts, but they produce infective conidia or zoospores that adhere to the surface of the host or are ingested by it. Each spore is capable of attacking only one animal, so one mycelial individual is limited to the resources of a single nematode, rotifer, or other invertebrate. Infective conidia germinate on or in the nematode, rotifer, or other host to form digestive hyphae within the body. Once the host is completely colonized and digested, hyphae reemerge to produce another crop of infective conidia. Species of Chytridiomycota and Oomycota produce infective spores that eventually form a thallus within the host animal's body; a new generation of zoospores forms within the thallus and eventually is released to the outside. A third group of microfauna-destroying fungi are the egg and cyst parasites that also attack their hosts by means of infective conidia or zoospores (Carris and Glawe 1989). Taxonomically, parasites of nematode eggs and cysts are a different group of fungi from the predators and parasites of adult nematodes, and specialized techniques are required for their recovery. The former group is not covered in this chapter, but an introduction to those fungi can be found in Carris and Glawe (1989).

FUNGAL ENDOPARASITES AND PREDATORS OF BDELLOID ROTIFERS

Bdelloid rotifers (Phylum Rotifera, Class Bdelloidea), or simply bdelloids, are microscopic aquatic animals that are ubiquitous in soil, manure piles, rotting wood, decaying leaves, compost piles, old bracket fungi, living and dead mosses or lichens, and organic debris of all types. They are also common in almost all freshwater habitats and in some brackish water and marine habitats. Unlike most rotifer species, bdelloids have the ability to retract and extend their bodies at will. In unfavorable circumstances, bdelloids retract their bodies to form globose to subglobose cysts. In that dormant state they can survive adverse conditions for prolonged periods. Within minutes of the return of favorable conditions, the rotifers extend their bodies from the encysted state and resume activity. Bdelloids consume bacteria, fungal spores, and probably other microscopic life forms. They are distributed globally and are available as prey or hosts for a multitude of predators and parasites (Figs. 19.1 and 19.2); more than 60 species of fungi are known to attack bdelloids as either endoparasites or predators. Few attempts have been made to catalogue the parasites of rotifers either on a regional or global scale; the true number of species is probably in the hundreds.

To date, the most effective technique for recovering parasites of bdelloids is to use a host as bait (Barron 1985a). A small sample of soil or organic debris is sprinkled over a plate containing a large number (100–10,000) of rotifers. If a parasitic fungus is present in the sprinkled sample, it will attack the bait. In 3–5 days, hyphae from the fungus causing an initial infection will break out through the cuticle of the host and sporulate. Those spores will initiate another cycle, and in 3–5 days numerous additional rotifers will be infected. At this time, the quality and quantity of the fungal material should be sufficient to make mounts for identification and for voucher specimens and to obtain the fungus in pure culture. Transmission and amplification of infection does not always occur, however, and it is sometimes necessary to find and recover a single infected rotifer in a Petri dish by inspection under the low (35–100×) power of a compound microscope. Although scanning an entire Petri dish under a compound microscope is time consuming and tedious, it may be the only way to find a rare species of fungus.

The initial steps in using an animal host as bait are to isolate the animal from its natural home and to develop culture methods that will maintain a stable stock of the host and produce large populations on demand.

OBTAINING ROTIFERS FOR BAIT

A gram of soil or debris is sprinkled on a water agar (2% agar) plate, and 1–2 ml of water is added. The material is mixed by swirling gently and is incubated for 5–7 days. Bdelloid rotifers are located by examining the plate under a dissecting microscope or the low power of a compound microscope. Because these rotifers reproduce parthenogenetically, one rotifer is enough to start a colony. When a rotifer is located, it is aspirated with a Pasteur pipette and placed in a drop of water on a fresh water-agar plate. The "rotifer" drop is irrigated with several drops of sterile water to separate it from debris and contaminating microfauna. The bdelloid then is transferred to a second drop of water on the plate and irrigated again. Irrigation and transfer are repeated several times. Finally, the washed rotifer is placed on a plate of water agar along with a drop of liquid 2% malt-extract agar or Czapek's agar minus sucrose (0% Cz) and enough sterile water to produce a thin film (1–3 mm) over the surface of the water agar. The plate is incubated at room temperature (20–25°C). If bacteria are present, the water becomes cloudy after a few days. Five to 10 ml of sterile distilled water can be added to the plate and mixed thoroughly with the water already present; excess liquid then is drawn off with a Pasteur pipette. This procedure is repeated until the water is clear. Enough nutrient energy is available in 0% Cz to support bacterial growth for a large population of bdelloids. The purpose

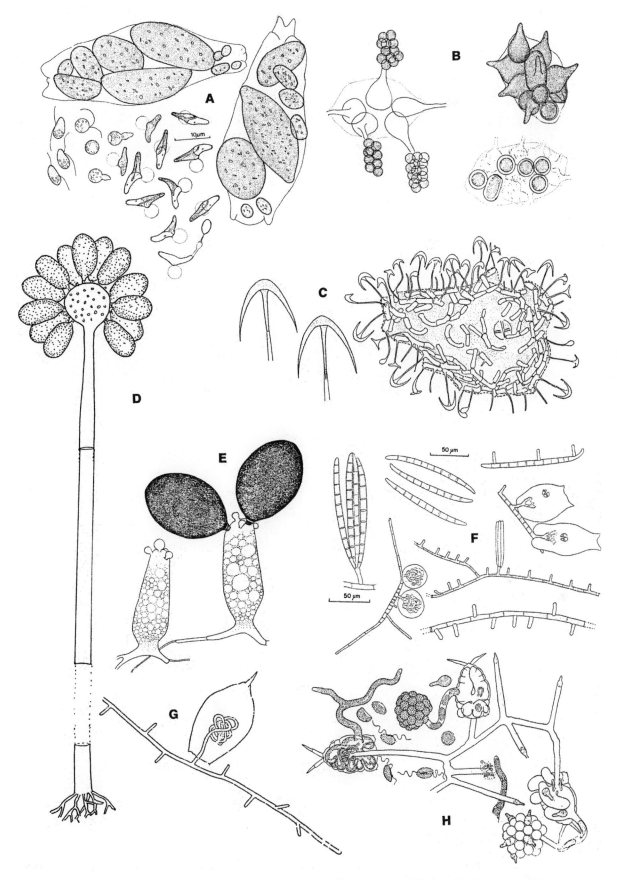

FIGURE 19.1 Fungal parasites and predators of rotifers. **A.** *Haptoglossa humicola* (from Barron 1981b). **B.** *Lagenidium oviparasiticum* (from Barron 1967). **C.** *Triacutus subcuticularis* (from Barron and Tzean 1981). **D.** *Rhopalomyces magnus* (modified from Drechsler 1955). **E.** *Brachymyces megalosporus* (from Barron 1980d). **F.** *Cephaliophora muscicola* (from Barron et al. 1990). **G.** *Zoophagus insidians* (from Prowse 1954). **H.** *Sommerstorffia spinosa* (from Karling 1952).

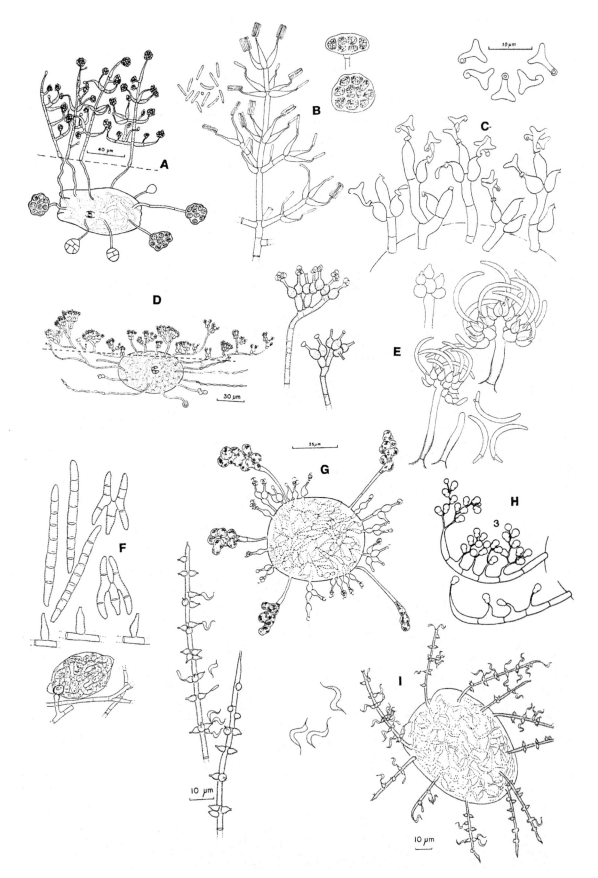

FIGURE 19.2 Fungal parasites and predators of rotifers. **A.** *Rotiferophthora ovispora* (from Barron 1980a). **B.** *Rotiferophthora cylindrospora* (from Barron 1985a). **C.** *Haptospora appendiculata* (from Barron 1991a). **D.** *Tolyplocladium lignicola* (from Barron 1983). **E.** *Medusamyces lunulosporus* (from Barron and Szijarto 1990). **F.** *Dwayaangam heterospora* (from Barron 1991b). **G.** *Culicinomyces parasiticus* (from Barron 1980b). **H.** *Pseudomeria mucosa* (from Barron 1980c). **I.** *Harposporium spirosporum* (from Barron 1986).

of irrigation in the early stages is to prevent bacterial buildup from overwhelming a solitary rotifer or a small population of them. After the rotifer population increases, its grazing will control the bacteria. Populations can be started from a single rotifer, but use of 10 or more is advantageous. The chance of a culture becoming contaminated, however, increases with the seed number. If the rotifers die or fail to multiply, the water agar is covered with a thin film of sterile pond water (rather than sterile plain water) for incubating rotifer cultures. Sterile pond water may simulate the ionic strength and pH of the rotifer's natural aquatic environment better.

The rotifer population builds up slowly at first. After several weeks, however, it is large enough to be divided among several plates. Within 1 or 2 months, flourishing cultures of rotifers should be available. Rotifers are maintained by weekly transfers to fresh plates of 0% Cz. If necessary, stock cultures can be prepared for long-term storage. Rotifers are poured onto sterile soil (autoclaved, cooled, and autoclaved again) in small culture bottles. The soil cultures are allowed to grow for a week or two then slowly dried. Bdelloid rotifers in dry sterile soil can be stored for long periods (years) at room temperature or under refrigeration.

Several single-rotifer cultures should be started at the same time because other microorganisms, including flagellates, ciliates, and amoebae, may be carried over with the initial rotifer in the pipette. Populations of those contaminating organisms can build up much faster than those of rotifers and can cause problems later. Amoebae consume the spores of endoparasites; flagellates and ciliates compete with rotifers for food and may interfere with their reproduction. Contaminated cultures, therefore, should be discarded immediately. One clean culture is enough to establish a rotifer stock for baiting purposes.

Rotifers also can be collected for culturing or baiting with a Baermann funnel system, which provides a sample comparatively free of debris. The system, which was developed initially for the collection of nematodes, is explained in detail later in this chapter (see "Fungal Endoparasites and Predators of Nematodes").

Fungal Endoparasites

Collection

About 10 ml of water is added to a stock culture of rotifers in a Petri dish. The mixture is swirled gently, and 2 ml of suspension is poured into each of five plates. Additional water may be added as water evaporates. Rotifer stocks are incubated for 3 days at room temperature to produce active populations for bait. The rotifer

bait cultures are seeded by sprinkling from 0.1 to 1.0 g of soil or organic debris over the plate. The mixture is swirled gently, but clear areas are left on the plate to facilitate future microscopic observations. Rotifer bait cultures also can be seeded by taking a larger sample of original material, preparing a slurry in sterile pond water, and adding a few drops of this slurry to the bait plates. Seeded plates are incubated at room temperature and examined for signs of parasitized rotifers, daily, under the low power of a compound microscope or a high-magnification dissecting microscope.

When bdelloid rotifers are attacked by fungi, they become immobilized and encyst quickly, assuming a spherical to ellipsoid shape. The parasite continues to grow inside the body of the encysted rotifer, and eventually hyphae break out at a number of points over its surface. At this stage, infected rotifers can be distinguished under the microscope from encysted, noninfected rotifers. A parasitized rotifer is recovered from a Petri plate (under the compound microscope) using a Pasteur pipette (carrying over as little debris as possible). The contents of the pipette (i.e., one to several drops of water or slurry), including the rotifer, are deposited onto water agar. The parasitized rotifer then must be located using a dissecting microscope. When located, the rotifer is picked up on a needle with a flattened tip and transferred to a drop of clean water. The rotifer then is irrigated with water and transferred to a fresh water drop.

If the infected rotifer is placed on a water-agar plate, conidia are produced within 1 or 2 days. Fresh rotifers now can be added to the plate, and an epidemic of infections will be initiated. Alternatively, the rotifer can be used to start a pure culture of the fungus (see "Culturing Parasites," later in this chapter).

The largest number of fungal species probably is recovered for the least investment of time and resources using the baiting technique. Most of the described species of parasites of bdelloid rotifers, however, have been recovered using a single strain of host. If different rotifer species or an assortment of species are used as bait, then the numbers of known species are likely to increase dramatically. Also, if a method were developed to recover rotifers directly from a substratum, it is more likely that host-specific or rare parasites would be recovered. Sometimes rotifers come through the Baermann funnel (see "Obtaining Nematodes," later in this chapter, and Fig. 19.3) in abundance and provide that opportunity.

Rotifers also can be obtained from substrata expected to support high populations of bdelloids (e.g., well-decomposed manure, compost). Substratum samples are used as the source of both the bait and the parasites. The sample material is mixed with water to make a slurry. Several drops of the slurry are transferred to a plate of 0% Czapek's or water agar and incubated. The method

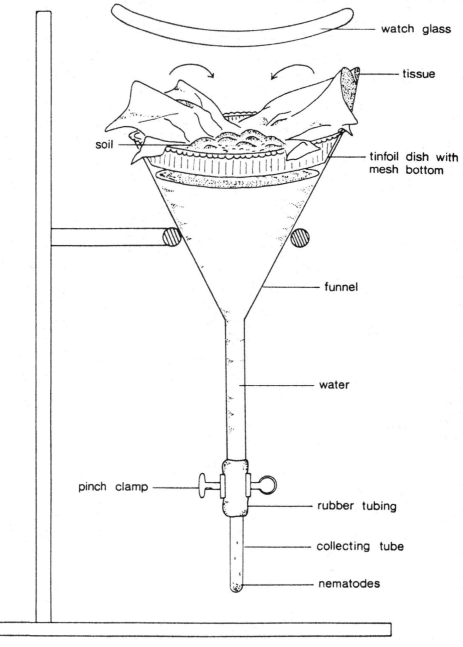

watch glass

tissue

soil

tinfoil dish with mesh bottom

funnel

water

pinch clamp

rubber tubing

collecting tube

nematodes

FIGURE 19.3   Diagram of a Baermann funnel apparatus (see "Obtaining Nematodes," in text).

depends on a natural increase in the population of rotifers already present in the sample and subsequent attack by parasites also present. It works much more slowly than the baiting technique. Preparing the slurry by homogenizing the sample in water improves the chances of recovery. Increasing the volume of material processed for a slurry or baiting suspension enhances recovery of endoparasites. With large samples, however, the common parasites obscure the rarer species. Small samples must be used, therefore, to recover rare endoparasites.

Karling (1946) discovered numerous zoosporic endoparasites of rotifers, nematodes, and other microfauna by simply covering a soil sample with a layer of water and incubating. This method depends on the natural buildup of populations of microorganisms followed by parasitism or predation of their members and eventual population decline. The method requires numerous inspections over long periods. It is slow, difficult to control, and somewhat unpredictable, but if methodically and intensively applied, it can lead to rewarding results.

Routine Identifications

In many cases, a parasite will develop and sporulate on the original baited plate. Infected rotifers can be picked up with a fine needle, mounted in acetic alcohol followed by lactophenol, examined microscopically, and tentatively identified. Lactophenol mounts can be saved and stored for future reference or as permanent records. Establishment of cultures will facilitate identifications further.

Culturing Parasites

An infected rotifer is transferred to clean agar using a fine needle or a micromanipulator. The rotifer is dragged or pushed over or under the agar surface with the tip of a sterile needle or micromanipulator probe to clean off all debris and contaminating organisms. The cleaned rotifer then is transferred to a plate with fresh potato-dextrose agar amended with streptomycin sulfate (50–100 mg/l) and incubated at room temperature (Aschner and Kohn 1958). Within a few days, hyphae from the rotifer grow into the surrounding agar. In an alternative culturing method, a drop of Grace's medium modified by adding fetal bovine serum, streptomycin, and penicillin (Appendix II) is placed on a sterile cover slip (Glockling and Shimazu 1997; Glockling 1998). The infected host animal is lifted from the isolation plate with a fine needle or micromanipulator and placed in the drop. The cover-slip is turned upside-down over a sterile cavity slide so that the drop hangs down from the cover slip. The slide with the hanging drop is incubated inside a humid chamber made from another Petri plate. Growth of the fungal parasite from the nematode is monitored daily with an inverted microscope or with the low-power objective of a compound microscope. Once a mycelium is visible, the slide culture is taken apart, and the mycelium is transferred to new plate of cornmeal agar or other medium (Appendix II).

Many rotifer parasites do not sporulate on nutrient-rich agar. Sporulation is induced by placing a cm² block of agar with mycelium from a pure culture onto water agar or water agar (Appendix II) supplemented with rotifer extract or debris from old rotifer cultures that has been filter-sterilized through a 0.2-μm Millipore filter. After sporulation begins, the plate is flooded with a fresh culture of rotifers. That results in massive simultaneous infections. Lactophenol mounts made at 6-hour intervals preserve the details of colonization of the infected animals as well as details of the parasite life cycle. Finding a dead animal with fungus inside does not necessarily mean that the fungus killed the animal. To prove the pathogenicity of an organism, the conditions of Koch's postulates must be satisfied. In other words, the suspected causal agent (the fungus in this case) (1) must be associated constantly with the disease; (2) must be isolated and grown in pure culture; (3) must result in the original disease when a healthy organisms is inoculated with it; and (4) must be reisolated from the experimentally infected plant. Such studies of the infection process should be a requisite when describing new species.

Species of the genus *Rotiferophthora* (Fig. 19.2 A and B) are extremely common parasites of bdelloid rotifers. Infected rotifers float at or near the surface of the water film. Verticillate conidiophores grow into the air above the body and produce abundant spore masses that are hydrophobic. The conidiophores with conidial masses act like tiny sails, and the infected rotifers float over the surface of the water for prolonged periods. Such infected rotifers can be located easily on the surface of a Petri dish under a dissecting microscope using a low angled light. With a steady hand and a fine needle, conidia can be picked off the aerial conidiophores and transferred to rotifer populations to start new cultures of the host and parasite.

Fungi that attack rotifers complete the parasitic cycle very quickly. A rotifer attacked by an endoparasite may encyst within minutes. In 24 hours much of the body is colonized, and within 48 hours the body is filled with hyphae, which are emerging through the cuticle. In 72 hours, sporulation is initiated on external hyphae. Following sporulation, the dead host and exhausted reproductive apparatus of the parasite degrade rapidly. Thus, cultures must be inspected at least every few days.

PREDATORY FUNGI

Predatory fungi develop an extensive hyphal system in the substratum; the hyphae give rise to trapping devices, either adhesive pegs, nets, or constricting rings, which are used to attract and capture rotifers and nematodes. Preliminary evidence indicates that several species of predatory fungi capture both bdelloid and loricate rotifers. The procedures used to locate and recover fungal predators are similar to those used for endoparasitic species. If a predator is present, it will capture rotifers and anchor itself to the plate by hyphal growth into the agar. Plates that have been used to collect endoparasites can be washed off under running tap water to remove loose debris and then can be inspected under a dissecting microscope for clusters of rotifer bodies. The clusters indicate zones where predation has occurred. Alternatively, debris from ponds, ditches, puddles, and so forth can be transferred to stackable, nested glass dishes and examined periodically under a dissecting scope for evidence of rotifer predators.

AQUATIC ROTIFERS

Few serious studies of parasites of aquatic rotifers have been carried out; not surprisingly, then, no special techniques for the recovery of parasites or predators of aquatic rotifers have been described. An obvious approach is to use a standardized technique for recovery of aquatic invertebrates, such as a sweep net, then to plate out the harvested populations in sterile pond water. Simple, unsophisticated techniques for the recovery of predators and parasites of aquatic rotifers also can be used. Living or dead plant material is harvested from ditches, ponds, and other aquatic habitats; transferred to sterile pond water in stackable, nested culture dishes; incubated; and periodically inspected for the presence of predators or parasites. The levels of parasitic infections in aquatic habitats, including infections by fungi, are likely to be low. It is also likely that fewer hyphomycetous and more zoosporic parasites will be found. In general, densities of all types of spores are higher in soils than in water.

DIVERSITY OF ENDOPARASITES AND PREDATORS

Although some 60 species of fungi are known to attack bdelloid rotifers (Figs. 19.1 and 19.2), few serious attempts have been made to recover and characterize members of this group. Using baiting techniques and a single host species, that number might be extended to 100 species. Developing methods to recover parasites of natural populations of a variety of rotifer species from soil or aquatic environments should increase the number to several hundred.

TAXONOMIC DISTRIBUTION OF GENERA

The genera of fungal predators and parasites of bdelloid rotifers are distributed among the fungal classes, as shown in Table 19.1. A key to those genera is provided in Appendix A at the end of this chapter.

DESCRIBING NEW OR RARE PARASITES

Parasites or predators that are new or rare should be properly described and catalogued for future reference. The following is the minimal protocol for describing such fungi.

1. Ten semipermanent slide mounts of infected rotifers should be prepared by fixing the animals in acetic alcohol and mounting in lactophenol.

TABLE 19.1
Fungal Predators and Parasites of Bdelloid Rotifers (Phylum: Genera)

Chytridiomycota: *Catenaria, Endochytrium, Olpidium, Rhizophydium*
Zygomycota: *Rhopalomyces, Brachymyces, Zoophagus*
Oomycota: *Atkinsiella, Lagenidium, Sommerstorffia, Haptoglossa*
Anamorph fungi: *Rotiferophthora, Harposporium, Haptospora, Pseudomeria, Lecophagus, Cephaliophora, Dwayaangam, Medusamyces, Tolyplocladium, Culicinomyce, Triacutus*

2. A pure culture of the fungus on potato-dextrose agar amended with streptomycin should be obtained, if possible (see "Culturing Parasites," earlier). New strains should be deposited in a public culture collection (see Appendix III).
3. Washed rotifers on which conidiophores (or equivalent spore-bearing structures) are developing are placed into a plate containing a population of bait rotifer to initiate an epidemic.
4. If an epidemic is initiated, then the entire population of parasitized rotifers is fixed in acidified alcohol in a small vial. An equal amount of lactophenol is added, the mixture is placed on a slide, and the cover glass is sealed. The preparation can serve as the type material or voucher for future studies.
5. During this initial period of study, the distinctive features of the fungus and its culture are recorded as diagrams, measurements, and photomicrographs. Publications should be illustrated with diagrams and photomicrographs.

FUNGAL ENDOPARASITES AND PREDATORS OF NEMATODES

Endoparasitic fungi (e.g., Figs. 19.4 and 19.5) attack nematodes by means of spores that adhere to the cuticle or are ingested by the animal as food. Once attached or ingested, the spores germinate, and hyphae proliferate inside the body of the host. All assimilative stages of the endoparasites are contained entirely within the body of the host. Only reproductive or fertile hyphae extend to the exterior of the animal and proliferate. Nematodes attacked by an endoparasite may be quite active for a day or two following initial infection, during which time they move and feed in a more or less normal fashion. At any given time, only a small percentage of the individuals in a population of nematodes will be infected by an endoparasitic fungus. The best approach for a collector, therefore, is to recover a large number of nematodes

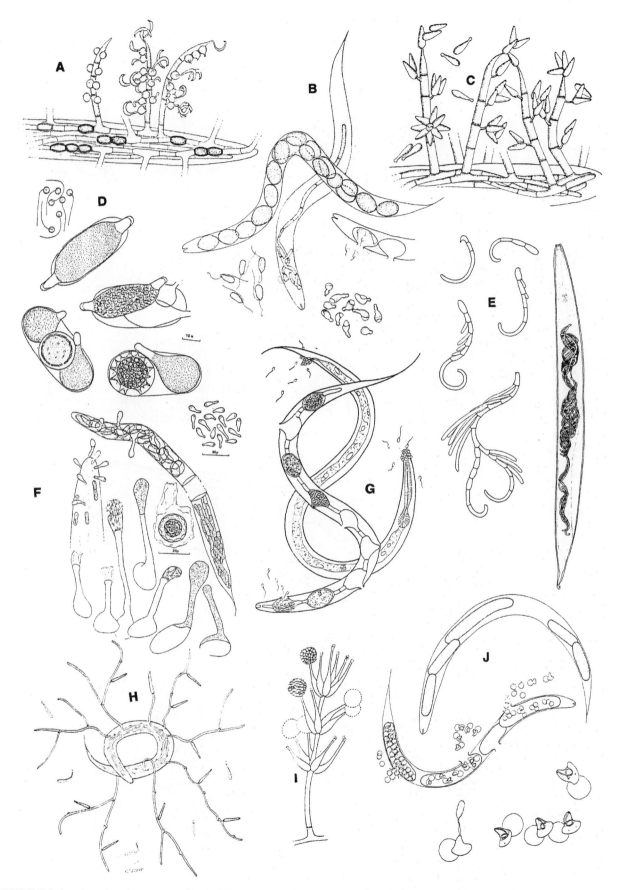

FIGURE 19.4 Fungal endoparasites of nematodes. **A.** *Harposporium anguillulae* (from Barron 1977). **B.** *Myzocytium humicola* (from Barron 1977). **C.** *Drechmeria coniospora* (from Barron 1977). **D.** *Olpidium vermicola* (from Barron and Szijarto 1986). **E.** *Spirogyromyces vermicola* (from Tzean and Barron 1981). **F.** *Gonimochaete lignicola* (from Barron 1985b). **G.** *Catenaria anguillulae* (from Barron 1977). **H.** *Nematoctonus leiosporus* (from Barron 1977). **I.** *Verticillium* species (from Barron 1977). **J.** *Haptoglossa heterospora* (from Barron 1977).

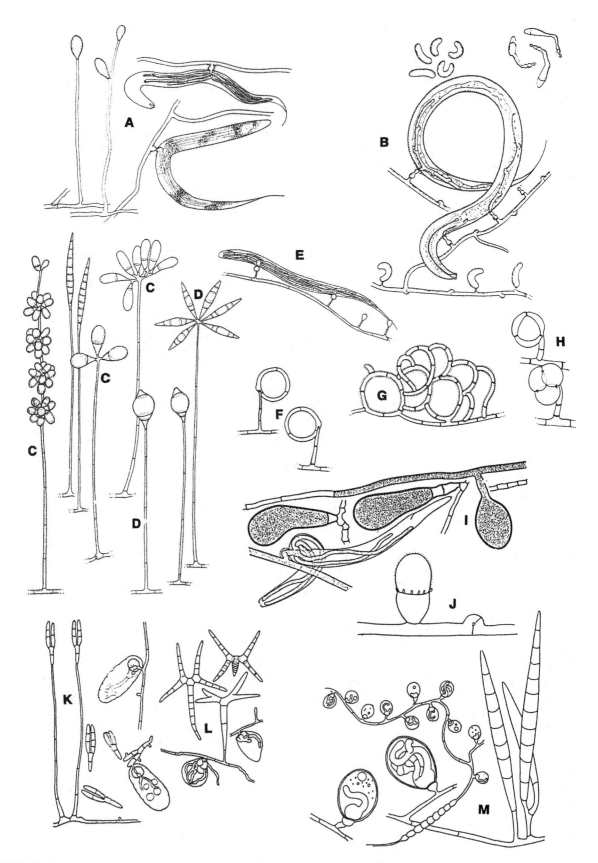

FIGURE 19.5 Fungal predators of nematodes and amoebae. **A.** *Stylopage hadra* (from Barron 1977). **B.** *Nematoctonus* species. (from Barron 1977). **C.** *Arthrobotrys* species. **D.** *Monacrosporium* sensu Subramanian (from Barron 1977). **E.** Adhesive knobs. **F.** Non-constricting rings. **G.** Adhesive net. **H.** Constricting rings. **I.** *Cystopage* (from Drechsler 1941c). **J.** Stephanocyst of *Hyphoderma* species. **K.** *Pedilospora dactylopaga* (from Drechsler 1934). **L.** *Triposporina quadridens* (from Drechsler 1962). **M.** *Dactylella passalopaga* (from Barron 1981a).

from a natural population and to screen them for parasites indigenous to the population.

Obtaining Nematodes

The Baermann funnel technique is an effective method for recovering large numbers of nematodes from soil or organic debris (Giuma and Cooke 1972; Barron 1977; Gray 1984). This technique depends on motile nematodes wriggling down through soil or organic material and passing through facial tissues. A Baermann funnel (Fig. 19.3) apparatus is prepared as follows:

1. The bottom is cut from a 15-cm diameter aluminum foil pie plate.
2. Nylon mesh is cut to fit the hole and is glued to the bottom of the dish.
3. The lip of the pie plate rests on the edge of a 15-cm glass funnel that is supported in a ring on a burette stand.
4. A piece of rubber tubing connects the base of the funnel and a 5-ml collecting tube.
5. From 50 to 250 g of soil or organic debris are placed on a double layer of facial tissue that covers the mesh in the pie plate.
6. Water is added to the funnel until it touches the bottom of the mesh.
7. The pie plate is covered with a large watch glass or an uncut, inverted pie plate to prevent dehydration.

Nematodes wriggle through the tissue and mesh screen and sink by gravity to the bottom of the collecting tube. After 6–12 hours, the rubber tube is closed with a pinch clamp, and the collecting tube is removed. About 4 ml of the supernatant carefully is removed with a Pasteur pipette without disturbing the nematodes at the bottom. The remaining liquid (0.5–1.0 ml) is shaken vigorously to disperse the nematodes and then poured onto the surface of a water-agar plate. Plates are inspected daily for up to 10 days for parasitized nematodes. Nematodes in a collecting tube very quickly become inactive, and if left overnight, they may die before they are harvested. Thus, it is best to process a large sample of soil or debris and empty the collecting tube every few hours. Occasionally, large numbers of bdelloid rotifers or tardigrades accumulate in the collecting tube. Investigators should take the opportunity to examine them for parasitic fungi.

Recovery of Parasites

Either freshly poured water-agar plates or plates that were poured 7 days previously can be used to recover

parasites. The 1 ml of water from the collecting tube forms a liquid film over the surface of a freshly poured plate and will remain liquid for many days. If necessary, water can be added to the plate to maintain the film, or it can be used to irrigate the preparation to control the buildup of bacteria. Free-water habitats favor zoosporic endoparasites such as *Myzocytium*. The 7-day-old agar, in contrast, rapidly absorbs the 1 ml of water from the collecting tube so that no water film remains on the surface. That condition favors hyphomycetous endoparasites such as *Harposporium* and *Verticillium* but prevents development of many of the flagellated species. Bacteria build up very quickly on the agar plates. Nematodes from the collecting tube can be washed several times by low-speed centrifugation before being added to the agar plates to control the bacteria. Alternatively, the nematodes can be added to water on the Petri plate and cleaned daily by irrigation. The first few days after recovery are critical for obtaining a good collection of infected nematodes.

Often, parasitized nematodes carrying fully developed fungal conidiophores and conidia are observed within 48 hours (Fig. 19.6). Transmission of the disease may result in numerous additional infections within a week or so. Semipermanent lactophenol mounts of infected nematodes can be made directly from the original plates for reference purposes.

Many parasites of nematodes are host specific. As a result, in the population of nematodes coming through a funnel, some nematodes may be parasitized by a host-specific fungus. Those fungi will not be transmitted to the surviving nematodes of other species and must be sought diligently because they are the most likely to be

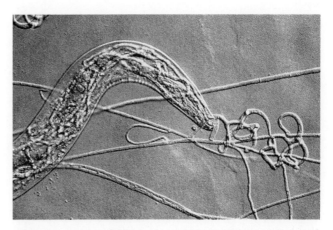

FIGURE 19.6 The body of a nematode that has been penetrated by *Arthrobotrys oligospora*, whose hyphae can be seen proliferating inside the host.

rare or new. That requires careful scrutiny of the nematode harvest for the first 48 hours. Rare or new endoparasites can be given special attention and with care can be obtained in pure culture.

CULTURING PARASITES

Methods to obtain pure cultures of hyphomycetous endoparasites are the same as those for rotifer parasites. Nematodes with early stages of infection, in which hyphae have penetrated the cuticle to the exterior but are not yet producing spores, can be cleaned by aqueous irrigation, dragged across water agar, then plated on a fresh plate of potato-dextrose agar amended with streptomycin (Aschner and Kohn 1958). Cultures can be assessed for growth under the low power of a compound microscope after 24–48 hours.

CULTURING NEMATODES

Some of the Baermann Funnel extracts will not contain parasitized nematodes. If left on water agar, most nematode species die quickly, but some rhabditoids adapt and multiply under Petri plate conditions. In a week or two a population of a single species of nematode will remain. That population can be used as a stock culture of nematodes. Nematodes can be grown easily on a host of suitable substrata, including water agar to which 0.5–1.0 g of oatmeal, cooked potato, soup powder, or peanut butter has been added. A few drops of nematode suspension are added and incubated at room temperature. A pure culture of a single nematode species also can be obtained by collecting a gravid female, washing it, and transferring it to a fresh plate by itself.

TAXONOMIC DISTRIBUTION OF GENERA

The genera of fungal endoparasites of nematodes are distributed among the fungal classes, as shown in Table 19.2. A key to those genera is provided in Appendix B at the end of this chapter. A useful key to the nematode-destroying fungi including both endoparasites and predators and covering all species known at that time can be found in Cooke and Godfrey (1964). The taxonomy of the biflagellate Oomycete parasites of nematodes has been reviewed by Glockling and Beakes (2000). Some of the hyphomycetous genera of endoparasites are diagnosed in Gams (1988, 1997).

TABLE 19.2

Fungal Endoparasites of Nematodes (Phyllum: Genera)

Chytridiomycota: *Catenaria, Endochytrium, Olpidium, Phlyctochytrium, Rhizophydium, Bicricium*
Oomycota: *Gonimochaete, Haptoglossa, Lagenidium, Myzocytium, Protascus, Pythium*
Zygomycota: *Euryancale, Helicocephalum, Meristacrum, Rhopalomyces, Zygnemomyces,* a possible *Conidiobolus* species
Anamorph Fungi: *Botryotrichum, Drechmeria, Haptocara, Harposporium, Hirsutella, Nematoctonus, Plesiospora, Verticillium, Cephalosporiopsis*

RECOVERY OF PREDATORS

A number of methods work well for the recovery of predatory fungi from soil and organic debris (Barron 1977; Gray 1984). For example, predatory fungi can be recovered from nematodes collected using the Baermann funnel technique described previously (see "Obtaining Nematodes," earlier). All species of *Pleurotus* and some species of *Hyphoderma* are predaceous on nematodes (Liou and Tzean 1992; Tzean and Liou 1993; Thorn et al. 2000); nevertheless, those fungi would not be recovered using methods described here. Species of *Pleurotus, Hyphoderma,* and other macroscopic fungi that are putative nematode predators should be collected and cultured using methods outlined in Chapter 8. Once a culture is established, the isolate can be challenged with nematodes by growing it on water agar and adding a few drops of nematode culture.

Drechsler's Technique

The earliest method used to collect predatory fungi was developed by Drechsler (1941b), who first discovered and described many of the predators. Drechsler observed that free-living nematodes multiply abundantly in agar cultures prepared for isolation of plant parasitic Oomycotina from plant roots. He also noted that those nematodes subsequently are attacked by a variety of predatory fungi. Based on those observations, he determined that a pinch of leaf mold, rotting plant tissue, or other organic debris added to a Petri dish of dilute cornmeal agar previously inoculated with a species of *Pythium* or *Phytophthora* would set off a natural chain of events resulting in large populations of all kinds of microscopic life forms including bacteria, nematodes, mites, amoebae, and so forth. He also observed that the dominant animal populations then were attacked by predators and parasites. In fact, this method, when properly applied, yields a greater diversity of parasites and predators of both nematodes and amoebae than any other technique. The drawback of the technique is that the sequence of natural events may take weeks or even months to run its course. Plates must be inspected many times over that incubation period, and contaminating organisms, such as fungus-feeding mites, can affect the results seriously. Thus, although this technique is highly productive for serious research workers, it is not the most appropriate for those who are attempting rapid surveys.

Nematode Baits

Free-living rhabditoid nematodes established in culture can be used as baits for predatory fungi (Eren and Pramer 1965) (Fig. 19.7). Plates containing low-nutrient agar (dilute cornmeal or water agar) are inoculated with a suspension containing several hundred nematodes obtained from a stock culture. At the same time, the plates are sprinkled with 0.1–1.0 g of soil or organic debris, incubated at room temperature, and inspected at intervals. Plates are scanned under a dissecting microscope for aggregations of captured nematodes, which are the first sign of predation and which begin to appear after about 5 days. The predatory fungi begin to sporulate, and erect conidiophores bearing one to many apical conidia can be spotted at various locations on the plate. The dissecting-scope light should be set at a very low angle. The improved contrast picks up even the tiniest hyphal structures growing above the agar. Low-angle light also minimizes radiant heat that will frizzle delicate structures. If possible, a fiberoptic (cold light) light source should be used. Lactophenol mounts can be made of conidiophores, conidia, and trapping

devices for future reference. Pure cultures can be obtained easily by picking conidia off the tips of the conidiophores with a fine needle and transferring them to new plates of nutrient agar. Some predatory fungi grow rapidly but produce few conidiophores. *Arthrobotrys flagrans*, for example, often is missed because of its early appearance and poor reproduction. Also, some predators with nonseptate hyphae cannot be grown in culture or may not sporulate under the conditions in a Petri plate (Fig. 19.8).

TAXONOMIC DISTRIBUTION OF PREDATOR GENERA

The predatory genera of fungi are listed in Table 19.3. A key to those genera is provided in Appendix C at the end of this chapter. Rubner (1996) recently revised the predatory fungi of the *Dactylella-Monacrosporium* complex. He described 39 species in some detail and provided a key to the species. Schenck and colleagues (1977) reviewed the taxonomy of *Arthrobotrys* and related genera and made a number of new combinations (Fig. 19.9). The discovery that fungi of the *Arthrobotrys* complex are anamorphs of species of discomycetes in the

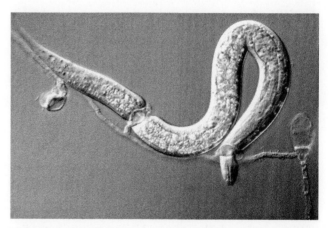

FIGURE 19.8 Nematode caught and held by constricting rings that have arisen on the hypha from a germinating spore (right) of *Arthrobotrys anchonia.*

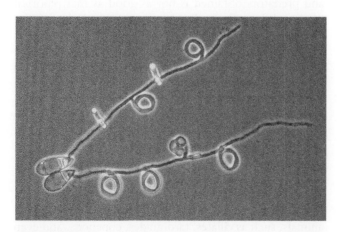

FIGURE 19.7 Nematode in which two spores of the fungus, *Arthrobotrys anchonia,* have germinated. Each spore has produced a germ tube bearing four constricting-ring trapping devices. One of the rings has been triggered and shows the three component cells.

TABLE 19.3
Predatory Fungi (Phylum: Genera)

Zygomycota: *Cystopage, Stylopage, Rhopalomyces*
Ascomycota and Anamorph Fungi: *Arthrobotrys, Dactylaria, Dactylella, Monacrosporium*
Basidiomycota: *Hohenbuehelia, Hyphoderma, Nematoctonus, Pleurotus*

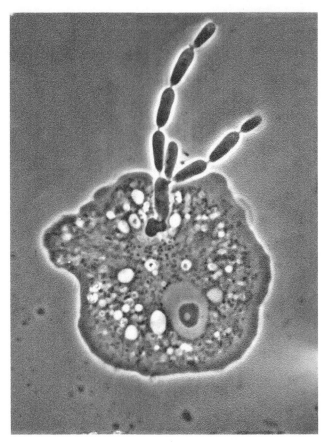

FIGURE 19.10 Nematode that has been attacked and killed by the endoparasitic fungus, *Verticillium* species. The conidiophores are now bursting out through the body of the victim to produce additional spores.

FIGURE 19.9 An amoeba that has been attacked by the fungus *Amoebophilus* species. Only the heart-shaped haustorium is found inside the membrane. Nutrients are extracted from the host through the haustorium, which supplies the branching chains of spores that are produced to the exterior.

family Orbiliaceae has led to an extensive reassessment of the taxonomy of those fungi (Rubner 1996; Scholler et al. 1999). The type of trapping organ has been found to be a reliable morphological criterion for delimiting more natural genera (Scholler et al. 1999). Thorn and Barron (1986) reviewed the genus *Nematoctonus*, providing keys, descriptions, and illustrations.

FUNGAL PARASITES AND PREDATORS OF AMOEBAE AND OTHER SOIL ORGANISMS

Charles Drechsler published descriptions of most of the species of fungi that parasitize or prey on amoebae in *Mycologia* between 1935 and 1961 (Dayal 1973). His

method for recovering nematode-destroying fungi (see "Drechsler's Technique," earlier) works equally well for amoebae-destroying species. In fact, Drechsler's technique yields a high diversity of fungal endoparasites and predators of a variety of microscopic animals. For the serious student, it is the best single method for recovery of such fungi. Baiting techniques can be used with amoebae-destroying fungi, but the diversity of parasites and predators obtained is not as good as that reported by Drechsler (G. L. Barron, unpublished data) (Fig. 19.10).

Most parasites and predators of amoebae are Oomycotina or Chytridiomycotina with nonseptate hyphae. A comprehensive key to those fungi can be found in Dayal (1973), along with an extensive bibliography of original descriptions and illustrations of species in that group. Only six species of hyphomycetous fungi are known to prey on amoebae (Barron 1981a): *Dactylella tylopaga*, *D. passelopaga*, *Tridentaria carnivora*, *T. glossopaga*, *Pedilospora dactylopaga*, and *Triposporina quadridens*.

Reports of fungal parasites and predators of other soil animals, such as ciliates, tardigrades, copepods, and even springtails, are scattered through the literature. A number of very interesting fungi have been recorded, but extensive studies have not been carried out on any of those groups.

APPENDIX A

KEY TO THE GENERA OF FUNGI THAT PREY ON OR PARASITIZE BDELLOID ROTIFERS

1. Parasitizing or preying on adult rotifers   2
1. Parasitizing or preying on eggs   18

2. Rotifers trapped on hyphal pegs or branches (predators)   3
2. Infections initiated by spores or modified spores (endoparasites)   5

3. Living hyphae and spores septate   *Lecophagus, Cephaliophora* (Fig. 19.1F)
3. Living hyphae and spores nonseptate   4

4. Rotifers captured on blunt adhesive hyphal pegs   *Zoophagus* (Fig. 19.1G)
4. Rotifers captured on pointed tips of branched hyphae   *Sommerstorffia* (Fig. 19.1H)

5. Parasite producing motile zoospores   6
5. Spores nonmotile   7

6. Infection thallus consisting of one large cell   *Atkinsiella, Haptoglossa* (Fig. 19.1A)
6. Infection thallus divided into segments   *Lagenidium, Myzocytium* (Fig. 19.1B)

7. Infected rotifers floating by means of aerial conidiophores   *Rotiferophthora* (Fig. 19.1A)
7. Sporulation on infected rotifers underwater   8

8. Hyphae extending from rotifer to produce large, terminal dictyospores   9
8. Dictyospores not obvious   10

9. Dictyospores elongate   *Culicinomyces* (Fig. 19.1G)
9. Dictyospores about as long as broad   *Rotiferophthora* (Fig. 19.1A)

10. Conidia with several elongate arms   11
10. Conidia not as above   12

11. Arms of conidia broad, blunt   *Dwayanngam* (Fig. 24.3F), *Haptospora* (Fig. 19.1C)
11. Arms of conidia thin, sharply pointed   *Triacutus* (Fig. 19.1C)

12. Spores serpentine or angular with very sharp points   *Harposporium* (Fig. 19.2I)
12. Spores not as above   13

13. Spores spherical, produced sympodially   *Pseudomeria* (Fig. 19.2H)
13. Spores not sympodial (i.e., phialidic, etc.)   14

14. Spores with tiny doughnut-shaped basal appendage   *Haptospora* (Fig. 19.2C)
14. Spores lacking basal appendage   15

15. Spores long, crescent-shaped, in apical cluster   *Medusamyces* (Fig. 19.2E)
15. Spores not crescent-shaped   16

16. Conidiogenous cells flask-shaped or tubular   *Rotiferophthora* (Fig. 19.2A)
16. Conidiogenous cells spherical with tubular mouth   17

17. Spores cylindrical   *Harposporium* (Fig. 19.2I)
17. Spores spherical to ellipsoid   *Tolypocladium* (Fig. 19.2D)

18. Conidia produced, no zoospores   19
18. Zoospores produced, no conidia   20

19. Conidia numerous in apical cluster on tall conidiophores   *Rhopalomyces* (Fig. 19.1D)
19. Conidia few on very short conidiophore   *Brachymyces* (Fig. 19.1E)

20. Zoospores biflagellate   *Lagenidium* (Fig. 19.1B)
20. Zoospores uniflagellate   21

21. Thallus epibiotic   *Rhizophydium*
21. Thallus endobiotic   22

22. Thallus lacking rhizoids   *Olpidium*
22. Rhizoids arising from thallus   *Endochytrium*

APPENDIX B

KEY TO THE GENERA OF FUNGI ENDOPARASITIC ON NEMATODES

1. Living hyphae or thallus nonseptate (Oomycota, Chytridiomycota, Zygomycota)   2
1. Living hyphae septate (Anamorph Fungi)   16

2. Sporangia produce motile zoospores   3
2. Zoospores absent   11

3. Zoospores with a single posterior flagellum   4
3. Zoospores biflagellate   8

4. Thallus epibiotic, attached to host by rhizoids   5
4. Thallus endobiotic   6

5. Rhizoids originating from penetration bulb   *Phlyctochytrium*
5. Rhizoids arising directly from thallus   *Rhizophydium*

6. Thallus lacking rhizoids   *Olpidium* (Fig. 19.4D)
6. Thallus with rhizoids   7

7. Thallus monocentric   *Endochytrium*
7. Thallus polycentric   *Catenaria* (Fig. 19.4G)
8. Infection resulting in a swollen, unicellular thallus
                              Haptoglossa (Fig. 19.4J)
8. Infection resulting in multicellular thallus   9
9. Thallus forming a linear series of swollen segments
                              Myzocytium (Fig. 19.4F)
9. Thallus hyphal   10
10. Hyphae less than 4 μm in diameter   *Pythium*
10. Hyphae more than 5 μm in diameter   *Lagenidium*
11. Mycelial infection forming chain of swollen
    segments   12
11. Fungus forms hyphae or hyphallike structures   14
12. Spores produced on erect, aerial
    sporangiophores   13
12. Spores produced in sporangia, expelled
    through exit tubes   *Protascus*
13. Spores borne in apical cluster on sporangiophore
                              Gonimochaete (Fig. 19.4F)
13. Spores borne in succession from corkscrew apex of
    sporangiophore   *Meristacrum*
14. Hyphae nonseptate   15
14. Hyphae septate   17
15. Spores small (±10 μm)   *Euryancale*
15. Spores very large (more than 40 μm)   16
16. Spores in apical cluster on vesicle   *Rhopalomyces*
                                          (Fig. 19.1D)
16. Spores in helicoid chain   *Helicocephalum*
17. Hyphae with clamps at septa
                              Nematoctonus (Fig. 19.4H)
17. Hyphae lacking clamps   18
18. Conidiophores lacking, conidia arise directly
    from conidiogenous cell   19
18. Conidiophores present   21
19. Conidia thick-walled, spherical, pigmented
                                          Botryotrichum
19. Conidia thin-walled, colorless   20
20. Conidia lens-shaped, from phialides on hyphae
    outside host   *Hirsutella*
20. Conidia spherical, borne on phialides produced
    inside host   *Plesiospora*
21. Conidia several-celled   *Haptocara*
21. Conidia one-celled   22

22. Conidiogenous cells aphanophialides, conidia
    teardrop-shaped   *Drechmeria* (Fig. 19.4C)
22. Conidiogenous cell globose to subglobose or
    flask-shaped   23
23. Conidia with odd, helical, or fusiform shapes;
    ingested by nematode   *Harposporium* (Fig. 19.4A)
23. Conidia globose, elliptical, or wedge-shaped;
    adhering to cuticle   *Verticillium* (Fig. 19.4H)

APPENDIX C

KEY TO THE GENERA OF FUNGI THAT PREY ON NEMATODES

1. Hyphae with clamps at the septa   2
1. Hyphae lacking clamps   3
2. Gill-forming teleomorph produced in culture
                              Hohenbuehelia (Fig. 19.5B)
2. No teleomorph produced   4
3. Hyphae with hourglass-shaped adhesive knobs
                              Nematoctonus (Fig. 19.5B)
3. Hyphae with adhesive stephanocysts   *Hyphoderma*
                                          (Fig. 19.5J)
4. Hyphae nonseptate   5
4. Hyphae septate   7
5. Conidiophore with apical cluster of large, dark
    conidia   *Rhopalomyces* (Fig. 19.1D)
5. Conidia colorless or absent   6
6. Spores chlamydospore-like cysts   *Cystopage*
                                          (Fig. 19.11)
6. Conidiophore with solitary, hyaline, nonseptate
    conidium   *Stylopage* (Fig. 19.5A)
7. Conidia nonseptate or uniseptate   *Arthrobotrys*
                                          (Fig. 19.5C)
7. Conidia multiseptate   8
8. Conidia cylindrical   *Dactylella, Dactylaria*
8. Conidia with enlarged central cell *Monacrosporium*
                                          (Fig. 19.5D)

20

FUNGI ASSOCIATED WITH VERTEBRATES

RICHARD C. SUMMERBELL

An important component of the fungal biodiversity of any given area occurs in habitats defined or conditioned primarily by vertebrates. Such habitats include the animals themselves, which are colonized by commensal and disease-causing fungi, as well as organic materials making up the dwelling places of those animals. Nominally, the fungal biota associated with vertebrates or anything that they have influenced include human diseases and the organisms growing in sites conditioned by human occupation and cultivation. Conservation-oriented ecology, however, may be limited to what is perceived to be natural, and nature may be seen as an ideal (although threatened) biological realm existing beyond the human sphere. Areas affected by human activity may be viewed as damaged or artificial, and thus unnatural, in direct proportion to the degree of modification. Those areas may be considered to be the research domain of the applied sciences. Similarly, diseases of humans and domesticated animals, although acknowledged as natural in origin, may seem irrelevant to most ecological study and may seem to fall entirely into the domain of the medical and veterinary discourses (Foucault 1972). Such boundaries between disciplines, however, are artificial. Natural and "unnatural" ecologies intergrade in studies of broad-scale habitat changes (e.g., polder-making in the Netherlands); species introductions; and the effects of agricultural, forestry, and conservation practices on ecosystems. Similarly, the medical and nonmedical arenas intergrade in studies of human diseases that are environmentally acquired, whether directly or indirectly via vectors or alternate hosts, and in studies of diseases that colonize not only humans and domestic animals but also animals indigenous to their ecosystems.

Because the present treatment of fungal biodiversity has a general biological orientation, I arbitrarily have limited the extent of my coverage along those continua and will consider only the following topics: (1) fungal diseases that are acquired more or less equally by humans or by other animals directly from relatively undisturbed environments; (2) communicable diseases and commensalisms of wild vertebrate animals; and (3) fungi characteristic of the dwelling places of wild vertebrate animals. I do not deal with the following topics: (1) mycotic diseases essentially found only in humans (e.g., onychomycosis), even when the causative agents are derived from environmental sources; (2) communicable diseases and commensal colonizations of humans, for which significant reservoirs of the causative organisms occur only in environments under strong human influence; (3) analogous communicable diseases and commensalisms of domestic animals; (4) fungal populations of human dwellings or other environments produced by human artifice such as haylofts, storage bins, refuse and compost mounds, motor vehicles, and so on, including resting places of domesticated animals significantly influenced by humans. Coprophilous fungi and rumen symbionts are treated in detail elsewhere (see Chapter 21 and Chapter 22).

In writing this chapter, I have made a number of assumptions about the investigators who will be consulting this work. The first is that the investigators are competent to detect and identify fungi from vertebrate sources. Therefore, I have summarized commonly used techniques for conducting such studies rather than treating them in detail. I also am assuming that the investigators are not physicians and are not directly involved in the procurement of human medical samples. Thus, I describe techniques related to medical sampling only to the extent necessary to permit understanding of their general nature. Nevertheless, any well-rounded survey of biodiversity in an ecosystem should include data from local medical and veterinary authorities on pathogens and their environmental prevalence. The material I present should increase the ability of biodiversity investigators to assess the strengths and weaknesses of biomedical sampling methodologies and evaluation criteria. It also will help them to distinguish the species isolated biomedically that derive from relatively undisturbed, as opposed to human-influenced, nature. I provide more detail about techniques used by investigators working in licensed biomedical or veterinary laboratories who may be directly involved in biodiversity studies. I also document techniques that may be used by investigators receiving subcultures from medical or veterinary sources or isolating pathogenic organisms directly from environmental material.

TAXONOMIC AND ECOLOGICAL SYNOPSIS OF PATHOGENIC FUNGI

Additional information on most of the topics discussed in this chapter (except the techniques themselves) can be obtained from Rippon (1988) or Kwon-Chung and Bennett (1992), as can basic medical and veterinary mycology information that I do not specifically credit to a source. Additional information on identification, classification, and biomedical status of the organisms mentioned can be obtained from de Hoog and colleagues (2000) and McGinnis (1980).

Because medically important fungi come from a wide variety of often relatively well-known fungal groups, there are no special criteria for valid taxonomic description of such fungi, nor are there dedicated culture collections. The University of Alberta Microfungus Collection and Herbarium, the Centraalbureau voor Schimmelcultures, and the American Type Culture Collection have outstanding collections. Herbarium specimens traditionally are prepared by drying cultures after fumigating the plates with formaldehyde vapor (add about 10 drops of commercial 37% formaldehyde to the medium, seal plate tightly, and retain at room temperature for 24 hours) and then demonstrating that attempted subcultures are uniformly nonviable.

Most of the information here assumes that the investigator will be dealing with cultures or direct detection of fungal structures in host materials. As an alternative to such direct detection techniques, skin prick serological testing provides the investigator with a rough idea of the prevalence of a few important pathogenic fungi by revealing previous exposure in healthy humans or animals. Such testing is primarily useful for detecting exposure to two virulent Onygenalean pathogens, *Histoplasma capsulatum* and *Coccidioides immitis*. Although the technique can be applied to any mammalian species (Ainsworth and Austwick 1973), testing of humans is most common. Sterile culture filtrates of the fungus are prepared and standardized using classic serological antigen preparation techniques (Ainsworth and Austwick 1973 and references therein). Those standardized extracts, such as histoplasmin from *Histoplasma,* then are injected subcutaneously in an area that has been shaved or clipped where necessary. Induration of the skin (hard swelling) over an area of more than 5 mm dia-meter at 48 hours is considered to be a positive test for prior exposure. The results from epidemiologically well-designed population surveys accurately delineate geographic regions where a fungus is endemic. A much less accurate but still useful approximation of such distributions may be obtained by compiling diagnostic mycotic immunology results from medical and veterinary laboratories, especially if data on the patients' travel histories are also available.

More sensitive and specific *in vitro* immunological tests have considerable potential for detecting exposure to pathogenic fungi or, in some cases, active disease in populations of humans and animals. The specifics of such indirect, complex tests, however, are beyond the scope of this volume. They are provided in Kaufman and Reiss (1992).

SOURCES OF MYCOTIC DISEASES

The proportion of mycotic diseases of humans and animals that are regularly communicable or contagious is small. Most mycotic diseases are acquired from non-animal sources in the environment; and most mycotic disease agents have other, usually saprobic, ecological roles through which they exert their main ecological impact. *Aspergillus fumigatus,* for example, although capable of causing disease in a variety of heavily exposed or immunocompromised humans and animals, is primarily a fungus of compost such as leaf litter or accumulated bird or bat dung.

Often, isolation of fungi from human and animal infections is of ecological interest because such infections reveal the presence in the environment of groups of fundamentally saprobic organisms that are overlooked or underrepresented by conventional sampling techniques for soil and plant material. Examples of such fungal saprobes or facultative saprobes, whose distributions mainly are elucidated with biomedical isolations, include *Scedosporium prolificans, Cladophialophora bantiana,* and *Acremonium falciforme,* as well as species in the genera *Blastomyces, Histoplasma, Coccidioides, Fonsecaea,* and *Onychocola.* In addition, disease agents participate in the regulation of animal population densities, influencing such factors as the availability of animals to predators. The enumeration of such agents is thus directly salient to conservation biology. Apart from any enumeration based on counting actual cases of infection, biomedically important organisms often can be enumerated directly from environmental sources using techniques specifically developed for detection of pathogens.

TYPES OF INTERACTIONS BETWEEN FUNGI AND VERTEBRATES

Many fungi isolated from vertebrate animals are neither commensals nor pathogens. They are contaminants, pathologically insignificant organisms of external origin, often isolated from environmentally exposed materials such as lung tissue or other parts of the respiratory tract, the gut, and the skin. Categories of association (commensalism, pathogenicity, contamination) overlap to some extent because a commensal or ordinarily harmless environmental fungus may become pathogenic with changes in the immunological or ecological status of the animal host. Isolation of an associated fungus is often easier than the correct evaluation of its status. Thus

medical and veterinary records in their raw state contain a mix of reports on the isolation of medically significant and insignificant fungi. Some evaluation criteria for determining significance are discussed here.

Weak and Fortuitous Associations

Fungi of external origin often are acquired as dormant contaminant propagules from aerial sediments, surfaces, or ingested materials. Moist or mucous surfaces exposed to the external environment, such as respiratory epithelium or amphibian skin, often exert a rapid immunological clearance of some organisms, leaving a distinctive assemblage of more immunoresistant organisms on the surface. For example, fungal propagules inhaled into human pulmonary systems appear to be cleared at different rates, with certain species such as *Aspergillus fumigatus* persisting longer than others even when not germinating or causing infection (Okudaira et al. 1977). Such species, although primarily saprobic, are often facultatively pathogenic. Gut passage also has a profoundly selective effect on dormant propagules, which is reflected in analyses of stool from humans and other animals. Investigations of coprophilous fungi also reflect this gastrointestinal selective process as well as the subsequent selective colonization and enrichment process of deposited dung. Accumulations of skin, scales, oils, and other such material on the body surface may facilitate the growth of some saprobes. Saprobes such as *Fusarium solani* and *Pseudallescheria boydii,* for example, may proliferate on wounds and in deep cracks in the skin, where serous exudate (pus) and dead skin are available nutrient sources (English 1968; R. C. Summerbell, personal observation).

Commensal Associations

Commensal organisms primarily colonize keratinized or mucous-covered animal surfaces and often use host secretions from the skin (e.g., sebum) or vagina, or gut materials such as oral or buccal cavity food residues, and fecal material. Yeasts in the genera *Candida, Malassezia,* and *Trichosporon* are the best known of those commensals (Rippon 1988). Those yeasts are essentially saprobic but are resistant to clearance by the immune system in certain body sites. *Malassezia* species are obligate commensals, occurring only on skin or within outer ear canals of mammals, including humans; six of the seven described species are so specialized that they will not grow on laboratory media without special lipid supplements substituting for skin lipids. Certain *Candida* species, particularly *C. albicans, C. tropicalis, C. glabrata, C. parapsilosis,* and *C. krusei,* also are associated strongly with mammals and birds, especially their gastrointestinal tracts and/or normal skin, and seldom are isolated from other sources except as transient inoculum or as active contaminants of fermented food and beverages. No such specialization yet has been shown with any *Trichosporon* species, but the species in this genus only recently have been clarified by molecular analysis (Guého et al. 1992).

Truly commensal, purely filamentous fungi have not been encountered. The filamentous growth habit involves substratum penetration, which may activate additional immune responses in vertebrates. Also, filamentous growth is spatially inappropriate for close packing within small surface niches and for the stabilization of nonpenetrative surface adhesion by breakdown of elongated structures into small, separate physical units.

Pathogenic Associations

Relatively few fungi are specialized animal pathogens; some of those are exclusively pathogenic, and others are also capable of saprobic existence. Certain other fungal pathogens appear to be saprobes but may cause disease in otherwise healthy vertebrates if suitably introduced into the body. A much greater number of species are opportunistically pathogenic. Ordinarily they are harmless, but they can invade animal tissues or extensively colonize body surfaces when a host becomes immunocompromised or, in some cases, is exposed to a heavy load of potentially immunodepressing fungal inoculum. Saprobes that do not act as opportunistic pathogens but that are connected with severe allergic reactions or adverse chemical exposures, such as mycotoxicoses, usually are not considered pathogenic because they do not establish populations in or on vertebrates. I do not discuss them in this chapter.

FUNGI ASSOCIATED WITH VERTEBRATE PATHOGENESIS

Virulent Systemic Pathogens and Phylogenetically Related Cutaneous Pathogens

The majority of specialized filamentous fungal pathogens of homeothermic vertebrates are members of, or anamorphs associated with, the Ascomycete order Onygenales. This group includes skin-infecting dermatophytes in the genus *Arthroderma* and its related anamorphs in *Trichophyton, Microsporum,* and *Epidermophyton,* as well as internally invasive (systemic) pathogens in the genus *Ajellomyces* and its related anamorphs in *Blastomyces, Histoplasma,* and *Paracoccidioide* (Figs. 20.1 and 20.2). Also related to this order is the anamorph species *Coccidioides immitis,* a particularly virulent species strongly resembling the anamorphic

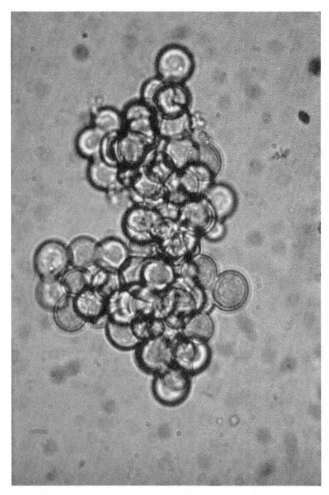

FIGURE 20.1   Yeast cells of *Blastomyces dermatitidis* as seen converted at 37°C on Blasto D medium (under oil immersion).

Malbranchea seen in the genus *Uncinocarpus* (Pan et al. 1994b). A few specialized systemic pathogens belong to other groups. *Penicillium marneffei,* for example, is of Eurotialean affinity; *Filobasidiella neoformans* is heterobasidiomycetous (Filobasidiales); and *Pneumocystis carinii* is a distant relative of the Endomycetales (Edman and Sogin 1994).

Dermatophytes are divided ecologically into three groups: (1) anthropophilic, specific to humans; (2) zoophilic, specific to nonhuman animals; and (3) geophilic, soil saprobes growing mainly on keratinous debris but occasionally causing infection in a human or other animal (Weitzman and Summerbell 1995). Zoophilic and geophilic dermatophytes have access to large numbers of habitable sites because they can colonize hair, feathers, and other dissociated keratinous materials. Most zoophiles and geophiles are distributed worldwide, but the former group may be limited to areas inhabited by particular types of animals.

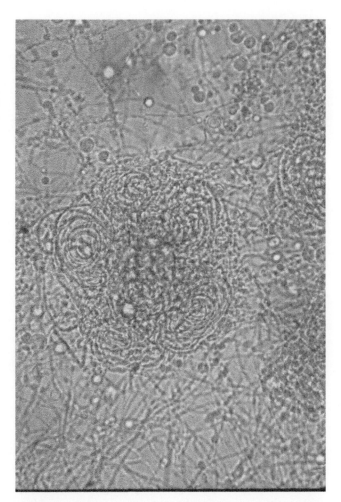

FIGURE 20.2 *Ajellomyces dermatitidis* (anamorph *Blastomyces dermatitidis*) conidia and infertile gymnothecia of the mold phase (10× objective). Note coiled appendages of the pseudogymnothecia.

Ajellomyces and other systemically pathogenic Onygenales can infect a wide range of terrestrial vertebrates. The life cycles of those organisms are poorly known, and it is not clear if any have a normal animal host. (A "normal animal host" is a single host with which a fungus constantly is associated ecologically, infecting but probably not killing it. Fortuitous hosts, in contrast, only occasionally are infected but may experience much more severe symptoms.) It seems most likely that these fungi persist in infected vertebrates until the animals die and then colonize conducive soils or plant debris where they produce conidia or ascospores, which may infect additional vertebrates (Maddy and Crecelius 1967). Infected animals apparently cannot transmit the disease directly to other animals. The animal hosts thus appear to function as stable sites for perennation rather than as loci of population increase and dispersal, although the latter functions also may be fulfilled to some extent. These fungi

usually are regarded as primarily saprobic, although little is known about their capabilities or significance as saprobes.

Systemically pathogenic Onygenales tend to be endemic to particular geographic areas for largely undetermined reasons. *Blastomyces dermatitidis* is specific to an enigmatic range of riparian and rodent-influenced habitats containing soil and plant debris, mostly in boreal eastern Canada and adjacent northern Wisconsin and Michigan and in a more southerly zone extending from Kentucky through Arkansas. It also is found rarely in other parts of the world, most notably in southern Africa, where an evolutionarily divergent lineage recognized as a separate serotype occurs. *Histoplasma capsulatum* var. *capsulatum* is indigenous to the valleys of the Mississippi River system where it occurs in soils, particularly those influenced by bird excretions and possibly other sources of organic nitrogen. It also occurs in limited parts of eastern Canada, the American southeast, Central America, South America, and Africa. It is associated strongly with depositions of bird or bat guano (hence also with caves). An African and Eurasian variety of this species, *H. capsulatum* var. *farciminosum*, causes epizootic outbreaks of a lymphocutaneous disease among equines, although the mechanism of transmission among animals is unclear (Ainsworth and Austwick 1973). It has been eliminated at least in some areas by "vigorous slaughter" of all affected animals, implying the absence of significant environmental (i.e., nonhost) reservoirs in some ecosystems where the disease is found (Ainsworth and Austwick 1973). *Paracoccidioides brasiliensis,* which seldom has been isolated except from infected humans, appears to be associated with moist soils in the humid, forested areas of southern Mexico, Central America, and South America east of the Andes and north of the Argentine pampas. Finally, *Coccidioides immitis* occurs primarily in desert soils of the Lower-Sonoran life-zone areas of the American southwest and northern Mexico (Rippon 1988), but it also is found in limited areas in mostly arid parts of Central and South America.

Emmonsia crescens, recently shown through mating to be the anamorph of *Ajellomyces crescens,* a species closely related to the virulent *Ajellomyces* species (Sigler 1996), occurs in temperate and tropical soils. *Emmonsia parva,* a species with no known teleomorph but confirmed by DNA-sequence analysis to be related closely to *E. crescens,* has a similar distribution (Peterson and Sigler 1998). The conidia of those fungi may swell after inhalation into mammalian lung passages and block small areas, but the fungus does not replicate in the host. That condition, called adiaspiromycosis, has been detected mostly in apparently healthy, small mammals (especially rodents and insectivores) during autopsy following trapping

(Ainsworth and Austwick 1973). A few cases of significant pulmonary obstruction have been reported in humans who have been exposed to heavy concentrations of airborne *Emmonsia* conidia in dust. In some cases obstruction evidently is related to immunosuppression inhibiting the clearance of the organism.

The life cycles and distributions of non-Onygenalean virulent systemic pathogens are also poorly known. *Penicillium marneffei* has a discrete endemic zone in the interior mountains of southern China and adjacent northern Indochina. It may be associated with the bamboo rat (*Rhizomys sinensis*). *Filobasidiella neoformans* var. *neoformans* (anamorph *Cryptococcus neoformans* var. *neoformans*) appears to be a cosmopolitan colonizer of bird and bat guano, but the closely related and equally animal-pathogenic *F. neoformans* var. *bacillispora* grows as an endophyte in the now globally dispersed Australian Red River gum tree (*Eucalyptus camaldulensis*) and a few related species (Ellis and Pfeiffer 1990). *Pneumocystis carinii* appears to be a contagious disease agent in a wide range of vertebrates including humans and rats, but the mechanism through which this marginally virulent opportunist maintains an epidemiologically significant population is not well understood (Laakkonen 1998). Infected humans, at least, are normally immunocompromised, and the great majority of patients in recent years have been infected with AIDS (acquired immunodeficiency syndrome). Molecular evidence suggests that the isolates taken from different animal groups are specific to those groups (Stringer 1996). *Pneumocystis carinii* cannot be grown in artificial culture but can be detected in the environment by a combination of filter air sampling and nucleic acid amplification (Olsson et al. 1996, 1998).

Sporothrix schenckii, an anamorphic species closely related to *Ophiostoma stenoceras* (Ophiostomataceae), mainly causes a slow-spreading lymphatic and subcutaneous ("lymphocutaneous") infection. It shows the same degree of specialization for pathogenesis as the Onygenalean systemic pathogens do, including conversion to a particulate vegetative phase (a budding yeast) within the host tissue and resistance to killing by the neutrophil and macrophage components of the vertebrate cellular immune system. It has a worldwide distribution and is associated mostly with finely divided, highly cellulosic plant debris such as peat moss and grass stalks. Some parts of Central and South America are hyperendemico—that is, the organism and its diseases are unusually common there (Rippon 1988; Summerbell et al. 1993).

Subcutaneous Pathogens

The virulent systemic, lymphocutaneous, and dermatophyte pathogens mentioned earlier (see "Virulent Systemic Pathogens and Phylogenetically Related Cutaneous Pathogens") are known or suspected to have a life cycle that regularly includes and evolutionarily reinforces vertebrate pathogenesis, leading to the development of attributes indicative of pathogenic specialization. In contrast, some fungi traditionally regarded as pathogenic appear to grow in the environment as saprobes but are capable of causing disease if traumatically implanted into the dermal layers of the skin. Those "subcutaneous pathogens" are distinguished roughly in medical mycology from traditionally defined "opportunistic pathogens" by their ability to cause disease in an immunologically normal host (a few fungi such as *Pseudallescheria boydii*, however, must be cited both as subcutaneous pathogens and as opportunists). Herpotrichiellaceous anamorphs in the genera *Fonsecaea*, *Phialophora* (pro parte), *Cladophialophora*, and *Exophiala* are prominent in this epidemiological category (Fig. 20.3), as are certain other Loculoascomycetous species and anamorphs such as *Curvularia lunata*, *Neotestudina rosatii*, and *Leptosphaeria senegalensis*. *Cladophialophora bantiana*, another fungus with Herpotrichiellaceous affinities, may cause a virulent, fatal brain cavitation, most likely subsequent to inhalation of inoculum. Hypocrealean anamorphs such as *Acremonium falciforme*, *A. kiliense*, and *Fusarium verticillioides* are also prominent subcutaneous pathogens, as are Microascalean fungi in the genera *Pseudallescheria* and *Scedosporium*. Finally, several *Phialophora* and *Phialemonium* species of unknown phylogenetic affinities frequently cause subcutaneous mycosis. Many

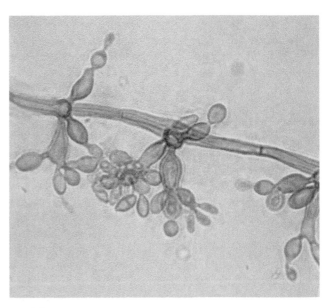

FIGURE 20.3 *Fonsecaea pedrosoi* conidiophores (under oil immersion, phase contrast).

subcutaneous pathogens are infrequently or never isolated except in this context, yet all must have fundamental adaptive niches elsewhere in nature. *Cladophialophora bantiana,* for example, appears to be associated with decaying material of certain Cupressaceous plants (Rippon 1988).

Ocular Pathogens

The mammalian eye presents a special habitat that can be invaded successfully by a large number of ordinarily harmless fungal species if they are implanted by thorns, slivers, or similar plant debris. *Fusarium solani* is a particularly common cause of blindness in tropical and subtropical areas. Reptiles may also be affected (Zwart et al. 1973).

Opportunistic Pathogens

A great variety of fungi cause infections opportunistically. Opportunistically pathogenic species are particularly common in certain fungal groups, including the Mucorales, Endomycetes, Eurotiales, Microascales, Dothideales (mainly Herpotrichiellaceae and Pleosporaceae), Hypocreales, and Filobasidiales. De Hoog and colleagues (2000) have compiled an extensive list of fungal pathogens and opportunists, along with their phylogenetic affinities, where known. Among the phylogenetically unrelated fungal analogues that are studied by mycologists, the Oomycetous order Saprolegniales frequently is associated with fish diseases, whereas species of the opportunistic prokaryotic genera *Nocardia* and *Actinomadura* cause infections of various vertebrates. The sole known Oomycetous mammalian pathogen, *Pythium insidiosum,* is in the order Peronosporales. Achlorophyllous green algae in the genus *Prototheca,* occasional agents of opportunistic infection, traditionally are included in the medical mycology diagnostic.

Some prominent opportunists are the normally or facultatively commensal *Candida* (Endomycetalean affinities), *Trichosporon* (Filobasidialean affinities), and *Malassezia* (basidiomycetous but otherwise of uncertain affinity) yeast species mentioned previously (see "Commensal Associations"). A large ecological group of opportunists consists of typical thermotolerant, facultative "compost fungi," such as *Aspergillus fumigatus, A. flavus,* and *A. terreus* (Eurotialean affinities); *Pseudallescheria boydii* (Microascalean affinities); and *Rhizopus* species, *Rhizomucor pusillus,* and *Absidia corymbifera* (Mucoralean affinities). Although any of those species can be isolated from composting materials such as warm, corrupted silage (R. C. Summerbell, unpublished data) and similar habitats, some also have other interesting ecological niches such as seed pathogenesis (*A. flavus*) or growth on materials submerged in tepid standing water (*Pseudallescheria*). If inoculum of some of those fungi reaches a critical level, it can cause massive epidemics in birds. *Aspergillus fumigatus* causes outbreaks of pulmonary disease and is particularly troublesome for domesticated fowl and birds in zoos and aviaries. It is extremely hazardous to penguins in captivity. *Dactylaria gallopava,* another fungus of warm rotting vegetation, can cause outbreaks of brain infection in flocks of turkeys and other gallinaceous fowl.

A third group of opportunists consists of thermotolerant, phytopathogenic *Cochliobolus* anamorphs (*Bipolaris, Curvularia, Exserohilum*) and *Alternaria alternata* (Pleosporaceous affinities). A fourth comprises primarily plant-pathogenic or rhizosphere-associated *Fusarium* species in the sections Liseola, Elegans, and Martiella, as well as certain members of *Acremonium* subgenus *Acremonium* (Hypocrealean affinities). Finally, a fifth prominent group includes black-yeast anamorphs genetically traceable to the family Herpotrichiellaceae (Masclaux et al. 1995); most opportunistic (as opposed to subcutaneous) infections ascribable to this group are caused by *Exophiala* species.

Some opportunistic fungi, besides causing serious diseases in immunocompromised animals, cause irritating, but not life-threatening, conditions through colonization of some internal surfaces of relatively immunocompetent animals. *Aspergillus fumigatus,* for example, can colonize ear canals, nasal sinuses, and bronchi of humans, progressing to an invasive disease only if the host's immunological condition deteriorates as a result of some independent cause. Aspergilli, *Pseudallescheria* species, and Dothidealean fungi are particularly likely to be involved in such colonizations; *Candida* is also important in colonizations of the ear canal, mouth, pharynx, and esophagus.

Two categories of opportunistic fungal infections are known primarily from cattle and, less commonly, from other ungulates and equines. The agents of those infections are aspergilli, Mucorales, and *Candida* species, which cause abortions, and *Cryptococcus neoformans, Candida* species, and aspergilli (also the aerobic actinomycete *Nocardia*), which cause mastitis (udder infection). *Mortierella wolfii,* a distinctive agent of mycotic abortion in cattle, seldom is isolated from any other type of mycosis. The agent may enter the affected site by inhalation or ingestion and subsequent hematogenous spread. Heavy environmental exposure to the causative agent seems to predispose an animal to site-specific (placenta and fetus in spontaneous abortion, udder in mastitis) opportunistic infections, at least with molds and cryptococci. Thus, the prevalence of those mycoses in domestic animals may be related to "unnaturally" high levels of inoculum in artificial shelters or in poorly stored feeds. Occurrence of those mycoses in wild animals is possible but has not been substantiated.

ABUNDANCE AND DIVERSITY OF MEDICALLY IMPORTANT FUNGI

de Hoog and associates (2000) listed more than 300 species of medically important fungi. Their list, which was based mainly on human cases, overlaps extensively with the list of significant fungal pathogens of other terrestrial mammals and birds, as well as the known pathogens of reptiles and amphibians and the non-zoosporic pathogens of fish. Although a few of the species are included on the basis of dubious or spurious records, the number of known potential fungal pathogens of humans and other homeothermic vertebrates is certainly well more than 300 and growing. Virulent systemic pathogens can be cosmopolitan (e.g., *Sporothrix schenckii, Cryptococcus neoformans* var. *neoformans*), or they can be confined to a variety of either temperate or tropical habitats (e.g., *Blastomyces, Histoplasma*) or tropical and subtropical habitats (*Coccidioides, Paracoccidioides, Penicillium marneffei, C. neoformans* var. *gattii*). Most opportunistic systemic pathogens (e.g., medically important Aspergilli) and zoophilic dermatophytes, as well as all commensal yeasts, are cosmopolitan. Fungal species that cause mycetoma (Mahgoub and Murray 1973), chromoblastomycosis, and other subcutaneous infections or are potential invaders of the traumatized cornea are particularly diverse in the tropics and subtropics (Rippon 1988). A 50,000-hectare tropical site likely would contain from one to four virulent fungal pathogens, 10–20 agents of subcutaneous mycosis, three to 10 noncommensal agents of cutaneous/superficial infection (e.g., dermatophytes, skin-infecting *Scytalidium* and *Phaeoannellomyces*), 15–20 commensal yeasts, 150–200 noncommensal potential opportunists, and 10–30 potential corneal invaders. Estimates of fungi at a 50,000-hectare temperate site would include from zero to four virulent pathogens, five to 15 agents of subcutaneous mycosis, three to 10 noncommensal agents of cutaneous/superficial infection, 15–20 commensal yeasts, 150–200 noncommensal potential opportunists, and two to 10 potential corneal invaders. Those estimates derive entirely from knowledge of the habitat range of various species and from general knowledge of the diversity of pathogenic fungal isolates reported in biomedical compilations from various areas. No survey of the abundance of animal pathogens in a defined area has been conducted directly. The virulent pathogens, certain subcutaneous and cutaneous pathogens, and the commensal yeasts likely would be detected only in investigations involving diseased animals (i.e., in medical and veterinary studies or investigations of trapped animals). The great majority of opportunists and corneal invaders would be detected in surveys of soil, litter, and vegeta-tion fungi; geophilic dermatophytes and some zoophiles specifically could be baited from soil using hair baits (see "Environmental Isolation of Medically Important Fungi" later in this chapter).

ISOLATION TECHNIQUES AND ASSESSMENT OF PATHOGENIC STATUS

IMPEDIMENTS TO ISOLATION STUDIES OF FUNGAL PATHOGENS

Isolation of pathogenic fungi often is facilitated by medical or veterinary inquiries. When an organism is both a potential contaminant and a potential opportunist, however, rigorous, "gold standard" criteria for verifying infections in particular cases may be difficult to meet. Demonstration of congruous fungal elements in a lung tissue biopsy, for example, may be necessary to ascribe invasive status to a normally saprobic mold isolated in culture from patient sputum or bronchial washing samples; biopsy, however, is a stressful, invasive medical procedure that may not be advisable for the patient in question. Such instances of hindered sampling have engendered the need to list certain fungi as suspected but not proven opportunistic pathogens and also have generated a vast literature of dubious or specious presumptions of pathogenicity (e.g., see analysis of atypical cryptococcoses by Krajden and colleagues 1991). Cases of infection by virulent fungi may be readily identified from medical and veterinary records, but verifying that isolations of potential opportunistic fungi represent instances of disease rather than contamination is frequently difficult. Even in medical or veterinary charts, the status of some isolations remains ambiguous. In general, any proposed infection by an opportunistic fungus should be verified by demonstrating histopathologically that infected tissue was penetrated by fungal elements (mycelium, yeast cells, etc.) compatible with the fungus isolated. Tissue confirmation of infection can be carried out either with formal histopathology (sampling, embedding, sectioning, and staining tissues) or simple direct hydroxide examination (see "Superficial or Cutaneous Infections," later in this chapter).

In general, any study involving isolation of fungi from humans (except in the course of medical procedures undertaken for diagnostic purposes) requires formal ethical sanction and explicit patient consent where appropriate. Animal studies are also stringently regulated in many countries. Statistical compilation of medical and veterinary laboratory records, however, is not regulated generally.

Although most pathogenic fungi grow readily in pure culture, a few species do not. The most prominent of the latter species is *Pneumocystis carinii*, an important opportunist in patients with AIDS. However, specific primers associated with that species are now available, permitting DNA-based identification of the organism in host material (Wakefield and Hopkin 1994) and in air filter samples and confirmation of its identity (Olsson et al. 1996, 1998). Other "unculturable" species include *Rhinosporidium seeberi*, a mesomycetozoan, and *Lacazia loboi*, a rare tropical pathogen of humans and dolphins.

A practical impediment to the study of pathogenic *Arthroderma* species and related *Trichophyton*, *Microsporum*, and *Epidermophyton* anamorphs is the strong tendency of isolates to degenerate in pure culture, losing the ability to form conidia and, in teleomorphic species, losing mating competence. *Paracoccidioides brasiliensis*, the agent of paracoccidioidomycosis, an important disease in South America, is difficult to maintain alive in pure culture, as are many Oomycetes and some *Malassezia* yeasts.

The biohazardous nature of some species is another potential impediment to study. *Coccidioides immitis*, *Histoplasma capsulatum*, and *Blastomyces dermatitidis* are considered to be in biohazard class 3, the second-highest hazard grade, because of their virulence and their easily aerosolized, respirable conidia. Study of such organisms requires specialized containment facilities. Infections of laboratory workers may be unusually severe because of the high inoculum loads in laboratory exposures. Shipping of cultures for analysis, especially across international borders, is costly and requires permits obtained only after navigating a complicated regulatory maze. Biohazard class 2 organisms, such as opportunistic aspergilli and yeasts, as well as dermatophytes, also require special facilities and shipping protocols.

Finally, environmental isolation of certain pathogenic fungi is notoriously difficult. This matter is discussed later in this chapter (see "Environmental Isolation of Medically Important Fungi").

ISOLATION OF DISEASE-CAUSING FUNGI FROM INFECTIONS

Strategies for isolating or otherwise detecting human- and animal-associated fungi vary according to the kinds of body sites infected by the fungi. The main categories of fungal diseases are as follows: (1) superficial or cutaneous infection (i.e., infection of the epidermal stratum corneum of the skin or of specialized skin structures, such as hair or nails, or colonization of skin or hair sur-

faces without penetration); (2) subcutaneous infection (i.e., infection of the dermal layers of the skin, sometimes progressing to local infection of underlying tissues); and (3) systemic infection, broadly defined as infection of internal body tissues and fluids or of the mucous membranes of the digestive, upper-respiratory, or genital tracts. Isolation strategies for these three disease categories as well as strategies for isolating causal disease agents from environmental substrata will be addressed.

Superficial or Cutaneous Infections

To isolate normally non-nonpathogenic, skin-surface yeasts, one draws a sterile cotton bacteriological swab moistened with sterile water or physiological saline across the skin and then streaks the swab onto appropriate growth medium, usually Sabouraud peptone-glucose agar with chloramphenicol (SABC; see Appendix II). Yeasts grow within 2–7 days at 30–35°C, a temperature range approximating that of the normal skin surface of homeotherms. Lower temperatures are used for isolation of fungi from poikilotherms. Skin mycota such as *Candida*, *Trichosporon*, and *Rhodotorula* species, as well as *Malassezia pachydermatis*, are readily isolated this way. Animals living in association with cold water may yield distinctive skin yeasts, such as members of the genera *Leucosporidium* (Bruce and Morris 1973) and *Itersonilia*.

Body surface mycota can be sampled with greater quantitative accuracy using contact plates. Those plates are designed so that even when the medium is poured to form a surface flush with the rim of the lower plate, an air space remains when the lid is closed. The surface of the medium is appressed directly to the skin, covered, and incubated. Use of medium such as Bandoni's sorbose yeast extract tetracycline medium (Bandoni 1981; Appendix II), which restricts colony diameter, may prevent overgrowth of the target mycota by spreading hyphomycetous and mucoralean contaminants (Summerbell et al. 1989a).

Superficial skin infections caused by dermatophytes, *Candida albicans*, and other tissue-penetrating skin pathogens usually are sampled by scraping flaky squames of infected epithelium from the skin surface. Typically those "skin scrapings" are collected with a sterile scalpel blade from areas of vigorous fungal growth at the margins of a skin lesion. Areas infected with *Microsporum canis* and closely related species can be visualized by their fluorescence when illuminated with a Wood's light (ultraviolet light filtered through cobalt glass) and accurately sampled. That trait can be very useful, for example, for detecting *M. canis* in felines. In animals with thick fur or hair, a sterilized brush with stiff bristles (e.g., a toothbrush) can be used to loosen and collect flakes of infected skin (Moriello and Deboer 1991). The skin scrapings are collected on black paper previously creased so that it can be refolded into a leak-proof packet. Scrapings are mounted on slides in 25% sodium or potassium hydroxide/5% glycerol, heated for 1 hour at 51–54°C to emulsify fats, and examined microscopically to determine the general categories of organisms growing in or adhering to the skin.

Scrapings showing presumed hyphomycete or yeast filaments, endomycetous type (blastic-multipolar budding) yeast cells, or no fungal elements are inoculated onto Sabouraud peptone-glucose agar (SAB) with cycloheximide, chloramphenicol, and gentamicin (CCG, to form SABCCG; Appendix II). Cycloheximide is a selective fungicide that inhibits growth of many nonpathogenic contaminants as well as some pathogenic fungi but facilitates isolation of certain pathogens and related organisms (e.g., dermatophytes and other Onygenalean fungi, certain *Candida* yeasts), many anamorphic Herpotrichiellaceous opportunistic pathogens (e.g., *Exophiala* species), and some opportunists (e.g., some species of *Aspergillus* and *Pseudallescheria*). If cycloheximide-sensitive species of medical significance are also likely to be present, particularly in reptiles and amphibians, SAB-chloramphenicol or Littman oxgall streptomycin agar is used in addition to SABCCG. If Oomycetous infection is possible, SAB is used without antibiotics.

The principal ringworm fungus of ungulates, *Trichophyton verrucosum*, poses a particular problem because of its slow growth. To detect it, skin scrapings or hair samples are inoculated onto Bromcresol purple-milk solids-yeast extract medium (BCP-MS-YE; Appendix II). If present, the fungus causes a distinctive halo around the colonies as a result of enzymatic clearance of casein solids (Kane and Smitka 1978).

Dermatophyte infections of areas covered with coarse hair or fur can be detected by microscopic examination and culture of the basal, "root" ends of hairs plucked from infected lesions. If hairs in the lesion are broken or too fragile to pluck, surface scrapings can be taken with a scalpel and a few hair roots can be extracted with the tip of the blade. If hair or fur is dense, scrapings can be taken with a sterile brush or by vigorously rubbing the underlying skin with a cotton swab. The swab technique does not allow for direct microscopic examination of the skin material for dermatophyte elements; thus, senescent or dead fungal material (e.g., drug-affected or from an older portion of a lesion) unable to grow in culture may not be detected. Cutaneous infections in reptiles and amphibians often are examined in small biopsies of affected areas.

In general, superficial cutaneous infections are distinguished from ubiquitous skin surface contamination by

the following: (1) isolation from a lesion compatible with a species that has been isolated only as an active pathogen (e.g., a dermatophyte); (2) isolation of a known pathogenic or opportunistic fungus from a lesion in which the hydroxide mount or biopsy tissue shows distinctive fungal elements unlikely to have been derived from another species; or (3) repeated isolation from a lesion of a fungus manifesting compatible fungal elements in direct microscopic or histopathological examination in the absence of a more likely, morphologically compatible pathogen. The presence of the latter pathogen must have been ruled out by unsuccessful attempted cultures on suitable stringent selective media from more than one independent sample (temporally independent in a living organism, spatially independent in autopsy).

Mucosal Infections

The moisture of mucous surfaces facilitates isolation of organisms, whereas the presence of immune components tends to diminish the burden of fungal contaminants. A sterile cotton swab generally is used to wipe oral, genital, and other mucous membrane surfaces investigated for *Candida* or other yeast colonization. In general bacteriology, the swab then is used to make a smear on a glass slide, which is gram-stained to examine for bacteria, yeasts, and protoctistans. After this procedure, or in its absence, the swab can be streaked across one or more culture plates (in a mycology laboratory) using SAB, Littman, or Inhibitory Mold Agar (Appendix II), with the possible addition of SABCCG as a second medium. In general medical microbiology laboratories, sheep blood agar (Appendix II) or any other common general bacteriological medium (e.g., chocolate agar) is used more commonly, and yeast will grow within 36 hours at 32–37°C. Infections may be distinguished from commensal colonizations by direct microscopic observation of an unusually heavy growth of the organism in question plus the presence of pseudohyphal or filamentous elements not found in purely commensal growth.

In cases in which opportunistic fungi have superficially colonized mucosal surfaces of the nasal cavity or sinuses, fungal crust is scraped from the surfaces and plated on the same medium set as for subcutaneous infection (see "Subcutaneous or Lymphocutaneous Infection," later in this chapter). It is especially important to observe the crust microscopically in a hydroxide preparation, characterizing the hyphae and noting pigmentation and presence of conidia. Often, much of the crust consists of dead material and therefore is difficult to culture. The filaments usually can be recognized as zygomycetous (usually *Rhizopus*, when cultured), dematiaceous (usually a Pleosporalean anamorph such as *Alternaria*, when cultured), or hyaline. Aspergilli and *Scedosporium* may conidiate *in situ*, although even the conidia are often not viable by the time samples are available.

Ear and Eye Infections

Ear Canals. The ears of many mammals other than humans can be colonized by a commensal yeast, *Malassezia pachydermatis*, which may proliferate and cause infection under certain circumstances (e.g., occlusion, immunosuppression). *Candida* species, certain *Aspergillus* species, *Pseudallescheria boydii*, and a few other organisms also can colonize human or, occasionally, animal external ear canals (an infection called otomycosis). A sterile cotton swab is used to collect ear secretions and inoculate plates containing SABC medium. A 10% potassium or sodium hydroxide mount is examined microscopically to verify that any filamentous fungus isolated is producing mycelium and conidiophores (where applicable) in the ear and is not merely a contaminant from an external source.

Eyes. No fungi colonize any portion of the eye primarily. An extensive and growing list of fungi, however, are known to infect the corneal surface subsequent to a traumatic eye injury, typically a scratch or puncture caused by plant material. After appropriate anesthesia is given (local or general, depending on the species of animal) corneal scrapings are collected with a suitable sterile scalpel blade and then cultured on SAB (Appendix II). Infection is verified by the presence of fungal filaments in the scrapings when examined in a hydroxide mount or gram smear.

Subcutaneous or Lymphocutaneous Infection

Fungi that cause subcutaneous infections usually enter the dermis through traumatic injury caused by a thorn, an arthropod bite, or a sliver of wood or dried herbaceous stalk. The same is true of *Sporothrix schenckii*, a fungus that causes lymphocutaneous systemic infection (i.e., an infection that progresses from a dermal primary inoculation site via lymph channels to more distal lymph nodes). Fungi are sampled from subcutaneous sites by surgical biopsy or attempted complete excision of the infected area (the latter is not possible for *Sporothrix*); in some human cases it is done by the less invasive method of punch biopsy performed by a dermatologist. Isolation from pus obtained from lesions by aspiration with a syringe or by lancing the lesions and swabbing the interiors is also possible, but not recommended, because in

many cases the etiologic agent is a fungal species that equally could be a mere surface contaminant. Such contaminants are not uncommonly isolated even from theoretically axenically drawn aspirates. In such cases, distinguishing the true infectious agent from skin contaminants becomes impossible, and only a biopsy can resolve the issue. The gold standard for identifying an etiologic agent in subcutaneous infection is usually isolation from internal tissue of a lesion shown histopathologically to have morphologically compatible fungal filaments invading tissue. An exception is *S. schenckii* because it essentially never is isolated as a biomedical contaminant. Any sort of isolation, even by the relatively nonspecific method of deep swabbing, is considered conclusive of etiology in suspected or compatible cases. The same is true for cases in which skin lesions are caused by other virulent, dimorphic, systemic pathogens such as *Blastomyces*, *Coccidioides*, or *Histoplasma*.

A sample of infected tissue can be ground with a tissue grinder (a sterile tubular glass mortar and pestle with matching sintered surfaces on both the pestle and the inside bottom of the tube) in sterile distilled water, and the slurry can be plated on the following (see Appendix II): (1) SABCCG; (2) sheep-blood CCG agar, brain-heart-infusion CCG agar, or an equivalent rich medium that supports growth of cultures requiring high levels of protein; and (3) a mycological medium, such as Littman oxgall streptomycin or Inhibitory Mold Agar (IMA), that permits growth of most hyphomycetous fungi but restricts colony diameter. If an Oomycetous infection is a possibility (as judged by compatible symptoms and direct microscopy), SAB without antibiotics may be needed. In the special case of mycetoma-type infections, sclerotium-like fungal grains issue from sinus channels within the lesion and emerge in pus at the skin surface. The grains at the surface are usually too degraded by bacteria to culture, but those extracted from biopsy material can be swished through agar to remove surface contamination and plated directly or pulverized in sterile water and then plated. The surfaces of a portion of the grains can be disinfected briefly with 70% ethanol, especially the hard, dense mycetoma grains, such as those found in *Madurella* infections (Mahgoub and Murray 1973).

Other Sources of Systemic Fungi

A standard set of growth media (Appendix II) for isolation of systemic fungi usually includes the following: (a) SABCCG; (b) a high-protein medium such as sheep-blood CCG agar, brain-heart-infusion CCG agar, or SABHI-CCG agar mainly for growth of virulent Onygenalean pathogens; and (c) a colony-growth-restricting medium such as Littman oxgall streptomycin or IMA. The partial redundancy among those media is important because it is impossible to control the quality of every batch of each individual medium to ensure that it supports every potentially important organism that it is designed to isolate. Generally, representative organisms should be checked explicitly for growth on new batches of medium. Representing all potentially important organisms is impracticable. However, the investigator should at least be certain that typical isolates of all significant species not explicitly represented in quality control testing are expected to have the physiological capability to grow on two media in the standard set. That way, if one medium is unsuitable as a result of a quality problem that is not readily evident, each species will have a fall-back medium on which to grow. Plates are maintained at from 28° to 32°C for isolation of fungi from homeotherms and at an appropriate ambient temperature for isolation from poikilotherms. Plates are retained for at least 3 weeks, and preferably for 5 weeks if *Histoplasma* infection is a possibility.

An increasing trend, in this time of human immunodeficiency virus (HIV) pandemic and other emerging types of immunosuppression, is to add blood agar amended with CCG and 2–3% egg albumen to the standard set of media for isolation of dimorphic Onygenalean pathogens from a background of heavy *Candida* yeast contamination (Kane et al. 1983; Chaturvedi et al. 1990). The majority of medically important *Candida* species is inhibited by the biotin-binding properties of egg white, and about half the yeast species potentially colonizing oropharyngeal sites in patients with HIV are inhibited by cycloheximide. The medium is recommended for any systemic isolation (e.g., respiratory isolations in cancer patients) in which yeast contamination is likely to be a problem.

Common systemic mycology samples include sputum (viscous material coughed from respiratory passages, with minimal salivary contamination), bronchoalveolar lavage fluid (bronchial washings) obtained by forcibly injecting sterile fluid into a portion of the respiratory tree using a tubular apparatus and immediately using suction to collect it again, biopsy tissue, urine, cerebrospinal fluid obtained by spinal puncture, and blood. All except blood usually are plated on the standard set of media; blood also can be cultured in this manner if freshly obtained or stabilized with anticoagulants such as heparin or citrate. In recent years, however, mechanized systems for blood culture have proved more rapid and efficient than plate culture (Merz and Roberts 1995). The patented Bactec system (Becton-Dickinson and Company, Appendix IV), for example, uses bottles of special medium designed to maximize growth of common pathogenic organisms in blood. Bottles with special fungal medium

are available for use when mycotic disease is suspected, but most fungi will grow in the general bacteriological media used as a default by most investigators. Medium constituents are prepared with a radioactive carbon isotope. The gas head in the bottle, sealed in with a rubber septum, is sampled automatically by a needle connected to a scintillation counter, which alerts investigators to any increase of radioactive carbon dioxide above background levels, indicating aerobic respiration. Microbiological staff then directly examine material aspirated from the bottle containing growth and plate it out on media appropriate for culturing the types of organisms seen. Bottles apparently negative for fungal growth can be cultured as a safeguard, especially if cryptococcosis is suspected.

Development of fully automated molecular chain reaction equipment that can identify hundreds of organisms directly from a minute clinical sample appears to be well under way. Culture is often still necessary, however, for confirming identifications, for testing drug susceptibilities, or for additional genetic testing during disease outbreaks. Automated isolation or genetic detection machinery relies on the general paucity of contaminating organisms in normally sterile bodily materials. It cannot be used generally to analyze samples from environmentally exposed body sites, including lungs and digestive tract, unless specific organisms can be targeted in some way and distinguished from contaminants.

Systemic fungi from nonmucosal sources can be distinguished from contaminants by (1) the presence in a normally sterile body site of fungal structures morphologically compatible with the fungus isolated from a biopsy or autopsy specimen; (2) repeated isolation of the organism from a normally sterile body site in the absence of direct microscopic detection (used mainly for unusual fungi from blood); (3) repeated isolation or, for known virulent opportunists, single isolation of an environmentally relatively uncommon organism from an immunologically influenced nonsterile site (e.g., isolation of *Aspergillus fumigatus* from lung) if symptoms and signs (e.g., infiltrates in X-ray) are compatible with infection; (4) isolation of an organism essentially never obtained in the laboratory except as a pathogen; or (5) isolation of a fungus verified as present by any of the following techniques: specific fluorescent antibody staining of a biopsy or autopsy specimen (Krajden et al. 1991); fungal culture corroborated by an independent direct chemical test (including molecular genetic detection) conducted using normally sterile material, a specific antibody titer considered indicative of active infection (Kaufman and Reiss 1992), a direct antigen test surpassing established critical values (Kaufman and Reiss 1992), or an overall clinical diagnosis definitively excluding all other reasonable possibilities.

ENVIRONMENTAL ISOLATION OF MEDICALLY IMPORTANT FUNGI

This section briefly reviews special techniques used in environmental isolations for biomedical investigations. I assume that the reader is aware of general techniques such as serial dilution (see other chapters in this volume). Such isolations most frequently are carried out in areas suspected of being foci of disease clusters or outbreaks but also can be carried out with ordinary samples from any environment. Some potential opportunists almost invariably will be isolated from any habitat, but the most highly virulent Onygenalean pathogens are usually specific to particular endemic regions (see "Virulent Systemic Pathogens and Phylogenetically Related Cutaneous Pathogens," earlier).

SYSTEMIC PATHOGENS

The systemic pathogens *Coccidioides immitis* and *Sporothrix schenckii* can be isolated by plating serial dilutions of appropriate bulk materials (soil, peat moss, pulverized plant material) on SABCCG (or its commercial equivalents, see Appendix II; Dixon et al. 1991). Isolation of *S. schenckii* can be improved by using SABCCG with enhanced cycloheximide and antibacterials (Zhang and Andrews 1993). Incubation at 37°C may favor growth of the former fungus, but temperature should not exceed 32°C for best detection of the latter. *Cryptococcus neoformans* can be isolated readily at 37°C on media such as Littman oxgall or SAB but is most efficiently detected on Staib's *Guizotia abyssinica* seed medium or its more defined counterpart, caffeic acid medium (Appendix II). Brown melanin pigment typically is produced in and beneath the colonies on these media, a reaction not occurring in common saprobic cryptococci or most other fungi.

Histoplasma capsulatum and *Blastomyces dermatitidis* seldom are isolated successfully from nonclinical materials, except by preparing slurries of bulk material (soil, dust, bird or bat guano) in physiological saline and injecting that material into rodents along with sufficient antibiotics to inhibit most bacteria (Ajello and Weeks 1983). All animals showing signs of disease are sacrificed immediately, and portions of their livers and spleens are plated out on the systemic mycology isolation media set (see "Other Sources of Systemic Fungi," earlier) to recover pathogens. At the end of 6 weeks, all remaining animals are sacrificed and similarly examined. *Blastomyces* appears to be ephemeral in the environment and most readily encountered during periods of relatively cool, damp weather. It is, therefore, seldom isolated in *post hoc*

studies of sites associated with outbreaks because the incubation period in the host before frank disease occurs is usually long enough (up to 100 days) for the weather to have changed substantially before the investigator arrives on the scene.

A report by Baumgardner and Paretsky (1999) documented an isolation of *B. dermatitidis* from environmental materials using a selective enrichment culturing technique. The overall reliability of this technique is unknown, but its development is of great interest.

DERMATOPHYTES

Dermatophytes and their nonpathogenic congeners generally are isolated from environmental materials using hair-baiting techniques (Vanbreuseghem 1949). In the simplest version, autoclaved horse mane or tail hair is cut into 1- to 3-cm lengths and placed on a Petri plate filled with the test soil (Dawson and Gentles 1961). Alternatively, hair of a child (ideally younger than 1 year old) may be cut into 1- to 3-cm lengths and placed on the surface of garden soil half filling a glass Petri dish (Stockdale 1961). The plates then are autoclaved at 20-lbs pressure for 20 minutes to sterilize the initial soil and hair, and then they are seeded with a quantity of the material to be tested. The plates are incubated for up to 4 weeks at room temperature. Many permutations of these protocols and alternate keratin sources are also possible.

Hairs are examined under a dissecting microscope for evidence of *Arthroderma* gymnothecia, pseudogymnothecia (ascoma-like structures that produce only conidia), macroaleurioconidia, or conspicuous tufts or strands of microaleurioconidia. Those structures may be removed with a fine needle and cultured; if that is not possible, hairs may be removed, cut into minute segments, and plated or streaked onto SABCCG. That test, of course, does not give a quantitative estimate of the amount of inoculum present in the soil or other material being tested. Carnivore dung containing hair, as well as finely divided keratinous materials such as feathers and quills retrieved from soil or animal dwellings (e.g., nests), also frequently will yield *Arthroderma* or its anamorphs if incubated in a moist chamber (Appendix I; Currah 1985; D. W. Malloch, personal communication; R. C. Summerbell, personal observation).

Quantitative isolation of dermatophyte inoculum from within human dwellings traditionally has involved swabbing surfaces with a cotton swab soaked in physiological saline and then streaking the swab on SABCCG medium. Sugimoto and associates (1995) improved that technique by adding another selective antifungal agent, 5-fluorocytosine, to the basic SABCCG base. The effectiveness of the technique with complex materials such as soils or litter has not yet been determined.

OPPORTUNISTIC PATHOGENS

When isolating species of readily culturable opportunistic fungal pathogens from bulk materials such as soil, selective factors can be used to increase isolation of specific species. Incubation of a general growth medium, such as Littman oxgall streptomycin, at 37°C may increase recovery of *Aspergillus fumigatus*, *Pseudallescheria boydii*, and other thermotolerant opportunists significantly (Summerbell et al. 1989b). Using a medium such as SABCCG amended with 0.04–0.10% cycloheximide facilitates isolation of Herpotrichiellaceous anamorphs (e.g., *Exophiala*, *Cladophialophora*; Dabrowa et al. 1964; Summerbell et al. 1989b). Benomyl (10 µg/ml) in any general fungal growth medium will select for endomycetous and many basidiomycetous yeasts, Mucorales, Microascaceae and anamorphs (e.g., *Pseudallescheria*, *Scopulariopsis*); Pleosporaceae; and Pyrenophoraceae (e.g., anamorphs *Exserohilum* and *Alternaria*; Baicu and Diaconu 1975; Summerbell 1993). Benomyl (trade name Benlate) generally has been available as 50% active ingredient in an inert clay carrier at garden shops. DuPont, its manufacturer, announced that for economic reasons it ceased selling Benlate after December 31, 2001. The compound could be added to any fungal growth medium by suspending the quantity appropriate for the desired dilution in a small volume of sterile distilled water and, just before the plates are poured, swirling the suspension vigorously into cooling medium to ensure uniform distribution (Hutchison 1990c). If the insoluble carrier (clay) was visible at low dilutions, mixing was inadequate. Most nonpathogenic Basidiomycetes and all Mucorales also will grow in media amended with 10 µg/ml benomyl, and those fungi, if they are abundant, may hinder the selective isolation of other target fungi.

At present, the molecular investigation of complex environmental materials for medically important fungi is an open research field. Investigators should keep in mind, however, that nonviable material frequently will give positive results in highly sensitive procedures such as the polymerase chain reaction and that some undescribed or poorly studied nonpathogenic relatives of medically important fungi may be present in the environment.

Air sampling frequently has been used to detect opportunistic *Aspergillus* species and other aerially disseminated pathogenic fungi. Although most procedures are the same as those for air sampling in general (Fradkin 1987), there are two important considerations when

pathogenic organisms are sought. First, for pathogens of humans and physiologically similar species, the medium (e.g., malt extract agar, Littman oxgall agar, SAB; Appendix II) is best incubated at 37°C. Second, an efficient volumetric device, such as an Andersen sampler, should be used because the species of concern usually have small, easily respirable conidia and, therefore, may not be recovered adequately with gravitational sedimentation sampling (Solomon 1975; Burge et al. 1987). In contrast to general air sampling, the focus species in biomedical air sampling are often relatively uncommon. Volumetric sampling increases the chances of detecting the rare pathogenic spores.

PREVALENCE OF MEDICALLY IMPORTANT FUNGI

The most common type of data analysis with regard to the prevalence of fungal infections is the single-site or collaborative multicenter compilation of the frequency of species occurrence in medical or veterinary isolations (e.g., Summerbell et al. 1989b). Although such compilations may reveal distinct and generally predictive trends, they are not based on random samples. Samples may be biased by factors such as which diseases the patients consider to be serious, which are covered by health insurance, which affect well-off as opposed to poor persons, the population of physicians or veterinarians with which a laboratory deals (e.g., generalists, particular specialists), which detection techniques a laboratory or its overarching regulatory system deems useful, the extent of travel and immigration in the population served, and so on. Generally only a subgroup of working laboratories is able or willing to participate in standardized data compilation efforts, leading to a strong reporter bias. Another critical variable with any studies involving opportunistic fungi is whether the reporting laboratories used adequate standards in distinguishing infections from contamination (McGinnis 1980). Such potential distorting influences must be considered when one reviews biomedical isolation data.

All diseases are theoretically susceptible to epidemiological analysis (e.g., numbers of cases per 100,000 individuals in a population). In practice, however, accuracy in such estimations requires either that the disease be relatively rare and dramatic or that the state make reporting of the disease mandatory. Fungal diseases of humans and domestic animals, which tend to be non–life–threatening if contagious and noncontagious, if life-threatening, seldom are placed on mandatory reporting lists. A substantial proportion of the infections diagnosed, therefore, tend to be lost in the unanalyzed records of laboratories not participating in data compilation efforts. Also, many physicians will diagnose certain diseases (especially dermatophytosis) purely symptomatically, without microscopy or culture. Such diagnoses are unproven and may be altered if the patient fails to respond to antifungal therapy. Patients, themselves, may elect to forego full diagnosis if they discover that a disease is nonthreatening or that treatment is expensive or troublesome. The same decision may be made for a domestic animal. In principle, common epidemiological devices, such as questionnaire surveys of a population, could accurately disclose the prevalence of fungal diseases. In practice, such surveys might be problematical and generally have not been done. Many patients are poorly informed about the nature of their mycotic diseases and only too readily forget any exotic Latinized disease name they have

been given. Also, few general-practice physicians are sufficiently knowledgeable about fungal diseases to give specific diagnostic information to patients.

In other arenas of investigation, the quantification of medically important fungi can be quite conventional and straightforward. The statistical analysis of data obtained from isolations of fungi from soil, litter, or air samples, for example, is exactly the same as for nonpathogenic fungi, and it will not be discussed here. Statistics books and Chapter 5 should be consulted, as necessary. In addition, if an investigator can collaborate with a zoologist to procure a statistically valid sample of wild vertebrates in an area, he or she then can proceed to determine the prevalence of disease or previous disease exposure (via serological testing) among those animals with equal validity, provided that legitimate veterinary or immunological diagnostic procedures are used.

21

COPROPHILOUS FUNGI

JOHN C. KRUG, GERALD L. BENNY, AND HAROLD W. KELLER

Coprophilous fungi are fungi that inhabit or are associated with the dung of animals, including soil contaminated with dung. Most of these fungi are found on mammalian dung from domesticated farm animals, such as cattle, horses, and sheep; from wild mammals, both herbivorous and carnivorous; and from birds. Development of the warm-blooded condition appears to be important in the evolution of the coprophilous habit (Webster 1970). Only a few fungal taxa are known from the droppings of groups of invertebrates and other vertebrates.

A fecal substratum may occur as a single deposit, or it may be scattered within a restricted area. The mycobiota appear specialized to some extent on particular kinds of dung. Dung itself includes various microhabitats; some fungi, for example, are found on the dung surface, whereas others are restricted to crevices within the substratum. Additional habitats develop as the dung fragments. The mycobiota found on dung from forested areas may differ from that found on dung from open areas because the surrounding habitat appears to influence diversity. The mycobiota on dung differ significantly between tropical and temperate latitudes. These differences may reflect the need for tropical species to initiate sporulation sooner after colonization than their temperate counterparts because dung decomposes more rapidly in tropical climates (Wicklow 1981). Fungi are poorly represented in the tropics on dung that has been devoured largely by scarabaeid (dung) beetles. Such dung consists mainly of an outer crust and lacks the moisture necessary for fungal growth.

Coprophilous fungi often grow and fruit only under a narrow range of conditions. Thus, different mycobiota may occur at different temperatures (e.g., during different seasons) or at contrasting moisture levels (e.g., in arid regions versus humid forests or during a wet season versus a dry season). The pH of a substratum, and therefore of its mycobiota, may vary with its stage of decay. Because most coprophilous fungi prefer a pH of around 7.0, the mycobiota found in strongly acidic conditions, such as those in pure coniferous forests or polluted sites, is rather restricted as compared to those in more alkaline conditions (e.g., in deciduous forests and arid regions). Some fungal taxa may be quite ephemeral and thus will be observed only on fresh dung, which is generally a poorer substratum for most Myxomycetes, Ascomycetes, and Basidiomycetes than aged dung but preferable for the isolation of Zygomycetes.

Many of the dung fungi are present on the substratum as spores. Thus, difficulties in eliciting fruiting, sporulation, or spore germination and in obtaining axenic cultures may limit the completeness of a study. Likewise, the lack of sporulation of anamorphs and production of a sexual stage, contamination by other fungi, competition by other microbes and predators, and age of the substratum may be limiting factors.

Coprophilous fungi are ideal organisms for both teaching and research in ecology and biodiversity and also provide useful systems for studies of community ecology (Wicklow 1981, 1992; Keller and Braun 1999). Succession is studied easily using moist chambers, which are simple to construct and are described in "Moist Chambers for the Development of Fungi," Appendix I. Many coprophilous taxa produce important chemical compounds that may inhibit competing and invading organisms or stimulate fungal growth. Wicklow (1988) has reviewed the role of such compounds in deterring predation. Gloer (1995, 1997) has shown that coprophilous fungi, especially those slow-growing taxa developing in middle or late succession, offer a rich and important source for new antifungal natural products. Many of those compounds possess novel ring systems (e.g., preussomerin A from *Preussia isomera*; Weber et al. 1990), a relatively rare occurrence in natural products chemistry (Gloer 1995). Thus, coprophilous fungi are an outstanding source of unknown compounds with diverse biogenetic origins and promising biological activity (Gloer 1997).

DIVERSITY AND DISTRIBUTION

Concentrated work on coprophilous fungi has been done largely in Europe and North America and to a lesser extent in southern South America. Recent studies have added some information for central and east Africa; Japan; and limited parts of Asia, New Zealand, and Venezuela. Still, large areas of the world are only poorly known or have not been studied at all.

A survey of the literature shows that the majority of the studies, collections, and published records of coprophilous Myxomycetes, like Zygomycetes, Ascomycetes and Basidiomycetes, come from Europe and North America, with fewer records from Africa and South America (Eliasson and Lundqvist 1979). Coprophilous fungi are the central focus of many of those papers, whereas reports of Myxomycetes appear more incidental. The majority of coprophilous Myxomycetes are presumed to be secondary inhabitants of dung. About 12 species are found exclusively on dung and 15 predominantly on dung. This represents about 2% of the approximately 700 described species of Myxomycetes. No obvious adaptations to endocoprophily have been found.

With our current state of knowledge, it is difficult to estimate patterns of species richness and endemism for coprophilous fungi with any precision. Certainly, some taxa are more prevalent at tropical as opposed to temperate latitudes, and others occur more frequently in arid than in humid regions. Certain species appear to be restricted to specific areas, and some fungi (e.g., *Wawelia regia*, known only from a site near Kracow in southern Poland), appear to be rare. Some genera (*Pleophragmia* and *Radiomyces*) are generally rare, although as Richardson (2001) illustrated, rarity may be apparent rather than real. Certain taxa (e.g., *Dictyocoprotus mexicanus* from Mexico and *Protomycocladus faisalabadensis* from Pakistan) are known from single collections. Other species (*Phaeotrichum hystricinum* on porcupine dung) are restricted to specific types of substrata. Nevertheless, in most instances, estimates of frequency and species richness are correlated with collecting intensity, geographic origin, and the expertise and interests of the mycologists.

Sampling has been, at best, haphazard, but in fact is probably nonrandom, being concentrated in areas of animal concentrations or in places where general collecting is being carried out. Consequently, it is difficult to estimate species richness in an area in any real way. Caves occupied by porcupines and certain wood rats in North America and by hyrax in Africa are an exception. Dung at those sites has been deposited in layers or pushed to the entrance of the cave for several generations, and some estimate of species richness should be possible.

Nevertheless, in most communities, a small number of species makes up the majority of the known mycobiota. Even in studies with large sample sizes, a modest increase in sampling will lead to a significant increase in numbers of species (Richardson 2001). Also, Richardson found a greater diversity of fungi in winter than in summer at northern latitudes. Thus, a large sample taken through different seasons is required for any realistic estimate of diversity.

In addition, because many fungi will not be isolated with incubation and culturing (Amann et al. 1995), a more complete estimate of a mycobiota can be obtained by the addition of PCR-RFLP (polymerase chain reaction–restriction-fragment-length polymorphism) (Viaud et al. 2000) using the ITS (Internal Transcribed Spacer) region of ribosomal DNA. The latter authors have shown that adding a molecular approach significantly increases any estimate of diversity.

On the basis of both animal and habitat diversities, one would expect a more diverse mycobiota in the tropics than in temperate regions. Cain (1934) recorded 112 taxa of coprophilous Sphaeriales from Ontario, Canada. At that time, Sphaeriales was used in the broad sense roughly equivalent to the current concept of pyrenomycetes. In East Africa and Venezuela, if one restricts the comparison to the genera covered by Cain, comparable numbers of taxa are 118 and 122, respectively. However, if one includes all coprophilous pyrenomycetes for these regions, the numbers of taxa are 151 and 159. Cailleux (1971) recorded 91 taxa from the Central African Republic. His data are not entirely comparable because he excluded the Sporormiaceae and a number of other pyrenomycetes. Richardson (2001) found 153 taxa in 50 samples from the zone 0-30° north and south of the equator. That figure drops off substantially as one proceeds beyond 40° north or south. In contrast, Bell (1983) recorded 66 taxa of pyrenomycetes from New Zealand, which support the existence of a latitudinal gradient.

In tropical savanna habitats, such as those in east Africa, one would expect a richer mycobiota than in temperate regions because of a greater concentration and diversity of substratum and climatic differences. About 100 taxa of coprophilous Ascomycetes were recorded in about 20 collections from the area of Lake Amboseli, with up to 30% overlap between collections. In comparison, collections from the forested and alpine areas of Mt. Kenya also yielded about 100 taxa, but the number of taxa per collection was never more than 20 as compared to 30 to 35 for Lake Amboseli. It is interesting that, 65–70% of the fungi from Mt. Kenya, which is more temperate, differed from those at Lake Amboseli. In comparison, Lundqvist (1972) recorded 129 taxa from the mycologically rich

Gotland Island in southern Sweden, but numbers of taxa from other parts of Sweden ranged from only 74 to 95. In a study published in 2001, however, Richardson confirmed statistically that species richness decreases with increasing latitude, both on dung in general and on specific substrata. It would appear then that even in fungi, diversity in tropical regions is greater than in temperate regions, even at higher elevations.

Although the number of coprophilous taxa decreases with increasing latitude, the number of taxa per sample does not decrease significantly, although the individual taxa are different (Richardson 2001). Lundqvist (1972) pointed out, however, that groups of north-temperate fungi and tropical–subtropical fungi appear to be distinct and are likely distinct from a corresponding group of south-temperate fungi. Within each group the distribution of taxa is fairly cosmopolitan.

Myxomycetes are not particularly common inhabitants of dung in temperate forests, and truly coprophilous species are either absent or at least rare. Only five species were recorded on dung collected from southwestern Virginia in the United States, and none of those was considered truly coprophilous. Values for species diversity (0.35) and average number of species per culture (0.15) were exceedingly low (Stephenson 1989). Eliasson and Lundqvist (1979) noted that *Perichaena corticalis* var. *liceoides* seems to prefer dung from domestic animals, whereas *Stemonitis fusca* has been found only on dung from forest animals.

The most extensive survey for coprophilous Zygomycetes was carried out in southern California and adjacent parts of Arizona and Mexico. Over a period of 35 years, 88 species distributed among 42 genera (exclusive of the large genera *Mucor* and *Rhizopus* because a monograph suitable for identifying the species was not available at that time) were recorded (R. K. Benjamin personal communication). In comparison, 78 species in 26 genera were reported for Pakistan after 5 years of extensive collecting (Mirza et al. 1979). In more temperate Switzerland, Lendner (1908) found 20 species in 10 genera.

ECOLOGY

Spores of many coprophilous taxa germinate only after passage through the mammalian digestive tract (Janczewski 1871; Massee and Salmon 1902). Such spores, especially among Ascomycetes, frequently are surrounded by some sort of gelatinous sheath or possess other kinds of gelatinous attachments on the sides or ends of the spore. The spores are discharged forcefully into the air and readily adhere to nearby vegetation.

When the animal whose dung they colonize feeds on the vegetation, the spores are ingested with it. Germination, growth, and sporulation occur on the freshly deposited dung. This cycle is illustrated in Larsen (1971: fig. 1) and Furuya (1990: fig. 14). Some myxomycete spores seem unable to survive digestion, which may explain why so few are found on freshly deposited dung (Marx 1992). Some coprophilous fungi are dispersed by rain, insects, other arthropods, or even mammals. Insects, especially, play an important role in dispersal, a fact overlooked by earlier workers (Stevenson and Dindal 1987). Mites also play a role in dispersal of a wide range of coprophilous fungi (Malloch and Blackwell 1992), ingesting and then disseminating certain myxomycete spores (Keller and Smith 1978). The fungal fruiting bodies of some cleistothecial Ascomycetes that are dispersed by mammals, particularly carnivores, possess modified appendages for attachment to fur. Such structures, which promote dispersal to different dung deposits, are particularly advantageous for fungi that inhabit dung of certain rodents. The rodents, which deposit dung in front of their burrows, act as a sort of vector. Under some arid conditions, fungi have cleistothecial ascocarps that are dispersed more frequently by animal vectors than by forceful discharge (Wicklow 1981).

Coprophilous fungi are very important in an ecosystem because they break down substrata and recycle micronutrients. A succession of fungi participates in that process. Frequently, Zygomycetes are observed early in succession (within 5–14 days), followed by Ascomycetes, then Basidiomycetes, and much later by Myxomycetes, over a period as long as 2 months. Lodha (1974) concluded that the timing of appearance of the different coprophilous fungi overlaps considerably. For example, certain Basidiomycetes that use lignin sometimes sporulate together with the Ascomycetes rather than later in succession (Wicklow 1981). The sequence of species is based to some extent on the types of substrata, carbohydrates, cellulose, and lignins used by each. Competition from other organisms and availability of specific nutrients and substrates also may modify the strict linear progression of the succession. Webster (1970) showed that the correlation between taxonomic groups and nutritional substrates is not exact and that depletion of other nutritionally important compounds is also important in the stages of succession. The presence of an essential nutrient is more important than external environmental factors (Wicklow 1981). Certain fungi are antagonistic to and suppress the fruiting of other fungi (Harper and Webster 1964) through hyphal interference, which apparently is widespread among coprophilous fungi (Ikediugwu and Webster 1970a, 1970b). Other fungi produce diffusible antibiotics that restrict the growth of competing and faster-growing fungi. Rapid spore germination and growth are certainly advantageous for the colonization of dung (Harper and Webster 1964), and thus speed of germination and rates of growth influence fungal succession. From an ecological viewpoint, succession certainly is influenced by competition.

The activities of nonfungal competitors also can influence the ability of particular fungi to survive. Certain arthropods and worms contribute to the fragmentation process of the dung and thus influence the mycobiota. Arthropod and fly larvae are known to prey on fungi, consuming both sporocarps and mycelium (Helsel and Wicklow 1979). That may allow some of the rarer fungal species to sporulate by decreasing the numbers of their more common fungal competitors (Wicklow 1981). Some fungi (e.g., *Chaetomium*) have evolved ornamented perithecial hairs that may act as a defense mechanism (Wicklow 1981). The same also may be true of the stiff, bristle-like hairs in *Coniochaeta* species. Members of the Chaetomiaceae and Sporormiaceae produce a number of diffusible compounds that also may act as defense mechanisms, whereas the dark melaninized pigments in the walls of some spores may provide protection against ultraviolet radiation.

Environmental factors also influence the composition of the mycobiota to some extent. Certain species of *Thelebolus* (Wicklow and Malloch 1971) and *Preussia* are dominant on leporid dung under colder temperatures of early spring, whereas other species dominate during the warmer summer months. Wicklow and Moore (1974) also demonstrated the effects of temperature on fungal succession. Light also can influence fungal succession (Morinaga et al. 1980); some taxa are positively phototropic (e.g., *Ascobolus* species and some *Podospora* species). The influence of moisture is self-evident (see Kuthubutheen and Webster 1986; Wicklow 1992).

Few investigators have attempted to isolate Myxomycetes systematically from large quantities of dung from herbivorous animals. Most studies have focused on dung samples from cattle (Kowalski 1972; Angel and Wicklow 1975; Keller and Anderson 1978; Cox 1981; Keller and Schoknecht 1989; Eliasson et al. 1991). In a notable exception, Eliasson and Lundqvist (1979) sampled dung from 25 herbivorous species including, in descending order of importance for number of myxomycete collections, cow, hares, horse, rabbit, elk and moose, roe deer, goat, and sheep. Based on a literature survey, they also located some 80 species in 23 genera of Myxomycetes that had been recorded from dung. Stephenson (1989) sampled dung from white-tailed deer and American bison. Published records of Myxomycetes from dung of carnivorous and omnivorous vertebrates are rare. Books by Martin and Alexopoulos (1969) and Stephenson and Stempen (1994) hardly mention those substrata. In part this reflects the tendency of myxomycologists to gravitate to the most productive habitats, which for Myxomycetes are the dung of herbivorous vertebrate mammals, especially cattle. Bird dung is generally a poor substratum for Myxomycetes, although they have been recorded from partridge dung (Schnittler and Novozhilov 1996), which also harbors a number of specialized Ascomycetes. Myxomycetes are isolated from dung almost exclusively in moist-chamber cultures, but some expert collectors have found specimens on cow dung in the field (Kowalski 1969a, 1969b).

Many mycologists have sampled Ascomycetes from animal substrata in different geographic areas; for example, Bell (1983) surveyed 17 substrata from New Zealand, and Cailleux (1971) surveyed the dung of 13 animals from central Africa. The frequencies of occurrence of Ascomycetes species on different substrata also have been compared in detail (e.g., Richardson 1972, for Scotland), with some species showing a certain degree of substratum specificity (Lundqvist 1972, for the Nordic region). Abdullah (1982) also noted frequencies of occurrence, including a few Zygomycetes and Basidiomycetes, in addition to Ascomycetes. Parker (1979) was able to group fungi from five substrata by substratum and habitat, whereas Angel and Wicklow (1975) suggested a relationship between certain herbivores and the mycobiota colonizing their dung. All of these studies suggested that a relationship exists between the type of substratum and the composition of the coprophilous mycobiota; ruminant dung, for example, appears to have a different mycobiota from that of lagomorph dung. Lundqvist (1972) suggested that fungi can be grouped with a series of hosts, but that the habitat of the host is likely more important than the substratum. Parker (1979) made a similar suggestion, and Wicklow (1981) showed that microclimate also influences species composition. Feeding habits of the host also can influence the species diversity on the substratum (Ebersohn and Eicker 1992). Most of the described studies were restricted to dung of herbivorous mammals with the exception of that of Lundqvist (1972), who also included some data for carnivores.

TAXONOMIC SYNOPSIS

In this review we deal separately with the Myxomycetes, Zygomycetes, Ascomycetes, and Basidiomycetes. The only other group containing coprophilous taxa, the Dictyosteliomycetes or cellular slime molds, are treated in Chapter 25, so we do not consider them here. Members of the Myxobacteriales, although not classified as fungi, frequently are encountered in studies of coprophilous fungi; we will mention them briefly here. Hawksworth and colleagues (1995) summarized the key papers needed for the identification of most fungal genera. Access to type material and voucher specimens is also essential for identification and systematics. The following herbaria contain important collections of coprophilous fungi: BPI, E, FH, FLAS, GB, L, LPS, NY, PC, S, SANK, TRTC, and UPS; cultures at ATCC, CBS, IMI, and NRRL (see Holmgren et al. 1990 for full details; International Mycological Institute [CABI], *Catalogue of the Culture Collections, Filamentous Fungi, Yeasts, and Plant Pathogenic Bacteria* 10th Edition [1992] for information on cultures; and Appendix III).

MYXOBACTERIA

The Myxobacteria, which originally were confused with Myxomycetes, include many species that produce fruiting bodies on dung (Pfister 1993). Thaxter (1892 [corrected in Thaxter 1893], 1897, 1904) described and illustrated many of these organisms. The most current classifications can be found in Staley and associates (1989) and Holt and colleagues (1994). McCurdy (1969) included color photographs of many species that might be encountered on dung. These papers should be consulted for the characterization and identification of these organisms.

Myxobacteria can be isolated from many organic substrata, including dung and soil. A common isolation technique is to place rabbit pellets on soil as bait; many Myxobacteria species will form fruiting bodies on the dung (Staley et al. 1989). Seven genera and 22 species of Myxobacteria are coprophilous or, at least, fruit on dung baits. They are classified in four families in the order Myxococcales (Staley et al. 1989), as shown in Table 21.1.

TABLE 21.1
Families and Genera of Coprophilous Myxobacteria, Order Myxococcales

Archangiaceae: *Archangium*
Cystobacteraceae: *Cystobacter, Melittangium, Stigmatella*
Myxococcaceae: *Myxococcus*
Polyangiaceae: *Polyangium, Chondromyces*

MYXOMYCETES

The Myxomycetes once were thought to be a class of fungi (e.g., Keller 1982), but recent molecular evidence (Patterson and Sogin 1992; Cavalier-Smith 1998) suggests a protozoan origin. This small class of organisms consists of three subclasses (Ceratiomyxomycetidae, Myxogastromycetidae, and Stemonitomycetidae), and six orders (Ceratiomyxales, Echinosteliales, Liceales, Trichiales, Physarales, and Stemonitales), with approximately 57 genera and 700 described species. They have been classified as animals (Mycetozoa), as plants (Myxomycophyta), as fungi (Myxomycetes), and as protists (Protista) by various taxonomists over the years, depending on the phase of the life cycle that the taxonomist emphasized. The class name Myxomycetes reflects the fungus-like reproductive phase (fruiting bodies and spores) of these organisms. They sometimes are classified as protozoans (Class Mycetozoa) because of their animal-like myxamoebae and plasmodial assimilative phase. Although their evolutionary origins are obscure and somewhat controversial, the group now is believed to have descended from protozoan ancestors (Olive 1975). Myxomycetes sometimes are referred to as the acellular, plasmodial, or true slime molds. The last name is probably the most acceptable when distinguishing the Myxomycetes from the dictyostelid cellular slime molds and the protostelids. The subclasses are distinguished by the type of sporophore development, type of plasmodium, and method of bearing spores. The various orders, families, genera, and species are distinguished by spore color and by characteristics of the fruiting bodies such as the presence or absence of a peridium, a capillitium, deposits of calcium carbonate, or a columella. Refer to Chapter 25, this volume, for additional information.

Noteworthy monographic publications on the Myxomycetes that should be consulted for keys, illustrations, and species descriptions include Lister (1894, 1911, 1925), Macbride (1899, 1922), Macbride and Martin (1934), Hagelstein (1944), Martin (1949), Martin and Alexopoulos (1969), Martin and associates (1983), and Nannenga-Bremekamp (1991). We follow the classification of Martin and associates (1983). A key and survey to the coprophilous genera and species are provided by Eliasson and Keller (1999).

Five orders of Myxomycetes include coprophilous species (Table 21.2). Those species may be exclusively coprophilous, known predominantly from dung and associated litter, or recorded on dung but found predominantly on other substrata.

ZYGOMYCETES

The Zygomycetes are a diverse group of fungi consisting of 10 orders, 18 families, and 122 genera (Benny et al. 2001), almost half of the genera being in the Mucorales. Recent phylogenetic studies clearly show that the Mucorales contains only a single family, the Mucoraceae (Voigt et al. 1999; O'Donnell et al. 2001; Voigt and

TABLE 21.2

Species of Coprophilous Myxomycetes and Their Associations with Dung

Echinosteliales
  Echinosteliaceae: *Echinostelium minutum**
Liceales
  Cribrariaceae: *Cribraria violacea**
  Liceaceae:
    Kelleromyxa fimicola[†] (= *Licea fimicola*)
    Licea alexopouli,[†] *L. belmontiana*,* *L. pusilla*,* *L. tenera*,*
      *L. variabilis**
Physarales
  Didymiaceae:
    Diderma applanatum,* *D. effusum*,* *D. globosum*,*
      D. cf. *niveum*,* *D. simplex*,* *D. testaceum*,* *Didymium
      anellus*, * *D. annulisporum*,[†] *D. clavus*,* *D. difforme*,*
      D. dubium,* *D. iridis*,* *D. karstensii*,* *D. ovoideum*,*
      D. melanospermum,* *D. nigripes*,* *D. nullifilum*,[†]
      D. ochroideum,* *D. quitense*,* *D. rugulosporum*,[†]
      D. saturnus,* *D. squamulosum*,* *D. trachysporum*,*
      D. vaccinum,* *D. verrucosporum**
    *Lepidoderma chailletii**
  Physaraceae:
    Badhamia apiculospora,[‡] *B. gracilis*,* *B. macrocarpa*,*
      B. spinispora[‡]
    *Craterium leucocephalum**.
    Fuligo cinerea,* *F. septica**
    *Leocarpus fragilis**
    Physarum bitectum,* *P. brunneolum*,* *P.compressum*,*
      P. confertum,* *P. contextum*,* *P. didermoides*,* *P. fimetarium*,[†]
      P. gyrosum,* *P. leucopus*,* *P. mucosum*,* *P. notabile*,*
      P. nucleatum,* *P. nutans*,* *P. ovisporum*,* *P. pusillum**
Stemonitales
  Stemonitaceae:
    *Colloderma oculatum**
    Comatricha mirabilis,[†] *C. nigra*,* *C. pulchella**
    *Lamproderma scintillans**
    Macbrideola coprophila,[†] *M. cornea**
    Stemonitis fusca,* *S. pallida**
Trichiales
  Dianemaceae: *Calomyxa metallica**
  Trichiaceae:
    *Arcyodes incarnata**
    Arcyria cinerea,* *A. elaterensis*,[†] *A. incarnata*,* *A. leiocarpa*,*
      A. pomiformis,* *A. stipata**
    *Oligonema schweinitzii**
    Perichaena chrysosperma,* *P. corticalis*,* *P. corticalis* var.
      liceoides,[‡] *P. depressa*,* *P. luteola*,[†] *P. minor*,* *P. pedata*,* *P.
      quadrata*,* *P. syncarpon*,* *P. vermicularis**
    Trichia brunnea,[†] *T. botrytis*,* *T. contorta*,* *T. fimicola*[†]

* Species recorded on dung but found predominantly on other substrates.

[†] Obligate coprophiles.

[‡] Species known predominantly from dung and associated litter.

Wöstemeyer 2001). Several taxa from the Dimargaritales, Entomophthorales, Kickxellales, Mucorales, and Zoopagales (e.g., Helicocephalidaceae: *Helicocephalum, Rhopalomyces*; Piptocephalidaceae: *Piptocephalis, Syncephalis*) can be isolated from dung. Zygomycetes have been discussed extensively by Benjamin (1958, 1959, 1960, 1961, 1962, 1963, 1965, 1966, 1979), Benjamin and Mehrotra (1963), Benny (1982), Humber (1989), Morton and Benny (1990), and Benny and colleagues (2001). Here we follow the classification scheme of Benjamin (1979) as emended by Humber (1989), Morton and Benny (1990), and Cavalier-Smith (1998). No key to all of the coprophilous genera of Zygomycetes exists. Keys to the mucoralean genera can be found in Hesseltine and Ellis (1973) and O'Donnell (1979).

Coprophilous Zygomycete genera are listed in Table 21.3. Their contained species can be obligate coprophiles, in which case dung extract or hemin must be present in the culture medium for sporulation, or nonobligate coprophiles. In the latter case the species in a genus can be predominantly coprophilous but with one or a few noncoprophilous species or records; they typically can include many coprophilous and noncoprophilous species; or the species may occur primarily on other substrata but with one or a few coprophilous species or records.

TABLE 21.3
Genera of Coprophilous Zygomycetes and Their Associations with Dung

---

Basidiobolales
  Basidiobolaceae: *Basidiobolus**
Dimargaritales
  Dimargaritaceae: *Dimargaris,*[†] *Dispira,*[†] *Tieghemiomyces*[†]
Entomophthorales
  Ancylistaceae: *Conidiobolus**
Kickxellales
  Kickxellaceae: *Coemansia,** *Kickxella,*[†] *Spirodactylon,*[†] *Spiromyces*[†]
Mortierellales
  Mortierellaceae: *Dissophora,*[‡] *Mortierella*[‡]
Mucorales
  Mucoraceae: *Absidia,** *Actinomucor,*[‡] *Backusella,** *Benjaminiella,*[†] *Blakeslea,*[‡] *Chaetocladium,*[†] *Choanephora,*[‡] *Circinella,** *Cokeromyces,*[†] *Cunninghamella,*[‡] *Dichotomocladium,*[†] *Ellisomyces,*[†] *Fennellomyces,*[†] *Gilbertella,*[‡] *Helicostylum,*[‡] *Micromucor,*[‡] *Mucor,** *Mycotypha,** *Parasitella,** *Phascolomyces,*[†] *Phycomyces,*[†] *Pilaira,*[†] *Pilobolus,*[§] *Pirella,** *Protomycocladus,*[‡] *Radiomyces,** *Rhizomucor,** *Rhizopus,** *Syncephalastrum,*[‡] *Thamnidium,** *Thamnostylum,*[†] *Utharomyces,*[†] *Zychaea*[†]
Zoopagales
  Helicocephalidaceae: *Helicocephalum,** *Rhopalomyces*[‡]
  Piptocephalidaceae: *Piptocephalis,** *Syncephalis**
  Sigmoideomycetaceae: *Reticulocephalis,*[†] *Sigmoideomyces,** *Thamnocephalis*[†]
  Zoopagaceae: *Stylopage*[‡]

** Genus typically with many coprophilous and noncoprophilous species.

†Nonobligate coprophiles: species predominantly coprophilous, but genus may include one or more noncoprophilous species or records.

‡Species primarily on other substrates, but genus may include one or more coprophilous species or records.

§Obligate coprophiles.*

ASCOMYCETES

The 46 orders of Ascomycetes (see Barr 2001, Barr and Huhndorf 2001, Geiser and LoBuglio 2001, Pfister and Kimbrough 2001, and Samuels and Blackwell 2001 for synopses of the major orders) include numerous families, many of which contain coprophilous species. Our synopsis of these forms reflects the range of genera with coprophilous taxa but is not necessarily all inclusive. We have followed the taxonomy of Eriksson and Hawksworth (1998), except when accepted revisions or data supporting alternative designations are available. The increasing availability of ultrastructural and molecular evidence is rapidly changing our views of the classification of the Ascomycetes (e.g., Landvik et al. 1997, 1998; Brummelen 1998; Suh and Blackwell 1999). Sugiyama and colleagues (1999) have shown based on molecular data that the Myxotrichaceae should be excluded from the Onygenales, as positioned below, and placed among the Leotiales and Erysiphales. No comprehensive key to the coprophilous genera is available, although Richardson and Watling (1997) published keys to some of the more common taxa, and Bell (1983) provided keys to the coprophilous mycobiota of New Zealand. Keys to the more common Hyphomycetes, many of which represent anamorphs of Ascomycetes, have been provided by Seifert and colleagues (1983) and Ellis and Ellis (1988).

Our synopsis of the coprophilous Ascomycete taxa is provided in Table 21.4, along with an indication of the degree of coprophily. The latter designation is likely somewhat arbitrary in those genera with few species or records.

BASIDIOMYCETES

Very few genera of Basidiomycetes contain coprophilous taxa, although the 32 orders contain numerous families. Our synopsis of the coprophilous taxa follows the arrangement of Moser (1978) for the Agaricales, Miller and Miller (1988) for the Nidulariales and Tulostomatales, and Ginns and Lefebvre (1993) for all other orders. As molecular evidence becomes available a clearer understanding of Basidiomycete phylogeny is emerging. For updated classifications, see Hibbett and Thorn (2001), Swan et al. (2001), and Wells and Bandoni (2001). The Platygloeales is currently placed with the Uredinomycetes (Swan et al. 2001), as is the anamorphic basidiomycetous yeast *Sporobolomyces*, which includes one coprophilous species (Fell et al. 2001; Swan et al. 2001). No comprehensive key to the coprophilous Basidiomycete genera is available, but there are keys to the more common coprophilous taxa (Richardson and

TABLE 21.4

Genera of Coprophilous Ascomycetes and Their Associations with Dung

Leotiales
    Orbiliaceae: *Caccobius,** *Orbilia*[†]
    Sclerotiniaceae: *Martininia,*[†] *Coprotinia*[†]
    Thelebolaceae: *Ascozonus,** *Coprobolus,** *Leptokalpion,** *Pseudascozonus,** *Ramgea,** *Thelebolus**
    Rutstroemiaceae: *Rutstroemia*[†]
Pezizales
    Ascobolaceae: *Ascobolus,** *Guilliermondia,** *Saccobolus,** *Thecotheus**
    Ascodesmidaceae: *Ascodesmis*[‡]
    Eoterfeziaceae: *Eoterfezia**
    Otideaceae: *Boubovia,*[‡] *Byssonectria,*[‡] *Chalazion,*[‡] *Cheilymenia,** *Cleistothelebolus,** *Coprobia,** *Dennisiopsis,** *Lasiobolidium,** *Lasiobolus,**
        *Mycoarctium,** *Ochotrichobolus,** *Orbicula,*[†] *Pseudombrophila,** *Scutellinia,*[†] *Trichobolus,** *Trichophaea*[†]
    Pezizaceae: *Cleistoiodophanus,** *Hapsidomyces,** *Iodophanus,** *Peziza*[†]
    Pyronemataceae: *Coprotiella,** *Coprotus* (including *Ascophanus*),** *Dictyocoprotus**
Diaporthales
    Valsaceae: *Phomatospora*[†]
Dothideales
    Eremomycetaceae: *Eremomyces**
    Microthyriaceae: *Microthyrium*[†]
    Phaeotrichaceae: *Chaetopreussia,** *Delitschia,** *Pleophragmia,** *Preussia* (including *Sporormiella*),** *Semidelitschia,** *Sporormia,** *Spororminula,**
        Westerdykella[†]
    Testudinaceae: *Faurelina**
    Venturiaceae: *Protoventuria*[†]
Eurotiales
    Monascaceae: *Monascus*[†]
    Pseudeurotiaceae: *Pleuroascus,** *Pseudeurotium*[†]
    Trichocomaceae: *Byssochlamys,*[†] *Emericella,*[†] *Emericellopsis,*[†] *Eupenicillium,*[†] *Eurotium,*[†] *Hamigera,*[†] *Neosartorya,*[†] *Talaromyces*[†]
Hypocreales
    Hypocreaceae: *Bulbithecium,** *Hapsidospora,** *Heleococcum,*[†] *Leucosphaerina,** *Mycoarchis,** *Nectria,*[†] *Neocosmospora,*[†] *Nigrosabulum,**
        *Roumegueriella,** *Selinia**
Laboulbeniales
    Pyxidiophoraceae: *Mycorhynchidium,** *Pyxidiophora*[‡]
Microascales
    Chadefaudiellaceae: *Chadefaudiella**
    Microascaceae: *Enterocarpus,** *Kernia,** *Leuconeurospora,** *Lophotrichus,*[†] *Microascus,*[†] *Petriella,*[†] *Pseudallescheria,*[†] *Sphaeronaemella*[†]
Onygenales
    Arthrodermataceae: *Arthroderma,*[†] *Ctenomyces,*[†] *Gymnoascoideus*[†]
    Gymnoascaceae: *Arachniotus,*[‡] *Arachnomyces,** *Gymnascella,** *Gymnoascus,** *Leucothecium**
    Myxotrichaceae: *Myxotrichum,*[†] *Pseudogymnoascus*[†]
    Onygenaceae: *Amauroascus,*[†] *Aphanoascus,** *Ascocalvatia,** *Auxarthron,** *Kuehniella,*[†] *Onygena,*[‡] *Polytolypa,** *Renispora,** *Shanorella,**
        *Uncinocarpus,** Xynophila**
Ophiostomatales
    Kathistaceae: *Kathistes**
    Ophiostomataceae: *Klasterskya*[†]
Sordariales
    Ceratostomataceae: *Melanospora* (including *Microthecium, Sphaerodes*),[†] *Rhytidospora,** *Pteridiosperma*[†]
    Chaetomiaceae: *Achaetomium,*[†] *Boothiella,*[†] *Chaetomidium,*[‡] *Chaetomium,*[‡] *Corynascus,*[†] *Farrowia,*[‡] *Subramaniula,** *Thielavia*[†]
    Coniochaetaceae: *Coniochaeta*[‡]
    Lasiosphaeriaceae: *Anopodium,** *Apiosordaria,*[†] *Apiospora,*[†] *Apodospora,** *Arniella,** *Arnium,** Bombardioidea,** *Camptosphaeria,**
        *Fimetariella,** *Jugulospora,*[†] *Lasiosphaeria* (including *Bombardia, Cercophora*),[‡] *Periamphispora,** *Podospora* (including *Schizothecium*),**
        *Strattonia,** *Triangularia,*[‡] *Tripterosporella,** *Zopfiella* (including *Tripterospora*),[‡] *Zygopleurage,** *Zygospermella**
    Sordariaceae: *Apodus,*[†] *Copromyces,** *Emblemospora,** *Gelasinospora,*[‡] *Neurospora,*[†] *Sordaria**
Trichosphaeriales
    Trichosphaeriaceae: *Collematospora**
Xylariales
    Xylariaceae: *Ascotricha,*[†] *Hypocopra,** *Podosordaria,** *Poronia,** *Wawelia,** *Xylaria*[†]

---

   * Species predominantly coprophilous or occupying dung-impregnated litter, but genus may include one or more noncoprophilous species or records.

   [†] Species primarily on other substrates, but genus may contain one or more coprophilous species or records.

   [‡] Genus typically with many coprophilous and noncoprophilous species.

TABLE 21.5
Genera of Coprophilous Basidiomycetes and Their Associations with Dung

Agaricales
    Agaricaceae: *Agaricus*,*
    Bolbitiaceae: *Agrocybe*,* *Bolbitius*,[†] *Conocybe* (including *Pholiotina*)*
    Coprinaceae: *Anellaria*,[‡] *Copelandia*,[‡] *Coprinus*,[†] *Panaeolus*,[†] *Psathyrella**
    Cortinariaceae: *Naucoria*,* *Phaeogalera**
    Galeropsidaceae: *Weraroa**
    Pluteaceae: *Volvariella**
    Strophariaceae: *Psilocybe*,[†] *Stropharia**
    Tricholomataceae: *Lepista*,* *Pseudoclitocybe*,* *Tephrocybe**
Atheliales
    Atheliaceae: *Byssocorticium**
Nidulariales
    Nidulariaceae: *Crucibulum*,* *Cyathus*[†]
    Sphaerobolaceae: *Sphaerobolus**
Tulostomatales
    Tulostomataceae: *Schizostoma*[‡]
Auriculariales
    Sebacinaceae: *Sebacina**
Platygloeales
    Cystobasidiaceae: *Cystobasidium*,* *Platygloea sensu lato* (the single coprophilous species, *P. fimetaria* cannot be accommodated in *Platygloea sensu stricto.*)*

* Species primarily on other substrates, but genus may include one or more coprophilous species or records.
[†] Genus typically with many coprophilous and noncoprophilous species.
[‡] Species predominately coprophilous or occupying dung-impregnated litter, but genus may include one or more noncoprophilous species or records.

Watling 1997) and to the coprophilous Agaricales (Largent and Baroni 1988).

The taxa of coprophilous Basidiomycetes are listed in Table 21.5. The degree to which each taxon is associated with dung is also noted.

CHARACTERIZATION OF TAXA

MYXOMYCETES

Life Cycle

Myxomycetes produce fruiting bodies that bear reproductive units called spores. The spores range from about 5–22 μm in diameter, but most are about 10 μm in diameter. Spores are released when the fruiting bodies are disturbed by animals, rain, or wind. Wind currents disperse the spores over wide geographic areas so that many myxomycete species have a cosmopolitan distribution. Spores that land where adequate moisture is available germinate,

producing one to four amoeboid, nonflagellated myxamoebae or unequally biflagellate, swimming swarm cells. Dry conditions favor cell division and growth of myxamoebae into large populations of haploid cells that serve as future gametes. Under adverse conditions, the myxamoebae may encyst by rounding up and developing a thin wall. That resting stage, called a microcyst, promotes survival until favorable conditions return. Free water favors conversion of myxamoebae into swarm cells, which do not divide and also may serve as gametes. Two haploid gametes usually fuse to form zygotes. The zygote nucleus and its descendants divide mitotically without cell division, giving rise to a free-living, multinucleate, acellular mass of protoplasm, the plasmodium, which exhibits rhythmic protoplasmic streaming. This assimilative (mobile, growing, feeding) phase represents the "slime" stage of the life cycle. Feeding is animal-like, with the plasmodium engulfing particulate matter, usually bacteria. Plasmodia may be microscopic and undetectable with the naked eye, or they may reach up to a meter across. Under unfavorable environmental conditions, a plasmodium may transform itself into a sclerotium, a flat, hardened sheet or kernel-like mass; that resting stage provides another means of survival. The plasmodium passes through a series of developmental stages, eventually transforming itself into various types of fruiting bodies. Detailed information on life cycle events is available in Gray and Alexopoulos (1968) and Alexopoulos and colleagues (1996).

Fruiting Body Types

There are four distinguishable types of mature myxomycete fruiting bodies: the sporangium (which is most common), plasmodiocarp, aethalium, and pseudoaethalium. Mycologists use these terms in species descriptions to designate particular myxomycete habits and morphology, as follows:

Sporangium. A sporangium is a small, stipitate (stalked) or nonstipitate (sessile) structure of restricted size (generally 1 mm in diameter) and definite shape, often roughly spherical, formed by the clumping of the plasmodium. In certain minute species (Echinosteliales) a plasmodium may form only one sporangium. In contrast, under ideal conditions, hundreds may be present in phaneroplasmodial species (Physarales) (Fig. 21.2).

Aethalium. An aethalium is a relatively large, sessile, round or mound-shaped fruiting body formed from one plasmodium, as in *Lycogala* species and *Fuligo* species (Fig. 21.3). This is the largest type of myxomycete fruiting body, sometimes greater than 20 cm across in *Fuligo septica*.

FIGURE 21.1 *Kelleromyxa fimicola.* Cluster of immature sporangia on dung in moist chamber culture (group of sporangia 1 mm across).

FIGURE 21.2 *Didymium saturnus.* Brown sporangia on oat straw (sporangia 0.8 mm in diameter).

Pseudoaethalium. A pseudoaethalium is many sporangia packed so closely together as to suggest an aethalium as in *Dictydiaethalium* species, or partially fused as in *Tubifera* species, in which individual sporangia are clearly distinguishable at maturity.

Plasmodiocarp. A plasmodiocarp is a sessile, elongated, simple, wormlike, branched, ring-shaped, or netted fruiting body formed when a plasmodium concentrates and

remains intact in the main veins. *Hemitrichia serpula* forms such a fruiting body.

An individual sporangium is usually of characteristic size, shape, and color and represents the most common and basic fruiting body type. The structural parts of a sporangium, going from exterior to interior, are the peridium, hypothallus, stalk, columella, capillitium, and spores. The peridium and spores are essential components, whereas the hypothallus, stalk, columella, and

FIGURE 21.3 *Fuligo cinerea*. Close-up of tightly compacted sporangial remnants of an aethalium broken open to show the black spore mass (image 6 cm across).

capillitium can be present or absent. These terms are used in keys to the orders, families, genera, and species of Myxomycetes to refer to basic morphological characters (Keller and Braun 1999).

Peridium. In the endosporous Myxomycetes, an acellular, outer enclosing envelope, or structural wall, surrounds the spore mass. Because of the peridium, the spores develop internally; such development is called endosporous. The peridium may persist at maturity as a single- (e.g., *Perichaena microspora*), double- (e.g., *P. depressa*), or triple-layered (e.g., *Physarum bogoriense*) wall. The peridium can have a predetermined lid, which opens to release the spores, as in *Perichaena corticalis*, or can be evanescent, as in *Stemonitis* species.

Hypothallus. A hypothallus is a layer at the base of the fruiting body deposited by the plasmodium during fruiting. The structure ranges from membranous, transparent, and inconspicuous to tough, spongy, and conspicuous.

Stalk. A stalk is a structure that elevates the spore case above the substratum; it sometimes is referred to as a stipe in species descriptions.

Columella. The columella is a sterile central structure inside the sporangium that may function to support the capillitium. It represents the continuation of the stipe or,

in forms with sessile fruiting bodies, arises from the base of the sporangium. Its shape (dome-shaped, spherical, elongated), size, and texture can vary. All species of Trichiales and Liceales lack this structure.

Capillitium. A capillitium is a system of sterile threads that forms within a spore mass.

Spores. Spores are microscopic reproductive units, formed inside the fruiting body. Spore color, shape, size, and ornamentation are among the more constant characteristics used in species descriptions. Spores generally are globose, free as single units or aggregated into loose or tight clusters (Fig. 21.1), and range from 5 to 22 μm in diameter. Mature spores are generally uninucleate and typically haploid. Spores are protected during airborne dissemination by walls. The spore walls may have smooth, spiny, or warty surfaces, marked (partially, evenly, or banded) with a reticulate pattern or variations thereof (Fig. 21.4).

Plasmodia

Three general developmental types of plasmodia are recognized (Alexopoulos et al. 1996). They characterize taxa at different taxonomic ranks.

Aphanoplasmodium. An aphanoplasmodium is a plasmodium characterized in its early stages of development

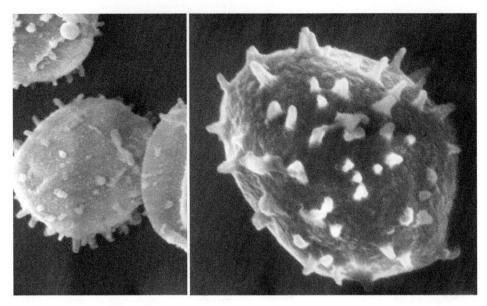

FIGURE 21.4 *Badhamia spinispora*. SEM of spores. Unique ornamentation showing conspicuous spines on one side of spore and smooth surface on opposite side (spores 12 μm in diameter).

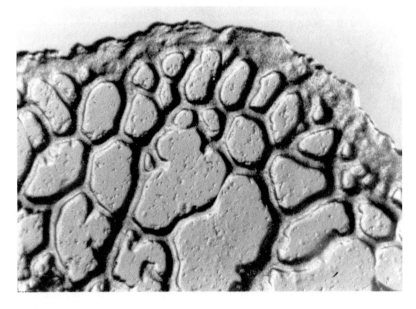

FIGURE 21.5 *Kelleromyxa fimicola*. Anterior portion of mature phaneroplasmodium. Note plasmodium with thickened, advancing, feeding fan with trailing veins, giving it a three-dimensional appearance (2 mm across).

by a network of flattened, threadlike, transparent, and nearly invisible strands that lack polarity and directional movement. The protoplasm is not coarsely granular, and the ectoplasmic and endoplasmic regions of young strands are indistinct. Such plasmodia tend to grow submerged in water or under very wet conditions. All members of the order Stemonitales have this type of plasmodium.

Phaneroplasmodium. This plasmodium is the largest, most common, and often most colorful type seen in the field. It typically gives rise to many fruiting bodies, which may cover an extensive area. In some species, the phaneroplasmodium heaps up to form a single, massive fruiting body. During initial phases of development, the phaneroplasmodium is a microscopic puddle of granular protoplasm similar to a protoplasmodium. It soon grows larger and is visible to the naked eye. At maturity it exhibits polarity and directional movement, with an advancing, fan-shaped feeding edge anteriorly, and a trailing network of veins posteriorly (Fig. 21.5). The veins deposit excreta along their margins, which frequently appears as "plasmodial tracks" on the substratum (Fig. 21.6).

Veins typically have two distinct regions: an inner, fluid sol layer, which undergoes rapid, rhythmic, reversible streaming, and an outer, thickened, nonstreaming gel

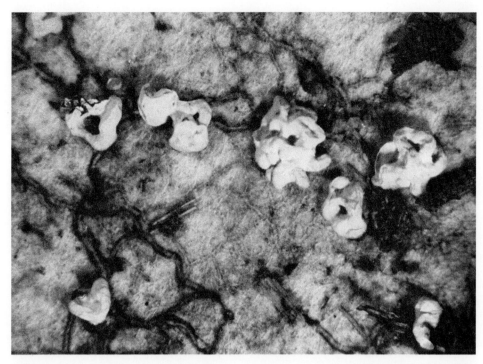

FIGURE 21.6 *Badhamia spinispora*. White, irregular sporangia in dung culture on filter paper. Note black plasmodial tracks on filter paper marking the former position of the plasmodial veins. Excreted matter is deposited along both sides of the veins as the plasmodium migrates over the surface (individual sporangia ca. 3-mm diameter).

layer. The entire phaneroplasmodium has a raised, three-dimensional aspect with definite margins. This type of plasmodium grows best in the absence of free water (Fig. 21.7). It is exemplified best by members of the order Physarales.

Protoplasmodium. The protoplasmodium is the smallest type of plasmodium and remains microscopic throughout its existence. The highly granular and homogeneous plate-shaped mass of protoplasm does not develop the veinlike strands, reticulum, or advancing fans seen in other types of plasmodia. The protoplasm streams slowly and irregularly. Each protoplasmodium gives rise to a single, tiny sporangium. A good example of this type of plasmodium is found in the myxomycete genus *Echinostelium*, order Echinosteliales.

Terminology Used in Keys

To use myxomycete keys found in monographic works, investigators must be familiar with the following terms and the structures that they represent. A comprehensive glossary of terms is available in Keller and Braun (1999).

- Badhamioid: Denoting a type of capillitium that consists of a network of calcareous tubules; found in species of the genus *Badhamia*.

- Calyculus: The lower portion of the sporangium, persisting as a cuplike or saucerlike base after spore release in some species of *Cribraria* and *Arcyria*.

- Calcareous bodies ("lime knots"): Deposits of calcium carbonate ($CaCO_3$) characteristically found on the peridium, stalk, or columella or in the capillitium of many species in the Physarales. Granules of $CaCO_3$ in expanded capillital threads are referred to as "lime knots" (see "Physaroid," later in this list). The word "lime," a term for calcium oxide (CaO), has been substituted incorrectly for calcium carbonate in some myxomycete literature. The capillitium in *Badhamia* species is a network of calcareous tubules.

- Clustered spores (spore balls): Varying numbers of spores (2–40) may adhere to each other to varying degrees, forming a cluster with (hollow) or without (solid) an interior cavity. Spore clusters are most notable in species of the genus *Badhamia*.

- Cortex: Thick, calcareous, outer covering of an aethalium, as in *Fuligo septica* and *Mucilago crustacea* of the Physarales; sometimes broadly defined as the wall of an aethalium as found in *Enteridium* and *Lycogala* in the Liceales and *Amaurochaete* and *Brefeldia* of the Stemonitales.

- Crystalline calcareous bodies: Crystals loosely scattered on the surface of the peridium or compacted to

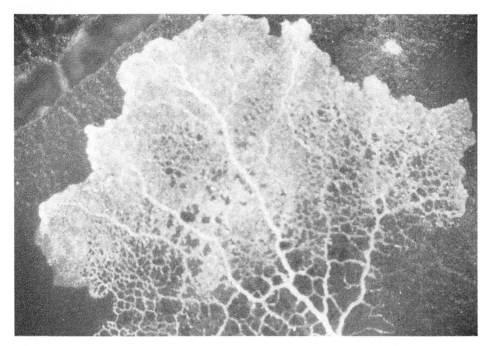

FIGURE 21.7   *Badhamia spinispora*. White phaneroplasmodium (2-cm diameter at widest point) on water agar culture medium.

form a smooth, eggshell-like outer crust, as in *Didymium* species.

- Dark-spored: Denoting spores that are dark-colored (black or brown) in mass; characteristic of the orders Physarales and Stemonitales (Fig. 21.8).
- Dictydine granules (plasmodic granules): Minute (0.5–3.0 μm in diameter), conspicuous, strongly refractive, and often dark-colored granules found on the peridium in species of *Cribraria*, *Dictydium*, and *Lindbladia*. The developmental origin of the granules is poorly known, in part because no species exhibiting them has been grown from spore to spore in culture.
- Elaters: Free, usually unbranched, elastic capillitial threads marked with spiral bands; characteristic of the genera *Oligonema* and *Trichia*. Genera such as *Arcyria*, *Hemitrichia*, and *Metatrichia* have elastic capillitial threads that branch and anastomose to varying degrees. The elastic threads are hygroscopic. They undergo twisting movements in response to changes in ambient humidity, thereby disseminating spores.
- Fruiting body: A spore-bearing structure, also known as a sporophore, sporocyst, spore case, or fructification. Four types of fruiting bodies are recognized: sporangium, plasmodiocarp, aethalium, and pseudoaethalium (see "Fruiting Bodies," earlier).
- Granular calcium carbonate: Amorphous (noncrystalline) or granular deposits of calcium carbonate ($CaCO_3$) usually found in the peridium, capillitium, or stalk of the fruiting body in the Family Physaraceae.

- Helical (spiral) bands: Thickened and raised bands, smooth or with ornamentation, that encircle the capillitial threads or elaters in a spiral pattern. Species of *Hemitrichia*, *Metatrichia*, and *Trichia* provide good examples of these bands.
- Light-spored: Denoting spores of bright colors (hyaline, gray, red, and yellow) in mass; characteristic of orders such as the Echinosteliales and Trichiales.
- Physaroid: Denoting a network of slender, noncalcareous, capillitial threads interconnected by enlarged calcareous nodes; found in *Physarum* species.
- Pseudocapillitium: An irregular structure found among the spores within the fruiting body in the order Liceales. In *Dictydiaethalium* it consists of slender threads, which are remnants of angles between adjoining peridial walls. In species of *Lycogala* it consists of noncalcareous tubules, usually greater than 6 μm in diameter; in *Enteridium*, it appears as perforated plates or membranes, and in *Tubifera* it appears as bristles. Its mode of development is unknown.
- Surface net: The system of branching, capillitial threads arising from the columella in *Stemonitis* species that eventually fuse at the surface into a net.
- True capillitium: Threadlike structures present in the fruiting bodies of Myxomycetes. Development occurs inside a system of preformed vacuoles that coalesce into channels, giving rise to threads of uniform diameter, often less than 6.0 μm. Examples are found in many taxa belonging to the endosporous orders Echinosteliales, Physarales, Stemonitales, and Trichiales.

FIGURE 21.8 *Didymium annulisporum* in dung culture on filter paper. Mature sporangia broken open to show black spore mass (individual sporangia 0.1 to 0.4 mm in diameter).

ZYGOMYCETES

Life Cycle

The majority of the coprophilous Zygomycetes, especially those in the Mucorales, are heterothallic, which means that they require two mating types (designated plus [+] and minus [−]) to reproduce sexually. In most cases only a single mating type (either the plus or the minus) is present on dung, and, therefore, reproduction is always asexual. In that case sporangiospores on dung germinate to produce a hyphal thallus that after 2 or more days forms sporangia or merosporangia. Those (mero)sporangia give rise to sporangiospores by cleavage plane formation (O'Donnell 1979). On dehiscence, the spores germinate if they land on a substratum with nutrients and moisture. This stage of the life cycle can be repeated numerous times. If the fungus is homothallic (mating types are not required, the fungus is self-fertile), then zygospores are formed along with sporangia.

In heterothallic species sexual hyphae grow together when the (+) and (−) mating types are in close proximity. The sexual hormone trisporic acid is produced from β-carotene by cross-feeding intermediates between the two mating types (Gooday 1994). That results in the formation of progametangia, which fuse at the apex when they contact one another. The apex of each progametangium is walled off by a septum, and the progametangia are then termed gametangia. The central septum dissolves, and the nuclei in the cell, now the prozygosporangium, which is supported by a suspensor, fuse. The prozygosporangium enlarges, its wall thickens and becomes ornamented, and it becomes the zygosporangium. An ultrastructural study of zygospore ontogeny is reported by O'Donnell (1979). The zygosporangium germinates to form a hypha or a germ sporangium. The life cycles of *Rhizopus stolonifer* and *Mucor piriformis* are covered in Alexopoulos and associates (1996) and Michailides and Spotts (1990).

The germ sporangium can contain sporangiospores that are all of the plus (+) or the minus (−) mating type, or both mating types can be present (Michailides and Spotts 1988). The zygospores of most species must undergo a resting period of from a few days to a few months before germinating. Reports of zygospore germination are rare in the literature. Zygospores of *Radiomyces* species must reach a certain age, and require a carbon source and scarification to germinate (Embree 1978); those of *Mucor piriformis* require drying and rewetting (Michailides and Spotts 1988), immersion in water, or placement on acidified water agar (Guo and Michailides 1998).

Zygomycetes form a sexual spore, the zygospore, which is rarely seen in most species. Consequently, species in this group are classified according to the types of asexual reproductive structures and hyphal septa they produce. These fungi form unicellular, nonmotile, asexual sporangiospores (aplanospores) by cleavage plane formation in sporangia or merosporangia. Some species produce chlamydospores, arthrospores, yeast cells, or a combination of these asexual spores. Most species produce aseptate hyphae, although some species form hyphal bodies or protoplastlike structures. Members of Dimargaritales and Kickxellales regularly form septa with plugs in lenticular cavities.

Asexual Structures

Sporangium. A sporangium (including sporangiolum; see Benny 1995b) is a structure borne on the apex of a sporangiophore or its branches, or on a pedicel or denticle that arises from the side of a sporangiophore, or from a vesicle in which 1 to 100,000 spores are produced. A sporangium can be globose to obovoid or flask- or dumbbell-shaped to somewhat cylindrical in a few taxa. Its outer wall can be smooth, have terminal spines, or be covered with calcium oxalate crystals or spines, and at maturity it can deliquesce, persist, or become evanescent, depending on the species. Sporangia in species of *Blakeslea*, *Choanephora*, *Poitrasia* (all Choanephoraceae), and *Gilbertella* (Gilbertellaceae) have persistent walls bearing calcium oxalate crystals and are characterized by a single (or more in *Poitrasia*) longitudinal suture that extends from the bottom to the top and then to the bottom on the opposite side. Sporangia can be nonapophysate (sporophore straight or constricted where it joins the sporangium), apophysate (sporophore apex widens where it joins the sporangium), or vesiculate (slightly constricted immediately below the sporangium but wider below the constriction). The sporangia in most Mucorales contain a relatively large, conspicuous columella (an extension of the sporophore) that is persistent, smooth-walled, and variously shaped (hemispherical, hourglass-shaped, globose, obpiriform to cylindrical). In some species it may bear one or more apical spinelike structures. Columella in some small, especially unisporate sporangia, if present, can be hemispherical (and often compressed when a spore is present) or discoid. In Mortierellaceae (Mucorales) the columella can be hemispherical to septum-like and is relatively small and inconspicuous. Sporangia may be produced alone or with chlamydospores, arthrospores, or yeast cells (Fig. 21.9 A and B).

Sporangiolum. We follow Benny (1995b) and consider the sporangiolum to be a small sporangium with a persistent wall and a few spores (see "Sporangium," earlier).

Merosporangium. A merosporangium is a cylindrical sporangium containing from one to 12 or 15 spores that usually are borne in a uniseriate arrangement; *Syncephalastrum racemosum* occasionally bears spores in a biseriate arrangement (Benjamin 1959). Merosporangia are produced by some or all species of Mucorales (*Syncephalastrum*, Syncephalastraceae), Dimargaritales, Kickxellales, and Zoopagales (*Piptocephalis*, *Syncephalis*, Piptocephalidaceae; *Rhopalomyces*, Helicocephalidaceae). The merosporangia are unispored in *Rhopalomyces* and the Kickxellales; two-spored in Dimargaritales; and one- to many-spored in *Piptocephalis*, *Syncephalastrum*, and *Syncephalis* (Fig. 21.9 C, F, and G).

Sporangiospore. A sporangiospore is a nonmotile, unicellular, and often hyaline and smooth-walled spore. Morphologically, it can be nearly globose, ovoid, ellipsoid, fusoid, or irregular and in exceptional cases, cylindrical, naviculate, or discoid; most commonly, spores are ovoid to ellipsoidal. *Gilbertella*, *Blakeslea*, *Choanephora*, and *Poitrasia* have fusiform spores with several thin, hyaline appendages attached apically; walls range from smooth and hyaline to brown and striate. Thicker appendages are found on the sporangiospores of *Aquamortierella elegans*, which is aquatic.

Conidium (conidia). A conidium is a spore originating from a special hypha (condiophore) by the formation of a delimiting septum. In Entomophorales the spore wall and the conidiophore wall have the same layers (Benny et al. 2001). The term often has been used for the unispored sporangia that occur in several genera of Mucorales (e.g., *Cunninghamella*, *Choanephora*). In a unispored sporangium, however, the wall of the sporangiospore is deposited inside the wall of the sporangium and is separate from it.

Chlamydospores. Chlamydospores are solitary, intercalary, and relatively thick-walled asexual spores whose primary function is perennation, not dissemination (Griffiths 1974). They are produced in swellings that form in young, aseptate mycelia. Powell and colleagues (1981) described the formation of these spores in *Gilbertella persicaria*. Chlamydospores are most common in members of the Mucorales, although they also have been reported from *Rhopalomyces* (Zoopagales). Certain taxa are characterized by the presence of conspicuous, pigmented chlamydospores in subaerial (*Phascolomyces articulosus*) or aerial (*Chlamydoabsidia padenii*) hyphae. Other taxa have relatively inconspicuous, hyaline spores that have thick, smooth walls. Those single cells or chains of spores form in subaerial or aerial hyphae or in sporophores. Some species commonly form chlamydospores, but many other taxa do so only under certain conditions, especially in nature.

Arthrospores. Arthrospores are doliform spores with relatively thin walls compared to chlamydospores, and

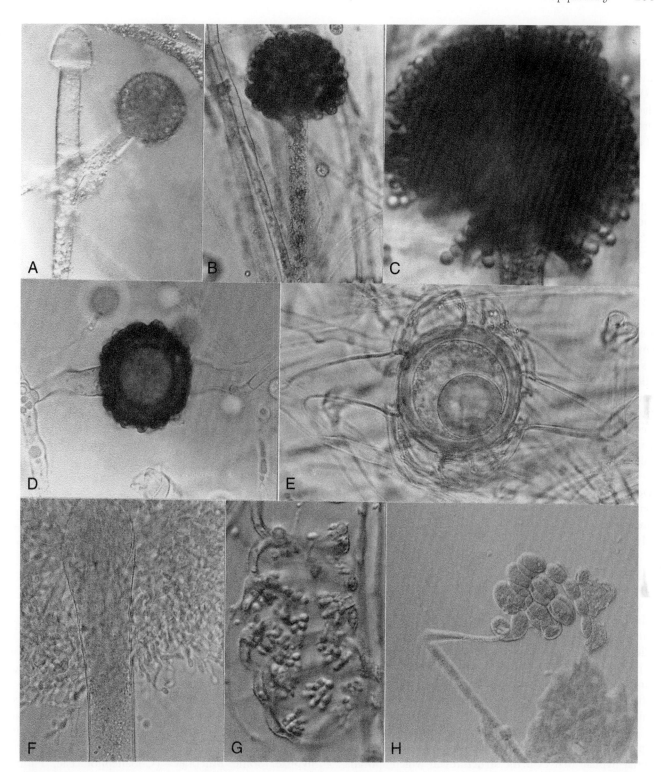

FIGURE 21.9   **A-E.** Mucorales: **A.** *Actinomucor elegans* sporangium and columella (upper branch), 50 μm = line 16 mm long. **B.** *Cunninghamella echinulata* with unispored sporangia, 50 μm = line 17.5 mm long. **C.** *Syncephalastrum racemosum* showing merosporangia, 20 μm = line 14 mm long. **D.** *Cunninghamella echinulata* zygospore, 50 μm = line 17.5 mm long. **E.** *Radiomyces mexicana* zygospore with appendaged suspensors, 50 μm = line 21.5 mm long. **F.** Dimargaritales: *Dimargaris cristalligena* vesicle bearing sporiferous branchlets and two-spored merosporangia, 20 μm = line 15 mm long. **G.** Kickxellales: *Spirodactylon aureum* coiled fertile hyphae bearing sporocladia, pseudophialides, and unispored merosporangia, 20 μm = line 14.5 mm long. **H.** Zoopagales: *Helicocephalum sarcophilum* coiled fertile hyphal-like apex consisting of a chain of arthrospores, 100 μm = line 14 mm long.

they are produced singly or in chains. Dehiscence is rhexolytic, schizolytic, or intermediate between those types. Arthrospore formation in *Ellisomyces anomalus* and *Mucor rouxii* is discussed by Beakes and colleagues (1984). The spores of *Helicocephalum* (Helicocephalidaceae, Zoopagales), which are relatively large and brown-pigmented, are considered to be arthrospores (Fig. 21.9 H).

Yeast Cells. Yeast cells (Cole et al. 1980) are round to long cells that reproduce vegetatively by budding or germinate to produce a mycelium. They are found in some members of the Mucorales (*Benjaminiella, Cokeromyces, Mycotypha;* Benny et al. 1985) under normal growth conditions on the surface of rich media (e.g., MEYE, see Appendix II). *Mucor rouxii* (Mucoraceae) also forms yeast cells but only below the surface of a thick layer (2.0–2.5 cm) of poured agar (Bartnicki-Garcia 1978).

Pedicels. Pedicels are relatively thin structures that bear small sporangia; they arise laterally from a sporophore or its branches or from a vesicle. Pedicels are tapered or parallel and are shorter than the diameter of the supported sporangium, or they may have parallel walls one to many times longer than the diameter of the sporangium. Long pedicels are straight, curved to recurved, or twisted and contorted. Pedicels are found in several taxa of Mucorales and Zoopagales.

Denticles. Denticles are tapered pedicellar remnants left on a fertile vesicle after sporangiolar dehiscence. They are present in *Benjaminiella, Mycotypha,* and *Radiomyces* (all Mucorales).

Hyphae. Thalli of most Zygomycetes consist of aerial and subaerial hyphae or modifications thereof; they are coenocytic (aseptate) in Endogonales, Glomales, and Mucorales except where old or injured hyphae or reproductive structures are delimited. In Dimargaritales and Kickxellales, hyphae are regularly septate, with each septum containing a septal plug in a lenticular cavity. Animal parasites in Zoopagales and Entomophthorales often produce coenocytic hyphal bodies after spore germination.

Sporophores (Sporangiophores). Sporophores are aerial hyphae that bear the structures in which sporangiospores are formed. They arise directly from the substratum or from stolons or opposite rhizoids.

Stolons and Rhizoids. Stolons and rhizoids are vegetative structures produced by some Mucorales and Entomophthorales. Stolons are aerial "runners" (Hawksworth et al. 1995) that terminate in rhizoids or sporangiophores. Rhizoids are rootlike hyphae that lack nuclei (Hawksworth et al. 1995); they attach to surfaces and support spore-forming structures. Stolons and rhizoids are characteristic of many genera of the Mucorales, such as *Absidia, Rhizopus,* and *Rhizomucor.* The formation of stolons and rhizoids is an important characteristic in the classification of some taxa.

Sexual Structures

Zygospores. Zygospores are sexual spores of Zygomycetes; they are rarely observed except in homothallic species. Occasionally, heterothallic species form zygospores on initial isolation, in which case it is advisable to make several isolates of the fungus to ensure that both mating types are acquired. Morphology of the zygospore, including the suspensor and zygosporangia, and its site of formation in the culture may be important key characters. Zygospores of Dimargaritales and Kickxellales are hyaline, have "smooth" or dimpled walls, and form in the substratum.

Zygosporangium. Characteristics of the zygosporangial wall such as color, light reaction, and chemical makeup (e.g., transparent or translucent versus opaque, noncarbonaceous versus hard and carbonaceous) may be important in classification but are usually not noted in descriptions. Of greater importance is wall ornamentation, which has been relatively well documented using scanning electron microscopy (SEM), especially for members of the Mucorales. Ornamentation can be quite variable, although most mucoralean species have warts that in some cases are described as pyramidal. Less commonly, zygosporangial walls are undulate, smooth, and transparent as opposed to opaque and translucent.

Suspensors. Suspensors are hyphae that support a zygospore (Hawksworth et al. 1995). Each zygospore forms on two suspensors, which, in most species are of more or less equal size and shape (isogamous type of some mycologists). A few species produce suspensors that are distinctly unequal in size and shape (heterogamous). In a few taxa, one or both suspensors have appendages (straight, branched, or dichotomous). Among Mucorales, suspensors most commonly are opposed (apex pointed toward one another, with zygospore between); the zygospores are formed in aerial hyphae. Some Mucorales (Choanephoraceae, Pilobolaceae) and Zoopagales have apposed (parallel and straight or entwined with the zygospore formed above) suspensors, which are found only at or in the substratum. Tong-type suspensors (at maturity they resemble ice-tongs) of *Phycomyces* are a modification of the apposed suspensor; they form as the huge zygospore matures, causing the suspensors to separate. Suspensors in Dimargaritales and Kickxellales are unmodified and resemble hyphae (Fig. 21.9 E).

ASCOMYCETES

Life Cycle

The life cycle of the Ascomycetes has many variations; no one is typical of the group. Several types of those life cycle are represented by Alexopoulos and associates (1996: figs. 7-7, 9-3, 12-18, 13-15) and Krug (1999: fig. 9.17). In general, following mitosis, ascospores are discharged from the ascus and germinate to form a mycelium into which the nuclear progeny from the spore are distributed. Asexual reproduction can be initiated by fragmentation of the mycelium, by formation of resting spores, or by conidia that arise directly from the mycelium or from a range of specialized structures on the mycelium. A synopsis of the various asexual states is presented by Seifert and Gams (2001). On discharge, conidia produced by the growing mycelium form mycelia similar to those formed by germinating ascospores.

Many Ascomycetes are homothallic, but in the species that are heterothallic, an opposite mating type, derived from a mycelium that originated from a separate spore, is required for sexual reproduction. Morphologically distinct multinucleate antheridia and uninucleate ascogonia, the male and female structures, respectively, differentiate from hyphae in the mycelium. On contact the nuclear contents from the antheridium pass into the ascogonium. In some Ascomycetes the conidia act as spermatia and on contact with the tip of the ascogonium empty their contents into the ascogonium. Following cytoplasmic fusion, the ascogonium forms ascogenous hyphae. Simultaneously the ascogonial stalk and the hyphae around the ascogonium divide and intertwine to form the ascocarp wall. The young ascus forms from the ascogenous hyphae. Nuclear fusion takes place in the young ascus, which then begins to elongate, and the zygote nucleus undergoes meiosis. Typically, the haploid products of meiosis undergo mitosis to form eight nuclei, which are incorporated into the ascospores during spore formation. At maturity the ascospores usually are discharged forcefully by various structures in the ascus tip, although disintegration and mechanical breakdown of the ascocarp are also important means of ascospore release.

Ascomycetes include a diverse assemblage of fungi; the coprophilous species are distributed among more than 150 genera (see, "Ascomycetes" under "Taxonomic Synopsis," earlier). A sexual spore (ascospore) is formed in all species of coprophilous taxa, and frequently an asexual stage (anamorph) is formed as part of the life cycle as well. Most species are wind dispersed, although insects or water play a role in some taxa. Mammals are also important agents in dispersal (see, "Ecology," earlier). The spores are produced within asci in several types of fruiting bodies. In some genera, a sterile tissue (see "Stroma" and "Sclerotium" under "Fruiting Body," later in this chapter) is associated with the fruiting body, which typically is a protective structure but can also signal formation of a resting stage. It does not function in spore release. Classification in the Ascomycetes has traditionally been based primarily on the morphology of the ascospore. Considering the degree of variability within the group, a considerable amount of information is required to characterize a taxon adequately.

Fruiting Body

Stroma. A stroma is a sterile tissue that contains ascocarps in some Ascomycetes; they can be stalked, sessile, or occasionally threadlike as in some *Xylaria* species. If stalked, then ascocarps are embedded in the upper portion. Sessile stromata are typically compound or occasionally even confluent; in *Hypocopra*, they are clypeate or shieldlike. Stromata sometimes are reduced to a small ring around the perithecial neck or replaced by a spreading, surface mycelium. They are usually discoid, globose, cushionlike, or elongated and are dark colored (but pale or flesh-colored in *Selinia*). Ascocarps usually protrude from the stroma (semi-immersed); in *Poronia* species, they are immersed in the discoid portion of the stroma. They sometimes are associated with an anamorph, as in some species of *Podosordaria* and *Xylaria*.

Sclerotium. A sclerotium is a hard, sterile, dark or brightly colored structure from which ascocarps arise in certain discoid fungi (e.g., Sclerotiniaceae) and some species of *Eupenicillium*. Frequently sclerotia appear well before the development of ascocarps and asci.

Ascocarp. An ascocarp is an open or closed structure containing the asci and ascospores. There are three morphological types of ascocarps: discoid (subglobose, saucer-shaped, or turbinate) (Fig. 21.10), with an open or closed hymenium (spore-bearing layer); flask-shaped (Fig. 21.11), with an enclosed hymenium; and cleistothecial (typically globose), in which the hymenium is enclosed. Shapes are helpful in grouping taxa, but they are not taxonomically significant. Other criteria to note are color (light or dark), discoid types of the external and hymenial surfaces, and presence or absence of an ostiole (opening for spore discharge). An ostiole is present in many pyrenomycetes and loculoascomycetes; it is not a significant single characteristic at the generic level.

Ascocarp Wall. The ascocarp wall is a tissue enclosing the hymenium; it usually is called the excipulum in discoid

FIGURE 21.10 Discoid or turbinate ascocarp (apothecium) of *Hapsidomyces venezuelensis in situ* (ascocarp ca. 1.5 mm diameter). (Reprinted from Krug and Jeng 1984; photo by R. S. Jeng)

FIGURE 21.11 Flask-shaped ascocarp (perithecium) of *Preussia minima* (perithecium 100–200 × 90–120 µm; ascus ca. 90–100 × 13–18 µm).

fungi (discomycetes) and the peridium in flask fungi (pyrenomycetes). The ascocarp wall sometimes is composed of a loose mass of hyphae (e.g., *Gymnoascella*). It occasionally is lacking (e.g., *Ascodesmis*), in which case the ascocarp is composed of a naked mass of asci and associated elements. In most genera the walls are differentiated into several layers comprised of a number of possible tissue types (Korf 1973). The wall in certain species of pyrenomycetes appears platelike or cephalothecoid in surface view, but this type of wall has no apparent evolutionary significance. *Bombardioidea* species have a diagnostic central peridial layer of gelatinous hyphae. Reactions of the wall with Melzer's reagent (IKI) and associated color

changes (usually dextrinoid, to reddish or red-brown), especially in discoid ascocarps, may be diagnostic.

Hairs. Some ascocarps have filamentous, erect, hooked, forked, wavy (e.g., *Kernia* species), coiled, or spiraled (e.g., *Lasiobolidium*) hairs or appendages. The hairs in some taxa of Lasiosphaeriaceae even may be agglutinated. Some hairs are characteristically ornamented or blistered (e.g., *Chaetomium*). Hair characteristics are taxonomically significant in some genera but more typically are important for distinguishing species.

Setae. Setae are stiff, erect, thick-walled hairs (as in *Coniochaeta*). They are rooted deeply within the wall and are enlarged or branched at the base in some discoid ascocarps. Presence or absence of septa in the setae is diagnostic for some genera (e.g., *Cheilymenia*).

Asci and Sterile Elements

Ascus. An ascus is a saclike structure enclosing the ascospores. The type of ascus is fundamentally important in the classification of ascomycetes. In discoid fungi the ascus may be operculate, opening with a lid or similar discharge mechanism, and relatively thin-walled, or it may be inoperculate and thick-walled. Pyrenomycetous fungi may be thin-walled or expansive because of an elastic inner membrane (loculoascomycetes; e.g., *Preussia*), in which case the wall appears thickened. Various structures for spore discharge are present at the apex. The reaction of the ascus wall and apical discharge apparatus with IKI is critical in both pyrenomycetous and discoid fungi and also with Congo Red in the latter. For example, the ascal wall of *Ascobolus*, an operculate discomycete, turns blue in contact with IKI. Ascus shape (cylindrical, clavate, globose) may be indicative of some species or even genera; in discomycetes and pyrenomycetes, it is typically cylindrical or clavate, and in many cleistothecial ascocarps, it is globose. In some species, asci are evanescent (disintegrate) and must be observed at an early stage. The number of spores in an ascus (typically eight) characterizes some species; if the number is large, it can be estimated, but usually it is a multiple of eight.

Ascal Apical Apparatus. In operculate discomycetes, the opening is usually a lid (*Ascobolus*) or sometimes a slit (*Thecotheus*). In pyrenomycetous fungi, the spore discharge apparatus may be a complex plug (*Xylaria*), a ring of varying complexity (*Podospora* species), or a globule (*Lasiosphaeria*). In IKI the apical apparatus may turn blue (Fig. 21.12) or sometimes reddish. In some discomycetes, the entire ascal wall colors. Because some of these reactions are ephemeral, the preparation should be

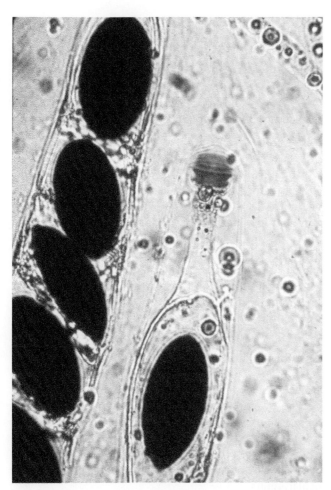

FIGURE 21.12 Ascal apical apparatus of *Hypocopra kansensis* stained in aqueous ink blue showing the complex ringlike nature of the apical plug and the gelatinous sheath surrounding the ascospores (apical apparatus 14–18 × 12–14 µm; spore ca. 50–62 × 25–28 µm).

observed carefully following addition of IKI. Baral (1987) showed that mounting conditions may affect the visibility of these reactions.

Paraphyses and Other Sterile Elements Located Among the Asci. Paraphyses and other sterile elements are typically filamentous, but sometimes they are tubular or reduced to groups of cells. Tips can be straight, swollen, hooked, or twisted and sometimes are glued together with, in most cases, a dark-colored substance. The paraphyses may contain globules and/or pigments.

Spores

Ascospores. Ascospores are the sexual propagules of ascomycetes. They are critical for describing most fungi. In ascocarp-forming ascomycetes, ascospores are only found in mature fruiting bodies. Thus, studying spore-bearing ascocarps is essential. Heterothallic species form

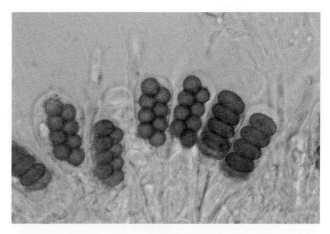

FIGURE 21.13 Asci of *Saccobolus infestans* showing the eight ascospores arranged in a cluster (asci ca. 60–90 × 12–18 µm; spore cluster ca. 21–28 × 10–14 µm; spores ca. 9–11 × 5–6 µm). (Photo by R. S. Jeng)

spores only when both mating types are present. The shape, symmetry, size, number of cells per spore, and pigmentation of ascospores, as well as the presence or absence of mucilaginous attachments, appendages (Fig. 21.14), or a sheath (Fig. 21.12), are critical for the characterization of many species. Usually ascospores are arranged in a row in the ascus, but they may be uniseriate or biseriate; in *Saccobolus* the spores are clustered (Fig. 21.13). In *Lasiosphaeria* young spores are hyaline and wormlike, becoming pigmented and ellipsoidal at maturity. The pigmented stage, together with the hyaline basal cell, sometimes appears only after discharge. Many spores are bilaterally or dorsiventrally flattened. In some two-celled spores the basal cell is small and disappears prior to maturity (*Hypocopra*); in other instances septation occurs late in spore development. Thus, investigators should look for septation both early and late in development. Mucilaginous attachments and sheaths should be observed in aqueous mounting media because they frequently disappear or break down under other mounting conditions. Attachments may be erect, lash-like, or united. Spores often are ornamented with spines, warts, ridges (Fig. 21.15), pits (Fig. 21.16), or reticulae (*Neurospora* species), which have been documented with SEM for a number of genera. Spores are copper-colored in some genera (e.g., *Microascus* and related genera), but more typically they are colorless or dark brown. Occasionally young spores are dextrinoid (Microascaceae).

Germ Pores. Germ pores are openings in the spore wall for exit of the germination tube. When present, the number (except in some species of *Gelasinospora*), position, size, and arrangement (for some species of *Gelasinospora*) are diagnostic of some species. Terminal pores

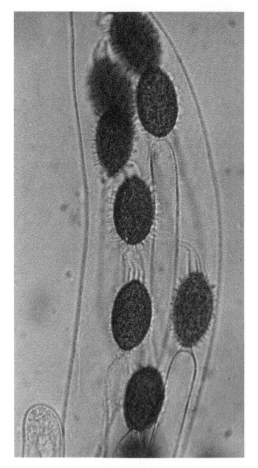

FIGURE 21.14 Ascus and ascospores of *Podospora* species showing the short lash-like gelatinous appendages attached to the sides of the upper cell of the ascospores, the apical striated gelatinous appendage, and the elongated basal cell of the ascospores (upper cell of ascospores ca. 30–39 × 18–24 μm; basal cell ca. 80–120 × 9–12 μm). (Photo by R. S. Jeng)

may be apical or subapical. The gelatinous sheath may invaginate around the pore (e.g., *Sordaria*).

Germ Slit. A germ slit is an elongated opening in the spore wall for exit of the germination tube (e.g., *Coniochaeta*). The position, direction, and sometimes the length of the slit are usually characteristic at the species level. In some species, especially those with spores flattened on one side, the slit may be difficult to observe.

De Bary Bubbles. A De Bary bubble is a conspicuous gaseous bubble in the center of a spore. Usually only one bubble is present. The bubbles form in spores mounted in certain media or in spores following desiccation as a result of contraction of the cytoplasm from the spore wall. De Bary bubbles are diagnostic of certain genera (e.g., *Coprotus*), but they also are known in a wide range of genera for which they have no taxonomic significance.

Asexual State

Anamorph. The anamorph is the asexual state in the fungal life cycle. The anamorphs of coprophilous fungi are generally similar to those of noncoprophilous fungi. Diagnostic criteria for anamorphs of noncoprophilous fungi are given in Chapter 11. For those anamorphs that have been associated with teleomorphs, determining the form genus is probably sufficient for identification. A particular anamorph is frequently characteristic of a particular genus.

BASIDIOMYCETES

Life Cycle

In most Basidiomycetes, except those in which the fructification remains closed, the basidiospores are dis-

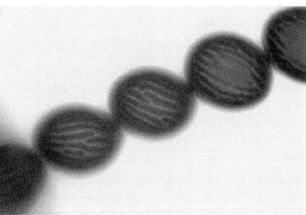

FIGURE 21.15 Ascospores of *Neurospora pannonica* showing the riblike ornamentation on the spore wall (spores ca. 30–38 × 20–24 μm). (Reprinted from Krug and Khan 1991.)

FIGURE 21.16 Ascospores of *Gelasinospora hapsidophora* showing pits in the spore wall (spores ca. 19–26 × 14–16 μm). (Reprinted from Khan and Krug 1989; photo by R. S. Khan)

charged forcefully from the basidium. Basidiospores usually germinate to form a mycelium of separate strands of hyphae with uninucleate cells. Most Basidiomycetes are heterothallic. The cytoplasm of cells from two compatible mycelia fuse to form a dicaryon. The resulting binucleate cell divides, giving rise to a binucleate mycelium. Fructifications develop from that mycelium. Fruiting bodies are composed of hyphae, which form a hymenium in the external region of the fructification. The basidia and basidiospores develop there by nuclear fusion and meiosis from binucleate hyphae. For a diagrammatic representation of the various stages in development, see Alexopoulos and associates (1996: fig. 16.6). The basidiospores form externally on the sterigmata of the developing basidium. In the heterobasidiomycetes (e.g., Auriculariales and Platygloeales), the young basidium undergoes further division to form four cells, each bearing a sterigma. Prior to spore discharge a droplet often forms adjacent to the basidiospore and sterigma and aids in spore release (see Alexopoulos et al.

1996: fig. 16-10). Spores germinate when they land on a suitable substratum. Although asexual reproduction is not as significant in the Basidiomycetes as it is in the Zygomycetes and Ascomycetes, some spores may bud to form conidia, or a mycelium may fragment into asexual spores.

In comparison with other major groups of fungi (Table 21.2, 21.3 and 21.4), considerably fewer genera of Basidiomycetes include coprophilous species, and most of those genera belong to the order Agaricales with a few representatives in other orders (Table 21.5). All of the species form basidia and basidiospores (sexual spores), which together with the type of fruiting body and presence of cystidia are important in classification. Wind is the primary dispersal agent for spores, except in the Nidulariales, for which water is important in dispersal. The criteria used to describe coprophilous Agaricales are similar to those used for terrestrial members of the order. Readers should refer to Chapter 8 for details on characterizing Agaricales and lignicolous macrofungi, and to Chapter 17 for information on the Atheliales. For characteristics of the Nidulariales, see Brodie (1975) and Miller and Miller (1988); for the Tulostomatales, see Miller and Miller (1988); and for the Auriculariales and Platygloeales, see Martin (1952).

Fruiting Body

Fructification. Most Agaricales form a stalked, enlarged, generally top-shaped fruiting body with basidia borne on gills. In the Nidulariales the fruiting body is oval to urn-shaped; in Tulostomatales, it is globose to oval and stalked; and in Atheliales it is spreading and adheres to the substratum. It has a rather gelatinous texture in Auriculariales and Platygloeales but becomes horny under dry conditions; it is irregularly cupulate to resupinate in Auriculariales and resupinate and spreading in Platygloeales.

Lamellae (Gills). Lamellae are leaves or flaps of tissue that protrude from the underside of the cap in Agaricales, bearing basidia, spores, and sometimes cystidia. The type of attachment of a gill to the stipe and characteristics of the gill margin are sometimes significant for defining genera.

Peridium. The peridium is a one- to several-layered wall of the fruiting body in the Nidulariales and Tulostomatales. In *Sphaerobolus*, one layer is gelatinous.

Exoperidium. The exoperidium is the outer layer of the peridium. In *Sphaerobolus*, it splits into stellate lobes exposing the inner peridium; in *Schizostoma* it has a sand-case texture and sloughs off apically.

Endoperidium. The endoperidium is the inner layer of the peridium. In *Schizostoma*, it can be smooth or pitted; it splits to create lacerate fragments.

Peridiole. Peridioles are thick, oval- or disc-shaped structures in the fruiting body of Nidulariales that contain the basidia and spores. They are dark-colored in *Cyathus* and *Sphaerobolus*, but white in *Crucibulum*.

Funicular Cord. The funicular cord is an expandable cord attached to the peridiole and the inner peridium in the Nidulariales, which functions in discharge of the peridioles. It is a single, simple strand in *Crucibulum*; it is more complex in other genera.

Gleba. The gleba is spore-bearing tissue within the fruiting body of Tulostomatales and related orders.

Capillitium. A capillitium is a thick-walled cell within the gleba; such cells form fragmented, septate hyphae in *Schizostoma*.

Basidium and Sterile Elements

Basidium. The basidium is a clublike structure with 2–4 apical sterigmata that bear basidiospores. It is usually entire but is divided longitudinally in Auriculariales and Platygloeales.

Phragmobasidium. A phragmobasidium is a basidium divided into separate cells following meiosis, as in Auriculariales and Platygloeales. In cross-section the metabasidium (part where meiosis occurs) appears as four cells.

Sterigma (pl. Sterigmata). A sterigma is an apical extension of the basidium that bears the basidiospores; it is elongated in Auriculariales and Platygloeales. There are usually four sterigmata on each basidium.

Cystidia. Cystidia are sterile structures located between the basidia in some Agaricales and Atheliales; they may extend beyond the hymenium and sometimes are encrusted with crystals of calcium oxalate near their apices (Keller 1985; Arnott 1995). Specialized types of cystidia are found on different parts of the basidiocarp in the Agaricales.

Spores

Basidiospores. Basidiospores can be colorless or pigmented. They are typically one-celled but are septate in a few genera. Basidiospore shape, size, color, and ornamentation are important diagnostic characters for genera and species.

Germ Pore. A germ pore is an opening in the wall at the spore apex; it is for exit of the germ tube.

Spore Print. A spore print is a deposit of spores, usually made on white paper or a glass slide, to indicate spore color; it is primarily used with Agaricales.

SAMPLING AND CULTURING MYXOMYCETES

SAMPLING

Dung from herbivorous animals such as cattle, horses, bison, and rabbits is the best source of Myxomycetes. Dung should be several days old with a crusted outer surface and hardened interior; soft, freshly defecated feces should be avoided. Most investigators who use moist chambers (see Appendix I) to isolate Myxomycetes record the length of time in storage but fail to note the approximate age of the dung when collected, which is important for aging specimens. Investigators should collect dung using forceps and rubber gloves to decrease the possibility of contracting diseases. Samples should be placed in paper bags, not in airtight containers (plastic bags, jars with screw-on lids) and stored in a dry environment. Dung from certain animals may harbor disease-causing organisms such as *Histoplasma capsulatum* (*Ajellomyces capsulatus*) and *Coccidioides immitis*. Investigators should exercise extreme caution when handling and examining moist chamber cultures of dung samples taken from areas where those diseases are endemic. Moist chamber cultures should be opened and examined only under a laminar flow hood. Additional information on sampling and isolation techniques is provided in Chapter 25.

ISOLATION

Myxomycetes are most commonly isolated using the moist chamber technique. Application of that technique is discussed in detail in Appendix I and will not be repeated here.

The most important characters for identification of a slime mold are the color and type of plasmodium and the fructification that develops from it. Little is known about the plasmodia of nearly 25% of the known myxomycete species. In cases in which plasmodia are feeding actively on the dung surface in moist chamber cultures, a thin slice of the substratum with a portion of the plasmodium can be removed and placed on a water agar medium (Appendix II), allowing the plasmodium to spread over the surface (Timnick 1947). The plasmodium can be fed sterile, finely ground, uncooked oat flakes to sustain growth on the agar medium and to promote eventual development of fruiting bodies. Fresh spores obtained in this way can be sown on a nutrient-poor agar medium, such as water agar or half-strength cornmeal agar (Gray and Alexopoulos 1968), in an attempt to cultivate the myxomycete from spore to spore.

Myxomycete fruiting bodies are harvested from cultures after they dry. Raising the lid of the Petri dish on one side allows the culture to dry gradually. If it dries too quickly, the sporangia tend to collapse, and the spores agglutinate; if a culture dries too slowly, filamentous fungi often overgrow the fruiting bodies. Cultures must be observed carefully and harvested in a timely fashion to avoid the development of filamentous fungi and infestation by mites and collembola, which may damage the fruiting bodies. Failure to detect the fructifications may be caused by a delay in observing the cultures or observation at too low a magnification (50–100× gives the best results). Some of the larger fruiting bodies will, of course, be clearly visible with the naked eye. A sharp scalpel is used to make an incision around the fruiting bodies and transfer them to a collection box.

Sometimes only one or a few fruiting bodies of a given species will develop in a moist chamber culture, in which case a slide preparation can serve as a numbered collection. Whenever a slide is made, great care should be taken to ensure that all characteristics of the fruiting body are preserved. Indeed, this may be the only way to obtain voucher specimens of some *Echinostelia*, given that the fruiting bodies of this genus are so tiny and fragile. Slides with frosted ends can be labeled with a sharp lead pencil or a thin-pointed permanent marker and covered with transparent tape to keep the writing from smudging. Such slides are made quickly and can be filed for future reference.

The moist chamber technique has several advantages. Many of the fruiting bodies of Myxomycetes are too small to be seen with the naked eye in the field and are found only under a dissecting microscope in the laboratory. Field time may not be sufficient for finding and collecting those fruiting bodies, or the weather may not have been suitable (too dry or cold) for fruiting. Use of a moist chamber not only extends the collection period but also enables the myxomycologist in the field to collect additional samples of dung that can be stored for later study in the laboratory (Braun and Keller 1977). The technique is ideal for Myxomycetes requiring a time for development (at least 1 month).

CULTURING

Decoctions of natural substrata enhance spore germination (Smart 1937) and are used routinely by myxomycologists. Formulae for different media appropriate for culturing Myxomycetes are available in Gray and Alexopoulos (1968:247–249) and Appendix II. Keller and Eliasson (1992) used an oat-dung extract plain agar medium (ODA) to culture *Perichaena depressa* from spore to spore. Details of that spore-isolation technique are provided by Spiegel and colleagues in Chapter 25.

Spores are obtained by crushing a single sporangium. The released spores are spread in a small area on the agar surface of a sterile plastic Petri dish (60 mm in diameter × 15 mm high) containing about 8 ml of medium. Some spores are submerged by slashing the agar, and others are left free on the agar surface. The area used is circled on the reverse side of the plate with a black marking pencil to facilitate the location of the spore mass under a compound microscope. Three drops of a dilute solution of *Escherichia coli* are added to the culture. Stock cultures of *E. coli* are maintained on a nutrient agar medium and transferred to fresh slants the day before use. The dilute bacterial inoculum is prepared by adding the pure bacterium to 10 ml of glass-distilled water and mixing thoroughly. Trichiaceous and stemonitaceous Myxomycetes require the presence of free water in early stages of their life histories, so the agar surface is flooded with 2 ml of sterile glass-distilled water. Physaraceous species grow best on a dry agar surface, and for isolation of those species, no free water is added. Culture dishes are stored in an incubator at 23°C in total darkness. Physaraceous species are cultured under a 12-hour dark/12-hour light cycle. The water, except for a thin layer to prevent desiccation, should be pipetted off on the seventh day, and the plates should be examined. Pulverized rolled oats then are sprinkled over the surface of the agar. Cultures are returned to the incubator and examined periodically for the presence of fruiting bodies.

SPECIMEN STORAGE

Specimens collected in the field or harvested from moist chambers should be mounted by gluing small pieces of the dung bearing the fruiting bodies to the inside lid of a small box so that the label and the specimen will be inseparable. Simple matchboxes may be convenient for this purpose, in which case the collection number should be recorded on both parts of the box as an identifier. Larger fructifications or thick pieces of substratum should be glued to the bottom of the box. Failure to secure specimens properly often results in shattered sporangia. Delicate species and fragile substrata should be handled carefully with needle-nosed forceps when gluing in boxes (Sundberg and Keller 1996).

Collecting boxes of uniform size and depth are most desirable. The National Fungus Collection (BPI) uses standard-size boxes measuring 4.5 × 10.5 × 2.0 cm or for larger specimens, 10.0 × 9.25 × 2.0 cm. These boxes have a fairly shallow lid, which facilitates examination of the specimen with a hand lens or dissecting microscope. Deeper boxes may be necessary, however, to protect and accommodate larger specimens.

The identification label for a field-collected specimen should be pasted to the lid's outer surface. It always should include the following information: name of species, locality of collection (country, state or province, county, city, name of street, park, or any other landmark or specific place), type of habitat, kind of substratum (dung and animal source), collector's name and collection number, date, and name of the person who identified the specimen.

MICROSCOPIC EXAMINATION

Some Myxomycetes can be identified to species on sight by the experienced myxomycologist. Most species, however, must be examined with a compound microscope for positive identification. Generally, magnifications of from 400× to 1000× are required to resolve delicate ornamentation of spores and capillitial threads. A calibrated ocular micrometer is necessary for measuring spore diameters.

General Procedures

Internal structures of a sporangium are best observed in glass-slide mounts if no spores are present. Thus, wind-blown sporangia, nearly void of spores, can be selected for mounts, or spores can be removed from sporangia by (1) firmly grasping the sporangium with jewelers' needle-nosed forceps and gently blowing on the sporangium; (2) gently blowing on the sporangium through a hand-held blower (a flexible rubber tube with an attached glass pipette having a very small terminal opening); or (3) placing broken sporangia in a water-filled test tube and shaking vigorously (Sundberg and Keller 1996).

The capillitia of *Arcyria*, *Comatricha*, and *Stemonitis* should be freed of spores without interfering with the attachments and arrangement of the capillitial threads. To do this, the investigator holds the sporangium with forceps over a drop of mounting fluid on a glass slide and gently blows on the sporangium until most of the spores are removed. Spores that fall into the mounting medium can be measured, and episporic ornamentation can be observed.

Three-dimensional aspects of sporangia; the structural patterns of the columella; and the attachment, branching, and surface net of the capillitium are observed in genera such as *Clastoderma*, *Comatricha*, *Lamproderma*,

and *Stemonitis* from glass-slide mounts of sporangia. The sporangium is centered in a small droplet of mounting medium, preferably water. Beginners tend to use too much mounting medium, which results in too much movement of the specimen and cover slip. Practice and experience are important. The investigator takes a no. 1 glass cover slip, breaking it to form four triangular-shaped fragments. Using needle-nose forceps or a dissecting needle, she or he positions a fragment on each side of the mounting droplet on the slide. Another no. 1 cover slip, angled at 45 degrees, is lowered gradually and gently over the mounting medium so that no air bubbles form. Because the cover slip is raised slightly, the specimen is not crushed, and the internal structure of the sporangium is preserved. Fruiting bodies of species in the Physarales that contain calcium carbonate deposits always should be mounted in water first so that structural features can be observed.

Temporary Water Mounts

Sporangia are mounted in water on a glass slide, and a cover slip is dropped into place. Such mounts are the best choice for the routine identification of Myxomycetes. Often, isolated air bubbles are trapped under the cover slip, producing a "lens effect" that increases magnification above that produced by the microscope and highlights the surface markings of spores and capillitial threads. In some cases it is desirable to immerse the fruiting body in a wetting agent (e.g., ethyl alcohol) first and then add water.

Semipermanent Mounts

Before mounting, the sporangium is placed in absolute ethyl alcohol that serves as a wetting agent. A drop of 3% potassium hydroxide (KOH) is added before the alcohol evaporates completely, causing capillitial threads and spores to swell and assume normal size. Excess KOH is removed with bits of filter paper; then a drop of 8% aqueous glycerin is added, and a cover slip is applied. Slides may be stored in dust-free containers such as Petri dishes; as the water evaporates more glycerin is added periodically. After the glycerin is concentrated, the cover slip is ringed with a cementing substance such as clear nail polish or Zut (a commercial slide-ringing compound). Drawbacks of this procedure are that KOH obscures details of spore ornamentation and that aqueous glycerin dissolves calcareous structures.

Permanent Mounts

Clear lactophenol is the mounting medium of choice when preparing glass-slide mounts for species in the

Echinosteliales, Liceales, Stemonitales, and Trichiales; it also is the routine mounting medium of choice for most fungi. It serves as a wetting agent and causes shriveled structures to swell and assume normal size. However, lactophenol also dissolves all calcareous structures, and the trapped carbon dioxide bubbles left behind make preservation of the slide extremely difficult. Physarales, often characterized by the presence of $CaCO_3$ in the peridium, capillitium, columella, or stalk, can be placed in lactophenol to test for the presence of calcareous deposits (release of bubbles indicates $CaCO_3$ was present). Lactophenol often is used with a dye such as cotton blue to stain hyaline structures and to add contrast. Cotton blue should not be used with dark-spored Myxomycetes because it deepens the colors and makes it more difficult to resolve ornamentation. In some instances polyvinyl lactophenol is used as an alternative mounting medium because it is viscous but not hydrophobic. It is ideal for embedding fruiting bodies of the Stemonitales, after blowing out the spore mass, to reveal the capillitial structures. It also dissolves calcareous structures (e.g., calcareous parts of the capillitium of *Badhamia* and *Physarum*) so that other internal parts (e.g., peridia) camouflaged with a calcareous crust become clearly visible in polyvinyl lactophenol. The latter material polymerizes slowly and hardens in several days.

SAMPLING AND CULTURING ZYGOMYCETES

SAMPLING

Zygomycetes, especially members of the genus *Mucor* (Mucorales), are encountered on all, or nearly all, collections of dung from wild animals (see Murie 1974 for information on North American animals; accompanying a mammalogist or herpetologist into the field to find places to collect is also useful). However, only a few, or sometimes one or no species of Zygomycetes, are present on dung of laboratory-reared animals. These fungi are not cellulosic, as far as is known, and therefore grow only on the water-soluble nutrients contained in dung. Suitable dung is dark, shiny, and as fresh as possible. Because the nutrients have been leeched from dung that appears rough or is whitish, such droppings are not a good source of Zygomycetes; spores may have been washed away or may be too old to germinate.

As mentioned earlier, at least one species of *Mucor* can be found on almost any dung collected. Some species of *Pilobolus*, an obligate coprophile, are found on the dung of only a single host species, although other members of

the genus have been recorded from the dung of a diverse array of animals, including monkeys, foxes, pigs, sheep, panda, bears, elephants, and chickens (Hu et al. 1989). Other coprophilous Mucorales (e.g., *Phycomyces blakesleeanus* and *Thamnidium elegans*) are encountered only during cooler times of the year in regions with mild winters, such as southern California and Florida. *Utharomyces epallocaulus* is known only from areas with summer rains (Kirk 1993). Dung collected in the Bahamas, Mexico, Taiwan (Kirk and Benny 1980), and the United States (Arizona, Florida; G. L. Benny unpublished data) has been a recent source of *U. epallocaulus*. The relatively large size of the feces of mammalian herbivores facilitates its collection. The dung of these large animals is not the best source for other genera of coprophilous Zygomycetes, however, because relatively few genera are encountered and the possibility of bringing mites into the laboratory increases (Fig. 21.17).

The best source of coprophilous Zygomycetes is the dung of small animals, especially rats and mice, although good results have been obtained from the feces of bats, small lizards, and frogs. Dung of other small animals may yield interesting or new taxa; *Dichotomocladium floridanum*, for example, is known only from pocket gopher (*Geomys pinetis*) dung collected in the vicinity of Gainesville, Florida (Benny and Benjamin 1993). Dung of these animals can be collected if the gophers are caught in traps, but it also can be found where they live or congregate to feed. In dry or desert areas, where vegetation is sparse, dung pellets may be deposited at irreg-

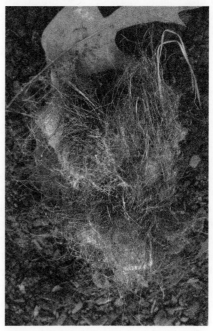

FIGURE 21.17 *Phycomyces* species growing on dog dung in nature (0.3×).

ular intervals along established animal runs. In rocky areas dung may be easier to find on uneven terrain because it may roll down to lower areas and accumulate where it is more readily collected. In the Mojave Desert of southern California, kangaroo rats dig shallow holes, and they often deposit one or two fecal pellets nearby. In the desert, ants also may collect pieces of small mammal dung and deposit them in the pile of debris that surrounds the entrance to the nest. During the late fall or early winter (before winter rains or snow), dung at the base of Joshua trees (*Yucca brevifolia*) at about 1,220-m elevation in the vicinity of Cajon Pass in the San Bernardino Mountains of California often is the best source of *Spirodactylon aureum*. Other good collecting sites for dung are storage areas (sheds, garage), wood piles, or any areas where small animals can feed or live. In Florida, small lizards often are found on window screens, especially at night, where they feed on insects attracted to the light. Frogs also may be found around windows. Both lizards and frogs deposit feces on walls or window ledges where it can be collected. The dung of insects or other small animals/invertebrates can be collected, but few data regarding isolation of fungi from that source are available. Indoh and Oyatsu (1965) isolated *Heterocephalum auranticum* from cockroach dung.

Dung should be collected using forceps and rubber gloves to decrease the possibility of contracting diseases. Most dung from small animals will be dry if the environment is also dry. Wet dung should be placed in paper bags; it should not be stored in airtight containers, including plastic bags, because fungal growth will commence before the dung can be plated out in the laboratory. When completely dry, dung pellets can be stored in airtight containers in a refrigerator for long periods.

ISOLATION

Moist chambers of sizes appropriate for the dung pellets to be examined are used to isolate Zygomycetes. For a discussion of techniques, see Appendix I.

Species in the Mucorales usually start to appear after 4 or 5 days when moist chambers are incubated at room temperature (about 22–25°C). It may be best, therefore, to start incubating on a Wednesday or Thursday so that the succession of fungi can be observed during the following week. Several taxa only grow at temperatures below 20°C, so it is beneficial to incubate some samples at 18°C; such cultures are initiated a day or two earlier (Monday or Tuesday) for the following week's isolations.

Coprophilous Zygomycetes are isolated most easily using stainless-steel, minuten insect pins mounted in wooden (long fireplace matches work well) or metal vise-

grip handles for wet spored species and watch-makers forceps (available at electron microscopy supply houses) for taxa with dry or wet spores. Forceps are sterilized in an open flame, cooled in 95% ethyl alcohol, passed quickly through the flame again to burn off the excess alcohol, and (when cool) used to transfer the spores to sterile media. Minuten insect pins are sterilized by passing the tips through the flame from an alcohol lamp or very briefly (excess heat can melt the pin tip) through a gas flame. The pin is cooled by inserting the tip into the sterile agar in a Petri dish. A wet sporangium or spore drop adheres when touched with the tip of the pin or forceps and can be transferred to the surface of the agar medium.

Following each examination of the dung collections, the tall sporophores of *Mucor* and other taxa should be pushed down (sweep across and close to the dung pellet) with a steel needle or inoculating loop. This is continued for several days until few, if any, of these tall sporophores are produced. This procedure will facilitate isolation of the zygomycetous fungal parasites (*Piptocephalis*, *Syncephalis*), which usually do not grow as high as the host fungi, and other late arrivals, including *Coemansia*.

CULTURING

Zygomycetes, especially members of Mucorales, can be grown on many culture media. Nutrient-rich media that are most useful in their culture include malt-extract yeast-extract agar, Emerson's YpSs agar, V8-juice agar, and potato-dextrose agar (PDA) (see Appendix II). Other culture media that have proved effective, especially if zygospore formation is desired, are Pablum agar, Pablum agar + dextrose (PAB-DEX; Benjamin 1959), Leonian's agar + yeast extract (LYE; Malloch and Cain 1971), acidic potato-dextrose agar (APDA), whey agar (WHEY), YpSs agar + dextrose (YPD; Benny and Benjamin 1975), Weitzman Silva-Hutner medium (WSH; Weitzman and Silva-Hutner 1967), tomato paste-oatmeal agar (TPO; Hesseltine 1960), and cherry agar (CA; Schipper 1969) (see Appendix II).

Nutrient-poor media that have proved useful in the culture of Zygomycetes include modified synthetic *Mucor* agar (MSMA; Benny and Benjamin 1975), wheat germ agar (WG; Benny 1972), cornmeal agar (CMA; Benjamin 1959), potato-carrot agar (PCA; modified from O'Donnell 1979), *Choanephora* agar (CH; after Barnett and Lilly 1955), and 0.2-strength YpSs agar (YpSs/5; after Benjamin 1985). For even more dilute media use 0.2-strength CM or WG agar (Appendix II).

Other culture media may work equally well; readers should refer to published reports, evaluations, and formulae, especially those dealing with fungi from local fruits or vegetables.

Basidiobolales, Entomophthorales, Mortierellales, and Mucorales

Most, if not all, species in the Mucorales and these other orders grow and sporulate well on most nutrient media. Most Mucorales grow well at 25°C although some taxa may do much better at 30°C. No coprophilous thermophilic Mucorales are known, but two genera (*Rhizomucor*, *Thermomucor*) are thermophilic and can be isolated from compost piles that may contain dung. Many species are psychrotolerant (e.g., *Chaetocladium brefeldii*, *Pirella circinans*, *Phycomyces* species, *Thamnidium elegans*) and can reproduce asexually at 25°C, or they are psychrophilic (e.g., *Chaetocladium jonesii*, *Helicostylum* species) and sporulate only at temperatures below 20°C. Psychrotolerant and psychrophilic species grow and sporulate well at 18°C, as do many mesophilic Zygomycetes (they just grow more slowly). Zygospore formation in *C. brefeldii*, *P. circinans*, and *Phycomyces* species proceeds quite well at 15°C, whereas in *T. elegans* and *Helicostylum* species, the sexual spores are produced only after 30–60 days at 7–10°C (Benjamin and Hesseltine 1959; Benny and Benjamin 1976; Benny and Schipper 1992; Benny 1992, 1995a). In areas with a warm, summer climate those latter psychrotolerant and psychrophilic species can be isolated only when or where it is cool (higher elevations, winter).

Zygospore formation may occur optimally in the dark (Hesseltine and Rogers 1987), but formation of asexual sporulating structures may or may not be influenced by light; some taxa are phototropic. In *Thermomucor* species zygospore formation is enhanced on APDA medium (Appendix II). Sealing Petri dishes with Parafilm or other types of tape, or placing growth containers in a closed incubating vessel may inhibit sporulation because of carbon dioxide accumulation.

Nutrient-poor culture media (CMA, WG, MSMA, PCA, CH) are used to simulate natural growth conditions and are optimal for the observation of branching patterns, stolon formation, and vegetative structures. They often promote spore formation.

Initial isolation of Zygomycetes often is conducted best using nutrient-rich culture media (Emerson's YpSs agar, MEYE, PDA, V8-juice agar). MEYE has been used with great success because it is nutrient-rich and relatively clear; most fungi grow rapidly on this medium, and the earliest stages of spore germination can be observed with a dissecting microscope using light reflected through the plate. It may be desirable to transfer the culture again to the same or different agar medium to ensure that the culture is clean and/or to promote sporulation.

Pilobolus species are obligate coprophiles and must be grown on media containing dung extract (DEA) or hemin (Levetin and Caroselli 1976). *Pilobolus* sporangia are transferred with forceps, preferably directly from the sporophore to the medium. They are incubated at 37°C overnight. The next day the plate is placed in a lighted incubator or next to an outside window. Sporangia should form in less than a week.

Dimargaritales, Kickxellales, and Zoopagales

Members of all of these taxa grow well on MEYE or other culture media. The Dimargaritales and Zoopagales (e.g., *Piptocephalis* and *Syncephalis*) require a host, but all culturable Kickxellales are saprobes. Many species will grow at 25°C, but other taxa only grow below 20°C. If facilities are available, maintenance of both the original samples and the subsequent isolation plates at 18°C will yield more organisms for isolation.

Obligate parasites (e.g., *Piptocephalis*, *Syncephalis*, both Piptocephalidaceae, Zoopagales; *Dimargaris*, *Dispira*, *Tieghemiomyces*, all Dimargaritaceae, Dimargaritales) require a host fungus, usually in the Mucorales, but some Dimargaritaceae (e.g., *Dispira implicata*, *D. simplex*; Benjamin 1961; Misra and Lata 1979) only grow on species of *Chaetomium*. *Cokeromyces recurvatus* is used routinely, if available, because the colony is dark and very low growing (Benjamin 1959); *Micromucor ramannianus* is the preferred host for a few species. Members of the Mucorales or a species of *Chaetomium* from the original isolation plate can be used as hosts. Use of the original host is often a requirement for the culture of some *Syncephalis* species, but mucoralean fungi often grow higher and faster than the parasite and can quickly overgrow it.

MICROSCOPIC EXAMINATION

Cultures from nutrient-rich media may enhance zygospore formation but are not the best for study or identification because of increased aerial hyphae and reduced asexual reproduction. In cultures from nutrient-poor media, asexual reproduction may be enhanced, but zygospore formation usually is reduced or completely inhibited. Such media are excellent for studying branching patterns, formation of stolons, rhizoids, and other vegetative structures.

Identification requires the preparation of microscope slides that show the required morphological characteristics without distorting the fungus. The best mounting agent in that case is KOH-phloxine (two solutions: 2% KOH and distilled water with enough phloxine B added to turn the solution pink; water without the dye can be used). A small amount of the culture is transferred to a drop of 95% ethyl alcohol near the center of a slide (to facilitate removal of air bubbles); the excess alcohol is drained, one drop of 2% KOH is added, and the specimen is manipulated to suit. One drop of either the phloxine solution or water is added, and the specimen is covered with a no. 1 cover slip. The cover slip is sealed with melted paraffin (transferred with a large paper clip, unbent to form a "U" with one side about 2 cm long and straight). If the seal is good (no air bubbles), the paraffin is coated with clear fingernail polish extending far enough on both sides of the paraffin for a thin layer to attach to both the slide and cover slip. Cover slip shards can be placed on opposite sides of large specimens to support the cover glass. Shards are made by holding two glass cover slips, one on top of the other, by one edge between the thumb and index finger; with the opposite edge against a paper towel, slightly forward and downward pressure is applied until the cover slips break. Care must be taken to avoid injury from the glass fragments. These mounts are good for 2–3 days. More permanent preparations lasting 1 year or more are made by mounting fungi in lactophenol cotton blue or other liquid mounting agents.

STORAGE

A fungus can be transferred from a culture plate to a test tube containing the same or another culture medium for storage in a refrigerator. V8-juice agar promotes sporulation, as do many nutrient-poor agars, and such media are ideal for refrigerator storage. The cap should be loose and wrapped together with the upper portion of the test tube in Parafilm or some other stretchable tape to prevent contamination by other fungi or mites. Most culturable Zygomycetes can be stored this way for 6–12 months; however, *Basidiobolus* and *Conidiobolus* may need to be transferred to new media every 2–3 weeks. Zygomycetes can be stored for much longer periods after lyophilization (in double-strength skim milk or in skim milk and sodium glutamate and either honey, raffinose, or trehalose; Berny and Hennebert 1991) or freezing at −85°C in cryovials in 1 ml of 50% glycerol; other cryoprotectants or strengths of glycerol may work better for particular species (see Chapter 3).

Initial culture isolates may be contaminated by other fungi, especially Deuteromycetes, or bacteria. Subsequent isolations may be required, therefore, before a pure culture is obtained. Such isolations, whether initial or subsequent, should be carried out using the freshest cultures possible to minimize contamination by fungi, bacteria, and mite infestations. Isolation plates should be discarded as soon as possible after use to prevent mites

from moving through a culture collection. Constant checking of cultures under a dissecting microscope for fungal contamination or mite tracks on the agar is a useful preventive measure.

SAMPLING AND CULTURING ASCOMYCETES AND BASIDIOMYCETES

SAMPLING

Traditionally, investigators have sampled fungi in a completely haphazard manner within a designated region while engaged in general field work. To obtain any idea of species richness, however, some sort of systematic sampling (e.g., with protocols based on transects or grids) is required. Such attempts with coprophilous fungi could limit the species recorded considerably because dung deposits of some species are close together, whereas those of other species are separated widely.

Because spring mycobiota are observably different from those of summer and autumn, at least in north temperate regions, sampling should be carried out both in spring and summer. That likely also applies to the south temperate biome. Richardson (2001) found that some fungi were recorded only in the winter; thus, some sampling in the winter is beneficial. In the tropics sampling should be undertaken in both dry and wet seasons. In areas of significant elevational range, samples should be taken at low, middle, and high elevations. Finally, both dry and moist habitats, both within and outside of wooded or forested areas, should be sampled.

Evidence indicates that droppings of different ages support communities of different composition (Wicklow 1992). Thus, sampling dung of different ages (both fresh and aged) will yield the maximum number of species. Some fungi are ephemeral and will be found only on fresh dung, which should be examined without drying. Generally more taxa will be found by incubating dry dung in moist chambers (see "Incubation," later in this chapter). Dung that has been largely devoured by dung beetles should be avoided because the mycobiota and resulting yield are poor.

Many fungi are substratum-specific. Thus, samples should include dung of animals representing the spectrum of feeding habits (herbivorous and carnivorous) and ecologies and should include both domesticated and wild animals. The samples should be separated by substrate type in the field (see Dorst and Dandelot 1970; Murie 1974). Deposits of dung in caves should be sampled at various depths because such layers represent several generations of animals.

Samples should be sufficient for subsequent incubation. If the fecal pellets are small, such as those deposited by deer and rabbit, at least a handful is necessary. More will increase the likelihood of obtaining an idea of the complete mycobiota. With larger animals (e.g., cow, horse), efforts should be made to obtain a #3 or #5 brown paper grocery bag of dung. Although large samples are preferable, an inadequate sample is better than no sample if one recognizes that only a portion of the mycobiota will be obtained. Gloves should be used, especially in tropical regions, to avoid contracting diseases.

Samples are examined under a dissecting microscope. Fruiting bodies of Ascomycetes are removed with a dissecting needle or fine-pointed probe and mounted in water on a slide for microscopic examination. In special instances, other mounting fluids (e.g., Shear's mounting medium; Malloch 1981) or water containing a drop of liquid detergent are used. Ephemeral fungi (e.g., *Pyxidiophora* species) are removed using fragments of glass cover slips, and both are mounted directly on the slide.

If fleshy Basidiomycetes are present on field collections, spore prints should be prepared, and notes should be taken on the macromorphological features of the fruiting bodies while they are still fresh. Additional information on preparation of notes is given in "Describing Macromorphological and Micromorphological Characters" under "Collecting and Describing Macrofungi" in Chapter 8. Following macroscopic study, the material is air-dried with a small piece of substratum and placed in a small specimen box.

INCUBATION

Because only a small portion of a coprophilous mycobiota will be obtained directly from observation of field samples, dung is incubated in moist chambers following its initial examination (Lundqvist 1972; Furuya 1990). Appropriate procedures for the use of moist chambers are provided in Appendix I.

The material in a moist chamber is observed every few days for 6–8 weeks. Developing fruiting bodies are studied microscopically as described for the field-collected material (see "Sampling," earlier). Providing different conditions in different moist chambers can increase the number of species recorded. Some fungi have narrow temperature latitudes, so maintaining some moist chambers at cooler temperatures may increase the number of species obtained. That practice is especially appropriate for, but not restricted to, samples from northern or montane regions. Likewise, certain taxa have narrow pH requirements, so adding ammonia or ammonium hydroxide, rather than water, to a few moist chambers may yield additional species. Finally, different fungi can have different moisture requirements, such that the composition of the mycobiota observed may vary under different conditions of wetness. Some chambers should be subjected to a wet/dry cycle. Basidiomycetes, certain Ascomycetes (e.g., *Bombardioidea* species), and some Myxomycetes develop in response to such cycles.

Permanent voucher specimens include dried pieces of dung that bear fruiting bodies (preferably at least 12) and all prepared slides. Any Basidiomycetes that develop are described and dried as noted earlier. If spore prints are not obtained, mature spores should be sought on the stipes or among spores deposited on the absorbent material in the moist chamber (Richardson and Watling 1997). As the dung in the moist chamber becomes depleted, additional substratum is moistened and inserted.

CULTURING

Warcup (1950) described a dilution technique for soil fungi that has been equally useful for coprophilous fungi, especially Ascomycetes. Some taxa isolated by this technique have not appeared in moist chambers. A small portion of dung is macerated by hand, placed in a 100- or 250-ml beaker with 50–70% ethyl alcohol sufficient to cover it, and left for about 5 minutes. The alcohol treatment eliminates some fast-growing molds, as well as some Zygomycetes, and should be omitted if isolation of those groups is desired. Bills and Polishook (1993) found 4 minutes in alcohol to be optimal. After alcohol treatment, the beaker is shaken to resuspend the macerated dung, and small portions of the suspension are added in sequence to sterile Petri plates until the suspension is exhausted. If the fluid portion of the suspension is exhausted before the dung particles, a minimal amount of additional alcohol can be added. Semi-cooled agar medium to which a pinch of powdered antibiotic (e.g., penicillin-G or streptomycin sulphate) has been added to reduce bacterial contamination is poured into the Petri plates to cover the bottom of each plate, and the plates are rotated carefully to distribute the suspension evenly. Agar should not touch the rim of the plate and should be allowed to cool until almost solidified before pouring to avoid killing dormant spores in the suspension. The plates are allowed to cool and examined as colonies begin to appear; those colonies are transferred to fresh agar plates. A natural medium (e.g., V8 or Weitzman and Silva-Hutner, see Appendix II) and a semisynthetic one (e.g., modified Leonian's medium) are used for isolation.

Bills and Polishook (1993) modified this technique by using dichloran-rose bengal-chloramphenicol (DRBC), a medium that restricts growth of rapidly growing fungi, to which 0.1 g of chloramphenicol was added after cooling. They added the ethyl-alcohol–treated dung samples directly to the molten, semicooled agar, agitated the mixture, and poured it directly into the Petri plates.

Frequently, cultures are used for detailed studies of selected taxa. Fructifications on dry dung, from moist chambers, or from colonies on dilution plates can provide fruiting bodies for establishing such cultures. The fruiting body or colony is removed with a sterilized needle or probe. The fruiting body also can be rolled in the agar to remove any adhering debris or can be sterilized by momentary immersion in a drop of fresh household bleach (5.25% sodium hypochlorite). Alternatively, a fruiting body can be broken open against the side of a Petri plate, and the spores can be removed. The cleaned fruiting body, colony, or spores are transferred to clean agar in a sterile Petri plate. Clean cultures also can be obtained for some fungi by placing a fruiting body, with the spore-bearing layer facing down, in petroleum jelly spread on the inside of a Petri plate lid. Spores that drop to the agar surface are transferred to a clean Petri plate. This technique is especially useful for discoid fruiting bodies. It also can be used with Basidiomycetes, especially Agaricales, in which a piece of gill tissue is stuck to the lid.

Antibiotics are frequently useful when isolating fungi, both Ascomycetes and Zygomycetes, from natural substrata. Sterilized filtrates of chlortetracycline hydrochloride and streptomycin sulphate in concentrations of 0.05 mg/l and 0.1 mg/l can be used when isolating fungi from substrata with a mixed fungal population and for cleaning fungal colonies infected with bacteria (Benny et al. 2001). Scott and colleagues (1993) supplemented Leonian's medium with 60 ppm each of chlortetracycline hydrochloride, streptomycin sulphate, and benzylpenicillin (penicillin-G) for direct plating of ascospores and perithecial contents. Some fungi are sensitive to these or other antibiotics, in which case alternative antibiotics should be selected.

Most Basidiomycetes grow slowly in culture. To eliminate the faster-growing molds and Ascomycetes, the media are supplemented (Hutchison and Summerbell 1990) with 2 mg/l of chlortetracycline hydrochloride, 30 mg/l of streptomycin sulphate, and 2 mg/l of benomyl (Benlate, 50% wettable powder; DuPont, Appendix IV). Benomyl is especially effective and can be added as a sole supplement to standard culture media. It is also useful at considerations of 5 to 20 ppm for isolating Zygomycetes from natural substrata.

Some coprophilous fungi are difficult to culture, possibly because of delayed spore germination. In addition, a number of factors, including the age of the spore, are correlated with probability of germination (Webster 1970). Various stimulants, such as acetate, hydroxide, phenol, heat, and rumen fluid, have been used to enhance spore germination. Furuya and Naito (1980) showed that among phenol derivatives, compounds with a substitution at the "para" position are most effective.

Another useful technique is to treat the spores with a 10% solution of mixed pancreatic enzymes buffered to the pH of the digestive tract of the species from whose dung the fungus was isolated. Generally, a pH of 6.8 with a 0.01 M buffer of $HPO_4^=$ and $H_2PO_4^-$ in a ratio of 3:1 is effective. Germination usually occurs if spores are treated for 20 minutes with the buffered solution and then streaked on a clean agar plate that is incubated for 24 hours at 37°C. Safar and Cooke (1988a) used a simpler version of this technique, with a pH of 9.5. Spore germination and fruiting body formation are sometimes enhanced by growth in cultures contaminated with molds and bacteria, which may release stimulatory compounds. In other instances, however, certain bacteria or bacterial filtrates inhibit germination (Safar and Cooke 1988b). Dung often contains coprogen, a growth compound that is produced by a broad spectrum of fungi and bacteria (Hesseltine et al. 1953; Webster 1970).

Growth and production of fruiting bodies in culture usually will occur under natural light conditions, but ultraviolet light can sometimes enhance fruiting body production and ostiole formation (in pyrenomycetes). Incubation at several temperatures is often beneficial because some fungi have narrow temperature requirements for fruiting. In addition to the media used for isolation, Czapek's (CZ) and dung-extract (Krug 1989) media (Appendix II) promote growth of some fungi. Other species grow better on a weak medium (e.g., cornmeal dextrose agar).

Frequently, vouchers are desired for deposition in herbaria. They can be prepared by drying the mature culture in a frostfree freezer as outlined by Wu in "Specimen Preparation," Chapter 2 and "Maintenance and Preservation of Cultures," Chapter 3.

MICROSCOPIC EXAMINATION

Asci and ascospores of Ascomycetes and basidia and basidiospores of Basidiomycetes are essential for identification. Consequently, culture media that promote the formation of such structures must be used. Usually, rich media such as Leonian's and Weitzman Silva-Hutner are satisfactory, but a poorer medium may be desirable to reduce growth of aerial hyphae. Identification of most Basidiomycetes should be possible from fructifications produced in moist chambers.

Diagnostic features are examined in material on microscope slides. A very small portion of a culture usually is mounted in water or Shear's mounting medium. Pretreatment with ethyl alcohol as described for Zygomycetes may be advantageous in some circumstances. For example, treatment of *Chaetomium* and similar taxa with ethyl alcohol dissipates trapped air bubbles and enables one to observe the hairs clearly. The mount is covered with a cover slip and stained with an aqueous solution of cotton blue, phloxine, Congo red, other stain, or Melzer's reagent. Some of the color reactions with ascal apical structures in Melzer's reagent or Congo red are transitory. Thus, it is best to place a single drop of the reagent or stain at the edge of the cover slip and to observe the preparation continually as the solution seeps toward the fungal material. The appendages and sheaths that occur on some spores can be elucidated by mounting in slightly diluted India ink. Semipermanent mounts are retained in Melzer's reagent or lactophenol.

STORAGE

Cultures can be preserved for some months in a refrigerator. Storage of a culture in sterilized screw-cap bottles containing sterile medium increases its longevity over storage in test tubes. As with Zygomycetes, it is best to wrap the caps with Parafilm. Both modified Leonian's and Weitzman Silva-Hutner media are satisfactory for storage. The simplest method of long-term storage is to cover a culture with a moderate layer of mineral oil and maintain it at room temperature (Smith 1990).

INTEGRATION OF INFORMATION

The data available on coprophilous fungi are still very limited. The taxonomy of many genera, for which no monographs are available, has not been studied in detail (or at all). The coprophilous mycobiota of only a few regions have been surveyed, and for many areas of the world, even the most elementary information is lacking. Even fewer data on the biology of coprophilous fungi are available, although some attention now is being directed to the ecology of these organisms and their roles in the ecosystem. Some information on their physiology is available, and Gloer (1995, 1997) is beginning to uncover the nature of their chemical metabolites. Still, our knowledge of their biology is almost negligible and is restricted to a limited number of species. If fungi are to be integrated more fully into programs for the enumeration and conservation of biodiversity, then additional comprehensive monographic treatments and mycobiotic surveys are needed. Such efforts should include more information on ecology and should focus on geographic areas that have not been surveyed adequately. Surveys are particularly important for regions where the native ecosystem is under pressure for commercial development.

With the present state of our knowledge, it is almost impossible to designate truly rare or threatened species that merit conservation efforts. Although it is true that many taxa are known from only single collections, in many instances that scarcity probably reflects low intensity of collecting rather than rarity. As Richardson (2001) pointed out, rarity may be an artifact of the specialized requirements of coprophilous fungi. Efforts for the preservation of coprophilous fungi should be directed toward the conservation of habitats that are threatened with extinction, especially those with threatened animals. Many coprophilous fungi have rather narrow host ranges; with every loss of habitat or host, the potential for loss or reduction in the mycobiota is high.

Hyde and Hawksworth (1997) indicated that obtaining realistic estimates of microfungal diversity as part of surveys to measure and monitor biodiversity may be impractical because of the enormous scope of the task. Coprophilous fungi, however, could be useful indicators of both biodiversity and habitat quality (Richardson 2001). Richardson (2001) also showed that those fungi might, in limited instances, allow for a quick, simple, inexpensive quantitative assessment of biodiversity and provide information relevant to the larger ecosystem.

ACKNOWLEDGMENTS. The research for this paper was supported in part by grants from the Natural Sciences and Engineering Research Council of Canada. We are especially grateful to the following colleagues for helpful comments, information, and/or discussions: Drs. R. Bandoni, T. Baroni, U. Eliasson, J. Ginns, J. Kimbrough, N. Lundqvist, D. Malloch, O. Miller, M. Schnittler, S. Udagawa, and D. Wicklow.

22

ANAEROBIC ZOOSPORIC FUNGI ASSOCIATED WITH ANIMALS

DANIEL A. WUBAH

Plant material is composed of structural polysaccharides, cellulose, hemicelluloses (including xylans), and pectic compounds that are coupled together in variable and complex ways. Because most mammalian herbivores do not produce cellulases and other enzymes required to degrade most of those plant materials, they depend on microorganisms inhabiting their gastrointestinal (GI) tract to digest the plant materials and provide easily usable forms of carbon, energy, some vitamins, and microbial protein. As a result, such herbivores can be considered as walking menageries containing whole communities of different species of microorganisms that are committed by evolution to live together (Attenbrough 1990).

Bacteria, protozoa, and fungi are the three major groups of microorganisms that inhabit the GI tract. Those microorganisms are suspended in the liquid phase of the rumen digesta, attached to plant fragments, or attached to the epithelial lining of the rumen (Trinci et al. 1994). Of the three groups of microbes, the fungi were discovered most recently (Orpin 1975). Three types of fungi occur in the GI tract of herbivores (Orpin and Joblin 1988). The first group includes facultatively anaerobic and aerobic fungi, which are considered to be transients because they continually enter the rumen in feed and may not be able to grow under the anaerobic conditions of the rumen environment (Clarke and DiMenna 1961; Sivers 1962; Lund 1974; Bauchop 1989). The second group consists of two species that parasitize ophryoscolecid ciliate protozoa: *Sagittospora cameroni* and *Sphaerita hoari* parasitize *Diplodinium minor* and *Eremoplaston bovis*, respectively (Lubinsky 1955a, 1955b). The third group, unlike all other known fungi, consists of the obligatorily anaerobic zoosporic fungi that are saprotrophic on plant material (Orpin 1975, 1976, 1977b; Bauchop 1989).

The GI tract of herbivores is made up of the foregut, which consists of the hind part of the esophagus and the stomach, and the hindgut, which includes the small intestines and the cecum. Herbivores are divided into two groups based on the part of the GI tract (hindgut or foregut) in which most of their microbial digestion occurs. In hindgut-fermenting herbivores, such as elephants (Proboscidae) and horses (Equidae), ingested plant material passes rapidly through the GI tract, and most of the nutrients obtained come from the contents of the plant cell rather than from its wall (Hume 1989). Microbial digestion in the hindgut-fermenting herbivores occurs mainly in the large intestine and cecum after minor gastric digestion. The ruminants, or foregut-fermenting herbivores, include the cloven-hoofed animals, kangaroos, and other mammals that depend on pregastric microbial fermentation in the rumen for their nutrients. They retain plant material in the rumen, where it is digested extensively, for prolonged periods (Church 1969).

RUMEN

The foregut of ruminants has four chambers (Fig. 22.1): the rumen, reticulum, omasum, and abomasum. The former three are pregastric chambers that are formed from modifications of the esophagus and the stomach. The rumen and reticulum together form a large fermentation vessel with a capacity of approximately 10 liters in sheep and from 100 to 150 liters in cows (Hobson and Wallace 1982). The rumen, which is the largest pregastric chamber, creates a highly specialized environment. It maintains heterogeneous plant material and microorganisms in a dynamic, amorphous fluid under conditions similar to those in a continuous culture chamber. Rumen contents may become stratified as a result of particle size and specific gravity, and a gas space

exists above the partially digested food. The pH of the rumen fluctuates between 5.8 and 6.8 depending on the frequency of feeding and food type (Trinci et al. 1994). Copious amounts of saliva, which contains bicarbonate and phosphate buffers that maintain the pH, constantly enter the rumen with the food. Heat from the animal's metabolism and from microbial fermentation maintains the temperature of the rumen between 38°C and 40°C depending on frequency of food or water intake (Orpin 1975). The composition of the gases in the rumen is approximately 62% CO_2, 30% CH_4, 7% N_2, 0.6% O_2, 0.2% H_2, and 0.1% H_2S (Hobson 1971). The redox potential of the rumen ranges from −250 to −450 mV because facultative anaerobic microbes such as bacteria and yeasts quickly utilize air that accompanies the feed into the rumen.

TROPHIC RELATIONSHIPS BETWEEN ANAEROBIC FUNGI AND RUMEN ORGANISMS

Interactions are known to occur between anaerobic fungi and methanogenic bacteria (Mountfort et al. 1982), between anaerobic fungi and nonmethanogenic bacteria, and between anaerobic fungi and protozoa (Romulo et al. 1989; Marvin-Sikkema et al. 1990). The synergistic relationships between anaerobic fungi and methanogens have been well established; however, the benefits of such interactions to the animal host are not completely understood. Cultures of an anaerobic fungus, *Neocallimastix frontalis*, degraded filter paper more extensively in the presence of methanogens than in their absence (Bauchop and Mountfort 1981). Fermentation shifted from the formation of electron sink products, such as lactate and H_2, to the production of more reduced end products, such as CO_2, CH_4, and acetate, in the co-culture. Mountfort and colleagues (1982) demonstrated that a monoculture of *N. frontalis* digested 53% of the dry-matter, whereas co-cultures of the fungus with *Methanosarcina barkeri* or *Methanobrevibacter* species converted 69% and 87%, respectively. A triculture of all three organisms converted 98% of filter paper dry weight to CO_2 and CH_4. That increased production of CO_2 and CH_4 may result in a decrease in net carbon available to the host animal.

Hydrogenosomes, enzymes involved in the production of H_2 and CO_2 in anaerobic fungi (Heath et al. 1983; Yarlett et al. 1986a; Munn et al. 1988; Webb and Theodorou 1988), are located near the surface of the plasma membrane. Methanogens attach to the surfaces of the zoospores or thalli of those fungi

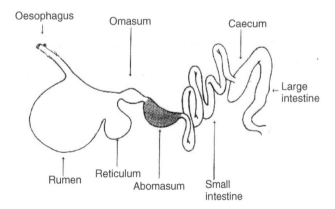

FIGURE 22.1   The digestive tract of a ruminant.

when grown with them. Methanogens may be readily able to use the products of the hydrogenosomes for growth (Yarlett et al. 1986a). Fonty and colleagues (1988) and Marvin-Sikkema and associates (1990) showed that when *Piromyces* species and *Caecomyces* species were grown in culture with methanogens, acetate production increased. *Methanobrevibacter smithii*, a methanogen isolated from sewage sludge, caused a shift toward acetate production when grown in culture on barley straw with either *Neocallimastix* species or *Piromyces* species (Joblin et al. 1989). However, the level of cellulose digestion was lower than that observed by Mountfort and associates (1982; see previous paragraph). Joblin and Naylor (1989) suggested that plant polymers such as lignin reduce cellulolytic activity in the co-cultures.

Interactions between nonmethanogenic bacteria and anaerobic fungi may be synergistic, symbiotic, or competitive. Such complex interactions occur in the rumen. Intracellular bacteria such as mycoplasmas have been associated with anaerobic fungi isolated from a steer (Kudo et al. 1990). Although the ecological significance of the association of mycoplasmas with rumen fungi is not known, Kudo and colleagues (1990) suggested that the axenic status of all cultures of anaerobic fungi isolated to date needs to be reexamined.

Growth of *Bacteroides ruminicola* in culture with *Neocallimastix frontalis* increased the degradation of xylan and barley straw (Richardson et al. 1986; Williams et al. 1991). The interaction between those two species may be synergistic given that degradation of xylan in the co-culture was more than three times that in monocultures of either the bacterium or the fungus (Williams et al. 1991). *Badteriudes succinogenes* had no effect on the degradation of straw by mixed cultures of fungi (Irvine and Stewart 1991). *Succinivibrio dextrinosolvens* also has a synergistic relationship with *N. frontalis* during xylan utilization. However, *Streptococcus bovis* and *Lachnospira multiparus* suppressed xylan utilization by that fungus (Williams et al. 1991).

Cellulose digestion by *Neocallimastix* species is increased by *Selenomonas ruminantium*, a bacterium that consumes H_2, succinate, and lactate (Marvin-Sikkema et al. 1990). That bacterium also decreased the cellulolytic activity and increased lactate production of *Piromyces communis* and increased cellulose digestion by *Caecomyces communis* (Bernalier et al. 1991). Xylan degradation by *N. frontalis* did not increase when the fungus was cultured with one of two strains of *S. ruminantium* (Williams et al. 1991). Digestion did increase, however, in a triculture of all three microorganisms.

An insignificant increase in the population of fungal zoospores was observed in animals from which the pro-

tozoa in the rumen were removed (Soetanto et al. 1985; Newbold and Hillman 1990). In defaunated sheep, however, zoospores and zoosporangia of anaerobic fungi showed twofold to fourfold increases (Romulo et al. 1989). Bauchop (1981) suggested that the increase in fungal populations following defaunation might be influenced by an animal's diet because populations of zoospores in animals fed high-fiber diets are larger than those in animals fed low-fiber or high-protein-concentrate diets; the opposite is true for ciliated protozoa. Orpin (1983) suggested that the interaction between anaerobic fungi and rumen protozoa may be complementary rather than competitive.

ROLE OF RUMEN FUNGI IN FIBER DEGRADATION

The role of rumen fungi in the degradation of plant fiber has been examined extensively (Bauchop 1981, 1983, 1986; Windham and Akin 1984; Akin and Rigsby 1987; Akin and Benner 1988; Akin et al. 1988; Joblin and Naylor 1989). The rhizoids of vegetative thalli are better at penetrating plant tissue than are bacteria and protozoa, so they gain access to plant material that is not available to other rumen microorganisms (Orpin 1977a; Orpin and Joblin 1988). Bauchop and Mountfort (1981) suggested that such penetration leads to faster and more complete degradation of forage that enters the rumen. An appresorium-like structure for penetration of stomata or cuticle was described in isolates from the rumen of swamp buffalo in Malaysia (Ho et al. 1988).

Degradation of lignin-containing walls of plant cells is an important characteristic of rumen fungi and has been discussed by various investigators (Bauchop 1979, 1981; Mountfort et al. 1982; Akin and Rigsby 1987; Akin and Benner 1988). Zoospores of many species appear to colonize the lignin-containing tissues preferentially and to establish colonies localized on sclerenchyma and xylem cells. Early observations indicated that lignified cell walls were degraded to a greater extent by rumen fungi than by rumen bacteria or protozoa. Those observations, coupled with the importance of lignin to forage quality (Akin 1986), prompted speculation about lignin degradation by those microorganisms. Orpin (1983) suggested that lignin degradation by rumen fungi is unlikely because they cannot ferment simple phenolic acids. In addition, fiber degradation is inhibited in the presence of 10 mM concentrations of phenolic acids (Akin and Rigsby 1985). Nonetheless, ultrastructural investigations indicate that fungal rhizoids or rhizomycelia penetrate lignified cell walls, with initial colonization of the secondary wall layer occurring from the lumen of the cell (Akin et al. 1989). Highly recalcitrant cell walls, such as the mestome sheath of leaf blades, are degraded partially, and the bundle integrity is weakened (Wubah et al. 1993). Fungal filaments often traverse the entire wall of lignified cells, including middle lamellar regions (Akin et al. 1989).

Some rumen fungi dissolve small amounts of phenolic compounds from plant cell walls (Orpin 1983; Gordon 1985). Research with specifically labeled 14-C polysaccharides or lignin indicated that the rumen fungi, in solubilizing phenolics, degrade lignocellulose, although they cannot use the lignin moiety (Akin and Benner 1988; Gordon and Phillips 1989). Further, a gel-permeation-chromatography study of fungal degradation of high molecular weight lignosulfates in the rumen demonstrated little decomposition of those polymeric phenolics (Wubah et al. 1993).

Ultrastructural studies have shown that anaerobic fungi penetrate the cuticle, which is a rigid structural barrier on the outside of the plant epidermis; other microorganisms cannot cross that barrier. Anaerobic fungi often enter the leaf interior through the stomata in the epidermal layer (Akin et al. 1983). That practice gives the fungi another advantage in degrading plant fiber by substantially increasing the area available for attack. Little is known about the ability of the fungi to attack cuticle *per se*. To test for solubilization or simple penetration of the plant cuticle by the fungi, D. E. Akin (unpublished data) sealed leaf blade sections of orchard grass (*Dactylis glomerata*) with dental wax to prevent attack of the cut ends and incubated them with anaerobic fungal cultures for 2 weeks. Microscopic evaluation indicated that the cuticles were sloughed from some sections but that large sheets of cuticle remained intact. Masses of fungal filaments were present on the cuticle surface near the ruptured sites. Nuclear magnetic resonance spectroscopy of orchard grass residue (lignified tissue and cuticle) that was incubated with anaerobic fungi did not show recognizable changes in the fatty acid esters in the residue. Results from those ultrastructural and chemical investigations suggest that rhizoids or rhizomycelia of the fungi penetrated the cuticle through stomata, cracks, and at particular sites perhaps by force arising from the extensive hydrolysis, and chemical degradation of the cuticle by the fungi was not apparent (Wubah et al. 1993).

Species of *Piromyces*, *Neocallimastix*, *Orpinomyces*, and *Ruminomyces* degrade fiber to a substantial degree. *Caecomyces* species degrade fiber less than the other genera (Gordon 1990), perhaps because they lack an extensive rhizoidal system. No advantage to an animal from fiber degradation by a monocentric versus a polycentric organism is obvious (Gordon and Akin, unpublished data). In a study of Bermuda grass (*Cynodon dactylon*) stems and leaves, *Neocallimastix* strain MC-2 and *Orpinomyces* strain PC-2 degraded the most plant material (Akin et al. 1990). *Ruminomyces* strain PC-1 eventually degraded as much dry matter as the *Neocallimastix* species but at a slower rate. *Piromyces* strain MC-1 grew poorly on Bermuda grass material, eventually degrading similar amounts of leaf material as the other species but degrading significantly less of the stem biomass after 9 days. Data suggest that the ability to degrade fiber varies within a genus of fungi and that plants differ in their ability to support fungal growth.

When rumen fungi attack recalcitrant plant cell walls, they weaken the textural strength of the residue (Akin et al. 1989, 1990). In a study of the ability of pure cultures to reduce textural strength of plants, species of Neocallimastix, Orpinomyces, and *Anaeromyces* weakened stem tissues of *Cynodon dactylon* about 40% after 9 days, but rates differed among genera and species. The greater ability of rumen fungi, compared to rumen bacteria, to weaken forage fiber may be important in enhancing forage utilization by the host animal (Borneman and Akin 1990). Sulfur fertilization of the warm-season grass, *Digitaria pentzii*, increases growth of fungal populations in the rumen, which results in increased feed intake by cattle (Akin et al. 1983). In addition, high fungal populations in the rumen and chewing characteristics related to greater feed use appear to be associated (Weston et al. 1988).

TAXONOMY, DIVERSITY, DISTRIBUTION, AND ECOLOGY

SYNOPSIS OF TAXA

Obligate anaerobic zoosporic fungi first were recognized in 1976 (Orpin 1976) and assigned to the class Chytridiomycetes (Heath et al. 1983). Li and Heath (1992) subsequently examined the phylogenetic relationship between the anaerobic zoosporic fungi and other chytrids with molecular techniques and ultimately described a new order, Neocallimasticales, to accommodate these fungi (Li et al. 1993). The currently accepted higher classification of those fungi according to Ho and Barr (1995) is as follows:

Division/Phylum: Eumycota
  Subdivision: Mastigomycotina
    Class: Chytridiomycetes
      Order: Neocallimasticales
        Family: Neocallimasticaceae

At the genus and species levels, however, various aspects of the taxonomic and nomenclature of the group remain controversial. Orpin (1976, 1977b) used the generic names *Sphaeromonas* and *Piromonas* for isolates with posteriorly uniflagellate zoospores. Gold and colleagues (1988) challenged the use of those generic names for fungi because the names had been used originally (Liebetanz 1910) to describe protozoa that are anteriorly flagellated, obtain nutrients by phagocytosis, and divide by binary fission. The anaerobic fungi that were assigned those names, in contrast, are posteriorly flagellated, obtain nutrients saprotrophically with rhizoids, and do not divide by binary fission. Gold and colleagues (1988) renamed *Sphaeromonas* and *Piromonas* as *Caecomyces* and *Piromyces*, respectively. The original isolate that was used to establish the anaerobic zoosporic fungi as a group was renamed *Neocallimastix patriciarum* (Orpin and Munn 1986) when an isolate that had been classified as *N. frontalis* (Heath et al. 1983) was found to be different from the original isolate.

To date, 17 species of anaerobic fungi in five (currently recognized) genera formally have been described in the literature (Trinci et al. 1994), but Ho and Barr (1995) have suggested reducing the number of species to 13, based on synonymy. They placed species in three monocentric genera, namely *Neocallimastix*, *Piromyces*, and *Caecomyces*, and two polycentric genera, *Anaeromyces* and *Orpinomyces*. A key to the genera and selected species is provided in the Appendix at the end of this chapter. Species whose taxonomy remains controversial are not included. Current studies using molecular techniques should clarify some of the uncertainties.

SPECIES DISTRIBUTIONS AND RICHNESS

Anaerobic zoosporic fungi have been isolated from the GI tract of a diverse array of domesticated and wild herbivores (Table 22.1) ranging in size from the small blue duiker (Dehority and Varga 1991) to the large African elephant (Li et al. 1990). They appear to be ubiquitous in ruminants (Artiodactyla), and they have been isolated from rodents (Rodentia), marsupials (Marsupialia), and certain species of the Perissodactyla, as well as the elephants (Proboscidae). They have not been isolated from certain other species of herbivores in which they might be expected to occur (Li and Heath 1993). Species from all genera of anaerobic zoosporic fungi have been isolated from various ruminants, but only *Piromyces* and *Caecomyces* have been isolated from horses and elephants (Orpin 1981; Gold et al. 1988; Li et al. 1990; Teunissen et al. 1991). The genera or species of these fungi may show host specificity, but that phenomenon

TABLE 22.1

Mammalian Herbivores from which Anaerobic Zoosporic Fungi Have Been Isolated

Species	Common name
Foregut/pregastric fermenters	
Order Artiodactyla	
Ovis aries	Sheep
Bos taurus	Cow
B. indicus	Kedah Kelantan cow
B. gaurus	Gaur
Bubalus bubalis	Swamp buffalo
Oryx leucoryx	Arabian oryx
O. dammah	Scimitar-horned oryx
Taurotragus eurycerus	Bongo
Tragelaphus strepsiceros	Greater kudu
Hippotragus equinus	Roan antelope
Cervus elaphus	Feral red deer
Rangifer tarandus	Reindeer
Camelus bactrianus	Bactrian camel
C. dromedarius	Dromedary camel
Vicugna vicugna	Vicuña
Llama species	Llama
Aepyceros species	Impala
Capra hircus	Domestic goat
Cephalus monticola	Blue duiker
Order Rodentia	
Dolichotis patagonum	Mara
Order Diprotodontia	
Macropus giganticus	Grey kangaroo
M. rufogriseus	Redneck wallaby
M. robustus	Wallaroo
Wallabia bicolor	Swamp wallaby
Hindgut fermenters	
Order Perrisodactyla	
Rhinoceros unicornis	Indian rhinoceros
Diceros bicornis	Black rhinoceros
Equus burcheli	Common zebra
E. caballus	Horse
Order Proboscidea	
Loxodonta africana	African elephant
Elephas maximus	Indian elephant

Based on Li and Heath (1993); taxonomy follows Wilson and Reeder (1993).

has not been established. Recently *Neocallimastix*, *Piromyces*, and *Orpinomyces* were isolated from the anoxic mud layers at the bottom of a pond on a dairy farm (Wubah and Kim 1994, 1995).

Despite the broad presence of the anaerobic zoosporic fungi among mammal groups, most detailed studies of the group have been carried out on fungi from domestic ruminants such as sheep and cattle (Bauchop 1989). Sources of the fungal isolates include saliva; partially digested plant material from the rumen, reticulum, omasum, abomasum, and cecum; and freshly voided and

TABLE 22.2

Estimated Numbers of Thallus-Forming Units of Fungi in the Digestive Tract of Cattle

Organ	Number of units*
Rumen and reticulum	7.6×10^8
Omasum	7.6×10^7
Abomasum	3.1×10^6
Small intestines	2.5×10^6
Cecum	3.1×10^6
Large intestines	6.7×10^6
Feces	4.2×10^4

* Estimates from Trinci et al. (1994).

dry feces (Lowe et al. 1987c; Gold et al. 1988; Milne et al. 1989; Wubah et al. 1991d; Barr et al. 1995). Population levels of thallus-forming units of these fungi have been estimated for individual animals (Table 22.2). They appear to vary among different parts of the digestive tract.

GEOGRAPHIC DISTRIBUTION

Orpin (1975) isolated the first pure culture of anaerobic zoosporic fungi from the rumen of sheep in Britain. Since then, they have been shown to occur on all the continents, having geographic distributions that coincide with those of their host animals. Isolates have been obtained from animals in temperate and subtemperate regions of Australia (Gordon and Phillips 1989), Canada (Kudo et al. 1990, Barr et al. 1995), Czechoslovakia (Marounek and Hodrova 1989), France (Fonty et al. 1988), Indonesia (Trinci et al. 1994), Japan (Ushida et al. 1989), Malaysia (Y. W. Ho et al. 1990), The Netherlands (Teunissen et al. 1991), New Zealand (Bauchop 1979a, 1979b), Norway (Orpin et al. 1985), Russia (Kostyukovsky et al. 1991), and the United States (Akin et al. 1988; Wubah et al. 1991d). These fungi also have been isolated from tropical and subtropical regions in Chile (Trinci et al. 1994), Ghana (D. A. Wubah, unpublished data), Ethiopia (Milne et al. 1989), Tanzania (Breton et al. 1991), and Tunisia (Breton et al. 1991).

LIFE CYCLES

Although zoospores and vegetative thalli with zoosporangia have been documented for most of the species,

their complete life cycles are yet to be reported. In fact, details of the life cycles of only two (*Neocallimastix* species and *Caecomyces equi*) of the five recognized genera of anaerobic zoosporic fungi have been described (Bauchop 1979; Gold et al. 1988). The life cycles of *Neocallimastix* species consist of alternating motile-zoosporic and vegetative-zoosporangial stages (Bauchop 1983, 1989; Citron et al. 1987; Lowe et al. 1987a; Mountfort 1987; Orpin and Joblin 1988) and a resting structure known as the resistant sporangial (RS) stage (Wubah et al. 1991a). *In vivo* and *in vitro* studies indicate that the alternation between the zoosporic and the zoosporangial stages lasts from 24 to 32 hours (Joblin 1981; Bauchop 1983; Lowe et al. 1987a; Wubah et al. 1991b), although zoosporogenesis of young sporangia may occur 8 hours after encystment under appropriate conditions (Orpin 1977a). The RS stage, which forms 36 hours after zoospores are inoculated into fresh media, is dark brown at maturity, and the pigments in its walls have been identified with histochemical tests as melanin (Wubah et al. 1991a).

The life cycle of *C. equi* also includes a zoosporic stage that alternates with a zoosporangial stage (Gold et al. 1988). No resting stage has been observed, however, even in a fungus isolated from dry feces (Wubah et al. 1991d). The occurrence of resting stages in the life cycles of other genera of zoosporic fungi first was reported by Wubah and colleagues (1991d), who noted that isolates of *P. communis* and *Orpinomyces* strain PC-2 also produce melanized sporangia in culture. Those melanized stages share characteristics with the RS stage of *Neocallimastix* Species. No RS stage has been described for the genus *Anaeromyces*, a situation that requires further investigation.

SURVIVAL AND TRANSFER OUTSIDE THE ANIMAL HOST

Fonty and colleagues (1988) investigated the means by which fungi in the genus *Neocallimastix* are transferred to the rumens of lambs. Those fungi are found in the rumen of 2-week-old lambs that are separated from other sheep at birth, which suggests that contact between a newborn lamb and its mother or other sheep is not necessary for the introduction of the fungi. When isolated lambs were maintained under aseptic conditions, however, no rumen fungi were observed in their rumens after an equal period.

The isolation of anaerobic zoosporic fungi from saliva and feces suggests that saliva and feces are possible routes of transfer (Lowe et al. 1987c; Trinci et al. 1988;

Milne et al. 1989; Theodorou et al. 1990). Wubah and associates (1991d) isolated species of *Neocallimastix*, *Piromyces*, *Orpinomyces*, and *Caecomyces* from both fresh and dry feces of a cow in Georgia; the same isolates were obtained from the rumen of the cow. Each of those isolates produced a melanized resting stage *in vitro*, and similar melanized sporangia were observed in the fecal smears from which they were isolated. That observation is consistent with the hypothesis that the anaerobic zoosporic fungi can survive outside the animal host, a phenomenon that has been investigated by several workers (Milne et al. 1989; Wubah et al. 1991a, 1991d). Wubah and Kim (1995) obtained an isolate of *Piromyces* from the anoxic layers of a pond in which cattle defecate. That observation supports the hypothesis that fecal matter also may be a route for the transfer of anaerobic zoosporic fungi from an adult to a young animal.

BIODIVERSITY INVENTORY

TECHNIQUES FOR COLLECTING SAMPLES

The methods used to sample anaerobic fungi are usually modifications of techniques used by microbiologists to sample anaerobic bacteria. The procedure selected depends on the site from which the samples are to be collected. Some of those techniques, by sampling location, are described in the following sections.

Saliva

Saliva samples are taken by fitting an animal with an esophageal fistula that is attached to a tube. Fluid samples are collected from the esophagus into a sterile tube while the animal is neither eating nor ruminating. Freshly collected saliva from drooling animals also can serve as a source of fungi. Saliva samples often must be enriched with the addition of antibiotics because the population of anaerobic fungi may be low.

Rumen

Partially digested plant material from the rumen is the most common source of anaerobic fungi. Animals from which digesta is collected usually are fitted permanently with a cannula or a fistula with an internal diameter ranging from 5 to 20 cm. Animals usually are fed a high-fiber diet prior to sample collection to stimulate growth of the fungi. Digesta is collected into a bucket by inserting a tube (sheep and goats) or a hand (cows) through the fistula. To obtain rumen fluid with zoospores, digesta is strained through three layers of cheesecloth into a container that is sealed quickly. To ensure their survival, fungi generally are transported from a collection site to the laboratory in a thermos flask. Prior to use, the flask is filled with warm water (50–60°C). Carbon dioxide may be bubbled through the water for approximately 20 minutes before the flask is sealed. The flask is emptied before the sample is added. Fungal thalli that are attached to plant particles can be obtained from the strained digesta. Fungi must be isolated within 30 minutes of collecting the digesta to maximize survival of the fungal species.

Feces

Fecal samples can be obtained from the ground in grazing pastures or from the floor in a zoo cage. In controlled experiments, however, the feces must be collected before they touch the ground to prevent contamination by extraneous microorganisms. A plastic bag or bucket can be tied to the rear end of the animal or held in place to receive the feces. Fecal samples can be stored at 4°C for an extended period before fungal isolation. Isolates of anaerobic zoosporic fungi have been obtained from feces that have been dry for up to 1 year (D. A. Wubah, unpublished data).

Whole Animal

In addition to the specific sites described earlier, samples can be obtained from the entire GI tract of an animal (Davies et al. 1993). Briefly, the GI tract is removed intact from a carcass immediately after slaughter and dissected into organs after tying off the various sections with twine to reduce cross-contamination. Samples of digesta (approximately 20–400 g fresh weight) are removed from the rumen, omasum, abomasum, small intestine, cecum, and large intestine; an approximately equivalent amount of feces is collected before slaughter. The pH of digesta and feces is determined using a portable pH meter. The total gut content from each organ is determined by one of two methods depending on the part of the tract being sampled. Either the entire content of an organ is emptied into a container and weighed directly, or the organ and its contents are weighed, the contents removed, and the empty organ washed and weighed. The latter procedure is used for the omasum because it is impossible to remove all of the digesta from that organ without washing. In the period between slaughter and fungal enumeration, samples are maintained at about 37°C.

ISOLATION, CULTURE, AND MAINTENANCE

In the laboratory, isolation and maintenance of anaerobic fungi routinely are performed in an anaerobic glove box that can be obtained commercially. The glove box is equipped with an incubator that is maintained at 39°C. The atmosphere within the glove box often is made up of 95% CO_2 and 5% H_2. Oxygen is excluded or taken up by a palladium catalyst. Moisture levels are kept low by placing desiccants inside the glove box. Under field conditions, isolations can be initiated in Hungate tubes with media only.

Several methods have been used to isolate and culture zoosporic fungi from the rumen (Orpin 1975, 1976, 1977a, 1977b; Bauchop 1979; Joblin 1981; Lowe et al. 1985; Borneman et al. 1989). One method involves overlaying partially molten agar with filtered rumen fluid and incubating it at 39°C for 48 hours, during which time zoospores settle and form individual thalli from which pure cultures can be obtained (Orpin 1975). Lowe and colleagues (1985) used a plate culture technique to isolate rumen fungi from rumen digesta of sheep and cattle. The roll-bottle method of Joblin (1981) involves inoculating a dilution series of molten agar medium with filtered rumen fluid. After a period of incubation, axenic cultures (probably not free of bacteria unless antibiotics have been added to the agar medium) are obtained from the individual colonies formed. Methanogenic bacteria are the most prevalent bacterial contaminants in pure cultures of anaerobic zoosporic fungi. Antibiotics such as penicillin G, streptomycin sulfate, neomycin, and chloramphenicol can be added to the isolation media to suppress bacterial growth (Lowe et al. 1985; Orpin and Joblin 1988; Borneman et al. 1989; Wubah et al. 1991a, b, c, d). If antibiotics are not added to the media in which pure cultures are maintained, contamination invariably occurs after two to three transfers.

Pure cultures are maintained in undefined, semidefined, or defined media. The most commonly used undefined medium consists of centrifuged rumen fluid, tryptone, yeast extract, a carbon source, a buffer, L-cysteine as a reducing agent, and vitamins (Orpin and Joblin 1988; Wubah et al. 1991c). Some isolates have been cultured on semidefined medium containing all of the previously listed components except rumen fluid (Orpin and Greenwood 1986; Lowe et al. 1985). Borneman and associates (1989) and Wubah and colleagues (1991a, 1991b, 1991c, 1991d) substituted a volatile fatty acid mixture and hemin for the rumen fluid. *Neocallimastix frontalis* has been maintained in a defined medium without yeast extract and rumen fluid (Lowe et al. 1985).

All the rumen fungi studied to date exhibit a mixed acid fermentation (Theodorou et al. 1988), the end products of which are formate, acetate, lactate, ethanol, carbon dioxide, and hydrogen (Bauchop and Mountfort 1981; Orpin and Munn 1986; Lowe et al. 1987b; Borneman et al. 1989). Traces of succinate also have been detected in some cultures (Phillips and Gordon 1988; Borneman et al. 1989). No isolates have been found yet to grow in media with pectin as the sole carbon source (Orpin and Joblin 1988), although various species of anaerobic zoosporic fungi produce low levels of pectin-hydrolyzing enzymes in some culture media (Williams and Orpin 1987). The fungi also can grow on a wide range of monosaccharides, disaccharides, and polysaccharides (Orpin 1975, 1976, 1977b; Orpin and Letcher 1979; Bauchop 1981, 1983; Mountfort and Asher 1985, 1989; Orpin and Munn 1986; Williams and Orpin 1987; Phillips and Gordon 1988; Borneman et al. 1989). These fungi can be archived by cryopreservation (Yarlett et al. 1986b).

For prolonged maintenance in the laboratory, pure cultures of anaerobic fungi must be transferred every 3–4 days into fresh basal anaerobic medium (BAM; see Wubah et al. 1991d and Appendix II). Joblin (1981), however, reported that cultures could be maintained for several months on plant tissues stored at 39°C without subculturing. Long-term storage is also possible by means of cryopreservation in liquid nitrogen at −70°C, with 5% dimethylsulfoxide or 10% (vol/vol) glycerol as a cryoprotectant (Yarlett et al. 1986b; Phillips and Gordon 1988).

IDENTIFICATION

Genera of anaerobic fungi are characterized on the basis of number of flagella per zoospore, thallus morphology (monocentric or polycentric), and rhizoid type (filamentous or a vegetative cell). Species are delimited on the basis of zoospore ultrastructure (Ho and Barr 1995) and source of the isolate. Using those characters, 17 species in six genera have been described in the literature; however, only five genera have been recognized since the synonymy of *Ruminomyces* (*Anaeromyces*) was established. The genus *Piromyces* has six species: *P. communis* (Gold et al. 1988), *P. dumbonica* (Li et al. 1991), *P. mae* (Li et al. 1991), *P. rhizinflata* (Breton et al. 1991), *P. minutus* (Ho et al. 1993b), and *P. spiralis* (Ho et al. 1993c). *Neocallimastix*, the most extensively studied genus, includes four species: *N. frontalis* (Orpin 1975), *N. patriciarum* (Orpin and Munn 1986), *N. hurleyensis* (Webb and Theodorou 1991), and *N. variabilis* (Ho et al. 1993a). The two species of *Caecomyces* are *C. communis* (Gold et al. 1988) and *C. equi* (Gold et al. 1988). The two polycentric genera are *Orpinomyces* and *Anaeromyces*. The former genus has three species:

Orpinomyces bovis (Barr et al. 1989), *O. joyonii* (Breton et al. 1989), and *O. intercalaris* (Ho et al. 1994). The latter genus has two species: *Anaeromyces elegans* (Y. W. Ho et al. 1990) and *Ruminomyces* (*Anaeromyces*; Ho et al. 1993d) *mucronatus* (Breton et al. 1990).

The species of *Piromyces* produce uniflagellate zoospores that develop into a monocentric thallus consisting of a zoosporangium and an extensively branched rhizoidal system (Orpin 1977b). Host species and ultrastructural characteristics have been used to distinguish the species in this genus. *Piromyces mae* frequently produces one or more papillae, whereas *P. dumbonica* and *P. rhizinflata* are nonpapillate. *Piromyces rhizinflata*, which was isolated from a Saharan ass, differs from its congeners based on types of carbohydrates making up the cell wall and its constricted neck (Breton et al. 1991; Barr et al. 1995). *Piromyces minutus* and *P. spiralis* were isolated from a deer and a goat, respectively, and their thalli are the smallest among the species of *Piromyces* (Ho et al. 1993b, c).

Neocallimastix patriciarum and *N. frontalis* have multiflagellate zoospores that develop into monocentric thalli. The two species are distinguished on the basis of zoospore characteristics that include number of flagellae, presence or absence of an equatorial constriction, distribution of intracellular organelles, arrangement of kinetosomes, major fermentation products, and method of carbohydrate metabolism (Orpin and Munn 1986). The third species of *Neocallimastix*, originally called strain R1, recently was renamed *N. hurleyensis* on the basis of the size, location, and complexity of a large organelle (presumptive hydrogenosome) present in the zoospore (Webb and Theodorou 1991). The fourth species, *N. variabilis*, was isolated from a cow and differs from the other three in its ability to develop endogenously or exogenously (Ho et al. 1993a).

Each species of *Caecomyces* has a posteriorly directed, uniflagellate zoospore that develops into a vegetative thallus consisting of zoosporangia attached to one or more large vegetative cells (Orpin 1976; Gold et al. 1988; Wubah et al. 1991c). *Caecomyces communis* differs from *C. equi* by producing more than one bulbous vegetative cell, whereas the latter species consistently produces a zoosporangium attached to a single vegetative cell (Gold et al. 1988).

Members of the genus *Anaeromyces* also produce uniflagellate zoospores; they develop into polycentric thalli with fusiform or ellipsoidal sporangia. *Anaeromyces elegans* and *Ruminomyces mucronatus* were isolated from the rumen of cattle and sheep, respectively. The genus *Orpinomyces* was created by Barr and colleagues (1989) to accommodate rumen zoosporic fungi that have multiflagellate zoospores and develop into polycentric thalli. Breton and colleagues (1989) described an isolate

with similar characteristics but described it as a new species, *Neocallimastix joyonii*, in a different genus. That organism later was renamed *Orpinomyces joyonii* (Li et al. 1991). The type species of the genus, *Orpinomyces bovis*, was isolated from a cow (Barr et al. 1989).

ENUMERATION OF ANAEROBIC FUNGI

Several methods have been designed to quantify anaerobic zoosporic fungi. Counts of individual zoospores and zoosporangia have been used to estimate fungal populations *in vitro* (Joblin 1981; Windham and Akin 1984) and *in vivo* (Grenet et al. 1989; Ushida et al. 1989). Breton and colleagues (1991) used colony-forming units per gram dry weight of feces as the basis for quantifying species of *Piromyces*.

An endpoint dilution procedure, based on the technique of most probable numbers, was developed to enumerate rumen fungi as thallus-forming units (Theodorou et al. 1990). Briefly, the procedure involves preparing a 10-fold dilution series of the sample in an antibiotic-containing BAM. Defined medium B (Theodorou et al. 1990) plus 10% (vol/vol) clarified ruminal fluid was used for the dilution series. The dilution medium contained freshly prepared penicillin G and streptomycin sulfate at final concentrations of 240 U/ml and 75 U/ml, respectively. The medium was added in 45-ml amounts to 60-ml serum bottles and warmed to 39°C before inoculation. The dilution series was initiated either by addition of 5 ml of a pooled culture supernatant to the 45 ml of medium in the serum bottle or by addition of 45 ml of medium to 5 g (fresh weight) of ruminal digesta, fresh feces, or air-dried feces in a plastic bag flushed with CO_2. Ruminal and fecal samples were mixed from 1 to 3 minutes to obtain a particulate suspension. The dilution series was continued down to 10^{-6} by transferring 5-ml amounts from one serum bottle to the next. Three 10-ml portions of the appropriate dilution (10^{-3} to 10^{-6}) then were transferred to enumeration tubes (three replicate tubes per dilution) containing approximately 100 mg of wheat straw. Enumeration tubes were incubated at 39°C and checked daily with an inverted microscope for the presence of anaerobic fungi. Tubes were scored positive if motile zoospores were present in the culture supernatant and if rhizoids and zoosporangia were associated with the plant fragments. The majority of tubes that contained anaerobic fungi were scored positive within 3–4 days of the onset of incubation. All negative tubes were investigated after 10 days to confirm the absence of anaerobic fungi. Fungal populations were expressed as thallus-forming units (TFU) per gram of dry matter (DM).

An indirect method based on zoospore concentrations and life cycle parameters has been used to quantify anaerobic fungi in the rumen. Using life history parameters and growth kinetics of these fungi, France and his colleagues (France et al. 1990) constructed and solved a mathematical model of the life cycle in a steady state so that the population of particle-attached fungal thalli could be calculated from the concentration of free-swimming zoospores in rumen fluid. The values obtained were consistent with observations of samples from the rumen and feces. Our limited knowledge of the life cycle of anaerobic fungi, however, makes it difficult to evaluate the reliability of this method.

CONCLUSIONS

The discovery of the anaerobic zoosporic fungi led to a change in the description of the fungal class Chytridiomycetes to include chytrids with more than one flagellum per zoospore. Furthermore, the physiological characteristics of this unique group of fungi indicate that it plays a critical role in the digestion of fiber in herbivores. In fact, it has been demonstrated that these fungi can be manipulated to enhance digestion in domestic ruminants. Enzymes produced by these fungi also may be valuable for the digestion of plant waste and the production of methane and hydrogen. Recently, these fungi have been isolated from a pond in a cow pasture and from sediments in a palustrine marsh, which suggests that they may play a role in nutrient recycling in such environments. The possible contribution of these fungi to the carbon and hydrogen budget in such anoxic habitats may be an interesting area for future research.

APPENDIX

KEY TO THE GENERA AND SELECTED SPECIES OF ANAEROBIC ZOOSPORIC FUNGI*

1. Zoospore polyflagellate	2
1. Zoospore uniflagellate	3
2. Thallus monocentric	*Neocallimastix* 5
2. Thallus polycentric and mycelioid	*Orpinomyces* 6
3. Thallus monocentric	4
3. Thallus polycentric and mycelioid	*Anaeromyces* 7
4. Thallus with filamentous rhizoids	*Piromyces* 8
4. Thallus with vegetative cell	*Caecomyces* 12
5. Zoospores released through apical pore followed by sporangial rupture	*N. frontalis*
5. Zoospores released through a distinctive apical pore	*N. hurleyensis*
6. Terminal globose sporangia on simple or branched sporangial stalks	*O. joyonii*
6. Intercalary globose sporangia	*O. intercalaris*
7. Hyphae with lobed or beadlike structures	*A. elegans*
7. Hyphae without lobed or beadlike structures	*A. mucronatus*
8. Zoospore release by dissolution of sporangial wall	9
8. Zoospore release through pores or papillae	10
9. Zoospore release accompanied by sporangial wall dissolution; rhizoids nonspiral	*P. communis*
9. Zoospore release preceded by sporangial wall dissolution; rhizoids spiral	*P. spiralis*
10. Sporangia less than 30 mm in diameter; main rhizoid slender	*P. minutus*
10. Sporangia more than 30 mm in diameter; main rhizoid tubular	11
11. Mature sporangia nonpapillated; neck constricted	*P. rhizinflata*
11. Mature sporangia papillated; neck broad	*P. mae*
12. Sporangium with one vegetative cell; from cecum of horse	*C. equi*
12. Sporangium with one or more vegetative cell; from hindgut and rumen of herbivore	*C. communis*

* Based on Ho and Barr (1995).

RECOMMENDED PROTOCOLS FOR SAMPLING PARTICULAR GROUPS OF FUNGI: COLLECTING AND ISOLATION PROTOCOLS FOR AQUATIC FUNGI AND FOR PROTOCTISTANS FORMERLY TREATED AS FUNGI

23

Fungi in Freshwater Habitats

CAROL A. SHEARER, DEBORAH M. LANGSAM, AND JOYCE E. LONGCORE

The fungi in freshwater habitats comprise a diverse assemblage of true fungi and fungus-like straminipiles and protists. Fungi are found in all types of freshwater habitats, including rivers, streams, drainage ditches, marshes, swamps, bogs, lakes, ponds, temporary ponds, and wetlands. Fungal saprobes play a major role in freshwater ecosystems as decomposers of plant and animal tissues, and parasitic fungi occur on planktonic and benthic algae, aquatic macrophytes, fish, amphibians, invertebrates, and animal-like protists. Aquatic habitats have disappeared at a rapid rate over the past 100 years, and many of them have been and continue to be altered by agrarian practices, land development, and pollution. Consequently, it is important that the species present in existing aquatic habitats be documented so that the effects of habitat alterations on species composition, and hence function, can be detected. It is also imperative that cultures of freshwater fungi be isolated and maintained for future use in habitat restoration, experimental approaches to aquatic ecosystem function, and phylogenetic studies.

Representatives of all fungal classes can be found in or isolated from freshwater habitats. This chapter, however, deals primarily with the groups of fungi that occur most frequently in freshwater habitats and play the most significant ecological roles in freshwater communities. The groups are (1) Chytridiomycetes and Hyphochytriomycetes; (2) Peronosporomycetes (formerly Oomycetes); (3) Ascomycetes; and (4) mitosporic fungi. It is highly likely that yeasts play significant roles in freshwater habitats, but studies documenting their occurrence and importance are rare. The reader is referred to Chapter 16 for methods regarding that group.

CHYTRIDIOMYCETES AND HYPHOCHYTRIOMYCETES

SUMMARY OF EXISTING KNOWLEDGE

The Chytridiomycetes, which are true Fungi (Förster et al. 1990; Bowman et al. 1992), and the Hyphochytriomycetes, which are allied with the straminipiles (Barr 1992; Van der Auwera et al. 1995), are treated together because species from both groups are outwardly similar, occupy the same habitats, and are studied with the same techniques. The hyphochytrids consist of a single order with three families and four genera (Karling 1977; Fuller 1990). Members characteristically produce zoospores with a single, anterior, tinsellated flagellum. The five orders that comprise the class Chytridiomycetes are defined on the basis of differences in ultrastructural characters of zoospores (D. J. S. Barr 1990, 2001). Members of all orders except some species in the specialized Neocallimastigales have zoospores with a single, posteriorly directed flagellum. The Chytridiales, Monoblepharidales, and Blastocladiales are found in water and in soils. The Spizellomycetales primarily inhabit soils but may be found at the margins of lentic and lotic aquatic habitats. Representatives of the Neocallimastigales are obligate anaerobes and, thus far, have been isolated only from the digestive systems and feces of herbivorous animals, with the exception of a single isolation from a farm pond (Wubah and Kim 1995).

The Blastocladiales contains five families and 13 genera. Some taxa, including the well-known, experimental organisms, *Allomyces* and *Blastocladiella,* are saprotrophic, but many others are specialized parasites of invertebrates (e.g., *Coelomomyces* species in mosquito larvae). *Physoderma,* a genus whose members parasitize aquatic and semiaquatic plants, also belongs to this order. Monoblepharidales contains four families and six genera, all saprotrophic (Forget et al. 2002). Members of the Chytridiales first were described in the 1850s; currently four families and about 80 genera containing more than 500 species are recognized. In 1980, Spizellomycetales, which contains four families and 10 genera, was separated from Chytridiales on the basis of ultrastructural characters of zoospores (Barr 1980).

Most of the diversity of the Chytridiomycetes lies within the Chytridiales, but unfortunately, studies of that group are severely limited by the lack of adequate species descriptions, mentors who can help with identifications, and recent identification guides. The most recent monograph that includes Chytridiomycetes and Hyphochytriomycetes is Sparrow's *Aquatic Phycomycetes* (1960), which he later followed with a key to genera (Sparrow 1973). More than 23 genera and 300 species have been described since Sparrow's monograph and are listed along with other changes in taxonomy in a bibliography by Longcore (1996). Taxa described since 1960, however, have not been incorporated into taxonomic keys. *Zarys hydromikologii* (Batko 1975) is dedicated exclusively to aquatic fungi, but its usefulness is limited in many countries because the keys are written in Polish and are not comprehensive in their coverage of species. *Chytridiomycetarum Iconographica* (Karling 1977) contains commentaries on genera of chytrids and hyphochytrids and many drawings. Because it is tempting to use Karling's drawings to "picture-key" aquatic fungi, we emphasize that his book portrays only a fraction of the described species. Investigators who wish to identify chytrids and hyphochytrids, particularly those described since 1960, need to refer to original descriptions.

Reviews of the ecology of freshwater fungi (Sparrow 1968; Dick 1976; Powell 1993) provide an overview of what is known about the diversity of zoosporic fungi. Knowledge of the diversity of zoosporic fungi in several geographic areas has accrued as a consequence of the career research of mycologists who specialize in those groups. Notable in this regard with respect to aquatic habitats are the papers of H. M. Canter and colleagues, which document chytrids associated with algae in the Lake District of England (see references in Karling 1977). Knowledge of the aquatic mycota of the Lake District was broadened further by L. G. Willoughby's studies of the saprotrophic chytrids of the margins and muds of several lakes in that area (see references in Karling 1977).

Similarly, a general knowledge of the diversity of chytrids and other aquatic fungi found in northern Michigan, in the United States, exists as a result of the research of F. K. Sparrow and colleagues. Over a span of about 20 years, those researchers published papers on zoosporic fungi from aquatic sites and soils throughout the northern counties of the lower peninsula of Michigan. Their studies emphasized the fungi, not the habitat.

The Chytridiomycetes and Hyphochytriomycetes are denoted as "aquatic" fungi because they disperse through water with motile spores. After rain or snow melt, most soils are transformed into an "aquatic" habitat in which chytrids and hyphochytrids are widespread. Consequently, many aquatic fungi have been studied from soil samples, which contain resting spores of zoosporic fungi, because they are relatively easy to collect and transport. For example, Karling based his reports of zoosporic fungi from India, Africa, New Zealand, Oceania, and various South American countries mostly on his studies of soil samples, although some of the soil samples came from dried aquatic habitats. Willoughby (1962a, 1962b) reported differences between the species found in lakes and those usually found in soils, but many species have been reported from both aquatic habitats and soils.

Chytrids associated with discrete, countable substrata such as algae have been quantified; for reviews see Masters (1976) and Powell (1993). In northern North America, conifers produce an annual shower of pollen, which falls in such abundance that yellow pollen lines surround lake shores. Ulken and Sparrow (1968) used a modification of the most probable number (MPN) method used by bacteriologists and found that the number of chytrid zoospores that attack pollen grains per liter of lake water peaked during the 2 weeks following the height of the pollen shower.

Endemism has not been known among chytrids and hyphochytrids; however, recent molecular evidence (Morehouse et al. 2000) suggests that *Batrachochytrium dendrobatidis*, a chytridialean pathogen of amphibians, may have recently spread to several continents. Sparrow's experience with Michigan fungi allowed him to observe that several of the fungi he found in a bog in the Hawaiian Islands belong to the same species that occur in bogs in northern Michigan (Sparrow 1965). Chytrid species with morphologies so distinctive as to preclude misidentification have been reported from different continents. The prevailing hypothesis is that chytrid species are distributed worldwide, with occurrence determined by local conditions rather than geography (Sparrow 1968).

COLLECTION, IDENTIFICATION, AND DEPOSITION OF SPECIMENS

Collection

Chytrids are microscopic, cannot be collected by unaided sight, and usually are not detected with standard microbiological methods such as dilution plating. They can be found by examining their natural substrata under a microscope, but this can be time-consuming and suitable natural substrata are not always easily found during a collecting trip. Therefore, "baits" are commonly used to provide chytrids and hyphochytrids with a fresh substratum to colonize. Baits are selected to represent naturally occurring organic debris in pieces that are thin enough to allow viewing with transmitted light on a microscope slide. Commonly used baits include chitin (bits of purified shrimp exoskeleton or insect wings); cellulose (cellophane, lens paper, white onion skin, and bleached corn straw); keratin (defatted, blond baby hair; pieces of snake skin; or defatted wool fibers); and pollen grains of various types, especially, but not limited to, pine (*Pinus* species), spruce (*Picea* species), and sweetgum (*Liquidambar* species). Those baits, except pollen, are boiled from 3 to 5 minutes before using. Larger baits, such as seeds, fruits, and twigs, are used to attract members of Blastocladiales

and Monoblepharidales. Collection and baiting need to be tailored to the group of fungi sought; detailed methods for use with members of the various orders of chytrids and hyphochytrids are found in Fuller and Jaworski (1987). Whisler (1987) offered advice regarding the Monoblepharidales and Blastocladiales, and W. W. Martin (1987) explained methods of collecting parasites of aquatic insects. Later in this chapter, we include methods that we have used successfully for baiting for Chytridiales, Spizellomycetales, and Hyphochytriales. Investigators should consult D. J. S. Barr (1987), Sparrow (1960), Emerson (1950), and Couch (1939) for additional insights.

Organic debris and water can be collected from a study site and baited in the laboratory. A small amount of debris (bottom organic matter amounting to one or two waterlogged leaves, a small aquatic plant, or more than 10 cc of detritus) and water from the same habitat are added to a deep, glass Petri plate or finger bowl. Two pieces (about $0.5-1$ cm $\times$ $0.5-1$ cm) each of cellophane, onion skin, and chitin, plus a thin shower of pollen, are added as bait. The baited collection is known as a gross culture and should be incubated at a temperature near that of the water at the collection site. The gross culture should consist mostly of water with a small amount of bait and organic debris because excess organic matter encourages growth of bacteria instead of chytrids. Chytrids usually appear on pollen grains after $1-2$ days and on the other baits after several days. Many cellulosic chytrids have generation times of $4-7$ days, and a longer time may elapse before sporangia appear. Gross cultures may continue to yield fungi for weeks and should be examined periodically.

If knowledge of the phenology of the mycota is important, chytrids are collected by placing baits enclosed in a weighted (sterilized glass marbles work well), nylon net bag into the habitat being studied. Pollen is wrapped in lens paper so that the grains become enmeshed among cellulose fibers and are retained in the bag. The net bag is fastened in an inconspicuous place and left at the site for about 1 week if temperatures are higher than 20°C and up to a month if temperatures are below 10°C. Bait bags and surrounding water are retrieved into a sterile container and transported to the laboratory, where baits and water are transferred to glass Petri plates or finger bowls. Baits then are examined for chytrids using a compound microscope. Additional baits can be added, if needed, to encourage development of populations of fungi.

Chytrids can be found and isolated directly from natural substrata. Such substrata include algae, senescent plants, waterlogged wood, and insect exuviae. Algae are carried to the laboratory in the water in which they are growing. A subsample (the amount that can be picked up with forceps) is removed, dipped in 50°C water for about 1 minute (D. J. S. Barr, personal communication), and returned to the gross culture at a location where it can be found later. The temperature shock kills the algae without completely disrupting membranes, and the newly dead algae attract chytrids that occur on naturally senescent algae.

Some aquatic plants have leaves that are a few cell layers thick. Such leaves easily can be examined for fungi by placing them on a microscope slide. If leaves are thicker, epidermal peels can be examined. Chytrids and peronosporomycetous fungi occur most frequently on senescent leaves, and hyphomycetes most commonly occur in more decomposed materials. If no newly senescent leaves are present, a small sample of leaves can be heat-treated as for algae. If fungi are not seen in collected aquatic plants, onion skin can be used as a substitute substratum. We have isolated chytrids directly from *Eriocaulon, Typha, Utricularia,* and *Potamogeton* (J. E. Longcore, unpublished data). Exuviae of aquatic insects also harbor many interesting chytrids and can be examined easily using a compound microscope.

Waterlogged wood supports the growth of members of the Monoblepharidales and Blastocladiales. Waterlogged twigs that still retain bark are collected and packed into jars with water from the collection site. Storage of these collections for weeks or months at 5°C before subsamples are brought to room temperature enhances the likelihood of finding monoblepharidalean fungi. The fungi can be detected by using a dissecting microscope to locate tufts of hyphae, which often grow as an understory beneath peronosporomycetes.

Identification

Before fungi are isolated, data on their substrata and growth habits should be recorded. After chytrids are in culture, isolates can be characterized by photographing developmental sequences of the organism growing on nutrient agar. Zoospores, germlings, intermediate stages, mature sporangia, method of zoospore discharge, and resting spores should be included. The rhizoidal system should be well documented, focusing on number of rhizoidal axes, branching pattern, nature of the rhizoidal system where it attaches to the sporangium, and rhizoidal tips. Other characters include generation time on nutrient agar, maximum temperature at which the isolate grows, and color of colonies in culture. Molecular sequence data and/or data on ultrastructure of the zoospore should be included in descriptions of new species to ensure that species are placed in the correct orders and genera (D. J. S. Barr 1980, 1990; James et al. 2000).

Historically, most chytrids were described and identified based on observations of the organisms on their natural substrata or baits. Sometimes the chytrid being studied was isolated into unifungal culture, and enough of the morphology was documented to ensure adequate descriptions. Some chytrids have unique morphologies and are identifiable despite incomplete descriptions. Many descriptions, however, especially those lacking detailed observations of the rhizoidal system, the mechanism of zoospore discharge, and a depiction of the zoospore, are inadequate to assure accurate identification. Many descriptions of species also rely on characters that have been shown to vary (Miller 1976). In previous years several centers existed where mycologists with special knowledge about chytrid, hyphochytrid, and peronosporomycete genera and species handed down that understanding to their students. Such centers are now rare, and those wishing to identify zoosporic fungi must rely primarily on published material.

For identification of saprotrophic Monoblepharidales and Blastocladiales, investigators can refer to keys and drawings in Sparrow (1960, 1973) and Karling (1977) and to original descriptions for species named since 1960 (Longcore 1996). Couch and Bland (1985) provided keys to species of *Coelomomyces,* and Barr (1984) provided a key to most of the saprobic spizellomycetalean species. For the Chytridiales and Hyphochytriales, Sparrow's keys (1960, 1973) and Karling's drawings (1977) must be supplemented with individual papers. See Longcore (1996) for a list of taxonomic changes in the Chytridiomycetes since Sparrow's monograph. Many species will not be identifiable, and those investigators interested in determining diversity first must become, or cooperate with, systematists for the group. The foregoing advice for methods to isolate, characterize, and identify chytrids and hyphochytrids is based on experience gained in an ongoing attempt to determine the diversity of Chytridiomycetes in two Maine lakes. The original purpose of the study was to compare the fungi in lakes with different pH; however, studies of the systematics of many of the isolates were necessary before names could be applied (Longcore 1992a, 1992b, 1993, 1995).

Deposition

In the past, chytrids on natural substrata have been preserved for documentation in lactophenol on glass microscope slides. Such microscope slides, however, are only semipermanent and require curatorial upkeep. Furthermore, they are not adequate substitutes for cultures and photographs of developmental sequences. Photographs of developmental sequences from pure cultures now are used as types when describing new species and to confirm identities (e.g., see Salkin 1970; Powell and Koch 1977a; Barr 1986; Barr and Désaulniers 1987; Barr et al. 1987; Longcore 1992a, 1992b, 1993, 1995). For examples of excellent documentation of chytrids parasitic on algae see the many, well-illustrated papers of Canter and colleagues (e.g., Canter 1968, 1985; Canter et al. 1992). Because making reliable identifications is difficult and researchers may have differing concepts of species, we urge investigators to deposit voucher cultures in a culture collection and to publish and archive photographs of the developmental morphology of those organisms.

ISOLATION AND CULTURING

Isolation

In any serious study of the diversity of zoosporic fungi the organisms must be isolated into pure culture. That is the best, and frequently the only, way to document sufficient taxonomic characters of a chytrid either to determine its identity or to describe it in such a way that others can later determine its identity. Historically, that practice has not been followed routinely; as a result, many species are characterized inadequately. Exceptions exist. Some parasites have been well documented after study in two-membered cultures, and some saprobes are so distinctive that they can be identified without culturing. Culturing, however, yields the most characters for use in identification and should be standard practice for saprobes.

Chytrids and hyphochytrids found in gross cultures can be isolated into pure culture whenever mature sporangia are present. Subcultures of species of interest are established by adding infected baits to sterile, dilute pond water with fresh baits. Subculturing, however, is not always successful. Consequently, when mature sporangia are present on baits or natural substrata, it is best to isolate them without further subculturing.

The piece of colonized substratum is placed on a 9-cm-diameter plate of nutrient agar. We commonly use PmTG (Appendix II) and weak PmTG (mPmTG) nutrient medium made with 1% agar and containing antibiotics. While the fungus is viewed with a dissecting microscope (40–80× with substage lighting), the sporangium is pushed through the agar from one side of the Petri plate to the other with the point of a fine needle to remove bacteria and other fungi. The cleansed sporangium is placed on a clean area of the plate or transferred to a fresh plate of nutrient agar. A square is drawn in the agar around the isolated sporangium with the sterile tip of a needle to facilitate its location under the compound microscope. After four to six sporangia, or more, are cleaned, the plate is sealed with a strip of laboratory film (e.g., Parafilm) and incubated at the same temperature as the gross culture. Isolation plates should be checked daily by inverting the sealed culture plate onto the stage of a compound microscope and scanning with the 10× objective.

It may be necessary to remove fungal "weeds" to prevent overgrowth of the target organism. Seeds, snake skin, and fruit baits attract peronosporomycetes, which produce many zoospores and rapidly can overgrow chytrids on isolation plates. Therefore, snake skin, fruit, and seeds are better placed in their own gross cultures when baiting for keratinophilic chytrids, Monoblepharidales, and Blastocladiales. The chance of isolating chytrids into pure culture increases with the number of sporangia cleaned. If the target organism does not grow on any of the standard isolation media, then investigators should experiment with other nutrient media. *Rhizoclosmatium globosum* and *Chytriomyces hyalinus* are good practice organisms when learning isolation techniques. They commonly occur on chitin bait and practically leap into culture on PmTG agar. Many species that colonize pollen also are easy to bring into culture.

An alternative isolation method, which is particularly valuable with tiny sporangia, involves washing the substratum that contains the mature zoosporangia with distilled water from a wash bottle. The washed substratum is placed in a depression slide with some distilled water, and the slide is maintained in a moist chamber (Appendix I). During the next hour or two, the water is examined using a compound microscope. When discharged zoospores appear, the liquid (and zoospores) is withdrawn with a sterile Pasteur pipette and spread on nutrient agar. Marking the area where the water was applied on the bottom of the culture plate facilitates later inspections for growing thalli. If the medium supports growth of the fungus, thalli will grow from the zoospores, frequently free of bacteria, which may have been present on the substratum.

Polycentric organisms (e.g., *Nowakowskiella, Cladochytrium, Hyphochytrium*) are isolated by the previously mentioned methods or by simply placing a very small piece of substratum containing the rhizomycelium on nutrient agar that contains antibiotics. Replicates will increase the chances that a colony will appear without being overgrown by other fungi. Pieces of mycelium from collections of *Allomyces, Monoblepharis, Monoblepharella,* and *Gonapodya* can be drag-cleaned through agar and allowed to develop on PmTG agar plates. Emerson (1950), Dogma (1973), and D. J. S. Barr (1987) also discuss isolation procedures.

Culturing

After a colony is established on an isolation plate, parts of it can be transferred to nutrient agar without antibiotics to check for bacterial contamination. Adding sterile water to colonies of monocentric species will spread zoospores and establish additional colonies. Bacteria-free colonies then can be transferred to agar slants in 20-mm

× 125-mm screw-cap culture tubes. Water collects from condensation when autoclaved agar slants cool. This water does not evaporate from screw-capped tubes and can be tipped over monocentric chytrids to spread zoospores over the surface of the slant. Polycentric species need not be tipped. Cultures are incubated at their optimum growing temperature until colonies are well established and then refrigerated (5°C) for storage. Most isolates from temperate regions that are stored as described remain viable for at least 3 months. Duplicate tubes should be kept in separate incubators as backup. Many chytrid cultures retain their viability after freezing and storage in liquid nitrogen (Barr and Babcock 1994, Boyle et al. 2003). Other methods for culturing can be found in Emerson (1958), D. J. S. Barr (1987), and Whisler (1987).

ASSESSING BIODIVERSITY AND ABUNDANCE

Attempts to thoroughly document the diversity of chytrids in aquatic habitats are not common. Willoughby's report of the chytrids of marginal and permanently submerged muds of Estwaite Water in the English Lake District remains one of the better examples of diversity studies of chytrids (Willoughby 1961). Although Willoughby did not isolate chytrids into pure culture, he sampled during all months of the year and documented fungi with drawings of their morphology throughout their development. Czeczuga and colleagues reported on the aquatic fungi (Chytridiomycetes, Hyphochytriomycetes, Peronosporomycetes, Ascomycetes, and mitosporic fungi) of many lakes and rivers in Poland (e.g., Czeczuga and Woronowicz 1992). They included limnologial data in their reports, but, in fact, they found few chytrids and did not document their identities sufficiently to allow comparisons with results from other studies. A report on some chytrids from aquatic sites in Taiwan (Chen and Chien 1995, 1998; Chen et al. 2000), although not based on the systematic sampling of a single wetland, is praiseworthy because the investigators brought the isolates into pure culture and documented the development of the chytrid thalli with photographs.

The difficulty in isolating and identifying chytrids in aquatic habitats makes studying their diversity challenging but not impossible. Thorough studies of limited areas with reliable species identifications are of greater scientific value than ambitious surveys for which identifications are dubious and cultures or documentation of identities are nonexistent. We recommend, therefore, that those wishing to assess the diversity of fungi in aquatic habitats start with a small area. We suggest that

investigators attempting to determine all species of chytrids and hyphochytrids in an area make bait collections from detritus and water, bait *in situ,* and examine natural substrata because different methods do not always reveal the same species. Sampling during all seasons of several years and sampling as many microhabitats as possible will increase the probability that most species will be found.

Quantitative methods have not been refined for this group. Numbers can be determined using bacteriological (e.g., MPN) and limnological collecting and counting methods when the natural substratum is particulate. For chytrids that only are recovered on baits, abundance is difficult to determine. Crude abundance, however, can be ranked by determining the number of bait bags that contain the target species out of the total number of bait bags placed in a study site.

PERONOSPOROMYCETES (FORMERLY OOMYCETES)

SUMMARY OF EXISTING KNOWLEDGE

The first mention of a peronosporomycetous "fungus" probably can be traced to a letter from Arderon (1748:321–322), in which he describes a "distemper" of fish resembling a "mouldiness" and involving "minute tubes filled with a brownish liquor." In the disease, "Mortification seizes upon the Tail itself, and gradually creeps along it until it reaches the Intestines, at which time the Fish immediately dies." The illustrations accompanying the letter do not reveal the genus of the organism observed by Arderon, but his description leaves little doubt that the causative agent of the disease was a pathogenic isolate of the Saprolegniaceae, a family of straminipilous fungi aligned for many years with the Oomycete "watermolds." Despite numerous references made after 1748 to the occurrence of aquatic "fungi" on various substrata, more than 70 years passed before Gruithuisen (1821) published the first account of zoosporangial structure and zoospore discharge in a watermold. Oogonia and antheridia were not observed and identified as sexual structures of those organisms until more than 20 years later (Schleiden 1842; Thuret 1850; Pringsheim 1855). Nevertheless, the early morphological studies set the stage for debates about the evolutionary history of those organisms that have persisted over the past 150 years.

Early on, Peronosporomycetes were aligned with the algae by Nees von Esenbeck (1823). Both he and Carus

(1823), however, recognized the difficulty in establishing a meaningful position for these organisms within the traditional plant and animal kingdoms. The motile spores and filamentous thalli of the organisms were representative of some algal groups, but the colorless nature of the filaments eventually led to their affiliation with the fungi (De Bary 1866) and then to membership in the Oomycetes. Although the fungal connection was contested by many investigators (e.g., Shaffer 1975; Dick 1976, 1990a; Cavalier-Smith 1986, 1987), the affiliation—largely based on tradition—persisted for many years (Barr 1992). An impressive body of evidence, grounded in part on zoospore ultrastructure (e.g., Barr and Allan 1985; Powell et al. 1985), metabolic pathways and storage products (e.g., Bartnicki-Garcia and Wang 1983; Warner and Domnas 1987; Griffith et al. 1992), and nucleic-acid sequencing (e.g., Gunderson et al. 1987; Förster et al. 1990; Bhattacharya et al. 1992; Karlovsky and Fartmann 1992), has taken the debate full circle by clearly linking those organisms with the chrysophyte algae, in the Kingdom Straminipila. At the same time, the fungus-like morphology and obligate osmotrophic physiology now is viewed as support for categorizing the group as chromistan or straminipilous fungi (Dick 1995, 1999). The redesignation of the group as Peronosporomycetes was largely because of the taxonomic imprecision of the term Oomycetes (Dick 1999).

Today the Peronosporomycetes comprise a group of heterotrophic organisms displaying the osmotrophic mode of nutrition traditionally associated with true fungi. The vast majority are saprotrophic or facultatively parasitic organisms occurring in well-aerated freshwater or moist soils. Stagnant waters provide a habitat for some Peronosporomycetes, including a few facultative anaerobes (Emerson and Natvig 1981); others inhabit estuarine and marine environments, and a number of representatives are obligate parasites of aquatic or terrestrial organisms. In aquatic habitats, parasites attack fish, algae, mosquitoes, and small invertebrates such as nematodes and rotifers.

According to Dick (1999), the entire group includes around 95 genera and somewhere between 900 and 1,500 species (depending on the criteria used for classification). The classification scheme currently favored has roots in Sparrow's (1976) division of the group (under its former Oomycete label) into "saprolegnian" and "peronosporacean" galaxies of organisms. Based on the structure of the oospheres and oospores as well as the structure and behavior of the zoospores, Sparrow's scheme was based on the theory that the group originated from two separate heterokont ancestors. The current classification scheme, proposed by Dick (1995, 1999) and supported by a number of nucleic-acid analy-

sis studies (e.g., Dick et al. 1999; Riethmüller et al. 1999; Hudspeth et al. 2000; Petersen and Rosendahl 2000), recognizes three subclasses. The Peronosporomycetidae are mycelial fungi that include the Peronosporales and Pythiales. The Saprolegniomycetidae are also mycelial and include the Saprolegniales and the Scerosporales. The Rhipidiomycetidae include the nonmycelial (monocentric and polar) Rhipidiales. A number of other orders are aligned with the Peronosporomycetes, but their taxonomic positions remain uncertain at this time. Those include the Leptomitales, Salilagenidiales, Myzocytiopsidales, and the Olpidiopsidales.

COLLECTION, IDENTIFICATION, AND DEPOSITION OF SPECIMENS

Collection

Although members of the Peronosporomycetes can be found on submerged twigs and decaying fruits, most early investigators relied on water samples of lakes, streams, and ponds to provide organisms for study. They transferred samples to Petri plates in the laboratory, diluted them with distilled water, and baited them with a wide range of nutrient substrata such as egg yolk, dead flies, snake skin, and mushroom grubs. Butler's (1907) discovery of *Pythium* isolates in soil samples led Harvey (1925), and those who followed him, to rely on soil, as well as water, as a source for saprolegniaceous fungi. The discoveries that charcoal-filtered distilled (CFD) water (Livingston et al. 1905) promoted the formation of motile spores and that hemp seed inhibited the growth of bacteria (Harvey 1925) allowed for the characterization of numerous water molds over the first half of the 20th century.

Those indirect methods of collection persist today (Sparrow 1960; Seymour and Fuller 1987; Dick 1990a). Moist and dry soils, as well as sediments from ponds, lakes, and streams can be collected in small plastic bags for transport to the laboratory. As with any sampling regimen, detailed information pertaining to the collection such as date, geographic location, habitat, type of material collected, air and water temperatures, and so forth, must be recorded. Ideally, samples should be kept cool (or at temperatures approximating those of the collection site) and cultured within a few hours of collection. Seymour and Fuller (1987) found, however, that samples could be kept at ambient temperatures for as long as a month without significantly changing the yield.

Once in the laboratory, a thick slurry is made by adding sterilized distilled, CFD, or deionized water to the bag. Enough is removed to coat one-third of the bottom of a deep (100-mm diameter × 20-mm high) Petri plate, and additional water is added to fill the plate one-third to one-half full. Alternatively, water may be added to soil placed directly in the bottom of a Petri plate. In either case, two or three autoclaved hemp seed halves or two or three small (1 cm × 1 cm) squares of shed snake skin are added to the mixture. The snake skin, soaked for 1 minute in water immediately after it has been heated to boiling, is floated on the surface of the diluted soil samples. Water samples are treated in a similar fashion; they may be used directly or diluted with sterilized distilled, CFD, or deionized water. Plates can be examined after 2–3 days of incubation at 10–20°C; additional incubation (up to 2 weeks) is advised, however, so that some of the slow-growing water molds may appear. Incubation at temperatures that approximate those of natural habitats has yielded a number of so-called "rare" Peronosporomycetes (Perrott 1960). That finding raises questions of whether the species are truly rare or whether the standard isolation and culture techniques used are appropriate for all geographic locations.

Attempts to improve collection techniques for Peronosporomycetes include the use of *in situ* baiting (e.g., Minden 1916; Höhnk and Bock 1954; Cooke and Bartsch 1959; Bandoni et al. 1975). Substrata (e.g., hemp seed, bits of fruit, nutrient agar) are placed in containers (e.g., mesh bags, aluminum tea balls, dialysis tubing), suspended in an aquatic habitat or buried in sediment, and examined after several days of exposure. *In situ* baiting techniques are essential for isolating members of the Rhipidiaceae (Dick 1990a); solanaceous or rosaceous fruits are placed in plastic-coated, wire-mesh cages and suspended in stagnant or slow-moving shallow water. After 10 days, fruit surfaces are examined for the presence of pustulelike outgrowths containing Peronosporomycetes. After vigorously rinsing, the pustules are removed and teased apart in preparation for microscopic examination. Baits placed in water containing high levels of carbon dioxide and low levels of oxygen (e.g., shady bogs, swamps) also may yield members of the Leptomitales (Stevens 1974).

Dilution plating (Dick 1990a) has been used successfully to isolate many pythiaceous taxa. In that technique, about 50 g fresh weight of soil is brought to 250 ml with the addition of 0.08% water agar. Sequential dilutions with the water agar are made to achieve final dilutions ranging between 1:100 and 1:500. Glass rods are used to spread 1-ml aliquots of desired dilutions on the surface of fresh (no more than 36 hours old) VP$_3$ agar (Appendix II). The plates are incubated at 20–22°C, and after 42–48 hours, they are rinsed gently with tap water and examined. The plates are examined again after

another 24 hours, and resulting hyphae are transferred to fresh agar plates.

Identification

Identification is based on characteristics of the vegetative hyphae and the reproductive structures, including patterns of zoosporangium discharge. The morphologies of antheridia, oogonia, and oospores are critical features, as are patterns of fertilization (i.e., where antheridia occur in relationship to the oogonia and how they attach to the female structures).

Young, axenic (see "Isolation and Culturing," earlier in this chapter) cultures should be examined periodically, generally between 3 and 21 days after baiting or initial incubation, for the development of vegetative as well as sexual and asexual reproductive structures. Alternatively, the development of reproductive structures, including the discharge of zoosporangia, can be stimulated in the laboratory through a variety of techniques (Stevens 1974; Fuller and Jaworski 1987). Many of the techniques involve the transfer of actively growing hyphae to fresh sterilized distilled, CFD, or deionized water; others include exposure of a colony to dilute salt solutions, particularly those containing calcium chloride, or to specific nutrient regimens. Development of reproductive structures in some species may require special light or temperature protocols.

In some cases, actively growing cultures can be examined directly by placing Petri plates on the stage of a dissecting or compound microscope. More detailed information, however, usually is acquired from wet mounts of hyphae cut from baited water cultures; razor blades, scalpels, or scissors used for microscopic surgical procedures are particularly effective in separating hyphae and attendant structures from baits. Alternatively, pustules of peronosporomycetous material can be removed from baited fruit with forceps or a probe. In either case, material is transferred to a glass slide and a cover slip is applied. An eyedropper is effective for transferring the hyphae produced by many species. Particularly delicate structures, however, can be floated above a submerged cover slip that then is lifted from the water and inverted over a glass slide. Although most stains distort the delicate structures of these organisms, in some instances the addition of black ink (e.g., Sheaffer Jet Black) can be useful in revealing surface ornamentation or sheaths on reproductive structures (Seymour 1987). Prolonged observations of a particular slide can be facilitated by ringing the coverslip with petroleum jelly or nail polish, which prevents drying of the wet mount. More sophisticated apparatuses that allow observations of developing structures while organisms are under less physiological stress can be purchased or fashioned (Salkin and Robertson 1970; Amon et al. 1987).

Although the taxonomic and phylogenetic relationships within some families of the Peronosporomycetes are being investigated actively (e.g., Powell and Blackwell 1995; Cooke et al. 2000; Leclerc et al. 2000), no recent comprehensive monograph of the group exists. Nonetheless, a number of investigators (Sparrow 1960; Ainsworth et al. 1973b; Karling 1981; Erwin et al. 1983; Dick 1990a) have provided useful overviews of those organisms. Waterhouse's (1970) monograph of *Phytophthora* and Van der Plaats-Niterink's (1981) work with *Pythium* are helpful in understanding aspects of the Pythiaceae biology. Monographs of the genera *Achlya* (Johnson 1956), *Aphanomyces* (Scott 1961), and *Saprolegnia* (Seymour 1970) remain valuable tools for identification of peronosporomycete fungi.

Deposition

Most standard dehydration techniques, preservatives (e.g., Amann's solution and formalin-acetic acid-alcohol; see Appendix II), and mounting media distort vegetative and reproductive features of Peronosporomycetes, thereby preventing long-term archiving of specimens. Langsam and Armbruster (1985) reported that a glutaraldeyde/osmium tetroxide fixation procedure minimizes distortion in saprolegniaceous Peronosporomycetes, thus preserving taxonomically important characters for permanent slide mounts. Nevertheless, type specimens of most Peronosporomycetes have been lost, preserved material (where it exists) is poor at best, and traditional "herbaria" do not exist for the Peronosporomycetes. The deposition of living cultures is, therefore, particularly important for members of this group. The World Data Centre for Microorganisms (Appendix III) maintains a comprehensive directory of collections where microorganisms and cell lines of all types may be deposited or obtained. The center maintains links to small, obscure collections as well as to those maintained through well-known organizations such as the American Type Culture Collection and the Centraalbureau voor Schimmelcultures (Appendix III). Although useful, the collections are incomplete; only a single specimen of some species is available, and many species are not represented at all. In addition, living cultures are subject to genetic change (Goos and Pollack 1965), can lose the ability to reproduce, and may be contaminated by other organisms. Thus, any culture obtained from such collections should be reisolated and its identification should be verified.

ISOLATION AND CULTURING

Although the techniques for collection are useful in obtaining Peronosporomycetes, they rarely yield the axenic cultures needed for their identification and for long-term survival in culture. Numerous methods for separating Peronosporomycetes from contaminating bacteria, fungi, and protists have been reviewed by others (Johnson 1956; Sparrow 1960; Seymour 1970; Stevens 1974; Fuller and Jaworski 1987). One of the most common methods is the removal of single hyphal filaments from actively growing colonies in gross cultures, using flame-sterilized forceps. Filaments are transferred to Petri plates containing agar-based nutrient media (e.g., cornmeal, peptone–yeast–glucose extract, potato dextrose, V8 juice; Appendix II) and incubated until colonies of actively growing mycelia form. To obtain single-spore isolates, water containing motile and encysted spores is streaked with a flame-sterilized bacteriological loop along the surface of Petri plates containing water agar. After 12–24 hours of incubation (usually at 20–22°C), germinating spores are located and cut from the agar plates. They then are "planted" on fresh agar-based nutrient medium and allowed to incubate for 3–5 days or until recognizable colonies form.

Bacterial contamination, which often persists through this stage, sometimes is eliminated through the use of antibiotics (Stevens 1974). In one regimen (Whisler 1987), 400 units of penicillin G and 0.2 mg streptomycin-SO$_4$ per millimeter of medium are added to molten media or water shortly after autoclaving. Antibiotics, such as germanium dioxide, also can be used to inhibit diatom growth, and pimaricin is useful in limiting the growth of contaminating fungi (Fuller and Jaworski 1987). In any case, antibiotics should be used with care because they also can inhibit the growth of certain peronosporomycetous fungi (Sorenson 1964).

Alternative methods of freeing cultures from bacterial contaminants rely on the ability of peronosporomycetous fungi to grow faster than their bacterial contaminants. Bacteria-free isolates can be propagated by cutting blocks of agar from the edge of an actively growing colony and positioning them on yet another agar plate. Occasionally, bacteria adhere to hyphal filaments and grow in tandem with them. The bacteria, however, nearly always are confined to the surface layer. To obtain bacteria-free hyphae, a plate containing the contaminated colony is inverted. A sterile probe is run between the agar and the plate to separate the agar from the plate so that it drops onto the inner surface of the Petri plate cover. In this technique (modified from Schmitthenner and Hilty 1962), a piece of agar containing hyphae is removed from the newly uncovered lower surface of the agar and planted on fresh nutrient medium. The cut to remove the agar block from the larger agar layer must be shallow to prevent contamination by bacteria growing on the surface (now the bottom of the agar layer).

Other investigators (Raper 1937; Seymour and Fuller 1987; Dick 1990a) have suggested using small heat-resistant glass rings to promote bacteria-free growth. In one technique, a 12-mm × 12-mm ring is dipped in 90% ethyl alcohol, flamed, and transferred to the surface of an agar plate so that the residual heat from the flaming causes the ring to partially embed itself into the agar. Hyphae then can be placed on the agar inside the ring. The ring blocks the growth of surface hyphae and forces hyphae to grow under the ring and emerge outside of it without contaminating bacteria.

Once isolates are axenic, members of the Peronosporomycetes can be maintained on nutrient agar plates or slants or grown on appropriate baits in water culture. Water cultures are established by cutting blocks of hyphal-colonized agar and placing them in Petri plates with sterilized distilled, CFD, or deionized water. Alternatively, baits can be applied to the edge of a colony actively growing on agar. The baits are removed after incubation for 24 hours and placed in Petri plates containing sterilized distilled, CFD, or deionized water and additional bait material.

Numerous attempts have been made to develop reliable techniques for prolonged storage and preservation of peronosporomycetous taxa. Dehydration (Coker 1923; Goldie-Smith 1956), submersion of agar slants in mineral oil (Buell and Weston 1947; Goldie-Smith 1956; Seymour 1970), and freezing with liquid nitrogen (Hohl and Iselin 1987) are successful but only with some species of Peronosporomycetes. As a result, the most reliable method remains to be storage of peronosporomycetous thalli on appropriate nutrient agar slants or in water culture grown on bait suitable for the particular species. Cultures generally can be maintained for 1–6 months at 20–22°C without being transferred.

Despite its reliability, one of the biggest disadvantages to maintaining Peronosporomycetes in water culture is the danger of mite infestation and subsequent fouling of the cultures by unwanted fungi and bacteria. One method of limiting mite infestation involves the use of glycerin moats. The moats are made by inverting an empty Petri plate and separating the two halves. A thin layer of glycerin is poured into the inverted cover of the Petri plate. The bottom half of the Petri plate is replaced, and the space between the inner periphery of the lip of the Petri plate cover and the outer periphery of the lip of the Petri plate bottom fills with glycerin to form the moat. The moat is placed below a stack of Petri plates. Mites crawling up the side of the moat fall into the glycerin, where they are trapped and eventually die. A number of other methods are summarized by Onions (1989).

ASSESSING BIODIVERSITY AND ABUNDANCE

For the most part, the assessment of peronosporomycete biodiversity and abundance relies heavily on many of the indirect baiting, trapping, and culturing techniques (Sparrow 1968; Dick 1976, 1990a) described earlier in this section. However, direct observations of materials removed from the environment (e.g., submerged twigs and fruits, phytoplankton and zooplankton, aquatic animals) also can be helpful (Stevens 1974), particularly in determining the abundance of peronosporomycetous parasites of other organisms (Unestam and Weiss 1970). Additional growth of organisms that have invaded substrata can be stimulated by submerging them in jars or shallow trays of sterile diluted pond, lake, or stream water and keeping them in the dark at 15–50°C for a week (Stevens 1974).

Techniques notwithstanding, the microscopic sizes of most Peronosporomycetes and their propagules make study of their ecology challenging. The difficulties, thoroughly outlined by Sparrow (1968) and Dick (1976), have yet to be overcome entirely and continue to interfere with a comprehensive understanding of the occurrence of Peronosporomycetes in most habitats. First and foremost among the difficulties are sampling errors that result in an inaccurate representation of the actual communities of organisms inhabiting particular ecosystems. Some Peronosporomycetes, for example, may "escape" occasional collections from particular habitats as a result of seasonal occurrence or patchy distributions; alternatively, artificially inflated estimates of abundance may occur as a result of a nonrepresentative appearance of organisms in samples. Estimates of biodiversity and abundance also may be skewed by the inclusion of transient Peronosporomycetes, which enter ecosystems by a variety of means (e.g., spring runoff, snow melt, animal carriers, plant debris). Such organisms may be isolated from collections and mistakenly included as part of a community although they are not sustained in that system by vegetative growth or reproduction over prolonged periods. Additional difficulties arise as a result of the delicate nature of peronosporomycete structures (e.g., zoospores) that may be destroyed during the rigors of sampling, transport, and culture. At the other end of the spectrum is the loss of resistant structures such as oospores; our limited knowledge of peronosporomycete physiology impedes our ability to recover and stimulate germination of many resistant propagules, causing them to be overlooked as viable members of a community. An overarching concern stems from the fact that conventional techniques tend to sample reproductive propagules but do not measure the occurrence of Peronosporomycetes in their mycelial form or provide any understanding of their activity, productivity, or impact on the ecosystem.

To some extent, the difficulties in assessing biodiversity and abundance can be mitigated by developing protocols that help reduce sampling errors. At the very least, such protocols involve making replicate collections across seasons to increase the probability that actual populations of organisms will be represented accurately (Hughes 1962; Dick 1966; Alabi 1971). It also involves a number of other techniques. Sampling, for example, must occur across a variety of microhabitats within a particular environment. Collections can be taken horizontally across transects (Dick 1971), vertically within individual soil cores, at various depths within a body of water (Paterson 1967), or intensively within a prescribed area of soil or water (Dick 1966). That type of microhabitat sampling, although useful, presents its own set of challenges. The process of taking mud cores, for example, can disturb adjacent sites, thereby making it difficult to pull multiple cores from the same area without disturbing the relative position of soil particles and peronosporomycete propagules. Dick (1966) addressed that issue, recommending that replicate samples be taken from a single core.

Samples also must be treated to promote the growth of a wide variety of organisms. That means trapping organisms, or exposing samples to a wide array of baits and nutrient agars, as well as subjecting cultures to conditions that closely mimic those of the environment (e.g., comparable temperature, light, hydrogen ion concentrations, degree of oxygenation). It also may involve techniques that maximize yields from individual samples. Fuller and Poyton (1964), for example, recommended centrifugation to concentrate reproductive propagules prior to plating. Other investigators have tried to separate and/or dilute samples to reduce the possibility that the presence of some organisms will be masked by others. Diluted samples can be plated on nutrient agar (Willoughby 1962a) or fractionated to increase yields. Dick (1966) diluted 50- to 150-g samples of soil with 250-ml sterile lake water and then fractionated the samples into three parts: supernatant, coarse plant material, and sedimented slurry. He found little difference between the yields from the latter two fractions, but at least one species (*Aphanomyces* species) was found only in the supernatant.

Dick's (1976) account of "phycomycete" ecology probably remains the most comprehensive assessment of the difficulties related to determining abundance of Peronosporomycetes in natural environments.

Not surprisingly perhaps, the major obstacles involve the elimination of sampling errors that confound collection results. To be sure, pioneer studies of peronosporomycete abundance have been undertaken by some investigators, but such studies are limited in number, scope, and usefulness. Sparrow (1968), for example, cited investigations by Suzuki in which numbers of Peronosporomycetes were estimated by counting the number of growths emerging from a hemp seed immersed in water; however, Sparrow's account points to the flaws in Suzuki's work and warns that the numbers it produced are probably inaccurate. Other avenues of investigation involve plating techniques in which aqueous samples, either diluted or undiluted, are combined with agar, in some instances containing bacteriostatic agents (Willoughby 1962a), and centers of mycelial growth are counted as an estimate of zoospores per unit of water (Collins and Willoughby 1962). In one refinement of the procedure (Dick 1966), a grid is drawn on the bottom of plates containing agar into which a sample has been incorporated; blocks of agar are removed from the grid, and each is placed in a drop of sterile water and topped with half a hemp seed. After 48 hours, blocks showing growth of saprolegniaceous organisms are counted and cultured for identification, and the grid then is resampled for new agar blocks. Other investigators have tried to establish relative measures of abundance of Peronosporomycetes by determining ratios between numbers of organisms and numbers of collections examined (Dick and Newby 1961; Apinis 1964; Dick 1971) or by estimating numbers of organisms on nonnumeric scales (Roberts 1963). Unfortunately, the basic issues related to sampling error prevent such studies from providing definitive information about abundance for most populations of Peronosporomycetes. Clearly, this is an area that remains problematic and largely unexplored.

ASCOMYCETES

SUMMARY OF EXISTING KNOWLEDGE

Although the distinctive freshwater ascomycota was discovered mostly within the past 50 years, its existence is now well recognized (Dudka 1985; Shearer 1993, 2001b; Goh and Hyde 1996; Hyde et al. 1997). Knowledge of that mycota has come primarily from a few studies of fungi that colonize selected submerged substrata in limited geographic areas rather than from more broadly based studies. Such studies include those by Ingold (1951, 1954, 1955), Ingold and Chapman (1952), Shearer (1972), Cavaliere (1975), Minoura and Muroi (1978), Shearer and von Bodman (1983), Dudka

(1985), Shearer and Crane (1986, 1995), Magnes and Hafellner (1991), Hyde (1992a, b, c), and Tsui and colleagues (2000). Preliminary studies indicated that freshwater ascomycetes might play important roles in aquatic habitats as saprobes, endophytes, and parasites of emergent aquatic macrophytes; as decomposers of submerged allochthonous woody debris; and as a food resource for invertebrates (Shearer 1993). Despite their putative importance, however, the identities, distribution patterns, and activities of freshwater ascomycetes remain largely unknown. With the exception of a study along a latitudinal gradient through eastern North America (Crane and Shearer 1994; Shearer 2001a), freshwater ascomycetes have not been studied on a broad geographic scale, and no comprehensive keys or monographs are available for their identification.

About 450 species of ascomycetes have been reported from freshwater habitats (Shearer 2001a). They occur in both lentic and lotic habitats, as parasites and endophytes of aquatic macrophytes and algae, and as saprobes on dead plant material. Based on the classification systems of D. J. S. Barr (1987, 1990), freshwater ascomycetes comprise 56 families and 17 orders. Orders with 20 or more representatives from freshwater are Helotiales (99 species), Pleosporales (90 species), Sordariales (82 species) Melanommatales (25 species), Eurotiales (25 species), and Halosphaeriales (20 species). Those orders represent distinctly different evolutionary lines clearly demonstrating that the freshwater ascomycetes are a polyphyletic group and that adaptation to the freshwater habitat has occurred numerous times.

COLLECTION, IDENTIFICATION, AND DEPOSITION OF SPECIMENS

Collection

Freshwater ascomycetes can be found by collecting and examining living and dead macrophytes around the edges of ponds, lakes, bogs, swamps, wetlands, drainage ditches, and small temporary water bodies. Submerged and partially submerged woody debris is also a good source of freshwater ascomycetes, especially in lotic habitats, and a few species occur on submerged deciduous tree leaves. Although fruiting bodies (ascomata) of some species can be seen on substrata in the field using a hand lens, many are undetectable until substrata are examined with a dissecting microscope. Substrata can be collected in the field by hand or with the aid of a sturdy rake. In some cases, a tree trimmer can be used to cut macrophytes rooted in deep water. For large woody debris, a saw, twig cutters, and an ax are useful for obtaining samples. Once material is collected, it should be placed in plastic bags containing several pieces of paper toweling to absorb excess water. Samples in plastic bags should be stored on ice in a cooler until transported to the laboratory to inhibit the growth and respiration of bacteria and other organisms. At the time of collection, detailed information concerning the specimen should be recorded, including date of collection, geographic location (including latitude and longitude and elevation), identity of the substrata, state of the substrata (living, dead, degree of decomposition, submerged or not), type of aquatic habitat, air and water temperatures, pH and other available water-chemistry parameters, and any other pertinent factors.

Substrata are returned to the laboratory and rinsed with sterile tap, distilled, or deionized water. Attached aufwuchs (i.e., growths such as algae, protozoa, bryozoans, etc.) and mud and sand should be removed from the surfaces of substrata by hand or by gentle scraping with a spatula. That cleaning allows fungi to sporulate and increases visibility of fungal fruiting bodies. Substrata then should be transferred to moist chambers for incubation. Glass or plastic Petri plates containing moistened filter paper; plastic boxes, such as those used for the refrigerator storage of food; or plastic bags containing moistened paper toweling can be used as incubation chambers (see Appendix I).

Material placed in a moist chamber should just cover the bottom surface. If substrata are densely layered, ascomata can be obscured and oxygen can be depleted, thereby inhibiting growth and sporulation of the ascomycetes. Standing water should be poured from the moist chamber before incubation to prevent buildup of bacterial populations and anaerobic conditions. Incubated samples should be kept at temperatures similar to those of the natural habitat, although ambient room temperatures may suffice, and in alternating light and dark conditions (12 hours/12 hours), which apparently enhance sexual reproduction (Shearer and von Bodman 1983). Substrata placed in moist chambers should be examined prior to incubation, and any freshwater ascomycetes present should be recorded because some fungal species deteriorate once they are removed from water and may not be recognizable after a period of incubation. Once incubation begins, samples should be examined periodically for about 6 months.

Because it is often difficult to determine the origin and identity of randomly collected samples or how long they have been submerged, investigators often use baits to follow the colonization and development of freshwater ascomycetes (Shearer 1972; Shearer and von Bodman 1983). Baits such as twigs or wood blocks cut from identified wood sources can be attached to masonry bricks or cement blocks and placed in the habitat for specified periods. They also can be assembled into packs with nylon twine and attached to a line tethered to a relatively permanent structure such as a pier or a tree (Shearer 1972). Substrata such as dead macrophyte leaves and stems can be placed in nylon mesh bags and submerged in the habitat, as explained earlier. Given the fluctuating water levels and often muddy conditions common in freshwater habitats, a good deal of thought and planning should go into the selection of baiting sites and the methods of placing and retrieving baits. Allowances should be made for both flood and drought conditions.

Indirect collection techniques such as plating water, mud, or homogenized substratum are not very efficient for detecting freshwater ascomycetes. Ascospores of some freshwater ascomycetes germinate poorly or not at all. Those that do germinate often are overgrown by fast-growing, weedy fungi, such as species of *Mucor*, *Penicillium*, *Trichoderma*, and *Sclerotium*. Although cultures of putative ascomycetes may be obtained, inducing reproduction to confirm the identities of the cultures often requires long periods and special culture procedures. In addition, some species do not reproduce in culture, and without reproductive structures, freshwater ascomycetes cannot be identified unless molecular techniques are used. Thus, testing the numerous isolates that result from plating is quite laborious and often futile.

The techniques of foam examination and membrane filtration of water that are so useful for the identification and quantification of aquatic hyphomycetes (Iqbal and Webster 1973b, 1977) are of little value for freshwater ascomycetes. Ascospores of freshwater ascomycetes occur at much lower densities than conidia of aquatic hyphomycetes and, thus, rarely are encountered on membrane filters. In addition, the ascospores of many freshwater ascomycetes are not very distinctive morphologically, and unless the investigator already knows the freshwater ascomycetes likely to be present, they cannot be identified. Those limitations make direct quantification of the spora of freshwater ascomycetes in water very difficult.

Identification

Freshwater ascomycetes are identified by the morphology and anatomy of the ascomata and the morphologies of the hamathecium, asci, and ascospores. For known freshwater ascomycetes, micromorphological examination of squash mounts of ascomata taken directly from the substratum is usually sufficient for identification. In the case of unknown or problematic specimens, ascomata must be sectioned so that anatomical details and centrum structure can be examined. Rough sections of ascomata may be made with a razor blade, but more definitive work requires thin sections. The freezing microtome is useful for sectioning fresh specimens, especially Dis-

comycetes and ascomata embedded in host or substratum. Both fresh and rehydrated dried specimens embedded in low-viscosity resin and sectioned with an ultra-microtome provide excellent material for examination (Huhndorf 1991). Making squash mounts of portions of the centrum in water and then adding a dilute suspension of India ink is also important. The procedure leads to hydration, revealing gelatinous coatings on physes and asci and ascospore sheaths and appendages, if present. If neither sheaths nor appendages are revealed by the India ink, the ink should be removed by pulling distilled water under the coverslip. In some freshwater ascomycetes, the ink stains the appendages, which are visible only after surrounding ink is removed. The ascus tips should be stained with Melzer's reagent and aqueous cotton blue so that the nature of the apical apparatus and/or apex can be determined. Important morphological features of species should be documented with photographs of fresh material as soon as possible after collection. Often, drying or chemically preserving material alters its appearance and may render ascospore appendages and sheaths invisible. Photographs should be deposited with voucher specimens.

The taxonomic diversity of freshwater ascomycetes and the absence of taxonomic keys make them difficult to identify, although there are several general works on Ascomycetes that can be useful (e.g., Dennis 1968; Korf 1973; Luttrell 1973; Müller and von Arx 1973; Kohlmeyer and Kohlmeyer 1979; Sivanesan 1984; Kohlmeyer and Volkmann-Kohlmeyer 1990), as well as monographs of specific groups (e.g., Müller 1950; Holm 1957; Munk 1957; Eriksson 1967; Hedjaroude 1969; Lundqvist 1972; M. E. Barr 1978, 1987, 1990a, 1990b, 1990c; Leuchtmann 1984; Shoemaker and Babcock 1989). For detailed studies of particular taxa, original descriptions must be used.

Deposition

It is critical that slides and/or dried material of both type and voucher specimens be deposited in recognized herbaria (Dennis 1968; Kohlmeyer and Kohlmeyer 1979). A published record of a species occurrence is useless unless it can be validated by later investigators. Such specimens also can prove useful in population studies. In addition, dried ascomata may serve as a source of genetic information for future taxonomic and population studies. Specimens should be dried while they are still in good condition—that is, as soon as cultures, sections, squash mounts, and photographs have been obtained. In dry climates, specimens can be dried at room temperature with no special procedures. In humid areas, a desiccant may be needed along with gentle heat; high drying temperatures should be avoided because they can degrade DNA. It is extremely important that detailed information about a collection accompany every herbarium specimen (see "Collection," under "Ascomycetes," earlier). Such information, when compiled, is invaluable for interpreting the ecologies and distribution patterns of freshwater ascomycetes.

ISOLATION AND CULTURING

Cultures of ascosocmycetes can be obtained from single ascospores and/or ascospores contained in single asci. The ascospores of most freshwater ascomycetes germinate relatively easily. Freshwater ascomycetes with closed ascomata can be isolated by crushing mature ascomata in 2–4 ml of sterile distilled water to form a spore suspension. Discharged ascospores that have accumulated around ostioles can be removed with a flamed, cooled needle and stirred into sterile water. Vigorous agitation of suspensions with a tube agitator will separate ascospores and dislodge attached bacteria. Spore suspensions are poured onto plates containing antibiotic distilled water agar (AWA, Appendix II) and are allowed to stand for 1–2 hours so that the ascospores can settle. Excess water is decanted gently from the plates, which then are incubated at a slight angle so that remaining water can drain. Ascomata also can be crushed on the surface of AWA plates with a fine needle, and the ascospores and asci can be spread on the agar surface. Ascospores that accumulate at the ostioles of ascomata can be streaked on AWA to separate them. Streaking techniques should be used for ascospores that contain large quantities of lipid and hence are likely to float if suspended in distilled water.

To isolate discomycetes, a small piece of filter paper is attached with a drop of water or petroleum jelly to the lid of an AWA plate about 1 cm from the edge of the lid. An apothecium is removed from the substratum with a sterile needle and placed on the filter paper so that its surface is facing down toward the agar surface. The lid is placed on the bottom plate and rotated about 45 degrees every 30 minutes so that ascospores are discharged in a circular path around the plate. That procedure is reversed for pyrenomycetous forms with forcible ascospore discharge. In the latter, a small piece of substratum with perithecia directed downward is mounted on the filter paper attached to the lid of a Petri plate, and the plate is inverted so that the agar receives the upwardly discharged ascospores.

After incubation at room temperature for 24 hours, ascospore-containing plates should be examined microscopically with transmitted light, and the position of germinated ascospores should be marked. It is extremely important that single, recognizable, uncontaminated ascospores be isolated. Examination of a plate and transfer of germinated ascospores within the next 24–48 hours is a critical step. If plates are left for longer periods, the ascospores may be overgrown by contaminants or become unrecognizable because of growth. If ascospores do not germinate within 3 days, they still can be transferred to a nutrient medium and be observed periodically. Transfer to a nutrient medium often stimulates germination. Because fruiting ability varies among strains and reproduction may be heterothallic, at least four to eight isolates of each species should be made. Single germinated ascospores should be transferred from AWA to a standard nutrient medium (Appendix II) such as corn-meal agar (CMA); malt extract agar (MEA); peptone-yeast extract-glucose agar (PYG); or yeast extract-soluble starch agar (YPSS). For long-term (6–12 months) storage of cultures, strips of moistened balsa or birch wood can be added to culture tubes of the previously mentioned media before autoclaving to provide a long-lasting substratum. To store cultures for several years, pieces of colonized agar are placed in small sterile vials containing a 16% glycerin solution and maintained at −40°C. Cultures of new, unusual, or rare species should be deposited in one or more internationally recognized culture collections such as those at the American Type Culture Collection or the International Mycological Institute (Appendix III).

To induce ascomata formation so that identities can be verified, cultures are grown on alfalfa stems, wood strips, or the substrata on which they occur (e.g., *Juncus*, *Typha*, willow roots). Six to eight pieces (4 cm long) of substratum are added to 50 ml of sterile distilled water (filtered water from the natural habitat also may be used) in 125-ml Erlenmeyer flasks and autoclaved for 1 hour. Flasks are inoculated with mycelium from stock tubes and placed on a rotary shaker (100 rpm) at 22°C until substrata are visibly well colonized. Cultures also may be incubated under stationary conditions at ambient temperatures and swirled once or twice a day. Substrata are transferred aseptically to Petri dish moist chambers (see "Collection" under "Ascomycetes," earlier), sealed with Parafilm, and incubated between 18–22°C in alternating light/dark (12 hours/12 hours) conditions. Daylight fluorescent light is acceptable, although an additional near-ultraviolet light may be useful in stimulating fruiting in some species. Times required for fruiting vary greatly among species, ranging from about 2 weeks to 6 months.

Because aquatic hyphomycete anamorphs have been reported for several freshwater ascomycetes (Webster and Descals 1979; Webster 1992), cultures should be tested for the production of such states (see "Collection,

Identification, and Deposition of Specimens" under "Mitosporic Fungi," later in this chapter). Numerous freshwater ascomycetes produce pycnidial anamorphs. Those stages usually form on standard laboratory media, so examination of stock cultures is sufficient for their detection.

ASSESSING BIODIVERSITY AND ABUNDANCE

The best technique for assessing biodiversity of freshwater ascomycetes is the direct examination of substrata. A range of known substrata can be collected from the habitat, and/or substrata baits can be placed in and then retrieved from the habitat. To obtain the most reliable information about biodiversity, we strongly recommend that both of those procedures be used. Collections should be made in all seasons because species composition can vary seasonally. In all aquatic habitats, gradients in habitat parameters and substrata occur from terrestrial to fully aquatic areas. Further, within any given aquatic habitat, environmental conditions can vary with depth, currents, and physical structure of the water body (e.g., riffles versus pools in streams). Such factors can affect species occurrences and must be taken into account in any sampling protocol. Using a comprehensive sampling protocol and a wide range of substrata, direct examination of substrata can be used to develop a species list for a particular habitat. The direct examination of substrata approach is limited, however, because (1) only taxa that are reproducing at the time of collection or during incubation can be detected; (2) the amount of substratum examined is usually quite small compared to the total amount in the habitat; and (3) the relationships between reproductive structures and biomass and activity are not known.

It is difficult to assess abundance of ascomycetes, primarily because no satisfactory techniques for quantifying them at the species level exist, and secondly, because their distributions on substrata are very patchy. The number of ascomycete species per twig, leaf, or bait can range from zero to five at most. Generally, two to three species occur on the same piece of substratum. One can estimate, however, the relative importance of species by determining the frequency of occurrence of individual species on a given number of substrata and dividing this value by the sum of the frequencies of occurrence of all species found on the same substrata. That value then can be used to rank species within the habitat in order of importance. The limitations of the direct-examination-of-substrata method also apply to that estimate of importance.

MITOSPORIC FUNGI

SUMMARY OF EXISTING KNOWLEDGE

Several groups of mitosporic fungi occur in freshwater, although the species present at particular sites vary with the types of habitat and substratum represented. The best-known and most studied group of mitosporic fungi is the "aquatic," or "Ingoldian," hyphomycete group, members of which are distinguished by their tetraradiate, branched, or sigmoid conidia that are released into and dispersed by water (Ingold 1975; Webster and Descals 1981; Bärlocher 1992). Although the first species was described in 1880 (Saccardo 1880), it was not until Ingold discovered a varied mycota on alder and willow leaves submerged in a stream (Ingold 1942) that the diversity of this distinctive group of fungi was recognized. About 300 species have been described thus far (Bärlocher 1992). Most aquatic hyphomycetes occur in lotic habitats on decaying deciduous leaves and woody debris of allocthonous origin. A few species, however, have been reported from lentic habitats (Suzuki and Nimura 1961), brackish waters (Jones and Oliver 1964; Müller-Haeckel and Marvanová 1979b), and terrestrial habitats (Park 1974; Webster 1977, 1981; Sridhar and Kaveriappa 1987; Iqbal et al. 1995).

The "aeroaquatic hyphomycetes," whose conidia are modified in a variety of ways to trap air for flotation, comprise a second group of mitosporic fungi (Fisher 1979; Michaelides and Kendrick 1982; Webster and Descals 1981; Premdas and Kendrick 1991). Mycelia of aeroaquatic fungi are thought to grow on submerged decaying leaves, woody debris, dead emergent macrophytes, and a variety of other decaying plant parts. Unlike most aquatic hyphomycetes, which require submersion for sporulation, the aeroaquatic fungi sporulate when they are exposed to air. Aeroaquatic fungi can be found in all types of aquatic habitats, but they occur most commonly in lentic habitats with fluctuating water levels, such as small ponds, marshes, swamps, and ditches.

Dematiaceous, and to a lesser extent, hyaline hyphomycetes, nematode-trapping hyphomycetes, and coelomycetes all are encountered regularly on a wide variety of submerged plant substrata in both lentic and lotic habitats. Those fungi usually go undetected, however, unless a substratum on which they occur is incubated in a moist chamber after collection (Shearer 1972; Lamore and Goos 1978; Shearer and Crane 1986; Révay and Gönczöl 1990; Sivichai et al. 2000; Tsui et al. 2000). We will not discuss those fungi or the aeroaquatic fungi in detail. Generally, however, the methods used to collect, isolate, and culture them are very similar to those used for the aquatic ascomycetes. Therefore, we refer the reader to the previous section on Ascomycetes and the papers cited therein as starting points for dealing with these fungi. No comprehensive keys to the freshwater mitotic fungi exist. The best strategy for identifying them is to use a general reference to arrive at a suitable genus (Barron 1968; Ellis 1971, 1976; Carmichael et al. 1980; Sutton 1980; Barnett and Hunter 1998; Kiffer and Morelet 2000) and then to consult the CABI Bioscience Database of Fungal Names (Appendix III) to find the species described in the genus. Literature citations provided in that database can be used to obtain the original descriptions of species needed for identification.

In the remainder of this section, we deal only with the aquatic hyphomycetes. Descals (1997) described detailed methods for the collection, identification, isolation, and deposition of aquatic hyphomycetes. Basic, and in some instances, different techniques are described in the following sections.

COLLECTION, IDENTIFICATION, AND DEPOSITION OF SPECIMENS

Collection

Aquatic hyphomycetes can be collected from submerged decaying vegetation, from foam, or directly from water. This group extensively colonizes deciduous leaves and wood of riparian origin in flowing water. Submerged leaves, twigs, and subsamples of logs are collected, placed in plastic bags, and labeled with the same information listed in the section "Collection" under "Ascomycetes," earlier. Until they are processed in the laboratory, samples should be placed on ice to retard bacterial growth. Leaves and woody samples are processed differently. Leaves should be washed gently in tap, distilled, or stream water. Depending on the purpose of the sampling, wash water and incubation water may be sterilized prior to use. Samples should be placed in deep Petri dishes (100-mm diameter × 20-mm high) in about 30 ml of sterile tap, distilled, or stream water. Entire leaves or fragments of leaves can be used but are often difficult to examine microscopically. An alternative is to cut 10–20 disks from a leaf using a handheld paper hole punch and letting the disks fall into the Petri dish. Parts of the petiole should be included with the leaf disks because that substratum is often a source of abundant conidia. Leaf disks are retrieved easily, mounted on a microscope slide, and examined for conidia. Disks also provide a mechanism for quantifying the presence of

species on a leaf as a proportion of disks sampled that carry a given species (Shearer and Lane 1983). Only a single leaf or 10–20 disks or fragments should be added to a dish. The presence of too much material can inhibit sporulation and lead to the buildup of bacteria and protozoans. In general, substrata should be incubated at the temperature of the stream at the collection site, but sometimes achieving the goals of the sampling will require use of a range of temperatures. Because aquatic hyphomycetes are temperature sensitive (Suberkropp 1984), it is advisable to work out the temperature range supporting sporulation of the greatest number of species for the habitat under study.

After incubation for 24–72 hours, dishes are examined using a dissecting microscope with an elevated stage and subsurface illumination for the initial location and identification of conidia. The conidia are opaque and show up well against a darkish background. Clusters of conidia can be pipetted from the bottom of the dish to a microscope slide, the drop covered with a coverslip, and conidia located and examined. Small fragments of leaves and leaf disks also can be mounted on a microscope slide in lactic acid with cotton blue (an intensely blue solution is required), and their surfaces and edges can be examined for conidiophores and conidia. Leaf disks can be fixed and stored in lactic acid until they can be examined.

Conidia of aquatic hyphomycetes are not detected easily on woody debris placed in dishes of water. Exceptionally good conidial production, however, is obtained by placing twigs in cylindrical tubes with water and passing a stream of air bubbles along the twig (Shearer and Webster 1991). The resulting conidial suspension is membrane filtered, the conidia are stained in lactic acid with cotton blue, and then they are examined for identification and quantification.

The membrane filtration of conidia from water (Iqbal and Webster 1973b) now is used widely to characterize the conidial pool of aquatic hyphomycetes in water (Iqbal and Webster 1977; Müller-Haeckel and Marvanová 1979a; Shearer and Webster 1985a, 1985b, 1985c; Iqbal 1993; Descals 1997). In that technique, 500–1,000 ml of stream water are passed through a cellulose membrane filter (47-mm diameter, 5–8 μm pore size) using a handheld vacuum pump. To analyze the resulting data statistically, a minimum of four water samples should be filtered at each site. Membrane filters are placed in plastic Petri dishes and immediately air-dried. Dishes are labeled (top and bottom, in case the tops are misplaced) and stored until sample examination. We have kept dried samples for up to 3 years without damage to conidia.

To examine fungi, a square (1 × 1 cm) is cut with a razor blade from the area of filtration of the membrane and placed on a glass slide. A drop of lactic acid containing cotton blue or trypan blue (0.1% w/v) is placed gently to one side of the square to avoid dislodging conidia and is allowed to hydrate the membrane filter completely before a coverslip is added (Iqbal and Webster 1973b). Slides may be warmed at 60°C for 1 hour to clear the filter, but this is not usually necessary. Conidia present on the membrane filter are identified and quantified under the compound microscope at 200×. A major limitation of this collection technique is the identification of species solely from spores, especially for species that have undergone convergent evolution in spore shape. For example, many species, including some ascomycetes, have sigmoidal spores; unless the method of spore production (i.e., type of conidiogenesis or ascosporogenesis) is known, many of the spores cannot be identified with certainty. Many spores on membrane filters have been in the water for some time, and their shapes and pigmentation may have changed, which also makes identification difficult.

Stream foam is a rich source of conidia of aquatic hyphomycetes because conidia are trapped on the surface of air bubbles (Iqbal and Webster 1973a, 1977). Foam is scooped from natural foam traps in the stream using small wide-mouth glass or plastic jars (baby food jars are particularly useful). The number of samples taken at each site depends on the purpose of the collection and whether statistical analyses will be applied to the data. A minimum of four samples from each site is recommended. Care should be taken not to take up water with the sample, and standing water in the jar should be decanted to avoid dilution of conidia (Descals 1997). Jars should be labeled with a collection number, and habitat information such as listed earlier (see "Collection," under "Ascomycetes") should be recorded in a data book with the corresponding number. Samples should be placed on ice immediately for transport to the lab. Samples may be stored for up to a month at 13°C or fixed with a preservative such as Formalin-acetic acid-alcohol (FAA) (Webster and Descals 1981; Descals 1997). For direct examination, a small drop of foam is placed on a glass slide with a coverslip. A drop of foam also can be spread on a slide, allowed to dry, covered with a drop of lactic acid with cotton blue and a coverslip, and then examined. It is important to remove sand grains and other debris before applying a coverslip to maximize the visibility of the conidia. Foam examination has the same limitations as membrane sampling of water and the additional limitation that slow-flowing streams and those with a neutral to basic pH usually produce little foam. Iqbal (1993) detailed a method for making artificial foam to entrap fungal spores that might be adaptable for such streams.

Identification

The shapes and sizes of the conidia of many aquatic hyphomycetes are distinctive enough to permit identification to species. As more species are described, however, the distinctions among congeneric species become less clear, and pure cultures of isolates must be compared carefully. At present, except for a small manual prepared by Ingold (1975), no monographs exist for this group, although one is in preparation (E. Descals, personal communication). One therefore must obtain the original descriptions of all aquatic hyphomycete taxa before serious study is initiated. A good starting point for identifying the literature required is the review by Webster and Descals (1981).

Deposition

Deposition of herbarium specimens for this group is problematic. The conidia are quite delicate and hyaline and do not preserve well on dried substrata. Traditionally, slides of material fixed in lactophenol with acid fuchsin or cotton blue and sealed with nail polish have been used as vouchers. Because phenol is a known carcinogen, many investigators have switched to using lactic acid as a preservative. Its efficacy over the long term is not known. We highly recommend use of the double coverslip technique (Volkmann-Kohlmeyer and Kohlmeyer 1996) to make permanent slides. Foam samples fixed with FAA also can be deposited in herbaria, but the FAA must be replaced periodically. Photographs of conidia and conidiogenesis deposited with microscope slides could provide useful information should the quality of the specimens decline as a result of mounting in lactic acid. Again the importance of providing detailed habitat information with herbarium specimens cannot be overemphasized.

ISOLATION AND CULTURING

Conidia of aquatic hyphomycetes germinate well and rapidly so that species are isolated easily. Conidia can be located in dishes of submerged substrata or foam when viewed under the dissecting microscope with illumination from below. Conidia then are picked up with a pipette drawn out to form a capillary tube or a mounted needle or hair and placed on a low-nutrient medium with antibiotics added (e.g., 0.1% MEA; Webster and Descals 1981; Descals 1997) or AWA (Appendix II). Conidia are moved along the surface of the agar to dislodge attached bacteria, and their final location is marked on the underside of the dish. Leaf disks or leaf fragments known to harbor target species can be shaken in sterile, distilled-water blanks, and a 1- to 2-ml aliquot of conidial suspension is plated onto AWA. Conidia are allowed to settle for several hours; the excess water is decanted; and the plate, tilted at a 45 angle, is incubated at ambient temperatures. Whichever technique is used, germinated conidia must be identified and transferred to nutrient media within 24–36 hours (see "Isolation and Culturing" under "Ascomycetes," earlier). The medium traditionally used for culturing aquatic hyphomycetes is MEA Webster and Descals 1981, Descals 1997). Our work indicates that many aquatic hyphomycetes grow as well or better on media such as CMA and YpSs (Appendix II; C. A. Shearer, unpublished data). Cultures can be stored on agar slants (MEA, CMA, YpSs; Appendix II) in screw-top test tubes at 4°C for about 6 months. Warm-water species should be stored at 15–17°C.

Once uncontaminated cultures are obtained, their identities should be confirmed with sporulation tests. Thin slices of colonized agar (YpSs and MEA agar are particularly good for this purpose) are submerged in sterile distilled water in Petri plates at 15–18°C for cold-water species and 22–25°C for warm-water species. Submerged cultures are examined periodically for conidia over several weeks using a dissecting microscope and transmitted light. If no conidia are produced in static-water culture, isolates then must be tested in aerated culture. Cultures are aerated by passing air forcibly through an aeration stone or Pasteur pipette (drawn out to a fine point) in flasks or culture tubes containing distilled water and thin slices of colonized agar. Water samples periodically are removed from the cultures with a sterile pipette and checked microscopically for conidia. As soon as the identity of an isolate has been confirmed, the isolate should be deposited in a major culture collection where cultures are lyophilized because some isolates lose their ability to sporulate after repeated transfers.

Many ascomycete and a few basidiomycete sexual states have been reported for the aquatic hyphomycetes (Webster 1992). To make connections between sexual (teleomorph) and asexual (anamorph) states, either colonies derived from single ascospores (teleomorph) must be induced to produce the conidial state (anamorph) or colonies derived from single conidia must be induced to produce ascomata with asci and ascospores. Colonies derived from the teleomorph can be tested for the production of an anamorphic state using the procedures to induce sporulation in the previous paragraph. To obtain the teleomorph from cultures of the anamorph, natural substrata can be inoculated with the test culture and incubated at temperatures of the habitat for relatively long periods (see "Isolation and Culturing" under "Ascomycetes," earlier). Detailed methods are available in Webster and Descals (1979).

ASSESSING BIODIVERSITY AND ABUNDANCE

Of all the freshwater fungi, the aquatic hyphomycetes are most suited to rapid, broad geographic surveys. Using both membrane filtration of conidia and collection of foam, an individual can collect and fix samples over a wide geographic area containing a variety of freshwater habitats and, at a later time, identify and quantify the conidia of species present. The two techniques are complementary. Foam naturally concentrates conidia and generally contains a large number of species, although species with large, branched conidia usually are overrepresented. Membrane filtration generally yields fewer species than foam but more of the species that are trapped less easily usually are present. It also can provide quantitative data (see later in this section). Membrane filtration can provide a reasonably complete picture of species composition in streams lacking foam if additional water samples are taken. Foam and membrane filtration samples represent the conidial pool, a conglomerate of the reproductive efforts of different communities occurring on different substrata (e.g., deciduous leaves, woody debris, fruits, and aquatic macrophytes), and should not be portrayed as representing a single community. Limitations to using these two techniques to assess biodiversity include the following: (1) difficulty in confirming identifications of "loose" conidia as described previously (see "Collection" under Mitosporic Fungi," earlier); (2) lack of information on the geographic origins of conidia in flowing water; (3) undetectability of species present but not sporulating; (4) absence of foam; and (5) turbidity that severely restricts the quantity of water that can be filtered.

For a comprehensive study of the biodiversity of a small number of habitats, examination of a wide range of substrata, membrane filtration of water, and foam collection, if coupled with a well-designed sampling protocol that incorporates seasonal components, should result in a complete and accurate species list. The reliability of that list is increased greatly when pure-culture studies are used to confirm the identities of problematic taxa.

Methods for directly assessing abundance by quantitative measurement of biomass at the species level are not yet sufficiently developed for routine use. In most studies of aquatic hyphomycetes, abundance is estimated indirectly to establish ranks or importance values for individual species (Suberkropp and Klug 1976; Shearer and Lane 1983; Gessner et al. 1993). All of those techniques are based on conidial production. Because conidia vary greatly in size (and hence biomass), and because the relationship between hyphal biomass and the conidial biomass it produces it is not known for most species, biomass measurements at the species level have not been made. Careful laboratory studies to determine the correlation between biomass of hyphae and that of conidia (Suberkropp 1991) for all the species in a community would allow the interpolation of hyphal biomass from the measurement of conidial biomass from a substratum. The use of immunological assays to estimate biomass of individual species shows promise, but high variance is still a problem (Newell 1992c).

Some investigators have used presence of conidia on substrata to determine species frequencies on leaves or leaf disks to determine the relative importance of the species in a community (Suberkropp and Klug 1976; Shearer and Lane 1983). Suberkropp and Klug (1976) also estimated species density and then calculated an importance index based on relative frequencies on leaves and relative density. Others have ranked species based on conidial production from particular substrata (Wood-Eggenschwiler and Bärlocher 1983). Suberkropp (1991) showed that hyphal biomass and conidial production are strongly correlated in aquatic hyphomycetes, thereby lending support to the use of conidia as an indicator of species importance. Species' conidial frequencies and relative frequencies also have been calculated from membrane filtration data (Iqbal and Webster 1977; Bärlocher and Rosset 1981; Shearer and Webster 1985a, b, c). Species also can be ranked by the frequencies with which their conidia are found in foam, if fields are randomly sampled until a predetermined total number of conidia (e.g., 250, 500, 1,000) is identified. Species can be ranked according to the frequencies of their respective conidia in the total number of conidia identified. Again caution must be used in interpreting membrane and foam data, given that they represent the reproductive output of several different communities.

CONCLUSIONS

The biodiversity of fungi in freshwater habitats is quite complex and includes all groups of fungi, although the Basidiomycetes are mostly absent. Globally, fungal biodiversity has been determined for only a minute number of freshwater habitats. Those habitats that have been studied usually have been surveyed for one taxonomic group at a time, rather than for all groups in concert. Thus, the complete fungal biodiversity for representative freshwater habitats is unknown.

A major barrier to biodiversity studies of freshwater fungi is the absence of up-to-date monographs and keys, which prevents all but a very few highly trained taxonomic experts from carrying out broad, reliable surveys. Well-illustrated monographs for the major groups of freshwater fungi would allow nonspecialists to collect and identify taxa. A barrier to quantitative studies is the absence of an easily used and accurate method for measuring abundance at the species level. That severely limits our ability to conduct rigorous, quantitative studies comparing fungal communities in different habitats and geographic locations. Those barriers must be overcome if we are to move forward.

Despite the foregoing limitations, the challenges and rewards of studying freshwater fungi are enormous. The sheer magnitude of freshwater habitats that have never been studied with respect to fungi is enticing. Many new species remain to be discovered, and many questions regarding roles; substratum specificities; and habitat, microhabitat, and broad-scale geographic distribution patterns remain to be answered.

24

Marine and Estuarine Mycelial Eumycota and Oomycota

Jan Kohlmeyer, Brigitte Volkmann-Kohlmeyer, and Steven Y. Newell

The existence of both indigenous and widely distributed fungi confined to the marine habitat was not recognized until about 40 years ago. Although some species of true marine fungi had been described in the 100 years prior to that (Kohlmeyer and Kohlmeyer 1979), the field of marine mycology was established only with the publication of Johnson and Sparrow's (1961) book on fungi in oceans and estuaries. Because marine fungi are not a taxonomic group and cannot be defined by nutritional or physiological requirements, Kohlmeyer and Kohlmeyer (1979) proposed the following ecological definitions: Obligate marine fungi are those that grow and sporulate exclusively in a marine or estuarine habitat and are permanently or intermittently submerged; facultative marine fungi are those that normally occupy freshwater or terrestrial milieus but are able to grow (and possibly also to sporulate) in the marine environment. Fungi restricted to immersed parts of estuarine or marine plants are considered terrestrial but halotolerant.

The definition given here for obligate marine fungi applies equally well to some members of the Oomycota, a phylum of protoctistan, eukaryotic, decomposers whose phylogenetic affinities are somewhat unclear. Several authors have suggested that the Oomycota is a group of nonfungal microbes whose fungus-like morphologies arose through convergent evolution (Dick 1990a; Brasier and Hansen 1992; Erwin and Ribeiro 1996). Other mycologists have provided alternative classification schemes for these organisms as Oomycetes in the Kingdom Chromista (Cavalier-Smith et al. 1994) or Peronosporomycetes in the Kingdom Straminipila (Beakes 1998; Dick et al. 1999; Dick 2001b). The oomycotes have diploid mycelial states, as opposed to the haploid mycelial condition of eumycotic fungi; form egglike sexual spores ("oospores") after fusion of haploid gametangia; and have biflagellate swimming zoospores as asexual propagules (Dick 1990a, 1995).

Obligate marine oomycotes grow and produce sporangia exclusively in marine or estuarine habitats. Nothing is known, however, about the extent to which characteristic marine mycelial oomycotes (e.g., *Halophytophthora* species) can be present and active upstream from estuaries, so the possibility remains that some halophytophthorans could be facultatively marine. Evidence of weak penetration of estuaries by freshwater mycelial oomycetes also exists, although it appears unlikely that those oomycetes (saprolegniaceans) play any active role downstream from freshwater courses (Harrison and Jones 1975; Padgett et al. 1988). No mycelial oomycetes characteristic of freshwater have been found as participants in any marine leaf-decomposition system (e.g., Newell and Fell 1995, 1997). However, *Aphanomyces* species, in the Saprolegniaceae, is secondarily active in ulcerative disease of estuarine fish (Noga 1993).

HABITATS AND MODES OF LIFE

EUMYCOTA

Marine and estuarine environments, namely the large oceans and their shores with sandy or rocky beaches, river mouths, tidal creeks, sounds and lagoons, and other bodies of saltwater or brackish water connected to the ocean, occupy 75% of the globe. Inland salt lakes probably should be included also because their mycota appears to be identical with that of the oceans

(Anastasiou 1961, 1963). Fungi occur mainly in the intertidal zone where most of their hosts or substrata are located, but some indigenous species have been collected in the deep sea also, with the deepest record at 5,315 m (Kohlmeyer 1977). The distribution of fungi in marine habitats is limited primarily by availability of dissolved oxygen; low levels of oxygen in the water or sediments inhibit or prevent fungal growth. Fungi did not settle on wood panels submerged off the California coast in a "minimum oxygen zone" with a dissolved oxygen content of 0.30 ml/liter; fungal growth did occur at another site with an oxygen content of 1.26 ml/liter (Kohlmeyer 1969a).

Marine fungi are saprotrophs, symbionts, or parasites on plants or animals. All are microscopic; the largest marine ascomycetes and basidiomycetes are only 4–5 mm in diameter. Saprobic fungi are important decomposers of cellulose, in the form of driftwood, pilings, mangrove roots, and marsh plants, and are found also on washed-up algae, seagrass leaves, and animal products (e.g., chitin, keratin, tunicin, calcium carbonate). Marine fungi and algae are able to form several types of symbiotic associations. Lichenoids have phototrophic partners, usually microscopic cyanobacteria or green algae, that also can occur in a free-living state. *Halographis runica* (Figs. 24.1AB, 24.2) is an endolith in submerged snail shells,

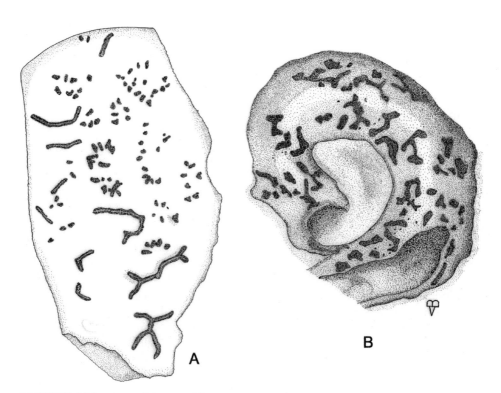

A

B

FIGURE 24.1 A. Lirelliform irelliform ascomata and smaller pycnidia of *Halographis runica* in the shell of *Vasum muricaturm*. **B.** Calcareous tube of a worm, strongly decomposed by the lichenoid *Halographis runica*, itself in the process of degeneration. (A, B reprinted with permission from Kohlmeyer and Volkmann-Kohlmeyer 1988.)

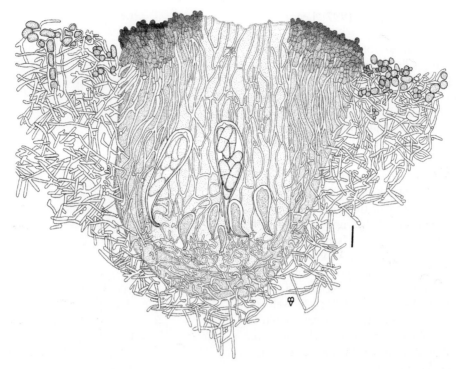

FIGURE 24.2 Cross-section through an ascoma of *Halographis runica* after dissolution of the calcareous shell of *Vasum muricaturm* by 5% HCl. Vegetative hyphae surround the ascoma. A thin layer of cyanobacterial cells grows at the surface. Scale bar = 15 µm. (Reprinted with permission from Kohlmeyer and Volkmann-Kohlmeyer 1988.)

worm tubes, and corals (Kohlmeyer and Volkmann-Kohlmeyer 1988). True submarine lichens, such as *Arthopyrenia halodytes*, which is found in barnacle tests and mollusk shells, and *Verrucaria serpuloides*, which occurs on submerged rocks, represent obligate associations between a mycobiont and a photobiont, forming a coherent structure, the thallus (see Chapter 9). Finally, mycophycobioses are obligate symbioses between systemic fungi and marine macroalgae in which the habit of the alga dominates (Kohlmeyer and Kohlmeyer 1979). Relationships between the intertidal brown alga *Ascophyllum nodosum* and the fungus *Mycophycias ascophylli* and between members of the red algal genus *Apophlaea* and their endosymbiont *M. apophlaeae* are examples of mycophycobioses (Kohlmeyer and Volkmann-Kohlmeyer 1998).

About 40 species of higher marine fungi are parasitic; most of them belong to the Ascomycotina. All but two of the known species parasitize algae (Kohlmeyer and Volkmann-Kohlmeyer 2003b); *Trichomaris* is found on the carapace of crabs, and *Cytospora rhizophorae* is found on the proproots of mangroves. *Mycaureola dilseae* is the only known filamentous basidiomycete that parasitizes algae. The number of affected host plants or animals is usually small, which makes collecting parasites in the field difficult and time-consuming.

OOMYCOTA

Although parasitoidal and algal-pathogenic species of marine oomycotes are known (see Karling 1981; Porter 1986; Noga 1990), some of them form hyphal thalli but not extensive mycelia (e.g., *Gonimochaete latitubus* in a marine nematode; see Newell et al. 1977; Glockling and Beakes 2000; and Chapter 19). Most species of mycelial marine oomycotes have been described from leaf-decomposition systems (e.g., Nakagiri et al. 1994; but see de Cock 1986; Kerwin et al. 1992). It appears that marine mycelial oomycotic decomposers (principally *Halophytophthora* species) are primarily decomposers of fallen leaves that enter the aquatic environment along the coast, as opposed to leaves produced on plants subject to tidal submergence (Newell et al. 1987; Newell 1992a, 1996). Species of *Pythium* may be exceptions (see de Cock 1986); in fact, *P. grandisporangium* originally was described from a marine intertidal grass (Fell and Master 1975). The most prominent contributors of fallen leaves along marine shorelines (12 t/ha/year litterfall at the equator; Saenger and Snedaker 1993) are the mangroves of subtropical to tropical zones (Tomlinson 1986), and most of the information on marine mycelial oomycotes has been gathered in mangrove ecosystems (Fell and Master 1975; Nakagiri 1993; Newell and Fell 1995;

Nakagiri et al. 1996; Tan and Pek 1997; Leaño et al. 1998).

Marine oomycotes are a part of the distinct decomposer community of submerged mangrove leaves (Somerfield et al. 1998). Studies of the distribution of fungal zoospores in mangrove systems indicate that the spores are released from submerged decaying leaves lying close to the mangrove stands and do not travel in large concentrations farther than a few meters from those stands (Newell and Fell 1992). Within as few as 2 hours after leaf fall, mycelial oomycotes begin to occupy mangrove leaves, and within 24–48 hours of submergence 80–100% of fallen leaves, and of the composite area of those leaves, often are occupied (Fell and Master 1975; Newell and Fell 1995). Preliminary samples of fallen leaves in temperate coastal marine water columns suggest that marine mycelial oomycotes are also prominent leaf decomposers outside of the mangrove ecosystem (Anastasiou and Churchland 1969; Newell 1992a; Newell and Fell 1995), so any temperate marine zones in which leaves of trees and shrubs may be deposited (e.g., river mouths, landward edges of saltmarshes) are likely habitats of marine mycelial oomycotes.

TAXONOMIC STATUS, DIVERSITY, AND DISTRIBUTION

EUMYCOTA

Compared with the number of terrestrial fungi, the number of higher marine fungi is small (Kohlmeyer and Kohlmeyer 1979). To date 444 species have been described from marine habitats (Hyde and Pointing 2000). Marine filamentous Eumycota can be characterized and identified by using descriptions, keys, and illustrations in Kohlmeyer and Kohlmeyer (1964–1969, 1979), as well as those in Kohlmeyer and Volkmann-Kohlmeyer (1991) and Hyde and Pointing (2000). So far, no cases of true endemic marine fungi have been recorded, unless host-specific fungi on endemic hosts are considered. *Mycophycias apophlaeae*, for example, occurs only on *Apophlaea* species in New Zealand, and numerous species of ascomycetes and coelomycetes are found only on *Juncus roemerianus* on the east coast of the United States and in the Gulf of Mexico.

Our experience based on collecting marine fungi in a variety of environments and oceans has shown that the greatest diversity of species is encountered in intertidal habitats such as sandy beaches, jetties (e.g., *Moana turbinulata*, Fig. 24.3), salt marshes, and the mangal. Beaches harbor arenicolous fungi, for example, members

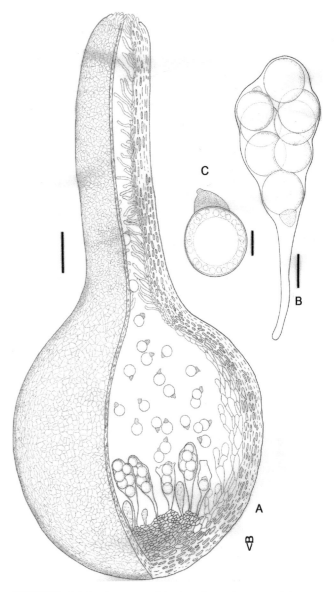

FIGURE 24.3 *Moana turbinulata* from the intertidal zone. **A.** Ascoma; partly shown in longitudinal section; scale bar = 50 µm. **B.** Ascus; scale bar = 10 µm. **C.** Appendaged ascospore; scale bar = 5 µm. (Reprinted with permission from Kohlmeyer and Volkmann-Kohlmeyer 1989a.)

of the large ascomycete genus *Corollospora*, that grow on buried dead algae or lignocellulose. Intermittently submerged parts of senescent marsh plants are inhabited by numerous caulicolous and foliicolous species of marine fungi, whereas the nonimmersed upper parts are occupied by a wide variety of terrestrial fungi. A similar pattern of distribution of lignicolous fungi is found on submersed and immersed roots and trunks of mangrove trees.

As far as we know, no thorough inventory of marine fungal biodiversity has been carried out in any area. We are in the process of an intensive survey of fungi in the salt marshes of the coastal United States, concentrating initially on inhabitants of the endemic needlerush, *Juncus roemerianus* (Juncaceae). So far we have found more than 100 species of ascomycetes and mitosporic fungi on that host alone, with marine species at the base, halotolerant terrestrial forms at the top, and facultative marine species in between. Almost all of those species are new to science (Kohlmeyer and Volkmann-Kohlmeyer 1993a, 1993b, 1996a, 1996b, 2000, 2001; Kohlmeyer et al. 1995b, 1995c, 1995d, 1995e, 1997, 1998). Species of cord grass (*Spartina* species, Poaceae) live in habitats similar to those of the needlerush and also harbor large numbers of filamentous higher fungi (Fig. 24.4) (Gessner and Goos 1973a, 1973b; Gessner and Kohlmeyer 1976; Gessner 1977).

Oceans in certain geographic areas of the world have been explored more extensively than others for marine fungi. Among them are those around the British Isles (Jones 1962, 1963a, 1963b, 1965), the Hawaiian Islands (Kohlmeyer and Volkmann-Kohlmeyer 1989a; Volkmann-Kohlmeyer and Kohlmeyer 1993), Hong Kong (Poon and Hyde 1998; Vrijmoed et al. 1982a, 1982b), the Seychelles (Hyde and Jones 1989), the Caribbean (Kohlmeyer 1980; Kohlmeyer and Volkmann-Kohlmeyer 1987a), and the South China Sea adjacent to Brunei (Hyde 1988a, 1988b, 1990a, 1991).

The main distribution areas of marine fungi are long narrow tracts of the intertidal zone, running parallel to coastlines or tidal creeks. Wider areas of occurrence can be found in extended salt marshes where *Juncus roemerianus* may cover thousands of hectares, such as in areas of Louisiana. The total marsh area covered by this plant in North Carolina, our main collecting site, is about 40,000 hectares (Eleuterius 1976).

OOMYCOTA

Far fewer species of marine mycelial oomycotes than species of true marine fungi are known. In part, this reflects the small number of investigators of marine oomycotic biodiversity and the use of methods in studies of distribution of decomposer microbes that prevent the observation of oomycotes (Newell and Fell 1994). Currently available methods selectively favor the detection of oomycotes (see "Oomycota" under "Collecting Methods," later in this chapter), but it also may be that some species of oomycotes do not produce the sporangia or oospores necessary for their detection (Newell and Fell 1994). At present, known mycelial oomycotes include 14 species described in the genus *Halophytophthora* (Nakagiri et al. 1996, 1998, 2001), and an approximately equal number of species in the other three genera, *Pythium*, *Haliphthoros*, and *Salilagenidium* (formerly *Lagenidium*), combined (see Bland et al. 1981, Dick 1998, and Chapter 19 for genera of nonmycelial oomycotes). *Halophytophthora* and *Pythium* are members of the family Pythiaceae of the subclass Peronosporomycetidae; the latter two fall into the Haliphthoraceae and Salilagenidiaceae, respectively, of the subclass Saprolegniomycetidae (Dick 1995). No modern monograph or comprehensive key comparable to that available for the marine eumycetes exits for the Oomycota. Ho and colleagues (1991) provided a key to species of *Halophytophthora*. That publication is now out of date (Nakagiri et al. 1994), but a new key to the halophytophthoras will soon be available (Nakagiri et al. 2001). When Johnson and Sparrow (1961) wrote their comprehensive tome on marine "fungi," no "marine phytophthoras" (Stamps et al. 1990), now halophytophthoras (Ho and Jong 1990), had been described. It may be, based on DNA evidence, that not all species of

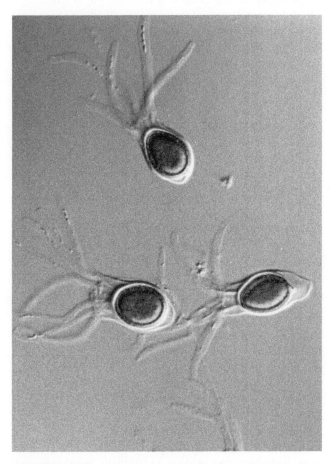

FIGURE 24.4 *Octopodotus stupendus* from *Spartina*. One-celled conidia surrounded by a gelatinous sheath bearing apical appendages.

Halophytophthora are congeners (A. W. A. M. de Cock, unpublished data).

The most commonly encountered species of marine mycelial oomycote is *Halophytophthora vesicula* (e.g., see Newell and Fell 1995, 1997), the first "marine" phytophthoras described (from British Columbia, Anastasiou and Churchland 1969). It occurs in both tropical and subtropical mangrove habitats and in temperate habitats at coastal sites around the world (Fell and Master 1975; Nakagiri et al. 1996) and decomposes fallen, submerged leaves. The species has an inverted cone-shaped structure within its sporangial apex (Newell 1996: fig. 1). The apex everts, producing a balloonlike vesicle that opens at its tip to release the zoospores. There are two distinct types of *H. vesicula*, one that produces "delicate vesicles," which are very thin-walled and not easily discerned, and one that produces "robust vesicles," which are thicker-walled and more easily seen (Nakagiri 1993: figs. 12,13; Nakagiri et al. 1996). The delicate type is by far the most common (Newell and Fell 1994), although available cultures derived from the type material (IFO [Institute for Fermentation, Osaka, Japan] 32216) belong to the "robust" type. The taxonomy of *H. vesicula* is unsettled—the two types may be different species, and *H. batemanensis* and *H. polymorphica* (Gerrettson-Cornell and Simpson 1984) are probably conspecific with the "delicate" *H. vesicula* (Newell and Nakagiri, unpublished data). Studies of halophytophthoran classification based on analyses of DNA (ribosomal DNA, mitochondrial DNA) and isoenzymes are under way (M. Hudspeth, D. Hudspeth, A. Nakagiri, S. Y. Newell, and D. Porter, unpublished data).

Salilagenidium callinectes and *Haliphthoros milfordensis* are examples of marine mycelial or mycelioid (minimal branching of hyphae) oomycotes that parasitize ova and larvae of marine crustacea and mollusks and are sometimes pathogenic in adults (Nakamura and Hatai 1995; Roza and Hatai 1999). The life stages of these species are illustrated in Fuller and Jaworski (1987: *Lagenidium* = *Salilagenidium*; Dick 1995). *Pythium marinum* is an example of a marine mycelial oomycote that is pathogenic in marine macroalgae (e.g., *Porphyra perforata*, a commercially important alga). Descriptions of interactions between algae and oomycotes and references are available in Kerwin and colleagues (1992) and de Cock (1986).

The species of *Halophytophthora* with the apparently most restricted geographic distribution (among those species that have been recorded more than once) is *H. operculata*, which has been found only in Australian and Singaporean mangroves (Nakagiri et al. 1996). One species, *H. masteri*, apparently is restricted to one species of substratum, the black mangrove, *Avicennia germinans* (Newell and Fell 1995). Other halophytophthoras have much wider distributions with respect to substratum (i.e., leaf type) and/or geography. *Halophytophthora exoprolifera*, for example, is found in subtropical mangroves in Bermuda and the Bahamas in the Atlantic and in the Ryukyus in the Pacific (Nakagiri 1993; Nakagiri et al. 1996; S. Y. Newell, unpublished data). Species that originally were named to match what appeared to be a limited distribution (e.g., *H. bahamensis*) now are known to occur much more widely (Fell and Master 1975; Newell and Fell 1995), but some species have been found only in the Atlantic (e.g., *H. masteri*) or the Pacific (e.g., *H. avicennae*) (Nakagiri et al. 1996; A. Nakagiri, unpublished data). Some species of *Halophytophthora* (e.g., *H. vesicula*, *H. bahamensis*, *H. kandeliae*) actually may include more than one species under their binomials (Nakagiri 1993; Nakagiri et al. 1996; S. Y. Newell and A. Nakagiri, unpublished data), which could mask some endemicity.

No thorough inventories of the biodiversity of marine mycelial oomycotes have been carried out. However, small-scale studies of the distribution and ecology of members of this group and searches for new species are in progress in the coastal waters of the southeastern United States, Bahamas, subtropical islands of Japan, and Singapore (Nakagiri et al. 1996; Newell and Fell 1997). All of those investigations are outgrowths of the pantropical search for marine oomycotes conducted by Fell and Master (1975 and references therein). Based on current taxonomic knowledge, we estimate that any 50,000-hectare coastal strip is likely to harbor between five and 15 known species of marine mycelial oomycotes.

BIODIVERSITY INVENTORY

COLLECTING METHODS

Eumycota

The main factor controlling the occurrence of marine fungi, besides dissolved oxygen and substratum, is water temperature. Therefore, in tropical zones where the fluctuations in water temperatures are small, samples can be taken year round. In temperate areas, collecting carried out in the summer months yields the highest diversity of species. Subtropical zones, where freezes may occur in January and February and tropical conditions prevail during the summer (e.g., mid-Atlantic states of the United States), should be sampled at least twice, in late winter and in summer. Tropical and temperate organisms, including marine algae and animals, overlap in such areas, and the same is true for the marine mycota. In our studies of the mycota on *Juncus roemerianus*, we sampled once monthly for more than 2 years in North Carolina and found only a few seasonal species. Many species, however, were recorded only once or twice during those years, and other species may have been missed because sampling was sporadic. At present, the presence or absence of a fungus can be determined solely by the occurrence of fruiting bodies and propagules. Certain species survive unfavorable conditions in the hyphal form; *Glomerobolus gelineus,* for example, does not form propagules during the summer months but rather in the fall when temperatures are lower; it continues to fruit until the following spring (Kohlmeyer and Volkmann-Kohlmeyer 1996a).

Sampling conditions and, therefore, collecting methods for marine fungi may vary depending on the substratum, host, and habitat. A relatively simple survey of arenicolous fungi can be made by collecting foam along the beach with a spoon-type skimmer. The foam is poured into a bottle and refrigerated for a few hours until the air bubbles have disappeared. Fungal propagules sink to the bottom and can be picked up with a pipette and examined under a compound microscope (Kohlmeyer and Kohlmeyer 1979). The ascospores or conidia present in the foam provide an indication of the diversity of species in a particular beach. Another way to obtain arenicolous fungi is to collect wet floatsam (e.g., pieces of algae, seagrasses, wood, feathers) at the water's edge, mix it with moist (not wet) sand, add some foam, if possible, and half-fill a sterile plastic bag (e.g., Whirl-Pak, 23 cm × 11 cm) with the mixture. The bags are stored at room temperature for some months and examined regularly under a dissecting microscope for fruiting bodies. Some slow-growing ascomycetes first appeared in our collections after 2 years of incubation.

Twenty-four species of Ascomycotina and mitosporic fungi have been reported from $CaCO_3$ in marine habitats, mostly in the tropics. Corallicolous fungi occur in and on dead corals. Five species of *Koralionastes* (Fig. 24.5) and also the very rare *Corallicola nana* are associated mostly with crustose sponges that all live on the lower side of dead coral slabs (Kohlmeyer and Volkmann-Kohlmeyer 1987b, 1990; Volkmann-Kohlmeyer and Kohlmeyer 1992). Ascomata of *Lulworthia calcicola* occupy the same substratum but never are surrounded by sponges; they also may develop on thin crusts of calcareous or filamentous algae (Kohlmeyer and Volkmann-Kohlmeyer 1989b). Corallicolous fungi are generally rare, and looking for them is both tedious and time-consuming (Kohlmeyer and Volkmann-Kohlmeyer 2003a). One may have to lift dozens of coral rocks to find a few ascomata.

Endolithic fungi occur inside corals, calcified algae, and other calcareous substrata. Examples include *Halographis runica* (Figs. 24.1AB, 24.2), which is found in

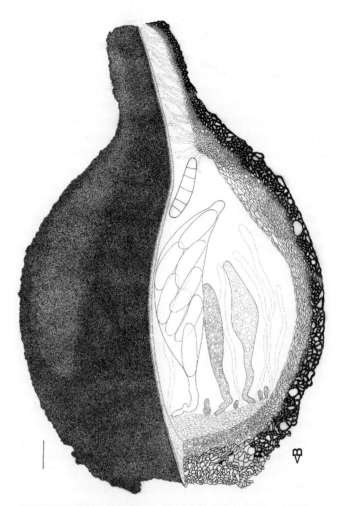

FIGURE 24.5 Subiculate ascoma of *Koralionastes violaceus* from dead coral; partly shown in longitudinal section. Scale bar = 50 μm. (Reprinted with permission from Kohlmeyer and Volkmann-Kohlmeyer 1990.)

coral slabs, worm tubes, and shells of the snail *Vasum* species (Kohlmeyer and Volkmann-Kohlmeyer 1988, 1992), always on the side turned toward the sediment; *Lulworthia kniepii* in coralline algae (Kohlmeyer 1963); *L. curalii* in conch shells and calcareous algae (Kohlmeyer 1984; Kohlmeyer and Volkmann-Kohlmeyer 1991); *Xenus lithophylli* from calcified algae attached to coral rock (Kohlmeyer and Volkmann-Kohlmeyer 1992); and the two lichenized species, *Chadefaudia corallinarum,* from coralline algae, and *Arthopyrenia halodytes,* from mollusk shells and barnacles (Kohlmeyer and Kohlmeyer 1979). Unidentifiable fungi also have been found in the skeletons of live corals (LeCampion-Alsumard et al. 1995). Tests of Foraminifera harbor ascomata of two species of *Arenariomyces, Corollospora maritima,* and two species of *Lindra* (Kohlmeyer 1984, 1985); ascomata of two species of *Halosphaeria, Corollospora pulchella,* and the conidia of *Cirrenalia pygmea*, *Trichocladium alopallonellum*, and *Periconia prolifica* occur in and on the lining of old shipworm tubes (Kohlmeyer 1969b, 1984).

Caulicolous and foliicolous fungi can be collected in salt marshes. Senescent leaves or culms of standing marsh plants (e.g., *Juncus roemerianus, Spartina* species) are cut at the base close to the rhizome and stored in plastic bags in a refrigerator or freezer until examined or airdried. Because the fungal structures usually are immersed in the plant tissues, the culms or leaves are split lengthwise with a razor blade and scanned under the dissecting microscope for embedded fruiting bodies of ascomycetes or coelomycetes. As mentioned earlier, marine, facultatively marine, and halotolerant terrestrial species occur in zones from the base to the tip of the plants, respectively (Gessner 1977; Kohlmeyer et al. 1995e).

Manglicolous marine fungi grow on permanently or intermittently submerged parts of mangrove trees. Except for a few species that are restricted to bark (e.g., *Etheirophora blepharospora*), most fungi in this group require wood for their development and, therefore, are found only on decorticated areas of the trees. Dead or damaged roots (e.g., proproots of *Rhizophora* species and pneumatophores of *Avicennia*) are cut close to the sediment and then examined under a dissecting microscope. A zonation of marine and terrestrial fungi, similar to the one on marsh plants, can be observed along the roots (Hyde 1990a, b, 1991; Kohlmeyer et al. 1995e). Marine fungi also develop on submerged branches and trunks of dead trees; pieces of wood containing fruiting bodies must be removed for examination. Such substrata often bear large colonies of carbonaceous ascomata that can be detected easily with the naked eye.

Oomycota

The methods discussed in this and the following subsection refer only to the decomposer mycelial oomycotes. Methods appropriate for collection of algal-pathogenic pythia are available in Kerwin and colleagues (1992); methods for animal–parasitic oomycotes can be found in Newell and associates (1977), Fuller and Jaworski (1987), and Barron (Chapter 19). Other, general references include Sparrow (1960), Waterhouse (1971), Park (1980), Van der Plaats-Niterink (1981), Buczacki (1983), Erwin and colleagues (1983), Dick (1990a), Lucas and colleagues (1991), Singleton and colleagues (1992), Erwin and Ribeiro (1996), Hall (1996), Weber and colleagues (1997), and Dick (2001a, 2001b).

The decomposer marine mycelial oomycotes (halophytophthoras and pythia) can be collected from submerged decaying leaves or from the water column of coastal intertidal and subtidal zones. Samples should be collected during both cold and warm seasons because some species may be temperature limited (e.g., *Halophytophthora exoprolifera* to colder intervals, *H. spinosa* (Fig. 24.6) to warmer ones; Nakagiri et al. 1996). Because oomycotes do not form complex-tissue fruiting bodies like ascomata (Dick 1990a), they must be surveyed using growth-stimulation methods. Standard-size leaf disks can be removed from decaying leaves with cork borers that are sterilized between samples to prevent cross-contamination (Newell and Fell 1995 and references therein). This method permits calculation of frequencies of occurrence for individual species of oomycotes, provided that leaves are sampled nonselectively from the environment. Leaves are collected by walking at low tide or snorkeling at high tide, especially in mangrove habitats where sediments are very soft. Leaves are rinsed in the field to remove surface clay or silt films and transported to the laboratory in plastic bags containing just enough water to prevent drying. They must be protected from heat and direct sunlight. Ideally the bags should be sterile, but this is not essential because bags are very unlikely to be contaminated with oomycotes.

Disks should be removed from leaves and agar-plated as soon as possible after return to the laboratory. Ideally,

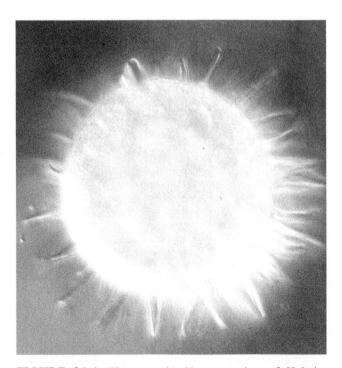

FIGURE 24.6 The sea-urchin-like sporangium of *Halophytophthora spinosa*. This species is commonly second to *H. vesicula* in frequency of occurrence on submerged decaying leaves of mangroves in the subtropics and tropics. An exit or dehiscence tube, through which zoospores swim away, is seen beginning to form. Diameter of the sporangium, about 70 µm.

leaves will be processed under a laminar-flow cleanhood to limit contamination by spoilage fungi (Newell and Fell 1994). Oomycotes are not commonly aerial contaminants. Leaves are drained and can be surface-treated with biocide (Newell 1996) if the collector wishes to record only (or principally) species present as internally active mycelia (Newell et al. 1987). Washing samples or simply dipping them into biocide solutions is not sufficient to prevent significant outgrowth of microbes present as surface forms (e.g., conidia or weak microcolonies); measures must be taken to enhance the likelihood that the biocide reaches its targets (Newell and Fell 1982). When more than one leaf-disk is allotted to an agar plate, the oomycote mycelium from one disk may merge into that of another, so that the number of positive disks cannot be recorded correctly. This problem can be ameliorated if plates are read under a dissecting microscope 3–4 days after disk plating so that disks yielding oomycotic mycelium and those negative for oomycotes can be recorded.

It is important that oomycote-selective agar medium be used when surveying for marine mycelial oomycotes. Use of inappropriate ingredients can reduce severely the accuracy of observed frequencies (e.g., tenfold for *H. vesicula* when high concentrations of chloramphenicol are present in the medium; Newell and Fell 1994). Competitors (bacteria, eumycetes) and predators (labyrinthulids) must be restrained from inhibiting or preventing oomycotic outgrowth from leaf-disk samples. A/T Oomycote Medium, a medium modified from Newell and Fell (1994), is particularly effective at selecting for halophytophthoras (Appendix II). Competition from eumycetes, however, is not fully eliminated with this medium, and labyrinthulids are not markedly suppressed (S. Y. Newell, unpublished data); additional modification of the composition of the medium is needed. Preliminarily reports (A. Statzell-Tallman, unpublished data) indicate that adding amphotericin-B to the medium at a concentration of 4 mg/liter will depress eumycetic growth more sharply without suppressing oomycote growth. Chloramphenicol, an antibacterial agent, is used only at low concentrations in A/T medium because it is toxic to oomycotes (Newell and Fell 1994). The antibacterial agent, streptomycin (which can be toxic to oomycotes) and the common antifungal agent, pimaricin (which is toxic to *H. spinosa*), should not be used (Newell and Fell 1992).

Plates of leaf disks should be incubated at approximate field temperatures for 3–4 days in the dark (to minimize loss of antieumycetic activity of the light-sensitive amphotericin) and then switched to incubation in alternating light and dark (12 hours/12 hours) to stimulate formation of zoosporangia. Plates should be read 10–20 days after sampling, depending on the temperature of incubation. If reading is delayed, sporangia may deteriorate through autolysis or be preyed on by labyrinthulids and other microbes, preventing identification of some species.

The surfaces of disks, the edges of disks (especially at the disk/agar interface), at least part of the mycelia around disks, and the areas of mycelial contact with the edges of the plates should be inspected. Sometimes sporangia form at only one of these sites. Both subplate transmitted and surface illumination should be used. A dissecting microscope (e.g., Wild M8) magnification of 25× is satisfactory for scanning, but zooming to 50 or 100× may be required for confident detection of small and/or translucent sporangia and for differentiation among types of sporangia with roughly similar shape. Until one is very familiar with the appearance of the sporangia of the marine mycelial oomycotes at the dissecting microscope, and whenever an identification is doubtful, zoosporangia should be examined under a compound microscope at 100–400× using interference contrast lighting to confirm conclusions drawn at the dissecting microscope.

Zoospore-release behavior and morphology of structures involved in zoospore release differ among species of marine mycelial oomycotes (e.g., Nakagiri et al. 1994); therefore, zoospore release must be induced for each type of sporangium encountered to confirm species identifications. Methods for inducing sporangium maturation, dehiscence-tube production, and zoospore release are discussed by Nakagiri (1993; Nakagiri et al. 1996; also, see methods references cited at the beginning of this section, especially Fuller and Jaworski 1987). Induction usually involves changes of salinity and temperature, as well as exposure to light, brief drying, or sterols or phospholipids (e.g., *H. exprolifera*, for which both temperature and available plant lipids are important; Ho et al. 1992). Other species (e.g., *H. vesicula*) produce sporangia and release zoospores after the agar containing their mycelium is submerged in dilute seawater (15 g l^{-1}). Autoclaving the seawater used may eliminate its stimulatory effect on some species (Fell and Master 1975; Harrison et al. 1980; Keller et al. 1988). Obtaining zoospore-release information requires long periods at the compound microscope and considerable patience. Sporangia that are compressed or allowed to dry on microscope slides will be damaged and lose the capacity to release zoospores. Slides containing sporangia in water can be maintained in damp chambers (e.g., supported above the bottoms of closed Petri dishes containing 1 ml of distilled water) and observed periodically until zoospore release is imminent. Dilution of the preparation by addition of freshwater at one side of the coverslip and drawing of liquid into absorbent paper at the other side may elicit zoospore release (but too much or too rapid dilution can cause sporangia to burst and be destroyed). Microcontinuous-flow chambers are ideal for observing zoospore release (Quesnel 1969; Falloon 1977; Amon et al. 1987; a good general reference for microtechnique is Descals 1997). Research is needed to define additional routine treatments that could be effective in inducing zoospore release by marine mycelial oomycotes.

Potentially useful references for identifying marine mycelial oomycotes include the following: Höhnk 1953 (drawings of marine pythia); Sparrow 1960 (drawings of and key to aquatic pythia and phytophthoras); Johnson and Sparrow 1961 (drawings and keys to marine pythia); Anastasiou and Churchland 1969 (drawings and photographs of *Halophytophthora vesicula*); Fell and Master 1975 (photographs of *H. bahamensis, H. epistomium, H. mycoparasitica, H. spinosa, Pythium grandisporangium*); Van der Plaats-Niterink 1981 (keys to and drawings of pythia); Pegg and Alcorn 1982 (drawings and photographs of *H. operculata*); de Cock 1986 (drawings of marine pythia and a phytophthora); Dick 1990b (keys to pythia); H. H. Ho and associates 1990 (oospores of *H. epistomium*); Stamps and colleagues 1990 (photographs and key to marine "phytophthoras"); Ho and associates 1991 (photographs of *H. kandeliae* and key to halophytophthoras); Ho and colleagues 1992 (photographs of *H. exoprolifera*); Nakagiri 1993 (photographs of *H. exoprolifera, H. spinosa*, and *H. vesicula* "delicate" and "robust"); Nakagiri and associates 1994 (photographs of *H. masteri* and *H. tartarea*); Newell and Fell 1994 (photographs of *H. masteri*); Pan and colleagues 1994a (scanning electron microscope images of seven species of *Halophytophthora*); Newell 1996 (photograph of *H. vesicula* "robust"); Fell and Newell 1998 (photograph of *H. vesicula* "delicate"); Nakagiri and associates 1998 (photographs of *H. mycoparasitica*); and Nakagiri and colleagues 2001 (photographs of *H. porrigovesica* and key to halophytophthoras).

Some species or strains of species of marine mycelial oomycotes may be present on leaf-sampling plates as mycelia but may be unable to form sporangia or oospores under the particular conditions present (e.g., presence of strong competitors such as *H. vesicula*; Newell and Fell 1997). Strains of *H. kandeliae*, for example, are unable to form sporangia on agar (Ho et al. 1991; S. Y. Newell and A. Nakagiri, unpublished data), so their frequencies could be underestimated, especially when *H. vesicula* is present (Newell and Fell 1995). Even strains of *H. vesicula* can be weak producers of sporangia on agar plates; sometimes they form sporangia only at the outer edges of plates, where the mycelium contacts the wall of the dish. Other species readily form oospores on agar but not sporangia (e.g., *H. exoprolifera*), so whenever oospores are encountered, one should search for associated sporangia among the oospores. One halophytophthora, *H.*

tartarea, occurs on leaf-sampling plates only as oosporic mycelium with sporangium-like structures, producing its characteristic sporangia only after subculture (Newell 1992b, as Species 3; Nakagiri et al. 1994).

Theoretically, one can sample the water column for zoospores or zoosporic cysts of marine mycelial oomycotes (Newell and Fell 1992; cf. Hallett and Dick 1981). Water can be sampled by baiting (e.g., Newell et al. 1987), but estimated concentrations of zoospores or cysts based on baiting seem overly conservative (Newell and Fell 1995). Although apparently accurate estimates of zoospore concentrations were obtained by selective culturing from gelled water samples in laboratory experiments with *H. vesicula* and *H. spinosa* (Celio and Padgett 1989), no zoospores were recovered from mangrove environments using this technique (S. Y. Newell, unpublished data). Only one set of samples was tested, however, so the technique should be reevaluated. Most-probable-number techniques (e.g., Cooper 1993) are not practical because of the apparently low concentration of halophytophthoran zoospores in nature. Vacuum filtration of water samples onto membrane filters followed by plating of the filters on oomycote-selective medium apparently renders zoospores or cysts nonviable and thus cannot be used to recover halophytophthoras (Newell et al. 1987; D. Porter, unpublished data). Gentler techniques for concentrating spores and cysts, such as continuous centrifugation (Hallett and Dick 1981), reverse-flow filtration (Sieburth 1979), and gravity filtration (Leadbetter 1993), have potential for use with marine mycelial oomycetes, but only continuous centrifugation has been sufficiently tested and the testing was performed in a freshwater system. Reverse-flow filtration led to a loss of cell viability in delicate marine protozoa (Gifford 1985).

ISOLATION METHODS

Eumycota

Isolation methods for marine fungi are similar to the ones adopted for terrestrial species, except for the use of a saltwater medium. The substratum has to be air-dried to facilitate isolation. We excise the top of an ascoma or conidioma with a sterile razor blade, remove the contents of the centrum with a fine needle, and transfer them into a drop of sterile distilled water in an inoculation loop. We then streak the loop on seawater agar with antibiotics (Appendix II). Other media and methods can be used to induce fruiting in some taxa (Hyde et al. 1987). Marine hyphomycetes are usually more difficult to isolate than ascomycetes or coelomycetes because the conidia develop on the surface of substrata also covered by bacteria and exuded propagules of other fungal species. Hyphae of sporulating colonies are picked off carefully with a fine sterile needle under a dissecting microscope, transferred into a drop of sterile water, and further processed as described for ascospores. Small blocks of agar bearing germinating spores or conidia are cut from the plate with a sterilized needle and transferred to plates or tubes of the same medium without antibiotics. They are stored in incubators and, depending on their origin, maintained at about 20°C for temperate species and at about 25°C for tropical and subtropical species. For further details, see *Marine Mycology—A Practical Approach* (Hyde and Pointing 2000).

Oomycota

Pure cultures of species of marine mycelial oomycetes can be prepared from leaf-sampling plates in several ways, as follows. We describe three procedures in the following paragraphs.

Hyphal-Tip Culturing. Hyphal tips are cut from mycelial edges in an area of the agar plate that appears to contain only one species, based on presence of sporangia and/or oospores. The tips are transferred aseptically under a laminar-flow cleanhood to new plates of oomycote-selective agar medium (A/T Oomycote Medium; Appendix II). As soon as the mycelia grow away from any bacteria or other contaminant, unintentionally carried with the hyphal tips, the hyphal tips are cut and transferred again, this time to medium without inhibitors (A/T medium minus amphotericin, etc.). About 1 mm^3 of autoclaved plant material (e.g., peanut, sesame, or safflower seed) is placed on top of the agar in the plate to provide any lipids that may be required for sporangium or oospore genesis (Fell and Master 1975; Ho et al. 1992; D. Porter, personal communication). Regardless of the medium used, low concentrations of agar (e.g., 0.5%) are likely to stimulate oospore production (Paternoster and Burns 1996). The plate is incubated in light and dark (12 hours/12 hours) until sporangia are developed sufficiently for identification. To ensure cultural purity, a single sporangium can be placed in sterile seawater and the zoospores can be baited onto autoclaved plant bits for transfer to new agar plates. Agar cultures can be maintained for 6–9 months before transfer, in plastic dishes (60 mm diam. × 15 mm high) with tightly fitting lids, at room temperature (about 22°C), or in sterile dilute seawater in sealed containers at 15°C, (Smith and Onions 1994; see Nakagiri and Jones 2000 for methods of long-term storage). Reducing the rate of water loss by placing highly nonpolar sealing film (e.g., Parafilm M) around the plates may be advantageous. It is not unusual for marine mycelial oomycotes to die suddenly in culture for no apparent reason, so multiple replicates of particularly valuable cultures should be

maintained (S. Y. Newell, unpublished data). Transferring cultures is more often successful if a piece of the plant from the original plate (e.g., the mycelial-occupied peanut bit) also is transferred (S. Y. Newell, unpublished data). Even when an agar culture appears to have died, oases of living material may be present in the plate. Rescue is sometimes successful if half of the agar is removed aseptically and the resultant depression is filled with sterile dilute seawater along with an autoclaved bit of peanut. The cut-out agar that was removed is submerged in a separate dish of sterile dilute seawater containing a bit of autoclaved peanut.

Sampling and Culturing Sporangia. A single sporangium or small groups of sporangia are cut from the original leaf sampling plates. The presence of sporangia from one species only can be confirmed using a compound microscope. The sporangia are placed onto a plate of A/T medium under the dissecting microscope and pushed across the agar surface with a fine, sterile nichrome inoculating needle. A spatulate point can be produced at the tip of a nichrome needle with a small hammer, or a fine point can be produced with a small file, for use in excavating agar or micromanipulating sporangia. The final destinations of isolated sporangia are marked on the bottom of the plate. Hyphae are produced either by direct sporangial germination (e.g., *Halophytophthora exoprolifera*) or from mature unreleased zoospores (e.g., *H. vesicula*). Hyphal tips are excised and transferred as described in method 1, earlier. For species that do not produce hyphae in this way (e.g., *H. bahamensis* and *H. epistomium*), isolated sporangia are placed on a slide in dilute seawater (15 g l^{-1} or less) until zoospores are released (naturally or by artificial stimulation; see "Oomycota" under "Collecting Methods," earlier). Immediately after zoospore release begins, sporangia should be placed on agar plates and manipulated as already described. A few zoospores usually are trapped in the sporangia and germinate from that position. Also, a drop of the seawater into which the zoospores were released can be transferred to the A/T medium and spread over the agar surface. Flagellate-isolation microtechniques also can be used (Caron 1993). Isolated zoospores will encyst and may germinate to form small mycelia, the hyphal tips of which can be transferred to new agar plates.

Baiting Zoospores. Sporangia are isolated and their identity is confirmed at 400× magnification. The separated sporangia are submerged in dilute sterile seawater (15 g l^{-1} or less). Released zoospores are baited onto autoclaved plant bits. This can be accomplished on plates or in miniaturized fashion in well slides. Plant bits are transferred to new plates of A/T medium with inhibitors soon after it is observed that zoospores have been released. Inhibitors are useful, even if cultures of origin on the leaf-sampling plates appear pure, because bacterial or fungal contamination is likely to occur during initial identification procedures or during baiting in well slides.

SPECIMEN REPOSITORIES

Eumycota

The only major repository of well-preserved specimens of marine fungi is at the Institute of Marine Sciences (IMS) of the University of North Carolina (Appendix III). The collection includes dried or preserved herbarium material and about 15,000 permanent microscope slides. Type material of some marine species is available at AHFH, B, BM, BPI, BRIP, C, CBS, CO, CP, DAOM, FH, H, HKU(M), ILLS, IMI, JE, K, LPS, MA, NY, P, PC, UC, UME, and UPS (Kohlmeyer and Volkmann-Kohlmeyer 1991). Culture collections of marine fungi are maintained at the University of Portsmouth (about 5,000 strains, the largest collection; Hyde and Pointing 2000) and at the IMS, with some additional cultures available at the American Type Culture Collection (ATCC) and the Centraalbureau voor Schimmelcultures (CBS); the Institute for Fermentation (IFO) at Osaka, Japan; and the Department of Biology and Chemistry at the City University of Hong Kong (see Appendix III).

Oomycota

At present, the most extensive repositories of living cultures of marine mycelial oomycotes are at the IFO, CBS, and ATCC and in the personal laboratories of S. Y. Newell (University of Georgia) and M. D. Coffey (University of California, Riverside) (see Appendix III). Because the taxonomy of marine mycelial oomycotes is in a state of flux, it would be wise to have an expert check the identities of cultures from large collections (e.g., ATCC). Investigators also should be aware that mycelial eukaryotes can undergo genetic change when kept on artificial media for long periods (e.g., Gramss 1991; Petersen et al. 1999).

DETERMINATION OF RELATIVE IMPORTANCE AND ABUNDANCE

EUMYCOTA

Newell (2000) reviewed methods for measuring fungal mass and provided instructions for taking direct microscopic measurements of hyphal biovolume and indirect measurements using biochemical indices of proxies for biomass (e.g., glucosamine/muramic-acid, ergosterol). The only practical indicator of occurrence of marine fungi is the presence of fruiting bodies or propagules. The vegetative mycelium in a substratum may extend much farther from a fruiting body and overlap colonies of other species. Therefore, quantification of the total standing crop of a particular species of fungus by direct observation is not possible at present. However, density of ascomata and rate of ascospore expulsion can be measured directly (e.g., Newell and Wasowski 1995). A formula (Hyde 1986) often used to demonstrate the frequency of species within collections made in a particular geographic area is as follows:

$$\text{percent occurrence} = \frac{\text{no. collections of a sporulating species}}{\text{no. all samples supporting sporulating fungi}} \times 100$$

Two facts must be kept in mind when this formula is applied. First, fungi that are present in the substratum in the nonfruiting, vegetative state will be missed. Second, the data must be comparable—that is, they must come from fungi occupying identical habitats (e.g., beaches) and identical substrata or hosts (e.g., roots of *Rhizophora* species; leaves of *Juncus* species). Mixed material does not provide meaningful frequency data.

We collected fruiting bodies of certain ascomycete species in almost every monthly sample taken in *Juncus roemerianus* marshes during a 2-year period and consider such species abundant (e.g., *Atrotorquata lineata, Papulosa amerospora*). Other species that we consider rare were found only once or twice. Ascomata of *Gaeumannomyces medullaris* were collected only six times, but its conidial anamorph, *Trichocladium medullare*, was present 33 times (Kohlmeyer and Volkmann-Kohlmeyer 1995; Kohlmeyer et al. 1995c and unpublished data). Clearly the occurrence of ascomata alone can provide misleading information about the presence and actual abundance of a particular species. We determined that 20 dead standing leaves or culms of *J. roemerianus* provide a representative sample of the common fungal taxa fruiting at a particular site (e.g., a tidal creek) at a certain time. A similar collection strategy probably can be applied to other salt-marsh plants, such as *Spartina* species. Adequate sample sizes for other habitats (e.g., sandy beaches, mangals) have not been established. However, recent publications discuss frequencies of manglicolous, lignicolous, and arenicolous fungi of the Pacific and Caribbean (Volkmann-Kohlmeyer and Kohlmeyer 1993) and of marine mangrove fungi (Hyde and Jones 1988).

OOMYCOTA

Frequencies of occurrence for species of marine mycelial oomycotes can be obtained by disk-sampling of naturally decomposing submerged leaves (see "Oomycota" under "Collecting Methods," earlier). By placing yellow (senescent) leaves from trees (which contain no oomycotes; Newell et al. 1987) into seawater sites, one can determine the frequencies of occurrence for zoospores or zoospore cysts of particular marine mycelial oomycotes after various periods (e.g., Newell and Fell 1995, 1997). The $R \times C$ G-Test (Sokal and Rohlf 1981) is a good (and easy) nonparametric statistical method for testing significance of differences among frequencies. Large numbers of sample replicates are desirable; with a sample size of 20, the 95% confidence interval is "21 percentage units at an observed frequency of 50%" (Snedecor and Cochran 1967:6). Currently no well-developed methods analogous to the biomass-index methods for eumycetes are available for measuring the biomass or productivity of marine oomycotes (Newell 1994; Gessner and Newell 1997; Fell and Newell 1998; but see Lee and Mullins [1994], Larsen et al. [2000] and Newell [2001]). Oomycotic storage carbohydrates and phospholipid fatty acids may be unique enough to serve as biomass indices.

CHARACTERIZATION OF ORGANISMS

EUMYCOTA

Adequate descriptions for taxa of marine Ascomycotina and mitosporic fungi always should be accompanied by satisfactory photomicrographs and/or camera lucida drawings. We would like to emphasize again (see also Kohlmeyer and Kohlmeyer 1979) that all data must be supported by good voucher material, preferably permanently mounted microscope slides (Volkmann-Kohlmeyer and Kohlmeyer 1996). Temporary slides ringed with fingernail polish are mostly useless because they dry out after a short time. Too many papers are published with long lists of fungal names but no indication of where vouchers are available for examination. We concur strongly with Dennis (1968, p. XXIX), who wrote that "Lists of records that cannot be verified are mere waste paper." The need for the deposition of voucher material also was emphasized in 2000 by Agerer and associates (2000) and is now a prerequisite for publication in *Mycological Research.*

OOMYCOTA

Fewer than ten investigators worldwide have surveyed marine mycelial oomycotes using methods that maximize the likelihood of observing all the species present at a given locale. Temperate locales have been neglected in this work. It is likely, therefore, that many species of marine mycelial oomycotes remain to be described. The requirements for new descriptions and for depositions of type materials are the same for oomycotes as they are for eumycetes (see Nakagiri et al. 1994, for an example of species descriptions for marine mycelial oomycotes). Four invaluable general references are Hawksworth (1974), Stearn (1995), Hawksworth and colleagues (1995), and Korf (1995). Permanent slides (Volkmann-Kohlmeyer and Kohlmeyer 1996) and dried type material must be prepared and deposited in a recognized repository (Agerer et al. 2000). However, neither type of preserved specimen is likely to be as valuable to subsequent investigators as are similar preparations of eumycetes because oomycote structures do not preserve well and stages of zoospore release cannot be preserved. Therefore, authors of new taxa must prepare subcultures from the type material and subcultures from paratype material and deposit them in recognized culture collections. In addition, any unpublished photographs (prints and negatives) and drawings should be deposited with the type specimens to enable subsequent examiners to see the full range of morphologies produced by the original culture.

ACKNOWLEDGMENTS. This work has supported in part by the U.S. National Science Foundation (Grants BSR9203915 to J. Kohlmeyer and B. Volkmann-Kohlmeyer and OCE-9115642 and 9521588 to S. Y. Newell).

25

MYCETOZOANS

FREDERICK W. SPIEGEL, STEVEN L. STEPHENSON, HAROLD W. KELLER,
DONNA L. MOORE, AND JAMES C. CAVENDER

Mycetozoa is a name applied to the polyphyletic assemblage of terrestrial amoeboid protists that produce aerial spore-bearing structures. Another term used for these organisms is slime molds. The categories to which mycetozoans can be assigned are the Eumycetozoa and the acrasids (Olive 1975). The taxon Eumycetozoa appears to be a monophyletic group (Olive 1975; Spiegel et al. 1995a; Baldauf and Doolittle 1997; Baldauf et al. 2000) that includes the Myxogastria (myxomycetes or plasmodial slime molds) (Fig. 25.1), the Dictyostelia (dictyostelid cellular slime molds) (Fig. 25.2), and the Protostelia (protostelids) (Fig. 25.3). The eumycetozoans may be a sister group to the Fungi and Animal clade of the crown eukaryotes. The acrasid cellular slime molds, called Acrasea by Olive (1975), appear to be a polyphyletic group (Olive 1975; Blanton 1990). The acrasids *Acrasis* (Fig. 25.4), *Pocheina*, and *Guttulinopsis* are members of the amoeboflagellate class Heterolobosea (Page and Blanton 1985; Blanton 1990; Baldauf et al. 2000), a group that also includes nonfruiting taxa such as *Naegleria*, *Vahlkampfia*, and *Tetramitus*. The other acrasids include the copromyxid genera *Copromyxa* and *Copromyxella* (Blanton 1990) and the unusual genus *Fonticula* (Blanton 1990). The phylogenetic affinities of the latter three genera are unknown.

The ability of mycetozoans to fruit is what distinguishes them from other amoeboid protists; it is their

FIGURE 25.1 *Diachea splendens* Peck. One of the most beautiful "jewels of nature," this species is perhaps the most brilliantly colored of all Myxomycetes, with its sparkling, metallic-iridescent blue peridium and white calcareous stalks (individual sporangium 0.5 mm in diameter). (Photo by Stephan Briere)

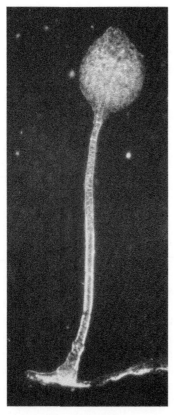

FIGURE 25.2 Sorocarp, or fruiting body, of *Dictyostelium discoideum* (height of individual sorocarp approximately 2 mm). (Photo by Frederick Spiegel)

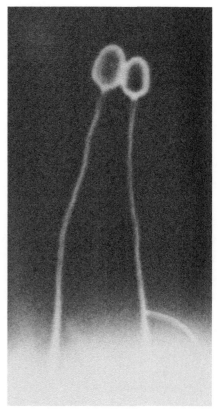

FIGURE 25.3 Two sporocarps of the protostelid *Nematostelium ovatum* (height of individual sporocarps approximately 150 μm). (Photo by Frederick Spiegel)

FIGURE 25.4 Sorocarps of *Acrasis rosea* and sporocarp of *Protostelium mycophaga* (total width of figure approximately 400 µm). (Photo by Frederick Spiegel)

fruiting bodies that are detected most commonly on a substratum and that are used to identify these organisms. Although we use the term fruiting body throughout this chapter, we should note some more specific terms. In myxomycetes and protostelids, the fruiting body develops from a single cell and is called a sporocarp (Olive 1975; Spiegel 1990). The sporocarps of most protostelids have a single spore, although there are a few species that regularly have two- to four-spored fruiting bodies. All myxomycete sporocarps have multiple spores. The morphology of myxomycete sporocarps is extremely variable; specialized terms used to describe those structures can be found in a number of publications (e.g., Martin and Alexopoulos 1969; Olive 1975; Keller 1982; Martin et al. 1983; Frederick 1990; Nannenga-Bremekamp 1991; Stephenson and Stempen 1994; Keller and Braun 1999; Krug et al., Chapter 21 of this volume). In cellular slime molds (dictyostelids and acrasids), in contrast, many amoebae aggregate to form a fruiting body, or sporocarp (Olive 1975; Raper 1984; Cavender 1990). The easily dispersed propagules of the fruiting bodies usually are called spores (Figs. 25.5 and 25.6). The propagules of the fruiting bodies of copromyxids, which are indistinguishable from individual cysts, are an exception. In that case, the propagules are called sorocysts (Olive 1975; Blanton 1990). If a spore or group of spores is borne above the substratum by a distinctive structure, that supporting structure is called a stalk. The stalk may be acellular, as in the protostelids, myxomycetes, *Fonticula*, and the dictyostelid genus *Acytostelium* (Martin and Alexopoulos 1969; Olive 1975; Martin et al. 1983; Raper 1984; Frederick 1990; Spiegel 1990; Nannenga-Bremekamp 1991; Stephenson and Stempen 1994), or it may be cellular (Fig. 25.7), including either dead stalk

cells, as in most dictyostelids (Olive 1975; Raper 1984; Cavender 1990), or living stalk cells, as in *Acrasis* and *Pocheina* (Olive 1975; Blanton 1990).

Fruiting bodies range in size from that of some myxomycete aethalia, which may approach 1 m in maximum extent, to the microscopic fruiting bodies of most protostelids, the smallest of which may be less than 10 µm tall. Fruiting bodies are not necessarily reproductive structures because their spores are not always the direct products of cell divisions. That is the case, for example, in dictyostelids (Raper 1984). However, all fruiting bodies produced by mycetozoans appear to function in the dispersal of those organisms.

In all mycetozoans, the trophic state consists of one or more types of amoeboid cells that feed phagotrophically on other microorganisms. Several terms are used for those trophic cells (Table 25.1). The terms are descriptive and denote nothing about phylogenetic relationships. For the ease of communication we use only the following terms. An amoeba is a cell that produces pseudopodia as it moves and feeds. Typically, the term amoeba is used unmodified for a cell that has one or a few nuclei. Amoebae in one group of mycetozoans may be distinctly different from amoebae in another group. A plasmodium is an amoeba with many nuclei and reticulated cytoplasmic strands. An amoeboflagellate is an amoeboid cell that produces one or more flagella when sufficient water is present. Some eumycetozoans have a life cycle in which an amoeboflagellate state is followed by a morphologically distinct amoeboid state that does not readily produce flagella. In that context, the latter state is called an obligate amoeba (Spiegel and Feldman 1985). Under harsh conditions, the amoebae of most mycetozoans can round up, produce cell walls, and

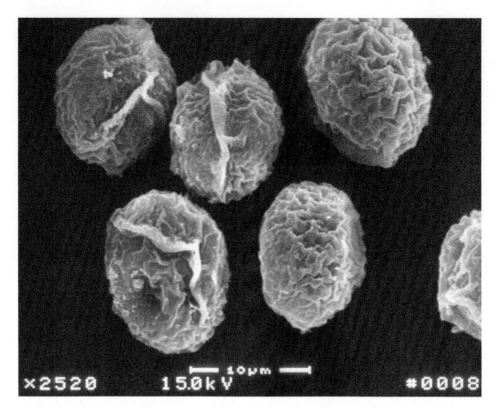

FIGURE 25.5 *Badhamia rhytidosperma* H.W. Keller and Schoknecht. SEM showing unique spore ornamentation: half of the spore surface with a wrinkled-reticulate pattern, and the remainder almost smooth with a ridge line marking the future site of germination. (Scale bar = 10 µm.)

TABLE 25.1

Terms Used for the Trophic Cells of Mycetozoans

Trophic state	Protostelids	Myxomycetes	Dictyostelids	Acrasids
Uninucleate amoeba	Amoeba	Amoeba	Amoeba	Amoeba
	Myxamoeba	Myxamoeba	Myxamoeba	Myxamoeba
	Amoeboflagellate	Amoeboflagellate		
	Obligate amoebae (some species)			
Flagellate cell	Amoeboflagellate	Amoeboflagellate	NA	Amoeboflagellate
	Swarm cell	Swarm cell		
	Myxoflagellate	Myxoflagellate		
Plurinucleate cell	Obligate amoeba	Protoplasmodium	NA	Plurinucleate amoeba (some species)
Multinucleate cell	Plasmodium	Protoplasmodium	NA	NA
	Obligate amoeba	Aphaneroplasmodium		
		Phaneroplasmodium		
Encysted trophic cell	Microcyst (amoeba; [plasmodium in some species])	Microcyst (amoeba) (plasmodium)	Microcyst	Microcyst
		Sclerotium of macrocysts		

NA, Not applicable.

become dormant microcysts. The plasmodia of myx-omycetes encyst to become sclerotia. The zygotic cysts of dictyostelids are called macrocysts.

Mycetozoans feed on bacteria, yeasts, spores, and hyphae of filamentous fungi, algae, and other protists in a variety of terrestrial habitats (Martin and Alexopoulos 1969; Olive 1975; Martin et al. 1983; Raper 1984; Cavender 1990; Frederick 1990; Spiegel 1990; Stephenson and Stempen 1994; Keller and Braun 1999; Krug et al., Chapter 21 of this book). Terrestrial habitats are defined as those that are not inundated with water and in which fruiting bodies are able to suspend spores in the air above the surface of the substratum. Mycetozoans have been found in various habitats worldwide. They are present in temperate, boreal, and tropical forests as well as in grasslands, deserts, and tundra (Martin and Alexopoulos 1969; Olive 1975; Keller 1982; Raper 1984; Stephenson and Stempen 1994). Mycetozoans are found in soil, leaf litter, herbivore dung, rotting logs (Figs. 25.8 and 25.9), the bark surface of living and dead trees, and dead aerial parts of plants, such as grass culms (see Fig. 25.4), old infructescences and inflorescences, and leaves (Table 25.2) (Martin and Alexopoulos 1969; Stevens 1974; Olive 1975; Raper 1984).

Mycetozoans usually are located on the outer surface of a substratum when their fruiting bodies are observed. The fruiting bodies of many myxomycetes and most species of the protostelid genus *Ceratiomyxa* (Fig. 25.10) are macroscopic, are often brightly colored, and can be seen readily in the field. Although anecdotal evidence suggests that heavy fruitings of some dictyostelids on dung occasionally may be observed in nature (Olive 1975), fruiting bodies of the rest of the mycetozoans are tiny and ephemeral. They are observed only when substratum samples are brought into the laboratory, maintained in moist chambers (Appendix I) or on agar plates, and examined with a stereomicroscope or compound microscope (Martin and Alexopoulos 1969; Olive 1975; Raper 1984).

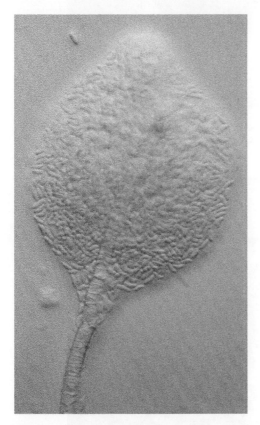

FIGURE 25.6   Sorus of *D. discoideum* (length of sorus approximately 0.5 mm). (Photo by Frederick Spiegel)

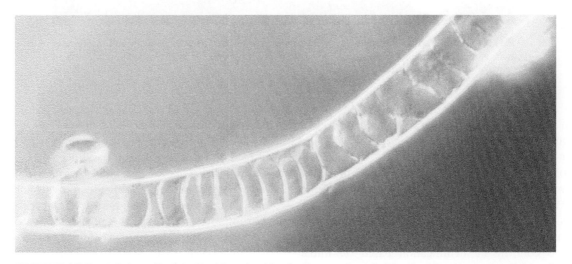

FIGURE 25.7   Cellular stalk of *D. discoideum* (width of stalk approximately 30 μm). (Photo by Frederick Spiegel)

FIGURE 25.8 *Physarum melleum* (Berkeley and Broome) Massee. Freshly formed yellow globose sporangia with white calcareous stalks. (Photo by Ray Simons)

FIGURE 25.9 *Physarum polycephalum* Schweinitz. Bright yellow phaneroplasmodium on water agar. Note the migrating, fan-shaped, feeding, anterior edge and the posterior trailing network of veins. (Photo by Samuel Ristich)

Moist chambers consist of Petri dishes or fingerbowls in which substrata are placed onto an appropriate source of moisture, either moistened paper or agar. The moist-chamber technique also works well for some larger species of myxomycetes when harsh environmental conditions prevent fruiting in the field (Braun and Keller 1977). Most species of mycetozoans can be identified on the basis of fruiting body morphology alone. However, it is often useful to culture protostelids, dictyostelids, and acrasids to confirm an identification based on initial observations of fruiting body morphology.

The presence of myxomycete plasmodia on a substratum or in moist chambers is often readily obvious. In the field, large plasmodia often are seen on rotting wood or feeding on wood-rotting basidiomycetes (Fig. 25.9). Plasmodia also may be located by the slime trails that they leave behind as they migrate. In moist chambers, plasmodia of all types are recognizable. Amoeboflagellates of myxomycetes swimming in a meniscus of water around a piece of substratum in moist chambers can be recognized when observed under the 10× objective of a compound microscope. However, species identification of myx-

FIGURE 25.10 The fruiting bodies of *Ceratiomyxa fruticulosa* on a rotting log (specimen approximately 1 cm long). (Photo by Frederick Spiegel)

TABLE 25.2
Substrata on which Mycetozoans Are Known to Occur

Group	ADPP	Litter	Soil	Bark	Dung	Rotting wood
Protostelids	+++	+++	+	+++	+	++
Myxomycetes	++	+++	+++	++	++	+++
Dictyostelids	*	++	+++	*	+++	*
Acrasis	+++	–	–	+	–	–
Pocheina	–	–	–	+++	–	*
Copromyxa	–	–	–	–	++	–
Copromyxella	–	–	+	–	+	–
Fonticula	–	+	–	–	+	–

ADPP, Attached dead plant parts; –, absent; *, very rare; +, occasional; ++, common; +++, abundant.

omycetes is usually impossible in the absence of fruiting bodies. The trophic cells of other mycetozoans are microscopic. Although, in principle, they should be recognizable with light microscopic examination of moist chambers, they are not usually distinct enough to be detected among the mixture of microorganisms present.

The presence of soil-inhabiting mycetozoans may be determined by plating suspensions of that substratum and a preferred food organism on agar and looking for plaques containing the mycetozoans of interest (Cavender and Raper 1965; Feest 1987; Stephenson and Cavender 1996). The rates at which protostelids and other small mycetozoans colonize substrata can be sampled with sterilized substrata that are introduced into a microhabitat. From one to several weeks later, the substrata are transferred to moist chambers and searched periodically over several weeks for fruiting bodies (Moore and Spiegel 1995). Any serious effort to sample the biodiversity of mycetozoans requires access to a laboratory equipped with compound and stereomicroscopes and microbiological culture facilities, including stock cultures of bacteria and yeasts that can be used as food (Stevens 1974).

TAXONOMY, DIVERSITY, AND DISTRIBUTION

EUMYCETOZOA

Olive (1975) postulated that the Eumycetozoa are a monophyletic group. Since then, numerous morphological studies have shown that the myxomycetes and most species described as protostelids are members of a monophyletic group (Spiegel and Feldman 1988; Spiegel 1990, 1991). In addition, molecular evidence from a portion of the sequence of the nuclear small subunit ribosomal RNA gene suggests that the dictyostelids, protostelids, and the myxomycetes may form a monophyletic group (Spiegel et al. 1995a) as does recent sequence data from the elongation factor 1-alpha (EF-1α) gene (Baldauf and Doolittle 1997; Baldauf et al. 2000). Although the phylogenetic relationships within the group are not entirely clear, mycologists generally agree that the Myxogastria and Dictyostelia each form a monophyletic group within Eumycetozoa and that the Protostelia form a paraphyletic group. F. W. Spiegel is in the process of revising the taxonomy of the group, but for convenience we will follow Olive's (1975) classification in this chapter. Under this system, Eumycetozoa is a class, and Protostelia, Dictyostelia, and Myxogastria are subclasses. The taxa of higher rank to which Olive assigned Eumycetozoa are artificial and do not need to be treated here. Except for particular examples of genera and species, we will not deal with lower level taxa within each group. Treatments of those taxa can be found in specialized publications listed in Table 25.3.

Protostelia

The protostelids include 37 described species in 17 genera (Spiegel 1990; Spiegel et al. 1995a). In addition, more than 50 undescribed species have been observed. All species of protostelids produce microscopic fruiting bodies characterized by a delicate, acellular stalk that supports one or a few spores. Their trophic cells are quite diverse (Spiegel and Feldman 1985; Spiegel 1990). All

TABLE 25.3
Important Recent Publications on Mycetozoans

Protostelids: Olive 1975; Spiegel 1990, 1991; Spiegel et al. 1995a*

Myxomycetes: Martin and Alexopolous 1969; Olive 1975; Martin et al. 1983; Frederick 1990; Keller and Braun 1999; Nannenga-Bremekamp 1991; Stephenson and Stempen 1994; Spiegel et al. 1995a*

Dictyostelids: Olive 1975; Raper 1984; Hagiwara 1989; Cavender 1990; Spiegel et al. 1995a*

Acrasids: Olive 1975; Raper 1984; Page and Blanton 1985*; Blanton 1990

* Publications dealing primarily with the phylogeny of mycetozoans.

species except the larger species of *Ceratiomyxa* must be detected by microscopy. The latter produce white or brightly colored slime columns that support the multiple sporocarps, and the entire structure is easily visible macroscopically.

There is some question whether all species so far described as protostelids are Eumycetozoa. Taxa in question include *Endostelium* species and *"Protostelium" pyriformis* (Spiegel 1990; F. W. Spiegel and J. Feldman, unpublished data), which we treat here as eumycetozoans for convenience. Phylogenetic analysis based on morphological characters suggests that there are two main lineages of Protostelia, one of which also includes the myxomycetes (Spiegel 1991; Spiegel et al. 1995a). Publications by Olive (1967, 1970, 1975, 1982) and Spiegel (1990) are useful for identifying protostelids. In addition, illustrations and up-to-date keys to the protostelids, based on fruiting body morphology and trophic-cell morphology, are available on the protostelid home page on the World Wide Web (Appendix III), compiled by F. W. Spiegel and D. D. Rhoads.

Most authenticated specimens are kept as cultures or as material embedded for electron microscopy. They are available from F. W. Spiegel, curator of the University of Arkansas Mycetozoan Culture Collection, which includes approximately 25 species. In addition, a few species are available from the American Type Culture Collection (ATCC; Appendix III).

Protostelids are located by detecting their fruiting bodies after examining substrata microscopically. Examination of substrata collected throughout the world indicates that the protostelids are probably ubiquitous (Olive 1975; Feest 1987; Spiegel 1990). The major collectors of protostelids from the 1960s through the early 1980s were L. Olive, C. Stoianovitch, and the students in Olive's laboratory at Columbia University, New York and the University of North Carolina, Chapel Hill. Subsequent collecting has been carried out by individuals from the laboratory of F. W. Spiegel. Protostelids have been collected from subarctic North America, temperate and desert areas of North America, the Caribbean, Brazil, Melanesia, Micronesia, Polynesia (including Hawaii), Japan, portions of southeastern Asia, England, and northern and central Africa. The most intensively collected areas are those within 240 km of New York City; in southwestern North Carolina, centering on the town of Highlands; within 32 km of the University of North Carolina, Chapel Hill; and in Tahiti, all by individuals from Olive's laboratory. Individuals from Spiegel's laboratory have collected extensively within 32 km of Oxford, Ohio; within 80 km of Fayetteville, Arkansas; and on the island of Hawaii. In addition, collecting has been carried out in a number of areas in Puerto Rico and Costa Rica (Moore et al. 1996; Stephenson et al. 1997b).

Although quantitative studies in which protostelids have been reported are limited (Best and Spiegel 1984; Feest 1987; Moore and Spiegel 1995, 2000a, 2000b, 2000c), certain conclusions can be drawn. Some species, such as *Protostelium mycophaga*, *Soliformovum irregularis*, *Nematostelium* species (see Fig. 25.3), *Ceratiomyxa fruticulosa*, and *Schizoplasmodiopsis pseudoendospora*, are very common; other species range from uncommon to rare. We hypothesize, however, that the great majority of species are cosmopolitan and that endemism does not exist. Our hypothesis is supported by the observation that almost all described species are present in all areas that have been collected intensively, although published collection records are few. Publication of information in the collecting notes of L. Olive and C. Stoianovitch would add greatly to our knowledge of the distribution of mycetozoans. Also, the wider use of a new technique that provides quantitative data (Moore and Spiegel 1995) will help to reveal differences in mycetozoan species assemblages that may exist among habitats and differences among mycetozoan species in biogeographic patterns of distribution.

Protostelids are found in a wide range of terrestrial habitats; however, assemblages of species in different microhabitats within the same habitat may be more dissimilar than the assemblages associated with similar microhabitats in different habitats (Olive 1975; Best and Spiegel 1984; Moore and Spiegel 1995, 2000a). Important microhabitats for protostelids include dead aerial parts of plants, litter, bark of living trees, bark and wood of dead trees and logs, soil, and herbivore dung (Baker 1975; Olive 1975; Best and Spiegel 1984; Feest 1987; Moore and Spiegel 1995). Olive (1975) was the first to note that certain species are present on bark or on dead aerial plant parts but tend not to occur in both microhabitats. Best and Spiegel (1984) supported this observation and noted that the species found on dead plant parts appear to be the same in different habitats. Likewise, in certain mesic and dry habitats, litter supports species that are uncommon on both dead aerial plant parts and/or bark but lacks some species that are common in those microhabitats (Moore and Spiegel 1995, 2000a). Ordination analysis (Moore and Spiegel 2000a) has shown that in northwestern Arkansas, differences in microhabitat have a greater influence on species distribution patterns than differences in habitat—that is, that species found in the litter microhabitat are distinct from those occurring in dead aerial plant-part microhabitat. Litter assemblages from grasslands are similar to litter assemblages from forests. The same pattern is true for the assemblages of species associated with dead aerial plant parts. However, in tropical rainforests in Puerto Rico, the distinction between litter and aerial microhabitats is less pronounced (Moore and Spiegel 2000b). In any event, investigators attempting to catalogue the biodiversity of protostelids in a habitat must sample each of the major microhabitats in which these organisms are found, preferably in proportion to their representation in the habitat.

Some qualitative predictions can be made concerning species presence in habitats in various ecosystems, although too few data are available to allow for quantitative predictions. If the microhabitats in forested habitats, both tropical and temperate, were sampled one would expect to find all or the great majority of the known species of protostelids in a single 50,000-hectare (500 km^2) plot. For example, several forest sites, a few hectares in size, on the island of Hawaii each have yielded more than 80% of the described species of protostelids in less than 40 collections (F. W. Spiegel and D. Hemmes, unpublished data). Similarly, all species except those that prefer bark or rotting wood are likely to be present in temperate grasslands. In arid grasslands and desert ecosystems, litter and bark species would be common, but fewer of the protostelids common on dead aerial plant parts would be found. It is also likely, based on an examination of the rate at which new species have been described, that numerous undescribed species will be found in any truly intensive survey for protostelids. From 1960 to 1984, Olive's group described 28 of the 37 known species, or 1.12 species per year. Although only three new species were described between 1985 and 1996, little effort was being directed toward collection and identification of specimens. During 1994 to 1996, Spiegel's group spent considerable time field collecting in the Fayetteville, Arkansas area; at least five undescribed species were observed (one has been isolated and named; Spiegel et al. 1995b). Recent surveys of Hawaiian protostelids have yielded evidence of more than 50 undescribed species (F. W. Spiegel and D. Hemmes, unpublished data). It is doubtful, therefore, that the full range of protostelid diversity presently is known.

Myxogastria

The myxomycetes, or plasmodial slime molds, have been known from their fruiting bodies since at least the mid-seventeenth century, and their life history has been understood for more than a century (Martin and Alexopoulos 1969). Around 800 species have been described (Hawksworth et al. 1995), approximately 100 of them in the last 25 years. The subclass typically is divided into five orders, excluding the Ceratiomyxida, which we treat as a protostelid (Martin and Alexopoulos 1969; Olive 1975). Neither the phylogenetic relationships among the orders nor those among the families in each order are known. Keller (1996) reviewed current taxonomic practices, importance of proper collecting and field observations, and spore–to-spore cultures as they relate to future studies in the myxomycetes.

All species of myxomycetes produce spores that germinate as amoeboflagellates. Feest (1987) suggested that the amoeboflagellate is probably the trophic state in which the majority of myxomycetes spend most of their lives. Eventually, the amoeboflagellate develops into a plasmodium (Fig. 25.11). This process can result from gametic fusion between compatible amoeboflagellates or it can be apomictic (Collins 1980, 1981). Some morphologically defined species encompass several cryptic, sexual species and a number of apomictic species (Clark 1995). Macroscopic plasmodia with prominent veins showing strong shuttle streaming are termed phaneroplasmodia and often can be seen in the field. Large reticulate plasmodia that have delicate, transparent veins and erratic streaming are referred to as aphanoplasmodia, and rounded, irregular to circular microscopic plasmodia without shuttle streaming are known as protoplasmodia (Alexopoulos 1960). All plasmodia, regardless of type, eventually develop into one or more fruiting bodies with multiple spores.

In sexual species, meiosis occurs in the spores after the spore wall matures (Aldrich 1967). The basic terminology used to describe the various types of fruiting bodies is available in Martin and Alexopoulos (1969), Keller (1982), Stephenson and Stempen (1994), and Keller and Braun (1999) or on The Internet Guide of Myxomycetes home page on the World Wide Web (Appendix III), compiled by S. L. Stephenson and D. Binion. In brief, small, discrete fruiting bodies are termed sporangia (Figs. 25.12 and 25.13). They may be stalked or sessile. Partially fused sporangia are referred to as pseudoaethalia; they are usually sessile, but in a few species they are stalked. Plasmodium-shaped fruiting bodies are called plasmodiocarps (Fig. 25.14), and massive fruiting bodies are known as aethalia (Figs. 25.15 and 25.16). A given species may produce more than one type of fruiting body in a single fructification. For example, *Physarum bivalve* can form both plasmodiocarps and sporangia. The type of fruiting body, the presence and type of capillitia, the

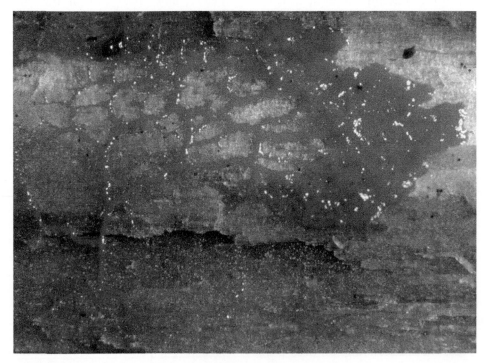

FIGURE 25.11 *Physarum roseum* Berkeley and Broome. Bright red phaneroplasmodium (specimen 1 cm across). (Photo by Ray Simons)

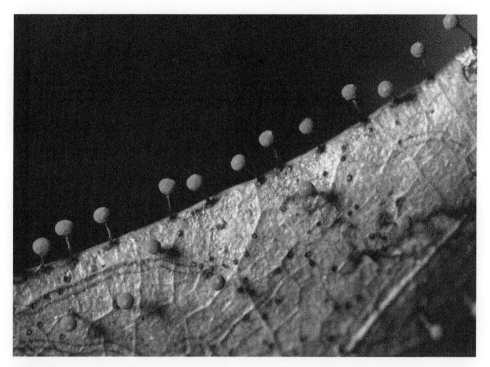

FIGURE 25.12 *Physarum roseum* Berkeley and Broome. Habit of stalked, bright red sporangia on decaying leaf (individual sporangium 0.4 mm in diameter). Note in the left foreground the black parallel lines of the plasmodial vein. Plasmodial tracks are often present on the surface of decaying leaves found in deep leaf litter. (Photo by Ray Simons)

FIGURE 25.13 *Badhamia rhytidosperma* H.W. Keller and Schoknecht. Habit on dung culture. Note chalky-white, calcareous fruiting bodies (1–2 mm across) forming widely spaced, irregular, gyrose-confluent lobes. (Photo by H.W. Keller)

FIGURE 25.14 *Didymium perforatum* Yamashiro. Habit of plasmodiocarp appearing labyrinthiform or closely reticulate. This rare species, only known from Japan, Pakistan, and Kansas, has an iridescent peridium sprinkled with pale yellow calcareous crystals. (Specimen 2 cm across.) (Photo by Ray Simons)

FIGURE 25.15 *Fuligo megaspora* Sturgis. Habit on twig. Chalky-white, calcareous aethalium (2 cm long). This species is frequently found in the Everglades National Park, Florida. (Photo by H.W. Keller)

color and structure of the spores, and the presence or absence of calcium carbonate in the fruiting body are characteristics useful for identifying myxomycetes to species (Martin and Alexopoulos 1969).

Most species are thought to be cosmopolitan, although some species may be restricted to the tropics and subtropics and some may be strictly temperate. Also, some species, although widespread, are more typical of some parts of the world than others (Schnittler and Stephenson 2000; Stephenson et al. 2000), and at least a few species are limited to particular types of habitats. Such is the case for *Barbeyella minutissima*, which

FIGURE 25.16 *Enteridium splendens* (Morgan) Macbride var. *juranum* (Meylan) Härkönen. Aethalium broken open to expose coppery, spore mass. (Specimen 2 cm in diameter.) (Photo by H.W. Keller)

appears to have a distribution restricted to montane spruce and spruce-fir forests of the Northern Hemisphere (Schnittler et al. 2000). An example of a species common in the Old World but not the New World is *Erionema aureum* (Hagiwara and Yamamoto 1995).

Fruiting bodies of myxomycetes are encountered in a number of microhabitats, including rotting wood, dead aerial plant parts, litter, soil, the bark surface of living trees, and dung, wherever these microhabitats are found. Each microhabitat appears to support a distinct assemblage of species (Stephenson 1989). The greatest diversity of myxomycetes is known from temperate forests, although these organisms also inhabit deserts, grasslands, alpine areas at high elevations, coniferous forests and tundra at high latitudes, and tropical forests (Alexopoulos 1963; Farr 1976; Blackwell and Gilbertson 1980; Martin et al. 1983; Stephenson et al. 1992; Stephenson and Laursen 1993, 1998; Stephenson and Stempen 1994; Novozhilov et al. 1999; Schnittler and Stephenson 2000; Stephenson et al. 2000). T. Brooks, H. W. Keller, and their coworkers have studied the distribution of corticolous myxomycetes in temperate North America extensively (Brooks 1967; Keller and Brooks 1973, 1975, 1976a, 1976b, 1977; Brooks et al. 1977; Keller 1980; Keller and Braun 1999).

It is not clear that the trophic stages of a given species are active in all the microhabitats in which its fruiting bodies are found because plasmodia can migrate some distance before fruiting (Feest 1987). For example, a species that fruits on litter or at the base of a tree may feed in the soil. However, if pieces of substratum are incubated in moist chambers for several days with no sign of plasmodial slime trails, and if fruiting then occurs, it is safe to conclude that trophic cells were active on the substratum because any migrating plasmodia present would have been expected to fruit in the moist chamber within a day of plating. That behavior is especially true for species with protoplasmodia or small phaneroplasmodia or aphanoplasmodia.

The large number of species known from temperate regions likely reflects the concentration of collector activity more than the actual abundance of species. The myxobiota of the temperate United States, for example, has been collected intensively throughout the past century by T. H. Macbride and G. W. Martin in Iowa; R. Hagelstein in New York and Pennsylvania; A. P. Morgan, K. L. Braun, and H. W. Keller in Ohio, Arkansas, and Kansas; T. E. Brooks in Arkansas, Kansas, and Kentucky; B. A. Branson in Kentucky; W. C. Sturgis and D. H. Mitchel in Colorado; C. J. Alexopoulos in Texas; D. T. Kowalski in California; S. L. Stephenson in Virginia, West Virginia, and Montana; and many others too numerous to mention. Even so, some states, such as Alabama, Missouri, and Oklahoma, still lack any published lists of myxomycetes. Some portions of Europe, particularly Britain, the Netherlands, Germany, France, Sweden, and Spain, also have been collected intensively. The New and Old World Tropics are still relatively poorly known, although studies in Puerto Rico in 2001 (Novozhilov et al. 2001), Costa Rica in 2000 (Schnittler and Stephenson 2000), and southern Mexico in 1999 (Lado et al. 1999) have provided a considerable body of information on the myxomycetes of at least some Neotropical regions. Myxomycetes have been reported to be less abundant and diverse in tropical forests than in temperate forests (Alexopoulos 1970), and the data accumulated thus far suggest that such is the case in at least some situations. For example, myxomycetes are certainly uncommon in tropical forests at the wet end of the moisture gradient (Schnittler and Stephenson 2000). It is possible that the delicate fruiting bodies of myxomycetes disappear more rapidly in the tropics than in temperate regions, perhaps because of high rainfall and the high incidence of myxomyceticolous fungi and myxomycetophagous invertebrates (Alexopoulos 1970; Farr 1976; Blackwell 1984; Wheeler 1984; Rogerson and Stephenson 1993). Other hypotheses suggest that tropical myxomycetes spend more time in the trophic state than do temperate myxomycetes so fruiting bodies occur with low frequency.

Most collections of myxomycetes are preserved in herbaria as specimens of dried fruiting bodies. The fruiting bodies usually are stored in sturdy cardboard boxes with the substratum glued in place to prevent movement that could damage the specimen. Fruiting bodies also may be stored as permanent slide mounts (Fig. 25.17) (Sundberg and Keller 1996). The U.S. National Fungus Collections (Appendix III) have the single largest collection of myxomycetes, with 50,000–60,000 specimens (Keller 1996). Other large collections include those at the New York Botanical Garden, the Academy of Natural Sciences in Philadelphia, the Fairmont State College herbarium (FWVA), the personal collection of H. W. Keller, and the cryptogamic herbarium of the University of Florida in Gainesville (Appendix III). Relatively few myxomycetes are kept in culture, and most of those are species in the orders Physarida, Stemonitida, and Echinosteliida. Many of those species are available from the ATCC (Appendix III) (Keller 1996). Cultures of some species are also available from a number of researchers, including J. Clark at the University of Kentucky, Louisville, and E. F. Haskins at the University of Washington, Seattle.

Collection data on myxomycetes to date reflect more the activity of collectors than the actual distribution of species. Relatively few quantitative studies exist on the distribution of these organisms in nature. Some studies, such as those of myxomycetes in soil by Feest and coworkers (Feest 1987), have provided data on the numbers of trophic cells in soil but not on the numbers of taxa present simply because most of the collections did not fruit. Nonetheless, it is clear that myxomycetes are major components of agricultural soils. Studies of myxomycete biodiversity are limited in scope and geographic range (Stephenson et al. 1993). Based on results from a 5-year study of the myxomycetes occurring in temperate forests in the Mountain Lake area of Virginia (Stephenson 1988, 1989), however, it seems likely that a 50,000-hectare plot would contain about 250 species, or approximately 80% of the species reported for eastern North America. Any sampling effort should focus on all the microhabitats in which myxomycetes are found and should extend over a period of several years to ensure that all rare species are found and that species common in one year and scarce the next are recorded. A 50,000-hectare plot in the tropics certainly would yield fewer species because some of the microhabitats (e.g., the bark surface of living trees) that support diverse assemblages of myxomycetes in temperate forests are characterized by relatively few species in the tropics (Schnittler and Stephenson 2000).

Dictyostelia

Dictyostelids are a relatively homogeneous group of cellular slime molds consisting of three well-known genera and 71 described species. The majority of the known species have been described since 1970. In addition, about a dozen undescribed species in culture (J. C. Cavender, unpublished data) currently are being described by H. Hagiwara, J. C. Cavender, and various coworkers. All dictyostelids are characterized by having uninucleate cells with a reticulate, peripheral nucleolus (Olive 1975; Raper 1984; Cavender 1990). Amoeboid trophic cells, with acutely pointed pseudopodia, differentiate into aggregating cells that migrate in streams to an aggregation center. Aggregation is stimulated by chemical attractants called acrasins. Examples of specific acrasins are cyclic adenosine monophosphate, in *Dictyostelium discoideum*, and the dipeptide, glorin, in *Polysphondylium violaceum*. The multicellular aggregation, or pseudoplasmodium, develops into one or more elongated slug-shaped structures (each of which is called a slug, or grex) (Figs. 25.18 and 25.19) that may migrate in some species or transform directly into a mature fruiting body. A migrating pseudoplasmodium secretes a protective slime sheath and, in some species, a stalk. The slug of the well-studied *D. discoideum* has been demonstrated to be sensitive to light, heat, relative humidity,

FIGURE 25.17 *Macbrideola synsporos* (Alexopoulos) Alexopoulos. Habit of holotype specimen. Sporangium stalked, tapering upward and continuing into the middle of the spore case as a columella. Capillitial threads arising from the tip and sides of the columella forming a globose surface network with few free ends. Spores clustered, adhering in groups of 8 to 16. (Total height = 0.4 mm.) (Photo by H.W. Keller)

and osmotic forces (Raper 1984). The fruiting body consists of a stalk that may display pseudobranching and one or more sori of spores. The spores of most species are elliptical or oblong, but a few species produce spherical spores. The stalk in members of the genus *Acytostelium* is acellular, whereas stalks in members of the other two genera are filled with dead stalk cells, which are polyhedral and cellulosic. *Dictyostelium* species have a stalk that is unbranched or irregularly branched, whereas *Polysphondylium* species have regularly spaced whorls of pseudobranches along the central main stalk. Spore walls and stalk sheaths also contain cellulose. Aggregation and fruiting represent an asexual dispersal process, but sex is known for many species (Raper 1984; Cavender 1990).

Zygotes form from the fusion of two compatible, gametic amoebae. The zygotes develop into giant cells, which become aggregation centers that attract several hundred amoebae. The centers lay down a thick, multilayered, cellulose cell wall and become macrocysts. The giant cell devours and digests all the amoebae in the macrocyst and enters a period of dormancy. Meiosis occurs prior to macrocyst germination, and the products of meiosis divide mitotically several times to produce a population of several tens of amoebae that escape from the germinating macrocyst. Species may be heterothallic or homothallic.

Dictyostelids are found in the soil worldwide, particularly in the surface humus layers (Cavender 1973, 1990;

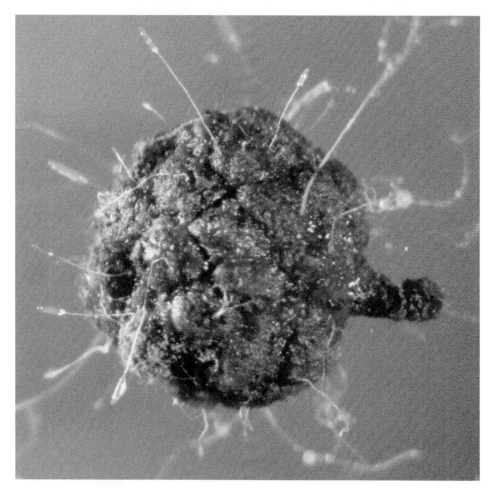

FIGURE 25.18   Numerous slugs of *Dictyostelium* species migrating onto agar from a flower bud. (Specimen approximately 1 mm long.) (Photo by Frederick Spiegel)

Raper 1984; Feest 1987; Hagiwara 1989; Landolt and Stephenson 1990). They also are found very commonly on dung, especially herbivore dung, although no species appear to be strictly coprophilic. Dictyostelids also are common in litter (Stephenson and Landolt 1995). F. W. Spiegel (unpublished data) has isolated *D. mucoroides* once from bark, and D. L. Moore (unpublished data) has found them on dead aerial plant parts on a few occasions. The canopy soil microhabitat of tropical rain forests also supports an assemblage of dictyostelids (Landolt and Stephenson 1997; Stephenson and Landolt 1998). Although dictyostelids often invade dung opportunistically from the soil, Suthers (1985) demonstrated that ground-feeding birds can transport spores of tropical species to temperate regions and deposit them in their feces. Ground-dwelling rodents and amphibians are known to disperse spores in temperate forests (Stephenson and Landolt 1992) and probably play the same role in tropical forests. In soil and litter, spores can be transported by invertebrates such as nematodes and small arthropods and/or by water (Feest 1987). Huss (1989) has shown that earthworms, in particular, are a source of dictyostelid propagules. Mites that contaminate cultures have been observed to carry dictyostelids from one culture plate to another (F. W. Spiegel, unpublished observation). Soil and litter arthropods that are allowed to migrate across plates of agar carry numerous dictyostelids, which fruit on the agar (G. Mecum and F. W. Spiegel, unpublished observation).

Dictyostelids are more common in forest soils than in agricultural soils, grassland soils, or deserts (Raper 1984; Feest 1987; Cavender 1990). In fact, the differences between forest soils and soils from a disturbed environment in diversity and dominant species of dictyostelids are striking (Hammer 1984; Cavender et al. 1993). Hammer (1984) found greater densities and frequencies of *Dictyostelium mucoroides* and *D. sphaerocephalum* in pastures and hay fields than in forests. More species are found at lower latitudes than at higher latitudes, and at a given latitude, more species are found at lower eleva-

FIGURE 25.19 Individual slug of *Dictyostelium* sp. (Slug approximately 1 mm long.) (Photo by Frederick Spiegel)

tions than at higher elevations. More clones of dictyostelids are present in moist soils than in dry soils, although they are rare in saturated soils. Some species appear to be strictly tropical; others are strictly temperate; and others, although cosmopolitan, are more common in either tropical or temperate regions (Raper 1984; Swanson et al. 1999; J. C. Cavender, unpublished data). Distribution patterns worldwide are not dissimilar from those of higher plants and probably are determined by the same factors (Swanson et al. 1999). The highest diversity of dictyostelids has been reported from Neotropical rain forest soils (Vadell et al. 1995). Most species are strictly bactivorous, but one species, *D. caveatum*, preys on other species of dictyostelids by invading their aggregates and ingesting their cells (Waddell 1982).

Much of the worldwide collecting of dictyostelids has been carried out by J. C. Cavender and coworkers or by S. L. Stephenson and J. C. Landolt. In addition, H. Hagiwara has collected extensively in Japan and Nepal, and A. Feest and H. Hodgson have collected dictyostelids in Britain. Dictyostelids are collected by culturing them from soil, litter, and dung. Proper identification requires that dictyostelids be cultured. They can be grown on weak nutrient agar in conjunction with bacteria, usually *Escherichia coli*, as a food source. Cultures are preserved by lyophilization and can be acquired from ATCC. J. C. Cavender, curator of the Kenneth B. Raper Cellular Slime Mold Culture Collection (The University of Ohio, Athens), and J. C. Landolt (Shepherd College in West Virginia) can supply cultures of species not available from ATCC (Appendix III).

Dictyostelids range in abundance from a few 10s of clones per gram of soil to several thousand (Raper 1984; Feest 1987; Cavender 1990). If investigators sampled a 50,000-hectare plot in a temperate deciduous forest in eastern North America adequately over several seasons, they could expect to find 15–20 species (Cavender 1990). In some Neotropical rain forests, one might expect to find as many as 30 or more species in a plot of similar size (Vadell et al. 1995). Smaller numbers of species would be expected from plots in grasslands, deserts, boreal and montane forests, and tundra. Dictyostelids are most abundant during spring and fall, with fewer clones isolated in summer and winter (Raper 1984; Feest 1987; Cavender 1990).

ACRASIDS

All acrasid cellular slime molds have amoebae that aggregate to form fruiting bodies directly. Some species also have a flagellated form that germinates from the spore (Olive and Stoianovitch 1974; Olive et al. 1983; Page and Blanton 1985; Blanton 1990; F. W. Spiegel, unpublished data). The genera *Acrasis*, *Pocheina* (*Guttulina*), and *Guttulinopsis* have limax (monopodial) amoebae, mitochondria with platelike cristae, and fruiting bodies with viable stalk cells as well as spores (Olive 1975; Dykstra 1977; Raper 1984; Blanton 1990). They are classified as fruiting members of the amoeboid taxon Heterolobosea, which also includes the nonfruiting limax amoebae and amoeboflagellate genera *Naegleria*, *Tetramitus*, and *Vahlkampfia* (Page and Blanton 1985; Blanton 1990; Baldauf et al. 2000). The genera *Copromyxa* and *Copromyxella* have limax amoebae and tubular mitochondrial cristae and form simple fruiting bodies consisting entirely of sorocysts (Olive 1975; Dykstra 1977; Spiegel and Olive 1978; Raper 1984; Blanton 1990). The phylogenetic affinities of these genera are not known. The unusual, monotypic genus *Fonticula* has amoebae with acutely pointed subpseudopodia, mitochondria with platelike cristae, and a fruiting body with a cellular stalk tube through which the cells of the aggregate move to form a sorus of spores (Dykstra 1977; Raper 1984; Blanton 1990).

The primary resources for identifying acrasids are the monographic works of Olive (1975) and Raper (1984). Additional material can be found in Blanton (1990).

Cultures of *Acrasis rosea* are available from the University of Arkansas Mycetozoan Culture Collection in Fayetteville. *Acrasis granulata* has not been seen since it was described (Olive 1975; Raper 1984), and no cultures of *Pocheina* species are available. Cultures of *Guttulinopsis nivea* are available from the Kenneth B. Raper Cellular Slime Mold Collection (Appendix III). Finally, *Fonticula alba* is available from ATCC.

Acrasis rosea is widespread and relatively common, although it is more common in some years than it is in others in a given area (Olive 1975). For instance, in northwest Arkansas, *A. rosea* was present on 10–15% of the collections of dead aerial plant parts made during the summer of 1993, but the species was found fewer than five times on similar collections in three subsequent summers (F. W. Spiegel and D. L. Moore, unpublished observation). It is found primarily on dead aerial plant parts and appears to be present wherever protostelids occur. An unnamed flagellated species of *Acrasis* (see Fig. 25.4) has been found once on hickory (*Carya ovata*) bark in central Iowa (F. W. Spiegel, unpublished data). *Pocheina* species are found primarily on conifer bark (Olive 1975; Olive et al. 1983; Raper 1984; Blanton 1990). *Guttulinopsis* species are relatively common on herbivore dung (Olive 1975; Raper 1984; Blanton 1990); *G. nivea* is also common in soil at Tikal in Guatemala, where it may be spread by monkeys (J. C. Cavender, unpublished observation). *Copromyxa* species, some species of *Copromyxella*, and *Fonticula alba* also are coprophilous, although they are much less common than species of *Guttulinopsis* (Olive 1975; Raper 1984; Blanton 1990). Other species of *Copromyxella* infrequently are isolated from soil (Raper 1984; Blanton 1990). As is evident from this paragraph, little information is available on the acrasids. It certainly should be possible to sample their diversity in greater detail, however, using the sampling techniques discussed in the next section.

ENSURING ADEQUATE BIODIVERSITY INVENTORY

Sampling the biodiversity of most mycetozoans is relatively inexpensive, inasmuch as little or no complex equipment is needed. The only costly aspect of such projects is travel if areas to be sampled are far from the investigator's laboratory. Maintenance of herbarium or culture collections of myxomycetes or mycetozoans involves additional costs.

As mentioned in the section on "Taxonomy, Diversity, and Distribution," earlier in this chapter, mycetozoans are found in a variety of habitats throughout the world and in a variety of microhabitats within each habitat (Table 25.2). To survey the mycetozoans in a given habitat, each of the microhabitats in that habitat must be sampled. If field sites are not near an adequate processing laboratory, samples can be stored and processed the following day. If collecting sites are located at greater distances, samples must be air-dried in the field before they are returned to the investigator's laboratory. In all instances a compound microscope, a stereomicroscope, and a laboratory equipped for basic mycological culturing are required for processing. In the following treatment we will discuss sampling, collecting, and isolation (where pertinent) protocols for the different microhabitats and the various mycetozoans found in each. These techniques can be used a number of ways. If one is interested primarily in a preliminary survey, numerous samples may be taken at one time and examined to get a rough estimate of the species present in an area. At the other extreme, if one is interested in monitoring the mycetozoan community in detail, one can establish a sampling protocol that allows rigorous statistical analysis of the data obtained. In either case, the basic methods of preparing and examining the samples are the same.

DEAD AERIAL PLANT PARTS

This microhabitat includes a broad range of substrata that are found on standing plants and consist of dead or moribund, nonwoody tissues. Prominent examples include herbaceous stems and leaves and old inflorescences and infructescences on both woody and herbaceous plants. In general, such substrata are collected when they are brown and show signs of fungal decay (Olive 1975; Spiegel 1990). They support many protostelids, *Acrasis rosea*, and many of the species of smaller myxomycetes (Martin and Alexopoulos 1969; Olive 1975; Best and Spiegel 1984; Blanton 1990; Spiegel 1990). In addition, some of the species of larger myxomycetes may migrate onto standing dead (and living) plants to fruit. It is unlikely, however, that the latter species are trophically active on such substrata. Dictyostelids, some protostelids, many myxomycetes, and all other acrasids are extremely rare in this microhabitat.

Dead aerial plant parts can be collected at any time of year and from any habitat. They can be plated out immediately or dried and stored in a cool dry place for periods of more than a year. If substrata are stored moist, they will be overgrown by fungi, and mycetozoans will be difficult to locate. Each collection should be made from a single plant; given an identification number; stored in a paper bag, plastic bag, or glassine envelope; and documented with geographic location, habitat, microhabitat, collecting date, and other pertinent information. Mycetozoans on these substrata are microscopic and cannot be seen directly. Even direct examination of fresh substrata with a stereomicroscope or a compound microscope may not reveal any mycetozoan fruiting bodies. Therefore, substrata should be plated out using the technique described in Olive (1975) or some variation of it (Spiegel 1990), as outlined as follows:

1. Pieces of substratum are cut into sections 1- to 3-cm long with a sterile blade and soaked in sterile distilled water for 10–30 minutes. This is the most commonly used method for finding protostelids and is most conducive for quantifying them. If one desires only to determine if species are present in a sample, then the substratum may be crushed into small pieces and suspended in water. Soaking may be omitted to reduce risk of displacing clones of mycetozoans. In that case, the substrata will absorb moisture from the agar when plated.

2. The pieces are placed in a Petri dish containing either water agar or a low-nutrient agar such as weak malt extract–yeast extract (wMY), hay infusion (HI), or oak-bark agar (OBA; see Appendix II). The lack of nutrients will prevent the substratum from being overgrown by fungi and bacteria. The pieces of substratum may be drained of excess water, or they may be placed directly on the agar. If crushed and suspended, aliquots of the substratum can be poured directly onto the plate.

3. After 2 or 3 days under ambient conditions, the pieces of substratum are examined for fruiting bodies with a compound microscope at low power. Once an investigator has developed some skill with these mycetozoans, samples can be screened with a good-quality stereomicroscope.

4. The plated substrata may be observed every 2 or 3 days for up to 3 weeks using a compound microscope or good stereomicroscope, and the species present are recorded. If large numbers of samples are being examined, they can be observed once after 5–7 days and again after 10–14 days with little chance of missing any species. Species of protostelids may be identified on the basis of fruiting body morphology by using keys in Olive (1975), illustrations in Spiegel (1990), or the keys and illustrations on the protostelid home page (Appendix III). Myxomycetes can be identified using Martin and Alexopoulos (1969), Martin and colleagues (1983), Stephenson and Stempen (1994), or Keller and Braun (1999). On occasion, species identification can be confirmed by careful examination of the site where each mycetozoan is fruiting to determine if trophic cells or prespore cells typical of the species in question are present.

If detailed enumeration of clones is necessary, each continuous patch of fruiting bodies of mycetozoans is treated as a single clone. As a rule, two patches of the same species that are separated by one or more fields of view as observed with a 10× objective (approximately 0.2 mm) are considered to be discontinuous and thus to represent separate clones. The location of each clone should be marked the first time it is observed so that it will not be subsequently counted as a new clone.

Some species have ephemeral sporocarps that disappear shortly after they mature. Many such species fruit only at night or for a short period early in the morning. Therefore, samples of substratum should be examined during the day and at night so that such species are not overlooked (Olive and Stoianovitch 1977, 1981; Spiegel 1984). Late-fruiting mycetozoans occasionally are found up to 6 weeks after plating.

New species of mycetozoans and those that are difficult to identify directly from their fruiting bodies should be isolated and examined in detail. The spore-touch technique is the most useful for isolating most mycetozoans that fruit on dead plant parts (Olive 1975; Spiegel 1990). In that technique, spores are located microscopically, retrieved with a sterile needle, and transferred to a plate of fresh agar that has been streaked with a potential food organism. Some of the bacteria and yeasts present in the isolation plate are likely to be transferred along with the spores. Dragging spores across the surface of the agar with a micromanipulator needle will remove surrounding microorganisms (Olive 1975). Initial isolations should include a number of spores; single-cell isolation should be carried out only after the mycetozoan is in culture. As a rule, mycetozoans from dead aerial plant parts and litter are relatively easy to isolate and culture on wMY or HI agar (Olive 1975; Spiegel 1990; Appendix II). Some of the smaller mycetozoans, particularly species of *Echinostelium*, have been recovered and then cultured with considerable success on plates of soft agar, with or without an infusion of plant parts (Haskins et al. 1983; Haskins and McGuinness 1986; Appendix II). Once a mycetozoan is isolated, its amoebae and spores should be flooded with sterile distilled water to determine whether amoeboflagellates are produced.

Myxomycetes may appear as plasmodia in moist-chamber agar plates. Those plasmodia can be transferred to water-agar plates with sterile oat flakes and then transferred again to fresh plates each week. Getting isolated plasmodia to fruit in culture is often difficult, which makes identification difficult (Feest 1987). It is sometimes possible to induce fruiting by providing a plasmodium with bits of sterile straw or filter paper to which it can migrate from the agar. Other methods that have been used to culture plasmodia or induce fruiting are described by Wollman (1966) and Clark and Stephenson (1994).

Two techniques can be used to quantify the rate of colonization of a substratum by mycetozoans. In a study of the succession of protostelids on a substratum, S. Miller (unpublished data) recorded the sequence of protostelids on the tassel florets of corn (*Zea mays*) from anthesis until the plant fell to the ground about 12 weeks later. He showed that protostelids were present from the time the trichomes on the florets disappeared 1–2 weeks postanthesis until the plant hit the ground. Species such as *Protostelium mycophaga* and *Soliformovum irregularis* appeared first and were followed by species such as *Schizoplasmodiopsis pseudoendospora* and *Schizoplasmodium cavostelioides* in later weeks. Clones of *P. mycophaga* were most abundant during the first 2 weeks it was present; fewer clones were found in each subsequent week. In contrast the number of clones of *S. irregularis* remained constant throughout the study. Other species were abundant for 2–3 weeks and then disappeared.

Colonization rates by mycetozoans also can be measured using a technique developed by Moore (Moore and Spiegel 1995). Sterilized wheat straws, approximately 5-cm long, are hung at various heights above the ground from monofilament fishing line strung between aluminum poles on vegetation. They are left in the field for periods of various lengths and then collected. The collected straws are plated and examined for the presence of mycetozoans clones. Comparison of these data with similar data obtained for naturally occurring substrata in the same microhabitat has demonstrated that all the species present on naturally occurring substrata appear to colonize the straw. Therefore, the straws replicate natural substrata in a given dead aerial plant microhabitat. Because the substratum being used is standardized and always available, similarities and differences in mycetozoan communities occupying the same type of microhabitat but in different habitats or occupying different microhabitats within a single habitat can be measured quantitatively. This method also allows for the comparison of colonization rates in different seasons (see "Protostelia" under "Determination of Relative Importance and Abundance," later in this chapter).

LITTER

Litter consists of recognizable dead plant parts that have fallen to the ground. Initially litter may be identical to dead aerial plant parts, but the two microhabitats are colonized by different assemblages of species (Moore and Spiegel 1995). Some protostelids are common in litter and rare on dead aerial plant parts. In temperate regions, many more myxomycetes are found in litter than on dead aerial plant parts, although the species fruiting on litter

may be trophically active in the soil (Feest 1987). In tropical forests, myxomycetes appear to be relatively more common on dead aerial plant parts (Edmunds and Stephenson 1996). Dictyostelids are not uncommon in decomposing litter (Raper 1984; Feest 1987), and there are a few reports of *Acrasis rosea* in litter (Raper 1984). The very rare acrasid, *Fonticula alba,* has been found on litter containing isopod droppings (G. Mecum and F. W. Spiegel, unpublished data).

Macroscopic myxomycetes can be observed in and collected directly from litter during warm, moist periods. In temperate North America, collecting is best between the last frost of spring and the first frost of autumn. Optimal conditions for collecting freshly matured fruiting bodies occur 24–72 hours after abundant rain, at temperatures ranging from 20°C to 35°C. Fruiting bodies of many species are visible from some distance; others are located only by systematic examination of bits of litter with a good quality, 10- to 25-× hand lens. Substrata with fruiting bodies attached should be dried, mounted with glue in small cardboard boxes, and stored in a cool dry place. Some practice, such as freezing the collection for several days before storing, should be used to control insect damage (Martin and Alexopoulos 1969; Sundberg and Keller 1996; Keller and Braun 1999). Older techniques, such as storing collections with insecticides, no longer are recommended.

Methods for sampling litter for microscopic mycetozoans are essentially identical to those used for dead aerial plant parts. Pieces of litter can be soaked and plated out, or sterilized straws may be introduced into the litter (see "Dead Aerial Plant Parts," earlier) and anchored into place with stainless-steel straight pins (Moore and Spiegel 1995). Pins with colored heads are easier to locate in the litter than pins with plain heads (G. Bryant, unpublished data). To characterize the biodiversity of a given area, all habitats and microhabitats in the area must be sampled. If the resulting data are to be analyzed statistically, then an appropriate sampling scheme must be used. Stephenson (1989) collected samples of forest-floor litter at regular intervals along a transect extending the entire length of a given study area. He placed each sample in a sterile plastic bag (a paper bag should be used if the sample is to be stored for any length of time) and returned it to the laboratory, where it was subdivided and used to prepare two or three moist-chamber cultures. It also is possible to transport Petri dishes already lined with filter paper into the field and add individual samples of litter to prepare each culture. In that way, the sample placed in each culture is obtained from a different microsite on the forest floor.

The so-called Cavender method for sampling dictyostelids may be applied to litter as well (see "Soil," next subsection). The major difference between sampling soil and sampling litter is that litter must be finely ground and suspended in water with the appropriate food microorganisms, whereas soil need not be.

SOIL

Soil, including the humus layer, is the primary microhabitat for dictyostelids (Raper 1984; Feest 1987; Cavender 1990) and many myxomycetes (Feest 1987). In addition, a few species of protostelids (Feest 1987) and some members of the genus *Copromyxella* (Raper 1984; Blanton 1990) can be isolated from soil. Investigators rarely detect mycetozoans directly in soil; rather, their presence is revealed only after they have been induced to fruit in the laboratory.

Investigators isolate mycetozoans from soil samples using a number of techniques. They can sprinkle soil onto a plate of water agar, wMY agar, HI agar, or some other weak nutrient agar that has been streaked with the appropriate food organisms (Appendix II). *E. coli* is used most often as food for cellular slime molds (Raper 1984; Feest 1987). More often, researchers use some variation of the "Cavender Method," which is detailed in Cavender and Raper (1965), Raper (1984), and Cavender (1990).

In brief, the Cavender method involves collecting soil from a number of sites in a given habitat and diluting and suspending a set mass of soil from each sample in a set volume of sterile distilled water. Two dilutions are made to rid the suspension of factors that may be present in some soils and inhibit mycetozoan growth. A set volume of this suspension is spread evenly on a plate of a weak nutrient agar such as HI or wMY agar (Appendix II) and then overlain with a turbid suspension of *E. coli* in water. Lids of these plates should be left open for several hours to let the excess liquid evaporate. Plates are incubated at ambient temperatures for 3 or 4 days and then examined for clones of cellular slime molds. For soils from cooler regions, plates should be incubated at a temperature below 20°C to allow temperature-sensitive species, such as *Dictyostelium septentrionalis*, to develop. If the suspension of soil is dilute enough, individual clones will form plaques in the bacterial lawn. Cellular slime molds, particularly the dictyostelids, appear within 3–4 days; are subcultured easily; and then can be identified to species, using keys and descriptions in Raper (1984), Hagiwara (1989), Cavender (1990), and Cavender and colleagues (1995). Subculturing first with 1.5% nonnutrient agar streaked with 12- to 24-hour pregrown *E. coli* will decrease contamination. Spores are removed from a sorus by touching it with a sterile glass needle. A 0.1% lactose-peptone medium streaked with a light suspension of *E. coli* is used for later subcultures (Appendix II). If the

original suspension plates are maintained for several weeks, plasmodia of myxomycetes often will appear. The plasmodia can be cultured, but inducing them to fruit is often difficult (Stephenson and Landolt 1992).

The Cavender method allows one to determine the species present in a given habitat but provides little information on their spatial organization within the site because the sample has been homogenized. In a modification of the technique, Eisenberg (1976) used soda straws to core soil to obtain subsamples (often side by side) whose exact position relative to other subsamples is known. Such sampling allows one to estimate the size of individual populations (clones) to a resolution of a few millimeters. The contents of each straw is plated out by the Cavender method. A species present in one subsample but absent from adjacent subsamples can be interpreted to represent an individual clone. The technique also can be used to determine the degree to which the distribution clones of different species overlap.

Feest (1987) modified the Cavender method for sampling myxomycetes from soil. Instead of *E. coli* he spread the yeast *Saccharomyces cerevisiae* over the soil mixture in the Petri plate. Myxomycetes plasmodia appeared several days after the samples were plated. Whether or not the yeast is the primary food for the mycetozoans that develop in the plate is not clear. Plasmodia do not appear immediately after plating, and we surmise that most of the myxomycetes present are initially amoeboflagellates. Amoeboflagellates tend to feed on bacteria (not on yeasts). Protostelids in the genus *Schizoplasmodiopsis* also are found in such samples (Feest 1987, personal communication), and *Schizoplasmodiopsis* species feed strictly on bacteria. It is likely that the yeast prevents the plates from being overrun by filamentous fungi while at the same time allowing the food bacteria to grow. Most of the myxomycetes that appear in these plates do not fruit readily. As a result, this technique is of limited value for researchers interested in cataloging diversity to the species level.

One category of "soil" that has received little attention as a source of mycetozoans is the material that collects around the bases of vascular epiphytes on the trunks and branches of trees in tropical and temperate-region rain forests. That "canopy soil" consists of decaying organic matter from the epiphytes, tree bark, and intercepted litter (Benzing 1983; Lesica and Antibus 1990). The microhabitat supports both dictyostelids and myxomycetes in the Neotropics (Landolt and Stephenson 1997, unpublished data).

BARK

Many species of myxomycetes and protostelids are corticolous. In addition, both *Pocheina* and *Acrasis* are found on bark, with the former being almost exclusively corticolous. On rare occasions, a dictyostelid may be isolated from bark. Olive (1975) noted that bark-inhabiting species of mycetozoans are, as a rule, much more difficult to culture than species of mycetozoans that grow in other microhabitats, even if the bark-inhabiting species is the same as a species that was collected from a nonbark microhabitat. There is as yet no satisfactory explanation for this phenomenon.

The protostelids, acrasids, and many of the smaller myxomycetes that grow on bark can be found only by using variations of the moist-chamber technique (Appendix I). Gilbert and Martin (1933) were the first to demonstrate the use of moist chambers for isolating corticolous mycetozoans. While using such chambers to show their botany classes the algae that grew on the bark of living trees, they discovered a number of tiny fructifications of two undescribed species of myxomycetes that had developed on the bark along with the algae. Preparing moist chambers is simple (Stevens 1974; Keller and Brooks 1976b; Stephenson 1985; Keller and Braun 1999). Plastic or glass Petri dishes (100 mm in diameter ×15-mm high) can be fitted with sterile discs of filter paper or absorbent paper toweling. The bottom of the dish is covered with a single, nonoverlapping layer of bark (Fig. 25.20). Enough sterile distilled water (10–15 ml) to wet the bark thoroughly is added. The amount of water will vary with the amount of bark making up the sample and differences in the absorptive capacity of bark from different species of trees. The plates are set aside to soak for 10–12 hours. After soaking, excess water is removed carefully or, if necessary, more water may be added. Too much water often interferes with development of *Echinostelium* species, which may appear in the first 24 hours, whereas lack of water will lead to abnormal development of members of the Physarida and Stemonitida that may appear later. Observation of cultures should begin after approximately 24 hours to ensure detection of the smaller mycetozoans and should be repeated daily for 7–10 days. At this point (if filamentous fungi have not overrun and contaminated the culture), allowing the bark culture to dry completely and then rewetting it appears to stimulate further development and fruiting in some myxomycetes, especially species of *Licea*. Longer periods of incubation under moist conditions favor aphanoplasmodial and phaneroplasmodial species of myxomycetes. Culture dishes should be kept at room temperature or at incubator temperatures of 23–25°C and under normal light conditions or a 12 hours/12 hours light/dark cycle. Each Petri dish should be labeled with the collection number of the sample and the date the sample was wetted. Labels should be applied to the side of the Petri dish so as not to obscure the field of vision through the lid.

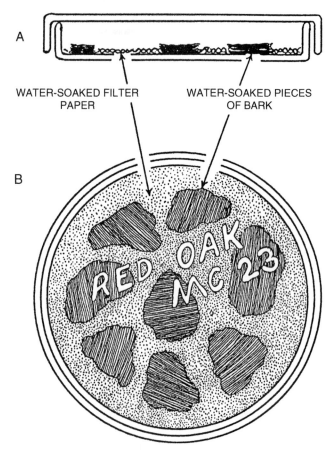

FIGURE 25.20 Petri dish set up as a moist chamber for culturing fungi. **A.** Side view. **B.** View from above. (Adapted with permission from Stephenson and Stempen 1994.)

With myxomycetes, it is very important to observe the color and type of plasmodium as well as all possible features of the fructifications that develop from it. Little is known about the plasmodia of nearly 25% of the known species of myxomycetes. In cases where plasmodia are feeding actively on the bark surface in moist-chamber cultures, a thin slice of the substratum with a portion of the plasmodium can be removed and placed on an appropriate agar medium such as CM/2 (half-strength cornmeal agar; Appendix II). The plasmodium is allowed to spread over the agar surface (Timnick 1947). If it continues to feed and grow, fruiting bodies eventually may form on the agar medium. Fresh spores obtained in this way can be sown on the proper agar medium in an attempt to cultivate the myxomycete from spore to spore. Some bark cultures produce plasmodia that never fruit. In such cases, one should remove the lid of the Petri dish and allow the bark to dry completely. As noted earlier, rewetting the bark after several days usually induces certain species to fruit. Apparently, some species must go through a sclerotal stage before fruiting is initiated.

To harvest myxomycete fruiting bodies, one should raise the lid of the Petri dish on one side, permitting the contents to dry gradually. If the cultures dry too quickly, the sporangia have a tendency to collapse and the spores agglutinate; if the cultures dry too slowly, filamentous fungi often overgrow the fruiting bodies. Cultures must be observed carefully to prevent the development of filamentous fungi and to check for the presence of mites and collembola, which may damage the fruiting bodies. Failure to detect the fructifications may be the result of a delay in observing the cultures or of the use of a stereomicroscope at too low a magnification (50–100× gives the best results). Fruiting bodies of some species are large and clearly visible to the naked eye. Bark fragments hosting only a single species occasionally are found, but usually a mixture of several species is present. Separating mixed species into individual collections is merely a matter of patience. The substratum is cut around the bases of the fruiting bodies with a sharp scalpel, which is used to transfer them to a collecting box. Sometimes, very small pieces of bark (less than 1 mm²) must be moved to separate species. The specimens should be placed on small spots of glue on the inner surface of the lid of the collection box. That procedure will ensure that the specimen and its label, which is attached to the outer surface of the lid, are inseparable. The label on the collection box should contain all the information on the moist chamber plus the harvesting date and eventually the scientific name of the specimen (Fig. 25.21).

Sometimes only one or two fruiting bodies of a given species develop in a moist-chamber culture. When that occurs, a slide preparation may serve as a numbered collection. Whenever a slide is made, great care should be taken to ensure that all characteristics of the fruiting body are preserved. Indeed, that may be the only way to obtain voucher specimens of *Echinostelium* species because the fruiting bodies in members of this genus are so tiny and fragile. The use of slides with frosted ends labeled with a sharp lead pencil represents a quick way to make a slide record that can be filed for future reference. The writing on the microscope slide should be covered with transparent scotch tape to keep it from smudging (Sundberg and Keller 1996).

The moist-chamber technique can be modified to promote growth and fruiting of particular taxa. Partially submerging the bark in standing water is especially useful for stimulating heavy fruiting of protostelids and acrasids (L. Olive and C. Stoianovitch, personal communication; F. W. Spiegel, unpublished data). In other cases, small bits of bark can be placed on weak nutrient agar or water agar instead of on moist paper (Olive 1975; Best and Spiegel 1984; Spiegel 1990). That practice makes fruiting bodies of protostelids that extend from the sides of the bark easier to view with a compound microscope.

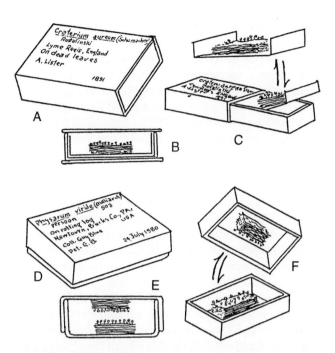

FIGURE 25.21 Two methods of storing specimens of myxomycetes in small, cardboard boxes. **A–C.** Box with a tray. **D–F.** Box with a lid. (Adapted with permission from Stephenson and Stempen 1994.)

Direct observation of the corticolous protostelids is important because those species are often difficult to isolate and subculture, and such observations may be the only way to identify them.

The moist-chamber technique has several advantages. The fruiting bodies of many corticolous mycetozoans are so small that they cannot be seen by the naked eye in the field and can be located only under a stereomicroscope in the laboratory. Frequently, the time required to hunt and collect fruiting bodies of myxomycetes in the field is prohibitive, or environmental conditions may have been unsuitable for fruitings (e.g., a prolonged dry spell or winter conditions).

The bark surface of both the main trunk and branches of trees may contain many myxomycetes; larger species often can be located directly in the field. In the central and southeastern United States rainy periods lasting several days, along with daytime temperatures of 21–32°C, stimulate fruiting by the greatest diversity of myxomycetes. During the months of July, August, and September, under optimal conditions, it is not uncommon to find as many as four species intermixed on the same piece of bark and up to ten species or even more on an individual tree (Braun and Keller 1977; Brooks et al. 1977). Collectors should allow time for the formation of mature fruiting bodies, however, before venturing to likely habitats. Plasmodial activity is greatest

24–48 hours after heavy rains; mature fruiting bodies appear after 3–7 days.

The species of tree, the general shape of the tree to some degree, surface topography of the bark, and the presence of an epiphytic cover of lichens and bryophytes are factors that influence the location of sporangia on the bark surface (Brooks et al. 1977). Individual trees have distinctive drainage patterns and watercourses along inclined lateral branches and the main trunk. Natural flowways; crotches; and the presence of algae, mosses, liverworts, ferns, and lichens increase the water-absorbing capacity of bark and its moisture retention. Certain species of corticolous myxomycetes are found repeatedly in such microhabitats, whereas others occur only on bare bark in more exposed areas. In eastern North America, the greatest diversity of corticolous myxomycetes occurs on red cedar (*Juniperus virginiana*), American elm (*Ulmus americana*), apple (*Malus* species), and grape (*Vitis* species). In some areas, certain species of pine (*Pinus*) and oak (*Quercus*) provide particularly good substrata. As already noted, the diversity of corticolous myxomycetes appears to be much lower in tropical forests than in temperate forests (Binion et al. 1997; Schnittler and Stephenson 2000).

Species with brightly colored sporangia that are readily visible to the naked eye can serve as "indicator species." The presence of their sporangia on the trunks of living trees indicates that rainfall has probably been sufficient for other, more difficult-to-see species of myxomycetes to be present. Many species belonging to the order Physarida (e.g., *Physarum crateriforme*) are good indicator species. Slime tracks left on a bark surface by migrating plasmodia are permanent, visible signs of plasmodial activity and another indication that myxomycetes may be found somewhere on the tree. Rare species or even undescribed species often turn up with careful collecting. Myxomycetes thought to be rare or uncommon may appear suddenly and sporadically in great numbers one season and then not be seen again for years. The lack of good geographic distribution records is one reason that distribution maps seldom are prepared for mycetozoans. Specimens collected in the field are treated as described in the section on "Litter," earlier.

DUNG

Most dictyostelids, some myxomycetes, a few protostelids, the copromyxids, *Guttulinopsis* species, and *Fonticula* species are at least facultatively coprophilous. Most of these mycetozoans are found on herbivore dung, but some also have been found on carnivore dung (Olive 1975; Raper 1984; Blanton 1990; Krug et al. Chapter 21 of this volume). Although dictyostelids have been known primarily from dung for a long time, it is likely that they exploit dung by migrating to it from the soil. That may be true for many of the other mycetozoans as well. *Guttulinopsis vulgare*, however, has been found on cow dung carefully collected within seconds of deposition to exclude any contact with soil (F. W. Spiegel, unpublished data), which suggests that this species can pass through the digestive tract of cattle. Other coprophilous mycetozoans may be carried from bolus to bolus by insects.

For most mycetozoans, it is best to collect dung that is less than a day old, although some dictyostelids and many myxomycetes can be found on older, dried material. Although heavy fruitings of dictyostelids can be seen in the field, collecting samples of dung and examining them in the laboratory is more useful. The best way to collect dung is with a trowel or other similar tool, placing each sample in a plastic bag. Another method is to invert a plastic bag around the hand and pick up a sample, which then is enclosed in the bag as it is turned right-side-out. Dung samples are placed on moistened paper towels in finger bowls and covered. If the dung is relatively dry, it may be moistened with water. After 1 or 2 days, the dung should be examined under a stereomicroscope for fruiting bodies. Examination should continue for several weeks; some myxomycetes and dictyostelids may appear during those later observations.

Species of *Copromyxa* and *Guttulinopsis* appear within 2 or 3 days on fresh dung. They are small, inconspicuous, and easily missed if sporulation by zygomycetes is heavy. One way to overcome that problem is to freeze the dung sample overnight (Spiegel and Olive 1978). Ideally, the surface of the sample will be frozen solid, but the center still will be slightly soft. That treatment delays the appearance of zygomycetes by 2–4 days, during which it is easier to locate the fruiting bodies of acrasids, protostelids, and dictyostelids. Nevertheless, we do not recommend this method for tropical species of dictyostelids, which may be sensitive to freezing (Stephenson and Cavender 1996).

Because the fruiting bodies of all coprophilous mycetozoans, except for some myxomycetes, are delicate and ephemeral, they often must be cultured for further characterization before identifications can be made or confirmed. Culturing is carried out most effectively using the spore-touch method. Spores are picked up with a sterile needle and transferred to a weak nutrient-agar plate that has been streaked with a bacterium, usually *E. coli*, or, to avoid contamination, to a nonnutrient agar heavily streaked with *E. coli*. Dictyostelids, which rarely have bacteria in their sori, usually transfer cleanly. Acrasids, in contrast, often have bacteria throughout their sori. The most common bacterial contaminants are motile *Pseudomonas* species. They can be eliminated by culturing on several-day-old, nonnutrient agar before culturing on a nutrient medium such as 0.1% lactose–peptone agar (Appendix II). If a monoxenic culture is desired, it is often possible to clean up a culture by allowing amoebae to migrate several times through a streak of the desired food bacterium. Four to five passes is usually sufficient to eliminate unwanted bacteria.

ROTTING WOOD

Most of the better-known species of large myxomycetes and many protostelids fruit on rotting wood, although cellular slime molds are rare in this microhabitat. The logs either may be standing or fallen, with or without bark. Relatively warm and moist environmental conditions are optimal for finding myxomycetes and members of the protostelid genus *Ceratiomyxa*. As with bark, it is best to collect mature fruiting bodies and to place them in carefully labeled collecting boxes in the field. Gluing specimens into the box will minimize specimen damage on the trip to the laboratory. Once in the laboratory the collections should be allowed to dry completely to prevent the growth of filamentous fungi and to deter invertebrate pests. Collections should be stored in a cool, dry place, preferably treated to kill insects and mites (Stevens 1974). Preliminary identifications can be made after examining the collections with a good hand lens or stereomicroscope; examination under a compound microscope is usually necessary for identification of species. Samples with more than one species present should be split carefully so that a herbarium specimen can be kept for each species. Permanent slides of fruiting bodies are stored with the dried specimens (Stephenson and Stempen 1994; Sundberg and Keller 1996).

Protostelids (other than *Ceratiomyxa* species) must be observed in the laboratory. Small pieces of wood are placed in standing water in a Petri dish or onto weak-nutrient-agar plates, which then are kept in moist chambers. The specimens are examined daily for up to 2 weeks with both a stereomicroscope and a compound microscope. Species are identified, recorded, and (if necessary) cultured as described in the section on "Dead Aerial Plant Parts," earlier.

DETERMINATION OF RELATIVE IMPORTANCE AND ABUNDANCE

All mycetozoans are predators of bacteria, fungi, and other microorganisms in decomposer systems. In many of those systems, eumycetozoans are the most common amoeboid organisms present (Raper 1984; Feest 1987).

Their abundance probably reflects the high level of dispersability conferred on them by their ability to fruit. Because they fruit, and because fruiting makes them readily identifiable, mycetozoans represent an excellent model for monitoring the dynamics of the predatory protists of the decomposer systems. However, quantitative studies of the biodiversity of mycetozoans are still in their infancy. Therefore, we will use this section to provide a set of suggestions for future work, with some comments on work in progress.

Studies of mycetozoan biodiversity have to focus on the particular mycetozoans to be studied and the specific microhabitats in which they occur, but that focus does not necessarily mean that such studies have to be limited to just one group of mycetozoans. For example, studies of protostelids on dead aerial plant parts also can be expected to yield information on small myxomycetes, such as *Echinostelium* species and *Acrasis rosea* (Olive 1975; Best and Spiegel 1984). Studies of dictyostelids in soil also will yield data on myxomycetes and protostelids (Feest 1987; Stephenson and Landolt 1995). Investigations of macroscopic myxomycetes, however, will not necessarily provide information about microscopic mycetozoans. A truly comprehensive study of all mycetozoans in an area would have to be a multiyear project using a number of researchers familiar with the techniques applicable to each group of mycetozoans and/or the types of substrata to be examined. Because sampling of macroscopic myxomycetes requires the use of the largest plots (see "Myxomycetes," later in this chapter), comprehensive studies should be confined to small subplots within those larger plots.

DICTYOSTELIA

Methods for sampling soil dictyostelids have been worked out in more detail than for any other group of mycetozoans. Cavender and his coworkers (see Cavender and Raper 1968; Cavender 1973, 1990; Raper 1984; Cavender et al. 1995; Vadell et al. 1995; Swanson et al. 1999) and Stephenson and Landolt (Landolt and Stephenson 1991, 1995; Stephenson et al. 1991, 1996, 1997a, 1997b; Landolt et al. 1992a, 1992b, 1994) have sampled soils from different habitats around the world. Their studies have provided a great deal of data on the biogeography of dictyostelids, and certain trends are evident (Swanson et al. 1999). Some species, such as the crampon-based Dictyostelia, appear to be strictly tropical (Raper 1984). Although those species sometimes are transported to temperate habitats by migrating birds (Suthers 1985), most do not appear to survive for long at higher latitudes. *Dictyostelium rhizopodium* and *D. vinaceo-fuscum* have been found in a few temperate areas but only in habitats where freezing does not occur. *Dictyostelium discoideum* is primarily a temperate species and only occasionally occurs in tropical environments. Other species such as *Polysphondylium violaceum* and *D. mucoroides* appear to be cosmopolitan. Some species appear to be much more restricted in their distribution and occur only in a single geographic region. Prominent examples are *D. deminutivum*, *D. fasciculatum*, and *Acytostelium ellipticum* (Swanson et al. 1999).

Spatial distribution of dictyostelids within a plot may be sampled using the soda-straw technique of Eisenberg (1976; see "Soil," earlier). Although it has not been used widely, any future sampling of biodiversity of dictyostelids can be improved by including that technique to determine the number of clones in a sample plot, with higher resolution. The soda-straw technique is more effective than the classical Cavender method. Although the latter method allows clones derived from individual cells to be counted, it is not possible to tell if all the cells of a given species in a sample are from one rich pocket of bacteria in the sample or from several independent feeding sites—that is, it is difficult to tell if one is dealing with one patch (clone) or many (Feest 1987). However, Holmes (1991), working at Tikal in Central America, found that the Eisenberg technique sampled fewer of the species present than did the Cavender method.

We recommend that sampling efforts to determine the biodiversity of dictyostelids in a study area include replicate plots in areas representing similar soil conditions and vegetation types and that all vegetation types and soil conditions (i.e., different moisture regimens and slope aspects) occurring in the study area be sampled in proportion to their occurrence. The relative amounts of dormant versus actively feeding cells can be determined by freezing a subset of each sample, which kills any active amoebae but does not harm dormant cells such as cysts and spores (Kuserk 1980). The frozen subset and a comparable unfrozen one are diluted and plated. After a suitable period, the clones produced by each sample are counted. The number of clones missing from the plates of the frozen sample, as a percentage of the number of clones from the unfrozen sample, approximates the percentage of cells in the sample that are active amoebae. In seasonal climates, study areas should be sampled throughout the year for several years because the occurrence of dictyostelids may vary between seasons and between years (see Raper 1984). In a comprehensive study of mycetozoans, the subplots used for sampling dictyostelids can be the same as those used to sample litter protostelids. The scope of a sampling effort is limited only by the number of dilution plates one can observe at a given time. Generally, most workers who have developed some skill at observing and recording mycetozoans can examine up to 100 or more plates per day. Sampling strategies must take this limitation into account.

If dung is a major component of the ecosystem being inventoried, sampling protocols can be designed for the systematic collection of dung samples of standard sizes and ages of decomposition (see "Acrasids," later in this chapter).

MYXOGASTRIA

In Wood, Litter, and Soil

Macroscopic myxomycetes can be sampled in wood, litter, and soil with the techniques described by Stephenson (1988, 1989). Species abundance can be estimated quantitatively in a temperate or boreal forest by collecting all the fruitings of myxomycetes that occur in one or more 20 m × 50 m (0.1-ha) plot located within the study area. A somewhat smaller plot (perhaps 20 m × 20 m) is probably more appropriate for tropical forests and grasslands. The four corners of the plot are marked with stake-wire flags if the plot is to be used for a single field season. More permanent markers (e.g., wooden stakes, sections of metal or PVC pipe) are used to mark plots that will be sampled for several field seasons. A fairly conspicuous, brightly colored marker also should be placed at the center of each long-term plot to aid in relocating the plot if the interval between visits is long and/or if marked seasonal changes (e.g., leaf fall) occur between visits. The route to plots situated some distance from a trail or road should be flagged as well. For each plot, latitude and longitude and/or Universal Transverse Mercator (UTM) grid coordinates can be determined by means of a portable global positioning system (GPS) unit. On each visit to the plot, as much of the plot as possible should be examined. Collectors should work from one end of the plot to the other along three or four transects parallel to the sides of the plot at intervals of no more than five meters. The actual sampling points along each transect, which should be approximately the same on each visit, are determined by the availability and location of potential substrata for myxomycetes. Because many species of myxomycetes seem to fruit at about the same time throughout a given locality, observations outside the plot may provide some insight into the species fruiting within the plot. Collectors should sample each plot throughout the year for several years. They can expect to find approximately 50–75% of the species present in the first year and declining numbers of the remainder in subsequent years. Sampling ends when no additional species are collected in a given year.

If all myxomycetes in a given plot are to be sampled, a technique for collecting standard-size samples of substrata should be devised for dead plant parts, dung, and bark to be examined in moist chambers. In addition, the activities and abundance of trophic cells can be assessed by plating suspensions of substratum with bacteria or yeasts on agar (Raper 1984; Feest and Madelin 1985; Feest 1987). That procedure will allow an investigator to estimate number of plasmodium-forming units present, but the abundances of individual species are unlikely to be recorded because few of the myxomycetes isolated in this manner will fruit.

On Bark, Dead Plant Parts, and Dung

Corticolous myxomycetes can be collected in the field if they produce macroscopic fruiting bodies. Because individual trees vary, and because myxomycetes are not randomly distributed, it would be difficult to derive data more quantitative than presence/absence of species. Again, the same trees should be sampled throughout the year for several years. As the incidence of new records levels off, one can assume that essentially all species that occur on a particular tree have been collected. Microscopic myxomycetes on bark must be sampled using the moist-chamber technique. Quantitative data concerning the abundance of microscopic corticolous myxomycetes, in principle, could be collected if standardized methods for establishing sample size and homogeneity were developed (see Stephenson 1989; Keller and Braun 1999).

Myxomycetes on dead aerial plant parts can be studied using the same protocols as those used for protostelids (see "Protostelia," later in this chapter). The protocols for sampling myxomycetes on dung are those used for sampling acrasids on dung (see "Acrasids," later in this chapter).

PROTOSTELIA

Quantitative studies of microscopic mycetozoans on dead plant parts and in litter are hindered by the fact that each of these microhabitats includes numerous heterogeneous elements of substratum of different sizes. Also, the appearance of mycetozoans from a piece of substratum that has been plated out in the laboratory gives no indication whether those organisms were active or dormant in the environment when the substratum was collected. At best, only presence/absence data can be collected for each species in each collection (Best and Spiegel 1984).

The development of a standardized technique for introducing sterilized substrata into a microhabitat has opened a number of possibilities for quantitative assessment of the biodiversity of mycetozoans in litter and on dead plant parts (Moore and Spiegel 1995). Examina-

tion of naturally occurring substrata suggests that most, if not all, mycetozoans that occur in those microhabitats show no preferences for particular substrata (Best and Spiegel 1984)—that is, a mycetozoan is no more or less likely to fruit on a dead leaf than on a grass culm. Moore and Spiegel (1995) showed that protostelids and other mycetozoans will colonize segments of sterilized wheat straws introduced into a given microhabitat (see "Dead Aerial Plant Parts" and "Litter," earlier). Hanging straws and litter straws are introduced into 1-m² plots that are established within larger sampling areas. Straws from each microhabitat are collected from the plots in random order at intervals and plated out using the standard techniques for protostelids. Sterilized straws that have not been put into the field should be plated out as controls to determine if contaminating protostelids were present on the straws. The presence/absence of species in a plot based on samples of naturally occurring substrata from each microhabitat also should be recorded. Comparisons between naturally occurring and introduced substrata have shown that qualitative species' abundances on naturally occurring substrata and on the straws do not differ (Moore and Spiegel 1995, 2000a, 2000b). More common species are also more likely to colonize the straws earlier than rarer species. The units measured are presumed to be individual clones determined as contiguous patches of fruiting bodies (see "Dead Aerial Plant Parts," earlier).

If straws are devoid of microorganisms when they are introduced into a microhabitat, the recovery of mycetozoans when the straws are collected and plated out indicates that the species present were active in the environment during the period of the study, and seasonal patterns of substratum colonization can be determined. Moore's results (Moore and Spiegel 2000c) show that the colonization rate of mycetozoans in litter is about eight times higher in summer than in winter. Straws in the aerial environment are colonized in the summer at a rate that is nearly 20 times greater than that in the winter. Further experiments are being designed to determine if this pattern of colonization results from seasonal differences in the activity of invertebrate vectors or in the rate at which mycetozoans produce dispersible propagules. Sampling with introduced substrata also can be used to compare aerial and litter microhabitats within a given habitat (e.g., temperate deciduous forest or grassland) or between habitats or to compare a particular type of microhabitats among habitats widely separated geographically. Species that prefer dead aerial plant parts are common in several different forests and grasslands in northwestern Arkansas. The same is true for species that prefer litter. The species composition of litter differs from the species composition of aerial microhabitats at all sites (Moore and Spiegel 2000a).

Generally, two straws are placed on a plate; and examination of each straw requires from 2 to 10 minutes. Straws first are observed between 3 and 5 days after plating and then at 3-day intervals for at least 2 weeks. About 100 plates can be examined per day by one observer. Sampling strategy should take such logistic factors into account.

The techniques for sampling protostelids on bark and dung are essentially the same as those used to sample microscopic corticolous myxomycetes (see "On Bark, Dead Plant Parts, and Dung," earlier, and "Acrasids," later in this chapter).

ACRASIDS

Acrasids are much less common than the other mycetozoans in most habitats and can be recorded as they are encountered as the result of sampling techniques being used for other mycetozoans in a given microhabitat. Acrasids on dead plant parts and litter are encountered when using the techniques designed to sample protostelids (see previous section on "Myxomycetes"). Acrasids in soil can be sampled with the Cavender method (see "Dictyostelids," earlier).

The diversity of coprophilous acrasids and other mycetozoans has not been measured in any quantitative fashion, although they may be common in some areas. Many acrasids are primarily coprophilous, however, and given the nature of this substratum, it should be possible to measure their diversity and that of other mycetozoans present both qualitatively and quantitatively. Dung from a given species of host tends to be of a characteristic consistency and size. With experience, a collector can select boli of the same freshness. By sampling dung from animals in a group or occupying the same habitat, many replicate samples of the same size and freshness could be obtained, allowing for quantitative estimates of acrasid species' abundances. In principle, one should be able to compare diversity between groups in different locations, between groups with different diets, within groups at different seasons, and between groups of animals of different species. One also could determine whether acrasids and other mycetozoans prefer boli at different stages of decomposition.

Certain predictions are possible with members of the acrasid group. F. W. Spiegel (unpublished observation) attempted to find as many sources of *Copromyxa protea* as possible in central and eastern North Carolina. Based on his collecting experience, he noted that the species is more common on cow dung than horse dung in a pasture shared by the two animals. The species has a patchy distribution, but when found in a given pasture, it can be collected repeatedly. Its species appear to be more common in pastures where grazing is the sole source of food for the animals. When food was supplemented with antibiotics, mycetozoans were never found. The validity of these and similar observations could be tested with carefully controlled and quantifiable collecting techniques.

SUMMARY

Relatively few effective, quantitative data on the biodiversity of mycetozoans have been collected, probably reflecting an inclination on the part of collectors to determine only whether species are present in an environment. With a little thought, it should be possible to develop techniques effective at providing quantitative data. Because mycetozoans can be collected and identified with a few inexpensive resources, it should be possible to train local collectors around the world to collect and compile data on mycetozoan biodiversity.

CHARACTERIZING MYCETOZOANS

Mycetozoans are recognized by their fruiting bodies, which means that any taxonomic survey of the group must focus on the presence of fruiting bodies on a given substratum. In addition, the presence of myxomycetes, although not the individual species, can be assessed by observing their distinctive plasmodia on a substratum. With the incorporation of the use of specific molecular markers (e.g., randomly amplified polymorphic DNA, or RAPD) into field-based research, it ultimately may be possible to determine the presence of species in the absence of visible fruiting bodies. Until then, however, we must continue to identify species visually. To some extent, the information necessary for characterization of mycetozoans depends on the taxa involved and the microhabitats being sampled.

MACROSCOPIC MYXOMYCETES

Macroscopic myxomycetes are analogous to the more familiar higher fungi (including mushrooms) because fruiting structures are usually discrete. Often, fruiting bodies of a given species will appear only once or a few times during a study period. Fruiting bodies of even the most common species will turn up only a few times in a 20- × 50-m plot on collecting trip. Therefore, it is possible to collect samples from all fructifications that are found and to store them as herbarium specimens. Identification of each specimen collected should be based on macro- and micromorphological examination of one or more fruiting bodies. Permanent slides showing the micromorphological characters of each species can be maintained along with the actual specimens. If molecular characters are to be examined, DNA can be extracted and preserved. The minimum information associated with each specimen should include the name of the species, the collector, the type of substratum on which fruiting occurred, and the date and location of the collection. With the availability of relatively inexpensive GPS receivers, it is advisable to record the coordinates of the collecting site rather than the vague descriptors that classically have been used to describe a location. The body of information relating to each collection should be available on a computer database for easy retrieval.

MICROSCOPIC MYCETOZOANS

It is difficult and impractical to collect every microscopic mycetozoan that may turn up in a survey. For instance, in a survey of protostelids, *Acrasis* species, *Echinostelium* species, or coprophilous acrasids, one can encounter several hundred individual patches (clones) of fruiting bodies in a collecting period. Some patches may consist of only a few fruiting bodies, and those may be intermixed with sporulating fungi or other mycetozoans, which makes the physical collection of the fruiting bodies difficult. Even when a fruiting body is collected, it often is not possible to preserve it intact. If a physical record of the mycetozoan is required, then the species must be isolated and cultured. Some species are relatively easy to isolate and culture, although the process may take several weeks. Others, such as many bark-inhabiting mycetozoans (Olive 1975), are difficult or even impossible to culture. Photomicrographic or digital images of each clone could, in principle, be taken, but this still would involve a tremendous cost in time and resources. The most practical approach for recording microscopic mycetozoans is with monochromatic digital images, more than a thousand of which can be stored on a compact disc. Still, it is not practical to collect a specimen or an image of every single clone of a mycetozoan that appears in a survey.

Because collecting permanent samples of most microscopic mycetozoans is so difficult, it is imperative that anyone involved in collecting data have the opportunity to study with someone experienced in the identification of species based on fruiting-body morphology. Fortunately, identification of almost all protostelids, *Acrasis* species, *Pocheina* species, and coprophilous acrasids is relatively straightforward. When the identification reached for a particular species is equivocal, the investigator should examine the substratum for trophic cells or fruit-

ing bodies in various stages of development or germinate spores to produce trophic stages. Characteristics of such structures can help to confirm (or refute) a preliminary identification (Olive 1975; Raper 1984; Spiegel 1990; Protostelid home page). The presence of flagella in *Planoprotostelium aurantium*, for example, definitively distinguishes it from *Protostelium mycophaga* (Olive 1975). Spores that have germinated in water will reveal that feature. Undescribed species must be followed through their life cycles so that their trophic states can be characterized. Ideally, such species should be cultured, although that is not always possible (e.g., *Ceratiomyxa hemisphaerica*; Olive and Stoianovitch 1979).

Records should be kept of the frequency of occurrence of a species on different substrata. If a voucher is required, a photographic or (better) a digital record of the fruiting bodies can be maintained. Cultures should be made if the researcher needs material for a monograph or for comparisons with specimens from other microhabitats, habitats, or geographic locations. Also, a set of cultures of species occurring in a study area should be maintained for future reference. Cultures of most species can be maintained on plates or slants of wMY agar (Appendix II) with the appropriate food organisms (Spiegel 1990). Transfers can be made monthly or the cultures can be stored at 15°C for 6–13 months. Cultures that are dried and stored at room temperature can be revived up to a year later by rewetting with sterile distilled water. Microscope slides of fruiting bodies and trophic cells can be prepared from cultured specimens. One preferred technique that preserves structure extremely well involves fixing the specimen for electron microscopy by flat embedding some material between a coverslip and a slide.

Regardless of whether specimens are collected, the records for each mycetozoan located at each site should include the name of the organism, the type of substratum on which it occurred, the collection site for substratum (GPS coordinates preferred) and elevation, the date the substratum was collected, the collector, the date the substratum was plated out, the date each species appeared, the name of person identifying the mycetozoans, whether any specimens were kept, and where they can be found. To the extent possible, that information should be made generally available on a computer database.

Dictyostelids from soil usually are located as clonal plaques in lawns of *E. coli*. If they are developing normally on these plates, they need not be cultured. Plaques that cannot be readily identified must be brought into culture, a process that is much quicker and easier with dictyostelids than with any other mycetozoans. Nonetheless, it is impractical to assume that a specimen or culture can be taken of each of the hundreds of clones that will appear in a survey. Therefore, the presence of a species should be recorded in a manner identical to that suggested for microscopic mycetozoans. Dictyostelid spore suspensions can be preserved for up to a year on silica gel. Lyophilized spores suspended in milk protein will persist for longer periods (50 years or more; Raper 1984).

Sampling for mycetozoans at a site usually stops when all species collected have been recorded earlier in the study. The time required to reach this point varies with the group under study. Stephenson (1988, 1989) suggested that 5 years of collecting in all seasons provides a reasonable estimate of the number of myxomycete species present in a temperate deciduous forest. At the other extreme, only 10–12 weeks of sampling may be required to estimate the number of protostelid species in the litter microhabitat of a similar forest community (Moore and Spiegel 1995). In general, we recommend that intensive sampling of all microhabitats of a habitat in proportion to their abundances be carried out in all seasons for at least 2 years. At the end of that time the rate of appearance of unrecorded species can be assessed and a decision can be made about whether to continue collecting data.

INTEGRATION OF INFORMATION

So little quantitative information is available on the biodiversity of mycetozoans that the most effective way to assemble a body of data would be to generate one or more monographic treatments of these organisms. Because some of the new collecting techniques generate large amounts of information, the monographs would have to be updated fairly frequently. Therefore, it would be advisable to have the data on which the monograph is based available on a computer database as well as in the classical published form. Such databases can be updated much more rapidly than can standard monographs. In addition, a computer database can include additional information such as records of species' distributions, substratum relationships, and population estimates.

Because so little is known about the true status of the less commonly encountered species of mycetozoans, it is inappropriate to suggest that any are endangered. As more information is accumulated, however, it may emerge that distributions and frequencies of certain more common mycetozoans may be indicators of environmental conditions that bear further attention. Dictyostelids, for example, are known to be very sensitive to soil disturbance and could serve as indicators of such disturbance. Mycetozoans on dead aerial plant parts and bark also may be useful for monitoring acid rain or air quality (Wrigley de Basanta 2000).

ACKNOWLEDGMENTS. We express our appreciation to E. F. Haskins and J. C. Landolt for reviewing this chapter and to the many students and colleagues who shared data and observations on the biology of mycetozoans with us.

26

FUNGI ASSOCIATED WITH AQUATIC ANIMALS

THOMAS G. RAND

Fungi associated with aquatic animals have been the focus of considerable research over the past 30 years. As a result, fungi belonging to a wide range of taxa now are recognized to be common and significant causes of disease in a wide diversity of aquatic microinvertebrates, macroinvertebrates, and vertebrates around the world. Important reviews of the literature on some of these mycopathogens and the diseases they cause include Alderman (1976, 1982), Alderman and Polglase (1986), Austwick and Keymer (1981), Cooper and Jackson (1981), Couch and Bland (1985), Elkan (1981), Fisher and colleagues (1978), Jangoux (1987), Johnson (1983), Johnson and Sparrow (1961), Kinne (1980, 1983, 1984), Kohlmeyer and Kohlmeyer (1979), Lauckner (1984), Lightner (1981), Lichtwardt (1986, 1996), McVicar (1977, 1979, 1981, 1982), Migaki and Jones (1983), Migaki and associates (1984), Neish and Hughes (1980), Pickering and Willoughby (1982), Polglase and colleagues (1986), Porter (1986), Rand (1996), Reichenbach-Klinke (1973), Roberts (1989), Samson and associates (1988), Sindermann (1970, 1977), Sindermann and Lightner (1988), Sparks (1985), Sparrow (1960), Unestam (1973) and Wolke (1975).

Despite significant advances in our understanding of fungal pathogens, especially of their functional and structural relationships with their hosts, the basic biology of the majority of those pathogens remains poorly known (Rand 1996). In addition, an increasing number of fungi is being isolated from cultured aquatic animals. Although such fungi may interfere with the successful husbandry of those animals, they remain largely unidentified owing to a lack of mycologists with expertise in this area. Relatively few investigators are trained in the methods used for investigating aquatic animal mycoses. Increased availability of methodological manuals and critical reviews (e.g., Rand 2000) to guide researchers who are unfamiliar with the area and the contemporary techniques used for such investigations may help to offset those problems. This chapter reviews some of the contemporary methods for the collection and isolation of fungi from aquatic animals and discusses their current applications and limitations.

TAXONOMIC STATUS, DISTRIBUTION, AND ABUNDANCE

More than 250 species, comprising 86 genera from all major fungal taxonomic groups except the Basidiomycotina, are recognized from aquatic animals worldwide. However, only some 50 of those genera include species considered to be significant aquatic-animal pathogens (Table 26.1). Mycopathogens that infect aquatic animals can be either obligate or facultative invaders (Rand 1996). The two known obligate pathogens are *Ichthyophonus hoferi*, a well pathogen of fish (Lauckner 1984), and *Trichomaris invadens*, an Ascomycete that invades Alaskan tanner crabs, *Chionoecetes bairdi* (Sparks and Hibbitts 1979; Hibbits et al. 1981; Alderman 1982; Sparks 1982a, 1982b). Until recently, *I. hoferi* was considered to be the best-known pathogenic fungus of fish. However, its phylogenetic position (including the phylum to which it belongs) is now in dispute (see Ragan et al. 1996; Spanggaard et al. 1996) and awaiting additional analysis (Ragan et al., unpublished data). Both of the pathogens have stringent nutritional requirements, and neither species has been grown *in vitro* for an extended period. Trichomycete species that are important in the intestinal tracts of arthropods appear to have formed obligate, but nonpathogenic, associations with their hosts. Many of those species have not been cultured *in vitro* either, and their characteristics in culture are unknown (Lichtwardt 1986, 1996).

The majority of fungal invaders of aquatic animals are considered to be facultative parasites (Alderman 1982). Many of those fungi can be isolated from their host tissues and are cultured easily *in vitro* on a variety of nonspecific, glucose-based substrata (see "Isolation and Culture," later in this chapter). In nature, some taxa with motile spores, such as various thraustochytrid species, *Aphanomyces* species, *Lagenidium* species, and *Saprolegnia* species also exhibit little specificity toward natural substrata. Those taxa can be collected from the water column and/or a variety of biological surfaces, including decaying algal, higher plant and animal tissues, and sediments, using baiting methods (see "Isolation and Culture," later in this chapter) (Fuller et al. 1964; Willoughby and Pickering 1977; Willoughby 1978; Moss 1986; Porter 1986; D. J. S. Barr 1987; Fraser et al. 1992; Rand 1992b). Anamorphic fungi, including *Aspergillus, Cladosporium, Exophalia, Fusarium, Paecilomyces*, and *Trichosporon*, also frequently have been isolated from sediments, water, and decaying plant and animal matter (Horter 1960; Miller and Whitney 1981; Porter 1987; Söderhäll et al. 1993; Rand et al. 2000a).

The majority of infectious mycopathogens of aquatic animals are species of the phyla Oomycetes (Chromista)

TABLE 26.1

Genera of Fungi Isolated from Aquatic Animals from Freshwater, Estuarine, and Marine Environments, Worldwide

Kingdom and phylum	Genus	Site of interaction
Chromista		
Labyrinthomorpha	*Labyrinthuloides*	C
	Schizochytrium	S, D
	Thraustochytrium	S, D
	Ulkenia	S, D
Oomycota	*Achlya*	S, C
	Aphanomyces	S, C
	Aphanomycopsis	D
	Atkinsiella	S, C
	Branchiomyces	C
	Couchia	D
	Dictyuchus	S
	Haliphthoros	S, C
	Hyphochytrium	S
	Lagenidium	S, C, D
	Leptolegnia	S
	Leptolegniella	S
	Leptomitus	S
	Plectospira	S
	Pythium	S, C, D
	Saprolegnia	S, C
	Sirolpidium	S
	Thaloassomyces	S
Fungi		
Chytridiomycota	*Allomyces*	S
	Catenaria	D
	Coelomomyces	D
	Nephromyces	D
Zygomycetes	*Basidiobolus*	S, D
	Rhizomucor	D
Ascomycetes (including anamorphic fungi)	*Aureobasidium*	D
	Aspergillus	D, M
	Candida	S, D
	Cladosporium	S, C, D
	Cryptococcus	S
	Didmaryia	S
	Didymella	S
	Exopthiala	D
	Fusarium	S, C, D, M
	Hormonconis	D
	Metschnikowia	S
	Paecilomyces	S
	Penicillium	S, D, M
	Phoma	D
	Ochroconis	D
	Ramularia	S
	Rhodotorula	S
	Septocylindricum	S
	Torula	S
	Trichomaris	S, C
	Trichosporon	S, C
Fungi incertae sedis	*Cycloptericola*	C
	Ostracoblabe	S

S, Surface; C, cutaneous; D, deep; M, mycotoxin producers.

and Chytridiomycota (Fungi) (Alderman 1982; Post 1983; Roberts 1989; Rand 1996). Within those groups, at least 12 genera (*Aphanomyces, Aphanomysopsis, Atkinsiella, Catenaria, Coelomomyces, Couchia, Haliphthoros, Ichthyophonus, Lagenidium, Leptolegnia, Saprolegnia*, and *Sirolpidium*) contain species that are notoriously destructive, aquatic-animal pathogens (Sproston 1944; Vishniac 1955, 1958; Dorier and Degrange 1961; Knittel 1966; Ruggieri et al. 1970; Egusa and Ueda 1972; Bootsma 1973; Ross et al. 1975; Armstrong et al. 1976; Hatai et al. 1977, 1980; Neish 1977; Ochiai et al. 1977; Sindermann 1977; Bahnweg and Bland 1980; Bahnweg and Gotelli 1980; Lightner 1981; Copland and Willoughby 1982; McVicar 1982; Pickering and Willoughby 1982; Johnson 1983; Persson and Söderhäll 1983; Ramos-Flores 1983; Burns 1985; Smith et al. 1985; Noga and Dykstra 1986; Paperna 1986; Bruno and Stamps 1987; W. W. Martin 1987; Hatai and Lawhavinit 1988; Samson et al. 1988; Rand 1990; Sitja-Bobadilla and Alvarez-Pellitero 1990; Hanke et al. 1991; Rand 1992b; Rand and Munden 1993; Söderhäll et al. 1993).

Some anamorphic fungi are also significant pathogens of aquatic animals. As pathogens, some taxa cause primary infections, whereas others are opportunistic, secondary invaders of host animals under stress of primary infection by other organisms (Alderman 1981, 1982; Miller and Flemming 1983; Rand et al. 2000a). Among the significant pathogens in this group are members of the genera *Aspergillus, Candida, Exopthalia, Fusarium, Paecilomyces, Phoma*, and *Ochroconis* (Doty and Slater 1946; Horter 1960; Carmichael 1966; Egusa and Ueda 1972; Ross and Yasutake 1973; Lightner 1975; Solangi and Lightner 1976; Ajello et al. 1977; Fisher et al. 1978; Richards et al. 1978; Blazer and Wolke 1979; Burns et al. 1979; Alderman 1981, 1982; Bain and Egusa 1981; Hose et al. 1984; Hatai et al. 1986a, 1986b; Turton and Wardlaw 1987; Raghukumar and Lande 1988; Muhvich et al. 1989; Geiser et al. 1998; Rand et al. 2000a).

Modest numbers of species from the Entomophthorales (Yang 1962; Tills 1977), Mucorales (Wolf and Smith 1999), Labyrinthomorpha (Sindermann 1970; Polglase 1981; McLean and Porter 1982; Jones and O'Dor 1983; Bower 1987a, 1987b), and meiotic Ascomycetes (Sparks and Hibbits 1979; Hibbits et al. 1981; Porter 1982) also have been recognized as pathogens of aquatic animals. Some of those species, such as *Trichomaris invadens*, have been implicated as causative agents in the massive dieoffs of Alaskan tanner crabs (Alderman 1982; Porter 1982; Sparks 1982a, 1982b). Labyrinthrulids and thraustochytrids appear to be important molluscan parasites, and at least one species, *Labyrinthuloides haliotidis*, causes significant mortality of juvenile abalone in British Columbian waters (Bower et

al. 1987a, 1987b). Additional information on those and other aquatic animal mycopathogens and on the diseases they cause can be found in Alderman (1976, 1982), Alderman and Polglase (1986), Bower (1987a, 1987b), Johnson and Sparrow (1961), Kohlmeyer and Kohlmeyer (1979), Polglase and colleagues (1986), Porter (1986), Lauckner (1984), Rand (1996), Neish and Hughes (1980), Sindermann (1970), and Wolke (1975).

Numerous cases of nutritional disorders and pathologies also have been reported in aquatic animals under intensive culture. Those disorders have been linked to the presence of specific toxins or contaminants produced by anamorphic fungi (especially *Aspergillus, Fusarium*, and *Penicillium*) in food (Rand 1996). Probably the most important fungal toxins, or mycotoxins, are those responsible for fish and crustacean mycotoxicosis (Horter 1960; Lightner 1988; Hendricks and Bailey 1989; Lovell 1989, 1991, 1992; Roberts 1989; Jantrarotai and Lovell 1990a, b; Jantrarotai et al. 1990; Galash and Marchenko 1991; Shigemori et al. 1991; Tacon 1992). More than 200 mycotoxins have been identified from fish feed (Tacon 1992). That cases of mycotoxicoses and populations of their causative species likely will increase is of considerable concern (Rand 1996) as the aquaculture industry expands to include new and exotic species and moves into new geographic areas.

It is also probable that many new species associated with aquatic animals will be uncovered as mycotic investigations intensify, especially in little-studied regions or biological communities. For example, fungi associated with animals from coral reef (see Ramos-Flores 1983; Geiser et al. 1998) and tropical rainforest regions of the world have been studied remarkably little. *Trichomyces* in the guts of aquatic arthropods are hardly known (Lichtwardt 1986, 1996). The areas of mycology dealing with fungi infecting many groups of aquatic invertebrates, including annelids, nematodes, branchiurans, tunicates, and ascidians, remain in their infancy.

Aquatic-animal pathogenic fungi, such as members of the genera *Aphanomyces, Coelomomyces, Lagenidium, Ichthyophonus*, and *Saprolegnia*, that have been studied intensively are cosmopolitan in their distributions and invade wide ranges of hosts (Rand 1996). Epidemiological data indicate, however, that those aquatic animal mycopathogens probably have restricted and patchy distributions within the host range (Ruggieri et al. 1970; McVicar 1979, 1981; Sparks and Hibbits 1979; Munro et al. 1983; Dykstra et al. 1986, 1989; Whisler 1987; Hatai and Lawhavinit 1988; Levine et al. 1990; Fraser et al. 1992; Rand 1992a, 1992b; Rand and Munden 1993). The patchy patterns of host–animal infections may reflect gene-based differences in susceptibilities of host populations to particular mycopathogens. Such patterns also could result from climatic and biotic factors that differ among locations within the host range and influence the probability of successful infection of susceptible host species (Rand 1996).

Furthermore, the few studies of the temporal-distribution patterns of zoosporic aquatic-animal mycopathogens indicate that the number of infected hosts varies seasonally (Neish and Hughes 1980; Wood and Willoughby 1986; Puckridge et al. 1989; Levine et al. 1990; Rand 1992a; Bly et al. 1993). In contrast, *Ichthyophonus hoferi* exhibits a secular (as opposed to annual) pattern of occurrence in Nova Scotian yellowtail flounder (*Limanda ferruginea*) and herring (*Clupea harengus*), with a periodicity of between 18 and 30 years (Sindermann 1970). Interestingly, however, Rand (1992a) showed that recruitment of this pathogen into *L. ferruginea* on the Nova Scotian shelf is synchronized seasonally and possibly related to warming of water temperatures in the spring. Those results suggest that attempts to collect mycopathogen-infected aquatic animals should be timed carefully to ensure the highest probability of successful recovery of infected animals.

RESOURCES NEEDED FOR ADEQUATE BIODIVERSITY INVENTORY

COLLECTING SAMPLES

A prerequisite to the study of fungi from aquatic animals is the collection of infected specimens. Most fungus-infected specimens have been collected as moribund or clinically diseased specimens and often are presented in the literature as case reports (Blazer and Wolke 1979; Hatai et al. 1980, 1986a; Hibbitts et al. 1981; Polglase 1981; McLean and Porter 1982; Sparks 1982a, 1982b; Jones and O'Dor 1983; Miller and Fleming 1983; Ramos-Flores 1983; Harrell et al. 1986; Paperna 1986; Polglase et al. 1986; Bower 1987a, 1987b; Wiles and Rand 1987; Hatai and Lawhavinit 1988; Noga et al. 1988; Raghukumar and Lande 1988; Dykstra et al.1989; Muhvich et al. 1989; Rand 1990; Strongman and Rand 1991; Rand and Munden 1993; Strongman et al. 1997; Wolf and Smith 1999; Rand et al. 2000a). The collection of infected animals often is prompted by the need for an accurate diagnosis of the etiological agent and a description of the disease for control and treatment purposes. Infected aquatic animals also are collected during surveys to determine the prevalence of a disease (i.e., number of host infected/number of host examined ×100) and the distribution of the fungal agent within a "normal" host population (Ruggieri et al. 1970; McVicar 1977, 1979, 1981, 1982; Agius 1978; Hicks

1982; Burns 1985; Paperna 1986; Banning 1987; Gartner and Zwerner 1988; Dykstra et al. 1989; Levine et al. 1990; Rand 1990).

The same collection techniques are used to obtain both infected and uninfected specimens. Choice of technique is governed by behavior of the species to be collected, season, sample site, and nature of working conditions. Investigators also must use proper sampling and testing procedures to ensure that a representative sample is obtained in a reproducible manner. According to Amos (1985) and Munro and colleagues (1983), a sample should consist of a statistically significant number of randomly collected specimens. The minimum sample size should provide 95% confidence that infected specimens will be included in the sample assuming a minimum prevalence of infection equal to or greater than 2%, 5%, or 10%, respectively. Tables showing the minimum sample sizes required to detect pathogens for populations varying from 50 to an infinite number of individuals are given in Amos (1985) and Munro and colleagues (1983). In addition, samples should be obtained from sites that have been randomly selected prior to sampling (McVicar 1982; Banning 1987; Levine et al. 1990). Aquatic insects and other microinvertebrates collected (100 specimens/week/site) for fungi in the field should be reared in Petri plates in the laboratory to allow time (generally 1–6 weeks) for infections that were not apparent in the field to appear (W. W. Martin 1987; Whisler 1987). It should be emphasized, however, that even one diseased specimen can provide information useful for diagnostic mycological and histopathological investigations (e.g., Porter 1982; Wiles and Rand 1987; Strongman and Rand 1991; Rand 1996). Specimens should be alive when collected, and, in most instances, processed as soon as possible thereafter.

Disease surveys are hampered by the limitations and selectivity of sampling gear (Munro et al. 1983; Banning 1987). Use of a variety of sampling techniques, or the consistent use of a single method, are good ways to mitigate such problems (Frimeth 1987). Other factors that can interfere with the collection of representative and random samples and with accurate estimation of disease prevalence in free-ranging host species have been discussed by Amos (1985), Banning (1987), Munro and colleagues (1983), and Rand (1990).

HANDLING SPECIMENS

Once collected, infected specimens should be processed before death or as soon after death as possible to avoid degenerative postmortem tissue changes, artifact formation, and specimen contamination from air- and water-borne microorganisms and the normal commensal biota within the host. Contamination makes diagnosis either unreliable or impossible. Specimens that cannot be processed immediately after capture should be refrigerated or kept on ice until they can be examined, but for no more than 12–24 hours (Porter 1982; Levine et al. 1990). If specimens cannot be chilled or examined within 24 hours, they should be preserved in chemical fixatives such as 10% buffered formalin (Humason 1976). Both small and large diseased specimens can be fixed whole, although body cavities of large animals should be opened to allow penetration of the fixative. Diseased-tissue samples (1–5 mm^3) also can be excised from infected specimens and prepared for light microscopy and/or electron microscopy using a suitable fixative (Humason 1976; McDowell and Trump 1976; Culling et al. 1985). Specimens with fungal infections also can be frozen (Amos 1985; Levine et al. 1990; Rand 1990). However, freezing is the least preferred way of storing infected material (Pritchard and Kruse 1982; Bullock 1989). Although brief freezing does not appear to destroy fungal spores, it does severely limit how fungus-infected tissues can be used in subsequent studies (Cooper and Jackson 1981).

Living or dead specimens with mycoses that are collected at remote sites where they cannot be examined can be sent to other laboratories for examination (Porter 1982). Living specimens and chilled, freshly dead material should be transported as quickly as possible under aseptic conditions (Needham 1981). Freshly dead material should be packaged in ethylene-oxide-sterilized plastic sampling bags (e.g., Whirl-paks or baggies) and separated from packing ice that may carry its own fungal community. Fixed materials also should be shipped as quickly as possible. Specimens must be well packed, with contents clearly indicated on the outside of the package. Good packaging in clean or sterile containers protects the specimens in transit as well as personnel that will be handling them (Pritchard and Kruse 1982). Useful guidelines for the packing and mailing of specimens are given in Needham (1981), Pritchard and Kruse (1982), and Bunkley-Williams and Williams (1994).

EXAMINATION OF HOST SPECIMENS

Before a study is initiated, study objectives should be clearly established because they invariably will dictate the procedures and techniques to be used when a specimen is examined. All examinations of both living and dead specimens should begin with the preparation of a case record that notes the following: (1) common and scientific names of the specimen; (2) collection date; (3) collection site with geographic coordinates and elevation, if possible; (4) collector's name and collection number (if the host is maintained); (5) collection methods used;

and (6) specimen length, width, weight, sex (Pritchard and Kruse 1982), and any other specimen-specific data that the investigator considers to be important. The importance of recording these data cannot be overemphasized. Case data can be used to determine disease prevalence in the host population; temporal and spatial distribution patterns of infected specimens (McVicar 1982; Munro et al. 1983; Rand 1990); and age-, sex-, and size-distribution profiles of infected and uninfected specimens (McVicar 1981, 1982; Gartner and Zwerner 1988; Rand 1990). The data also can be used to estimate "wasting," as indicated by reduced condition indices (CI), which are calculated as follows:

$$CI = \frac{\text{animal weight g}}{\text{length in cm}^3} \times 100$$

(Möller 1974; Rand 1990; Rand and Cone 1990), or

$$CI = \frac{\text{mollusc-meat dry weight g}}{\text{volume between valves cm}^3} \times 100$$

(Alderman and Jones 1971). Unpublished case data also can provide comprehensive background information on aquatic animal mycoses that can be a useful guide for subsequent investigators (Austwick and Keymer 1981).

Once those data have been noted, any diagnostic signs of the disease associated with the external surfaces of the specimen should be recorded. If external signs are present, the specimen should be photographed digitally or with color slide film (ASA less than 100), which can be used to produce color slides and prints for presentations as well as high-quality black-and-white negatives for publication. Subsequent steps in the examination should include inspection of smears and squashes of lesion material; thorough inspection of internal structures, using aseptic techniques (Elkan 1981; Needham 1981; Frye 1984; Amos 1985; Klontz 1985); and collection of tissue samples for light and electron microscopy and diagnostic mycological studies. The sequence in which the steps of a postmortem examination are carried out can be varied according to the commercial or aesthetic value of living specimens, specimen "freshness" and size, nature of the working conditions, individual preference, and study objectives. Suggested procedures for postmortem examination are provided in Amos (1985), Elkan (1981), Frye (1984), Pritchard and Kruse (1982), and Bullock (1989).

TISSUE SMEARS AND SQUASH PREPARATIONS

Microscopic examination of smears and/or squashes of lesion material is a common procedure for diagnosing mycotic diseases (Hibbits et al. 1981; Lightner 1981;

Miller and Fleming 1983; Ramos-Flores 1983; Noga and Dykstra 1986; Polglase et al. 1986; Bower, 1987a, 1987b; W. W. Martin 1987; Wiles and Rand 1987; Rand and Cone 1990). Tissue smears and squashes are made by taking exudate and/or tissue samples from the edge and center of a lesion and placing them in a drop of sterile seawater, physiological saline, or staining solution (e.g., lactophenol blue) on a clean glass microscope slide using a sterilized dissecting needle or scalpel. Tissue smears are made by spreading the droplet with a circular motion until it forms a thin film over the surface of the slide. Tissues to be squashed are covered with a glass coverslip or microscope slide, which is pressed gently until the cells are flattened enough to permit the passage of transmitted light. Temporary preparations can be mounted in water or lactic acid and examined directly. Semipermanent and permanent preparations should be fixed in a suitable fixative such as Schaudinn's fluid (Polglase 1981); 10% neutral buffered formalin (Gartner and Zwerner 1988; Rand 1990); alcohol-acetic acid 10% formalin (W. W. Martin 1987), or 5–10% formaldehyde (W. W. Martin 1987). Preparations can be stained, if desired, and examined using a compound microscope equipped with bright-field, phase-contrast, dark-field, or Nomarski differential interference objectives (10×, 40×, 100×).

Stains used to reveal fungi in smeared or squashed lesion material include: lactophenol cotton blue (W. W. Martin 1987; Gartner and Zwerner 1988; Levine et al. 1990); modified Wright stain (Cabisco Chemicals Wright Stain Solution, Carolina Biological Supply, Appendix IV) (Harrell et al. 1986); Gram's stains (Harrell et al. 1986; Wiles and Rand 1987; Schiewe et al. 1988); Semichron's acetocarmine (Gartner and Zwerner 1988); and Roques' strong silver proteinate (Polglase 1981). Stain composition and staining procedures can be found in Humason (1976), Culling and colleagues (1985), and Appendix II.

ISOLATION AND CULTURE

It is usually necessary to isolate and culture fungi from lesions on living or freshly killed specimens before they can be identified or used for experiments. Alderman (1982) suggested that a major problem in investigations of aquatic-animal mycoses is the difficulty of isolating the fungi, especially those from surface lesions. Surface lesions, particularly advanced ones, usually contain mixed infections of both the primary pathogen(s) and secondary invaders (Alderman 1982; Wiles and Rand 1987; Noga et al. 1988; Rand and Wiles 1988). Attempts to isolate fungal pathogens from lesion tissue may fail when those organisms are overgrown by other species present.

In addition, the physiological and nutritional requirements of fungal pathogens are generally unknown and may not be met in culture, explaining why some of the fungi either cannot be cultured or are difficult to maintain (Bahnweg and Bland 1980; Hibbits et al. 1981; Alderman 1982; McLean and Porter 1982; Rand 1990).

In some aquatic animals with translucent bodies, the presence of fungi can be detected by staining the hosts with 0.01% methylene blue or lactophenol (with or without cotton blue). Stained specimens are mounted on glass slides and, using bright-field or phase-contrast microscopy, are examined for evidence of resting spores or mycelia (W. W. Martin 1987; Whisler 1987).

Infected material can be excised using aseptic methods, or it can be scraped from the center or edge of the lesion, where active growth and multiplication of fungi occurs, with a sterilized wire loop, needle, or swab. Sampling lesion material in these ways risks contamination from the commensal mycotic or bacterial community that may be associated with uninfected tissues. That risk can be reduced if lesion material is surface sterilized with 70% ethanol prior to sampling (Wiles and Rand 1987). Another sampling problem concerns the selection and removal of representative material from diseased specimens. As far as I am aware, no one has evaluated the efficacy of the various sampling methods that are used to recover fungi from diseased animal tissues. Tretsven (1963) showed, however, that tissue excision is more efficacious than scraping and swabbing for the recovery of bacteria from fish tissues. For that reason, excision should be considered the sampling technique of choice for investigation of animals with mycotic diseases.

Sampled lesion material should be transferred to a Petri dish so that associated fungal taxa can be isolated by either baiting or plating techniques. A variety of fungi with motile spores have been isolated from diseased lesion material by baiting (Fig. 26.1). This technique involves the transfer of small pieces (less than 5 mm^3) of lesion material into sterilized Erlenmeyer flasks containing 100–150 ml of sterile seawater or physiological saline and a few hemp seeds, pollen grains, algae, blades of grass, defatted hairs, insect wings, flies or other insects, small pieces (less than 1 mm^2) of snake skin, or some other organic material as bait. Alternatively, inoculum can be placed in one section of a four-section Petri dish (available from Fisher Scientific Co.) and bait can be added to the remaining three sections. After 2–7 days of incubation, the bait is removed aseptically and examined microscopically for evidence of fungal colonization. Colonized baits can be removed and plated onto an isolation medium supplemented with antibiotics (0.25–0.50 g/liter of both penicillin G and streptomycin sulphate or chloramphenicol) to suppress bacterial overgrowth, or they can be left to grow on the baits (D. J.

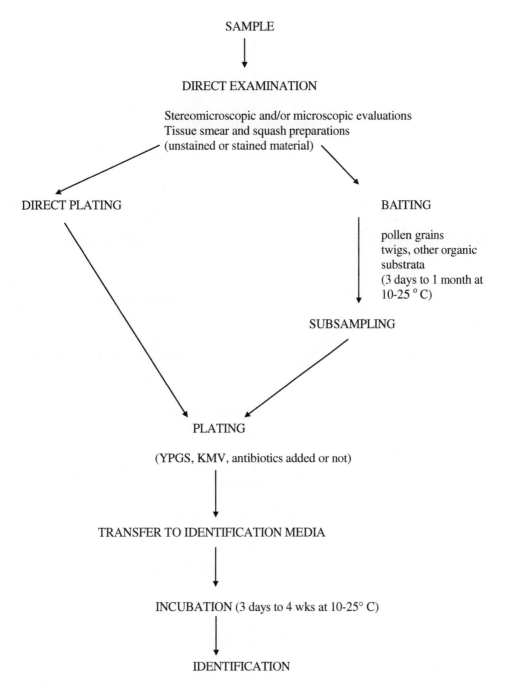

Figure 26.1 Methods for the examination, isolation, and cultivation of zoosporic fungi from aquatic animals.

S. Barr 1987). Pimaricin (Sigma Chemical Company, Appendix IV) can be added (0.1 g/liter) to a medium after autoclaving to inhibit fungi other than Oomycetes (D. J. S. Barr 1987). The purpose of baiting is to limit bacterial overgrowth and at the same time provide a minimally enriched nutritional medium favored by the fungi associated with the lesion material. The bait selected determines which taxa of fungi will grow (D. J. S. Barr

1987; Porter 1987; Whisler 1987). Additional information on baiting technique is available in Sparrow (1960).

Many taxa of fungi, especially species of anamorphic fungi, have been isolated from animal tissues by plating infected lesion material directly onto nutrient medium (Fig. 26.2). Important factors in the isolation of such fungi are the composition, pH, and salt concentration of the medium and the incubation temperature (Johnson

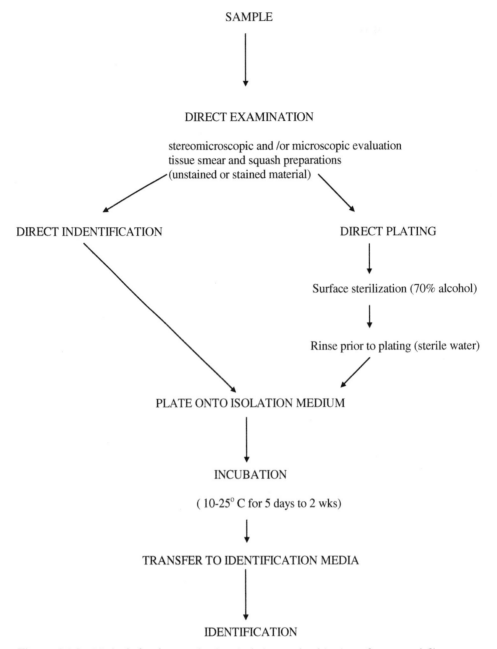

SAMPLE

DIRECT EXAMINATION

stereomicroscopic and /or microscopic evaluation
tissue smear and squash preparations
(unstained or stained material)

DIRECT INDENTIFICATION   DIRECT PLATING

Surface sterilization (70% alcohol)

Rinse prior to plating (sterile water)

PLATE ONTO ISOLATION MEDIUM

INCUBATION

(10-25° C for 5 days to 2 wks)

TRANSFER TO IDENTIFICATION MEDIA

IDENTIFICATION

Figure 26.2 Methods for the examination, isolation, and cultivation of yeasts and filamentous nonzoosporic fungi from aquatic animals.

and Sparrow 1961; Alderman 1982; Okamoto et al. 1985; D. J. S. Barr 1987; Rand 1990). Generally, nonspecific media supplemented with antibiotics, glucose, yeast extract, peptone, and/or salts are used, and samples are incubated at temperatures approximating those of the aquatic environment from which the samples were obtained (about 10–25°C). Media appropriate for the isolation and cultivation of zoosporic fungi and chromista from aquatic animals include KMV (keto methylvalerate), serum seawater agar, yeast protein-glucose agar, yeast protein-soluble starch agar, and water

agar (see Appendix II). For nonzoosporic fungi, appropriate isolation and cultivation media include malt extract agar, potato dextrose agar, dilute potato dextrose agar, and water agar. Formulae for these media are provided in Appendix II. If nonspecific media do not support the isolation and growth of the etiological agent, the medium can be modified. Agarose, for example, can be used instead of agar. Powdered activated charcoal can be added (15 g/liter) to the medium to absorb factors inimical to the growth of fungi (Rand 1990). Fresh seawater, which has some antimycotic properties and may reduce growth of some aquatic fungi, is best avoided (Rand 1990). If seawater is required, it should be stored in the dark for 1–2 months before being used to reduce the mycostatic effect it has on the spores of some fungal species (Kirk 1980). Storage in the dark for an extended period reduces the effect and allows for germination of spores that will not germinate in fresh, natural seawater. An alternative is to include sodium chloride (10.0–25.0 g/liter) or artificial sea salts (1.0–3.6%/liter) in media when isolating and growing marine or estuarine fungi (D. J. S. Barr 1987). Certain fungi can be isolated and maintained on living cell lines or organisms in culture. Harrell and colleagues (1986) and Schiewe and associates (1988) used chinook-salmon-embryo cell cultures for isolating and maintaining a chytridlike pathogen from kidney and splenic tissues of chinook salmon (*Oncorhynchus twhawytscha*). *Ichthyophonus* species can be maintained in the laboratory by force feeding or injecting viable resting spores and other propagules into the peritoneal cavity of uninfected, naive fish hosts (Rand 1990). Although the infected fish die 3–4 months after inoculation, high numbers of infective propagules can be harvested and used to infect other experimentally maintained animals. A similar method is used to maintain *Coelomomyces* from insect larvae in laboratory-maintained colonies of aquatic dipterans (Whisler 1987).

Some fungi from animal tissues cannot be isolated and/or grown in culture but can be collected from lesion material and used for taxonomic or other studies. The structures of ascospores and perithecia of *Trichomaris invadens* were studied after they were removed from carapace surfaces of infected tanner crabs and processed for light and electron microscopy (Porter 1982).

IDENTIFICATION

The generic status of most fungi isolated from lesions can be determined by comparing their taxonomically distinguishing characteristics (including both gross and fine structures as observed *in situ* and *in vitro* (e.g., see Pickering et al. 1979) with the descriptions in taxonomic keys. Methods for the identification of some fungal species have been treated in Alderman (1982), Pickering and Willoughby (1982), Roberts (1989), and Sindermann and Lightner (1988). Useful taxonomic keys for the identification of fungal species that have been recovered from animal tissues include Barron (1983), C. Booth (1971), Couch and Bland (1985), Domsch and colleagues (1993), Ellis (1971, 1976), Johnson and Sparrow (1961), Karling (1977, 1981), Kohlmeyer and Kohlmeyer (1979), Scott (1961), Seymour (1970), and Sparrow (1960). Specific fungal groups in the genera *Aspergillus, Fusarium*, and *Penicillium* are treated in Pitt

(1979, 1991), Klitch and Pitt (1988), Nelson and colleagues (1983), Domsch and colleagues (1993), and Samson and colleagues. (1995). In some cases, taxonomic difficulties or inadequate keys may preclude the ready identification of species (e.g., Beakes 1983; Bower 1987a, 1987b; Rand 1990). For example, the validity of characters used to distinguish some members of the Thraustochytridales is uncertain (Moss 1986; Porter 1986). In addition, some species, especially saprolegniaceous fungi infesting fishes and crayfish, do not produce in culture the reproductive structures that permit their ready identification (Pickering and Willoughby 1982; Bullis et al. 1990; Fraser et al. 1992; Rand 1996). *Ichthyophonus hoferi* is also difficult to identify based on morphology alone (Rand 1996; Rand et al. 2000b). Identification of fungal isolates at the species level may require comparison with original species descriptions available in the literature or with representative specimens on loan from laboratory or museum collections (e.g., The American Type Culture Collection; International Mycological Institute; Canadian Collection of Fungal Cultures; Fusarium Research Center; Herbarium, Department of Fish Diseases, Nippon Veterinary and Animal Science University; National Collection of Yeast Cultures, Food Research Institute; see Appendix III), if they are available. In cases where specific identifications remain intractable, living cultures, along with pertinent collection and background information, can be sent to experts who study the group for assistance.

Many of the taxonomic problems associated with difficult fungal taxa soon may be resolved using molecular techniques. Isozyme profiles and polyclonal-antibody specificities have been used to discriminate among certain strains of pathogenic fungi of fish and aquatic insects (Peduzzi and Bizzozero 1977; Beakes and Ford 1983; Bullis et al. 1990; Khachatourians 1996). The fluorescent-antibody, indirect fluorescent-antibody, and Ouchterlony tests, which use antisera that cause agglutination in the presence of a specific antigen(s), have proved useful for the identification of certain mycopathogens from molluscs and fishes (McVicar 1982; Bower et al. 1989). The tests are relatively easy to use once the antisera have been made (see Cuello 1983; Amos 1985; Culling et al. 1985; Bullock 1989, and references therein). Restriction fragment-length polymorphisms, which are useful in separating and biotyping strains of medically important species of fungi (Pincus et al. 1988; Croft et al. 1990), likely will prove invaluable for distinguishing among aquatic animal mycopathogens (Rand 1996; Geiser et al. 1998). Likewise, the development of single ribosomal DNA probes for highly conserved genetic sequences in aquatic animal mycopathogens should help to reveal systematic relationships and differentiate species in enigmatic genera such as

Branchiomyces, Ichthyophonus, Nephromyces, and *Ostracoblabe* (Rand et al. 1994; Ragan et al. 1996; Rand 1996; Rand et al. 2000b). Those highly sensitive and specific nuclear probes unquestionably will play important roles in the rapid diagnosis and identification of fungal pathogens in both wild and captive aquatic animal populations (Rand 1996). Serological techniques also can be used to identify certain taxa of fungi. Although use of the procedure in studies of aquatic animal mycoses is still in its infancy, it has been used widely in the study of other aquatic animal diseases, especially those of bacterial etiology. Serological techniques provide sensitive, rapid, and inexpensive methods for assessing the immune status of organisms, evaluating vaccination programs, and studying epidemiological aspects of disease (Austin and Austin 1987; Bullock 1989). They also have great potential for investigations of aquatic animal mycoses.

INTEGRATION

On June 11, 1992, Canada became the first of 153 nations to sign the United Nations Draft Convention on Biological Diversity at the Earth Summit in Rio de Janeiro, Brazil (Morrow 1993). The importance of that convention cannot be overstated. In principal, it stands to preserve the ecological, genetic, social, economic, scientific, cultural, recreational, and aesthetic values of biological diversity and its components, which are considered important for evolution and for maintaining life-sustaining systems of the biosphere (Morrow 1993). Despite those very noble convention objectives, we have, in fact, little understanding of how the fungi now known interact with the abiotic and biotic components of their environments. That deficiency is as true for the fungi associated with aquatic animals as it is for almost every other group of fungi known. Our ignorance derives largely from a lack of manpower and financial resources, a point so aptly recognized by Hawksworth (1991). At a time when there is renewed interest in conservation, biodiversity, the environment, and biotechnology and when the need for individuals with the taxonomic skills to document biodiversity is increasing, infrastructural and financial support for systematic mycology has declined (Hawksworth 1991). The implications of the Convention for aquatic-animal-mycopathogen research and for our understanding of biodiversity are immense.

Reported cases of mycotoxicoses and infectious species likely will increase as the aquaculture industry expands to include new and exotic species and moves into new geographic locations (Rand 1996). If that expanded economic activity is to be sustained, then it is imperative that animal health be ensured. That requires that crucial pathogens be collected properly, identified correctly for risk assessment, and properly maintained (Miller 1995). The knowledge obtained, and the pathogens, then can be used for the management of aquatic animal diseases through prevention and the development of pathogen-resistant hosts. Pathogen cultures also can be used in the development of gene probes that will facilitate rapid diagnoses and enable managers to monitor pathogen abundances and movements in the field. The latter information, especially, will lead to a greater understanding of the diversity and dynamics of those pathogens as well as of their ecological role in natural systems. Moreover, the isolation and maintenance of some pathogens may lead to the development of commercial microbial products, such as pesticides, fertilizers, and drugs, and technologies (Miller 1995). All of those activities demand individuals highly knowledgeable in the area of fungal taxonomy and trained in the isolation and cultivation of fungi from aquatic animals.

Focusing more attention on aquatic-animal mycoses will lead to an increase in the number of recognized pathogenic and mycotoxin-producing species reported from that group of potential hosts. Unfortunately, much of that information probably will be scattered widely in the scientific literature, making it difficult for many workers to access it readily. Establishment of a database to provide an up-to date, easily accessible list of the fungal species, their taxonomic affiliations, the scientific and common names of host species, site(s) of infection, habitat type, geographic distribution, and relevant scientific references would help to alleviate this problem. Such data banks already have been started at some of the larger mycological herbaria (Hawksworth 1991; Miller 1995). The Biotechnology Section, Commercial Chemicals Branch, Environment Canada already has developed such a database that provides information on protistan and microbial pathogens, including some 209 fungal species from aquatic animals and plants, worldwide. Greater financial support is needed to investigate and document the pathogenic and toxin-producing fungi from aquatic-animal hosts to meet the objectives set out in the Earth Summit Conference.

III

APPENDICES, GLOSSARY, LITERATURE CITED, AND INDEX

I

MOIST CHAMBERS FOR THE DEVELOPMENT OF FUNGI

JOHN C. KRUG

Genera and/or species in various groups of fungi can be distinguished only on the basis of characteristics of their fruiting bodies or spores. For other genera and species, such characters enhance the ease and accuracy of identification. Consequently, obtaining fruiting bodies and spores of fungi located during inventories and monitoring surveys is critical (or, at least, highly desirable) if species richness of an area and geographic and habitat distributions of particular species are to be determined. Obtaining such material is not always easy because many species of fungi do not fruit on a regular basis but only under highly specific conditions. The fruiting bodies of other species are so small and/or rare that they can be difficult to locate. To increase the likelihood of recording nonfruiting, as well as rare, tiny, or inconspicuous forms, mycologists have developed a laboratory method to increase detection—the moist-chamber technique. This technique involves keeping a substratum moist and maintaining it over long periods in a chamber in which ambient conditions may be manipulated; during that period, the researcher observes and records the sequence of fungi as they develop and fruit (see "Taxonomy, Diversity, and Distribution" in Chapter 25). The fungi also can be removed for additional culturing and for study.

The appropriateness of the technique depends on the types of fungi likely to be present on the substratum, but frequently it is the only means of detecting or stimulating development of minute fruiting bodies. Dormant spores and/or mycelia immersed in substrata collected under dry weather conditions often can be induced to form fruiting bodies, which facilitates a better understanding of a component mycobiota. In addition, it is possible to stimulate maturation of immature fruiting bodies, rehydrate desiccated fruiting bodies, or sometimes break the dormancy of resting spores and sclerotia in such chambers. This appendix offers suggestions of materials appropriate for the construction of moist chambers, reviews conditions under which such chambers might be maintained, considers the use of moist chambers for different types of fungi, and discusses potential problems with the technique.

STANDARD CHAMBER

A chamber consists of a large glass or plastic container lined on the bottom with some absorbent material (e.g., Perlite) overlaid with sterile filter paper. The chamber is moistened with sterilized, glass-distilled water, which is applied with an atomizer. The choice of chamber depends on the group of fungi, the size of the substratum, and the nature of the research. If the chamber is to be reused, it must be able to withstand autoclaving or harsh disinfection. Dry heat sterilization can be used for chambers that may not withstand autoclaving (e.g., certain plastic containers). See Figure I.1.

EQUIPMENT

Containers

A wide range of containers can be used as chambers, including, for example, glass Petri plates, glass crystallizing dishes, glass baking dishes, and glass dishes or bowls

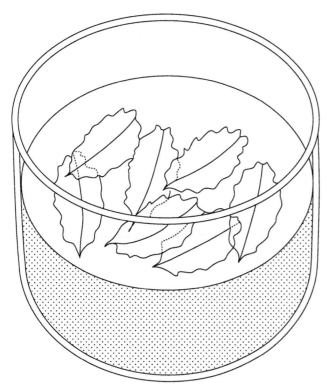

FIGURE I.1 Diagram of typical moist chamber, with cover removed, composed of glass container with substratum placed on sterilized filter paper lying above moisture-absorbing material.

with straight sides. The latter two can be covered with square pieces of glass. Heavy "stacking" glass culture dishes are another good choice. Plastic Petri plates; plastic sandwich boxes; plastic bakery, yogurt, or margarine containers; salad bar bowls; boxes; and dishes also can be used, as can plastic bottles (e.g., beverage and vinegar bottles), polycarbonate and metal storage boxes, and plastic bags with and without a zipper closing.

Labeling Devices

Enough essential information (e.g., moist-chamber or collection number, initiation date, substratum, locality) must be included on the individual substrata, the containers in a chamber, or the chamber itself to ensure that samples do not become confused. Information should be written with a permanent ink magic marker, Staedtler waterproof marker, wax pencil, or other device that is persistent and resistant to water. Full collection data are recorded in the field or laboratory record book.

Absorbing Materials

Filter paper (coarse grade), paper towels, *Sphagnum* or comparable absorbing mosses, vermiculite, Perlite, sand, and wood chips can be used to line moist chambers. Except for Perlite, which is essentially inert, and filter paper, which may be more or less sterile, materials should be heat sterilized. Certainly *Sphagnum* moss can possess foreign fungi, and Actinomycetes may be present in vermiculite. Some materials (e.g., vermiculite, wood chips) should be rinsed lightly in water before sterilization to remove any dirt and adhering debris and to aid moisture penetration. Placing the filter paper on the lid of the chamber can help to control excess moisture; placing it on the floor of the chamber frequently leads to excessive moisture. Small and delicate substrata (e.g., small leaves or stems) can be supported on top of glass rods or tooth picks so they avoid direct contact with the wet chamber floor.

Wetting Agents and Devices

Water usually is used to wet the substratum and to maintain moisture in the chamber. In cases in which a change in pH is desirable (e.g., to approach a neutral pH or to approximate the conditions preferred by certain fungi more closely) agents such as acetic acid or ammonium hydroxide can be substituted. Sterilized, glass-distilled water is preferred to avoid introduction of foreign microbes. I do not recommend use of tap water with chlorine additives. Water can be applied directly from a tap, from a fine pipette, or with a squirt gun or atomizer. The wetting device should have a volume regulator, however, so that excess moisture will not be dispensed.

Insecticides

Appropriate insecticides for use in moist chambers include methyl benzoate, naphthalene, and para-dichlorobenzene (PDB). The latter two substances are applied as crystals and used only under conditions of extreme infection.

INCUBATION

Ambient temperature and normal daylight are adequate for incubation under most circumstances. Slight air movement is also generally desirable, especially if the air is heavy or humid. Placing the chamber near an open window or slow-moving fan usually ensures sufficient air movement. Generally, it is preferable to use a relatively cool incubation temperature (not higher than 18°C). Incubation of some chambers at warmer temperatures, however, or even in the refrigerator, may stimulate additional species adapted to such temperatures to develop, depending on the source of the substratum (e.g., subtropical or arctic sites). Likewise, varying other standard conditions may stimulate fruiting of additional species with more specific fruiting requirements. Incubation under ultraviolet light, for example, may stimulate additional species to produce fruiting bodies, and subjecting some chambers to a wet/dry moisture cycle may encourage development of species adapted to different moisture levels or a dormancy period. A standard 12 hours/12 hours light/dark cycle also may be substituted for normal day length if a standardized day length is desirable.

CHAMBERS FOR USE WITH PARTICULAR FUNGAL GROUPS OR SITUATIONS

CORTICOLOUS AND HERBACEOUS MYXOMYCETES

The bottom of a plastic or glass Petri dish (100 mm in diameter × 20-mm high) is lined with sterile filter-paper discs or absorbent paper toweling. The sample is placed on the filter paper so that the bottom of the dish is covered but the individual pieces of substratum do not overlap. Enough sterile, glass-distilled water is added (10–15 ml) to wet the sample thoroughly. The correct amount of water will vary with the type and size of sample and its absorptive capacity; different species of bark, for example, have different absorptive capacities. Alternatively, the sample may be soaked in water before placement in the Petri dish. If desired, the pH may be adjusted to 7.0 by adding dilute potassium hydroxide or acetic acid to ensure that the water does not influence which fungi appear in the chamber. The plates then are set aside to soak for 10–12 hours to thoroughly saturate the sample(s). After soaking, excess water is removed carefully with a small pipette or, if necessary, more water is added. Too much water can interfere with development of the tiny *Echinostelia*, which may appear in the first 24 hours; not enough water can impede normal development of physaraceous and stemonitaceous species, which may develop later. Allowing the substratum (e.g., bark culture) to dry after the development of the first series of organisms and then rewetting it appears to stimulate development and fruiting of additional species, perhaps because some species go through an encystment stage before fruiting is initiated. Long periods of incubation under moist conditions favor the development of aphanoplasmodial and phaneroplasmodial species (Keller and Braun 1999).

MATURATION OF LARGE ASCOCARPS AND STROMATA

The maturation of ascocarps and stromata often can be induced by wrapping perithecia and stromata in moistened paper toweling, placing them in heavy glass bowls, and leaving them to incubate for an appropriate number of days. Lining the bowls with moistened Perlite or wood chips will increase the moisture level in the chamber. Placing perithecia and stromata in closed plastic bags also can be effective.

DEVELOPMENT OF MICROSCOPIC FUNGI

Leaves, bark, wood, fruits, dung, seaweed, and similar substrata can be placed in plastic or glass containers with appropriate covers. For large pieces of wood, transparent polycarbonate boxes with snap-on lids are especially useful. The bottom of the chamber is lined with Perlite or similar absorbent material and covered with moistened, sterile filter paper. The substratum is presoaked or placed directly on the filter paper. Excess water is removed, although periodic remoistening may be required. A succession of fungi should appear, requiring that the chamber be examined frequently. As a substratum becomes depleted from the removal of developing species for study, additional substratum can be moistened and placed in the chamber.

A plastic salad-type container with the bowl inverted over the lid also can be used as a chamber. The lid, which is lined with a moistened filter paper, can be snapped shut or left partially open for aeration; the substratum is placed in the lid. That setup, which is especially good for wood and seaweed, allows for easy manipulation of material and fits under a dissecting microscope.

SUCCESSION OF COPROPHILOUS FUNGI

Petri plates (100 mm in diameter × 20 mm high) make ideal chambers for the dung of small animals, which can be an excellent source of Zygomycetes and some Ascomycetes. Filter paper is fitted into the bottom of the plate and moistened, and excess water is removed with a pipette. The dung is placed on the filter paper, with pellets at least 1.0–1.5 mm apart.

Sterilizable glass containers with lids are most appropriate for large pieces of substratum. Such substrata yield some Zygomycetes, a variety of Ascomycetes and Basidiomycetes, and some Hyphomycetes and Myxomycetes. Both Furuya (1990) and Lundqvist (1972) described such chambers. The dung is soaked for several hours and then placed in a chamber containing a suitable absorptive material, such as *Sphagnum* moss or Perlite, usually overlain with filter paper. Excess water is removed. As the chamber dries, water is added (usually every few days) with an atomizer. Maintaining some chambers with excess moisture may encourage development of additional species of Basidiomycetes because the sequence in appearance of species in that group varies with the wetness of the dung (Richardson and Watling 1997).

Some additional fungi, especially Basidiomycetes, may be obtained if a thick layer of a sterilized, general-purpose isolation medium (e.g., Leonian's agar medium [ML]; Appendix II) is poured into a sterilized chamber. The dung is placed on the agar and incubated. The moisture from the agar is sufficient; supplementary water is not required.

Richardson and Watling (1997) suggested placing the top one-third of a plastic bottle (e.g., from lemonade or mineral water) onto a Petri-plate chamber. The cap of the bottle is replaced with a cotton plug, thus ensuring aeration. The bottle provides the chamber height required by developing Basidiomycetes, especially Agaricales.

INITIATION OF FUNGAL GROWTH IN CULTURE

A deep, sterilized, plastic or glass Petri dish containing a layer of water agar (lacking nutrients) or half-strength cornmeal agar provides an especially good chamber environment for organisms that do not complete their life cycle in culture (e.g., Zoopagales and their microscopic animal associates). The substratum is placed on the agar and incubated. The outgrowth of hyphae can be observed under a compound microscope through the lower side of the Petri dish.

Plasmodia from species of plasmodial slime molds, which actively feed on bark surfaces in moist-chamber cultures, sometimes do not fruit. If a thin slice of the substratum with a portion of the plasmodium is removed and placed on an appropriate agar medium, the plasmodium will spread over the agar surface (Timnick 1947). If the plasmodium continues to feed and grow, fruiting bodies eventually will form on the agar.

SPORE PRINT CHAMBERS

A spore print is an accumulation of basidiospores obtained during their natural ballistic discharge from the fruiting body. A setup to capture the print is prepared by separating the cap (or a section of the cap) of a fleshy Basidiomycete (e.g., Agaricales) from the stipe and placing it lamellae- or pore-side down on a piece of white paper. Care must be taken not to damage the lamellae or pores and to maintain them in a vertical position. The paper and the cap then are placed in a deep container, relative to the height of the fruiting body, and covered with a sheet of paper toweling. The chamber is closed or covered with a lid and maintained at room temperature. Maintenance of the spore-print setup in a reclosable plastic bag at room temperature also works well for many fungi (e.g., *Cortinarius* species). Sterile 60- or 100-mm Petri dishes are very good spore-print chambers for small fruiting bodies when the spore print is to be saved or transported for subsequent culturing of the basidiospores. Temperature extremes should be avoided. This technique is ideal for small to medium-size Agaricales such as *Clitocybe* species and *Lactarius* species.

POTENTIAL PROBLEMS

One problem sometimes encountered in moist chambers is overgrowth of the substratum by aggressive fungal invaders, frequently Hyphomycetes and sometimes Zygomycetes. Species of *Trichoderma* are, on occasion, especially aggressive in chambers with coprophilous fungi. Incubation at a relatively cool temperature (e.g., ≤18°C) may help to prevent that problem. In addition, the water in the chamber should be limited so that the chamber is not saturated but retains adequate moisture to induce fungal development. The formation of glossy moisture films on the surface of the substratum indicates that the chamber is too wet. Excessive moisture can be reduced by placing salt solutions in the chamber. When remoistening a chamber, as little moisture as possible should be added. Small pipettes and fine atomizers are effective at controlling the amount of moisture released. Use of a secondary moist chamber is another way to maintain a humid atmosphere while preventing direct contact of the substrata with excessive moisture. A Styrofoam ice chest or large plastic box is lined with moist towels or a layer of moist vermiculite, and the small moist chambers are placed inside it.

The appearance in a chamber of mites, fly larvae, collembola, and parasitic worms, sometimes originating in the substratum, can be detrimental because insects and mites eat fungal mycelia and fruiting bodies. Such animals also may spread undesirable contaminating fungi (e.g., *Trichoderma* species, *Verticillium lecanii*). Placing a few crystals of paradichlorobenzene or naphthalene in the chamber usually will solve the problem. However, careful examination of certain substrata, such as dung, for eggs, pupae, and cysts of invertebrates and their removal also will help to prevent introduction of such pests. Preferably, a piece of substratum lacking detectable eggs, pupae, or cysts should be selected. Because contamination of cultures with mites and fly larvae is possible, moist chambers should be incubated away from sterile culturing areas and culture collections. Also, once a substratum is exhausted or a chamber is old and unproductive, chamber contents should be sterilized and discarded to prevent any mites present from moving to other chambers and cultures. If insects or other arthropods persist despite such precautions, the chamber can be sterilized again, or under extreme circumstances,

increased concentrations of insecticides (e.g., methyl benzoate) can be added. The latter option is rarely used because some insecticides interfere with sporulation and development of some fungi. Because some insects and other arthropods act as fungal vectors, especially for some coprophilous fungi, the application of insecticides also may interfere with or prevent further fungal succession. Chambers suspected of containing mites can be isolated on a raised pedestal, or similar object, to prevent the spread of mites to other chambers while retaining the potential as vectors. Richardson and Watling (1997) suggested that mites can be controlled by placing a chamber on a glass plate smeared with petroleum jelly to which an acaricide (e.g., methyl benzoate) has been added. Another alternative is to place the chamber on top of a Petri dish, resting in a tray with a thin layer of mineral oil.

In some tropical and subtropical regions, substrata may disintegrate rapidly because of microbial activity, a process frequently enhanced by foreign invaders. In the case of coprophilous fungi, dung beetles often are also a potential problem. Placing the substratum or, where appropriate, the chamber in a balloon or closed mesh bag may help to exclude such pests.

ACKNOWLEDGMENTS. I with to thank the many individuals who provided information, suggestions, and editorial comments. Special thanks go to Drs. J. Ammirati, G. L. Benny, G. Bills, M. Blackwell, W. Gams, H. Keller, D. Malloch, D. Pfister, M. Richardson, and G. Samuels.

FORMULAE FOR SELECTED MATERIALS USED TO ISOLATE AND STUDY FUNGI AND FUNGAL ALLIES

COMPILED BY GERALD F. BILLS AND MERCEDES S. FOSTER

INTRODUCTION

Laboratory studies of fungi are an essential complement to their observation in the field. In this appendix, we provide formulae for media and reagents used in the laboratory for isolation of fungi into pure culture and for their cultivation, tissue preservation, microscopical examination, and biochemical and physiological characterization. Laboratory and microscopy techniques used to study fungi are nearly as varied as the fungi themselves. The material we provide here focuses on techniques, media, and reagents mentioned in the handbook volume. It is not meant to be exhaustive, and additional recipes and procedures can be found among previous compilations of mycological methods (Tuite 1969; C. Booth 1971; Largent et al. 1977; Watling 1980; Stevens 1981; Fuller and Jaworski 1987; Singleton et al. 1992; Gams et al. 1998; Kurtzman and Fell 1998).

Isolation and cultivation of fungi in pure culture require preparation of sterilized culture media. Most culture media contain at least water, a simple or complex source of carbon (carbohydrates, lipids, peptides, amino acids, organic acids) and nitrogen (peptides, amino acids, ammonium, nitrate), phosphate, sulfur, essential metal cations, and sometimes vitamins and essential amino acids. Nutrients can be supplied as pure reagents, when all media components need to be defined and reproducible, or, as is often the case, as complex plant- or animal-derived additives (such as oatmeal, hay infusion, peptone, sheep blood). Agar is added to solidify media. Fungal growth and development are extremely sensitive to the composition of the medium, to laboratory conditions, and to the physical environment (light, temperature, humidity). Those factors often are varied to influence the growth form, life cycle, and metabolic products of fungi in culture. For example, low-nutrient, or dilute, media or media with only polysaccharide carbon sources often favor sporulation and suppress mycelial growth. Media with high concentrations of monosaccharides or disaccharides tend to favor vegetative growth, whereas extremely high carbohydrate or salt concentrations may inhibit growth by reducing water activity.

A pressurized steam chamber (i.e., an autoclave or pressure cooker) is the most common means of sterilizing media; the heat also solubilizes agar and other media components. Autoclaves and pressure cookers are potentially dangerous if used improperly. Anyone using them must understand fully their operation, capacity, and safety features. Solutions also can be sterilized by filtration through membranes with 0.2-μm-size pores and designed to exclude bacterial and fungal cells. Filter sterilization is necessary when heating decomposes or deactivates components of the solution.

Except where indicated otherwise, media components are weighed on a precision balance and added to a 2,000-ml flask or Pyrex bottle. About half the water and agar are added to the flask along with the other ingredients. The mixture is heated and stirred constantly until the ingredients dissolve. The remaining water and agar then are added, and the mixture is heated again until completely dissolved. The liquid mixture is poured into a number of smaller (about 500 ml) flasks or bottles with caps until the bottles are about half filled. If flasks or bottles are filled completely, the mixture will boil over during sterilization. The flasks or bottles are capped loosely or covered with aluminum foil and then sterilized in an autoclave or pressure cooker for 15–20 minutes at 121°C (7–12 kg [15–25 lb] pressure), after which the caps or foils are tightened and the bottles are allowed to cool. Sterilization times may need to be extended to 30–45 minutes for especially viscous or dirty natural materials, such as soil or oatmeal, or carefully controlled for components that may suffer from heat degradation, such as high glucose concentrations. Heat-sensitive components, such as vitamins or antibiotics, are added after sterilization but before the medium cools completely and solidifies. Warm media (45–50°C) are dispensed into sterile Petri dishes. Warm melted media also can be measured and dispensed into tubes (to about one-fourth to one-third the tube volume) with a pipette or peristaltic pump before autoclaving. Tubes then are capped loosely and autoclaved. After autoclaving, but before the agar solidifies, the tubes are inclined or slanted to increase the growing surface of the solidified agar, and the caps are tightened.

The prescribed amount of powdered agar should be increased by about 5 g per liter whenever media are prepared in a very warm climate or if an especially solid medium, such as that used for single-spore isolations, is desired. If the medium requires a specific pH, then a pH meter probe is placed into the solubilized medium and an acid or base solution slowly is added while the medium is stirred with a magnetic bar.

Direct microscopic observation of fungi is fundamental for their study and identification, and stereomicro-scopes and compound microscopes are essential laboratory tools. Fungi are observed directly by stereomicroscopy or are dissected and mounted on microscope slides for observation by compound microscopy. Cells and tissues can be observed unstained with bright-field, phase-contract, or differential-interference contrast microscopy. Staining improves contrast and visibility of certain fungal structures, differentiates fungal cells from surrounding host tissues, and helps the investigator to visualize and diagnose subcellular organelles. Mounting media are used for long-term storage or permanent archiving of fungal cells or tissues mounted on microscope slides. Cell and tissue fixation may be needed to kill fungal cells, improve penetration by straining agents, or prevent degradation of cellular structures after cell death for long-term storage.

CULTURE AND ISOLATION MEDIA

ALPHA-CELLULOSE (ALPHACEL) AGAR

(Sloan et al. 1960; Weitzman and Silva-Hutner 1967)

Alpha-cellulose	20.0 g
Coconut milk (from a ripe coconut)	50.0 ml
Oatmeal cereal for babies	10.0 g
Tomato paste	10.0 g
KH_2PO_4	1.5 g
$MgSO_4 + 7H_2O$	1.0 g
$NaNO_3$	1.0 g
Yeast extract	3.0 g
Agar	20.0 g
Distilled water	1,000.0 ml

Final pH, unadjusted. Medium appropriate for characterization of nonsporulating isolates from soil.

A/T OOMYCETE MEDIUM

(modified from Newall and Fell 1994)

V8 juice	20.0 ml
Amphotericin-B	2.5 mg
Thiabendazole	10.0 mg
Chloramphenicol	10.0 mg
Ampicillin	250.0 mg
Rifampicin	10.0 mg
Agar	20.0 g
Dilute seawater (15 g salts/liter)	1,000.0 ml

Bring seawater, V8 juice, and agar to boil. Autoclave for 20 minutes. Cool to 40–50°C. Add inhibitors as mixed powder, and agitate thoroughly to mix. Some ingredients are highly toxic; laboratory safety-apparel and gloves are required when handling. This is particularly effective at selecting for *Halophytophthora* species from submerged organic debris.

BANDONI'S SORBOSE-YEAST EXTRACT-TETRACYCLINE MEDIUM

(Bandoni 1981)

L(-) sorbose	4.0 g
Yeast extract	0.5 g
Distilled water	1,000.0 ml

After autoclaving, cool to 50°C and add the following:

Tetracycline (in 95% ethanol)	100.0 mg

Other antibacterial antibiotics may be substituted for isolations from natural substrata.

BASAL CONTROL MEDIUM—BAM

(Wubah et al. 1991d)

Ingredients per 100-ml batch include the following:

Mineral solution A (Caldwell and Bryant 1966)	15.0 ml
Mineral solution B (Caldwell and Bryant 1966)	15.0 ml
0.1% (w/v) resazurin	0.1 ml
Volatile fatty acids solution (Caldwell and Bryant 1966)	0.2 ml
0.05% (w/v) hemin in 1:1 ethanol-0.05M NaOH solution	0.2 ml
Cellobiose	2.0 g
Yeast extract	3.5 mg
Trypticase	14.0 mg
Cysteine hydrochloride	3.5 mg
Sodium carbonate	21.0 mg
Deionized distilled water	50.0 ml

BLOOD AGAR, SHEEP

(modified from Atlas and Parks 1993)

Heart infusion agar*	120.0 g
Distilled water	3,000.0 ml

Autoclave at 15 psi for 25 minutes. Cool and add the following:

Sterile, defibrinated sheep blood	160.0 ml

Optional supplements:

(a) Chloramphenicol, gentamicin, and cycloheximide, alone or in combination (see "Antibacterial Antibiotics" under "Antibiotics," later in this appendix). Add to medium just before pouring. Dispense aseptically, and slant in sterile tubes.

(b) Antibiotic solution and egg albumen (Kane et al. 1983). Add 90.0 ml albumen with the antibiotics and blood. Sterilize 250-ml blender cylinder. Clean eggs, then soak in 95% ethanol for 15 minutes. Clean hands with alcohol. Crack eggs with sterile knife; separate egg whites into sterile blender. Quickly switch blender on and off to beat egg whites evenly (if overbeaten, too much froth forms; too much oxygenation destroys activity). Pour egg whites into sterile flask. Store any excess at 4°C.

BRAIN-HEART-INFUSION AGAR*

(Atlas and Parks 1993)

Calf-brain infusion	200.0 g
Beef-heart infusion	250.0 g
Peptone	10.0 g
Glucose	2.0 g
NaCl	5.0 g
Na_2HPO_4	2.5 g
Agar	15.0 g
Deionized water	1,000.0 ml

Mix ingredients and bring to boil to dissolve completely. Optional additions (alone or in combination) include chloramphenicol, gentamicin, or cycloheximide. Add to medium just before pouring; mix well, and dispense.

BROMCRESOL PURPLE-MILK SOLIDS-YEAST EXTRACT MEDIUM—BCP-MS-YE

(Kane and Smitka 1978)

Milk solution:

Skim milk powder	80.0 g
Bromcresol purple (1.6% in ethanol)	2.0 ml
Yeast extract	4.0 g
Distilled water	1,000.0 ml

Put on magnetic stirrer; mix well, and autoclave at 10 psi for 8 minutes.

Agar solution:

Agar	30.0 g
Distilled water	900.0 ml

Soak mixture for 15 minutes; autoclave at 15 psi for 15 minutes.

Mix milk and agar solutions and cool to 50°C. Optional additions (alone or in combination) include chloramphenicol, cycloheximide, and gentamicin. Add to medium just before pouring. Mix well, dispense, and slant in sterile tubes.

Brain-Heart Infusion Agar, Dilute (1/10 BHIv)

Agar	15.0 g
Brain-heart infusion*	3.7 g
Thiamine HCl	200.0 µg
Biotin	50.0 µg
Glass-distilled water	1,000.0 ml

This medium is prepared by using 10% of the recipe for commercially prepared, dried brain-heart infusion. The two vitamins may not be essential, but some evidence suggests that they are stimulatory to some species of fungi. Medium also can be used for storing refrigerated cultures.

Caffeic Acid Agar

(Salkin 1988; also known as Phenol Oxidase Medium, Pincus and Salkin 1988)

Glucose	5.0 g
Ammonium sulfate	5.0 g
Yeast extract	2.0 g
KH$_2$PO$_4$	0.8 g
MgSO$_4$ + 7H$_2$O	0.7 g
Caffeic acid	180.0 mg
Ferric citrate	2.0 mg
Purified agar (Noble)	20.0 g
Distilled water	1,000.0 ml

While mixing, bring agar and distilled water to a boil. Add remaining ingredients, and mix until dissolved. Autoclave for 12 minutes at 15 psi. Final pH, 5.5 + 0.2. *Cryptococcus neoformans* and *C. laurentii* can be used as positive and negative quality control organisms, respectively.

Carboxymethyl Cellulose Agar—CMC

(Gams 1960; Söderström and Bååth 1978; Bååth 1988)

Sodium carboxymethyl cellulose	10.0 g
Yeast extract	0.5 g
(NH$_4$)$_2$SO$_4$	1.0 g
NaNO$_3$	2.0 g
KH$_2$PO$_4$	1.0 g
MgSO$_4$ + 7H$_2$O	0.5 g
KCl	0.5 g
CaCl$_2$	50.0 mg
FeSO$_4$ + 7H$_2$O	10.0 mg
CuSO$_4$	10.0 mg
MnSO$_4$ + 4H$_2$O	5.0 mg
ZnSO$_4$ + 7H$_2$O	1.0 mg
Agar	15.0 g
Water	1,000.0 ml

For convenience, the last five salts can be combined as a stock solution. Sodium carboxymethyl cellulose forms a highly insoluble gel, and some brands are easier to dissolve than others. Combine all dry ingredients in a flask. Slowly add hot water, stirring constantly. Stir mixture for 15–30 minutes. Medium probably will be lumpy; most lumps disperse after autoclaving. Add antibacterial antibiotics after autoclaving (see "Antibacterial Antibiotics" under "Antibiotics," later). This is a low-nutrient source; it supports only limited colony development. Because the medium is nearly transparent, however, minute hyphal growth is seen easily with a dissecting scope. It is appropriate for isolating and cultivating soil fungi.

Cellulose Azure Broth Tubes

(Smith 1977; Thorn 1993)

Basal medium:

KH$_2$PO$_4$	1.25 g
MgSO$_4$ + 7H$_2$O	625.0 mg
Agar	15.0 g
Distilled water	1,000.0 ml

Overlay:

Yeast extract	1.0 g
Malt extract	5.0 g
Cellulose azure	5.0 g
Agar	15.0 g
Distilled water	1,000.0 ml

Dissolve basal medium by heating; then dispense 15 ml each into 25-ml square flint-glass bottles; autoclave, and allow to stand vertically. Prepare a small batch of overlay medium (20% more than apparently required) in a flask with magnetic stir bar. Heat to dissolve, autoclave, and then, while stirring constantly, dispense 0.8 ml aseptically over the surface of the basal medium in each bottle using a broad-bore pipette (e.g., 1-ml automatic pipette). Cellulose azure must be washed repeatedly in distilled water to remove soluble dye and then dried before use. It is appropriate for detection of cellulolytic activity and characterization of nonsporulating isolates.

Choanephora Agar—CH

(Barnett and Lilly 1955)

Glucose	3.0 g
Casamino acids*	2.0 g
KH$_2$PO$_4$	1.0 g
MgSO$_4$ + 7H$_2$O	0.5 g
Thiamine HCl	25.0 mg
Distilled water	1,000.0 ml
Agar	20.0 g

Mix and adjust pH to 6.0 before autoclaving.

CHOCOLATE AGAR*

(Balows et al. 1985)

Casein	7.5 g
Peptone	7.5 g
Cornstarch	1.0 g
K₂HPO₄	4.0 g
KH₂PO₄	1.0 g
NaCl	5.0 g
Agar	15.0 g
Distilled water	1,000.0 ml

Prepare and autoclave the sterile basal medium listed earlier. Aseptically add 5–10% v/v defibrinated blood (sheep, horse, rabbit) and heat at 80°C for 15 minutes or until the color is chocolate brown. Other basal media, such as dehydrated casein-peptone agar[a], GC agar base[a], or GC medium base[a], may be substituted. This medium is appropriate for culture and diagnosis of fungi from clinical samples.

CHRISTENSEN'S UREA BROTH, MODIFIED

(Kane and Fischer 1971)

(a)	KH₂PO₄	2.0 g
	Yeast extract	1.0 g
	Phenol red (0.2%, in ethanol)	8.0 ml
	Distilled water	900.0 ml

Adjust to pH 6.9 with 1.0 M KOH, and autoclave.

(b)	Urea	20.0 g
	Glucose	1.0 g
	Distilled water	100.0 ml

Sterilize (b) with a 0.2-μm filter and add aseptically to (a). Dispense 8 ml of mixture aseptically into autoclaved 25-ml square flint bottles with a repeating syringe or pipette. This medium is appropriate for characterization of nonsporulating isolates.

CORNMEAL AGAR—CMA

Yellow cornmeal	40.0 g
Agar	15.0 g
Distilled water	1,000.0 ml

Boil water and add cornmeal. Simmer for 15 minutes. Filter through three or four layers of gauze. Allow the infusion to rest 15 minutes. Slowly decant all but the bottom 200 ml. Bring volume back to 1,000 ml and add agar. Dehydrated cornmeal agar[a] is available from various distributors. Antibacterial antibiotics should be added after autoclaving if the medium is used for isolations from natural substrata. This medium is appropriate for isolating and cultivating endophytes, microfungi, soil fungi, and others.

CORNMEAL AGAR, ALKALINE—ACMA

(Nagai et al. 1995, 1998)

(a)	Dehydrated cornmeal agar*	17.0 g
	Distilled water	900.0 ml
(b)	NaH₂PO₄ + 2H₂O	3.0 g
	Na₂CO₃	3.0 g
	Distilled water	100.0 ml

Autoclave solutions (a) and (b) separately, then mix. Final pH should be adjusted with 1N NaOH to about 9.7. This is an appropriate medium for isolating and cultivating alkalophilic fungi.

CORNMEAL AGAR—CM+

Cornmeal agar*	17.0 g
Glucose	2.0 g
Yeast extract	1.0 g
Distilled water	1,000.0 ml

Medium appropriate for culture of myxomycetes, some heartier species of protostelids and acrasids, as well as other filamentous fungi.

CORNMEAL AGAR—CM/2

Dehydrated cornmeal agar*	8.5 g
Agar	7.5 g
Glucose	1.0 g
Yeast extract	0.5 g
Distilled water	1,000.0 ml

This medium is appropriate for culture of myxomycetes and some heartier species of protostelids and acrasids.

CORNMEAL DEXTROSE AGAR

Cornmeal agar*	17.0 g
Glucose	2.0 g
Agar	10.0 g

Antibacterial antibiotics, alone or in combination, should be added if medium is used for isolations. This medium is appropriate for isolation and culture of microfungi and coprophilous fungi.

CREATINE AGAR

(Frisvad 1981)

Creatine	3.0 g
Sucrose	30.0 g

$K_2HPO_4 + 3H_2O$	1.3 g
Bromcresol purple	50.0 mg
Mineral solution	10.0 ml
Agar	20.0 g
Distilled water	1,000.0 ml

Mineral stock solution:

KCl	5.0 g
$MgSO_4 + 7H_2O$	5.0 g
$FeSO_4 + 7H_2O$	0.1 g
Distilled water	100.0 ml

Adjust pH of medium with sterile 1N NaOH to pH 8, and autoclave. This is appropriate for characterization of nonsporulating isolates and certain species of *Penicillium* and *Aspergillus*.

CZAPEK AGAR*—CZ

(Cooney and Emerson 1964)

Sucrose	30.0 g
$NaNO_3$	3.0 g
K_2HPO_4	1.0 g
$MgSO_4 + 7H_2O$	0.05 g
KCl	0.5 g
$FeSO_4 + 7H_2O$	10.0 mg
Agar	20.0 g
Tap water	1,000.0 ml

This is an appropriate medium for characterization and maintenance of a wide variety of filamentous fungi and yeasts. A variation of Czapek-Dox agar, later in this appendix.

CZAPEK'S AGAR MINUS SUCROSE—0% CZ

$NaNO_3$	3.0 g
K_2HPO_4	1.0 g
$MgSO_4 + 7H_2O$	50.0 mg
KCl	0.5 g
$FeSO_4 + 7H_2O$	0.01 g
Agar	20.0 g
Tap water	1,000.0 ml

CZAPEK'S AGAR, GLUCOSE-SUBSTITUTED

(Abdel-Hafez et al. 1989)

Glucose	10.0 g
$NaNO_3$	3.0 g
K_2HPO_4	1.0 g
$MgSO_4 + 7H_2O$	50.0 mg
KCl	0.5 g
$FeSO_4 + 7H_2O$	10.0 mg

Yeast extract	0.5 g
Rose bengal	60.0 mg
Agar	15.0 g
NaCl	150.0–200.0 g
Tap water	1,000.0 ml

Antibacterial antibiotics should be added when used for isolations from natural substrata. This is an appropriate medium for isolating and cultivating halophilic fungi; incubate isolation plates in the dark at 28°C for 6–8 weeks.

CZAPEK-DOX AGAR*

Sucrose	30.0 g
$NaNO_3$	2.0 g
K_2HPO_4	1.0 g
$MgSO_4 + 7H_2O$	0.5 g
KCl	0.5 g
$FeSO_4 + 7H_2O$	0.01 g
Agar	15.0 g
Distilled water	1,000.0 ml

Antibacterial antibiotics should be added when used for isolations from natural substrata. This medium is appropriate for isolation and cultivation of microfungi, animal pathogens, and fungi from extreme environments.

CZAPEK-DOX AGAR, DILUTED

Sucrose	3.0 g
$NaNO_3$	2.0 g
K_2HPO_4	1.0 g
$MgSO_4 + 7H_2O$	0.5 g
KCl	0.5 g
$FeSO_4 + 7H_2O$	10.0 mg
Agar	15.0 g
Distilled water	1,000.0 ml

This medium is appropriate for isolating and cultivating fungi from extreme environments.

DICHLORAN-GLYCEROL-18 MEDIUM*—DG18

(Pitt and Hocking 1997)

Glucose	10.0 g
Bacteriological peptone	5.0 g
KH_2PO_4	1.0 g
$MgSO_4 + 7H_2O$	0.5 g
Dichloran (0.2%, in 95% ethanol)	1.0 ml
Glycerol (analytical reagent grade)	220.0 g
Agar	15.0 g
Distilled water	1,000.0 ml
Chlortetracycline (0.1%, aqueous)	To give 50 µg/ml

Add all ingredients except glycerol, dichloran, and chlortetracycline to 800 ml distilled water. After heating to dissolve agar, add glycerol and dichloran solution, giving final concentrations of 18% (w/w) glycerol and 2 g of dichloran/ml. Adjust final volume with distilled water to 1 liter. Add filter-sterilized 0.1% aqueous solution of chlortetracycline to medium just before pouring to give a concentration of 50 µg/ml. Final pH is 5.6; a_w is 0.955. Dichloran is highly toxic and should be used with extreme care and under proper ventilation. This is an appropriate medium for isolating and cultivating osmotolerant, xerotolerant, and xerophilic fungi.

DICHLORAN-ROSE BENGAL-CHLORAMPHENICOL AGAR*—DRBC

(Anonymous 1990)

Glucose	10.0 g
Peptone	5.0 g
KH_2PO_4	1.0 g
$MgSO_4 + 7H_2O$	0.5 g
Dichloran	2.0 mg
Rose bengal	25.0 mg
Chloramphenicol	0.1 g
Agar	15.0 g
Distilled water	1,000.0 ml

Dichloran has low water solubility. A convenient stock solution can be made by dissolving 0.2 g of dichloran in 100 ml ethanol; 1 ml is added to each liter of medium to attain a concentration of 2 mg/liter. Dichloran also can be dissolved in a few milliliters of acetone and added to the medium prior to autoclaving. Chloramphenicol can be purchased as a supplement in ampoules, or it can be weighed and added directly to the medium. The manufacturer of DRBC recommends use of chloramphenicol, but other combinations of antibacterial antibiotics can be added after autoclaving. This medium is appropriate for isolation of soil fungi and fungi from dried seeds and plant materials. Originally developed for isolation of fungi from food products and osmotolerant, xerotolerant, and xerophilic fungi.

DUNG EXTRACT AGAR—DEA

Dung extract:

Air-dried dung (cow or horse)	200.0 g
Tap water	1,000.0 ml

Boil mixture 20 minutes and filter through several layers of gauze or a coarse filter paper. Supernatant can be autoclaved and saved for future use.

Dung-extract agar:

Bring filtered dung extract up to 1,000 ml with tap water. Add 20 g agar and autoclave.

A variation of dung extract agar involves replacing 500 ml of the water in the recipe of another medium, such as cornmeal agar, with 500 ml of the diluted dung extract. This is for use in isolating and cultivating coprophilous fungi.

EMERSON YEAST STARCH AGAR*

(Cooney and Emerson 1964; Tansey and Jack 1976)

Yeast extract	4.0 g
K_2HPO_4	1.0 g
$MgSO_4 + 7H_2O$	0.5 g
Soluble starch	15.0 g
Agar	20.0 g
Rose bengal	50.0 mg
Tap water	250.0 ml
Distilled water	750.0 ml
Gentamicin	50.0 mg/liter
Penicillin G	100 units

Antibiotics are added just before pouring. Penicillin G, conversion of units to mg: 1,600 units of K^+ salt = 1 mg; 1,650 units of Na^+ salt = 1 mg. This is an appropriate medium for isolating and cultivating thermotolerant and thermophilic fungi and others.

GLUCOSE AMMONIUM NITRATE AGAR—GAN

(Gochenaur 1978)

Glucose	10.0 g
NH_4NO_3	1.0 g
KH_2PO_4	1.0 g
$MgSO_4+7H_2O$	0.5 g
Rose bengal	30.0 mg
Agar	20.0 g
Water	1,000.0 ml

Add antibacterial antibiotics after autoclaving, and adjust pH if desired. Substitution of 2–10 mg cyclosporin A^b for the rose bengal is a useful variation of GAN. This medium is appropriate for isolation of soil fungi; incubate in the dark.

GRACE'S INSECT-TISSUE CULTURE MEDIUM

Grace's Insect Tissue Culture Medium (Sigma Chemical #G 9771)*	44.5 g
Distilled water	1,000.0 ml

Mix solution and filter sterilize.

The medium can be modified by adding fetal bovine serum, 50–100 µl/ml; streptomycin, 100 mg/liter; or penicillin, 60 mg/liter (Glockling and Shimazu 1997; Glockling 1998). This medium is appropriate for use with fungal parasites of insects and invertebrates.

HAY INFUSION AGAR—HI

(Stevens 1981)
Hay infusion:

Dried hay or grass	50.0 g
Tap water	1,000.0 ml

Autoclave for 20 minutes at 15 psi and filter.

Hay infusion agar:

Hay infustion	1,000.0 ml
K$_2$HPO$_4$	2.0 g
Agar	15.0 g

Adjust pH to 6.2 with 1 N KOH or 5% lactic acid. This medium is appropriate for moist-chamber plates; all-purpose culture medium for most protostelids and many myxomycetes, dictyostelids, and acrasids.

INHIBITORY MOLD AGAR*—IMA

(modified from Atlas and Parks 1993)

Trypticase peptone (tryptone)	3.0 g
Thiotone peptone	2.0 g
Yeast extract	5.0 g
Glucose	5.0 g
Soluble starch	2.0 g
Dextrin	1.0 g
Chloramphenicol	0.125 g
Na$_2$HPO$_4$	2.0 g
MgSO$_4$+7H$_2$O	0.8 g
FeSO$_4$	40.0 mg
MnSO$_4$	40.0 mg
NaCl	40.0 mg
Agar	20.0 g
Distilled water	1,000.0 ml

Mix ingredients and bring to boil to dissolve completely. Chloramphenicol is autoclaved along with other ingredients.

HORIKOSHI I—BASAL MEDIUM

(Horikoshi 1991)

Glucose	10.0 g
Polypeptone	5.0 g
Yeast extract	5.0 g
KH$_2$PO$_4$	1.0 g
MgSO4+7H$_2$O	0.2 g
Na$_2$CO$_3$	10.0 g
Agar	20.0 g
Distilled water	1,000.0 ml

Add antibacterial antibiotics after autoclaving. This is an appropriate medium for isolation and cultivation of alkalophilic fungi.

HORIKOSHI II—BASAL MEDIUM

(Horikoshi 1991)

Soluble starch	10.0 g
Yeast extract	5.0 g
KH$_2$PO$_4$	1.0 g
MgSO$_4$+7H$_2$O	0.2 g
Na$_2$CO$_3$	10.0 g
Agar	20.0 g
Distilled water	1,000.0 ml

Add antibacterial antibiotics after autoclaving. This is an appropriate medium for isolating and cultivating alkalophilic fungi.

HYPOMYCES FRUITING AGAR—HFA

(Carey and Rogerson 1981)

Preparation of the medium requires the mixture of two media prepared simultaneously.

Potato infusion:

Potato (scrubbed and sliced)	200.0 g
Glucose	15.0 g
Deionized water	500.0 ml

Boil potatoes 30 minutes in water and strain. Bring the potato infusion up to 500 ml and add glucose.

Agar medium:

Malt extract (Difco)	20.0 g
Agar	20.0 g
Deionized water	1,000.0 ml

Autoclave potato infusion and agar medium separately. Mix them while both are still hot, and pour into Petri dishes. The potato infusion can be stored at 4°C. HFA has been used to stimulate development of perithecia of *Hypomyces polyporinus*. It is appropriate for isolating and cultivating fungicolous fungi.

KMV

Glucose	1.0 g
Gelatin hydrolysate	1.0 g

Peptone	10.0 g
Agar	20.0 g
Seawater	1,000.0 ml

Ingredients can be dissolved in distilled, estuarine, or marine water. Add antibacterial antibiotics after autoclaving. This is an appropriate medium for isolation, cultivation, and maintenance of zoosporic fungi and Chromista from aquatic animals.

LACTOSE-PEPTONE AGAR—LPA

Agar	15.0 g
Lactose	0.1–1.0 g
Peptone	0.1–1.0 g
Distilled water	1,000.0 ml

This medium is appropriate for isolation and cultivation of dictyostelids.

LACTOSE-YEAST EXTRACT AGAR—LYA

Agar	15.0 g
Lactose	0.1–1.0 g
Yeast extract	0.1–0.5 g
Distilled water	1,000.0 ml

This medium is appropriate for isolation and cultivation of dictyostelids.

LEONIAN'S AGAR MEDIUM—ML

(Malloch 1981)

Peptone	0.625 g
Maltose	6.25 g
Malt extract	6.25 g
KH_2PO_4	1.25 g
$MgSO_4+7H_2O$	0.625 g
Agar	20.00 g
Distilled water	1,000.0 ml

This is a general-purpose culture and isolation medium; it is especially good for basidiomycetes.

LITTMAN OXGALL AGAR*

(Atlas and Parks 1993)

Peptone	10.0 g
Glucose	10.0 g
Oxgall	15.0 g
Agar	20.0 g

Crystal violet	10.0 mg
Distilled water	1,000.0 ml

Final pH adjusted to 7.0 + 0.2 at 25°C. Optional addition: Streptomycin, 100 mg/liter, to make Littman Oxgall-Streptomycin Agar. This medium is appropriate for isolation of dermatophytic fungi.

MALT EXTRACT AGAR—MEA

(Dreyfuss 1986; Bills and Polishook 1994)

Malt extract	2.0–20.0 g
Yeast extract (optional)	0.2–2.0 g
Agar	20.0 g
Water	1,000.0 ml
Cyclosporin A[†]	2.0–10.0 mg

The quantity of malt extract often is varied. Yeast extract is optional, depending on fungi to be isolated; if included, it produces malt yeast agar (MYA). Antibacterial antibiotics and cyclosporin A (if control of colony expansion is desired) are added after autoclaving. Addition of cyclosporin A produces malt cyclosporin A agar. This medium is appropriate for isolation of soil and leaf-litter fungi.

MALT EXTRACT BROTH*

Malt extract	6.0 g
Maltose	1.8 g
Glucose	6.0 g
Yeast extract	1.2 g
Distilled water	1,000.0 ml

Mix, dispense in tubes or flasks, and autoclave. This is appropriate for liquid cultures of filamentous fungi and yeasts and growth of cells for DNA extraction.

MALT EXTRACT AGAR WITH PEPTONE

Malt extract	20.0 g
Peptone	1.0 g
Glucose	20.0 g
Agar	20.0 g
Distilled water	1,000.0 ml

This is an appropriate medium for isolation, cultivation, and maintenance of nonzoosporic fungi from aquatic animals. Ingredients can be dissolved in distilled, estuarine, or marine water. Antibiotics are added after autoclaving (e.g., 200 ppm cycloheximide; 100 ppm streptomycin sulfate).

MALT EXTRACT AGAR PLUS COPPER—MEC

(Arnebrant et al. 1987)

Malt extract	20.0 g
Agar	15.0 g
$CuSO_4 + 5 H_2O$	0–400.0 mg
Distilled water	1,000.0 ml
Chlortetracycline	30.0 mg

Chlortetracycline is added after autoclaving. To enrich isolations for zygomycetes, add 1.0 ml of benomyl stock solution to yield a final concentration of 1.0 mg/liter. To prepare stock solution, dissolve 2.0 g of benomyl wettable powder in a small amount of DMSO in a 100-ml volumetric flask. Gradually add 100 ml sterile distilled water to bring to volume. This is an appropriate medium for isolating and cultivating heavy-metal-tolerant fungi.

MALT EXTRACT–YEAST EXTRACT AGAR— MEYE

Malt extract	3.0 g
Yeast extract	3.0 g
Peptone	5.0 g
Glucose	10.0 g
Distilled water	1,000.0 ml
Agar	15.0 g

This medium is appropriate for isolating and cultivating coprophilous fungi.

MALT EXTRACT–YEAST EXTRACT 5–12 AGAR—MY5–12 AGAR

(Pitt and Hocking 1987)

Malt extract	20.0 g
Yeast extract	5.0 g
NaCl	50.0 g
Glucose	120.0 g
Agar	20.0 g
Distilled water	To make 1 liter

Dissolve ingredients in water, and autoclave at 121°C for 10 minutes. Overheating will inhibit gelling of agar. The final a_w is 0.96. This is an appropriate medium for isolation and cultivation of halophilic fungi.

MALT EXTRACT–YEAST EXTRACT 10–12 AGAR—MY10–12 AGAR

(Pitt and Hocking 1987)

Malt extract	20.0 g
Yeast extract	5.0 g
NaCl	100.0 g
Glucose	120.0 g
Agar	20.0 g
Distilled water	To make 1 liter

Combine ingredients; bring mixture to a boil to melt agar. *Do not autoclave.* Final a_w is 0.93. This is an appropriate medium for halophilic fungi.

MALT EXTRACT–YEAST EXTRACT AGAR, WEAK—wMYA

Agar	5.0 g
Malt extract	20.0 mg
Yeast extract	20.0 mg
K_2HPO_4	0.75 g

This medium is appropriate for isolation plates and induction of sporulation of Ascomycetes and anamorphic fungi. It also is an all-purpose culture medium for most mycetozoans.

MALT-YEAST-PEPTONE AGAR—MYPA

Malt extract	7.0 g
Yeast extract	0.5 g
Peptone	1.0 g
Agar	15.0 g
Distilled water	1,000.0 ml

Penicillin G (0.5 g/liter) and streptomycin sulfate (0.5 g/liter) can be added just before pouring to produce MYP-ps agar (Carreiro and Koske 1992). This is an appropriate medium for cultivation and isolation of psychrotolerant and psychrophilic fungi.

MILK SOLIDS-GLUCOSE AGAR

(Summerbell et al. 1988)

(a) Skim milk powder		40.0 g
Distilled water		500.0 ml
(b) Glucose		20.0 g
Agar		20.0 g
Distilled water		500.0 ml
Bromcresol purple (1.6%, in ethanol)		1.0 ml

Dissolve solution (b) by heating, then autoclave both parts separately. When cooled to about 50°C, combine (a) and (b) and add bromcresol purple. Adjust the pH to 6.5 so that agar is pale blue (near "cadet blue" of Kornerup and Wanscher 1978). This is an appropriate medium for characterization of basidiomycetes and other fungi.

PEPTONIZED MILK-TRYPTONE-GLUCOSE AGAR—PMTG

Peptonized milk	1.0 g
Tryptone	1.0 g
Glucose	5.0 g
Agar	10.0 g
Distilled water	1,000.0 ml

MODIFIED SYNTHETIC *MUCOR* AGAR— MSMA

(Benny and Benjamin 1975)

Glucose	10.0 g
NaNO₃	4.0 g
KH₂PO₄	0.5 g
MgSO₄+7H₂O	0.5 g
Thiamine HCl	25.0 mg
Distilled water	1,000.0 ml
Agar	15.0 g

MYCOBIOTIC AGAR

Soytone*	10.0 g
Glucose	10.0g
Agar	15.0 g
Cycloheximide	0.5 g
Chloramphenicol	50.0 mg
Distilled water	1,000.0 ml

This is useful for isolating dermatophytes and other pathogenic fungi from clinical samples. Commercial preparations are available from Difco (Bacto MycoBiotic Agar) and BBL (Mycosel Agar).

OAK-BARK AGAR—OBA

(Blanton 1990)

Oak bark (*Quercus* species) infusion:

Oak bark	2.5–7.5 g
Tap water	1,000.0 ml

Autoclave and filter. Readjust volume to 1 liter.

Oak bark agar:

Oak bark infusion	1,000.0 ml
Agar	15.0–20.0 g

Adjust pH to 6.0 with 1 N KOH or 5% lactic acid and autoclave. Small pieces of sterile bark may be added. The hardness of the agar and strength of the infusion can be varied. This medium is appropriate for moist-chamber plates and for culture of many mycetozoans

OAT-DUNG EXTRACT AGAR—ODA

(Keller and Eliasson 1992)

Oat-dung extract:

Fresh horse dung	50.0 g
Rolled oats	5.0 g
Distilled water	1,000.0 ml

Add dung and rolled oats to a flask containing water. Mix contents thoroughly and allow to stand 24 hours. Boil for 30 minutes, pour through filter paper, and allow to cool.

Oat-dung extract agar:

Dilute oat-dung extract 1:5 with distilled water and add enough agar to yield a 2.0% agar medium. This is appropriate for characterization of coprophilous fungi.

OATMEAL AGAR I*—OA

(Cooney and Emerson 1964)

Oatmeal	60.0 g
Agar	20.0 g
Tap water	To make 1 liter

Cook oatmeal with 700 ml tap water by steaming for 1 hour. Strain through cheesecloth to remove solids, and bring volume to 1 liter. Add agar. Prepare this medium in an oversized flask because it usually produces excessive foaming and can boil over when autoclaved. Autoclave 30–40 minutes at 15 psi. This is appropriate for induction of sporulation in nonsporulating isolates and for characterization of a wide variety of fungi.

OATMEAL AGAR II*—OA

Rolled oats	30.0 g
Agar	20.0 g
Tap water	To make 1 liter

Boil oats for 15–30 minutes. Filter through cheesecloth or gauze. Bring filtrate volume to 1 liter. Add agar. Autoclave 30–40 minutes at 15 psi. Alternatively, grind uncooked oats in a blender, combine with agar and water, and autoclave for 45 minutes. Prepare this medium in an oversized flask because the medium usually produces excessive foaming when autoclaved. This is appropriate for induction of sporulation in nonsporulating isolates and characterization of a wide variety of fungi.

PEPTONIZED MILK-TRYPTONE-GLUCOSE AGAR—PMTG

(See entry following "Milk Solids-Glucose Agar," earlier.)

PEPTONE-YEAST-GLUCOSE LIQUID ISOLATION MEDIUM—PYG

(See entry following "Yeast-Glucose Agar," later).

PIMARICIN-VANCOMYCIN-PENTACHLORONITROBENZENE-CORNMEAL AGAR—P10VP

(Tsao and Ocana 1969)

Cornmeal agar*	17.0 g
Water	1,000.0 ml
Pimaricin	10.0 mg
Vancomycin HCl	0.2 g
Pentachloronitrobenzene (PCNB)	0.1 g

Add the last three ingredients to cooled agar immediately before pouring plates. This medium is appropriate for isolation of soil fungi.

VANCOMYCIN-PIMARICIN-PENTACHLORONITOBENZENE-PENICILLIN-CORNMEAL AGAR—VP₃ AGAR

Dehydrated cornmeal agar	17.0 g
Agar	23.0 g
Sucrose	20.0 g
$MgSO_4+7H_2O$	10.0 mg
$CaCl_2$	10.0 mg
Water	989.0 ml
Microelement stock solution‡	1.0 ml.

Autoclave for 30 minutes. When cooled to 50°C, the following antibiotic combination is added aseptically:

Vancomycin	75.0 mg
Pentachloronitrobenzene (PCNB)	100.0 mg
Penicillin G	50.0 mg
Pimaricin	5.0 mg
Rose bengal	2.5 mg

‡ Microelement stock solution:

$ZnCl_2$	100.0 mg
$FeSO_4+7H_2O$	2.0 mg
MoO_3	2.0 mg
$CuSO_4+5H_2O$	2.0 mg
Distilled water	100.0 ml

Mix the antibiotics into a paste with 5 ml of sterile water; the container should be rinsed with an additional 5 ml of sterile water. The concentrated solution is dispersed in the liquid agar by gentle agitation. Plates are poured immediately, stored in the dark, and used within 36 hours. This medium is appropriate for the isolation and cultivation of Oomycetes and freshwater fungi.

POLYPORE EXTRACT-INFUSION AGAR

Polypore flour	200.0 g
Distilled water	1,000.0 ml

This medium is appropriate for characterization of some fungicolous fungi.

POTATO CARROT AGAR—PCA

(Gams et al. 1998)
Potato-carrot extract:

Peeled carrots	40.0 g
Peeled potatoes	40.0 g
Tap water	1,000.0 ml

Shred carrots and chop potatoes into small pieces; add to water and boil for 15 minutes; filter the mixture through a layer of gauze. The extract maybe autoclaved for future use.

Potato-carrot agar:

Potato-carrot extract	500.0 ml
Tap water	500.0 ml
Agar	20.0 g

This medium is very useful for stimulation of sporulation in culture. It may be used as an isolation medium for soil and litter fungi by adding antibacterial antibiotics after autoclaving.

POTATO-DEXTROSE AGAR—PDA

(Stevens 1981)

PDA is one of the most commonly used culture media for fungi because of its simple formulation and its ability to support mycelial growth of a wide range of fungi. Several formulations are in the literature; one is provided here. The recipe for commercially prepared PDA is provided as well.

(a) PDA from fresh potatoes:

Potatoes	200.0 g
Glucose	20.0 g

Agar	20.0 g
Distilled water	1,000.0 ml

Chop potatoes and place in water. Autoclave for 20 minutes at 15 psi. Strain autoclaved potatoes and liquid through several layers of gauze, squeezing the potatoes to extract all the liquid. Add enough water to bring the potato extract to 1,000 ml. Add glucose and agar, and autoclave for 15–20 minutes at 15 psi.

(b) PDA from commercial preparation:

Potato-dextrose agar*	39.0 g
Distilled water	1,000.0 ml

Antibiotics (e.g., 200 mg cycloheximide; 100 mg streptomycin sulfate) can be added to medium after autoclaving. This is an appropriate medium for isolation, cultivation, and maintenance of a wide variety of fungi. For cultivation and maintenance of nonzoosporic fungi from aquatic animals or marine fungi; distilled, estuarine, or marine water can be used.

POTATO-DEXTROSE AGAR, ACIDIC—APDA

Dehydrated potato-dextrose agar*	39.0 g
Distilled water	1,000.0 ml

When cooled enough to pour plates, add 6 drops 50% lactic acid. This medium is appropriate for isolating and cultivating coprophilous fungi.

POTATO-DEXTROSE AGAR, DILUTE—dPDA

Potato-dextrose agar*	3.9 g
Difco special agar*	14.0 g
Distilled water	1,000.0 ml

Antibiotics are added to the medium after autoclaving. This is an appropriate medium for isolation, cultivation, and maintenance of nonzoosporic fungi from aquatic animals; distilled, estuarine, or marine water can be used.

POTATO-PEPTONE-GLUCOSE AGAR—PPGA

Potatoes	200.0 g
Peptone	5.0 g
Glucose	5.0 g
$Na_2HPO_4+12H_2O$	3.0 g
KH_2PO_4	0.5 g
Agar	15.0 g
Distilled water	To make 1 liter

Prepare a potato extract from fresh potatoes following formulation for potato-dextrose agar listed earlier. Mix ingredients and autoclave.

SABOURAUD PEPTONE-GLUCOSE AGAR*—SAB

Neopeptone	10.0 g
Glucose	40.0 g
Agar	15.0 g
Distilled water	1,000.0 ml

The medium may be modified by reducing the glucose to 20.0 or 10.0 g. This medium is used for characterization of a wide variety of filamentous fungi and yeasts.

SABOURAUD AGAR WITH CYCLOHEXIMIDE—SABC

Peptone	10.0 g
Glucose	40.0 g
Agar	15.0 g
Cycloheximide	0.1 g
Chloramphenicol	0.1 g
Deionized water	1,000.0 ml

Mix ingredients and bring to boil to dissolve completely; autoclave. Optional addition: cycloheximide, 0.4 g.

SABOURAUD BRAIN-HEART-INFUSION AGAR—SABHI

(Atlas and Parks 1993)

Calf brain infusion*	100.0 g
Beef heart infusion*	125.0 g
Glucose	21.0 g
Agar	15.0 g
Neopeptone	5.0 g
Proteose peptone	5.0 g
NaCl	2.5 g
Na_2HPO_4	1.25 g
Distilled water	1,000.0 ml

Optional additions (alone or in combination): chloramphenicol, cycloheximide, gentamicin, or other antibiotics.

SABOURAUD DEXTROSE-YEAST AGAR—SABDY

Neopeptone*	10.0 g
Glucose	40.0 g
Yeast extract	10.0 g
Agar	15.0 g
Distilled water	1,000.0 ml

Mix ingredients and bring to a boil to dissolve completely. Autoclave at 121°C for 15 minutes at 15 psi.

Optional additions (alone or in combination): chloramphenicol, cycloheximide, and gentamicin to make SABCCG (see "Antibacterial Antibiotics" under "Antibiotics," later in this appendix). Media equivalent to SAB (with or without CCG) are commercially available from many suppliers (e.g., from BBL as Mycosel; Difco as Mycobiotic, etc.; Appendix IV). This medium is appropriate for insect-inhabiting fungi.

SABOURAUD MALTOSE AGAR

Neopeptone	10.0 g
Maltose	40.0 g
Agar	15.0 g
Distilled water	1,000.0 ml

The medium may be modified by reducing the maltose to 20.0 g. This medium is used for characterization of a wide variety of filamentous fungi and yeasts.

SEAWATER AGAR—SA

Glucose	1.0 g
Yeast extract	1.0 g
Peptone	0.5 g
Agar	18.0 g
Distilled water	1,000.0 ml
Instant Ocean salt	38.0 g
Penicillin G (optional)	100.0 mg
Streptomycin sulfate (optional)	100.0 mg

Antibiotics may be added to medium after autoclaving, to suppress bacteria.

SERUM SEAWATER AGAR—SSA

Agar	12.0 g
Seawater	1,000.0 ml
Horse serum (calf serum)	10.0 ml

This is an appropriate medium for isolation, cultivation, and maintenance of zoosporic fungi and Chromista from aquatic animals; ingredients can be dissolved in distilled, estuarine, or marine water. Antibiotics may be added to medium after autoclaving, to suppress bacteria.

SILICA-GEL MEDIUM

(Parkinson et al. 1989)

Silicic acid	12.0 g
KOH	200.0 ml
NaNO₃	0.8 g
KH₂PO₄	0.4 g
KCl	0.2 g
Double-distilled water	170.0 ml
Orthophosphoric acid (20% v/v)	30.0 ml
MgSO4+H₂O (10% solution)	2.0 ml
FeSO₄+7H₂O (0.4% solution w/v).	1.0 ml

Add silicic acid to KOH (84 g in 200 ml double-distilled water) to produce potassium silicate; dissolve mixture by heating, place in 400-ml bottle, and autoclave. Mix and autoclave remaining components (sterile salts solutions), except the FeSO₄ solution, which is filter-sterilized and added last. Chill salts solution on ice and aseptically add to bottle containing sterile potassium silicate. Mix contents, and pour rapidly into ultracleaned glass Petri dishes. Medium is sufficient for 10–16 plates. Gel sets in 15 minutes. Leave plates out overnight, and then pour off the water on the surface. Streptomycin sulfate (1 g/liter) can be added to the medium to prevent bacterial contamination.

For ultra-pure studies, all glassware is washed with chromic acid, rinsed in double-distilled water, and heated to 500°C for 1 hour in a muffle furnace to remove all traces of organic carbon. Foil used to cover glassware and silicic acid also should be heated to 500°C, for 1 hour and 2 hours, respectively. This is an appropriate medium for isolation and cultivation of oligotrophic fungi.

SOFT AGAR—SA

Agar	7.5 g
Infusion of dead plant parts, autoclaved in distilled water	2.5 g in 1,000.0 ml

Dead plant parts of any type (e.g., dried infructescences) can be used. This medium is appropriate for culture of some of the smaller myxomycetes, particularly *Echinostelium* species.

SOIL EXTRACT AGAR—SEA

(Apinis 1963)

Soil extract:

Steam 1 kg of soil (preferably sandy loam) in 1 liter distilled water at 100°C for 1 hour. Filter first through cheesecloth; then centrifuge to clear. Autoclave extract for 1 hour on each of 2 consecutive days before preparing medium.

Soil extract agar:

Soil extract	100.0 ml
Glucose	0.5 g
K₂HPO₄	0.5 g
MgSO₄+7H₂O	0.5 g

Agar	20.0 g
Distilled water	900.0 ml

This is an appropriate medium for isolating and cultivating soil fungi and thermotolerant and thermophilic fungi.

SOIL EXTRACT-ROSE BENGAL AGAR— SERBA

Soil extract:

Collect 75–200 g soil, preferably from the study site. If an inadequate quantity of soil is received from a remote site, substitute a rich garden soil or a soil with high organic matter. Place soil in filtering funnel and pour boiling water through until 500 ml of extract is obtained.

Soil extract-rose bengal agar:

Soil extract	500.0 ml
Rose bengal	0.03–0.05 g
Lactic acid (optional)	Few drops, to adjust pH to 4.5
Agar	20.0 g
Distilled water	500.0 ml

Antibacterial antibiotics (see "Antibiotics," later) can be added after autoclaving. This is appropriate for isolation of soil fungi; incubate medium in the dark.

SOIL BASIDIOMYCETE ISOLATION MEDIUM WITH LIGNIN, GUAIACOL, AND BENOMYL

(Thorn et al. 1996)

Basal medium:

KH_2PO_4	0.5 g
$MgSO_4+7H_2O$	0.2 g
NH_4NO_3	0.1 g
KCl	0.1 g
$FeSO_4+7H_2O$	20.0 mg
$Ca(NO_3)_2+4H_2O$	50.0 mg
Malt extract	2.0 g
Agar	15.0 g
Distilled water	1,000.0 ml

The basal medium is autoclaved in flasks with magnetic stir bars and is cooled to about 55°C.

Soil basidiomycete isolation medium:

After the basal medium has cooled, add the following aseptically:

1 M KOH	Up to 5.0 ml
Guaiacol	0.4 ml
Indulin AT, alkali lignin (suspended, partially dissolved in 10 ml dioxane)	1.0 g
Chlortetracycline HCl	60.0 mg

Streptomycin sulfate	30.0 mg
Penicillin G (Na salt)	30.0 mg
Benomyl[‡]	4.0 mg

[‡]Benomyl wettable powder in 2 ml of 1:1 acetone:ethanol (70 %) to yield 2 mg/ml. Adjust pH to 6.0–6.5 with KOH; amount will depend on source and acidity of alkali lignin. Stir medium; pour approximately 20 ml/plate into 100-mm diameter Petri dishes. This is appropriate for isolation of basidiomycetes from soil.

STARCH-MILK AGAR

Nonfat powdered milk	1.0 g
Soluble starch (food grade)	2.0 g
Agar	2.0 g
Murashige and Skoog basal salt solution[a]	200.0 ml

Murashige and Skoog basal salt solution (Sigma Chemical #M0529) is purchased as a sterile 10× solution and diluted to 1×. Autoclave and pour with agitation. This medium is appropriate for isolating and cultivating endophytes.

SYNTHETIC NUTRIENT AGAR—SNA

(Gerlach and Nirenberg 1982)

KH_2PO_4	1.0 g
KNO_3	1.0 g
$MgSO_4+7H_2O$	0.5 g
KCl	0.5 g
Glucose	0.2 g
Sucrose	0.2 g
Agar	20.0 g
Distilled H_2O	1,000.0 ml

After autoclaving and cooling, add small squares (about 1 cm^2) of sterile filter paper to medium. This medium is recommended for observing sporulation of slimy-spored mitosporic fungi (e.g., *Clonostachys, Fusarium, Cylindrocarpon, Verticillium*). If antibacterial antibiotics are added, it is also an excellent low-nutrient medium for isolation of soil fungi.

TRICHODERMA ISOLATING MEDIUM

(Elad and Chet 1983; Askew and Laing 1993)

Glucose	3.0 g
Agar	20.0 g
$MgSO_4+7H_2O$	0.2 g
K_2HPO_4	0.9 g
KCl	0.15 g
NH_4NO_3	1.0 g
Chloramphenicol	0.25 g
Fenaminosulf (Lesan, formerly Dexon, 60% wettable powder)	0.3 g

Pentachloronitrobenzene (PCNB, Terraclor 75% w.p.)	0.2 g
Rose bengal	0.15 g
Distilled water	1,000.0 ml

Captan (20 mg/liter) can be added after autoclaving. Fenaminosulf can be replaced with propamocarb or metalaxyl to increase suppression of oomycetes.

TWEEN-20 AGAR, MODIFIED

(Hankin and Anagnostakis 1975)

Peptone	10.0 g
$CaCl_2+H_2O$	0.1 g
Agar	20.0 g
Distilled water	1,000.0 ml
Tween-20 (polyoxyethylenesorbitan monolaurate)	10.0 ml

Final pH, unadjusted. Autoclave Tween-20 separately and add to agar cooled to about 50°C. The original medium is modified by deletion of 5.0 g NaCl, which can be toxic to many basidiomycetes. This medium is appropriate for characterization of nonsporulating isolates.

V8-JUICE AGAR

$CaCO_3$ (powder)	3.0 g
V8 juice	180.0 ml
Distilled water	820.0 ml
Agar	20.0 g

Combine ingredients and autoclave. Agitate molten medium to maintain $CaCO_3$ in suspension while pouring plates. This medium is appropriate for characterization and stimulation of sporulation.

VOGEL'S MEDIUM N

(Davis and de Serres 1970)

Concentrated solution N[‡]	20.0 ml
Sucrose	15.0 g
Agar	20.0 g
Distilled water	980.0 ml
HCl (1 N)	As necessary to acidify medium to pH 4.0

Stock solutions

Concentrated solution N[‡]:

$Na_3Citrate + 5H_2O$	150.0 g
KH_2PO_4	250.0 g
$MgSO_4 + 7H_2O$	10.0 g
$CaCl_2 + 2H_2O$	5.0 g

Biotin solution[§]	5.0 ml
Trace element solution[‖]	5.0 ml
Distilled water	To make 1 liter

Mix and store in refrigerator. Prepare fresh every 6 months. $CaCl_2 + 2H_2O$ should be predissolved in a small volume of water to prevent precipitation.

Biotin solution[§]

Biotin	5.0 mg
Ethanol (50%)	100.0 ml

Trace-element stock solution[‖]

Citric acid + $1H_2O$	5.00 g
$ZnSO_4 + 7H_2O$	5.00 g
$Fe(NH_4)_2(SO_4)_2 + 6H_2O$	1.00 g
$CuSO_4 + 5H_2O$	0.25 g
$MnSO_4 + 1H_2$	50.0 mg
H_3BO_3	50.0 mg
$Na_2MO_4 + 2H_2O$	50.0 mg
Distilled water	100.00 ml

This is an appropriate medium for Phoenicoid fungi. For selective isolation of *Neurospora crassa* (Maheshwari and Antony 1974), 3–5 g soil is suspended in 10 ml of 10^{-3} M furural (2-furaldehyde) and heated to 60°C for 30 minutes. Furfural solution is removed by gentle centrifugation. Plate small amounts of soil on acidified Vogel's Medium N; Petri dishes should be incubated at room temperature.

VP₃ AGAR

(See "Vancomycin–Pimaricin–Pentachloronitrobenzene–Penicillin–Cornmeal Agar," earlier).

WATER AGAR—WA

Agar	15.0–20.0 g
Distilled water	1,000.0 ml

Various combinations of antibiotics (e.g., 200 ppm cycloheximide; 100 ppm streptomycin sulfate) are added to the medium after autoclaving to form antibiotic water agar (AWA). It is appropriate for isolation, cultivation, and maintenance of zoosporic fungi, nonzoosporic fungi, and Chromista from aquatic animals; distilled, estuarine, or marine water can be used. It also is used for isolation of endophytes, plant pathogens, and soil fungi.

WEITZMAN AND SILVA-HUTNER AGAR

(Weitzman and Silva-Hutner 1967)

Alpha-cellulose	20.0 g
Oatmeal cereal for babies	10.0 g
Tomato paste	10.0 g

KH$_2$PO$_4$	1.5 g
MgSO$_4$ + H$_2$O	1.0 g
NaNO$_3$	1.0 g
Yeast extract	3.0 g
Agar	18.0 g
Distilled water	1,000.0 ml

Adjust pH to 5.6 with 5 M NaOH before autoclaving. The medium was derived from alpha-cellulose agar (see earlier) when it was discovered that coconut milk was not needed to promote sporulation in certain ascomycetes. This medium promotes sporulation of many soil-borne and coprophilous ascomycetes. It also is appropriate for characterization of nonsporulating isolates.

WHEAT GERM AGAR—WG

(Benny 1972)

Wheat germ extract:

Boil 15 g wheat germ in 700 ml distilled water; filter, and bring supernatant volume to 1,000 ml.

Wheat germ agar:

Glucose	5.0 g
Agar	15.0 g
Wheat germ extract	1,000.0 ml

This medium is appropriate for cultivation of certain Zygomycetes.

WICKERHAM'S YEAST EXTRACT-MALT EXTRACT MEDIUM

Yeast extract	3.0 g
Malt extract	3.0 g
Peptone	5.0 g
Glucose	10.0 g
Agar	20.0–30.0 g
Distilled water	1,000.0 ml

Various carbon sources can be substituted for glucose. Freshwater or filtered sea water can be used (for ocean sampling, use sea water at 37 ‰; inshore and estuarine sampling, use brackish water at 15 ‰). Antibiotics can be added to the medium: (1) chloramphenicol, 200 mg/liter (0.02%), added prior to autoclaving; or (2) penicillin G and streptomycin sulfate, each 150–500 mg/liter, added dry to autoclaved, cooled medium.

YEAST-GLUCOSE AGAR

(Cooney and Emerson 1964)

Yeast extract	5.0 g
Glucose	10.0 g
Agar	20.0 g
Tap water	1,000.0 ml

This is an appropriate medium for various filamentous fungi and yeasts and thermotolerant and thermophilic fungi.

YEAST-GLUCOSE-PHOSPHATE AGAR—YPG (ALSO KNOWN AS YEAST PROTEIN-GLUCOSE AGAR)

Yeast extract	8.0 g
Glucose	20.0 g
K$_2$HPO$_4$	2.0 g
MgSO$_4$ + 7H$_2$O	1.0 g
Agar	15.0 g
Distilled water	1,000.0 ml

Antibiotics can be added to the medium after autoclaving, if desired. This is an appropriate medium for isolation, cultivation, and maintenance of zoosporic fungi and Chromista from aquatic animals; distilled, estuarine, or marine water can be used.

PEPTONE-YEAST-GLUCOSE LIQUID ISOLATION MEDIUM—PYG (ALSO KNOWN AS PEPTONE–YEAST EXTRACT–GLUCOSE AGAR)

Yeast extract	5.0 g
Peptone	10.0 g
Glucose	20.0 g
Agar	20.0–30.0 g
Distilled water	1,000.0 ml

Various carbon sources can be substituted for glucose. Freshwater or filtered sea water can be used (for ocean sampling, use sea water at 37 ‰; inshore and estuarine sampling, use brackish water at 15 ‰). Optional additions: (a) chloramphenicol, 200 mg/liter (0.02‰), added prior to autoclaving; or (b) penicillin G and streptomycin sulfate, each 150–500 mg/liter, added dry to autoclaved, cooled medium.

YEAST-PEPTONE-DEXTROSE AGAR—YPD

Glucose	20.0 g
Peptone	5.0 g
Yeast extract	10.0 g
Agar	20.0 g
Tap water	1,000.0 ml

YEAST-SOLUBLE STARCH AGAR*—YpSs (ALSO KNOWN AS YEAST PROTEIN-SOLUBLE STARCH AGAR)

Soluble starch	20.0 g
Yeast extract	1.0 g
K$_2$HPO$_4$	1.0 g

MgSO₄ + 7H₂O   1.0 g
Distilled water   1,000.0 ml

Antibiotics can be added to the medium after autoclaving, if desired. This is an appropriate medium for isolation, cultivation, and maintenance of zoosporic fungi and Chromista from aquatic animals; distilled, estuarine, or marine water can be used.

ANTIBIOTICS

ANTIMICROBIAL ANTIBIOTICS

Addition of antibacterial antibiotics to fungal isolation media often is essential to prevent bacterial growth. Streptomycin sulfate, chlortetracycline, oxytetracycline, ampicillin, chloramphenicol, and vancomycin are frequently used. Most antibiotics are heat labile and are added to cooled media (45–50°C) just as plates are poured. An exception is chloramphenicol, which can be autoclaved and usually is used at a concentration of 100 mg/liter medium. Most antibiotics are prepared in sterile water or filter-sterilized and stored as concentrated stock solutions. Concentrate is added to media to yield a final concentration of 30–100 mg/liter of medium. Acidification of media to pH 4.5 with lactic acid also can help to limit bacterial growth. Addition of rose bengal to isolation media also inhibits growth of many bacteria and reduces radial extension of fungal colonies. Rose bengal is a singlet oxygen-generating photosensitizer. Singlet oxygen damages membranes and cytoplasmic contents of cells (Daub et al. 1992). Therefore, rose bengal–containing media should be incubated in the dark. A useful guide for selection of antimicrobial antibiotics can be found in Glasby (1992).

Chlortetracycline (50.0 mg/liter) and streptomycin sulfate (50.0 mg/liter) are a good combination of antibiotics to use in general surveys of filamentous soil fungi. Those antibiotics interfere specifically with bacterial-ribosome function and never have been observed to adversely affect establishment of basidiomycete or ascomycete cultures from direct plating of spores. Some antibacterial antibiotics may be selectively toxic toward certain fungi, particularly oomycetes and basidiomycetes (Tuite 1969; Singleton et al. 1992). If selective toxicity is suspected, other formulations of antibiotics should be substituted.

ANTIBACTERIAL ANTIBIOTICS

Most antibacterial antibiotics are heat sensitive; therefore, concentrated stock solutions should be prepared beforehand, filter-sterilized, and stored refrigerated or frozen. Stock solutions are added to partially cooled media after autoclaving to obtain the recommended final concentrations indicated in the following list. Other antibacterials can be used, but these are among the most commonly used.

(a) Chloramphenicol, prepared in ethanol (95%); final concentration in medium, 100 mg/liter. Chloramphenicol can be added to media before autoclaving, if desired.
(b) Gentamicin, prepared in distilled water; final concentration in medium, 50 mg/liter.
(c) Streptomycin sulfate, prepared in distilled water; final concentration in medium, 50 mg/liter.
(d) Tetracycline or chlortetracycline, prepared in distilled water; final concentration in medium, 50 mg/liter.

ANTIFUNGAL ANTIBIOTICS

Antifungal antibiotics often have a spectrum of action skewed toward other fungal taxa and, therefore, can be used to isolate certain types of fungi preferentially. At sublethal concentrations they can slow hyphal extension and can aid isolation by permitting longer incubation periods for isolation plates. The following antifungal antibiotics often are used in the range of concentrations indicated.

(a) Cycloheximide, prepared in acetone solution; final concentration in medium, 100–500 mg/liter. Cycloheximide is toxic; investigators should take appropriate safety precautions when using it.
(b) Cyclosporin A[†], final concentration in medium, 4–10 mg/liter.
(c) Pentachloronitrobenzene (PCNB), 100–750 mg/liter. PCNB is toxic; investigators should take appropriate safety precautions when using it.
(d) Pimaricin (natamycin), 5–100 mg/liter.

DIELDRIN MEDIUM SUPPLEMENT TO ELIMINATE MITES

As a last resort, when mites have infested or threaten to destroy cultures, media can be made lethal to mites by adding dieldrin at a concentration of 20 μg/ml. Dieldrin is extremely toxic to animals but seems to have little effect on fungi. Investigators should avoid inhalation and contact with skin. Glassware and used media need to be handled as toxic waste.

Dieldrin	200.0 mg
Acetone	10.0 ml

Add 1 ml of the acetone stock solution to medium before or after autoclaving. Gams and associates (1998) suggested a similar medium treatment using lindane at a concentration of 75 μg/ml.

FIXATIVES

ACIDIFIED ALCOHOL

Glacial acetic acid	250.0 ml
Ethanol, absolute	750.0 ml

2X CTAB (HEXADECYLTRIMETHYLAMMONIUM BROMIDE) BUFFER

Ingredients per 100-ml batch:

Tris-HCl	1.21 g
NaCl	8.182 g
EDTA	74.0 mg
CTAB (Sigma Chemical #H5882 or M7635)	2.0 g
Polyvinylpyrolidone	2.0 ml
2-mercaptoethanol	0.2 ml
Distilled water	To make 100.0 ml

Pipette 500-μl aliquots into sterile vials. Take proper precautions when handling 2-mercaptoethanol, which has a very disagreeable odor and can irritate the eyes and nose. A single batch has a shelf-life of 2–3 months. Fresh fungal tissues (up to 200–300 mg) are placed directly into 500 μl of 2X CTAB buffer, which prevents degradation of sporocarp DNA during long-term storage.

DMSO, SUPERSATURATED SOLUTION

(Wu et al. 2000)

Ingredients per 100-ml batch:

DMSO (dimethyl sulfoxide)	20.0 ml
Distilled water	80.0 ml
EDTA (ethylenediaminetetraacetic acid disodium salt dihydrate)	8.4 g
NaCl	~36.0 g (bring solution to saturation)

This formula yields a solution of 20% DMSO and 250 mM EDTA in saturated NaCl. It prevents degradation of sporocarp DNA during long-term storage. Pipette 500-μl aliquots into sterile vials. Fresh fungal tissues (up to 200–300 mg) are placed directly into 500 μl of solution.

FORMALIN, NEUTRAL, BUFFERED

Formalin (37% aqueous formaldehyde)	100.0 ml
Distilled water	900.0 ml
$CaCO_3$ or $MgCO_3$	saturated

Prepared neutral buffered formalin solution can be purchased from some reagent suppliers (e.g., Sigma Chemical).

FORMALIN-ACETIC ACID-ALCOHOL FIXATIVE—FAA

Formalin (37% aqueous formaldehyde)	5.0 ml
Acetic acid	5.0 ml
Ethanol (50%)	90.0 ml

KEW MIXTURE

Ethanol (95%)	50.0 ml
Formalin (37% aqueous formaldehyde)	5.0 ml
Glycerol	5.0 ml
Water	40.0 ml

SCHAUDINN'S FLUID

(Polglase 1981)

$HgCl_2$ (saturated aqueous)	66.0 ml
Ethanol (95%)	33.0 ml
Glacial acetic acid	5.0 ml

Prepare a saturated aqueous solution of mercuric chloride with distilled water and store in a glass bottle in a cool, dark place. Some precipitate should remain as evidence of saturation. Schaudinn's Fluid does not store for more than a few days; however, the mercuric chloride and ethanol may be mixed as a stock solution that will keep for several months. Add 5 ml of acetic acid to 100 ml of stock solution to prepare a working solution for use within 48 hours. It can be used with fungi of aquatic animals. Take appropriate precautions when handling and disposing of mecuric chloride.

MOUNTING MEDIA

Amann's Solution

Phenol crystals	20.0 g
Lactic acid	16.5 ml
Glycerol	32.0 ml
Distilled water	20.0 ml

A trace of acid fuchsin or cotton blue can be added, if desired. It serves as both a preservative and a mounting medium and can be substituted for aqueous glycerol.

AQUEOUS GLYCEROL

Glycerol	100.0 ml
Chloral hydrate	5.0 g
Distilled water	5.0 ml

A trace of dye (e.g., cotton blue or acid fuchsin) can be added, if desired. This is an excellent medium for general morphological studies.

HOYER'S MEDIUM

Gum arabic	30.0 g
Chloral hydrate	200.0 g
Glycerol	16.0 ml
Distilled water	50.0 ml

The gum arabic and chloral hydrate, in turn, are dissolved in the water without heat, added to the glycerol, and mixed thoroughly. It is used sparingly only as a medium in which to position fungi on slides or cover glasses before adding mountant.

KOH-PHLOXINE

Two solutions

(a) KOH (2%)

(b) Phloxine

Distilled water with enough phloxine B added to turn the solution pink (approximately 0.5 g in 100 ml water).

A small amount of culture or tissue is transferred to a drop of 95% ethanol near the center of a slide; the excess alcohol is drained; 1 drop of 2% KOH is added, and the specimen is manipulated to suit. One drop of the phloxine solution is added, and the specimen is covered with a cover slip. This is an excellent mounting agent for revealing morphological characteristics of a fungus without distorting it.

LACTIC-ACID MOUNTING MEDIUM

Specimens are mounted in reagent-grade lactic acid. Acid fuchsin, cotton blue, or trypan blue can be added to the lactic acid solution (0.1% w/v) for microscopic examinations of spores and other fungal structures.

LACTOPHENOL MOUNTING MEDIUM

Phenol	30.0 g
Glycerol	31.8 ml
Lactic acid	24.6 ml
Distilled water	30.6 ml

LACTOPHENOL-ANILINE BLUE

Phenol	20.0 g
Glycerol	31.0 ml
Lactic acid	16.0 ml
Aniline blue	0.05–1.0 g
Distilled water	20.0 ml

The amount of aniline blue used depends on the amount of staining desired. This is appropriate for fungi on insects.

LACTOPHENOL-COTTON BLUE

(Stevens 1981)

Phenol	20.0 g
Glycerol	31.0 ml
Lactic acid	16.0 ml
Cotton blue	0.05–1.0 g
Distilled water	20.0 ml

The amount of cotton blue used depends on the amount of staining desired.

MELZER'S REAGENT

Iodine	1.5 g
Potassium iodide	5.0 g
Chloral hydrate	100.0 g
Distilled water	100.0 ml

This medium is used to visualize amyloid and dextrinoid cell-wall materials in fungal spores and hyphae.

POLYVINYL ALCOHOL-LACTIC ACID-GLYCEROL MOUNTANT—PVLG

Polyvinyl alcohol	16.0 g
Glycerol	10.0 ml
Lactic acid	100.0 ml
Distilled water	100.0 ml

Add alcohol to mixture of other ingredients. Place solution in a hot water bath (70–80°C) for 4 hours to dissolve. Mixture should be 50–75% hydrolyzed, with a viscosity of 20–25 centipoise in a 4% aqueous solution at 20°C. It can be stored in dark bottles for about 1 year. It is a permanent mountant.

SHEAR'S MOUNTING MEDIUM

Potasium acetate	6.0 g
Glycerol	120.0 ml
Ethanol (95%)	180.0 ml
Distilled water	300.0 ml

WHITE'S SOLUTION

(modified from Barras 1972)

$HgCl_2$	1.0 g
NaCl	6.5 g
HCl	1.25 ml
Ethanol (95%)	250.0 ml
Sterile distilled water	750.0 ml

STAINS

ACETO-CARMINE STAIN

Glacial acetic acid	45.0 ml
Distilled water	55.0 ml
Powdered carmine	1.0 g to saturate solution
5% $FeCl_3$ + $6H_2O$	5 ml/100 ml aceto-carmine stock solution

Boil carmine in water and acetic acid in chemical hood, filter, and store in a dark bottle at 4°C to make stock solution. To use, add ferric hydrate to given amount of stock solution until precipitation barely begins; then add an equal amount of stock solution. This is a nuclear stain.

SEMICHRON'S ACETOCARMINE

Glacial acetic acid	250.0 ml
Distilled water	250.0 ml
Carmine alum lake	5.0 g (to saturate solution)
Ethanol (70%)	500.0 ml

Combine water and acid in a flask, and then add carmine to make a saturated solution. Bring contents of the flask to about 95°C over a boiling water bath, but do not boil. Cool and filter. Measure stain volume, and add an equal volume of 70% ethanol. The stain will keep for years if stored in a cool dark place. Discard stock stain when stained cytoplasm begins to appear muddy or when stained specimens appear bluish-purple.

ACETO-ORCEIN STAIN

Orcein	1.0–2.0 g
Glacial acetic acid, at 100°C	45.0 ml
Distilled water	55.0 ml

Dissolve stain in boiling acid; add water, cool, and filter. This is a nuclear stain; it is an outstanding routine mounting medium for insect pathogens.

ACRIFLAVIN FLUORESCENT STAIN

(Raju 1986)

Prepare:

4N HCl
Concentrated HCl in 70% ethanol (2:98 v/v)
Acriflavin (100–200 µg/ml) in 0.1N HCl
Potassium metabisulphite ($K_2O_5S_2$) (5 mg/ml) in 0.1N HCl

Incubate all solutions at 30°C. Wash cells, hyphae, or tissue in 4N HCl for 20–30 minutes at 30°C, then wash once with distilled water. Stain for 20–30 minutes in acriflavin solution. Wash with HCl-ethanol solution three times, and then twice with distilled water. Mount cells in a drop of 25% glycerol. Morphology of the cells' nuclei is observed using a fluorescence microscope at an excitation wavelength of 450 nm.

AMMONIACAL CONGO RED

Congo red	0.5 g
Aqueous ammonia solution (reagent grade 27–31%)	33.0 ml
Water	66.0 ml

This solution is good for staining tissues and spores of macrofungi and microfungi.

ANILINE BLUE STAIN, ACIDIFIED

Aniline blue powder	0.1 g
Sterile distilled water	100.0 ml
Lactic acid (85%)	50.0 ml

Mix vigorously until powder is dissolved; add acid, and mix again.

ANILINE BLUE STAIN, NONACIDIFIED

Aniline blue powder	0.05 g
Sterile distilled water	100.0 ml

Agitate mixture vigorously to dissolve powder.

BISMARK BROWN

Bismark brown	2.0 g
Ethanol (95%)	100.0 ml

This medium is good for staining fungi on insects.

CALCOFLUOR WHITE M2R

Prepare 0.05–0.1% calcofluor white (ACS Chemical Index No. 40622) in 0.2 M tris buffer, pH 8.0. Stain cells for 30–60 seconds. Wash and mount cells in water. Morphology of the cell walls is observed using a fluorescence microscope at an excitation wavelength of 350 nm. Wall materials will glow green–yellow. This medium is good for observing endophytic fungi immersed in plant tissue, and conidiogenesis.

CHLORAZAL BLACK E STAIN IN LACTOGLYCEROL

Chlorazol black E‡	0.1 g
Lactic acid	33.3 ml
Glycerol	33.3 ml
Distilled water	33.3 ml

This medium is suitable for examination of vesicular–arbuscular mycorrhizae in roots and endophytes in other tissues. (‡ACS Chemical Index No. 30235)

COOMASSIE BRILLIANT BLUE R-250 STAIN

This stain should be prepared 24 hours in advance of intended use. It contains equal volumes of the following:

Trichloroacetic acid (0.15%)
Coomassie brilliant blue R-250‡ (0.6% in 99% methanol)

Clear tissue for 10–60 minutes in a mixture of ethanol-chloroform (3 : 1 v/v) with 0.15% trichloroacetic acid, using several solution changes. This is a protein-specific stain useful for cytoplasm-rich structures such as germ hyphae and haustoria. (‡ACS Chemical Index No. 42660)

DAPI FLUORESCENT STAIN

DAPI (4′,6-diamidino-2-phenylindole dihydrochloride) forms fluorescent complexes with natural double-stranded DNA and shows fluorescence specificity for certain base-pair clusters. Because of that property, DAPI is a useful tool in various cytochemical investigations, especially those requiring visualization of nuclei. Care should be taken when handling DAPI because it labels A-T rich sites of all DNA.

(1) Prepare a stock solution of DAPI at 10 mg/ml in distilled water; protect from light, and store at 4°C. Prepare a 1,000–5,000-fold dilution in 0.1 M phosphate buffer or distilled water for labeling.
(2) Prepare a fresh 4% formaldehyde solution in phosphate buffer (or one you select) for fixation. A 0.2% Triton-X100 solution also may be required to increase permeability of cells, depending on tissue complexity.
(3) Fix cells for 10–30 minutes at room temperature.
(4) Rinse cells three times in buffer.
(5) If necessary, immerse cells in Triton-X solution for about 5 minutes to increase permeability.
(6) Repeat step (4).
(7) Incubate cells at room temperature for 1–5 minutes in DAPI labeling solution.
(8) Repeat step (4).
(9) Mount cells in 90% glycerol/10% phosphate buffer. N-propyl gallate (0.1–1.0 mg/ml) can be added to mounting solution to prevent fading.
(10) View samples on epifluorescent microscope with ultraviolet excitation (approximately 359 nm) and blue emission (approximately 461 nm).

DIAZONIUM BLUE B STAIN—DBB

(Summerbell 1985)

Prepare diazonium blue B at 1 mg/ml in 0.1 M Tris-HCL buffer (pH 7). Refrigerate at 4°C. Mycelia or yeast cells are harvested while actively growing, rinsed three times with distilled water, incubated 12 hours in 1 M KOH at 4°C, rinsed thoroughly (three or more times) with distilled water, transferred to cold (4°C) 0.1 M Tris-HCl buffer at pH 7.0, and maintained on ice while fresh DBB (1 mg/ml) is prepared in the same, cold buffer. This is an appropriate stain for saprobes, yeasts, and basidiomycetes.

GRAM'S STAIN

Gram's stain is a double-staining method used in most examinations and preliminary identifications of bacteria. Differential staining involves the use of several sequential stains and shows the differences among bacteria with different cell wall structures. Gram-"positive" bacteria appear violet or blue in color, whereas those that are

Gram-"negative" appear red to pink in color. Fungi and yeasts exhibit a Gram-positive reaction. Commercial Gram stain kits are widely available. The method here uses a carbol-fuchsin counterstain, but safranin counter staining is often used.

Preparation of Gram's Stain reagents first requires preparation of the following solutions:

Crystal violet solution:

Crystal violet	2.0 g
Ammonium oxalate	0.8 g
Ethanol	20.0 ml
Distilled water	80.0 ml

Gram's iodine solution:

Iodine	1.0 g
KI 2.0 g	
Distilled water	300.0 ml

Dilute carbol-fuchsin solution

Basic fuchsin	1.0 g
Absolute ethanol	10.0 ml
Phenol	5.0 g
Distilled water	1,600.0 ml

Smear a thin layer of cells onto a glass slide with a cotton swab. Flood the fixed smear with Gram's crystal violet stain for 1.5 minutes. Pour off the dye and wash slide with water. Flood the slide with Gram's iodine for 1 minute. Iodine is a mordant that fixes the Gram-positive cells so that they will not lose the violet color. Pour off the iodine and wash. Decolorize the slide with 95% ethanol. Add the ethanol a drop at a time until it runs colorless off the slide. Remove the alcohol at once by washing in cold water. Counterstain by flooding the slide with basic fuchsin solution for 1.5 minutes. Pour off basic fuchsin, wash slide, and blot dry. Dark-blue or violet staining indicates a Gram-positive reaction; if cells are red or pink, they are Gram negative.

MALACHITE-GREEN ACID-FUCHSIN

(Alexander 1980)

Malachite green (1%, in 95% ethanol)	2.0 ml
Acid fuchsin (1%, in distilled water)	10.0 ml
Glycerol	40.0 ml
Ethanol (95%)	20.0 ml
Phenol	5.0 g
Lactic acid	1.0–6.0 ml
Distilled water	50.0 ml

AQUEOUS METHYL GREEN

Distilled water	50.0 ml
Methyl green	0.5 g

Let the mixture sit overnight to dissolve completely.

METHYLENE BLUE

Methylene blue	1.0 mg
Sodium citrate	0.2 g
Distilled water	10.0 ml

Mix 0.1 ml of cell suspension with 0.1 ml methylene blue-sodium citrate gently for 1 minute. Live cells remain clear; dead cells are blue. Stain for use with aquatic animals.

PERIODIC ACID-SCHIFF STAIN

(Dring 1955; Farris 1966; Nair 1976)

Periodic acid, 1% solution:

Periodic acid (50%)	2.0 ml
Distilled water	98.0 ml

Schiff's reagent:

Basic fuchsin	1.0 g
Potassium metabisulphite ($K_2O_5S_2$)	2.0 g
Distilled water	200.0 ml
Concentrated HCl	2.0 ml
Deactivated charcoal	1.0–2.0 g

Boil distilled water; add basic fuchsin slowly; mix and cool to 50°C. Add potassium metabisulphite. Mix and cool to room temperature before adding HCl. Keep in the dark overnight for bleaching to occur. Add charcoal and filter through coarse paper and then a fine filter paper. Store in the refrigerator.

This staining procedure can be used on a variety of plant and animal materials to determine the presence and intracellular localization of starch, glycogen, and cellulose. Fix plant tissue in formalin-acetic acid-alcohol fixative (see "Fixatives," earlier). Tissue is immersed in 1% aqueous periodic acid for 5 minutes, rinsed in tap water for 10 minutes, immersed in Schiff's reagent for 5 minutes, washed again in tap water for 10 minutes, and then immersed in a solution containing 5 ml 10% aqueous $K_2O_5S_2$, 5 ml 1 M HCl, and 90 ml distilled water for 5 minutes. The solution is changed, and the tissue is immersed for another 5 minutes; then it is washed in tap water for 10 minutes, completely dehydrated through an absolute ethanol to xylol series, and mounted in Permount.

ROSE BENGAL STAIN

Rose bengal powder	0.5 g
Ethanol (absolute)	5.0 ml
Distilled water	95.0 ml

Agitate mixture vigorously to dissolve powder. This medium may replace aniline blue for viewing of endophytic mycelia.

TRYPAN BLUE IN LACTOGLYCEROL

Trypan blue[‡]	0.05 g
Lactic acid	33.3 ml
Glycerol	33.3 ml
Distilled water	33.3 ml

This medium is suitable for transmitted light microscopy. ([‡]ACS Chemical Index No. 23850)

WRIGHT STAIN, MODIFIED[a]

Wright stain (powdered)	0.3 g
Glycerol	3.0 ml
Methanol	97.0 ml

Mix Wright stain and glycerol in a mortar. Slowly add methanol and mix. Let mixture stand overnight, and then filter. Store in a dark bottle. Buffered modified Wright stain is available in prepared form from several suppliers.

STORAGE SOLUTION

COPENHAGEN SOLUTION

Ethanol	52.6 ml
Glycerol	5.3 ml
Water	42.1 ml

---

* Available dehydrated or in powdered form, or a medium equivalent is available from commercial suppliers such as BBL (Becton-Dickinson), Difco, Oxoid (Unipath), Sigma Chemical Company, and others (see Appendix IV).

[†] Cyclosporin A can be purchased from a pharmacy in the form of ampoules of Sandimmune I.V. or from Sigma-Aldrich, Inc. (see Appendix IV).

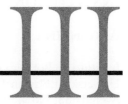

INSTITUTIONS WITH SIGNIFICANT COLLECTIONS OF FUNGI OR FUNGAL ALLIES AND FUNGUS-RELATED WEBSITES

COMPILED BY FIONA A. WILKINSON AND MERCEDES S. FOSTER

Throughout the text of this volume, authors refer to herbaria, museums, universities, and other institutions that house important collections of preserved specimens or live cultures of fungi and fungal allies. To facilitate location of and access to those institutions, we provide contact information for each in the first part of this appendix (see "Institutions," later in this chapter). In most instances we provide the URL address for the institution website, which provides more detailed information such as mail and e-mail addresses and names of contacts. Often, the websites also include detailed information about specimen holdings, procedures for requesting loans, and investigator visits. If a URL address was not available, we provide mailing addresses, telephone, and fax numbers. We list only those institutions referred to by name or herbarium code in the text of this volume. Individuals interested in other institutions should consult one of several websites that list institutional names and contact information (e.g., see web addresses for *The Internet Directory for Botany: Botanical Museums, Herbaria, Natural History Museums* and for *The Index Herbariorum*, later in this chapter). In several instances, authors refer to private collections held by individuals. We provide the web addresses for the institutions with which those individuals are associated. Contact information for the individuals can be obtained by accessing those sites. Information is arranged alphabetically by country and includes herbarium codes.

In the second part of this appendix, we provide a list of URLs for websites (arranged alphabetically) mentioned in the text of this volume. Websites with information on fungi and fungal allies are numerous and extensive and beyond the scope of this appendix. Most sites mentioned, however, provide links to many other sites. In addition, individuals should consult directories to appropriate websites, such as *The Internet Directory for Botany: Lower Plants, Algae—Bryophytes—Fungi* (see "Websites," later in this chapter).

INSTITUTIONS

ARGENTINA

Instituto de Botánica Carlos Spegazzini, Herbario (LPS)
Universidad Nacional de La Plata
Avenida 53, No. 477
1900 La Plata, Buenos Aires
Telephone: [54] 221/421-9845

AUSTRALIA

Australian National Collection of Fungi: made up of the collections of three of the major mycological herbaria in Australia: BRIP, DAR, and VPRI
CSIRO, Australian National Herbarium, Research Collections and Services (QRS)
http://www.tfrc.csiro.au/

Plant Pathology Herbarium (BRIP)
Department of Primary Industries
80 Meiers Road
Indooroopilly, Queensland 4068
Telephone: [61] 7/3896-9340.
Fax: [61] 7/3896-9533.

Plant Pathology Herbarium (DAR)
Orange Agricultural Institute
Forest Road
Orange, New South Wales 2800
Telephone: [61] 2/6391-3800.
Fax: [61] 2/6391-3899.

Institute for Horticultural Development, Herbarium (VPRI)
Department of Agriculture, Victoria
http://www.nre.vic.gov.au/agvic/ihd/index.htm

Royal Botanic Gardens Melbourne
http://www.rbgmelb.org.au/

State Herbarium of South Australia, Plant Biodiversity Centre (AD)
http://www.flora.sa.gov.au/

Tasmanian Herbarium, University of Tasmania (HO)
http://www.tmag.tas.gov.au/herb.htm

Queensland Herbarium, Indooroopilly
http://www.env.qld.gov.au/environment/science/her barium/

AUSTRIA

Karl–Franzens–Universität Graz, Institut für Botanik, Herbarium (GZU)
http://www.kfunigraz.ac.at/botwww/home.htm

BELGIUM

Mycothèque de l'Université Catholique de Louvain (MUCL)
http://www.belspo.be/bccm/mucl.htm

CANADA

Canadian Collection of Fungal Cultures (CCFC), Agriculture and Agri-Food Canada
http://sis.agr.gc.ca/brd/ccc

Department of Agriculture, Ottawa, Mycology (DAOM)
http://res2.agr.ca/ecorc/daom/daom_e.htm

Royal Ontario Museum, Cryptogamic Herbarium (TRTC)
http://www.rom.on.ca/

University of Alberta Microfungus Collection and Herbarium (UAMH)
http://www.devonian.ualberta.ca/uamh/

DENMARK

Royal Veterinary and Agricultural University, Centre for Advanced Food Studies
http://www.adm.kvl.dk/english/research/instcent.htm

Royal Veterinary and Agricultural University, Plant Biology Department, Herbarium (CP)
http://www.ibt.dtu.dk/lmc/research/res98-00/mycolo.htm

Technical University of Denmark, Mycology Group, BioCentrum, Culture Collection (IBT)
http://www.ibt.dtu.dk/mycology/index.htm

University of Copenhagen, Herbarium (C)
http://www.botanicalmuseum.dk/bot

ESTONIA

Institute of Zoology and Botany, Department of Biological Collections, Herbarium (TAA), Tartu
http://www.zbi.ee/

FINLAND

University of Helsinki, Herbarium (H)
Finnish Museum of Natural History
http://www.helsinki.fi/kmus/

FRANCE

Université Montpellier II
Institut de Botanique, Herbier (MPU)
163 rue Auguste-Broussonet
F–34090 Montpellier

Laboratoire de Cryptogamie (PC), Paris
http://www.mnhn.fr/

Musée National d'Histoire Naturelle
Department of Marine Biology, Herbier Crouan (CO)
B.P. 225
F-29125 Concarneau
Telephone: [33] 2/98 97 0659.
Fax: [33] 2/98 97 8124.

Musée National d'Histoire Naturelle, Herbier (P)
16 rue Buffon
F-75005 Paris
Telephone: [33] 1/ 40 79 33 53.
Fax: [33] 1/ 40 79 33 42.

HUNGARY

Hungarian Natural History Museum, Department of Botany, Herbarium (BP)
http://www.bot.nhmus.hu/enoveny.html

ITALY

Industrial Yeasts Collection
http://www.agr.unipg.it/dbvpg/

JAPAN

Institute for Fermentation (IFO)
http://wwwsoc.nii.ac.jp/ifo/

Japan Collection of Microorganisms (JCM)
http://www.jcm.riken.go.jp/

National Institute of Sericultural and Entomological Science (NISES), Owashi
http://ss.nises.affrc.go.jp/index-e.html

Nippon Veterinary and Animal Science University, Department of Fish Diseases, Herbarium
http://www.nvau.ac.jp/english/index.htm

Sanko Co., Ltd. (SANK)
Laboratory for Biological Resources Research
http://www.sankyo.co.jp/english/index.html

LUXEMBOURG

Musée national d'histoire naturelle, Herbarium (LUX)
http://www.mnhn.lu/recherche/botan.htm

THE NETHERLANDS

Centraalbureau voor Schimmelcultures (CBS)
http://www.cbs.knaw.nl/

National Herbarium Nederland, Leiden University branch (L)
http://nhncml.leidenuniv.nl/rhb/

NEW ZEALAND

Landcare Research (PDD)
http://www.landcare.cri.nz/

PEOPLE'S REPUBLIC OF CHINA

Center for Collection of General Microbiological Cultures (CCGMC)
Institute of Microbiology, Academia Sinica
P.O. Box 2714, Beijing

University of Hong Kong Culture Collection—HKUCC
http://www.hku.hk/ecology/mycology/cc/

City University of Hong Kong
http://www.cityu.edu.hk/cityu/dpt-acad/fse-bch.htm
(and http://ccsql1.cityu.edu.hk/bhdbapp/links/default.htm)

PORTUGAL

Portuguese Yeast Culture Collection (PYCC)
http://wdcm.nig.ac.jp/CCINFO/CCINFO.xml?595

REPUBLIC OF GERMANY

Berlin-Dahlem, Botanischer Garten und Botanisches Museum, Herbarium (B)
http://www.bgbm.org

Botanisches Staatssammlung, Munich (M)
http://www.botanik.biologie.uni-muenchen.de/botsamml/home.html

Centraalbureau voor Schimmelcultures (CBS)—Fungal Biodiversity Center
http://www.cbs.knaw.nl/

Deutsche Sammlung von Mikroorganismen und
Zellkulturen GmbH (DSMZ)
[German Collection of Microorganisms and Cell
Cultures]
http://www.dsmz.de/

Friedrich–Schiller–Universität Jena, Herbarium
Haussknecht (JE)
http://www.uni-jena.de/biologie/spezbot/

Philipps-Universität Marburg
http://www.uni-marburg.de/

Universität Regensburg, Regensburgische Botanische
Gesellschaft, Herbarium (REG)
http://www.biologie.uni-regensburg.de/Botanik/
Poschlod/index.html

RUSSIA

All-Russian Collection of Microorganisms
http://www.vkm.ru/

SCOTLAND

Royal Botanic Garden (Edinburgh), Herbarium (E)
http://www.rbge.org.uk

SLOVAKIA

Culture Collection of Yeasts (CCY), Institute of Chemistry, Slovak Academy of Sciences
http://nic.savba.sk/sav/inst/chem/intro.html

Research Institute for Viticulture and Enology
http://wdcm.nig.ac.jp/CCINFO/CCINFO.
xml?28#ch1

SPAIN

Real Jardín Botánico, Herbario (MA)
http://www.rjb.csic.es

SWEDEN

Bergius Foundation, Herbarium (SBT)
http://www.bergianska.se/stift/

Göteborg University, Systematic Botany and Plant
Ecology, Herbarium (GB)
http://www.systbot.gu.se/

Swedish Museum of Natural History, Botany Departments, Herbarium (S)
http://www.nrm.se/

Umeå University, Herbariet (UME)
http://www.eg.umu.se

Uppsala University, Museum of Evolution, Botany
Section (Fytoteket), Herbarium (UPS)
http://www-hotel.uu.se/evolmuseum/fytotek/

SWITZERLAND

Eidgenössische Technische Hochschule Zürich,
Herbarium (ZT)
http://www.geobot.umnw.ethz.ch/

UNITED KINGDOM

Commonwealth Agricultural Bureau International—
CABI
http://www.cabi.org/
Includes the International Mycological Institute,
Herbarium (IMI)

National Collection of Yeast Cultures (NCYC)
http://www.ifrn.bbsrc.ac.uk/ncyc/Default.html
(includes a directory to other UK culture collections)

Natural History Museum [British Museum of Natural
History] (BM)
http://nhm.ac.uk

Royal Botanic Gardens, Kew (K)
http://www.rbgkew.org.uk

University of Portsmouth
http://www.port.ac.uk

UNITED STATES

Academy of Natural Sciences in Philadelphia
http://www.acnatsci.org/

Agricultural Research Service Culture Collection
(NRRL)
http://nrrl.ncaur.usda.gov/the_collection3.htm

Agricultural Research Service Collection of
Entomopathogenic Fungal Cultures (ARSEF)
http:www.ars.usda.gov/is/np/systematics/fungibact.htm

American Type Culture Collection (ATCC)
http://www.atcc.org/SearchCatalogs/
SpecialCollections.cfm

Central Washington University, Biological Science
Department, Herbarium (ELRG)
Ellensburg, Washington 98926-7537

Cornell University, Herbarium (BH)
http://www.cwu.edu/

Cornell University, Wiegand Herbarium (CU)
(CU transferred to Cornell University Herbarium)

Cornell University, Plant Pathology Herbarium (CUP)
http://ppathw3.cals.cornell.edu/CUPpages/CUP.
html

Deaver Herbarium, Northern Arizona University
(ASC)
http://art.artsci.nau.edu/~it/ascs/ascs98/ascs.990326
.html

Fairmont State College, Herbarium (FWVA)
Biology Department
Fairmont State College
Locust Avenue
Fairmont, West Virginia 26554-2470
Telephone: [1] 304-367-4158.
Fax: [1] 304-366-4870.

Farlow Herbarium, Harvard University (FH)
http://www.herbaria.harvard.edu/Collections/farlow/
farlow.html

Field Museum of Natural History (F)
http://www.fieldmuseum.org

Fungal Genetics Stock Center
http://www.fgsc.net/

Fusarium Research Center
http://frc.cas.psu.edu/

Herman J. Phaff Yeast Culture Collection, University
of California, Davis
http://foodscience.ucdavis.edu/CIFAR/home_pag.htm

Illinois Natural History Survey, Herbarium (ILLS)
http://www.inhs.uiuc.edu/~kenr/herbarium.html

Institute of Marine Sciences (IMS), University of
North Carolina
http://www.marine.unc.edu/

Natural History Museum of Los Angeles County,
Herbarium (LAM)
http://www.nhm.org/research/home.html

New York Botanical Garden (NY)
http://www.nybg.org/

NRRL Culture Collection (see Agricultural Research
Service Culture Collection, earlier)

Oregon State University, Department of Botany and
Plant Pathology, Mycological Collections
http://ocid.nacse.org/research/herbarium/myco/

Rancho Santa Ana Botanic Garden, Herbarium (RSA)
http://www.cgu.edu/inst/rsa/

Rutgers University, Mycological Herbarium (RUTPP)
http://www.rci.rutgers.edu/~white/herbarium

San Francisco State University, Harry D. Thiers
Herbarium (SFSU)
http://www.mycena.sfsu.edu/

Smithsonian Institution (US)
http://www.si.edu/

United States Department of Agriculture, Agricultural
Research Service
Collection of Entomopathogenic Fungi
http://www.ars.usda.gov/is/np/systematics/
fungibact.htm

United States Department of Agriculture, Forest
Service, Madison
Center for Forest Mycology Research
http://www.fpl.fs.fed.us/rwus/rwu4501.htm

United States Department of Agriculture, Forest
Service, Forest Sciences Laboratory, Corvallis
http://mgd.nacse.org/cgi-bin/qml2.0/fslmyco/
fslherb.qml

United States Department of Agriculture, Forest
Service, Pacific Northwest Research Station, Corvallis,
Forest Mycology and Mycorrhiza Research
http://www.fsl.orst.edu/mycology/truffs.html

United States Department of Energy, Joint Genome
Institute
http://www.jgi.doe.gov

United States National Fungus Collections (BPI)
Systematic Botany and Mycology Laboratory
http://nt.ars-grin.gov/SBMLweb/
SystematicsResources/FungusCol/NFCIndex.htm
(and http://nt.ars-grin.gov/SBMLweb/Databases/
NFCDataBases.htm)

University of California, Berkeley, University and
Jepson Herbaria (UC)
http://ucjeps.berkeley.edu/

University of California, Riverside
Botany and Plant Sciences Department, Herbarium
(UCR)
http://cnas.ucr.edu/

University of Florida and Florida Museum of Natural
History, Herbarium (FLAS)
Mycological Collection, Bryophyte and Lichen
Collections
http://www.flmnh.ufl.edu/natsci/herbarium/
flasfung.htm

University of Georgia, Plant Pathology Department,
Herbarium (GAM)
http://mars.cropsoil.uga.edu/ppath/index.htm

University of Michigan, Herbarium (MICH)
http://www.herb.lsa.umich.edu/

University of North Carolina and North Carolina
Botanical Garden, Herbarium (NCU)
(including the J. N. Couch Collection)
http://www.herbarium.unc.edu

University of Southern California, Herbarium (AHFH)
AHFH transferred to Rancho Santa Ana Botanic
Garden and Natural History Museum of Los Angeles
County

INSTITUTIONS HOUSING PERSONAL COLLECTIONS MENTIONED IN THIS VOLUME

Central Missouri State University
http://www.cmsu.edu/biology/

Columbia University
http://www.columbia.edu/

Cornell University, Department of Entomology
http://www.nysaes.cornell.edu/ent/

Instituto de Ecología y Sistemática, La Habana, Cuba
http://www.cuba.cu/ciencia/citma/ama/ecologia

Oregon State University (OSC)
http://www.orst.edu/

Shepherd College
http://www.shepherd.wvnet.edu/

Smithsonian Institution (including National Museum
of Natural History)
http://www.mnh.si.edu/

Taiwan Agricultural Research Institute, National Plant
Genetic Resources Center, Wufeng, Taichung, Taiwan
http://192.192.196.1/indexe.html

University of California, Berkeley, Department of
Environmental Science, Policy, and Management
http://www.berkeley.edu/

University of Arkansas, Fayetteville
http://www.uark.edu/

University of Kentucky, Louisville Organismal and
Integrative Biology
http://biology.uky.edu/oib/

University of Washington, Seattle
http://www.washington.edu/

University of Western Australia
http://www.uwa.edu.au/

WEBSITES

American Bryological and Lichenological Society
http://www.unomaha.edu/~abls/

American Phytopathological Society (APS)
http://www.apsnet.org/visitors/top.asp

Ascomycete Species from Freshwater Habitats
http://fm5web.life.uiuc.edu:23523/ascomycete/

The Bank of European Glomales (BEG)
http://www.ukc.ac.uk/bio/beg/

Basidiomycete Phylogeny website
http://www.biology.duke.edu/fungi/phylogeny

The British Lichen Society
http://www.argonet.co.uk/users/jmgray

The British Mycological Society
http://www.ulst.ac.uk/faculty/science/bms/

The Bryologist (with searchable reference database)
http://www.toyen.uio.no/botanisk/lavherb.htm

Centre for Agriculture and Biosciences International
(CABI); includes the International Mycological Institute
(IMI)
http://www.cabi.org/

CABI Bioscience and CBS Database of Fungal Names
http://indexfungorum.org/Names/Names.asp

Centro Internacional de Agricultura Tropical (CIAT)—
Cali, Colombia
http://www.ciat.cgiar.org/

Common Data Structure for European Floristic Databases, Biodiversity and Biological Collections Web Server
http://www.keil.ukans.edu/

A Conservation Overview of Australian Non-marine
Lichens, Bryophytes, Algae, and Fungi
http://www.ea.gov.au/biodiversity/threatened/action
/cryptogams/6a.html

Cornell Center for Fungal Biology, Resources
http://ccfb.cornell.edu/links.html

Costa Rican National Biodiversity Inventory—Fungal
Component
http://www.inbio.ac.cr/papers/gt_Hongos/en/index.
htm

EMBL, see Genebank for European Molecular Biology
Laboratory

Fungal Genetics Stock Center
http://www.fgsc.net/

Fusarium Research Center
http://frc.cas.psu.edu/

FusKey (computer assisted diagnostic aids for
Penicillium and *Fusarium*)
http://sis.agr.gc.ca/brd/fusarium/

GenBank: NCBI = National Center for Biotechnology
Information
http://www.ncbi.nlm.nih.gov

Genebank for European Molecular Biology Laboratory
(EMBL)—EMBL Nucleotide Sequence Database
http://www.ebi.ac.uk/embl/

The German Mycological Society (Die Deutsche
Gesellschaft für Mykologie)
http://www.dgfm-ev.de/

Gray Herbarium Card Index, see International Plant
Name Index

Great Smoky Mountains National Park—All Taxa
Biodiversity Inventory (ATBI)
http://www.discoverlife.org/

Index Fugorum (funindex)
http:www.indexfungorum.org/

Index Herbariorum online (Holmgren et al. 1990)
http://www.nybg.org/bsci/ih/ih.html

Integrated Taxonomic Information System (ITIS)
http://www.itis.usda.gov/

International Association for Plant Taxonomy
http://www.bgbm.org/iapt/default.htm

International Code of Botanical Nomenclature
http://www.bgbm.org/iapt/nomenclature/code/
default/html

International Culture Collection of (Vesicular)
Arbuscular Mycorrhizal Fungi (INVAM)
(extensive website on the collection and other
mycologic information)
http://invam.caf.wvu.edu/

International Mycological Institute (IMI), see CABI

International Plant Names Index (IPNI)
http://www.ipni.org/index.html

The Internet Directory for Botany: Botanical
Museums, Herbaria, Natural History Museums
http://www.botany.net/IDB/subject/botmus.html

Internet Directory for Botany: Lower Plants, Algae—
Bryophytes—Fungi
http://www.botany.net/IDB/subject/botcryp.html#
fungi

The Internet Guide to Myxomycetes
http://www.wvonline.com/myxo

Kenneth B. Raper Cellular Slime Mold Collection,
Department of Environmental and Plant Biology, Ohio
University, Athens
http://www.plantbio.ohiou.edu/index.htm

LIAS—A Global Information System for Lichenized
and Non-Lichenized Ascomycetes
http://www.lias.net

Lichen Determination Keys Available on INTERNET
(H. J. M. Sipman)
http://www.bgbm.org/sipman/keys/

Mycology Resources 2 (list of mycology-related
websites, directories, online publications, etc.)
http://www.edae.gr/myco2.html

The Mycological Society of America
http://msafungi.org

Myconet
http://www.umu.se/myconet/Myconet.html

Natural Science Collections Alliance (formerly the Association for Systematics Collections)
http://nscalliance.org

Plant Pathology Internet Guide Book
http://www.bspp.org.uk/ppigb/

Protostelid Home Page
http://www.comp.uark.edu/~fspiegel/protist.html

Society for the Preservation of Natural History Collections
http://www.spnhc.org

TAXACOM
http://usobi.org/archives/taxacom.html

The Trichomycetes: Fungal Associates of Arthropods (Lichtwardt et al. 2001)
http://www.nhm.ukans.edu/~fungi

United States Department of Agriculture (USDA), Animal Plant Health Inspection Service
National Center for Import and Export (regulations and guidelines regarding research with soils from outside the United States)
http://www.aphis.usda.gov/NCIE/

LIAS—A Global Information System for Lichenized and Non-Lichenized Ascomycete
http://www.lias.net/

Lichen Determination Keys Available on INTERNET (H. J. M. Sipman)
http://www.bgbm.org/sipman/keys/

United States Department of Agriculture (USDA) Forest Service, Forestry Sciences Laboratory, Mycological Herbarium Database
http://mgd.nacse.org/cgi-bin/qml2.0/fslmyco/fslherb.qml

United States Department of Transportation (regulations for safety aspects of domestic transportation of infectious substances)
http://hazmat.dot.gov/rules.htm

United States National Science Foundation (NSF)
http://www.nsf.gov/

WWW Virtual Library: See Mycologic Resources

The World Data Centre for Microorganisms (WDCM) (Comprehensive directory of collections where microorganisms and cell lines can be deposited or obtained. Links to small, obscure and to well known collections)
http://wdcm.nig.ac.jp/

World Federation for Culture Collections (WFCC), Committee on Postal, Quarantine, and Safety Regulations
http://wdcm.nig.ac.jp/wfcc/postal_committee.html

VENDORS

The list of vendors is provided to aid investigators in locating equipment, materials, and supplies necessary to carry out studies of fungal diversity in the field and in the laboratory. The list reflects the experiences of the authors and is not exhaustive. Mention of vendors should not be construed as an endorsement of those companies over others offering comparable products.

EQUIPMENT AND SUPPLIES FOR FIELD AND LABORATORY

Akteingesellschaft Sigg
Household Appliances
CH-8500 Frauenfeld
Switzerland
Telephone: 041-54-26-31-31
Fax: 041-54-21-66-05
[portable food dehydrators]

Amerex Instruments, Inc.
P.O. Box 787
Lafayette, CA 94549
USA
Telephone: 925-299-0743
Fax: 925-299-0745
http://www.amerexinst.com/
[top-loading, portable autoclaves]

Becton-Dickinson and Co.
BBL Branch
1 Becton Drive
Franklin Lakes, NJ 07417
USA
Telephone: 201-847-6800
http://www.bd.com
[Bactec mechanized system for blood cultures]

Bel-Art Products
6 Industrial Road
Pequannock, NJ 07440
USA
Telephone: 973-694-0500
Fax: 973-694-7199
http://www.bel-art.com/
[general laboratory supplies, microsieve sets, plasticware]

Bellco Glass, Inc.
340 Edrudo Road
Vineland, NJ 08360-3493
USA
Telephone: 800-257-7043; 609-691-1075
Fax: 609-691-3247
http://www.bellcoglass.com/

[glassware, incubators, laminar flow products, freeze-dry ampoules]

Ben Meadows Co.
P.O. Box 80549
3589 Broad Street
Chamblee, GA 30341
USA
Telephone: 800-241-6401; 404-455-0907
Fax: 800-628-2068; 404-457-1841
http://www.benmeadows.com/
[drawing and drafting supplies, global positioning system (GPS) receivers, hand tools, insect pins, magnifiers, measuring tapes, soil analysis equipment, soil collection bags, tree-marking supplies, waterproof field books, weather instruments]

BioQuip Products
17803 La Salle Avenue
Gardena, CA 90248-3602
USA
Telephone: 310-324-0620
Fax: 310-324-7931
E-mail: bioquip@aol.com
http://www.bioquip.com/
[equipment, supplies, and books for entomology and related sciences; nets, pins, black lights, storage cases]

Carolina Biological Supply Co.
2700 York Road
Burlington, NC 27215
USA
Telephone: 800-334-5551
Fax: 919-584-3399; 800-227-7112
http://www.carolina.com
[entomologic supplies, general laboratory supplies, non-viable hemp seeds]

Cole-Parmer Instrument Co.
625 East Bunker Court
Vernon Hills, IL 60061
USA
Telephone: 800-323-4340
Fax: 847-247-2929
E-mail: info@coleparmer.com
http://www.coleparmer.com/
Branch locations throughout the world; contact information available on the website.
[pumps, tubing, fittings, plasticware, glassware, equipment for biotechnology, environmental control, balances, thermometers, accessories]

Consolidated Plastics Co., Inc.
8181 Darrow Road
Twinsburg, OH 44807

USA
Telephone: 216-425-3900
Fax: 216-425-3333
[plastic boxes, plasticware, shipping supplies, tubing]

CUDA Fiberoptics
6025 Chester Avenue
Jacksonville, FL 32217-2242
USA
Telephone: 904-737-7611
Fax: 904-733-4832
http://www.cuda.com/
[fiberoptic fibers]

Deakin Equipment
1361 Powell Street
Vancouver, British Columbia V5L 1G8
Canada
Telephone: 604-253-2685; 800-663-3735
Fax: 604-253-4639; 800-634-8388
http://www.deakin.com/
[forestry tools, GPS units and field data collection devices, gear and safety equipment for remote field camps]

Dynal Biotech, Inc.
5 Delaware Drive
Lake Success, NY 11042
USA
Telephone: 800-638-9416
E-mail: ustechserv@dynalbiotech.com
http://www.dynal.no/
[Dynabeads]
Dynal Biotech (headquarters)
P.O. Box 114, Smestad
N-0309 Oslo
Norway
Telephone: + 47 22 06 10 00
Fax: + 47 22 50 70 15
E-mail: dynal@dynalbiotech.com
[Dynabeads]

Fine Science Tools (USA), Inc.
373-G Vintage Park Drive
Foster City, CA 94404-1139
USA
Telephone: 800-521-2109; 650-349-1636
Fax: 800-523-2109; 650-349-3729
http://www.finescience.com/
[fine forceps, scalpels, stage micrometers]

Fisher Scientific
Branch locations throughout the United States, and worldwide export service; contact information available on the website.
http://www.fishersci.com/

[complete line of scientific supplies, equipment, and reagents]

Forestry Suppliers, Inc.
205 W. Rankin Street
P.O. Box 8397
Jackson, MS 39284-8397
USA
Telephone: 800-647-5368; international 601-354-3565
Fax: 800-813-2439; international 601-355-5126
http://www.forestry-suppliers.com/
[drawing and drafting supplies, global positioning systems (GPS) receivers, hand tools, insect pins, magnifiers, measuring tapes, soil analysis equipment, soil collection bags, tree-marking supplies, waterproof field books, weather instruments]

Forma Scientific, Inc.
P.O. Box 649
Marietta, OH 45750
USA
Telephone: 740-373-4763; 800-848-3080
Fax: 740-374-1817
http://www.forma.com
[biosafety cabinets, cryopreservation systems, incubators, shakers]

Gallard-Schlesinger Industries, Inc.
777 Zeckendorf Boulevard
Garden City, NY 11530
Telephone: 516-229-4000
Fax: 516-229-4015
E-mail: info@gallard.com
http://www.gallard.com/
[laboratory chemicals, equipment, and other laboratory products]

General Oceanics, Inc.
1295 N.W. 163rd Street
Miami, FL 33169
USA
Telephone: 305-621-2882
Fax: 305-621-1710
http://www.generaloceanics.com
[sterile bag sampler]

Gilson Company, Inc.
P.O. Box 200
Lewis Center, OH 43035-0200
USA
Telephone: 740-548-7298, 800-444-1508
Fax: 740-548-5314
http://www.globalgilson.com/
[environmental testing supplies, sieving equipment]

Grainger Worldwide
Branch locations throughout the United States, and worldwide export service; contact information available on the website.
http://www.grainger.com
[electrical equipment, fluorescent lamps, hand and power tools, hardware, handling material, paper and plastic bags, safety equipment]

Haake, Inc.
c/o Thermo Haake
53 W. Century Road
Paramus, NJ 07652
USA
Telephone: 201-265-7865
Fax: 201-265-1977
http://www.thermo.com
[waterbath—Haake D3 circulator]

Hubco, Inc. (Hutchinson Bag Co.)
P.O. Box 1286
215 South Poplar
Hutchinson, KS 67504-1286
USA
Telephone: 620-663-8301; 800-563-1867
Fax: 620-663-5053
http://www.hubcoinc.com
[Hubco sand sample bags]

Kimble/Kontes
Vineland, NJ
USA
Telephone: 888-546-2531; 856-692-3600
Fax: 856-794-9762
International distributor contact information available on the website.
http://www.kimble-kontes.com/html/Home.html
[glassware, Kimracks, Petri dishes, plasticware, vacuum pumps]

L. L. Bean, Inc.
Freeport, ME 04033-0001
USA
Telephone: 800-221-4221; 207-552-6878
Fax: 207-552-3080
http://www.llbean.com/
[basket packs, camping gear]

Millipore Corp.
80 Ashby Road
Bedford, MA 01730
USA
Telephone: 800-MILLIPORE (800-645-547-673)
Fax: 781-533-3110
http://www.millipore.com/

[cell culture, filtration products, membranes, water purification]

Nalgene Nunc International
75 Panorama Creek Drive
P.O. Box 20365
Rochester, NY 14602-0365
USA
Telephone: 800-625-4327
Fax: 716-586-8987
http://www.nalgenunc.com/
[filtration products, plasticware, racks, tubing]

Neville Crosby, Inc.
445 Terminal Street
Vancouver, British Columbia V6A 2L7
Canada
Telephone: 604-662-7272; 800-663-6733
Fax: 604-662-8133; 800-873-8166
http://www.esupplyshop.com/
[forestry tools, GPS units, and field data]

New Brunswick Scientific Co., Inc.
P.O. Box 4005
44 Talmadge Road
Edison, NJ 08818-4005
USA
Telephone: 732-287-1200
Fax: 732-287-4222
http://www.nbsc.com/
[fermenters, incubators, media dispensers, gyratory shaker G-25]

NuAire, Inc.
2100 Fernbrook Lane
Plymouth, MN 55447
USA
Telephone: 800-328-3352
Fax: 763-553-0459
http://www.nuaire.com/
[biosafety cabinets and laminar flow products]

Ott-Light True Color Lamps
1214 West Cass Street
Tampa, FL 33606
USA
Telephone: 813-621-0058; 800-842-8848
Fax: 813-626-8790
http://www.ottlight.com
[lights that replicate natural light indoors]

Philip Harris Scientific
618 Western Avenue
Park Royal
London W3 0TE
United Kingdom
Telephone: +44 (0) 181 992 555
Fax: +44 (0) 181 993 8020
http://www.philipharris.co.uk
[complete lines of scientific supplies and equipment]

Pye Unicam, Ltd.
York Street
Cambridge CB1 2PX
United Kingdom
Products sold through distributors; see website for contact information.
[pye unicam SP 1800-UV spectrophotometer, scientific instruments]

Spectrum Medical Industries, Inc.
1100 Rankin Road
Houston, TX 77073-4716
USA
Telephone: 800-634-3300; 713-44-2900
Fax: 713-443-3100
[filtration and separation products, Spectramesh, Trans-fertubes]

Stuewe and Sons, Inc.
2290 SE Kiger Island Drive
Corvallis, OR 97333-9461
USA
Telephone: 503-553-5331
Fax: 541-754-6617
http://www.stuewe.com
[Cone-tainers, nursery supplies]

Techni-Tool, Inc.
1547 N. Trooper Road
P.O. Box 1117
Worcester, PA 19490-1117
USA
Telephone: 800-832-4866
Fax: 800-854-8665
http://www.techni-tool.com/
[electrical hardware, equipment cases, forceps, hand tools, pin vises, soldering equipment]

Tekmar-Dohrmann
7143 E. Kemper Road
Cincinnati, OH 45249
USA
Telephone: 513-247-7000; 800-543-4461
Fax: 513-247-7050
http://www.tekmar.com/
[specialized laboratory blenders, e.g., "Stomacher Blender"]

Tetko, Inc.
111 Calumet Street
Depew, NY 14043
USA
Telephone: 716-683-4050
Fax: 716-683-4053
[nylon sieves]

VirTis Co., Inc.
815 Route 208
Gardiner, NY 12525-9989
USA
Telephone: 800-765-6198; 845-255-5000
Fax: 845-255-5338
http://www.virtis.com/default.html
[freeze-drying equipment]

Wheaton Science Products
1501 N. 10th Street
Millville, NJ 08332-2093
USA
Telephone: 800-225-1437; international 856-825-1100
Fax: 856-825-1368, international 856-825-4568
http://www.wheatonsci.com/
[alcohol lamps, cell culture equipment, glassware, self-refilling syringes, vials]

HERBARIUM AND ARCHIVES SUPPLIES

Carr McLean, Ltd.
461 Horner Avenue
Toronto, Ontario M8W 4X2
Canada
Telephone: 800-268-2123
Fax: 800-871-2397
http://www.carrmclean.ca/
[archival specimen storage: acid-free boxes and paper, plastic bags, photograph and slide storage systems]

Delta Designs, Ltd.
P.O. Box 1733
Topeka, KS 66601
USA
Telephone: 785-234-2244; 800-656-7426
Fax: 785-233-1021
http://www.deltaltd.com/
[herbarium cabinets]

Herbarium Supply Co.
3483 Edison Way
Menlo Park, CA 94025
USA
Telephone: 650-366-8868
Fax: 650-365-5492
http://www.herbariumsupply.com/
[glue, hand tools, labels, mounting paper, packets, plant presses, specimen cases]

Pacific Papers
P.O. Box 606
Cotati, CA 94931
USA
Telephone: 800-876-1151
Fax: 425-482-0534
http://www.pacific-papers.com/
[herbarium paper supplies, plant presses]

University Products
Telephone: 800-336-4847
Fax: 800-532-9281
http://www.universityproducts.com
[archival-quality boxes and other products for herbaria]

MICROSCOPY AND MICROMANIPULATORS

Carl Zeiss
P.O. Box 1380
73444 Oberkochen
Germany
Telephone: 49-7364-2 00
Fax: 49-7364-68 08
http://www.zeiss.com/
[image analysis, microscopes, micromanipulators, photomicroscopy]

Carl Zeiss North America
One Zeiss Drive
Thornwood, NY 10594
USA
Telephone: 914-747-1800
Fax: 914-682-8296
http://www.zeiss.com/
[image analysis, microscopes, micromanipulators, photomicroscopy]

Graticules, Ltd.
(Division of Pyser-SGI)
Morley Road, Tonbridge
Kent, TN9 1RN
United Kingdom
Telephone: 1732-864111
Fax: 1732-770217
http://www.pyser-sgi.com/graticules/graticules.htm
[England finder, stage graticules, eyepiece micrometers]

Nikon, Inc.
Instrument Group
1300 Walt Whitman Road
Melville, NY 11747-3064
USA
Telephone: 516-547-8500
Fax: 516-547-0306
http://www.nikonusa.com/
[microscopes, photomicroscopy, survey equipment]

ProSciTech
P.O. Box 111
Thuringowa, Queensland 4817
Australia
Telephone: (07) 4774 0370
Fax: National (07) 4789 2313, international +61 7 4789 2313
http://www.proscitech.com/
[light and electron microscopy, microprobe and histology supplies, England finders, scientific instruments]

Soma Scientific Instruments
5319 University Drive
PMB #366
Irvine, CA 92612-2938
USA
Telephone: 949-854-0220
Fax: 949-854-0223
http://somascientific.com/
[hydraulic and mechanical micromanipulators]

Toyo Rikki Co., Ltd.
1-10-3-307 Shouwa-cho
Akishima-shi, Tokyo 196
Japan
Fax: 81-425-45-8955
[Skerman micromanipulator and microforge]

REAGENTS AND CULTURE MEDIA

Calbiochem-Novabiochem Corp.
P.O. Box 12087
La Jolla, CA 92039-2087
USA
Telephone: 800-854-3417; 858-450-9600
Fax: 800-776-0999; 858-453-3552
http://www.calbiochem.com/
[antibiotics, fine chemicals, mycotoxins]

Difco Laboratories
P.O. Box: 331058
Irvine, CA 92714

USA
Telephone: 714-260-3980; 800-521-0851
Fax: 201-847-5757
http://www.difco.com/
[culture media, diagnostic kits, reagents]

Fluka Holding AG
Buchs
Switzerland
Telephone: 41-81-755-2511
Fax: 41-81-756-5449
http://www.sigmaaldrich.com/
[antibiotics, culture media, fine chemicals, mycotoxins]

Oxoid Division—see Unipath Company
Q-Biogene
2251 Rutherford Road
Carlsbad, CA 92008
Telephone: 800-424-6101
Fax: 760-918-9313
http://www.qbiogene.com/
[glass beads, polymerase chain reaction (PCR) reagents, fluorescent transfection reagents]

Quest International
Bioproducts Group
Woods Corner
Norwich, NY 13815
USA
Telephone: 800-833-8308
http://www.sheffield-products.com/pharma_ingredients/
[culture media, protein and yeast hydrolysates]

Unipath Co. (Oxoid Division)
P.O. Box 691
800 Proctor Avenue
Ogdensburg, NY 13669
USA
Telephone: 800-567-8378
Fax: 613-226-3728
[cryopreservation supplies, culture media, diagnostic kits, Oxoid culture media, reagents]

Unipath Ltd.
Wade Road
Basingstoke
Hampshire, RG24 8PW
United Kingdom
Telephone: 0256-841144
Fax: 0256-463388
[cryopreservation supplies, culture media, diagnostic kits, Oxoid culture media, reagents]

Sigma-Aldrich, Inc.
St. Louis, MO
USA
Telephone: 800-325-3010; 314-771-5765
Fax: 314-771-5757
http://www.sigmaaldrich.com/
[antibiotics, chemicals, culture media, electrophoresis reagents, immunochemicals, media components, mycotoxins, stains]

Sigma Chemical Co.
6050 Spruce Street
St. Louis, MO 63103
USA
Telephone: 314-771-5750; 800-325-3010
Fax: 800-325-5052
http://www.sigma.sial.com

MOLECULAR TECHNIQUES

All chemicals, biochemicals, and enzymes can be obtained from the following sources. Refer to catalogues for specific materials.

Amersham Pharmacia LKB Biotechnology, Inc.
800 Centennial Avenue
P.O. Box 1327
Piscataway, NJ 08855-1327
USA
Telephone: 800-526-3593
http://www.apbiotech.com/na/
[LKB products; sequencing set, LKB 2010 MacroPhor unit equipped with a water-jacketed cooling plate and external cooling unit]

Barnstead Thermolyne
2555 Kerper Boulevard
Dubuque, IA 52001-1478
USA
Telephone: 800-446-6060
Fax: 563-589-0516
[thermocyclers]

BDH Laboratory Supplies
Poole, Dorset BH15 ITD
United Kingdom
Telephone: +44 1 202-660-444
Fax: +44 1 202-666-856
http://www.bdh.com/
[instrumentation, laboratory consumables; laboratory, electrophoresis, molecular biology reagents; cell diagnostic stains, HPLC solvents, BDH buffers]

Beckman Instruments, Inc.
P.O. Box 3100
2500 Harbor Boulevard
Fullerton, CA 92634-3100
USA
Telephone: 714-871-4848; 800-742-2345
Fax: 714-773-8898; 800-643-4366
http://www.beckman.com/
[ultracentrifuge beckman L5-50E and type 50 rotor]

Bio-Rad Laboratories
1000 Alfred Nobel Drive
Hercules, CA 94547
USA
Telephone: 510-724-7000
Fax: 510-741-5817
http://www.bio-rad.com/
[electrophoresis reagents, cloning material, and chromatography supplies]

Boehringer Mannheim Corp.
Biochemical Products
9115 Hague Road
P.O. Box 504
Indianapolis, IN 46250-0414
USA
http://www.roche-applied-science.com/
[nucleic acid isolation and purification products, PCR products, reagents, etc.]

DuPont
DuPont Corporate Information Center
Barley Mill Plaza, P10
Wilmington, DE 19880-0010
Telephone: 800-441-7515 (U.S. callers only); 1-302-774-1000 (worldwide)
http://www.dupont.com
[genescreen plus nylon membranes; Benlate]

Isco, Inc.
4700 Superior Street
P.O. Box 82531
Lincoln, NE 68504
USA
Telephone: 402-464-0231; 800-228-4373
Fax: 402-465-3064
http://www.isco.com/
[gradient peristaltic pump, ISCO Model 184, ISCO Fraction collector, Model 6 UV-unit and ISCO UV-5 absorbence monitor]

Eastman Kodak Co.
http://www.kodak.com/
[Kodak x-ray film and Kodak X-omatic x-ray cassette]

Godax Laboratories, Inc.
720-B Erie Avenue
Takoma Park, MD 20912
USA
Telephone: 301-588-2825
Fax: 301-589-6023
http://www.godax.com/
[nochromix]

Invitrogen Corp.
1600 Faraday Avenue
P.O. Box 6482
Carlsbad, CA 92008
Telephone: 760-603-7200
Fax: 760-602-6500
http://www.invitrogen.com
[gene primers, molecular biology products]

Polaroid Corp.
784 Memorial Drive
Cambridge, MA 02139
USA
Telephone: 781-386-2000
http://www.polaroid.com
[Polaroid MP4 camera equipped with a Wratten No. 9 red filter]

Promega
2800 Woods Hollow Road
Madison, WI 53711-5399
USA
Telephone: 800-356-9526; 608-274-4330
Fax: 800-356-1970
http://www.promega.com
[taq track sequencing system]

MJ Research, Inc.
Waltham, MA 02451
Telephone: 888-735-8437
Fax: 617-923-8080
http://www.mjr.com
[thermal controller, PTC-100 programmable]

Novagen, Inc.
601 Science Drive
Madison, WI 53711
USA
Telephone: 608-238-6110
Fax: 608-238-1388
http://www.novagen.com
[reagents for the isolation, analysis, and purification of genes and protein products]

Stratagene
La Jolla, CA 92037
USA
Telephone: 800-424-5444
Fax: 512-321-3128
http://www.stratagene.com/
[thermocyclers]

Ultra-Violet Products, Inc.
2066 W. 11th Street
Upland, CA 91786
USA
Telephone: 800-452-6788; 909-946-3197
Fax: 909-946-3597
http://www.uvp.com/html/ultraviolet_products.html
[transilluminator ultraviolet products]

Ultra-Violet Products, Ltd.
Unit 1, Trinity Hall Farm Estate
Nuffield Road
Cambridge CB4 1TG
United Kingdom
Telephone: +44 (0) 1223-420022
Fax: +44 (0) 1223-420561
http://www.uvp.com/html/ultraviolet_products.html
[transilluminator ultraviolet products]

COMPUTER PROGRAMS AND SOFTWARE

Environmental Systems Research Institute, Inc.
380 New York Street
Redlands, CA 92373-8100
USA
Telephone: 909-793-2853
Fax: 909-307-3025
http://www.esri.com/
[ARC/INFO Geographic Information Systems (GIS)]

NCSS Statistical Software
329 North 1000 East
Kaysville, UT 84037
USA
Telephone: 800-898-6109
Fax: 801-546-3907
http://www.ncss.com/

NIH *Image* Software
National Institutes of Health
http://rsb.info.nih.gov/nih-image/

SciTech International, Inc.
2525 N. Elston Avenue
Chicago, IL 60647-2003
USA
Telephone: 800-622-3345; 773-486-9191
Fax: 773-486-9234
http://www.scitechint.com/
[software packages for chemistry, drawing, graphics, mapping, statistics]

SPSS, Inc.
233 S. Wacker Drive
11th floor
Chicago, IL 60606

USA
Telephone: 800-543-2185
Fax: 800-841-0064
http://www.spss.com/

Statistical Solutions
Stonehill Corporate Center
Suite 104
999 Broadway
Saugus, MA 01906
USA
Telephone: 781-231-7680; 800-262-1171
Fax: 781-231-7684
http://www.statsolUSA.com/

GLOSSARY

F. M. DUGAN

Mycological literature uses hundreds of terms foreign to the nonspecialist. With this glossary, I have tried to render such literature more accessible, although the scope and detail of the terminology pertaining to fungal ecology and morphology vastly exceed the terms presented here. Additional terms are defined in the text; the reader should consult the "Index" for their locations. Readers requiring more precise and extended terminology should consult pertinent dictionaries, glossaries, and texts. A number of references are listed at the end of the chapter.

TERMS

Acanthophysis (pl. **ancanthophyses**)—a **hyphoid** with numerous branches (arrows), resembling a bottle brush (Fig. 1).

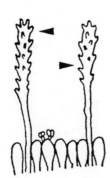

FIGURE 1   Acanthophysis (arrows)

Acerose—needlelike (Fig. 2). Cf. **acicular**, **filiform**, **fusiform**.

FIGURE 2   Acerose

Acervulus (pl. **acervuli**)—a mat of conidiogenous cells that erupts through the epidermis of the host (Fig. 3b) to release **conidia** (Fig. 3a). The Melanconiales produce acervuli when on their hosts but usually form **sporodochia** or **stromata** in culture.

FIGURE 3   Acervulus (a. conidia; b. host epidermis)

Acicular—having a sharply pointed apex (Fig. 4). Cf. **acerose**, **subulate**.

FIGURE 4   Acicular

Acrasid—type of **slime mold** in which aggregated **myxamoebae** retain cell walls. Cf. **dictyostelid**.

Adelophialide—a reduced **phialide** (Fig. 5b) not separated from the subtending **hypha** (Fig. 5c) by a **septum**, that produces conidia (Fig. 5a).

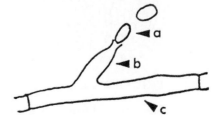

FIGURE 5   Adelophialide (a. conidium; b. adelophialide; c. hypha)

Adnate—of the **gills** (Fig. 6a) of an **agaric**: broadly attached to the **stipe** (Fig. 6b). Cf. **adnexed**, **subdecurrent**.

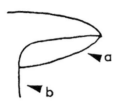

FIGURE 6   Adnate

Adnexed—of the **gills** (Fig. 7a) of an **agaric**: narrowly attached to the **stipe** (Fig. 7b). Cf. **adnate**, **free**.

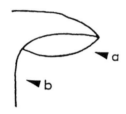

FIGURE 7   Adnexed

Aeciospore—nonseptate, nonrepeating spore in the **rusts** (Fig. 8).

FIGURE 8   Aeciospore

Aecium (pl. **aecia**)—in the **rusts**, the fruiting body producing aeciospores (Fig. 9).

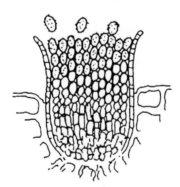

FIGURE 9   Aecium

Agar—1. A substance extracted from red algae and used to solidify nutrient media. 2. The resultant solidified growth medium. See **culture medium**.

Agaric—a mushroom with a spore-bearing surface composed of **gills** (Fig. 10, arrow) or **lamellae**. Cf. **bolete**.

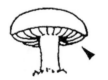

FIGURE 10   Agaric

Allantoid—sausage-shaped (Fig. 11). Cf. **lunate**, **reniform**.

FIGURE 11   Allantoid

Amoeba (pl. **amoebae**)—a nonflagellated, motile, single cell capable of engulfing and digesting bacteria or other food items. See **myxamoeba**.

Amerospore—a single-celled spore (Fig. 12, various spores).

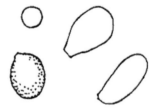

FIGURE 12   Amerospore

Amphigynous—condition in which the **antheridium** (Fig. 13b) surrounds the basal **hypha** (Fig. 13c) of the **oogonium** (Fig. 13a). Cf. **paragynous**, **hypogynous**.

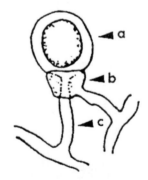

FIGURE 13   Amphigynous

Ampulliform—flasklike (Fig. 14). Cf. **lacrymoid**, **pyriform**.

FIGURE 14   Ampulliform

Amyloid—staining dark blue with Meltzer's reagent. For example, an amyloid ring at the apex of an **ascus** (Fig. 15a) or amyloid reticulations on a **basidiospore** (Fig. 15b). The indication for a positive amyloid reaction is often written as "I+" or "J+."

FIGURE 15   Amyloid

Anamorph—phase in the life cycle in which the fungus propagates asexually; generally characterized by production of **conidia**. See **mitosporic fungi**.

Annellide—a conidiogenous (see **conidium**) cell that produces a series of ringlike scars at its apex (Fig. 16, arrow).

FIGURE 16   Annellide

Annulus (pl. **annuli**)—a ring of tissue on the **stipe** of certain mushrooms (Fig. 17, arrow). See partial veil.

FIGURE 17   Annulus

Antheridium (pl. **antheridia**)—male sex organ. It attaches to the female **oogonium** during mating of oomycetes (Fig. 18a). The same term is applied to a male sex organ in ascomycetes (Fig. 18b). See **ascogonium**, **trichogyne**.

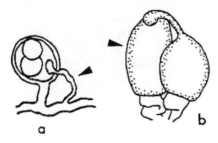

FIGURE 18   Antheridium

Anthracnose—a plant disease characterized by limited necrotic lesions on leaves or stems.

Apiculus (pl. **apiculi**)—a short projection on a spore (Fig. 19, arrow) where the spore was attached to the sporogenous cell.

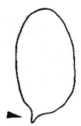

FIGURE 19   Apiculus

Aplerotic—condition in which the **oospore** (Fig. 20a) does not fill the **oogonium** (Fig. 20b). Cf. **pleurotic**.

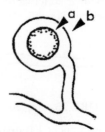

FIGURE 20   Aplerotic

Apothecium (pl. **apothecia**)—the fruiting body of a cup fungus or **discomycete** (Fig. 21).

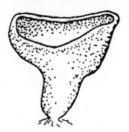

FIGURE 21   Apothecium

Appendiculate—of an agaric: the fringed appearance of the rim (arrow) of the **pileus** (Fig. 22a). See also **veil**. Of a spore: possessing one or more setulae (Fig. 22b, arrows).

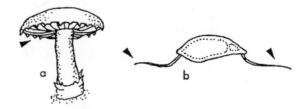

FIGURE 22   Appendiculate

Applanate—flat or nearly so. Usually pertaining to the superior surface of a **pileus** (Fig. 23, arrow).

FIGURE 23   Applanate

Appressed—flattened down; for example, the hairs or scales on the surface of a **pileus** (Fig. 24). Cf. **squarrose**.

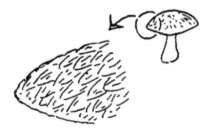

FIGURE 24   Appressed

Appressorium (pl. **appressoria**)—expanded or flattened tip of a **hypha,** which attaches to the epidermis of a host or other surface (Fig. 25, arrow).

FIGURE 25   Appressorium

Arbuscule—a finely branched **haustorium** characterizing **vesicular–arbuscular mycorrhizae (endomycorrhizae).**

Archicarp—see **ascogonium.**

Areolate—divided into small, more or less angular spaces by fissures or cracks (Fig. 26). Cf. **rimose-areolate.**

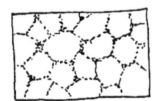

FIGURE 26   Areolate

Arthrospore—an asexual spore resulting from fragmentation of a **hypha** (Fig. 27).

FIGURE 27   Arthrospore

Ascocarp—fruiting body of an ascomycete (Fig. 28).

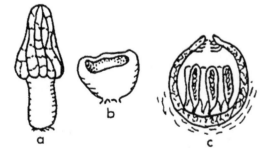

FIGURE 28   Ascocarp (a. morel; b. cup fungus; c. pyrenomycete)

Ascogonium (pl. **ascogonia**)—female sex cell (arrow) in the **ascomycetes** (Fig. 29). See **antheridium, trichogene.**

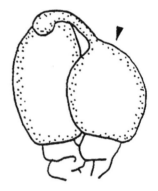

FIGURE 29   Ascogonium

Ascomycete—fungus whose **meiotic** products are **ascospores** enclosed in an **ascus.**

Ascospore—spore produced inside an ascus (Fig. 30).

Ascus (pl. **asci**)—the "spore sack" of an **ascomycete** (Fig. 30). An ascus usually contains 8 **ascospores.**

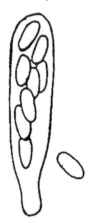

FIGURE 30   Ascospore Ascus

Aspergillum (pl. **aspergilla**)—fruiting structure of *Aspergillus* spp. (Fig. 31), consisting of a **foot cell** (e), **stipe** (d), **vesicle** (c), **phialides** (b), and chains of conidia (a). **Biseriate** aspergilla possess **metulae,** but **uniseriate** aspergilla do not.

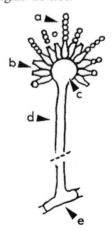

FIGURE 31   Aspergillum

Axenic—a **culture** consisting of a single organism growing on an inert nutrient medium.

Bait—an item placed in a given environment to sample the fungi in that environment. The bait attracts fungi that are later isolated from it.

Basidiocarp—fruiting body of a **basidiomycete** (Fig. 32a, polypore; b, agaric).

FIGURE 32 Basidiocarp

Basidiole—an immature **basidium** (Fig. 33, arrow).

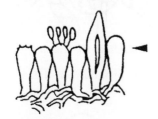

FIGURE 33 Basidiole

Basidiomycete—a fungus whose **meiotic** products are **basidiospores** produced on a **basidium**.

Basidiospore—spore produced on a **basidium** (Fig. 34).

Basidium (pl. **basidia**)—spore-producing structure of a **basidiomycete**. It normally produces four **basidiospores**. Undivided **holobasidia** lack internal **septa** (Fig. 34a). **Phragmobasidia** are divided into several cells (Fig. 34b).

 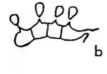

FIGURE 34 Basidia

Biotroph (adj. **biotrophic**)—an organism that grows only in intimate association with the cytoplasm of another living organism.

Biseriate—indicating an **aspergillum** possessing **metulae** (Fig. 35a) in addition to **phialides** (35b). Or, of an **ascus**: containing two rows of **ascospores**. Cf. **uniseriate**.

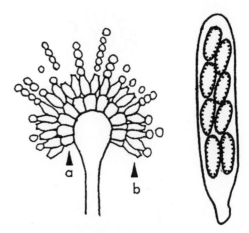

FIGURE 35 Biseriate

Bitunicate ascus—having two walls (Fig. 36a, b). The term is applied to **asci** of the **Loculoascomycetes**. See **jack-in-the-box**.

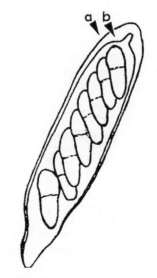

FIGURE 36 Bitunicate ascus

Biverticillate—indicating a **penicillus** that has two branch points; the first branch point produces the **metulae** (Fig. 37a), and the second produces the **phialides** (Fig. 37b). Cf. **monoverticillate**, **terverticillate**.

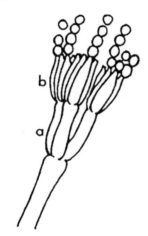

FIGURE 37   Biverticillate

Black mildew—member of the Meliolales, a group of fungi **biotrophic** on tropical vegetation.

Blight—a plant disease characterized by sudden wilting and discoloration.

Bluestain—blue–gray discoloration of wood caused by fungal growth.

Bolete—mushroom with a spore-bearing surface composed of **tubes**, the openings of which are called **pores** (Fig. 38, arrow).

FIGURE 38   Bolete

Brown rot—1. Fruit decay caused by fungal species in the genus *Monilinia*. 2. A type of wood decay in which cellulose is degraded preferentially, leaving a residue of lignin. Cf. **white rot**.

Bullate—possessing a robust, knoblike apex (Fig. 39). Cf. **umbonate**.

FIGURE 39   Bullate

Campanulate—of the **pileus** of an **agaric**: bell-shaped (Fig. 52).

FIGURE 52   Companulate

Canker—a plant disease characterized by a sharply defined necrosis of bark tissues.

Capillitium—threadlike structures interspersed among the spores of gasteromycetes and myxomycetes (Fig. 40).

FIGURE 40   Capillitium

Capitate—with a pronounced head (Fig. 41). Cf. **spathulate**.

FIGURE 41   Capitate

Cheilocystidium (pl. **cheilocystidia**)—a **cystidium** (Fig. 42, arrow) positioned on the end of a **lamella**. Cf. **pleurocystidium**.

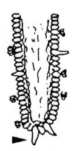

FIGURE 42   Cheilocystidium

Chitin—primary component of the cell walls of true fungi.

Chlamydospore—thick-walled spore, functioning as a resting spore, and formed inside a preexisting cell (Fig. 43). Sometimes referred to as a **gemma**.

FIGURE 43 Chlamydospore

Chrysocystidium (pl. **chrysocystidia**)—a **gloeocystidium** (arrows) that turns yellow in alkaline solutions (Fig. 44). See also **cystidium**.

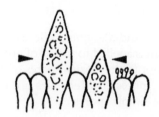

FIGURE 44 Chrysocystidium

Chytrid—type of microscopic, generally aquatic fungus whose **zoospore** has a single, posterior **flagellum**.

Chytridiomycete—see **chytrid**.

Circinate—coiled at the tip (Fig. 45). Cf. **hamate**.

FIGURE 45 Circinate

Clamp connection—buckle-shaped **septum** that characterizes **hyphae** of many **basidiomycetes** (Fig. 46).

FIGURE 46 Clamp connection

Clavate—club-shaped (Fig. 47). Cf. **capitate**.

FIGURE 47 Clavate

Cleistothecium (pl. **cleistothecia**)—small fruiting body that encloses one or more **asci**, and that lacks an **ostiole** (Fig. 48). Cf. **perithecium**.

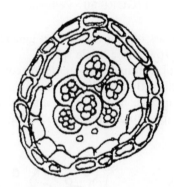

FIGURE 48 Cleistothecium

Clypeus—shield-shaped **stroma** (arrow) around the mouth of a fruiting body (Fig. 49).

FIGURE 49 Clypeus

Coelomycete—**mitosporic fungus** producing **conidia** within a fruiting body such as an **acervulus** or a **pycnidium**. Cf. **hyphomycete**.

Coenocyte (adj. **coenocytic**)—fungus in which the nuclei are not separated from each other by **septa**; characterized by absence of **septa**.

Collarette—cup-shaped collar (Fig. 50, arrow) at the apex of a **phialide**.

FIGURE 50   Collarette

Colony—a mass of cells or **hyphae**, generally resulting from the growth of a single spore.

Columella (pl. **columellae**)—sterile structure located on a central axis within a fruiting body and projecting into it from the base; frequently an internal extension of the stalk supporting a fruiting body (Fig. 51, arrows).

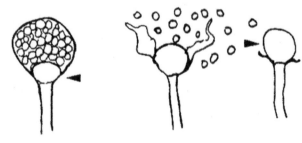

FIGURE 51   Columella

Conchate—of the **pileus** of a basidiomycete; shaped like one half of a bivalve shell (Fig. 53).

FIGURE 53   Conchate

Conidiogenesis—act of producing **conidia**.

Conidiogenous—producing **conidia**.

Conidiophore—1. A **hypha** that bears a **conidiogenous** cell. 2. The conidiogenous cell itself. (Fig. 54, arrow).

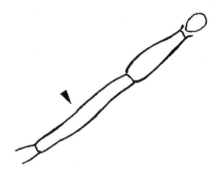

FIGURE 54   Conidiophore

Conidium (pl. **conidia**)—nonmotile, asexual spore (Fig. 55). The term is applied to spores of **mitosporic fungi** but usually not to the asexual spores of the Mucorales. The asexual spores of Mucorales are called **sporangiospores** or simply spores.

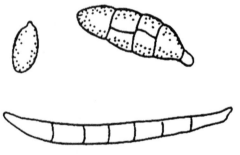

FIGURE 55   Conidium

Convex—broadly rounded (Fig. 56). Cf. **plano-convex**, **pulvinate**.

FIGURE 56   Convex

Coprophilic—adapted to growth on dung.

Coremium (pl. **coremia**)—a type of **synnema** with a broomlike apex.

Corniform—bent at the apex, horn-shaped (Fig. 57). Cf. **hamate**.

FIGURE 57   Corniform

Corticolous—adapted to growth on bark.

Costate—ribbed, veined, fluted (Fig. 58).

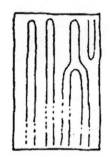

FIGURE 58   Costate

Crenate—with rounded teeth, scalloped (Fig. 59).

FIGURE 59   Crenate

Crenulate—with small, rounded teeth; finely scalloped (Fig. 60).

FIGURE 60   Crenulate

Crozier—the initial hook-shaped stage of a developing **ascus** (Fig. 61).

FIGURE 61   Crozier

Cryoprotectant—substance enabling cell survival during preservation at low temperature.

Culture—*n.* growth of one to several organisms on a **culture medium**; *v.* to grow such organisms on **culture media** in the laboratory.

Culture medium—an inert medium used to grow organisms in the laboratory. See **agar**.

Cymbiform—see **navicular**.

Cyst—thick-walled structure enabling survival of the enclosed cell(s) during adverse conditions.

Cystidiole—immature or undifferentiated **cystidium** (Fig. 62, arrow).

FIGURE 62   Cystidiole

Cystidium (pl. **cystidia**)—a sterile element (arrows) on a surface of a **basidiocarp**, especially the **hymenium** (Fig. 63). Cf. **paraphysoid, seta**.

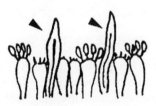

FIGURE 63   Cystidium

Cystosorus (pl. **cystosori**)—compact aggregation of **cysts** produced by some **chytrids**.

Decurrent—of the **gills** (Fig. 64a) of an **agaric**: running down the **stipe** (Fig. 64b). Cf. **subdecurrent**.

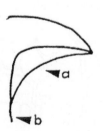

FIGURE 64   Decurrent

Dendrophysis (pl. **dendrophyses**)—a **hyphoid** with treelike branching (Fig. 65, arrows). Cf. **dichophysis**.

FIGURE 65   Dendrophysis

Dentate—with toothlike projections (Fig. 66). Cf. **incised**, **serrate**.

FIGURE 66   Dentate

Denticulate—with small, toothlike projections (Fig. 67).

FIGURE 67   Denticulate

Depressed—1. Of a **pileus**: sunken in the center (Fig. 68a). 2. Of **gills** or **tubes**: indented toward the point of attachment with the **stipe**. (Fig. 68b).

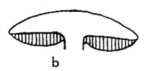

FIGURE 68   Depressed

Deuteromycete—see **mitosporic fungi**.
Dichophysis (pl. **dichophyses**)—a **hyphoid** with dichotomous branching (Fig. 69, arrow).

FIGURE 69   Dichophysis

Diclinous—condition in which the **antheridium** (Fig. 70a) originates on a different **hypha** (Fig. 70c) from the **oogonium** (Fig. 70b). Cf. **monoclinous**.

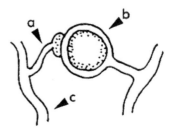

FIGURE 70   Diclinous

Dictyospore—a spore with both longitudinal and transverse **septa** (Fig. 71).

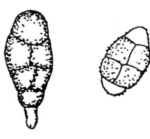

FIGURE 71   Dictyospore

Dictyostelid—type of **slime mold** in which **amoebae** aggregate to form the **sorocarp**. Distinguished from **acrasids** by the well-differentiated, stalked sorocarp.
Didymospore—a spore with a single **septum** (Fig. 72).

FIGURE 72   Didymospores

Dimidiate—of a **pileus**: semicircular (Fig. 73).

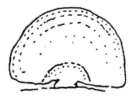

FIGURE 73   Dimidiate

Discomycete—type of fungus in which the **ascocarp** is generally cup-shaped or convoluted.

Doliform—barrel-shaped (Fig. 74).

FIGURE 74   Doliform

Downy mildews—group of obligate plant pathogens in the Peronosporales.

Echinulate—with spines (Fig. 75).

FIGURE 75   Echinulate

Ectal—of the exterior.

Ectomycorrhiza—type of **mycorrhiza** in which the plant root is ensheathed by the fungus and penetrated by the **Hartig net**. Cf. **endomycorrhiza**.

Effused-reflexed—of a pileus **resupinate** except for an upturned upper portion (Fig. 76).

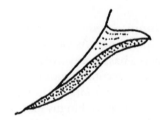

FIGURE 76   Effused-reflexed

Emarginate—of the **gills** (Fig. 77a) of an **agaric**: notched by the **stipe** (Fig. 77b). Cf. **seceding**, **sinuate**.

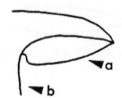

FIGURE 77   Emarginate

Endoconidium (pl. **endocondia**)—a **conidium** produced within a **hypha** (Fig. 78, arrow).

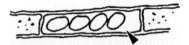

FIGURE 78   Endoconidium

Endomycorrhiza—a type of **mycorrhiza** in which cells of the plant root are penetrated by **vesicles** and **arbuscules**; also called **vesicular–arbuscular mycorrhiza**. Cf. ectomycorrhiza.

Endophyte—type of fungus growing partially or completely inside host plant tissues; *sensu stricto*, a non-pathogenic fungus whose vegetative growth is entirely within host tissues.

Enteroblastic—form of **conidigenesis** in which an inner wall or neither wall of the **conidiogenous** cell is used in formation of conidia . Enteroblastic **phialidic** (Fig. 79a). Cf. enteroblastic **tretic** (Fig. 79b).

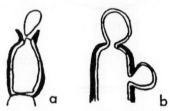

FIGURE 79   Enteroblastic

Epithecium (pl. **epithecia**)—in **discomycetes**, a tissue formed over the top of the **asci** by the interwoven apices of the **paraphyses** (Fig. 80, arrow). Cf. **pseudothecium**.

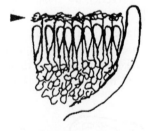

FIGURE 80   Epithecium

Excipulum—a layer of an **apothecium**. **Ectal** excipulum is toward the exterior (Fig. 81a); **medullary** excipulum is interior (Fig. 81b).

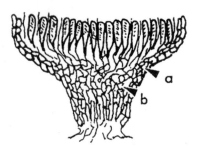

FIGURE 81   Excipulum

Falcate—sickle-shaped (Fig. 82). Cf. **lunate**.

FIGURE 82   Falcate

Filiform—threadlike (Fig. 83). Cf. **acerose**.

FIGURE 83   Filiform

Fimbriate—with a finely torn margin; fringed (Fig. 84). Cf. **lacerate**.

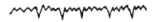

FIGURE 84   Fimbriate

Flagellum (pl. **flagella**)—a hairlike, whiplike, or tinsel-like organelle used to propel a motile cell. A **whiplash-flagellum** is naked (Fig. 85a); a **tinsel-flagellum** has small hairs called **mastigonemes** (Fig. 85b).

FIGURE 85   Flagellum

Fly speck—disease of pome fruits caused by *Schizothyrium pomi;* the skin of the fruit is discolored by specks of fungal growth.

Foot cell—In *Aspergillus,* the basal cell of the **aspergillum** (Fig. 86a). In *Fusarium,* the basal cell of the **conidium** (Fig. 86b).

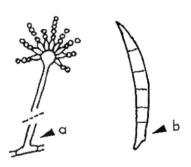

FIGURE 86   Foot cell

Free—of the **gills** (Fig. 87a) of an **agaric**: not attached to the **stipe** (Fig. 87b). Cf. **adnexed, seceding**.

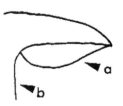

FIGURE 87   Free

Fungus (pl. **fungi**)—*sensu stricto,* a member of the Eumycota, including Ascomycota (**ascomycetes**), Basidiomycota (**basidiomycetes**), Chytridiomycota (**chytrids**), Zygomycota (**zygomycetes**), and associated **anamorphs**; *sensu lato,* including the previously mentioned groups plus fungus-like organisms more closely related to protozoans or algae.

Fungicolous—adapted to parasitism, commensalism, or mutualism with other **fungi**.

Fusiform—spindle-shaped (Fig. 88). Cf. **acerose**.

FIGURE 88   Fusiform

Gall—a plant part that swells in response to the action of a parasite.

Gasteromycete—**basidiomycete** fungus in which the spores are enclosed in a peridium and are not forcibly discharged.

Gemma (pl. **gemmae**)—see **chlamydospore**.

Gill—see **lamella.**

Gleba—the mass of spore-bearing tissue within a puffball or truffle (Fig. 89, arrow).

FIGURE 89   Gleba

Gloeocystidium (pl. **gloeocystidia**)—a **cystidium** with conspicuous contents or one that stains readily (Fig. 90, arrows). Cf. **lamprocystidium, leptocystidium**.

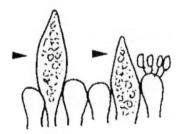

FIGURE 90   Gloeocystidium

Graminicolous—adapted to parasitism, commensalism, or mutualism with grasses.

Hamate—hooked at the apex (Fig. 91). Cf. **circinate, corniform**.

FIGURE 91   Hamate

Hartig net—in an **ectomycorrhiza**, the network of intercellular **hyphae** in the cortex of the host rootlets.

Haustorium (pl. **haustoria**)—fungal organ that penetrates host tissue and absorbs nutrients (Fig. 92, arrow).

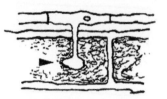

FIGURE 92   Haustorium

Helicospore—spore with a helical structure (Fig. 93).

FIGURE 93   Helicospore

Hepaticolous—adapted to parasitism, commensalism, or mutualism with liverworts (Hepaticae).

Heterobasidium (pl. **heterobasidia**)—a septate **basidium** (Fig. 94); any basidium other than the **holobasidium** characteristic of the higher **basidiomycetes**. Equivalent to **phragmobasidium**.

FIGURE 94   Heterobasidium

Heterotrophic—living by assimilation of organic compounds originally synthesized by other organisms.

Hirsute—with long hairs (Fig. 95). Cf. **pubescent**, **strigose**.

FIGURE 95   Hirsute

Hispid—with bristles or short, stiff (coarse) hairs (Fig. 96). Cf. **strigose**.

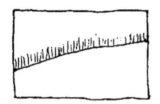

FIGURE 96   Hispid

Holdfast—an organ that attaches the **thallus** to a **substratum**.

Holobasidium (pl. **holobasidia**)—see **basidium**.

Holoblastic—a form of **conidiogenesis** in which cell walls of the **conidiogenous** cell are incorporated into the new conidium (Fig. 97, arrow).

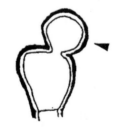

FIGURE 97   Holoblastic

Holomorph—the entire fungal organism, including **anamorph**, **synanamorph** (if any), and **teleomorph**.

Hyalo-—a prefix indicating a colorless condition. For example, hyalophragmiae (Fig. 98) are colorless spores with multiple transverse septa. Cf. **phaeo-**.

FIGURE 98   Hyalophragmiae

Hymenium (pl. **hymenia**)—the fruiting surface of an **ascocarp** (Fig. 99a) or a **basidiocarp**. (Fig. 99b).

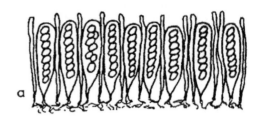

FIGURE 99   Hymenium

Hypha (pl. **hyphae**, adj. **hyphal**)—basic, threadlike element that forms the **thallus** of most fungi (Fig. 100). Cf. **mycelium**.

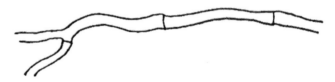

FIGURE 100   Hypha

Hyphal body—short, thick **hypha** that enlarges by fission and budding and subsequently produces **conidiophores**.

Hyphidium (pl. **hyphidia**)—sterile element in the **hymenium** of a basidiomycete. See **cystidium**, **hyphoid**.

Hyphoid—branching **cystidium**. See **acanthophysis**, **dendrophysis**, **dichophysis**.

Hyphomycete—**mitosporic fungus** producing **conidia** that are not enclosed within a fruiting body. Cf. **coelomycete**.

Hypogynous—condition in which the **antheridium** (Fig. 101b) originates inside the **hypha** (Fig. 101c) basal to the **oogonium** (Fig. 101a).

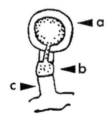

FIGURE 101   Hypogynous

Hysterothecium (pl. **hysterothecia**)—oblong, cleft **ascocarp** (Fig. 102) characteristic of the Hysteriales.

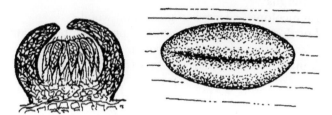

FIGURE 102   Hysterothecium

I+—see **amyloid**.

Imbricate—of the **pilei** of **agarics** or **polypores**: in a series of overlapping shelf- or shinglelike layers (Fig. 103).

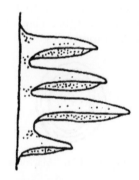

FIGURE 103   Imbricate

Incised—as if cut into (Fig. 104). Cf. **dentate**, **lacerate**.

FIGURE 104   Incised

Inflexed—of the margin of a **pileus**: curving down (Fig. 105, arrow). Cf. **involute**, **reflexed**.

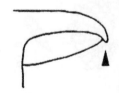

FIGURE 105   Inflexed

Infundibuliform—funnel-shaped (Fig. 106).

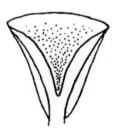

FIGURE 106   Infundibuliform

Ingoldian—of aquatic **hyphomycetes**: producing elongate or branched **conidia**.

Involute—of the margin of a **pileus**: rolled inward (Fig. 107, arrow). Cf. **inflexed**.

FIGURE 107   Involute

J+—see **amyloid**.

Jack-in-the-box—type of discharge mechanism found in **bitunicate** asci, in which the inner wall of the **ascus** is extruded (Fig. 108).

FIGURE 108   Jack-in-the-box

Jelly fungi—colloquial name for the Tremellales, a group of fungi, with a gelatinous texture, that produce **basidiospores**.

Kinetospore—see **zoospore**.

Lacerate—jagged and roughly torn (Fig. 109). Cf. **fimbriate**, **incised**, **lacinate**.

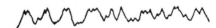

FIGURE 109   Lacerate

Lacinate—roughly torn into lobes (Fig. 110). Cf. **incised, lacerate**.

FIGURE 110   Lacinate

Lacrymoid—tear-shaped (Fig. 111). Cf. **ampulliform**.

FIGURE 111   Lacrymoid

Lamella (pl. **lamellae**)—one of the **gills** of an agaric (Fig. 112, arrow).

FIGURE 112   Lamellae

Lamprocystidium (pl. **lamprocystidia**)—thick-walled **cystidium**; a **seta** (Fig. 113, arrows). Cf. **leptocystidium, gloeocystidium**.

FIGURE 113   Lamprocystidium

Leptocystidium (pl. **leptocystidia**)—thin-walled **cystidium** without conspicuous contents (Fig. 114, arrows). Cf. **lamprocystidium, gloeocystidium**.

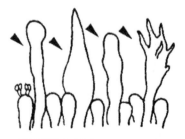

FIGURE 114   Leptocystidia

Lichen—entity resulting from the stable and characteristic association of an alga (or cyanobacterium) and a **fungus**.

Lignicolous—adapted to growth on wood.

Loculoascomycete—fungus that produces **bitunicate asci** within a cavity in a **stroma**.

Lunate—crescent-shaped (Fig. 115). Cf. **falcate**.

FIGURE 115   Lunate

Macrocyclic—of **rust** fungi: possessing a life cycle producing all the spore states. Cf. **microcyclic**.

Macrocyst—thick-walled spore (Fig. 116) that is the sexual stage of Acrasiomycetes.

FIGURE 116   Macrocyst

Macrofungus (pl. **macrofungi**)—**fungi** having large fruit bodies.

Mastigoneme—see **flagellum**.

Medullary—toward the interior.

Meiotic—denoting a product of meiosis—that is, haploid cells resulting from sex and recombination.

Merosporangium (pl. **merosporangia**)—cylindrical sporangium (Fig. 117, arrow) characteristic of several genera in the Mucorales.

FIGURE 117   Merosporangium

Metula (pl. **metulae**)—a structure subtending a **phialide** in a **biseriate aspergillum** or in a **penicillus**. See **biseriate**.

Microcyclic—of **rust** fungi: possessing a life cycle producing only some of the spore states. Cf. **macrocyclic**.

Microfungus (pl. **microfungi**)—fungi having small or microscopic fruit bodies.

Mitosporic fungi—fungi producing spores by mitosis; fungi producing **conidia**. Generally reserved for fungi whose **teleomorphs**, if discovered, would be classified as **ascomycetes** or **basidiomycetes**.

Monoblastic—with only one **conidiogenous** site (arrow) per **holoblastic** conidiogenous cell (Fig. 118).

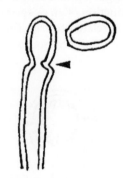

FIGURE 118   Monoblastic

Monoclinous—the condition in which the **antheridium** (Fig. 119a) originates on the same **hypha** (Fig. 119c) as the **oogonium** (Fig. 119b).

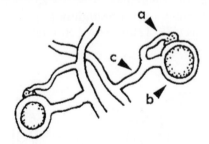

FIGURE 119   Monoclinous

Monotretic—with only one **conidiogenous** site (arrow) per **tretic** conidiogenous cell (Fig. 120).

FIGURE 120   Monotretic

Monoverticillate—indicating a **penicillus** with a single branch point (Fig. 121). Cf. **biverticillate**.

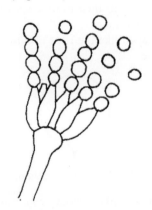

FIGURE 121   Monoverticillate

Muriform—characterized by possession of both longitudinal and transverse septa. See **dictyospore**.

Muscicolous—adapted to parasitism, commensalism, or mutualism with mosses (Musci).

Mushroom—**macrofungus** with fleshy texture, usually an **agaric**, **bolete,** or one of the larger **ascomycetes**. Cf. **toadstool**.

Mutualism—a particular type of symbiosis; an intimate association between individuals of two different species where both partners benefit.

Mycelium (pl. **mycelia**)—an aggregation of **hyphae** (Fig. 122).

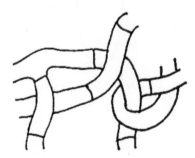

FIGURE 122   Mycelium

Mycorrhiza (pl. **mycorrhizae**)—a symbiotic association between a **fungus** and a plant root. See **ectomycorrhiza, endomycorrhiza**.

Myxamoeba (pl. **myxamoebae**)—an amoeboid cell of the **Myxomycetes** (Fig. 123).

FIGURE 123   Myxamoeba

Myxomycete—one of several types of **slime molds**. Myxomycetes aggregate into **plasmodia** in which true cell walls are absent; their minute **sporangia** are often colorful and elaborate.

Napiform—shaped like the root of a turnip (Fig. 124). Cf. **turbinate**.

FIGURE 124   Napiform

Navicular—shaped like a boat seen from above (Fig. 125). Cf. **fusiform**.

FIGURE 125   Navicular

Necrotroph (adj. **necrotrophic**)—an organism that derives nutrition from dead host tissues. Cf. **biotroph**.

Obovoid—egg-shaped, with the broad end at the apex (Fig. 126). Cf. **ovoid**.

FIGURE 126   Obovoid

Obpyriform—pear-shaped, with the broad end at the apex (Fig. 127). Cf. **pyriform**.

FIGURE 127   Obpyriform

Oogonium (pl. **oogonia**)—female sex organ of an **oomycete** (Fig. 128, arrow).

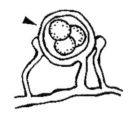

FIGURE 128   Oogonium

Oomycete—a "fungus" whose sexual stage is characterized by the production of an **oospore**. Such organisms are only distantly related to true **fungi**.

Oospore—sexually produced spore of the **oomycetes** (Fig. 129, arrow).

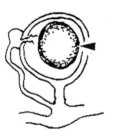

FIGURE 129   Oospore

Operculum (pl. **opercula**)—a lid ("trap door") covering the opening on a **sporangium** (Fig. 130a, arrow) or an **ascus** (Fig. 130b, arrow) through which spores are discharged.

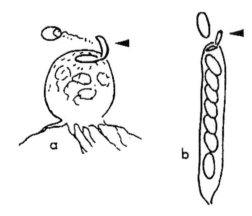

FIGURE 130   Operculum

Ostiole—the opening of a **pycnidium**, **perithecium**, or **pseudothecium** (Fig. 131).

FIGURE 131 Ostiole

Ovoid—egg-shaped with the narrow end at the apex (Fig. 132). Cf. **obovoid**.

FIGURE 132 Ovoid

Papillate—possessing one or more nipplelike swellings (Fig. 133). Cf. **umbonate**.

FIGURE 133 Papillate

Paragynous—condition in which the **antheridium** (Fig. 134a) penetrates the **oogonium** (Fig. 134b) laterally. Cf. **amphigynous**.

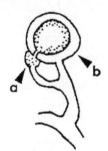

FIGURE 134 Paragynous

Paraphysis (pl. **paraphyses**)—a sterile threadlike element inside a **perithecium** (Fig. 135, arrows). Cf. **paraphysoid**, **periphysis**, **pseudoparaphysis**. Sometimes used with reference to **basidiomycetes**; see **hyphidium**.

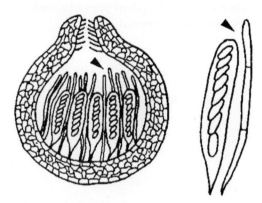

FIGURE 135 Paraphysis

Paraphysoid—narrow, anastomosing, sterile, threadlike element in a **pseudothecium** (Fig. 136, arrow). Cf. **paraphysis**, **pseudoparaphysis**.

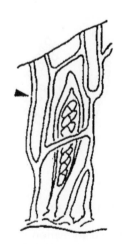

FIGURE 136 Paraphysoid

Parasite (adj. **parasitic**)—organism living in or on another organism from which it obtains nutrients. See **biotroph**, **necrotroph**.

Partial veil—a tissue that joins the margin of the **pileus** to the **stipe** during development of an **agaric** (Fig. 137, arrow). Its remnants may appear as an **appendiculate** margin on the pileus or as an **annulus** on the stipe. See **veil**. Cf. **universal veil**.

FIGURE 137   Partial veil

Pellis—skin or cortex of a **sporocarp**. See **pileipellis**, **stipitipellis**.

Penicillus (pl. **penicilli**)—brushlike structure producing **conidia** in *Penicillium* species and related **fungi** (Fig. 138). Penicilli with higher orders of branching are composed of **rami** and **metulae**, in addition to **phialides**. See **monoverticillate**, **biverticillate**, **terverticillate**.

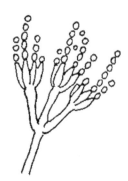

FIGURE 138   Penicillus

Percurrent—repeated formation of **conidia** at a single locus, so that a series of ringlike scars form on the **conidiogenous** cell (Fig. 139). See **annellide**.

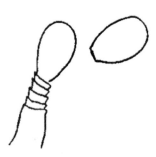

FIGURE 139   Percurrent

Peridiole—walled unit of the **gleba** in **gasteromycetes**; an "egg" in the bird's nest fungi, Nidulariales (Fig. 140, arrow).

FIGURE 140   Peridiole

Periphysis (pl. **periphyses**)—short, threadlike element (Fig. 141, arrow) inside the neck, near the **ostiole**, of a **perithecium** or **pseudothecium**.

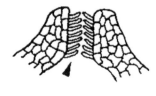

FIGURE 141   Periphyses

Perithecium (pl. **perithecia**)—small, sacklike fruiting body of several types of **ascomycetes** (Fig. 142). Perithecia possess **ostioles** and walls. Cf. **cleistothecium**, **pseudothecium**.

FIGURE 142   Perithecium

Phaeo-—prefix indicating dark coloration (e.g., phaeophragmiae are dark spores with multiple, transverse **septa**). Cf. **hyalo-**. (Fig. 143).

FIGURE 143   Phaeo-

Phialide (adj. **phialidic**)—an **enteroblastic conidiogenous** cell, often more or less bottle-shaped, that produces **conidia** in basipetal fashion (Fig. 144, arrow).

FIGURE 144   Phialide

Phoenicioid fungi—fungi growing in burnt areas.
Phragmobasidium (pl. **phragmobasidia**)—see **basidium**.
Phragmospore—a spore with more than one crosswall or **septum** (Fig. 145).

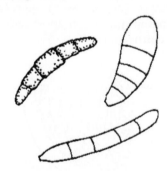

FIGURE 145   Phragmosporae

Pileipellis—skin or cortical layer of the **pileus**.
Pileus (pl. **pilei**)—the cap of a **mushroom** (Fig. 146, arrow).

FIGURE 146   Pileus

Plane—see **applanate**.
Plano-convex—broadly convex (Fig. 147). Cf. **applanate, convex**.

FIGURE 147   Plano-convex

Plasmodiocarp—branched, veinlike fruiting structure of many myxomycetes (Fig. 148).

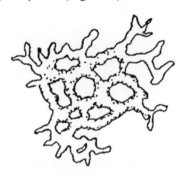

FIGURE 148   Plasmodiocarp

Plasmodium (pl. **plasmodia**)—unwalled (naked), multinucleate mass of protoplasm of **myxomycetes** (Fig. 149). It feeds on bacteria in an amoeboid fashion.

FIGURE 149   Plasmodium

Plectomycete—*sensu lato*, a **fungus** whose **asci** are enclosed in **cleistothecia**; *sensu stricto*, a member of the Eurotiales.
Plerotic—condition in which the **oospore** (Fig. 150a) fills the **oogonium** (Fig. 150b). Cf. **aplerotic**.

FIGURE 150   Plerotic

Pleurocystidium (pl. **pleurocystidia**)—a **cystidium** positioned on the side of a **lamella** (Fig. 151, arrow). Cf. **cheilocystidium**.

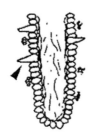

FIGURE 151   Pleurocystidium

Plurivorus—adapted to numerous substrata or hosts.
Polyblastic—with several **conidigenous** sites per **holoblastic** conidiogenous cell (Fig. 152).

FIGURE 152   Polyblastic

Polyphialide—a **phialide** that has more than one **conidiogenous** locus (Fig. 153, arrow).

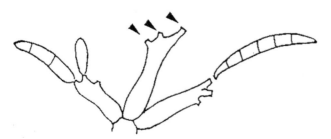

FIGURE 153   Polyphialide

Polypore—fungus whose **basidiocarp** is hard, often shelflike, and whose spore-bearing surface is composed of **tubes** or pores (Fig. 154).

FIGURE 154   Polypore

Polytretic—with several **conidiogenous** sites (arrows) per **tretic** conidiogenous cell (Fig. 155).

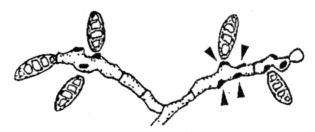

FIGURE 155   Polytretic

Pore—the opening of a **tube**. See **bolete**, **polypore**.
Powdery mildew—member of the Erysiphales, a group of **biotrophic**, plant **parasitic fungi** producing their **asci** in **cleistothecia**.
Pseudoepithecium (pl. **pseudoepithecia**)—an amorphous substance (not a tissue) aggregated around the tips of **paraphyses** in an **apothecium** (Fig. 156, arrow). Cf. **epithecium**.

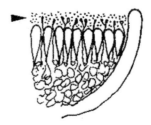

FIGURE 156   Pseudoepithecium

Pseudomycelium—chainlike series of cells resembling **mycelium**; especially common in **yeasts** (Fig. 157).

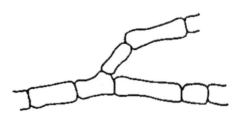

FIGURE 157   Pseudomycelium

Pseudoparaphysis (pl. **pseudoparaphyses**)—a sterile thread inside a **pseudothecium** (Fig. 158, arrow). Cf. **paraphysis, paraphysoid, periphysis**.

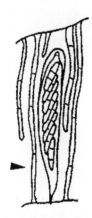

FIGURE 158   Pseudoparaphysis

Pseudothecium (pl. **pseudothecia**)—saclike fruiting structure of a **loculoascomycete** (Fig. 159). Outwardly resembles a **perithecium**.

FIGURE 159   Pseudothecium

Pubescent—with short, soft hairs (Fig. 160). Cf. **hirsute, tomentose**.

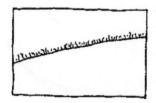

FIGURE 160   Pubescent

Puffball—a mushroomlike **basidiocarp** lacking **gills** or **pores**, with an outer wall completely enclosing the spores, and opening by a **stoma** or by rupture. Characteristic of the Lycoperdales.

Pulvinate—cushion-shaped (Fig. 161).

FIGURE 161   Pulvinate

Pycnidium (pl. **pycnidia**)—a saclike body containing **conidia** (Fig. 162).

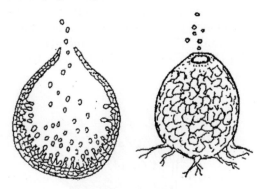

FIGURE 162   Pycnidium

Pycniospore—a **spermatium** (fertilizing spore) in the **rust** fungi.

Pycnium (pl. **pycnia**)—the fruiting body producing **pycniospores** in the **rust** fungi (Fig. 163).

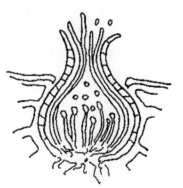

FIGURE 163   Pycnium

Pyrenomycete—colloquial term for fungi producing **perithecia** or **pseudothecia**. Modern usage sometimes excludes fungi producing the latter.

Pyriform—pear-shaped (Fig. 164). Cf. **ampulliform**, **lacrymoid**.

FIGURE 164   Pyriform

Quaterverticillate—indicating a **penicillus** that has one degree of branching more than **terverticillate**—that is, with branch points of the stipe (Fig. 165f) subtending **rami** (Fig. 165e), **ramuli** (Fig. 165d), **metulae** (Fig. 165c), and **phialides** (Fig. 165b) with conidia (Fig. 165a).

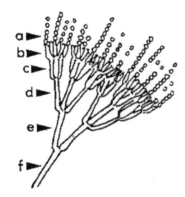

FIGURE 165   Quaterverticillate

Rachiform—condition in which the **conidiogenous** loci in a cell are arranged as a rachis (Fig. 166). Cf. **raduliform**, **sympodial**.

FIGURE 166   Rachiform

Raduliform—condition in which the **conidiogenous** loci consist of fine protrusions distributed on the upper portion of a clavate cell (Fig. 167).

FIGURE 167   Raduliform

Ramus (pl. **rami**)—on a **terverticillate** or **quaterverticillate penicillus**, the branch subtended by the **stipe**.

Ramulus (pl. **ramuli**)—on a **quaterverticillate penicillus**, the branch subtended by the **ramus** and supporting the **metulae**.

Reflexed—of the margin of a **pileus**: curved upward (Fig. 168, arrow). Cf. **inflexed**, **revolute**.

FIGURE 168   Reflexed

Reniform—kidney-shaped (Fig. 169). Cf. **allantoid**.

FIGURE 169   Reniform

Resupinate—of a sporocap of a macrofungus: directly in contact with the substratum (Fig. 170).

FIGURE 170   Resupinate

Reticulate—netlike (Fig. 171).

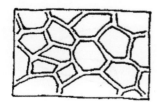

FIGURE 171   Reticulate

Revolute—of the margin of a **pileus**: coiled upward (Fig. 172, arrow). Cf. **reflexed**.

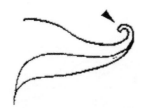

FIGURE 172   Revolute

Rhizoid—a small, rootlike structure (Fig. 173a) at the base of a sporangium that is connected to other sporangia by stolons (Fig. 173b). Cf. **rhizomorph**.

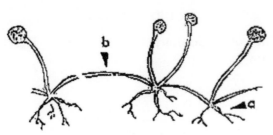

FIGURE 173   Rhizoid

Rhizomorph—a rootlet- or shoestringlike aggregation of **hyphae** possessing an apical meristem and rind. Characteristic of *Armillaria mellea*.

Rimose—cracked or fissured in multiple directions (Fig. 174). Cf. **rugulose**.

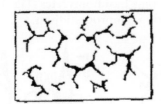

FIGURE 174   Rimose

Rimose-areolate—possessing divisions created by cracks or fissures and grading into a **rimose** margin (Fig. 175). Cf. **areolate, rimose**.

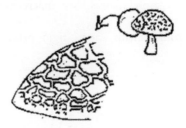

FIGURE 175   Rimose-areolate

Rivulose—resembling a system of streams or rivers (Fig. 176).

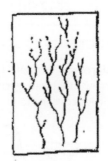

FIGURE 176   Rivulose

Rugose—with wrinkles (Fig. 177). Cf. **rugulose**.

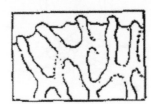

FIGURE 177   Rugose

Rugulose—with small wrinkles (Fig. 178). Cf. **rugulose**.

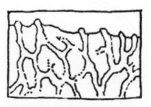

FIGURE 178   Rugulose

Rust—1. Member of the Urediniales, one of the groups of **microfungi** producing phragmobasidia. 2. Plant disease caused by such a **fungus** and characterized by rust-colored spore masses. See **macrocyclic**, **microcyclic**.

Scabrous—rough, with short, rigid projections (Fig. 179).

FIGURE 179   Scabrous

Sclerotium (pl. **sclerotia**)—a small, hard, multicellular, resting body (arrow) that can germinate to produce vegetative or reproductive structures (Fig. 180).

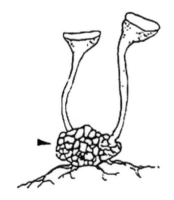

FIGURE 180   Sclerotium

Scolecospore—threadlike spore (Fig. 181).

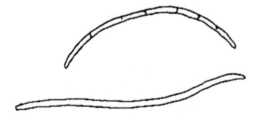

FIGURE 181   Scolecospore

Seceding—of the **gills** of an **agaric**: abruptly separated from the **stipe** (Fig. 182). Cf. **emarginate**, **free**, **sinuate**.

FIGURE 182   Seceding

Septum (pl. **septa**, adj. **septal**)—cell wall in a **hypha** (Fig. 183 left, arrow) or spore (Fig. 183 right, arrow).

FIGURE 183   Septum

Serrate—notched on one side, like a saw (Fig. 184). Cf. **dentate**.

FIGURE 184   Serrate

Sessile—of a **sporocarp**: without a **stipe** (Fig. 185, arrow).

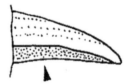

FIGURE 185   Sessile

Sequestrate fungi—**macrofungi**, such as truffles, whose fruit bodies are produced underground.

Seta (pl. **setae**)—hairlike bristle on a fruiting body (Fig. 186a); a thick-walled bristle such as occurs in the **hymenia** of some **basidiomycetes** (Fig. 186b). Sometimes used for a **lamprocystidium**.

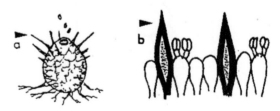

FIGURE 186   Setae

Setula (pl. **setulae**)—thin hairlike appendage on a spore (Fig. 187, arrow). See **appendiculate**. In basidiomycetes, a small **cystidium** darkening in potassium hydroxide.

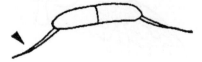

FIGURE 187   Setula

Sinuate—of the **gills** (Fig. 188a) of an **agaric**: with a curved indentation near the **stipe** (Fig. 188b). Cf. **emarginate, seceding, subdecurrent**.

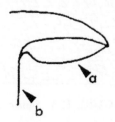

FIGURE 188   Sinuate

Slime mold—one of several "fungi" related to protozoans and producing an **amoeba** stage in the life cycle; many slime molds produce **plasmodia** and/or **plasmodiocarps**. See also **acrasid, dictyostelid, myxomycete**.

Smut—1. Member of the Ustilaginales, one of the groups of **microfungi** producing **phragmobasidia**. 2. Plant disease caused by such a fungus and characterized by dark, powdery spore masses.

Sooty molds—any of several generally saprotrophic fungi forming dark masses of **hyphae** on plant leaves. Cf. **black mildew**.

Sorocarp—fruiting body in Acrasiomycetes (Fig. 189). See **acrasid, dictyostelid, slime mold**.

FIGURE 189   Sorocarp

Sorus (pl. **sori**)—spore-producing structure in one of several groups of **fungi**, especially **rust** and **smut** fungi.

Spathulate—spoonlike (Fig. 190). Cf. **capitate, clavate**.

FIGURE 190   Spathulate

Spermatium (pl. **spermatia**)—spore whose function is fertilization. See **pycniospore**.

Spermogonium (pl. **spermogonia**)—fruiting body that produces **spermatia**. A **pycnium** is a kind of spermogonium present in the **rust** fungi.

Sphaerocyst—globose cells in the tissues of *Russula* species and *Lacterius* species (Fig. 191).

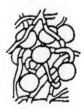

FIGURE 191   Sphaerocyst

Sporangiolum (pl. **sporangiola**)—small **sporangium** containing few spores (Fig. 192).

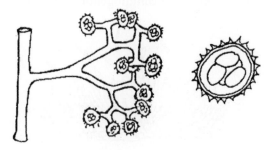

FIGURE 192   Sporangiolum

Sporangiophore—stalklike **hypha** (arrow) that bears a **sporangium** (Fig. 193).

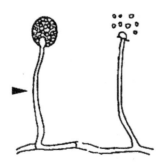

FIGURE 193   Sporangiophore

Sporangiospore—asexual spore of a member of the Mucorales; borne inside a **sporangium**.

Sporangium (pl. **sporangia**)—a spore-containing, saclike body (arrow) on the tip of a **sporangiophore** (Fig. 194).

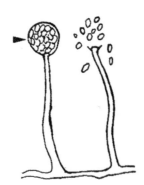

FIGURE 194   Sporangium sphaerocyst

Sporodochium (pl. **sporodochia**)—cushion-shaped **stroma** covered with **conidiogenous** cells (Fig. 195).

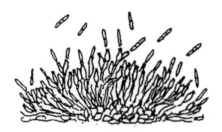

FIGURE 195   Sporodochium

Squama (pl. **squamae**)—a scale (Fig. 196).

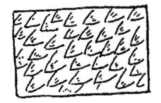

FIGURE 196   Squama

Squamose (**squamous**)—of a **pileus** or **stipe**: with **squamae**; scaled (Fig. 197). Cf. **squamulose**.

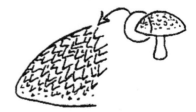

FIGURE 197   Squamose

Squamule—a small scale (Fig. 198). Cf. **squama**.

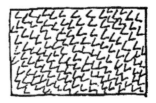

FIGURE 198   Squamule

Squamulose—of a **pileus** or **stipe**: with **squamules**. Cf. **squamose**.

Squarrose—with recurved scales, as of the surface of a **pileus** (Fig. 199).

FIGURE 199   Squarrose

Staurospore—a spore with radiating appendages (Fig. 200).

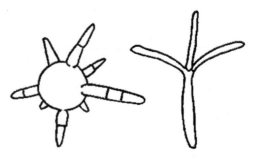

FIGURE 200   Staurosporae

Sterigma (pl. **sterigmata**)—a small branch or extension (arrow) that bears a **conidium**, **sporangium**, or especially a **basidiospore** (Fig. 201).

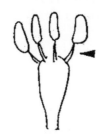

FIGURE 201   Sterigma

Stipe—portion of the **sporocarp** that supports the spore-producing structures; for example, the stalk of a mushroom (Fig. 202, arrow).

FIGURE 202   Stipe

Stipitipellis—the skin or cortical layer of the **stipe**.

Stolon—an aerial **hypha** connecting fascicles of **rhizoids**. Characteristic of several genera of the Mucorales.

Stoma (pl. **stomata**)—term used for the **ostiole** of **basidiocarps** of certain **puffballs** and their allies.

Strain—1. *Sensu stricto,* a genetically uniform collection of cells or **hyphae**. 2. A serially propagated **culture** of a single species.

Striate—lined (Fig. 203 a. spore; b. pileus).

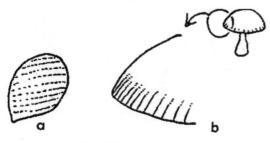

FIGURE 203   Striate

Strigose—with long, coarse hairs (Fig. 204). Cf. **hirsute**, **hispid**.

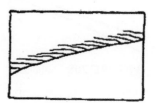

FIGURE 204   Strigose

Stroma (pl. **stromata**)—compact structure (arrows) on (or in) which fruiting bodies are produced (Fig. 205). Cf. **subiculum**.

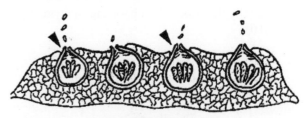

FIGURE 205   Stroma

Subdecurrent—of the **gills** of an **agaric**: tending to extend down the **stipe** (Fig. 206). Cf. **adnate**, **decurrent**.

FIGURE 206   Subdecurrent

Subiculum (pl. **subicula**)—a loose **hyphal** network in which fruiting bodies (arrows) are formed (Fig. 207). Cf. **stroma**.

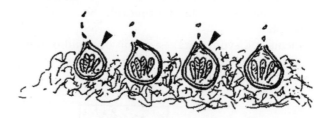

FIGURE 207   Subiculum

Substrate—1. The nutrient source on which an organism feeds. 2. The material (**substratum**) to which an organism is attached. 3. The substance acted on by an enzymatic or chemical process.

Substratum (pl. **substrata**)—1. The material to which an organism is attached (see also **substrate**). 2. The layer beneath the soil surface.

Subulate—awl-shaped (Fig. 208). Cf. **acicular**.

FIGURE 208   Subulate

Subumbonate—with a broad, low umbo (Fig. 209). Cf. **bullate, convex, umbonate.**

FIGURE 209   Subumbonate

Suspensor—**hypha** (arrow) supporting a **zygospore** (Fig. 210).

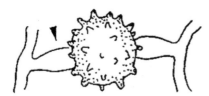

FIGURE 210   Suspensor

Sympodial—of a **conidiophore**: with the main axis elongating in a zigzag manner (Fig. 211). Cf. **rachiform**.

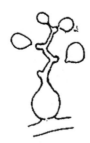

FIGURE 211   Sympodial

Synanamorph—one of two or more **anamorphs** belonging to the same **teleomorph**. See also **holomorph**.

Synnema (pl. **synnemata**)—a group of **conidiophores** closely packed together to form a single, spore-bearing stalk (Fig. 212). See **coremium**.

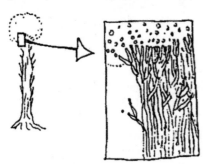

FIGURE 212   Synnema

Teleomorph—stage of the fungus that produces sexual spores via meiosis; applied to **ascomycetes** and **basidiomycetes.**

Teliospore—sexual spore of a **rust** (Fig. 213a) or a **smut** (Fig. 213b) fungus.

FIGURE 213   Teliospore

Telium (pl. **telia**)—fruiting body producing **teliospores** in **rusts** or **smuts** (Fig. 214).

FIGURE 214   Telium

Terverticillate—indicating a **penicillus** that has three levels of branching points (Fig. 215 a. rami; b. metulae; c. phialides). Cf. **biverticillate, quaterverticillate.**

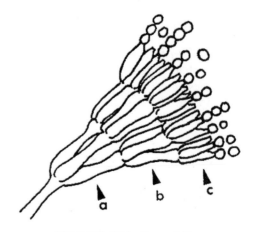

FIGURE 215   Terverticillate

Textura—tissue composed of **hyphae**. Several characteristic types are recognizable:

Textura angularis: hyphae polygonal in cross section (Fig. 216).

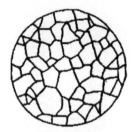

FIGURE 216   Textura angularis

Textura epidermoidea: hyphae in cross section resemble jigsaw puzzle pieces (Fig. 217).

FIGURE 217   Textura epidermoidea

Textura globosa: hyphae nearly circular in cross section (Fig. 218).

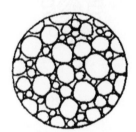

FIGURE 218   Textura globosa

Textura intricata: hyphae appear interwoven (Fig. 219).

FIGURE 219   Textura intricata

Textura oblita: hyphae appear more or less parallel, with thick walls (Fig. 220).

FIGURE 220   Textura oblita

Textura porrecta: hyphae appear more or less parallel, with long cells (Fig. 221).

FIGURE 221   Textura porrecta

Textura prismatica: hyphae appear more or less parallel, with short cells (Fig. 222).

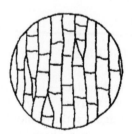

FIGURE 222   Textura prismatica

Thallic—referring to the type of **conidiogenesis** in which **conidia** are produced by fragmentation of a **hypha** (Fig. 223). See **arthrospore**.

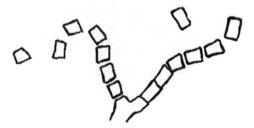

FIGURE 223   Thallic

Thallus (pl. **thalli**)—1. Vegetative body of a **fungus**. 2. Also used to refer to the entire body of a fungus. (Fig. 224).

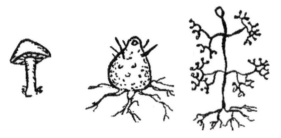

FIGURE 224   Thallus

Tinsel-flagellum—see **flagellum**.

Tissue types—see **textura**.

Toadstool—an **agaric** or **bolete**, especially those that are inedible or poisonous. A more pejorative term than **mushroom**.

Tomentose—downy; with soft, matted hairs (Fig. 225). Cf. **pubescent**, **villose**.

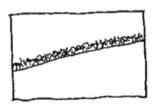

FIGURE 225   Tomentose

Trabecula (pl. **trabeculae**)—1. In **gasteromycetes**, plates of primordial tissue attached to a **columella** (Fig. 226). 2. In **agarics**, the primordium of a **lamella**. 3. In **ascomycetes**, a **paraphysoid**.

FIGURE 226   Trabecula

Trama—tissue internal to the **hymenium** of **agarics** and other **basidiomycetes**. The principal types of trama based on their arrangement in the hymenium are: **Convergent trama** (Fig. 227).

FIGURE 227   Convergent trama

Divergent trama (Fig. 228).

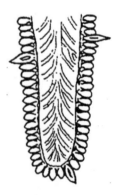

FIGURE 228   Divergent trama

Interwoven trama (Fig. 229).

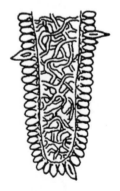

FIGURE 229   Interwoven trama

Parallel trama (Fig. 230).

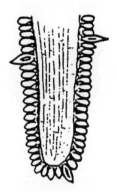

FIGURE 230 Parallel trama

Tretic—of conidia produced by protrusion of the inner wall of the **conidiogenous** cell through pores in the outer wall (Fig. 231).

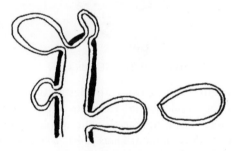

FIGURE 231 Tretic

Trichogyne—receptive **hypha** (arrow) of a female organ in **ascomycetes** (Fig. 232). See **antheridium, ascogonium**.

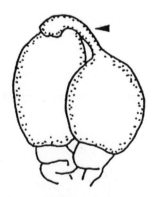

FIGURE 232 Trichogyne

Tuberculate—with massive warts (Fig. 233). Cf. **verrucose**.

FIGURE 233 Tuberculate

Tubes—spore-bearing surfaces of a **bolete** or **polypore** (Fig. 234).

FIGURE 234 Tubes

Turbinate—shaped like a top (Fig. 235). Cf. **napiform, pyriform**.

FIGURE 235 Turbinate

Umbilicate—of a **pileus**: deeply and sharply depressed in the center (Fig. 236).

FIGURE 236 Umbilicate

Umbonate—possessing a raised swelling or umbo (Fig. 237). Cf. **bullate, papillate**.

FIGURE 237 Umbonate

Ungulate—hoof-shaped sporocarp (Fig. 238).

FIGURE 238   Ungulate

Uniseriate—indicating an **aspergillum** with **phialides** but lacking **metulae** (Fig. 239a). Or, of an **ascus** containing only one row of ascospores (Fig. 239b). Cf. **biseriate**.

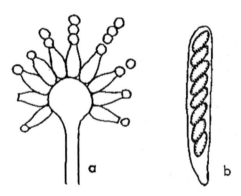

FIGURE 239   Uniseriate

Universal veil—a tissue enclosing the whole of a developing **agaric** (Fig. 240, arrow). See **veil**. Remnants of a universal veil may appear as an **appendiculate** margin of the **pileus** (Fig. 241a), patches on the pileus, or a **volva** (Fig. 241b). Cf. **partial veil**.

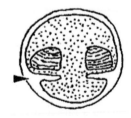

FIGURE 240   Universal veil

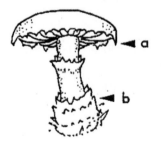

FIGURE 241   Veil

Uredinium (pl. **uredinia**)—fruiting body that produces **urediniospores** in the **rust** fungi (Fig. 242).

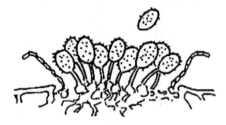

FIGURE 242   Uredinium

Urediniospore (**uredospore, urediospore**)—the repeating spore in the **rust** fungi (Fig. 243).

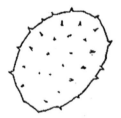

FIGURE 243   Urediospore

Vascular wilt—wilting caused by destruction of the vascular (water conducting) elements of the plant.

Veil—a tissue covering parts or the whole of an **agaric** during its development (Figs. 240, 241). See **partial veil, universal veil**.

Verrucose—with warts (Fig. 244). Cf. **tuberculate, verruculose**.

FIGURE 244   Verrucose

Verruculose—with small warts (Fig. 245). Cf. **verrucose**.

FIGURE 245   Verruculose

Vesicle—a swollen hyphal structure. For example, in *Aspergillus*, the swelling at the apex of the **stipe** (Fig. 246a). In *Pythium*, the swelling into which the **zoospores** are discharged from the **zoosporangium** (Fig. 246b). In **endomycorrhizae**, a swollen fungal structure inside a host cell.

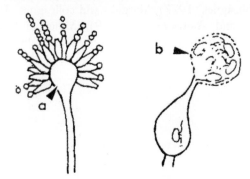

FIGURE 246   Vesicle

Vesicular-arbuscular mycorrhiza—see **endomycorrhiza**.

Villose—with long, soft hairs (Fig. 247). Cf. **tomentose**.

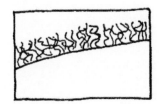

FIGURE 247   Villose

Volva—of an **agaric**: remnants of a **universal veil** at the base of a **stipe** (Fig. 243c). See **veil, universal veil**.

Whip-lash flagellum—see **flagellum**.

White rot—a type of decay in which lignin is decomposed more completely than cellulose, leaving a light-colored residue of cellulose. Cf. **brown rot**.

Witch's broom—any of several plant diseases resulting in a localized, dense proliferation of leafy tissue.

Woronin body—a small, refractive body associated with the **septal** pores in **ascomycetes** (Fig. 248, arrow).

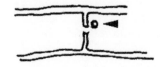

FIGURE 248   Woronin body

Xiphoid—swordlike (Fig. 249).

FIGURE 249   Xiphoid

Yeast—a unicellular, budding **fungus** (Fig. 250).

FIGURE 250   Yeast

Zoosporangium—a **sporangium** that contains **zoospores** (Fig. 251, arrow).

FIGURE 251   Zoosporangium

Zoospore—a motile spore with one or more **flagella** (Fig. 252).

FIGURE 252   Zoospore

Zygomycete—a **fungus** whose sexual state is characterized by the production of a **zygospore** (Fig. 253, arrow).

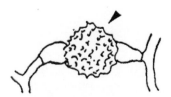

FIGURE 253   Zygospore

Zygospore—the sexual spore of a **zygomycete**.

672   *F. M. Dugan*

REFERENCES

The publications cited here provide direct entry into terminology applicable to morphology and other aspects of eufungi, oomycetes, acrasiomycetes, myxomycetes, and protostelids. Some of the references are useful for identification of isolates or specimens. I did not, however, choose them for that purpose but rather to illustrate variation in morphological structures or to explain terminology beyond the scope of this introductory glossary.

The publications encompass both popular and scholarly literature and do not cover all taxa with equal emphasis. The references are Alexopoulos and colleagues (1996); Barnett and Hunter (1998); Breitenbach and Kränzlin (1984, 1986); Carmichael and colleagues (1980); Cummins and Hiratsuka (1983); Hanlin (1990, 1998a, 1998b); Hawksworth and colleagues (1995); Kirk and colleagues (2001); Largent (1986); Largent and colleagues (1977); Margulis and colleagues (1990); Martin and Alexopoulos (1969); Miller and Miller (1988); O'Donnell (1979); Snell and Dick (1957); Stern (1992); Ulloa and Hanlin (2000); Vánky (1987); and von Arx (1981).

LITERATURE CITED

Abbott, L.K. 1982. Comparative anatomy of vesicular-arbuscular mycorrhizas formed on subterranean clover. Australian Journal of Botany 30:485–499.

Abbott, L.K., and C. Gazey. 1994. An ecological view of the formation of VA mycorrhizas. Plant and Soil 159:69–78.

Abdel-Hafez, S.I.I. 1982a. Thermophilic and thermotolerant fungi in the desert soils of Saudia Arabia. Mycopathologia 80:15–20.

——. 1982b. Osmophilic fungi of desert soils in Saudi Arabia. Mycopathologia 80:9–14.

Abdel-Hafez, S.I.I., M.A. Zidan, M.M.K. Bagy, and M.A. Abdel-Sater. 1989. Distribution of two halophilic fungi in the Egyptian soils and glycerol accumulation. Cryptogamie: Mycologie 10:125–133.

Abdullah, S.K. 1982. Coprophilous mycoflora on different dung types in southern desert of Iraq. Sydowia 35:1–5.

Abdullah, S.K., and S.M. Al-Bader. 1990. On the thermophilic and thermotolerant mycoflora of Iraqi soils. Sydowia 42:1–7.

Abe, J-i. P., and K. Katsuya. 1995. Vesicular-arbuscular mycorrhizal fungi in coastal dune plant communities. II. Spore formation of *Glomus* spp. predominates under geographically separated patches of *Elymus mollis*. Mycoscience 36:113–116.

Achenbach-Richter, L., R. Gupta, W. Zillig, and C.R. Woese. 1988. Rooting the archaebacterial tree: the pivotal role of *Thermococcus celer* in archaebacterial evolution. Systematic and Applied Microbiology 10:231–240.

Adams, G.C., and B.R. Kropp. 1996. *Athelia arachnoidea*, the sexual state of *Rhizoctonia carotae*, a pathogen of carrot in cold storage. Mycologia 88:459–472.

Adams, J.E., and E.D. McCune. 1979. Application of the generalized jackknife to Shannon's measure of information used as an index of diversity. Pp. 117–132. *In* J.G.F. Grassle, G.P. Patil, W. Smith, and C. Taille (eds.), Ecological Diversity in Theory and Practice. International Co-operative Publishing House, Fairland, Maryland.

Adams, P.B. 1979. A rapid method for quantitative isolation of sclerotia of *Sclerotinia minor* and *Sclerotium cepivorum* from soil. Plant Disease Reporter 63:349–351.

——. 1990. The potential of mycoparasites for biological control of plant diseases. Annual Review of Phytopathology 28:59–72.

Adams, P.B., and W.A. Ayers. 1985. The world distribution of the mycoparasites *Sporidesmium sclerotivorum*, *Teratosperma oligocladum*, and *Laterispora brevirama*. Soil Biology and Biochemistry 17:583–584.

Adams, P.B., W.A. Ayers, and J. Marois. 1985. Energy efficiency of the mycoparasite *Sporidesmium sclerotivorum* in vitro and soil. Soil Biology and Biochemistry 17:155–158.

Adams, P.B., J.J. Marois, and W.A. Ayers. 1984. Population dynamics of the mycoparasite, *Sporidesmium sclerotivorum*, and its host, *Sclerotinia minor*, in soil. Soil Biology and Biochemistry 16:627–633.

Agerer, R. 1991. Studies on ectomycorrhizae XXXIV. Mycorrhizae of *Gomphidius glutinosus* and of *G. roseus*, with some remarks on Gomphidiaceae (Basidiomycetes). Nova Hedwigia 53:127–170.

Agerer, R., J. Ammirati, T.J. Baroni, P. Blanz, R. Courtecuisse, D.E. Desjardin, W. Gams, N. Hallenberg, R. Halling, D.L. Hawksworth, E. Horak, R.P. Korf, G.M. Mueller, F. Oberwinkler, G. Rambold, R.C. Summerbell, D. Triebel, and R. Watling. 2000. Always deposit vouchers. Mycological Research 104:642–644.

Agius, C. 1978. Infection by an *Ichthyophonus*-like fungus in the deep-sea scabbard fish *Aphanopus carbo* (Lowe) (Trichiuridae) in the North East Atlantic. Journal of Fish Diseases 1:191–193.

Agrios, G.N. 1988. Plant Pathology. 3rd ed. Academic Press, San Diego, California.

Ahlich, K., and T.N. Sieber. 1996. The profusion of dark septate endophytic fungi in non-mycorrhizal fine roots of forest trees and shrubs. New Phytologist 132:259–270.

Ahmad, J.S., and R. Baker. 1987. Rhizosphere competence of *Trichoderma harzianum*. Phytopathology 77:182–189.

Ahmad, S. 1956. Ustilaginales of West Pakistan. Commonwealth Mycological Institute Mycological Papers. 64:1–17.

Ahmadjian, V. 1993. The Lichen Symbiosis. 2nd ed. J. Wiley, New York.

Ainsworth, A.M., and A.D.M. Rayner. 1991. Ontogenetic stages from coenocyte to basidiome and their relation to phenoloxidase activity and colonization processes in *Phanerochaete magnoliae*. Mycological Research 95:1414–1422.

Ainsworth, G.C. 1976. Introduction to the History of Mycology. Cambridge University Press, Cambridge, England.

Ainsworth, G.C., and P.K.C. Austwick. 1973. Fungal Diseases of Animals. 2nd ed. Commonwealth Agricultural Bureaux, Bucks, England.

Ainsworth, G.C., F.K. Sparrow, and A.S. Sussman. 1973a. The Fungi. An Advanced Treatise. IVA. A Taxonomic Review with Keys: Ascomycetes and Fungi Imperfecti. Academic Press, New York.

——. 1973b. The Fungi. An Advanced Treatise. IVB. A Taxonomic Review with Keys: Basidiomycetes and Lower Fungi. Academic Press, New York.

Ajello, L., and R.J. Weeks. 1983. Soil decontamination and other control measures. Pp. 229–238. *In* A.F. DiSalvo (ed.), Occupational Mycoses. Lea and Febiger, Philadelphia, Pennsylvania.

Ajello, L., M.R. McGinnis, and J. Camper. 1977. An outbreak of phaeohyphomycoses in rainbow trout caused by *Scolecobasidium humicola*. Mycopathologia 62:15–22.

Akin, D.E. 1986. Chemical and biological structures in plants as related to microbial degradation of forage cell walls. Pp. 139–157. *In* L.P. Milligan, W.L. Grovum, and A. Dobson (eds.), Control of Digestion and Metabolism in Ruminants: Proceedings of the Sixth International Symposium on Ruminant Physiology. Prentice Hall, Englewood Cliffs, New Jersey.

Akin, D.E., and R. Benner. 1988. Degradation of polysaccharides and lignin by ruminal bacteria and fungi. Applied and Environmental Microbiology 54:1117–1125.

Akin, D.E., and L.L. Rigsby. 1985. Influence of phenolic acids on rumen fungi. Agronomy Journal 77:180–182.

——. 1987. Mixed fungal populations and lignocellulosic tissue degradation in the bovine rumen. Applied and Environmental Microbiology 53:1987–1995.

Akin, D.E., W.S. Borneman, and C.E. Lyon. 1990. Degradation of leaf blades and stems by monocentric and polycentric isolates of ruminal fungi. Animal Feed Science and Technology 31:205–221.

Akin, D.E., W.S. Borneman, and W.R. Windham. 1988. Rumen fungi: morphological types from Georgia cattle and their attack on forage cell walls. BioSystems 21:385–391.

Akin, D.E., G.L.R. Gordon, and J.P. Hogan. 1983. Rumen bacterial and fungal degradation of *Digitaria pentzii* grown with or without sulfur. Applied and Environmental Microbiology 46:738–748.

Akin, D.E., C.E. Lyon, W.R. Windham, and L.L. Rigsby. 1989. Physical degradation of lignified stem tissues by ruminal fungi. Applied and Environmental Microbiology 55:611–616.

Alabi, R.O. 1971. Seasonal periodicity of Saprolegniaceae at Ibadan, Nigeria. Transactions of the British Mycological Society 56:337–341.

Al-Delaimy, K.S. 1977. Protein and amino acid composition of truffle. Journal of the Canadian Institute of Food Science and Technology 10:221–222.

Alderman, D.J. 1976. Fungal diseases in marine animals. Pp. 223–260. *In* E.B. Gareth Jones (ed.), Recent Advances in Marine Mycology. Elek Science, London, England.

——. 1981. *Fusarium solani* causing an exoskeletal pathology in cultured lobsters, *Homarus volgaris*. Transactions of the British Mycological Society 76:25–27.

——. 1982. Fungal diseases of aquatic animals. Pp. 189–242. *In* R.J. Roberts (ed.), Microbial Diseases of Fish. Academic Press, London, England.

Alderman, D.J., and E.B.G. Jones. 1971. Shell disease of oysters. Fishery Investigations, London (Series 2) 26:1–19.

Alderman, D.J., and J.L. Polglase. 1986. Are fungal diseases significant in the marine environment? Pp. 189–198. *In* S.T. Moss (ed.), The

Biology of Marine Fungi. Cambridge University Press, Cambridge, England.

Aldrich, H.C. 1967. The ultrastructure of meiosis in three species of *Physarum*. Mycologia 59:127–148.

Alexander, M.P. 1980. A versatile stain for pollen fungi, yeast and bacteria. Stain Technology 55:13–18.

Alexopoulos, C.J. 1960. Gross morphology of the plasmodium and its possible significance in the relationships among the myxomycetes. Mycologia 52:1–20.

——. 1963. The myxomycetes II. Botanical Review 29:1–78.

——. 1970. Rain forest Myxomycetes. Pp. 21–23. *In* H.T. Odum (ed.), A Tropical Rain Forest. U. S. Atomic Energy Commission, Washington, D.C.

Alexopoulos, C.J., C.W. Mims, and M. Blackwell. 1996. Introductory Mycology. 4th ed., John Wiley, New York.

Al-Hamdani, A.M., and R.C. Cooke. 1983. Effects of the mycoparasite *Pythium oligandrum* on cellulolysis and sclerotium production by *Rhizoctonia solani*. Transactions of the British Mycological Society 81:619–670.

Allen, E.B., and M.F. Allen. 1984. Competition between plants of different successional stages: mycorrhizae as regulators. Canadian Journal of Botany 62:2625–2629.

Allen, E.B., M.F. Allen, D.J. Helm, J.M. Trappe, R. Molina, and E. Rincon. 1995. Patterns and regulation of mycorrhizal plant and fungal diversity. Plant and Soil 170:47–62.

Allen, M.F. 1991. The Ecology of Mycorrhizae. Cambridge University Press, Cambridge, England.

Allen, T.F.H., and T.W. Hoekstra. 1990. The confusion between scale-defined levels and conventional levels of organization in ecology. Journal of Vegetation Science 1:5–12.

——. 1992. Toward a Unified Ecology. Columbia University Press, New York.

Allen, W. 2001. Green Phoenix: Restoring the Tropical Forests of Guanacaste, Costa Rica. Oxford University Press, New York.

Allendorf, F.W., and R.F. Leary. 1986. Heterozygosity and fitness in natural populations of animals. Pp. 57–76. *In* M.E. Soulé (ed.) Conservation Biology: The Science of Scarcity and Diversity. Sinauer Associates, Sunderland, Massachusetts.

Almasan, A., and N.C. Mishra. 1991. Recombination by sequence repeats with formation of suppressive or residual mitochondrial DNA in *Neurospora*. Proceedings of the National Academy of Sciences USA 88:7684–7688.

Almeida, R.T., and N.C. Schenck. 1990. A revision of the genus *Sclerocystis* (Glomaceae, Glomales). Mycologia 82:703–714.

Alsheikh, A.M. 1994. Taxonomy and mycorrhizal ecology of the desert truffles in the genus *Terfezia*. Unpubl. Ph.D. dissert., Oregon State University, Corvallis, Oregon.

Alsheikh, A.M., and J.M. Trappe. 1983a. Taxonomy of *Phaeangium lefebvrei*, a desert truffle eaten by birds. Canadian Journal of Botany 61:1919–1925.

——. 1983b. Desert truffles: the genus *Tirmania*. Transactions of the British Mycological Society 81:83–90.

Alstrup, V., and D.L. Hawksworth. 1990. The lichenicolous fungi of Greenland. Meddelelser om Grønland Bioscience 31:1–90.

Amann, R.I., W. Ludwig, and K.H. Schleifer. 1995. Phylogenetic identification and in situ detection of individual microbial cells without cultivation. Microbiology Reviews 59:143–169.

Ammirati, J. 1979. Chemical studies of mushrooms: the need for voucher specimens. Mycologia 71:437–441.

Ammirati, J., S. Ammirati, L. Norvell, T. O'Dell, M. Puccio, M. Seidl, G. Walker, The Puget Sound Mycological Society, S. Redhead, J. Ginns, H. Burdsall, T. Volk, and K. Nakasone. 1994. A preliminary report on the fungi of Barlow Pass, Washington. McIlvainea 11:10–33.

Amon, J.P., D. Jennings, and D.R. Cool. 1987. Use of a continuous culture chamber for physiological, ecological and morphological studies of monocentric fungi. Pp. 132–136. *In* M.S. Fuller and A. Jaworski (eds.), Zoosporic Fungi in Teaching and Research. Southeastern Publishing, Athens, Georgia.

Amos, K.H. (ed.). 1985. Procedures for the Detection and Identification of Certain Fish Pathogens. 3rd ed. Fish Health Section, American Fisheries Society, Corvallis, Oregon.

An, Z.-Q., B.Z. Gao, and J.W. Hendrix. 1993. Populations of spores and propagules of mycorrhizal fungi in relation to the life cycles of tall fescue and tobacco. Soil Biology and Biochemistry 25:813–817.

An, Z.-Q., J.H. Grove, J.W. Hendrix, D.E. Hershman, and G.T. Henson. 1990. Vertical distribution of endogonaceous mycorrhizal fungi associated with soybean, as affected by soil fumigation. Soil Biology and Biochemistry 22:715–719.

Anastasiou, C.J. 1961. Fungi from salt lakes. I. A new species of *Clavariopsis*. Mycologia 53:11–16.

——. 1963. Fungi from salt lakes. II. Ascomycetes and fungi imperfecti from the Salton Sea. Nova Hedwigia 6:243–276.

Anastasiou, C.J., and L.M. Churchland. 1969. Fungi on decaying leaves in marine habitats. Canadian Journal of Botany 47:251–257.

Anderson, D.C., K.T. Harper, and R.C. Holmgren. 1982. Factors influencing development of cryptogamic soil crusts in Utah deserts (Effects of grazing, management of rangelands). Journal of Range Management 35:180–185.

Anderson, J.B., and L.M. Kohn. 1995. Clonality in soilborne, plant-pathogenic fungi. Annual Review of Phytopathology 33:369–391.

Anderson, J.B., M.L. Smith, and J.N. Bruhn. 1994. Relatedness and spatial distribution of *Armillaria* genets infecting red pine seedlings. Phytopathology 84:822–829.

Anderson, J.M., and J.M. Swift. 1983. Decomposition in tropical forests. Pp. 287–309. *In* S.L. Sutton, T.C. Whitmore, and A.C. Chadwick (eds.), Tropical Rain Forest: Ecology and Management. Blackwell Scientific, Oxford, England.

Anderson, R.C., A.E. Liberta, L.A. Dickman, and J.A. Katz. 1983. Spatial variation in vesicular-arbuscular mycorrhiza spore density. Bulletin of the Torrey Botanical Club 110:519–525.

Ando, K. 1992. A study of terrestrial aquatic fungi. Transactions of the Mycological Society of Japan 33:415–425.

——. 1993a. *Kodonospora*, a new staurosporous hyphomycete genus from Japan. Mycological Research 97:506–508.

——. 1993b. Three new species of staurosporous hyphomycetes from Japan. Transactions of the Mycological Society of Japan 34:399–408.

Ando, K., and K. Tubaki. 1984a. Some undescribed hyphomycetes in rainwater draining from intact trees. Transactions of the Mycological Society of Japan 25:39–47.

——. 1984b. Some undescribed hyphomycetes in the rain drops from intact leaf-surface. Transactions of the Mycological Society of Japan 25:21–37.

Ando, K., K. Tubaki, and M. Arai. 1987. *Trifurcospora*: a new generic name for *Flabellospora irregularis*. Transactions of the Mycological Society of Japan 28:469–473.

Andrews, J. 1995. Fungi and the evolution of growth form. Canadian Journal of Botany 73 (Suppl. 1):S1206–S1212.

Andrews, J.H. 1991. Comparative Ecology of Microorganisms and Macroorganisms. Springer-Verlag, New York.

——. 1992. Fungal life-history strategies. Pp. 119–145. *In* G.C. Carroll and D.T. Wicklow (eds.), The Fungal Community: Its Organization and Role in the Ecosystem. 2nd ed. Marcel Dekker, New York.

Andrews, J.H., and C.M. Kenerley. 1978. The effects of a pesticide program on non-target epiphytic microbial populations of apple leaves. Canadian Journal of Microbiology 24:1058–1072.

Andrews, S., and J.I. Pitt. 1986. Selective medium for isolation of *Fusarium* species and dematiaceous hyphomycetes from cereals. Applied and Environmental Microbiology 51:1235–1238.

——. 1987. Further studies on the water relations of xerophilic fungi, including some halophiles. Journal of General Microbiology 133:233–238.

Angel, K., and D.T. Wicklow. 1975. Relationships between coprophilous fungi and fecal substrates in a Colorado grassland. Mycologia 67:63–74.

Anke, H., and O. Sterner. 1988. Transformation of isovelleral by the parasitic fungus *Calcarisporium arbuscula*. Phytochemistry 27:2765–2767.

Anonymous. 1990. Food Mycology. A Guide to the Use of Oxoid Culture Media. Unipath, Basingstoke, United Kingdom.

——. 1996. Laboratory Biosafety Guidelines. 2nd ed. Health Canada, Health Protection Branch, Laboratory Centre for Disease Control, Ottawa, Canada.

——. 2001. CBS List of Cultures: Fungi and Yeasts. 34th ed. Institute of the Royal Academy of Arts and Sciences, Baarn, The Netherlands.

Aoki, T., S. Tokumasu, and K. Tubaki. 1990. Fungal succession on momi fir needles. Transactions of the Mycological Society of Japan 31:355–374.

Aoki, T., S. Tokumasu, and F. Oberwinkler. 1992. Fungal succession on fir needles in Germany. Transactions of the Mycological Society of Japan 33:359–374.

Apinis, A.E. 1963. Occurrence of thermophilous microfungi in certain alluvial soils near Nottingham. Nova Hedwigia 5:57–78.

——. 1964. Concerning occurrence of phycomycetes in alluvial soils of certain pastures, marshes, and swamps. Nova Hedwigia 8:103–126.

——. 1972. Thermophilous fungi in certain grasslands. Mycopathologia et Mycologia Applicata 48:63–74.

Apinis, A.E., and C.G.C. Chesters. 1964. Ascomycetes of some salt marshes and sand dunes. Transactions of the British Mycological Society 47:419–435.

Apinis, A.E., and G.J.F. Pugh. 1967. Thermophilous fungi of birds' nests. Mycopathologia et Mycologia Applicata 33:1–9.

Aptroot, A. 1991. A monograph of the Pyrenulaceae (excluding *Anthracothecium* and *Pyrenula*) and the Requienellaceae, with notes on the Pleomassariaceae, the Trypetheliaceae and Mycomicrothelia (Lichenized and non-lichenized Ascomycetes). Bibliotheca Lichenologica 44:1–178.

——. 1995. A monograph of *Didymosphaeria*. Studies in Mycology 37:1–160.

——. 1997. Lichen biodiversity in Papua New Guinea, with the report of 173 species on one tree. Bibliotheca Lichenologica 68:203–213.

Aptroot, A., and M.R.D. Seaward. 1999. Annotated checklist of Hong Kong lichens. Tropical Bryology 17:51–101.

Aptroot, A., and H.J.M. Sipman. 1997. Diversity of lichenized fungi in the tropics. Pp. 93–106. *In* K.D. Hyde (ed.), Biodiversity of Tropical Microfungi. Hong Kong University Press, Hong Kong, China.

Aptroot, A., P. Diederich, E. Sérusiaux, and H. Sipman. 1997. Lichens and lichenicolous fungi from New Guinea. Bibliotheca Lichenologica 64:1–220.

Aptroot, A., C.M. van Herk, L.B. Sparrius, and P.P.G. van den Boom. 1999. Checklist van de Nederlandse lichenen en lichenicole fungi. Buxbaumiella 50:4–64.

Archibold, E.E.A. 1949. The specific character of plant communities. II. A quantitative approach. Journal of Ecology 37:274–288.

Arderon, W. 1748. The substance of a letter from Mr. *William Arderon* F.R.S. to Mr. *Henry Baker* F.R.S. Philosophical Transactions 45:321–323.

Armstrong, D.A., D.U. Buchanon, and R.S. Caldwell. 1976. A mycosis caused by *Lagenidium* sp. in laboratory reared larvae of the

Dungeness crab, *Cancer magister*, and possible chemical treatments. Journal of Invertebrate Pathology 28:329–336.

Armstrong, R.A. 1988. Substrate colonization, growth, and competition. Pp. 3–16. *In* M. Galun (ed.), CRC Handbook of Lichenology. Vol. II. CRC Press, Boca Raton, Florida.

Arnebrant, K., E. Bååth, and A. Nordgren. 1987. Copper tolerance of microfungi isolated from polluted and unpolluted forest soil. Mycologia 79:890–895.

Arnebrant, K., E. Bååth, and B.E. Söderström. 1990. Changes in microfungal community structure after fertilization of Scots pine forest soil with ammonium nitrate or urea. Soil Biology and Biochemistry 22:309–312.

Arnold, A.E., Z. Maynard, G.S. Gilbert, P.D. Coley, and A. Thomas. 2000. Are tropical endophytes hyperdiverse? Ecology Letters 3:267–274.

Arnold, G.R.W. von. 1969. Bestimmungsschlüssel für die wichtigsten und häufigsten mykophilen ascomyceten und hyphomycetes. Zeitschrift für Pilzkunde 35:41–45.

——. 1970. Über Sibirina und Sympodiophora, zwei neue Gattungen der Moniliales. Nova Hedwigia 19:299–304.

——. 1976. Internationale Bibliographie der hypomycetaceae (mycophyta, ascomycotina). Bibliographische Mitteilungen der Universitätsbibliothek, Jena No. 25.

——. 1986. Beitrag zur Kenntnis der Pilzflora Cubas. Feddes Repertorium 97:59–65.

——. 1987. Beitrag zur Kenntnis der Pilzflora Kubas. III. Feddes Repertorium 98:351–355.

——. 1988. Beitrag zur Kenntnis der Pilzflora Kubas. IV. Feddes Repertorium 99:27–31.

Arnold, G.R.W. von, and R.F. Castañeda. 1987. Neue Hyphomyzeten–Arten aus Kuba. II. *Verticillium antillanum, Nakataea curvularioides* und *Cladobotryum cubitense*. Feddes Repertorium 98:411–417.

Arnolds, E. 1981. Ecology and Coenology of Macrofungi in Grasslands and Moist Heathlands in Drenthe, The Netherlands. Lubrecht and Cramer, Port Jervis, New York.

——. 1988. The changing macromycete flora in The Netherlands. Transactions of the British Mycological Society 90:391–406.

——. 1991. Mycologists and nature conservation. Pp. 243–264. *In* D.L. Hawksworth (ed.). Frontiers in Mycology. CAB International, Kew, Surrey, United Kingdom.

——. 1992. The analysis and classification of fungal communities with special reference to macrofungi. Pp. 7–47. *In* W. Winterhoff (ed.), Fungi in Vegetation Science. Kluwer Academic, Dordrecht, The Netherlands.

Arnott, H.J. 1995. Calcium oxalate in fungi. Pp. 73–111. *In* S.R. Khan (ed.). Calcium Oxalate in Biological Systems. CRC Press, Boca Raton, Florida.

Arora, D. 1987. Mushrooms Demystified. A Comprehensive Guide to Fleshy Fungi. 2nd ed. 10 Speed Press, Berkeley, California.

Arrhenius, O. 1921. Species and area. Journal of Ecology 9:95–99.

Arrhenius, S.P., and J.H. Langenheim. 1986. The association of Pestalotia species with members of the leguminous tree genera *Hymenaea* and *Copaifera* in the neotropics. Mycologia 78:673–676.

Arthur, J.C. 1934. Manual of the Rusts in United States and Canada. Hafner, New York.

Articus, K., J.-E. Mattsson, L. Tibell, M. Grube, and M. Wedin. 2002. Ribosomal DNA and β-tubulin data do not support the separation of the lichens *Usnea florida* and *U. subfloridana*. Mycological Research 106:412–418.

Aruguete, D.M., J.H. Aldstadt, and G.M. Mueller. 1998. Accumulation of several heavy metals and lanthanides in mushrooms (Agaricales) from the Chicago region. Science of the Total Environment 224:43–56.

Arup, U., S. Ekman, L. Lindblom, and J.E. Mattsson. 1993. High performance thin layer chromatography (HPTLC), an improved technique for screening lichen substances. Lichenologist 25:61–71.

Arvidsson, L. 1976. *Athelia arachnoidea* (Berk.) Jül. and its influence on epiphytic cryptogams in urban areas. Göteborgs Svampklubb Årsskrift 1975–1976:4–10.

——. 1978. Svampangrepp på lavar—en orsak till lavöken. Svensk Botanisk Tidskrift 72:285–292.

Aschner, M., and S. Kohn. 1958. The biology of *Harposporium anguillulae*. Journal of General Microbiology 19:182–189.

Ashida, J. 1965. Adaptation of fungi to metal toxicants. Annual Review of Phytopathology 3:153–174.

Askew, D.J., and M.D. Laing. 1993. An adapted selective medium for the quantitative isolation of *Trichoderma* species. Plant Pathology 42:686–690.

Assigbetse, K.B., D. Fernandez, M.P. Dubois, and J.P. Geiger. 1994. Differentiation of *Fusarium oxysporum* f.sp. *vasinfectum* races on cotton by random amplified polymorphic DNA (RAPD) analysis. Phytopathology 84:622–626.

Atienza, V., and D.L. Hawksworth. 1994. *Minutoexcipula tuckerae* gen. et sp. nov., a new lichenicolous deuteromycete on *Pertusaira texana* in the United States. Mycological Research 98:587–592.

Atlas, R.M., and L.C. Parks. 1993. Handbook of Microbiological Media. CRC Press, Boca Raton, Florida.

Attenborough, D. 1990. Living together. Pp. 163–184. *In* D. Attenborough (ed.), The Trials of Life. Collins, London, England.

Attili, D.S., and R.A.P. Grandi. 1995. Filamentous fungi from soil and leaf litter in the Ecological Reserve of Juréia-Ititans (Atlantic Rain Forest), São Paulo, Brazil. Abstracts of VII International Symposium on Microbial Ecology, Santos. Brazilian Society of Microbiology, São Paulo, Brazil.

Attili, D.S., J.A. Zavatini, and I.M.D.A. Pinto. 1996. Características da Estação ecológica de Juréia-Itantins, Mata Atlántica, São Paulo, Brazil. Geografía 20: in press.

Augspurger, C.K. 1983. Seed dispersal of the tropical tree, *Platypodium elegans*, and the escape of its seedlings from fungal pathogens. Journal of Ecology 71:759–771.

——. 1990. Spatial patterns of damping-off disease during seedling recruitment in tropical forests. Pp.131–144. *In* J.J. Burdon and S.R. Leather (eds.), Pests, Pathogens and Plant Communities. Blackwell Scientific, London, England.

Austin, B., and D.A. Austin. 1987. Bacterial Fish Pathogens: Disease in Farmed and Wild Fish. John Wiley, New York.

Austwick, P.K.C., and I.F. Keymer. 1981. Fungi and actinomycetes. Pp. 193–231. *In* J.E. Cooper and O.F. Jackson (eds.), Diseases of the Reptilia. Vol. 1. Academic Press, London, England.

Avise, J. 1989. Gene trees and organismal histories: a phylogenetic approach to population biology. Evolution 43:1192–1208.

Awamah, M.S., A. Alsheikh, and S. Al-Ghawas. 1979. Mycorrhizal synthesis between *Helianthemum ledifolium*, *H. salicifolium* and four species of the genera *Terfezia* and *Tirmania* using ascospores and mycelial cultures obtained from ascospore germination. P. 23. *In* C.P.P. Reid (ed.), Abstracts of the 4th North American Conference on Mycorrhizae. Fort Collins, Colorado.

Awasthi, D.D. 1988. A key to the macrolichens of India and Nepal. Journal of the Hattori Botanical Laboratory 65:207–302.

——. 1991. A Key to the Microlichens of India, Nepal and Sri Lanka. J. Cramer, Berlin, Germany.

Ayers, T.T. 1935. Parasitism of *Dispira cornuta*. Mycologia 27:235–261.

Ayers, W.A., and P.B. Adams. 1979. Mycoparasitism of sclerotia of *Sclerotinia* and *Sclerotium* species by *Sporidesmium sclerotivorum*. Canadian Journal of Microbiology 25:17–23.

——. 1981. Mycoparasitism of sclerotial fungi by *Teratosperma oligocladum*. Canadian Journal of Microbiology 27:886–892.

——. 1983. Improved media for growth and sporulation of *Sporidesmium sclerotivorum* (new species parasitic on sclerotia of plant-pathogenic *Sclerotinia* spp.). Canadian Journal of Microbiology 29:325–330.

Ayers, W.A., and R.D. Lumsden. 1977. Mycoparasitism of oospores of *Pythium* and *Aphanomyces* species by *Hyphochytrium catenoides*. Canadian Journal of Microbiology 23:38–44.

Ayers, W.A., S.P. Lee, A. Tsuneda, and Y. Hiratsuka. 1980. The isolation, identification and bioassay of the antifungal metabolites produced by *Monocillium nordinii*. Canadian Journal of Microbiology 26:766–773.

Azcón, R. 1993. Growth and nutrition of nodulated mycorrhizal and non-mycorrhizal *Hedysarum coronarium* as a result of treatment with fractions from plant growth-promoting rhizobacteria. Soil Biology and Biochemistry 25:1037–1042.

Azcón, R., R. Rubio, and J.M. Barea. 1991. Selective interactions between different species of mycorrhizal fungi and *Rhizobium meliloti* strains, and their effects on growth, N_2-fixation (^{15}N) and nutrition of *Medicago sativa* L. New Phytologist 117:399–404.

Azema, R.C. 1975. Le gènre *Septobasidium* Pat. Documents Mycologiques 6:1–24.

Bååth, E. 1980. Microfungi in a clear-cut pine forest soil in central Sweden. Canadian Journal of Botany 59:1331–1337.

——. 1988. A critical examination of the soil washing technique with special reference to the effect of the size of the soil particles. Canadian Journal of Botany 66:1566–1569.

Bååth, E., B. Lundgren, and B.E. Söderström. 1984. Fungal populations in podzolic soils experimentally acidified to simulate acid rain. Microbial Ecology 10:197–203.

Bachofen, R. 1986. Microorganisms in extreme environments. Introduction. Experientia 42:1179–1182.

Bacon, C.W. 1990. Isolation, culture, and maintenance of endophytic fungi of grasses. Pp. 259–282. *In* D.P. Labeda (ed.), Isolation of Biotechnological Organisms from Nature. McGraw-Hill, New York.

Bacon, C.W., and D.M. Hinton. 1996. Symptomless endophytic colonization of maize by *Fusarium moniliforme*. Canadian Journal of Botany 74:1195–1202.

Bacon, C.W., and J.F. White, Jr. 1994. Stains, media, and procedures for analyzing endophytes. Pp. 47–56. *In* C.W. Bacon and J.F. White, Jr. (eds.), Biotechnology of Endophytic Fungi of Grasses. CRC Press, Boca Raton, Florida.

Bagyaraj, D.J. 1984. Biological interactions with VA mycorrhizal fungi. Pp. 131–153. *In* C. Ll. Powell and D.J. Bagyaraj (eds.), VA Mycorrhiza. CRC Press, Boca Raton, Florida.

Bahnweg, G., and C.E. Bland. 1980. Comparative physiology and nutrition of *Lagenidium callinectes* and *Haliphthoros milfordensis*, fungal parasites of marine crustaceans. Botanica Marina 23:689–698.

Bahnweg, G., and D. Gotelli. 1980. Physiology and nutrition of *Lagenidium callinectes*, a fungal parasite of the blue crab (*Callinectes sapidus*). Botanica Marina 23:219–225.

Baicu, T., and V. Diaconu. 1975. Mediu selectiv pentru izolarea speciilor de *Alternaria*. Studii şi Cercetǎri de Biologie 27:233–234.

Bailey, A.D., and D.D. Wynn-Williams. 1982. Soil microbiological studies at Signey Island, South Orkney Islands. British Antarctic Survey Bulletin 51:167–191.

Bain, B.Z., and S. Egusa. 1981. Histopathology of black gill disease caused by *Fusarium solani* (Martius) infection in the Kuruma prawn *Panaeus japonicus* Bate. Journal of Fish Diseases 4:195–201.

Baker, J.G. 1975. Annotated checklist of the simpler mycetozoans of North Carolina. Journal of the Elisha Mitchell Scientific Society 90:135–136.

Baker, K.F., and R.J. Cook. 1974. Biological Control of Plant Pathogens. American Phytopathological Society Press, St. Paul, Minnesota.

Baker, R. 1987. Mycoparasitism: ecology and physiology. Canadian Journal of Plant Pathology 9:370–379.

——. 1988. *Trichoderma* spp. as plant-growth stimulants. CRC Critical Reviews in Biotechnology 7:97–106.

——. 1991. Molecular biology in control of fungal pathogens. Pp. 259–271. *In* D. Arora, B. Rai, K.G. Mukerji, and G.R. Knudsen (eds.), Handbook of Applied Mycology. Vol. 1. Marcel Dekker, New York.

Balazy, S. 1993. Entomophthorales. Flora of Poland (Flora Polska) Fungi (Mycota). Polish Academy of Sciences, W. Szafer Institute of Botany, Krakow, Poland.

Baldauf, S.L., and W.F. Doolittle. 1997. Origin and evolution of the slime molds (Mycetozoa) Proceedings of the National Academy of Sciences USA 94:12007–12012.

Baldauf, S.L., and J.D. Palmer. 1993. Animals and fungi are each other's closest relatives: Congruent evidence from multiple proteins. Proceedings of the National Academy of Sciences USA 90: 11558–11562.

Baldauf, S.L., A.J. Roger, I. Wenk-Siefert, and W.F. Doolittle. 2000. A kingdom-level phylogeny of eukaryotes based on combined protein data. Science 290:972–977.

Balows, A., W.J. Hausler, Jr., and H.J. Shadomy. 1985. Manual of Clinical Microbiology. 4th ed. American Society for Microbiology, Washington, D.C.

Bandoni, R.J. 1981. Aquatic hyphomycetes from terrestrial litter. Pp. 693–708. *In* D.T. Wicklow and G.C. Carroll (eds.), The Fungal Community, Its Organization and Role in the Ecosystem. Vol 2. Marcel Dekker, New York.

——. 1984. The Tremellales and Auriculariales: an alternative classification. Transactions of the Mycological Society of Japan 25:489–530.

——. 1995. Dimorphic Heterobasidiomycetes: taxonomy and parasitism. Studies in Mycology 38:13–27.

Bandoni, R.J., and F. Oberwinkler. 1981. *Hyalopycnis blepharistoma*, a pycnidial basidiomycete. Canadian Journal of Botany 59:1613–1620.

Bandoni, R.J., J.D. Parsons, and S.A. Redhead. 1975. Agar "baits" for the collection of aquatic fungi. Mycologia 67:1020–1024.

Banerjee, P., and R.C. Anderson. 1992. Long-term effects of soil fumigation and inorganic nutrient addition on the rhizoplane mycoflora of little bluestem (*Schizachyrium scoparium*). Mycologia 84:843–848.

Banerjee, S.K., and V.N. Iyer. 1995. Quick and easy spreading technique for electron microscopy of DNA. BioTechniques 18:946–947.

Banning, P. van. 1987. Long-term recording of some fish diseases using general fishery research surveys in the south-east part of the North Sea. Diseases of Aquatic Organisms 3:1–11.

Banno, I. 1967. Studies on the sexuality of *Rhodotorula*. Journal of General and Applied Microbiology 13:167–196.

Barak, R., Y. Elad, D. Mirelman, and I. Chet. 1985. Lectins: a possible basis for specific recognition in the interaction of *Trichoderma* and *Sclerotium rolfsii*. Phytopathology 75:458–462.

Baral, H.O. 1984. Taxonomische und ökologische Studien über die Koniferen bewohnenden europäischen Arten der Gattung Lachnellula Karsten. Beiträge zur Kenntnis der Pilze Mitteleuropas 1:143–156.

——. 1987. Lugol's solution/IKI versus Melzer's reagent: hemiamyloidity, a universal feature of the ascus wall. Mycotaxon 29:399–450.

——. 1992. Vital versus herbarium taxonomy: morphological differences between living and dead cells of ascomycetes, and their taxonomic implications. Mycotaxon 44:333–390.

——. 1999. A monograph of *Helicogonium* (= *Myriogonium*, Leotiales), a group of non-ascocarpous intrahymenial mycoparasites. Nova Hedwigia 69:1–71.

Baral, H.O., and G.J. Krieglsteiner. 1985. Bausteine zu einer Askomyzeten-Flora der Bundesrepublik Deutschland: In Süddeutschland gefundene inoperculate Diskomyzeten—mit taxonomischen, ökologischen, chorologischen, Hinweisen und einer Farbtafel. Zeitschrift für Mykologie (Beihefte) 6:1–160.

Barbour, M.G., J.H. Burk, and W.D. Pitts. 1987. Terrestrial Plant Ecology. 2nd ed. Benjamin/Cummings, Menlo Park, California.

Barker, S.M., and H.L. Barnett. 1973. Nitrogen and vitamin requirements for axenic growth of the haustorial mycoparasite *Dispira cornuta*. Mycologia 65:21–27.

Barklund, P., and T. Kowalski. 1996. Endophytic fungi in branches of Norway spruce with particular reference to *Tryblidiopsis pinastri*. Canadian Journal of Botany 74:673–678.

Barkman, J.J., J. Moravec, and S. Rauschert. 1986. Code of phytosociological nomenclature. Vegetatio 67:145–158.

Bärlocher, F. (ed.). 1992. Ecology of Aquatic Hyphomycetes. Springer-Verlag, Berlin, Germany.

Bärlocher, F., and J. Rosset. 1981. Aquatic hyphomycete spora of two Black Forest and two Swiss Jura streams. Transactions of the British Mycological Society 76:351–355.

Barnett, H.L. 1963. The nature of mycoparasitism by fungi. Annual Review of Microbiology 17:1–14.

——. 1964. Mycoparasitism. Mycologia 56:1–19.

——. 1968. The effects of light, pyridoxine and biotin on the development of the mycoparasite, *Gonatobotryum fuscum*. Mycologia 60:244–251.

——. 1970. Nutritional requirements for axenic growth of some haustorial mycoparasites. Mycologia 62:750–760.

Barnett, H.L., and F.L. Binder. 1973. The fungal host-parasite relationship. Annual Review of Phytopathology 11:273–292.

Barnett, H.L., and B.B. Hunter. 1998. Illustrated Genera of Imperfect Fungi. 4th ed., American Phytopathological Society Press, St. Paul, Minnesota.

Barnett, H.L., and V.G. Lilly. 1955. The effects of humidity, temperature and carbon dioxide on sporulation of *Choanephora cucurbitarum*. Mycologia 47:26–29.

——. 1958. Parasitism of *Calcarisporium parasiticum* on species of *Physalospora* and related fungi. Bulletin of the West Virginia University Agricultural Experimental Station 420T:1–37.

——. 1962. A destructive mycoparasite, *Gliocladium roseum*. Mycologia 54:72–77.

Barnett, J.A., R.W. Payne, and D. Yarrow. 2000. Yeasts: Characteristics and Identification. 3rd ed. Cambridge University Press, Cambridge, England.

Barns, S.M., D.J. Lane, M.L. Sogin, C. Bibeau, and W.G. Weisburg. 1991. Evolutionary relationships among pathogenic *Candida* species and relatives. Journal of Bacteriology 173:2250–2255.

Barr, D.J.S. 1980. An outline for the reclassification of the Chytridiales, and for a new order, the Spizellomycetales. Canadian Journal of Botany 58:2380–2394.

——. 1984. The classification of *Spizellomyces, Gaertneriomyces, Triparticalcar,* and *Kochiomyces* (Spizellomycetales, Chytridiomycetes). Canadian Journal of Botany 62:1171–1201.

——. 1986. *Allochytridium expandens* rediscovered: morphology, physiology and zoospore ultrastructure. Mycologia 78:439–448.

——. 1987. Isolation, culture and identification of Chytridiales, Spizellomycetales, and Hyphochytriales. Pp. 118–120. *In* M.S. Fuller and A. Jaworski (eds.), Zoosporic Fungi in Teaching and Research. Southeastern Publishing, Athens, Georgia.

——. 1990. Phylum Chytridiomycota. Pp.454–466. *In* L. Margulis, J.O. Corliss, M. Melkonian, and D. Chapman (eds.), Handbook of Protoctista. Jones and Bartlett, Boston, Massachusetts.

——. 1992. Evolution and kingdoms of organisms from the perspective of a mycologist. Mycologia 84:1–11.

——. 2001. Chytridiomycota. Pp. 93–112. *In* D.J. McLaughlin, E.G. McLaughlin, and P.A. Lempke (eds.), The Mycota. Vol. VII. Systematics and Evolution. Part A. Springer-Verlag, Berlin, Germany.

Barr, D.J.S., and P.M.E. Allan. 1985. A comparison of the flagellar apparatus of *Phytophthora, Saprolegnia, Thraustochytrium,* and *Rhizidiomyces*. Canadian Journal of Botany 63:138–154.

Barr, D.J.S., and C.E. Babcock. 1994. Cryopreservation of unicellular, zoosporic fungi, a simple method. U.S. Federal Culture and Collections Newsletter 24:6.

Barr, D.J.S., and R. Bandoni. 1979. A new species of *Rozella* on a basidiomycete. Mycologia 71:1261–1264.

Barr, D.J.S., and N.L. Désaulniers. 1987. *Allochytridium luteum* n. sp.: morphology, physiology and zoospore ultrastructure. Mycologia 79:193–199.

——. 1988. Precise configuration of organisms of the chytrid zoospore. Canadian Journal Botany 66:869–876.

Barr, D.J.S., N.L. Désaulniers, and J.S. Knox. 1987. *Catenochytridium hemicysti* n. sp.: morphology, physiology and zoospore ultrastructure. Mycologia 79:587–594.

Barr, D.J.S., H. Kudo, K.D. Jakober, and K-J. Cheng. 1989. Morphology and development of rumen fungi: *Neocallimastix* sp., *Piromyces communis,* and *Orpinomyces bovis* gen. nov., sp. nov. Canadian Journal of Botany 67:2815–2824.

Barr, D.J.S., L.J. Yanke, H.D. Bae, T.A. McAllister, and K-J. Cheng. 1995. Contributions on the morphology and taxonomy of some rumen fungi from Canada. Mycotaxon 54:203–214.

Barr, M.E. 1972. Preliminary studies on the Dothideales in temperate North America. Contributions to the University of Michigan Herbarium 9:523–638.

——. 1978. The Diaporthales in North America, with emphasis on *Gnomonia* and its segregates. Mycologia Memoirs 7:1–232.

——. 1979. A classification of Loculoascomycetes. Mycologia 71: 935–957.

——. 1980. On the family Tubeufiaceae (Pleosporales). Mycotaxon 12:137–167.

——. 1987. Prodromus to class Loculoascomycetes. Published by the author, Amherst, Massachusetts.

——. 1990a. Prodromus to non-lichenized, pyrenomycetous members of class Hymenoascomycetes. Mycotaxon 39:43–184.

——. 1990b. Melanommatales (Loculoascomycetes). North American Flora II 13:1–129.

——. 1990c. Some dictyosporous genera and species of Pleosporales in North America. Memoirs of the New York Botanical Gardens 62:1–92.

Barr, M.E., and S.M. Huhndorf. 2001. Loculoascomycetes. Pp. 283–305. *In* D.J. McLaughlin, E.G. McLaughlin, and P.A. Lempke (eds.), The Mycota. Vol. VII. Systematics and Evolution. Part A. Springer-Verlag, Berlin, Germany.

Barras, S.J. 1970. Antagonism between *Dendroctonus frontalis* and the fungus *Ceratocystis minor*. Annals of the Entomological Society of America 63:1187–1190.

——. 1972. Improved White's solution for surface sterilization of pupae of *Dendroctonus frontalis*. Journal of Economic Entomology 65:1504.

Barras, S.J., and T.J. Perry. 1972. Fungal symbionts in the prothoracic mycangium of *Dendroctonus frontalis* (Coleopt., Scolytidae). Zeitschrift für Angewandte Entomologie 71:95–104.

——. 1975. Interrelationships among microorganisms, bark or ambrosia beetles, and woody host tissue: An annotated bibliography 1965–1974. USDA Forest Service, General Technical Report 5–10.

Barrett, J. 1987. Molecular variation and evolution. Pp. 83–95. *In* A.D.M. Rayner, C.M. Brasier, and D. Moore (eds.), Evolutionary Biology of the Fungi. Cambridge University Press, Cambridge, England.

Barron, G.L. 1962. New species and new records of *Oidiodendron*. Canadian Journal of Botany 40:589–607.

——. 1967. New species and new records of fungi that attack microscopic animals. Canadian Journal of Botany 67:267–271.

——. 1968. The Genera of Hyphomycetes from Soil. Williams and Wilkins, Baltimore, Maryland.

——. 1971. Soil fungi. Pp. 405–427. *In* C. Booth (ed.), Methods in Microbiology. Vol. 4. Academic Press, London, England.

——. 1977. The Nematode-Destroying Fungi. Canadian Biological Publications, Guelph, Ontario, Canada.

——. 1980a. Fungal parasites of rotifers: two new verticillate endoparasites with aerial conidiophores. Canadian Journal of Botany 58:432–438.

——. 1980b. Fungal parasites of rotifers: a new *Tolypocladium* with underwater conidiation. Canadian Journal of Botany 58:439–442.

——. 1980c. Fungal parasites of rotifers: a new genus of Hyphomycetes endoparasitic on *Adineta* [*Pseudomeria mucosa*]. Canadian Journal of Botany 58:443–446.

——. 1980d. A new genus of the Zygomycetes. Canadian Journal of Botany 58:2450–2453.

——. 1981a. Parasites and predators of microscopic animals. Pp. 167–199. *In* G. Cole and W.B. Kendrick (eds.), Biology of Conidial Fungi. Vol. 2. Academic Press, New York.

——. 1981b. Two new fungal parasites of bdelloid rotifers. Canadian Journal of Botany 59:1449–1455.

——. 1983. Structure and biology of a new *Tolypocladium* attacking bdelloid rotifers. Canadian Journal of Botany 61:2566–2569.

——. 1985a. Fungal parasites of bdelloid rotifers: *Diheterospora*. Canadian Journal of Botany 63:211–222.

——. 1985b. A new *Gonimochaete* with an oospore state. Mycologia 77:17–23.

——. 1986. A new *Harposporium* parasitic in bdelloid rotifers. Canadian Journal of Botany 64:2379–2382.

——. 1991a. A new genus of the Hyphomycetes endoparasitic in bdelloid rotifers with conidia that lodge in the mastax. Canadian Journal of Botany 69:503–506.

——. 1991b. A new species of *Dwayaangam* parasitic on eggs of rotifers and nematodes. Canadian Journal of Botany 69:1402–1406.

——. 1992. Lignolytic and cellulolytic fungi as predators and parasites. Pp. 311–326. *In* G.C. Carroll and D.T. Wicklow (eds.), The Fungal Community: Its Organization and Role in the Ecosystem. 2nd ed. Marcel Dekker, New York.

Barron, G.L., and J.T. Fletcher. 1970. *Verticillium albo-atrum* and *V. dahliae* as mycoparasites. Canadian Journal of Botany 48:1137–1139.

——. 1972. *Rhopalomyces elegans* Corda, a host of *Mycogone perniciosa*. Mushroom Science 8:383–386.

Barron, G.L., and E. Szijarto. 1986. A new species of *Olpidium* parasitic in nematode eggs. Mycologia 78:972–975.

——. 1990. A new genus of the Hyphomycetes endoparasitic in rotifers. Mycologia 82:134–137.

Barron, G.L., and S.S. Tzean. 1981. A subcuticular endoparasite impaling bdelloid rotifers using three-pronged spores. Canadian Journal of Botany 59:1207–1212.

Barron, G.L., C. Morikawa, and M. Saikawa. 1990. New *Cephaliophora* species capturing rotifers and tardigrades. Canadian Journal of Botany 68:685–690.

Bartnicki-Garcia, S. 1970. Cell wall composition and other biochemical markers in fungal phylogeny. Pp. 81–103. *In* J.B. Harborne (ed.), Phytochemical Phylogeny. Academic Press, London, England.

——. 1978. *Mucor rouxii*: mold-yeast dimorphism. Pp. 146–148. *In* M.S. Fuller (ed.), Lower Fungi in the Laboratory. Palfrey Contributions in Botany no.1. Department of Botany, University of Georgia, Athens, Georgia.

Bartnicki-Garcia, S., and M.C. Wang. 1983. Biochemical aspects of morphogenesis in *Phytophthora*. Pp. 121–137. *In* D.C. Erwin, S. Bartnicki-Garcia, and P.H. Tsao (eds.), Phytophthora, Its Biology, Ecology and Taxonomy. American Phytopathological Society Press, St. Paul, Minnesota.

Bas, C., T.W. Kuyper, M.E. Noordeloos, and E.C. Vellinga (eds.). 1988–1995. Flora Agaricina Neerlandica: Critical Monographs on Families of Agarics and Boleti Occurring in the Netherlands. Vol. 1–3. Rotterdam, The Netherlands.

Bassett, E.N., and P. Fenn. 1984. Latent colonization and pathogenicity of *Hypoxylon atropunctatum* on oaks. Plant Disease 68:317–319.

Bastos, C.N. 1996. Mycoparasitic nature of the antagonism between *Trichoderma viride* and *Crinipellis perniciosa*. Fitopatologia Brasileira 21:50–54.

Bastos, C.N., H.C. Evans, and R.A. Samson. 1982. A new hyperparasitic fungus, *Cladobotryum amazonense*, with potential for control of fungal pathogens of cocoa. Transactions of the British Mycological Society 77:273–278.

Bastos, C.N., S.J. Neill, and R. Horgan. 1986. A metabolite from *Cladobotryum amazonense* with antibiotic activity. Transactions of the British Mycological Society 86:571–578.

Bates, J.W., and A.M. Farmer (eds.). 1992. Bryophytes and Lichens in a Changing Environment. Clarendon Press, Oxford, United Kingdom.

Batista, A.C. 1959. Monografia dos Fungos Micropeltaceae. Publicações Instituto de Micologia da Universidade do Recife. 56:1–519.

Batista, A.C., and R. Ciferri. 1963. Capnodiales. Saccardoa, Monographie Mycologicae 2:1–298.

Batko, A. 1975. Zarys hydromikologii. Paßstwowe Wydawnictwo Naukowe. Warsaw, Poland.

Batra, L.R. 1967. Ambrosia fungi: a taxonomic revision, and nutritional studies of some species. Mycologia 59:976–1017.

——. (ed.). 1979. Insect-Fungus Symbiosis, Nutrition, Mutualism, and Commensalism. Allanheld, Osmun, Montclair, New Jersey.

——. 1985. Ambrosia beetles and their associated fungi: research trends and techniques. Proceedings of the Indian Academy of Sciences 94:137–148.

——. 1987. Insect-associated, filamentous endomycetes: their growth and strategies for survival. Pp. 415–428. *In* G.S. de Hoog, M. Th. Smith, and A.C.M. Weijman (eds.), The Expanding Realm of Yeast-like Fungi. Centraalbureau voor Schimmelcultures, Baarn and Elsevier Science, Amsterdam, The Netherlands.

——. 1991. World species of *Monilinia* (Fungi): their ecology, biosystematics and control. Mycologia Memoir No.16.

Batra, L.R., and S.W.T. Batra. 1977. Termite-fungus mutualism. Pp. 117–163. *In* L.R. Batra (ed.), Insect-Fungus Symbiosis, Nutrition, Mutualism, and Commensalism. Allanheld, Osmun, Montclair, New Jersey.

——. 1979. Termite-fungus mutualism. Pp. 117–163. *In* L.R. Batra (ed.), Insect-Fungus Symbiosis. Nutrition, Mutualism, and Commensalism. Allanheld, Osmun, Montclair, New Jersey.

Bauchop, T. 1979a. Rumen anaerobic fungi of cattle and sheep. Applied and Environmental Microbiology 38:148–158.

——. 1979b. The rumen anaerobic fungi: Colonizers of plant fibre. Annales de Recherches Vétérinaires 10:246–248.

——. 1981. The anaerobic fungi in rumen fiber digestion. Agriculture and Environment 6:339–348.

——. 1983. The gut anaerobic fungi: colonizers of dietary fiber. Pp. 143–148. *In* G. Wallace and L. Bell (eds.), Fiber in Human and Animal Nutrition. Royal Society of New Zealand, Wellington, New Zealand.

——. 1986. Rumen anaerobic fungi and the utilization of fibrous feeds. Reviews in Rural Science 6:118–123.

——. 1989. Biology of gut anaerobic fungi. BioSystems 23:53–64.

Bauchop, T., and D.O. Mountfort. 1981. Cellulose fermentation by a rumen anaerobic fungus in both the absence and presence of rumen methanogens. Applied and Environmental Microbiology 42:1103–1110.

Bauer, R., and F. Oberwinkler. 1990a. Direct cytoplasm-cytoplasm connection: an unusual host-parasite interaction of the tremelloid mycoparasite *Tetragoniomyces uliginosus*. Protoplasma 154:157–160.

——. 1990b. Haustoria of the mycoparasitic heterobasidiomycete *Christiansenia pallida*. Cytologia 55:419–424.

——. 1991. The colacosomes: new structures at the host-parasite interface of a mycoparasitic basidiomycete. Botanica Acta 104:53–57.

Bauer, R., F. Oberwinkler, M. Piepenbring, and M.L. Berbee. 2001. Basidiomycota: Ustilaginales. Pp. 57–83. *In* D.J. McLaughlin, E.G. McLaughlin, and P.A. Lempke (eds.), The Mycota. Vol. VII. Systematics and Evolution. Part B. Springer-Verlag, Berlin, Germany.

Bauer, R., F. Oberwinkler, and K. Vánky. 1997. Ultrastructural markers and systematics in smut fungi and allied taxa. Canadian Journal Botany 75:1273–1314.

Baumgardner, D.J., and D.P. Paretsky. 1999. The *in vitro* isolation of *Blastomyces dermatitidis* from a woodpile in north central Wisconsin, USA. Medical Mycology 37:163–168.

Baxter, A.P., and G.C.A. Westhuizen. 1984. A synoptic key to South African isolates of *Colletotrichum*. South African Journal of Botany 3:265–266.

Beakes, G. 1983. A comparative account of cyst coat ontogeny in saprophytic and fish-lesion (pathogenic) isolates of the *Saprolegnia diclina-parasitica* complex. Canadian Journal of Botany 61:603–622.

——. 1998. Relationships between lower fungi and protozoa. Pp. 351–373. *In* G.H. Coombs, K. Vickerman, M.A. Sleigh, and A. Warren (eds.), Evolutionary Relationships Among Protozoa. Chapman Hall, London, England.

Beakes, G., and H. Ford. 1983. Esterase isoenzyme variation in the genus *Saprolegnia*, with particular reference to the fish pathogenic *S. diclina-parasitica* complex. Journal of General Microbiology 129:2605–2619.

Beakes, G.W., G.M. Campos-Takaki, and G. Takaki. 1984. The effect of glucose concentration and light on thallospore ultrastructure in *Ellisomyces anomalus* (Thamnidiaceae, Mucorales). Canadian Journal of Botany 62:2677–2687.

Beare, M.H., D.C. Coleman, D.A. Crossley, Jr., P.F. Hendrix, and E.P. Odum. 1995. A hierarchical approach to evaluating the significance of soil biodiversity to biogeochemical cycling. Plant and Soil 170:5–22.

Beaton, G., and G. Weste. 1982. A new species of *Scleroderma* from Victoria, Australia. Transactions of the British Mycological Society 79:41–43.

Beaver, R.A. 1989. Insect-fungus relationships in the bark and ambrosia beetles. Pp. 121–143. *In* N. Wilding, N.M. Collins, P.M. Hammond, and J.F. Webber (eds.), Insect-Fungus Interactions. Academic Press, New York.

Bécard, G., and Y. Piché. 1989. New aspects on the acquisition of biotrophic status by a vesicular-arbuscular mycorrhizal fungus, *Gigaspora margarita*. New Phytologist 112:77–83.

——. 1994. Establishment of vesicular-arbuscular mycorrhiza in root organ culture: review and proposed methodology. Pp. 549–568. *In* J.R. Norris, D. Read, and A.K. Varma (eds.), Techniques for Mycorrhizal Research. Academic Press, London, England.

Beiswenger, J.M., and M. Christensen. 1989. Fungi as indicators of past environments. Current Research in the Pleistocene 6:54–56.

Bell, A. 1983. Dung Fungi: An Illustrated Guide to Coprophilous Fungi in New Zealand. Victoria University Press, Wellington, New Zealand.

Bell, A.A., and M.H. Wheeler. 1986. Biosynthesis and functions of fungal melanins. Annual Review of Phytopathology 24:411–451.

Belloch, C., A. Querol, M.D. García, and E. Barrio. 2000. Phylogeny of the genus *Kluyveromyces* inferred from the mitochondrial cytochrome-c oxidase II gene. International Journal of Systematic and Evolutionary Microbiology 50:405–416.

Benjamin, C.R., and C.W. Hesseltine. 1959. Studies on the genus *Phycomyces*. Mycologia 51:751–771.

Benjamin, P.K., R.C. Anderson, and A.E. Liberta. 1988. Vesicular-arbuscular mycorrhizal ecology of little bluestem across a prairie-forest gradient. Canadian Journal of Botany 67:2678–2685.

Benjamin, R.K. 1958. Sexuality in the Kickxellaceae. Aliso 4:149–169.

——. 1959. The merosporangiferous Mucorales. Aliso 4:321–433.

——. 1960. Two new members of the Mucorales. Aliso 4:523–530.

——. 1961. Addenda to "the merosporangiferous Mucorales." Aliso 5:11–19.

——. 1962. A new *Basidiobolu*s that forms microspores. Aliso 5:223–233.

——. 1963. Addenda to "the merosporangiferous Mucorales." II. Aliso 5:273–288.

——. 1965. Addenda to "the merosporangiferous Mucorales." III. *Dimargaris*. Aliso 6:1–10.

——. 1966. The merosporangium. Mycologia 58:1–42.

——. 1971. Introduction and Supplement to Roland Thaxter's Contribution Towards a Monograph of the Laboulbeniaceae. J. Cramer, Lehre, Germany.

——. 1979. Zygomycetes and their spores. Pp. 573–616. *In* W.B. Kendrick (ed.), The Whole Fungus, The Sexual-Asexual Synthesis. Vol. 1. National Museum of Natural Sciences, Ottawa, Canada.

——. 1985. A novel new species of *Syncephalis* (Zoopagales: Piptocephalidaceae) from California that forms hypogenous merosporangia. Aliso 11:1–15.

——. 1986. Laboulbeniales on semiaquatic Hemiptera. V. *Triceromyces*: with a description of monoecious-dioecious dimorphism in the genus. Aliso 11:245–278.

——. 1989. Taxonomy and morphology of *Aporomyces* (Laboulbeniales). Aliso 12:335–367.

——. 1992a. *Cupulomyces*, a new genus of Laboulbeniales (Ascomycetes) based on *Stigmatomyces lasiochili*. Aliso 13:355–364.

——. 1992b. A new genus of Laboulbeniales (Ascomycetes) on a species of *Phalacrichus* (Coleoptera: Dryopoidea; Limnichidae), with a note on mirror-image asymmetry in the order. Aliso 13:427–446.

——. 1995. *Corylophomyces*, a new dioecious genus of Laboulbeniales on Corylophidae (Coleoptera). Aliso 14:41–57.

——. 1999. Laboulbeniales on semiaquatic Heteroptera. VIII. *Monandromyces*, a new genus based on *Autophagomyces microveliae* (Laboulbeniales). Aliso 18:71–91.

Benjamin, R.K., and B.S. Mehrotra. 1963. Obligate azygospore formation in two species of *Mucor* (Mucorales). Aliso 5:235–245.

Benny, G.L. 1972. Histochemistry of the cell wall and septum of vegetative cells of selected species of Dimargaritaceae and Kickxellaceae. Mycologia 64:854–862.

——. 1982. Zygomycetes. Pp. 184–195. *In* S.P. Parker (ed.), Synopsis and Classification of Living Organisms. Vol. I. McGraw-Hill, New York.

——. 1992. Observations on Thamnidiaceae (Mucorales). V. *Thamnidium*. Mycologia 84:834–842.

——. 1995a. Observations on Thamnidiaceae. VII. *Helicostylum* and a new genus, *Kirkia*. Mycologia 87:253–264.

——. 1995b. Classical morphology in zygomycete taxonomy. Canadian Journal of Botany 73 (1): S725-S730.

Benny, G.L., and R.K. Benjamin. 1975. Observations on Thamnidiaceae (Mucorales). New taxa, new combinations, and notes on selected species. Aliso 8:301–351.

——. 1976. Observations on Thamnidiaceae (Mucorales). II. *Chaetocladium, Cokeromyces, Mycotypha,* and *Phascolomyces.* Aliso 8:391–424.

——. 1993. Observations on Thamnidiaceae (Mucorales). VI. Two new species of *Dichotomocladium* and the zygospores of *D. hesseltinei* (Chaetocladiaceae). Mycologia 85:660–671.

Benny, G.L., and K. O'Donnell. 2000. *Amoebidium parasiticum* is a protozoan, not a trichomycete. Mycologia 92:1133–1137.

Benny, G.L., and M.A.A. Schipper. 1992. Observations on Thamnidiaceae (Mucorales). IV. *Pirella.* Mycologia 84:52–63.

Benny, G.L., R.A. Humber, and J.B. Morton. 2001. Zygomycota: Zygomycetes. Pp. 113–146. *In* D.J. McLaughlin, E.G. McLaughlin, and P.A. Lempke (eds.), The Mycota. Vol. VII. Systematics and Evolution. Part A. Springer-Verlag, Berlin, Germany.

Benny, G.L., P.M. Kirk, and R.A. Samson. 1985. Observations on Thamnidiaceae (Mucorales). III. Mycotyphaceae fam. nov. and a re-evaluation of *Mycotypha* sensu Benny & Benjamin illustrated by two new species. Mycotaxon 22:119–148.

Bentivenga, S.P., and B.A.D. Hetrick. 1992. Seasonal and temperature effects on mycorrhizal activity and dependence of cool- and warm-season tallgrass prairie grasses. Canadian Journal of Botany 70:1596–1602.

Bentivenga, S.P., and J.B. Morton. 1995. A monograph of the genus *Gigaspora,* incorporating developmental patterns of morphological characters. Mycologia 87:719–731.

Benyagoub, M., S.H. Jabaji-Hare, G. Banville, and P.M. Charest. 1994. *Stachybotrys elegans:* a destructive mycoparasite of *Rhizoctonia solani.* Mycological Research 98:493–505.

Benyagoub, M., S.H. Jabaji-Hare, H. Chamberland, and P.M. Charest. 1996. Cytochemical and immunocytochemical investigation of the mycoparasitic interaction between *Stachybotrys elegans* and its host *Rhizoctonia solani* (AG-3). Mycological Research 100:79–86.

Benzing, D.H. 1983. Vascular epiphytes: a survey with special reference to their interactions with other organisms. Pp. 11–24. *In* S.L. Sutton, T.C. Whitmore, and A.C. Chadwick (eds.), Tropical Rain Forest: Ecology and Management. Blackwell Scientific, Oxford, England.

Berbee, M.L. 1996. Loculoascomycete origins and evolution of filamentous ascomycete morphology based on 18S rRNA gene sequence data. Molecular Biology and Evolution 13:462–470.

Berbee, M.L., and J.W. Taylor. 1992a. Convergence in ascospore discharge mechanism among pyrenomycetous fungi based on 18S ribosomal RNA gene sequence. Molecular Phylogenetics and Evolution 1:59–71.

——. 1992b. Two ascomycete classes based on fruiting-body characters and ribosomal DNA sequence. Molecular Biology and Evolution 9:278–284.

——. 1993a. Dating the evolutionary radiations of the true fungi. Canadian Journal of Botany 71:1114–1127.

——. 1993b. Ascomycete relationships: dating the origin of asexual lineages with 18S ribosomal RNA gene sequence data. Pp. 67–78. *In* D.R. Reynolds and J.W. Taylor (eds.), The Fungal Holomorph: Mitotic, Meiotic and Pleomorphic Speciation in Fungal Systematics. CAB International, Wallingford, Oxon, United Kingdom.

——. 2001. Fungal molecular evolution: gene trees and geologic time. Pp. 229–245. *In* D.J. McLaughlin, E.G. McLaughlin, and P.A. Lempke (eds.), The Mycota. Vol. VII. Systematics and Evolution. Part B. Springer-Verlag, Berlin, Germany.

Berbee, M.L., D.A. Carmean, and K. Winka. 2000. Ribosomal DNA and resolution of branching order among the ascomycota: how many nucleotides are enough? Molecular Phylogenetics and Evolution 17:337–344.

Berch, S.M., and M.A. Castellano. 1986. Sporulation of *Endogone pisiformis* in axenic and monoxenic culture. Mycologia 78:292–295.

Bergero, R., M. Girlanda, G.C. Varese, D. Intili, and A.M. Luppi. 1999. Psychrooligotrophic fungi from Arctic soils of Franz Joseph Land. Polar Biology 21:361–368.

Bernalier, A., G. Fonty, and P. Gouet. 1991. Cellulose degradation by two rumen anaerobic fungi in monoculture or in coculture with rumen bacteria. Animal Feed Science and Technology 32:131–136.

Bernstein, M., H. Howard, and G. Carroll. 1973. Fluorescence microscopy of Douglas fir foliage epiflora. Canadian Journal of Botany 19:1129–1130.

Berny, J.-F., and G.L. Hennebert. 1991. Viability and stability of yeast cells and filamentous fungus spores during freeze-drying: effects of protectants and cooling rates. Mycologia 83:805–815.

Berry, C.R. 1959. Factors affecting parasitism of *Piptocephalis virginiana* on other Mucorales. Mycologia 51:824–832.

Berry, L.A., and J.W. Deacon. 1992. Video-analysis of *Gliocladium roseum* in relation to mechanism of antagonism of plant pathogens. International Organization of Biological Control/WPRS Bulletin 15:64–66.

Berry, L.A., E.E. Jones, and J.W. Deacon. 1993. Interaction of the mycoparasite *Pythium oligandrum* with other *Pythium* species. Biocontrol Science and Technology 3:247–260.

Besl, H. von, W. Helfer, and N. Luschka. 1989. Basidiomyceten auf alten Porlingsfruchtkörpern. Berichte der Bayerischen Botanischen Gesellschaft 60:133–145.

Bessette, A.E., A.R. Bessette, and D.W. Fischer. 1997. Mushrooms of Northeastern North America. Syracuse University Press, Syracuse, New York.

Bessette, A.E., W.C. Roody, and A.R. Bessette. 2000. North American Boletes. A Color Guide to the Fleshy Pored Mushrooms. Syracuse University Press, Syracuse, New York.

Best, S., and F. Spiegel. 1984. Protostelids and other simple mycetozoans of Hueston Woods State Park and Nature Preserve. Pp. 116–121. *In* G.B. Willeke (ed.), Hueston Woods State Park and Nature Preserve, Proceedings of a Symposium April 16–18, 1982. Miami University, Oxford, Ohio.

Bethlenfalvay, G.J., S. Dakessian, and R.S. Pacovsky. 1984. Mycorrhizae in a southern California desert: ecological implications. Canadian Journal of Botany 62:519–524.

Bettucci, L., and M.-F. Roquebert. 1995. Microfungi from a tropical rain forest litter and soil: a preliminary study. Nova Hedwigia 61:111–118.

Bettucci, L., and M. Saravay. 1993. Endophytic fungi of *Eucalyptus globulus:* a preliminary study. Mycological Research 97:679–682.

Bever, J.D., J.B. Morton, J. Antonovics, and P.A. Schultz. 1995. Host specificity and diversity of arbuscular fungi in Glomales: an experimental approach using an old field soil. Journal of Ecology 84:71–82.

Bharat, R., R.S. Upadhyay, and A.K. Srivastava. 1988. Utilization of cellulose and gallic acid by litter inhabiting fungi and its possible implication in litter decomposition of a tropical deciduous forest. Pedobiologia 32:157–165.

Bhattacharjee, M., K.G. Mukerji, J.P. Tewari, and W.P. Skoropad. 1982. Structure and hyperparasitism of a new species of *Gigaspora.* Transactions of the British Mycological Society 78:184–188.

Bhattacharya, D., L. Medlin, P.O. Wainright, E.V. Ariztia, C. Bibeau, S.K. Stickel, and M.L. Sogin. 1992. Algae containing chlorophylls *a + c* are paraphyletic: molecular evolutionary analysis of the Chromophyta. Evolution 46:1801–1817.

Bicknell, J.N., and H.C. Douglas. 1970. Nucleic acid homologies among species of *Saccharomyces.* Journal of Bacteriology 101:505–512.

Bidartondo, M.J., A.M. Kretzer, E.M. Pine, and T.D. Bruns. 2000. High root concentration and uneven ectomycorrhizal diversity near *Sarcodes sanguinnea* (Ericaceae): a cheater that stimulates its victims? American Journal of Botany 87:1783–1788.

Biermann, B., and R.G. Linderman. 1983. Use of vesicular-arbuscular mycorrhizal roots, intraradical vesicles and extraradical vesicles as inoculum. New Phytologist 95:97–105.

Bills, G.F. 1995. Analyses of microfungal diversity from a user's perspective. Canadian Journal of Botany 73(Suppl. 1):S33–S41.

——. 1996. Isolation and analysis of endophytic fungal communities from woody plants. Pp. 31–65. *In* S.C. Redlin and L.M. Carris (eds.), Endophytic Fungi in Grasses and Woody Plants, Systematics, Ecology and Evolution. American Phytopathological Society Press, St. Paul, Minnesota.

Bills, G.F., and F. Peláez. 1996. Endophytic isolates of *Creosphaeria sassafras.* Mycotaxon 57:471–477.

Bills, G.F., and J.D. Polishook. 1991. Microfungi from *Carpinus caroliana.* Canadian Journal of Botany 69:1477–1482.

——. 1992. Recovery of endophytic fungi from *Chamaecyparis thyoides.* Sydowia 44:1–12.

——. 1993. Selective isolation of fungi from dung of *Odocoileus hemionus* (mule deer). Nova Hedwigia 57:195–206.

——. 1994. Abundance and diversity of microfungi in leaf litter of a lowland rain forest in Costa Rica. Mycologia 86:187–198.

Bills, G.F., G.I. Holtzman, and O.K. Miller, Jr. 1986. Comparison of ectomycorrhizal-basidiomycete communities in red spruce versus northern hardwood forests of West Virginia. Canadian Journal of Botany 64:760–768.

Binion, D., M. Jenkins, N.D. Missel, and S.L. Stephenson. 1997. Corticolous myxomycetes of the Luquillo Experimental Forest, Puerto Rico. Proceedings of the West Virginia Academy of Science 69:23.

Bisby, F.A. 1994. Plant Names in Botanical Databases. Plant Taxonomic Database Standards No. 3. Hunt Institute for Botanical Documentation, Pittsburgh, Pennsylvania.

Bisby, G.R., M.I. Timonin, and N. James. 1935. Fungi isolated from soil profiles in Manitoba. Canadian Journal of Research 13 (Ser. C):47–65.

Bissegger, M., and T.N. Sieber. 1994. Assemblages of endophytic fungi in coppice shoots of *Castanea sativa.* Mycologia 86:648–655.

Bissett, J. 1991a. A revision of the genus *Trichoderma.* II. Infrageneric classification. Canadian Journal of Botany 69:2357–2372.

——. 1991b. A revision of the genus *Trichoderma.* III. Section *Pachybasium.* Canadian Journal of Botany 69:2373–2417.

Bissett, J., and A. Borkent. 1988. Ambrosia galls: the significance of fungal nutrition in the evolution of the Cecidomyiidae (Diptera). Pp. 203–225. *In* K.A. Pirozynski and D.L. Hawksworth (eds.), Coevolution of Fungi with Plants and Animals. Academic Press, London, England.

Bissett, J., and D. Parkinson. 1979a. The distribution of fungi in some alpine soils. Canadian Journal of Botany 57:1609–1629.

——. 1979b. Functional relationships between soil fungi and environment in alpine tundra. Canadian Journal of Botany 57:1642–1659.

——. 1979c. Fungal community structure in some alpine soils. Canadian Journal of Botany 57:1630–1641.

——. 1980. Long-term effects of fire on the composition and activity of the soil microflora of a subalpine coniferous forest. Canadian Journal of Botany 58:1704–1721.

Bissett, J., and P. Widden. 1972. An automatic, multichamber soil-washing apparatus for removing fungal spores from soil. Canadian Journal of Microbiology 18:1399–1404.

Black, C.A., D.D. Evans, J.L. White, L.E. Ensminger, and F.E. Clark (eds.). 1965. Methods of Soil Analysis. Part 1. Physical and Mineralogical Properties, Including Statistics of Measurement and Sampling. American Society of Agronomy, Madison, Wisconsin.

Blackwell, M. 1984. Myxomycetes and their arthropod associates. Pp. 67–90. *In* Q. Wheeler and M. Blackwell (eds.), Fungus-Insect Rela-

tionships, Perspectives in Ecology and Evolution. Columbia University Press, New York.

——. 1994. Minute mycological mysteries: the influence of arthropods on the lives of fungi. Mycologia 86:1–17.

——. 2000. Terrestrial life—fungal from the start? Science 289:1884–1885.

Blackwell, M., and R.L. Chapman. 1993. Collection and storage of fungal and algal samples. Methods in Enzymology 224:65–77.

Blackwell, M., and R.L. Gilbertson. 1980. Sonoran Desert myxomycetes. Mycotaxon 11:139–149.

Blackwell, M., and J.W. Kimbrough. 1976a. A developmental study of the termite-associated fungus *Coreomycetopsis oedipus.* Mycologia 68:551–558.

——. 1976b. Ultrastructure of the termite-associated fungus *Laboulbeniopsis termitarius.* Mycologia 68:541–550.

Blackwell, M., and D. Malloch. 1989. *Pyxidiophora*: life histories and arthropod associations of two species. Canadian Journal of Botany 67:2552–2562.

Blackwell, M., and W. Rossi. 1986. Biogeography of fungal ectoparasites of termites. Mycotaxon 25:581–601.

Blackwell, M., J.R. Bridges, J.C. Moser, and T.J. Perry. 1986. Hyperphoretic dispersal of a *Pyxidiophora* anamorph. Science 232:993–995.

Blackwell, W.H., and M.J. Powell. 1991. A nomenclatural synopsis of *Septosperma* (Chytridiomycetes) and emendation of *S. rhizophydii.* Mycotaxon 42:43–52.

Bland, C.E., J.N. Couch, and S.Y. Newell. 1981. Identification of *Coelomomyces, Saprolegniales* and Lagenidiales. Pp. 129–162. *In* H.D. Burges (ed.), Microbial Control of Pests and Plant Diseases 1970–1980. Academic Press, London, England.

Blanton, R.L. 1990. Phylum Acrasea. Pp. 75–87. *In* L. Margulis, J.O. Corliss, M. Melkonian, and D. Chapman (eds.), Handbook of Protoctista. Jones and Bartlett, Boston, Massachusetts.

Blanz, P. 1978. Uber die systematische Stellung der Exobasidiales. Zeitschrift für Mykologie 44:91–107.

Blanz, P.A., and M.G. Unseld. 1987. Ribosomal RNA as a taxonomic tool in mycology. Studies in Mycology 30:247–258.

Blanz, P.A., M.G. Unseld, and I. Rauh. 1989. Group-specific differences in the secondary structure of the 28S ribosomal RNA of yeasts. Yeast 5 (Special issue):S399–S404.

Blazer, V.S., and R.E. Wolke. 1979. An *Exophiala*-like fungus as the cause of a systemic mycosis of marine fish. Journal of Fish Diseases 2:145–152.

Bledsoe, C., P. Klein, and L.C. Bliss. 1990. A survey of mycorrhizal plants on Truelove Lowland, Devon Island, N.W.T., Canada. Canadian Journal of Botany 68:1848–1856.

Bliss, D.E. 1951. The destruction of *Armillaria mellea* in citrus soils. Phytopathology 41:665–683.

Blomberg, A., and L. Adler. 1993. Tolerance of fungi to NaCl. Pp. 209–232. *In* D.H. Jennings (ed.), Stress Tolerance of Fungi. Marcel Dekker, New York.

Bly, J.E., L.A. Lawson, A.J. Szalai, and L.W. Clem. 1993. Environmental factors affecting outbreaks of winter saprolegniosis in channel catfish, *Ictalurus punctatus* (Rafinesque). Journal of Fish Diseases 16:541–549.

Boddy, L. 1992. Development and function of fungal communities in decomposing wood. Pp. 749–782. *In* G.C. Carroll and D.T. Wicklow (eds.), The Fungal Community: Its Organization and Role in the Ecosystem. 2nd ed. Marcel Dekker, New York.

Boddy, L., and G.S. Griffith. 1989. Role of endophytes and latent invasion in the development of decay communities in sapwood of angiospermous trees. Sydowia 41:41–73.

Boddy, L., and J.W. T. Wimpenny. 1992. Ecological concepts in food microbiology. Journal of Applied Bacteriology 73:23–38.

Boddy, L., D.W. Bardsley, and O.M. Gibbon. 1987. Fungal communities in attached ash branches. New Phytologist 107:143–154.

Boekhout, T. 1991. A revision of ballistoconidia-forming yeasts and fungi. Studies in Mycology 33.

——. 1995. *Pseudozyma* Bandoni emend. Boekhout, a genus for yeast-like anamorphs of Ustilaginales. Journal of General and Applied Microbiology 41:359–366.

Boekhout, T., R.J. Bandoni, J.W. Fell, and K.J. Kwon-Chung. 1998. Discussion of teleomorphic and anamorphic genera of basidiomycetous yeasts. Pp. 609–634. *In* C.P. Kurtzman and J.W. Fell (eds.), The Yeasts, A Taxonomic Study. 4th ed. Elsevier, Amsterdam, The Netherlands.

Boekhout, T., A. Fonseca, and W.H. Batenburg-van der Vegte. 1991. *Bulleromyces* genus novum (*Tremellales*), a teleomorph for *Bullera alba*, and the occurrence of mating in *Bullera variabilis*. Antonie van Leeuwenhoek 59:81–93.

Boekhout, T., A. Fonseca, J.-P. Sampaio, and W.I. Golubev. 1993. Classification of heterobasidiomycetous yeasts: characteristics and affiliation of genera to higher taxa of heterobasidiomycetes. Canadian Journal of Microbiology 39:276–290.

Boekhout, T., C.P. Kurtzman, K. O'Donnell, and M.T. Smith. 1994. Phylogeny of the yeast genera *Hanseniaspora* (Anamorph *Kloeckera*), *Dekkera* (Anamorph *Brettanomyces*), and *Eeniella* as inferred from partial 26S ribosomal DNA nucleotide sequences. International Journal of Systematic Bacteriology 44:781–786.

Boekhout, T., Y. Yamada, A.C.M. Weijman, H.J. Roeijmans, and W.H. Batenburg-van der Vegte. 1992. The significance of coenzyme Q, carbohydrate composition and septal ultrastructure for the taxonomy of ballistoconidia-forming yeasts and fungi. Systematic and Applied Microbiology 15:1–10.

Boerema, G.H. 1993. Contributions towards a monograph of *Phoma* (Coelomycetes)—II. Section *Peyronellaea*. Persoonia 15:197–221.

Boerema, G.H., J. de Gruyter, and H.A. van Kesteren. 1994. Contributions towards a monograph of *Phoma* (Coelomycetes)—III. 1. Section *Plenodomus*: taxa often with a *Leptosphaeria* teleomorph. Persoonia 15:431–487.

Boerner, R.E.J. 1988. Relative mycorrhizal dependency and responsiveness of congeneric, sympatric species: implications for secondary succession. Ohio Journal of Science 88(2):12.

Boesewinkel, H.J. 1976. Storage of fungal cultures in water. Transactions of the British Mycological Society 66:183–185.

Bogan, B.W., B. Schoenike, R.T. Lamar, and D. Cullen. 1996a. Manganese peroxidase mRNA and enzyme activity levels during bioremediation of polycyclic aromatic hydrocarbon-contaminated soil with *Phanerochaete chrysosporium*. Applied and Environmental Microbiology 62:2381–2386.

——. 1996b. Expression of *lip* genes during growth in soil and oxidation of anthracene by *Phanerochaete chrysosporium*. Applied and Environmental Microbiology 62:3697–3703.

Bohannan, B.J.M., and J. Hughes. 2003. New approaches to analyzing microbial diversity data. Current Opinion in Microbiology 6:282–287.

Bokhary, H.A., A.M. Sabek, A.H. Abu-Zinada, and M.O. Fallatah Jow. 1984. Thermophilic and thermotolerant fungi of arid regions of Saudi Arabia: occurrence, seasonal variation and temperature relationships. Journal of Arid Environments 7:263–274.

Bolan, N.S. 1991. A critical review on the role of mycorrhizal fungi in the uptake of phosphorus by plants. Plant and Soil 134:189–207.

Bollen, G.J. 1979. Side-effects of pesticides on microbial interactions. Pp. 451–481. *In* B. Schippers and W. Gams (eds.), Soil-borne Plant Pathogens. Academic Press, London, England.

——. 1993. Mechanisms involved in nontarget effects of pesticides on soil-borne pathogens. Pp. 281–301. *In* J. Altman (ed.), Pesticide Interactions in Crop Production, Beneficial and Deleterious Effects. CRC Press, Boca Raton, Florida.

Bollen, G.J., and A. Fuchs. 1970. On the specificity of the in vitro and in vivo antifungal activity of benomyl. Netherlands Journal of Plant Pathology 76:299–312.

Bonner, T.I., D.J. Brenner, B.R. Neufeld, and R.J. Britten. 1973. Reduction in the rate of DNA reassociation by sequence divergence. Journal of Molecular Biology 81:123–135.

Boosalis, M.G. 1956. Effect of soil temperature and green-manure amendment of unsterilzed soil on parasitism of *Rhizoctonia solani* by *Penicillium vermiculatum* and *Trichoderma* sp. Phytopathology 46:473–478.

——. 1964. Hyperparasitism. Annual Review of Phytopathology 2:363–376.

Booth, C. 1959. Studies of Pyrenomycetes. IV. *Nectria*. Mycological Papers 73:1–115.

——. 1966. The genus *Cylindrocarpon*. Mycological Papers 104:1–56.

——. 1971. The Genus *Fusarium*. Commonwealth Mycological Institute, Kew, United Kingdom.

Booth, T. 1969. Marine fungi from British Columbia: monocentic chytrids and chytridiaceous species from coastal and interior halomorphic soils. Syesis 2:141–161.

——. 1971. Distribution of certain soil inhabiting chytrid and chytridiaceous species related to some physical and chemical factors. Canadian Journal of Botany 49:1743–1755.

Booth, T., and P. Barrett. 1976. Taxonomic and ecologic observations of zoosporic fungi in soils of high-arctic ecosystem. Canadian Journal of Botany 54:533–538.

Bootsma, R. 1973. Infections with *Saprolegnia* in pike culture (*Esox lucius* L.). Aquaculture 2:385–394.

Borneman, W.S., and D.E. Akin. 1990. Lignocellulose degradation by rumen fungi and bacteria: ultrastructure and cell wall degrading enzymes. Pp. 325–339. *In* D.E. Akin, L.G. Ljungdahl, J.R. Wilson, and P.J. Harris (eds.), Microbial and Plant Opportunities to Improve Lignocellulose Utilization by Ruminants. Elsevier, New York.

Borneman, W.S., D.E. Akin, and L.G. Ljungdahl. 1989. Fermentation products and plant cell wall-degrading enzymes produced by monocentric and polycentric anaerobic ruminal fungi. Applied and Environmental Microbiology 55:1066–1073.

Borror, D.J., C.A. Triplehorn, and N.F. Johnson. 1992. An Introduction to the Study of Insects. 6th ed. Harcourt Brace, Fort Worth, Texas.

Bose, S.R. 1947. Hereditary (seed-borne) symbiosis in *Casuarina equisetifolia* Forest. Nature (London) 159:512–514.

Bougher, N.L., and M.A. Castellano. 1993. Delimitation of *Hymenogaster sensu stricto* and four new segregate genera. Mycologia 85:273–293.

Bougher, N.L., I.C. Tommerup, and N. Malajczuk. 1993. Broad variation in developmental and mature basidiome morphology of the ectomycorrhizal fungus *Hydnangium sublamellatum* sp. nov. bridges morphologically based generic concepts of *Hydnangium*, *Podohydnangium* and *Laccaria*. Mycological Research 97:613–619.

Boullard, B. 1957. La mycotrophie chez les ptéridophytes. Sa fréquence, ses caractères, sa signification. Le Botaniste Sér. 41, pp. 5–185.

——. 1979. Considerations sur la symbiose fongique chez les pteridophytes. Syllogeus 19:1–59.

Bourdot, H., and A. Galzin. 1928. Hyménomycètes de France: Hétéro-basi-diés-homobasidiés gymnocarpes. Société mycologique de France; M. Bry, Sceaux, France.

Boursnell, J.G. 1950. The symbiotic seed-borne fungus in the Cistaceae I. Distribution and function of the fungus in the seeding and in the tissues of the mature plant. Annals of Botany 24:217–243.

Bowen, A.R., J.L. Chen-Wu, M. Momany, R. Young, P.J. Szaniszlo, and P.W. Robbins. 1992. Classification of fungal chitin synthases.

Proceedings of the National Academy of Sciences USA 89:519–523.

Bower, S.M. 1987a. *Labyrinthuloides haliotidis* n. sp. (Protozoa: Labyrinthomorpha), a pathogenic parasite of small juvenile alabone in a British Columbia mariculture facility. Canadian Journal of Zoology 65:1996–2007.

——. 1987b. Pathogenicity and host specificity of *Labyrinthuloides haliotidis* (Protozoa: Labyrinthomorpha), a parasite of juvenile alabone. Canadian Journal of Zoology 65:2008–2012.

Bower, S.M., D.J. Whitaker, and R.A. Elston. 1989. Detection of the abalone parasite *Labyrinthuloides haliotidis* by a direct fluorescent antibody technique. Journal of Invertebrate Pathology 53:281–283.

Bowman, B.H., J.W. Taylor, A.G. Brownlee, J. Lee, S.-D. Lu, and T.J. White. 1992. Molecular evolution of the fungi: relationship of the Basidiomycetes, Ascomycetes and Chytridiomycetes. Molecular Biology and Evolution 9:285–296.

Boyetchko, S.M., and J.P. Tewari. 1991. Parasitism of spores of the vesicular-arbuscular mycorrhizal fungus, *Glomus dimorphicum*. Phytoprotection 72:27–32.

Boyle, D.G., A.D. Hyatt, P. Daszak, L. Berger, J.E. Longcore, D. Porter, S.G. Hengstberger, and V. Olsen. 2003. Cryo-archiving of *Batrachochytrium dendrobatidis* and other chytridiomycetes. Diseases of Aquatic Organisms 56:59–64.

Brady, B.L. 1960. Occurrence of *Itersonilia* and *Tilletiopsis* on lesions caused by *Entyloma*. Transactions of the British Mycological Society 43:31–50.

Brady, L.R. 1962. Phylogenetic distribution of parasitism by *Claviceps* species. Lloydia 25:1–36.

Braselton, J.P. 2001. Plasmodiophoromycota. Pp. 81–91. *In* D.J. McLaughlin, E.G. McLaughlin, and P.A. Lempke (eds.), The Mycota. Vol. VII. Systematics and Evolution. Part A. Springer-Verlag, Berlin, Germany.

Brasier, C.M. 1987. The dynamics of fungal speciation. Pp. 231–260. *In* A.D.M. Rayner, C.M. Brasier, and D. Moore (eds.), Evolutionary Biology of the Fungi. Cambridge University Press, Cambridge, England.

——. 1988. *Ophiostoma ulmi*, cause of Dutch elm disease. Advances in Plant Pathology 6:207–223.

Brasier, C.M., and E.M. Hansen. 1992. Evolutionary biology of *Phytophthora*. Part II: Phylogeny, speciation, and population structure. Annual Review of Phytopathology 30:173–200.

Braun, K.L., and H.W. Keller. 1977. The collection of plasmodial slime molds as a winter hobby. McIlvania 3:18–20.

Braun, U. 1981. Taxonomic studies in the genus *Erysiphe* I. Generic delimitation and position in the system of the *Erysiphaceae*. Nova Hedwigia 34:679–719.

——. 1995. A Monograph of *Cercosporella*, *Ramularia* and allied genera (phytopathogenic Hyphomycetes), Vol. 1. IHW-Verlag, Berlin, Germany.

——. 1998. A Monograph of *Cercosporella*, *Ramularia* and allied genera (phytopathogenic hyphomycetes), Vol. 2. IHW-Verlag, Berlin, Germany.

Bray, J.R., and C.T. Curtis. 1957. An ordination of upland forest communities of southern Wisconsin. Ecological Monographs 27:325–349.

Brayford, D. 1992. *Cylindrocarpon*. Pp. 103–106. *In* L.L. Singleton, J.D. Mihail, and C.M. Rush (eds.), Methods for Research on Soilborne Phytopathogenic Fungi. American Phytopathological Society Press, St. Paul, Minnesota.

Breitenbach, J., and F. Kränzlin. 1984. Fungi of Switzerland. Vol. 1. Ascomycetes. Verlag Mykologia, Lucerne, Switzerland.

——. 1986. Fungi of Switzerland. Vol. 2. Non-gilled Fungi. Verlag Mykologia, Lucerne, Switzerland.

——. 1991. Fungi of Switzerland. Vol. 3. Boletes and Agaricales. Part 1. Verlag Mykologia, Lucerne, Switzerland.

——. 1995. Fungi of Switzerland. Vol. 4. Agarics. Part 2. Verlag Mykologia, Lucerene, Switzerland.

——. 2000. Fungi of Switzerland. Vol. 5. Agarics. Part 3. Verlag Mykologia, Lucerene, Switzerland.

Breton, A., A. Bernalier, F. Bonnemoy, G. Fonty, B. Gaillard, and P. Gouet. 1989. Morphological and metabolic characterization of a new species of strictly anaerobic rumen fungus: *Neocallimastix joyonii*. FEMS Microbiology Letters 49:309–314.

Breton, A., A. Bernalier, M. Dusser, G. Fonty, B. Gaillard-Martinie, and J. Guillot. 1990. *Anaeromyces mucronatus* nov. gen., nov. sp. A new strictly anaerobic rumen fungus with polycentric thallus. FEMS Microbiology Letters 58:177–182.

Breton, A., M. Dusser, B. Gaillard-Martinie, J. Guillot, L. Millet, and G. Prensier. 1991. *Piromyces rhizinflata* nov. sp., a strictly anaerobic fungus from the faeces of the Saharan ass: a morphological, metabolic and ultrastructural study. FEMS Microbiology Letters 66:1–8.

Bricaud, O., C. Coste, D. Le Coeur, M. Glenn, T. Ménard, and C. Roux. 1992. Champignons lichénisés et lichénicoles de la France méridionale: espèces nouvelles et intéressantes (VI). Bulletin de la Société Linnéenne de Provence 43:81–96.

Bridge, P.D., Z. Kozakiewicz, and R.R.M. Paterson. 1992. PENIMAT: a computer assisted identification scheme for terverticillate *Penicillium* species. Mycological Papers 165:1–59.

Bridges, J.R. 1983. Mycangial fungi of *Dendroctonus frontalis* (Coleoptera: Scolytidae) and their relationship to beetle population trends. Environmental Entomology 12:858–861.

Bridges, J.R., and J.C. Moser. 1983. Role of two phoretic mites in transmission of bluestain fungus, *Ceratocystis minor*. Ecological Entomology 8:9–12.

Bridges, J.R., and T.J. Perry. 1985. Effects of mycangial fungi on gallery construction and distribution of bluestain in southern pine beetle-infested pine bolts. Journal of Entomological Science 20:271–275.

Bridges, J.R., J.E. Marler, and B.H. McSparrin. 1984. A quantitative study of the yeasts and bacteria associated with laboratory-reared *Dendroctonus frontalis* Zimm. (Coleopt., Scolytidae). Zeitschrift für Angewandte Entomologie 97:261–267.

Bridson, D., and L. Forman (eds). 1992. The Herbarium Handbook. Royal Botanic Gardens, Kew, England.

Britten, R.J. 1986. Rates of DNA sequence evolution differ between taxonomic groups. Science 231:1393–1398.

Britten, R.J., D.E. Graham, and B.R. Neufeld. 1974. Analysis of repeating DNA sequences by reassociation. Pp. 363–418. *In* L. Grossman and K. Moldave (eds.), Nucleic Acids and Protein Synthesis Part E. Methods in Enzymology. Vol. 29. Academic Press, New York.

Brock, T.D. 1978. Thermophilic Microorganisms and Life at High Temperatures. Springer-Verlag, New York.

Brodie, H.J. 1975. The Bird's Nest Fungi. University of Toronto Press, Toronto, Ontario, Canada.

Brodie, H.J., and R.W.G. Dennis. 1954. The Nidulariaceae of the West Indies. Transactions of the British Mycological Society 37:151–160.

Brodo, I.M., S.D. Sharnoff, and S. Sharnoff. 2001. Lichens of North America. Yale University Press, New Haven, Connecticut.

Brooks, T.E. 1967. A study of corticolous Myxomycetes. Unpubl. Ph.D. dissert., University of Kansas, Lawrence, Kansas.

Brooks, T.E., H.W. Keller, and M. Chassain. 1977. Corticolous *Myxomycetes* IV: a new species of *Diderma*. Mycologia 69:179–184.

Brotherson, J.D., S.R. Rushforth, and J.R. Johansen. 1983. Effects of long-term grazing on cryptogam crust cover in Navajo National Monument, Arizona. Journal of Range Management 36:579–581.

Brown, J.H. 1995. Macroecology. University of Chicago Press, Chicago, Illinois.

Brown, K.B., K.D. Hyde, and D.I. Guest. 1998. Preliminary studies on endophytic fungal communities of *Musa acuminata* species complex in Hong Kong and Australia. Fungal Diversity 1:27–51.

Brown, M.T., and I.R. Hall. 1989. Metal tolerance in fungi. Pp. 95–104. *In* A.J. Shaw (ed.), Heavy Metal Tolerance in Plants: Evolutionary Aspects. CRC Press, Boca Raton, Florida.

Bruce, J., and E.O. Morris. 1973. Psychrophilic yeasts isolated from marine fish. Antonie van Leeuwenhoek 39:331–339.

Brückner, H., and M. Przybylski. 1984. Isolation and structural characterization of polypeptide antibiotics of the peptaibol class by high-performance liquid chromatography with field desorption and fast atom bombardment mass spectrometry. Journal of Chromatography 296:263–275.

Brückner, H., P. Wunsch, and C. Kussin. 1989. Production of polypeptide antibiotics by molds of the genus *Gliocladium*. Pp. 103–106. *In* A. Aubry, M. Marraud, and B. Vitoux (eds.), Second Forum on Peptides. Inserm, London, England.

Brummelen, J. van. 1998. Reconsideration of relationships within the Thelebolaceae based on ascus ultrastructure. Persoonia 16:425–469.

Brummitt, R.K., and C.E. Powell. (eds.). 1992. Authors of Plant Names. Royal Botanic Gardens, Kew, England.

Brundrett, M. 1991. Mycorrhizas in natural ecosystems. Advances in Ecological Research. 21:171–313.

Brundrett, M., and S. Juniper. 1995. Non-destructive assessment of spore germination of VAM fungi and production of pot cultures from single spores. Soil Biology and Biochemistry 27:85–91.

Brundrett, M., and B. Kendrick. 1990. The roots and mycorrhizas of herbaceous woodland plants. II. Structural aspects of morphology. New Phytologist 114:469–479.

Brundrett, M.C., L. Melville, and R.L. Peterson (eds.). 1994. Practical Methods in Mycorrhiza Research. Mycologue Publications, Waterloo, Ontario, Canada.

Brundrett, M.C., Y. Piché, and R.L. Peterson. 1984. A new method for observing the morphology of vesicular-arbuscular mycorrhizae. Canadian Journal of Botany 62:2128–2134.

Brunk, M., and H.L. Barnett. 1966. Mycoparasitism of *Dispira simplex* and *D. parvispora*. Mycologia 58:518–528.

Brunner, F., and O. Petrini. 1992. Taxonomy of some *Xylaria* species and xylariaceous endophytes by isozyme electrophoresis. Mycological Research 96:723–733.

Bruno, D.W., and D.J. Stamps. 1987. Saprolegniasis of Atlantic salmon, *Salmo salar* L., fry. Journal of Fish Diseases 10:513–517.

Bruns, T.D., R. Fogel, T.J. White, and J.D. Palmer. 1989. Accelerated evolution of a false-truffle from a mushroom ancestor. Nature (London) 339:140–142.

Bruns, T.D., R. Vilgalys, S.M. Barns, D. Gonzalez, D.S. Hibbett, D.J. Lane, L. Simon, S. Stickel, T.M. Szaro, W.G. Weisburg, and M.L. Sogin. 1993. Evolutionary relationships within the fungi: analysis of nuclear small subunit rRNA sequences. Molecular Phylogenetics and Evolution 1:231–241.

Bruns, T.D., T.J. White, and J.W. Taylor. 1991. Fungal molecular systematics. Annual Review of Ecology and Systematics 22:525–564.

Buczacki, S.T. 1983. Zoosporic Plant Pathogens: a Modern Perspective. Academic Press, London, England.

Buell, C.B., and W.H. Weston. 1947. Application of the mineral oil conservation method to maintaining collections of fungus cultures. American Journal of Botany 34:555–561.

Buller, A.H.R. 1924. Researches on Fungi. Vol. 3. The production and liberation of spores in Hymenomycetes and Uredineae. Longmans, Green, London, England.

Bullis, R.A., E.J. Noga, and M.G. Levy. 1990. Immunological relationship of the fish-pathogenic oomycete *Saprolegnia parasitica* to other oomycetous and unrelated fungi. Journal of Aquatic Animal Health 2:223–227.

Bullock, A.M. 1989. Laboratory methods. Pp. 374–405. *In* R.J. Roberts (ed.), Fish Pathology. 2nd ed. Baillière Tindall, London, England.

Bullock, S., P.B. Adams, H.J. Willetts, and W.A. Ayers. 1986. Production of haustoria by *Sporidesmium sclerotivorum* in sclerotia of *Sclerotinia minor*. Phytopathology 76:101–103.

Bultman, T.L., and N.S. Conrad. 1998. Effects of endophytic fungus, nutrient level, and plant damage on performance of fall armyworm (Lepidoptera: Noctuidae). Environmental Entomology 27:631–635.

Bunkley-Williams, L., and E.H. Williams, Jr. 1994. Parasites of Puerto Rican Freshwater Sport Fishes. Puerto Rico Department of Natural and Environmental Resources, San Juan, Puerto Rico, and Department of Marine Sciences, University of Puerto Rico, Mayaguez, Puerto Rico.

Bunse, T., and G.K. Steigleder. 1991. The preservation of fungal cultures by lyophilization. Mycoses 34:173–176.

Burdon, J.J., A.H.D. Brown, and A.M. Jarosz. 1990. The spatial scale of genetic interactions in host-pathogen coevolved systems. Pp. 233–247. *In* J.J. Burdon and S.R. Leather (eds.), Pests, Pathogens and Plant Communities. Blackwell, London, England.

Burdsall, H.H., Jr. 1968. A revision of *Hydnocystis* (Tuberales) and of the hypogeous species of *Geopora* (Pezizales). Mycologia 60:496–525.

Burdsall, H.H., Jr., and E.B. Dorworth. 1994. Preserving cultures of wood-decaying Basidiomycotina using sterile distilled water in cryovials. Mycologia 86:275–280.

Burdsall, H.H., Jr., and R.L. Gilbertson. 1974. A new species of *Platygloea* occurring on *Peniophora tamaricicola* in Arizona. Mycologia 66:702–706.

Burdsall, H.H., Jr., H.C. Hoch, M.G. Boosalis, and E.C. Setliff. 1980. *Laetisaria arvalis* (Aphyllophorales, Corticiaceae): a possible biological control agent for *Rhizoctonia solani* and *Pythium species*. Mycologia 72:728–736.

Bureau of Land Management (BLM). 1996. Sampling Vegetation Attributes. U.S. Department of the Interior, BLM, National Applied Resource Sciences Center, Denver, Colorado. Interagency Technical Reference 1734–4.

Burge, H.A., M. Chatigny, J. Feeley, K. Kreiss, P. Morey, J. Otten, and K. Petersen. 1987. Guidelines for assessment and sampling of saprophytic bioaerosols in the indoor environment. Applied Industrial Hygiene 9:R10–R16.

Burge, M.N. (ed.). 1988. Fungi in Biological Control Systems. Manchester University Press, Manchester, England.

Burgeff, H. von. 1924. Untersuchungen über Sexualität und Parasitismus bei Mucorineen 1. Botanische Abhandlungen 4:1–135.

Burges, A. 1958. Micro-Organisms in the Soil. Hutchinson, London, United Kingdom.

Buriticá, P., and J.F. Hennen. 1980. Pucciniosireae (Uredinales, Pucciniaceae). Flora Neotropica. 24:1–50.

Burns, C.D., M.E. Berrigan, and G.E. Henderson. 1979. *Fusarium* sp. infections in the freshwater prawn *Macrobranchium rosenbergii* (De Man). Aquaculture 16:193–198.

Burns, C.W. 1985. Fungal parasitism in a copepod population: the effects of *Aphanomyces* on the population dynamics of *Boeckella dilatata* Sars. Journal of Plankton Research 7:201–205.

Burt, E.A. 1966. The Thelephoraceae of North America I-XV. Hafner, New York.

Butin, H. 1981. *Keissleriella bavarica* sp. nov., a hyperparasite on *Ascodichaena rugosa* Butin. Phytopathologische Zeitschrift 100:186–190.

———. 1992. Effect of endophytic fungi from oak (*Quercus robur* L.) on the mortality of leaf inhabiting gall insects. European Journal of Forest Pathology 22:237–246.

Butler, E.E. 1957. *Rhizoctonia solani* as a parasite of fungi. Mycologia 49:354–373.

Butler, E.E., and A.H. McCain. 1968. A new species of *Stephanoma*. Mycologia 60:955–959.

Butler, E.J. 1907. An account of the genus *Pythium* and some Chytrideaceae. Memoirs of the Department of Agriculture in India, Botanical Series no. 5, 1:1–161.

Cabello, M., L. Gaspar, and R. Pollero. 1994. *Glomus antarcticum* sp. nov., a new vesicular-arbuscular mycorrhizal fungus from Antarctica. Mycotaxon 51:123–128.

Cabral, D. 1985. Phyllosphere of *Eucalyptus viminalis*: dynamics of fungal populations. Transactions of the British Mycological Society 85:501–511.

Cabral, D., J.K. Stone, and G.C. Carroll. 1993. The internal mycobiota of *Juncus* spp.: microscopic and cultural observations of infection patterns. Mycological Research 97:367–376.

Cailleux, R. 1971. Recherches sur la Mycoflore coprophile Centrafricaine. Les genres *Sordaria, Gelasinospora, Bombardia* (Biologie—Morphologie—Systématique—Écologie). Bulletin de la Société Mycologique de la France 87:461–626.

Cain, R.F. 1934. Studies of coprophilous Sphaeriales of Ontario. University of Toronto Studies, Biological Series 38:1–126.

——. 1948. *Myriogonium*, a new genus among simplified ascomycetes. Mycologia 40:158–167.

——. 1972. Evolution of the fungi. Mycologia 64:1–14.

Cain, S.A., and G.M. de Castro Oliveira. 1959. Manual of Vegetation Analysis. Harper, New York.

Cal, A., de., E.M.-Sagasta, and M. Malgarejo. 1988. Antifungal substances produced by *Penicillium frequentans* and their relationship to the biocontrol of *Monilinia laxa*. Phytopathology 78:888–893.

——. 1990. Biological control of peach twig blight (*Monilinia laxa*) with *Penicillium frequentans*. Plant Pathology 39:612–618.

Calderone, R.A., and H.L. Barnett. 1972. Axenic growth and nutrition of *Gonatobotryum fuscum*. Mycologia 64:153–160.

Caldwell, D.R., and M.P. Bryant. 1966. Medium without rumen fluid for nonselective enumeration and isolation of rumen bacteria. Applied Microbiology 14:794–801.

Callaham, R.Z., and M. Shifrine. 1960. The yeasts associated with bark beetles. Forest Science 6:146–154.

Calpouzos, L., T. Thies, and C.M. R. Battle. 1957. Culture of the rust parasite, *Darluca filum*. Phytopathology 47:108–109.

Camacho, F.J., D.S. Gernandt, A. Liston, J.K. Stone, and A.S. Klein. 1997. Molecular taxonomy of spruce endophytic fungi. Inoculum: 46:7.

Cameron, R.E., R.C. Honour, and F.A. Morelli. 1976. Antarctic microbiology—preparation for Mars life detection, quarantine, and back contamination. Pp. 57–84. *In* M.R. Heinrich (ed.), Extreme Environments: Mechanisms of Microbial Adaptation. Academic Press, New York.

Campbell, D.G. 1989. Quantitative inventory of tropical forests. Pp. 523–533. *In* D.G. Campbell and H.D. Hammond (eds.), Floristic Inventory of Tropical Countries: The Status of Plant Systematics, Collections, and Vegetation, plus Recommendations for the Future. New York Botanical Gardens, Bronx, New York.

Candoussau, F., and J.-F. Magni. 1995. New French records. Mycologist 9:12–14.

Cannon, P.F. 1991. A revision of *Phyllachora* and some similar genera on the host family Leguminosae. Mycological Papers 163:1–302.

——. 1996. Filamentous fungi. Pp. 125–143. *In* G.S. Hall (ed.), Methods for the Examination of Organismal Biodiversity in Soils and Sediments. CAB International, Wallingford, Oxon, United Kingdom.

——. 1997a. Diversity of the Phyllachoraceae with special reference to the tropics. Pp. 255–278. *In* K.D. Hyde (ed.), Biodiversity of Tropical Microfungi. Hong Kong University Press, Hong Kong, China.

——. 1997b. Strategies for rapid assessment of fungal diversity. Biodiversity and Conservation 6:669–680.

——. 1999. Options and constraints in rapid diversity analysis of fungi in natural ecosystems. Fungal Diversity 2:1–15.

Cannon, P.F., and D.L. Hawksworth. 1982. A re-evaluation of *Melanospora* Corda and similar Pyrenomycetes, with a revision of the British species. Botanical Journal of the Linnean Society (London) 84:115–160.

——. 1995. The diversity of fungi associated with vascular plants: the known, the unknown, and the need to bridge the knowledge gap. Advances in Plant Pathology 11:277–302.

Cannon, P.F., and D.W. Minter. 1986. The Rhytismataceae of the Indian subcontinent. Commonwealth Mycological Institute, Mycological Papers 155:1–123.

Cannon, P.F., D.L. Hawksworth, and M.A. Sherwood-Pike. 1985. The British Ascomycotina: An Annotated Checklist. Commonwealth Mycological Institute, Kew, United Kingdom.

Cannon, P.F., S.K. Abdullah, and B.A. Abbas. 1995. Two new species of *Monascus* from Iraq, with a key to the known species of the genus. Mycological Research 99:659–662.

Canter, H.M. 1968. Studies on British chytrids XXVIII. *Rhizophydium nobile* sp. nov., parasitic on the resting spore of *Ceratium hirundinella* O.F. Müll. from the plankton. Proceedings of the Linnean Society of London 179:197–201.

——. 1985. Observations on the chytrid *Endocoenobium eudorinae* Ingold. Botanical Journal of the Linnean Society (London) 91: 95–116.

Canter, H.M., and C.T. Ingold. 1984. A chytrid on *Dacrymyces*. Transactions of the British Mycological Society 82:739–742.

Canter, H.M., G.H.M. Jaworski, and G.W. Beakes. 1992. Formae speciales differentiation in the chytrid *Zygorhizidium planktonicum* Canter, a parasite of the diatoms *Asterionella* and *Synedra*. Nova Hedwigia 55:437–455.

Cappellano, A., B. deQuartre, G. Valla, and A. Moiroud. 1987. Root-nodules formation by *Penicillium* sp. on *Alnus glutinosa* and *Alnus incana*. Plant and Soil 104:45–51.

Carbone, I., and L.M. Kohn. 1993. Ribosomal DNA sequence divergence within internal transcribed spacer 1 of the Sclerotiniaceae. Mycologia 85:415–427.

Carey, A.B. 1991. The biology of arboreal rodents in Douglas-fir forests.U.S.D.A., Forest Service, Pacific Northwest Research Station, Portland, Oregon. General Technical Report PNW-GTR-276.

Carey, P.D., A.H. Fitter, and A.R. Watkinson. 1992. A field study using the fungicide benomyl to investigate the effect of mycorrhizal fungi on plant fitness. Oecologia 90:550–555.

Carey, S.T., and C.T. Rogerson. 1976. Taxonomy and morphology of a new species of *Hypocrea* on *Marasmius*. Brittonia 28:381–389.

——. 1981. Morphology and cytology of *Hypomyces polyporinus* and its *Sympodiophora* anamorph. Bulletin of the Torrey Botanical Club 108:13–24.

Carling, D.E., M.F. Brown, and D.F. Millikan. 1976. Ultrastructural examination of the *Puccinia graminis-Darluca filum* host-parasite relationship. Phytopathology 66:419–422.

Carmichael, J.W. 1956. Frozen storage for stock cultures of fungi. Mycologia 48:378–381.

——. 1962. Viability of mold cultures stored at −20°C. Mycologia 54:432–436.

——. 1966. Cerebral mycetoma of trout due to a *Phialophora*-like fungus. Sabouraudia 5:120–123.

Carmichael, J.W., W.B. Kendrick, L. Conners, and L. Sigler. 1980. Genera of Hyphomycetes. University of Alberta Press, Edmonton, Alberta, Canada.

Caron, D.A. 1993. Enrichment, isolation, and culture of free-living heterotrophic flagellates. Pp. 77–89. *In* P.F. Kemp, B.F. Sherr, E.B. Sherr, and J.J. Cole (eds.), Handbook of Methods in Aquatic Microbial Ecology. Lewis Publishers, Boca Raton, Florida.

Carpenter, S.E., and J.M. Trappe. 1985. Phoenicoid fungi: a proposed term for fungi that fruit after heat treatment of substrates. Mycotaxon 23:203–206.

Carpenter, S.E., J.M. Trappe, and J. Ammirati, Jr. 1987. Observations of fungal succession in the Mount St. Helens devastation zone, 1980–1983. Canadian Journal of Botany 65:716–728.

Carreiro, M.M., and R.E. Koske. 1992. Room temperature isolations can bias against selection of low temperature microfungi in temperate forest soils. Mycologia 84:886–900.

Carreiro, M.M., R.L. Sinsabaugh, D.A. Repert, and D.F. Parkhust. 2000. Microbial enzyme shifts explain litter decay responses to simulated nitrogen deposition. Ecology 81:2359–2365.

Carris, L.M. 1992. *Vaccinium* fungi: *Pseudotracylla falcata* sp. nov. Mycologia 84:534–540.

———. 1994. Observations on *Pseudotracylla* species. Mycotaxon 50: 93–98.

Carris, L.M., and D.A. Glawe. 1989. Fungi colonizing cysts of *Heterodera glycines*. Bulletin of the University of Illinois Urbana-Champaign Agriculture Experiment Station 786:1–93.

Carroll, G.C. 1986. The biology of endophytism in plants with particular reference to woody perennials. Pp. 205–222. *In* N.J. Fokkema and J. van den Heuvel (eds.), Microbiology of the Phyllosphere. Cambridge University Press, Cambridge, England.

———. 1988. Fungal endophytes in stems and leaves: from latent pathogen to mutualistic symbiont. Ecology 69:2–9.

———. 1991. Fungal associates of woody plants as insect antagonists in leaves and stems. Pp. 253–271. *In* P. Barbosa, V.A. Krischik, and C.G. Jones (eds.), Microbial Mediation of Plant-Herbivore Interactions. John Wiley, New York.

———. 1995. Forest endophytes: pattern and process. Canadian Journal of Botany 73 (Suppl. 1):S1316–S1324.

Carroll, G.C., and F.E. Carroll. 1978. Studies on the incidence of coniferous needle endophytes in the Pacific Northwest. Canadian Journal of Botany 56:3034–3043.

Carroll, G.C., and D.T. Wicklow. 1992. The Fungal Community: Its Organization and Role in the Ecosystem. 2nd ed. Marcel Dekker, New York.

Carter, M.R. (ed.). 1993. Soil Sampling and Methods of Analysis. Lewis Publishers. Boca Raton, Florida.

Carus, C.G. 1823. Beitrag zur Geschichte der unter Wasser an verwesenden *Thierkörpen* sich erzeugenden Schimmel-oder Algen-Gattungen. Nova Acta Physico-Medica Academiae Caesareae Leopoldino-Carolinae Naturae Curiosorum 11:493–507.

Cassar, S.C., and M. Blackwell. 1996. Non-monophyly of ambrosia fungi in *Ambrosiella*. Mycologia 88:596–601.

Castañeda, R.F. 1986. Fungi Cubense. Instituto de Investigaciones Fundamentales en Agricultura Tropical, Habana, Cuba.

Castellano, M.A., and J.M. Trappe. 1985. Ectomycorrhizal formation and plantation performance of Douglas-fir nursery stock inoculated with *Rhizopogon* spores. Canadian Journal of Forest Research 15:613–617.

Castellano, M.A., J.E. Smith, T. O'Dell, E. Cázares, and S. Nugent. 1999. Handbook to Strategy 1, Fungal Taxa from the Northwest Forest Plan. U.S. Department of Agriculture, Forest Service, Pacific Northwest Research Station, Portland, Oregon. General Technical Report PNW-GTR-476.

Castellano, M.A., J.M. Trappe, Z. Maser, and C. Maser. 1989. Key to Spores of the Genera of Hypogeous Fungi of North Temperate Forests with Special Reference to Animal Mycophagy. Mad River Press, Arcata, California.

Castlebury, L.A., and L.L. Domier. 1998. Small subunit ribosomal RNA gene phylogeny of *Plasmodiophora brassicae*. Mycologia 90:102–107.

Cauchon, R., and G. Ouellette. 1964. Association of *Stromatocrea cerebriforme* and *Hypocreopsis* species with *Hymenochaete* species. Mycologia 56:453–455.

Cavaliere, A.R. 1975. Aquatic ascomycetes from Lake Itasca, Minnesota. Journal of the Minnesota Academy of Science 41:32–35.

Cavalier-Smith, T. 1986. The Kingdom Chromista: origin and systematics. Pp. 309–347. *In* F.E. Round and D.J. Chapman (eds), Progress in Phycological Research. Biopress, Bristol, England.

———. 1987. The origin of fungi and pseudofungi. Pp. 339–353 *In* A.D.M. Rayner, C.M. Brasier, and D. Moore (eds.), Evolutionary Biology of the Fungi. Cambridge University Press, Cambridge, England.

———. 1998. A revised 6-kingdom system of life. Biological Review 73:203–266.

———. 2001. What are fungi? Pp. 3–37. *In* D.J. McLaughlin, E.G. McLaughlin, and P.A. Lempke (eds.), The Mycota. Vol. VII. Systematics and Evolution. Part A. Springer-Verlag, Berlin, Germany.

Cavalier-Smith, T., M.T.E.P. Allsopp, and E.E. Chao. 1994. Thraustochytrids are chromists, not fungi:18S rRNA signatures of heterokonta. Philosophical Transactions of the Royal Society of London, Series B 346:387–397.

Cavender, J.C. 1973. Geographical distribution of Acrasieae. Mycologia 65:1044–1054.

———. 1990. Phylum Dictyostelida. Pp. 88–101. *In* L. Margulis, J.O. Corliss, M. Melkonian, and D. Chapman (eds.), Handbook of Protoctista. Jones and Bartlett, Boston, Massachusetts.

Cavender, J.C., and K.B. Raper. 1965. The Acrasieae in nature. I. Isolation. American Journal of Botany 52:294–296.

———. 1968. The occurrence and distribution of Acrasieae in forests of subtropical and tropical America. American Journal of Botany 55: 504–513.

Cavender, J.C., R.A. Bradshaw, J.P. Regner, and T. Damio. 1993. Response of soil dictyostelid slime molds to agricultural disturbance in a tropical environment. Biotropica 25:245–248.

Cavender, J.C., J. Cavender-Bares, and H.R. Hohl. 1995. Ecological distribution of cellular slime molds in forest soils of Germany. Botanica Helvetica 105:199–219.

Cázares, E., and J.M. Trappe. 1990. Alpine and subalpine fungi of the Cascade Mountains. I. *Hymenogaster glacialis* sp. nov. Mycotaxon 38:245–249.

———. 1991. Alpine and subalpine fungi of the Cascade and Olympic Mountains. 3. *Gastroboletus rubur* comb. nov. Mycotaxon 42: 339–345.

Cazin, J., Jr., D.F. Wiemer, and J.J. Howard. 1989. Isolation, growth characteristics, and long-term storage of fungi cultivated by attine ants. Applied and Environmental Microbiology 55:1346–1350.

Celio, D.A., and D.E. Padgett. 1989. An improved method of quantifying water mold spores in natural water columns. Mycologia 81:459–460.

Cerrato, F.R., R.E. De la Cruz, and D.H. Hubbell. 1976. Further studies on a mycoparasitic basidiomycete species. Applied and Environmental Microbiology 31:60–62.

Chand, T., and C. Logan. 1984. Antagonists and parasites of *Rhizoctonia solani* and their efficacy in reducing stem canker of potato under controlled conditions. Transactions of the British Mycological Society 83:107–112.

Chang, Y. 1967. The fungi of wheat straw compost. II. Biochemical and physiological studies. Transactions of the British Mycological Society 50:667–677.

Chang, Y., and H.J. Hudson. 1967. The fungi of wheat straw compost. I. Ecological studies. Transactions of the British Mycological Society 50:649–666.

Chapela, I.H. 1989. Fungi in healthy stems and branches of American beech and aspen: a comparative study. New Phytologist 113:65–75.

Chapela, I.H., and L. Boddy. 1988a. Fungal colonization of attached beech branches. I. Early stages of development of fungal communities. New Phytologist 110:39–45.

——. 1988b. Fungal colonization of attached beech branches. II. Spatial and temporal organization of communities arising from latent invaders in bark and functional sapwood, under different moisture regimes. New Phytologist 110:47–57.

Chapela, I.H., O. Petrini, and G. Bielser. 1993. The physiology of ascospore eclosion in *Hypoxylon fragiforme*: mechanisms in the early recognition and establishment of an endophytic symbiosis. Mycological Research 97:157–162.

Chapela, I.H., O. Petrini, and L. Hagmann. 1991. Monolignol glucosides as specific recognition messengers in fungus-plant symbioses. Physiological and Molecular Plant Pathology 39:289–298.

Chapela, I.H., O. Petrini, and L.E. Petrini. 1990. Unusual ascospore germination in *Hypoxylon fragiforme*: first steps in the establishment of an endophytic symbiosis. Canadian Journal of Botany 68:2571–2575.

Chapela, I.H., S.A. Rehner, T.R. Schultz, and U.G. Mueller. 1994. Evolutionary history of the symbiosis between fungus-growing ants and their fungi. Science 266:1691–1694.

Chaturvedi, S., H.S. Randhawa, V.P. Chaturvedi and Z.U. Khan. 1990. Efficacy of brain heart infusion-egg albumen agar, yeast extract phosphate agar and peptone glucose agar media for isolation of *Blastomyces dermatitidis* from sputum. Mycopathologia 112:105–112.

Chelly, J., J.-C. Kaplan, P. Maire, S. Gautron, and A. Kahn. 1988. Transcription of the dystrophin gene in human muscle and non-muscle tissues. Nature 333:858–860.

Chen, K.Y., and Z.C. Chen. 1996. A new species of *Thermoascus* with a *Paecilomyces* anamorph and other thermophilic species from Taiwan. Mycotaxon 60:225–240.

Chen, S.F., and C.Y. Chien. 1995. Some chytrids of Taiwan (I). Botanical Bulletin of Academia Sinica 36:235–241.

——. 1998. Some chytrids of Taiwan (II). Botanical Bulletin of Academia Sinica 39:47–56.

Chen, S.F., M.L. Hsu, and C.Y. Chien. 2000. Some chytrids of Taiwan (III). Botanical Bulletin of Academia Sinica 41:73–80.

Cherrett, J.M., R.J. Powell, and D.J. Stradling. 1989. The mutualism between leaf-cutting ants and their fungus. Pp. 93–120. *In* N. Wilding, N.M. Collins, P.M. Hammond, and J.F. Webber (eds.), Insect-Fungus Interactions. Academic Press, New York.

Chesters, C.G.C. 1949. Concerning fungi inhabiting soil. Transactions of the British Mycological Society 32:197–216.

Chet, I. 1987. *Trichoderma*—application, mode of action, and potential as a biocontrol agent of soilborne plant pathogenic fungi. Pp. 137–160. *In* I. Chet (ed.), Innovative Approaches to Plant Disease Control. J. Wiley, New York.

Chet, I., G.E. Harman, and R. Baker. 1981. *Trichoderma hamatum*: its hyphal interactions with *Rhizoctonia solani* and *Pythium* spp. Microbial Ecology 7:29–38.

Chirgwin, J.M., A.E. Przybyla, R.J. MacDonald, and W.J. Rutter 1979. Isolation of biologically active ribonucleic acid from sources enriched in ribonuclease. Biochemistry 18:5294–5299.

Chiu, S.W., M.J. Chen, and S.T. Chang. 1995. Differentiating homothallic mushroom *Volvariella* by AP-PCR and RFLPs. Mycological Research 99:333–336.

Christensen, M. 1969. Soil microfungi of dry to mesic conifer-hardwood forests in northern Wisconsin. Ecology 50:9–27.

——. 1981a. Species diversity and dominance in fungal communities. Pp. 201–232. *In* D.T. Wicklow and G.C. Carroll (eds.), The Fungal Community, Its Organization and Role in the Ecosystem. Marcel Dekker, New York.

——. 1981b. A synoptic key and evaluation of species in the *Aspergillus flavus* group. Mycologia 73:1056–1084.

——. 1982. The *Aspergillus ochraceus* group: Two new species from western soils and a synoptic key. Mycologia 74:210–225.

——. 1989. A view of fungal ecology. Mycologia 81:1–19.

Christensen, M., and J.S. States. 1982. *Aspergillus nidulans* group: *Aspergillus navahoensis*, and a revised synoptic key. Mycologia 74:226–235.

Christensen, M., and D.E. Tuthill. 1985. *Aspergillus*: an overview. Pp. 195–209. *In* R.A. Samson and J.I. Pitt (eds.), Advances in *Penicillium* and *Aspergillus* Systematics. Plenum, New York.

Christensen, M., and W.F. Whittingham. 1965. The soil microfungi of open bogs and conifer swamps in Wisconsin. Mycologia 57:882–896.

Christensen, M., W.F. Whittingham, and R.O. Novak. 1962. The soil microfungi of wet-mesic forests in southern Wisconsin. Mycologia 54:374–388.

Chung, K., A. Leuchtmann, and C.L. Schardl. 1996. Inheritance of mitochondrial DNA and plasmids in the ascomycetous fungus *Epichloe typhina*. Genetics 142:259–265.

Chupp, C. 1940. Further notes on double cover-glass mounts. Mycologia 32:269–270.

Church, D.C. 1969. Digestive Physiology and Nutrition of Ruminants. Vol I. Oregon State University Press, Corvallis, Oregon.

Citron, A., A. Breton, and G. Fonty. 1987. Rumen anaerobic fungi. Bulletin de L'Institut Pasteur 85:329–343.

Clapp, J.P., J.P.W. Young, J.W. Merryweather, and A.H. Fitter. 1995. Diversity of fungal symbionts in arbuscular mycorrhizas from a natural community. New Phytologist 130:259–265.

Claridge, A.W., and T.W. May. 1994. Mycophagy among Australian mammals. Australian Journal of Ecology 19:251–275.

Claridge, A.W., S.J. Cork, and J.M. Trappe. 2000. Diversity and habitat relationships of hypogeous fungi. I. Study design, sampling techniques and general survey results. Biodiversity and Conservation 9:151–173.

Claridge, A.W., A.P. Robinson, M.T. Tanton, and R.B. Cunningham. 1993. Seasonal production of hypogeal fungal sporocarps in a mixed-species eucalypt forest stand in South-eastern Australia. Australian Journal of Botany 41:145–167.

Clark, A.G., and C.M.S. Lanigan. 1993. Prospects for estimating nucleotide divergence with RAPDs. Molecular Biology and Evolution 10:1096–1111.

Clark, E.M., J.F. White, Jr., and R.M. Patterson. 1983. Improved histochemical techniques for the detection of *Acremonium coenophialum* in tall fescue and methods of in vitro culture of the fungus. Journal of Microbiological Methods 1:149–155.

Clark, G., and M.W. Dick. 1974. Long-term storage and viability of aquatic Oomycetes. Transactions of the British Mycological Society 63:611–612.

Clark, J. 1995. Myxomycete reproductive systems: additional information. Mycologia 87:779–786.

Clark, J., and S.L. Stephenson. 1994. *Didymium ovoideum* culture and mating system. Mycologia 86:392–396.

Clark, M.C. (ed.). 1980. A Fungus Flora of Warwickshire. British Mycological Society, London, England.

Clarke, D.C., and M. Christensen. 1981. The soil microfungal community of a South Dakota grassland. Canadian Journal of Botany 59:1950–1960.

Clarke, R.T.J., and M.E. DiMenna. 1961. Yeasts from the bovine rumen. Journal of General Microbiology 25:113–117.

Clark-Walker, G.D. 1992. Evolution of mitochondrial genomes in fungi. International Review of Cytology 141:89–127.

Clark-Walker, G.D., C.R. McArthur, and K.S. Sriprakash. 1985. Location of transcriptional control signals and transfer RNA sequences in *Torulopsis glabrata* mitochondrial DNA. EMBO Journal 4:465–473.

Claus, R., H.O. Hoppen, and H. Karg. 1981. The secret of truffles: a steroidal pheromone? Experientia 37:1178–1179.

Clauzade, G., and C. Roux. 1985. Likenoj de okcidenta Eǔropo. Illustra determinlibro. Bulletin de la Société Botanique du Centre-Ouest, n.s., num. spéc. 7:1–893.

Clauzade, G., P. Diederich, and C. Roux. 1989. Nelikeniĝintaj fungoj likenologaj. Ilustrita determinlibro. Bulletin de la Société Linnéenne de Provence 1:1–142.

Clay, K. 1988. Fungal endophytes of grasses: a defensive mutualism between plants and fungi. Ecology 69:10–16.

———. 1990. Fungal endophytes of grasses. Annual Review of Ecology and Systematics 21:275–297.

———. 1994. The potential role of endophytes in ecosystems. Pp. 73–86. *In* C.W. Bacon and J.W. White, Jr. (eds.), Biotechnology of Endophytic Fungi of Grasses. CRC Press, Boca Raton, Florida.

Claydon, N., M. Allen, J.R. Hanson, and A.G. Avent. 1987. Antifungal alkyl pyrones of *Trichoderma harzianum*. Transactions of the British Mycological Society 88:503–513.

Clements, F.E., and C.L. Shear. 1931. The Genera of Fungi. H.W. Wilson, New York.

Clifford, H.T., and W. Stephenson. 1975. An Introduction to Numerical Classification. Academic Press, London, England.

Cohen, J. 1988. Statistical power analysis for the behavioral sciences. 2nd ed. Lawrence Erlbaum Associates, Hilsdale, New Jersey.

Coker, W.C. 1923. The Saprolegniaceae, with Notes on Other Water Molds. University of North Carolina Press, Chapel Hill, North Carolina.

Coker, W.C., and J.N. Couch. 1928. The Gasteromycetes of the Eastern United States and Canada. University of North Carolina Press, Chapel Hill, North Carolina. [reprinted 1974, Dover Publications, Mineola, New York]

Cole, G.T., T. Sekiya, R. Kasai, and Y. Nozawa. 1980. Morphogenesis and wall chemistry of the yeast, "intermediate," and hyphal phases of the dimorphic fungus, *Mycotypha poitrasii*. Canadian Journal of Microbiology 26:36–49.

Coley-Smith, J.R., and R.C. Cooke. 1971. Survival and germination of fungal sclerotia. Annual Review of Phytopathology 9:65–92.

Coley-Smith, J.R., K. Verhoeff, and W.R. Jarvis. 1980. The Biology of *Botrytis*. Academic Press, London, England.

Colgan, W., M.A. Castellano, and J.W. Spatafora. 1997. Systematics of the Hysterangiaceae. Inoculum 48:7.

Collado, J., G. Platas, I. González, and F. Peláez. 2000. Geographical and seasonal influences on the distribution of fungal endophytes in *Quercus ilex*. New Phytologist. 144:525–532.

Collins, O.R. 1980. Apomictic-heterothallic conversion in a myxomycete, *Didymium iridis*. Mycologia 72:1109–1116.

———. 1981. Myxomycete genetics, 1960–1981. Journal of the Elisha Mitchell Scientific Society 97:101–125.

Collins, R.A., and A.M. Lambowitz. 1983. Structural variations and optional introns in the mitochondrial DNAs of *Neurospora* strains isolated from nature. Plasmid. 9:53–70.

Collins, V.G., and L.G. Willoughby. 1962. The distribution of bacteria and fungal spores in Blelham Tarn with particular reference to an experimental overturn. Archiv für Mikrobiologie 43:294–307.

Colwell, R.K., and J.A. Coddington. 1994. Estimating terrestrial biodiversity through extrapolation. Philosophical Transactions of the Royal Society London, B 345:101–118.

Colwell, R.R., R.A. Clayton, B.A. Ortiz-Conde, D. Jacobs, and E. Russek-Cohen. 1995. The microbial species concept and biodiversity. Pp. 3–15. *In* D. Allsopp, R.R. Colwell, and D.L. Hawksworth (eds.), Microbial Diversity and Ecosystem Function. CAB International, Wallingford, Oxon, United Kingdom.

Constantinescu, O., and S. Ryman. 1989. A new Ophiostoma on polypores. Mycotaxon 34:637–642.

Cook, A.A. 1975. Diseases of Tropical and Subtropical Fruits and Nuts. Hafner Press, New York.

———. 1978. Diseases of Tropical and Subtropical Vegetable and Other Plants. Hafner Press, New York.

Cook, L.M., K.D. Rigby, and M.R.D. Seaward. 1990. Melanic moths and changes in epiphytic vegetation in north-west England and north Wales. Biological Journal of the Linnean Society 39:343–354.

Cooke, D.E.L., A. Drenth, J.M. Duncan, G. Wagels, and C.M. Brasier. 2000. A molecular phylogeny of *Phytophthora* and related Oomycetes. Fungal Genetics and Biology 30:17–32.

Cooke, J.C., J.N. Gemma, and R.E. Koske. 1987. Observations of nuclei in vesicular-arbuscular mycorrhizal fungi. Mycologia 79:331–333.

Cooke, R. 1977. The Biology of Symbiotic Fungi. John Wiley, London, England.

Cooke, R.C., and B.E.S. Godfrey. 1964. A key to the nematode-destroying fungi. Transactions of the British Mycological Society 47:61–74.

Cooke, R.C., and A.D.M. Rayner. 1984. Ecology of Saprotrophic Fungi. Longman, London, England.

Cooke, R.C., and J.M. Whipps. 1980. The evolution of modes of nutrition in fungi parasitic on terrestrial plants. Biological Review 55:341–362.

———. 1993. Ecophysiology of Fungi. Blackwell Scientific, Oxford, England.

Cooke, W.B. 1955. Fungi, lichens and mosses in relation to vascular plant communities in eastern Washington and adjacent Idaho. Ecological Monographs 25:119–180.

———. 1979. The Ecology of Fungi. CRC Press, Boca Raton, Florida.

Cooke, W.B., and A.F. Bartsch. 1959. Aquatic fungi in water with high waste loads. Sewage and Industrial Wastes 31:1316–1322.

Cooksey, K.E. (ed.). 1997. Molecular Approaches to the Study of the Ocean. Kluwer Academic, Dordrecht, The Netherlands.

Cooney, D.G., and R. Emerson. 1964. Thermophilic Fungi. An Account of Their Biology, Activites, and Classification. W.H. Freeman, San Francisco, California.

Cooper, J.A. 1993. Estimation of zoospore density by dilution assay. Mycologist 7:113–115.

Cooper, J.E., and O.F. Jackson (eds.). 1981. Diseases of the Reptilia. Vol. 1. Academic Press, London, England.

Copland, J.W., and L.G. Willoughby. 1982. The pathology of *Saprolegnia* infections of *Anguilla anguilla* L. elvers. Journal of Fish Diseases 5:421–428.

Coppins, B.J. 2002. Checklist of Lichens of Great Britain and Ireland. 2nd ed. British Lichen Society, London, England.

Corda, A.C.J. 1839. Pracht-Flora europaeischer Schimmelbildungen. G. Fleischer, Leipzig, Germany.

Corn, P.S. 1993. Straight-line drift fences and pitfall traps. Pp. 109–117. *In* W.R. Heyer, M.A. Donnelly, R.W. McDiarmid, L.C. Hayek, and M.S. Foster. Measuring and Monitoring Biological Diversity: Standard Methods for Amphibians. Smithsonian Institution Press, Washington, D.C.

Cornejo, F.H., A. Varela, and S.J. Wright. 1994. Tropical forest litter decomposition under seasonal drought: nutrient release, fungi and bacteria. Oikos 70:183–190.

Cornelissen, J.H.C., and H. ter Steege. 1989. Distribution and ecology of epiphytic bryophytes and lichens in dry evergreen forests of Guyana. Journal of Tropical Ecology 5:131–150.

Corner, E.J.H. 1981. The agaric genera *Lentinus*, *Panus* and *Pleurotus* with particular reference to Malaysian species. Nova Hedwigia 69:1–169.

———. 1994. Agarics in Malaysia. I. Tricholomatoid. II. Mycenoid. Nova Hedwigia 109:1–271.

Corry, J.E.L. 1987. Relationships of water activity to fungal growth. Pp. 51–99. *In* L.R. Beuchat (ed.), Food and Beverage Mycology. 2nd ed. Van Nostrand Reinhold, New York.

Cottam, G., and J.T. Curtis. 1956. The use of distance measures in phytosociological sampling. Ecology 37:451–460.

Cotter, H.V.T., and G.F. Bills. 1985. Comparison of spatial patterns of sexual and vegetative states of *Boletinellus merulioides.* Transactions of the British Mycological Society 85:520–524.

Cotter, H.V.T., and R.O. Blanchard. 1982. The fungal flora of bark of *Fagus grandifolia.* Mycologia. 74:836–843.

Couch, J.N. 1935. *Septobasidium* in the United States. Journal of the Elisha Mitchell Scientific Society 51:1–77.

———. 1938. The Genus *Septobasidium.* University of North Carolina Press, Chapel Hill, North Carolina.

———. 1939. Technic for collection, isolation and culture of chytrids. Journal of the Elisha Mitchell Scientific Society 55:208–214.

Couch, J.N., and C.E. Bland (eds.). 1985. The Genus *Coelomomyces.* Academic Press, New York.

Courtecuisse, R., and B. Duhem. 1995. Mushrooms and Toadstools of Britain and Europe. Harper Collins, London, England.

Cox, G.W. 1996. Laboratory Manual of General Ecology. 7th ed. W.C. Brown, Dubuque, Iowa.

Cox, J.J. 1981. Notes on coprophilous Myxomycetes from the western United States. Mycologia 73:741–747.

Crane, J.L., and C.A. Shearer. 1994. Preliminary Observations on Latitudinal Distribution Patterns of Freshwater Ascomycetes. Abstracts of the Fifth International Mycological Congress, Vancouver, British Columbia.

Crespo, A., M.C. Molina, O. Blanco, B. Schroeter, L.G. Sancho, and D.L. Hawksworth. 2002. rDNA ITS and β-tubulin gene sequence analysis reveal two monophyletic groups within the cosmopolitan lichen *Parmelia saxatilis.* Mycological Research 106:788–795.

Crisan, E.V. 1973. Current concepts of thermophilism and the thermophilic fungi. Mycologia 65:1171–1198.

Crittenden, P.D., J.C. David, D.L. Hawksworth, and F.S. Campbell. 1995. Attempted isolation and success in the culturing of a broad spectrum of lichen-forming and lichenicolous fungi. New Phytologist 130:267–297.

Croan, S.C. 2000. Lyophilization of hypha-forming tropical wood-inhabiting Basidiomycotina. Mycologia 92:810–817.

Croan, S.C., H.H. Burdsall, Jr., and R.M. Rentmeester. 1999. Preservation of tropical wood-inhabiting basidiomycetes. Mycologia 91:908–916.

Croft, J.H., V. Bhattacherjee, and K.E. Chapman. 1990. RFLP analysis of nuclear and mitochondrial DNA and its use in *Aspergillus* systematics. Pp. 309–320. *In* R.A. Samson and J.I. Pitt, (eds.), Modern Concepts in P*enicillium* and *Aspergillus* Classification. Plenum, New York.

Cromack, K., Jr., and B.A. Caldwell. 1992. The role of fungi in litter decomposition and nutrient cycling. Pp. 653–668. *In* G.C. Carroll and D.T. Wicklow (eds.), The Fungal Community, Its Organization and Role in the Ecosystem. 2nd ed. Marcel Dekker, New York.

Crous, P.W., and M.J. Wingfield. 1994. A monograph of *Cylindrocladium,* including anamorphs of *Calonectria.* Mycotaxon 51:341–435.

Cuello, A.C. 1983. Immunohistochemistry. John Wiley, Chichester, Great Britain.

Culberson, C.F. 1969. Chemical and Botanical Guide to Lichen Products. University of North Carolina Press, Chapel Hill, North Carolina.

———. 1970. Supplement to "Chemical and Botanical Guide to Lichen Products". Bryologist 73:177–377.

Culberson, C.F., W.L. Culberson, and A. Johnson. 1977. Second Supplement to "Chemical and Botanical Guide to Lichen Products". American Bryological and Lichenological Society, Saint Louis, Missouri.

———. 1988. Gene flow in lichens. American Journal of Botany 75:1135–1139.

Cullen, D. 1997. Recent advances on the molecular genetics of ligninolytic fungi. Journal of Biotechnology 53:273–289.

Cullen, D., and J.H. Andrews. 1984. Epiphytic microbes as biological control agents. Pp. 381–399. *In* T. Kosuge and E.W. Nester (eds.), Plant-Microbe Interactions. MacMillan, New York.

Culling, C.F.A., R.T. Allison, and W.T. Barr. 1985. Cellular Pathology Technique. 4th ed. Butterworths, London, England.

Cummings, D.J., K.L. McNally, J.M. Domenico, and E.T. Matsuura. 1990. The complete DNA sequence of the mitochondrial genome of *Podospora anserina.* Current Genetics 17:375–402.

Cummins, G.B. 1950. An index of the plant rusts recorded for continental China and Manchuria. Plant Disease Reporter Supplement 196:520–556.

———. 1962. Supplement to Arthur's manual of the rusts in United States and Canada. Hafner, New York.

———. 1971. The Rust Fungi of Cereals, Grasses and Bamboos. Springer-Verlag, New York.

———. 1978. Rust Fungi on Legumes and Composites in North America. University of Arizona Press, Tucson, Arizona.

Cummins, G.B., and Y. Hiratsuka. 1983. Illustrated Genera of Rust Fungi. Rev. ed. American Phytopathological Society Press, St. Paul, Minnesota.

Cunningham, J., and L. Bakshi. 1976. Two new genera of leaf-parasitic fungi (Basidiomycetidae, Brachybasidiaceae). Mycologia 68: 640–654.

Cunningham, J.E., and M.A. Pickard. 1985. Maltol, a metabolite of *Scytalidium uredinicola* which inhibits spore germination of *Endocronartium harknessii,* the western gall rust. Canadian Journal of Microbiology 31:1051–1055.

Cunningham, P.C. 1981. Isolation and culture. Pp. 103–123. *In* M.J.C. Asher and P.J. Shipton (eds.), Biology and Control of Take-All. Academic Press, London, England.

Currah, R.S. 1985. Taxonomy of the Onygenales: Arthrodermataceae, Gymnoascaceae, Myxotrichaceae and Onygenaceae. Mycotaxon 24:1–216.

Currie, C.R. 1995. Dissemination of the mycoparasite *Scytalidium uredinicola* by *Epuraea obliquus* (Coleoptera: Nitidulidae). Canadian Journal of Botany 73:1338–1344.

Curtis, F.C., G.H. Evans, V. Lillis, D.H. Lewis, and R.C. Cooke. 1978. Studies on Mucoralean mycoparasites: I. Some effects of *Piptocephalis* species on host growth. New Phytologist 80:157–166.

Curtis, J.T. 1959. The Vegetation of Wisconsin: An Ordination of Plant Communities. University of Wisconsin Press, Madison, Wisconsin.

Czeczuga, B., and M. Orłowska. 1997. Hyphomycetes fungi in rainwater falling from building roofs. Mycoscience 38:447–450.

Czeczuga, B., and L. Woronowicz. 1992. Studies on aquatic fungi. XXI. The Lake Mamry complex. Acta Mycologica 27:93–103.

Czederpiltz, D.L.L., G.R. Stanosz, and H.H. Burdsall, Jr. 1999. Forest management and the diversity of wood-inhabiting fungi. McIlvainea 14:34–45.

Dabrowa, N., J.W. Landau, V.D. Newcomer, and O.A. Plunkett. 1964. A survey of tide-washed coastal areas of southern California for fungi potentially pathogenic to man. Mycopathologia 24:137–150.

Daft, G.C., and P.H. Tsao. 1983. Susceptibility of *Phytophthora cinnamomi* and *P. parasitica* to fungi known to parasitize other Oomycetes. Transactions of the British Mycological Society 81: 71–76.

———. 1984. Parasitism of *Phytophthora cinnamomi* and *P. parasitica* spores by *Catenaria anguillulae* in a soil environment. Transactions of the British Mycological Society 82:485–490.

Dahlberg, A., and J. Stenlid. 1994. Size, distribution and biomass of genets in populations of *Suillus bovinus* (L.: Fr.) Roussel revealed by somatic incompatibility. New Phytologist 128:225–234.

Dahmen, H., T. Staub, and F.J. Schwinn. 1983. Technique for long-term preservation of phytopathogenic fungi in liquid nitrogen. Phytopathology 73:241–246.

Dallyn, H., and A. Fox. 1980. Spoilage of materials of reduced water activity by xerophilic fungi. Pp. 304–307. *In* G.W. Gould and J.E.L. Corry (eds.), Microbial Growth and Survival in Extremes of Environment. Academic Press, New York.

Dalpe, Y., W. Litton, and L. Sigler. 1989. *Scytalidium vaccinii* sp. nov., an ericoid endophyte of *Vaccinium angustifolium* roots. Mycotaxon 35:371–377.

Daniel, J.W., and H.H. Baldwin. 1964. Methods of culture for plasmodial myxomycetes. Pp. 9–41. *In* D.M. Prescott (ed.), Methods in Cell Physiology. Vol. I. Academic Press, New York.

Daniels, B.A. 1981. The influence of hyperparasites of vesicular-arbuscular mycorrhizal fungi on growth of citrus. Phytopathology 71:212–213.

Daniels, B.A., and J.A. Menge. 1980. Hyperparasitization. Phytopathology 70:584–588.

Daniels, B.A., and H.D. Skipper. 1982. Methods for the recovery and quantitative estimation of propagules from soil. Pp. 29–35. *In* N.C. Schenck (ed.), Methods and Principles of Mycorrhizal Research. American Phytopathological Society Press, St. Paul, Minnesota.

Danielsen, I.C., and H. Wolffhechel. 1991. Biological control of *Pythium ultimum* with the antagonist *Geomyces pannorum*. Pp. 194–198. *In* A.B.R. Beemster, G.J. Bollen, M. Gerlagh, M.A. Ruissen, B. Schippers, and A. Tempel (eds.), Biotic Interactions and Soil-borne Diseases. Elsevier, Amsterdam, The Netherlands.

da Silva, M., and D.W. Minter. 1995. Fungi from Brazil, recorded by Batista and co-workers. Mycological Papers 169:1–585.

Daszak P., L. Berger, A.A. Cunningham, A.D. Hyatt, D.E. Green, and R. Speare. 1999. Emerging infectious diseases and amphibian population declines. Emerging Infectious Diseases 5:735–748.

Daszak, P., A.A. Cunningham, and A.D. Hyatt. 2000. Emerging infectious diseases of wildlife—threats to biodiversity and human health. Science 287:443–449.

Daub, M.E., G.B. Leisman, R.A. Clark, and E.F. Bowden. 1992. Reductive detoxification as a mechanism of fungal resistance to singlet oxygen-generating photosensitizers. Proceedings of the National Academy of Sciences USA 89:9588–9592.

David, J.C., and D.L. Hawksworth. 1989. *Lauderlindsaya*, a new genus in the Verrucariales for *Sphaerulina chlorococca* (Leighton) R. Sant. Sydowia 41:108–121.

Davies, D.R., M.K. Theodorou, M.I.G. Lawrence, and A.P.J. Trinci. 1993. Distribution of anaerobic fungi in the digestive tract of cattle and their survival in faeces. Journal of General Microbiology 139:1395–1400.

Davis, E.E. 1956. Preservation of myxomycetes. Mycologia 57:986–989.

Davis, R.H., and F.J. de Serres. 1970. Genetic and microbiological research techniques for *Neurospora crassa*. Methods in Enzymology 17A:79–143.

Dawson, C.O., and J.C. Gentles. 1961. The perfect states of *Keratinomyces ajelloi* Vanbreuseghem, *Trichophyton terrestre* Durie & Frey and *Microsporum nanum* Fuentes. Sabouraudia 1:49–57.

Dayal, R. 1973. Key to phycomycetes predaceous or parasitic in nematodes or amoebae I. Zoopagales. Sydowia 27:293–301.

Dayal, R., and G.L. Barron. 1970. *Verticillium psalliotae* as a parasite of *Rhopalomyces*. Mycologia 62:826–830.

Deacon, J.W. 1976. Studies on *Pythium oligandrum*, an aggressive parasite of other fungi. Transactions of the British Mycological Society 66:383–391.

——. 1988a. Behavioural responses of fungal zoospores. Microbiological Sciences 5:249–252.

——. 1988b. Biocontrol of soil-borne plant pathogens with introduced inocula. Philosophical Transactions of the Royal Society, London, B 318:249–264.

Deacon, J.W., and C.M. Henry. 1978. Mycoparasitism by *Pythium oligandrum* and *P. acanthicum*. Soil Biology and Biochemistry 10:409–415.

Deacon, J.W., and L.V. Fleming. 1992. Interactions of ectomycorrhizal fungi. Pp. 249–300. *In* M.F. Allen (ed.), Mycorrhizal Functioning: An Integrative Plant-Fungal Process. Chapman Hall, New York.

Deacon, J.W., S.A.K. Laing, and L.A. Berry. 1991. *Pythium mycoparasiticum* sp.nov., an aggressive mycoparasite from British soils. Mycotaxon 42:1–8.

De Bary, A. 1866. De la génération sexuelle dans les champignons. Annales des Sciences Naturelle, Botanique (5ᵉ sér.) 5:343–367.

de Cock, A.W.A.M. 1986. Marine Pythiaceae from decaying seaweeds in the Netherlands. Mycotaxon, 25:101–110.

de Hoog, G.S. 1978. Notes on some fungicolous hyphomycetes and their relatives. Persoonia 10:33–81.

——. (ed.). 1999. Ecology and evolution of black yeasts and their relatives. Studies in Mycology. 43:1–208.

de Hoog, G.S., T. Hijwegen, and W.H. Batenburg-van der Vegte. 1991. A new species of *Dissoconium*. Mycological Research 95:679–682.

de Hoog, G.S., J. Guarro, J. Gené, and M.J. Figueras. 2000. Atlas of Clinical Fungi. 2nd ed. Centraalbureau voor Schimmelcultures, Baarn, The Netherlands.

de Hoog, G.S., M.T. Smith, and E. Guého. 1986. A revision of the genus *Geotrichum* and its teleomorphs. Studies in Mycology 29:1–131.

Dehority, B.A., and G.A. Varga. 1991. Bacterial and fungal numbers in ruminal and cecal contents of the blue duiker (*Cephalophus monticola*). Applied and Environmental Microbiology 57:469–472.

Deighton, F.C. 1960. Collecting fungi in the tropics. pp. 78–83 *In* Herb. I.M.I. Handbook, Methods in Use at the Commonwealth Mycological Institute. Commonwealth Mycological Institute, Kew, Surrey, England.

——. 1969. Microfungi. IV: Some hyperparasitic hyphomycetes, and a note on *Cercosporella uredinophila* Sacc. Mycological Papers 118:1–41.

Deighton, F.C., and K.A. Pirozynski. 1972. Microfungi. V. More hyperparasitic Hyphomycetes. Mycological Papers 128:1–112.

Delatour, C. 1991. A very simple method for long-term storage of fungal cultures. European Journal of Forest Pathology 21:444–445.

DeLay, J. 1970. Reexamination of the association between melting point, buoyant density, and chemical base composition of deoxyribonucleic acid. Journal of Bacteriology 101:738–754.

De Leo, F., C. Urzì, and G.S. de Hoog. 1999. Two *Coniosporium* species from rock surfaces. Studies in Mycology 43:70–79.

Delmotte, F., E. Bucheli, and J.A. Shykoff. 1999. Host and parasite population structure in a natural plant-pathogen system. Heredity 82:300–308.

De Mars, B.G., and R.E.J. Boerner. 1995. Mycorrhizal status of *Deschampsia antarctica* in the Palmer Station area, Antarctica. Mycologia 87:451–453.

den Boer, P.J. 1977. Dispersal power and survival. Carabids in a cultivated countryside. Miscellaneous Papers Landbouwhogeschool Wageningen 14:1–190.

Denison, W.C., D.M. Tracy, F.M. Rhoades, and M. Sherwood. 1972. Direct, non-destructive measurement of biomass and structure in living old-growth Douglas-fir. Pp. 147–158. *In* J.F. Franklin, L.J. Dempster, and R.H. Waring (eds.), Research on Coniferous Forest Ecosystems, US/IBP Proceedings Northwest Scientific Association 45th Annual Meeting. USDA Forest Service, Pacific Northwest Forest and Range Experiment Station, Portland, Oregon.

Dennis, C., and J. Webster. 1971a. Antagonistic properties of species-groups of *Trichoderma*: I. Production of non-volatile antibiotics. Transactions of the British Mycological Society 57:25–39.

——. 1971b. Antagonistic properties of species-groups of *Trichoderma*. II. Production of volatile antibiotics. Transactions of the British Mycological Society 57:41–48.

——. 1971c. Antagonistic properties of species-groups of *Trichoderma*. III. Hyphal interaction. Transactions of the British Mycological Society 57:363–369.

Dennis, R.W.G. 1956. Some Xylarias of tropical America. Kew Bulletin 11:401–444.

——. 1957. Further notes on tropical American Xylariaceae. Kew Bulletin. 12:297–332.

——. 1958. Some Xylosphaeras of tropical Africa. Revista Biologica 1:175–208.

——. 1968. British Ascomycetes. J. Cramer, Lehre, Germany.

——. 1970. Fungus Flora of Venezuela and Adjacent Countries. Kew Bulletin Additional Series No.3. Lubrecht and Cramer, Port Jervis, New York.

——. 1978. The British Ascomycetes. J. Cramer, Vaduz, Liechtenstein.

——. 1986. Fungi of the Hebrides. Royal Botanic Gardens, Kew, Surrey, England.

——. 1995. The Fungi of South East England. Royal Botanic Gardens, Kew, Surrey, England.

DePriest, P.T. 1993. Variation in the *Cladonia chlorophaea* complex I: Morphological and chemical variation in southern Appalachian populations. Bryologist 96:555–563.

——. 1994. Variation in the *Cladonia chlorophaea* complex II: Ribosomal DNA variation in a southern Appalachian population. Bryologist 97:117–126.

Descals, E. 1997. Ingoldian fungi: some field and laboratory techniques. Bolletí de la Societat d'Història Natural de les Balears 40:169–221.

Desjardins, A.E., and T.M. Hohn. 1997. Mycotoxins in plant pathogenesis. Molecular Plant-Microbe Interactions 10:147–152.

de Trogoff, H., and J.L. Ricard. 1976. Biological control of *Verticillium malthousei* by *Trichoderma viride* spray on casing soil in commercial mushroom production. Plant Disease Reporter 60:677–680.

De Wachter, R., J.-M. Neefs, A. Goris, and Y. Van de Peer. 1992. The gene coding for small ribosomal subunit RNA in the basidiomycete *Ustilago maydis* contains a group I intron. Nucleic Acids Research 20:1251–1257.

Dewan, M.M., and K. Sivasithamparam. 1988. A plant-growth-promoting sterile fungus from wheat and rye-grass roots with potential for suppressing take-all. Transactions of the British Mycological Society 91:687–717.

——. 1989. Efficacy of treatment with a sterile red fungus for control of take-all in wheat. New Zealand Journal of Crop and Horticultural Science 17:333–336.

Dhillion, S.S., and J.C. Zak. 1993. Microbial dynamics in arid ecosystems: desertification and the potential of mycorrhizas. Revista Chilena de Historia Natural 665:253–270.

Dhingra, O.D., and N.M. Khare. 1973. Biological control of *Rhizoctonia bataticola* on urid bean. Phytopathologische Zeitschrift 76:23–29.

Di Bonito, R., M.L. Elliot, and E.A. Des Jardin. 1995. Detection of an arbuscular mycorrhizal fungus in roots of different plant species with the PCR. Applied and Environmental Microbiology 61:2809–2810.

Di Menna, M.E. 1959. Yeasts from the leaves of pasture plants. New Zealand Journal of Agricultural Research 2:394–405.

Dick, M.W. 1966. The Saprolegniaceae of the environs of Blelham Tarn: sampling techniques and the estimation of propagule numbers. Journal of General Microbiology 42:257–282.

——. 1971. The ecology of Saprolegniaceae in lentic and littoral muds with a general theory of fungi in the lake ecosystem. Journal of General Microbiology 65:325–337.

——. 1976. The ecology of aquatic phycomycetes. Pp. 513–542. *In* E.B. Gareth Jones (ed.), Recent Advances in Aquatic Mycology. Elek Science, London, England.

——. 1990a. Phylum Oomycota. Pp. 681–685. *In* L. Margulis, J.O. Corliss, M. Melkonian, and D. Chapman (eds.), Handbook of Protoctista. Jones and Bartlett, Boston, Massachusetts.

——. 1990b. Keys to *Pythium*. University of Reading Press, Reading, United Kingdom.

——. 1995. Sexual reproduction in the *Peronosporomycetes* (chromistan fungi). Canadian Journal of Botany 73(Suppl. 1):S712–S724.

——. 1998. The species and systematic position of *Crypticola* in the peronosporomycetes, and new names for *Halocrusticida* and species therein. Mycological Research 102:1062–1066.

——. 1999. Fungi: Classification of the Peronosporomycetes. *In* R.K. Robinson, C.A. Batt, and P.D. Patel (eds.), Encyclopedia of Food Microbiology (on-line version). Academic Press, San Diego, California.

——. 2001a. The Peronosporomycetes. Pp. 39–72. *In* D.J. McLaughlin, E.G. McLaughlin, and P.A. Lempke (eds.), The Mycota. Vol. VII. Systematics and Evolution. Part A. Springer-Verlag, Berlin, Germany.

——. 2001b. Straminipilous Fungi: Systematics of the Peronosporomycetes Including Accounts of the Marine Straminipilous Protists, the Plasmodiophorids and Similar Organisms. Kluwer Academic, Dordrecht, The Netherlands.

Dick, M.W., and D.L. Hawksworth. 1985. A synopsis of the biology of the Ascomycotina. Botanical Journal of the Linnean Society (London) 91:175–179.

Dick, M.W., and H.V. Newby. 1961. The occurrence and distribution of Saprolegniaceae in certain soils of south-east England. I. Occurrence. Journal of Ecology 49:403–419.

Dick, M.W., M.C. Vick, J.G. Gibbings, T.A. Hedderson, and C.C. Lopez Lastra. 1999. 18S rDNA for species of *Leptolegnia* and other Peronosporomycetes: justification for the subclass taxa Saprolegniomycetidae and Peronosporomycetidae and division of the Saprolegniaceae sensu lato into the Leptolegniaceae and Saprolegniaceae. Mycological Research 103:1119–1125.

Dickman, L.A., A.E. Liberta, and R.C. Anderson. 1984. Ecological interaction of little bluestem and vesicular-arbuscular mycorrhizal fungi. Canadian Journal of Botany 62:2272–2277.

Diederich, P. 1992. *Stromatopogon baldwinii* Zahlbr., a lichenicolous Coelomycete with dimorphic conidia. Lichenologist 24:371–375.

——. 1996. The lichenicolous Heterobasidiomycetes. Bibliotheca Lichenologica 61:1–198.

Diederich, P., and M.S. Christiansen. 1994. *Biatoropsis usnearum* Räsänen, and other heterobasidiomycetes on *Usnea*. Lichenologist 26:47–66.

Diederich, P., and J. Etayo. 1994. Taxonomic notes on the genus *Plectocarpon* (lichenicolous Ascomycotina). Nordic Journal of Botany 14:589–600.

——. 2000. A synopsis of the genera *Skyttea*, *Llimoniella* and *Rhymbocarpus* (lichenicolous Ascomycota, Leotiales). Lichenologist 32:423–485.

Diederich, P., and E. Sérusiaux. 2000. The lichens and lichenicolous fungi of Belgium and Luxembourg: An Annotated Checklist. Musée national d'Historie naturelle de Luxembourg.

Diederich, P., and M. Wedin. 2000. The species of *Hemigrapha* (lichenicolous Ascomycetes, Dothideales) on Peltigerales. Nordic Journal of Botany 20:203–214.

Diederich, P., M. Zhurbenko, and J. Etayo. 2002. The lichenicolous species of *Odontotrema* (syn. *Lethariicola*) (Ascomycota, Ostropales). Lichenologist 34:479–501.

Diehl, W.W. 1950. *Balansia* and the Balansieae in America. U.S. Department of Agriculture, Washington, D.C. Agriculture Monograph 4.

Dighton, J., J.M. Poskitt, and D.M. Howard. 1986. Changes in occurrence of basidiomycete fruit bodies during forest stand development with specific reference to mycorrhizal species. Transactions of the British Mycological Society 87:163–171.

Dik, A.J., and J.A. van Pelt. 1992. Interaction between phyllosphere yeasts, aphid honeydew and fungicide effectiveness in wheat under field conditions. Plant Pathology 41:661–675.

Dix, N.J. 1964. Colonization and decay of bean roots. Transactions of the British Mycological Society 47:285–292.

Dix, N.J., and J. Webster. 1995. Fungal Ecology. Chapman Hall, London, England.

Dixon, D.M., I.F. Salkin, R.A. Duncan, N.J. Hurd, J.H. Haines, M.E. Kemna, and F.B. Coles. 1991. Isolation and characterization of *Sporothrix schenckii* from clinical and environmental sources associated with the largest U.S. epidemic of sporotrichosis. Journal of Clinical Microbiology 29:1106–1113.

Döbbler, P. 1979. Untersuchungen an moosparasitischen Pezizales aus der Verwandtschaft von *Octospora*. Nova Hedwigia 31:817–864.

Dobbs, C.G., and M.P. English. 1954. *Piptocephalis xenophila* sp. nov. parasitic on non-mucorine hosts. Transactions of the British Mycological Society 37:375–389.

Dodd, J.C., A. Estrada, and M.J. Jeger. 1992. Epidemiology of *Colletotrichum gloeosporioides* in the tropics. Pp. 308–325. *In* J.A. Bailey and M.J. Jeger (eds.), *Colletotrichum*: Biology, Pathology and Control. CAB International, Wallingford, Oxon, United Kingdom.

Dodd, J.C., V. Gianinazzi-Pearson, S. Rosendahl, and C. Walker. 1994. European Bank of Glomales—an essential tool for efficient international and interdisciplinary collaboration. Pp. 41–45. *In* S. Gianinazzi and H. Schüepp (eds.), Impact of Arbuscular Mycorrhizas on Sustainable Agriculture and Natural Ecosystems. Birhaüser Verlag, Basel, Switzerland.

Dogma, I.J., Jr. 1973. Isolation and pure culture techniques for monocentric Chytridiomycetes. Nova Hedwigia 24:379–392.

Doi, Y. 1972. Revision of the Hypocreales with cultural observations. IV. The genus *Hypocrea* and its allies in Japan (2). Enumeration of the species. Bulletin of the National Science Museum 15:649–751.

——. 1978. Revision of the Hypocreales with cultural observations. XI. Additional notes on *Hypocrea* and its allies in Japan (1). Bulletin of the National Science Museum, Series B (Botany) 4:19–26.

Doidge, E.M. 1939. Preliminary study of the South African rust fungi. Bothalia 4:895–937.

Doll, R. 1973. Zwei bemerkenswerte Arten der Sphaeriales. Westfälische Pilzbriefe 9:64–67.

Domanski, S. 1972. Fungi: Polyporaceae I, Mucronoporaceae I. Foreign Scientific Publications, Warsaw, Poland.

Domanski, S., H. Ortos, and A. Skirgietto. 1973. Fungi. Foreign Scientific Publications, Warsaw, Poland.

Domsch, K.H., W. Gams, and T.-H. Anderson. 1980. Compendium of Soil Fungi. Vols. I, II. Academic Press, London, United Kingdom.

Domsch, K.H., W. Gams, and T. Anderson. 1993. Compendium of Soil Fungi. IHW-Verlag, Eching, Germany.

Donegan, K.K., D.L. Schaller, J.K. Stone, L.M. Ganio, G. Reed, P.B. Hamm, and R.J. Seidler. 1996. Microbial populations, fungal species diversity and plant pathogen levels in field plots of potato plants expressing the *Bacillus thuringiensis* var. *tenebrionis* endotoxin. Transgenic Research 5:25–35.

Donk, M.A. 1966. Check list of European hymenomycetous Heterobasidiae. Persoonia 4:145–335.

——. 1972. The heterobasidiomycetes: A reconnaissance—I. A restricted emendation. Proceedings of the Koninklijke Nederlandse Akademie van Wetenschappen. Ser. C. 75:365–375.

Dons, J.J.M., and J.G.H. Wessels 1980. Sequence organization of the nuclear DNA of *Schizophyllum commune*. Biochimica et Biophysica Acta 607:385–396.

Dorenbosch, M.M.J. 1970. Key to nine ubiquitous soil-borne Phoma-like fungi. Persoonia 19:173–186.

Dorier, A., and C. Degrange. 1961. L'evolution de l'*Ichthyosporidium* (*Ichthyophonus*) *hoferi* (Plehn and Mulsow) chez les salmonides d'elevage (truite arc en ceil et saumon de fontaine). Travaux du Laboratoire d'Hydrobiologie et de Pisciculture de l'Université de Grenoble 1960/1961:7–44

Dorst, J., and P. Dandelot. 1970. A Field Guide to the Larger Mammals of Africa. Collins, London, England.

Dos Santos, A.F., and O.D. Dhingra. 1982. Pathogenicity of *Trichoderma* spp. on the sclerotia of *Sclerotinia sclerotiorum*. Canadian Journal of Botany 60:472–475.

Doty, M.S., and D.W. Slater. 1946. A new species of *Heterosporium* pathogenic on young chinook salmon. American Midland Naturalist 36:663–665.

Douds, D.D., Jr., and N.C. Schenck. 1990. Cryopreservation of spores of vesicular-arbuscular mycorrhizal fungi. New Phytologist 115:667–674.

Dowding, P., and P. Widden. 1974. Some relationships between fungi and their environment in tundra regions. Pp. 123–150. *In* A.J. Holding, O.W. Heal, J.S.F. Maclean, Jr., and P.W. Flanagan (eds.), Soil organisms and decomposition in tundra. IBP Tundra Biome Steering Committee. Stockholm, Sweden.

Drechsler, C. 1933. Morphological diversity among fungi capturing and destroying nematodes. Journal of the Washington Academy of Sciences 23:138–141.

——. 1934. *Pedilospora dactylopaga* n. sp., a fungus capturing and consuming testaceous rhizopods. Journal of the Washington Academy of Sciences 24:395–402.

——. 1938. Two hyphomycetes parasitic on oospores of root-rotting Oomycetes. Phytopathology 28:81–103.

——. 1941a. Predaceous fungi. Biological Reviews of the Cambridge Philosophical Society 16:265–290.

——. 1941b. Some hyphomycetes parasitic on free-living terricolous nematodes. Phytopathology 31:773–802.

——. 1941c. Four phycomycetes destructive to nematodes and rhizopods. Mycologia 33:248–269.

——. 1943. Another hyphomycetous fungus parasitic on *Pythium* oospores. Phytopathology 33:227–333.

——. 1952. Another nematode-strangulating *Dactylella* and some related hyphomycetes. Mycologia 44:533–556.

——. 1955. A new species of *Rhopalomyces* occurring in Florida. Bulletin of the Torrey Botanical Club 82:473–479.

——. 1962. Two additional species of *Dactylella* parasitic on *Pythium* oospores. Sydowia 15:92–97.

——. 1963. A slender-spored *Dactylella* parasitic on *Pythium* oospores. Phytopathology 53:1050–1052.

Dreyfuss, M.M. 1986. Neue erkenntnisse aus einem pharmakologischen pilz-screening. Sydowia 39:22–36.

Dreyfuss, M.M., and I.H. Chapela. 1994. Potential of fungi in the discovery of novel, low-molecular weight pharmaceuticals. Pp. 49–80. *In* V.P. Gullo (ed.), The Discovery of Natural Products with Therapeutic Potential. Butterworth-Heinmann, Boston, Massachusetts.

Dreyfuss, M., and O. Petrini. 1984. Further investigations on the occurrence and distribution of endophytic fungi in tropical plants. Botanica Helvetica 94:33–40.

Dring, D.M. 1955. A periodic acid-schiff technique for staining fungi in higher plants. New Phytologist 54:277–279.

——. 1980. Contributions towards a rational arrangement of the *Clathraceae*. Kew Bulletin 35:1–96.

Droby, S., E. Chalutz, C.L. Wilson, and M. Wisniewski. 1989. Characterization of the biocontrol activity of *Debaryomyces hansenii* in the control of *Penicillium digitatum* on grapefruit. Canadian Journal of Microbiology 35:794–800.

Drooz, A.T. 1985. Insects of Eastern Forests. U.S. Department of Agriculture, Forest Service. Miscellaneous Publication No. 1426. Washington, D.C.

Dubos, B. 1987. Fungal antagonism in aerial agrobiocenoses. Pp. 107–135. *In* I. Chet (ed.), Innovative Approaches to Plant Disease Control. J. Wiley, New York.

Dudka, I.A. 1985. Ascomycetes-components of freshwater biocoenosis. Ukraeins'kyéí Botanichnyéí Zhurnal 42:86–95. [in Russian]

Durán, R. 1973. Ustilaginales. Pp. 281–300. *In* G.C. Ainsworth, F.K. Sparrow, and A.S. Sussman (eds.), The Fungi. An Advanced Treatise. IVB. A Taxonomic Review with Keys: Basidiomycetes and Lower Fungi. Academic Press, New York.

——. 1987. Ustilaginales of Mexico: Taxonomy, Symptomatology, Spore Germination, and Basidial Cytology. Washington State University Press, Pullman, Washington

Durán, R., and G.W. Fischer. 1961. The Genus *Tilletia*. Washington State University Press, Pullman, Washington.

Durrell, L.W., and L.M. Shields. 1960. Fungi isolated in culture from soils of the Nevada test site. Mycologia 52:636–641.

Dutta, B.K. 1981. Studies on some fungi isolated from the rhizosphere of tomato plants and the consequent prospect for the control of *Verticillium* wilt. Plant and Soil 63:209–216.

Dykstra, M.J. 1977. The possible phylogenetic significance of mitochondrial configurations in the acrasid cellular slime molds with reference to members of the Eumycetozoa and the fungi. Mycologia 69:579–591.

Dykstra, M.J., J.F. Levine, E.J. Noga, J.H. Hawkins, P. Gerdes, W.J. Hargis, Jr., H.J. Grier, and D. Te Strake. 1989. Ulcerative mycosis: a serious menhaden disease of the southeastern coastal fisheries of the United States. Journal of Fish Diseases 12:175–178.

Dyskstra, M.J., E.J. Noga, J.F. Levine, D.W. Moye, and J.H. Hawkins. 1986. Characterization of the *Aphanomyces* species involved with ulcerative mycosis (um) in menhaden. Mycologia 78:664–672.

Dylewski, D.P. 1990. Phylum Plasmodiophoromycota. Pp. 399–416. *In* L. Margulis, J.O. Corliss, M. Melkonian, and D. Chapman (eds.), Handbook of Protoctista. Jones and Bartlett, Boston, Massachusetts.

Eberhart, J.L., D.L. Luoma, and M.P. Amaranthus. 1996. Response of ectomycorrhizal fungi to forest management treatments—a new method for quantifying morphotypes. Pp. 96–99. *In* C. Azcon-Aguilar and J.M. Barea (eds.). Mycorrhizas in Integrated Systems from Genes to Plant Development. Office for Official Publications of the European Communities, Luxembourg.

Ebersohn, C., and A. Eicker. 1992. Coprophilous fungal species composition and species diversity on various dung substrates of African game animals. Botanical Bulletin Academia Sinica 33:85–95.

Eboh, D.O. 1986. A taxonomic survey of Nigerian rust fungi: *Uredinales nigerianensis*. IV. Mycologia 78:577–586.

Eckblad, F.-E., and A.-E. Torkelsen. 1974. Contributions to the Hypocreaceae and fungicolous Nectriaceae of Norway. Norwegian Journal of Botany 21:5–15.

Edgington, L.V., K.L. Khew, and G.L. Barron. 1971. Fungitoxic spectrum of benzimidazole compounds. Phytopathology 61:42–44.

Edman, J.C., and M.L. Sogin. 1994. Molecular phylogeny of *Pneumocystis carinii*. Pp. 91–105. *In* P.D. Walzer (ed.), *Pneumocystis carinii* Pneumonia. 2nd ed., Marcel Dekker, New York.

Edmunds, A.D., and S.L. Stephenson. 1996. Myxomycetes associated with the litter microhabitat in tropical forests of Costa Rica. Proceedings of the West Virginia Academy of Science 68:23–24.

Egea, J.M., and P. Torrente. 1994. El género de hongos liquenizados *Lecanactis* (Ascomycotina). Biblotheca Lichenologica 54.

Egger, K. 1995. Molecular analysis of ectomycorrhizal communities. Canadian Journal of Botany 73(Suppl.1):S1415-S1422.

Egger, K., R.M. Danielson, and J.A. Fortin. 1991. Taxonomy and population structure of E-strain mycorrhizal fungi inferred from ribosomal and mitochondrial DNA polymorphisms. Mycological Research 95:866–872.

Egger, K.N., and J.A. Fortin 1990. Identification of taxa of E-strain mycorrhizal fungi by restriction fragment analysis. Canadian Journal of Botany 68:1482–1488.

Egusa, S., and T. Ueda. 1972. A *Fusarium* sp. associated with black gill disease of the Kuruma prawn, *Penaeus japonicus* Bate. Bulletin of the Japanese Society of Scientific Fisheries 38:1253–1260.

Egli, S., F. Ayer, and F. Chatelain. 1990. Der Einfluss des Pilzsammelns auf die Pilzflora. Mycologica Helvetica 3:417–428.

——. 1997. Die Beschreibung der diversität von macromyzeten. Erfahrungen aus pilzökologischen langzeitstudien im pilzreeservat La Chanéaz, FR. Mycologia Helvetica 9:19–32.

Eisenberg, R.M. 1976. Two-dimensional microdistribution of cellular slime molds in forest soil. Ecology 57:380–384.

El Shafie, A.E., and J. Webster. 1979. *Curvularia* species as parasites of *Rhizopus* and other fungi. Transactions of the British Mycological Society 73:352–353.

Elad, Y., and I. Chet. 1983. Improved selective media for isolation of *Trichoderma* spp. or *Fusarium* spp. Phytoparasitica 11:55–58.

Elad, Y., R. Barak, and I. Chet. 1983a. Possible role of lectins in mycoparasitism. Journal of Bacteriology 154:1431–1435.

——. 1984. Parasitism of sclerotia of *Sclerotium rolfsii* by *Trichoderma harzianum*. Soil Biology and Biochemistry 16:381–386.

Elad, Y., I. Chet, P. Boyle, and Y. Henis. 1983b. Parasitism of *Trichoderma* spp. on *Rhizoctonia solani* and *Sclerotium rolfsii*—scanning electron microscopy and fluorescence microscopy. Phytopathology 73:85–88.

Elad, Y., I. Chet, and Y. Henis. 1981. A selective medium for improving quantitative isolation of *Trichoderma* spp. from soil. Phytoparasitica 9:59–67.

——. 1982. Degradation of plant pathogenic fungi by *Trichoderma harzianum*. Canadian Journal of Microbiology 28:719–725.

Elad, Y., I. Chet, and J. Katan. 1980. *Trichoderma harzianum*: a biocontrol agent effective against *Sclerotium rolfsii* and *Rhizoctonia solani*. Phytopathology 70:119–121.

Elad, Y., R. Lifshitz, and R. Baker. 1985. Enzymatic activity of the mycoparasite *Pythium nunn* during interaction with host and nonhost fungi. Physiological Plant Pathology 27:131–148.

Elad, Y., Z. Sadowski, and I. Chet. 1983c. Detection of mycoparasitism by infrared photomicrography. Microbial Ecology 9:185–187.

——. 1987. Scanning electron microscopical observations of early stages of interaction of *Trichoderma harzianum* and *Rhizoctonia solani*. Transactions of the British Mycological Society 88:259–263.

Elamo, P., M.L. Helander, and I. Saloniemi.1999. Birch family and environmental conditions affect endophytic fungi in leaves. Oecologia 118:151–156.

Eldridge, D.J., and R. Rosentreter. 1999. Morphological groups: a framework for monitoring microphytic crusts in arid landscapes. Journal of Arid Environments 41:11–25.

Eldridge, D.J., and M.E. Tozer. 1996. Soil crust *Xanthoparmeliae*: key indicators of rangeland health in semi-arid New South Wales. Australasian Lichenological Newsletter 38:19–20.

Eleuterius, L.N. 1976. The distribution of *Juncus roemerianus* in the salt marshes of North America. Chesapeake Science 17:289–292.

Eliasson, U., and N. Lundqvist. 1979. Fimicolous myxomycetes. Botaniska Notiser 132:551–568.

Eliasson, U.H., and H.W. Keller. 1999. Coprophilous myxomycetes: updated summary, key to species, and taxonomic observations on *Trichia brunnea*, *Arcyria elaterensis*, and *Arcyria stipata*. Karstenis 39:1–10.

Eliasson, U.H., H.W. Keller, and J.D. Schoknecht. 1991. *Kelleromyxa*, a new generic name for *Licea fimicola* (Myxomycetes). Mycological Research 95:1201–1207.

Elkan, E. 1981. Pathology and histopathological techniques. Pp. 75–91. *In* J.E. Cooper and O.F. Jackson (eds.), Diseases of the Reptilia. Vol. 1. Academic Press, London, Great Britain.

Elliot, S., M.W. Sabelis, A. Janssen, C.P.S. van der Geest, E.A.M. Berling, and J. Fransen. 2000. Can plants use entomopathogens as bodyguards? Ecology Letters 3:228–235.

Ellis, D., and T.J. Pfeiffer. 1990. Natural habitat of *Cryptococcus neoformans* var. *gattii*. Journal of Clinical Microbiology 28:1642–1644.

Ellis, J.J. 1966. On growing *Syncephalis* in pure culture. Mycologia 58:465–469.

——. 1979. Preserving fungus strains in sterile water. Mycologia 71:1072–1075.

Ellis, M.B. 1968. Dematiaceous Hyphomycetes. IX. *Spiropes* and *Pleurophragmium*. Mycological Papers 114:1–44.

——. 1971. Dematiaceous Hyphomycetes. Commonwealth Mycological Institute, Kew, Surrey, United Kingdom.

——. 1976. More dematiaceous hyphomycetes. Commonwealth Mycological Institute, Kew, Surrey, United Kingdom.

——. 1993a. Dematiaceous hyphomycetes. CAB International, Wallingford, Oxon, United Kingdom.

——. 1993b. More dematiaceous hyphomycetes. CAB International, Wallingford, Oxon, United Kingdom.

Ellis, M.B., and J.P. Ellis. 1985. Microfungi on Land Plants. An Identification Guide. Croom Helm, London, England.

——. 1988. Microfungi on Miscellaneous Substrates. Croom Helm, London, England.

——. 1997. Microfungi on Land Plants: An Identification Guide. Richmond, London, England.

Ellsworth, D.L., K.D. Rittenhouse, and R.L. Honeycutt. 1993. Artifactual variation in randomly amplified polymorphic DNA banding patterns. BioTechniques 14:214–217.

Embree, R.W. 1978. *Radiomyces*. Pp. 138–140. *In* M.S. Fuller (ed.), Lower Fungi in the Laboratory. Palfrey Contributions in Botany. No. 1. Department of Botany, University of Georgia, Athens, Georgia.

Emerson, R. 1950. Current trends of experimental research on the aquatic phycomycetes. Annual Review of Microbiology 4:169–200.

——. 1958. Mycological organization. Mycologia 50:589–621.

Emerson, R., and D.O. Natvig. 1981. Adaptation of fungi to stagnant waters. Pp. 109–128. *In* D.T. Wicklow and G.C. Carrol (eds.), The Fungal Community: Its Organization and Role in the Ecosystem. Vol. 2. Marcel Dekker, New York.

Emmons, C.W. 1930. *Cicinnobolus cesatii*, a study in host-parasite relationships. Bulletin of the Torrey Botanical Club 57:421–441.

English, M.P. 1968. Invasion of the skin by filamentous non-dermatophyte fungi. British Journal of Dermatology 80:282–286.

Epplen, J.T. 1988. On simple repeated GAT/CA sequences in animal genomes: a critical reappraisal. Journal of Heredity 79:409–417.

Eren, J., and D. Pramer. 1965. The most probable number of nematode-trapping fungi in soil. Soil Science 99:285.

Eriksson, J., and L. Ryvarden. 1973. The Corticiaceae of North Europe. Vol. 2. Fungiflora, Oslo, Norway.

——. 1975. The Corticiaceae of North Europe. Vol. 3. Fungiflora, Oslo, Norway.

——. 1976. The Corticiaceae of North Europe. Vol 4. Fungiflora, Oslo, Norway.

Eriksson, J., K. Hjortstam, and L. Ryvarden. 1978. The Corticiaceae of North Europe. Vol. 5. Fungiflora, Oslo, Norway.

——. 1984. The Corticiaceae of North Europe. Vol. 7. Fungiflora, Oslo, Norway.

Eriksson, O. 1966. On *Eudarluca caricis* (Fr.) O. Erikss., comb. nov., a cosmopolitan uredinicolous pyrenomycete. Botaniska Notiser 119:33–69.

——. 1967. On graminicolous pyrenomycetes from Fennoscandia. 2. Phragmosporous and scolecosporous species. Archiv för Botanik 6:381–440.

——. 1992. The Non-Lichenized Pyrenomycetes of Sweden. Sbt.-Forlaget, Lund, Sweden.

Eriksson, O.E., and D.L. Hawksworth. 1993. Outline of the Ascomycetes-1993. Systema Ascomycetum 12:51–257.

——. 1998. Outline of the Ascomycetes—1998. Systema Ascomycetum 16:83–296.

Eriksson, O.E., and K. Winka. 1998. Families and higher taxa of Ascomycota. Myconet 1:17–24. Available on the web at http://www.umu.se/myconet/publ.html

Eriksson, O.E., H.-O. Baral, R.S. Currah, K. Hansen, C.P. Kurtzman, G. Rambold, and T. Laessøe (eds.). 2002. Outline of Ascomycota—2003. Myconet 9:1–89.

Eriksson, O.E., A. Svedskog, and S. Landvik. 1993. Molecular evidence for the evolutionary hiatus between *Saccharomyces cerevisiae* and *Schizosaccharomyces pombe*. Systema Ascomycetum 11:119–162.

Erlich, H.A. (ed.). 1992. PCR Technology: Principles and Applications for DNA Amplification. W.H. Freeman, New York.

Erlich, H.A., and T.L. Bugawan. 1992. HLA class II gene polymorphism: DNA typing, evolution, and relationship to disease susceptibility. Pp 193–208. *In* H.A. Erlich (ed.), PCR Technology: Principles and Applications for DNA Amplification. W.H. Freeman, New York

Erwin, D.C., and O.K. Ribeiro. 1996. *Phytophthora* Diseases Worldwide. American Phytopathological Society Press, St. Paul, Minnesota.

Erwin, D.C., S. Bartnicki-Garcia, and P.H. Tsao (eds.). 1983. Phytophthora: Its Biology, Taxonomy, Ecology and Pathology. American Phytopathological Society Press, St. Paul, Minnesota.

Espinosa-Garcia, F.J., and J.H. Langenheim. 1990. The endophytic fungal community in leaves of a coastal redwood population–diversity and spatial patterns. New Phytologist 116:89–97.

Esquivel-R., E.A. 1984. *Cladosporiella cercosporicola* Deighton, hyperparásito del agente causal de la mancha de la hoja de la caña de azucar, *Cercosporidium koepkei* Esquivel. en Panama. Phytopathology 74:1014.

Esseen, P.-A., K.-E. Renhorn, and R.B. Pettersson. 1996. Epiphytic lichen biomass in managed and old-growth boreal forests: effect of branch quality. Ecological Applications 6:228–238.

Esslinger, T.L. 2003. A Cumulative Checklist of the Lichen-Forming, Lichenicolous, and Allied Fungi of the Continental United States and Canada. On the web at www.ndsu.nodak.edu/instruct/esslinge/chcklst/chcklst7.htm

Esslinger, T.L., and R.S. Egan. 1995. A sixth checklist of the lichen-forming, lichenicolous, and allied fungi of the continental United States and Canada. Bryologist 98:467–549.

Etayo, J. 1998. Some hypocrealean lichenicolous fungi from Southwest Europe. Nova Hedwigia 67:499–509.

Etayo, J., and P. Diederich. 1996. Lichenicolous fungi from the western Pyrenees, France and Spain. III. Species on *Lobaria pulmonaria*. Bulletin de la Société des Naturalistes Luxembourgeois 97:93–118.

Etter, B. 1929. New media for developing sporophores of wood-rot fungi. Mycologia 21:197–203.

Evans, G.H., and R.C. Cooke. 1981. *Piptocephalis*-induced changes in organization of marginal hyphae in *Mycotypha microspora* colonies. Transactions of the British Mycological Society 76:343–345.

——. 1982. Studies on mucoralean mycoparasites: III. Diffusible factors from *Mortierella vinacea* Dixon-Stewart that direct germ tube growth in *Piptocephalis fimbriata* Richardson and Leadbeater. New Phytologist 91:245–253.

Evans, G.H., F.C. Curtis, and R.C. Cooke. 1981. Host carbohydrate composition and trehalase activity in relation to mycoparasitism in

Piptocephalis species. Transactions of the British Mycological Society 77:21–26.

Evans, H.C. 1971. Thermophilous fungi of coal spoil tips. II. Occurrence, distribution and temperature relationships. Transactions of the British Mycological Society 57:255–266.

Fahima, T., and Y. Henis. 1995. Quantitative assessment of the interaction between the antagonistic fungus *Talaromyces flavus* and the wilt pathogen *Verticillium dahliae* on eggplant roots. Plant and Soil 176:129–137.

Falk, S.P., D.M. Gadoury, P. Cortesi, R.C. Pearson, and R.C. Seem. 1995a. Parasitism of *Uncinula necator* cleistothecia by the mycoparasite *Ampelomyces quisqualis*. Phytopathology 85:794–800.

Falk, S.P., D.M. Gadoury, R.C. Pearson, and R.C. Seem. 1995b. Partial control of grape powdery mildew by the mycoparasite *Ampelomyces quisqualis*. Plant Disease 79:483–490.

Falloon, R.E. 1977. Chamber for continuous microscopic observation of fungal growth. Transactions of the British Mycological Society 68:469–472.

Fani, R., G. Damiani, C. Di Serio, E. Gallori, A. Grifoni, and M. Bazzicalupo. 1993. Use of random amplified polymorphic DNA (RAPD) for generating specific DNA probes for microorganisms. Molecular Ecology 2:243–250.

Farnet, A.M., M. Roux, and J. Le Petit. 1999. Genotypic variations among isolates of *Marasmius quercophilus*, a white-rot fungus isolated from evergreen oak litter. Canadian Journal of Botany 77:884–890.

Farr, D.F., G.F. Bills, G.P. Chamuris, and A.Y. Rossman. 1989. Fungi on Plants and Plant Products in the United States. American Phytopathological Society Press, St. Paul, Minnesota.

Farr, E.R., J.A. Leussink, and F.A. Stafleu. 1979. Index Nominum Genericorum (Plantarum). [Generic names of fungi and vascular plants.] Regnum Vegetabile 100–102.

Farr, E.R., J.A. Leussink, and G. Zijlstra. 1986. Index Nominum Genericorum (Plantarum) Supplementum I. Regnum Vegetabile 113.

Farr, M.L. 1976. Myxomycetes. Flora Neotropica Monograph No. 16. New York Botanical Garden, New York.

Farris, S.H. 1966. A staining method for mycelium of *Rhabdocline* in Douglas-fir needles. Canadian Journal of Botany 44:1106–1107.

Fatzinger, C.W. 1985. Attraction of the black turpentine beetle (Coleoptera: Scolytidae) and other forest Coleoptera to turpentine-baited traps. Environmental Entomology 14:768–775.

Federal Geographic Data Committee. 1998. Content Standard for Digital Geospatial Metadata, Version 2. FGDC-STD-001. http://biology.usgs.gov/fgdc.metadata/version2/3_D.htm.

Feest, A. 1987. The quantitative ecology of soil mycetozoa. Progress in Protistology 2:331–361.

Feest, A., and M.F. Madelin. 1985. A method for the enumeration of myxomycetes in soils and its application to a wide range of soils. FEMS Microbiology Ecology 31:103–109.

Fell, J.W. 1976. Yeasts in oceanic regions. Pp. 93–124. *In* E.B. Gareth Jones (ed.), Recent Advances in Aquatic Mycology. Elek Science, London, England.

——. 1993. Rapid identification of yeast species using three primers in a polymerase chain reaction. Molecular Marine Biology and Biotechnology 2:174–180.

——. 1995. rDNA targeted oligonucleotide primers for the identification of pathogenic yeasts in a polymerase chain reaction. Journal of Industrial Microbiology 14:475–477.

Fell, J.W., and C.P. Kurtzman. 1990. Nucleotide sequence analysis of a variable region of the large subunit rRNA for identification of marine-occurring yeasts. Current Microbiology 21:295–300.

Fell, J.W., and I.M. Master. 1975. Phycomycetes (*Phytophthora* spp. nov. and *Pythium* sp. nov.) associated with degrading mangrove (*Rhizophora mangle*) leaves. Canadian Journal of Botany 53:2908–2922.

Fell, J.W., and S.Y. Newell. 1998. Biochemical and molecular methods for the study of marine fungi. Pp. 259–283. *In* K.E. Cooksey (ed.), Molecular Approaches to the Study of the Oceans. Chapman Hall, London, England.

Fell, J.W., T. Boekhout, A. Fonseca, and J.P. Sampaio. 2001. Basidiomycetous Yeasts. Pp. 5–35. *In* D.J. McLaughlin, E.G. McLaughlin, and P.A. Lempke (eds.), The Mycota. Vol. VII. Systematics and Evolution. Part B. Springer-Verlag, Berlin, Germany.

Fell, J.W., T. Boekhout, A. Fonseca, G. Scorzetti, and A. Statzell-Tallman. 2000. Biodiversity and systematics of basidiomycetous yeasts as determined by large-subunit rDNA D1/D2 domain sequence analysis. International Journal of Systematic and Evolutionary Microbiology 50:1351–1371.

Fell, J.W., I.L. Hunter, and A.S. Tallman. 1973. Marine basidiomycetous yeasts (*Rhodosporidium* spp. n.) with tetrapolar and multiple allelic bipolar mating systems. Canadian Journal of Microbiology 19:643–657.

Fell, J.W., A.C. Statzell, I.L. Hunter, and H.J. Phaff. 1969. *Leucosporidium* gen. n. the heterobasidiomycetous stage of several yeasts of the genus *Candida*. Antonie van Leeuwenhoek 35:433–462.

Fell, J.W., A. Statzell-Tallman, M.J. Lutz, and C.P. Kurtzman. 1992. Partial rRNA sequences in marine yeasts: a model for identification of marine eukaryotes. Molecular Marine Biology and Biotechnology 1:175–186.

Fellner, R. 1993. Air pollution and mycorrhizal fungi in Central Europe. Pp. 239–250. *In* D.N. Pegler, L. Boddy, B. Ing, and P.M. Kirk (eds.), Fungi of Europe: Investigation, Recording and Conservation. Royal Botanic Gardens, Kew, Surrey, England.

Felsenstein, J. 1988. Phylogenies from molecular sequences: inference and reliability. Annual Review of Genetics 22:521–565.

Fennell, D.I. 1960. Conservation of fungus cultures. Botanical Review 26:79–141.

——. 1973. Plectomycetes; Eurotiales. Pp. 45–68. *In* G.C. Ainsworth, F.K. Sparrow, and A.S. Sussman (eds.), The Fungi: An Advanced Treatise. Vol. 4A. Academic Press, New York.

Fergus, C.L. 1964. Thermophilic and thermotolerant molds and actinomycetes of mushroom compost during peak heating. Mycologia 56:267–284.

Ferguson, J.J., and S.H. Woodhead. 1982. Production of endomycorrhizal inoculum. A. Increase and maintenance of vesicular-arbuscular mycorrhizal fungi. Pp. 47–54. *In* N. Schenck (ed.), Methods and Principles of Mycorrhizal Research. American Phytopathological Society Press, St. Paul, Minnesota.

Ferry, B.W., M.S. Baddeley, and D.L. Hawksworth. 1973. Air Pollution and Lichens. University of Toronto Press, Toronto, Canada.

Fidalgo, M.E.P.K. 1968. The genus *Hexagona*. Memoirs of the New York Botanical Garden 17:35–108.

Fidalgo, O., and M.E.P.K. Fidalgo. 1968. Polyporaceae from Venezuela I. Memoirs of the New York Botanical Garden 17:1–34.

Field, K.G., G.J. Olsen, D.J. Lane, S.J. Giovannoni, M.T. Ghiselin, E.C. Raff, N.R. Pace, and R.A. Raff. 1988. Molecular phylogeny of the animal kingdom. Science 239:748–753.

Filson, R.B. 1996. Checklist of Australian Lichens and Allied Fungi. Flora of Australia Supplementary Series Number 7. Australian Biological Resources Study, Canberra, Australia.

Fischer, C.R., D.P. Janos, D.A. Perry, R.G. Linderman, and P. Sollins. 1994. Mycorrhiza inoculum potentials in tropical secondary succession. Biotropica 26:369–377.

Fischer, G.W., and C.S. Holton. 1957. Biology and control of the smut fungi. Ronald Press, New York.

Fischer, J.B., and J. Kane. 1971. The detection of contamination in *Trichophyton rubrum* and *Trichophyton mentagrophytes*. Mycopathologia et Mycologia Applicata 43:169–180.

Fisher, P.J. 1979. Colonization of freshly abscissed and decaying leaves by aero-aquatic hyphomycetes. Transactions of the British Mycological Society 73:99–102.

Fisher, P.J., and O. Petrini. 1987. Location of fungal endophytes in tissues of *Suaeda fruticosa*: A preliminary study. Transactions of the British Mycological. Society 89:246–249.

——. 1990. A comparative study of fungal endophytes in xylem and bark of *Alnus* species in England and Switzerland. Mycological Research 94:313–319.

——. 1992. Fungal saprobes and pathogens as endophytes of rice (*Oryza sativa* L.). New Phytologist 120:137–143.

Fisher, P.J., A.E. Anson, and O. Petrini. 1984a. Novel antibiotic activity of an endophytic *Cryptosporiopsis* sp. isolated from *Vaccinium myrtillus*. Transactions of the British Mycological Society 83:145–187.

——. 1984b. Antibiotic activity of some endophytic fungi from ericaceous plants. Botanica Helvetica 94:249–253.

——. 1986. Fungal endophytes in *Ulex europaeus* and *Ulex gallii*. Transactions of the British Mycological Society 86:153–193.

Fisher, P.J., F. Graf, L.E. Petrini, B.C. Sutton, and P.A. Wookey. 1995. Fungal endophytes of *Dryas octopetala* from a high polar semidesert and from the Swiss Alps. Mycologia 87:319–323.

Fisher, P.J., O. Petrini, and H.M. Lappin Scott. 1992. The distribution of some fungal and bacterial endophytes in maize (*Zea mays* L.). New Phytologist 122:299–305.

Fisher, P.J., O. Petrini, and J. Webster. 1991. Aquatic hyphomycetes and other fungi in living aquatic and terrestrial roots of *Alnus glutinosa*. Mycological Research 95:543–547.

Fisher, P.J., D.J. Stradling, and D.N. Pegler. 1994. *Leucoagaricus* basidiomata from a live nest of the leaf-cutting ant *Atta cephalotes*. Mycological Research 98:884–888.

Fisher, R.A., A.S. Corbet, and C.B. Williams. 1943. The relation between the number of species and the number of individuals in a random sample of an animal population. Journal of Animal Ecology 12:42–58.

Fisher, W.S., E.H. Nilson, J.F. Steenbergen, and D.V. Lightner. 1978. Microbial disease of cultured lobsters: a review. Aquaculture 14:115–140.

Fletcher, J.T., and G.W. Ganney. 1968. Experiments on the biology and control of *Mycogone perniciosa* Magn. Mushroom Science 7:221–237.

Fletcher, J.T., B.J. Smewin, and A. O'Brien. 1990. *Pythium oligandrum* associated with a cropping disorder of *Agaricus bisporus*. Plant Pathology 39:603–605.

Fletcher, L.D., E.J. Kerry and G.M. Weste. 1985. Microfungi of Mac. Robertson and Enderby Lands, Antarctica. Polar Biology 4:81–88.

Flynn, T.M., and W.G. Niehaus. 1997. Improved method for isolation of DNA from slow growing basidiomycetes such as *Armillaria mellea*. BioTechniques 22:47–52.

Fogel, R.M. 1976. Ecological studies of hypogeous fungi. II. Sporocarp phenology in a western Oregon Douglas fir stand. Canadian Journal of Botany 54:1152–1162.

——. 1980. Mycorrhizae and nutrient cycling in natural forest ecosystems. New Phytologist 86:199–212.

——. 1981. Quantification of sporocarps produced by hypogeous fungi. Pp. 553–568. *In* D.T. Wicklow and G.C. Carroll (eds.), The Fungal Community, Its Organization and Role in the Ecosystem. Marcel Dekker, New York.

Fogel, R.M., and G.A. Hunt. 1979. Fungal and arboreal biomass in a western Oregon Douglas-fir ecosystem: distribution patterns and turnover. Canadian Journal of Forest Research 9:245–256.

Fogel, R.M., and J.M. Trappe. 1978. Fungus consumption (mycophagy) by small animals. Northwest Science 52:1–31.

Fokkema, N.J. 1978. Fungal antagonisms in the phyllosphere. Annals of Applied Biology 89:115–119.

Foley, M.F., and J.W. Deacon. 1985. Isolation of *Pythium oligandrum* and other necrotrophic mycoparasites from soil. Transactions of the British Mycological Society 85:631–639.

——. 1986. Physiological differences between mycoparasitic and plant-pathogenic *Pythium* spp. Transactions of the British Mycological Society 86:225–231.

Fontana, A. 1960. Sopra un parassita di *Rhodopaxillus nudus* (Fr.) Maire nuovo per l'Italia: *Harziella capitata* Cost. et Matr. Allionia 6:35–41.

Fontana, A., and P. Fasolo Bonfante. 1971. Una nuova specie di "*Sporothrix*". Allionia 17:5–13.

Fonty, G., P. Gouet, and V. Sante. 1988. Influence d'une bactérie méthanogène sur l'activité cellulolytique et al métabolisme de deux espèces de champignons du rumen, *in vitro*. Résultats préliminaries. Reproduction Nutrition Development 28:133–134.

Forest Ecosystem Management Assessment Team (FEMAT). 1993. Forest Ecosystem Management: An Ecological, Economic and Social Assessment. U.S. Department of Agriculture, U.S. Department of the Interior, and others, Portland, Oregon.

Forget, L., J. Ustinova, Z. Wang, V.A.R. Huss, and B.F. Lang. 2002. *Hyaloraphidium curvatum*: A linear mitochondrial genome, tRNA editing and an evolutionary link to lower fungi. Molecular Biology and Evolution 19:310–319.

Förster, H., M.D. Coffey, H. Elwood, and M.L. Sogin. 1990. Sequence analysis of the small subunit ribosomal RNAs of three zoosporic fungi and implications for fungal evolution. Mycologia 82:306–312.

Förster, H., T.G. Kinscherf, S.A. Leong, and D.P. Maxwell. 1987. Molecular analysis of the mitochondrial genome of *Phytopthora*. Current Genetics 12:215–218.

Fosberg, F.R., and M.-H. Sachet. 1965. Manual for Tropical Herbaria. Regnum Vegetabile no. 39. Utrecht, The Netherlands.

Foster, J.W. 1949. Chemical Activities of Fungi. Academic Press, New York.

Foucard, T. 1990. Svenksa Skorplavs Flora. Interpublishing, Stockholm, Sweden.

Foucault, M. 1972. The Archaeology of Knowledge and the Discourse of Language. Pantheon Books, New York.

Fox, J.L., and C.A. Smith. 1988. Winter mountain goat diets in southeast Alaska. Journal of Wildlife Management 52:362–365.

Fox, R.T.V., J.J.W. Obanya Obore, and A.M. McQue. 1991. Prospects for the integrated control of *Armillaria* root rot of trees. Pp. 154–159. *In* A.B.R. Beemster, G.J. Bollen, M. Gerlagh, M.A. Ruissen, B. Schippers, and A. Tempel (eds.), Biotic Interactions and Soil-borne Diseases. Elsevier, Amsterdam, The Netherlands.

Fradkin, A. 1987. Sampling of microbiological contaminants in indoor air. Pp. 66–77. *In* J.K. Taylor (ed.), Sampling and Calibration for Atmospheric Measurements. ASTM, Philadelphia, Pennsylvania.

France, J., M.K. Theodorou, and D. Davies. 1990. Use of zoospore concentrations and life cycle parameters in determining the population of anaerobic fungi in the rumen ecosystem. Journal of Theoretical Biology 147:413–422.

Francke-Grossman, H. 1967. Ectosymbiosis in wood-inhabiting insects. Pp. 141–205. *In* S.M. Henry (ed.), Symbiosis. Vol. II. Academic Press, New York.

Franke, M., and J.B. Morton. 1994. Ontogenetic comparisons of arbuscular mycorrhizal fungi *Scutellospora heterogama* and *Scutellospora pellucida*: revision of taxonomic character concepts, species descriptions, and phylogenetic hypotheses. Canadian Journal of Botany 72:122–134.

Frankland, J.C. 1966. Succession of fungi on decaying petioles of *Pteridium aquilinum.* Journal of Ecology 54:41–63.

——. 1982. Biomass and nutrient cycling by decomposer basidiomycetes. Pp. 241–261. *In* J.C. Frankland, J.N. Hedger, and M.J. Swift (eds.), Decomposer Basidiomycetes: Their Biology and Ecology. Cambridge University Press, Cambridge, United Kingdom.

Frankland, J.C., J. Dighton, and L. Boddy. 1990. Methods for studying fungi in soil and forest litter. Pp. 343–404. *In* R. Grigorova and J.R. Norris (eds.), Methods in Microbiology: Techniques in Microbial Ecology. Vol. 22. Academic Press, London, United Kingdom.

Franz, F., R. Grotjahn, and G. Acker. 1993. Identification of *Naemacyclus minor* hyphae within needle tissues of *Pinus sylvestris* by immunoelectron microscopy. Archives of Microbiology 160:265–272.

Fraser, G.C., R.B. Callinan, and L.M. Calder. 1992. *Aphanomyces* species associated with red spot disease: an ulcerative disease of estuarine fish from eastern Australia. Journal of Fish Diseases 15:173–181.

Frederick, L. 1990. Phylum Plasmodial Slime Molds, Class Myxomycota. Pp. 467–483. *In* L. Margulis, J.O. Corliss, M. Melkonian, and D. Chapman (eds.), Handbook of Protoctista. Jones and Bartlett, Boston, Massachusetts.

Freeman, S., and R.J. Rodriguez. 1993. Genetic conversion of a fungal plant pathogen to a nonpathogenic, endophytic mutualist. Science 260:75–78.

——. 1995. Differentiation of *Colletotrichum* species responsible for anthracnose of strawberry by arbitrarily primed PCR. Mycological Research 99:501–504.

Freeman, S., M. Pham, and R.J. Rodriguez. 1993. Molecular genotyping of *Colletotrichum* species based on arbitrarily primed PCR, A + T-rich DNA, and nuclear DNA analyses. Experimental Mycology 17:309–322.

Freeman, W.M., S.J. Walker, and K.E. Vrana. 1999. Quantitative RT–PCR: pitfalls and potential. BioTechniques 26:112–122, 124–125.

Fries, N., and G. Swedjemark. 1985. Sporophagy in Hymenomycetes. Experimental Mycology 9:74–79.

Friese, C.F., and M.F. Allen. 1991. The spread of VA mycorrhizal fungal hyphae in the soil: inoculum types and external hyphal architecture. Mycologia 83:409–418.

Friese, C.F., and R.E. Koske. 1991. The spatial dispersion of spores of vesicular-arbuscular mycorrhizal fungi in a sand dune: microscale patterns associated with the root architecture of American beachgrass. Mycological Research 95:952–957.

Frimeth, J.P. 1987. A survey of the parasites of nonanadromous and anadromous brook charr (*Salvelinus fontinalis*) in the Tabusintac River, New Brunswick, Canada. Canadian Journal of Zoology 65:1354–1362.

Frisvad, J.C. 1981. Physiological criteria and mycotoxin production as aids in identification of common asymmetric penicillia. Applied and Environmental Microbiology 41:568–579.

Fritze, H., and E. Bååth. 1993. Microfungal species composition and fungal biomass in a coniferous forest soil polluted by alkaline deposition. Microbial Ecology 25:83–92.

Fröhlich, J., and K.D. Hyde. 2000. Palm Microfungi. University of Hong Kong Press, Hong Kong, China.

Frye, F.L. 1984. Euthanasia, necropsy technique and comparative histology of reptiles. Pp. 703–755. *In* G.L. Hoff, F.L. Frye, and E.R. Jacobson (eds.), Diseases of Amphibians and Reptiles. Plenum, New York.

Fukatsu, T., and H. Ishikawa. 1992. A novel eukaryotic extracellular symbiont in an aphid, *Astegopteryx styraci* (Homoptera, Aphididae, Hormaphidinae). Journal of Insect Physiology 38:765–773.

Fuller, M.S. 1990. Phylum Hyphochytriomycota. Pp. 380–387. *In* L. Margulis, J.O. Corliss, M. Melkonian, and D. Chapman (eds.), Handbook of Protoctista. Jones and Bartlett, Boston, Massachusetts.

——. 2001. Hyphochytriomycota. Pp. 73–80. *In* D.J. McLaughlin, E.G. McLaughlin, and P.A. Lempke (eds.), The Mycota. Vol. VII. Systematics and Evolution. Part A. Springer-Verlag, Berlin, Germany.

Fuller, M.S., and A. Jaworski. 1987. Zoosporic Fungi in Teaching and Research. Southeastern Publishing, Athens, Georgia.

Fuller, M.S., and R.O. Poyton. 1964. A new technique for the isolation of aquatic fungi. Bioscience 14:45–46.

Fuller, M.S., B.E. Fowles, and D.J. McLaughlin. 1964. Isolation and pure culture study of marine phycomycetes. Mycologia 56:745–756.

Funk, A. 1981. Parasitic microfungi of western trees. Canadian Forestry Service, Pacific Forest Research Centre, Victoria, British Columbia. Information Report BC-X-222.

——. 1985. Foliar fungi of western trees. Canadian Forestry Service, Pacific Forest Research Centre, Victoria, British Columbia. Information Report BC-X-265.

Furtado, J.S. 1981. Taxonomy of *Amauroderma* (Basidiomycetes, Polyporaceae). Memoirs of the New York Botanical Garden 34:1–109.

Furuya, K. 1990. Coprophilous fungi as microbial resources. Sankyo Kenkyusho Nempo (Annual Report of the Sankyo Research Laboratories, Ibaraki, Japan) 42:1–31.

Furuya, K., and A. Naito. 1979. An effective method for isolation of *Boothiella tetraspora* from soil. Transactions of the Mycological Society of Japan 20:309–311.

——. 1980. Stimulation of ascospore germination by phenolic compounds in members of the Sordariaceae. Transactions of the Mycological Society of Japan 21:77–85.

Gaertner, A. 1954. Über das Vorkommen niederer Erdphycomyceten in Afrika, Schweden und an einigen mittleuropaischen Standorten. Archiv für Mikrobiologie 21:4–56.

Gain, R.E., and H.L. Barnett. 1970. Parasitism and axenic growth of the mycoparasite *Gonatorhodiella highlei.* Mycologia 62:1122–1129.

Galash, V.T., and A.M. Marchenko. 1991. The effects of trichothecene mycotoxins on carp. Bulletin of the European Association of Fish Pathologists 11:63–64.

Galloway, D.J. (ed.). 1991. Tropical Lichens: Their Systematics, Conservation and Ecology. Clarendon Press, Oxford, Great Britain.

——. 1992a. Biodiversity: a lichenological perspective. Biodiversity and Conservation 1:312–323.

——. 1992b. Checklist of New Zealand lichens. [DSIR Land Research Scientific Report 26:1–58.] Manaaki Whenua Press, Lincoln, New Zealand.

——. 1994. Biogeography and ancestry of lichens and other ascomycetes. Pp. 175–184. *In* D.L. Hawksworth (ed.), Ascomycete Systematics: Problems and Perspectives in the Nineties. Plenum, New York.

——. 1995. Lichens in southern temperate rainforest and their role in maintenance of biodiversity. Pp. 125–135. *In* D. Allsopp, R.R. Colwell, and D.L. Hawksworth (eds.), Microbial Diversity and Ecosystem Function. CAB International, Wallingford, Oxon, United Kingdom.

Galloway, D.J., and A. Aptroot. 1995. Bipolar lichens: a review. Cryptogamic Botany 5:184–191.

Galun, M. (ed.). 1988. CRC Handbook of Lichenology. Volumes I, II, III. CRC Press, Boca Raton, Florida.

Gams, W. 1960. Studium zellulolytischer Bodenpilze mit Hilfe der Zellophanstreifen-Methode und mit Carboxymethyl-Zellulose. Sydowia 14:295–307.

——. 1971. *Cephalosporium*-artige Schimmelpilze (Hyphomycetes). Gustav Fischer Verlag, Stuttgart, Germany.

——. 1974. *Tympanosporium parasiticum* gen. et sp. nov., a hyperparasitic hyphomycete on *Tubercularia vulgaris* with pleomorphic conidiogenesis. Antonie van Leeuwenhoek 40:471–479.

——. 1975a. The perfect state of *Tilachlidium brachiatum*. Persoonia 8:329–331.

——. 1975b. *Nematogonium niveum*, a new hyperparasitic species of Hyphomycetes. Revue de Mycologie 39:271–278.

——. 1977. A key to the species of *Mortierella*. Persoonia 9:381–391.

——. 1983. Two species of mycoparasitic fungi. Sydowia 36:46–52.

——. 1986. The fungicolous Hyphomycete *Zakatoshia erikssonii* n.sp. in vivo and in vitro. Windahlia 16:59–64.

——. 1988. A contribution to the knowledge of nematophagous species of *Verticillium*. Netherlands Journal of Plant Pathology 94:123–148.

——. 1992. The analysis of communities of saprophytic microfungi with special reference to soil fungi. Pp. 183–223. *In* W. Winterhoff (ed.), Fungi in Vegetation Science. Kluwer Academic, Dordrecht, The Netherlands.

——. 1993. Supplement and Corrigendum to the "Compendium of Soil Fungi." IHW-Verlag, Eching, Germany.

——. 1997. *Cephalosporium*-like Hyphomycetes. Extracted, Translated, and Supplemented from *Cephalosporium*-artige Schimmelplize (Hyphomycetes). Gustav Fischer Verlag, Stuttgart, Germany, and W. Gams, Baarn, The Netherlands.

Gams, W., and J. Bissett. 1999. Morphology and identification of *Trichoderma*. Pp. 3–34 *In* C.P. Kubicek and G.E. Harman (eds.), *Trichoderma* and *Gliocladium*. Vol. 1. Taylor Francis, London, England.

Gams, W., and K.H. Domsch. 1960. Bemerkungen zu einigen schwer bestimmbaren Bodenpilzen. Nova Hedwigia 18:1–29.

——. 1967. Beiträge zur Anwendung der Bodenwaschtechnik für die Isolierung von Bodenpilzen. Archiv für Mikrobiologie 58:134–144.

——. 1969. The spatial and seasonal distribution of microscopic fungi in arable soils. Transactions of the British Mycological Society 52:301–308.

Gams, W., and A.C.M. Hoozemans. 1970. *Cladobotryum*-Konidienformen von *Hypomyces*-Arten. Persoonia 6:95–110.

Gams, W., and A. van Zaayen. 1982. Contribution to the taxonomy and pathogenicity of fungicolous *Verticillium* species. I. Taxonomy. Netherlands Journal of Plant Pathology 88:52–78.

Gams, W., and R. Zare. 2001. A revision of *Verticillium* sect. *Prostrata*. III. Generic classification. Nova Hedwigia 72:329–337.

Gams, W., E.S. Hoekstra, and A. Aptroot. 1998. CBS Course of Mycology. 4th ed. Centraalbureau voor Schimmelcultures, Baarn, The Netherlands.

Gamundi, I.J. 1979. Discomycetes of Argentina. Flora Cryptogamica de Tierra de Fuego 11:1–524.

Gandy, D.G. 1979. Inhibition of *Mycogone perniciosa* growth by *Acremonium strictum*. Transactions of the British Mycological Society 72:151–155.

Garbaye, J. 1994. Helper bacteria: A new dimension to the mycorrhizal symbiosis. New Phytologist 128:197–210.

Garber, R.C., and O.C. Yoder. 1983. Isolation of DNA from filamentous fungi and separation into nuclear, mitochondrial, ribosomal and plasmid components. Analytical Biochemistry 135:416–422.

——. 1984. Mitochondrial DNA of the filamentous ascomycete *Cochliobolus heterostrophus*—Characterization of the mitochondrial chromosome and population genetics of a restriction enzyme polymorphism. Current Genetics 8:621–628.

Garber, R.C., B.G. Turgeon, E.U. Selker, and O.C. Yoder. 1988. Organization of ribosomal RNA genes in the fungus *Cochliobolus heterostrophus*. Current Genetics 14:573–582.

Gardes, M., and T.D. Bruns. 1996. Community structure of ectomycorrhizal fungi in a *Pinus muricata* forest: above- and below-ground views. Canadian Journal of Botany 74:1572–1583.

Gargas, A., P.T. DePriest, M. Grube, and A. Tehler. 1995. Multiple origins of lichen symbioses in fungi suggested by SSU rDNA phylogeny. Science 268:1492–1495.

Garrett, S.D. 1963. Soil Fungi and Soil Fertility. Pergamon Press, Oxford, England.

Gartner, Jr., J.V., and D.E. Zwerner. 1988. An *Ichthyophonus*-type fungus in the deep-sea pelagic fish *Scopelogadus beanii* (Günther) (Pisces: Melamphaidae): pathology, geographic distribution and ecological implications. Journal of Fish Biology 32:459–470.

Gaskell, J., P. Stewart, P. Kersten, S.F. Covert, J. Reiser, and D. Cullen. 1994. Establishment of genetic linkage by allele-specific polymerase chain reaction: application to the lignin peroxidase gene family of *Phanerochaete chrysosporium*. BioTechnology 12:1372–1375.

Gaugh, Jr., H.G. 1982. Multivariate Analysis in Community Ecology. Cambridge University Press, Cambridge, England.

Gäumann, E. 1922. Uber die gattung Kordyana Rac. Annales Mycologici 20:257–271.

——. 1959. Die Rostpilze Mittleuropas. Buchdruckerei Büchler, Bern, Switzerland.

Gauslaa, Y. 1995. The *Lobarion*, an epiphyte community of ancient forests threatened by acid rain. Lichenologist 27:59–76.

Gazey, C., L.K. Abbott, and A.D. Robson. 1992. The rate of development of mycorrhizas affects the onset of sporulation and production of external hyphae by two species of *Acaulospora*. Mycological Research 96:643–650.

Gehrig, H., A. Schüssler, and M. Kluge. 1996. *Geosiphon pyriforme*, a fungus forming endocytobiosis with Nostoc (Cyanobacteria), is an ancestral member of the Glomales: evidence by SSU rRNA analysis. Journal of Molecular Evolution 43:71–81.

Geiser, D.M., and K.F. LoBuglio. 2001. The monophyletic plectomycetes: ascosphaerales, onygenales, eurotiales. Pp. 201–219. *In* D.J. McLaughlin, E.G. McLaughlin, and P.A. Lempke (eds.), The Mycota. Vol. VII. Systematics and Evolution. Part A. Springer-Verlag, Berlin, Germany.

Geiser, D.M., J.W. Taylor, K.B. Richie, and G.W. Smith. 1998. Cause of sea fan death in the West Indies. Nature 394:137–138.

Geiser, L.H., C.C. Derr, and K.L. Dillman. 1994a. Air Quality Monitoring on the Tongass National Forest. Methods and Baselines Using Lichens. U.S. Department of Agriculture, Forest Service, Alaska Region, Technical Bulletin R10-TB-46.

Geiser, L.H., K.L. Dillman, C.C. Derr, and M.C. Stensvold. 1994b. Lichens of Southeastern Alaska. U.S. Department of Agriculture, Forest Service, Alaska Region, Technical Bulletin R10-TB-45.

Gemma, J.N., and R.E. Koske. 1995. Mycorrhizae in Hawaiian epiphytes. Pacific Science 49:175–180.

Gemma, J.N., R.E. Koske, and M.M. Carreiro. 1989. Seasonal dynamics of five species of V-A mycorrhizal fungi in a sand dune. Mycological Research 92:317–321.

Gentry, A.H. 1992. Tropical forest biodiversity: distributional patterns and their conservational significance. Oikos 63:19–28.

Gerdemann, J.W., and J.M. Trappe. 1974. The endogonaceae in the Pacific Northwest. Mycologia Memoirs 5:1–76.

Gerlach, W., and H. Nirenberg. 1982. The genus *Fusarium* -a pictorial atlas. Milleilungen aus der Biologischen Bundesanstalt für Land- und Forstwirtschaft. Berlin-Dahlem 209:1–406.

Gerlagh, M., and I. Vos. 1991. Enrichment of soil with sclerotia to isolate antagonists of *Sclerotinia sclerotiorum*. Pp. 165–171. *In* A.B.R. Beemster, G.J. Bollen, M. Gerlagh, M.A. Ruissen, B. Schippers, and A. Tempel (eds.), Biotic Interactions and Soil-borne Diseases. Elsevier, Amsterdam, The Netherlands.

Gerlagh, M., H.M. Goossen-van de Geijn, B. Verdam, and N.J. Fokkema. 1995. Biological control of white mould (*Sclerotinia sclerotiorum*) in various crops by application of *Coniothyrium minitans*. Pp. 13–17. *In* J.M. Whipps and M. Gerlagh (eds.), Biological Control of Sclerotium-forming Pathogens. IOBC/WPRS (International

Organization for Biological and Integrated Control of Noxious Animals and Plants/West Palearctic Regional Section) Bulletin 18.

Gernandt, D.S., F.J. Camacho, and J.K. Stone. 1997. *Meria laricis*, an anamorph of *Rhabdocline*. Mycologia 89:735–744.

Gernandt, D.S., J.L. Platt, J.K. Stone, J.W. Spatafora, A. Holst-Jensen, R.C. Hamelin, and L.M. Kohn. 2001. Phylogenetics of the Helotiales and Rhytismatales based on partial small subunit nuclear ribosomal DNA sequences. Mycologia 93:915–933.

Gerrettson-Cornell, L., and J. Simpson. 1984. Three new marine *Phytophthora* species from New South Wales. Mycotaxon 19:453–470.

Gessner, M.O., and S.Y. Newell. 1997. Bulk quantitative methods for the examination of eukaryotic organoopsmotrophs in plant litter. Pp. 295–308. *In* M. McInerney, L. Stetzenbach, C.J. Hurst, G. Knudsen, and M. Walter (eds.), Manual of Environmental Microbiology. ASM (American Society for Microbiology) Press, Washington, D.C.

Gessner, M.O., M. Thomas, A.-M. Jean-Louis and E. Chauvet. 1993. Stable successional patterns of aquatic hyphomycetes on leaves decaying in a summer cool stream. Mycological Research 97:163–172.

Gessner, R.V. 1977. Seasonal occurrence and distribution of fungi associated with *Spartina alterniflora* from a Rhode Island estuary. Mycologia 69:477–491.

Gessner, R.V., and R.D. Goos. 1973a. Fungi from decomposing *Spartina alterniflora*. Canadian Journal of Botany 51:51–55.

——. 1973b. Fungi from *Spartina alterniflora* in Rhode Island. Mycologia 65:1296–1301.

Gessner, R.V., and J. Kohlmeyer. 1976. Geographical distribution and taxonomy of fungi from salt marsh *Spartina*. Canadian Journal of Botany 54:2023–2037.

Ghisalberti, E.L., and K. Sivasithamparam. 1991. Antifungal antibiotics produced by *Trichoderma* spp. Soil Biology and Biochemistry 23:1011–1020.

Gianinazzi, S. 1991. Vesicular-arbuscular (endo-) mycorrhizas: cellular, biochemical and genetic aspects. Agriculture, Ecosystems and Environment 35:105–119.

Gibbs, J.N. 1993. The biology of ophiostomatoid fungi causing sap stain in trees and freshly cut logs. Pp. 153–160. *In* M.J. Wingfield, K.A. Seifert, and J.F. Webber (eds.), *Ceratocystis* and *Ophiostoma*: Taxonomy, Ecology, and Pathogenicity. American Phytopathological Society Press, St. Paul, Minnesota.

Gifford, D.J. 1985. Laboratory culture of marine planktonic oligotrichs (*Ciliophora, Oligotrichida*). Marine Ecology Progress Series 23:257–267.

Gilbert, H.C., and G.W. Martin. 1933. Myxomycetes found on the bark of living trees. University of Iowa Studies in Natural History 15(3):3–8.

Gilbert, O.L. 2000. Lichens. New Naturalist. Vol. 86. Harper Collins, London, England.

Gilbertson, R.L. 1974. Fungi that Decay Ponderosa Pine. University of Arizona Press, Tucson, Arizona.

——. 1984. Relationships between insects and wood-rotting basidiomycetes. Pp. 130–165. *In* Q. Wheeler, and M. Blackwell (eds.), Fungus-Insect Relationships, Perspectives in Ecology and Evolution. Columbia University Press, New York.

Gilbertson, R.L., and L. Ryvarden. 1986. North American Polypores. Vol. 1 Fungiflora, Oslo, Norway.

——. 1987. North American Polypores. Vol. 2. Fungiflora, Oslo, Norway.

Gilkey, H.M. 1939. Tuberales of North America. Oregon State Monographs 1:1–63.

——. 1954. Taxonomic notes on Tuberales. Mycologia 46:783–793.

Gilliland, G., S. Perrin, and H. Bunn. 1990. Competitive PCR for quantitation of mRNA. Pp. 60–69. *In* M.A. Innis, D.H. Gelfand, J.J. Sninsky, and T.J. White (eds.), PCR Protocols, A Guide to Methods and Applications. Academic Press, New York.

Gilman, J.C. 1957. Soil fungi. Iowa State College Press, Ames, Iowa.

Gilman, J.C., and L.H. Tiffany. 1952. Fungicolous fungi from Iowa. Proceedings of the Iowa Academy of Science 59:99–110.

Giménez-Jurado, G. 1992. *Metschnikowia gruessii* sp. nov., the teleomorph of *Nectaromyces reukaufii* but not of *Candida reukaufii*. Systematic and Applied Microbiology 15:432–438.

Ginns, J. 1986. The genus *Syzygospora* (Heterobasidiomycetes: Syzygosporaceae). Mycologia 78:619–636.

Ginns, J., and M.N.L. Lefebvre. 1993. Lignicolous corticioid fungi (Basidiomycota) of North America. Systematics, Distribution, and Ecology. Mycologia Memoir 19:1–247, American Phytopathological Society Press, St. Paul, Minnesota.

Ginns, J., and S. Sunhede. 1978. Three species of *Christiansenia* (Corticiaceae) and the teratological galls on *Collybia dryophila*. Botaniska Notiser 131:167–173.

Girlanda, M., D. Isocrono, C. Bianco, and A.M. Luppi-Mosca. 1997. Two foliose lichens as microfungal ecological niches. Mycologia 89:531–536.

Girlanda, M., G.C. Varese, and A.M. Luppi-Mosca. 1995. *In vitro* interactions between saprotrophic microfungi and ectomycorrhizal symbionts. Allionia 33:81–86.

Giuma, A.Y., and R.C. Cooke. 1972. Some endozoic parasites on soil nematodes. Transactions of the British Mycological Society 59:213–218.

Gjaerum, H.B. 1986. East African rusts (Uredinales), mainly from Uganda 5. On families belonging to Gamnopetalae. Mycotaxon 27:507–550.

Glasby, J.S. 1992. Encyclopedia of Antibiotics. 3rd ed. John Wiley, Chichester, United Kingdom.

Glasel, J.A. 1995. Validity of nucleic acid purities monitored by 260 nm/280 nm absorbance ratios. BioTechniques 18:62–63.

Gleason, H.A. 1922. On the relation between species and area. Ecology 3:158–162.

——. 1925. Species and area. Ecology 6:66–74.

Glen-Lewin, D.C. 1977. Species diversity in North American temperate forests. Vegetatio 33:153–162.

Glenn, A.E., C.W. Bacon, R. Price, and R.T. Hanlin. 1996. Molecular phylogeny of *Acremonium* endophytes and its taxonomic implications. Mycologia 88:369–383.

Glockling, S.L. 1998. Two new species of rotifer-attacking fungi, *Rotiferophthora* from Japan and records of *Verticillium bactrosporum* in rotifer hosts. Mycological Research 102:145–150.

Glockling, S.L., and G.W. Beakes. 2000. A review of the taxonomy, biology and infection strategies of "biflagellate holocarpic" parasites of nematodes. Fungal Diversity 4:1–20.

Glockling, S.L., and M. Shimazu. 1997. Culturing of three species of endoparasitic fungi infecting nematodes. Mycological Research 101:55–60.

Gloer, J.B. 1995. The chemistry of fungal antagonism and defense. Canadian Journal of Botany 73(1):S1265–S1274.

——. 1997. Applications of fungal ecology in the search for new bioactive natural products. Pp. 249–268. *In* D.T. Wicklow and B.E. Söderström (eds.), The Mycota. Vol. IV. Environmental and Microbial Relationships. Springer-Verlag, Berlin, Germany.

Gochenaur, S.E. 1964. A modification of the immersion tube method for isolating soil fungi. Mycologia 56:921–923.

——. 1970. Soil mycoflora of Peru. Mycopathologia et Mycologia Applicata 42:259–272.

——. 1975. Distributional patterns of mesophilous and thermophilous microfungi in two Bahamian soils. Mycopathologia et Mycologia Applicata 57:155–164.

——. 1978. Fungi of Long Island oak-birch forest I. Community organization and seasonal occurrence of the opportunistic decomposers of the A horizon. Mycologia 70:975–994.

——. 1981. Response of soil fungal communities to disturbance. Pp. 459–479. *In* D.T. Wicklow and G.C. Carroll (eds.), The Fungal

Community, Its Organization and Role in the Ecosystem. Marcel Dekker, New York.

Gochenaur, S.E., and M.P. Backus. 1967. Mycoecology of willow and cottonwood lowland communities in southern Wisconsin. II. Soil microfungi in the sandbar willow stands. Mycologia 59:893–901.

Gochenaur, S.E., and W.F. Whittingham. 1967. Mycoecology of willow and cottonwood lowland communities in southern Wisconsin. I. Soil microfungi in the willow-cottonwood forests. Mycopathologia et Mycologia Applicata 33:125–139.

Gochenaur, S.E., and G.M. Woodall. 1974. The soil microfungi of a chronically irradiated oak-pine forest. Ecology 55:1004–1016.

Goetsch, W., and R. Stoppel. 1940. Die Pilze der Blattschneider Ameisen. Biologisches Zentralblatt 60:393–398.

Goh, T.K., and K.D. Hyde. 1996. Biodiversity of freshwater fungi. Journal of Industrial Microbiology 17:328–345.

Gold, J.J., I.B. Heath, and T. Bauchop. 1988. Ultrastructural description of a new chytrid genus of caecum anaerobe, *Caecomyces equi* gen. nov. sp. nov., assigned to the Neocallimasticaceae. BioSystems 21:403–415.

Gold, S.E., J.W. Duick, R.S. Redman, and R.J. Rodriguez. 2001. Molecular transformation, gene cloning, and gene expression systems for filamentous fungi. Pp. 199–238. *In* G.G. Khachatourians and D.K. Arora (eds.), Applied Mycology and Biotechnology Vol. I: Agriculture and Food Production. Elsevier, Amsterdam, The Netherlands.

Goldhammer, D.S., F.M. Stephen, and T.D. Paine. 1990. The effect of the fungi *Ceratocystis minor* (Hedgecock) Hunt, *Ceratocystis minor* (Hedgecock) Hunt var. *barassii* Taylor, and SJB 122 on reproduction of the southern pine beetle, *Dendroctonus frontalis* Zimmerman (Coleoptera: Scolytidae). Canadian Entomologist 122:407–418.

Goldie-Smith, E.K. 1956. Maintenance of stock cultures of aquatic fungi. Journal of the Elisha Mitchell Scientific Society 72:158–166.

Goldsmith, F.B., C.M. Harrison, and A.J. Morton. 1986. Description and analysis of vegetation. Pp. 437–524. *In* P.D. Moore and S.B. Chapman (eds.), Methods in Plant Ecology. Oxford, United Kingdom.

Gómez, L.D.P. 1994. Una nueva especie neotropical de *Termitomyces* (Agaricales: Termitomycetaceae). Revista de Biología Tropical 42:439–441.

Gooday, G.W. 1994. Hormones in mycelial fungi. Pp. 401–411. *In* J.G.H. Wessels and F. Meinhardt (eds.), The Mycota. Vol. I. Growth, Differentiation and Sexuality. Springer-Verlag, Berlin, Germany.

Goodwin, S.B., D.E. Legard, C.D. Smart, M. Levy, and W.E. Fry. 1999. Gene flow analysis of molecular markers confirms that *Phytophthora mirabilis* and *P. infestans* are separate species. Mycologia 91:796–810.

Goos, R.D. 1960. Soil fungi from Costa Rica and Panama. Mycologia 52:877–883.

———. 1963. Further observations on soil fungi in Honduras. Mycologia 55:142–150.

Goos, R.D., and F.G. Pollack. 1965. An illustration of the need for multiple techniques for the preservation of fungus cultures. Mycologia 57:975–977.

Gordon, G.L.R. 1985. The potential for manipulation of rumen fungi. Pp. 124–128. *In* R.A. Leng, J.S.F. Barker, D.B. Adams, and K.J. Hutchinson (eds.), Biotechnology and Recombinant DNA Technology in the Animal Production Industries. Reviews in Rural Science. Vol. 6. University of New England, Armidale, Australia.

———. 1990. Selection of anaerobic fungi for better fiber degradation in the rumen. Pp. 301–309. *In* D.E. Akin, L.G. Ljungdahl, J.R. Wilson, and P.J. Harris (eds.), Microbial and Plant Opportunities to Improve Lignocellulose Utilization by Ruminants. Elsevier, New York.

Gordon, G.L.R., and M.W. Phillips. 1989. Degradation and utilization of cellulose and straw by three different anaerobic fungi from the ovine rumen. Applied and Environmental Microbiology 55:1703–1710.

Gosden, J.R. 1996. PRINS and *In Situ* PCR Protocols. Methods in Molecular Biology 71. Humana Press, Totowa, New Jersey.

Goto, S., R. Ano, J. Sugiyama, and K. Horikoshi. 1981. *Exophiala alcalophila*, a new black yeast-like hyphomycete with an accompanying *Phaeococcomyces alcalophilus* morph, and its physiological characteristics. Transactions of the Mycological Society of Japan 22:429–439.

Gounot, A.-M. 1986. Psychrophilic and psychrotrophic microorganisms. Experientia 42:1192–1197.

Gouy, M., and W.-H. Li. 1989. Phylogenetic analysis based on rRNA sequences supports the archaebacterial rather than the eocyte tree. Nature 339:145–147.

Goward, T., B. McCune, and D. Meidinger. 1994. The Lichens of British Columbia. Part 1. Foliose and Squamulose Species. Ministry of Forests. Crown Publications, Victoria, British Columbia, Canada.

Graafland, W. 1953. Four species of *Exobasidium* in pure culture. Acta Botanica Neerlandica 1:516–522.

Graf, F., and E. Horak. 1993. Taxonomy and ecology of a new hypogeous Basiodiomycete, *Hymenogaster saliciphilus* sp. nov., from the alpine zone of the Grisons (Switzerland). Arctic and Alpine Mycology 3. Bibliotheca Mycologica 150:39–51.

Gramss, G. 1991. "Definitive senescence" in stock cultures of basidiomycetous wood-decay fungi. Journal of Basic Microbiology 31:107–112.

Grant, W.D., W.E. Mwatha, and B.E. Jones. 1990. Alkaliphiles: ecology, diversity and applications. FEMS Microbiology Reviews 75:255–270.

Gray, D.J., and G. Morgan-Jones. 1981. Host-parasite relationships of *Agaricus brunnescens* and a number of mycoparasitic hyphomycetes. Mycopathologia 75:55–59.

Gray, M.W. 1982. Mitochondrial genome diversity and the evolution of mitochondrial DNA. Canadian Journal of Biochemistry 60:157–171.

Gray, N.F. 1984. Ecology of nematophagous fungi: methods of collection, isolation and maintenance of predatory and endoparasitic fungi. Mycopathologia 86:143–153.

———. 1987. Nematophagous fungi with particular reference to their ecology. Biological Reviews of the Cambridge Philosophical Society 62:245–304.

Gray, W.D., and C.J. Alexopoulos. 1968. Biology of the Myxomycetes. Ronald Press, New York.

Grenet, E., G. Fonty, J. Jamot, and F. Bonnemoy. 1989. Influence of diet and monensin on development of anaerobic fungi in the rumen, duodenum, cecum, and feces of cows. Applied and Environmental Microbiology 55:2360–2364.

Greuter, W., J. McNeill, F.R. Barrie, H.M. Burdet, V. Demoulin, T.S. Filgueiras, D.H. Nicolson, P.C. Silva, J.E. Skog, P. Trehane, N.J. Turland, and D.L. Hawksworth. (eds. and comps.). 2000. International Code of Botanical Nomenclature (Saint Louis Code). Regnum Vegetabile 138:1–474.

Grieg-Smith, P. 1983. Quantitative Plant Ecology. 3rd ed. Blackwell Scientific, Oxford, Great Britain.

Griffin, D.M. 1972. Ecology of Soil Fungi. Syracuse University Press, Syracuse, New York.

———. 1981. Water and microbial stress. Pp. 91–136. *In* M.A. Alexander (ed.), Advances in Microbial Ecology. Vol. 5. Plenum, New York.

Griffin, G.J., and K.H. Garren. 1974. Population levels of *Aspergillus flavus* and the *Aspergillus niger* group in Virginia peanut field soils. Phytopathology 64:322–325.

Griffith, G.S., and L. Boddy. 1988. Fungal communities in attached ash (*Fraxinus excelsior*) twigs. Transactions of the British Mycological Society 91:599–606.

Griffith, J.M., A.J. Davis, and B.R. Grant. 1992. Target sites of fungicides to control Oomycetes. Pp. 69–100. *In* W. Köller (ed.), Target Sites of Fungicide Action. CRC Press, Boca Raton, Florida.

Griffith, N.T., and H.L. Barnett. 1967. Mycoparasitism by basidiomycetes in culture. Mycologia 59:149–154.

Griffiths, D.A. 1974. The origin, structure and function of chlamydospores in fungi. Nova Hedwigia 25:503–547.

Griffiths, R.P., E.R. Ingham, B.A. Caldwell, M.A. Castellano, and K. Cromack, Jr. 1991. Microbial characteristics of ectomycorrhizal mat communities in Oregon and California. Biology and Fertility of Soils 11:196–202.

Grime, J.P. 1977. Evidence for the existence of three primary strategies in plants and its relevance to ecological and evolutionary theory. American Naturalist 111:1169–1194.

———. 1979. Plant Strategies and Vegetation Processes. John Wiley, Chichester, England.

Grime, J.P., M.L. Mackey, S.H. Hillier, and D.J. Read. 1987. Floristic diversity in a model system using experimental microcosms. Nature 328:420–422.

Gross, K.L., M.R. Willig, L. Gough, R. Inouye, and S.B. Cox. 2000. Patterns of species diversity and productivity at different spatial scales in herbaceous plant communities. Oikos 89:417–427.

Grossman, L.I., and M.E.S. Hudspeth. 1985. Fungal mitochondrial genomes. Pp 65–103. *In* J.W. Bennett and L.L.Lasure (eds.), Gene Manipulations in Fungi. Academic Press, Orlando, Florida.

Grube, M., and J. Hafellner. 1990. Studien an flechtenbewohnenden Pilzen der Sammelgattung *Didymella* (Ascomycetes, Dothideales). Nova Hedwigia 51:283–360.

Grube, M., and S. Kroken. 2000. Molecular approaches and the concept of species and species complexes in lichenized fungi. Mycological Research 104:1284–1294.

Grube, M., M. Matzer, and J. Hafellner. 1995. A preliminary account of the lichenicolous *Arthonia* species with reddish, K⁺ reactive pigments. Lichenologist 27:25–42.

Gruithuisen, F. von P. 1821. Die Brachienschnecke und eine aus ihren ueberresten hervorwachsende lebendig-gebaehrende conferve. Nova Acta Physico-Medica Academiae Caesareae Leopoldino-Carolinae Naturae Curiosorum 10:437–452.

Gruyter, J. de, and M.E. Noordeloos. 1992. Contributions towards a monograph of *Phoma* (Coelomycetes)—I. 1. Section *Phoma*: taxa with very small conidia in vitro. Persoonia 15:71–92.

Gruyter, J. de, M.E. Noordeloos, and G.H. Boerema. 1993. Contributions towards a monograph of *Phoma* (Coelomycetes)—I. 2. Section *Phoma*: Additional taxa with very small conidia and taxa with conidia up to 7 μm long. Persoonia 15:369–400.

Guadet, J., J. Julien, J.F. Lafey, and Y. Brygoo. 1989. Phylogeny of some *Fusarium* species, as determined by large subunit rRNA sequence comparison. Molecular Biology and Evolution 6:227–242.

Guarro, J., S.K. Abdullah, S.M. Al-Bader, M.J. Figueras, and J. Gené. 1996. The genus *Melanocarpus*. Mycological Research 100:75–78.

Guého, E., C.P. Kurtzman, and S.W. Peterson. 1989. Evolutionary affinities of heterobasidiomycetous yeasts estimated from 18S and 25S ribosomal RNA sequence divergence. Systematic and Applied Microbiology 12:230–236.

———. 1990. Phylogenetic relationships among species of *Sterigmatomyces* and *Fellomyces* as determined from partial rRNA sequences. International Journal of Systematic Bacteriology 40:60–65.

Guého, E., M.T. Smith, G.S. de Hoog, G. Billon-Grand, R. Christen, and W.H. Batenburg-van der Vegte. 1992. Contributions to a revision of the genus *Trichosporon*. Antonie Van Leeuwenhoek 61:289–316.

Guilliermond, A. 1912. Les Levures. O. Doin, Paris, France.

Guiraud, P., R. Steiman, F. Seigle-Murandi, and L. Sage. 1995. Mycoflora of soil around the Dead Sea II—Deuteromycetes (except *Aspergillus* and *Penicillium*). Systematic and Applied Microbiology 18:318–322.

Gulden, G., K. Høiland, K. Bendiksen, T.E. Brandrud, B.S. Foss, H.B. Jenssen, and D. Laber. 1992. Macromycetes and air pollution: mycocoenological studies in three oligotrophic spruce forests in Europe. Bibliotheca Mycologica 144:1–81.

Gulya, T.J., S. Masirevic, and C.E. Thomas. 1993. Preservation of air-dried downy mildew sporangia in liquid nitrogen without cryoprotectants or controlled freezing. Mycological Research 97:240–244.

Gunde-Cimerman, N., P. Zalar, S. de Hoog, and A. Plemenitaš 2000. Hypersaline waters in salterns—natural ecological niches for halophilic black yeasts. FEMS Microbiology Ecology 32:235–240.

Gunderson, J.H., H. Elwood, A. Ingold, K. Kindle, and M.L. Sogin. 1987. Phylogenetic relationships between chlorophytes, chrysophytes, and oomycetes. Proceedings of the National Academy of Sciences USA 84:5823–5827.

Guo, L.-Y., and T.J. Michailides. 1998. Factors affecting the rate and mode of germination of *Mucor piriformis* zygospores. Mycological Research 102:815–819.

Guo, L.D., K.D. Hyde, and C.Y. Liew. 2000. Identification of endophytic fungi from *Livistona chinensis* based on morphology and rDNA sequences. New Phytologist 147:617–630.

Gupta, M., and P. Filner. 1991. Microsatellites amplify highly polymorphic DNA bands in spar of plant DNA. The International Society for Plant Molecular Biology, Third International Congress. Abstract 1705.

Gupta, R.C., R.L. Fotedar, and K.N. Pandey. 1983. *Drechslera rostrata*, a destructive mycoparasite of *Rhizopus nigricans*. Acta Phytopathologica Academiae Scientiarum Hungaricae 18:49–56.

Gupta, R.C., R.S. Upadhyay, and B. Rai. 1979. Hyphal parasitism and chlamydospore formation by *Fusarium oxysporum* on *Rhizoctonia solani*. Mycopathologia 67:147–152.

Gutell, R.R., and G.E. Fox. 1988. Compilation of large subunit RNA sequences presented in a structural format. Nucleic Acids Research 16:175–269.

Haase, G., L. Sonntag, B. Melzer-Krick, and G.S. de Hoog. 1999. Phylogenetic inference by SSU-gene analysis of members of the Herpotrichiellaceae with special reference to human pathogenic species. Studies in Mycology 43:80–97.

Hadrys, H., M. Balick, and B. Schierwater. 1992. Applications of random amplified polymorphic DNA (RAPD) in molecular ecology. Molecular Ecology 1:55–63.

Hafellner, J. von. 1979. Karschia, revision einer Sammelgattung an der Grenze von lichenisierten und nichtlichenisierten Ascomyceten. Beihefte zur Nova Hedwigia 62:1–248. J. Cramer, Vaduz, Liechtenstein.

———. 1984. Studien in Richtung einer natürlicheren Gliederung der Sammelfamilien Lecanoraceae und Lecideaceae. Beihefte zür Nova Hedwigia 79:241–371.

———. 1985. Studien über lichenicole Pilze und Flechten IV. Die auf *Brigantiaea*-Arten beobachteten Ascomyceten. Herzogia 7:163–180.

———. 1988. Principles of classification and main taxonomic groups. Pp. 41–52. *In* M. Galun (ed.), CRC Handbook of Lichenology. Vol. III. CRC Press. Boca Raton, Florida.

———. 1989. Studien über lichenicole Pilze und Flechten. VII. Über die neue Gattung *Lichenochora* (Ascomycetes, Phyllachorales). Nova Hedwigia 48:357–370.

———. 1994. Beiträge zu einem Prodromus der lichenicolen Pilze Österreichs und angrenzender Gebiete. I. Einige neue oder seltene Arten. Herzogia 10:1–28.

Hafellner, J., and P. Navarro-Rosinés. 1993. *Llimoniella* gen. nov.- eine weitere Gattung lichenicoler Discomyceten (Ascomycotina, Leotiales). Herzogia 9:769–778.

Hafellner, J., and W. Obermayer. 1995. *Cercidospora trypetheliza* und einige weitere lichenicole Ascomyceten auf *Arthrorhapis*. Cryptogamie, Bryologie, Lichénologie 16:177–190.

Hafellner, J., and R. Türk. 2001. Die lichenisierten Pilze Österreichs—eine Checkliste der bisher nachgewiesenen Arten mit Verbreitungsangaben. Stapfia 76:1–167.

Haff, L.A. 1994. Improved Quantitative PCR using nested primers. PCR Methods and Applications 3:332–337.

Hagelstein, R. 1944. The Mycetozoa of North America. Published by the author, Mineola, New York.

Hagiwara, H. 1989. The Taxonomic Study of Japanese Dictyostelid Cellular Slime Molds. National Science Museum, Tokyo, Japan.

Hagiwara, H., and Y. Yamamoto. 1995. Myxomycetes of Japan. Heibonsha, Tokyo, Japan.

Hagler, A.N., and D.G. Ahearn. 1987. Ecology of aquatic yeasts. Pp. 181–205. *In* A.H. Rose and J.S. Harrison (eds.), The Yeasts: Yeasts and the Environment. Academic Press, London, England.

Haines, J.H., and C.R. Cooper, Jr. 1993. DNA and mycological herbaria. Pp. 305–315. *In* D.R. Reynolds and J.W. Taylor (eds.), The Fungal Holomorph: Mitotic, Meiotic and Pleomorphic Speciation in Fungal Systematics. CAB International, Wallingford, Oxon, United Kingdom.

Hajlaoui, M.R., N. Benhamou, and R.R. Bélanger. 1992. Cytochemical study of the antagonistic activity of *Sporothrix flocculosa* on rose powdery mildew, *Sphaerotheca pannosa* var. *rosae*. Phytopathology 82:583–589.

Hale, M.E., Jr. 1979. How to Know the Lichens. 2nd ed. W.C. Brown, Dubuque, Iowa.

——. 1982. Lichens as bioindicators and monitors of air pollution in the Flat Tops Wilderness Area, Colorado. Final Report, U.S. Department of Agriculture, Forest Service Contract No. OM RFP R2–81-SP35, Smithsonian Institution, Washington, D.C.

——. 1983. The Biology of Lichens. 3rd ed. Edward Arnold, Baltimore, Maryland.

Hale, M.E., Jr., and M. Cole. 1988. Lichens of California. University of California Press, Berkeley, California.

Hall, A.V. 1988. Pest control in herbaria. Taxon 37:885–907.

Hall, G.S. 1996. Zoosporic fungi. Pp. 109–123. *In* Methods for the Examination of Organismal Diversity in Soils and Sediments. CAB International, Wallingford, Oxon, United Kingdom.

Hall, G.S., and D.W. Minter. 1994. International Mycological Directory. 3rd ed. CAB International, Wallingford, Oxon, United Kingdom.

Hall, R.A. 1980. Laboratory infection of insects by *Verticillium lecanii* strains isolated from phytopathogenic fungi. Transactions of the British Mycological Society 74:445–446.

Hallett, I.C., and M.W. Dick. 1981. Seasonal and diurnal fluctuations of oomycete propagule numbers in the free water of a freshwater lake. Journal of Ecology 69:671–692.

Halmschlager, E., H. Butin, and E. Donaubauer. 1993. Endophytische pilze in blättern un zwiegen von *Quercus petraea*. European Journal of Forest Pathology 23:51–63.

Hamelin, R.C., J. Beaulieu, and A. Plourde. 1995. Genetic diversity in populations of *Cronartium ribicola* in plantations and natural stands of *Pinus strobus*. Theoretical and Applied Genetics 91:1214–1221.

Hammer, C.A., II. 1984. Dictyostelids in Agricultural Soils. Unpubl Master's thesis, Ohio University, Athens, Ohio.

Hammond, P.M. 1990. Insect abundance and diversity in the Dumoga-Bone National Park, N. Sulawesi, with special reference to the beetle fauna of lowland rain forest in the Toraut region. Pp. 197–254. *In* W.J. Knight and J.D. Holloway (eds.), Insects and the Rain Forests of South East Asia (Wallacea). Royal Entomological Society of London, London, England.

——. 1995. Described and estimated species numbers: an objective assessment of current knowledge. Pp. 29–71. *In* D. Allsopp, R.R. Colwell, and D.L. Hawksworth (eds.), Microbial Diversity and Ecosystem Function. CAB International, Wallingford, Oxon, United Kingdom.

Hammond, P.M., and J.F. Lawrence. 1989. Mycophagy in insects: a summary. Pp. 275–324. *In* N. Wilding, N.M. Collins, P.M. Hammond, and J.F. Webber (eds.), Insect-Fungus Interactions. Academic Press, New York.

Hanke, A.R., S. Backman, D.J. Speare, and G.W. Friars. 1991. An uncommon presentation of fungal infection in Atlantic salmon fry. Journal of Aquatic Animal Health 3:192–197.

Hankin, L., and S.L. Anagnostakis. 1975. The use of solid media for detection of enzyme production by fungi. Mycologia 67:597–607.

Hanlin, R.T. 1998a. Illustrated Genera of Ascomycetes. Volume II. American Phytopathological Society Press, St. Paul, Minnesota.

——. 1998b. Combined Keys to Illustrated Genera of Ascomycetes I, II. American Phytopathological Society Press, St. Paul, Minnesota.

——. 1990. Illustrated Genera of ascomycetes. American Phytopathological Society Press, St. Paul, Minnesota.

Hansford, C.G. 1946. The follicolous ascomycetes, their parasites and associated fungi, especially as illustrated by Uganda specimens. Mycological Papers 15:1–240.

——. 1961. The Meliolineae, a monograph. Sydowia, Ser II., Beihefte 2:1–806.

——. 1963. Iconographia Meliolinearum. Sydowia, Ser. II, Beihefte 5:1–285.

Hanson, A.M. 1945. A morphological, developmental, and cytological study of four saprophytic chytrids. I. *Catenomyces persicinus* Hanson. American Journal of Botany 32:431–438.

Happ, G.M., C.M. Happ, and S.J. Barras. 1971. Fine structure of the prothoracic mycangium, a chamber for the culture of symbiotic fungi, in the southern pine beetle, *Dendroctonus frontalis*. Tissue and Cell 3:295–308.

Harkin, J.M., and J.R. Obst. 1973. Syringaldazine, an effective reagent for detecting laccase and peroxidase in fungi. Experientia 29:381–387.

Harley, J.L., and S.E. Smith. 1983. Mycorrihizal Symbiosis. Academic Press, London, England.

Harley, J.L., and J.S. Waid. 1955. A method of studying active mycelia on living roots and other surfaces in the soil. Transactions of the British Mycological Society 38:104–118.

Harman, G.E. 1991. Seed treatments for biological control of plant disease. Crop Protection 10:166–171.

Harmsen, M.C., F.H.J. Schuren, S.M. Moukha, C.M. Van Zuilen, P.J. Punt, and J.G.H. Wessels. 1992. Sequence analysis of the glyceraldehyde-3-phosphate dehydrogenase genes from the basidiomycetes *Schizophyllum commune, Phanerochaete chrysosporium*, and *Agaricus bisporus*. Current Genetics 22:447–454.

Harper, J.E., and J. Webster. 1964. An experimental analysis of the coprophilous fungus succession. Transactions of the British Mycological Society 47:511–530.

Harper, J.L. 1977. Population Biology of Plants. Academic Press, New York.

——. 1990. Pests, pathogens, and plant communities: an introduction. Pp. 3–14. *In* J.J. Burdon and S.R. Leather (eds.), Pests, Pathogens and Plant Communities. Blackwell Scientific, London, England.

Harrell, L.W., R.A. Elston, T.M. Scott, and M.T. Wilkinson. 1986. A significant new systemic disease of net-pen reared chinook salmon (*Oncorhynchus tshawytscha*) brood stock. Aquaculture 55:249–262.

Harrington, T.C. 1988. *Leptographium* species, their distributions, hosts and insect vectors. Pp. 1–39. *In* T.C. Harrington and F.W. Cobb, Jr. (eds.), *Leptographium* Root Diseases on Conifers. American Phytopathological Society Press, St. Paul, Minnesota.

——. 1992. *Leptographium*. Pp. 129–133. *In* L.L. Singleton, J.D. Mihail, and C.M. Rush (eds.), Methods for Research on Soilborne Phytopathogenic Fungi. American Phytopathological Society Press, St. Paul, Minnesota.

——. 1993. Diseases of conifers caused by species of *Ophiostoma* and *Leptographium*. Pp. 161–172. *In* M.J. Wingfield, K.A. Seifert, and

J.F. Webber (eds.), *Ceratocystis* and *Ophiostoma*: Taxonomy, Ecology, and Pathogenicity. American Phytopathological Society Press, St. Paul, Minnesota.

Harrington, T.C., and F.J. Cobb, Jr. (eds.). 1988. *Leptographium* Root Diseases on Conifers. American Phytopathological Society Press, St. Paul, Minnesota.

Harrington, T.C., F.W. Cobb, Jr., and J.W. Lownsberry. 1985. Activity of *Hylastes nigrinus*, a vector of *Verticicladiella wageneri*, in thinned stands of Douglas-fir. Canadian Journal of Forestry Research 15:519–523.

Harris, D.C. 1985. *Microdochium fusarioides* sp. nov. from oospores of *Phytophthora syringae*. Transactions of the British Mycological Society 84:358–361.

Harris, R.C. 1995. More Florida Lichens. Published by the author, Bronx, New York.

Harrison, J.L., and E.B.G. Jones. 1975. The effect of salinity on sexual and asexual sporulation of members of the Saprolegniaceae. Transactions of the British Mycological Society 65:389–394.

Harrison, P.J., R.E. Waters, and F.J.R. Taylor. 1980. A broad spectrum artificial seawater medium for coastal and open ocean phytoplankton. Journal of Phycology 16:28–35.

Harvey, A.E., M.F. Jurgensen, and M.J. Larsen. 1978. Seasonal distribution of ectomycorrhizae in a mature Douglas-fir/Larch forest soil in western Montana. Forest Science 24:203–208.

Harvey, J.V. 1925. A study of the water molds and Pythiums occurring in the soils of Chapel Hill. Journal of the Elisha Mitchell Scientific Society 41:151–164.

Hashioka, Y., and T. Fukita. 1969. Ultrastructural observations on mycoparasitism of *Trichoderma*, *Gliocladium* and *Acremonium* to phytopathogenic fungi. Reports of the Tottori Mycological Institute 7:8–18.

Hashioka, Y., and M. Komatsu. 1964. *Trichoderma viride*, as an antagonist of the wood-inhabiting Hymenomycetes, III. Species of *Hypocrea* occurring on log-woods of *Lentinus edodes*. Report of the Tottori Mycological Institute 4:1–5.

Hashioka, Y., H. Ishikawa, M. Komatsu, and I. Arita. 1961. *Trichoderma viride*, as an antagonist of the wood-inhabiting Hymenomycetes. II. A metabolic product of *Trichoderma fungistatic* to the Hymenomycetes. Reports of the Tottori Mycological Institute 1:9–18.

Haskins, E.F., and M.D. McGuinness. 1986. Comparative ultrastructural observations of spore wall structure in six species of *Echinostelium* and three species of Eumycetozoa. Mycologia 78:613–618.

Haskins, E.F., M.D. McGuinness, and C.S. Berry. 1983. *Semimorula*: new genus with myxomycete and protostelid affinities. Mycologia 75:153–158.

Haskins, R.J. 1939. Cellulose as a substratum for saprophytic chytrids. American Journal of Botany 26:635–639.

Hass, H., T.N. Taylor, and W. Remy. 1994. Fungi from the lower Devonian Rhynie chert: mycoparasitism. American Journal of Botany 81:29–37.

Hata, K., and K. Futai. 1995. Endophytic fungi associated with healthy pine needles and needles infested by the pine needle gall midge, *Thecodiopsis japonensis*. Canadian Journal of Botany 73:384–390.

——. 1996. Variation in fungal endophyte populations in needles of the genus *Pinus*. Canadian Journal of Botany 74:103–114.

Hatai, K., and O. Lawhavinit. 1988. *Lagenidium myophilum* sp. nov., a new parasite on adult northern shrimp (*Pandalus borealis* Krøyer). Transactions of the Mycological Society of Japan 29:175–184.

Hatai, K., B.Z. Bian, M.C.L. Baticados, and S. Egusa. 1980. Studies on the fungal diseases in crustaceans. II. *Haliphthoros philippinesis* sp. nov. isolated from cultivated larvae of the jumbo tiger prawn (*Penaeus monodon*). Transactions of the Mycological Society of Japan 21:47–55.

Hatai, K., S. Egusa, S. Takahashi, and K. Ooe. 1977. Study on the pathogenic fungus of mycotic granulomatosis I. Isolation and pathogenicity of the fungus from cultured agar infected with the disease. Fish Pathology 12:129–133.

Hatai, K., Y. Fujimaka, and S. Egusa. 1986a. A visceral mycosis in ayu fry, *Plecoglossus altivelis* Temminck and Schlegel, caused by a species of *Phoma*. Journal of Fish Diseases 9:4–9.

Hatai, K., S.S. Kubota, N. Kida, and S. Udagawa. 1986b. *Fusarium oxysporum* in red sea bream (*Pagrus* sp.). Journal of Wildlife Diseases 22:570–571.

Hauerslev, K. 1969. *Christiansenia pallida* gen. nov., sp. nov., a new parasitic homobasidiomycete from Denmark. Friesia 9:43–45.

——. 1977. New and rare Tremellaceae on record from Denmark. Friesia 11:94–115.

——. 1979. New or rare resupinate fungi from Denmark. Friesia 11:272–280.

Hausner, G., A. Belkhiri, and G.R. Klassen. 2000. Phylogenetic analysis of the small subunit ribosomal RNA gene of the hyphochytrid *Rhizidiomyces apophysatus*. Canadian Journal of Botany 78:124–128.

Hausner, G., J. Reid, and G.R. Klassen. 1992. Do galeate-ascospore members of the Cephaloascaceae, Endomycetaceae and Ophiostomataceae share a common phylogeny? Mycologia 84:870–881.

Hawksworth, D.L. 1972. Lichens 1961–1969. Index of Fungi Supplement. Commonwealth Mycological Institute, Kew, Surrey, United Kingdom.

——. 1974. Mycologist's Handbook. Commonwealth Mycological Institute, Kew, Surrey, United Kingdom.

——. 1977. Taxonomic and biological observations on the genus *Lichenoconium* (Sphaeropsidales). Persoonia 9:159–198.

——. 1979a. The lichenicolous Hyphomycetes. Bulletin of the British Museum (Natural History) 6:183–300.

——. 1979b. Studies in the genus *Endococcus* (Ascomycotina, Dothideales). Botaniska Notiser 132:283–290.

——. 1980. Notes on some fungi occurring on *Peltigera*, with a key to accepted species. Transactions of the British Mycological Society 74:363–386.

——. 1981a. A survey of the fungicolous conidial fungi. Pp. 171–244. *In* G.T. Cole and B. Kendrick (eds.), The Biology of Conidial Fungi. Vol. 1. Academic Press, New York.

——. 1981b. The lichenicolous Coelomycetes. Bulletin of the British Museum (Natural History) 9:1–98.

——. 1982. Secondary fungi in lichen symbioses: parasites, saprophytes and parasymbionts. Journal of the Hattori Botanical Laboratory 52:357–366.

——. 1983. A key to the lichen-forming, parasitic, parasymbiotic and saprophytic fungi occurring on lichens in the British Isles. Lichenologist 15:1–44.

——. 1988a. Effects of algae and lichen-forming fungi on tropical trees. Pp. 76–83. *In* V.P. Agnihotri, A.K. Sarbhoy, and D. Kumar (eds.), Perspectives in Mycology and Plant Pathology. Malhotra, New Delhi, India.

——. 1988b. The variety of fungal-algal symbioses, their evolutionary significance, and the nature of lichens. Botanical Journal of the Linnean Society (London) 96:3–20.

——. 1991. The fungal dimension of biodiversity: magnitude, significance, and conservation. Mycological Research 95:641–655.

——. 1993. The tropical fungal biota: census, pertinence, prophylaxis, and prognosis. Pp. 265–293. *In* S. Isaac, J.C. Frankland, R. Watling, and A.J.S. Whalley (eds.), Aspects of Tropical Mycology. Cambridge University Press, Cambridge, United Kingdom.

——. 1994. The recent evolution of lichenology: A science for our times. Cryptogamic Botany 4:117–129.

——. 1998 ["1997"]. Raffaele Ciferri, the crisis precipitated in the naming of lichen-forming fungi, and why whole lichens have no names. Archivio Geobotanica 3:3–9.

Hawksworth, D.L., and T. Ahti. 1990. A bibliographic guide to the lichen floras of the world. 2nd ed. Lichenologist 22:1–78.

Hawksworth, D.L., and P. Diederich. 1988. A synopsis of the genus *Polycoccum* (Dothideales), with a key to accepted species. Transactions of the British Mycological Society 90:293–312.

Hawksworth, D.L., and D.J. Hill. 1984. The Lichen-forming Fungi. Blackie, Glasgow, United Kingdom.

Hawksworth, D.L., and D. Jones. 1981. *Sclerococcum sphaerale* obtained in pure culture. Transactions of the British Mycological Society 77:485–489.

Hawksworth, D.L., and M.T. Kalin-Arroyo. 1995. Magnitude and distribution of biodiversity. Pp. 107–191. *In* V.H. Heywood (ed.), Global Biodiversity Assessment. Cambridge University Press, Cambridge, Great Britain.

Hawksworth, D.L., and J. Miadlikowska. 1997. New species of lichenicolous fungi occurring on *Peltigera* in Ecuador and Europe. Mycological Research 101:1127–1134.

Hawksworth, D.L., and J. Poelt. 1986. Five additional genera of conidial lichen-forming fungi from Europe. Plant Systematics and Evolution 154:195–211.

Hawksworth, D.L., and R. Santesson. 1990. A revision of the lichenicolous fungi previously referred to *Phragmonaevia*. Bibliotheca Lichenologica 38:121–143.

Hawksworth, D.L., and S. Udagawa. 1977. Contributions to a monograph of *Microthecium*. Transactions of the Mycological Society of Japan 18:143–154.

Hawksworth, D.L., and H. Wells. 1973. Ornamentation on the terminal hairs in *Chaetomium* Kunze ex Fr. and some allied genera. Mycological Papers 134:1–24.

Hawksworth, D.L., B.J. Coppins, and P.W. James. 1979. *Blarneya*, a lichenized hyphomycete from southern Ireland. Botanical Journal of the Linnean Society (London) 79:357–367.

Hawksworth, D.L., P.W. James, and B.J. Coppins. 1980. Checklist of British lichen-forming, lichenicolous and allied fungi. Lichenologist 12:1–115.

Hawksworth, D.L., P.M. Kirk, B.C. Sutton, and D.N. Pegler. 1995. Ainsworth & Bisby's Dictionary of the Fungi. 8th ed. CAB International, Wallingford, Oxon, United Kingdom.

Hawksworth, D.L., D.W. Minter, G.C. Kinsey, and P.F. Cannon. 1998. Inventorying a tropical fungal biota: intensive and extensive approaches. Pp. 29–50. *In* K.K. Janardhanan, C. Rajendran, K. Natarajan, and D.L. Hawksworth (eds.), Tropical Mycology. Science Publishers, Enfield, New Hampshire.

Hawksworth, D.L., R.R.M. Paterson, and N. Vote. 1993. An investigation into the occurrence of metabolites in obligately lichenicolous fungi from thirty genera. Bibliotheca Lichenologica 53:101–108.

Hayes, A.J. 1965. Some microfungi from Scots pine litter. Transactions of the British Mycological Society 48:179–185.

Hayes, A.S. 1979. The microbiology of plant litter decomposition. Scientific Progress (London) 66:25–42.

Hayes, J.L., and B.L. Strom. 1994. 4-allylanisole as an inhibitor of bark beetle (Coleoptera: Scolytidae) aggregation. Journal of Economic Entomology 87:1586–1594.

Haynes, K.A., T.J. Westerneng, J.W. Fell, and W. Moens. 1995. Rapid detection and identification of pathogenic fungi by polymerase chain reaction amplification of large subunit ribosomal DNA. Journal of Medical and Veterinary Mycology 33:319–325.

Hayward, G.D., and R. Rosentreter. 1994. Lichens as nesting material for northern flying squirrels in the northern Rocky Mountains. Journal of Mammalogy 75:663–673.

He, F., and P. Legendre. 1996. On species-area relations. American Naturalist 148:719–737.

Heath, I.B., T. Bauchop, and R.A. Skipp. 1983. Assignment of the rumen anaerobe *Neocallimastix frontalis* to the *Spizellomycetales* (Chytridiomycetes) on the basis of its polyflagellate zoospore ultrastructure. Canadian Journal of Botany 61:295–307.

Heckman, J.E., J. Sarnoff, B. Alzner-DeWeerd, S. Yin, and U.L. RajBhandary. 1980. Novel features in the genetic code and codon reading patterns in *Neurospora crassa* mitochondria based on sequences of six mitochondrial tRNAs. Proceedings of the National Academy of Sciences USA 77:3159–3163.

Hedger, J. 1985. Tropical agarics: resource relations and fruiting periodicity. Pp. 41–86. *In* D. Moore, L.A. Casselton, D.A. Wood, and J.C. Frankland (eds.), Developmental Biology of Higher Fungi. Cambridge University Press, Cambridge, England.

Hedjaroude, G.A. 1969. Études taxonomiques sur les *Phaeosphaeria* Miyake et leurs formes voisines (Ascomycetes). Sydowia 22:57–107.

Hegedus, D.D., and G.G. Khachatourians. 1993a. Construction of cloned DNA probes for the specific detection of the entomopathogenic fungus *Beauveria bassiana* in grasshoppers. Journal of Invertebrate Pathology 62:233–240.

——. 1993b. Identification of molecular variants in mitochondrial DNAs of members of the genera *Beauveria*, *Verticillium*, *Paecilomyces*, *Tolypocladium*, and *Metarhizium*. Applied and Environmental Microbiology 59:4283–4288.

——. 1996a. Detection of the entomopathogenic fungus *Beauveria bassiana* within infected migratory grasshoppers (*Melanoplus sanguinipes*) using polymerase chain reaction and DNA probe. Journal of Invertebrate Pathology 67:21–27.

——. 1996b. Identification and differentiation of the entomopathogenic fungus *Beauveria bassiana* using polymerase chain reaction and single strand conformation polymorphism analysis. Journal of Invertebrate Pathology 67:289–299.

——. 2001. Comparative study of mitochondrial DNA heterogeneity amongst isolates of the entomopathogenic fungi *Beauveria brongniartii* and *B. bassiana*. (unpubl. ms.).

Hegedus, D.D., T.A. Pfeifer, J.M. MacPherson, and G.G. Khachatourians. 1991. Cloning and analysis of five mitochondrial tRNA-encoding genes from the fungus *Beauveria bassiana*. Genetics 109:149–154.

Hegedus D.D., T.A. Pfiefer, D.S. Mulyk, and G.G. Khachatourians. 1997a. Nucleotide sequence and secondary structure prediction of the mitochondrial small rRNA of the entomopathogenic fungus *Beauveria bassiana*. (unpubl. ms.)

——. 1997b. Characterization and structure of the mitochondrial small rRNA gene of the entomopathogenic fungus *Beauveria bassiana*. Genome 41:471–476.

——. 1998. Characterization and structure of the mitochondrial small rRNA gene of the entomopathogenic fungus *Beauveria bassiana*. Genome 41:471–476.

Heid, C.A., J. Stevens, K.J. Livak, and P.M. Williams. 1996. Real time quantitative PCR. Genome Research 6:986–994.

Heilmann-Clausen, J. 2001. A gradient analysis of communities of macrofungi and slime moulds on decaying beech logs. Mycological Research 105:575–596.

Heinrich, Z., and W. Wojewoda. 1976. The effect of fertilization on a pine forest ecosystem in an industrial region. IV. Macromycetes. Ecologia Polska 24:319–330.

Heintz, C., and R. Blaich. 1990. *Verticillium lecanii* als Hyperparasit des Rebmehltaus (*Uncinula necator*). Vitis 29:229–232.

Helander, M.L., T.N. Sieber, O. Petrini, and S. Neuvonen. 1994. Endophytic fungi in Scots pine needles: spatial variation and consequences of simulated acid rain. Canadian Journal of Botany 72:1108–1113.

Held, A.A. 1972. Host-parasite relations between *Allomyces* and *Rozella*. Parasite penetration depends on growth response of host cell-wall. Archiv für Mikrobiologie 82:128–139.

——. 1980. Development of *Rozella* in *Allomyces*: a single zoospore produces numerous zoosporangia and resistant sporangia. Canadian Journal of Botany 58:959–979.

——. 1981. *Rozella* and *Rozellopsis*: naked endoparasitic fungi which dress-up as their hosts. Botanical Review 47:451–515.

Helfer, W. 1991. Pilze auf Pilzfruchtkörpern: Untersuchungen zur Ökologie, Systematik und Chemie. IHW-Verlag, Eching, Germany.

Helsel, E.D., and D.T. Wicklow. 1979. Decomposition of rabbit feces: role of the sciarid fly *Lycoriella mali* Fitch in energy transformations. Canadian Entomologist 111:213–217.

Hendricks, J.D., and G.S. Bailey. 1989. Adventitious toxins. Pp. 605–651. *In* J.E. Halver (ed.), Fish Nutrition. 2nd ed. Academic Press, New York.

Hendriks, L., A. Goris, Y. Van de Peer, J.-M. Neefs, M. Vancanneyt, K. Kersters, J.-F. Berny, G.L. Hennebert, and R. De Wachter. 1992. Phylogenetic relationships among ascomycetes and ascomycete-like yeasts as deduced from small ribosomal subunit RNA sequences. Systematic and Applied Microbiology 15:98–104.

Hensens, O. D., J.G. Ondeyka, A.W. Dombrowski, D.A. Ostlind, and D.L. Zink. 1999. Isolation and structure of nodulisporic acid A_1 and A_2, novel insecticides from a *Nodulisporium* sp. Tetrahedron Letters 40:5455–5458.

Henssen, A., 1995. Studies on the biology and structure of *Dacampia* (Dothideales), a genus with lichenized and lichenicolous species. Cryptogamic Botany 5:149–158.

Henssen, A., and H.M. Jahns. 1974. Lichenes: eine Einführung in die Flechtenkunde. Georg Thieme Verlag, Stuttgart, Germany.

Hepperly, P.R. 1986. Hansfordia sp.: a parasitic pathogen of dematiaceous plant pathogenic fungi in Puerto Rico. Journal of Agriculture of the University of Puerto Rico 70:113–119.

Heredia, G. 1993. Mycoflora associated with green leaves and leaf litter of *Quercus germana*, *Quercus sartorrii* and *Liquidambar styraciflua* in a Mexican cloud forest. Cryptogamie: Mycologie 14:171–183.

Herman, S.G. 1986. The Naturalist's Field Journal: a manual of instruction based on a system established by Joseph Grinnell. Buteo Books, Vermillon, South Dakota.

Herr, L. 1988. Biocontrol of Rhizoctonia crown end root rot of sugar beet by binucleate *Rhizoctonia* spp. and *Laetisaria arvalis*. Annals of Applied Biology 113:107–118.

Hervey, A., C.T. Rogerson, and I. Leong. 1977. Studies on fungi cultivated by ants. Brittonia 29:226–236.

Hesseltine, C.W. 1957. The genus *Syzygites* (Mucoraceae). Lloydia 20:228–237.

——. 1960. The zygosporic stage of the genus *Pirella* (Mucoraceae). American Journal of Botany 47:225–230.

Hesseltine, C.W., and J.J. Ellis. 1973. Mucorales. Pp. 187–217. *In* G.C. Ainsworth, F.K. Sparrow, and A.S. Sussman (eds.), The Fungi. An Advanced Treatise. IVB. A Taxonomic Review with Keys: Basidiomycetes and Lower Fungi. Academic Press, New York.

Hesseltine, C.W., and R. Rogers. 1987. Dark-period induction of zygospores in *Mucor*. Mycologia 79:289–297.

Hesseltine, C.W., A.R. Whitehill, C. Pidacks, M. Ten Hagen, N. Bohonos, B.L. Hutchings, and J.H. Williams. 1953. Coprogen, a new growth factor present in dung, required by *Pilobolus* species. Mycologia 45:7–19.

Hestbjerg, H., S. Elmholt, U. Thrane, and U.B. Jensen. 1999. A resource-saving method for isolation of *Fusarium* and other fungi from individual soil particles. Mycological Research 103:1545–1548.

Hetrick, B.A.D., and J. Bloom. 1983. Vesicular-arbuscular mycorrhizal fungi associated with native tall grass prairie and cultivated winter wheat. Canadian Journal of Botany 61:2140–2146.

Heywood, V.H. (ed.). 1995. Global Biodiversity Assessment. Cambridge University Press, Cambridge, Great Britain.

Hibbett, D.S., and M.J. Donoghue. 1995. Progress toward a phylogenetic classification of the Polyporaceae through parsimony analyses of mitochondrial ribosomal DNA sequences. Canadian Journal of Botany 73(Suppl. 1):S853–S861.

——. 1998. Integrating phylogenetic analysis and classification in fungi. Mycologia 90:347–356.

Hibbett, D.S., and R.G. Thorn. 2001. Basidiomycota: Homobasidiomycetes. Pp. 121–168 *In* D.J. McLaughlin, E.G. McLaughlin, and P.A. Lempke (eds.), The Mycota. Vol. VII. Systematics and Evolution. Part B. Springer-Verlag, Berlin, Germany.

Hibbett, D.S., E.M. Pine, E. Langer, G. Langer, and M.J. Donoghue. 1997. Evolution of gilled mushrooms and puffballs inferred from ribosomal DNA sequences. Proceedings of the National Academy of Sciences USA 94:12002–12006.

Hibbits, J. 1978. Marine Eccrinales (Trichomycetes) found in crustaceans of the San Juan Archipelago, Washington. Syesis 11:213–261.

Hibbitts, J., G.C. Hughes, and A.K. Sparks. 1981. *Trichomaris invadens* gen. et sp. nov. an ascomycete parasite of the tanner crab (*Chionoecetes bairdi*) Rathbun (Crustacea; Brachyura). Canadian Journal of Botany 59:2121–2128.

Hicks, D.M. 1982. Abundance and distribution of black mat syndrome on stocks of tanner crabs, *Chionoecetes bairdi*, in the northwestern Gulf of Alaska. Pp. 563–579. *In* B. Melteff (ed.), Proceedings of the International Symposium on the Genus *Chionoecetes*, Fairbanks, Alaska, 1982. Lowell Wakefield Symposia Series, University of Alaska, Fairbanks, Alaska. Alaska Sea Grant Report 82–10.

Hijwegen, T. 1992a. Glycolytic activities in some fungicolous fungi. Netherlands Journal of Plant Pathology 98:77–80.

——. 1992b. Biological control of cucumber powdery mildew with *Tilletiopsis minor* under greenhouse conditions. Netherlands Journal of Plant Pathology 98:221–225.

Hijwegen, T., and H. Buchenauer. 1984. Isolation and identification of hyperparasitic fungi associated with Erysiphaceae. Netherlands Journal of Plant Pathology 90:79–84.

Hill, M.O. 1973. Diversity and evenness: a unifying notation and its consequences. Ecology 54:427–432.

Hillis, D.M., and C. Moritz. (eds.) 1990. Molecular Systematics. Sinauer, Sunderland, Massachusetts.

Hillis, D.M., J.P. Huelsenbeck, and C.W. Cunningham. 1994. Application and accuracy of molecular phylogenies. Science 264:671–677.

Hintz, W.E., M. Mohan, J.B. Anderson, and P.A. Horgen. 1985. The mitochondrial DNAs of *Agaricus*: heterogeneity in *A. bitorquis* and homogeneity in *A. brunnescens*. Current Genetics 9:127–132.

Hintz, W.E.A., J.B. Anderson, and P.A. Horgan. 1988. Relatedness of three species of *Agaricus* inferred from restriction fragment length polymorphism analysis of the ribosomal DNA repeat and mitochondrial DNA. Genome 32:173–178.

Hirata, K.A. 1986. Host Range and Geographical Distribution of the Powdery Mildew Fungi. Japan Scientific Societies Press, Tokyo, Japan.

Hiratsuka, Y., and S. Sato. 1982. Morphology and taxonomy of rust fungi. Pp. 1–36. *In* K.J. Scott and A.K. Chakravarty (eds.), The Rust Fungi. Academic Press, London, England.

Hirsch, G., and U. Braun. 1992. Communities of parasitic microfungi. Pp. 225–250. *In* W. Winterhoff (ed.), Fungi in Vegetation Science. Kluwer Academic, Dordrecht, The Netherlands.

Hirschhorn, E. 1986. Las Ustilaginales de la Flora Argentina. Comisión de Investigaciones Científicas, La Plata, Argentina.

Hjortstam, K., and K.-H. Larsson. 1994. Annotated check-list to genera and species of corticioid fungi (Aphyllophorales, Basidiomycotina) with special regards to tropical and subtropical areas. Windahlia 21:1–75.

Ho, H.H. 1981. Synoptic keys to the species of *Phytophthora*. Mycologia 73:705–714.

Ho, H.H., and S.C. Jong. 1990. *Halophytophthora*, gen. nov., a new member of the family Pythiaceae. Mycotaxon 36:377–382.

Ho, H.H., H.S. Chang, and S.Y. Hsieh. 1991. *Halophytophthora kandeliae*, a new marine fungus from Taiwan. Mycologia 83:419–424.

Ho, H.H., S.Y. Hsieh, and H.S. Chang. 1990. *Halophytophthora epistomium* from mangrove habitats in Taiwan. Mycologia 82:659–662.

Ho, H.H., A. Nakagiri, and S.Y. Newell. 1992. A new species of *Halophytophthora* from Atlantic and Pacific subtropical islands. Mycologia 84:548–554.

Ho, I., and J.M. Trappe. 1987. Enzymes and growth substances of *Rhizopogon* species in relation to mycorrhizal hosts and infrageneric taxonomy. Mycologia 79:553–558.

Ho, Y.W., and D.J.S. Barr. 1995. Classification of anaerobic gut fungi from herbivores with emphasis on rumen fungi from Malaysia. Mycologia 87:655–677.

Ho, Y.W., N. Abdullah, and S. Jalaludin. 1988. Penetrating structures of anaerobic rumen fungi in cattle and swamp buffalo. Journal General Microbiology 134:177–181.

——. 1994. *Orpinomyces intercalaris*, a new species of polycentric anaerobic rumen fungus from cattle. Mycotaxon 50:139–150.

Ho, Y.W., D.J.S. Barr, N. Abdullah, S. Jalaludin, and H. Kudo. 1993a. *Neocallimastix variabilis*, a new species of anaerobic rumen fungus from cattle. Mycotaxon 46:241–258.

——. 1993b. A new species of *Piromyces* from the rumen of deer in Malaysia. Mycotaxon 47:285–293.

——. 1993c. *Piromyces spiralis*, a new species of anaerobic fungus from the rumen of goat. Mycotaxon 48:59–68.

——. 1993d. *Anaeromyces*, an earlier name for *Ruminomyces*. Mycotaxon 47:283–284.

Ho, Y.W., T. Bauchop, N. Abdullah, and S. Jalaludin. 1990. *Ruminomyces elegans* gen. et sp. nov., a polycentric anaerobic rumen fungus from cattle. Mycotaxon 38:397–405.

Hobson, P.N. 1971. Rumen micro-organisms. Progress in Industrial Microbiology 9:42–77.

Hobson, P.N., and J.R. Wallace. 1982. Microbial ecology and activities in the rumen. Part I. Critical Reviews in Microbiology 9:165–225.

Hoch, H.C. 1977a. Mycoparasitic relationships: *Gonatobotrys simplex* parasitic on *Alternaria tenuis*. Phytopathology 67:309–314.

——. 1977b. Mycoparasitic relationships. III. Parasitism of *Physalospora obtusa* by *Calcarisporium parasiticum*. Canadian Journal of Botany 55:198–207.

——. 1978. Mycoparasitic relationships. IV. *Stephanoma phaeospora* parasitic on a species of *Fusarium*. Mycologia 70:370–379.

Hoch, H.C., and G.S. Abawi. 1979a. Mycoparasitism of oospores of *Pythium ultimum* by *Fusarium merismoides*. Mycologia 71:621–625.

——. 1979b. Biological control of *Pythium* root rot of table beet with *Corticium* sp. Phytopathology 69:417–419.

Hoch, H.C., and M.S. Fuller. 1977. Mycoparasitic relationships. I. Morphological features of interactions between *Pythium acanthicum* and several fungal hosts. Archive of Microbiology 111:207–224.

Hoch, H.C., and P. Provvidenti. 1979. Mycoparasitic relationships: cytology of the *Sphaerotheca fuliginea-Tilletiopsis* sp. interaction. Phytopathology 69:359–362.

Hocking, A.D. 1991. Xerophilic fungi in intermediate and low moisture foods. Pp. 69–97. *In* D.K. Arora, K.G. Mukerji, and E.H. Marth (eds.), Handbook of Applied Mycology. Vol. 111. Foods and Feeds. Marcel Dekker, New York.

——. 1993. Responses of xerophilic fungi to changes in water activity. Pp. 233–256. *In* D.H. Jennings (ed.), Stress Tolerance of Fungi. Marcel Dekker, New York.

Hocking, A.D., and J.I. Pitt. 1980. Dichloran-glycerol medium for enumeration of xerophilic fungi from low-moisture foods. Applied and Environmental Microbiology 39:488–492.

Hoefnagles, M.H., S.W. Broome, and S.R. Shafer. 1993. Vesicular-arbuscular mycorrhizae in salt marshes in North Carolina. Estuaries 16:851–858.

Hofman, T.W., and G.J. Bollen. 1987. Effects of granular nematicides on growth and microbial antagonism to *Rhizoctonia solani*. Netherlands Journal of Plant Pathology 93:201–214.

Högberg, N., and J. Stenlid 1999. Population genetics of *Fomitopsis rosea*—a wood-decay fungus of the old-growth European taiga. Molecular Ecology 8:703–710.

Högberg, N., O. Holdenrieder, and J. Stenlid. 1999. Population structure of the wood decay fungus *Fomitopsis pinicola*. Heredity 83:354–360.

Hogg, B.M., and H.J. Hudson. 1966. Micro-fungi on leaves of *Fagus sylvatica*. I. The micro-fungal succession. Transactions of the British Mycological Society 49:185–192.

Hohl, H.R., and K. Iselin. 1987. Liquid nitrogen preservation of zoosporic fungi. Pp. 143–145. *In* M.S. Fuller and A. Jaworski (eds.), Zoosporic Fungi in Teaching and Research. Southeastern Publishing, Athens, Georgia.

Höhnk, W. von. 1953. Studien zur Brack und Seewassermykologie. III. Oomycetes: Zweiter Teil. Veröffentlichungen des Instituts für Meeresforschung in Bremerhaven 2:52–108.

Höhnk, W. von, and K.J. Bock. 1954. Ein Beitrag zur Ökologie der saprophytischen Wasserpilze. Veröffentlichungen des Instituts für Meeresforschung in Bremerhaven 3:9–26.

Holdenrieder, O., and T.N. Sieber. 1992. Fungal associations of serially washed healthy non-mycorrhizal roots of *Picea abies*. Mycological Research 96:151–156.

Holland, D.M., and R.C. Cooke. 1990. Activation of dormant conidia of the wet bubble pathogen *Mycogone perniciosa* by Basidiomycotina. Mycological Research 94:789–792.

Holland, D.M., M.L. Parker, R.C. Cooke, and G.H. Evans. 1985. Germination of bicellular conidia of *Mycogone perniciosa*, the wet bubble pathogen of the cultivated mushroom. Transactions of the British Mycological Society 85:730–735.

Hölldobler, B., and E.O. Wilson. 1990. The Ants. Harvard University Press, Cambridge, Massachusetts.

Holliday, P. 1980. Fungus Diseases of Tropical Crops. Cambridge University Press, Cambridge, England.

Hollis, S., and R.K. Brummitt. 1992. World Geographical Scheme for Recording Plant Distributions. Hunt Institute for Botanical Documentation, Carnegie Mellon University, Pittsburgh, Pennsylvania.

Holm, L. 1957. Études taxonomiques sur les Pléosporacées. Symbolae Botanicae Upsalienses. 14:1–188.

Holmes, M.T. 1991. Seasonal Variations in the Distribution of Dictyostelid Cellular Slime Molds at Three Sites at Tikal, Guatemala and Their Taxonomy. Unpubl. Master's thesis, Ohio University, Athens, Ohio.

Holmgren, P.K., N.H. Holmgren, and L.C. Barnett. 1990. Index Herbariorum. Part 1: The Herbaria of the World. 8th ed. New York Botanical Garden, Bronx, New York. [also Regnum Vegetabile 120:1–693.]

Holmquist, R., M.M. Miyamoto, and M. Goodman. 1988. Analysis of higher-primate phylogeny from transversion differences in nuclear and mitochondrial DNA by Lake's method of evolutionary parsimony and operator metrics. Molecular Biology and Evolution 5:217–236.

Holst-Jensen, A., L.M. Kohn, and T. Schumacher. 1997. Nuclear rDNA phylogeny of the Sclerotiniaceae. Mycologia 89:885–899.

Holt, J.G., N.R. Krieg, P.H.A. Sneath, J.T. Staley, and S.T. Williams (eds.). 1994. The fruiting, gliding bacteria: the myxobacteria. Pp. 515–525. *In* Bergey's Manual of Determinative Bacteriology, 9th ed. Lippincott Williams & Wilkins, Baltimore, Maryland.

Honda, D., T. Yokochi, T. Nakahara, S. Raghukumar, A. Nakagiri, K. Schaumann, and T. Higashihara. 1999. Molecular phylogeny of labyrinthulids and thraustochytrids based on the sequencing of 18S ribosomal RNA gene. Journal of Eukaryotic Microbiology 46:637–647.

Hora, F.B. 1958. Effect of lime on the production of a toadstool (*Omphalina maura* (Fr.) Gill.). Nature 181:1668–1669.

Horak, E. 1979. Fungi, Basidiomycetes, Agaricales y Gasteromycetes secotioides. Flora Criptogamica de Tierra del Fuego (Buenos Aires) 11:1–524.

Hori, H., and S. Osawa 1979. Evolutionary change in 5S RNA secondary structure and a phylogenetic tree of 54 5S RNA species. Proceedings of the National Academy of Sciences USA 76:381–385.

Horikoshi, K. 1991. Microorganisms in Alkaline Environments. Kodansha, Ltd., Tokyo, Japan.

Horikoshi, T., K.D. Dannenberg, T.H. Stadlbauer, M. Volkenandt, L.C. Shea, K. Aigner, B. Gustavsson, L. Leichman, R. Frosing, M. Ray, N.W. Gibson, C.P. Spears, and P.V. Dannenberg. 1992. Quantitation of thymidylate synthase, dihydrofolate reductase and DT-diaphoroase gene expression in human tumors using the polymerase chain reaction. Cancer Research 52:108–116.

Horn, H.S. 1966. Measurement of "overlap" in comparative ecological studies. American Naturalist 100:419–424.

Hornby, D. 1990. Biological Control of Soil-borne Plant Pathogens. CAB International, Wallingford, Oxon, United Kingdom.

Hornok, L., and I. Walcz. 1983. *Fusarium heterosporum*, a highly specialised hyperparasite of *Claviceps purpurea*. Transactions of the British Mycological Society 80:377–380.

Horowitz, N.H., M. Fling, H.A. Macleod, and N. Sueoka. 1960. Genetic determination and enzymatic induction of tyrosinase in *Neurospora*. Journal of Molecular Biology 2:96–104.

Horré, R., G.S. de Hoog., C. Kluczny, G. Marklein, and K.P. Schaal. 1999. rDNA diversity and physiology of *Ochroconis* and *Scolecobasidium* species reported from humans and other vertebrates. Studies in Mycology 43:194–205.

Horter, H. von. 1960. *Fusarium* als Erreger einer Hautmykose bei karpfen. Zeitschrift für Parasitenkunde 20:355–358.

Hose, J.E., D.V. Lightner, R.M. Redman, and D.V. Donald. 1984. Observations on the pathogenesis of the Imperfect fungus, *Fusarium solani*, in the California brown shrimp, *Penaeus californiensis*. Journal of Invertebrate Pathology 44:292–303.

Hosoya, T., and Y. Otani. 1995. *Gelatinopulvinella astraeicola* gen. et sp. nov., a fungicolous discomycete and its anamorph. Mycologia 87:689–696.

Hoßfeld, M.K.W. 1990. Suche nach neuen Wirkstoffen aus mycophilen Pilzen und Pilzen der Phylloplane. Unpubl. Ph.D. dissert. Universität Kaiserslautern, Kaiserslautern, Germany.

Howe, M.J., and K. Suberkropp. 1993. Effects of mycoparasitism on an aquatic hyphomycete growing on leaf litter. Mycologia 85:898–901.

Howell, C.R. 1991. Biological control of *Pythium* damping-off of cotton with seed-coating preparations of *Gliocladium virens*. Phytopathology 81:738–741.

——. 1999. Selective isolation from soil and separation in vitro of P and Q strains of *Trichoderma virens* with differential media. Mycologia 91:930–934.

Howell, C.R., and R.D. Stipanovic. 1983. Gliovirin, a new antibiotic from *Gliocladium virens*, and its role in the biological control of *Pythium ultimum*. Canadian Journal of Microbiology 29:321–324.

——. 1995. Mechanisms in the biocontrol of *Rhizoctonia solani*-induced cotton seedling disease by *Gliocladium virens*: antibiosis. Phytopathology 85:469–472.

Hsiang, T., N. Matsumoto, and S. Millett. 1999. Biology and management of *Typhula* snow molds of turfgrass. Plant Disease 83:788–798.

Hu, F.-M., R.-Y. Zheng, and G.-Q. Chen. 1989. A redelimitation of the species of *Pilobolus*. Mycosystema 2:111–133.

Huang, H.C., and H.J. Hoes. 1976. Penetration and infection of *Sclerotinia sclerotiorum* by *Coniothyrium minitans*. Canadian Journal of Botany 54:406–410.

Huang, H.C., and E.G. Kokko. 1988. Penetration of hyphae of *Sclerotinia sclerotiorum* by *Coniothyrium minitans* without the formation of appressoria. Journal of Phytopathology 123:133–139.

——. 1993. *Trichothecium roseum*, a mycoparasite of *Sclerotinia sclerotiorum*. Canadian Journal of Botany 71:1631–1638.

Huang, L.H. 1971. Studies on Soil Microfungi of Nigeria and Dominica. Unpubl. Ph.D. dissert., University of Wisconsin, Madison, Wisconsin.

——. 1973. *Zopfiella flammifera*, a new species from Nigerian soil. Mycologia 65:690–694.

——. 1978. *Gliocladium catenulatum*: hyperparasite of *Sclerotinia sclerotiorum* and *Fusarium* species. Canadian Journal of Botany 56:2243–2246.

Huang, L.H., and J.A. Schmitt. 1975. Soil microfungi of central and southern Ohio. Mycotaxon 3:55–80.

Hubbell, S.P. 1979. Tree dispersion, abundance, and diversity in a tropical dry forest. Science 203:1299–1309.

Hubert, E.E. 1935. Observations on *Tuberculina maxima*, a parasite of *Cronartium ribicola*. Phytopathology 25:253–261.

Hudson, H.J. 1968. The ecology of fungi on plant remains above the soil. New Phytologist 67:837–874.

Hudson, H.J., and J. Webster. 1958. Succession of fungi on decaying stems of *Agropyron repens*. Transactions of the British Mycological Society 41:165–177.

Hudspeth, D.S.S., S.A. Nadler, and M.E.S. Hudspeth. 2000. A COX2 molecular phylogeny of the Peronosporomycetes. Mycologia 94:674–684.

Hudspeth, M.E.S. 1992. The fungal mitochondrial genome—a broader perspective. Pp. 213–241. *In* D.K. Arora, R.P. Elander, and K.G. Mukerji (eds.), Handbook of Applied Mycology. Vol. IV. Fungal Biotechnology. Marcel Dekker, New York.

Hudspeth, M.E.S., D.S. Shmard, K.M. Tatti, and L.I. Grossman 1980. Rapid purification of yeast mitochondrial DNA in high yield. Biochimica et Biophysica Acta 610:221–228.

Hughes, C. 1998. Integrating molecular techniques with field methods in studies of social behavior: a revolution results. Ecology 79:383–399.

Hughes, G.C. 1962. Seasonal periodicity of the Saprolegniaceae in the south-eastern United States. Transactions of the British Mycological Society 45:519–531.

Hughes, S.J. 1976. Sooty molds. Mycologia 68:693–820.

——. 1993. *Meliolina* and its excluded species. Mycological Papers 166:1–255.

Huhndorf, S.M. 1991. A method for sectioning ascomycete herbarium specimens for light microscopy. Mycologia 83:520–524.

Huhtinen, S. 1989. A monograph of *Hyaloscypha* and allied genera. Karstenia 29:45–252.

Huldén, L. 1983. Laboulbeniales (Ascomycetes) of Finland and adjacent parts of the U.S.S.R. Karstenia 23:31–136.

Humason, G.L. 1976. Animal Tissue Techniques. Freeman, San Francisco, California.

Humber, R.A. 1989. Synopsis of a revised classification for the Entomophthorales (Zygomycotina). Mycotaxon 34:441–460.

——. 1992. Collection of Entomopathogenic Fungi: Catalog of Strains. U.S. Department of Agriculture, Agricultural Research Service, Beltsville, Maryland. Publication 110.

——. 1994. Special considerations for operating a culture collection of fastidious fungal pathogens. Journal of Industrial Microbiology 13:195–196.

Humble, S.J., and J.L. Lockwood. 1981. Hyperparasitism of oospores of *Phytophthora megasperma* var. *sojae*. Soil Biology and Biochemistry 13:355–360.

Hume, I. 1989. Reading the entrails of evolution. New Scientist 1160:43–47.

Humpert, A.J., E.L. Muench, A.J. Giachini, M.A. Castellano, and J.W. Spatafora. 2001. Molecular phylogenetics of *Ramaria* and related genera: evidence from nuclear large subunit and mitochondrial small subunit rDNA sequences. Mycologia 93:465–477.

Hunt, D.W.A., and K.F. Raffa. 1989. Attraction of *Hylobius radicis* and *Pachylobius picivorus* (Coleoptera: Curculionidae) to ethanol and turpentine in pitfall traps. Environmental Entomology 18:351–355.

Hunt, G.A., and J.M. Trappe. 1987. Seasonal hypogeous sporocarp production in a western Oregon Douglas-fir stand. Canadian Journal of Botany 65:438–445.

Hunt, R.S., and W.G. Ziller. 1978. Host-genus keys to the Hypodermataceae of conifer leaves. Mycotaxon 6:481–496.

Hunter, W.E., J.M. Duniway, and E.E. Butler. 1977. Influence of nutrition, temperature, moisture, and gas composition on parasitism of *Rhizopus oryzae* by *Syncephalis californica*. Phytopathology 67:664–669.

Hunter-Cervera, J.C., and A. Belt. 1992. Isolation. Pp. 561–570. *In* J. Lederberg (ed.), Encyclopedia of Microbiology. Vol. 2. Academic Press, San Diego, California.

Hurlbert, S.H. 1971. The non-concept of species diversity: A critique and alternative parameters. Ecology 52:577–586.

——. 1984. Pseudoreplication and the design of ecological field experiments. Ecological Monographs 54:187–211.

Huse, K.J. 1981. The distribution of fungi in sound-looking stems of *Picea abies* in Norway. European Journal of Forest Pathology 11:1–6.

Huss, M.J. 1989. Dispersal of cellular slime molds by two soil invertebrates. Mycologia 81:677–682.

Hutchison, L.J. 1990a. Studies on the systematics of ectomycorrhizal fungi in axenic culture. II. The enzymatic degradation of selected carbon and nitrogen compounds. Canadian Journal of Botany 68:1522–1530.

——. 1990b. Studies on the systematics of ectomycorrhizal fungi in axenic culture. III. Patterns of polyphenol oxidase activity. Mycologia 82:424–435.

——. 1990c. Studies on the systematics of ectomycorrhizal fungi in axenic culture. IV. The effect of some selected fungitoxic compounds upon linear growth. Canadian Journal of Botany 68:2172–2178.

——. 1991. Description and identification of cultures of ectomycorrhizal fungi found in North America. Mycotaxon 42:387–504.

Hutchison, L.J., and R.C. Summerbell. 1990. Studies on the systematics of ectomycorrhizal fungi in axenic culture. Reactions of mycelia to diazonium blue B staining. Mycologia 82:36–42.

Hwang, K., D.A. Stelzig, H.L. Barnett, P.P. Roller, and M.I. Kelsey. 1985. Partial purification of the growth factor mycotrophein. Mycologia 77:109–113.

Hyde, K.D. 1986. Frequency of occurrence of lignicolous marine fungi in the tropics. Pp. 311–322. *In* S.T. Moss (ed.), The Biology of Marine Fungi. Cambridge University Press, Cambridge, Great Britain.

——. 1988a. Studies on the tropical marine fungi of Brunei. Botanical Journal of the Linnean Society (London) 98:135–151.

——. 1988b. Studies on the tropical marine fungi of Brunei. II. Notes on five interesting species. Transactions of the Mycological Society of Japan 29:161–171.

——. 1990a. A study of the vertical zonation of intertidal fungi on *Rhizophora apiculata* at Kampong Kapok Mangrove, Brunei. Aquatic Botany 36:255–262.

——. 1990b. Vertical zonation of intertidal mangrove fungi. Pp. 302–306. *In* T. Hatori, Y. Ishida, Y. Maruyama, R.Y. Myota, A.

Uchida (eds.), Recent Advances in Microbial Ecology. Japanese Scientific Society Press, Tokyo, Japan.

——. 1991. Fungal colonization of *Rhizophora apiculata* and *Xylocarpus granatum* poles in Kampong Kapok mangrove, Brunei. Sydowia 43:31–38.

——. 1992a. Tropical Australian freshwater fungi. I. Some Ascomycetes. Australian Systematic Botany 5:109–116.

——. 1992b. Tropical Australian freshwater fungi. II. *Annulatascus velatispora* gen. et sp. nov., *A. bipolaris* sp. nov. and *Nais aquatica* sp. nov. (Ascomycetes). Australian Systematic Botany 5:117–124.

——. 1992c. Tropical Australian freshwater fungi IV. *Halosarpheia aquatica* sp. nov., *Garethjonesia lacunospora* gen.& sp. nov. and *Ophioceras dolichostomum* (Ascomycetes). Australian Systematic Botany 5:407–414.

Hyde, K.D., and S.A. Alias. 2000. Biodiversity and distribution of fungi associated with decomposing *Nypa fruticans*. Biodiversity and Conservation 9:393–402.

Hyde, K.D., and D.L. Hawksworth. 1997. Measuring and monitoring the biodiversity of microfungi. Pp. 11–28. *In* K.D. Hyde (ed.), Biodiversity of Tropical Microfungi. Hong Kong University Press, Hong Kong, China.

Hyde, K.D., and E.B.G. Jones. 1988. Marine mangrove fungi. Marine Ecology (Pubblicazioni della Stazione Zoologica di Napoli) 9:15–33.

——. 1989. Ecological observations on marine fungi from the Seychelles. Botanical Journal of the Linnean Society (London) 100: 237–254.

Hyde, K.D., and S.B. Pointing (eds). 2000. Marine Mycology, A Practical Approach. Fungal Diversity Press, Hong Kong, China.

Hyde, K.D., C.A. Farrant, and E.B.G. Jones. 1987. Isolation and culture of marine fungi. Botanica Marina 30:291–303.

Hyde, K.D., J.E. Taylor, and J. Fröhlich. 2000. Genera of Ascomycetes from Palms. University of Hong Kong Press, Hong Kong, China.

Hyde, K.D., S.W. Wong, and E.B.G. Jones. 1997. Freshwater Ascomycetes. Pp. 179–187. *In* K.D. Hyde (ed.), Biodiversity of Tropical Microfungi. Hong Kong University Press, Hong Kong, China.

Ichielevich-Auster, M., B. Sneh, Y. Koltin, and I. Barash. 1985. Suppression of damping-off caused by *Rhizoctonia* species by a nonpathogenic isolate of *R. solani*. Phytopathology 75:1080–1084.

Ihlen, P.G. 1995. The lichenicolous fungi on *Thamnolia vermicularis* in Norway. Graphis Scripta 7:17–24.

——. 1998. The lichenicolous fungi on species of the genera *Baeomyces, Dibaeis,* and *Icmadophila* in Norway. Lichenologist 30:27–57.

Ikediugwu, F.E.O. 1976a. Ultrastructure of hyphal interference between *Coprinus heptemerus* and *Ascobolus crenulatus*. Transactions of the British Mycological Society 66:281–290.

——. 1976b. The interface in hyphal interference by *Peniophora gigantea* against *Heterobasidion annosum*. Transactions of the British Mycological Society 66:291–296.

Ikediugwu, F.E.O., and J. Webster. 1970a. Antagonism between *Coprinus heptemerus* and other coprophilous fungi. Transactions of the British Mycological Society 54:181–204.

——. 1970b. Hyphal interference in a range of coprophilous fungi. Transactions of the British Mycological Society 54:205–210.

Indoh, H., and A. Oyatsu. 1965. On *Heterocephalum aurantiacum* Thaxter newly found in Okinawa. Transactions of the Mycological Society of Japan 5:79–82. [in Japanese]

Ing, B. 1974. Mouldy myxomycetes. Bulletin of the British Mycological Society 8:25–30.

Ingham, R.E. 1992. Interactions between invertebrates and fungi: effects on nutrient availability. Pp. 669–690. *In* G.C. Carroll and D.T. Wicklow (eds.), The Fungal Community: Its Organization and Role in the Ecosystem. 2nd ed. Marcel Dekker, New York.

Ingold, C.T. 1942. Aquatic hyphomycetes of decaying alder leaves. Transactions of the British Mycological Society 25:339–417.

——. 1951. Aquatic ascomycetes: *Ceriospora caudae suis* n. sp. and *Ophiobolus typhae*. Transactions of the British Mycological Society 34:210–215.

——. 1954. Aquatic ascomycetes: discomycetes from lakes. Transactions of the British Mycological Society 37:1–18.

——. 1955. Aquatic ascomycetes: further species from the English Lake District. Transactions of the British Mycological Society 38:157–168.

——. 1975. An Illustrated Guide to Aquatic and Water-Borne Hyphomycetes (Fungi Imperfecti) with Notes on their Biology. Freshwater Biological Association, Scientific Publication No.30.

Ingold, C.T., and B. Chapman. 1952. Aquatic ascomycetes: *Loramyces juncicola* Weston and *L. macrospora* n.sp. Transactions of the British Mycological Society 35:268–272.

Innis, N., J. Gelfand, and T.J. White. 1990. PCR-Protocols: A Guide to Methods and Applications. Academic Press, San Diego, California.

International Mycological Institute. 1992. Catalogue of the Culture Collection, Filamentous Fungi, Yeasts and Plant Pathogenic Bacteria. 10th ed. CAB International, Wallingford, Oxon, United Kingdom.

Iqbal, S.H. 1993. Efficiency of artificial foam in trapping conidia of Ingoldian fungi. Annales Botanici Fennici 30:153–160.

Iqbal, S.H., and J. Webster. 1973a. The trapping of aquatic hyphomycete spores by air bubbles. Transactions of the British Mycological Society 60:37–48.

——. 1973b. Aquatic hyphomycete spora of the River Exe and its tributaries. Transactions of the British Mycological Society 61:331–346.

——. 1977. Aquatic hyphomycete spora of some Dartmoor streams. Transactions of the British Mycological Society 69:233–241.

Iqbal, S.H., G. Akhtar, and F.-e-Bareen. 1995. Endophytic freshwater hyphomycetes of submerged leaves of some plants lining a canal bank. Pakistan Journal of Plant Science 1:239–254.

Irvine, H.L., and C.S. Stewart. 1991. Interactions between anaerobic cellulolytic bacteria and fungi in the presence of *Methanobrevibacter smithii*. Letters on Applied Microbiology 12:62–64.

Iscove, N.N., M. Barbara, M. Gu, M. Gibson, C. Modi, and N. Winegarden. 2002. Representation is faithfully preserved in global cDNA amplified exponentially from sub-picogram quantities of mRNA. Nature Biotechnology 20:940–943.

Ito, S. 1950. Mycological Flora of Japan. Vol. II. Yokenda, Tokyo, Japan.

Ito, T. 1991. Frozen storage of fungal cultures deposited in the IFO culture collection. IFO [Institute for Fermentation, Osaka, Japan] Research Communication 15:119–128.

Ito, T., and T. Yokoyama. 1983. Preservation of basidiomycete cultures by freezing. IFO [Institute for Fermentation, Osaka, Japan] Research Communication 11:60–70.

Jackson, A.M., J.M. Whipps, and J.M. Lynch. 1991. *In vitro* screening for the identification of potential biocontrol agents of *Allium* white rot. Mycological Research 95:430–434.

Jacobsen, C.S. 1995. Microscale detection of specific bacterial DNA in soil with magnetic capture-hybridization and PCR amplification assay. Applied and Environmental Microbiology 61:3347–3352.

Jacobson, K.M., O.K. Miller, and B.J. Turner. 1993. Randomly amplified polymorphic DNA markers are superior to somatic incompatibility tests for discriminating genotypes in natural populations of the ectomycorrhizal fungus *Suillus granulatus*. Proceedings of the National Academy of Sciences USA 90:9159–9163.

Jager, G., and H. Velvis. 1985. Biological control of *Rhizoctonia solani* on potatoes by antagonists. 4. Inoculation of seed tubers with *Verticillium biguttatum* and other antagonists in field experiments. Netherlands Journal of Plant Pathology 91:49–63.

Jager, G., H. Velvis, J.G. Lamers, A. Mulder, and J. Roosjen. 1991. Biological, chemical and integrated control of *Rhizoctonia solani* in potato. Pp. 187–193. *In* A.B.R. Beemster, G.J. Bollen, M. Gerlagh, M.A. Ruissen, B. Schippers, and A. Tempel (eds.), Biotic Interactions and Soil-borne Diseases. Elsevier, Amsterdam, The Netherlands.

Jahn, H. 1963. Mitteleuropäische Porlinge (Polyporaceae s. lato) und ihr Vorkommen in Westfalen. Westfälische Pilzbriefe 4:1–143.

——. 1968. Das Bisporetum antennatae, Pilzgesellschaft auf den Schnittflächen von Buchenholz. Westfälische Pilzbriefe 7:41–47.

Jalkanen, R., and R. Laakso. 1986. *Hendersonia acicola* in an epidemic caused by *Lophodermella sulcigena*, with special reference to biological control. Karstenia 26:49–56.

James, P.W., and P.A. Wolseley. 1992. Acidification and the Lobarion: a case for biological monitoring. British Lichen Society Bulletin 71:4–12.

James, T.Y., D. Porter, C.A. Leander, R. Vigalys, and J.E. Longcore. 2000. Molecular phyogenetics of the Chytridiomycota supports the utility of ultrastructural data in chytrid systematics. Canadian Journal of Botany 78:336–350.

Janczewski, E. von G. 1871. Morphologische Untersuchungen über *Ascobolus furfuraceus*. Botanische Zeitung 29:257–262.

Jangoux, M. 1987. Diseases of Echinodermata. I. Agents microorganisms and protistans. Diseases of Aquatic Organisms 2:147–162.

Janos, D.P. 1980. Mycorrhizae influence tropical succession. Biotropica 12(Suppl):56–64.

——. 1992. Heterogeneity and scale in tropical vesicular-arbuscular mycorrhiza formation. Pp. 276–282. *In* D.J. Read, D.H. Lewis, A.H. Fitter, and I.J. Alexander (eds.), Mycorrhizas in Ecosystems. CAB International, Wallingford, Oxon, United Kingdom.

Janse, B.J.H., J. Gaskell, M. Akhtar, and D. Cullen. 1998. Expression of *Phanerochaete chrysosporium* genes encoding lignin peroxidases, manganese peroxidases, and glyoxal oxidase in wood. Applied and Environmental Microbiology 64:3536–3538.

Jansen, A.E. 1984. Vegetation and macrofungi of acid oakwoods in the north-east of The Netherlands. Agricultural Research Reports (Wageningen) 923:1–162.

Jantrarotai, W., and R.T. Lovell. 1990a. Subchronic toxicity of dietary aflatoxin B_1 to channel catfish. Journal of Aquatic Animal Health 2:248–254.

——. 1990b. Acute and subchronic toxicity of cyclopiazonic acid to channel catfish. Journal of Aquatic Animal Health 2:255–260.

Jantrarotai, W., R.T. Lovell, J. and J.M. Grizzle. 1990. Acute toxicity of aflatoxin B_1 to channel catfish. Journal of Aquatic Animal Health 2:237–247.

Janzen, D.H., W. Hallwachs, J. Jimenez, and R. Gómez. 1993. The role of the parataxonomists, inventory managers, and taxonomists in Costa Rica's national biodiversity inventory. Pp. 223–254. *In* W.V. Reid, S.A. Laird, C.A. Meyer, R. Gómez, A. Sittenfeld, D.H. Janzen, M.A. Gollin, and C. Juma (eds.), Biodiversity Prospecting. World Resources Institute, Washington, D. C.

Jarfster, A.G., and D.M. Sylvia. 1993. Inoculum production and inoculation strategies for vesicular-arbuscular mycorrhizal fungi. Pp. 349–377. *In* F.B. Metting (ed.), Soil Microbial Technologies: Applications in Agriculture, Forestry, and Environmental Management. Marcel Dekker, New York.

Jarvis, B., and A.P. Williams. 1987. Methods for detecting fungi in foods and beverages. Pp. 599–636. *In* L.R. Beuchat (ed.), Food and Beverage Mycology. 2nd ed. Van Nostrand Reinhold, New York.

Jarvis, W.R., L.A. Shaw, and J.A. Traquair. 1989. Factors affecting antagonism of cucumber powdery mildew by *Stephanoascus flocculosus* and *S. rugulosus*. Mycological Research 92:162–165.

Jeffries, P. 1985. Mycoparasitism within the Zygomycetes. Botanical Journal of the Linnean Society (London) 91:135–150.

——. 1995. Biology and ecology of mycoparasitism. Canadian Journal of Botany 73(Suppl. 1):S1284–S1290.

Jeffries, P., and J.C. Dodd. 1991. The use of mycorrhizal inoculants in forestry and agriculture. Pp. 155–185. *In* D.K. Arora, B. Rai, K.G. Mukerji, and G. Knudsen (eds.), Handbook of Applied Mycology. Vol. I. Soil and Plants. Marcel Dekker, New York.

Jeffries, P., and P.M. Kirk. 1976. New technique for the isolation of mycoparasitic Mucorales. Transactions of the British Mycological Society 66:541–543.

Jeffries, P., and T.W.K. Young. 1976. Ultrastructure of infection of *Cokeromyces recurvatus* by *Piptocephalis unispora* (Mucorales). Archive of Microbiology 109:277–288.

——. 1978. Mycoparasitism by *Piptocephalis unispora* (Mucorales): host range and reaction with *Phascolomyces articulosus*. Canadian Journal of Botany 56:2449–2459.

——. 1994. Interfungal Parasitic Relationships. CAB International, Wallingford, Oxon, United Kingdom.

Jeffries, P., T. Spyropoulos, and E. Vardavarkis. 1988. Vesicular-arbuscular mycorrhizal status of various crops in different agricultural soils of northern Greece. Biology and Fertility of Soils 5:333–337.

Jennings, D.H. 1984. Polyol metabolism in fungi. Advances in Microbial Physiology 25:149–193.

——. 1993. Understanding tolerance to stress: laboratory versus environmental actuality. Pp. 1–12. *In* D.H. Jennings (ed.), Stress Tolerance of Fungi. Marcel Dekker, New York.

Jensen, A.B., A. Gargas, J. Eilenberg, and S. Rosendahl. 1998. Relationships of the insect-pathogenic order entomophthorales (Zygomycota, Fungi) based on phylogenetic analyses of nuclear small subunit ribosomal DNA sequences (SSU rDNA). Fungal Genetics and Biology 24:325–334.

Jessen, R.J. 1955. Determining the fruit count on a tree by randomized branch sampling. Biometrics 11:99–109.

Jloba, N.M., I.I. Sidorova, and G.G. Shumskaya. 1980. The antibiotic and hydrolytic activity of some species of mycophilous fungi. Vestnik Moskovskogo Universiteta, Seriya XVI, Biologiya 1980(1):64–68.

Joblin, K.N. 1981. Isolation, enumeration, and maintenance of rumen fungi in roll tubes. Applied and Environmental Microbiology 42:1119–1122.

Joblin, K.N., and G.E. Naylor. 1989. Fermentation of woods by rumen anaerobic fungi. FEMS Microbiology Letters. 65:119–125.

Joblin, K.N., G.P. Campbell, A.J. Richardson, and C.S. Stewart. 1989. Fermentation of barley straw by anaerobic rumen bacteria and fungi in axenic culture and in co-culture with methanogens. Letters in Applied Microbiology 9:195–197.

Johansson, D. 1974. Ecology of vascular epiphytes in West African rain forest. Acta Phytogeographica Suecica 59:1–129.

Johnson, G.C., and A.K. Martin. 1992. Survival of wood-inhabiting fungi stored for 10 years in water and under oil. Canadian Journal of Microbiology 38:861–864.

Johnson, H.W. 1923. Relationships between hydrogen ion, hydroxyl ion, and salt concentrations and the growth of seven molds. Iowa Agricultural Experiment Station, Iowa State College for Agriculture and the Mechanic Arts, Research Bulletin 76:307–344.

Johnson, J.A., and N.J. Whitney. 1989. An investigation of needle endophyte colonization patterns with respect to height and compass direction in a single crown of balsam fir (*Abies balsamea*). Canadian Journal of Botany 67:723–725.

——. 1994. Cytotoxicity and insecticidal activity of endophytic fungi from black spruce (*Picea mariana*) needles. Canadian Journal of Microbiology 40:24–27.

Johnson, J.L. 1985. Determination of DNA base composition. Pp. 1–31. *In* G. Gottschalk (ed.) Methods in Microbiology. Academic Press, New York.

Johnson, N.C., and D.A. Wedin. 1997. Soil carbon, nutrients, and mycorrhizae during conversion of dry tropical forest to grassland. Ecological Applications 7:171–182.

Johnson, N.C., T.E. O'Dell, and C.S. Bledsoe. 2000. Methods for ecological studies of mycorrhizae. Pp. 378–412. *In* P. Robertson, P. Sollins, and C.S. Bledsoe (eds.), Standardized Soil Methods for Long-Term Ecological Research. Cambridge University Press, Cambridge, England.

Johnson, N.C., D. Tilman, and D. Wedin. 1992. Plant and soil controls on mycorrhizal fungal communities. Ecology 73:2034–2042.

Johnson, N.C., D.R. Zak, D. Tilman, and F.L. Pfleger. 1991. Dynamics of vesicular-arbuscular mycorrhizae during old field succession. Oecologia 86:349–358.

Johnson, P.T. 1983. Diseases caused by viruses, rickettsiae, bacteria and fungi. Pp. 2–78. *In* A.J. Provenzano (ed.), The Biology of the Crustacea. Vol. 6. Pathobiology. Academic Press, New York.

Johnson, T.W., Jr. 1956. The Genus *Achlya*: Morphology and Taxonomy. University of Michigan Press, Ann Arbor, Michigan.

——. 1973. Aquatic fungi of Iceland: uniflagellate species. Acta Naturalia Islandica 22:1–77.

——. 1975. Aquatic fungi of Scandinavia: some ornamented chytrids from southern Norway. Norwegian Journal of Botany 22:249–257.

——. 1977. Resting spore germination in three chytrids. Mycologia 69:34–45.

Johnson, T.W., Jr., and F.K. Sparrow, Jr. 1961. Fungi in Oceans and Estuaries. J. Cramer, Weinheim, Germany.

Jones, E.B.G. 1962. Marine fungi. Transactions of the British Mycological Society 45:93–114.

——. 1963a. Marine fungi. II. Ascomycetes and Deuteromycetes from submerged wood and drift *Spartina*. Transactions of the British Mycological Society 46:135–144.

——. 1963b. Observations on the fungal succession on wood test blocks submerged in the sea. Journal of the Institute of Wood Science 11:14–23.

——. 1965. *Halonectria milfordensis* gen. et sp. nov., a marine Pyrenomycete on submerged wood. Transactions of the British Mycological Society 48:287–290.

Jones, E.B.G., and A.C. Oliver. 1964. Occurrence of aquatic hyphomycetes on wood submerged in fresh and brackish water. Transactions of the British Mycological Society 47:45–48.

Jones, E.E., and J.W. Deacon. 1995. Comparative physiology and behaviour of the mycoparasites *Pythium acanthophoron*, *P. oligandrum* and *P. mycoparasiticum*. Biocontrol Science and Technology 5:27–39.

Jones, E.P., R. Mahendran, M.R. Spottswood, Y.C. Yang, and D.L. Miller. 1990. Mitochondrial DNA of *Physarum polycephalum*: physical mapping, cloning and transcription mapping. Current Genetics 17:331–337.

Jones, G.M. and R.K. O'Dor. 1983. Ultrastructural observations on a thraustochytrid fungus parasitic in the gills of squid (*Illex illecebrosus* Lesueur). Journal of Parasitology 69:903–911.

Jones, K.G., and M. Blackwell. 1996. Ribosomal DNA sequence analysis places the yeast-like genus *Symbiotaphrina* within filamentous ascomycetes. Mycologia 88:212–218.

——. 1998. Phylogenetic analysis of ambrosial species in the genus *Raffaelea* based on 18S rDNA sequences. Mycological Research 102:661–665.

Jordan, E.G., and H.L. Barnett. 1978. Nutrition and parasitism of *Melanospora zamiae*. Mycologia 70:300–312.

Jørstad, I. 1956. Uredinales from South America and tropical North America, chiefly collected by Swedish botanists. I. Arkiv för Botanik, Ser. 2, 3:443–490.

——. 1959. Uredinales from South America and tropical North America, chiefly collected by Swedish botanists II. Arkiv för Botanik, Ser. 2, 4:59–103.

Joshi, S.R., G.D. Sharma, and R.R. Mishra. 1994. Effect of disturbance in microbial population and their activities in forest soils at higher altitudes of Meghalaya. Pp. 298–309. *In* Y.P.S. Pangtey and R.S. Rawal (eds.), High Altitudes of the Himalaya: Biogeography, Ecology & Conservation. Gyanodaya Prakashan, Nainital, India.

Ju, Y.-M., and J.D. Rogers. 1996. A Revision of the Genus *Hypoxylon*. American Phytopathological Society Press, St. Paul, Minnesota.

Jülich, W. 1981. Higher Taxa of Basidiomycetes. Bibliotheca Mycologica 85:1–485.

———. 1983. Parasitic heterobasidiomycetes on other fungi. A key to the European taxa. International Journal of Mycology and Lichenology 1:189–203.

Jülich, W., and J.A. Stalpers. 1980. The Resupinate Non-Poroid Aphyllophorales of the Temperate Northern Hemisphere. North-Holland, Amsterdam, The Netherlands.

Jump, J.A. 1954. Studies on sclerotization in *Physarum polycephalum*. American Journal of Botany 41:561–567.

Jumpponen, A. 2003. Soil fungal community assembly in a primary successional glacier forefront ecosystem as inferred from rDNA sequence analyses. New Phytologist 158:569–578.

Jurgensen, M.F., M.J. Larsen, G.D. Mroz, and A.E. Harvey. 1986. Timber harvesting, soil organic matter and site productivity. Pp. 43–53. *In* C. Smith, C. Tattersall, C.W. Martin, and L.M. Tritton (eds.), Proceedings, 1986 Symposium on the Productivity of Northern Forests Following Biomass Harvesting. Durham, New Hampshire. U. S. Department of Agriculture, Forest Service, Northeastern Forest Experiment Station, Broomall, Pennsylvania. General Technical Report NE-115a.

Jurzitza, G. 1979. The fungi symbiotic with anobiid beetles. Pp. 65–76. *In* L.R. Batra (ed.), Insect-Fungus Symbiosis, Nutrition, Mutualism, and Commensalism. Allanheld Osmun, Montclair, New Jersey.

Käärik, A. 1965. The identification of the mycelia of wood-decay fungi by their oxidation reactions with phenolic compounds. Studia Forestalia Suecica 31:1–38.

Kajimura, H., and N. Hijli. 1992. Dynamics of the fungal symbionts in the gallery system and the mycangia of the ambrosia beetle, *Xylosandrus mutilatus* (Blandford) (Coleoptera: Scolytidae) in relation to its life history. Ecological Research 7:107–117.

Kalb, K., J. Hafellner, and B. Staiger. 1995. *Haematomma*-Studien. II. Lichenicole pilze auf arten der flechtengattung *Haematomma*. Bibliotheca Lichenologica 59:199–222.

Kaltenboeck, B., J.W. Spatafora, X. Zhang, K.G. Kousalas, M. Blackwell, and J. Storz. 1992. Efficient production of single-stranded DNA as long as 2 kb for sequencing of PCR-amplified DNA. BioTechniques 12:164–171.

Kaltenecker, J.H., M. Wicklow-Howard, and R. Rosentreter. 1999. Biological soil crusts in three sagebrush communities recovering from a century of livestock trampling. Pp. 222–226. *In* E.D. McArthur, W.K. Ostler, and C.L. Bambolt (comps.), Proceedings: Shrubland Ecotones, Ehpraim, Utah. August 12–14, 1998. U.S. Department of Agriculture, Forest Service, Rocky Mountain Research Station, Ogden, Utah. RMRS-P-11.

Kane, J., and J.B. Fischer. 1971. The differentiation of *Trichophyton rubrum* and *T. mentagrophytes* by use of Christensen's urea broth. Canadian Journal of Microbiology 17:911–913.

Kane, J., and C.M. Smitka. 1978. The early detection and identification of *Trichophyton verrucosum*. Journal of Clinical Microbiology 8:740–747.

Kane, J., J. Righter, S. Krajden, and R.S. Lester. 1983. Blastomycosis: a new endemic focus in Canada. Canadian Medical Association Journal 129:728–731.

Kane, M., and J.T. Mullins. 1973. Thermophilic fungi in a municipal waste compost system. Mycologia 65:1087–1100.

Kappen, L. 1988. Ecophysiological relationships in different climatic regions. Pp. 37–100. *In* M. Galun (ed.), CRC Handbook of Lichenology. Vol. II. CRC Press, Boca Raton, Florida.

Karatygin, I.V. 1999. Problems in the macrosystematics of fungi. Mikologiya i Fitopatologiya 33:150–165.

Karling, J.S. 1942. Parasitism among the chytrids. American Journal of Botany 29:24–35.

———. 1945. Brazilian chytrids. VI. Rhopalophlyctis and Chytriomyces, two new chitinophyllic operculate genera. American Journal of Botany 32:362–369.

———. 1946. Brazilian chytrids. VIII. Additional parasites of rotifers and nematodes. Lloydia 9:1–12.

———. 1952. *Sommerstorffia spinosa* Arnaudow. Mycologia 44:387–412.

———. 1960. Parasitism among chytrids. II *Chytriomyces verrucosus* sp. Nov. and *Phlyctochytrium synchytrii*. Bulletin of the Torrey Botanical Club 87:326–336.

———. 1964. *Synchytrium*. Academic Press, New York.

———. 1968. The Plasmodiophorales. 2nd ed. Hafner, New York.

———. 1977. *Chytridiomycetarum Iconographia*. Lubrecht & Cramer, Monticello, New York.

———. 1981. Predominantly Holocarpic and Eucarpic Simple Biflagellate Phycomycetes. 2nd ed. Cramer, Vaduz, Liechtenstein.

Karlovsky, P., and B. Fartmann. 1992. Genetic code and phylogenetic origin of oomycetous mitochondria. Journal of Molecular Evolution 34:254–258.

Katumoto, K. 1983. Hyperparasitic species of *Spiropes* in Japan. Transactions of the Mycological Society of Japan 24:249–258.

Kaufman, L., and E. Reiss. 1992. Serodiagnosis of fungal diseases. Pp. 506–528 *In* N.R. Rose, E.C. DeMacario, J.L. Fahey, H. Friedman, and G.M. Penn (eds.), Manual of Clinical Laboratory Immunology. American Society for Microbiology Press, Washington, D.C.

Kauserud, H. 1995. The Diversity and Ecology of Wood-Inhabiting Fungi Colonizing *Alnus incana*. Unpubl. Master's thesis. University of Oslo, Oslo, Norway.

Kays, S., and J.L. Harper. 1974. The regulation of plant and tiller density in a grass sward. Journal of Ecology 62:97–105.

Keeling, P.J., M.A. Luker, and J.D. Palmer. 2000. Evidence from beta-tubulin phylogeny that microsporidia evolved from within the fungi. Molecular Biology and Evolution 17:23–31.

Keinath, A.P., D.R. Fravel, and G.C. Papavizas. 1991. Potential of *Gliocladium roseum* for biocontrol of *Verticillium dahliae*. Phytopathology 81:644–648.

Keissler, K. Von. 1930. Die Flechtenparasiten. Rabenhorsts Kryptogamenfloraq, Leipzig, Germany.

Keller, H.W. 1980. Corticolous Myxomycetes VIII: *Trabrooksia*, a new genus. Mycologia 72:395–403.

———. 1982. Myxomycota, Myxomycetes. Pp. 165–172. *In* S.P. Parker (ed.), Synopsis and Classification of Living Organisms. Vol. 1. McGraw Hill, New York.

———. 1996. Biosystematics of Myxomycetes: A Futuristic View. Plenary Address. Pp. 23–37. *In* Second International Congress on the Systematics and Ecology of Myxomycetes (ICSEM₂). Real Jardín Botánico. Consejo Superior de Investigaciones Científicas (CSIC), Madrid, Spain.

Keller, H.W., and L.L. Anderson. 1978. Some coprophilous species of Myxomycetes. ASB (Association of Southeastern Biologists) Bulletin 25:67.

Keller, H.W., and K.L. Braun. 1999. Myxomycetes of Ohio: Their Systematics, Biology, and Use in Teaching. Ohio Biological Survey, Columbus, Ohio.

Keller, H.W., and T.E. Brooks. 1973. Corticolous Myxomycetes I: two new species of *Dydimium*. Mycologia 65:286–294.

———. 1975. Corticolous Myxomycetes III: a new species of *Badhamia*. Mycologia 67:1218–1222.

———. 1976a. Corticolous Myxomycetes IV: *Badhamiopsis*, a new genus for *Badhamia ainoae*. Mycologia 68:834–841.

———. 1976b. Corticolous Myxomycetes V: observations on the genus *Echinostelium*. Mycologia 68:1204–1220.

———. 1977. Corticolous Myxomycetes VII: Contribution toward a monograph of *Licea*, five new species. Mycologia 69:667–684.

Keller, H.W., and U.H. Eliasson. 1992. Taxonomic evaluation of *Perichaena depressa* and *P. quadrata* based on controlled cultivation, with additional observations on the genus. Mycological Research 96:1085–1097.

Keller, H.W., and J.D. Schoknecht. 1989. Life cycle of a new annulate-spored *Didymium*. Mycologia 81:248–265.

Keller, H.W., and D.M. Smith. 1978. Dissemination of myxomycete spores through feeding activities (ingestion-defecation) of an acarid mite. Mycologia 70:1239–1241.

Keller, J. 1985. Les Cystides Cristallifères des Aphyllophorales. Mycologia Helvetica 1:277.

Keller, M.D., W.K. Bellows, and R.L.L. Guillard. 1988. Microwave treatment for sterilization of phytoplankton culture media. Journal of Experimental Marine Biology and Ecology 117:279–283.

Keller, S. 1987. Arthropod-pathogenic entomophthorales of Switzerland. I. *Conidiobolus*, *Entomophaga* and *Entomophthora*. Sydowia 40:122–167.

———. 1991. Arthropod-pathogenic entomophthorales of Switzerland. II. *Erynia*, *Eryniopsis*, *Neozygites*, *Zoophthora* and *Tarichium*. Sydowia 43:39–122.

Keller, S., and G. Zimmermann. 1989. Mycopathogens of soil insects. Pp. 238–270. *In* N. Wilding, N.M. Collins, P.M. Hammond, and J.F. Webber (eds.), Insect-Fungus Interactions. Academic Press, New York.

Kellner, M., A. Burmester, A. Wöstemeyer, and J. Wöstemeyer. 1993. Transfer of genetic information from the mycoparasite *Parasitella parasitica* to its host *Absidia glauca*. Current Genetics 23:334–337.

Kellock, L.M., and N.J. Dix. 1984. Antagonism by *Hypomyces aurantius*. II. Ultrastructural studies of hyphal disruption. Transactions of the British Mycological Society 82:335–338.

Kelly, A., A.R. Alcalá-Jiménez, B.W. Bainbridge, J.B. Heale, E. Pérez-Artés, and R.M. Jiménez Díaz. 1994. Use of genetic fingerprints and random amplified polymorphic DNA to characterize pathotypes of *Fusarium oxysporum* f. sp. *ciceris* infecting chickpea. Phytopathology 84:1293–1298.

Kelly, K.L. 1965. ISCC-NBS Color-Name Charts Illustrated with Centroid Colors. Standard Sample #2106. U.S. Department of Commerce, National Bureau of Standards Circular 553 (Supplement). U. S. Government Printing Office, Washington, D.C.

Kelly, K.L., and D.B. Judd. 1955. The ISCC-NBS Method of Designating Colors and a Dictionary of Color Names. U.S. Department of Commerce, National Bureau of Standards Circular 553, U.S. Government Printing Office, Washington, D.C.

Kempton, R.A. 1979. The structure of species abundance and measurement of diversity. Biometrics 35:307–322.

Kendrick, B. 1992. The Fifth Kingdom. 2nd ed. Mycologue, Waterloo, Ontario, Canada.

Kendrick, B., and F. Di Cosmo. 1979. Teleomorph-anamorph connections in ascomycetes. Pp. 283–410. *In* B. Kendrick (ed.), The Whole Fungus: The Sexual-Asexual Synthesis. National Museums of Canada, Ottawa, Ontario, Canada.

Kendrick, W.B., and A. Burges. 1962. Biological aspects of the decay of *Pinus sylvestris* leaf litter. Nova Hedwigia 4:313–342.

Kenerley, C.M., and J.P. Stack. 1987. Influence of assessment methods on selection of fungal antagonists of the sclerotium-forming fungus *Phymatotrichum omnivorum*. Canadian Journal of Microbiology 33:632–635.

Kenkel, N.C., and T. Booth. 1992. Multivariate analysis in fungal ecology. Pp. 209–227. *In* G.C. Carroll and D.T. Wicklow (eds.), The Fungal Community: Its Organization and Role in the Ecosystem. 2nd ed. Marcel Dekker, New York.

Kerrigan, R.W., P. Callac, J. Xu, and R. Noble. 1999. Population and phylogenetic structure within the *Agaricus subfloccosus* complex. Mycological Research 103:1515–1523.

Kerwin, J.L., L.M. Johnson, H.C. Whisler, and A.R. Tuininga. 1992. Infection and morphogenesis of *Pythium marinum* in species of *Porphyra* and other red algae. Canadian Journal of Botany 70:1017–1024.

Kettlewell, H.B.D. 1973. The Evolution of Melanism. Clarendon Press, Oxford, United Kingdom.

Khachatourians, G.G. 1996. Biochemistry and molecular biology of entomopathogenic Fungi. Pp. 331–363. *In* D.H. Howard and J.D. Miller (eds.), The Mycota. Vol. VI. Animal and Human Relationships. Springer-Verlag, Berlin, Germany.

Khachatourians, G.G., E.P. Valencia, and G.S. Miranpuri. 2001. *Beauveria bassiana* and other entomopathogenic fungi in the management of insect pests. Pp. 239–275. *In* O. Koul and G.S. Dhaliwal (eds.), Microbial Biopesticides. Taylor and Francis, London, United Kingdom.

Khan, R.S., and J.C. Krug. 1989. New species of *Gelasinospora*. Mycologia 81:226–233.

Kiffer, E., and M. Morelet. 2000. The Deuteromycetes, Mitosporic Fungi: Classification and Generic Keys. Science Publishers, Enfield, New Hampshire.

Kim, K.K., D.R. Fravel, and G.C. Papavizas. 1990. Production, purification, and properties of glucose oxidase from the biocontrol fungus *Talaromyces flavus*. Canadian Journal of Microbiology 36:199–205.

Kinne, O. (ed.). 1980. Diseases of Marine Animals. Vol. I. General Aspects, Protozoa to Gastropoda. John Wiley, Chichester, Great Britain.

———. 1983. Diseases of Marine Animals. Vol. II. Introduction. Bivalvia to Scaphopoda. Biologische Anstalt Helgoland, Hamburg, Germany.

———. 1984. Diseases of Marine Animals. Vol. IV, Pt. 1. Introduction, Pisces. Biologische Anstalt Helgoland, Hamburg, Germany.

Kinoshita, T., J. Imamura, H. Nagai, and K. Shimotohno. 1992. Quantification of gene expression over a wide range by the polymerase chain reaction. Analytical Biochemistry 206:231–235.

Kirby, J.J.H., and A.D.M. Rayner. 1988. Disturbance, decomposition and patchiness in thatch. Proceedings of the Royal Society of Edinburgh 94B:145–153.

Kirby, J.J.H., J. Webster, and J.H. Baker. 1990. A particle plating method for analysis of fungal community composition and structure. Mycological Research 94:621–626.

Kirk, P.M. 1981. New or interesting microfungi III. A preliminary account of microfungi colonizing *Laurus nobilis* leaf litter. Transactions of the British Mycological Society 77:457–473.

———. 1982. New or interesting microfungi V. Microfungi colonizing *Laurus nobilis* leaf litter. Transactions of the British Mycological Society 78:293–303.

———. 1983. New or interesting microfungi X. Hyphomycetes on *Laurus nobilis* leaf litter. Mycotaxon 18:259–298.

———. 1984. New or interesting microfungi XIII. Ascomycetes on *Laurus nobilis* leaf litter. Mycotaxon 19:307–322.

Kirk, P.M. 1993. Distribution of Zygomycetes—the tropical connection. Pp. 91–102. *In* S. Isaac, J.C. Frankland, R. Watling, and A.J.S. Whalley (eds.), Aspects of Tropical Mycology. Cambridge University Press, Cambridge, United Kingdom.

Kirk, P.M., and G.L. Benny. 1980. The genus *Utharomyces* Boedijn (Pilobolaceae: Zygomycetes). Transactions of the British Mycological Society 75:123–131.

Kirk, P.M., P.F. Cannon, J.C. David, and J.A. Stalpers (eds.). 2001. Ainsworth and Bisby's Dictionary of the Fungi, 9th ed. CAB International, Wallingford, Oxon, United Kingdom.

Kirk, P.W., Jr. 1980. The mycostatic effect of seawater on spores of terrestrial and marine higher fungi. Botanica Marina 23:233–238.

Kirsop, B.E. 1991. Maintenance of yeasts. Pp. 161–182. *In* B.E. Kirsop and A. Doyle (eds.), Maintenance of Microorganisms and Cultured Cells: A Manual of Laboratory Methods. 2nd ed. Academic Press, London, England.

Kistler, H.C., and V.P.W. Miao. 1992. New modes of genetic change in filamentous fungi. Annual Review of Phytopathology 30:131–152.

Kjøller, A., and S. Struwe. 1982. Microfungi in ecosystems: fungal occurrence and activity in litter and soil. Oikos 39:389–422.

Klecan, A.L., S. Hippe, and S.C. Somerville. 1990. Reduced growth of *Erysiphe graminis* f.sp. *hordei* induced by *Tilletiopsis pallescens*. Phytopathology 80:325–331.

Klein, A.S., and D.E. Smith. 1996. Erratum: phylogenetic inferences on the relationships of North American and European *Picea* species based on nuclear ribosomal 18S sequences and the internal transcribed spacer 1 region. Molecular Phylogenetics and Evolution 5:286–287.

Kleinschmidt, A.K. 1968. Monolayer techniques in electron microscopy of nucleic acid molecules. Pp. 361–377. *In* L. Grossman and K. Moldave (eds.), Nucleic Acids. Part B. Methods in Enzymology, Vol. 12. Academic Press, New York.

Klepzig, K.D., K.F. Raffa, and E.B. Smalley. 1991. Association of an insect-fungal complex with red pine decline in Wisconsin. Forest Science 37:1119–1139.

Klingner, A., B. Hundeshagen, H. Kernebeck, and H. Bothe. 1995. Localization of the yellow pigment formed in roots of gramineous plants colonized by arbuscular fungi. Protoplasma 185:50–57.

Klingström, A., and S.M. Johansson. 1973. Antagonism of *Scytalidium* isolates against decay fungi. Phytopathology 63:473–479.

Klitch, M.A., and J.I. Pitt. 1988. A Laboratory Guide to Common *Aspergillus* Species and Their Teleomorphs. Commonwealth Scientific and Industrial Research Organization (CSIRO) Division of Food Processing, North Ryde, New South Wales, Australia.

Klontz, P. 1985. Diagnostic methods in fish diseases: Present status and needs. Pp. 3–10. *In* A.E. Ellis (ed.), Fish and Shellfish Pathology. Academic Press, London, England.

Knight, C., and B.E.J. Wheeler. 1977. Perennation of *Diplocarpon rosae* on rose leaves. Transactions of the British Mycological Society 69:385–389.

Knittel, M.D. 1966. Topical application of malachite green for control of common fungus infections in adult spring salmon. Progress in Fish Culture 28:51–53.

Knudsen, S. 2002. A Biologist's Guide to Analysis of DNA Microarray Data. Wiley-Liss, New York.

Kobayasi, Y. 1982. Keys to the taxa of the genera *Cordyceps and Torrubiella*. Transactions of the Mycology Society of Japan 23:329–364.

Koç, N.K., and G. Défago. 1983. Studies on the host range of the hyperparasite *Aphanocladium album*. Phytopathologische Zeitschrift 107:214–218.

Koç, N.K., H.R. Forrer, and H. Kern. 1981. Studies on the relationship between *Puccinia graminis* and the hyperparasite *Aphanocladium album*. Phytopathologische Zeitschrift 101:131–135.

Koch, J. 1994. Growth of *Coniophora puteana* modified by a gall-inducing mycoparasite. Mycological Research 98:1263–1271.

Kohlmeyer, J. 1963. Parasitische und epiphytische Pilze auf Meeresalgen. Nova Hedwigia 6:127–146.

——. 1969a. Deterioration of wood by marine fungi in the deep sea. American Society for Testing of Materials, Special Technical Publication 445:20–29.

——. 1969b. The role of marine fungi in the penetration of calcareous substances. American Zoologist 9:741–746.

——. 1977. New genera and species of higher fungi from the deep sea (1615–5315 m). Revue de Mycologie 41:189–206.

——. 1980. Tropical and subtropical filamentous fungi of the Western Atlantic Ocean. Botanica Marina 23:529–544.

——. 1984. Tropical marine fungi. Marine Ecology (Pubblicazioni della Stazione Zoologica di Napoli) 5:329–378.

——. 1985. Marine fungi (Ascomycetes) within and on tests of Foraminifera. Marine Biology 90:147–149.

Kohlmeyer, J., and E. Kohlmeyer. 1964–1969. Icones Fungorum Maris. J. Cramer, Weinheim, Germany. 7 fascicles.

——. 1972. Permanent microscopic mounts. Mycologia 64:666–669.

——. 1979. Marine Mycology: The Higher Fungi. Academic Press, New York.

Kohlmeyer, J., and B. Volkmann-Kohlmeyer. 1987a. Marine fungi from Belize with a description of two new genera of Ascomycetes. Botanica Marina 30:195–204.

——. 1987b. Koralionastetaceae fam. nov. (Ascomycetes) from coral rock. Mycologia 79:764–778.

——. 1988. *Halographis* (Opegraphales), a new endolithic lichenoid from corals and snails. Canadian Journal of Botany 66:1138–1141.

——. 1989a. Hawaiian marine fungi, including two new genera of Ascomycotina. Mycological Research 92:410–421.

——. 1989b. A new *Lulworthia* (Ascomycotina) from corals. Mycologia 81:289–292.

——. 1990. New species of *Koralionastes* (Ascomycotina) from the Caribbean and Australia. Canadian Journal of Botany 68:1554–1559.

——. 1991. Illustrated key to the filamentous higher marine fungi. Botanica Marina 34:1–61.

——. 1992. Two Ascomycotina from coral reefs in the Caribbean and Australia. Cryptogamic Botany 2:367–374.

——. 1993a. *Atrotorquata* and *Loratospora*: new ascomycete genera on *Juncus roemerianus*. Systema Ascomycetum 12:7–22.

——. 1993b. Two new genera of *Ascomycotina* from saltmarsh *Juncus*. Systema Ascomycetum 11:95–106.

——. 1995. Fungi on *Juncus roemerianus*. 1. *Trichocladium medullare* sp. nov. Mycotaxon 53:349–353.

——. 1996a. Fungi on *Juncus roemerianus*. 6. *Glomerobolus* gen. nov., the first ballistic member of Agonomycetales. Mycologia 88:328–337.

——. 1996b. Fungi on *Juncus roemerianus*. 7. *Tiarosporella halmyra* sp. nov. Mycotaxon 59:79–83.

——. 1998. *Mycophycias*, a new genus for the mycobionts of *Apophlaea*, *Ascophyllum* and *Pelvetia*. Systema Ascomycetum 16:1–7.

——. 2000. Fungi on *Juncus roemerianus*. 14. Three new coelomycetes, including *Floricola*, anam.-gen. nov. Botanica Marina 43:385–392.

——. 2001. The biodiversity of fungi on *Juncus roemerianus*. Mycological Research 105:1411–1412.

——. 2003a. Fungi from coral reefs: A commentary. Mycological Research 107:386–387.

——. 2003b. Marine ascomycetes from algae and animal hosts. Botanica Marina 46:285–306.

Kohlmeyer, J., H.-O. Baral, and B. Volkmann-Kohlmeyer. 1998. Fungi on *Juncus roemerianus*. 10. A new *Orbilia* with ingoldian anamorph. Mycologia 90:303–309.

Kohlmeyer, J., B. Bebout, and B. Volkmann-Kohlmeyer. 1995a. Decomposition of mangrove wood by marine fungi and teredinids in Belize. Marine Ecology (Pubblicazioni della Stazione Zoologica di Napoli) 16:27–39.

Kohlmeyer, J., J.W. Spatafora, and B. Volkmann-Kohlmeyer. 2000. Lulworthiales, a new order of marine Ascomycota. Mycologia 92:453–458.

Kohlmeyer, J., B. Volkmann-Kohlmeyer, and O.E. Eriksson. 1995b. Fungi on *Juncus roemerianus*. 2. New dictyosporous ascomycetes. Botanica Marina 38:165–174.

———. 1995c. Fungi on *Juncus roemerianus*. 3. New ascomycetes. Botanica Marina 38:175–186.

———. 1995d. Fungi on *Juncus roemerianus*. 4. New marine ascomycetes. Mycologia 87:532–542.

———. 1995e. Fungi on *Juncus roemerianus*. New marine and terrestrial ascomycetes. Mycological Research l00:393–404.

———. 1997. Fungi on *Juncus roemerianus*. 9. New obligate and facultative marine Ascomycotina. Botanica Marina 40:291–300.

Kohn, L.M., and E. Stasovski. 1990. The mycorrhizal status of plants at Alexandra Fiord, Ellesmere Island, Canada, a high arctic site. Mycologia 82:23–35.

Koide, R.T., D.L. Shumway, and S.A. Mabon. 1994. Mycorrhizal fungi and reproduction of field populations of *Abutilon theophrasti* Medic (Malvaceae). New Phytologist 126:123–130.

Kojima, Y., Y. Tsukuda, Y. Kawai, A. Tsukamoto, J. Sugiura, M. Sakaino, and Y. Kita. 1990. Cloning, sequence analysis, and expression of ligninolytic phenoloxidase genes of the white-rot basidiomycete *Coriolus hirsutus*. Journal of Biological Chemistry 265:15224–15230.

Kolasa, J., and S.T.A. Pickett (eds.). 1991. Ecological Heterogeneity. Ecological Studies 86.

Komatsu, M. 1976. Studies on *Hypocrea*, *Trichoderma* and allied fungi antagonistic to shiitake, *Lentinus edodes*, (Berk.) Sing. Report of the Tottori Mycological Institute 13:1–113.

Komatsu, M., and Y. Hashioka. 1964. *Trichoderma viride*, as an antagonist of the wood-inhabiting Hymenomycetes, V. Lethal effect of the different *Trichoderma* forms on *Lentinus edodes* inside logwoods. Report of the Tottori Mycological Institute 4:11–18.

Komiya, H., M. Miyazaki, and S. Takemura. 1981. The nucleotide sequence of 5S ribosomal RNA from *Schizosaccharomyces pombe*. Journal of Biochemistry 89:1663–1666.

Kondratyuk, S.Y., and B.J. Coppins (eds.). 1998. Lobarion Lichens as Indicators of the Primeval Forests of the Eastern Carpathians, Darwin International Workshop, 25–30 May 1998, Kostrino, Ukraine. M.H. Kholodny Institute of Botany, Ukrainian Phytosociological Centre, Kiev, Ukraine.

Korf, R.P. 1973. Discomycetes and Tuberales. Pp. 249–319. *In* G.C. Ainsworth, F.K. Sparrow, and A.S. Sussman (eds.), The Fungi. An Advanced Treatise. IVA. A Taxonomic Review with Keys: Ascomycetes and Fungi Imperfecti. Academic Press, New York.

———. 1983. *Cyttaria* (Cyttariales): coevolution with Nothofagus, and evolutionary relationship to the Boedijnopezizeae (Pezizales, Sarcoscyphaceae), parasitic fungi which form stem and twig galls. Australian Journal of Botany, Supplemental Series 10:77–87.

———. 1995. Authors, reviewers, and editors of articles proposing new names: A few guidelines. Mycotaxon 54:413–419.

Korf, R.P., and P. Lizon. 2000. Validation of Nannfeldt's ordinal name Helotiales. Mycotaxon 75:501–502.

Kormanik, P.P., W.C. Bryan, and R.C. Schultz. 1980. Procedures and equipment for staining large numbers of plant root samples for endomycorrhizal assay. Canadian Journal of Microbiology 26:536–538.

Kornerup, A., and J.H. Wanscher. 1978. Methuen Handbook of Colour. 3rd ed. Eyre Methuen, London, United Kingdom.

Koske, R.E. 1981. *Labyrinthula* inside the spores of a vesicular-arbuscular mycorrhizal fungus. Mycologia 73:1175–1180.

———. 1987. Distribution of VA mycorrhizal fungi along a latitudinal temperature gradient. Mycologia 79:55–68.

Koske, R.E., and J.N. Gemma. 1989. A modified procedure for staining roots to detect VA mycorrhizas. Mycological Research 92:486–505.

———. 1990. VA mycorrhizae in strand vegetation of Hawaii: Evidence for long-distance codispersal of plants and fungi. American Journal of Botany 77:466–474.

———. 1992. Restoration of early and late successional dune communities at Province Lands, Cape Cod National Seashore. Cooperative National Park Service Studies Unit, University of Rhode Island, Narragansett, Rhode Island, Technical Report NPS/NARURI/NRTR-92/03.

Koske, R.E., and B. Tessier. 1983. A convenient, permanent slide mounting medium. Mycology Society of America News Letter 34:59.

Kostyukovsky, V.A., O.N. Okunev, and B.V. Tarakanov. 1991. Description of two anaerobic fungal strains from the bovine rumen and influence of diet on the fungal population in vivo. Journal of General Microbiology 137:1759–1764.

Kouvelis, V.N., R. Zare, P.D. Bridge, and M.A. Typas. 1999. Differentiation of mitochondrial subgroups in the *Verticillium lecanii* species complex. Letters in Applied Microbiology 28:263–268.

Kowalski, D.T. 1969a. A new coprophilous species of *Calonema* (Myxomycetes). Madrono 20:229–231.

———. 1969b. A new coprophilous species of *Didymium*. Mycologia 61:635–639.

———. 1972. *Squamuloderma*: a new genus of myxomycetes. Mycologia 64:1282–1289.

Kowalski, T., and R.D. Kehr. 1992. Endophytic fungal colonization of branch bases in several forest tree species. Sydowia 44:137–168.

Kozakiewicz, Z. 1989. *Aspergillus* species on stored products. Mycological Papers 161:1–188.

Krajden, S., R.C. Summerbell, J. Kane, I.F. Salkin, M.E. Kemna, M.G. Rinaldi, M. Fuksa, E. Spratt, C. Rodrigues, and J. Choe. 1991. Normally saprobic cryptococci isolated from *Cryptococcus neoformans* infections. Journal of Clinical Microbiology 29:1883–1887.

Kramer, C.L. 1973. Protomycetales and taphrinales. Pp. 33–41. *In* G.C. Ainsworth, F.K. Sparrow, and A.S. Sussman (eds.), The Fungi. An Advanced Treatise. IVA. A Taxonomic Review with Keys: Ascomycetes and Fungi Imperfecti. Academic Press, New York.

———. 1987. The Taphrinales. Pp. 151–166. *In* G.S. de Hoog, M.T. Smith, and A.C.M. Weijman (eds.), The Expanding Realm of Yeastlike Fungi. Elsevier, Amsterdam, The Netherlands.

Krantz, G.W. 1978. A Manual of Acarology. 2nd ed. Oregon State University, Corvallis, Oregon.

Kranz, J. 1974. A host list of the rust parasite *Eudarluca caricis* (Fr.) O. Eriks. Nova Hedwigia 24:169–180.

———. 1988. Measuring plant disease. Pp. 36–50. *In* J. Kranz and J. Rotem (eds.), Experimental Techniques in Plant Disease Epidemiology. Springer-Verlag, Berlin, Germany.

———. 1990. Fungal diseases in multispecies plant communities. New Phytologist 116:383–405.

Kranz, J., and W. Brandenburger. 1981. An amended host list of the rust parasite *Eudarluca caricis*. Zeitschrift für Pflanzenkrankheiten und Pflanzenschutz 88:682–702.

Krebs, C.J. 1989. Ecological Methodology. Harper Row, New York.

Kreger-van Rij, N.J.W., and M. Veenhuis. 1971. Some features of yeasts of the genus *Sporidiobolus* observed by electron microscopy. Antonie van Leeuwenhoek 37:253–255.

Kretzer, A.M., M.I. Bidartondo, and L.C. Grubisha. 2000. Regional specialization of *Sarcodes sanguinea* (Ericaceae) on a single fungal symbiont from the *Rhizopogon ellenae* (Rhizopogonaceae) species complex. American Journal of Botany 87:1778–1783.

Kristiansen, R., and A.-E. Torkelsen. 1994. Utbredelsen av seljepute (*Hypocreopsis lichenoides*) i Norge. Blekksoppen 63:23–24, 43–44.

Kroken, S., and J.W. Taylor. 2001. A gene genealogical approach to recognize phylogenetic species boundaries in the lichenized fungus *Letharia*. Mycologia 93:38–53.

Krug, J.C. 1989. *Periamphispora*, a new genus of the Sordariaceae. Mycologia 81:475–479.

——. 1999. Plant disease fungi. Pp. 140–149. *In* B.I. Roots, D.A. Chant, and C.E. Heidenreich (eds.), Special Places: The Changing Ecosystem of the Toronto Region. University of British Columbia Press, Vancouver, British Columbia, Canada.

Krug, J.C., and R.S. Jeng. 1989. *Hapsidomyces*, a new genus of the Pezizaceae with ornamented ascospores. Mycologia 76:748–751.

Krug, J.C., and R.S. Khan. 1991. A new homothallic species of *Neurospora* from Hungary. Mycologia 8:829–832.

Krumbien, W.E., and K. Jens. 1981. Biogenic rock varnishes of the Negev Desert (Israel), an ecological study of iron and manganese transformation by cyanobacteria and fungi. Oecologia 50:25–38.

Kück, U., H.D. Osiewacz, U. Schmidt, B. Kappelhoff, E. Schulte, U. Stahl, and K. Esser. 1985. The onset of senescence is affected by DNA rearrangements of a discontinuous mitochondrial gene in *Podospora anserina*. Current Genetics 9:373–382.

Kudo, H., K.D. Jakober, R.C. Phillipe, K.J. Cheng, D.J.S. Barr, and J.W. Costerton. 1990. Isolation and characterization of cellulolytic anaerobic fungi on associated mycoplasmas from the rumen of a steer fed a roughage diet. Canadian Journal of Microbiology 36:513–517.

Kuhlman, E.G., and F.R. Matthews. 1976. Occurrence of *Darluca filum* on *Cronartium strobilinum* and *C. fusiforme* infecting oak. Phytopathology 66:1195–1197.

Kuhlman, E.G., and T. Miller. 1976. Occurrence of *Tuberculina maxima* on fusiform rust galls in the southeastern United States. Plant Disease Reporter 60:627–629.

Kuhlman, E.G., J.W. Carmichael, and T. Miller. 1976. *Scytalidium uredinicola*, a new mycoparasite of *Cronartium fusiforme* on *Pinus*. Mycologia 68:1188–1194.

Kuhlman, E.G., F.R. Matthews, and H.P. Tillerson. 1978. Efficacy of *Darluca filum* for biological control of *Cronartium fusiforme* and *C. strobilinum*. Phytopathology 68:507–511.

Kuhls, K., E. Lieckfeldt, G.J. Samuels, W. Kovacs, W. Meyer, O. Petrini, W. Gams, T. Börner, and C.P. Kubicek. 1996. Molecular evidence that the asexual industrial fungus *Trichoderma reesei* is a clonal derivative of the ascomycete *Hypocrea jecorina*. Proceedings of the National Academy of Sciences USA 93:7755–7760.

Kühne, B., H.-P. Hanssen, W.-R. Abraham, and V. Wray. 1991. A phytotoxic eremophilane ether from *Hypomyces odoratus*. Phytochemistry 30:1463–1465.

Kumar, J., R.J. Nelson, and R.S. Zeigler. 1999. Population structure and dynamics of *Magnaporthe grisea* in the Indian Himalayas. Genetics 152:971–984.

Kümmerling, H., D. Triebel, and G. Rambold. 1993. *Lepraria neglecta* and its lichenicolous fungi. Bibliotheca Lichenologica 53:147–160.

Kuprevich, V.F., and V.G. Transchel'. 1957. Cryptogamic Plants of the USSR. Fungi IV. Rust Fungi, No. 1, Family Melampsoraceae. Academy of Sciences of the USSR, Komarov Institute of Botany, Moscow-Leningrad, USSR. [Translation of "Flora sporovych rastenii SSSR. Rzhavchinnye Griby," Moskva-Lenningrad, Izdatel'stvo Akademii Nauk Sssr, by E. Rabinovitz. 1970. Israel Program for Scientific Translations, Jerusalem, Israel.]

Kurokawa, S. (ed.). 2003. Checklist of Japanese Lichens. National Science Museum, Tokyo, Japan.

Kurtzman, C.P. 1968. Parasitism and axenic growth of *Dispira cornuta*. Mycologia 60:915–923.

——. 1987. Prediction of biological relatedness among yeasts from comparisons of nuclear DNA complementarity. Studies in Mycology 30:459–468.

——. 1993. Systematics of the ascomycetous yeasts assessed from ribosomal RNA sequence divergence. Antonie van Leeuwenhoek 63:165–174.

——. 1994. Molecular taxonomy of the yeasts. Yeast 10:1727–1740.

——. 1998. Discussion of teleomorphic and anamorphic ascomycetous yeasts and a key to genera. Pp. 111–121. *In* C.P. Kurtzman and J.W. Fell (eds.), The Yeasts, A Taxonomic Study. 4th ed. Elsevier, Amsterdam, The Netherlands.

Kurtzman, C.P., and J.W. Fell. 1998. The Yeasts, A Taxonomic Study. 4th ed. Elsevier, Amsterdam, The Netherlands.

Kurtzman, C.P., and Z. Liu. Evolutionary affinities of species assigned to *Lipomyces* and *Myxozyma* estimated from ribosomal RNA sequence divergence. Current Microbiology 21:387–393.

Kurtzman, C.P., and C.J. Robnett. 1991. Phylogenetic relationships among species of *Saccharomyces, Schizosaccharomyces, Debaryomyces* and *Schwanniomyces* determined from partial ribosomal RNA sequences. Yeast 7:61–72.

——. 1994a. Orders and families of ascosporogenous yeasts and yeast-like taxa compared from ribosomal RNA sequence similarities. Pp. 249–258. *In* D.L. Hawksworth (ed.), Ascomycete Systematics: Problems and Perspectives in the Nineties. Plenum, New York.

——. 1994b. Synonymy of the yeast genera *Wingea* and *Debaryomyces*. Antonie van Leeuwenhoek 66:337–342.

——. 1995. Molecular relationships among hyphal ascomycetous yeasts and yeastlike taxa. Canadian Journal of Botany 73(Suppl. 1):S824–S830.

——. 1997. Identification of clinically important ascomycetous yeasts based on nucleotide divergence in the 5' end of the large-subunit (26S) ribosomal DNA gene. Journal of Clinical Microbiology 35:1216–1223.

——. 1998. Identification and phylogeny of ascomycetous yeasts from analysis of nuclear large subunit (26S) ribosomal DNA partial sequences. Antonie van Leeuwenhoek 73:331–371.

Kuserk, F.T. 1980. The relationship between cellular slime molds and bacteria in forest soil. Ecology 61:1474–1485.

Kushner, D.J. 1978. Microbial Life in Extreme Environments. Academic Press, London, England.

Kuter, G.A. 1984. Hyphal interactions between *Rhizoctonia solani* and some *Verticillium* species. Mycologia 76:936–940.

Kuthubutheen, A.J. 1982. Thermophilous fungi from Malaysia. Transactions of the British Mycological Society 79:548–552.

Kuthubutheen, A.J., and J. Webster. 1986. Water availability and the coprophilous fungus succession. Transactions of the British Mycological Society 86:63–76.

Kuykendall, W.R., D.F. Hindal, and H.L. Barnett. 1983. Parasitism and nutrition of the contact mycoparasite *Acladium tenellum*. Mycologia 75:656–665.

Kwon-Chung, K.J. 1975. A new genus *Filobasidiella*, the perfect state of *Cryptococcus neoformans*. Mycologia 67:1197–1200.

Kwon-Chung, K.J., and J.E. Bennett. 1992. Medical Mycology. Lea and Febiger, Philadelphia, Pennsylvania.

Laakkonen, J. 1998. *Pneumocystis carinii* in wildlife. International Journal for Parasitology 28:241–252.

Laaser, G., K.-D. Jahnke, H. Prillinger, R. Bauer, P. Hoffmann, G. Deml, and F. Oberwinkler. 1988. A new tremelloid yeast isolated from *Asterophora lycoperdoides* (Bull.: Fr.) Ditm. Antonie van Leeuwenhoek 54:57–74.

Laaser, G., E. Möller, K.-D. Jahnke, G. Bahnweg, H. Prillinger, and H.H. Prell. 1989. Ribosomal DNA restriction fragment analysis as a taxonomic tool in separating physiologically similar basidiomycetous yeasts. Systematic and Applied Microbiology 11:170–175.

Lachance, M.-A. 1990. Ribosomal DNA spacer variation in the cactophilic yeast *Clavispora opuntiae*. Molecular Biology and Evolution 7:178–193.

Lachance, M.-A., C.A. Rosa, W.T. Starmer, B. Schlag-Edler, J.S.F. Barker, and J.M. Bowles. 1998. *Metschnikowia continentalis* var. *borealis*, *Metschnikowia continentalis* var. *continentalis*, and *Metschnikowia hibisci*, new heterothallic haploid yeasts from ephemeral flowers and associated insects. Canadian Journal of Microbiology 44:279–288.

Lado, C., M. Rodríguez-Palma, and A. Estrada-Torres. 1999. Myxomycetes from a seasonal tropical forest on the Pacific coast of Mexico. Mycotaxon 71:307–321.

Læssøe, T. 1989. Pilfinger (*Hypocreopsis lichenoides*) i Danmark. Svampe 20:56–58.

Læssøe, T., and D.J. Lodge. 1994. Three host-specific *Xylaria* species. Mycologia 86:436–446.

Læssøe, T., and B.M. Spooner. 1994. *Rosellinia* & *Astrocystis* (Xylariaceae): New species and generic concepts. Kew Bulletin 49:1–70.

Læssøe, T., J.D. Rogers, and A.J.S. Whalley. 1989. *Camillea, Jongiella* and light-spored species of *Hypoxylon*. Mycological Research 93:121–155.

Laing, S.A.K., and J.W. Deacon. 1990. Aggressiveness and fungal host ranges of mycoparasitic *Pythium* spp. Soil Biology and Biochemistry 22:905–911.

——. 1991. Video microscopical comparison of mycoparasitism by *Pythium oligandrum*, *P. nunn* and an unnamed *Pythium* species. Mycological Research 95:469–479.

Lake, J.A. 1987. A rate-independent technique for analysis of nucleic acid sequences: evolutionary parsimony. Molecular Biology and Evolution 4:167–191.

Lamar, R.T., M.W. Davis, D.M. Dietrich, and J.A. Glaser. 1994. Treatment of a pentachlorophenol- and creosote-contaminated soil using the lignin-degrading fungus *phanerochaete*(sic) *sordid a* (sic) : a field demonstration. Soil Biology and Biochemistry 26:1603–1611.

Lamar, R.T., B. Schoenike, A. Vanden Wymelenberg, P. Stewart, D. Dietrich, and D. Cullen. 1995. Quantitation of fungal mRNAs in complex substrates by reverse transcription PCR and its application to *Phanerochaete chrysosporium*-colonized soil. Applied and Environmental Microbiology 61:2122–2126.

Lamb, I.M. 1963. Index Nominum Lichenum Inter Annos 1932 et 1960 Divulgatorum. Ronald Press, New York.

Lambowitz, A.M. 1989. Infectious introns. Cell 56:323–326.

Lamore, B.J., and R.D. Goos. 1978. Wood-inhabiting fungi of a freshwater stream in Rhode Island. Mycologia 70:1025–1034.

Landeweert, R., P. Leeflang, T.W. Kuyper, E. Hoffland, A. Rosling, K. Wernars, and E. Smit. 2003. Molecular identification of ectomycorrhizal mycelium in soil horizons. Applied and Environmental Microbiology 69:327–333.

Landolt, J.C., and S.L. Stephenson. 1990. Cellular slime molds in forest soils of West Virginia. Mycologia 82:114–119.

——. 1991. Cellular slime molds from tropical rain forests of eastern Peru. Cryptogamic Botany 2/3:258–260.

——. 1995. Recent collections of dictyostelid cellular slime molds from the Pacific Rim and Oceania. Proceedings of the West Virginia Academy of Science 67:23.

——. 1997. Dictyostelid cellular slime molds in canopy soils of tropical forests. Proceedings of the West Virginia Academy of Science 69:13.

Landolt, J.C., S.L. Stephenson, D.F. Home, and J.C. Cavender. 1994. Recent collections of dictyostelid cellular slime molds from Central and South America. Abstracts of the Fifth International Mycological Congress, Vancouver, British Columbia 1:118.

Landolt, J.C., S.L. Stephenson, G.A. Laursen, and R. Densmore. 1992a. Distribution patterns of cellular slime molds in the Kantishna Hills, Denali National Park and Preserve, Alaska. Arctic and Alpine Research 24:244–248.

Landolt, J.C., S.L. Stephenson, and C.W. Stihler. 1992b. Cellular slime molds from West Virginia caves including notes on the occurrence and distribution of *Dictyostelium rosarium*. Mycologia 84:399–405.

Landvik, S. 1996. *Neolecta*, a fruit-body-producing genus of the basal ascomycetes, as shown by SSU and LSU rDNA sequences. Mycological Research 100:199–202.

Landvik, S., K.N. Egger, and T. Schumacher. 1997. Towards a subordinal classification of the Pezizales (Ascomycota): phylogenetic analyses of SSU rDNA sequences. Nordic Journal of Botany 17:403–418.

Landvik, S., R. Kristiansen, and T. Schumacher. 1998. Phylogenetic and structural studies in the Thelebolaceae (Ascomycota). Mycoscience 39:49–56.

Lane, D.J., B. Pace, G.J. Olsen, D.A. Stahl, M.L. Sogin, and N.R. Pace. 1985. Rapid determination of 16S ribosomal RNA sequences for phylogenetic analyses. Proceedings of the National Academy of Sciences USA 82:6955–6959.

Lang, G.E., W.A. Reiners, and L.H. Pike. 1980. Structure and biomass dynamics of epiphytic lichen communities of balsam fir. Ecology 6:541–550.

Langdon, R.F.N. 1954. The origin and differentiation of *Claviceps* species. University of Queensland Papers (Botany) 3:61–68.

Lange, M. 1978. Fungus flora in August: ten years observations in a Danish beech wood district. Botanisk Tidskrift 73:21–54.

——. 1992. Sequence of macromycetes on decaying beech logs. Persoonia 14:449–456.

Langen, G., B. Beissmann, H.J. Reisener, and K. Kogel. 1992. A β-1, 3-D-endo-mannanase from culture filtrates of the hyperparasites *Verticillium lecanii* and *Aphanocladium album* that specifically lyses the germ pore plug from uredospores of *Puccinia graminis* f.sp. *tritici*. Canadian Journal of Botany 70:853–860.

Langerak, C.J. 1977. The role of antagonists in the chemical control of *Fusarium oxysporum* f.sp. *narcissi*. Netherlands Journal of Plant Pathology 83(Suppl.1):365–381.

Langsam, D.M., and B.A. Armbruster. 1985. Preservation of saprolegniaceous fungi for light microscopy. Mycologia 77:829–832.

Langvad, F. 1980. A simple and rapid method for qualitative and quantitative study of the fungal flora of leaves. Canadian Journal of Microbiology 26:666–670.

Lapeyrie, F.F., and G.A. Chilvers. 1985. An endomycorrhiza-ectomycorrhiza succession associated with enhanced growth of *Eucalyptus dumosa* seedlings planted in calcareous soil. New Phytologist 100:93–104.

Large, E.C. 1940. The Advance of the Fungi. J. Cape, London, England.

Largent, D.L. 1977. How to Identify Mushrooms to Genus I: Macroscopic Features. Mad River Press, Eureka, California.

——. 1986. How to Identify Mushrooms to Genus I. Macroscopic Features. Revised ed. Mad River Press, Eureka, California.

Largent, D.L., and T.J. Baroni. 1988. How to Identify Mushrooms to Genus VI: Modern Genera. Mad River Press, Eureka, California.

Largent, D.L., D. Johnson, and R. Watling. 1977. How to Identify Mushrooms to Genus III: Microscopic Features. Mad River Press, Eureka, California.

Larsen, J., K. Mansfield-Giese, and L. Bødker. 2000. Quantification of *Aphanomyces euteiches* in pea roots using specific fatty acids. Mycological Research 104:858–864.

Larsen, K. 1971. Danish endocoprophilous fungi, and their sequence of occurrence. Botanisk Tidsskrift 66:1–32.

Larsen, M.J., M.F. Jurgensen, and A.E. Harvey. 1982. N₂ fixation in brown-rotted soil wood in an intermountain cedar-hemlock ecosystem. Forest Science 28:292–296.

Last, F.T., K. Natarajan, V. Mohan, and P.A. Mason. 1992. Sequences of sheathing (ecto-) mycorrhizal fungi associated with man-made forests, temperate and tropical. Pp. 214–219. *In* D.J. Read, D.H. Lewis, A.H. Fitter, and I.J. Alexander (eds.), Mycorrhizas in Ecosystems. CAB International, Wallingford, Oxon, United Kingdom.

Latch, G.C.M., M.J. Christensen, and G.J. Samuels. 1984. Five endophytes of *Lolium* and *Festuca* in New Zealand. Mycotaxon 20:535–550.

Lauckner, G. 1984. Diseases caused by microorganisms: agents: fungi. Pp. 83–113 *In* O. Kinne (ed.), Diseases of Marine Animals. Volume IV, Part 1. Introduction, Pisces. Biologische Anstalt Helgoland, Hamburg, Germany.

Laundon, G.F. 1973. Uredinales. Pp. 247–279. *In* G.C. Ainsworth, F.K. Sparrow, and A.S. Sussman (eds.), The Fungi. An Advanced Treatise. IVB. A Taxonomic Review with Keys: Basidiomycetes and Lower Fungi. Academic Press, New York.

Law, R., and D.H. Lewis. 1983. Biotic environments and the maintenance of sex–some evidence from mutualistic symbioses. Biological Journal of the Linnean Society (London) 20:249–276.

Lawrey, J.D. 1984. Biology of Lichenized Fungi. Praeger, New York.

——. 1991. The species-area curve as an index of disturbance in saxicolous lichen communities. Bryologist 94:377–382.

——. 1992. Natural and randomly-assembled lichen communities compared using the species-area curve. Bryologist 95:137–141.

——. 1995. The chemical ecology of lichen mycoparasites: a review. Canadian Journal of Botany 73(Suppl. 1):S603–S608.

Lawrey, J.D., and P. Diederich. 2003. Lichenicolous fungi: Interactions, evolution, and biodiversity. Bryologist 106:80–120.

Lawrey, J.D., A.Y. Rossman, and R. Lowen. 1994. Inhibition of selected hypocrealean fungi by lichen secondary metabolites. Mycologia 86:502–506.

Le Campion-Alsumard, T., S. Golubic, and K. Priess. 1995. Fungi in corals: symbiosis or disease? Interaction between polyps and fungi causes pearl-like skeleton biomineralization. Marine Ecology Progress Series ll7:137–147.

Le Gal, M. 1953. Les discomycètes de Madagascar. Prodrome à une Flore Mycologique de Madagascar et Dépendances. Vol. 4. Académie des Sciences, Paris, France.

——. 1959. Discomycètes du Congo Belge d'après les recoltes de Madame Coossens-Fontana. Bulletin du Jardin Botanique de Bruxelles 29:73–132.

Le Picard, Y.T., and B. Trique. 1987. Antagonistes et hyperparasites du *Fulvia fulva* (Cooke) Ciferri. Cryptogamie Mycologie 8:43–50.

Leach, C.M. 1971. A practical guide to the effects of visible light and ultraviolet light on fungi. Methods in Microbiology 4:609–664.

Leacock, P.R. 1997. Diversity of Ectomycorrhizal Fungi in Minnesota's Ancient and Younger Stands of Red Pine and Northern Hardwood-Conifer Forests. Unpubl. Ph.D. dissert., University of Minnesota, St. Paul, Minnesota.

Leadbetter, B.S.C. 1993. Preparation of pelagic protists for electron microscopy. Pp. 241–251. *In* P.F. Kemp, B.F. Sherr, E.B. Sherr, and J.J. Cole (eds.), Handbook of Methods in Aquatic Microbial Ecology. Lewis Publishers, Boca Raton, Florida.

Leander, C.A., and D. Porter. 2001. The Labyrinthulomycota is comprised of three distinct lineages. Mycologia 93:459–464.

Leaño, E.M., L.L.P. Vrijmoed, E.B.G. Jones. 1998. Physiological studies on *Halophytophthora vesicula* (straminipilous fungi) isolated from fallen mangrove leaves from Mai Po, Hong Kong. Botanica Marina 41:411–419.

Lebeau, J.B., and C.E. Logsdon. 1958. Snow mold of forage crops in Alaska and Yukon. Phytopathology 48:148–150.

Lebeck, L.M. 1989. Extracellular symbiosis of a yeast-like microorganism within *Comperia merceti* (Hymenoptera: Encyrtidae). Symbiosis 7:51–66.

Leclerc, M.C., J. Guillot, and M. Deville. 2000. Taxonomic and phylogenetic analysis of Saprolegniaceae (Oomycetes) inferred from LSU rDNA and ITS sequence comparisons. Antonie van Leeuwenhoek 77:369–377.

Lee, J., and J.T. Mullins. 1994. Cytoplasmic water-soluble β-glucans in *Achlya*: response to nutrient limitation. Mycologia 86:235–241.

Lee, P.-J., and R.E. Koske. 1994. *Gigaspora gigantea*: parasitism of spores by fungi and actinomycetes. Mycological Research 98:458–466.

Lee, Y.-B., and K. Sugiyama. 1986. On a new genus of the Laboulbeniales: *Majewskia*. Mycologia 78:289–292.

Legault, D., M. Dessureault, and G. Laflamme. 1989. Mycoflore des aguilles de *Pinus banksiana* et *Pinus resinosa*. I. Champignons endophytes. Canadian Journal of Botany 67:2052–2060.

Legendre, P., and M.-J. Fortin. 1989. Spatial pattern and ecological analysis. Vegetatio 80:107–138.

Legendre, P., and L. Legendre. 1998. Numerical Ecology. 2nd ed. Elsevier, Amsterdam, The Netherlands.

Leinhos, G.M.E., and H. Buchenauer. 1992. Hyperparasitism of selected fungi on rust fungi of cereal. Zeitschrift für Pflanzenkrankheiten und Pflanzenschutz 99:482–498.

Lendner, A. 1908. Les Mucorinées de la Suisse. Matériaux pour la Flore Cryptogamique Suisse 3:1–182.

Léon-Gallegos, H.M., and G.B. Cummins. 1981. Uredinales (Royas) de México. Vols. I, II. Secretaría de Agricultura y Recursos Hidráulicos (SARH), Culiacan, Sinaloa, Mexico.

Lerner, L.E., and I.I. Sidorova. 1978. Interaction of mycoparasites with some fungi in a culture. Vestnik Moskovskogo Universiteta, Seriya XVI, Biologiya:81–85.

Lesica, P., and R.K. Antibus. 1990. The occurrence of mycorrhizae in vascular epiphytes of two Costa Rican rain forests. Biotropica 22:250–258.

Lesica, P., B. McCune, S.V. Cooper, and W.S. Hong. 1991. Differences in lichen and bryophyte communities between old-growth and managed second-growth forests in the Swan Valley, Montana. Canadian Journal of Botany 69:1745–1755.

Leslie, J.F., C.A.S. Pearson, P.E. Nelson, and T.A. Tousson. 1990. *Fusarium* spp. from corn, sorghum, and soybean fields in the central and eastern United States. Phytopathology 80:343–350.

Leuchtmann, A. 1984. Über *Phaeosphaeria* Miyake und andere bitunicate ascomyceten mit mehrfach querseptierten ascosporen. Sydowia 37:79–195.

Leuchtmann, A., C.L. Schardl, and M.R. Siegel. 1994. Sexual compatibility and taxonomy of a new species of *Epichloë* symbiotic with fine fescue grasses. Mycologia 86:802–812.

Leufven, A., and L. Nehls. 1986. Quantification of different yeasts associated with the bark beetle, *Ips typographus*, during its attack on a spruce tree. Microbial Ecology 12:237–243.

Levetin, E., and N.E. Caroselli. 1976. A simplified medium for growth and sporulation of *Pilobolus* species. Mycologia 68:1254–1258.

Levine, J.F., J.H. Hawkins, M.J. Dykstra, E.J. Noga, D.W. Moye, and R.S. Cone. 1990. Species distribution of ulcerative lesions on finfish in the Tar-Pamlico River estuary, North Carolina. Diseases of Aquatic Organisms 8:1–5.

Lewinsohn, D., E. Lewinsohn, C.L. Bertagnolli, and A.D. Partridge. 1994. Blue-stain fungi and their transport structures on the Douglas-fir beetle. Canadian Journal of Forest Research 24:2275–2283.

Lewis, D.H., J.M. Whipps, and R.C. Cooke. 1995. Mechanisms of biological disease control with special reference to the case study of *Pythium oligandrum* as an antagonist. Pp. 191–217. *In* J.M. Whipps and R.D. Lumsden (eds.), Biotechnology of Fungi for Improving Plant Growth. Cambridge University Press, Cambridge, Great Britain.

Lewis, J.A., and G.C. Papavizas. 1985. Effect of mycelial preparations of *Trichoderma* and *Gliocladium* on populations of *Rhizoctonia solani* and the incidence of damping-off. Phytopathology 75:812–817.

——. 1992. Potential of *Laetisaria arvalis* for the biocontrol of *Rhizoctonia solani*. Soil Biology and Biochemistry 24:1075–1079.

Lewis, K.J., and S.A. Alexander. 1986. Insects associated with the transmission of *Verticicladiella procera*. Canadian Journal of Forest Research 16:1330–1333.

Lewis, K., J.M. Whipps, and R.C. Cooke. 1990. Mechanisms of biological disease control with special reference to the case study of *Pythium oligandrum* as an antagonist. Pp. 191–217. *In* J.M. Whipps and R.D. Lumsden (eds.), Biotechnology of Fungi for

Improving Plant Growth. Cambridge University Press, Cambridge, England.

Li, J., and I.B. Heath. 1992. The phylogenetic relationships of anaerobic chytridiomycetous gut fungi (Neocallimasticaceae) and the Chytridiomycota. I. Cladisitc analysis of rRNA sequences. Canadian Journal of Botany 70:1738–1746.

———. 1993. Chytridiomycetous gut fungi, oft overlooked contributors to herbivore digestion. Canadian Journal of Microbiology 39:1003–1013.

Li, J., I.B. Heath, and T. Bauchop. 1990. *Piromyces mae* and *Piromyces dumbonica*, two new species of uniflagellate anaerobic chytridiomycete fungi from the hindgut of the horse and elephant. Canadian Journal of Botany 68:1021–1033.

Li, J., I.B. Heath, and K.-J. Cheng. 1991. The development and zoospore ultrastructure of a polycentric chytridiomycete gut fungus, *Orpinomyces joyonii* comb. nov. Canadian Journal of Botany 69:580–589.

Li, J., I.B. Heath, and L. Packer. 1993. The phylogenetic relationships of the anaerobic chytridiomycetous gut fungi (Neocallimasticaceae) and the Chytridiomycota II. Cladisitc analysis of structural data and description of Neocallimasticales ord. nov. Canadian Journal of Botany 71:393–407.

Lichtwardt, R.W. 1986. The Trichomycetes: Fungal Associates of Arthropods. Springer-Verlag, New York.

———. 1996. Trichomycetes and the arthropod gut. Pp. 315–330. *In* D.H. Howard and J.D. Miller (eds.), The Mycota. Vol. VI. Human and Animal Relationships. Springer-Verlag, Berlin, Germany.

Lichtwardt, R.W., M.J. Cafaro, and M.M. White. 2001. The Trichomycetes: Fungal Associates of Arthropods, Rev. ed. Department of Ecology & Evolutionary Biology, University of Kansas, Lawrence, Kansas. Published on the web: http://www.nhm.ku.edu/~fungi

Liebetanz, E. 1910. Die parasitischen Protozoen der Wiederkäuermagens. Archiv für Protistenkunde 19:19–80.

Lifshitz, R., M. Dupler, Y. Elad, and R. Baker. 1984. Hyphal interactions between a mycoparasite, *Pythium nunn*, and several soil fungi. Canadian Journal of Microbiology 30:1482–1487.

Lightner, D.V. 1975. Some potentially serious disease problems in the culture of penaeid shrimp in North America. Pp. 75–97. *In* Proceedings of the Third U.S.–Japan Meeting on Aquaculture. Tokyo, Japan. Special Publication, Fishery Agency, Japanese Government and Japan Sea Regional Fisheries Research Laboratory, Nigata, Japan.

———. 1981. Fungal diseases of marine crustacea. Pp. 451–484. *In* E.W. Davidson (ed.), Pathogenesis of Invertebrate Microbial Disease. Allandeld Osmun Publishers, Totowa, New Jersey.

———. 1988. Aflatoxicosis of panaeid shrimp. Pp. 96–99. *In* C.J. Sindermann and D.V. Lightner (eds.), Disease Diagnosis and Control in North American Marine Aquaculture. Elsevier, New York.

Ligrone, R. 1988. Ultrastructure of a fungal endophyte in *Phaeoceros laevis* (L.) Prosk. (Anthocerotophyta). Botanical Gazette 149:92–100.

Ligrone, R., and C. Lopes. 1989. Cytology and development of a mycorrhiza-like infection in the gametophyte of *Conocephalum conicum* (L.) Dum. (Marchantiales, Hepatophyta). New Phytologist 111:423–433.

Ligrone, R., K. Pocock, and J.G. Duckett. 1993. A comparative ultrastructural study of endophytic basidiomycetes in the parasitic achlorophyllous hepatic *Cryptothallus mirabilis* and the closely allied photosynthetic species *Aneura pinguis* (Metzgeriales). Canadian Journal of Botany 71:666–679.

Lim, T.K., and K.G. Rohrbach. 1980. Role of *Penicillium funiculosum* strains in the development of pineapple fruit diseases. Phytopathology 70:663–665.

Lim, T.-K., and Z.W.N. Wan. 1983. Mycoparasitism of the coffee rust pathogen, *Hemileia vastatrix*, by *Verticillium psalliotae* in Malaysia. Pertanika 6:23–25.

Lincoff, G.H. 1981. The Audubon Society Field Guide to North American Mushrooms. Alfred A. Knopf, New York.

Lindblad, I. 1998. Wood-inhabiting fungi on fallen logs of Norway spruce: relations to forest management and substrate quality. Nordic Journal of Botany 18:243–255.

———. 2000. Host specificity of some wood-inhabiting fungi in a tropical forest. Mycologia 92:399–405.

———. 2001. Diversity of poroid and some corticioid fungi along the rainfall gradient in tropical forest, Costa Rica. Journal of Tropical Ecology 17:353–369.

Linderman, R.G. 1988. Mycorrhizal interactions with the rhizosphere microflora: the mycorrhizosphere effect. Phytopathology 78:366–371.

Lindgren, B.S. 1983. A multiple funnel trap for scolytid beetles (Coleoptera). Canadian Entomologist 115:299–302.

Linnemann, G. 1941. Die Mucorineen-Gattung Mortierella Coemans. Pflanzenforschung 23:1–64.

Liou, J.Y., and S.S. Tzean. 1992. Stephanocysts as nematode-trapping and infecting propagules. Mycologia 84:786–790.

Lipka, J. 1985. Weiteres über *Buchwaldoboletus lignicola* (Kallenbach) Pilát und *Phaeolus schweinitzii* (Fries) Patouillard. Zeitschrift für Mykologie 51:47–50.

Lipsey, M.W. 1990. Design Sensitivity: Statistical Power for Experimental Research. Sage Publishing, Newbury Park, California.

Liskauskas, A.P., and M.M. Ferguson. 1990. Enzyme heterozygosity and fecundity in a naturalized population of brook trout (*Salvelinus fontinalis*). Canadian Journal of Fisheries and Aquatic Sciences 47:2010–2015.

Lister, A. 1894. A Monograph of the Mycetozoa being a descriptive catalogue of the species in the Herbarium of the British Museum. British Museum (Natural History), London, England.

———. 1911. A Monograph of the Mycetozoa. 2nd ed. (revised by G. Lister). British Museum of Natural History, London, England.

———. 1925. A monograph of the Mycetozoa. 3rd ed. (revised by G. Lister). British Museum of Natural History, London, England.

Liston, A., and E. Alvarez-Buylla. 1995. Internal transcribed spacer sequences of conifers: "there is a fungus among us." Inoculum 46:26.

Liston, A., W.A. Robinson, J.M. Oliphant, and E.R. Alvarez-Buylla. 1996. Length variation in the nuclear ribosomal DNA internal transcribed spacer region of non-flowering seed plants. Systematic Botany 21:109–120.

Liu, Y., S. Whelen, and B.D. Hall. 1999. Phylogenetic relationships among ascomycetes: evidence from an RNA polymerase II subunit. Molecular Biology and Evolution 16:1799–1808.

Liu, Z., and C.P. Kurtzman. 1991. Phylogenetic relationships among species of *Williopsis* and *Saturnospora* gen. nov. as determined from partial rRNA sequences. Antonie van Leeuwenhoek 60:21–30.

Livingston, B.E., J.C. Britton, and F.R. Reid. 1905. Studies on the properties of an unproductive soil. U.S. Department of Agriculture, Bureau of Soils Bulletin 28:5–39.

Livsey, S., and D.W. Minter. 1994. The taxonomy and biology of *Tryblidiopsis pinastri*. Canadian Journal of Botany 72:549–557.

Llimona, X., and N.L. Hladun. 2001. Checklist of the lichens and lichenicolous fungi of the Iberian Peninsula and the Balearic Islands. Bocconea 14:1–581.

LoBuglio, K.F., J.I. Pitt, and J.W. Taylor. 1993. Phylogenetic analysis of two ribosomal DNA regions indicates multiple independent losses of a sexual *Talaromyces* state among asexual *Penicillium* species in subgenus *Biverticillium*. Mycologia 85:592–604.

Lodge, D.J. 1993. Nutrient cycling by fungi in wet tropical forests. Pp. 37–57. *In* S. Isaac, J.C. Frankland, R. Watling, and A.J.S. Whalley (eds.), Aspects of Tropical Mycology. Cambridge University Press, Cambridge, United Kingdom.

——. 1996. Microorganisms. Pp. 53–108. *In* D.P. Reagan and R.B. Waide (eds.), The Food Web of a Tropical Forest. University of Chicago Press, Chicago, Illinois.

Lodge, D.J., and S. Cantrell. 1995a. Diversity of litter agarics at Cuyabeno, Ecuador: calibrating sampling efforts in tropical rainforest. Mycologist 9:149–151.

——. 1995b. Fungal communities in wet tropical forests: variation in time and space. Canadian Journal of Botany 73(Suppl. 1):S1391–S1398.

Lodge, D.J., I. Chapela, G. Samuels, F.A. Uecker, D. Desjardin, E. Horak, O.K. Miller, Jr., G.L. Hennbert, C.A. Decock, J. Ammirati, H.H. Burdsall, Jr., P.M. Kirk, D.W. Minter, R. Halling, T. Læssøe, G. Mueller, F. Oberwinkler, D.N. Pegler, B. Spooner, R.H. Petersen, J.D. Rogers, L. Ryvarden, R. Watling, E. Tunbull, and A.J.S. Whalley. 1995. A survey of patterns in fungal diversity in non-lichenized fungi. Mitteilungen der Eidgenössischen Forschungs-anstalt für Wald, Schnee und Landschaft 70:157–173.

Lodge, D.J., P.J. Fisher, and B.C. Sutton. 1996a. Endophytic fungi of *Manilkara bidentata* leaves in Puerto Rico. Mycologia 88:733–738.

Lodge, D.J., D.L. Hawksworth, and B.J. Ritchie. 1996b. Microbial diversity and tropical forest functioning. Pp. 69–100. *In* G. Orians, R. Dirzo, and J.H. Cushman (eds.), Biodiversity and Ecosystem Processes in Tropical Forests. Springer-Verlag, Berlin, Germany.

Lodge, D.J., W.H. McDowell, and C.P. McSwiney. 1994. The importance of nutrient pulses in tropical forests. Trends in Ecology and Evolution 9:384–387.

Lodge, D.J., L. Ryvarden, and O. Perdomo. 2001. Studies in neotropical polypores 11. *Antrodia aurantia*, a new species from the Dominican Republic, Greater Antilles. Mycotaxon 80:261–266.

Lodha, B.C. 1974. Decomposition of digested litter. Pp. 213–241. *In* C.H. Dickinson and G.J.F. Pugh (eds.), Biology of Plant Litter Decomposition. Vol. I. Academic Press, London, England.

Lohman, M.L. 1938. Notes on Indiana fungi—1937. Proceedings of the Indiana Academy of Science 47:88–92.

Lohman, M.L., and A.J. Watson. 1943. Identity and host relations of *Nectria* species associated with disease of hardwoods in the eastern states. Lloydia 6:77–108.

Longcore, J.E. 1992a. Morphology, occurrence, and zoospore ultra-structure of *Podochytrium dentatum* sp. nov. (Chytridiales). Mycologia 84:183–192.

——. 1992b. Morphology and zoospore ultrastructure of *Chytriomyces angularis* sp. nov. (Chytridiales). Mycologia 84:442–451.

——. 1993. Morphology and zoospore ultrastructure of *Lacustromyces hiemalis* gen. et sp. nov. (Chytridiales). Canadian Journal of Botany 71:414–425.

——. 1995. Morphology and zoospore ultrastructure of *Entophlyctis luteolus* sp. nov. (Chytridiales): Implications for chytrid taxonomy. Mycologia 87:25–33.

——. 1996. Chytridiomycete taxonomy since 1960. Mycotaxon 60:149–174.

Longcore J.E., A.P. Pessier, and D.K. Nichols. 1999. *Batrachochytrium dendrobatidis* gen. et sp. nov., a chytrid pathogenic to amphibians. Mycologia 91:219–227.

Longman, K.A., and J. Jenik. 1987. Tropical Forest and its Environment. 2nd ed. Longman Scientific and Technical, Essex, United Kingdom.

Louis, I., and G. Lim. 1987. Spore density and root colonization of vesicular-arbuscular mycorrhizas in tropical soil. Transactions of the British Mycological Society 88:207–212.

Lovell, R.T. 1989. Nutrition and Feeding in Fish. Van Nostrand Reinhold, New York.

——. 1991. Cyclopiazonic acid, a potentially serious mould toxin. Aquaculture Magazine 17(July/August): 66–68.

——. 1992. Mycotoxins: hazardous to farmed fish. Feed International 13:24–28.

Lowe, S.E., G.G. Griffiths, A. Milne, M.K. Theodorou, and A.P.J. Trinci. 1987a. The life cycle and growth kinetics of an anaerobic rumen fungus. Journal of General Microbiology 133:1815–1827.

Lowe, S.E, M.K. Theodorou, and A.P.J. Trinci. 1987b. Growth and fermentation of an anaerobic fungus on various carbon sources and effect of temperature on development. Applied and Environmental Microbiology 53:1210–1215.

——. 1987c. Isolation of anaerobic fungi from saliva and faeces of sheep. Journal of General Microbiology 133:1829–1834.

Lowe, S.E., M.K. Theodorou, A.P.J. Trinci, and R.B. Hespell. 1985. Growth of anaerobic rumen fungi on defined and semi-defined media lacking rumen fluid. Journal of General Microbiology 131:2225–2229.

Lowen, R. 1995. Acremonium section Lichenoidea section nov. and Pronectria oligospora species nov. Mycotaxon 53:81–95.

Lowen, R., B.L. Brady, D.L. Hawksworth, and R.R.M. Paterson. 1986. Two new lichenicolous species of *Hobsonia*. Mycologia 78:842–846.

Lowy, B. 1952. The genus *Auricularia*. Mycologia 44:656–692.

——. 1968. Taxonomic problems in the Heterobasidiomycetes. Taxon 17:118–127.

——. 1971. Tremellales. Flora Neotropica Monograph 6:1–153.

Lubinsky, G. 1955a. On some parasites of parasitic protozoa. I. *Sphaerita hoari* sp. n.—a chytrid parasitizing *Eremoplastron bovis*. Canadian Journal of Microbiology 1:440–450.

——. 1955b. On some parasites of parasitic protozoa. II. *Sagittospora cameroni* gen. n., sp. n.—a phycomycete parasitizing Ophryoscolecidae. Canadian Journal of Microbiology 1:675–684.

Lucas, J.A., R.C. Shattock, D.S. Shaw, and L. Cooke (eds.). 1991. Phytophthora. Cambridge University Press, Cambridge, England.

Lücking, R. 1995. Biodiversity and conservation of foliicolous lichens in Costa Rica. Mitteilungen der Eidgenössischen Forschungsanstalt für Wald, Schnee und Landschaft 70:63–92.

Lücking, R., and A. Lücking. 1996. Foliicolous bryophytes and lichens. Pp. 67–72. *In* S.R. Gradstein, P. Hietz, R. Lücking, A. Lücking, H.J.M. Sipman, H.F.M. Vester, J.H.D. Wolf, and E. Gardette, How to Sample the Epiphytic Diversity of Tropical Rain Forests. Ecotropica 2:59–72.

Lücking, R., P. Diederich, and J. Etayo, 1999. Checklist of foliicolous lichens and their lichenicolous fungi. Part II: Lichenicolous fungi (extended version). http://www.uni-bayreuth.de/departments/planta2/ass/robert/lichens/homepage.html

Ludwig, J.A., and J.F. Reynolds. 1988. Statistical Ecology: A Primer on Methods and Computing. John Wiley, New York.

Luff, M.L., M.D. Eyre, and S.P. Rushton. 1989. Classification and ordination of habitats of ground beetles (Coleoptera, Carabidae) in North-East England. Journal of Biogeography 16:121–130.

Lumbsch, H.T. 1998. The use of metabolic data in lichenology at the species and subspecific levels. Lichenologist 30:357–367.

Lumsden, R.D. 1981. Ecology of mycoparasitism. Pp. 295–318. *In* D.T. Wicklow and G.C. Carroll (eds.), The Fungal Community, Its Organization and Role in the Ecosystem. Marcel Dekker, New York.

——. 1992. Mycoparasitism of soilborne plant pathogens. Pp. 275–293. *In* G.C. Carroll and D.T. Wicklow (eds.), The Fungal Community: Its Organization and Role in the Ecosystem. 2nd ed. Marcel Dekker, New York.

Lumsden, R.D., and J.A. Lewis. 1989. Selection, production, formulation and commercial use of plant disease biocontrol fungi: problems and progress. Pp. 171–190. *In* J.M. Whipps and R.D. Lumsden (eds.), Biotechnology of Fungi for Improving Plant Growth. Cambridge University Press, Cambridge, England.

Lund, A. 1974. Yeasts and moulds in the bovine rumen. Journal of General Microbiology 81:453–462.

Lundborg, A., and T. Unestam. 1980. Antagonism against *Fomes annosus*. Comparison between different test methods in vitro and in vivo. Mycopathologia 70:107–115.

Lundqvist, N. 1972. Nordic Sordariaceae s. lat. Symbolae Botanicae Upsalienses 20:1–374.

——. 1980. On the genus *Pyxidiophora* sensu lato (Pyrenomycetes). Botaniska Notiser 133:121–144.

Luoma, D.L. 1988. Biomass and Community Structure of Sporocarps Formed by Hypogeous Ectomycorrhizal Fungi within Selected Forest Habitats of the H.J. Andrews Experimental Forest, Oregon. Unpubl. Ph.D. dissert., Oregon State University, Corvallis, Oregon.

——. 1991. Annual changes in seasonal production of hypogeous sporocarps in Oregon Douglas-fir forests. Pp. 83–89. *In* L.F. Ruggiero, K.B. Aubry, A.B. Carey, and M.H. Huff (eds.), Wildlife Habitat Relationships in Old-Growth Douglas-fir Forests. U.S. Department of Agriculture, Forest Service, Pacific Northwest Research Station, Portland, Oregon. General Technical Report PNW-BGTR-285.

Luoma, D.L., J.L. Eberhart, and M.P. Amaranthus. 1996a. Biodiversity of ectomycorrhizal types from southwest Oregon. Pp. 249–253. *In* T.N. Kaye, A. Liston, R.M. Love, D.L. Luoma, and M.V. Wilson (eds.), Conservation and Management of Native Plants and Fungi. Native Plant Society of Oregon, Corvallis, Oregon.

——. 1996b. Response of ectomycorrhizal fungi to forest management treatments—sporocarp production. Pp. 553–556. *In* C. Azcon-Aguilar and J.M. Barea (eds.), Mycorrhizas in Integrated Systems from Genes to Plant Development. Office for Official Publications of the European Communities, Luxembourg.

——. 1996c. Response of ectomycorrhizal fungi to forest management treatments: implications for long-term ecosystem productivity. Pp. 23–26. *In* D. Pilz and R. Molina (eds.), Managing Forest Ecosystems to Conserve Fungus Diversity and Sustain Wild Mushroom Harvests. U.S. Department of Agriculture, Forest Service, Pacific Northwest Research Station, Portland, Oregon. General Technical Report PNW-GTRB-371.

——. 1996d. Community structure and dynamics of ectomycorrhizal fungi in managed forest stands: demonstration of ecosystem management options (DEMO) program. Pp. 27–31. *In* D. Pilz and R. Molina (eds.), Managing Forest Ecosystems to Conserve Fungus Diversity and Sustain Wild Mushroom Harvests. U.S. Department of Agriculture, Forest Service, Pacific Northwest Research Station, Portland, Oregon. General Technical Report PNW-GTRB-371.

——. 1997. Biodiversity of ectomycorrhizal types from Southwest Oregon. Pp. 249–253. *In* T.N. Kaye, A. Liston, R.M. Love, D.L. Luoma, R.J. Meinke, and M.V. Wilson (eds.), Conservation and Management of Native Plants and Fungi. Native Plant Society of Oregon, Corvallis, Oregon.

Luoma, D.L., R.E. Frenkel, and J.M. Trappe. 1991. Fruiting of hypogeous fungi in Oregon Douglas-fir forests: seasonal and habitat variation. Mycologia 83:335–353.

Lussenhop, J. 1981. Analysis of microfungal component communities. Pp. 37–45. *In* D.T. Wicklow and G.C. Carroll (eds.), The Fungal Community, Its Organization and Role in the Ecosystem. Marcel Dekker, New York.

Lutchmeah, R.S., and R.C. Cooke. 1984. Aspects of antagonism by the mycoparasite *Pythium oligandrum*. Transactions of the British Mycological Society 83:696–700.

——. 1985. Pelleting of seed with the antagonist *Pythium oligandrum* for biological control of damping-off. Plant Pathology 34:528–531.

Lüth, P. 1998. Entwicklüng, Zulassung und Markteinführung eines biologischen Pflanzenschutzmittels am Beispiel von Contans (R)WG. Pp. 145–160. *In* A. Dowe (ed.), Symposium 50 Jahre Phytomedizin an der Universität Rostock. Universität Rostock, Rostock, Germany.

Luttrell, E.S. 1951. Taxonomy of the Pyrenomycetes. University of Missouri Studies 24:1–120.

——. 1955. The ascostromatic Ascomycetes. Mycologia 47:511–532.

——. 1965. Paraphysoids, pseudoparaphyses, and apical paraphyses. Transactions of the British Mycological Society 48:135–144.

——. 1973. Loculoascomycetes. Pp. 135–219. *In* G.C. Ainsworth, F.K. Sparrow, and A.S. Sussman (eds.), The Fungi. An Advanced Treatise. IVA. A Taxonomic Review with Keys: Ascomycetes and Fungi Imperfecti. Academic Press, New York.

Lutzoni, F., M. Pagel, and V. Reeb. 2001. Major fungal lineages are derived from lichen symbiotic ancestors. Nature 411:937–940.

Lynch, M., and B.G. Milligan. 1994. Analysis of population genetic structure with RAPD markers. Molecular Ecology 3:91–99.

MacArthur, R.H. 1965. Patterns of species diversity. Biological Review 40:510–533.

MacArthur, R.H., and E.O. Wilson. 1967. The Theory of Island Biogeography. Monographs in Population Biology No. 1.

Macauley, B.J., and L.B. Thrower. 1966. Succession of fungi in leaf litter of *Eucalyptus regnans*. Transactions of the British Mycological Society 49:509–520.

Macbride, T.H. 1899. The North American Slime-Moulds, being a list (descriptions) of myxomycetes hitherto described from North America, including Central America. Macmillan, New York.

——. 1922. The North American Slime-Moulds. Macmillan, New York.

Macbride, T.H., and G.W. Martin. 1934. The Myxomycetes. Macmillan, New York.

Maclean, D.J., K.S. Braithwaite, J.A.G. Irwin, J.M. Manners, and J.V. Groth. 1995. Random amplified polymorphic DNA reveals relationships among diverse genotypes in Australian and American collections of *Uromyces appendiculatus*. Phytopathology 85:757–765.

Madden, J.L., and M.P. Coutts. 1979. The role of fungi in the biology and ecology of woodwasps. Pp. 165–174. *In* L.R. Batra (ed.), Insect-Fungus Symbiosis, Nutrition, Mutualism, and Commensalism. Allanheld Osmun, Montclair, New Jersey.

Maddy, K.T., and H.G. Crecelius. 1967. Establishment of *Coccidioides immitis* in negative soils following burial of infected animals and animal tissues. Pp. 309–312. *In* L. Ajello (ed.), Coccidioidomycosis. University of Arizona Press, Tucson, Arizona.

Madelin, M.F. 1968. Fungi parasitic on other fungi and lichens. Pp. 253–269. *In* G.C. Ainsworth and A.S. Sussman (eds.), The Fungi. An Advanced Treatise. III. The Fungal Population. Academic Press, New York.

——. 1984. Myxomycetes, microorganisms and animals: A model of diversity in animal-microbial interactions. Pp. 1–33. *In* J.M. Anderson, A.D.M. Rayner, and D.W.H. Walton (eds.), Invertebrate-Microbial Interactions. Cambridge University Press, Cambridge, United Kingdom.

Madelin, M.F., and A. Feest. 1982. *Dipodascus macrosporus* sp. nov. (Hemiascomycetes), associated with plasmodia of *Badhamia utricularis*. Transactions of the British Mycological Society 79:331–335.

Madi, L., T. Katan, and Y. Henis. 1992. Inheritance of antagonistic properties and lytic enzyme activities in sexual crosses of *Talaromyces flavus*. Annals of Applied Biology 121:565–576.

Maerz, A., and M.R. Paul. 1950. A Dictionary of Color. 2nd ed. McGraw-Hill, New York.

Magan, N., and J. Lacey. 1984a. Effect of temperature and pH on water relations of field and storage fungi. Transactions of the British Mycological Society 82:71–81.

——. 1984b. Effect of water activity, temperature and substrate on interactions between field and storage fungi. Transactions of the British Mycological Society 82:83–93.

Magan, N., and J.M. Whipps. 1988. Growth of *Coniothyrium minitans*, *Gliocladium roseum*, *Trichoderma harzianum* and *T. viride* from alginate pellets and interactions with water availability. EPPO [European and Mediterranean Plant Protection Organization] Bulletin 18:37–45.

Magee, B.B., T.M. D'Souza, and P.T. Magee. 1987. Strain and species identification by restriction fragment length polymorphisms in the ribosomal DNA repeat of *Candida* species. Journal of Bacteriology 169:1639–1643.

Maggi, O., and A.M. Persiani. 1992. Études comparatives sur les microchampignons en écosystèmes tropicaux. Rapport final sur les recherches mycologiques du sol. Mycologia Helvetica 5:79–98.

Maggi, O., A.M. Persiani, M.A. Casada, and F.D. Pineda. 1990. Edaphic mycoflora recovery in tropical forests after shifting cultivation. Acta Oecologica 11:337–350.

Magnes, M., and J. Hafellner. 1991. Ascomyceten auf Gefäßpflanzen an Ufern von Gebirgsseen in den Ostalpen. Bibliotheca Mycologica 139:1–182.

Magurran, A.E. 1988. Ecological Diversity and Its Measurement. Croom Helm, London, England.

Maheshwari, R., and A. Antony. 1974. A selective technique for the isolation of *Neurospora crassa* from soil. Journal of General Microbiology 81:505–507.

Mahgoub, E.S., and I.G. Murray. 1973. Mycetoma. W. Heinemann Medical Books, London, Great Britain.

Mahoney, D.P. 1972. Soil and Litter Microfungi of the Galapagos Islands. Unpubl. Ph.D. dissert., University of Wisconsin, Madison, Wisconsin.

Mahoney, D.P., L.H. Huang, and M.P. Backus. 1969. New homothallic Neurosporas from tropical soils. Mycologia 61:264–272.

Majerus, M.E.N. 1989. Melanic polymorphism in the peppered moth, *Biston betularia*, and other Lepidoptera. Journal of Biological Education 23:267–284.

Majewski, T. 1994. The Laboulbeniales of Poland. Polish Botanical Studies Vol. 7. Pp. 3–466. Polish Academy of Sciences, Krakow, Poland.

Malathrakis, N.E. 1985. The fungus *Acremonium alternatum* Linc: Fr., a hyperparasite of the cucurbits powdery mildew *Sphaerotheca fuliginia*. Zeitschrift für Pflanzenkrankheiten und Pflanzenschutz 92:509–515.

Malcolm, W.M., and D.J. Galloway. 1997. New Zealand Lichens. Checklist, Key, and Glossary. Museum of New Zealand, Wellington, New Zealand.

Malençon, M.G. 1942. Études de parasitisme mycopathologique. I. Sur une propriété mycétophage du *Claudopus byssisedus*. II. Un Hyphomycète parasite de *l'Endogone microcarpa* Tul. Revue de Mycologie 7(N.S.):27–52.

Malik, K.A., K. Sultana, F. Wajid, and Farooq-e-Azam. 1982. Taxonomy of cellulolytic fungi isolated from salt affected soils. Biology of Saline Soils, Final Technical Report. Soil Biology Division, Nuclear Institute for Agriculture and Biology. Faisalabad, Pakistan.

Malloch, D. 1981. Moulds. Their Isolation, Cultivation, and Identification. University of Toronto Press, Toronto, Ontario, Canada.

——. 1995. Fungi with heteroxenous life histories. Canadian Journal of Botany 73 (Suppl. 1):S1334–S1342.

Malloch, D., and M. Blackwell. 1992. Dispersal of fungal diaspores. Pp. 147–171. *In* G.C. Carroll and D.T. Wicklow (eds.), The Fungal Community: Its Organization and Role in the Ecosystem. 2nd ed. Marcel Dekker, New York.

——. 1993a. Life histories of three undescribed species of *Pyxidiophora* occurring on beached marine algae. Inoculum 44:48.

——. 1993b. Dispersal biology of Ophiostomatoid fungi. Pp. 195–206. *In* M.J. Wingfield, K.A. Seifert, and J.F. Webber (eds.), *Ceratocystis* and *Ophiostoma*: Taxonomy, Ecology and Pathogenicity. American Phytopathological Society, St. Paul, Minnesota.

Malloch, D., and R.F. Cain. 1971. New genera of Onygenaceae. Canadian Journal of Botany 49:839–846.

Malloch, D., and C.T. Rogerson. 1978. Fungi of the Canadian boreal forest region: *Catulus aquilonius* gen. et sp. nov., a hyperparasite on *Seuratia millardetii*. Canadian Journal of Botany 56:2344–2347.

Mandel, M., and J. Marmur 1968. Use of ultraviolet absorbance-temperature profile for determining the guanine plus cytosine content of DNA. Pp. 195–206. *In* L. Grossman and K. Moldave (eds.), Nucleic Acids Part B. Methods in Enzymology, Vol. 12. Academic Press, New York.

Manion, P.D. 1981. Tree Disease Concepts. Prentice-Hall, Englewood Cliffs, New Jersey.

Manion, P.D., and D.H. Griffin. 1986. Sixty-five years of research on *Hypoxylon* canker of aspen. Plant Disease 70:803–808.

Mannarelli, B.M., and C.P. Kurtzman. 1998. Rapid identification of *Candida albicans* and other human pathogenic yeasts by using short oligonucleotides in a PCR. Journal of Clinical Microbiology 36:1634–1641.

Manocha, M.S. 1981. Host specificity and mechanism of resistance in a mycoparasitic system. Physiological Plant Pathology 18:257–265.

——. 1987. Cellular and molecular aspects of fungal host-mycoparasite interaction. Zeitschrift für Pflanzenkrankheiten und Pflanzenschutz 94:431–444.

——. 1991. Physiology and biochemistry of biotrophic mycoparasitism. Pp. 273–300. *In* D.K. Arora, B. Rai, K.G. Mukerji, and G. Knudsen (eds.), Handbook of Applied Mycology. Vol. I. Soil and Plants. Marcel Dekker, New York.

Manocha, M.S., and J.M. Deven. 1975. Host-parasite relations in a mycoparasite IV. A correlation between the levels of γ-linolenic acid and parasitism of *Piptocephalis virginiana*. Mycologia 67:1148–1157.

Manocha, M.S., and R. Golesorkhi. 1979. Host-parasite relations in a mycoparasite. V. Electron microscopy of *Piptocephalis virginiana* infection in compatible and incompatible hosts. Mycologia 71:565–576.

——. 1981. Host-parasite relations in a mycoparasite. 7. Light and scanning electron microscopy of interactions of *Piptocephalis virginiana* with host and nonhost species. Mycologia 73:976–987.

Manocha, M.S., and L.L. Graham. 1982. Host cell wall synthesis and its role in resistance to a mycoparasite. Physiological Plant Pathology 20:157–164.

Manocha, M.S., and A.S. Sahai. 1991. Cell surface characterization of five haustorial mycoparasites with lectin probes. Mycologia 83:643–649.

——. 1993. Mechanisms of recognition in necrotrophic and biotrophic mycoparasites. Canadian Journal of Microbiology 39:269–275.

Mao, J., B. Appel, J. Schaack, S. Sharp, H. Yamada, and D. Soll. 1982. The 5S RNA genes of *Schizosaccharomyces pombe*. Nucleic Acids Research 10:487–500.

Marcelli, M.P. 1992. Ecologia Liquenica nos Manguezais do Sul-Sudeste Brasileiro. Bibliotheca Lichenologica 47:1–310.

Margalef, R. 1958. Information theory in ecology. General Systematics 3:36–71.

——. 1972. Homage to Evelyn Hutchinson, or why there is an upper limit to diversity. Pp. 211–235. *In* E.S. Deevey (ed.), Growth by Intussusception; Ecological Essays in Honor of G. Evelyn Hutchinson. Transactions. Connecticut Academy of Arts and Sciences Vol. 44:1–443.

Margulis, L., J.O. Corliss, M. Melkonian, and D. Chapman (eds.). 1990. Handbook of Protoctista. Jones and Bartlett, Boston, Massachusetts.

Marin, A.B., L.M. Libbey, and M.E. Morgan. 1984. Truffles: on the scent of buried treasure. McIlvainea 6:34–38.

Marois, J.J., D.R. Fravel, and G.C. Papavizas. 1984. Ability of *Talaromyces flavus* to occupy the rhizosphere and its interaction with *Verticillium dahliae.* Soil Biology and Biochemistry 16:387–390.

Marounek, M., and B. Hodrova. 1989. Susceptibility and resistance of anaerobic rumen fungi to antimicrobial feed additives. Letters in Applied Microbiology 9:173–175.

Marr, C.D. 1979. Laccase and tyrosinase oxidation of spot test reagents. Mycotaxon 9:244–276.

Marr, C.D., D.W. Grund, and K.A. Harrison. 1986. The taxonomic potential of laccase and tyrosinase spot tests. Mycologia 78:169–184.

Marschner, H. 1998. Role of root growth, arbuscular mycorrhiza, and root exudates for the efficiency in nutrient acquisition. Field Crops Research 56:203–207.

Martin, F.N. 1992. *Pythium.* Pp. 39–49. *In* L.L. Singleton, J.D. Mihail, and C.M. Rush (eds.), Methods for Research on Soilborne Phytopathogenic Fungi. American Phytopathological Society Press, St. Paul, Minnesota.

Martin, F.N., and J.G. Hancock. 1987. The use of *Pythium oligandrum* for biological control of preemergence damping-off caused by *P. ultimum.* Phytopathology 77:1013–1020.

Martin, G.W. 1949. Fungi. Myxomycetes. North American Flora 1:1–151.

——. 1952. Revision of the North Central tremellales. University of Iowa Studies in Natural History 19:1–122.

Martin, G.W., and C.J. Alexopoulos. 1969. The Myxomycetes. University of Iowa Press, Iowa City, Iowa.

Martin, G.W., C.J. Alexopoulos, and M.L. Farr. 1983. The Genera of Myxomycetes. University of Iowa Press, Iowa City, Iowa.

Martin, M.M. 1987. Invertebrate-Microbial Interactions: Ingested Fungal Enzymes in Arthropod Biology. Cornell University Press, Ithaca, New York.

Martin, W.W. 1987. Zoosporic parasites of aquatic insects: collection, identification, and culture. Pp. 137–142. *In* M.S. Fuller and A. Jaworski (eds.), Zoosporic Fungi in Teaching and Research. Southeastern Publishing, Athens, Georgia.

Martínez, I., and J. Hafellner. 1998. Lichens and lichenicolous fungi on Peltigerales in the Iberian Peninsula and the Canary Islands. Mycotaxon 69:271–310.

Marvanová, L. 1977. A contact biotrophic mycoparasite on aquatic hyphomycete conidia. Transactions of the British Mycological Society 68:485–488.

Marvanová, L., and R.J. Bandoni. 1987. *Naiadella fluitans* gen. et sp. nov.: a conidial basidiomycete. Mycologia 79:578–586.

Marvanová, L., and P.J. Fisher. 1991. A new endophytic hyphomycete from alder roots. Nova Hedwigia. 52:33–37.

Marvanová, L., and K. Suberkropp. 1990. *Camptobasidium hydrophilum* and its anamorph, *Crucella subtilis*: a new Heterobasidiomycete from streams. Mycologia 82:208–217.

Marvin-Sikkema, F.D., A.J. Richardson, C.S. Stewart, J.C. Gottschal, and R.A. Prins. 1990. Influence of hydrogen-consuming bacteria on cellulose degradation by anaerobic fungi. Applied and Environmental Microbiology 56:3793–3797.

Marx, D.H., and W.J. Daniel. 1976. Maintaining cultures of ectomycorrhizal and plant pathogenic fungi in sterile water cold storage. Canadian Journal of Microbiology 22:338–341.

Marx, H. 1992. Myxomyceten auf Dung—ein Versuch zur Erklärung der Sporenherkunft. Boletus 16:111–114.

Masclaux, F., E. Gueho, G.S. de Hoog, and R. Christen. 1995. Phylogenetic relationships of human-pathogenic *Cladosporium* (*Xylohypha*) species inferred from partial LS rRNA sequences. Journal of Medical and Veterinary Mycology 33:327–338.

Maser, C., Z. Maser, J.W. Witt, and G. Hunt. 1986. The northern flying squirrel: a mycophagist in southwestern Oregon. Canadian Journal of Zoology 64:2086–2089.

Maser, C., J.M. Trappe, and R.A. Nussbaum. 1978. Fungal-small mammal interrelationships with emphasis on Oregon coniferous forests. Ecology 59:799–809.

Maser, Z., C. Maser, and J.M. Trappe. 1985. Food habits of the northern flying squirrel (*Glaucomys sabrinus*) in Oregon. Canadian Journal of Zoology 63:1084–1088.

Mason, P.A., M.O. Musoko, and F.T. Last. 1992. Short-term changes in vesicular-arbuscular mycorrhizal spore populations in *Terminalia* plantations in Cameroon. Pp. 261–267. *In* D.J. Read, D.H. Lewis, A.H. Fitter, and I.J. Alexander (eds.), Mycorrhizas in Ecosystems. CAB International, Wallingford, Oxon, United Kingdom.

Mason, P.A., J. Wilson, F.T. Last, and C. Walker. 1983. The concept of succession in relation to the spread of sheathing mycorrhizal fungi on inoculated tree seedlings growing in unsterile soils. Plant and Soil 71:247–256.

Massee, G., and E.S. Salmon. 1902. Researches on coprophilous fungi. II. Annals of Botany 16:57–93.

Masters, M.J. 1976. Freshwater phycomycetes on algae. Pp. 489–512. *In* E.B. Gareth Jones (ed.), Recent Advances in Aquatic Mycology. Elek Science, London, England.

Mathiassen, G. 1993. Corticolous and lignicolous Pyrenomycetes s. lat. (Ascomycetes) on *Salix* along a mid-Scandinavian transect. Sommerfeltia 20:1–180.

Matsushima, T. 1971. Microfungi of the Solomon Islands and Papua-New Guinea. Published by the author, Kobe, Japan.

——. 1975. Icones microfungorum a Matushima lectorum. Published by the author, Kobe, Japan.

——. 1995. Saprophytic microfungi from Taiwan. Hyphomycetes. Matsushima's Mycological Memoirs Vol. 8. Published by the author, Kobe, Japan.

Mattirolo, O. 1907. Gli autoptici di Carlo Vittadini e la loro importanza nello studio della idnologia. Società Italiana di Scienze Naturali VS:1–396.

Mattsson, J-E., and H.T. Lumbsch. 1989. The use of the species pair concept in lichen taxonomy. Taxon 38:238–241.

Matzer, M. 1996. Lichenicolous ascomycetes with fissitunicate asci on foliicolous lichens. Mycological Papers 171:1–202.

Matzer, M., and J. Hafellner. 1990. Eine Revision der lichenicolen Arten der Sammelgattung Rosellina (Ascomycetes). Bibliotheca Lichenologica 37:1–138.

Mavridou, A., and M.A. Typas. 1998. Intraspecific polymorphism in *Metarhizium anisopliae* var. *anisopliae* revealed by analysis of rRNA gene complex and mtDNA RFLPs. Mycological Research 102:1233–1241.

May, R.M. 1975. Patterns of species abundances and diversity. Pp. 81–120. *In* M.L. Cody and J.M. Diamond (eds.), Ecology and Evolution of Communities. Harvard University Press, Cambridge, Massachusetts.

——. 1981. Theoretical Ecology: Principles and Applications. 2nd ed. Blackwell Scientific, Boston, Massachusetts.

——. 1991. A fondness for fungi. Nature 352:475–476.

McCallum, S.H. 1986. American National Standard for Information Sciences–Bibliographic Information Interchange. Publication ANSI Z39.2–1985. American National Standards Institute, New York.

McCarroll, R., G.J. Olsen, Y.D. Stahl, C.R. Woese, and M.L. Sogin. 1983. Nucleotide sequence of the *Dictyostelium discoideum* small-subunit ribosomal ribonucleic acid inferred from the gene sequence: evolutionary implications. Biochemistry 22:5858–5868.

McCredie, T.A., and K. Sivasithamparam. 1985. Fungi mycoparasitic on sclerotia of *Sclerotinia sclerotiorum* in some Western Australian soils. Transactions of the British Mycological Society 84:736–739.

McCune, B. 1990. Rapid estimation of abundance of epiphytes on branches. Bryologist 93:39–43.

——. 1993. Gradients in epiphyte biomass in three *Pseudotsuga-Tsuga* forests of different ages in western Oregon and Washington. Bryologist 96:405–411.

——. 1994. Using epiphyte litter to estimate epiphyte biomass. Bryologist 97:396–401.

——. 2000. Lichen communities as indicators of forest health. Bryologist 103:353–356.

McCune, B., and J.A. Antos. 1981a. Diversity relationships of forest layers in the Swan Valley, Montana. Bulletin of the Torrey Botanical Club 108:354–361.

——. 1981b. Correlations between forest layers in the Swan Valley, Montana. Ecology 62:1196–1204.

McCune, B., and P. Lesica. 1992. The trade-off between species capture and quantitative accuracy in ecological inventory of lichens and bryophytes in forests in Montana. Bryologist 95:296–304.

McCune, B., J.P. Dey, J.E. Peck, D. Cassell, K. Heiman, S. Will-Wolf, and P.N. Neitlich. 1997a. Repeatability of community data: species richness versus gradient scores in large-scale lichen studies. Bryologist 100:40–46.

McCune, B., J. Dey, J. Peck, K. Heiman, and S. Will-Wolf. 1997b. Regional gradients in lichen communities of the southeast United States. Bryologist 100:145–158.

McCune, B., R. Rosentreter, J.M. Ponzetti, and D.C. Shaw. 2000. Epiphyte Habitats in an Old Conifer Forest in Western Washington, U.S.A. Bryologist 103:417–427.

McCurdy, H.D. 1969. Study on the taxonomy of the Myxobacteriales. I. Record of Canadian isolates and survey of methods. Canadian Journal of Microbiology 15:1453–1461.

McCutcheon, T.L., G.C. Carroll, and S. Schwab. 1993. Genotypic diversity in populations of a fungal endophyte from Douglas fir. Mycologia 85:180–186.

McDowell, E.M., and B.F. Trump. 1976. Histologic fixatives suitable for diagnostic light and electron microscopy. Archives of Pathology and Laboratory Medicine 100:405–414.

McFadden, G.I., I. Bonig, E.C. Cornish, and A.E. Clarke. 1988. A simple fixation and embedding method for use in hybridisation histochemistry on plant tissues. Histochemical Journal 20:575–586.

McGee, P.A. 1989. Variation in propagule numbers of vesicular-arbuscular mycorrhizal fungi in a semi-arid soil. Mycological Research 92:28–33.

McGinnis, M.R. 1980. Laboratory Handbook of Medical Mycology. Academic Press, New York.

McGinnis, M.R., A.A. Padhye, and L. Ajello. 1974. Storage of stock cultures of filamentous fungi, yeasts, and some aerobic actinomycetes in sterile distilled water. Applied and Environmental Microbiology 28:218–222.

McGonigle, T.P., and A.H. Fitter. 1990. Ecological specificity of vesicular-arbuscular mycorrhizal associations. Mycological Research 94:120–122.

McKay, G.J., D. Egan, E. Morris, C. Scott, and A.E. Brown. 1999. Genetic and morphological characterization of *Cladobotryum* species causing cobweb disease of mushrooms. Applied and Environmental Microbiology 65:606–610.

McKenney, M.C., and D.L. Lindsey. 1987. Improved method for quantifying endomycorrhizal fungi spores from soil. Mycologia 79:779–782.

McKerracher, L.J., and I.B. Heath. 1985. The structure and cycle of the nucleus-associated organelle in two species of *Basidiobolus*. Mycologia 77:412–417.

McKnight, K.H. 1985. The small-spored species of *Podaxis*. Mycologia 77:24–35.

McKnight, K.H., and V.B. McKnight. 1987. A Field Guide to Mushrooms: North America. Houghton Mifflin, Boston, Massachusetts.

McLaren, D.L., H.C. Huang, and S.R. Rimmer. 1986. Hyperparasitism of *Sclerotinia sclerotiorum* by *Talaromyces flavus*. Canadian Journal of Plant Pathology 8:43–48.

McLaren, D.L., H.C. Huang, G.C. Kozub, and S.R. Rimmer. 1994. Biological control of Sclerotinia wilt of sunflower with *Talaromyces flavus* and *Coniothyrium minitans*. Plant Disease 78:231–235.

McLaren, D.L., H.C. Huang, S.R. Rimmer, and E.G. Kokko. 1989. Ultrastructural studies on infection of sclerotia of *Sclerotinia sclerotiorum* by *Talaromyces flavus*. Canadian Journal of Botany 67:2199–2205.

McLaughlin, D.J., M.E. Berres, and L.J. Szabo. 1995. Molecules and morphology in basidiomycete phylogeny. Canadian Journal of Botany 73(Suppl. 1):S684–S692.

McLaughlin, D.J., E.G. McLaughlin, and P.A. Lempke. 2001. The Mycota. Vol. VII. Sytematics and Evolution. Parts A and B. Springer-Verlag, Berlin, Germany.

McLean, J.A. 1985. Ambrosia beetles: A multimillion dollar degrade problem of sawlogs in coastal British Columbia. Forestry Chronicles 61:295–298.

McLean, N., and D. Porter. 1982. The yellow-spot disease of *Tritonia diomeda* Bergh, 1894 (Mollusca: Gastropoda: Nudibranchia): encapsulation of the thraustochytriaceous parasite by host amoebocytes. Journal of Parasitology 68:243–252.

McNabb, R.F.R. 1964a. New Zealand Tremellales–I. New Zealand Journal of Botany 2:403–414.

——. 1964b. Taxonomic studies in the Dacrymycetaceae. I. *Cerinomyces* Martin. New Zealand Journal of Botany 2:415–424.

——. 1965a. Taxonomic studies in the Dacrymycetaceae. II. *Calocera* (Fries) Fries. New Zealand Journal of Botany 3:31–58.

——. 1965b. Taxonomic studies in the Dacrymycetaceae. III. *Dacryopinax* Martin. New Zealand Journal of Botany 3:59–72.

——. 1965c. Taxonomic studies in the Dacrymycetaceae. IV. *Guepiniopsis* Patouillard. New Zealand Journal of Botany 3:159–167.

——. 1965d. Taxonomic studies in the Dacrymycetaceae. V. *Heterotextus* Lloyd. New Zealand Journal of Botany 3:215–222.

——. 1965e. Taxonomic studies in the Dacrymycetaceae. VI. *Femsjonia* Fries. New Zealand Journal of Botany 3:223–228.

——. 1966a. New Zealand Tremellales–II. New Zealand Journal of Botany 4:533–545.

——. 1966b. Taxonomic studies in the Dacrymycetaceae. VII. *Ditiola* Fries. New Zealand Journal of Botany 4:546–558.

——. 1969. New Zealand Tremellales–III. New Zealand Journal of Botany 7:241–261.

——. 1973. Taxonomic studies in the Dacrymycetaceae. VIII. *Dacrymyces* Nees ex Fries. New Zealand Journal of Botany 11:461–524.

McQuilken, M.P., J.M. Whipps, and R.C. Cooke. 1990. Oospores of the biocontrol agent *Pythium oligandrum* bulk-produced in liquid culture. Mycological Research 94:613–616.

McVicar, A.H. 1977. *Ichthyophonus* as a pathogen in farmed and wild fish. Bulletin de l'Office International des Épizooties 87:517–519.

——. 1979. *Ichthyophonus* in haddock and plaice in Scottish waters. ICES 1979/G:48, ICES [International Council for the Exploration of the Sea], Copenhagen, Denmark.

——. 1980. The effects of *Ichthyophonus* infection in haddock *Melanogrammus aeglefinus* and plaice *Pleuronectes platessa* in Scottish waters. International Council for the Exploration of the Sea, Special Meeting on Diseases of Commercially Important Marine Fish and Shellfish, Copenhagen, No 16.

——. 1981. An assessment of *Ichthyophonus* disease as a component of natural mortality in plaice populations in Scottish waters. ICES 1981/G:49, ICES [International Council for the Exploration of the Sea] Copenhagen, Denmark.

——. 1982. *Ichthyophonus* infections of fish. Pp. 243–269. *In* R.J. Roberts (ed.), Microbial Diseases of Fish. Academic Press, London, England.

Mehus, H. 1986. Fruit body production of macro-fungi in some north Norwegian forest types. Norwegian Journal of Botany 6:679–701.

Meier, R., and I. Charvat. 1993. Reassessment of tetrazolium bromide as a viability stain for spores of vesicular-arbuscular mycorrhizal fungi. American Journal of Botany 80:1007–1015.

Meijer, G., B. Megnegneau, and E.G.A. Linders. 1994. Variability for isozyme, vegetative compatibility and RAPD markers in natural populations of *Phomopsis subordinaria*. Mycological Research 98:267–276.

Mendgen, K. 1981. Growth of *Verticillium lecanii* in pustules of stripe rust (*Puccinia striiformis*). Phytopathologische Zeitschrift 102:301–309.

Mendgen, K., and R. Casper. 1980. Detection of *Verticillium lecanii* in pustules of bean rust (*Uromyces phaseoli*) by immunofluorescence. Phytopathologische Zeitschrift 99:362–364.

Mendonça-Hagler, L.C., A.N. Hagler, and C.P. Kurtzman. 1993. Phylogeny of *Metschnikowia* species estimated from partial rRNA sequences. International Journal of Systematic Bacteriology 43:368–373.

Menhinick, E.F. 1964. A comparison of some species-individuals diversity indices applied to samples of field insects. Ecology 45:859–861.

Merz, W.G., and G.D. Roberts. 1995. Detection and recovery of fungi from clinical specimens. Pp. 709–722. *In* P.R. Murray, E.J. Baron, M.A. Pfaller, F.C. Tenover, and R.H. Yolken (eds.), Manual of Clinical Microbiology. 6th ed. American Society for Microbiology Press, Washington, D.C.

Metzler, B., F. Oberwinkler, and H. Petzold. 1989. *Rhynchogastrema* gen. nov. and *Rhynchogastremaceae* fam. nov. Systematic and Applied Microbiology 12:280–287.

Meyer, F.H. 1973. Distribution of ectomycorrhizae in native and man-made forests. Pp. 79–105. *In* G.C. Marks and T.T. Kozlowski (eds.), Ectomycorrhizae: Their Ecology and Physiology. Academic Press, New York.

Micales, J.A., and R.J. Stipes. 1987. A reexamination of the fungal genera *Cyrphonectria* and *Endothia*. Phytopathology 77:650–654.

Michaelides, J., and B. Kendrick. 1982. The bubble-trap propagules of *Beverwykella*, *Helicoon* and other aero-aquatic fungi. Mycotaxon 14:247–260.

Michailides, T.J., and R.A. Spotts. 1988. Germination of zygospores of *Mucor piriformis* (on the life history of *Mucor piriformis*). Mycologia 80:837–844.

——. 1990. Postharvest diseases of stone and pome fruits caused by *Mucor piriformis* in the Pacific Northwest and California. Plant Disease 74:537–543.

Middelhoven, W.J. 1993. Catabolism of benzene compounds by ascomycetous and basidiomycetous yeasts and yeastlike fungi: a literature review and an experimental approach. Antonie van Leeuwenhoek 63:125–144.

Mietzsch, E., H.T. Lumbsch, and J.A. Elix. 1993. Notice: a new computer program for the identification of lichen substances. Mycotaxon 47:475–479.

Migaki, G., and S.R. Jones. 1983. Mycotic diseases in marine mammals. Pp. 1–27. *In* E.B. Howard (ed.), Pathobiology of Marine Mammal Diseases. Vol. 2. CRC Press, Boca Raton, Florida.

Migaki, G., E.R. Jacobson, and H.W. Casey. 1984. Fungal diseases in reptiles. Pp. 183–204. *In* G.L. Hoff, F.L. Frye, and E.R. Jacobson (eds.), Diseases of Amphibians and Reptiles. Plenum, New York.

Milanez, A.I. 1967. Resting spores of *Phlyctochytrium planicorne* on Saprolegniaceae. Transactions of the British Mycological Society 50:679–681.

Milgroom, M.G. 1996. Recombination and the multilocus structure of fungal populations. Annual Review of Phytopathology 34:457–477.

Milgroom, M.G., and P. Cortesi. 1999. Analysis of population structure of the chestnut blight fungus based on vegetative incompatibility genotypes. Proceedings of the National Academy of Sciences USA 96:10518–10523.

Miller, C.E. 1976. Substrate-influenced morphological variations and taxonomic problems in freshwater, posteriorly uniflagellate phycomycetes. Pp. 469–487. *In* E.B. Gareth Jones (ed.), Recent Advances in Aquatic Mycology. Elek Science, London, England.

Miller, D.D., P.A. Domoto, and C. Walker. 1985. Mycorrhizal fungi at eighteen apple rootstock plantings in the United States. New Phytologist 100:379–391.

Miller, J.D. 1995. Microbial germplasm conservation working group. Pp 54–57. *In* G. Baillargeon, D. Bardeau, Y. Bélanger, D. Leger, E.E. Lester, and D. Miller (eds.), Proceedings of the National Workshop on a Canadian Germplasm Network. Research Branch, Agriculture and Agri-Foods Canada, Ottawa, Ontario, Canada.

Miller, J.D., and L.C. Fleming. 1983. Fungi associated with an infestation of *Pseudocarcinonemertes homari* on *Homarus americanus*. Transactions of the British Mycological Society 80:9–12.

Miller, J.D., and N.J. Whitney. 1981. Fungi from the Bay of Fundy. III. Geofungi in the marine environment. Marine Biology 65:61–68.

Miller, J.H., J.E. Giddens, and A.A. Foster. 1957. A survey of the fungi of forest and cultivated soils of Georgia. Mycologia 49:779–808.

Miller, O.K., Jr. 1972. Mushrooms of North America. E.P. Dutton, New York.

——. 1983. Ectomycorrhizae in the Agaricales and Gasteromycetes. Canadian Journal of Botany 61:909–916.

Miller, O.K., Jr., and G.A. Laursen. 1978. Ecto- and endomycorrhizae of Arctic plants at Barrow, Alaska. Pp. 229–237. *In* L.L. Tieszen (ed.), Vegetation and Production Ecology of an Alaskan Arctic Tundra. Ecological Studies 29. Springer-Verlag, New York.

Miller, O.K., Jr., and H.H. Miller. 1988. Gasteromycetes. Morphological and Developmental Features with Keys to the Orders, Families and Genera. Mad River Press, Eureka, California.

Miller, R.M., and D.J. Lodge. 1997. Fungal responses to disturbance—Agriculture and Forestry. Pp. 65–84. *In* K. Esser, P.A. Lemke, and D.T. Wicklow (eds.), The Mycota. Vol. V. Environmental and Microbial Relationships. Springer-Verlag, Berlin, Germany.

Miller, R.M., and J.D. Jastrow. 1994. Vesicular-arbuscular mycorrhizae and biogeochemical cycling. Pp. 189–212. *In* F.L. Pfleger and R.G. Linderman (eds.), Mycorrhizae and Plant Health. American Phytopathological Society Press, St. Paul, Minnesota.

Miller, R.M., D.R. Reinhardt, and J.D. Jastrow. 1995. External hyphal production of vesicular-arbuscular mycorrhizal fungi in pasture and tallgrass prairie communities. Oecologia 103:17–23.

Miller, S.L. 1986. Hypogeous fungi from the southeastern United States. I. The genus *Rhizopogon*. Mycotaxon 27:193–218.

Miller, S.L., and E.B. Allen. 1992. Mycorrhizae, nutrient translocation, and interactions between plants. Pp. 301–332. *In* M.F. Allen (ed.), Mycorrhizal Functioning: An Integrative Plant-Fungal Process. Chapman Hall, New York.

Miller, S.L., T.M. McClean, J.F. Walker, and B. Buyck. 2001. A molecular phylogeny of the Russulales including agaricoid, gasteroid and pleurotoid taxa. Mycologia 93:344–354.

Miller, S.L., P. Torres, and T.M. McClean. 1994. Persistence of basidiospores and sclerotia of ectomycorrhizal fungi and *Morchella* in soil. Mycologia 86:89–95.

Millner, P.D., and D.G. Kitt. 1992. The Beltsville method for soilless production of vesicular-arbuscular mycorrhizal fungi. Mycorrhiza 2:9–15.

Milne, A., M.K. Theodorou, M.G.C. Jordan, C. King-Spooner, and A.P.J. Trinci. 1989. Survival of anaerobic fungi in feces, in saliva, and in pure culture. Experimental Mycology 13:27–37.

Minden, M. von. 1916. Beiträge zur Biologie und Systematik einheimischer submerser Phycomyceten. Pp. 146–255. *In* R. Falck, Mykologische Untersuchungen und Berichte, Heft 2. G. Fischer, Jena, Germany.

Minoura, K., and T. Muroi. 1978. Some freshwater Ascomycetes from Japan. Transactions of the Mycological Society of Japan 19:129–134.

Minter, D.W. 1981. *Lophodermium* on pines. Mycology Papers 147:1–54.

——. 1996. Recording and mapping fungi. Pp. 321–382. *In* B.C. Sutton (ed.), A Century of Mycology. Cambridge University Press, Cambridge, United Kingdom.

Minter, D.W., R. Lowen, and S. Diamandis. 1987. *Zeus olympius* gen. et sp. nov. and *Necttria ganymede* sp. nov. from Mount Olympus, Greece. Transactions of the British Mycological Society 88:55–61.

Mirza, J.H., S.M. Khan, S. Begum, and S. Shagufta. 1979. Mucorales of Pakistan. University of Agriculture, Faisalabad, Pakistan.

Misra, J.K., and R.W. Lichtwardt. 2000. Illustrated genera of Trichomycetes: fungal symbionts of insects and other arthropods. Science Publishers, Enfield, New Hampshire.

Misra, P.C., and K. Lata. 1979. Studies on Dimargaritaceae (Mucorales). II. A new *Dispira* parasitic on ascomycetous hosts. Mycotaxon 8:372–376.

Mitchell, C.P., C.S. Millar, and D.W. Minter. 1978. Studies on decomposition of Scots pine needles. Transactions of the British Mycological Society 71:343–348.

Mitchell, J.K., D.H. Smith, and R.A. Taber. 1987. Potential for biological control of *Cercosporidium personatum* leafspot of peanuts by *Dicyma pulvinata*. Canadian Journal of Botany 65:2263–2269.

Mitchell, J.K., R.A. Taber, and R.E. Pettit. 1986. Establishment of *Dicyma pulvinata* in *Cercosporidium personatum* leaf spot of peanuts: effect of spray formulation, inoculation time, and hours of leaf wetness. Phytopathology 76:1168–1171.

Mix, A.J. 1949. A monograph of the genus *Taphrina*. University of Kansas Science Bulletin 33:3–167.

——. 1954. Additions and emendations to a monograph of the genus Taphrina. Transactions of the Kansas Academy of Science 57:55–65.

Modjo, H.S., and J.W. Hendrix. 1986. The mycorrhizal fungus *Glomus macrocarpum* as a cause of Tobacco Stunt Disease. Phytopathology 76:688–691.

Molina, R., and J.G. Palmer. 1982. Isolation, maintenance, and pure culture manipulation of ectomycorrhizal fungi. Pp. 115–129. *In* N.C. Schenck (ed.), Methods and Principles of Mycorrhizal Research. American Phytopathology Society Press, St. Paul, Minnesota.

Molina, R., and J.M. Trappe. 1982. Patterns of ectomycorrhizal host specificity and potential among Pacific Northwest conifers and fungi. Forest Science 28:423–258.

Molina, R., H. Massicotte, and J.M. Trappe. 1992. Specificity phenomena in mycorrhizal symbiosis: community-ecological consequences and practical implications. Pp. 357–423. *In* M.F. Allen (ed.), Mycorrhizal Functioning: An Integrative Plant-Fungal Process. Chapman Hall, New York.

Molina, R., T. O'Dell, D. Luoma, M. Amaranthus, M. Castellano, and K. Russell. 1993. Biology, ecology, and social aspects of wild edible mushrooms in the forests of the Pacific Northwest: A preface to managing commercial harvest. U.S. Department of Agriculture, Forest Service, Pacific Northwest Research Station, Portland, Oregon. General Technical Report PNW-GTR-309.

Molina, R., D. Pilz, J. Smith, S. Dunham, T. Dreisbach, T. O'Dell, and M. Castellano. 2001. Conservation and management of forest fungi in the Pacific Northwestern United States: an integrated ecosystem approach. Pp. 19–63. *In* D. Moore, M.M. Nauta, S.E. Evans, and M. Rotheroe (eds.), Fungal Conservation, Issues and Solutions. Cambridge University Press, Cambridge, England.

Molitor, F., and P. Diederich. 1997. Les pyrénolichens aquatiques du Luxembourg et leurs champignons lichénicoles. Bulletin de la Société de Naturalistes de Luxembourg 98:69–92.

Möller, A. 1893. Die Pilzgärten einiger südamerikanischer Ameisen. G. Fischer, Jena, Germany.

Möller, H. 1974. *Ichthyosporidium hoferi* (Plehn et Muslow) (Fungi) as parasite in the Baltic cod (*Gadus morhua* L.). Kieler Meeresforschungen 30:37–41.

Moncalvo, J.-M., F.M. Lutzoni, S.A. Rehner, J. Johnson, and R. Vilgalys. 2000. Phylogenetic relationships of agaric fungi based on nuclear large subunit ribosomal DNA sequences. Systematic Biology 49:278–305.

Moncalvo, J.-M., R. Vilgalys, S.A. Redhead, J.E. Johnson, T.Y. James, M.C. Aime, V. Hofstetter, S.J.W. Verduin, E. Larsson, T.J. Baroni, R.G. Thorn, S. Jacobsson, H. Clémençon, and O.K. Miller. 2002. One hundred and seventeen clades of euagarics. Molecular Phylogenetics and Evolution 23:357–400.

Monreal, M., S.M. Berch, and M. Berbee. 1999. Molecular diversity of ericoid mycorrhizal fungi. Canadian Journal of Botany 77:1580–1594.

Montacchini, F., and R.C. Lomagno. 1977. Researches on *Tuber melanosporum* ecology II. Inhibitory action on wild erbaceous species. Allionia 22:81–85.

Montfoort, D., and R.C. Ek. 1990. Vertical Distribution and Ecology of Epiphytic Bryophytes and Lichens in a Lowland Rain Forest in French Guiana. Herbarium, Institute of Systematic Botany, Utrecht, The Netherlands.

Montrocher, R., M.-C. Verner, J. Briolay, C. Gautier, and R. Marmeisse. 1998. Phylogenetic analysis of the *Saccharomyces cerevisiae* group based on polymorphisms of rDNA spacer sequences. International Journal of Systematic Bacteriology 48:295–303.

Moody, A.R., and D. Gindrat. 1977. Biological control of cucumber black root rot by *Gliocladium roseum*. Phytopathology 67:1159–1162.

Moon, C.D., B. Scott, C.L. Schardl, and M.J. Christensen. 2000. The evolutionary origins of *Epichloë* endophytes from annual ryegrasses. Mycologia 92:1103–1118.

Moore, D., and F.W. Spiegel. 1995. A new technique for sampling protostelids. Mycologia 87:414–418.

——. 2000a. Microhabitat distribution of protostelids in temperate habitats in Northwest Arkansas. Canadian Journal of Botany 78:985–994.

——. 2000b. Microhabitat distribution of protostelids in tropical forests of the Caribbean National Forest, Puerto Rico. Mycologia 92:616–625.

——. 2000c. The effect of season on protostelid communities. Mycologia 92:599–608.

Moore, D., S.L. Stephenson, and F.W. Spiegel. 1996. Protostelids from tropical forests of Costa Rica and Puerto Rico. Inoculum 47:22.

Moore, J.C., D.E. Walter, and H.W. Hunt. 1988. Arthropod regulation of micro- and mesobiota in below-ground detrital food webs. Annual Review of Entomology 33:419–439.

Moore, R.T. 1980. Taxonomic proposals for the classification of marine yeasts and other yeastlike fungi including the smuts. Botanica Marina 23:361–373.

——. 1985. The challenge of the dolipore/parenthesome septum. Pp. 175–212. *In* D. Moore, L.A. Casselton, D.A. Wood, and J.C. Franklin (eds.), Developmental Biology of Higher Fungi. Cambridge University Press, Cambridge, England.

——. 1996. An inventory of the phylum Ustomycota. Mycotaxon 59:1–31.

Moore-Landecker, E. 1990. Fundamentals of the Fungi. 3rd ed. Prentice Hall, Englewood Cliffs, New Jersey.

Mordue, J.E.M., and G.C. Ainsworth. 1984. Ustilaginales of the British Isles. Mycological Papers 154:1–96.

Moreau, F. 1939. Sur deux champignons parasites des Saprolégniacées. Bulletin Trimestriel de la Société Mycologique de France 55:95–98.

Morehouse, E.A., T.Y. James, A.R.D. Ganley, R. Vilgalys, L. Berger, P.J. Murphy, and J.E. Longcore. 2003. Multilocus sequence typing suggests the chytrid pathogen of amphibians is a recently emerged clone. Molecular Ecology 12:395–403.

Morgan-Jones, G., and W. Gams. 1982. Notes on hyphomycetes. XLI. An endophyte of *Festuca arundinacea* and the anamorph of *Epichloe typhina*, new taxa in one of two new sections of *Acremonium*. Mycotaxon 15:311–318.

Moriello, K.A., and D.J. Deboer. 1991. Fungal flora of the haircoat of cats with and without dermatophytosis. Journal of Medical and Veterinary Mycology 29:285–292.

Morinaga, T., S. Ito, and K. Minoura. 1980. Fungal succession on deer dung. The effects of light on the development of certain fungi. Transactions of the Mycological Society of Japan 21:351–357.

Morrall, R.A.A. 1974. Soil microfungi associated with aspen in Saskatchewan: synecology and quantitative analysis. Canadian Journal of Botany 52:1803–1817.

Morris, R.A.C., J.R. Coley-Smith, and J.M. Whipps. 1995a. Quantitative isolation of the mycoparasite *Verticillium biguttatum* from soil. Soil Biology and Biochemistry 27:793–799.

——. 1995b. The ability of the mycoparasite *Verticillium biguttatum* to infect *Rhizoctonia solani* and other plant pathogenic fungi. Mycological Research 99:997–1003.

Morrow, J.D. 1993. Intellectual property and biodiversity. Pp. 26–34. *In* G. Baillargeon, D. Barbeau, Y. Bélanger, D. Leger, E.E. Lister, and D. Miller (eds.), Proceedings of the National Workshop on a Canadian Germplasm Network. Research Branch, Agriculture and Agri-Foods Canada, Ottawa, Ontario, Canada.

Morton, J.B. 1986. Three new species of *Acaulospora* (Endogonaceae) from high aluminum, low pH soils in West Virginia. Mycologia 78:641–648.

——. 1988. Taxonomy of VA mycorrhizal fungi: classification, nomenclature, and identification. Mycotaxon 32:267–324.

——. 1990a. Evolutionary relationships among arbuscular mycorrhizal fungi in the Endogonaceae. Mycologia 82:192–207.

——. 1990b. Species and clones of arbuscular mycorrhizal fungi (Glomales, Zygomycetes): their role in macro- and microevolutionary processes. Mycotaxon 37:493–515.

——. 1993. Problems and solutions for the integration of glomalean taxonomy, systematic biology, and the study of mycorrhizal phenomena. Mycorrhiza 2:97–109.

——. 1999. Evolution of endophytism in arbuscular mycorrhizal fungi of Glomales. Pp.121–140. *In* C.W. Bacon and J.H. White (eds.), Microbial Endophytes. Marcel Dekker, New York.

Morton, J.B., and G.L. Benny. 1990. Revised classification of arbuscular mycorrhizal fungi (Zygomycetes): a new order, Glomales, two new suborders, Glomineae and Gigasporineae, and two new families, Acaulosporaceae and Gigasporaceae, with an emendation of Glomaceae. Mycotaxon 37:471–491.

Morton., J.B., and D. Redecker. 2001. Two new families of Glomales, Archaeosporaceae and Paraglomaceae, with two new genera *Archaeospora* and *Paraglomus*, based on concordant molecular and morphological characters. Mycologia 93:181–195.

Morton, J.B., S.P. Bentivenga, and J.D. Bever. 1995. Discovery, measurement, and interpretation of diversity in arbuscular endomycorrhizal fungi (Glomales, Zygomycetes). Canadian Journal of Botany 73(Suppl. 1):S25–S32.

Morton, J.B., S.P. Bentivenga, and W.W. Wheeler. 1993. Germ plasm in the International Collection of Arbuscular and Vesicular-arbuscular Mycorrhizal Fungi (INVAM) and procedures for culture development, documentation, and storage. Mycotaxon 48:491–528.

Moser, J.C., T.J. Perry, J.R. Bridges, and H.-F. Yin. 1995. Ascospore dispersal of *Ceratocystiopsis ranaculosus*, a mycangial fungus of the southern pine beetle. Mycologia 87:84–86.

Moser, M. 1978. Die Röhrlinge und Blätterpilze (Polyporales, Boletales, Agaricales, Russulales). Kleine Kryptogamenflora IIb/2. Basidiomyceten. 2 Teil, 4 Auflage. G. Fischer, Jena, Germany.

Moss, S.T. 1986. Biology and phylogeny of Labyrinthulales and Thraustochytriales. Pp. 105–129. *In* S.T. Moss (ed.), The Biology of Marine Fungi. Cambridge University Press, Cambridge, Great Britain.

Mouchacca, J. 1995. Thermophilic fungi in desert soils: a neglected extreme environment. Pp. 265–288. *In* D. Allsopp, R.R. Colwell, and D.L. Hawksworth (eds.), Microbial Diversity and Ecosystem Function. CAB International, Wallingford, Oxon, United Kingdom.

Mountfort, D.O. 1987. The rumen anaerobic fungi. FEMS Microbiology Reviews 46:401–408.

Mountfort, D.O., and R.A. Asher. 1985. Production and regulation of cellulase by two strains of the rumen anaerobic fungus *Neocallimastix frontalis*. Applied and Environmental Microbiology 49:1314–1322.

——. 1989. Production of xylanase by the ruminal anaerobic fungus *Neocallimastix frontalis*. Applied and Environmental Microbiology 55:1016–1022.

Mountfort, D.O., R.A. Asher, and T. Bauchop. 1982. Fermentation of cellulose to methane and carbon dioxide by a rumen anaerobic fungus in a triculture with *Methanobrevibacter* sp. strain RA1 and *Methanosarcinia barkeri*. Applied and Environmental Microbiology 44:128–134.

Moustafa, A.F., M.S. Sharkas, and S.M. Kamel. 1976. Thermophilic and thermotolerant fungi in the desert and salt-marsh soils of Kuwait. Norwegian Journal of Botany 23:213–220.

Moustert, L., P.W. Crous, and O. Petrini. 2000. Endophytic fungi associated with shoots and leaves of *Vitis vinifera*, with specific reference to the *Phomopsis viticola* complex. Sydowia 52:46–58.

Mower, R.L., W.C. Snyder, and J.G. Hancock. 1975. Biological control of ergot by fusarium. Phytopathology 65:5–10.

Mowll, J.L., and G.M. Gadd. 1985. Effect of vehicular lead pollution on phylloplane mycoflora. Transactions of the British Mycological Society 84:685–689.

Mueller, G.M. 1992. Systematics of Laccaria (Agaricales) in the continental United States and Canada, with discussions on extralimital taxa and descriptions of extant types. Fieldiana: Botany, New Series 30:1–158.

——. 1999. A new challenge for mycological herbaria: destructive sampling of specimens for molecular data. Pp. 287–300. *In* D.A. Metsger and S.C. Byers (eds.), Managing the Modern Herbarium, An Inter-Disciplinary Approach. Society for the Preservation of Natural History Collections, Washington, D.C., and The Royal Ontario Museum, Toronto, Ontario, Canada.

Mueller, G.M., and J.F. Ammirati. 1993. Cytological studies in *Laccaria* (Agaricales). II. Assessing phylogenetic relationships among *Laccaria*, *Hydnangium*, and other agaricles. American Journal of Botany 80:322–329.

Mueller, G.M., and M.M. Mata. 2000. Inventory of Costa Rican Fungi. Web site: http://www.inbio.eas.ualberta.ca/papers/gt_Hongos/en/index.html.

Mueller, J.D., M. Niedbalski Cline, J.B. Sinclair, and B.J. Jacobsen. 1985. An in vitro test for evaluating efficacy of mycoparasites on sclerotia of *Sclerotinia sclerotiorum*. Plant Disease 69:584–587.

Mueller-Dombois, D., and H. Ellenberg. 1974. Aims and Methods of Vegetation Ecology. John Wiley, New York.

Muhvich, A.G., R. Reimschuessel, M.M. Lipsky, and R.O. Bennett. 1989. *Fusarium solani* isolated from newborn bonnethead sharks, *Sphyrna tiburo* (L.). Journal of Fish Diseases 12:57–62.

Mulder, A., L.J. Turkensteen, and A. Bouman. 1992. Perspectives of green-crop-harvesting to control soil-borne and storage diseases of seed potatoes. Netherlands Journal of Plant Pathology 98(Suppl. 2):103–114.

Müller, E. 1950. Die schweizerischen arten der gattung *Leptosphaeria* und ihrer Verwandten. Sydowia 4:185–319.

Müller, E., and J.A. von Arx. 1962. Die gattungen der didymosporen Pyrenomyceten. Beiträge zur Kryptogamenflora der Schweiz 11:1–922.

———. 1973. Pyrenomycetes: meliolales, coronophorales, sphaeriales. Pp. 87–132. *In* G.C. Ainsworth, F.K. Sparrow, and A.S. Sussman (eds.), The Fungi. An Advanced Treatise. IVA. A Taxonomic Review with Keys: Ascomycetes and Fungi Imperfecti. Academic Press, New York.

Muller, K.E., and V.A. Benignus. 1992. Increasing scientific power with statistical power. Neurotoxicology and Teratology 14:211–219.

Müller-Haeckel, A., and L. Marvanová. 1979a. Periodicity of aquatic hyphomycetes in the subarctic. Transactions of the British Mycological Society 73:109–116.

———. 1979b. Freshwater hyphomycetes in brackish and sea water. Botanica Marina 22:421–424.

Mulligan, D.F.C., and J.W. Deacon. 1992. Detection of presumptive mycoparasites in soil placed on host-colonized agar plates. Mycological Research 96:605–608.

Mullis, K., and F.A. Faloona. 1987. Specific synthesis of DNA in vitro via a polymerase catalyzed chain reaction. Methods in Enzymology 155:335–350.

Mundkar, B.B., and M.J. Thirumalachar. 1952. Ustilaginales of India. Commonwealth Mycological Institute, Kew, Surrey, England.

Munk, A. 1954. Notes on some *Hypocreales* recently found in Denmark. Botanisk Tidsskrift 51:220–229.

———. 1957. Danish pyrenomycetes: a preliminary flora. Dansk Botanisk Arkiv 17:1–491.

Munn, E.A., C.G. Orpin, and C.A. Greenwood. 1988. The ultrastructure and possible relationships of four obligate anaerobic chytridiomycete fungi from the rumen of sheep. BioSystems 22:67–81.

Munro, A.L.S., A.H. McVicar, and R. Jones. 1983. The epidemiology of infectious disease in commercially important wild marine fish. Pp. 21–32. *In* J.E. Stewart (ed.), Diseases of Commercially Important Marine Fish and Shellfish. ICES Marine Science Symposium. Rapports et Procès-Verbaux des V Réunion, Conseil International pour l'Exploration de la Mer Vol. 182. Copenhagen, Denmark.

Munsell, A.H. 1966. Munsell Book of Color. MacBeth Color and Photometry, Kollmorgen Corp., Newburgh, New York.

Murakami, Y. 1987. Spatial distribution of *Russula* species in *Castanopsis cuspidata* forest. Transactions of the British Mycology Society 89:187–193.

Murie, O.J. 1974. A Field Guide to Animal Tracks. 2nd ed. The Peterson Field Guide Series. Vol. 9. Houghton Mifflin, Boston, Massachusetts.

Murrill, W.A. 1915. Tropical Polypores. Published by the Author, New York.

Murrin, F., J. Holtby, R.A. Nolan, and W.S. Davidson. 1986. The genome of *Entomophaga aulicae* (Entomophthorales, Zygomycetes): base composition and size. Experimental Mycology 10:67–75.

Mushin, T.M., and T. Booth. 1987. Fungi associated with halophytes of an inland salt marsh, Manitoba, Canada. Canadian Journal of Botany 65:1137–1151.

Mushin T.M., T. Booth, and K.H. Zwain. 1989. A fungal endophyte associated with a desert parasitic plant. Kavaka 17:1–5.

Muthumeenakshi, S., P.R. Mills, A.E. Brown, and D.A. Seaby. 1994. Intraspecific molecular variation among *Trichoderma harzianum* isolates colonizing mushroom compost in the British Isles. Microbiology 140:769–777.

Naeem, S., L.J. Thompson, S.P. Lawler, J.H. Lawton, and R.M. Woodfin. 1994. Declining biodiversity can alter the performance of ecosystems. Nature (London) 368:734–737.

Nag Raj, T.R. 1993. Coelomycetous Anamorphs with Appendage-Bearing Conidia. Mycologue, Waterloo, Ontario, Canada.

Nagahama, T., H. Sato, M. Shimazu, and J. Sugiyama. 1995. Phylogenetic divergence of the entomophthoralean fungi: evidence from nuclear 18S ribosomal RNA gene sequences. Mycologia 87:203–209.

Nagai, K., K. Suzuki, and G. Okada. 1998. Studies on the distribution of alkalophilic and alkali-tolerant soil fungi II: fungal flora in two limestone caves in Japan. Mycoscience 39:293–298.

Nagai, K., T. Sakai, R.M. Rantiatmodjo, K. Suzuki, W. Gams, and G. Okada. 1995. Studies on the distribution of alkalophilic and alkali-tolerant soil fungi I. Mycoscience 36:247–256.

Nagtzaam, M.P.M. 1998. Biological control of *Verticillium dahliae* by *Talaromyces flavus*. Unpubl. Ph.D. dissert., Agricultural University Wageningen, Wageningen, The Netherlands.

Nagtzaam, M.P.M., G.J. Bollen, and A.J. Termorshuizen. 1998. Efficacy of *Talaromyces flavus* alone or in combination with other antagonists in controlling *Verticillium dahliae* in growth chamber experiments. Journal of Phytopathology 146:165–173.

Nair, J. 1976. A method for staining infection hyphae in pine leaves. Stain Technology 51:47–49.

Nakagiri, A. 1993. Growth and reproduction of *Halophytophthora* species. Transactions of the Mycological Society of Japan 34:87–99.

Nakagiri, A., and E.B.G. Jones. 2000. Maintenance of cultures of marine fungi. Pp. 62–68. *In* K.D. Hyde and S.B. Pointing (eds.), Marine Mycology, A Practical Approach. Fungal Diversity Press, Honk Kong, China.

Nakagiri, A., T. Ito, L. Manoch, and M. Tanticharoen. 2001. A new *Halophytophthora* species, *H. porrigovesica*, from subtropical and tropical mangroves. Mycoscience 42:33–41.

Nakagiri, A., S.Y. Newell, and T. Ito. 1994. Two new *Halophytophthora* species, *H. tartarea* and *H. master*, from intertidal decomposing leaves in saltmarsh and mangrove regions. Mycoscience 35:223–232.

Nakagiri, A., S.Y. Newell, T. Ito, T.K. Tan, and C.L. Pek. 1996. Biodiversity and ecology of the oomycetous fungus *Halophytophthora*. Pp. 273–280. *In* I.M. Turner, C.H. Diong, S.S.L. Lim, and P.K.L. Ng (eds.), Biodiversity and the Dynamics of Ecosystems. Diversitas in the Western Pacific and Asia (DIWPA) Series. Vol. 1. DIWPA, Singapore.

Nakagiri, A., I. Okane, and T. Ito. 1998. Zoosporangium development, zoospore release and culture properties of *Halophytophthora mycoparasitica*. Mycoscience 39:87–95.

Nakamura, K., and K. Hatai. 1995. Three species of Lagenidiales isolated from the eggs and zoeae of the marine crab *Portunus pelagicus*. Mycoscience 36:87–95.

Nakase, T., M. Hamamoto, and J. Sugiyama. 1991. Recent progress in the systematics of basidiomycetous yeasts. Japanese Journal of Medical Mycology 32(Suppl.):21–30.

Nakase, T., A. Takematsu, and Y. Yamada. 1993. Molecular approaches to the taxonomy of ballistosporus yeasts based on the analysis of the partial nucleotide sequences of 18S ribosomal ribonucleic acids. Journal General and Applied Microbiology 39:107–134.

Nakasone, K.K. 1990. Cultural studies and identification of wood-inhabiting Corticiaceae and selected Hymenomycetes from North America. Mycologia Memoirs No. 15.

Nannenga-Bremekamp, N.E. 1991. A Guide to Temperate Myxomycetes. Biopress, Bristol, United Kingdom.

Nannfeldt, J.A. 1932. Studien über die Morphologie und Systematik der Nicht-Lichenisierten Inoperculaten Discomyceten. Nova Acta Regiae Societatis Scientiarum Upsaliensis. Ser. IV. 8:1–368.

Nantel, P., and P. Neuman. 1992. Ecology of ectomycorrhizal-basidiomycete communities on a local vegetation gradient. Ecology 73:99–117.

Nardon, P., and A.M. Grenier. 1989. Endocytobiosis in coleoptera: Biological, biochemical, and genetic aspects. Pp. 175–216. *In* W. Schwemmler (ed.), Insect Endocytobiosis: Morphology, Physiology, Genetics, Evolution. CRC Press, Boca Raton, Florida.

Nash, T.H., III (ed.). 1996. Lichen Biology. Cambridge University Press, Cambridge, United Kingdom.

Nash, T.H., III, and R.S. Egan. 1988. The biology of lichens and bryophytes. Bibliotheca Lichenologica 30:11–22.

Nash, T.H., III, and V. Wirth (eds.). 1988. Lichens, Bryophytes and Air Quality. Cramer, Stuttgart, Germany.

Nash, T.H., III, B.D. Ryan, C. Gries, and F. Bungartz (eds.). 2002. Lichen flora of the greater Sonoran Desert region. Vol. 1. Lichens Unlimited, Arizona State University, Tempe, Arizona.

Nauta, M.M., and E.C. Vellinga. 1995. Atlas van Nederlandse Paddestoelen. A.A. Balkema, Rotterdam, The Netherlands.

Navarro-Rosinés, P., and N.L. Hladun. 1990. El género *Sacropyrenia* Nyl. (ascomicetes liquenícolas) en Europa y norte de Africa. Candollea 45:469–489.

——. 1994. Datos sobre el género *Rhagadostoma* (ascomicetes lichenícolas, Sordariales). Bulletin de la Société Linnéenne de Provence 45:431–442.

Navarro-Rosinés, P., and C. Roux. 1995. Le genre *Weddellomyces* (Dothideales, Dacampiaceae) en Catalogne et en Provence. Mycotaxon 53:161–187.

Navarro-Rosinés, P., M. Boqueras, and C. Roux. 1998. Nuevos datos para el género *Lichenochora* (Phyllachorales, Ascomicetes liquenícolas). Bulletin de la Société Linnéenne de Provence 49:107–124.

Nawawi, A., J. Webster, and R.A. Davey. 1977. *Dendrosporomyces prolifer* gen. et sp. nov., a basidiomycete with branched conidia. Transactions of the British Mycological Society 68:59–63.

Needham, J.R. 1981. Microbiology and laboratory techniques. Pp. 9–132. *In* J.E. Cooper and O.F. Jackson (eds.), Diseases of the Reptilia. Vol. 1. Academic Press, London, Great Britain.

Neer, A.J., K.G. Mukerji, B.C. Sharma, and A.K. Varma. 1993. A new species of *Gigaspora* from desert soils of Rajasthan, India. World Journal of Microbiology and Biotechnology 9:291–294.

Nees von Esenbeck, C.G.D. 1823. Zusatz. Pp. 507–522. *In* C.G. Carus, Beitrag Zur Geschichte der unter Wasser an verwesenden Thierkörpern sich erzeugenden Schimmel—oder Algengattungen. Nova Acta Physico-Medica Academiae Caesareae Leopoldino-Carolinae Naturae Curiosorum 11.

Neish, G.A. 1977. Observations on saprolegniasis of adult sockeye salmon, *Oncorhynchus nerka* (Walbaum). Journal of Fish Biology 10:513–522

Neish, G.A., and G.C. Hughes. 1980. Fungal Diseases of Fishes. T.F.H. Publications, Neptune City, New Jersey.

Nelson, A.C., R.O. Novak, and M.P. Backus. 1964. A new species of *Neurospora* from soil. Mycologia 56:384–392.

Nelson, D.L., and D.L. Sturges. 1982. A snowmold disease of mountain big sagebrush *Artemisia tridentata vaseyana*. Phytopathology 72:965.

Nelson, E.E., and H.A. Fay. 1985. Maintaining cultures of wood-rotting fungi. U.S. Department of Agriculture, Forest Service, Pacific Northwest. Forest and Range Experiment Station, Portland, Oregon, Research Note PNW 428:1–3.

Nelson, P.E., T.A. Tousson, and W.F.O. Marasas. 1983. *Fusarium* species, an illustrated manual for identification. Pennsylvania State University Press, University Park, Pennsylvania.

Newbold, C.J., and K. Hillman. 1990. The effect of ciliate protozoa on the turnover of bacterial and fungal protein in the rumen of sheep. Letters in Applied Microbiology 11:100–102.

Newell, K. 1984a. Interaction between two decomposer basidiomycetes and a collembolan under Sitka spruce: distribution, abundance and selective grazing. Soil Biology and Biochemistry 16: 227–233.

——. 1984b. Interaction between two decomposer basidiomycetes and a collembolan under Sitka spruce: grazing and its potential effects on fungal distribution and litter decomposition. Soil Biology and Biochemistry 16:235–239.

Newell, S.Y. 1992a. Autumn distribution of marine Pythiaceae across a mangrove-saltmarsh boundary. Canadian Journal of Botany 70:1912–1916.

——. 1992b. Ecomethodology for organoosmotrophs: Prokaryotic unicellular versus eukaryotic mycelial. Microbial Ecology 28:151–157.

——. 1992c. Estimating fungal biomass and production in decomposing litter. Pp. 521–561. *In* G.C. Carroll and D.T. Wicklow (eds.), The Fungal Community: Its Organization and Role in the Ecosystem. 2nd ed. Marcel Dekker, New York.

——. 1994. Ecomethodology for organoosmotrophs: prokaryotic, unicellular versus eukaryotic mycelial. Microbial Ecology 28:151–157.

——. 1996. Established and potential impacts of eukaryotic mycelial decomposers in marine/terrestrial ecotones. Journal of Experimental Marine Biology and Ecology 200:187–206.

——. 2000. Methods for determining biomass and productivity of mycelial marine fungi. Pp. 69–91. *In* K.D. Hyde and S.B. Pointing (eds.), Marine Mycology, A Practical Approach. Fungal Diversity Press, Hong Kong, China.

——. 2001. Fungal biomass and productivity. Pp. 357–372. *In* J.H. Paul (ed.), Marine Microbiology. Methods in Microbiology. Vol. 30. Academic Press, London, England.

Newell, S.Y., and J.W. Fell. 1982. Surface sterilization and the active mycoflora of leaves of a seagrass. Botanica Marina 25:339–346.

——. 1992. Distribution and experimental responses to substrate of marine oomycetes (*Halophytophthora* spp.) in mangrove ecosystems. Mycological Research 96:851–856.

——. 1994. Parallel testing of media for measuring frequencies of occurrence for *Halophytophthora* spp. (Oomycota) from decomposing mangrove leaves. Canadian Journal of Microbiology 40:250–256.

——. 1995. Do halophytophthoras (marine Pythiaceae) rapidly occupy fallen leaves by intraleaf mycelial growth? Canadian Journal of Botany 73:761–765.

——. 1997. Competition among mangrove oomycetes, and between oomycetes and other microbes. Aquatic Microbial Ecology 12:21–28.

Newell, S.Y., and J. Wasowski. 1995. Sexual productivity and spring intramarsh distribution of a key salt-marsh microbial secondary producer. Estuaries 18:241–249.

Newell, S.Y., J.D. Miller, and J.W. Fell. 1987. Rapid and pervasive occupation of fallen mangrove leaves by a marine zoosporic fungus. Applied and Environmental Microbiology 53:2464–2469.

Newell, S.Y., R. Cefalu, and J.W. Fell. 1977. *Myzocytium, Haptoglossa*, and *Gonimochaete* (fungi) in littoral marine nematodes. Bulletin of Marine Science 27:177–207.

Nicot, J. 1960. Some characteristics of the microflora in desert sands. Pp. 94–97. *In* D. Parkinson and J.S. Waid (eds.), The Ecology of Soil Fungi. Liverpool University Press, Liverpool, Great Britain.

——. 1968. Sur le mycoparasitisme de *Calcarisporium arbuscula* Preuss. Bulletin Trimestriel de la Société Mycologique de France 84:85–92.

Nicot, J., and F. Durand. 1965. Remarques sur la moisissure fongicole *Amblyosporium botrytis* Fres. Bulletin Trimestriel de la Société Mycologique de France 81:623–649.

Niemelä, T., and I. Nordin. 1985. *Hypocreopsis lichenoides* (Ascomycetes) in North Europe. Karstenia 25:75–80.

Niemelä, T., P. Renvall, and R. Penttilä. 1995. Interactions of fungi at late stages of wood decomposition. Annales Botanici Fennici 32: 141–152.

Nikolcheva, L.G., A.M. Cockshutt, and F. Bärlocher. 2003. Determining diversity of freshwater fungi on decaying leaves: Comparison of traditional and molecular approaches. Applied and Environmental Microbiology 69:2548–2554.

Niku-Paavola, M.L., L. Raaska, and M. Itavaara. 1990. Detection of white-rot fungi by a non-toxic stain. Mycological Research 94:27–31.

Nimis, P.L. 1993. The Lichens of Italy, an Annotated Catalogue. Monografia XII. Museo Regionale di Scienze Naturali, Torino, Italy.

Nimis, P.L., C. Scheidegger, and P.A. Wolseley (eds.). 2002. Monitoring with Lichens—Monitoring Lichens. NATO Science Series 4, Vol. 7. Kluwer Academic, The Hague, The Netherlands.

Nishida, F.H. 1989. Review of mycological studies in the Neotropics. Pp. 494–522. *In* D.G. Campbell and H.D. Hammond (eds.), Floristic Inventory of Tropical Countries: The Status of Plant Systematics, Collections, and Vegetation. New York Botanical Gardens, Bronx, New York.

Nishida, H., and J. Sugiyama. 1993. Phylogenetic relationships among *Taphrina, Saitoella,* and other higher fungi. Molecular Biology and Evolution 10:431–436.

———. 1994. Archiascomycetes: detection of a major new lineage within the Ascomycota. Mycoscience 35:361–366.

Nishida, H., K. Ando, Y. Ando, A. Hirata, and J. Sugiyama. 1995. *Mixia osmundae*: transfer from the Ascomycota to the Basidiomycota based on evidence from molecules and morphology. Canadian Journal of Botany 73(Suppl. 1): S660–S666.

Nishii, T., and A. Nakagiri. 1991. Cyropreservation of oomycetous fungi in liquid nitrogen. IFO [Institute for Fermentation, Osaka, Japan] Research Communication 15:105–118.

Nisken, S.J. 1962. A water sampler for microbiological studies. Deep Sea Research 9:501–503.

Noble, H.M., D. Langley, P.J. Sidebottom, S.J. Lane, and P.J. Fisher. 1991. An echinocandin from an endophytic *Cryptosporiopsis* sp. and *Pezicula* sp. in *Pinus sylvestris* and *Fagus sylvatica*. Mycological Research 95:1439–1440.

Nobles, M.K. 1965. Identification of cultures of wood-inhabiting Hymenomycetes. Canadian Journal of Botany 43:1097–1139.

Noda, H., and N. Kawahara. 1995. Electrophoretic karyotype of intracellular yeast-like symbiotes in rice planthoppers and anobiid beetles. Journal of Invertebrate Pathology 65:118–124.

Noda, H., and K. Kodama. 1996. Phylogenetic position of yeastlike endosymbionts of anobiid beetles. Applied and Environmental Microbiology 62:162–167.

Noda, H., and T. Omura. 1992. Purification of yeast-like symbiotes of planthoppers. Journal of Invertebrate Pathology 59:104–105.

Noda, H., N. Nakashima, and M. Koizumi. 1995. Phylogenetic position of yeast-like symbiotes of rice planthoppers based on partial 18S rDNA sequences. Insect Biochemistry and Molecular Biology 25:639–646.

Noga, E.J. 1990. A synopsis of mycotic diseases of marine fishes and invertebrates. Pp. 143–160. *In* F.O. Perkins and T.C. Cheng (eds.), Pathology in Marine Science. Academic Press, San Diego, California.

———. 1993. Fungal diseases of marine and estuarine fishes. Pp. 85–109. *In* J.A. Couch and J.W. Fournie (eds.), Pathobiology of Marine and Estuarine Organisms. CRC Press, Boca Raton, Florida.

Noga, E.J., and M.J. Dykstra. 1986. Oomycete fungi associated with ulcerative mycosis in menhaden, *Brevoortia tyrannus* (Latrobe). Journal of Fish Diseases 9:47–53.

Noga, E.J., J.F. Levine, M.J. Dykstra, and J.H. Hawkins. 1988. Pathology of ulcerative mycosis in Atlantic menhaden *Brevoortia tyrannus*. Diseases of Aquatic Organisms 4:189–197.

Nograsek, A. 1990. Ascomyceten auf Gefäßpflanzen der Polsterseggenrasen in den Ostalpen. Bibliotheca Mycologica 133:1–271.

Noordeloos, M.E. 1992. Entoloma s.l. *In* M. Canatusso (ed.), Funghi Europaei, Vol. 5, Biella Giovanna, Saronno, Italy.

Nordgren, A., E. Bååth, and B. Söderström. 1985. Soil microfungi in an area polluted by heavy metals. Canadian Journal of Botany 63:448–455.

Norris, D.M. 1979. The mutualistic fungi of Xyleborini beetles. Pp. 53–63. *In* L.R. Batra (ed.), Insect-Fungus Symbiosis, Nutrition, Mutualism, and Commensalism. Allanheld, Osmun, Montclair, New Jersey.

North, M. 1993. Stand Structure and Truffle Abundance Associated with Northern Spotted Owl Habitat. Unpubl. Ph.D. dissert. University of Washington, Seattle, Washington.

North, M., and J. Trappe. 1994. Small mammal exclosures for studies of hypogeous fungi. Mycologia 86:586–587.

Norvell, L.L., and J. Roger. 1998. The Oregon *Cantharellus* study project: pacific golden chanterelle, preliminary observations and productivity data (1986–1997). Inoculum 49:40.

Novak, R.O., and M.P. Backus. 1963. A new species of *Mycotypha* with a zygosporic stage. Mycologia 55:790–798.

Novak, R.O., and W.F. Whittingham. 1968. Soil and litter microfungi of a maple-elm-ash floodplain community. Mycologia 60:776–787.

Novozhilov, Y.K., M. Schnittler, A.W. Rollins, and S.L. Stephenson. 2001. Myxomycetes in different forest types of Puerto Rico. Mycotaxon 77:285–299.

Novozhilov, Y.K., M. Schnittler, and S.L. Stephenson. 1999. Myxomycetes of the Taimyr Peninsula (north-central Siberia): taxonomy and distribution. Karstenia 39:77–97.

Nuzum, C. 1989. A simple method for the preservation of some nonsporing fungi. Australasian Plant Pathology 18:104–105.

Oberwinkler, F. 1990. New genera of auricularioid heterobasidiomycetes. Report of the Tottori Mycological Institute 28:113–127.

———. 1992. Biodiversity amongst filamentous fungi. Biodiversity and Conservation 1:293–311.

———. 1993. Diversity and phylogenetic importance of tropical heterobasidiomycetes. Pp. 121–147. *In* S. Isaac, J.C. Frankland, R. Watling, and A.J.S. Whalley (eds.), Aspects of Tropical Mycology. Cambridge University Press, Cambridge, United Kingdom.

Oberwinkler, F., and R.J. Bandoni. 1981. *Tetragoniomyces* gen. nov. and Tetragoniomycetaceae fam. nov. (Tremellales). Canadian Journal of Botany 59:1034–1040.

———. 1982. Carcinomycetaceae: a new family in the Heterobasidiomycetes. Nordic Journal of Botany 2:501–516.

———. 1983. *Trimorphomyces*: a new genus in the Tremellaceae. Systematic and Applied Microbiology 4:105–113.

———. 1984. *Herpobasidium* and allied genera. Transactions of the British Mycological Society 83:639–658.

Oberwinkler, F., and R. Bauer. 1990. *Cryptomycocolax*: a new mycoparasitic heterobasidiomycete. Mycologia 82:671–692.

Oberwinkler, F., and B. Lowy. 1981. *Syzygospora alba*, a mycoparasitic heterobasidiomycete. Mycologia 73:1108–1115.

Oberwinkler, F., R.J. Bandoni, P. Blanz, G. Deml, and L. Kisimova-Horovitz. 1982. Graphiolales: Basidiomycetes parasitic on palms. Plant Systematics and Evolution 140:251–277.

Oberwinkler, F., R.J. Bandoni, P. Blanz, and L. Kisimova-Horovitz. 1983. *Cystofilobasidium*: a new genus in the Filobasidiaceae. Systematic and Applied Microbiology 4:114–122.

Oberwinkler, F., R.J. Bandoni, R. Bauer, G. Deml, and L. Kisimova-Horovitz. 1984. The life-history of *Christiansenia pallida*, a dimorphic, mycoparasitic Heterobasidiomycete. Mycologia 76:9–22.

Oberwinkler, F., R. Bauer, and R. Bandoni. 1990a. Heterogastridiales: a new order of basidiomycetes. Mycologia 82:48–58.

Oberwinkler, F., R. Bauer, and J. Schneller. 1990b. *Phragmoxenidium mycophilum*, an unusual mycoparasitic heterobasidiomycete. Systematic and Applied Microbiology 13:186–191.

Ochiai, T., K. Kodera, T. Kon, T. Miyazaki, and S.S. Kubota. 1977. Studies on disease owing to erroneous-swallowing in ayu fry. Fish Pathology 12:135–139.

Ochman, H., and A.C. Wilson. 1987. Evolution in bacteria: evidence for a universal substitution rate in cellular genomes. Journal of Molecular Evolution 26:74–86.

O'Dell, T.E., and J.F. Ammirati. 1994. Diversity and sporocarp productivity of ectomycorrhizal fungi in old growth *Pseudotsuga menziesii*—*Tsuga heterophylla* forests with different understory vegetation. Abstracts of the Fifth International Mycological Congress, Vancouver, British Columbia 1:160.

O'Dell, T.E., J.F. Ammirati, and E.G. Schreiner. 1999. Species richness and abundance of ectomycorrhizal basidiomycete sporocarps on a moisture gradient in the *Tsuga heterophylla* zone. Canadian Journal of Botany 77:1699–1711.

O'Dell, T.E., M.A. Castellano, and R.J. Molina. Conservation status of fungi in the Pacific Northwest: The Northwest Forest Plan. Canadian Journal of Botany. Unpubl. ms.

O'Dell, T.E., D.L. Luoma, and R.J. Molina. 1992a. Ectomycorrhizal fungal communities in young, managed, and old-growth douglas-fir stands. Northwest Environmental Journal 8:166–168.

O'Dell, T.E., H.B. Massicote, and J.M. Trappe. 1993. Root colonization of *Lupinus latifolius* Agardh. and *Pinus contorta* Dougl. by *Phialocephala fortinii* Wang & Wilcox. New Phytologist 124:93–100.

O'Dell, T.E., J.E. Smith, M.A. Castellano, and D.L. Luoma. 1996. Diversity and Conservation of Forest Fungi. Pp. 5–18. *In* R. Molina and D. Pilze (eds.), Managing Forest Ecosystems to Conserve Fungus Diversity and Sustain Wild Mushroom Harvests. U.S. Department of Agriculture, Forest Service, Pacific Northwest Research Station, Portland, Oregon. General Technical Report PNW-GTR-371.

O'Dell, T. E, J.M. Trappe, N.J. Weber, and E.G. Schreiner. 1992b. Fungal diversity in Olympic National Park. Northwest Environmental Journal 8:170–172.

O'Donnell, J., and C.H. Dickinson. 1980. Pathogenicity of *Alternaria* and *Cladosporium* isolates on *Phaseolus*. Transactions of the British Mycological Society 74:335–342.

O'Donnell, K., and E. Cigelnik. 1997. Two divergent intragenomic rDNA ITS2 types within a monophyletic lineage of the fungus *Fusarium* are nonorthologous. Molecular Phylogenetics and Evolution 7:103–116.

O'Donnell, K., E. Cigelnik, and G. L Benny. 1998. Phylogenetic relationships among the *Harpellales* and *Kickxellales*. Mycologia 90:624–639.

O'Donnell, K., E. Cigelnik, N.S. Weber, and J.M. Trappe. 1996. Phylogenetic relationships among ascomycetous truffles and the true and false morels inferred from 18S and 28S ribosomal DNA sequence analysis. Mycologia 89:48–65.

O'Donnell, K., F. Lutzoni, T. Ward, and G. Benny. 2001. Evolutionary relationships among mucoralean fungi (Zygomycota): Evidence for family polyphyly on a large scale. Mycologia 93:286–296.

O'Donnell, K.L. 1979. Zygomycetes in Culture. Palfrey Contributions in Botany. No. 2. Department of Botany, University of Georgia, Athens, Georgia.

Ohenoja, E. 1993. Effect of weather conditions on the larger fungi at different forest sites in northern Finland in 1976–1988. Acta Universitatis Ouluensis, Series A 243:1–69.

Ohenoja E., and K. Metsänheimo. 1982. Phenology and fruiting body production of macro-fungi in subarctic Finnish Lapland. Pp 371–389 *In* G.A. Laursen and J.F. Ammirati (eds.), Arctic and Alpine Mycology. University of Washington Press, Seattle, Washington.

Ohmasa, M., Y. Abe, K. Babasaki, H. Hiraide, and K. Okabe. 1992. Preservation of cultures of mushrooms by freezing. Transactions of the Mycology Society of Japan 33:467–479.

Ohr, H.D., and D.E. Munnecke. 1974. Effects of methyl bromide on antibiotic production by *Armillaria mellea*. Transactions of the British Mycological Society 62:65–72.

Okamoto, N., K. Nakase, H. Suzuki, J. Nakai, K. Fujii, and T. Sano. 1985. Life history and morphology of *Ichthyophonus hoferi in vitro*. Fish Pathology 20:273–285.

Oksanen, J. 1988. Impact of habitat, substrate and microsite classes on the epiphyte vegetation: interpretation using exploratory and canonical correspondence analysis. Annales Botanica Fennici 25:59–71.

Okuda, T., M. Yanagisawa, F. Fujimori, Y. Nishizuka, Y. Takehana, and M. Sugiyama. 1995. New isolation methods and polymerase chain reaction strain discrimination techniques for natural products screening programs. Canadian Journal of Botany 73 (Suppl. 1): S946–S954.

Okudaira, M., H. Kurata, and F. Sakabe. 1977. Studies on the fungal flora in the lung of human necropsy cases. A critical survey in connection with the pathogenesis of opportunistic fungus infections. Mycopathologia 61:3–18.

Olive, L.S. 1967. The Protostelida—a new order of the Mycetozoa. Mycologia 59:1–29.

——. 1968. An unusual new heterobasidiomycete with *Tilletia*-like basidia. Journal of the Elisha Mitchell Scientific Society 84:261–266.

——. 1970. The mycetozoa: a revised classification. Botanical Review 36:59–87.

——. 1975. The Mycetozoans. Academic Press, New York.

——. 1982. Eumycetozoa. Pp. 521–525. *In* S.P. Parker (ed.), Synopsis and Classification of Living Organisms. McGraw-Hill, New York.

Olive, L.S., and C. Stoianovitch. 1974. A cellular slime mold with flagellate cells. Mycologia 66:685–690.

——. 1977. *Clastostelium*, a new ballistosporous protostelid (mycetozoa) with flagellate cells. Transactions of the British Mycological Society 69:83–88.

——. 1979. Observations of the mycetozoan genus *Ceratiomyxa*: description of a new species. Mycologia 71:546–555.

——. 1981. *Protostelium expulsum*, sp. nov., a simple mycetozoan with a unique method of spore discharge. Transactions of the British Mycological Society 76:303–309.

Olive, L.S., C. Stoianovitch, and W.E. Bennett. 1983. Descriptions of acrasid cellular slime molds: *Pocheina rosea* and a new species, *Pocheina flagellata*. Mycologia 75:1019–1029.

Oliver, I., and A.J. Beattie. 1993. A possible method for the rapid assessment of biodiversity. Conservation Biology 7:562–568.

Olivieri, S.T., J. Harrison, and J.R. Busby. 1995. Data and information management and communication. Pp. 607–670. *In* V.H. Heywood (ed.), Global Biodiversity Assessment. Cambridge University Press, Cambridge, England.

Olsson, M., A. Sukura, L.A. Lindberg, and E. Linder. 1996. Detection of *Pneumocystis carinii* DNA by filtration of air. Scandinavian Journal of Infectious Diseases 28:279–282.

Olsson, M., C. Lidman, S. Latouche, A. Bjorkman, P. Roux, E. Linder, and M. Wahlgren. 1998. Identification of *Pneumocystis carinii* f. sp. *hominis* gene sequences in filtered air in hospital environments. Journal of Clinical Microbiology 36:1737–1740.

Olsson, S., and Y. Persson. 1994. Transfer of phosphorus from *Rhizoctonia solani* to the mycoparasite *Arthrobotrys oligospora*. Mycological Research 98:1065–1068.

Omacini, M., E.J. Chaneton, and C.M. Ghersa. 2001. Symbiotic fungal endophytes control insect host-parasite interaction webs. Nature 409:78–81.

O'Neill, E.G., R.V. O'Neill, and R.J. Norby. 1991. Hierarchy theory as a guide to mycorrhizal research on large-scale problems. Environmental Pollution 73:271–284.

Onions, A.H.S. 1989. Prevention of mites in cultures. World Federation for Culture Collections. UNESCO/WFCC Education Committee. Web site: http://www2.cbs.knaw.nl/publications/download/wfcc01.pdf

Orange, A., P.W. James, and F.J. White. 2001. Microchemical Methods for the Identification of Lichens. British Lichen Society, London, England.

Orpin, C.G. 1975. Studies on the rumen flagellate *Neocallimastix frontalis*. Journal of General Microbiology 91:249–262.

——. 1976. Studies on the rumen flagellate *Sphaeromonas communis*. Journal of General Microbiology 94:270–280.

——. 1977a. Invasion of plant tissue by the rumen flagellate *Neocallimastix frontalis*. Journal of General Microbiology 98:423–430.

——. 1977b. The rumen flagellate *Piromonas communis*: its life cycle and invasion of plant material in the rumen. Journal of General Microbiology 99:107–117.

——. 1981. Isolation of cellulolytic phycomycete fungi from the caecum of the horse. Journal of General Microbiology 123:287–296.

——. 1983. The role of ciliate protozoa and fungi in the rumen digestion of plant cell walls. Animal Feed Science and Technology 10:121–143.

Orpin, C.G., and Y. Greenwood. 1986. Nutritional and germination requirements of the rumen chytridiomycete *Neocallimastix patriciarum*. Transactions of the British Mycological Society 86:103–109.

Orpin, C.G., and K.N. Joblin. 1988. The rumen anaerobic fungi. Pp. 129–150. *In* P.N. Hobson (ed.), The Rumen Microbial Ecosystem. Elsevier Applied Science, London, England.

Orpin, C.G., and A.J. Letcher. 1979. Utilization of cellulose, starch, xylan and other hemicelluloses for growth by the rumen phycomycete *Neocallimastix frontalis*. Current Microbiology 3:121–124.

Orpin, C.G., and E.A. Munn. 1986. *Neocallimastix patriciarum* sp. nov., a new member of the Neocallimasticaceae inhabiting the rumen of sheep. Transactions of the British Mycological Society 86:178–181.

Orpin, C.G., S.D. Mathiesen, Y. Greenwood, and A.S. Blix. 1985. Seasonal changes in the ruminal microflora of the high-arctic Svalbard reindeer (*Rangifer tarandus platyrhynchus*). Applied and Environmental Microbiology 50:144–151.

Orpurt, P.A., and J.T. Curtis. 1957. Soil microfungi in relation to the prairie continuum in Wisconsin. Ecology 38:628–637.

Otrosina, W.J., T.E. Chase, F.W. Cobb, Jr., and K. Korhonen. 1993. Population structure of *Heterobasidion annosum* from North America and Europe. Canadian Journal of Botany 71:1064–1071.

Ou, S.H. 1985. Rice Diseases. 2nd ed. Commonwealth Mycological Institute, Kew, Surrey, England.

Øvstedal, D.O., and R.I.L. Smith. 2001. Lichens of Antarctica and South Georgia. Cambridge University Press, Cambridge, United Kingdom.

Owens, J.L., and C.M. Uyeda. 1991. Single primer amplification of avian DNA detects polymorphic loci. Animal Biotechnology 2:107–122.

Pace, N.R., G.J. Olson, and C.R. Woese. 1986. Ribosomal RNA phylogeny and the primary lines of evolutionary descent. Cell 45:325–326.

Pacioni, G. 1994. Wet-sieving and decanting techniques for the extraction of spores of vesicular-arbuscular fungi. Pp. 777–782. *In* J.R. Norris, D. Read, and A.K. Varma (eds.), Techniques for Mycorrhizal Research. Academic Press, London, England.

Paden, J.W. 1967. A centrifugation technique for separating ascospores from soil. Mycopathologia et Mycologia Applicata 33:382–384.

Padgett, D.E., A.S. Kendrick, J.H. Hearth, and W.D. Webster. 1988. Influence of salinity, temperature, and nutrient availability on the respiration of saprolegniaceous fungi (Oomycetes). Holarctic Ecology 11:119–126.

Page, F.C., and R.L. Blanton. 1985. The Heterolobosea (Sarcodina: Rhizopoda), a new class uniting the Schizopyrenida and the Acrasidae (Acrasea). Protistologica 21:121–132.

Palm, M.E., and I.H. Chapela. 1998. Mycology in Sustainable Development: Expanding Concepts, Vanishing Borders. Parkway, Boone, North Carolina.

Palmer, M.W., and P.S. White. 1994. Scale dependence and the species-area relationship. The American Naturalist 144:717–740.

Pan, J., H.H. Ho, and S.C. Jong. 1994a. A scanning electron microscopy study of *Phytophthora* and *Halophytophthora* species. Mycotaxon 51:257–279.

Pan, S., L. Sigler, and G.T. Cole. 1994b. Evidence for a phylogenetic connection between *Coccidioides immitis* and *Uncinocarpus reesei* (Onygenaceae). Microbiology 140:1481–1494.

Pandit, A., and R. Maheshwari. 1996. Life-history of *Neurospora intermedia* in a sugar cane field. Journal of Biosciences 21:57–79.

Pannetier, C., S. Delassus, S. Darche, C. Saucier, and P. Kourilsky. 1993. Quantitative titration of nucleic acids by enzymatic amplification reactions run to saturation. Nucleic Acids Research 21:577–583.

Papavizas, G.C. 1985. *Trichoderma* and *Gliocladium*: biology, ecology and potential for biocontrol. Annual Review of Phytopathology 23:23–54.

Papavizas, G.C., B.B. Morris, and J.J. Marois. 1983. Selective isolation and enumeration of *Laetisaria arvalis* from soil. Phytopathology 73:220–223.

Paperna, I. 1986. *Ichthyophonus* infection in grey mullets from Southern Africa: histopathogical and ultrastructural study. Diseases of Aquatic Organisms 1:89–97.

Paquin, B., M.-J. Laforest, L. Forget, I. Roewer, W. Zhang, J. Longcore, and B.F. Lang. 1997. The fungal mitochondrial genome project: evolution of fungal mitochondrial genomes and their gene expression. Current Genetics 31:380–395.

Parbery, D.G. 1978. *Phyllachora*, *Linochora* and hyperparasites. Pp. 263–277. *In* C.V. Subramanian (ed.), Taxonomy of Fungi. Pt. 1. University of Madras, Madras, India.

Parbery, I.H., and J.F. Brown. 1986. Sooty moulds and black mildews in extra-tropical rainforests. Pp. 101–120. *In* N.J. Fokkema and J. van den Heuvel (eds.), Microbiology of the Phyllosphere. Cambridge University Press, Cambridge, England.

Park, D. 1974. Aquatic hyphomycetes in non-aquatic habitats. Transactions of the British Mycological Society 63:183–189.

——. 1980. A method for isolating pigmented cellulolytic *Pythium* from soil. Transactions of the British Mycological Society 75:491–492.

Park, D., and P.M. Robinson. 1964. Isolation and bioassay of a fungal morphogen. Nature 203:988–989.

Park, Y.-H., J.P. Stack, and C.M. Kenerley. 1992. Selective isolation and enumeration of *Gliocladium virens* and *G. roseum* from soil. Plant Disease 76:230–235.

Parker, A.D. 1979. Associations between coprophilous Ascomycetes and fecal substrates in Illinois. Mycologia 71:1206–1214.

Parkinson, D. 1994. Filamentous fungi. Pp. 329–350. *In* R.W. Weaver, S. Angle, P. Bottomley, D. Bezdicek, S. Smith, A. Tabatabai, and A.

Wollum (eds.), Methods of Soil Analysis, Part 2. Microbiological and Biochemical Properties. Soil Science Society of America, Madison, Wisconsin.

Parkinson, D., and A. Thomas. 1965. A comparison of methods for the isolation of fungi from rhizospheres. Canadian Journal of Microbiology 11:1001–1007.

Parkinson, D., and J.S. Waid (eds.). 1960. The Ecology of Soil Fungi. Liverpool University Press, Liverpool, Great Britain.

Parkinson, D., and S.T. Williams. 1961. A method for isolating fungi from soil microhabitats. Plant and Soil 13:347–355.

Parkinson, D., T.R.G. Gray, and S.T. Williams. 1971. Methods for Studying the Ecology of Soil Micro-organisms. IBP Handbook No. 19. Blackwell Scientific, Oxford, England.

Parkinson, S.M., M. Wainwright, and K. Killham. 1989. Observations on oligotrophic growth of fungi on silica gel. Mycological Research 93:529–534.

Parmasto, E. 1974. *Helicogonium jacksonii* (Dipodascales) found in Caucasus. Folia Cryptogamica Estonica 6:41–43.

———. 1998. *Athelia arachnoidea*, a lichenicolous basidiomycete in Estonia. Folia Cryptogamica Estonica 32:63–66.

Pasarell, L., and M.R. McGinnis. 1992. Viability of fungal cultures maintained at –70°C. Journal of Clinical Microbiology 30:1000–1004.

Pasricha, R., R.N. Kumar, and K.G. Mukerji. 1994. Effect of water stress on growth and sporulation of coprophilous fungi. Nova Hedwigia 59:157–162.

Paternoster, M.P., and R.G. Burns. 1996. A novel medium for the oosporogenesis of *Aphanomyces cochlioides*. Mycological Research 100:936–938.

Paterson, R.A. 1967. Benthic and planktonic phycomycetes from Northern Michigan. Mycologica 59:405–416.

Paterson, R.R.M., and P.D. Bridge. 1994. Biochemical Techniques for Filamentous Fungi. CAB International, Wallingford, Oxon, United Kingdom.

Patouilliard, N. 1900. Essai taxonomique sur les families et les genres des hyménomycètes. Lucien Declume, Lons-le-Saunier, France.

Patterson, D.J., and M.L. Sogin. 1992. Eukaryote origins and protistan diversity. Pp 13–46. *In* H. Hartman and K. Matsuno (eds.), The Origin and Evolution of the Cell. World Scientific, Singapore.

Paulitz, T.C., and R. Baker. 1988. Interactions between *Pythium nunn* and *Pythium ultimum* on bean leaves. Canadian Journal of Microbiology 34:947–951.

Paulitz, T.C., and R.G. Linderman. 1991a. Lack of antagonism between the biocontrol agent *Gliocladium virens* and vesicular arbuscular mycorrhizal fungi. New Phytologist 117:303–308.

———. 1991b. Mycorrhizal interactions with soil organisms. Pp. 77–129. *In* D.K. Arora, B. Rai, K.G. Mukerji, and G. Knudsen (eds.), Handbook of Applied Mycology. Vol. I. Soil and Plants. Marcel Dekker, New York.

Paulitz, T.C., and J.A. Menge. 1984. Is *Spizellomyces punctatum* a parasite or saprophyte of vesicular-arbuscular mycorrhizal fungi? Mycologia 76:99–107.

———. 1986. The effects of a mycoparasite on the mycorrhizal fungus, *Glomus deserticola*. Phytopathology 76:351–354.

Payne, T.L., J.C. Dickens, and J.C. Richerson. 1984. Insect predator-prey coevolution via enantiomeric specificity in a kairomone-pheromone system. Journal of Chemical Ecology 10:487–492.

Pearce, R.B. 1984. Staining fungal hyphae in wood. Transactions of the British Mycological Society 82:564–567.

Pearson, D.L. 1994. Selecting indicator taxa for the quantitative assessment of biodiversity. Philosophical Transactions of the Royal Society, London. Series B, 345:75–79.

Peck, J.E., and B. McCune. 1997. Effects of green tree retention on epiphytic lichen communities: A retrospective approach. Ecological Applications 7:1181–1187.

Peduzzi, R., and S. Bizzozero. 1977. Immunochemical investigation of four *Saprolegnia* species with parasitic activity in fish: serological and kinetic characterization of a chymotrypsin-like activity. Microbial Ecology 3:107–118.

Pegg, K.G., and J.L. Alcorn. 1982. *Phytophthora operculata* sp. nov., a new marine fungus. Mycotaxon 16:99–102.

Pegler, D.N. 1977. A Preliminary Agaric Flora of East Africa. Royal Botanic Gardens, Kew, Surrey, England.

———. 1983a. Agaric Flora of the Lesser Antilles. Royal Botanic Gardens, Kew, Surrey, England.

———. 1983b. The Genus *Lentinus*. A World Monograph. Royal Botanic Gardens, Kew, Surrey, England.

———. 1986. Agaric Flora of Sri Lanka. Royal Botanic Gardens, Kew, Surrey, England.

Pegler, D.N., L. Boddy, B. Ing, and P.M. Kirk. 1993. Fungi of Europe: Investigation, Recording, and Mapping. Royal Botanic Gardens, Kew, Surrey, England.

Pegler, D.N., T. Læssøe, and B.M. Spooner. 1995. British Puffballs, Earthstars and Stinkhorns: An Account of the British Gasteroid Fungi. Royal Botanic Gardens, Kew, Surrey, England.

Pehl, L., and H. Butin. 1994. Endophytische Pilze in Blättern von Laubbäumen und ihre Beziehungen zu Blattgallen (Zoocecidien). Mitteilungen aus der Biologishen Bundesanstalt für Land- und Forstwirtschaft 297:1–56. Berlin-Dahlem, Germany.

Peláez, F., A. Cabello, G. Platas, M.T. Díez, A. González del Val, A. Basilio, I. Martán, F. Vicente, G.F. Bills, R.A. Giacobbe, R.E. Schwartz, J.C. Onishi, M.S. Meinz, G.K. Arbruzzo, A.M. Flattery, L. Kong, and M.B. Kurtz. 2000. The discovery of enfumafungin, a novel antifungal compound produced by endophytic *Hormonema* species, biological activity and taxonomy of the producing organisms. Systematic and Applied Microbiology 23:333–343.

Peláez, F., J. Collado, F. Arenal, A. Basilio, A. Cabello, M.T. Díez-Matas, J.B. García, A. González del Val, V. González, J. Gorrochategui, P. Hernández, I. Martín, G. Platas, and F. Vicente. 1998. Endophytic fungi from plants living on gypsum soils as source of secondary metabolites with antimicrobial activity. Mycological Research 102:755–761.

Pemberton, C.M., R.A. Davey, J. Webster, M.W. Dick, and G. Clark. 1990. Infection of *Pythium* and *Phytophthora* species by *Olpidiopsis gracilis* (Oomycetes). Mycological Research 94:1081–1085.

Peresse, M., and D. Le Picard. 1980. *Hansfordia pulvinata*, mycoparasite destructeur du *Cladosporium fulvum*. Mycopathologia 71:23–30.

Perkins, D.D. 1962. Preservation of *Neurospora* stock cultures with anhydrous silica gel. Canadian Journal of Microbiology 8:591–594.

———. 1977. Details for preparing silica gel stocks. *Neurospora* Newsletter 24:16–17.

Perkins, D.D., B.C. Turner, and E.G. Barry. 1976. Strains of *Neurospora* collected from nature. Evolution 30:281–313.

Perring, T.M., A.D. Cooper, R.J. Rodriguez, C.A. Farrar, and T.S. Bellows, Jr. 1993. Identification of a whitefly species by genomic and behavioral studies. Science 259:74–77.

Perrott, P.E. 1960. The ecology of some aquatic Phycomycetes. Transactions of the British Mycological Society 43:19–30.

Perry, T.J. 1991. A synopsis of the taxonomic revisions in the genus *Ceratocystis* including a review of blue-staining species associated with *Dendroctonus* bark beetles. U.S. Department of Agriculture, Forest Service, Southern Forest Experiment Station, New Orleans, Louisiana. General Technical Report SO-86.

Persiani, A.M., and O. Maggi. 1988. Fungal communities in the rhizosphere of *Coffea arabica* L. in Mexico. Micologia Italiana 17:21–37.

Persson, M., and K. Söderhäll. 1983. *Pacifastacus leniusculus* Dana and its resistance to the parasitic fungus *Aphanomyces astaci* Shikora. Pp.

292–298. *In* C.R. Goldman (ed.), Freshwater Crayfish. AVI Publishing, Westport, Connecticut.

Persson, Y. 1991. Mycoparasitism by the Nematode-trapping Fungus *Arthrobotrys oligospora*. Unpubl. Ph.D. dissert., Lund University, Lund, Sweden.

Persson, Y., and E. Bååth. 1992. Quantification of mycoparasitism by the nematode-trapping fungus *Arthrobotrys oligospora* on *Rhizoctonia solani* and the influence of nutrient levels. FEMS Microbiology Ecology 101:11–16.

Persson, Y., and E. Friman. 1993. Intracellular proteolytic activity in mycelia of *Arthrobotrys oligospora* bearing mycoparasitic or nematode trapping structures. Experimental Mycology 17:182–190.

Persson, Y., M. Veenhuis, and B. Nordbring-Hertz. 1985. Morphogenesis and significance of hyphal coiling by nematode-trapping fungi in mycoparasitic relationships. FEMS Microbiology Ecology 31:283–291.

Peter, M., F. Ayer, and S. Egli. 2001. Nitrogen addition in a Norway spruce stand altered macromycete sporocarp production and belowground ectomycorrhizal species composition. New Phytologist 149: 311–325.

Petersen, A.B., and S. Rosendahl. 2000. Phylogeny of the Peronosporomycetes (Oomycota) based on partial sequences of the large ribosomal subunit (LSU rDNA). Mycological Research 104: 1295–1303.

Petersen, P.M. 1970. Danish fireplace fungi: An ecological investigation of fungi on burns. Dansk Botanisk Arkiv 27:1–97.

——. 1971. The macromycetes in a burnt forest area in Denmark. Botanisk Tiddskrift 65:228–248.

Petersen, R.H., K.W. Hughes, S.A. Redhead, N. Psurtseva, and A.S. Methven. 1999. Mating systems in the Xerulaceae (Agaricales, Basidiomycetes): *Flammulina*. Mycoscience 40:411–426.

Peterson, R.L., and S.M. Bradbury. 1995. Use of plant mutants, intraspecific variants, and non-hosts in studying mycorrhiza formation and function. Pp. 157–180. *In* A. Varma and B. Hock (eds.), Mycorrhiza: Structure, Function, Molecular Biology and Biotechnology. Springer Verlag, Berlin, Germany.

Peterson, S.W., and C.P. Kurtzman 1991. Ribosomal RNA sequence divergence among sibling species of yeasts. Systematic and Applied Microbiology 14:124–129.

Peterson, S.W., and L. Sigler. 1998. Molecular genetic variation in *Emmonsia crescens* and *Emmonsia parva*, etiologic agents of adiaspiromycosis, and their phylogenetic relationship to *Blastomyces dermatitidis* (*Ajellomyces dermatitidis*) and other systemic fungal pathogens. Journal of Clinical Microbiology 36:2918–2925.

Petrini, L., and O. Petrini. 1985. Xylariaceous fungi as endophytes. Sydowia 38:216–234.

Petrini, L.E. 1992. *Rosellinia* species of the temperate zones. Sydowia 44:169–281.

Petrini, O. 1985. Wirtsspezifität endophytischer Pilze bei einheimischen Ericaceae. Botanica Helvetica 95:213–238.

——. 1986. Taxonomy of endophytic fungi of aerial plant tissues. Pp. 175–187. *In* N.J. Fokkema and J. van den Heuvel (eds.), Microbiology of the Phyllosphere. Cambridge University Press, Cambridge, England.

——. 1987. Endophytic fungi of alpine Ericaceae. The endophytes of *Loiseleuria procumbens*. Pp. 71–77. *In* G.A. Laursen, J.F. Ammirati, and S.A. Redhead (eds.), Arctic and Alpine Mycology II. Plenum, New York.

——. 1991. Fungal endophytes of tree leaves. Pp. 179–197. *In* J.H. Andrews and S.S. Hirano (eds.), Microbial Ecology of Leaves. Springer-Verlag, New York.

Petrini, O., and G. Carroll. 1981. Endophytic fungi in foliage of some Cupressaceae in Oregon. Canadian Journal of Botany 59:629–636.

Petrini, O., and M. Dreyfuss. 1981. Endophytische Pilze in epiphytischen Araceae, Bromeliaceae, und Orchidaceae. Sydowia 34:135–148.

Petrini, O., and P.J. Fisher. 1986. Fungal endophytes of *Salicornia perennis*. Transactions of the British Mycological Society 87:647–651.

——. 1988. A comparative study of fungal endophytes in xylem and whole stem of *Pinus sylvestris* and *Fagus sylvatica*. Transactions of the British Mycological Society 91:233–238.

——. 1990. Occurrence of fungal endophytes in twigs of *Salix fragilis* and *Quercus robur*. Mycological Research 94:1077–1080.

Petrini, O., and E. Müller. 1979. Pilzliche endophyten, am beispiel von *Juniperus communis* L. Sydowia 32:224–251.

Petrini, O., P.J. Fisher, and L.E. Petrini. 1992a. Fungal endophytes of bracken (*Pteridium aquilinium*) with some reflections on biological control. Sydowia 44:282–293.

Petrini, O., U. Hake, and M.M. Dreyfuss. 1990. An analysis of fungal communities isolated from fruticose lichens. Mycologia 82:444–451.

Petrini, O., L.E. Petrini, and M.M. Dreyfuss. 1992b. Psychrophylic deuteromycetes from alpine habitats. Mycologia Helvetica 5:9–20.

Petrini, O., L.E. Petrini, and K.F. Rodrigues. 1995. Xylariaceous endophytes: An exercise in biodiversity. Fitopatologia Brasiliera 20:531–539.

Petrini, O., J. Stone, and F.E. Carroll. 1982. Endophytic fungi in evergreen shrubs in western Oregon: a preliminary study. Canadian Journal of Botany 60:789–796.

Pfeifer, T.A., and G.G. Khachatourians. 1987. The formation of protoplasts from *Beauveria bassiana*. Applied Microbiology and Biotechnology 26:248–253.

——. 1989. Isolation and characterization of DNA from the entomopathogen *Beauveria bassiana*. Experimental Mycology 13:392–402.

——. 1993. Isolation of DNA from entomopathogenic fungi grown in liquid cultures. Journal of Invertebrate Pathology 61:113–116.

Pfeifer, T.A., D.D. Hegedus, and G.G. Khachatourians. 1993. The mitochondrial genome of the entomopathogenic fungus *Beauveria bassiana*: Analysis of the ribosomal RNA region. Canadian Journal of Microbiology 39:25–31.

Pfender, W.F., and S.L. Wootke. 1988. Microbial communities of *Pyrenophora*-infested wheat straw as examined by multivariate analysis. Microbial Ecology 15:95–113.

Pfenning, L. 1993. Mikroskopische Bodenpilze des Ostamazonischen Regenwaldes (Brasilien). Unpubl. Ph. D. dissert., Universität Tübingen, Tübingen, Germany.

Pfister, D.H. 1993. Roland Thaxter and the myxobacteria. Pp. 1–11. *In* M. Dworkin and D. Kaiser (eds.), Myxobacteria II. American Society for Microbiology, Washington, D.C.

——. 1997. Castor, Pollux and life histories of fungi. Mycologia 89:1–23.

Pfister, D.H., and J.W. Kimbrough. 2001. Discomycetes. Pp. 257–281. *In* D.J. McLaughlin, E.G. McLaughlin, and P.A. Lempke (eds.), The Mycota. Vol. VII. Systematics and Evolution. Part A. Springer-Verlag, Berlin, Germany.

Pfister, D.H., J.R. Boise, and M.A. Eifler. 1990. A Bibliography of Taxonomic Mycological Literature 1753–1821. Mycologica Memoirs No. 17.

Phaff, H.J., and W.T. Starmer. 1987. Yeasts associated with plants, insects and soil. Pp. 123–180. *In* A.H. Rose and J.S. Harrison (eds.), The Yeasts: Biology of Yeasts. 2nd ed. Academic Press, London, England.

Phaff, H.J., M.W. Miller, and E.M. Mrak. 1966. The Life of Yeasts. Their Nature, Activity, Ecology, and Relation to Mankind. Harvard University Press, Cambridge, Massachusetts.

Phelps, J.W. 1973. Microfungi in two Wisconsin sand blows. Transactions of the British Mycological Society 61:386–390.

Philipp, W.-D. 1985. Extracellular enzymes and nutritional physiology of *Ampelomyces quisqualis* Ces., hyperparasite of powdery mildew, *in vitro*. Phytopathologische Zeitschrift 114:274–283.

——. 1988. Biologische Bekämpfung von Pflanzenkrankheiten. E. Ulmer, Stuttgart, Germany.

Philipp, W.-D., E. Beuther, D. Hermann, F. Klinkert, C. Oberwalder, M. Schmidtke, and B. Straub. 1990. Zur Formulierung des Mehltauhyperparasiten *Ampelomyces quisqualis* Ces. Zeitschrift für Pflanzenkrankheiten und Pflanzenschutz 97:120–132.

Phillips, A.J.L. 1986. Factors affecting the parasitic activity of *Gliocladium virens* on sclerotia of *Sclerotinia sclerotiorum* and a note on its host range. Journal of Phytopathology 116:212–220.

Phillips, A.J.L., and K. Price. 1983. Structural aspects of the parasitism of sclerotia of *Sclerotinia sclerotiorum* (Lib.) de Bary by *Coniothyrium minitans* Campb. Phytopathologische Zeitschrift 107:193–201.

Phillips, J.M., and D.S. Hayman. 1970. Improved procedures for clearing roots and staining parasitic and vesicular-arbuscular mycorrhizal fungi for rapid assessment of infection. Transactions of the British Mycological Society 55:158–161.

Phillips, M.W., and G.L.R. Gordon. 1988. Sugar and polysaccharide fermentation by rumen anaerobic fungi from Australia, Britain and New Zealand. BioSystems 21:377–383.

Phillips, R. 1991. Mushrooms of North America. Little Brown, Boston, Massachusetts.

Phillips, T.W., A.J. Wilkening, T.H. Atkinson, J.L. Nation, R.C. Wilkinson, and J.L. Foltz. 1988. Synergism of turpentine and ethanol as attractants for certain pine-infesting beetles (Coleoptera). Environmental Entomology 17:456–462.

Pickering, A.D., and L.G. Willoughby. 1982. *Saprolegnia* infections of salmonid fish. Pp. 271–297. *In* R.J. Roberts (ed.), Microbial Diseases of Fish. Academic Press, London, England.

Pickering, A.D., L.G. Willoughby, C.B. McGrory. 1979. Fine structure of secondary zoospore cyst cases of *Saprolegnia* isolates from infected fish. Transactions of the British Mycological Society 72:427–436.

Pielou, E.C. 1975. Ecological Diversity. Wiley, New York.

——. 1984. The Interpretation of Ecological Data. A Primer on Classification and Ordination. Wiley, New York.

Piepenbring, M., and R. Bauer. 1997. *Erratomyces*, a new genus of Tilletiales with species on Leguminosae. Mycologia 89:924–936.

Pike, L.H. 1978. The importance of epiphytic lichens in mineral cycling. Bryologist 81:247–257.

Pike, L.H., W.C. Denison, D.M. Tracy, M.A. Sherwood, and F.M. Rhoades. 1975. Floristic survey of epiphytic lichens and bryophytes growing on living, old-growth conifers in western Oregon. The Bryologist 78:389–402.

Pike, L.H., R.A. Rydell, and W.C. Denison. 1977. A 400-year-old Douglas fir tree and its epiphytes: biomass, surface area, and their distributions. Canadian Journal of Forest Research 7:680–699.

Pilz, D., and R. Molina (eds). 1996. Managing forest ecosystems to conserve fungus diversity and sustain wild mushroom harvests. U. S. Department of Agriculture, Forest Service, Pacific Northwest Research Station, Portland, Oregon. General Technical Report PNW-GTR-371.

Pimentel, G., L.M. Carris, and T.L. Peever. 2000. Characterization of interspecific hybrids between *Tilletia controversa* and *T. bromi*. Mycologia 92:411–420.

Pincus, D.H., and I.F. Salkin. 1988. Human infections caused by yeast-like fungi. Pp. 239–269. *In* B.B. Wentworth (Coord. ed.), Diagnostic Procedures for Mycotic and Parasitic Infections, 7th ed. American Public Health Association, Washington D.C.

Pincus, D.H., I.F. Salkin, and M.R. McGinnis. 1988. Rapid methods in medical mycology. Laboratory Medicine 19:315–320.

Pirozynski, K.A. 1969. Reassessment of the genus *Amblyosporium*. Canadian Journal of Botany 47:325–334.

——. 1974. *Meliolina mollis* and two hyperparasites in India. Kavaka 2:33–41.

——. 1976. Notes on hyperparasitic Sphaeriales, Hypocreales and "hypocreoid Dothideales." Kew Bulletin 31:595–610.

Pirozynski, K.A., and D.L. Hawksworth. 1988. Coevolution of Fungi with Plants and Animals. Academic Press, New York.

Pirozynski, K.A., and R.A. Shoemaker. 1972. *Vestigium*, a new genus of Coelomycetes. Canadian Journal of Botany 50:1163–1165.

Pirozynski, K.A., and L.K. Weresub. 1979. A biogeographic view of the history of ascomycetes and the development of their pleomorphism. Pp. 93–123. *In* B. Kendrick (ed.), The Whole Fungus: The Sexual-Asexual Synthesis. National Museums of Canada, Ottawa, Ontario, Canada.

Pitt, J.I. 1975. Xerophilic fungi and the spoilage of foods of plant origin. Pp. 273–307. *In* R.B. Duckworth (ed.), Water Relations of Foods. Academic Press, London, England.

——. 1979. The Genus *Penicillium* and its Teleomorphic States *Eupenicillium* and *Talaromyces*. Academic Press, London, United Kingdom.

——. 1981. Food spoilage and biodeterioration. Pp. 111–142. *In* G.T. Cole and B. Kendrick (eds.), Biology of Conidial Fungi. Vol. 2. Academic Press, New York.

——. 1990. Penname, a new computer key to common *Penicillium* species. Pp. 279–287. *In* R.A. Samson and J.I. Pitt (eds.), Modern Concepts in *Penicillium* and *Aspergillus* Classification. Plenum Press, New York.

——. 1991. A Laboratory Guide to Common *Penicillium* Species. 2nd ed. Commonwealth Scientific and Industrial Research Organization (CSIRO), Division of Food Processing. North Ryde, New South Wales, Australia.

Pitt, J.I., and A.D. Hocking. 1997. Fungi and Food Spoilage. 2nd ed. Blackie Academic Professional, London, United Kingdom.

Planta, R.J. 1997. Regulation of ribosome synthesis in yeast. Yeast 13:1505–1518.

Platt, J.L. 2000. Lichens, Earth Tongues, and Endophytes: Evolutionary Patterns Inferred from Phylogenetic Analyses of Multiple Loci. Unpubl. Ph.D. dissert. Oregon State University, Corvallis, Oregon.

Platt, J.L., and J.W. Spatafora. 2000. Evolutionary relationships of nonsexual lichenized fungi: molecular phylogenetic hypotheses for the genera *Siphula* and *Thamnolia* from SSU and LSU rDNA. Mycologia 92:475–487.

Pocock, K., and J.G. Duckett. 1984. A comparative ultrastructural analysis of the fungal endophytes in *Cryptothallus mirabilis* Malm. and other British thalloid hepatics. Journal of Bryology 13:227–233.

——. 1985a. Fungi in hepatics. Bryological Times 31:2–3.

——. 1985b. On the occurrence of branched and swollen rhizoids in British hepatics: their relationships with the substratum and associations with fungi. New Phytologist 99:281–304.

Poelt, J. 1969. Bestimmungsschlüssel europäischer Flechten. J. Cramer, Lehre, Germany.

Poelt, J. von, and A. Vězda. 1977. Bestimmungsschlüssel europäischer Flechten. I. J. Cramer, Vaduz, Liechtenstein.

——. 1981. Bestimmungsschlüssel europäischer Flechten. Bibliotheca Lichenologica 16:1–390.

Poindexter, J.S. 1981. Oligotrophy: feast and famine existence. Advances in Microbial Ecology 5:63–90.

Põldmaa, K. 1996. A new species of *Hypomyces* and three of *Cladobotryum* from Estonia. Mycotaxon 59:389–405.

——. 1999. The genus *Hypomyces* and allied fungicolous fungi in Estonia. I. Species growing on aphyllophoralean basidiomycetes. Folia Cryptogamica Estonica 34:27–43.

——. 2000. Generic delimitation of the fungicolous Hypocreaceae. Studies in Mycology 45:83–94.

Põldmaa, K., and G.J. Samuels. 1999. Aphyllophoricolous species of *Hypomyces* with KOH-negative perithecia. Mycologia 91:177–199.

Põldmaa, K., E. Larsson, and U. Kõljalg. 1999. Phylogenetic relationships in *Hypomyces* and allied genera, with emphasis on species growing on wood-decaying homobasidiomycetes. Canadian Journal of Botany 77:1756–1768.

Põldmaa, K., G.J. Samuels, and D.J. Lodge. 1997. Three new polyporicolous species of *Hypomyces* and their *Cladobotryum* anamorphs. Sydowia 49:80–93.

Põldmaa, P. 1966. Data on some Deuteromycetes occurring with uredinales and Erysiphales. Eesti NSV Teaduste Akadeemia Toimetised, Bioloogiline [Proceedings of the Academy of Sciences of Estonia, Biology] Seeria 3:374–396.

Polglase, J.L. 1981. A preliminary report on the Thraustochytrid(s) and Labyrinthrulid(s) associated with a pathological condition in the lesser octopus *Eledone cirrhosa*. Botanica Marina 23:699–706.

Polglase, J.L., D.J. Alderman, and R.H. Richards. 1986. Aspects of the progress of mycotic infections in marine animals. Pp. 155–164. *In* S.T. Moss (ed.), The Biology of Marine Fungi. Cambridge University Press, Cambridge, Great Britain.

Polishook, J.D., G.F. Bills, and D.J. Lodge. 1996. Microfungi from decaying leaves of two rain forest trees in Puerto Rico. Journal of Industrial Microbiology 17:284–294.

Polishook, J.D., A.W. Dombrowski, N.N. Tsou, G.M. Salituro, and J.E. Curotto. 1993. Preussomerin D from the endophyte *Hormonema dematioides*. Mycologia 85:62–64.

Pomerleau, R. 1980. Flore des Champignons au Quebec et regiones limitrophes. Les Editions la Presse, Montreal, Quebec, Canada.

Poon, M.O.K., and K.D. Hyde. 1998. Biodiversity of intertidal estuarine fungi on *Phragmites* in Mai Po marshes, Hong Kong. Botanica Marina 41:141–155.

Porter, D. 1982. The appendaged ascospores of *Trichomaris invadens* (Halosphaeriaceae), a marine ascomycetous parasite of the tanner crab, *Chionoecetes bairdi*. Mycologia 74:363–375.

——. 1986. Mycoses of marine organisms: an overview of pathogenic fungi. Pp. 141–153. *In* S.T. Moss (ed.), The Biology of Marine Fungi. Cambridge University Press, Cambridge, Great Britain.

——. 1987. Isolation of zoosporic marine fungi. Pp. 128–129. *In* M.S. Fuller and A. Jaworski (eds.), Zoosporic Fungi in Teaching and Research. Southeastern Publishing, Athens, Georgia.

Porter, D.M., and J.L. Steele. 1983. Quantitative assay by elutriation of peanut field soil for sclerotia of *Sclerotinia minor*. Phytopathology 73:636–640.

Post, F.J. 1977. Microbial ecology of the Great Salt Lake. Microbial Ecology 3:143–165.

Post, G. 1983. Textbook of Fish Health. T.F.H. Publications, Neptune City, New Jersey.

Powell, M.J. 1981. Structure of the interface between the haustorium of *Caulochytrium protostelioides* and the hyphal cytoplasm of *Cladosporium cladosporioides*. Journal of the Elisha Mitchell Scientific Society 97:171–182.

——. 1982. Ultrastructure of the host-parasite interface between *Allomyces javanicus* and its endoparasite *Catenaria allomycis*. Botanical Gazette 143:176–187.

——. 1984. Fine structure of the unwalled thallus of *Rozella polyphagi* in its host *Polyphagus euglenae*. Mycologia 76:1039–1048.

——. 1993. Looking at mycology with a Janus face: a glimpse at Chytridiomycetes active in the environment. Mycologia 85:1–20.

Powell, M.J., and W.H. Blackwell. 1995. Searching for homologous ultrastructural characters in zoosporic fungi. Canadian Journal of Botany 73 (Suppl.): S693–S700.

Powell, M.J., and W.J. Koch. 1977a. Morphological variations in a new species of *Entophlyctis*. I. The species concept. Canadian Journal of Botany 55:1668–1685.

——. 1977b. Morphological variations in a new species of *Entophlyctis*. II. Influence of growth conditions on morphology. Canadian Journal of Botany 55:1686–1695.

Powell, MJ., C.E. Bracker, and D.J. Sternshein. 1981. Formation of chlamydospores in *Gilbertella persicaria*. Canadian Journal of Botany 59:908–928.

Powell, M.J., L.P. Lehnen, Jr., and R.N. Bortnick. 1985. Microbody-like organelles as taxonomic markers among Oomycetes. BioSystems 18:321–334.

Premdas, P.D., and B. Kendrick. 1991. Colonization of autumn-shed leaves by four aero-aquatic fungi. Mycologia 83:317–321.

Preston, F.W. 1948. The commonness and rarity of species. Ecology 29:254–283.

Prillinger, H., J. Altenbuchner, G. Laaser, and C. Dörfler. 1993. Yeasts isolated from Homobasidiomycetes (*Asterophora, Collybia*): new aspects for sexuality, taxonomy, and speciation. Experimental Mycology 17:24–45.

Prillinger, H., J. Altenbuchner, B. Schulz, Ch. Dörfler, Th. Forst, G. Laaser, and U. Stahl. 1989. *Ustilago maydis* isolated from Homobasidiomycetes. Pp. 408–419. *In* Proceedings Braunschweig Symposium on Applied Plant Molecular Biology. Braunschweig, Germany.

Pringsheim, N. 1855. Über die Befruchtung der Algen. Verhandlungen der Königlich Preussischen Akademie der Wissenschaften zu Berlin, pp. 133–165.

Pritchard, M.H., and G.O.W. Kruse. 1982. The Collection and Preservation of Animal Parasites. University of Nebraska Press, Lincoln, Nebraska.

Prowse, G.A. 1954. *Sommerstorffia spinosa* and *Zoophagus insidians* predacious on rotifers, and *Rozellopsis inflata* the endoparasite of *Zoophagus*. Transactions of the British Mycological Society 37:134–150.

Puckridge, J.T., K.F. Walker, J.S. Langdon, C. Daley, and G.W. Beakes. 1989. Mycotic dermatitis in a freshwater gizzard shad, the bony bream, *Nematalosa erebi* (Günther), in the River Murray, South Australia. Journal of Fish Diseases 12:205–221.

Pugh, G.J.F. 1958. Leaf litter fungi found on *Carex paniculata* L. Transactions of the British Mycological Society 41:185–195.

——. 1980. Strategies in fungal ecology. Transactions of the British Mycological Society 75:1–14.

Pugh, G.J.F., and D. Allsopp. 1982. Microfungi on Signy Island, South Orkney Islands. British Antarctic Survey Bulletin 57:55–67.

Pugh, G.J.F., and L. Boddy. 1988. A view of disturbance and life strategies in fungi. Proceedings of the Royal Society of Edinburgh 94B:3–11.

Pugh, G.J.F., and N.G. Buckley. 1971. The leaf surface as a substrate for colonization by fungi. Pp. 431–445. *In* T.F. Preece and C.H. Dickinson (eds.), Ecology of Leaf Surface Micro-organisms. Academic Press, London, England.

Punsola, L., and J. Guarro. 1984. *Keratinomyces ceretanicus* sp. nov., a psychrophilic dermatophyte from soil. Mycopathologia 85:185–190.

Purvis, O.W., B.J. Coppins, D.L. Hawksworth, P.W. James, and D.M. Moore (eds.). 1992. The Lichen Flora of Great Britain and Ireland. Natural History Museum Publications, London, United Kingdom.

Putman, R.J. 1994. Community Ecology. Chapman Hall, London, England.

Quesnel, L.B. 1969. Methods of microculture. Pp. 365–425. *In* J.R. Norris and D.W. Ribbons (eds.), Methods in Microbiology. Vol. 1. Academic Press, London, England.

Rabatin, S.C., and B.R. Stinner. 1988. Indirect effects of interactions between VAM fungi and soil-inhabiting invertebrates on plant processes. Agriculture, Ecosystems and Environment 24:135–146.

Radford, A. 1993. A fungal phylogeny based upon orotidine 5'-monophosphate decarboxylase. Journal of Molecular Evolution 36:389–395.

Raffa, K.F., and K.D. Klepzig. 1989. Chiral escape of bark beetles from predators responding to a bark beetle pheromone. Oecologia 80:566–569.

Ragan, M.A., C.L. Goggin, R.J. Cawthorn, L. Cerenius, A.V.C. Jamieson, S.M. Plourd, T.G. Rand, K. Söderhäll, and R.R. Gutell. 1996. A novel clade of protistan parasites near the animal-fungal divergence. Proceedings of the National Academy of Sciences USA 93:11907–11912.

Raghukumar, C., and V. Lande. 1988. Shell disease of rock oyster *Crassostrea cucullata*. Diseases of Aquatic Organisms 4:77–81.

Rai, J.N., J.P. Tewari, and S.C. Agarwal. 1974. The genus *Aspergillus* in India. Beihefte zur Nova Hedwigia 47:487–499.

Raina, K., and J.M. Chandler. 1996. Recovery of genomic DNA from a fungus (*Sclerotinia homoeocarpa*) with high polysaccharide content. BioTechniques 21:1030–1032.

Rajendren, R.B. 1968. *Muribasidiospora*—a new genus of the Exobasidiaceae. Mycopathologia et Mycologia Applicata 36:218–222.

Raju, N.B. 1986. A simple fluorescent staining method for meiotic chromosomes of *Neurospora*. Mycologia 78:901–906.

Raju, N.B., and D. Newmeyer. 1977. Giant ascospores and abnormal croziers in a mutant of *Neurospora crassa*. Experimental Mycology 1:152–165.

Rakvidhyasastra, V., and E.E. Butler. 1973. Mycoparasitism by *Stephanoma phaeospora*. Mycologia 65:580–583.

Ralls, K., and J. Ballou 1983. Extinction: Lessons from zoos. Pp. 164–184. *In* C.M. Schonewald-Cox, S.M. Chambers, B. MacBryde, and W.L. Thomas (eds.), Genetics and Conservation. Benjamin Cummings, Menlo Park, California.

Rambelli, A. 1973. The rhizosphere of mycorrhizae. Pp. 299–349. *In* G.L. Marks and T.T. Koslowski (eds.), Ectomycorrhizae: Their Ecology and Physiology. Academic Press, New York.

Rambelli, A., A.M. Persiani, O. Maggi, D. Lunghini, S. Onofri, S. Riess, G. Dowgiallo, and G. Puppi. 1983. Comparative Studies on Microfungi in Tropical Ecosystems. UNESCO, Rome, Italy.

Rambelli, A., A.M. Persiani, O. Maggi, S. Onofri, S. Riess, G. Dowgiallo, and L. Zucconi. 1984. Comparative studies on microfungi in tropical ecosystems. Further mycological studies in South Western Ivory Coast forest. Report no. 2. Giornale Botanico Italiano 118:201–243.

Rambelli, A., L. Zucconi, S. Onofri, and L. Quadraccia. 1991. Comparative studies on microfungi in tropical ecosystems. Further mycological studies in South Western Ivory Coast forest. Report no. 2. Giornale Botanico Italiano 125:779–796.

Rambold, G. 1993. Further species of the genus *Tephromela* (Lecanorales). Sendtnera 1:281–288.

Rambold, G., and D. Triebel. 1990. *Gelatinopsis, Geltingia* and *Phaeopyxis*: three Helotialean genera with lichenicolous species. Notes from the Royal Botanical Garden, Edinburgh 46:375–389.

Ramirez, C. 1982. Manual and Atlas of the Penicillia. Elsevier Biomedical Press, Amsterdam, The Netherlands.

Ramírez Gómez, C. 1957. Contribución al estudio de la ecología de las levaduras. I. Estudio de levaduras aisladas de hongos carnosos. Microbiología Española 10:215–247.

Ramos-Flores, T. 1983. Lower marine fungus associated with black line disease in star corals (*Montastrea annularis*, E. & S.). Biological Bulletin 165:429–435.

Ramsbottom J. 1953. Mushrooms and Toadstools: A Study of the Activities of Fungi. Collins, London, England.

Rand, T.G. 1990. Studies on the Biology of *Ichthyophonus hoferi* Plehn and Mulsow, 1911 from Nova Scotian Yellowtail Flounder, *Limanda ferruginea* (Storer). Unpubl. Ph.D. dissert., University of New Brunswick, Fredericton, New Brunswick, Canada.

——. 1992a. Seasonal and geographical distribution of *Ichthyophonus hoferi* Plehn and Mulsow, 1911 in yellowtail flounder, *Limanda ferruginea* on the Nova Scotia shelf, Canada. Journal of the Marine Biological Association of the United Kingdom 72:669–674.

——. 1992b. Microbial analysis of Atlantic salmon incubation facilities in Newfoundland. Department of Fisheries and Oceans, Canada/Newfoundland Inshore Fisheries Development Agreement # 4086. St. Johns, Newfoundland, Canada.

——. 1996. Fungal diseases of fish and shellfish. Pp. 297–313. *In* H. Howard and J.D. Miller (eds.), The Mycota. Vol. VI. Human and Animal Relationships. Springer-Verlag, Berlin, Germany.

——. 2000. Diseases of marine animals. Pp. 21–48. *In* K.D. Hyde and S.B. Pointing (eds.), Marine Mycology: A Practical Approach. Fungal Diversity Press, Hong Kong, China.

Rand, T.G., and D.K. Cone. 1990. Effects of *Ichthyophonus hoferi* on condition indices and blood chemistry of experimentally infected rainbow trout (*Oncorhynchus mykiss*). Journal of Wildlife Disease 26:323–328.

Rand, T.G., and D. Munden. 1993. Involvement of zoospores of *Saprolegnia diclina* (Oomycotina: Saprolegniaceae) in the attachment and invasion of eggs of brook trout under experimental conditions. Journal of Aquatic Animal Health 5:233–239.

Rand, T.G., and M. Wiles. 1988. Bacterial involvement in saddleback disease of reef silversides *Atherina harringtonensis* Goode, in Bermuda. Journal of Fish Biology 32:805–816.

Rand, T.G., L. Bunkley-Williams, and E.H. Williams. 2000a. A hyphomycete fungus, *Paecilomyces lilacinus*, associated with wasting disease in two species of tilapia from Puerto Rico. Journal of Aquatic Animal Health 12:149–156.

Rand, T.G., M. Ragan, R. Singh, and A. Jamieson. 1994. Systematics of *Ichthyophonus* based on evidence from ribosomal DNA sequences. *In* Abstracts, 10th Biennial Meeting of the International Society of Evolutionary Protistology, Halifax, Nova Scotia, Canada.

Rand, T.G., K. White, J.J. Cannone, R.R. Gutell, C.A. Murphy, and M.A. Ragan. 2000b. *Ichthyophonus irregularis* sp. nov. from the yellowtail flounder *Limanda ferruginea* from the Nova Scotia shelf. Diseases of Aquatic Organisms 41:31–36.

Ranjard, L., D.P.H. Lejon, C. Mougel, L. Schehrer, D. Merdinoglu, and R. Chaussod. 2003. Sampling strategy in molecular microbial ecology: Influence of soil sample size on DNA fingerprinting analysis of fungal and bacterial communities. Environmental Microbiology 5:1111–1120.

Ranta, H., S. Neuvonen, and A. Ylimartimo. 1995. Interactions of *Gremmeniella abietina* and endophytic fungi in shoots of Scots pine trees treated with simulated acid rain. Journal of Applied Ecology 32:67–75.

Rao, N.N.R., and M.S. Pavgi. 1976. A mycoparasite on *Sclerospora graminicola*. Canadian Journal of Botany 54:220–223.

Raper, J.R. 1937. A method of freeing fungi from bacterial contamination. Science 85:342.

Raper, K.B. 1984. The Dictyostelids. Princeton University Press, Princeton, New Jersey.

Raper, K.B., and D.I. Fennell. 1965. The Genus *Aspergillus*. Williams and Wilkins, Baltimore, Maryland.

Rappaz, F. 1995. *Anthostomella* and related Xylariaceous fungi on hard wood from Europe and North America. Mycology Helvetica 7:99–168.

Raps, A., and S. Vidal. 1998. Indirect effects of an unspecialized endophytic fungus on specialized plant-herbivorous insect interactions. Oecologia 114:541–547.

Raven, P.H. 1988. Tropical floristics tomorrow. Taxon 37:549–560.

Rayner, A.D.M. 1994. Evolutionary processes affecting adaptation to saprotrophic life styles in ascomycete populations. Pp. 261–271. *In* D.L. Hawksworth (ed.), Ascomycete Systematics: Problems and Perspectives in the Nineties. Plenum Press, New York.

Rayner, A.D.M., and L. Boddy. 1988. Fungal Decomposition of Wood, Its Biology and Ecology. John Wiley, Chichester, England.

Rayner, A.D.M., L. Boddy, and C.G. Dowson. 1987. Temporary parasitism of *Coriolus* spp. by *Lenzites betulina*: A strategy for domain capture in wood decay fungi. FEMS Microbiology Ecology 45:53–58.

Rayner, A.D.M., R. Watling, and J.C. Frankland. 1985. Resource relations—an overview. Pp. 1–40. *In* D. Moore, L.A. Casselton, D.A. Wood, and J.C. Frankland (eds.), Developmental Biology of Higher Fungi. Cambridge University Press, Cambridge, England.

Rayner, M.C. 1915. Obligate symbiosis in *Calluna vulgaris*. Annals of Botany 29:96–131.

——. 1929. The biology of fungus infection in the genus *Vaccinium*. Annals of Botany 43:55–70.

Rayner, R.W. 1970. A Mycological Colour Chart. Commonwealth Mycological Institute, Kew, Surrey, England.

Read, D.J., J.R. Leake, and A.R. Langdale. 1989. The nitrogen nutrition of mycorrhizal fungi and their host plants. Pp. 181–204. *In* L. Boddy, R. Marchant, and D.J. Read (eds.), Nitrogen, Phosphorus and Sulfur Utilization by Fungi. Cambridge University Press, Cambridge, England.

Reddy, M.S., and C.L. Kramer. 1975. A taxonomic revision of the Protomycetales. Mycotaxon 3:1–50.

Redecker, D., J.B. Morton, and T.D. Bruns. 2000a. Ancestral lineages of arbuscular mycorrhizal fungi. Molecular Phylogenetics and Evolution 14: 276–284.

——. 2000b. Molecular phylogeny of the arbuscular mycorrhizal fungi *Glomus sinuosum* and *Sclerocystis coremioides*. Mycologia 92:282–285.

Redfern, D.B. 1989. The roles of the bark beetle *Ips cembrae*, the woodwasp *Urocerus gigas* and associated fungi in dieback and death of larches. Pp. 195–204. *In* N. Wilding, N.M. Collins, P.M. Hammond, and J.F. Webber (eds.), Insect-Fungus Interactions. Academic Press, New York.

Redhead, S.A., and D. Malloch. 1977. The Endomycetaceae: new concepts, new taxa. Canadian Journal of Botany 55:1701–1711.

Redhead, S.A., J.F. Ammirati, G.R. Walker, L.L. Norvell, and M.B. Puccio. 1994. *Squamanita contortipes*, the rosetta stone of a mycoparasitic agaric genus. Canadian Journal of Botany 72:1812–1824.

Redman, R.S., and R.J. Rodriguez. 1994. Factors affecting the efficient transformation of *Colletotrichum* species. Experimental Mycology 18:230–246.

Redman, R.S., J.C. Ranson, and R.J. Rodriguez. 1999. Conversion of the pathogenic fungus *Colletotrichum magna* to a nonpathogenic endophytic mutualist by gene disruption. Molecular Plant-Microbe Interactions 12:969–975.

Reed, C., and D.F. Farr. 1993. Index to Saccardo's Sylloge Fungorum. Rose Printing, Tallahassee, Florida.

Rehner, S.A., and G.J. Samuels. 1994. Taxonomy and phylogeny of *Gliocladium* analysed from nuclear large subunit ribosomal DNA sequences. Mycological Research 98:625–634.

Rehner, S.A., and F.A. Uecker. 1994. Nuclear ribosomal internal transcribed spacer phylogeny and host diversity in the coelomycete *Phomopsis*. Canadian Journal of Botany 72:1666–1674.

Reichenbach-Klinke, H.-H. 1973. Fish Pathology. T. F. H. Publications, Neptune City, New Jersey.

Reid, D.A. 1970. New or interesting records of British Hymenomycetes, IV. Transactions of the British Mycological Society 55:413–441.

Reinecke, P., and N.J. Fokkema. 1979. *Pseudocercosporella herpotrichoides*: storage and mass production of conidia. Transactions of the British Mycological Society 72:329–330.

Renkonen, O. 1938. Statistisch-ökologische Untersuchungen über die terrestrische käferwelt der finnischen Bruchmoore. Annales Zoologici Societatis Zoologicæ-Botanicæ Fennicæ Vanamo 6:1–231.

Renvall, P. 1995. Community structure and dynamics of wood-rotting Basidiomycetes on decomposing conifer trunks in northern Finland. Karstenia 35:1–51.

Reper, C., and M.J. Pennincky. 1987. Inhibition of *Pleurotus ostreatus* growth and fructification by a diffusible toxin from *Trichoderma hamatum*. Lebensmittel-Wissenschaft und-Technologie 20:291–292.

Révay, À., and J. Gönczöl. 1990. Longitudinal distribution and colonization patterns of wood-inhabiting fungi in a mountain stream in Hungary. Nova Hedwigia 51:505–520.

Reynolds, D.R. 1971. Notes on capnodiaceous fungi II: *Leptocapnodium*. Bulletin of the Torrey Botanical Club 98:151–154.

——. 1975. Observations on growth forms of sooty mold fungi. Nova Hedwigia 26:179–193.

——. 1978a. Foliicolous ascomycetes. 1. The capnodiaceous genus *Scorias* reproduction. Natural History Museum of Los Angeles County Contributions in Science 288:1–16.

——. 1978b. Foliicolous ascomycetes. 2. *Capnodium salicinum* Montagne emend. Mycotaxon 7:501–507.

——. 1979. Foliicolous ascomycetes. 3. The stipitate capnodiaceous species. Mycotaxon 8:417–445.

——. 1982. Foliicolous ascomycetes. 4. The capnodiaceous genus *Trichomerium*. Mycotaxon 14:189–220.

——. 1983. Foliicolous ascomycetes. 5. The capnodiaceous clypeate genus *Treubiomyces*. Mycotaxon 17:349–360.

——. 1985. Foliicolous ascomycetes. 6. The capnodiaceous genus *Limacinia*. Mycotaxon 23:153–168.

——. 1989. The bitunicate ascus paradigm. Botanical Review 55:1–52.

Ribeiro, W.R.C., and E.E. Butler. 1992. Isolation of mycoparasitic species of *Pythium* with spiny oogonia from soil in California. Mycological Research 96:857–862.

——. 1995. Comparison of the mycoparasites *Pythium periplocum*, *P. acanthicum* and *P. oligandrum*. Mycological Research 99:963–969.

Richards, R.H., A. Holliman, and S. Helgason. 1978. *Exophiala salmonis* infection in Atlantic salmon *Salmo salar* L. Journal of Fish Diseases 1:357–368.

Richardson, A.J., C.S. Stewart, G.P. Campbell, A.B. Wilson, and K.N. Joblin. 1986. Influence of co-culture with rumen bacteria on the lignolytic activity of phycomycetous fungi from the rumen. Abstracts of the XIV International Congress of Microbiology, Manchester, England. PC 2–24, 233.

Richardson, K.A., and R.S. Currah. 1995. The fungal community associated with the roots of some rainforest epiphytes of Costa Rica. Selbyana 16:49–73.

Richardson, M.J. 1970. Studies on *Russula emetica* and other agarics in a Scots pine plantation. Transactions of the British Mycological Society 55:217–229.

——. 1972. Coprophilous Ascomycetes on different dung types. Transactions of the British Mycological Society 58:37–48.

——. 2001. Diversity and occurrence of coprophilous fungi. Mycological Research 105:387–402.

Richardson, M.J., and G. Leadbeater. 1972. *Piptocephalis fimbriata* sp. nov., and observations on the occurrence of *Piptocephalis* and *Syncephalis*. Transactions of the British Mycological Society 58: 205–215.

Richardson, M.J., and R. Watling. 1969. Keys to Fungi on Dung. British Mycological Society, E. and E. Plumridge, Cambridge, England.

——. 1997. Keys to Fungi on Dung. 2nd ed. British Mycological Society, Stourbridge, United Kingdom.

Richmond, J.Y., and R.W. McKinney (eds.). 1999. Biosafety in Microbiological and Biomedical Laboratories. 4th ed. U. S. Department of Health and Human Services, Public Health Service, Centers for Disease Control and Prevention, and National Institutes of Health. U.S. Government Printing Office, Washington D. C. and web site: http://www.cdc.gov/od/ohs/biosfty/bmbl4/bubl4toc.html

Richter, D.L., and J.N. Bruhn. 1989. Revival of saprotrophic and mycorrhizal basidiomycete cultures from cold storage in sterile water. Canadian Journal of Microbiology 35:1055–1060.

Ridgway, R. 1912. Color Standards and Color Nomenclature. Published by the Author. Washington, D. C.

Riesen, T.K., and R.C. Close. 1987. Endophytic fungi in propiconazole-treated and untreated barley leaves. Mycologia 79:546–552.

Riesen, T.K., and T.N. Sieber. 1985. Endophytic Fungi in Winter Wheat (*Triticum aestivum* L.). Administration and Druck AG (ADAG), Zürich, Switzerland.

Riethmüller, A., M. Weiss, and F. Oberwinkler. 1999. Phylogenetic studies of Saprolegniomycetidae and related groups based on nuclear large subunit ribosomal DNA sequences. Canadian Journal of Botany 77:1790–1800.

Rifai, M.A. 1968. The Australasian Pezizales in the herbarium of the Royal Botanic Gardens, Kew. Verhandelingen van de Koninklijke Nederlandsche Akademie van Wetenschappen, Afdeling Natuurkunde, Tweede Sect. 57:1–295.

——. 1969. A revision of the genus *Trichoderma*. Mycological Papers 116:1–56.

Rippon, J.W. 1988. Medical Mycology: The Pathogenic Fungi and the Pathogenic Actinomycetes. 3rd ed. W. B. Saunders, Philadelphia, Pennsylvania.

Rishbeth, J. 1963. Stump protection against *Fomes annosus*. III. Inoculation with *Peniophora gigantea*. Annals of Applied Biology 52:63–77.

Roberson, R.W., E.S. Luttrell, and M.S. Fuller. 1990. Mycoparasitism of teliospores of *Ustilago bullata* by an oomycete. Canadian Journal of Botany 68:2415–2421.

Roberts, P. 1997. New heterobasidiomycetes from Great Britain. Mycotaxon 63:195–216.

Roberts, R.E. 1963. A study of the distribution of certain members of the Saprolegniales. Transactions of the British Mycological Society 46:213–224.

Roberts, R.G. 1990. Postharvest biological control of gray mold of apple by *Cryptococcus laurentii*. Phytopathology 80:526–530.

Roberts, R.G., S.T. Reymond, and B. Andersen 2000. RAPD fragment pattern analysis and morphological segregation of small-spored *Alternaria* species and species groups. Mycological Research 104:51–160.

Roberts, R.G., J.A. Robertson, and R.T. Hanlin. 1986. Fungi occurring in achenes of sunflower (*Helianthus annuus*). Canadian Journal of Botany 64:1964–1971.

Roberts, R.J. 1989. The mycology of teleosts. Pp. 320–336. *In* R.J. Roberts (ed.), Fish Pathology. 2nd ed. Baillière Tindall, London, England.

Robinson, P.M., and D. Park. 1965. The production and quantitative estimation of a fungal morphogen. Transactions of the British Mycological Society 48:561–571.

Rodrigues, K.F. 1994. The foliar fungal endophytes of the Amazonian palm *Euterpe oleracea*. Mycologia 86:376–385.

Rodrigues, K.F., and G.J. Samuels 1992. *Idriella* species endophytic in palms. Mycotaxon 43:271–276.

Rodrigues, K.F., O. Petrini, and A. Leuchtmann. 1995. Variability among isolates of *Xylaria cubensis* as determined by isozyme analysis and somatic incompatibility tests. Mycologia 87:592–596.

Rodriguez, R.J. 1993. Polyphosphate present in DNA preparations from filamentous fungal species of *Colletotrichum* inhibits restriction endonucleases and other enzymes. Analytical Biochemistry 209:291–297.

Rodriguez, R.J., and J.L. Owen. 1992. Isolation of *Glomerella musae* [Teleomorph of *Colletotrichum musae* (Berk. & Curt.) Arx] and segregation analysis of ascospore progeny. Experimental Mycology 16:291–301.

Roeper, R.A. 1988. Interaction between the ambrosia beetle *Corthylus punctatissimus* and its mutualistic fungi. Proceedings of the XVIII International Congress of Entomology, p. 410.

——. 1995. Patterns of mycetophagy in Michigan ambrosia beetles (Coleoptera: Scolytidae). Michigan Academician 26 (sic, = vol. 27):153–161.

Roeper, R.A., C.R. Hazen, D. Helsel, and M.A. Bunce. 1980. Studies on Michigan ambrosia fungi. Michigan Botanist 19:69–73.

Roger, A.J., M.W. Smith, R.F. Doolittle, and W.F. Doolittle. 1996. Evidence for the Heterolobosea from phylogenetic analysis of genes encoding glyceraldehyde-3-phosphate dehydrogenase. Journal of Eukaryotic Microbiology 43:475–485.

Rogers, J.D. 2000. Thoughts and musings on tropical Xylariaceae. Mycological Research 104:1412–1420.

Rogers, J.D., B.E. Callan, and G.J. Samuels. 1987. The Xylariaceae of the rain forests of North Sulawesi (Indonesia). Mycotaxon 29:113–172.

Rogers, J.D., A.Y. Rossman, and G.J. Samuels. 1988. *Xylaria* (Sphaeriales, Xylariaceae) from Cerro de la Neblina, Venezuela. Mycotaxon 31:103–153.

Rogerson, C.T. 1970. The hypocrealean fungi (Ascomycetes, Hypocreales). Mycologia 62:865–910.

Rogerson, C.T., and G.J. Samuels. 1985. Species of *Hypomyces* and *Nectria* occurring on Discomycetes. Mycologia 77:763–783.

——. 1989. Boleticolous species of *Hypomyces*. Mycologia 81:413–432.

——. 1993. Polyporicolous species of *Hypomyces*. Mycologia 85:231–272.

——. 1994. Agaricicolous species of *Hypomyces*. Mycologia 86:839–866.

Rogerson, C.T., and S.L. Stephenson. 1993. Myxomyceticolous fungi. Mycologia 85:456–469.

Rohde, C., and D. Claus (eds.). 1999. Shipping of infectious, noninfectious and genetically modified biological materials: International regulations. ECCO (available from Deutsche Sammlung von Mikroorganismen und Zellkulturen GmbH [DSMZ], Mascheroder Weg 1b, D-38124 Braunschweig, Germany. Website: http://www.gbf.de/dsmz/shipping/shipping.html.

Rohde, C., D. Claus, and K.A. Malik. 1995. Packing and shipping of biological materials: some instructions, legal requirements and international regulations. (Technical information sheet no.14). World Journal of Microbiology and Biotechnology 11:706–710. Web site: http://wdcm.nig.ac.jp/wfcc/info14.html.

Rohde, K. 1992. Latitudinal gradients in species diversity: the search for the primary cause. Oikos 65:514–527.

Rohlf, F.J., and R.R. Sokal. 1995. Statistical Tables. 3rd ed. Freeman, New York.

Roling, M.P., and W.H. Kearby. 1975. Seasonal flight and vertical distribution of Scolytidae attracted to ethanol in an oak-hickory forest in Missouri. Canadian Entomologist 107:1315–1320.

Roll-Hansen, F., and H. Roll-Hansen. 1979. Microflora of sound-looking wood in *Picea abies* stems. European Journal of Forest Pathology 9:308–316.

——. 1980a. Microorganisms which invade *Picea abies* in seasonal stem wounds. II. Ascomycetes, fungi imperfecti, and bacteria. European Journal of Forest Pathology 10:396–410.

——. 1980b. Microorganisms which invade seasonal stem wounds. I. General aspects. Hymenomycetes. European Journal of Forest Pathology 10:321–339.

Rollinger, J.L., and J.H. Langenheim, 1993. Geographic survey of fungal endophyte community composition in leaves of coastal redwood. Mycologia 85:149–156.

Romagnesi, H. 1967. Les Russules d'Europe et d'Afrique du Nord. Gantener, Vaduz, Liechtenstein.

Rombach, M.C., and D.W. Roberts. 1987. *Calcarisporium ovalisporum*, symbiotic with the insect pathogen *Hirsutella citriformis*. Mycologia 79:153–155.

Romulo, B., S.H. Bird, and R.A. Leng. 1989. Effects of defaunation and protein supplementation on intake, digestibility, N retention and fungal numbers in sheep fed straw-based diets. Pp. 285–288. *In* J.V. Nolan, R.A. Leng, and D.I. Demeyer (eds.), The Roles of Protozoa and Fungi in Ruminant Digestion. Penambul, Armidale, New South Wales, Australia.

Roos, U.P., and B. Guhl. 1996. A novel type of unorthodox mitosis in amoebae of the cellular slime mold (Mycetozoan) *Acrasis rosea*. European Journal of Protistology 32:171–189.

Rose, F. 1992. Temperate forest management: its effects on bryophyte and lichen floras and habitats. Pp. 210–233. *In* J.W. Bates and A.M. Farmer (eds.), Bryophytes and Lichens in a Changing Environment. Clarendon Press, Oxford, United Kingdom.

Rose, S.L. 1981. Vesicular-arbuscular endomycorrhizal associations of some desert plants of Baja California. Canadian Journal of Botany 59:1056–1060.

———. 1988. Above and below ground community development in a marine sand dune ecosystem. Plant and Soil 109:215–226.

Rosenberg, S.L. 1978. Cellulose and lignocellulose degradation by thermophilic and thermotolerant fungi. Mycologia 70:1–13.

Rosendahl, S., and J.W. Taylor. 1997. Development of multiple genetic markers for studies of genetic variation in arbuscular mycorrhizal fungi using AFLP. Molecular Ecology 6:821–829.

Rosentreter, R. 1986. Compositional patterns within a rabbitbrush (*Chrysothamnus*) community of the Idaho Snake River Plain. Pp. 273–277. *In* E.D. McArthur and B.L. Welch (comp.), Proceedings—Symposium on the Biology of *Artemisia* and *Chrysothamnus*, July 9–13, 1984, Provo, Utah. U. S. Department of Agriculture, Forest Service, Intermountain Research Station, Ogden, Utah. General Technical Report INT-200.

———. 1993. Vagrant lichens in North America. The Bryologist 96:333–338.

———. 1995. Lichen diversity in managed forests of the Pacific Northwest, USA. Pp. 103–124. *In* C. Scheidegger, P.A. Wolseley, and G. Thor. (eds.), Conservation Biology of Lichenised Fungi. Mitteilungen der Eidgenössischen Forschungsanstalt für Wald, Schnee und Landschaft 70:1–173.

Rosentreter, R., A. DeBolt, and C.C. Bratt. 1988. Curation of soil lichens. Evansia 5:23–25.

Rosenzweig, M.L. 1995. Species Diversity in Space and Time. Cambridge University Press, Cambridge, England.

Ross, A.J., and W.T. Yasutake. 1973. *Scolecobasidium humicola*, a fungal pathogen of fish. Journal of the Fisheries Research Board of Canada 30:994–995.

Ross, A.J., W.T. Yasutake, and S. Leek. 1975. *Phoma herbarum*, a plant saprophyte as a fish pathogen. Journal of the Fisheries Research Board of Canada 32:1648–1652.

Ross, D.W., P. Fenn, and F.M. Stephen. 1992. Growth of southern pine beetle associated fungi in relation to the induced wound response in loblolly pine. Canadian Journal of Forestry Research 22:1851–1859.

Ross, J.P., and R. Ruttencutter. 1977. Population dynamics of two vesicular-arbuscular endomycorrhizal fungi and the role of hyperparasitic fungi. Phytopathology 67:490–494.

Rossi, W., and A. Weir. 1998. *Triainomyces*, a new genus of Laboulbeniales on the pill-millipede, *Procyliosoma tuberculatum* from New Zealand. Mycologia 90:282–289.

Rossman, A.Y. 1987. The Tubeufiaceae and similar loculoascomycetes. Mycological Papers 157:1–71.

———. 1994. A strategy for an all-taxa inventory of fungal biodiversity. Pp. 169–194. *In* C.I. Peng and C.H. Chou (eds.), Biodiversity and Terrestrial Ecosystems. Academia Sinica Monograph Series No. 14, Taipei, Taiwan.

———. 1996. Morphological and molecular perspectives on systematics of the Hypocreales. Mycologia 88:1–19.

Rossman, A.Y., and C.T. Rogerson. 1981. A new species of *Hypomyces* (Hypocreaceae) with phragmosporous ascospores. Brittonia 33:382–384.

Rossman, A.Y., M.E. Palm, and L.J. Spielman. 1987. A Literature Guide for the Identification of Plant Pathogenic Fungi. American Phytopathological Society Press, St. Paul, Minnesota.

Rossman, A.Y., G.J. Samuels, C.T. Rogerson, and R. Lowen. 1999. Genera of Bionectriaceae, Hypocreaceae and Nectriaceae (Hypocreales, Ascomycetes). Studies in Mycology 43:1–248.

Rossman, A.Y., R.E. Tulloss, T.E. O'Dell, and R.G. Thorn. 1998. Protocols for an All Taxa Biodiversity Inventory of Fungi in a Costa Rican Conservation Area. Parkway Publishers, Boone, North Carolina.

Rosso, A.L., and R. Rosentreter. 1999. Lichen biodiversity and biomass in relation to management practices in forests of northern Idaho. Evansia 16:97–104.

Rousseau, A., N. Benhamou, I. Chet, and Y. Piché. 1996. Mycoparasitism of the extramatrical phase of *Glomus intraradices* by *Trichoderma harzianum*. Phytopathology 86:434–443.

Roux, C., and D. Triebel. 1994. Révision des espèces de *Stigmidium* et de *Sphaerellothecium* (champignons lichénicoles non lichénisés, Ascomycetes) correspondant à *Pharcidia epicymatia* sensu Keissler ou à *Stigmidium schaereri* auct. Bulletin de la Société Linnéenne de Provence 45:451–542.

Roux, C., O. Bricaud, E. Sérusiaux, and C. Coste. 1994. *Wentiomyces lichenicola* subsp. nov. *bouteillei* champignon lichénicole non lichénisé (Dothideales, Dimeriaceae). Mycotaxon 50:459–474.

Roux, C., J. Etayo, O. Bricaud, and D. Le Coeur. 1997. Les *Refractohilum* (Hyphomycètes, Moliniacés [sic]) à conidies pluriseptées en Europe et au Canada. Canadian Journal of Botany 75:1592–1600.

Roy, B.A. 1993. Floral mimicry by a plant pathogen. Nature 362:56–58.

———. 2001. Patterns of association between crucifers and their flower-mimic pathogens: Host jumps are more common than coevolution or cospeciation. Evolution 55:41–53.

Roy, B.A., D.R. Vogler, T.D. Bruns and T.M. Szaro. 1998. Cryptic species in the *Puccinia monoica* complex. Mycologia 90:846–853.

Royal Botanic Gardens. 1997. Index Kewensis 2.0 on Computer Disc. Oxford University Press, Oxford, England.

Royse, D.L., and S.M. Ries. 1978. The influence of fungi isolated from peach twigs on the pathogenicity of *Cytospora cincta*. Phytopathology 68:603–607.

Roza, D., and K. Hatai. 1999. Pathogenicity of fungi isolated from the larvae of the mangrove crab, *Scylla serrata*, in Indonesia. Mycoscience 40:427–431.

Rubner, A. 1996. Revision of predacious hyphomycetes in the *Dactylella-Monacrosporium* complex. Studies in Mycology 39:1–134.

Rudakov, O.L. 1978. Physiological groups in mycophilic fungi. Mycologia 70:150–159.

———. 1981. Mikofil'nye griby, ikh biologiya i prakticheskoe znachenie. Izd-vo Akademii Nauka SSSR Institut Mikrobiologii, Moskva, USSR.

Rudinsky, J.A., and G.E. Daterman. 1964. Field studies on flight patterns and olfactory responses of ambrosia beetles in Douglas-fir forests of western Oregon. Canadian Entomologist 96:1339–1352.

Ruggieri, G.D., R.F. Nigrelli, P.M. Powles, and D.G. Garnett. 1970. Epizootics in yellowtail flounder, *Limanda ferruginea* Storer, in the western North Atlantic caused by *Ichthyophonus*, a ubiquitous parasitic fungus. Zoologica (New York) 55:57–72.

Ruhling, A., E. Bååth, and B. Söderström. 1984. Fungi in metal-contaminated soils. Ambio 13:34–36.

Ruscoe, Q.W. 1971. Mycoflora of living and dead leaves of *Nothofagus truncata*. Transactions of the British Mycological Society 56:463–474.

Russell, G.F. 1999. An overview of bar code applications and issues in systematics collections. Pp. 253–261. *In* D.A. Metsger and S.C. Byers (eds.), Managing the Modern Herbarium, An Inter-Disciplinary Approach. Society for the Preservation of Natural History Collections, Washington, D. C., and The Royal Ontario Museum, Toronto, Ontario, Canada.

Ryan, B.D., and F.P. McWhorter. 1986. Processing lichen colonies growing on soil and moss, with glue to facilitate sectioning. Evansia 3:14–16.

Rykard, D.M., E.S. Luttrell, and C.W. Bacon. 1984. Conidiogenesis and conidiomata in the Clavicipitoideae. Mycologia 76:1095–1103.

Ryman, N. 1970. A genetic analysis of recapture frequencies of released young of salmon (*Salmo salar* L). Hereditas 65:159–162.

Ryvarden, L. 2000. Studies in neotropical polypores 2: a preliminary key to neotropical species of *Ganoderma* with a laccate pileus. Mycologia 92:180–191.

Ryvarden, L., and I. Johansen. 1980. A Preliminary Polypore Flora of East Africa. Fungiflora, Oslo, Norway.

Ryvarden, L., and M. Nuñez. 1992. Basidiomycetes in the canopy of an African rain forest. Pp. 116–118. *In* F. Hallé, and O. Pascal (eds.), Biologie d'une Canopée de Forêt Équatoriale -II., Lyon, France.

Saccardo, P.A. 1880. Conspectus generum fungorum Italiæ inferorium. Michelia II:1–38.

——. 1882–1972. Sylloge Fungorum Omnium Hucusque Cognitorum. Vols. 1–26 + suppl., Padua, Italy. R. Friedländer & Sohn, Berlin, Germany. [reprinted 1966, Johnson Reprint Corp., New York.]

Saenger, P., and S.C. Snedaker. 1993. Pantropical trends in mangrove above-ground biomass and annual litterfall. Oecologia 96:293–299.

Safar, H.M., and R.C. Cooke. 1988a. Exploitation of faecal resource units by coprophilous Ascomycotina. Transactions of the British Mycological Society 90:593–599.

——. 1988b. Interactions between bacteria and coprophilous Ascomycotina and a *Coprinus* species on agar and in copromes. Transactions of the British Mycological Society 91:73–80.

Safir, G.R. (ed.). 1987. Ecophysiology of VA Mycorrhizal Plants. CRC Press, Boca Raton, Florida.

Saha, D.C., M.A. Jackson, and J.M. Johnson-Cicalese. 1988. A rapid staining method for detection of endophytic fungi in turf and forage grasses. Phytopathology 78:237–239.

Sahashi, N., T. Kubono, and Y. Miyasawa. 1999. Temporal variations in isolation frequency of endophytic fungi of Japanese beech. Canadian Journal of Botany 77:197–202.

Sahr, T., H. Ammer, H. Besl, and M. Fischer. 1999. Infrageneric classification of the boleticolous genus *Sepedonium*: species delimitation and phylogenetic relationships. Mycologia 91:935–943.

Saiki, R.K., S. Scharf, F. Faloona, K.B. Mullis, G.T. Horn, H.A. Erlich, and N. Arnheim. 1985. Enzymatic amplification of β-globin genomic sequences and restriction site analysis for diagnosis of sickle cell anemia. Science 230:1350–1354.

Saikkonen, K., S.H. Faeth, M. Helander, and T.J. Sullivan. 1998. Fungal endophytes: a continuum of interactions with host plants. Annual Review of Ecology and Systematics. 29:319–343.

Saito, I. 1998. Non-gramineous hosts of *Myriosclerotinia borealis*. Mycoscience 39:145–153.

Saitou, N., and T. Imanishi. 1989. Relative efficiencies of the Fitch-Margoliash, maximum parsimony, maximum-likelihood, minimum-evolution, and neighbor-joining methods of phylogenetic tree construction in obtaining the correct tree. Molecular Biology and Evolution 6:514–525.

Saitou, N., and M. Nei. 1987. The neighbor-joining method: a new method for reconstructing phylogenetic trees. Molecular Biology and Evolution 4:406–425.

Saksirirat, W., and H.-H. Hoppe. 1990a. Light and scanning electron microscopic studies on the development of the mycoparasite *Verti-cillium psalliotae* Treschow on uredospores of the soybean rust (*Phakopsora pachyrhizi* Syd.). Journal of Phytopathology 128:340–344.

——. 1990b. *Verticillium psalliotae*, an effective mycoparasite of the soybean rust fungus *Phakopsora pachyrhizi* Syd. Zeitschrift für Pflanzenkrankheiten und Pflanzenschutz 97:622–633.

Salkin, I.F. 1970. *Allochytridium expandens*, gen. et sp.n.: Growth and morphology in continuous culture. American Journal of Botany 57:649–658.

——. 1988. Media and stains for mycology. Pp. 379–411. *In* B.B. Wentworth (Coord. ed.), Diagnostic Procedures for Mycotic and Parasitic Infections. 7th ed. American Public Health Association, Washington. D.C.

Salkin, I.F., and J.A. Robertson. 1970. Use of a tissue culture chamber for developmental studies of aquatic phycomycetes. Archiv für Microbiologie 70:157–160.

Salo, K. 1993. The composition and structure of macrofungus communities in boreal upland type forests and peatlands in North Karelia, Finland. Karstenia 33:61–99.

Sambrook, J., E.F. Fritsch, and T. Maniatis. 1989. Molecular Cloning: A Laboratory Manual. 2nd ed. Cold Spring Harbor Laboratory Press, Cold Spring Harbor, New York.

Samson, R.A. 1974. *Paecilomyces* and some allied hyphomycetes. Studies in Mycology 6:1–119.

Samson, R.A., M.J. Crisman, and M.R. Tansey. 1977. Observations on the thermophilous ascomycete *Thielavia terrestris*. Transactions of the British Mycological Society 69:417–423.

Samson, R.A., H.C. Evans, and J.-P. Latge. 1988. Atlas of Entomopathogenic Fungi. Springer-Verlag, Berlin, Germany.

Samson, R.A., E.S. Hoekstra, J.C. Frisvad, and O. Filtenborg. 1995. Introduction to Food-Borne Fungi. Centraalbureau voor Schimmelcultures, Baarn, The Netherlands.

Samuels, G.J. 1976. A revision of the fungi formerly classified as *Nectria* subgenus *Hyphonectria*. Memoirs of the New York Botanical Garden 26:1–126.

——. 1988. Fungicolous, lichenicolous, and myxomyceticolous species of *Hypocreopsis, Nectriopsis, Nectria, Peristomialis*, and *Trichonectria*. Memoirs of the New York Botanical Garden 48:1–78.

——. 1989. *Nectria* and *Sesquicillium*. Memoirs of the New York Botanical Garden 49:266–285.

Samuels, G.J., and M.E. Barr. 1997. Notes on and additions to the Niessliaceae (Hypocreales). Canadian Journal of Botany 75:2165–2176.

Samuels, G.J., and M. Blackwell. 2001. Pyrenomycetes—fungi with perithecia. Pp. 221–255. *In* D.J. McLaughlin, E.G. McLaughlin, and P.A. Lempke (eds.), The Mycota. Vol. VII. Systematics and Evolution. Part A. Springer-Verlag, Berlin, Germany.

Samuels, G.J., and K.A. Seifert. 1995. The impact of molecular characters on systematics of filamentous ascomycetes. Annual Review of Phytopathology 33:37–67.

Samuels, G.J., F. Candoussau, and J.-F. Magni. 1997. Fungicolous pyrenomycetes 1. *Helminthosphaeria* and the new family Helminthosphaeriaceae. Mycologia 89:141–155.

Samuels, G.J., S.L. Dodd, W. Gams, L.A. Castlebury, and O. Petrini. 2002. *Trichoderma* species associated with the green mold epidemic of commercially grown *Agaricus bisporus*. Mycologia 94:146–170.

Samuels, G.J., Y. Doi, and C.T. Rogerson. 1990. Hypocreales. Memoirs of the New York Botanical Gardens 59:6–108.

Samuels, G.J., R. Pardo-Schultheiss, K.P. Hebbar, R.D. Lumsden, C.N. Bastos, J.C. Costa, and J.L. Bezerra. 2000. *Trichoderma stromaticum*, sp. nov., a parasite of the cacao witches broom pathogen. Mycological Research 104:760–764.

Samuels, G.J., O. Petrini, K. Kuhls, E. Lieckfeldt, and C.P. Kubicek. 1998. The *Hypocrea schweinitzii* complex and *Trichoderma* sect. *Longibrachiatum*. Studies in Mycology 41:1–54.

Samuels, G.J., O. Petrini, and S. Manguin. 1994. Morphological and macromolecular characterization of *Hypocrea schweinitzii* and its *Trichoderma* anamorph. Mycologia 86:421–435.

Samuels, G.J., A.Y. Rossman, R. Lowen, and C.T. Rogerson. 1991. A synopsis of *Nectria* subgen. *Dialonectria*. Mycological Papers 164:1–48.

San Martín-González, F., and J.D. Rogers. 1989. A preliminary account of *Xylaria* of Mexico. Mycotaxon 34:283–373.

———. 1993. *Biscogniauxia* and *Camillea* in Mexico. Mycotaxon 47:229–258.

Sands, W.A. 1969. The association of termites and fungi. Pp. 495–524. *In* K. Krishna and F.M. Weesner (eds.), Biology of Termites. Academic Press, New York.

Sandys-Winsch, C., J.M. Whipps, M. Gerlagh, and M. Kruse. 1993. World distribution of the sclerotial mycoparasite *Coniothyrium minitans*. Mycological Research 97:1175–1178.

Sangalang, A.E., L.W. Burgess, D. Backhouse, J. Duff, and M. Wurst. 1995. Mycogeography of *Fusarium* species in soils from tropical, arid and mediterranean regions of Australia. Mycological Research 99:523–528.

Sanger, F., S. Nicklen, and A.R. Coulson. 1977. DNA sequencing with chain-terminating inhibitors. Proceedings of the National Academy of Sciences USA 74:5463–5467.

Santamaría, S. 1989. El orden Laboulbeniales (Fungi, Ascomycotina) en la Peninsula Iberica e Islas Baleares. Edicions especials de la Societat Catalana de Micologia, Barcelona, Spain.

———. 1996. *Parvomyces*, a new genus of Laboulbeniales from Spain. Mycological Research 99:1071–1077.

Santesson, R. 1952. Foliicolous lichens I. A revision of the taxonomy of the obligately foliicolous, lichenized fungi. Symbolae Botanicae Upsalienses. 12:1–590.

———. 1989. Parasymbiotic fungi on the lichen-forming basidiomycete *Omphalina foliacea*. Nordic Journal of Botany 9:97–99.

———. 1993. The Lichens and Lichenicolous Fungi of Sweden and Norway. STB-Forlaget, Lund, Sweden.

Sarkar, F.H., Y.-W. Li, and J.D. Crissman. 1993. A method for PCR sequencing of the p53 gene from a single 10-μm frozen or paraffin-embedded tissue section. BioTechniques 15:36–38.

Savile, D.B.O. 1959. Notes on *Exobasidium*. Canadian Journal of Botany 37:641–656.

———. 1968. Possible interrelationships between fungal groups. Pp. 649–675. *In* G.C. Ainsworth and A.S. Sussman (eds.), The Fungi. An Advanced Treatise. Vol. III. The Fungal Population. Academic Press, New York.

———. 1972. Arctic Adaptations in Plants. Canadian Department of Agriculture Monograph No. 6.

———. 1976. Evolution of the rust fungi (Uredinales) as reflected by their ecological problems. Evolutionary Biology 9:137–207.

———. 1980. Ecology, convergent evolution, and classification in Uredinales. Reports of the Tottori Mycological Institute 18:275–281.

Savulescu, T. 1953. Monografia Uredinalelor din Republica Populara Romana. Vols. 1 and 2. Editura Academiei Republicii Populare Romane, Bucuresti, Romania.

Scazzocchio, C. 1987. The natural history of fungal mitochondrial genomes. Pp. 53–73. *In* A.D.M. Rayner, C.M. Brasier, and D. Moore (eds.), Evolutionary Biology of the Fungi. Cambridge University Press, Cambridge, England.

Schardl, C.L., and A. Leuchtmann. 1999. Three new species of *Epichloë* symbiotic with North American grasses. Mycologia 91:95–107.

Schardl, C.L., and H.H. Wilkinson. 2000. Hybridization and cospeciation hypotheses for the evolution of grass endophytes. Pp. 63–83. *In* C.W. Bacon and J.W. White, Jr. (eds.), Microbial Endophytes. Marcel Dekker, New York.

Schardl, C.L., A. Leuchtmann, H.-F. Tsai, M.A. Collett, D.M. Watt, and D.B. Scott. 1994. Origin of a fungal symbiont of perennial ryegrass by interspecific hybridization of a mutualist with the ryegrass choke pathogen, *Epichloë typhina*. Genetics 136:1307–1317.

Schardl, C.L., J.-S. Liu, J.F. White, Jr., R.A. Finkel, Z.Q. An, and M.R. Siegel. 1991. Molecular phylogenetic relationships of nonpathogenic grass mycosymbionts and clavicipitaceous plant pathogens. Plant Systematics and Evolution 178:27–41.

Schenck, N.C., and Y. Pérez. 1990. Manual for the Identification of VA Mycorrhizal Fungi. 3rd ed. Synergistic, Gainesville, Florida.

Schenck, S., W.B. Kendrick, and D. Pramer. 1977. A new nematode-trapping hyphomycete and a reevaluation of *Dactylaria* and *Arthrobotrys*. Canadian Journal of Botany 55:977–985.

Schiewe, M.H., A.J. Novotny, and L.W. Harrell. 1988. Systemic fungal disease of salmonids. Pp. 344–346. *In* C.J. Sindermann and D.V. Lightner (eds.), Disease Diagnosis and Control in North American Marine Aquaculture. 2nd ed. Elsevier, Amsterdam, The Netherlands.

Schipper, M.A.A. 1969. Zygosporic stages in heterothallic *Mucor*. Antonie van Leeuwenhoek 35:189–208.

———. 1978. 1. On certain species of *Mucor* with a key to all accepted species. 2. On the genera *Rhizomucor* and *Parasitella*. Studies in Mycology 17:1–71.

Schleiden, J.M. 1842. Grundzüge der wissenschaftlichen Botanik, nebst einer methodologischen. Einleitung als anleitung zum Studium der Pflanze. Wilhelm Engelmann, Leipzig, Germany.

Schmid, E., and F. Oberwinkler. 1993. Mycorrhiza-like interaction between the achlorophyllous gametophyte of *Lycopodium clavatum* L. and its fungal endophyte studied by light and electron microscopy. New Phytologist 124:69–81.

Schmid, E., F. Oberwinkler, and L.D. Gomez. 1995. Light and electron microscopy of a host-fungus interaction in the roots of some epiphytic ferns from Costa Rica. Canadian Journal of Botany 73:991–996.

Schmit, J.P., J.F. Murphy, and G.M. Mueller. 1999. Macrofungal diversity in a temperate oak forest: a test of species richness estimators. Canadian Journal of Botany 77:1014–1027.

Schmitt, C.K., and N.G. Slack. 1990. Host specificity of epiphytic lichens and bryophytes: a comparison of the Adirondack Mountains (New York) and the southern Blue Ridge Mountains (North Carolina). Bryologist 93:257–274.

Schmitthenner, A.F., and J.W. Hilty. 1962. A modified dilution technique for obtaining single-spore isolates of fungi from contaminated material. Phytopathology 52:582–583.

Schnittler, M., and Y.K. Novozhilov. 1996. The myxomycetes of boreal woodlands in northern Karelia: a preliminary report. Karstenia 36:19–40.

Schnittler, M., and S.L. Stephenson. 2000. Myxomycete biodiversity in four different forest types in Costa Rica. Mycologia 92:626–637.

Schnittler, M., S.L. Stephenson, and Y.K. Novozhilov. 2000. Ecology and world distribution of *Barbeyella minutissima* (Myxomycetes). Mycological Research 104:1518–1523.

Schnürer, J., M. Clarholm, and T. Rosswall. 1985. Microbial biomass and activity in an agricultural soil with different organic matter contents. Soil Biology and Biochemistry 17:611–618.

Schoenlein-Crusius, I.H., and A.I. Milanez. 1995. Fungi of submerged leaves and stream water in the Atlantic rainforest in São Paulo state, Brazil. P. 65. *In* Abstracts. Seventh International Symposium on Microbial Ecology. Santos, Brazil, 1995.

Scholler, M., G. Hagedorn, and A. Rubner. 1999. A reevaluation of predatory orbiliaceous fungi. II. A new generic concept. Sydowia 51:89–113.

Scholz, P. 2000. Katalog der Flechten und flechtenbewohnenden Pilze Deutschlands. Schriftenreihe für Vegetationskunde 31:1–298.

Schoor, O., T. Weinschenk, J. Hennenlotter, S. Corvin, A. Stenzl, H.-G. Rammensee, and S. Stevanovic. 2003. Moderate degradation does not preclude microarray analysis of small amounts of RNA. BioTechniques 35:1192–1201.

Schouten, S.P., and M.H. Waandrager. 1979. Problems in obtaining pure cultures of *Cantharellus cibarius*. Mushroom Science 10:885–890.

Schowalter, T.D., and G.M. Filip (eds.). 1993. Beetle-pathogen Interactions in Conifer Forests. Academic Press, London, England.

Schroers, H.-J. 2000. Generic delimitation of *Bionectria* (Bionectriaceae, Hypocreales) based on holomorph characters and rDNA sequences. Studies in Mycology 45:63–82.

——. 2001. A monograph of *Bionectria* (Ascomycota, Hypocreales, Bionectriaceae) and its *Clonostachys* anamorphs. Studies in Mycology 46:1–214.

Schroers, H.-J., G.J. Samuels, K.A. Seifert, and W. Gams. 1999. Classification of the mycoparasite *Gliocladium roseum* in *Clonostachys* as *C. rosea*, its relationships to *Bionectria ochroleuca*, and notes on other *Gliocladium*-like fungi. Mycologia 91:365–385.

Schulthess, F., and S.H. Faeth. 1998. Distribution, abundances, and associations of the endophytic fungal community of Arizona fescue (*Festuca arizonica*). Mycologia 90:569–578.

Schultz, T.R., and R. Meier. 1995. A phylogenetic analysis of the fungus-growing ants (Hymenoptera: Formicidae: Attini) based on morphological characters of the larvae. Systematic Entomology 20:337–370.

Schulz, B., J. Sucker, H.J. Aust, K. Krohn, K. Ludewig, P.G. Jones, and D. Doring. 1995. Biologically active secondary metabolites of endophytic *Pezicula* species. Mycological Research 99:1007–1015.

Schulz, B., U. Wanke, S. Draeger, and H.-J. Aust. 1993. Endophytes from herbaceous plants and shrubs: effectiveness of surface-sterilization methods. Mycological Research 97:1447–1450.

Schumacher, T., and L. Ryvarden. 1981. *Dipodascus polyporicola* nov. sp., a parasitic Hemiascomycete on *Piptoporus soloniensis*. Mycotaxon 12:525–530.

Schüßler, A., P. Bonfante, E. Schnepf, D. Mollenhauer, and M. Kluge. 1996. Characterization of the *Geosiphon pyriforme* symbiosome by affinity techniques: confocal laser scanning microscopy (CLSM) and electron microscopy. Protoplasma 190:53–67.

Schüßler, A., D. Schwarzott, and C. Walker. 2001. A new fungal phylum, the Glomeromycota: Phylogeny and evolution. Mycological Research 105:1413–1421.

Scott, J.A., D.W. Malloch, and J.B. Gloer. 1993. *Polytolypa*, an undescribed genus in the Onygenales. Mycologia 85:503–508.

Scott, W.W. 1961. A monograph of the genus *Aphanomyces*. Virginia Agricultural Experiment Station, Technical Bulletin 151:1–95.

Seaby, D. 1987. Infection of mushroom compost by *Trichoderma* species. Bulletin of the Mushroom Growers Association [Mushroom Journal] 179:355–361.

Seaver, F.J. 1942. The North American Cup Fungi (Operculates). Published by the author, New York.

——. 1951. The North American Cup Fungi (Inoperculates). Published by the author, New York.

Seaward, M.R.D. 1988. Contribution of lichens to ecosystems. Pp. 107–129. *In* M. Galun (ed.), CRC Handbook of Lichenology. Vol. II. CRC Press, Boca Raton, Florida.

Seeler, E.V. 1943. Several fungicolous fungi. Farlowia 1:119–133.

Seifert, K.A. 1985. A monograph of *Stilbella* and some allied Hyphomycetes. Studies in Mycology 27:1–238.

——. 1990. Isolation of filamentous fungi. Pp. 21–51. *In* D.P. Labeda (ed.), Isolation of Biotechnological Organisms from Nature. McGraw-Hill, New York.

Seifert, K.A., and W. Gams. 2001. The taxonomy of anamorphic fungi. Pp. 307–347. *In* D.J. McLaughlin, E.G. McLaughlin, and P.A. Lempke (eds.), The Mycota. Vol. VII. Systematics and Evolution. Part A. Springer-Verlag, Berlin, Germany.

Seifert, K.A., W. Gams, P.W. Crous, and G.J. Samuels (eds). 2000. Molecules, morphology and classification: towards monophyletic genera in the Ascomycetes. Studies in Mycology 45:1–230.

Seifert, K.A., W.B. Kendrick, and G. Murase. 1983. A key to hyphomycetes on dung. University of Waterloo, Biological Series 27:1–62. [Waterloo, Ontario, Canada].

Seifert, K.A., R.A. Samson, and I.H. Chapela. 1995. *Escovopsis aspergilloides*, a rediscovered hyphomycete from leaf-cutter ant nests. Mycologia 87:407–413.

Seifert, K.A., J.F. Webber, and M.J. Wingfield. 1993. Methods for studying species of *Ophiostoma* and *Ceratocystis*. Pp. 255–259. *In* M.J. Wingfield, K.A. Seifert, and J.F. Webber (eds.), *Ceratocystis* and *Ophiostoma*: Taxonomy, Ecology, and Pathogenicity. American Phytopathological Society Press, St. Paul, Minnesota.

Selva, S.B. 1994. Lichen diversity and stand continuity in the northern hardwoods and spruce-fir forests of northern New England and western New Brunswick. Bryologist 97:424–429.

Sérusiaux, E., P. Diederich, A.M. Brand, and P. van den Boom. 1999. New or interesting lichens and lichenicolous fungi from Belgium and Luxembourg. VIII. Lejeunia, Revue de Botanique, N.S. 162:1–95.

Setliff, E.C. 1984. *Aporpium*, an example of horizontal gene transfer? Mycotaxon 18:19–21.

Seymour, R. 1970. The genus *Saprolegnia*. Nova Hedwigia 19:1–124.

——. 1987. *Geolegnia inflata*. Pp. 70. *In* M.S. Fuller and A. Jaworski (eds.), Zoosporic Fungi in Teaching and Research. Southeastern Publishing, Athens, Georgia.

Seymour, R., and M.S. Fuller. 1987. Collection and isolation of water molds (Saprolegniaceae) from water and soil. Pp. 125–127. *In* M.S. Fuller and A. Jaworski (eds.), Zoosporic Fungi in Teaching and Research. Southeastern Publishing, Athens, Georgia.

Shaffer, R.L. 1975. The major groups of Basidiomycetes. Mycologia 67:1–18.

Shannon, C.E., and W. Weaver (eds.). 1949. The Mathematical Theory of Communication. University of Illinois Press, Urbana, Illinois.

Sharma, I.K., and W.A. Heather. 1978. Parasitism of uredospores of *Melampsora larici-populina* Kleb. by *Cladosporium* sp. European Journal of Forest Pathology 8:48–54.

Sharma, I.K., and W.A. Heather. 1987. Light and electron microscope studies on *Cladosporium tenuissimum*, mycoparasitic on poplar leaf rust, *Melampsora larici-populina*. Transactions of the British Mycological Society 90:125–131.

Sharp, A.J. 1957. Vascular epiphytes in the Great Smoky Mountains. Ecology 38:654–655.

Shaw, D.E. 1990. Blooms of *Neurospora* in Australia. Mycologist 4:6–13.

——. 1998. Species of *Neurospora* recorded in Australia, and the collection of *Neurospora* conidia by honey bees in lieu of pollen. Mycologist 12:154–158.

Shaw, J. 1987. John Shaw's Closeups in Nature. Billboard Publications, New York.

Shaw, P.J.A. 1992. Fungi, fungivores, and fungal food webs. Pp. 295–310. *In* G.C. Carroll and D.T. Wicklow (eds.), The Fungal Community: It's Organization and Role in the Ecosystem. 2nd ed. Marcel Dekker, New York.

Shearer, B.L., R.J. Zeyen, and J.J. Ooka. 1974. Storage and behavior in soil of *Septoria* species isolated from cereals. Phytopathology 64:163–167.

Shearer, C.A. 1972. Fungi of the Chesapeake Bay and its tributaries. III. The distribution of wood-inhabiting ascomycetes and fungi imperfecti in the Patuxent River. American Journal of Botany 59:961–969.

——. 1993. The freshwater ascomycetes. Nova Hedwigia 56:1–33.

——. 2001a. The distribution of freshwater filamentous ascomycetes. Pp. 223–390. *In* J.K. Misra and B.W. Horn (eds.), Mycology: Trichomycetes, Other Fungal Groups and Mushrooms. Science Publishers, Enfield, New Hampshire.

——. 2001b. Freshwater Ascomycetes and Their Anamorphs. Web site: http://fm5web.life.uiuc.edu:23523/ascomycete/

Shearer, C.A., and J.L. Crane. 1986. Illinois fungi XII. Fungi and myxomycetes from wood and leaves submerged in southern Illinois swamps. Mycotaxon 25:527–538.

——. 1995. *Boerlagiomyces websteri*, a new ascomycete from fresh water. Mycologia 87:876–879.

Shearer, C.A., and L.C. Lane. 1983. Comparison of three techniques for the study of aquatic hyphomycete communities. Mycologia 75:498–508.

Shearer, C.A., and S.B. Von Bodman. 1983. Patterns of occurrence of ascomycetes associated with decomposing twigs in a midwestern stream. Mycologia 75:518–530.

Shearer, C.A., and J. Webster. 1985a. Aquatic hyphomycete communities in the River Teign.I. Longitudinal distribution patterns. Transactions of the British Mycological Society 84:489–501.

——. 1985b. Aquatic hyphomycete communities in the River Teign. II. Temporal distribution patterns. Transactions of the British Mycological Society 84:503–507.

——. 1985c. Aquatic hyphomycete communities in the River Teign. III. Comparison of sampling techniques. Transactions of the British Mycological Society 84:509–518.

——. 1991. Aquatic hyphomycetes in the River Teign. IV. Twig colonization. Mycological Research 95:413–420.

Shigemitsu, H., H. Kunoh, and S. Akai. 1978. Scanning electron microscopic studies of *Corticium rolfsii* sclerotia parasitized by *Aspergillus terreus*. Mycologia 70:935–943.

Shigemori, H., S. Wakuri, K. Yazawa, T. Nakamura, T. Sasaki, and J. Kobayashi. 1991. Fellutamides A and B, cytotoxic peptides from a marine fish-possessing fungus *Penicillium fellutanum*. Tetrahedron 47:8529–8534.

Shigo, A.L. 1960a. Parasitism of *Gonatobotryum fuscum* on species of *Ceratocystis*. Mycologia 52:584–598.

——. 1960b. Mycoparasitism of *Gonatobotryum fuscum* and *Piptocephalis xenophila*. Transactions of the New York Academy of Sciences, Ser. II, 22:365–372.

Shoemaker, R.A., and C.E. Babcock. 1989. *Phaeosphaeria*. Canadian Journal of Botany 67:1500–1599.

Shumway, D.L., and R.T. Koide. 1994. Reproductive responses of mycorrhizal colonization of *Abutilon theophrasti* Medic. Plants grown for two generations in the field. New Phytologist 128:219–224.

Sieber, T., and C. Hugentobler. 1987. Endophytishche Pilze in Blättern und Asten gesunder und geschädigter Buchen (*Fagus sylvatica* L.). European Journal of Forest Pathology 17:411–425.

Sieber, T.N. 1989. Endophytic fungi in twigs of healthy and diseased Norway spruce and white fir. Mycological Research 92:322–326.

Sieber, T.N., and C.E. Dorworth. 1994. An ecological study about assemblages of endophytic fungi in *Acer macrophyllum* in British Columbia: in search of candidate mycoherbicides. Canadian Journal of Botany 72:1397–1402.

Sieber, T.N., F. Sieber-Canavesi, and C.E. Dorworth. 1991. Endophytic fungi of red alder (*Alnus rubra* Bong.) leaves and twigs in British Columbia. Canadian Journal of Botany 69:407–411.

Sieber, T.N., F. Sieber-Canavesi, O. Petrini, A.K.M. Ekramoddoullah, and C.E. Dorworth. 1991. Characterization of Canadian and European *Melanconium* from some *Alnus* species by morphological, cultural and biochemical studies. Canadian Journal of Botany 69:2170–2176.

Sieber-Canavesi, F., and T.N. Sieber. 1993. Successional patterns of fungal communities in needles of European silver fir (*Abies alba* Mill.). New Phytologist 125:149–161.

Sieburth, J.M. 1979. Sea Microbes. Oxford University Press, New York.

Siegel, M.R., C.L. Schardl, and T.D. Phillips. 1995. Incidence and compatibility of nonclavicipitaceous fungal endophytes in *Festuca* and *Lolium* grass species. Mycologia 87:196–202.

Sieverding, E. 1990. Ecology of VAM fungi in tropical agrosystems. Agriculture, Ecosystems, and Environment 29:369–390.

——. 1991. Vesicular-Arbuscular Mycorrhiza Management in Tropical Agrosystems. Deutsche Gesellschaft für Technische Zusammenarbeit, Eschborn, Germany.

Sigler, L. 1996. *Ajellomyces crescens* sp. nov., taxonomy of *Emmonsia* spp., and relatedness with *Blastomyces dermatitidis* (teleomorph *Ajellomyces dermatitidis*). Journal of Medical Veterinary Mycology 34:303–314.

Sillett, S.C., and P.N. Neitlich. 1996. Emerging themes in epiphyte research in westside forests with special reference to cyanolichens. Northwest Science 70:54–60.

Sillett, S.C., S.R. Gradstein, and D. Griffin. 1995. Bryophyte diversity of *Ficus* tree crowns from cloud forest and pasture in Costa Rica. Bryologist 98:251–260.

Sillet, S.C., B. McCune, J.E. Peck, T.R. Rambo, and A. Ruchty. 2000. Dispersal limitations of epiphytic lichens result in species dependent on old-growth forests. Ecological Applications 10:789–799.

Simione, F.P., and E.M. Brown (eds.). 1991. ATCC Preservation Methods: Freezing and Freeze-drying. 2nd ed. American Type Culture Collection, Rockville, Maryland.

Simmons, E.G. 1981. *Halysiomyces*, a new dematiaceous genus from Arizona's Sonoran desert. Mycotaxon 13:407–411.

Simon, C. 1991. Molecular systematics at the species boundary: exploiting conserved and variable regions of the mitochondrial genome of animals via direct sequence from amplified DNA. Pp. 33–71. *In* G.M. Hewitt, A.W.B. Johnston, and J.P.W. Young (eds.), Molecular Techniques in Taxonomy. NATO ASI (Advanced Study Institutes) Sub-series H. Cell Biology Vol. 57. Springer-Verlag, Berlin, Germany.

Simon, L., J. Bousquet, R.C. Lévesque, and M. Lalonde. 1993. Origin and diversification of endomycorrhizal fungi and coincidence with vascular land plants. Nature 363:67–69.

Simpson, E.H. 1949. Measurements of diversity. Nature 163:688.

Sinclair, W.A., H.H. Lyon, and W.T. Johnson. 1987. Diseases of Trees and Shrubs. Cornell University Press, Ithaca, New York.

Sindermann, C.J. 1970. Principal Diseases of Marine Fish and Shellfish. Academic Press, New York.

——. (ed.). 1977. Disease Diagnosis and Control in North American Marine Aquaculture. Elsevier, Amsterdam, The Netherlands.

Sindermann, C.J., and D.V. Lightner (eds.). 1988. Disease Diagnosis and Control in North American Marine Aquaculture. 2nd ed. Elsevier, Amsterdam, The Netherlands.

Singer, R. 1970. Omphalinae (Clitocybeae—Tricholomataceae Basidiomycetes). Hafner, New York.

——. 1973. The Genera *Marasmiellus, Crepidotus* and *Simocybe* in the Neotropics. Nova Hedwigia (Beihefte) 44:1–517.

——. 1974. A monograph of *Favolaschia*. Nova Hedwigia (Beihefte) 50:1–108.

——. 1977. Keys for the identification of the species of agaricales I. Sydowia 30:192–279.

——. 1978. Keys for identification of the species of agaricales II. Sydowia 31:193–237.

——. 1982. *Hydropus* (Basidiomycetes–Tricholomataceae–Mycenae). Hafner, New York.

——. 1986a. How to become a mycologist. I. The conservation of material and notes. McIlvainea 7:18–23.

——. 1986b. The Agaricales in Modern Taxonomy. 4th ed. Koeltz Scientific, Koenigstein, Germany.

——. 1989. New taxa and new combinations of agaricales (Diagnose Fungorum Novorum Agaricalium IV). Field Museum of Natural History, Botany, New Series no. 21.

Singer, R., and A.H. Smith. 1958. Studies on secotiaceous fungi-II: *Endoptychum depressum*. Brittonia 10:216–221.

——. 1959. Studies on secotiaceous fungi-V. *Nivatogastrium* gen. nov. Brittonia 11:224–228.

——. 1960. Studies on secotiaceous fungi IX. The astrogastraceous series. Memoirs of the Torrey Botanical Club 21:1–112.

Singh, K., J.C. Frisvad, U. Thrane, and S.B. Mathur. 1991. An Illustrated Manual on Identification of Some Seed-Borne *Aspergilli*, *Fusaria*, and *Penicillia*. Danish Government Institute of Seed Pathology for Developing Countries, Hellerup, Denmark.

Singh, N. 1975. Nutritional requirements for growth of *Dispira cornuta* in axenic culture. Journal of General Microbiology 88:372–376.

Singh, N., and B.E. Plunkett. 1967. Cyanide tolerance and growth stimulation by the carbon of organic nitrogen compounds in hymenomycete-inhabiting fungi. Transactions of the British Mycological Society 50:359–376.

Singh, S., and D.K. Sandhu. 1986. Thermophilous fungi in Port Blair soils. Canadian Journal of Botany 64:1018–1026.

Singh, U.P., S.N. Vishwakarma, and K.C. Basuchaudhury. 1978. *Acremonium sordidulum* mycoparasitic on *Colletotrichum dematium* f. *truncata* in India. Mycologia 70:453–455.

Singleton, L.L., J.D. Mihail, and C.M. Rush (eds.). 1992. Methods for Research on Soilborne Phytopathogenic Fungi. American Phytopathological Society Press, St. Paul, Minnesota.

Sipman, H.J.M. 1992. Results of a lichenological and bryological exploration of Cerro Guaiquinima (Guyana Highland, Venezuela). Tropical Biology 6:1–31.

——. 1995. Preliminary review of the lichen biodiversity of the Colombian montane forests. Pp. 313–320. *In* S.P. Churchill, H. Balslev, E. Forero, and J.L. Luteyn (eds.). Biodiversity and Conservation of Neotropical Montane Forests. New York Botanical Garden, Bronx, New York.

——. 1996a. Lichenized and lichenicolous fungi 1993–1994. Progress in Botany 57:312–335.

——. 1996b. Corticolous lichens. Pp. 66–67. *In* S.R. Gradstein, P. Hietz, R. Lücking, A. Lücking, H.J.M. Sipman, H.F.M. Vester, J.H.D. Wolf, and E. Gardette, How to Sample the Epiphytic Diversity of Tropical Rainforests. Ecotropica 2:59–72.

——. 1997. Observations on the foliicolous lichen and bryophyte flora in the canopy of a semi-deciduous tropical forest. Abstracta Botanica 21:153–161.

Sipman, H.J.M., and A. Aptroot. 2001. Where are the missing lichens? Mycological Research 105:1433–1439.

Sipman, H.J.M., and T. Raus. 2002. An inventory of the lichen flora of Kalimnos and parts of Kos (Dodecanisos, Greece). Willdenowia 32:351–392.

Siqueira, J.O., D.H. Hubbell, J.M. Kimbrough, and N.C. Schenck. 1984. *Stachybotrys chartarum* antagonistic to azygospores of *Gigaspora margarita* in vitro. Soil Biology and Biochemistry 16:679–681.

Sitja-Bobadilla, A., and P. Alvarez-Pellitero. 1990. First report of *Ichthyophonus* disease in wild and cultured sea bass *Dicentrarchus labrax* from the Spanish Mediterranean area. Diseases of Aquatic Organisms 8:145–150.

Sivan, A., and G.E. Harman. 1991. Improved rhizosphere competence in a protoplast fusion progeny of *Trichoderma harzianum*. Journal of General Microbiology 137:23–30.

Sivanesan, A. 1977. The taxonomy and pathology of *Venturia* species. Bibliotheca Mycologica 59:1–139.

——. 1984. The Bitunicate Ascomycetes and Their Anamorphs. J. Cramer, Vaduz, Liechtenstein.

Sivers, V.S. 1962. Fungi of the order Mucorales in the rumen of cattle. Mikrobiologicheskii Zhurnal (Kiev) 24:14–19.

Sivichai, S., E.B.G. Jones, and N.L. Hywel-Jones. 2000. Fungal colonisation of wood in a freshwater stream at Khao Yai National Park, Thailand. Fungal Diversity 5:71–88.

Slack, N.G. 1988. The ecological importance of lichens and bryophytes. Bibliotheca Lichenologica 30:23–54.

Slifkin, M.K. 1961. Parasitism of *Olpidiopsis incrassata* on members of the Saprolegniaceae. I. Host range and effects of light, temperature, and stage of host on infectivity. Mycologia 53:183–193.

——. 1963. Parasitism of *Olpidiopsis incrassata* on members of the Saprolegniaceae. II. Effect of pH and host nutrition. Mycologia 55:172–182.

Sloan, B.J., J.B. Routien, and V.P. Miller. 1960. Increased sporulation in fungi. Mycologia 52:47–63.

Smart, R.F. 1937. Influence of certain external factors on spore germination in the Myxomycetes. American Journal of Botany 24:145–159.

Smith, A.H., and R. Singer. 1959. Studies on secotiaceous fungi-IV. *Gastroboletus*, *Truncocolumella* and *Chamonixia*. Brittonia 11:205–223.

Smith, B. 1986. Evaluation of Different Similarity Indices Applied to Data from the Rothamsted Insect Survey. Unpubl. Master's thesis, University of York, York, England.

Smith, B.J., and K. Sivasithamparam. 2000. Isozymes of *Ganoderma* species from Australia. Mycological Research 104:952–961.

Smith, C., L. Geiser, L. Gough, B. McCune, B. Ryan, and R. Showman. 1993. Species and communities. Pp. 41–66. *In* K. Stolte, D. Mangis, R. Doty, and K. Tonnessen (eds.), Lichens as Bioindicators of Air Quality. U.S. Department of Agriculture, Forest Service, Rocky Mountain Forest and Range Experiment Station, Fort Collins, Colorado. General Technical Report RM-224.

Smith, C.W. 1991. Lichen conservation in Hawaii. Pp. 35–45. *In* D.J. Galloway (ed.), Tropical Lichens: Their Systematics, Conservation and Ecology. Clarendon Press, Oxford, United Kingdom.

Smith, D. 1990. Notes on the Preservation of Fungi. CAB International, International Mycological Institute, Kew, Surrey, England.

——. 1993. Tolerance to freezing and thawing. Pp. 145–171. *In* D.H. Jennings (ed.), Stress Tolerance of Fungi. Marcel Dekker, New York.

Smith, D., and A.H.S. Onions. 1983. A comparison of some preservation techniques for fungi. Transactions of the British Mycological Society 81:535–540.

——. 1994. The Preservation and Maintenance of Living Fungi. 2nd ed. IMI Technical Handbook 2. CAB International, International Mycological Institute, Kew, Surrey, England.

Smith, F.E.V. 1924. Three diseases of cultivated mushrooms. Transactions of the British Mycological Society 10:81–97.

Smith, H.V., and A.H. Smith. 1973. How to Know the Non-Gilled Fleshy Fungi. W.C. Brown, Dubuque, Iowa.

Smith, J.E., D. McKay, and R. Molina. 1994. Survival of mycorrhizal fungal isolates in sterile water at two temperatures and retrieved on solid and liquid nutrient media. Canadian Journal of Microbiology 40:736–742.

Smith, J.E., R. Molina, D. McKay, M. Castellano, and D. Luoma. 1996. Measuring fungal succession in Douglas-fir forests. Pp. 32–35. *In* D. Pilz and R. Molina (eds.), Managing Forest Ecosystems to Conserve Fungal Diversity and Sustain Wild Mushroom Harvests. U.S. Department of Agriculture, Forest Service, Pacific Northwest Research Station, Portland, Oregon. General Technical Report PNW-GTR-371.

Smith, M.L., J.N. Bruhn, and J.B. Anderson. 1992. The fungus *Armillaria bulbosa* is among the largest and oldest living organism. Nature 356:428–431.

Smith, R.E. 1977. Rapid tube test for detecting fungal cellulase production. Applied and Environmental Microbiology 33:980–981.

Smith, S.E., and V. Gianinazzi-Pearson. 1988. Physiological interactions between symbionts in vesicular-arbuscular mycorrhizal plants. Annual Review of Plant Physiology and Plant Molecular Biology 39:221–244.

Smith, S.E., and D.J. Read. 1997. Mycorrhizal Symbiosis. Academic Press, New York.

Smith, S.N, R.A. Armstrong, J. Springate, and G. Barber. 1985. Infection and colonization of trout eggs by Saprolegniaceae. Transactions of the British Mycological Society 85:719–723.

Sneath, P.H.A., and R.R. Sokal. 1973. Numerical Taxonomy, The Principles and Practice of Numerical Classification. Freeman, San Francisco, California.

Snedecor, G.W., and W.G. Cochran. 1967. Statistical Methods. 6th ed. Iowa State University Press, Ames, Iowa.

Sneh, B. 1977. A method for observation and study of living fungal propagules in soil. Soil Biology and Biochemistry 9:65–66.

Sneh, B., L. Burpee, and A. Ogoshi. 1991. Identification of *Rhizoctonia* species. American Phytopathological Society Press, St. Paul, Minnesota.

Sneh, B., S.J. Humble, and J.L. Lockwood. 1977. Parasitism of oospores of *Phytophthora megasperma* var. *sojae*, *P. cactorum*, *Pythium* sp. and *Aphanomyces euteiches* in soil by Oomycetes, Chytridiomycetes, Hyphomycetes, Actinomycetes and bacteria. Phytopathology 67:622–628.

Snell, W.H., and E.A. Dick. 1957. A Glossary of Mycology. Harvard University Press, Cambridge, Massachusetts.

Söderhäll, K., J. Rantamäki, and O. Constantinescu. 1993. Isolation of *Trichosporon beigelii* from the freshwater crayfish *Astacus astacus*. Aquaculture 116:25–31.

Söderström, B.E., and E. Bååth. 1978. Soil microfungi in three Swedish coniferous forests. Holarctic Ecology 1:62–72.

Soetanto, H., G.L.R. Gordon, I.D. Hume, and R.A. Leng. 1985. The role of protozoa and fungi in fibre digestion in the rumen of sheep. Proceedings of the 3rd AAAP Animal Science Congress 2:805–807.

Sokal, R.R., and F.J. Rohlf. 1962. The comparison of dendrograms by objective methods. Taxon 11:33–40.

——. 1981. Biometry: The Principles and Practice of Statistics in Biological Research. 2nd ed. W.H. Freeman, New York.

——. 1995. Biometry. 3rd ed. W.H. Freeman, New York.

Solangi, M.A., and D.V. Lightner. 1976. Cellular inflammatory response of *Penaeus aztecus* and *P. setiferus* to the pathogenic fungus, *Fusarium* sp., isolated from the California brown shrimp, *P. californiensis*. Journal of Invertebrate Pathology 27:77–86.

Solheim, H. 1995. Early stages of blue-stain fungus invasion of lodgepole pine sapwood following mountain pine beetle attack. Canadian Journal of Botany 73:70–74.

Solheim, H., B. Långström, and C. Hellqvist. 1993. Pathogenicity of the blue-stain fungi *Leptographium wingfieldii* and *Ophiostoma minus* to Scots pine: effect of tree pruning and inoculum density. Canadian Journal of Forestry Research 23:1438–1443.

Solomon, W.R. 1975. Assessing fungus prevalence in domestic interiors. Journal of Allergy and Clinical Immunology 56:235–242.

Somerfield, P.J., J.M. Gee, and C. Aryuthaka. 1998. Meiofaunal communities in a Malaysian mangrove forest. Journal of the Marine Biological Association of the United Kingdom 78:717–732.

Sorenson, W.G. 1964. A Nutritional Study in the *Thraustotheca*-Complex of the Saprolegniaceae with Special Reference to its Taxonomic Implications. Unpubl. Ph.D. dissert. University of Texas, Austin, Texas.

Southern, E.M. 1975. Detection of specific sequences among DNA fragments separated by gel electrophoresis. Journal of Molecular Biology 98:503–517.

Southwood, T.R.E. 1978. Ecological Methods, with Particular Reference to the Study of Insect Populations. Chapman Hall, London, United Kingdom.

Spanggaard, B., P. Skouboe, L. Rossen, and J.W. Taylor. 1996. Phylogenetic relationships of the intercellular fish pathogen *Ichthyophonus hoferi* and fungi, choanoflagellates and the rosette agent. Marine Biology 126:109–115.

Sparks, A.K. 1982a. The histopathology and possible role in the population dynamics of the tanner crab, *Chionoecetes bairdi*, of the fungus disease (black mat syndrome) caused by *Trichomaris invadens*. Pp. 539–546. *In* B. Melteff (ed.), Proceedings of the International Symposium on the Genus *Chionoecetes*, Anchorage, Alaska, 1982. Alaska Sea Grant College Program, University of Alaska, Fairbanks, Alaska. Alaska Sea Grant Report 82–10.

——. 1982b. Observations on the histopathology and possible progression of the disease caused by *Trichomaris invadens*, an invasive ascomycete, in the tanner crab, *Chionoecetes bairdi*. Journal of Invertebrate Pathology 40:242–254

——. 1985. Synopsis of Invertebrate Pathology Exclusive of Insects. Elsevier, Amsterdam, The Netherlands.

Sparks, A.K., and J. Hibbitts. 1979. Black mat syndrome, an invasive mycotic disease of the Tanner crab, *Chionoecetes bairdi*. Journal of Invertebrate Pathology 34:184–191.

Sparrow, F.K. 1965. The occurrence of *Physoderma* in Hawaii, with notes on other Hawaiian phycomycetes. Mycopathologia et Mycologia Applicata 25:119–143.

——. 1973. Chytridiomycetes, Hyphochytridiomycetes. Pp. 85–110. *In* G.C. Ainsworth, F.K. Sparrow, and A.S. Sussman (eds.), The Fungi. An Advanced Treatise. IVB. A Taxonomic Review with Keys: Basidiomycetes and Lower Fungi. Academic Press, New York.

——. 1976. The present status of classification in biflagellate fungi. Pp. 213–222. *In* E.B. Gareth Jones (ed.), Recent Advances in Aquatic Mycology. Elek Science, London, England.

——. 1977. A *Rhizidiomycopsis* on azygospores of *Gigaspora margarita*. Mycologia 69:1053–1058.

Sparrow, F.K., Jr. 1960. Aquatic Phycomycetes. 2nd ed. University of Michigan Press, Ann Arbor, Michigan.

——. 1968. Ecology of freshwater fungi. Pp. 41–93. *In* G.C. Ainsworth and A.S. Sussman (eds.), The Fungi. An Advanced Treatise. III. The Fungal Population. Academic Press, New York.

Spatafora, J.W. 1995. Ascomal evolution among filamentous ascomycetes: evidence from molecular data. Canadian Journal of Botany 73 (Suppl. 1):S811–S815.

Spatafora, J.W., and M. Blackwell. 1993. Molecular systematics of unitunicate perithecial ascomycetes: the Hypocreales-Clavicipitales connection. Mycologia 85:912–922.

——. 1994a. The polyphyletic origins of ophiostomatoid fungi. Mycological Research 98:1–9.

——. 1994b. Cladistic analysis of partial SSU rDNA sequences among unitunicate perithecial ascomycetes and its implication on the evolution of centrum development. Pp. 231–241. *In* D.L. Hawksworth (ed.), Ascomycete Systematics: Problems and Perspectives in the Nineties. NATO ASI (Advanced Study Institutes) Sub-series A, Life Sciences. Vol. 269. Plenum Press, New York.

Spatafora, J.W., B. Volkmann-Kohlmeyer, and J. Kohlmeyer. 1998. Independent terrestrial origins of the Halosphaeriales (marine Ascomycota). American Journal of Botany 85:1569–1580.

Spencer, D.M. (ed.). 1978. The Powdery Mildews. Academic Press, London, England.

——. 1980. Parasitism of carnation rust (*Uromyces dianthi*) by *Verticillium lecanii*. Transactions of the British Mycological Society 74:191–194.

Spencer, D.M., and P.T. Atkey. 1981. Parasitic effects of *Verticillium lecanii* on two rust fungi. Transactions of the British Mycological Society 77:535–542.

Spiegel, F.W. 1984. *Protostelium nocturnum*, a new, minute, ballistosporous protostelid. Mycologia 76:443–447.

——. 1990. Phylum Plasmodial Slime Molds, Class Protostelids. Pp. 484–497. *In* L. Margulis, J.O. Corliss, M. Melkonian, and D. Chapman (eds.), Handbook of Protoctista. Jones Bartlett, Boston, Massachusetts.

———. 1991. A proposed phylogeny of the flagellated protostelids. BioSystems 25:113–120.

Spiegel, F.W., and J. Feldman. 1985. Obligate amoebae of the protostelids: significance for the concept of Eumycetozoa. BioSystems 18:377–386.

———. 1988. The trophic cells of *Clastostelium recurvatum*, a third member of the myxomycete-like protostelids. Mycologia 80:525–535.

Spiegel, F.W., and L.S. Olive. 1978. New evidence for the validity of *Copromyxa protea*. Mycologia 70:843–847.

Spiegel, F.W., S.B. Lee, and S.A. Rusk. 1995a. Eumycetozoans and molecular systematics. Canadian Journal of Botany 73(Suppl. 1):S738–S746.

Spiegel, F.W., D.L. Moore, and J. Feldman. 1995b. *Tychosporium acutostipes*, a new protostelid which modifies the concept of the Protosteliidae. Mycologia 87:265–270.

Spielman, L.J. 1985. A monograph of *Valsa* on hardwoods in North America. Canadian Journal of Botany 63:1355–1378.

Sproston, N.G. 1944. *Ichthyosporidium hoferi* (Plehn and Mulsow, 1911), an internal fungoid parasite of the mackerel. Journal of the Marine Biological Association of the United Kingdom 26:72–98.

SPSS. 1999. SPSS Base System User's Guide. SPSS, Chicago, Illinois.

Sridhar, K.R., and F. Bärlocher. 1992a. Endophytic aquatic hyphomycetes of roots of spruce, birch, and maple. Mycological Research 96:305–308.

———. 1992b. Aquatic hyphomycetes in spruce roots. Mycologia 84:580–584.

Sridhar, K.R., and K.M. Kaveriappa. 1987. Occurrence and survival of aquatic hyphomycetes under terrestrial conditions. Transactions of the British Mycological Society 89:606–609.

Srivastava, A.K., G. Défago, and T. Boller.1985a. Secretion of chitinase by *Aphanocladium album*, a hyperparasite of wheat rust. Experientia 41:1612–1613.

Srivastava, A.K., G. Défago, and H. Kern. 1985b. Hyperparasitism of *Puccinia horiana* and other microcyclic rusts. Phytopathologische Zeitschrift 114:73–78.

St. John, T.V., and R.E. Koske. 1988. Statistical treatment of endogonaceous spore counts. Transactions of the British Mycological Society 91:117–121.

Stafford, S. 1985. A statistics primer for foresters, the science of using data isn't just for researchers. Journal of Forestry 83:148–157.

Stahl, S.A., J.D. Rogers, and M.J. Adams. 1988. Observations on *Hendersonia pinicola* and the needle blight of *Pinus contorta*. Mycotaxon 31:323–337.

Stähle, U., and J. Kranz. 1984. Interactions between *Puccinia recondita* and *Eudarluca caricis* during germination. Transactions of the British Mycological Society 82:562–563.

Staley, J.T., J.B. Adams, and F.E. Palmer. 1992. Desert varnish: a biological perspective. Pp. 173–195. *In* G. Stotzky and J.M. Bollag (eds.), Soil Biochemistry. Vol. 7. Marcel Dekker, New York.

Staley, J.T., M.P. Bryant, N. Pfennig, and J.G. Holt (eds.). 1989. Fruiting gliding bacteria: the myxobacteria. Pp. 2139–2170. *In* Bergey's Manual of Systematic Bacteriology. Vol. 3. Williams and Wilkins, Baltimore, Maryland.

Stalpers, J.A. 1978. Identification of wood-inhabiting Aphyllophorales in pure culture. Studies in Mycology 16:1–248.

Stalpers, J.A., A. de Hoog, and I. Vlug. 1987. Improvement of the straw technique for the preservation of fungi in liquid nitrogen. Mycologia 79:82–89.

Stamets, P., and J.S. Chilton. 1983. The Mushroom Cultivator: A practical Guide to Growing Mushrooms at Home. Agarikon Press, Olympia, Washington.

Stamps, D.J., G.M. Waterhouse, F.J. Newhook, and G.S. Hall. 1990. Revised tabular key to the species of *Phytophthora*. 2nd ed. Mycological Papers 162:1–28.

States, J. 1985. Hypogeous, mycorrhizal fungi associated with ponderosa pine: sporocarp phenology. P. 271. *In* R. Molina (ed.), Proceedings of the 6th North American Conference on Mycorrhizae. U.S. Department of Agriculture, Forest Service, Forestry Research Laboratory, Oregon State University, Corvallis, Oregon.

States, J.S. 1978. Soil fungi of cool-desert plant communities in northern Arizona and southern Utah. Journal of the Arizona-Nevada Academy of Science 13:13–17.

Stearn, W.T. 1995. Botanical Latin. David Charles, London, Great Britain.

Steinkraus, D.C., G.O. Boys, and P.H. Slaymaker. 1993. Culture, storage, and incubation period of *Neozygites fresenii* (Entomophthorales: Neozygitaceae) a pathogen of the cotton aphid. Southwestern Entomology 18:197–202.

Stenton, H. 1953. The soil fungi of Wicken Fen. Transactions of the British Mycological Society 36:304–314.

Stephenson, S.L. 1985. Myxomycetes in the laboratory II: moist chamber cultures. American Biology Teacher 47:487–489.

———. 1988. Distribution and ecology of myxomycetes in temperate forests. I. Patterns of occurrence in the upland forests of southwestern Virginia. Canadian Journal of Botany 66:2187–2207.

———. 1989. Distribution and ecology of myxomycetes in temperate forests. II. Patterns of occurrence on bark surface of living trees, leaf litter, and dung. Mycologia 81:608–621.

Stephenson, S.L., and J.C. Cavender. 1996. Dictyostelids and myxomycetes. Pp. 91–101. *In* G.S. Hall (ed.), Methods for the Examination of Organismal Diversity in Soils and Sediments. CAB International, Wallingford, Oxon, United Kingdom.

Stephenson, S.L., and J.C. Landolt. 1992. Vertebrates as vectors of cellular slime moulds in temperate forests. Mycological Research 96:670–672.

———. 1995. The vertical distribution of dictyostelids and myxomycetes in the soil/litter microhabitat. Nova Hedwigia 62:105–117.

———. 1998. Dictyostelid cellular slime molds in canopy soils of tropical forests. Biotropica 30:657–661.

Stephenson, S.L., and G.A. Laursen. 1993. A preliminary report on the distribution and ecology of myxomycetes in Alaskan tundra. Bibliotheca Mycologica 150:251–257.

———. 1998. Myxomycetes from Alaska. Nova Hedwigia 66:425–434.

Stephenson, S.L., and H. Stempen. 1994. Myxomycetes: A Handbook of Slime Molds. Timber Press, Portland, Oregon.

Stephenson, S.L., I. Kalyanasundaram, and T.N. Lakhanpal. 1993. A comparative biogeographical study of Myxomycetes in the mid-Appalachians of eastern North America and two regions of India. Journal of Biogeography 20:645–657.

Stephenson, S.L., J.C. Landolt, and G.A. Laursen. 1991. Cellular slime molds in soils of Alaskan tundra, U.S.A. Arctic and Alpine Research 23:104–107.

———. 1997a. Dictyostelid cellular slime molds from western Alaska, U.S.A., and the Russian Far East. Arctic and Alpine Research 29:222–225.

Stephenson, S.L., J.C. Landolt, D. Mitchell, and W. Roody. 1996. Dictyostelid cellular slime molds of New Zealand. Proceedings of the 2nd International Congress on the Systematics and Ecology of Myxomycetes. 1:120. Real Jardín Botánico. Madrid, Spain.

Stephenson, S.L., J.C. Landolt, and D.L. Moore. 1997b. Protostelids, dictyostelids, and myxomycetes in the litter microhabitat of the Luquillo Experimental Forest, Puerto Rico. Proceedings of the West Virginia Academy of Science 69:11–12.

Stephenson, S.L., Y. Novozhilov, and M. Schnittler. 2000. Distribution and ecology of myxomycetes in high-latitude regions of the northern hemisphere. Journal of Biogeography 27:741–754.

Stephenson, S.L., R.L. Seppelt, and G.A. Laursen. 1992. The first record of a myxomycete from subantarctic Macquarie Island. Antarctic Science 4:431–432.

Sterflinger, K., and W.E. Krumbein. 1997. Dematiaceous fungi as a major agent for biopitting on Mediterranean marbles and limestone. Geomicrobiology Journal 14:219–230.

Sterflinger, K., R. de Baere, G.S. de Hoog, R. de Wachter, W.E. Krumbein, and G. Haase. 1997. *Coniosporium perforans* and *C. apollinis*, two new rock-inhabiting fungi isolated from marble in the Sanctuary of Delos (Cyclades, Greece). Antonie van Leeuwenhoek 72:349–363.

Sterflinger, K., G.S. de Hoog, and G. Haase. 1999. Phylogeny and ecology of meristematic ascomycetes. Studies in Mycology 43:5–22.

Stern, W.T. 1992. Botanical Latin. 4th ed. David Charles, London, England.

Stevens, R.B. (ed.). 1974. Mycology Guidebook. University of Washington Press, Seattle, Washington.

——. 1981. Mycology Guidebook. University of Washington Press, Seattle, Washington.

Stevenson, B.G., and D.L. Dindal. 1987. Functional ecology of coprophagous insects: a review. Pedobiologia 30:285–298.

Stevenson, S.K., and J.A. Rochelle. 1984. Lichen litterfall—its availability and utilization by black-tailed deer. Pp. 391–396. *In* W.R. Meehan, T.R. Merrell, Jr., and T.A. Hanley (eds.), Proceedings of a Symposium on Fish and Wildlife Relationships in Old-Growth Forests. Alaska District, American Institute of Fisheries Research Biologists, Juneau, Alaska.

Stewart, P., and D. Cullen. 1999. Organization and differential regulation of a cluster of lignin peroxidase genes of *Phanerochaete chrysosporium*. Journal of Bacteriology 181:3427–3432.

Stewart, P., P. Kersten, A. Vanden Wymelenberg, J. Gaskell, and D. Cullen. 1992. Lignin peroxidase gene family of *Phanerochaete chrysosporium*: complex regulation by carbon and nitrogen limitation and identification of a second dimorphic chromosome. Journal of Bacteriology 174:5036–5042.

Stockdale, P.M. 1961. *Nannizzia incurvata* gen. nov., sp. nov., a perfect state of *Microsporum gypseum* (Bodin) Guiart et Grigorakis. Sabouraudia 1:41–48.

Stolk, A.C., and R.A. Samson. 1972. The genus *Talaromyces*. Studies on *Talaromyces* and related genera II. Studies in Mycology 2:1–65.

Stolte, K., D. Mangis, R. Doty, and K. Tonnessen (eds.). 1993. Lichens as Bioindicators of Air Quality. U.S. Department of Agriculture, Forest Service, Rocky Mountain Forest and Range Experiment Station, Fort Collins, Colorado. General Technical Report RM-224.

Stoms, D.M., and J.E. Estes. 1993. A remote sensing research agenda for mapping and monitoring biodiversity. International Journal of Remote Sensing 14:1839–1860.

Stone, J.K. 1987. Initiation and development of latent infections by *Rhabdocline parkeri* on Douglas-fir. Canadian Journal of Botany 65:2614–2621.

——. 1988. Fine structure of latent infections by *Rhabdocline parkeri* on Douglas-fir, with observations on unidentified epiderman cells. Canadian Journal of Botany 66:45–54.

Stone, J.K., C.W. Bacon, and J.F. White, Jr. 2000. An overview of endophytic microbes: endophytism defined. Pp. 3–30. *In* C.W. Bacon and J.W. White, Jr. (eds.), Microbial Endophytes. Marcel Dekker, New York.

Stone, J.K., M.A. Sherwood, and G.C. Carroll. 1996. Canopy microfungi: function and diversity. Northwest Science 70:37–45.

Stone, J.K., O. Viret, O. Petrini, and I.H. Chapela. 1994. Histological studies on host penetration and colonization by endophytic fungi. Pp. 115–126. *In* O. Petrini and G.B. Ouellette (eds.), Alteration of Host Walls by Fungi. American Phytopathological Society Press, St. Paul, Minnesota.

Stotzky, G., R.D. Goos, and M.I. Timonin. 1962. Microbial changes occurring in soil as a result of storage. Plant and Soil 16:1–18.

Stoyke, G., and R.S. Currah. 1991. Endophytic fungi from the mycorrhizae of alpine ericoid plants. Canadian Journal of Botany 69:347–352.

Stoyke, G., K. Egger, R.S. Currah. 1992. Characterization of sterile endophytic fungi from the mycorrhizae of subalpine plants. Canadian Journal of Botany 70:2009–2016.

Straatsma, G., and R.A. Samson. 1993. Taxonomy of *Scytalidium thermophilum*, an important thermophilic fungus in mushroom compost. Mycological Research 97:321–328.

Straatsma, G., F. Ayer, and S. Egli. 2001. Species richness, abundance, and phenology of fungal fruit bodies over 21 years in a Swiss forest plot. Mycological Research 105:515–523.

Straatsma, G., J.P.G. Gerrits, M.P.A.M. Augustijn, H.J.M. Op den Camp, G.D. Vogels, and L.J.L.D. van Griensven. 1989. Population dynamics of *Scytalidium thermophilum* in mushroom compost and stimulatory effect on growth rate and yield of *Agaricus bisporus*. Journal of General Microbiology 135:751–759.

Straatsma, G., R.A. Samson, T.W. Olijnsma, H.J.M. Op den Camp, J.P.G. Gerrits, and L.J.L.D. van Griensven. 1994. Ecology of thermophilic fungi in mushroom compost, with emphasis on *Scytalidium thermophilum* and growth stimulation of *Agaricus bisporus* mycelium. Applied and Environmental Microbiology 60:454–458.

Streimann, H. 1986. Catalogue of the lichens of Papua New Guinea and Irian Jaya. Bibliotheca Lichenologica 22:1–45.

Stringer, J.R. 1996. *Pneumocystis carinii*: what is it, exactly? Clinical Microbiology Reviews 9:489–498.

Strong, D.R., Jr., and D.A. Levin. 1979. Species richness of plant parasites and growth form of their hosts. American Naturalist 114:1–22.

Strongman, D.B., and T.G. Rand. 1991. *Hirsutella nodulosa* from a rhodacarid mite on driftwood collected from an inland marine lake in Bermuda. Mycological Research 95:372.

Strongman, D.B., C. Morrison, and G. MacClelland. 1997. Lesions in the musculature of captive American plaice *Hippoglossoides platessoides* caused by the fungus *Hormonconis resinae* (Deuteromycetes). Diseases of Aquatic Organisms 28:107–113.

Studer, M., K. Flück, and W. Zimmermann. 1992. Production of chitinases by *Aphanocladium album* grown on crystalline and colloidal chitin. FEMS Microbiology Letters 99:213–216.

Stutz, J.C., and J.B. Morton. 1996. Successive pot cultures reveal high species richness of arbuscular endomycorrhizal fungi in arid ecosystems. Canadian Journal of Botany 74:1883–1889.

Stutz, J.C., R. Copeman, C.A. Martin, and J.B. Morton. 2000. Patterns of species composition and distribution of arbuscular mycorrhizal fungi in arid regions of southwestern North America and Namibia, Africa. Canadian Journal of Botany 78:237–245.

Suberkropp, K. 1984. Effect of temperature on seasonal occurrence of aquatic hyphomycetes. Transactions of the British Mycological Society 82:53–62.

——. 1991. Relationships between growth and sporulation of aquatic hyphomycetes on decomposing leaf litter. Mycological Research 95:843–850.

Suberkropp, K., and M.J. Klug. 1976. Fungi and bacteria associated with leaves during processing in a woodland stream. Ecology 57:707–719.

Subramanian, C.V., and B.P.R. Vittal. 1979a. Studies on litter fungi II. Fungal colonization of *Atlantia monophylla* Corr. leaves and litter. Nova Hedwigia 63:361–369.

——. 1979b. Studies on litter fungi III. Quantitative studies of the mycoflora of *Atlantia monophylla* Corr. litter. Nova Hedwigia 63:371–378.

——. 1980. Studies on litter fungi. IV. Fungal colonization of *Gymnosporia emarginata* leaves and litter. Transactions of the Mycological Society of Japan 21:339–344.

Sugimoto, R., T. Katoh, and K. Nishioka. 1995. Isolation of dermatophytes from house dust on a medium containing gentamicin and flucytosine. Mycoses 38:405–410.

Sugiyama, J. 1998. Relatedness, phylogeny and evolution of the fungi. Mycoscience 39:487–511.

Sugiyama, J., and S.-O. Suh. 1993. Phylogenetic analysis of basidiomycetous yeasts by means of 18S ribosomal RNA sequences: relationship of *Erythrobasidium hasegawianum* and other basidiomycetous yeast taxa. Antonie van Leeuwenhoek 63:201–209.

Sugiyama, M., A. Ohara, and T. Mikawa. 1999. Molecular phylogeny of onygenalean fungi based on small subunit ribosomal DNA (SSU rDNA) sequences. Mycoscience 40:251–258.

Suh, S-O., and M. Blackwell. 1999. Molecular phylogeny of the cleistothecial fungi placed in Cephalothecaceae and Pseudeurotiaceae. Mycologia 91:836–848.

Suh, S.-O., and J. Sugiyama. 1993. Phylogeny among the basidiomycetous yeasts inferred from small subunit ribosomal DNA sequence. Journal of General Microbiology 139:1595–1598.

Suh, S.-O., H. Noda, and M. Blackwell. 2001. Insect symbiosis: Derivation of yeast-like endosymbionts within an entomopathogenic filamentous lineage. Molecular Biology and Evolution 18:995–1000.

Summerbell, R.C. 1985. The staining of filamentous fungi with diazonium blue B. Mycologia 77:587–593.

——. 1987. The inhibitory effect of *Trichoderma* species and other soil microfungi on formation of mycorrhiza by *Laccaria bicolor* in vitro. New Phytologist 105:437–448.

——. 1989. Microfungi associated with the mycorrhizal mantle and adjacent microhabitats within the rhizosphere of black spruce. Canadian Journal of Botany 67:1085–1095.

——. 1993. The benomyl test as a fundamental diagnostic method for medical mycology. Journal of Clinical Microbiology 31:572–577.

Summerbell, R.C., J. Kane, and S. Krajden. 1989a. Onychomycosis, tinea pedis, and tinea manuum caused by non-dermatophytic filamentous fungi. Mycoses 32:609–619.

Summerbell, R.C., J. Kane, S. Krajden, and E.E. Duke. 1993. Medically important *Sporothrix* species and related Ophiostomatoid fungi. Pp. 185–192. *In* M.J. Wingfield, K.A. Seifert, and J.F. Webber (eds.), *Ceratocystis* and *Ophiostoma*: Taxonomy, Ecology and Pathogenicity. American Phytopathological Society Press, St. Paul, Minnesota.

Summerbell, R.C., S. Krajden, and J. Kane. 1989b. Potted plants in hospitals as reservoirs of pathogenic fungi. Mycopathologia 106:13–22.

Summerbell, R.C., S.A. Rosenthal, and J. Kane. 1988. Rapid method for the differentiation of *Trichophyton rubrum*, *Trichophyton mentagrophytes*, and related dermatophyte species. Journal of Clinical Microbiology 26:2279–2282.

Sundari, S.S., and M.S. Manocha. 1991. Interaction between biotrophic haustorial mycoparasite and protoplasts of host and nonhost fungi. Botanica Helvetica 101:141–148.

Sundberg, W.J., and H.W. Keller. 1996. Myxomycetes: some tools and tips on collection, care, and use of specimens. Inoculum 47:12–14.

Sundheim, L. 1982. Control of cucumber powdery mildew by the hyperparasite *Ampelomyces quisqualis* and fungicides. Plant Pathology 31:209–214.

——. 1986. Use of hyperparasites in biological control of biotrophic plant pathogens. Pp. 333–347. *In* N.J. Fokkema and J. van den Heuvel (eds.), Microbiology of the Phyllosphere. Cambridge University Press, Cambridge, England.

Sundheim, L., and T. Krekling. 1982. Host-parasite relationships of the hyperparasite *Ampelomyces quisqualis* and its powdery mildew host *Sphaerotheca fuliginea*. I. Scanning electron microscopy. Phytopathologische Zeitschrift 104:202–210.

Sundström, K.R. 1964. Studies on the physiology, morphology, and serology of *Exobasidium*. Symbolae Botanicae Upsalienses 18:1–89.

Suryanarayanan, T.S., G. Senthilarasu, and V. Muruganandam. 2000. Endophytic fungi from *Cuscuta reflexa* and its host plants. Fungal Diversity 4:117–123.

Suske, J., and G. Acker. 1987. Internal hyphae in young, symptomless needles of *Picea abies*: electron microscopic and cultural investigation. Canadian Journal of Botany 65:2098–2103.

Suthers, H.B. 1985. Ground-feeding migratory songbirds as cellular slime mold distribution vectors. Oecologia 65:526–530.

Sutton, B.C. 1973a. Hyphomycetes from Manitoba and Saskatchewan, Canada. Mycological Papers 132:1–143.

——. 1973b. Some hyphomycetes with holoblastic sympodial conidiogenous cells. Transactions of the British Mycological Society 61:417–429.

——. 1975. Hyphomycetes on cupules of *Castanea sativa*. Transactions of the British Mycological Society 64:405–426.

——. 1980. The Coelomycetes: Fungi Imperfecti with Pycnidia, Acervuli, and Stromata. Commonwealth Mycological Institute, Kew, Surrey, United Kingdom.

——. 1986. Improvisations on conidial themes. Transactions of the British Mycological Society 86:1–38.

——. 1992. The genus *Glomerella* and its anamorph *Colletotrichum*. Pp. 1–26. *In* J.A. Bailey and M.J. Jeger (eds.), *Colletotrichum*: Biology, Pathology and Control. CAB International, Wallingford, Oxon, United Kingdom.

——. 1996. Conidiogenesis, classification and correlation. Pp. 135–160. *In* B.C. Sutton (ed.), A Century of Mycology. Cambridge University Press, Cambridge, United Kingdom.

Sutton, B.C., and G.L. Hennebert. 1994. Interconnections amongst anamorphs and their possible contribution to ascomycete systematics. Pp. 77–100. *In* D.L. Hawksworth (ed.), Ascomycete Systematics: Problems and Perspectives in the Nineties. Plenum Press, New York.

Sutton, B.C., and C.S. Hodges, Jr. 1976. Eucalyptus microfungi: *Mycoleptodiscus* species and *Pseudotracylla* gen. nov. Nova Hedwigia 27:693–700.

Suzuki, S., and N. Nimura. 1961. Relation between the distribution of aquatic hyphomycetes in Japanese lakes and lake types. Botanical Magazine (Tokyo) 74:51–55.

Swai, I.S., and D.F. Hindal. 1981. Selective medium for recovering *Verticicladiella procera* from soils and symptomatic white pines. Plant Disease 65:963–965.

Swann, E.C., and J.W. Taylor. 1993. Higher taxa of Basidiomycetes: an 18S rRNA gene perspective. Mycologia 85:923–936.

——. 1995a. Phylogenetic perspectives on basidiomycete systematics: evidence from the 18S rRNA gene. Canadian Journal of Botany 73 (Suppl. 1):S862–S868.

——. 1995b. Toward a phylogenetic systematics of the Basidiomycota: integrating yeast and filamentous basidiomycetes using 18S rRNA gene sequences. Studies in Mycology 38:147–161.

——. 1995c. Phylogenetic diversity of yeast-producing basidiomycetes. Mycological Research 99:1205–1210.

Swann, E.C., E.M. Frieders, and D.J. McLaughlin. 1999. *Microbotryum*, *Kriegeria* and the changing paradigm in basidiomycete classification. Mycologia 91:51–66.

——. 2001. Urediniomycetes. Pp. 37–56. *In* D.J. McLaughlin, E.G. McLaughlin, and P.A. Lempke (eds.), The Mycota. Vol. VII. Systematics and Evolution. Part B. Springer-Verlag, Berlin, Germany.

Swanson, A.R., E.M. Vadell, and J.C. Cavender. 1999. Global distribution of forest soil dictyostelids. Journal of Biogeography 26:133–148.

Swart, H.J. 1975. Callosities in fungi. Transactions of the British Mycological Society 64:511–515.

Swift, M.J. 1976. Species diversity and the structure of microbial communities in terrestrial habitats. Pp. 185–222. *In* J.M. Anderson and A. Macfadyen (eds.), The Role of Terrestrial and Aquatic Organisms in Decomposition Processes. Blackwell Scientific, Oxford, England.

———. 1984. Microbial diversity and decomposer niches. Pp. 8–16. *In* M.J. Klug and C.A. Reddy (eds.), Current Perspectives in Microbiology. Proceedings of the 3rd International Symposium on Microbial Ecology. American Society of Microbiology, Washington, D.C.

Swinscow, T.D.V., and H. Krog. 1988. Macrolichens of East Africa. British Museum (Natural History), London, United Kingdom.

Swofford, D.L. 1989. PAUP Phylogenetic Analysis Using Parsimony. Version 3.0 D.L. Swofford, Illinois Natural History Survey, Champaign, Illinois.

Sykes, E.E., and D. Porter. 1980. Infection and development of the obligate parasite *Catenaria allomycis* on *Allomyces arbuscula*. Mycologia 72:288–300.

Sylvia, D.M. 1986. Spatial and temporal distribution of vesicular-arbuscular mycorrhizal fungi associated with *Uniola paniculata* in Florida foredunes. Mycologia 78:728–733.

———. 1994. Vesicular-arbuscular mycorrhizal fungi. Pp. 351–378. *In* R.W. Weaver, S. Angle, P. Bottomley, D. Bezdicek, S. Smith, A. Tabatabai, and A. Wollum (eds.), Methods of Soil Analysis, Part 2. Microbiological and Biochemical Properties. Soil Science Society of America, Madison, Wisconsin.

Sylvia, D.M., and N.C. Schenck. 1983. Germination of chlamydospores of three *Glomus* species as affected by soil matric potential and fungal contamination. Mycologia 75:30–35.

Sylvia, D.M., D.O. Wilson, J.M. Graham, J.J. Maddox, P. Millner, J.B. Morton, H.D. Skipper, S.F. Wright, and A. Jarstfer. 1993. Evaluation of vesicular-arbuscular mycorrhizal fungi in diverse plants and soils. Soil Biology and Biochemistry 25:705–713.

Systematics Agenda 2000. 1994. Systematics Agenda 2000: Charting the Biosphere. Technical Report. Systematics Agenda 2000, American Museum of Natural History, New York.

Szczepka, M.Z., and S. Sokól. 1984. *Buchwaldoboletus lignicola* (Kallenbach) Pilát und *Phaeolus schweinitzii* (Fries) Patouillard—das Problem ihres gemeinsamen Auftretens. Zeitschrift für Mykologie 50:95–99.

Sztejnberg, A., S. Galper, and N. Lisker. 1990. Conditions for pycnidial production and spore formation by *Ampelomyces quisqualis*. Canadian Journal of Microbiology 36:193–198.

Taber, R.A., and R.E. Pettit. 1975. Occurrence of thermophilic microorganisms in peanuts and peanut soil. Mycologia 67:157–161.

Tacon, A.G.J. 1992. Nutritional Fish Pathology: Morphological Signs of Nutritional Deficiency and Toxicity in Farmed Fish. FAO Fish Technical Paper, no. 330, FAO, Rome, Italy.

Talbot, P.H.B. 1968. Fossilized pre-Patouillardian taxonomy? Taxon 17:620–628.

Tan, T.K., and C.L. Pek. 1997. Tropical leaf litter fungi in Singapore with an emphasis on *Halophytophthora*. Mycological Research 101:165–168.

Tanabe, Y., K. O'Donnell, M. Saikawa, and J. Sugiyama. 2000. Molecular phylogeny of parasitic Zygomycota (Dimargaritales, Zoopagales) based on nuclear small subunit ribosomal DNA sequences. Molecular Phylogenetics and Evolution 16:253–262.

Tansey, M.R. 1971. Isolation of thermophilic fungi from self-heated, industrial wood chip piles. Mycologia 63:537–547.

———. 1973. Isolation of thermophilic fungi from alligator nesting material. Mycologia 65:594–601.

Tansey, M.R., and T.D. Brock. 1973. *Dactylaria gallopava*, a cause of avian encephalitis, in hot spring effluents, thermal soils and self-heated coal waste piles. Nature 242:202–203.

———. 1978. Life at high temperatures: ecological aspects. Pp. 159–216. *In* D.J. Kushner (ed.), Microbial Life in Extreme Environments. Academic Press, London, England.

Tansey, M.R., and M.A. Jack. 1976. Thermophilic fungi in sun-heated soils. Mycologia 68:1061–1075.

Tavares, I.I. 1985. Laboulbeniales (Fungi, Ascomycetes). J. Cramer, Braunschweig, Germany.

Tavares, I.I., and J. Balazuc. 1989. *Sugiyamaemyces*, a new genus of Laboulbeniales (Ascomycetes) on *Clidicus* (Scydmaenidae). Mycotaxon 34:565–576.

Taylor, B.L., and T. Gerrodette. 1993. The uses of statistical power in conservation biology: the Vaquita and Northern Spotted Owl. Conservation Biology 7:489–500.

Taylor, J.W. 1986. Fungal evolutionary biology and mitochondrial DNA. Experimental Mycology 10:259–269.

Taylor, J.W., E.C. Swann, and M.L. Berbee. 1994. Molecular evolution of ascomycete fungi: phylogeny and conflict. Pp. 201–212. *In* D.L. Hawksworth (ed.), Ascomycete Systematics: Problems and Perspectives in the Nineties. Plenum Press, New York.

Taylor, J.W., D.M. Geiser, A. Burt, and V. Koufopanou. 1999. The evolutionary biology and population genetics underlying fungal strain typing. Clinical Microbiology Reviews 12:126–146.

Taylor, L.R. 1978. Bates, Williams, Hutchinson—a variety of diversities. Pp. 1–18. *In* L.A. Mound and N. Waloff (eds.), Diversity of Insect Faunas. Symposia of the Royal Entomological Society of London no. 9. Blackwell Scientific, Oxford, England.

Taylor, T.N., W. Remy, H. Hass, and H. Kerp. 1995. Fossil arbuscular mycorrhizae from the early Devonian. Mycologia 87:560–573.

Tebbe, C.C., and W. Vahjen. 1993. Interference of humic acids and DNA extracted directly from soil in detection and transformation of recombinant DNA from bacteria and yeast. Applied and Environmental Microbiology 59:2657–2665.

Tebbe, C.C., D.F. Wenderoth, W. Vahjen, K. Lübke, and J.C. Munch. 1995. Direct detection of recombinant gene expression by two genetically engineered yeasts in soil on the transcriptional and translational levels. Applied and Environmental Microbiology 61:4296–4303.

Tedersoo, L., U. Kõljalg, N. Hallenberg, and K.-H. Larsson. 2003. Fine scale distribution of ectomycorrhizal fungi and roots across substrate layers including coarse woody debris in a mixed forest. New Phytologist 159:153–165.

Tegelström, H. 1986. Mitochondrial DNA in natural populations: An improved routine for the screening of genetic variation based on sensitive silver staining. Electrophoresis 7:226–229.

Tehler, A., J.S. Farris, D.L. Lipscomb, and M. Källersjö. 2000. Phylogenetic analyses of the fungi based on large rDNA data sets. Mycologia 92:459–474.

Templeton, A.R., and B. Read. 1983. The elimination of inbreeding depression in a captive herd of Speke's gazelle. Pp. 241–262. *In* C.M. Schonewald-Cox, S.M. Chambers, B. MacBryde, and W.L. Thomas (eds.). Genetics and Conservation. Benjamin/Cummings, Menlo Park, California.

Teng, S.C. 1996. Fungi of China. Mycotaxon, Ithaca, New York.

ter Steege, H., and J.H.C. Cornelissen. 1988. Collecting and studying bryophytes in the canopy of standing rainforest trees. Pp. 285–290. *In* J.M. Glime (ed.), Methods in Bryology. The Hattori Botanical Laboratory, Nichinan, Japan.

Tester, M., S.E. Smith, and F.A. Smith. 1987. The phenomenon of "nonmycorrhizal" plants. Canadian Journal of Botany 65:419–431.

Teunissen, M.J., H.J. Op den Camp, C.G. Orpin, J.H.J. Huis in't Veld, and G.D. Vogels. 1991. Comparison of growth characteristics of anaerobic fungi isolated from ruminant and non-ruminant herbivores during cultivation in a defined medium. Journal of General Microbiology 137:1401–1408.

Tews, L.L., and R.E. Koske. 1986. Toward a sampling strategy for vesicular-arbuscular mycorrhizas. Transactions of the British Mycological Society 87:353–358.

Thaxter, R. 1892. Contributions from the cryptogamic laboratory of Harvard University. XVIII. On the Myxobacteriaceae, a new order of Schizomycetes. Botanical Gazette 17:389–406.

——. 1893. A new order of Schizomycetes. Botanical Gazette 18:29–30.

——. 1896. Contribution towards a monograph of the Laboulbeniaceae. Part I. Memoirs of the American Academy of Arts and Sciences 12:187–429.

——. 1897. Contributions from the cryptogamic laboratory of Harvard University. XXXIX. Further observations on the Myxobacteriaceae. Botanical Gazette 23:395–411.

——. 1904. Contributions from the cryptogamic laboratory of Harvard University. LVI. Notes on the Myxobacteriaceae. Botanical Gazette 37:405–416.

——. 1908. Contribution towards a monograph of the Laboulbeniaceae. Part II. Memoirs of the American Academy of Arts and Sciences 13:217–469.

——. 1924. Contribution towards a monograph of the Laboulbeniaceae. Part III. Memoirs of the American Academy of Arts and Sciences 14:309–426.

——. 1926. Contribution towards a monograph of the Laboulbeniaceae. Part IV. Memoirs of the American Academy of Arts and Sciences 15:427–580.

——. 1931. Contribution towards a monograph of the Laboulbeniaceae. Part V. Memoirs of the American Academy of Arts and Sciences 16:1–435.

——. 1932. Contributions towards a monograph of the Laboulbeniaceae. Part V. Memoirs of the American Academy of Arts and Sciences 16:1–435.

Theodore, M.L., T.W. Stevenson, G.C. Johnson, J.D. Thornton, and A.C. Lawrie. 1995. Comparison of *Serpula lacrymans* isolates using RAPD PCR. Mycological Research 99:447–450.

Theodorou, M.K., M. Gill, C. King-Spooner, and D.E. Beever. 1990. Enumeration of anaerobic chytridiomycetes as thallus-forming units: novel method for quantification of fibrolytic fungal populations from the digestive tract ecosystem. Applied and Environmental Microbiology 56:1073–1078.

Theodorou, M.K., S.E. Lowe, and A.P.J. Trinci. 1988. The fermentative characteristics of anaerobic rumen fungi. BioSystems 21:371–376.

Theron, J., and T.E. Cloete. 2000. Molecular techniques for determining microbial diversity and community structure in natural environments. Critical Reviews in Microbiology 26:37–57.

Thiers, H.D. 1984. The secotioid syndrome. Mycologia 76:1–8.

——. 1989. *Gastroboletus* revisited. Memoirs of the New York Botanical Garden 49:355–359.

——. 1997. The Agaricales (gilled fungi) of California 9. Russulaceae I. *Russula*. Mad River Press, Eureka, California.

Thiers, H.D., and A.H. Smith. 1969. Hypogeous cortinarii. Mycologia 61:526–536.

Thomas, A., and R. Rosentreter. 1992. Antelope utilization of lichens in the Birch Creek Valley of Idaho. Pp. 58–66. *In* P. Riddle (Chr.), Proceedings of the 15th Biennial Pronghorn Antelope Workshop, Rock Springs, Wyoming. Wyoming Fish and Game Department, Cheyenne, Wyoming.

Thomas, L., and F. Juanes. 1996. The importance of statistical power analysis: An example from *Animal Behaviour*. Animal Behaviour 52:856–859.

Thomas, L., and C.J. Krebs. 1997. A review of statistical power analysis software. Bulletin of the Ecological Society of America 78:126–139.

Thomas, M.R., and R.C. Shattock. 1986. Filamentous fungal associations in the phylloplane of *Lolium perenne*. Transactions of the British Mycological Society 87:255–268.

Thomas, W.K., J. Maa, and A.C. Wilson. 1989. Shifting constraints on tRNA genes during mitochondrial evolution in animals. New Biologist 1:93–100.

Thompson, S.K. 1992. Sampling. J. Wiley, New York.

Thompson, S.K., and G.A.F. Seber. 1996. Adaptive Sampling. J. Wiley, New York.

Thomson, J.W. 1984. American Arctic Lichens. I. The Macrolichens. Columbia University Press, New York.

Thomson, J.W. 1997. American Arctic Lichens. II. The Microlichens. University of Wisconsin Press, Madison, Wisconsin.

Thorn, R.G. 1991. Taxonomic Studies of Cultures of Selected Corticiaceae (Basidiomycota). Unpubl. Ph.D. dissert. University of Toronto, Toronto, Ontario, Canada.

——. 1993. The use of cellulose azure agar as a crude assay of both cellulolytic and ligninolytic abilities of wood-inhabiting fungi. Proceedings of the Japan Academy B 69:29–34.

Thorn, R.G., and G.L. Barron. 1986. *Nematoctonus* and the tribe Resupinateae in Ontario, Canada. Mycotaxon 25:321–453.

Thorn, R.G., J-M. Moncalvo, C.A. Reddy, and R. Vilgalys. 2000. Phylogenetic analyses and distribution of nematophagy support a monophyletic Pleurotaceae within the polyphyletic pleurotoid-lentinoid fungi. Mycologia 92:241–252.

Thorn, R.G., C.A. Reddy, D. Harris, and E.A. Paul. 1996. Isolation of saprophytic basidiomycetes from soils. Applied and Environmental Microbiology 62:4288–4292.

Thuret, G. 1850. Recherches sur les zoospores des algues et les anthéridies des cryptogames. Annales des Sciences Naturelle, 3e. sér., Botanique 14:214–260.

Tibell, L. 1984. A reappraisal of the taxonomy of Caliciales. Nova Hedwigia (Beihefte) 79:597–713.

——. 1992. Crustose lichens as indicators of forest continuity in boreal coniferous forests. Nordic Journal of Botany 12:427–450.

Tien, M., and T.K. Kirk. 1984. Lignin-degrading enzyme from *Phanerochaete chrysosporium*: purification, characterization, and catalytic properties of a unique HO-requiring oxidase. Proceedings of the National Academy of Sciences USA 81:2280–2284.

Tills, D.W. 1977. The distribution of the fungus *Basidiobolus ranarum* Eidem, in fish, amphibians and reptiles of the southern Appalachian region. Transactions of the Kansas Academy of Science 80:75–77.

Timnick, M.B. 1947. Culturing myxomycete plasmodia for classroom use. Proceedings of the Iowa Academy of Science 53:191–193.

Tirilly, Y., F. Lambert, and D. Thouvenot. 1991. Bioproduction of [^{14}C] deoxyphomenone, a fungistatic metabolite of the hyperparasite *Dicyma pulvinata*. Phytochemistry 30:3963–3965.

Tisdall, J.M, S.E. Smith, and P. Rengasamy. 1997. Aggregation of soil by fungal hyphae. Australian Journal of Soil Research 35:55–60.

Todd, D.D. 1988. The effects of host genotype, growth rate, and needle age on the distribution of a mutualistic, endophytic fungus in Douglas-fir plantations. Canadian Journal of Forest Research 18:601–605.

Tokumasu, S., T. Aoki, and F. Oberwinkler. 1994. Fungal succession on pine needles in Germany. Mycoscience 35:29–37.

Tokumasu, S., K. Tubaki, and L. Manoch. 1997. Microfungal communities on decaying pine needles in Thailand. Pp. 93–106. *In* K.K. Janardhanan, C. Rajendran, K. Natarajan, and D.L. Hawksworth (eds.), Tropical Mycology. Science Publishers, Enfield, New Hampshire.

Toledo, L.S.D., and M.A. Castellano. 1996. A revision of the genera *Radiigera* and *Pyrenogaster*. Mycologia 88:863–884.

Tomasi, R. 1977. Funghi "predatori" di funghi. Micologia Italiana 6:3–12.

Tomlinson, P.B. 1986. The Botany of Mangroves. Cambridge University Press, Cambridge, England.

Tommerup, I.C. 1994. Methods for the study of the population biology of vesicular-arbuscular mycorrhizal fungi. Pp. 486–511. *In* J.R. Norris, D. Read, and A.K. Varma (eds.), Techniques for Mycorrhizal Research. Methods in Microbiology. Vol. 24. Academic Press, London, England.

Tommerup, I.C., and K. Sivasithamparam. 1990. Zygospores and asexual spores of *Gigaspora decipiens*, an arbuscular mycorrhizal fungus. Mycological Research 94:897–900.

Tommerup, I.C., J.E. Barton, and P.A. O'Brien. 1995. Reliability of RAPD fingerprinting of three basidiomycete fungi, *Laccaria, Hydnangium* and *Rhizoctonia*. Mycological Research 99:179–186.

Torkelsen, A.-E. 1996. Slekten syzygospora i norge. Agarica 14:81–89.

Toth, R., and D. Toth. 1982. Quantifying vesicular-arbuscular mycorrhizae using a morphometric technique. Mycologia 74:182–187.

Toti, L., O. Viret, G. Horat, and O. Petrini. 1993. Detection of the endophyte *Discula umbrinella* in buds and twigs of *Fagus sylvatica*. European Journal of Forest Pathology 23:147–152.

Touw, M., and P. Kores. 1984. Compactorization in herbaria: planning factors and four case studies. Taxon 33:276–287.

Trappe, J.M. 1969. Studies of *Cenococcum graniforme*. I. An efficient method for isolation from sclerotia. Canadian Journal of Botany 47:1389–1390.

——. 1972. Parasitism of *Helvella lacunosa* by *Clitocybe sclerotoidea*. Mycologia 64:1337–1340.

——. 1975. A revision of the genus *Alpova* with notes on *Rhizopogon* and the Melanogastraceae. Nova Hedwigia (Beihefte) 51:279–309.

——. 1979. The orders, families, and genera of hypogeous Ascomycotina (truffles and their relatives). Mycotaxon 9:297–340.

——. 1987. Phylogenetic and ecological aspects of mycotrophy in the angiosperms from an evolutionary standpoint. Pp. 5–25. *In* G. Safir (ed.), Ecophysiology of VA Mycorrhizal Plants. CRC Press, Boca Raton, Florida.

——. 1989. *Cazia flexiascus* gen. et sp. nov., a hypogeous fungus in the Helvellaceae. Memoirs of the New York Botanical Garden 49:336–338.

Trappe, J.M., and M.A. Castellano. 1991. Keys to the genera of truffles (Ascomycetes). McIlvainea 10:47–65.

Trappe, J.M., and J.W. Gerdemann. 1979. A neotype of *Endogone pisiformis*. Mycologia 71:206–209.

Trappe, J.M., and C. Maser. 1977. Ectomycorrhizal fungi: Interactions of mushrooms and truffles with beasts and trees. Pp. 65–179. *In* T. Walters (ed.), Mushrooms and Man, an Interdisciplinary Approach to Mycology. Linn-Benton Community College, Albany, Oregon.

Traquair, J.A., and W.E. McKeen. 1978. Necrotrophic mycoparasitism of *Ceratocystis fimbriata* by *Hirschioporus pargamenus* (Polyporaceae). Canadian Journal of Microbiology 24:869–874.

Traquair, J.A., D.A. Gaudet, and E.G. Kokko. 1987. Ultrastructure and influences of temperature on the *in vitro* production of *Coprinus psychromorbidus* sclerotia. Canadian Journal of Botany 65:124–130.

Traquair, J.A., R.B. Meloche, W.R. Jarvis, and K.W. Baker. 1984. Hyperparasitism of *Puccinia violae* by *Cladosporium uredinicola*. Canadian Journal of Botany 62:181–184.

Traquair, J.A., L.A. Shaw, and W.R. Jarvis. 1988. New species of *Stephanoascus* with *Sporothrix* anamorphs. Canadian Journal of Botany 66:926–933.

Tresner, H.D., M.P. Backus, and J.T. Curtis. 1954. Soil microfungi in relation to the hardwood forest continuum in southern Wisconsin. Mycologia 46:314–333.

Tretsven, W.I. 1963. Bacteriological survey of filleting processes in the Pacific Northwest. I. Comparison of methods of sampling fish for bacterial counts. Journal of Milk and Food Technology 26:602–603.

Tribe, H.T. 1960. Aspects of decomposition of cellulose in Canadian soils. I. Observations with the microscope. Canadian Journal of Microbiology 6:309–316.

Triebel, D. 1989. Lecideicole Ascomyceten. Eine Revision der obligat lichenicolen Ascomyceten auf lecideioiden Flechten. Bibliotheca Lichenologica 35:1–278.

——. 1993. Notes on the genus *Sagediopsis* (Verrucariales, Adelococcaceae). Sendtnera 1:273–280.

Triebel, D., G. Rambold, and J.A. Elix. 1995. A conspectus of the genus *Phacopsis* (Lecanorales). Bryologist 98:71–83.

Trinci, A.P.J., D.R. Davies, K. Gull, M.I. Lawrence, B.B. Nielsen, A. Rickers, and M.K. Theodorou. 1994. Anaerobic fungi in herbivorous animals. Mycological Research 98:129–152.

Trinci, A.P.J., S.E. Lowe, A. Milne, and M.K. Theodorou. 1988. Growth and survival of rumen fungi. BioSystems 21:357–363.

Tronsmo, A., and J. Ystaas. 1980. Biological control of *Botrytis cinerea* on apple. Plant Disease 64:1009.

Trüper, H.G., and E.A. Galinski. 1986. Concentrated brines as habitats for microorganisms. Experienta 42:1182–1187.

Trutmann, P., P.J. Keane, and P.R. Merriman. 1982. Biological control of *Sclerotinia sclerotiorum* on aerial parts of plants by the hyperparasite *Coniothyrium minitans*. Transactions of the British Mycological Society 78:521–529.

Tsai, H.F., J.-S. Liu, C. Staben, M.J. Christensen, G.C.M. Latch, M.R. Siegel, and C.L. Schardl. 1994. Evolutionary diversification of fungal endophytes of tall fescue grass by hybridization with *Epichloë* species. Proceedings of the National Academy of Sciences USA 91:2542–2546.

Tsao, P.H. 1970. Selective media for isolation of pathogenic fungi. Annual Review of Phytopathology 8:157–186.

Tsao, P.H., and G. Ocana. 1969. Selective isolation of species of *Phytophthora* from natural soils on an improved antibiotic medium. Nature 223:636–638.

Tschermak-Woess, E. 1988. The algal partner. Pp. 39–92. *In* M. Galun (ed.), CRC Handbook of Lichenology. Vol. I. CRC Press, Boca Raton, Florida.

Tsui, C.K., D.D. Hyde, and I.J. Hodgkiss. 2000. Biodiversity of fungi on submerged wood in Hong Kong streams. Aquatic Microbial Ecology 21:289–298.

Tsuneda, A., and Y. Hiratsuka. 1979. Mode of parasitism of a mycoparasite, *Cladosporium gallicola*, on western gall rust *Endocronartium harknessii*. Canadian Journal of Plant Pathology 1:31–36.

——. 1980. Parasitization of pine stem rust fungi by *Monocillium nordinii*. Phytopathology 70:1101–1103.

——. 1981. *Scopinella gallicola*, a new species from rust galls of *Endocronartium harknessii* on *Pinus contorta*. Canadian Journal of Botany 59:1192–1195.

Tsuneda, A., and W.P. Skoropad. 1978. Behaviour of *Alternaria brassicae* and its mycoparasite *Nectria inventa* on intact and on excised leaves of rapeseed. Canadian Journal of Botany 56:1333–1340.

——. 1980. Interactions between *Nectria inventa*, a destructive mycoparasite, and fourteen fungi associated with rapeseed. Transactions of the British Mycological Society 74:501–507.

Tsuneda, A., and R.G. Thorn. 1995. Interactions of wood decay fungi with other microorganisms, with emphasis on the degradation of cell walls. Canadian Journal of Botany 73(Suppl. 1):S1325–S1333.

Tsuneda, A., Y. Hiratsuka, and P.J. Maruyama. 1980. Hyperparasitism of *Scytalidium uredinicola* on western gall rust, *Endocronartium harknessii*. Canadian Journal of Botany 58:1154–1159.

Tsuneda, A., S. Murakami, W.M. Gill, and N. Maekawa. 1997. Black spot disease of *Lentinula edodes* caused by the *Hyphozyma* synanamorph of *Eleutheromyces subulatus*. Mycologia 89:867–875.

Tsuneda, A., W.P. Skoropad, and J.P. Tewari. 1976. Mode of parasitism of *Alternaria brassicae* by *Nectria inventa*. Phytopathology 66:1056–1064.

Tu, J.C. 1980. *Gliocladium virens*, a destructive mycoparasite of *Sclerotinia sclerotiorum*. Phytopathology 70:670–674.

——. 1984. Mycoparasitism by *Coniothyrium minitans* on *Sclerotinia sclerotiorum* and its effect on sclerotial germination. Phytopathologische Zeitschrift 109:261–268.

Tu, J.C., and O. Vaartaja. 1981. The effect of the hyperparasite (*Gliocladium virens*) on *Rhizoctonia solani* and on *Rhizoctonia* root rot of white beans. Canadian Journal of Botany 59:22–27.

Tubaki, K. 1955. Studies on the Japanese Hyphomycetes. (II) Fungicolous group. Nagaoa 5:11–40.

——. 1975. Hypomyces and the conidial states in Japan. Report of the Tottori Mycological Institute 12:161–169.

Tubaki, K., and T. Yokoyama. 1973. Successive fungal flora on sterilized leaves in the litter of forest III. IFO [Institute for Fermentation, Osaka, Japan] Research Communication 6:27–49.

Tuite, J. 1969. Plant Pathological Methods. Burgess, Minneapolis, Minnesota.

Tulloch, M. 1972. The genus *Myrothecium* Tode ex Fr. Mycological Papers 130:1–42.

Tulloss, R.E. 1997. Assessment of similarity indices for undesirable properties and a new tripartite similarity index based on cost functions. Pp. 122–137. *In* M.E. Palm and I.H. Chapela (eds.), Mycology in Sustainable Development: Expanding Concepts, Vanishing Borders. Parkway Publishers, Boone, North Carolina.

Tunset, K., A.C. Nilssen, and J. Andersen. 1988. A new trap design for primary attraction of bark beetles and bark weevils (Col., Scolytidae and Curculionidae). Journal of Applied Entomology 106:266–269.

——. 1993. Primary attraction in host recognition of coniferous bark beetles and bark weevils (Col., Scolytidae and Curculionidae). Journal of Applied Entomology 115:155–169.

Turhan, G. 1990. Further hyperparasites of *Rhizoctonia solani* Kühn as promising candidates for biological control. Zeitschrift für Pflanzenkrankheiten und Pflanzenschutz 97:208–215.

——. 1993. Mycoparasitism of *Alternaria alternata* by an additional eight fungi indicating the existence of further unknown candidates for biological control. Journal of Phytopathology 138:283–292.

Turhan, G., and K. Turhan. 1989. Suppression of damping-off of pepper caused by *Pythium ultimum* Trow and *Rhizoctonia solani* Kühn by some new antagonists in comparison with *Trichoderma harzianum* Rifai. Journal of Phytopathology 126:175–182.

Turnbow, R.H., Jr., and R.T. Franklin. 1980. Flight activity by Scolytidae in the northern Georgia piedmont. Journal of the Georgia Entomology Society 15:26–37.

Turner, G.J., and H.T. Tribe. 1976. On *Coniothyrium minitans* and its parasitism of *Sclerotinia* species. Transactions of the British Mycological Society 66:97–105.

Turner, W.B., and D.C. Aldridge. 1983. Fungal Metabolites II. Academic Press, London, England.

Turton, G., and A.C. Wardlaw. 1987. Pathogenicity of the marine yeasts *Metschnikowia zobellia* and *Rhodotorula rubra* for the sea urchin *Echinus esculentus*. Aquaculture 67:199–202.

Tuthill, D.E. 1985. The Soil Microfungi of a Native Prairie and Loess of the Midcontinental United States. Unpubl. Master's thesis. University of Wyoming, Cheyenne, Wyoming.

Tweddell, R.J., S.H. Jabaji-Hare, and P.M. Charest. 1994. Production of chitinases and β-1, 3 glucanases by *Stachybotrys elegans*, a mycoparasite of *Rhizoctonia solani*. Applied and Environmental Microbiology 60:489–495.

Tyler, G. 1985. Macrofungal flora of Swedish beech forest related to soil organic matter and acidity characteristics. Forest Ecology and Management 10:13–29.

——. 1989. Edaphical distribution patterns of macrofungal species in deciduous forest of south Sweden. Acta Oecologica, Oecologia Generalis 10:309–326.

Typas, M.A., A. Mavridou, and V.N. Kouvelis. 1998. Mitochondrial DNA differences provide maximum intraspecific polymorphism in the entomopathogenic fungi *Verticillium lecanii* and *Metarhizium anisopliae*, and allow isolate detection/identification. Pp. 227–238. *In* P. Bridges, Y. Couteaudier, and J. Clarkson (eds.), Molecular Variability of Fungal Pathogens. CAB International, Wallingford, Oxon, United Kingdom.

Tzean, S.S. 1994. Taxonomic study of Deuteromycotina and related teleomorphs in Taiwan. Pp. 203–213. *In* C.I. Peng and C.H. Chou (eds.), Proceedings of the International Symposium on Biodiversity and Terrestrial Ecosystems. Institute of Botany, Academica Sinica, Taipei, Taiwan. Monograph Series No. 14.

Tzean, S.S., and G.L. Barron. 1981. A new and unusual fungus living in the gut of free-living nematodes. Canadian Journal of Botany 59:1861-1866.

Tzean, S.S., and R.H. Estey. 1978a. Nematode-trapping fungi as mycopathogens. Phytopathology 68:1266–1270.

——. 1978b. *Schizophyllum commune* Fr. as a destructive mycoparasite. Canadian Journal of Microbiology 24:780–784.

——. 1991. *Geotrichopsis mycoparasitica* gen. et sp. nov. (Hyphomycetes), a new mycoparasite. Mycological Research 95:1350–1354.

——. 1992. *Geotrichopsis mycoparasitica* as a destructive mycoparasite. Mycological Research 96:263–269.

Tzean, S.S., and J.Y. Liou. 1993. Nematophagous resupinate basidiomycetous fungi. Phytopathology 83:1015–1020.

Udagawa, S., and Y. Horie. 1971. Taxonomical notes on mycogenous fungi. I. Journal of General and Applied Microbiology (Tokyo) 17:141–159.

Udagawa, S., and S. Uchiyama. 1998. Three new hyphomycetes isolated from soil and feather debris. Canadian Journal of Botany 76:1637–1646.

Udagawa, S.I., T. Awao, and S.K. Abdulllah. 1986. *Thermophymatospora* a new thermophilic genus of basidiomycetous hyphomycetes. Mycotaxon 27:99–106.

Udagawa, S., S. Uchiyama, and S. Kamiya. 1994. A new species of *Myxotrichum* with an *Oidiodendron* anamorph. Mycotaxon 52:197–205.

Uecker, F.A., W.A. Ayers, and P.B. Adams. 1978. A new hyphomycete on sclerotia of *Sclerotinia sclerotiorum*. Mycotaxon 7:275–282.

——. 1980. *Teratosperma oligocladum*, a new hyphomycetous mycoparasite on sclerotia of *Sclerotinia sclerotiorum*, *S. trifoliorum*, and *S. minor*. Mycotaxon 10:421–427.

——. 1982. *Laterispora brevirama*, a new hyphomycete on sclerotia of *Sclerotinia minor*. Mycotaxon 14:491–496.

Ueda, S., and S. Udagawa. 1983. *Thermoascus aegyptiacus*, a new thermophilic ascomycete. Transactions of the Mycological Society of Japan 24:135–142.

Ulken, A., and F.K. Sparrow, Jr. 1968. Estimation of chytrid propagules in Douglas Lake by the MPN//pollen grain method. Veröffentlichungen des Instituts für Meeresforschung in Bremerhaven 11:83–88.

Ulloa, M., and R.T. Hanlin. 2000. Illustrated Dictionary of Mycology. American Phytopathological Society Press, St. Paul, Minnesota.

Uma, N.U., and G.S. Taylor. 1987. Parasitism of leek rust urediniospores by four fungi. Transactions of the British Mycological Society 88:335–340.

Unestam, T. 1973. Fungal diseases of crustacea. Review of Medical and Veterinary Mycology 8:1–20.

Unestam, T., and D.W. Weiss. 1970. The host-parasite relationship between freshwater crayfish and the crayfish disease fungus

Aphanomyces astaci: responses to infection by a susceptible and a resistant species. Journal of General Microbiology 60:77–90.

United Nations Economic and Social Council Committee of Experts on the Transport of Dangerous Goods. 2001. Recommendations on the Transport of Dangerous Goods: Model Regulations. 12th ed.

United States Public Health Service, Centers for Disease Control, Office of Biosafety, Ad Hoc Committee on the Safe Shipment and Handling of Etiologic Agents. 1974. Classification of Etiologic Agents on the Basis of Hazard. 4th ed. U.S. Department of Health, Education, and Welfare, Washington, D.C.

Untereiner, W.A., N.A. Straus, and D. Malloch. 1995. A molecular-morphotaxonomic approach to the systematics of the Herpotrichiellaceae and allied black yeasts. Mycological Research 99:897–913.

Upadhyay, H.P. 1981. A monograph of *Ceratocystis* and *Ceratocystiopsis*. University of Georgia Press, Athens, Georgia.

——. 1993. Classification of the Ophiostomatoid fungi. Pp. 7–13. *In* M.J. Wingfield, K.A. Seifert, and J.F. Webber (eds.), *Ceratocystis* and *Ophiostoma*: Taxonomy, Ecology, and Pathogenicity. American Phytopathological Society Press, St. Paul, Minnesota.

Upadhyay, R.S., B. Rai, and R.C. Gupta. 1981. *Fusarium udum* parasitic on *Aspergillus luchuensis* and *Syncephalastrum racemosum*. Plant and Soil 63:407–413.

Urbasch, I. 1986. *Harzia acremonioides* als biotropher kontaktmykoparasit an *Stemphylium botryosum*. Zeitschrift für Pflanzenkrankheiten und Pflanzenschutz 93:392–396.

Urquhart, E.J., J.G. Menzies, and Z.K. Punja. 1994. Growth and biological control activity of *Tilletiopsis* species against powdery mildew (*Sphaerotheca fuliginea*) on greenhouse cucumber. Phytopathology 84:341–351.

Urzì, C., U. Wollenzien, G. Criseo, and W.E. Krumbein. 1995. Biodiversity of the rock inhabiting microbiota with special reference to black fungi and black yeasts. Pp. 289–302. *In* D. Allsopp, R.R. Colwell, and D.L. Hawksworth (eds.), Microbial Diversity and Ecosystem Function. CAB International, Wallingford, Oxon, United Kingdom.

USDA Forest Service and USDI Bureau of Land Management. 1994. Record of Decision for Amendments to Forest Service and Bureau of Land Management Planning Documents within the Range of the Northern Spotted Owl. Standards and Guidelines for Management of Habitat for Late-Successional and Old-Growth Forest Related Species. USDA-FS and USDI-BLM, Portland, Oregon. [U.S. Government Printing Office, 1994-589-11/0001, Washington, D.C.]

——. 2000. Final Supplemental Environmental Impact Statement for Amendments to the Survey and Manage, Protection Buffer, and other Mitigating Measures Standards and Guidelines. Vols. I, II. USDA-FS and USDI-BLM, Portland, Oregon Publication BLM/OR/WA/PT-00/065+1792.

Ushida, K., H. Tanaka, and Y. Kojima. 1989. A simple in situ method for estimating fungal population size in the rumen. Letters in Applied Microbiology 9:109–111.

Vadell, E.M., M.T. Holmes, and J.C. Cavender. 1995. *Dictyostelium citrinum*, *D. medusoides* and *D. granulophorum*: three new members of Dictyosteliaceae from forest soils of Tikal, Guatemala. Mycologia 87:551–559.

Vakili, N.G. 1985. Mycoparasitic fungi associated with potential stalk rot pathogens of corn. Phytopathology 75:1201–1207.

——. 1989. *Gonatobotrys simplex* and its teleomorph, *Melanospora damnosa*. Mycological Research 93:67–74.

Valentine, H.T., and S.J. Hilton. 1977. Sampling oak foliage by the randomized-branch method. Canadian Journal of Forest Research 7:295–298.

Valla, G., A. Cappellano, R. Hugueney, and A. Moiroud. 1989. *Penicillium nodositatum* Valla, a new species inducing myconodules on *Alnus* roots. Plant and Soil 114:142–146.

Vallim, M.A., B.J.H. Janse, J. Gaskell, A.A. Pizzirani-Kleiner, and D. Cullen. 1998. *Phanerochaete chrysosporium* cellobiohydrolase and cellobiose dehydrogenase transcripts in wood. Applied and Environmental Microbiology 64:1924–1928.

Van de Peer, Y., and R. De Wachter. 1997. Evolutionary relationships among the eukaryotic crown taxa taking into account site-to-site rate variation in 18S rRNA. Journal of Molecular Evolution 45:619–630.

Van de Peer, Y., L. Hendriks, A. Goris, J.-M. Neefs, M. Vancanneyt, K. Kersters, J.-F. Berny, G.L. Hennebert, and R. De Wachter. 1992. Evolution of basidiomycetous yeasts as deduced from small ribosomal subunit RNA sequences. Systematic and Applied Microbiology 15:250–258.

Van de Peer, Y., J.-M. Neefs, and R. De Wachter. 1990. Small ribosomal subunit RNA sequences, evolutionary relationships among different life forms, and mitochondrial origins. Journal of Molecular Evolution 30:463–476.

van den Boogert, P.H.J.F. 1989. Nutritional requirements of the mycoparasitic fungus *Verticillium biguttatum*. Netherlands Journal of Plant Pathology 95:149–156.

van den Boogert, P.H.J.F., and J.W. Deacon. 1994. Biotrophic mycoparasitism by *Verticillium biguttatum* on *Rhizoctonia solani*. European Journal of Plant Pathology 100:137–156.

van den Boogert, P.H.J.F., and W. Gams. 1988. Assessment of *Verticillium biguttatum* in agricultural soils. Soil Biology and Biochemistry 20:899–905.

van den Boogert, P.H.J.F., and G. Jager. 1983. Accumulation of hyperparasites of *Rhizoctonia solani* by addition of live mycelium of *R. solani* to soil. Netherlands Journal of Plant Pathology 89:223–228.

van den Boogert, P.H.J.F., and T.A.W.M. Saat. 1991. Growth of the mycoparasitic fungus *Verticillium biguttatum* from different geographical origins at near-minimum temperatures. Netherlands Journal of Plant Pathology 97:115–124.

van den Boogert, P.H.J.F., and H. Velvis. 1992. Population dynamics of the mycoparasite *Verticillium biguttatum* and its host, *Rhizoctonia solani*. Soil Biology and Biochemistry 24:157–164.

van den Boogert, P.H.J.F., G. Jager, and H. Velvis. 1990. *Verticillium biguttatum*, an important mycoparasite for the control of *Rhizoctonia solani* in potato. Pp. 77–91. *In* D. Hornby (ed.), Biological Control of Soil-borne Plant Pathogens. CAB International, Wallingford, Oxon, United Kingdom.

van den Boogert, P.H.J.F., H. Reinartz, K.A. Sjollema, and M. Veenhuis. 1989. Microscopic observations on the interaction of the mycoparasite *Verticillium biguttatum* with *Rhizoctonia solani* and other soil-borne fungi. Antonie van Leeuwenhoek 56:161–174.

Van der Auwera, G., R. De Baere, Y. Van de Peer, P. De Rijk, I. Van den Broeck, and R. De Wachter. 1995. The phylogeny of the Hyphochytriomycota as deduced from ribosomal RNA sequences of *Hyphochytrium catenoides*. Molecular Biology and Evolution 12:671–678.

Van der Knaap, E., R.J. Rodriguez, and D.W. Freckman. 1993. Differentiation of bacterial-feeding nematodes in soil ecological studies by means of arbitrarily-primed PCR. Soil Biology and Biochemistry 25:1141–1151.

Van der Plaats-Niterink, A.J. 1981. Monograph of the genus *Pythium*. Studies in Mycology 21:1–242.

van der Walt, J.P., and D. Yarrow. 1984. Methods for the isolation, maintenance, classification and identification of yeasts. Pp. 45–103. *In* N.J.W. Kreger-van Rij (ed.), The Yeasts: A Taxonomic Study. 3rd ed. Elsevier, Amsterdam, The Netherlands.

Van Gelderen de Komaid, A. 1988. Viability of fungal cultures after ten years of storage in sterile distilled water at room temperature. Revista Latinoamericana de Microbiología 30:219–221.

Van Oorschot, C.A.N. 1980. A revision of *Chrysosporium* and allied genera. Studies in Mycology 20:1–89.

Van Tongeren, O.F.R. 1995. Cluster analysis. Pp. 174–212. *In* R.H.G. Jongman, C.J.F. Ter Braak, and O.F.R. van Tongeren (eds.), Data Analysis in Community and Landscape Ecology. Cambridge University Press, Cambridge, England.

van Zaayen, A. 1981. *Verticillium* sp., a pathogen of *Agaricus bitorquis.* Mushroom Science 11:591–595.

van Zaayen, A., and W. Gams. 1982. Contribution to the taxonomy and pathogenicity of fungicolous *Verticillium* species. II. Pathogenicity. Netherlands Journal of Plant Pathology 88:143–154.

van Zaayen, A., and B. van der Pol-Luiten. 1977. Heat resistance, biology and prevention of *Diehliomyces microsporus* in crops of *Agaricus* species. Netherlands Journal of Plant Pathology 83:221–240.

Vanbreuseghem, R. 1949. La culture des dermatophytes in vitro sur cheveux isolés. Annales de Parasitologie Humaine et Comparée 24:559–573.

Vandenkoornhuyse, P., R. Husband, T.J. Daniell, I.J. Watson, J.M. Duck, A.H. Fitter, and J.P.W. Young. 2002. Arbuscular mycorrhizal community composition associated with two plant species in a grassland ecosystem. Molecular Ecology 11:1555–1564.

Vánky, K. 1987. Illustrated Genera of Smut Fungi. Lubrecht Cramer, Forestburgh, New York.

——. 1994. European Smut Fungi. G. Fischer, Stuttgart, Germany.

——. 1998. A survey of the spore-ball-forming smut fungi. Mycological Research 102:513–526.

Vánky, K., and L. Guo. 1987. Ustilaginales from China. Acta Mycologica Sinica (Suppl.) 1:227–250.

Vánky, K., J. Gönczöl, and S. Tóth. 1982. Review of the Ustilaginales of Hungary. Acta Botanica Academiae Scientiarum Hungaricae 28:255–277.

Varga, J., F. Kevei, A. Vriesema, F. Debets, Z. Kozakiewicz, and J.H. Croft. 1994. Mitochondrial DNA fragment length polymorphism in field isolates of the *Aspergillus niger* aggregate. Canadian Journal of Microbiology 40:612–621.

Vaughan Martini, A., and C.P. Kurtzman. 1985. Deoxyribonucleic acid relatedness among species of the genus *Saccharomyces* sensu stricto. International Journal of Systematic Bacteriology 35:508–511.

Venema, J., and D. Tollervey. 1995. Processing of pre-ribosomal RNA in *Saccharomyces cerevisiae*. Yeast 11:1629–1650.

Vesely, D. 1977. Potential biological control of damping-off pathogens in emerging sugar beet by *Pythium oligandrum* Drechsler. Phytopathologische Zeitschrift 90:113–115.

Viaud, M., A. Pasquier, and Y. Brygoo. 2000. Diversity of soil fungi studied by PCR-RFLP of ITS. Mycological Research 104:1027–1032.

Vilgalys, R., and M. Hester. 1990. Rapid genetic identification and mapping of enzymatically amplified ribosomal DNA from several *Cryptococcus* species. Journal of Bacteriology 172:4238–4246.

Villeneuve, N., M.M. Grandtner, and J.A. Fortin. 1989. Frequency and diversity of ectomycorrhizal and saprophytic macrofungi in the Laurentide Mountains of Quebec. Canadian Journal of Botany 67:2616–2629.

Viret, O., and O. Petrini. 1994. Colonisation of beech leaves (*Fagus sylvatica*) by the endophyte *Discula umbrinella* (Teleomorph: *Apiognomonia errabunda*). Mycological Research 98:423–432.

Vishniac, H.S. 1955. The morphology and nutrition of a new species of *Sirolpidium*. Mycologia 47:633–645.

——. 1958. A new phycomycete. Mycologia 50:66–79.

Visser, S., and D. Parkinson. 1975. Fungal succession on aspen poplar leaf litter. Canadian Journal of Botany 53:1640–1651.

Vitt, D.H., and R.J. Belland. 1997. Attributes of rarity among Alberta mosses: patterns and prediction of species diversity. Bryologist 100:1–12.

Vittadini, C. 1831. Monographia Tuberaceum. Felicis Rusconi, Milan, Italy.

——. 1842. Monographia Lycoperdineorum. Auguste Taurinorum, Torino, Italy.

Vobis, G., and D.L. Hawksworth. 1981. Conidial lichen-forming fungi. Pp. 245–273. *In* G.T. Cole and B. Kendrick (eds.), The Biology of Conidial Fungi. Vol. 1. Academic Press, New York.

Voglmayr, H., and I. Krisai-Greilhuber. 1996. *Dicranophora fulva*, a rare mucoraceous fungus growing on boletes. Mycological Research 100:583–590.

Vogt, K.A., J. Bloomfield, J.F. Ammirati, and S.R. Ammirati. 1992. Sporocarp production by Basidiomycetes, with emphasis on forest ecosystems. Pp. 563–581. *In* G.C. Carroll and D.T. Wicklow (eds.), The Fungal Community: Its Organization and Role in the Ecosystem. 2nd ed. Marcel Dekker, New York.

Vogt, K.A., R.L. Edmonds, and C.C. Grier. 1981. Biomass and nutrient concentrations of sporocarps produced by mycorrhizal and decomposer fungi in *Abies amabilis* stands. Oecologia 50:170–175.

Vogt, K.A., C.C. Grier, C.E. Meier, and R.L. Edmonds. 1982. Mycorrhizal role in net primary production and nutrient cycling in *Abies amabilis* ecosystems in Western Washington. Ecology 63:370–380.

Vogt, K.A., E.E. Moore, D.J. Vogt, M.J. Redlin, and R.L. Edmonds. 1983. Conifer fine root and mycorrhizal root biomass within the forest floors of Douglas-fir stands of different ages and site productivities. Journal of Forest Research 13:429–437.

Voigt, K., and J. Wöstemeyer. 2001. Phylogeny and origin of 82 zygomycetes from all 54 genera of the Mucorales and Mortierellales based on combined analysis of actin and translation elongation factor EF-1α genes. Gene 270:113–120.

Voigt, K., E. Cigelnik, and K. O'Donnell. 1999. Phylogeny and PCR identification of clinically important Zygomycetes based on nuclear ribosomal-DNA sequence data. Journal of Clinical Microbiology 37:3957–3964.

Volkmann-Kohlmeyer, B., and J. Kohlmeyer. 1992. *Corallicola nana* gen. & sp. nov. and other ascomycetes from coral reefs. Mycotaxon 44:417–424.

——. 1993. Biogeographic observations on Pacific marine fungi. Mycologia 85:337–346.

——. 1996. How to prepare truly permanent microscope slides. Mycologist 10:107–108.

von Arx, J.A. 1957. Die Arten der Gattung *Colletotrichum* Cda. Phytopathologische Zeitschrift 29:413–468.

——. 1963. Die Gattungen der Myriangiales. Persoonia 2:421–475.

——. 1981. The Genera of Fungi Sporulating in Pure Culture. J. Cramer, Vaduz, Liechtenstein.

——. 1987. Plant pathogenic fungi. Nova Hedwigia (Beihefte) 87:1–288.

von Arx, J.A., and E. Müller. 1954. Die Gattungen der amerosporen Pyrenomyceten. Beiträge zur Kryptogamenflora der Schweiz 11:1–434.

——. 1975. A re-evaluation of the bitunicate ascomycetes with keys to families and genera. Studies in Mycology 9:1–159.

von Arx, J.A., and M.A.A. Schipper. 1978. The CBS fungus collection. Advances in Applied Microbiology 24:215–236.

von Arx, J.A., and J.P. van der Walt. 1987. Ophiostomatales and Endomycetales. Pp.167–176. *In* G.S. De Hoog, M.T. Smith, and A.C.M. Weijman (eds.), The Expanding Realm of Yeast-like Fungi. Elsevier, Amsterdam, The Netherlands.

von Arx, J.A., M.J. Figueras, and J. Guarro. 1988. Sordariaceous ascomycetes without ascospore ejaculation. Nova Hedwigia (Beihefte) 94:1–104.

von Arx, J.A., J. Guarro, and M.J. Figueras. 1986. The Ascomycete Genus *Chaetomium*. J. Cramer, Berlin, Germany.

von Arx, J.A., J.P. van der Walt, and N.V.D.M. Liebenberg. 1982. The classification of *Taphrina* and other fungi with yeast-like cultural states. Mycologia 74:285–296.

Vouaux, L. 1912–14. Synopsis des champignons parasites des lichens. Bulletin Trimestriel de la Société Mycologique de France 28:177–256; 29:33–128, 399–494; 30:135–198, 281–329.

Vrijmoed, L.L.P., I.J. Hodgkiss, and L.B. Thrower. 1982a. Factors affecting the distribution of lignicolous marine fungi in Hong Kong. Hydrobiologia 87:143–160.

——. 1982b. Seasonal patterns of primary colonization by lignicolous marine fungi in Hong Kong. Hydrobiologia 89:253–262.

Vrijmoed, L.L.P., R.B. Sadaba, E.B.G. Jones, and I.J.H. Hodgkiss. 1995. An ecological study of fungi colonising decayed *Acanthus ilicifolius* at Mai Po mangroves, Hong Kong. P. 65. *In* Abstracts. 7th International Symposium on Microbial Ecology, Santos, São Paulo, Brazil, 1995. Brazilian Society of Microbiology, São Paulo, Brazil.

Waddell, D.R. 1982. A predatory slime mould. Nature 298:464–466.

Waide, R.B., M.R. Willig, C.F. Steiner, G. Mittelbach, L. Gough, S.I. Dodson, G.P. Juday, and R. Parmenter. 1999. The relationship between productivity and species richness. Annual Review of Ecology and Systematics 30:257–300.

Wainright, P.O., G. Hinkle, M.L. Sogin, and S.K. Stickel. 1993. Monophyletic origins of the Metazoa: An evolutionary link with fungi. Science 260:340–342.

Wainwright, M. 1992. The impact of fungi on environmental biogeochemistry. Pp. 601–618. *In* G.C. Carroll and D.T. Wicklow (eds.), The Fungal Community, Its Organization and Role in the Ecosystem, 2nd ed. Marcel Dekker, New York.

——. 1993. Oligotrophic growth of fungi. Pp. 127–144. *In* D.H. Jennings (ed.), Stress Tolerance of Fungi. Marcel Dekker, New York.

Wakefield, A.E., and J. Hopkin. 1994. A molecular approach to the diagnosis of *Pneumocystis carinii* pneumonia. Pp. 403–414. *In* P.D. Walzer (ed.), *Pneumocystis carinii* Pneumonia. 2nd ed., Marcel Dekker, New York.

Walker, C. 1992. Systematics and taxonomy of the arbuscular endomycorrhizal fungi (Glomales)—a possible way forward. Agronomie 12:887–897.

Walker, C., C.W. Mize, and H.S. McNabb, Jr. 1982. Populations of endogonaceous fungi at two locations in central Iowa. Canadian Journal of Botany 60:2518–2529.

Walker, J.A., and R.B. Maude. 1975. Natural occurrence and growth of *Gliocladium roseum* on the mycelium and sclerotia of *Botrytis allii*. Transactions of the British Mycological Society 65:335–339.

Walker, J.C., and D.W. Minter. 1981. Taxonomy of *Nematogonum, Gonatobotrys, Gonatobotryum* and *Gonatorrhodiella*. Transactions of the British Mycological Society 77:299–319.

Walker, W.F. 1984. 5S rRNA sequences from Atractiellales, and basidiomycetous yeasts and fungi imperfecti. Systematic and Applied Microbiology 5:352–359.

——. 1985a. 5S ribosomal RNA sequences from ascomycetes and evolutionary implications. Systematic and Applied Microbiology 6:48–53.

——. 1985b. 5S and 5.8S ribosomal RNA sequences and protist phylogenetics. Biological Systems 18:269–278.

Walker, W.F., and W.F. Doolittle. 1982. Redividing the basidiomycetes on the basis of 5S rRNA sequences. Nature (London) 299:723–724.

Walley, F.L., and J.J. Germida. 1995. Estimating the viability of vesicular-arbuscular mycorrhizae fungal spores using tetrazolium salts as vital stains. Mycologia 87:273–279.

Wang, A.M., and D.F. Mark. 1990. Quantitative PCR. Pp. 70–75. *In* M.A. Innis, D.H. Gelfand, J.J. Sninsky, and T.J. White (eds.), PCR Protocols: A Guide to Methods and Applications. Academic Press, New York.

Wang, A.M., M.V. Doyle, and D.F. Mark. 1989. Quantitation of mRNA by the polymerase chain reaction. Proceedings of the National Academy of Sciences USA 86:9717–9721.

Wang, C.J.K., and H.E. Wilcox. 1985. New species of ectendomycorrhizal and pseudomycorrhizal fungi: *Phialocephala finlandia, Chloridium paucisporum,* and *Phialocephala fortinii.* Mycologia 77:951–958.

Wang, J., L. Hu, S.R. Hamilton, K.R. Coombes, and W. Zhang. 2003. RNA amplification strategies for cDNA microarray experiments. BioTechniques 34:394–400.

Warcup, J.H. 1950. The soil-plate method for isolation of fungi from soil. Nature 166:117–118.

——. 1951a. Soil-steaming: a selective method for the isolation of ascomycetes from soil. Transactions of the British Mycological Society 34:515–518.

——. 1951b. Effect of partial sterilization by steam or formalin on the fungus flora of an old forest nursery soil. Transactions of the British Mycological Society 34:519–532.

——. 1955. Isolation of fungi from hyphae present in soil. Nature 175:953–954.

——. 1957. Studies on the occurrence and activity of fungi in a wheatfield soil. Transactions of the British Mycological Society 40:237–262.

——. 1959. Studies on basidiomycetes in soil. Transactions of the British Mycological Society 42:45–52.

Warcup, J.H., and K.F. Baker. 1963. Occurrence of dormant ascospores in soil. Nature 197:1317–1318.

Warcup, J.H., and P.H.B. Talbot. 1962. Ecology and identity of mycelia isolated from soil. Transactions of the British Mycological Society 45:495–518.

Ward, J.E., and G.T. Cowley. 1972. Thermophilic fungi of some central South Carolina forest soils. Mycologia 64:200–205.

Ward, M., L.J. Wilson, C.L. Carmona, and G. Turner. 1988. The *oli*C3 gene of *Aspergillus niger*: Isolation, sequence and use as a selectable marker for transformation. Current Genetics 14:37–42.

Warner, S.A., and A.J. Domnas. 1987. Biochemical characterization of zoosporic fungi: the utility of sterol metabolism as an indicator of taxonomic affinity. Pp. 202–208. *In* M.S. Fuller and A. Jaworski (eds.), Zoosporic Fungi in Teaching and Research. Southeastern Publishing, Athens, Georgia.

Watanabe, T. 1994. Pictorial Atlas of Soil and Seed Fungi. Lewis Publishers, Boca Raton, Florida.

Waterhouse, G.M. 1968. The genus *Pythium* Pringsheim. Mycological Papers 110:1–71.

——. 1970. The genus *Phytophthora* de Bray. Mycological Papers 122:1–59.

——. 1971. Phycomycetes. Pp. 183–192. *In* C. Booth (ed.), Methods in Microbiology. Academic Press, London, England.

——. 1973. Peronosporales. Pp. 165–183. *In* G.C. Ainsworth, F.K. Sparrow, and A.S. Sussman (eds.), The Fungi. An Advanced Treatise. IVB. A Taxonomic Review with Keys: Basidiomycetes and Lower Fungi. Academic Press, New York.

Watling, R. 1968. Hints on the microscopical examination of the agaric fruit-body. Microscopy 31:95–105.

——. 1974. Dimorphism in *Entoloma abortivum.* Pp. 449–470. *In* Travaux Mycologiques Dedies a R. Kuhner. Bulletin de la Société Linnéenne de Lyon. Numéro spécial 43.

——. 1980. How to Identify Mushrooms to Genus V: Cultural and Developmental Features. Mad River Press, Eureka, California.

Watling, R., and A.E. Watling. 1980. A Literature Guide for Identifying Mushrooms. Mad River Press, Eureka, California.

Watson, A.K., and J.E. Miltimore. 1975. Parasitism of the sclerotia of *Sclerotinia sclerotiorum* by *Microsphaeropsis centaureae.* Canadian Journal of Botany 53:2458–2461.

Watson, E.S., D.C. McClurkin, and M.B. Huneycutt. 1974. Fungal succession on loblolly pine and upland hardwood foliage and litter in North Mississippi. Ecology 55:1128–1134.

Watson, P. 1955. *Calcarisporium arbuscula* living as an endophyte on apparently healthy sporophores of *Russula* and *Lactarius*. Transactions of the British Mycological Society 38:409–414.

——. 1962. Culture of *Spinellus*. Nature (London) 195:1018.

——. 1965. Nutrition of *Spinellus macrocarpus*. Transactions of the British Mycological Society 48:73–80.

Waugh, R., and W. Powell. 1992. Using RAPD markers in crop improvement. Tibtech 10:186–191.

Webb, J., and M.K. Theodorou. 1988. A rumen anaerobic fungus of the genus *Neocallimastix*: Utrastructure of the polyflagellate zoospore and young thallus. BioSystems 21:393–401.

——. 1991. *Neocallimastix hurleyensis* sp. nov., an anaerobic fungus from the ovine rumen. Canadian Journal of Botany 69:1220–1224.

Weber, B.C., and J.E. McPherson. 1991. Seasonal flight patterns of Scolytidae (Coleoptera) in black walnut plantations in North Carolina and Illinois. The Coleopterist Bulletin 45:45–56.

Weber, C.A., M.E. Hudspeth, G.P. Moore, and L.I. Grossman. 1986. Analysis of the mitochondrial and nuclear genomes of two basidiomycetes, *Coprinus cinereus* and *Coprinus stercorarius*. Current Genetics 10:515–525.

Weber, H.A., N.C. Baenziger, and J.B. Gloer. 1990. Structure of preussomerin A: an unusual antifungal metabolite from the coprophilous fungus *Preussia isomera*. Journal of the American Chemical Society 112:6718–6719.

Weber, N.S. 1995a. Western American Pezizales. *Selenaspora guernisacii*, new to North America. Mycologia 87:90–95.

Weber, N.S. 1995b. A Morel Hunter's Companion: Guide to True and False Morels. Reprint ed. Thunder Bay Press, Lansing, Michigan.

Weber, N.S., J.M. Trappe, and W.C. Denison. 1997. Studies on Western American Pezizales. Collecting and describing ascomata—macroscopic features. Mycotaxon 61:153–176.

Webster, J. 1957. Succession of fungi on decaying cocksfoot culms. II. Journal of Ecology 45:1–30.

——. 1970. Coprophilous fungi. Transactions of the British Mycological Society 54:161–180.

——. 1977. Seasonal observations on "aquatic" hyphomycetes on oak leaves on the ground. Transactions of the British Mycological Society 68:108–111.

——. 1981. Biology and ecology of aquatic hyphomycetes. Pp. 681–691. *In* D.T. Wicklow and G.C. Carroll (eds.), The Fungal Community, Its Organization and Role in the Ecosystem. Marcel Dekker, New York.

——. 1992. Anamorph-teleomorph relationships. Pp. 100–117. *In* F. Bärlocher (ed.), Studies in Ecology, The Ecology of Aquatic Hyphomycetes. Springer-Verlag, New York.

Webster, J., and E. Descals. 1979. The teleomorphs of water-borne hyphomycetes from fresh water. Pp. 419–451. *In* W.B. Kendrick (ed.), The Whole Fungus, The Sexual-Asexual Synthesis Vol. 1. National Museum of Natural Sciences, Ottawa, Ontario, Canada.

——. 1981. Morphology, distribution, and ecology of conidial fungi in freshwater habitats. Pp. 295–355. *In* G.T. Cole and B. Kendrick (eds.), The Biology of Conidial Fungi. Vol. 1. Academic Press, New York.

Webster, J., and N.J. Dix. 1960. Succession of fungi on decaying cocksfoot culms. III. A comparison of the sporulation and growth of some primary saprophytes on stem, leaf blade and leaf sheath. Transactions of the British Mycological Society 43:85–99.

Webster, J., and N. Lomas. 1964. Does *Trichoderma viride* produce gliotoxin and viridin? Transactions of the British Mycological Society 47:535–549.

Wedin, M., and J. Hafellner. 1998. Lichenicolous species of *Arthonia* on Lobariaceae with notes on excluded taxa. Lichenologist 30:59–91.

Wedin, M., and L. Tibell. 1997. Phylogeny and evolution of Caliciaceae, Mycocaliciaceae and Sphinctrinaceae (Ascomycota), with notes on the evolution of the prototunicate ascus. Canadian Journal of Botany 75:1236–1242.

Wedin, M., H. Döring, A. Nordin, and L. Tibell. 2000. Small subunit rDNA phylogeny shows the lichen families Caliciaceae and Physciaceae (Lecanorales, Ascomycotina) to form a monophyletic group. Canadian Journal of Botany 78:246–254.

Wehmeyer, L.E. 1941. A revision of *Melanconis*, *Pseudovalsa*, *Prosthecium*, and *Titania*. University of Michigan Studies, Scientific Series 14:1–161.

Wei, J.-C. 1991. An Enumeration of Lichens in China. International Academic, Beijing, China.

Weindling, R. 1932. *Trichoderma lignorum* as a parasite of other soil fungi. Phytopathology 22:837–845.

——. 1938. Association effects of fungi. Botanical Review 4:475–496.

——. 1941. Experimental consideration of the mold toxins of *Gliocladium* and *Trichoderma*. Phytopathology 31:991–1003.

Weir, A., and G.W. Beakes. 1996. Correlative light- and scanning electron microscope studies on the developmental morphology of *Hesperomyces virescens*. Mycologia 88:677–693.

Weir, A., and P.M. Hammond. 1997a. A preliminary assessment of species-richness patterns of tropical, beetle-associated Laboulbeniales (Ascomycetes). Pp. 121–139. *In* K.D. Hyde (ed.), Biodiversity of Tropical Microfungi. Hong Kong University Press, Hong Kong.

——. 1997b. Laboulbeniales on beetles: host utilization patterns and species richness of the parasites. Biodiversity and Conservation 6:701–719.

Weir, A., and W. Rossi. 1995. Laboulbeniales parasitic on British Diptera. Mycological Research 99:841–849.

Weisburg, W.G., J.G. Tully, D.L. Rose, J.P. Petzel, H. Oyaizu, D. Yang, L. Mandelco, J. Sechrest, T.G. Lawrence, J. Van Etten, J. Maniloff, and C.R. Woese. 1989. A phylogenetic analysis of the mycoplasmas: Basis for their classification. Journal of Bacteriology 171:6455–6467.

Weising, K., H. Nybom, K. Wolff, and W. Meyer. 1995. DNA Fingerprinting in Plants and Fungi. CRC Press, Boca Raton, Florida.

Weising, K., F. Weigand, A.J. Driesel, G. Kahl, H. Zischler, and J.T. Epplen. 1989. Polymorphic simple GATA/GACA repeats in plant genomes. Nucleic Acids Research 17:10128.

Weissenhorn, I., A. Glashoff, C. Leyval, and J. Berthelin. 1994. Differential tolerance to Cd and Zn of arbuscular mycorrhizal (AM) fungal spores isolated from heavy metal-polluted and unpolluted soils. Plant and Soil 167:189–196.

Weitzman, I., and M. Silva-Hutner. 1967. Non-keratinous agar media as substrates for the ascigerous state in certain members of the Gymnoascaceae pathogenic for man and animal. Sabouraudia 5:335–340.

Weitzman, I., and R.C. Summerbell. 1995. The dermatophytes. Clinical Microbiology Reviews 8:240–259.

Wells, H.D., D.K. Bell, and C.A. Jaworski. 1972. Efficacy of *Trichoderma harzianum* as a biocontrol for *Sclerotium rolfsii*. Phytopathology 62:442–447.

Wells, K. 1994. Jelly fungi, then and now. Mycologia 86:18–48.

Wells, K., and R.J. Bandoni. 2001. Heterobasidiomycetes. Pp. 85–120. *In* D.J. McLaughlin, E.G. McLaughlin, and P.A. Lempke (eds.), The Mycota. Vol. VII. Systematics and Evolution. Part B. Springer-Verlag, Berlin, Germany.

Welsch, J., and M. McClelland. 1990. Fingerprinting genomes using PCR with arbitrary primers. Nucleic Acids Research 18:7213–7218.

Wennström, A. 1994. Systemic diseases on hosts with different growth patterns. Oikos 69:535–538.

Weston, R.H., J.R. Lindsay, D.B. Purser, G.L.R. Gordon, and P. Davis. 1988. Feed intake and digestion responses in sheep to the addition of inorganic sulfur to a herbage diet of low sulfur content. Australian Journal of Agricultural Research 39:1107–1119.

Wetmore, C.M. 1967 [1968]. Lichens of the Black Hills of South Dakota and Wyoming. Publications of the Museum, Michigan State University, Biological Series 3:209–464.

———. 1993. Lichens and Air Quality in Rainbow Lake Wilderness of The Chequamegon National Forest. Final Report for U. S. Department of Agriculture, Forest Service, Chequamegon National Forest and Northeastern Areas State and Private Forestry Forest Health Protection Contract 42–649, St. Paul, Minnesota. [http://ocid.nac.se.org/airlichenPDF/Rainbow.pdf]

———. 1995. Lichens and air quality in Lye Brook Wilderness of the Green Mountain National Forest. Final Report for U. S. Department of Agriculture, Forest Service, Green Mountain National Forest and Northeastern Areas State and Private Forestry Forest Health Protection Contract 42–649. St. Paul, Minnesota. [http://ocid.nac.se.org/airlichenPDF/LyeBrook.pdf]

Whaley, J.W., and H.L. Barnett. 1963. Parasitism and nutrition of *Gonatobotrys simplex*. Mycologia 55:199–210.

Whalley, A.J.S. 1987. *Xylaria* inhabiting fallen fruits. Agarica 8:68–72.

———. 1993. Tropical Xylariaceae: their distribution and ecological characteristics. Pp. 103–119. *In* S. Isaac, J.C. Frankland, R. Watling, and A.J.S. Whalley (eds.), Aspects of Tropical Mycology. Cambridge University Press, Cambridge, United Kingdom.

Wheeler, K.A., A.D. Hocking, and J.I. Pitt. 1988. Influence of temperature and water relations of *Polypaecilium pisce* and *Basipetospora halophila*, two halophilic fungi. Journal of General Microbiology 134:2255–2260.

Wheeler, Q.D. 1984. Evolution of slime mold feeding in leiodid beetles. Pp. 446–477. *In* Q. Wheeler and M. Blackwell (eds.), Fungus-Insect Relationships: Perspectives in Ecology and Evolution. Columbia University Press, New York.

Wheeler, Q., and M. Blackwell (eds.). 1984. Fungus-Insect Relationships: Perspectives in Ecology and Evolution. Columbia University Press, New York.

Whipps, J.M. 1987. Effect of media on growth and interactions between a range of soil-borne glasshouse pathogens and antagonistic fungi. New Phytologist 107:127–142.

———. 1991. Effects of mycoparasites on sclerotia-forming fungi. Pp. 129–140. *In* A.B.R. Beemster, G.J. Bollen, M. Gerlagh, M.A. Ruissen, B. Schippers, and A. Tempel (eds.), Biotic Interactions and Soil-borne Diseases. Elsevier, Amsterdam, The Netherlands.

———. 1992. Concepts in mycoparasitism and biological control of plant diseases. Pp. 54–59. *In* D.F. Jensen, J. Hockenhull, and N.J. Fokkema (eds.), New Approaches in Biological Control of Soil-Borne Diseases. IOBC/WPRS [International Organisation for Biological Control/West Palearctic Regional Section] Bulletin 15.

Whipps, J.M., and S.P. Budge. 1990. Screening for sclerotial mycoparasites of *Sclerotinia sclerotiorum*. Mycological Research 94:607–612.

Whipps, J.M., and M. Gerlagh. 1992. Biology of *Coniothyrium minitans* and its potential for use in disease biocontrol. Mycological Research 96:897–907.

Whipps, J.M., and R.D. Lumsden (eds.). 1989. Biotechnology of Fungi for Improving Plant Growth. Cambridge University Press, Cambridge, England.

Whipps, J.M., S.K. Grewal, and P. van der Goes. 1991. Interactions between *Coniothyrium minitans* and sclerotia. Mycological Research 95:295–299.

Whipps, J.M., S.P. Budge, and S.J. Mitchell. 1993a. Observations on sclerotial mycoparasites of *Sclerotinia sclerotiorum*. Mycological Research 97:697–700.

Whipps, J.M., M.P. McQuilken, and S.P. Budge. 1993b. Use of fungal antagonists for biocontrol of damping-off and Sclerotinia disease. Pesticide Science 37:309–313.

Whisler, H.C. 1968. Experimental studies with a new species of *Stigmatomyces* (Laboulbeniales). Mycologia 60:65–75.

———. 1987. On the isolation and culture of water molds: the blastocladiales and monoblepharidales. Pp. 121–124. *In* M.S. Fuller and A. Jaworski (eds.), Zoosporic Fungi in Teaching and Research. Southeastern Publishing, Athens, Georgia.

White, J.F., Jr. 1987. Widespread distribution of endophytes in the Poaceae. Plant Disease 71:340–342.

———. 1992. Endophyte-host associations in grasses. XVII. Ecological and physiological features characterizing *Epichloë typhina* and some anamorphic varieties in England. Mycologia 84:431–441.

———. 1993. Endophyte-host associations in grasses. XIX. A systematic study of some sympatric species of *Epichloë* in England. Mycologia 85:444–455.

———. 1994a. Endophyte-host associations in grasses. XX. Structural and reproductive studies of *Epichloë amarillans* sp. nov. and comparisons to *E. typhina*. Mycologia 86:571–580.

———. 1994b. Taxonomic relationships among the members of the Balansieae (Clavicipitales). Pp. 1–20. *In* C.W. Bacon and J.F. White, Jr. (eds.), Biotechnology of Endophytic Fungi of Grasses. CRC Press, Boca Raton, Florida.

White, J.F., Jr., and P.M. Halisky. 1992. Association of systemic fungal endophytes with stock-poisoning grasses. Pp. 574–578. *In* L.F. James, R.F. Keeler, E.M. Bailey, Jr., P.R. Cheeke, and M.P. Hegarty (eds.), Poisonous Plants: Proceedings of the 3rd International Symposium. Iowa State University Press, Ames, Iowa.

White, J.F., Jr., and G. Morgan-Jones. 1987. Endophyte-host associations in forage grasses. VII. *Acremonium chisosum*, a new species isolated from *Stipa eminens* in Texas. Mycotaxon 28:179–189.

White, J.F., Jr., and J.R. Owens. 1992. Stromal development and mating system of *Balansia epichloë*, a leaf-colonizing endophyte of warm-season grasses. Applied and Environmental Microbiology 58:513–519.

White, J.F., Jr., G.T. Cole, and G. Morgan-Jones. 1987. Endophyte-host associations in forage grasses. VI. A new species of *Acremonium* isolated from *Festuca arizonica*. Mycologia 79:148–152.

White, J.F., Jr., A.C. Morrow, and G. Morgan-Jones. 1990. Endophyte-host associations in forage grasses. XII. A fungal endophyte of *Trichachne insularis* belonging to *Pseudocercosporella*. Mycologia 82:218–226.

White, T.J., T. Bruns, S. Lee, and J. Taylor. 1990. Amplification and direct sequencing of fungal ribosomal RNA genes for phylogenetics. Pp. 315–322. *In* M.A. Innis, D.H. Gelfand, J.J. Sninsky, and T.J. White (eds.), PCR Protocols: A Guide to Methods and Applications. Academic Press, New York.

White, W.L. 1942. A new hemiascomycete. Canadian Journal of Research 20:389–395.

Whittaker, R.H. 1956. Vegetation of the Great Smoky Mountains. Ecological Monographs 26:1–80.

———. 1965. Dominance and diversity in land plant communities. Science 147:250–260.

———. 1972. Evolution and measurement of species diversity. Taxon 21:213–251.

———. 1975. Communities and Ecosystems. MacMillan, New York.

———. 1977. Evolution of species diversity in land communities. Evolutionary Biology 10:1–67.

Wicker, E.F. 1981. Natural control of white pine blister rust by *Tuberculina maxima*. Phytopathology 71:997–1000.

Wicker, E.F., and J.Y. Woo. 1973. Histology of blister rust cankers parasitized by *Tuberculina maxima*. Phytopathologische Zeitschrift 76:356–366.

Wicklow, D.T. 1973. Microfungal populations in surface soils of manipulated prairie stands. Ecology 54:1302–1310.

——. 1975. Fire as an environmental cue initiating ascomycete development in a tall grass prairie. Mycologia 67:852–862.

——. 1981. Biogeography and conidial fungi. Pp. 417–447. *In* G.T. Cole and B. Kendrick (eds.), The Biology of Conidial Fungi. Vol. 1. Academic Press, New York.

——. 1988. Metabolites in the coevolution of fungal chemical defense systems. Pp. 174–201. *In* K.A. Pirozynski and D. Hawksworth (eds.), Coevolution of Fungi with Plants and Animals, Academic Press, New York.

——. 1992. The coprophilous fungal community: an experimental system. Pp. 715–728. *In* G.C. Carroll and D.T. Wicklow (eds.), The Fungal Community: Its Organization and Role in the Ecosystem. 2nd ed. Marcel Dekker, New York.

Wicklow, D.T., and D. Malloch. 1971. Studies in the genus *Thelebolus*: temperature optima for growth and ascocarp development. Mycologia 63:118–131.

Wicklow, D.T., and V. Moore. 1974. Effect of incubation temperature on the coprophilous fungal succession. Transactions of the British Mycological Society 62:411–415.

Wicklow, D.T., and W.F. Whittingham. 1974. Soil microfungal changes among the profiles of disturbed conifer-hardwood forests. Ecology 55:3–16.

——. 1978. Comparison of soil microfungal populations in disturbed and undisturbed forests in northern Wisconsin. Canadian Journal of Botany 56:1702–1709.

Wicklow, D.T., and D.M. Wilson. 1990. *Paecilomyces lilacinus*, a colonist of *Aspergillus flavus* sclerotia buried in soil in Illinois and Georgia. Mycologia 82:393–395.

Widden, P. 1986a. Functional relationships between Quebec forest soil microfungi and their environment. Canadian Journal of Botany 64:1424–32.

——. 1986b. Microfungal community structure from forest soils in southern Quebec, using discriminant function and factor analysis. Canadian Journal of Botany 64:1402–1412.

——. 1986c. Seasonality of forest soil microfungi in southern Quebec. Canadian Journal of Botany 64:1413–1423.

Widler, B., and E. Müller. 1984. Untersuchungen über endophytische Pilze von *Arctostaphylos uva-ursi* (L.) Sprengel (Ericaceae). Botanica Helvetica 94:307–337.

Wilcox, H.E., and C.J.K. Wang. 1987. Mycorrhizal and pathological associations of dematiaceous fungi in roots of 7-month-old tree seedlings. Canadian Journal of Forest Research 17:884–889.

Wilding, N., N.M. Collins, P.M. Hammond, and J.F. Webber. 1989. Insect-Fungus Interactions. Academic Press, New York.

Wildman, H.G. 1991. Lithium chloride as a selective inhibitor of *Trichoderma* species on soil isolation plates. Mycological Research 95:1364–1368.

Wildman, H.G., and D. Parkinson. 1979. Microfungal succession on living leaves of *Populus tremuloides*. Canadian Journal of Botany 57:2800–2811.

Wiles, M., and T.G. Rand. 1987. Integumental ulcerative disease in a loggerhead turtle *Caretta caretta* at the Bermuda Aquarium: microbiology and histopathology. Diseases of Aquatic Organisms 3:85–90.

Williams, A.G., and G.C. Orpin. 1987. Polysaccharide-degrading enzymes formed by three species of anaerobic rumen fungi grown on a range of carbohydrate substrates. Canadian Journal of Microbiology 33:418–426.

Williams, A.G., S.E. Withers, and K.N. Joblin. 1991. Xylanolysis by cocultures of the rumen fungus *Neocallimastix frontalis* and ruminal bacteria. Letters in Applied Microbiology 12:232–235.

Williams, J.G.K., A.R. Kubelik, K.J. Livak, J.A. Rafalski, and S.V. Tingey 1990. DNA polymorphisms amplified by arbitrary primers are useful as genetic markers. Nucleic Acids Research 18:6531–6535.

Williams, P.H., and C.J. Humphries. 1994. Biodiversity, taxonomic relatedness, and endemism in conservation. Pp. 269–287. *In* P.L. Forey, C.J. Humphries, and R.I. Vane-Wright (eds.), Systematics and Conservation Evaluation. Clarendon Press, Oxford, England.

Williamson, B. 1994. Latency and quiescence in survival and success of fungal plant pathogens. Pp. 187–207. *In* J.P. Blakeman and B. Williamson (eds.), Ecology of Plant Pathogens. CAB International, Wallingford, Oxon, United Kingdom.

Williamson, D.H., and D.J. Fennell. 1975. The use of fluorescent DNA-binding agent for detecting and separating yeast mitochondrial DNA. Methods in Cell Biology 12:335–351.

Williamson, M.A., and N.J. Fokkema. 1985. Phyllosphere yeasts antagonize penetration from appressoria and subsequent infection of maize leaves by *Colletotrichum graminicola*. Netherlands Journal of Plant Pathology 91:265–276.

Willoughby, L.G. 1956. Studies on soil chytrids. I. *Rhizidium richmondense* sp. nov. and its parasites. Transactions of the British Mycological Society 39:125–141.

——. 1961. The ecology of some lower fungi at Esthwaite Water. Transactions of the British Mycological Society 44:305–332.

——. 1962a. The occurrence and distribution of reproductive spores of Saprolegniales in fresh water. Journal of Ecology 50:733–759.

——. 1962b. The ecology of some lower fungi in the English Lake District. Transactions of the British Mycological Society 45:121–136.

——. 1978. Saprolegnias of salmonid fish in Windermere: a critical analysis. Journal of Fish Diseases 1:51–67.

——. 1984. Viability of *Allomyces* in a dry soil, investigated by poly-cell-gel analysis. Transactions of the British Mycological Society 82:581–587.

——. 1988. *Saprolegnia* parasitized by *Mortierella alpina*. Transactions of the British Mycological Society 90:496–499.

Willoughby, L.G., and A.D. Pickering. 1977. Viable saprolegniaceae spores on the epidermis of the salmonid fish *Salmo trutta* and *Salvelinus alpinus*. Transactions of the British Mycological Society 68:91–95.

Will-Wolf, S. 1980. Structure of corticolous lichen communities before and after exposure to emissions from a "clean" coal-fired power generating station. Bryologist 83:281–295.

——. 1988. Quantitative approaches to air quality studies. Bibliotheca Lichenologica 30:109–140.

——. 1998. Lichens of Badlands National Park, South Dakota, USA. Pp. 323–336. *In* M.G. Glenn, R.C. Harris, R. Dirig, and M.S. Cole (eds.), Lichenographia Thomsoniana: North American Lichenology in Honor of John W. Thomson. Mycotaxon, Ltd., Ithaca, New York.

Will-Wolf, S., C. Scheidegger, and B. McCune. 2002. Methods for monitoring biodiversity and ecosystem function. Pp. 147–162. *In* P.L. Nimis, C. Scheidegger, and P. Wolseley. Monitoring with Lichens—Monitoring Lichens. NATO Science Series. Kluwer Academic, Dordrecht, The Netherlands.

Wilmotte, A., Y. Van de Peer, A. Goris, S. Chapelle, R. De Baere, B. Nelissen, J.-M. Neefs, G.L. Hennebert, and R. De Wachter. 1993. Evolutionary relationships among higher fungi inferred from small ribosomal subunit RNA sequence analysis. Systematic and Applied Microbiology 16:436–444.

Wilson, C., A. Ragnini, and H. Fukuhara. 1989. Analysis of the regions coding for transfer RNAs in *Kluyveromyces lactis* mitochondrial DNA. Nucleic Acids Research 17:4485–4491.

Wilson, D. 1995a. Fungal endophytes which invade insect galls: insect pathogens, benign saprophytes, or fungal inquilines? Oecologia 103:255–260.

——. 1995b. Endophyte—the evolution of a term, and clarification of its use and definition. Oikos 73:274–276.

Wilson, D., and G.C. Carroll. 1994. Infection studies of *Discula quercina*, an endophyte of *Quercus garryana*. Mycologia 86:635–647.

——. 1997. Avoidance of high-endophyte space by gall-forming insects. Ecology 78:2153–2163.

Wilson, D., and S.H. Faeth. 2001. Do fungal endophytes result in selection for leafminer ovipositional preference? Ecology 82:1097–1111.

Wilson, D.E., and D.M. Reeder. 1993. Mammal Species of the World: A Taxonomic and Geographic Reference. 2nd ed. Smithsonian Institution Press, Washington, D.C.

Wilson, J., K. Ingleby, P.A. Mason, K. Ibrahim, and G.J. Lawson. 1992. Long-term changes in vesicular-arbuscular mycorrhizal spore populations in *Terminalia* plantations in Côte d'Ivoire. Pp. 268–275. *In* D.J. Read, D.H. Lewis, A.H. Fitter, and I.J. Alexander (eds.), Mycorrhizas in Ecosystems. CAB International, Wallingford, Oxon, United Kingdom.

Wilson, M., and D.M. Henderson. 1966. British Rust Fungi. Cambridge University Press, Cambridge, England.

Wilson, M.V., and C.L. Mohler. 1983. Measuring compositional change along gradients. Vegetatio 54:129–141.

Wilson, M.V., and A. Shmida. 1984. Measuring beta diversity with presence-absence data. Journal of Ecology 72:1055–1064.

Windham, W.R., and D.E. Akin. 1984. Rumen fungi and forage fiber degradation. Applied and Environmental Microbiology 48:473–476.

Wingfield, M.J., K.A. Seifert, and J.F. Webber (eds.). 1993. *Ceratocystis* and *Ophiostoma*: Taxonomy, Ecology, and Pathogenicity. American Phytopathological Society Press, St. Paul, Minnesota.

Winka, K., O.E. Eriksson, and Å. Bång. 1998. Molecular evidence for recognizing the Chaetothyriales. Mycologia 90:822–830.

Winterhoff, W. 1992. Fungi in Vegetation Science. Handbook of Vegetation Science Vol. 19. Kluwer Academic, Dordrecht, The Netherlands.

Wirth, V. 1994. Checkliste der Flechten und flechtenbewohnenden Pilze Deutschlands—eine Arbeitshilfe. Beitrage zur Naturkunde, Serie A, 517:1–63.

——. 1995a. Flechtenflora: Bestimmung und Ökologische Kennzeichnung der Flechten Südwestdeutschlands und angrenzender Gebiete. 2. Auflage. E. Ulmer, Stuttgart, Germany.

——. 1995b. Die Flechten Baden–Württembergs. 2nd ed. 2 vols. E. Ulmer, Stuttgart, Germany.

Witcosky, J.J., T.D. Schowalter, and E.M. Hansen. 1987. Host-derived attractants for the beetles *Hylastes nigrinus* (Coleoptera: Scolytidae) and *Steremnius carinatus* (Coleoptera: Curculionidae). Environmental Entomology 16:1310–1313.

Woese, C.R. 1987. Bacterial evolution. Microbiological Reviews 51:221–271.

Wolda, H. 1981. Similarity indices, sample size and diversity. Oecologia (Berlin) 50:296–302.

Wolf, D.C., and H.D. Skipper. 1994. Soil sterilization. Pp. 41–51. *In* R.W. Weaver, S. Angle, P. Bottomley, D. Bezdicek, S. Smith, A. Tabatabai, and A. Wollum (eds.), Methods of Soil Analysis, Part 2. Microbiological and Biochemical Properties. Soil Science Society of America, Madison, Wisconsin.

Wolf, G., and F. Frič. 1981. A rapid method for staining *Erisyphe graminis* f. sp. *hordei* in and on whole barley leaves with a protein-specific dye. Phytopathology 71:596–598.

Wolf, J.C., and S.A. Smith. 1999. Systemic zygomycosis in farmed tilapia fish. Journal of Comparative Pathology 121:301–306.

Wolf, J.H.D. 1993. Ecology of Epiphytes and Epiphyte Communities in Montane Rain Forests, Colombia. Unpubl. Ph.D. dissert. University of Amsterdam. Elinkwijk B.V., Utrecht, The Netherlands.

Wolf, K., and L. Del Giudice. 1988. The variable mitochondrial genome of ascomycetes: organization, mutational alterations and expression. Advances in Genetics 25:185–308.

Wolke, R.E. 1975. Pathology of bacterial and fungal diseases affecting fish. Pp. 33–116. *In* W.E. Ribelin and G. Migaki (eds.), The Pathology of Fishes. University of Wisconsin Press, Madison, Wisconsin.

Wollenweber, H.W., and O.A. Reinking. 1935. Die Fusarien, ihre Beschreibung, Schadwirkung und Bekämpfung. Paul Parey, Berlin, Germany.

Wollenzien, U., G.S. de Hoog, W. Krumbein, and J.M.J. Uijthof. 1997. *Sarcinomyces petricola*, a new microcolonial fungus from marble in the Mediterranean basin. Antonie van Leeuwenhoek 71:281–288.

Wollman, C.E. 1966. Laboratory Culture of Selected Species of Myxomycetes with Special Reference to the Gross Morphology of Their Plasmodia. Unpubl. Ph.D. dissert., University of Texas, Austin, Texas.

Wolseley, P.A., and B. Aguirre-Hudson. 1991. Lichens as indicators of environmental change in the tropical forests of Thailand. Global Ecology and Biogeography Letters 1:170–175.

Wolseley, P.A., B. Aguirre-Hudson, and P.M. McCarthy. 2002. Catalog of the lichens of Thailand. Bulletin of the Natural History Museum (London), Botany 32:13–59.

Wolseley, P.A., C. Moncreiff, and B. Aguirre-Hudson. 1995. Lichens as indicators of environmental stability and change in the tropical forests of Thailand. Global Ecology and Biogeography Letters 4:116–123.

Wood, D.L. 1982. The role of pheromones, kairomones, and allomones in the host selection and colonization behavior of bark beetles. Annual Review of Entomology 27:411–446.

Wood, S.E., and L.G. Willoughby. 1986. Ecological observations on the fungal colonization of fish by Saprolegniaceae in Windermere. Journal of Applied Ecology 23:737–747.

Wood, S.L. 1982. The bark and ambrosia beetles of North and Central America (Coleoptera: Scolytidae), a taxonomic monograph. Great Basin Naturalist Memoirs no.6.

Wood, S.L., and D.E. Bright, Jr. 1987. A Catalog of Scolytidae and Platypodidae (Coleoptera). Part 1: Bibliography. Great Basin Naturalist Memoirs no.11.

——. 1992. A Catalog of Scolytidae and Platypodidae (Coleoptera). Part 2, A and B: Bibliography and Taxonomic Index. Great Basin Naturalist Memoirs no.13.

Wood, T.G., and R.J. Thomas. 1989. The mutualistic association between Macrotermitinae and *Termitomyces*. Pp. 69–92. *In* N. Wilding, N.M. Collins, P.M. Hammond, and J.F. Webber (eds.), Insect-Fungus Interactions. Academic Press, New York.

Woodbridge, B., J.R. Coley-Smith, and D.A. Reid. 1988. A new species of *Cylindrobasidium* parasitic on sclerotia of *Typhula incarnata*. Transactions of the British Mycological Society 91:166–169.

Wood-Eggenschwiler, S., and F. Bärlocher. 1983. Aquatic hyphomycetes in sixteen streams in France, Germany and Switzerland. Transactions of the British Mycological Society 81:371–379.

Worthen, W.B., and T.R. McGuire. 1990. Predictability of ephemeral mushrooms and implications for mycophagous fly communities. American Midland Naturalist 124:12–21.

Wöstemeyer, J., A. Wöstemeyer, A. Burmester, and K. Czempinski. 1995. Relationships between sexual processes and parasitic interactions in the host-pathogen system *Absidia glauca-Parasitella parasitica*. Canadian Journal of Botany 73(Suppl. 1):S243–S250.

Wright, D. 1996. An introduction to lichen spot tests. Bulletin of the California Lichen Society 3:9–12.

Wright, D.K., and M.M. Manos. 1990. Sample preparation from paraffin-embedded tissues. Pp. 153–158. *In* M.A. Innis, D.H. Gelfand, J.J. Sninsky, and T.J. White (eds.), PCR Protocols: A Guide to Methods and Applications. Academic Press, San Diego, California.

Wright, S.F., and A. Upadhyaya. 1998. A survey of soils for aggregate stability and glomalin, a glycoprotein produced by hyphae of arbuscular mycorrhizal fungi. Plant and Soil 198:97–107.

Wright, S.F., M. Franke-Snyder, J.B. Morton, and A. Upadhyaya. 1996. Time-course study and partial characterization of a protein on arbuscular mycorrhizal fungi during active colonization of roots. Plant and Soil 181:193–203.

Wright, S.F., A. Upadhyaya, and J.S. Buyer. 1998. Comparison of N-linked oligosaccharides of Glomalin from arbuscular mycorrhizal fungi and soils by capillary electrophoresis. Soil Biology and Biochemistry 30:1853–1857.

Wrigley de Basanta, D. 2000. Acid deposition in Madrid and corticolous myxomycetes. Stapfia 73:113–120.

Wu, Q.-X., G.M. Mueller, F.M. Lutzoni, Y.-Q. Huang, and S.-Y. Guo. 2000. Phylogenetic and biogeographic relationships of eastern Asian and eastern North American disjunct *Suillus* species (Fungi) as inferred from nuclear ribosomal RNA ITS sequences. Molecular Phylogenetics and Evolution 17:37–47.

Wubah, D.A., and S.H. Kim. 1994. Morphology and development of two new isolates of anaerobic zoosporic fungi from a pond. In Abstracts. 5th International Mycological Congress, Vancouver, British Columbia, Canada.

——. 1995. Isolation and characterization of a free-living species of *Piromyces* from a pond. Inoculum 46:44.

Wubah, D.A., D.E. Akin, and W.S. Borneman. 1993. Biology, fiber-degradation, and enzymology of anaerobic zoosporic fungi. Critical Reviews in Microbiology 19:99–115.

Wubah, D.A., M.S. Fuller, and D.E. Akin. 1991a. Resistant body formation in *Neocallimastix* sp., an anaerobic fungus from the rumen of a cow. Mycologia 83:40–48.

——. 1991b. *Neocallimastix*: a comparative morphological study. Canadian Journal of Botany 69:835–843.

——. 1991c. Studies on *Caecomyces communis*: morphology and development. Mycologia 83:303–310.

——. 1991d. Isolation of monocentric and polycentric fungi from the rumen and feces of cows in Georgia. Canadian Journal of Botany 69:1232–1236.

Wymelenberg, A.V., S. Denman, D. Dietrich, J. Bassett, X. Yu, R. Atalla, P. Predki, U. Rudsander, T.T. Teeri, and D. Cullen. 2002. Transcript analysis of genes encoding a family 61 endoglucanase and a putative membrane-anchored family 9 glycosyl hydrolase from *Phanerochaete chrysosporium*. Applied and Environmental Microbiology 68:5765–5768.

Wynn, A.R., and H.A.S. Epton. 1979. Parasitism of oospores of *Phytophthora erythroseptica* in soil. Transactions of the British Mycological Society 73:255–259.

Yadav, A.S. 1966. The ecology of microfungi on decaying stems of *Heracleum sphondylium*. Transactions of the British Mycological Society 49:471–485.

Yamada, Y., and H. Kawasaki. 1989. The molecular phylogeny of the Q$_8$-equipped basidiomycetous yeast genera *Mrakia* Yamada et Komagata and *Cystofilobasidium* Oberwinkler et Bandoni based on the partial sequences of 18S and 26S ribosomal ribonucleic acids. Journal of General and Applied Microbiology 35:173–183.

Yamada, Y., and H. Nakagawa. 1990. The molecular phylogeny of the basidiomycetous yeast species, *Leucosporidium scottii* based on the partial sequences of 18S and 26S ribosomal nucleic acids. Journal of General and Applied Microbiology 36:63–68.

Yamada, Y., H. Kawasaki, T. Nakase, and I. Banno. 1989a. The phylogenetic relationship of the conidium-forming anamorphic yeast genera *Sterigmatomyces*, *Kurtzmanomyces*, *Tsuchiyaea* and *Fellomyces*, and the telemorphic yeast genus *Sterigmatosporidium* on the basis of the partial sequences of 18S and 26S ribosomal ribonucleic acids. Agricultural Biological Chemistry 53:2993–3001.

Yamada, Y., K. Maede, I. Banno, and J.P. van der Walt. 1992. An emendation of the genus *Debaryomyces* Lodder et Kreger-van Rij and the proposals of two new combinations, *Debaryomyces carsonii* and *Debaryomyces etchellsii* (Saccharomycetaceae). Journal of General and Applied Microbiology 38:623–626.

Yamada, Y., Y. Nakagawa, and I. Banno. 1989b. The phylogenetic relationship of the Q$_9$-equipped species of the heterobasidiomycetous yeast genera *Rhodosporidium* and *Leucosporidium* based on the partial sequences of 18S and 26S ribosomal nucleic acids: the proposal of a new genus *Kondoa*. Journal of General and Applied Microbiology 35:377–385.

——. 1990b. The molecular phylogeny of the Q$_{10}$-equipped species of the heterobasidiomycetous yeast genus *Rhodosporidium* Banno based on the partial sequences of 18S and 26S ribosomal nucleic acids. Journal of General and Applied Microbiology 36:435–444.

Yamada, Y., T. Nagahama, H. Kawasaki, and I. Banno. 1990a. The phylogenetic relationship of the genera *Phaffia* Miller, Yoneyama et Soneda and *Cryptococcus* Kützing emend. Phaff et Spencer (Cryptococcaceae) based on the partial sequences of 18S and 26S ribosomal nucleic acids. Journal of General and Applied Microbiology 36:403–414.

Yamamoto, H., T. Kadzunori, and T. Uchiwa. 1985. Fungal flora of soil polluted with copper. Soil Biology and Biochemistry 17:785–790.

Yang, B.Y. 1962. *Basidiobolus meristosporus* of Taiwan. Taiwania 8:17–27.

Yarlett, N., C.G. Orpin, E.A. Munn, N.C. Yarlett, and C.A. Greenwood. 1986a. Hydrogenosomes in the rumen fungus, *Neocallimastix patriciarum*. Biochemical Journal 236:729–739.

Yarlett, N.C., N. Yarlett, C.G. Orpin, and D. Lloyd. 1986b. Cryopreservation of the anaerobic rumen fungus *Neocallimastix patriciarum*. Letters of Applied Microbiology 3:1–3.

Yarranton, G.A. 1972. Distribution and succession of epiphytic lichens on black spruce near Cochrane, Ontario. Bryologist 75:462–480.

Yarrow, D. 1998. Methods for the isolation, maintenance, and identification of yeasts. Pp. 77–100. *In* C.P. Kurtzman and J.W. Fell (eds.), The Yeasts, A Taxonomic Study. 4th ed. Elsevier Science, Amsterdam, The Netherlands.

Yarwood, C.E. 1973. Pyrenomycetes: erysiphales. Pp. 71–86. *In* G.C. Ainsworth, F.K. Sparrow, and A.S. Sussman (eds.), The Fungi. An Advanced Treatise. IVA. A Taxonomic Review with Keys: Ascomycetes and Fungi Imperfecti. Academic Press, New York.

Young, P.A. 1926. Penetration phenomena and facultative parasitism in *Alternaria*, *Diplodia*, and other fungi. Botanical Gazette 81:258–279.

Yu, K., and K.P. Pauls. 1992. Optimization of the PCR program for RAPD analysis. Nucleic Acids Research 20:2606.

Zahl, S. 1977. Jackknifing an index of diversity. Ecology 58:907–913.

Zahlbruckner, A. 1921–1940. Catalogus Lichenum Universalis. 10 vols. Borntraeger, Berlin, Germany.

Zak, J.C. 1988. Redevelopment of biological activity in strip-mine spoils: saprotrophic fungal assemblages of grass roots. Proceedings of the Royal Society of Edinburgh 94B:73–83.

——. 1992. Response of soil fungal communities to disturbance. Pp. 403–425. *In* G.C. Carroll and D.T. Wicklow (eds.), The Fungal Community: Its Organization and Role in the Ecosystem. 2nd ed. Marcel Dekker, New York.

——. 1993. The enigma of desert ecosystems: the importance of interactions among the soil biota to fungal biodiversity. Pp. 59–71. *In* S. Isaac, J. C Frankland, R. Watling, and A.J.S. Whalley (eds.), Aspects of Tropical Mycology. Cambridge University Press, Cambridge, United Kingdom.

Zak, J.C., and S. Visser. 1996. An appraisal of soil fungal biodiversity: the crossroads between taxonomic and functional biodiversity. Biodiversity and Conservation 5:169–183.

Zak, J.C., and W.G. Whitford. 1986. The occurrence of a hypogeous Ascomycete in the northern Chihuahuan desert. Mycologia 78:840–841.

Zak, J.C., and D.T. Wicklow. 1978. Response of carbonicolous ascomycetes to aerated steam temperatures and treatment intervals. Canadian Journal of Botany 56:2313–2318.

Zak, J.C., R. Sinsabaugh, and W.P. Mackay. 1995. Windows of opportunity in desert ecosystems: their implications to fungal community development. Canadian Journal of Botany 73 (Suppl. 1): S1407–S1414.

Zalar, P., G.S. de Hoog, and N. Gunde-Cimerman. 1999a. Ecology of halotolerant dothideaceous black yeasts. Studies in Mycology 43:38–48.

———. 1999b. *Trimmatostroma salinum*, a new species from hypersaline water. Studies in Mycology 43:57–62.

Zambino, P.J., and L.J. Szabo. 1993. Phylogenetic relationships of selected cereal and grass rusts based on rDNA sequence analysis. Mycologia 85:401–414.

Zar, J.H. 1996. Biostatistical analysis. 3rd ed. Prentice-Hall, Upper Saddle River, New Jersey.

Zare, R., and W. Gams. 2001. A revision of *Verticillium* section *Prostrata*. IV. The genera *Lecanicillium* and *Simplicillium* gen. nov. Nova Hedwigia 73:1–50.

Zare, R., W. Gams, and A. Culham. 2000. A revision of *Verticillium* sect. *Prostrata*. I. Phylogenetic studies using ITS sequences. Nova Hedwigia 71:465–480.

Zeller, S.M. 1922. Contributions to our knowledge of Oregon fungi-I. Mycologia 14:173–199.

———. 1943. North American species of *Galeropsis*, *Gyrophragmium*, *Longia*, and *Montagnea*. Mycologia 35:409–421.

———. 1947. More notes on Gastromycetes. Mycologia 39:282–312.

Zhang, Q. 1996. Fungal Community Structure and Microbial Biomass in a Semi-Arid Environment: Roles in Root Decomposition, Root Growth, and Soil Nitrogen Dynamics. Unpubl. Ph.D. dissert., Texas Tech University, Lubbock, Texas.

Zhang, X., and J.H. Andrews. 1993. Evidence for growth of *Sporothrix schenckii* on dead but not on living sphagnum moss. Mycopathologia 123:87–94.

Zhang, and Zak. 1996. Cited in figure legends, fig. 33 (chap.5). Citation requested from author.

Zheng, R.-Y. 1985. Genera of the Erysiphaceae. Mycotaxon 22: 209–263.

Zhuang, W. 1988. A new species of *Dencoeliopsis* and a synoptic key to the genera of the Encoelioideae (Leotiaceae). Mycotaxon 32:97–104.

Ziller, W.G. 1974. The Tree Rusts of Western Canada. Canadian Forest Service Publication No. 1329, Victoria, British Columbia.

Zolan, M.E. 1995. Chromosome-length polymorphism in fungi. Microbiological Reviews 59:686–698.

Zugmaier, W., R. Bauer, and F. Oberwinkler. 1994. Studies in Heterobasidiomycetes. 104. Mycoparasitism of some *Tremella* species. Mycologia 86:49–56.

Zundel, G.L. 1953. The Ustilaginales of the World. Pennsylvania State College School of Agriculture, State College, Pennsylvania. Department of Botany, Contributions, no.176.

Zwart, P., M.A. Verwer, G.A. De Vries, and E.J. Hermanides-Nijhof. 1973. Fungal infection of the eyes of the snake *Epicrates chenchria maurus*: Enucleation under halothane narcosis. Journal of Small Animal Practice 14:773–779.

Zycha, H., R. Siepmann, and G. Linnemann. 1969. Mucorales. J. Cramer, Lehre, Germany.

INDEX

Printed and bound by CPI Group (UK) Ltd, Croydon, CR0 4YY

08/05/2025

01865029-0001